Human Biochemistry and Disease

Human Biochemistry and Disease

Gerald Litwack

Former Chair of Biochemistry and Molecular Pharmacology
Thomas Jefferson University Medical College
Philadelphia, Pennsylvania

Former Visiting Scholar
Department of Biological Chemistry
Geffen School of Medicine at UCLA
Los Angeles, California

AMSTERDAM • BOSTON • HEIDELBERG • LONDON
NEW YORK • OXFORD • PARIS • SAN DIEGO
SAN FRANCISCO • SINGAPORE • SYDNEY • TOKYO

Academic Press is an imprint of Elsevier

Academic Press is an imprint of Elsevier
30 Corporate Drive, Suite 400, Burlington, MA 01803, USA
525 B Street, Suite 1900, San Diego, California 92101-4495, USA
84 Theobald's Road, London WC1X 8RR, UK

This book is printed on acid-free paper. ∞

Copyright © 2008, Elsevier Inc. All rights reserved.

Notice

No part of this publication may be reproduced or transmitted in any form or by any means, electronic or mechanical, including photocopying, recording, or any information storage and retrieval system, without permission in writing from the publisher.

Permissions may be sought directly from Elsevier's Science & Technology Rights Department in Oxford, UK: phone: (+44) 1865 843830, fax: (+44) 1865 853333, E-mail: permissions@elsevier.com. You may also complete your request online via the Elsevier homepage (http://elsevier.com), by selecting "Support & Contact" then "Copyright and Permissions" and then "Obtaining Permissions".

The Publisher

Library of Congress Cataloging-in-Publication Data
2007926493

British Library Cataloguing in Publication Data
A catalogue record for this book is available from the British Library.

ISBN 978-0-12-452815-4

For information on all Elsevier Academic Press publications
visit our Web site at www.books.elsevier.com

Printed in China

07 08 09 10 11 12   9 8 7 6 5 4 3 2 1

Working together to grow
libraries in developing countries

www.elsevier.com | www.bookaid.org | www.sabre.org

ELSEVIER   BOOK AID International   Sabre Foundation

Contents

Preface xiii
Dedication xv

CHAPTER 1: Introduction and General Considerations 1

Introduction 1
Integration of Biochemistry with Clinical Medicine 2
General Considerations 2
    The Human Body and Organ Systems 2
    The Cell 5
    Cell Membrane 6
    Nucleus 9
    Cytoplasm 18
    Receptors and Their Cellular Locations 26
    Biological Roles of Water 27
    Ion Channels 28
    pH 30
Further Reading 32

CHAPTER 2: Proteins 33

Prion Disease: A Fatal Protein Conformation 33
    Propagation of PrP^{Sc} from PrP^{c} in the Cell 35
Amino Acids 39
    Chirality 44
    Amino Acids Have Two or More Potential Charges 46
    Synthesis and Degradation of Amino Acids 50
Proteins 54
    Amino Acid Sequence 56
    Secondary Structure 57
Further Reading 92

CHAPTER 3: Enzymes   93

Clinical Enzymology in Diagnosis of Disease   93
Enzymes Are Catalytic Proteins   96
Kinetics   97
   The Michaelis-Menten Equation   99
   Enzyme Inhibition   101
   Allosterism   105
Classification   107
Coenzymes   112
Prosthetic Groups   119
Drugs and Enzymes   125
Further Reading   129

CHAPTER 4: Carbohydrates   131

Diabetes: A Prevalent Disease That Disrupts Glucose Utilization   131
   Insulin   138
   The Pancreatic Beta Cell   142
   Effects of Diabetes   145
Simple Sugars   146
Starch   153
Glycogen   154
   Breakdown of Glycogen for Energy Use (Glycogenolysis)   156
   Glycogen Synthesis   162
   Effects of Hormones on Glycogen Breakdown and Synthesis   163
   Glycogen Storage Diseases   169
Is Type 2 Diabetes a Disease of Protein Aggregation?   170
Use of Glucose for Energy   173
Glycerol Can Be Converted to Glucose   182
Glycoproteins   182
Blood Group Proteins   185
Lactose Intolerance   188
Glycobiology   188
Further Reading   188

CHAPTER 5: Lipids   189

Hypercholesterolemia: A Disease in Which Serum Cholesterol Is Not Properly Imported at the Cellular Level   189
   Biosynthesis of Cholesterol   193
   Synthesis of Bile Acids   193
   Prognosis   198
Fatty Acids and Fat   198
Fatty Acid Oxidation   202

Activation and Transport of Fatty Acids into
   Mitochondria   206
Lipid Metabolism and Hormonal Control   207
Phospholipids   220
Glycosphingolipids   226
Lipoproteins   233
Lipid Anchoring of Proteins to Membranes   236
Further Reading   238

CHAPTER 6: Nucleic Acids and Molecular Genetics   239

Huntington's Disease: A Trinucleotide Repeat Mutation   239
Purines and Pyrimidines   244
   Base-Pairing   249
   Biosynthesis of Purines and Pyrimidines and Their Catabolism   252
   Purine Interconversions   268
   Purine and Pyrimidine Nucleotide Catabolism   271
   Disorders of Purine and Pyrimidine Metabolism   283
Biosynthesis of Deoxyribonucleic Acids   288
   Mutations and Damage to DNA   298
   Specific Nucleases: Restriction Enzymes   300
   Natural Genomic DNA   305
   Sequencing DNA   308
   Inhibition of DNA Synthesis   311
Functional Genomics   312
   Gene Therapy   314
Ribonucleic Acids   316
Further Reading   322

CHAPTER 7: Transcription   323

Asbestosis: A Disease of Aberrant Transcription   323
Transcription Factors and Transcription Complex   329
Coactivators and Corepressors   337
The Glucocorticoid Receptor as a Model Transcription
   Factor   349
Chromatin   357
Further Reading   363

CHAPTER 8: Polypeptide Hormones   365

Panhypopituitarism: A Malfunction of the Hypothalamus–
   Pituitary–End Organ Axis   365
Humoral Mechanism   367

Posterior Pituitary   380
Actions of Releasing Hormones and Anterior Pituitary Hormones   387
   CRH–ACTH–Cortisol Pathway   387
   Growth Hormone–Releasing Hormone—Growth Hormone—Bodily Growth Path   392
Gonadotropins   406
Thyrotropin   417
Prolactin   430
Gastrointestinal Hormones   435
Further Reading   444

CHAPTER 9: Steroid Hormones   445

Stress: A State That Can Have Serious Pathological Consequences   445
   Adrenal Medulla   448
Adrenal Cortex   452
   Aldosterone   452
   Cortisol   466
   Dehydroepiandrosterone   474
Steroid Hormone Structures   476
Unliganded Forms of Receptors and Mechanism of Activation   478
Ligands and Receptor Conformation: The Sex Hormones   483
Peroxisome Proliferators and Orphan Receptors   488
Programmed Cell Death (Apoptosis) Induced by Glucocorticoids   491
Further Reading   496

CHAPTER 10: Metabolism   497

Hyperammonemia and Disruptions of the Urea Cycle   497
   Excess Ammonium Ion and Urea in the Blood Can Be Lethal   497
The Urea Cycle   499
Nitrogen Flow, Amino, and Amide Group Transfers in Amino Acid Metabolism   503
Transamination   508
Transamidation   512
Deamination   513
Amino Acid Oxidation   514
Amino Acid Decarboxylation   516
Metabolism of Specific Amino Acids to Key Substances   517
   Methionine   517

    Phenylalanine and Tyrosine   521
    Formation of Catecholamines   524
    Formation of Melanin   527
    Tryptophan   532
    Arginine   536
    Histidine   541
    Glutamate   544
    Serine   545

Catabolism of Amino Acids   546

Metabolism of Lipids   554

Glucagon   559

Fatty Acid Degradation   560

Fat as Storage Energy   561

Lipid and Carbohydrate Metabolism Are Jointly Regulated   565

Steroid Hormone Metabolism   566

Nucleic Acid Metabolism   570

DNA Damage and Repair   573

Cell Death   575

Carbohydrate Metabolism   577

Regulation of Blood Glucose Level   581

Overview   585

Further Reading   586

CHAPTER 11: Growth Factors and Cytokines   587

New Approaches to Ovarian Cancer, Such as the Action of TRAIL (TNF-Related Apoptosis-Inducing Ligand) Might Form the Basis of a Treatment   587

TNF Superfamily   594

Growth Factors   600
    Epidermal Growth Factor   604
    Transforming Growth Factor   612
    Fibroblast Growth Factor   616
    Nerve Growth Factor   623
    Colony-Stimulating Factor   627
    Erythropoietin   633
    Interferon-γ 638
    Insulin-like Growth Factors   643

Interleukins   654

Further Reading   683

CHAPTER 12: Membrane Transport 685

Cystic Fibrosis: A Genetic Disease Involving Aberrant Ion Transport 685
Types of Membrane Transport 695
  Absorption of Large Molecules 695
  Exocytosis 695
  Passive Diffusion or Osmosis 697
Energy-Requiring Transport: Active Transport 702
Simple and Coupled Transporters 704
Ions and Gradients 706
How Do Magnesium and Other Divalent Ions Enter Cells? 713
Proton (H^+) Transport 714
Amino Acid Transporters 720
Fatty Acid Uptake 723
Sodium Conductance and Voltage-Gated Sodium Channels 729
Multidrug Resistance Channel (MDR) of the ABC Transporter Superfamily 732
Blood-Brain Barrier 734
Further Reading 737

CHAPTER 13: Dietary Metals, Iron, Micronutrients, and Nutrition 739

Iron-Deficiency Anemia 739
Ingestion and Uptake of Iron 743
Heme Synthesis 753
Formation of Hemoglobin 760
Dietary Metals 764
  Copper 764
  Selenium 773
  Zinc 776
  Magnesium 780
  Calcium: A Micronutrient 783
  Molybdenum 788
  Iodine: A Micronutrient 795
Vitamins 801
  Water-Soluble Vitamins 801
  Fat-Soluble Vitamins 835
The Diet 849
  Protein Nutrition 849
Herbs and Nutraceuticals 852
Further Reading 852

CHAPTER 14: Blood and Lymphatic System   853

Deep Vein Thrombosis: A Major Health Problem   853
The Blood-Clotting Mechanism   864
Blood   871
    Transport of Oxygen   871
    Carbon Dioxide   878
    Blood Cells   879
    Blood Proteins   884
    Blood Type and Rh   888
Lymphatic System   894
Further Reading   898

CHAPTER 15: Immunobiochemistry   899

The Surveillance System and Cancer   899
Types of Antibodies   910
    Polyclonal and Monoclonal Antibodies   915
Opsonization   917
Antibody Formation   917
Autoimmunity   920
    Graves' Disease   920
    MHC Involvement   922
    Theory on the Development of Type I Diabetes (Insulin-Dependent Diabetes Mellitus)   930
Complement System   933
    Regulators of Complement Pathways   942
Properdin   949
C-Reactive Protein   951
Further Reading   953

CHAPTER 16: Neurobiochemistry   955

Pain: A Constant Health Problem   955
Substance P   962
Opioids and Morphine   968
Anandamide   974
Excitatory Amino Acids   988
The Classical Neurotransmitters   1003
Catecholamines and Monoamines   1027
Characteristics of the Brain   1044
Further Reading   1046

CHAPTER 17: Microbial Biochemistry 1047

AIDS: A Deadly Viral Disease 1047
Other Viruses of Current Interest 1068
- Human Rhinoviruses 1072
- Influenza Viruses 1072
- West Nile Virus 1076
- Severe Acute Respiratory Syndrome 1078

Bacteriophage 1089
A Bacterial Cell, *E. coli* 1107
- Diseases Caused by *E. coli* 1147

Further Reading 1151

Appendices

Appendix 1: Abbreviations of the Common Amino Acids 1153
Appendix 2: The Genetic Code 1155
- Base Pairing 1155

Appendix 3: Weights and Measures 1157

Glossary 1159
Index 1177

Preface

This is a different kind of a biochemistry textbook. The book is centered on human biochemistry and does not dwell on comparative biochemistry, except in a few cases to enhance meaning. This text is directed to medical students, graduate students, and undergraduate students, particularly those majoring in biochemistry or biology and those who are pre-medical students. The content is fairly concentrated, but there are many figures, making this a satisfying experience for visual learners. I have always felt that a picture to support the word is the best way to learn. Since I love to set ideas down in pictures, I have slanted the entire book in this direction. In addition, there are several pictures of structures, especially of proteins. Because so much information is now available on protein structure, students should become used to looking at three-dimensional structures that may resemble the actual protein in solution. Sometimes, little will be conveyed through the structure about its function; other times, especially when there is another macromolecule or small molecule with which the protein is reacting, the picture will impart a great feeling for how the protein is working, surpassing the verbal explanation. Also, there are no distinctions made between biochemistry, molecular biology, and cell biology; in my view, they are related seamlessly.

The impetus for creating this book came from many years of experience in planning for and teaching biochemistry to medical students. The majority of medical students, in my opinion, found biochemistry to be a grueling experience because they had a difficult time understanding how biochemistry relates to medicine or to disease. Part of this perception came from the way in which biochemistry is taught. Biochemists usually know rather little about disease, and clinicians know little about biochemistry. I have tried to make the relationship of biochemistry to medicine evident by introducing each biochemical topic with a study of a disease that represents the biochemical principles to be conveyed. For example, the subject of carbohydrate biochemistry is introduced by a discussion of diabetes, proteins by a discussion of prion disease, microbial biochemistry, by a discussion of HIV, and so on, with an introductory discussion of a relevant disease or clinical relationship in each chapter. This should make the study of biochemistry more meaningful for the medical student and not something to be avoided by the undergraduate or graduate student. After all, in many cases disease stems from abnormal biochemistry, and normalizing it may be the way to treat the disease. One needs to understand aberrant biochemistry and certainly normal biochemistry because this is the way in which cells in the body function.

Figures and tables are, for the most part, taken from the literature. Many citations to the sources for the data shown appear and these references will be useful to those readers who wish to pursue the literature beyond what is presented. For this reason, I have not appended a list of published papers at the end of each chapter, as is the usual custom, but rather I mention one or more specialized books for further reading.

Ten years ago, it might have taken me twice the time it actually took to prepare this book. Now with powerful search engines and availability of the literature on the Internet, writing this book was a pleasant experience. In particular, I need to give credit to the search engines and people who have helped me. Google search engine and to a lesser extent Google Scholar were very powerful tools. PubMed was especially helpful. Academic Press/Elsevier, through the courtesy of Jeremy Hayhurst, provided Science Direct, which allowed my entry into the current literature in many journals. Two university libraries were made available to me online: Dr. Thomas Nasca made it possible for me to utilize the Thomas Jefferson University library of my former institution. Dr. Elizabeth Neufeld, Chair of the Department of Biological Chemistry, David Geffen School of Medicine at UCLA, invited me to be a Visiting Scholar and at the same time made the library of the institution available to me. Because of this kind of assistance, I was able to generate most of the information I needed directly from my computer.

The Publisher, Academic Press/Elsevier, is one I have been associated with for many years. The Publishing Editor, Jeremy Hayhurst, was helpful and very supportive during the process and agreed with my idea for this text from the beginning. In the later stages of the completion of the book and its publication, Tari Broderick and Renske Van Dijk of Academic Press facilitated the final steps and production.

Gerald Litwack

For the people who worked with me in research over the years

Technicians, graduate students, post-doctoral fellows, and sabbatical visitors. A few of a great many are: Ann Trowbridge, Kris Morey, Nora Lichtash, Peter Bodine, Emad Alnemri, Sandy Singer, George Tryfiates, Tom Diamondstone, Emerich Fiala, Teresa Fernandes, Ilga Winicov, Tom Schmidt, Noreen Robertson, Sonia Lobo Planey, Andrea Miller, Violet Daniel, Costas Sekeris, Bob Baldridge, Gary Smith, Max Cake, Virginia Ohl, and David Phelps.

For the teachers, mentors, collaborators, and friends who inspired me or helped in some way

Kathryn Cook, Conrad Elvehjem, Jesse (Jerry) Williams, Jr., Edwin Bret Hart, Moe Cleland, Joe Nielands, Mavis Brandt, Charity Crocker, Roger Monier, Vern Schramm, Gordon Tomkins, Carlo Croce, Kay Huebner, Gary Stein, Brian Ketterer, Joe Gonnella, Tom Nasca, Sidney Weinhouse, Mannie Rubin, Marge Foti, Alan Kelly, Darwin and Ellie Prockop, Tony and Helen Norman, and many others.

For my family

Ellie, Geoff, Kate, Claudia, Debbie, and David.

CHAPTER 1

Introduction and General Considerations

Introduction

This is a text of human biochemistry interwoven with clinical aspects. It is a text that can be used by medical students, graduate students, undergraduate biochemistry majors, and other undergraduate majors interested in the interplay between biochemistry and disease.

Over the years, through my experience in teaching freshman medical students, it became clear to me that most students were eager to obtain relevant clinical experience, their reason for coming to medical school. As the increasing detail of biochemistry was presented, they lost sight of the role that biochemistry plays in medicine. Standing back and inspecting the scientific disciplines, it becomes obvious that medicine is based largely in biochemistry and physiology. One problem in making a basic science relevant is that most scientists are absorbed in their fields of endeavor and know little about medicine. On the other hand, few physicians know enough about basic science to make the coupling. My purpose here is to begin each chapter with the presentation of a disease illustrative of a biochemical subject. If I successfully start with a disease process, describing the molecular basis of the disease in biochemical terms, the relevance factor should be satisfied. Knowing the abnormal biochemistry explaining the disease process automatically leads to a discussion of the normal processes; thus, human biochemistry can be introduced in the context of disease. This should peak the interest of the medical student and illustrate the roles of biochemistry in clinical problems.

Most biochemistry graduate students know little about disease. This approach should be an interesting introduction to aspects of medicine. The material presented here can be mastered easily by the instructor in biochemistry and will allow biochemistry graduate students to learn more about disease processes.

Undergraduates, especially those contemplating a career in medicine or in medically related research, will find this a useful book. The instructor who wants to base a course on this book will undoubtedly attract an excited body of students. The text is not daunting; any interested instructor should be able to organize lectures of interest to undergraduates.

Integration of Biochemistry with Clinical Medicine

In general, each case of a disease is an example of a disruption of a normal biochemical process. So, I begin each biochemical topic with a disease that represents an abnormality of that process. Thus, the subject of proteins, for example, can be launched with the introduction of prion disease, an abnormal process of protein conformation. The medical explanation of the disease will be made first, progressing to the molecular events. This should raise the question, If this process is abnormal, what is the normal process? This automatically leads into the basic human biochemistry of proteins.

Usually, biochemistry textbooks start with proteins or carbohydrates. I chose to begin with proteins because virtually all enzymes are proteins, which make it possible to achieve complex chemical reactions under the restricted conditions of temperature, pressure, and so on, imposed by the human body. Thus, the reactions in carbohydrate metabolism, lipid metabolism, and those of proteins, nucleic acids, and so on, are all catalyzed by enzyme proteins. Then the study of enzymology becomes important to understand biological catalysis. And although the basic information is provided on the principles of enzymology, the chapter in this book discusses clinical enzymology to illustrate the usefulness of the basic information as it applies to the diagnosis of disease. In selecting disease entities to reflect specific biochemical topics, I tried to choose examples that medical students may eventually see in the hospital or in their practice or those of importance in world health. In a few cases, I resorted to a disease that may not be widely prevalent but may be the best representative of the biochemical topic under discussion. In some cases, there are prevalent diseases about which the biochemical and molecular understanding may be rudimentary still. I avoided their discussion, but they are mentioned in some context, although this is not meant to be a textbook on the introduction to medicine.

There are many figures in this book, which are rendered for the purpose of explaining pathways and to give an overall picture of what is happening. Some figures are highly simplified to serve as an introduction and to provide a snapshot of the "forest," introducing the concept before proceeding to the discussion of individual steps in a pathway or process. Sometimes detailed figures tie together a lot of information and give it some perspective. These figures may serve the research-oriented student.

General Considerations

The Human Body and Organ Systems

Figure 1-1 shows Leonardo da Vinci's famous Vitruvian man drawn in 1492. It would seem difficult to improve much on this drawing. From here we can break down the human organism into organ systems. The human body is extremely complex when we consider the one-celled

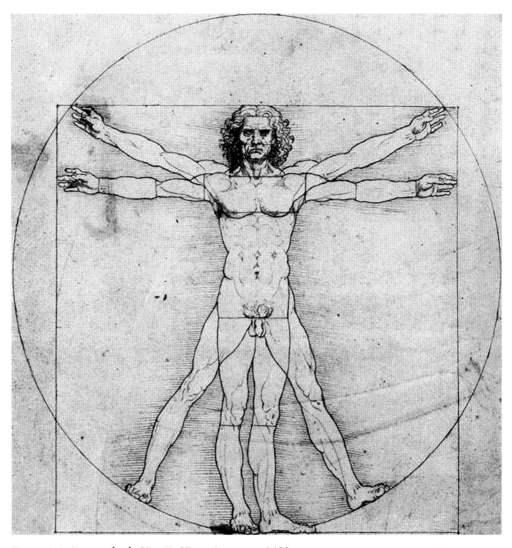

Figure 1-1. Leonardo da Vinci's Vitruvian man, 1492.

organism; in this unicellular form of life, all changes externally and internally are immediately sensed by the cell and it contains all information necessary to react or adapt to changes in these environments. Humans, on the other hand, has developed separate organ systems to carry out specific functions. Because these systems are spatially separated from one another, a large and complex system of communications has evolved so that one part of the body may sense what is going on in another part of the body and various mechanisms can be called into play when needed. All organs are composed of individual cells, and the communications, metabolic responses, etc., are all occurring at the cellular level. Those individual cellular responses summate to produce the organ responses and, in turn, the whole organism's response.

In all, there are 10 organ systems: integumentary (enclosure of the body as skin and its layers), cardiovascular, nervous, endocrine, blood and lymphatics, respiratory, skeletal-muscular, alimentary, urinary, and reproductive. All of these systems are included in the context of human biochemistry and disease in this text. Some of these systems are pictured in Figure 1-2 just to give a sense of them in the bodily context, information you already know intuitively.

In embryonic life, about 2 weeks after conception, cell division is occurring rapidly and the cells arrange into three germ layers: the outer layer, or **ectoderm;** the inner layer, or **endoderm;** and the middle layer, or **mesoderm.** All organs in the body derive from these three layers of cells. Tissues deriving from the ectoderm are skin (epidermis), the linings of all hollow organs whose surfaces are covered by epidermis, nerve tissue, salivary glands of the nose and mouth, and some epithelial tissues. The epithelial tissues derived from the endoderm (beside

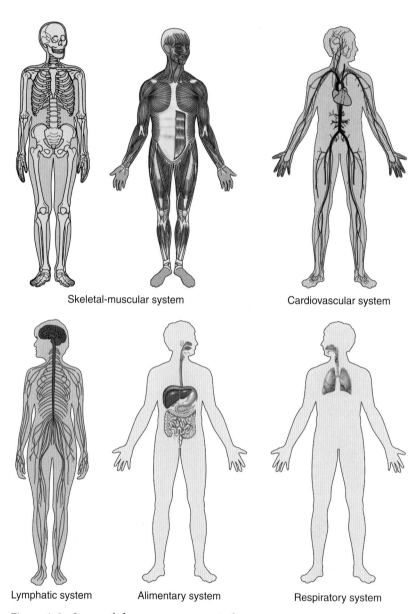

Figure 1-2. Some of the organ systems in humans.

those epithelial tissues deriving from the ectoderm) are the epithelial lining of the digestive tract (except for the open ends and the linings of the hollow structures that are outpockets of the digestive tract); the liver parenchyma, including connecting ducts; the linings of the lungs and the connection between the pharynx and the lungs; the bladder and urethra epithelia; and secretory glands of the digestive tract. From the mesodermal layer are derived blood cells, vessels for blood and lymph, adipose tissue, bone and cartilage, fibrous tissues, and muscle (see Figure 1-3). Most biochemistry takes place at the cellular level, and we need a clear understanding of the structure and functions of various parts of the cell. At the outset, biochemistry moves into the realm of the individual cell and away from the totality of the organ system; to understand the whole, we have to understand its parts.

The Cell

Most biochemical reactions take place inside a cell. Human cells have been highly differentiated to carry out specific functions, and these functions differ in cells in one organ compared to those in another.

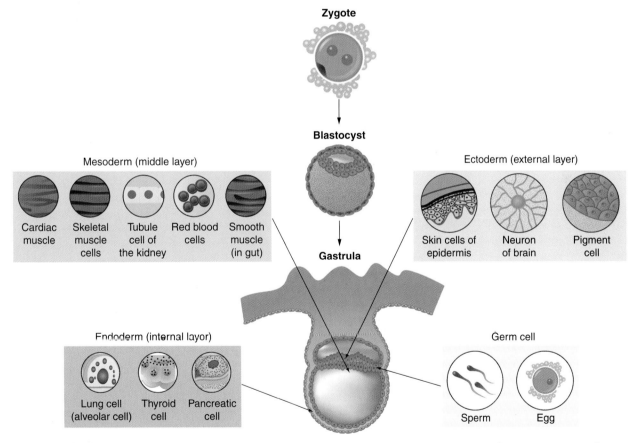

Figure 1-3. Illustration of the three embryonic layers of cells destined to give rise to specific tissues, some of which are shown. Redrawn from http://www.nhci.nlm.nih.gov/About/primer/genetics_cell.html.

Differentiation is achieved by the expression, during development, of specific genes and the repression of other genes that are not needed in the specialized cell. There are more than 200 kinds of cells in the body. A model of a typical single cell is shown in Figure 1-4.

Cells that have a nucleus are distinguished from unicellular organisms that do not have a nucleus (although they have organized deoxyribonucleic acid, or DNA). Thus, all cells in a human have a nucleus—with exception of the red blood cell, which functions without a nucleus (more about this later). Although all cells contain the same genes, they can be different in their functions because of the specific array of genes, of the total genome, that are expressed. They mostly share many structures.

Cell Membrane

The cell membrane (sometimes called the plasma membrane) envelops the cell and consists of a double layer of lipids with polar head groups facing outside. The bilayer is punctuated with transmembrane proteins that extend from the cell cytoplasm on the inside, through the membrane bilayer, and extend again on the outside. At either end of the membrane (looking through cross section), the internal lipids end on the outside and on the inside with charged polar-head groups; the lipid material between the polar head groups is nonpolar (uncharged).

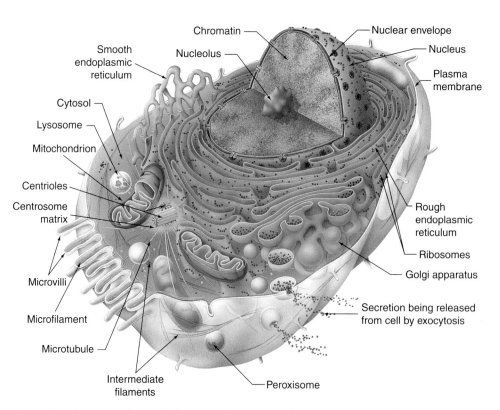

Figure 1-4. Cartoon of a typical liver cell. Redrawn from Department of Genetics, University of Minnesota, http://www.cbc.umn.edu/~mwd/cell.

Figure 1-5 provides a diagram of a section of a typical membrane. The polar head group usually consists (from the outside inward) of an ammonium group (NH_3^+), a hydrocarbon portion, and a phosphate group. The phosphate is connected (going toward the interior of the cell membrane) to glycerol, for example. This polar group connecting to the nonpolar chain constitutes one member of the two inner layers of the membrane and is referred to as a leaflet. This is envisioned in Figure 1-6. The bottom leaflet extending toward the cell interior would be a mirror image, thus constituting the double cell membrane shown in Figure 1-5. The membrane also contains cholesterol, a key ingredient, and it is present in an approximately one-to-one ratio with each phospholipid molecule (polar lipid molecule). Cholesterol is tucked between the hydrocarbon portions of the cell membrane. Its presence decreases the extent to which the bilayer can be deformed, inasmuch as the membrane itself is flexible and the presence of cholesterol lends some rigidity to the wall. Cholesterol also allows the "solubility" of nonpolar (uncharged) compounds as they pass from the outside of the cell, across the membrane to the cell interior. Otherwise, in its absence, nonpolar compounds might come out of solution in the form of damaging crystals within the membrane structure. Membranes also contain glycolipids that have sugar groups extending to the outside of the membrane. These groups serve specific biological functions and can be involved in the binding of extracellular proteins to the cell. Figure 1-7 presents this more thorough view of the membrane, including the various components discussed earlier. The cell membrane is not just a static structure to envelop the cell; rather, it contains many functional moieties, including those already mentioned and **receptors** of all kinds, mediating extracellular chemical messages, recognizing antigens, signaling infection by viruses, transporting small molecules (like glucose and amino acids, depending upon the tissue), and containing channels for the regulated transport of ions either by unidirectional import or by exchange for an internal ion or small molecule (called an antiporter). Later in this chapter there is a brief discussion of the cellular locations of many of these receptors.

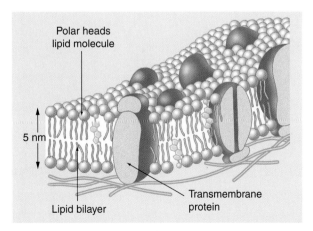

Figure 1-5. Drawing of a section of a cell membrane showing the lipid bilayer punctuated by transmembrane proteins. Redrawn from http://www.cytochemistry.net/cell-biology/membrane_intro.htm#Architecture, page 4.

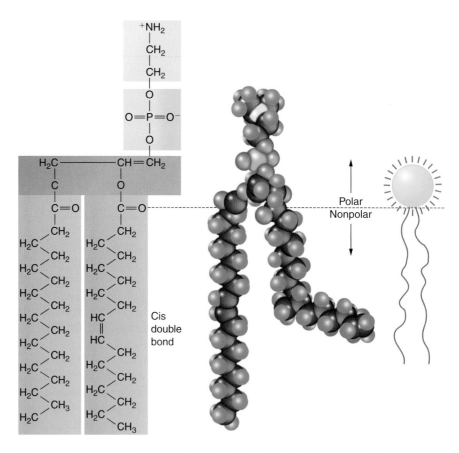

Figure 1-6. Structure of half of a cell membrane leaflet showing a polar group (outside) and extending to the center of the cell membrane bilayer through a nonpolar hydrocarbon structure. The bottom leaflet would be a mirror image, completing the double layer membrane.

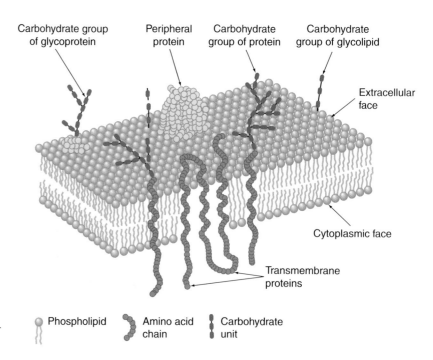

Figure 1-7. Depiction of a portion of the cell membrane showing inclusions of transmembrane proteins and substituent groups.

8   CHAPTER 1   Introduction and General Considerations

Nucleus

The nucleus resides somewhere in the middle of the cell cytoplasm (see Figure 1-4) and it contains the **genome** (the genes that make up the organism). The human genome consists of up to 3 billion base pairs (see Chapter 6 on nucleic acids) and up to 32,000 genes. Some individual genes may give rise to more than one protein, so a cell could contain up to 40,000 different proteins, although that number is still a guess. The nucleus is the site of gene expression, where signals coming into the nucleus through the cytoplasm evoke the expression (encoded by those signals) of certain genes or the repression of certain genes, depending on the signal. The result of a specific signal can be both expression in the form of **messenger ribonucleic acid (mRNA)** (produced from each gene expressed) and repression of specific genes. So the nucleus is really the command center for the cell, and it is under stimulation or inhibition by many factors, usually those that are capable of binding directly to the gene's DNA and causing the transcription of a specific gene to occur. These are called transcription factors. Mechanisms for these activities will be discussed later in chapters on proteins and nucleic acids. However, the **mitochondrion** also contains genetic material, which determines the expression of many of the components of that organelle. The mitochondria are visited later in this chapter when the cell cytoplasm is considered.

The nucleus is contained by a nuclear membrane that resembles the cell membrane. The nuclear membrane has many **nucleopores** (or **nuclear pores**) through which materials from the cytoplasm are transported, principally transcription factors (which are proteins) whose actions depend on their ability to interact with genomic DNA (specifically with gene promoters). Materials from the nucleus are also transported out into the cytoplasm by way of these nucleopores. Shortly, the discussion of nucleopores is expanded; this topic is of major importance because these structures are required to permit signals from the cytoplasm to reach the interior of the nucleus. A good view of the nucleus showing nucleopores and structures within the nucleus is afforded by Figure 1-8. The nucleus also contains its own cytoplasm called **nucleoplasm** (shortly, there is a discussion of the cytoplasm). The nucleus contains **chromatin** and

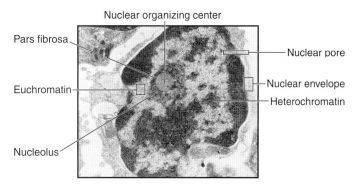

Figure 1-8. Structure of the cell nucleus. Reproduced from H. Busch, *The Cell Nucleus*, Academic Press, 1974.

nucleoli (one of which is called a **nucleolus**), which constitute the main structures within the nucleus.

Chromatin

Chromatin is made up of chromosomes that contain genetic information (genes). **Heterochromatin** describes condensed chromatin, which looks fairly dense under the electron microscope. Because chromatin needs to be unwound or opened to receive signals (such as transcriptional activation factors) from the cytoplasm, heterochromatin is transcriptionally inactive. On the other hand, **euchromatin** is the opened form of chromatin, is more stringy and less dense, and is characteristic of active transcription (producing mRNA from genes). A good picture of these two forms of chromatin is shown in Figure 1-9A, and the stringy or opened chromatin (obtained artificially by removal of histones) is magnified in Figure 1-9B. Although this picture is obtained by treating artificially dense chromatin, it gives an idea of what euchromatin looks like when it is actively transcribing mRNA from genes. The structure of the chromosome in its more native condition, with histones bound to it, is

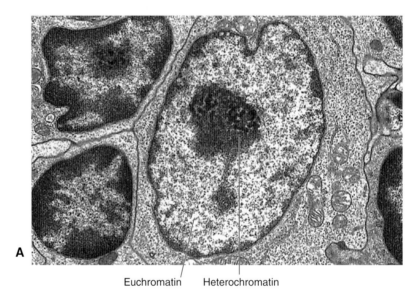

Figure 1-9. **A.** A portion of a cell nucleus visualized in the electron microscope using fixed ultrathin sections of tissues. Two forms of chromatin are evident: the dense chromatin characteristic of nontranscriptionally active genetic material (heterochromatin) and the more fiberlike, opened chromatin (euchromatin) characteristic of transcriptionally active material. **B.** A view of chromatin uncoiled by removal of histones (dextran sulphate and heparin) as seen under the electron microscope. Unraveled chromatin is shown in the *upper right inset.* Although artificially derived, euchromatin will resemble these open fibrous structures. Part A reproduced from http://www.bu.edu/histology/p/20104ooa.htm. Part B reproduced from http://cellbio.utmb.edu/cellbio/nucleus2.htm as modified from Bloom and Fawcett, *A Textbook of Histology,* 12th ed., Chapman and Hall.

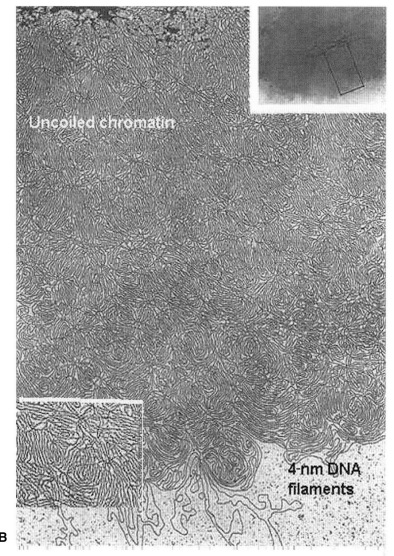

Figure 1–9, cont'd

visualized in Figure 1-10. Nucleosomes periodically occur on the DNA. Recent evidence suggests that there may be information in the DNA to locate the nucleosomes.

The Nucleolus

This nuclear structure is visualized in Figure 1-11. A nucleus may contain a single nucleolus or as many as four nucleoli. Although the nucleolus is a dense structure containing proteins, ribosomal particles, and lipids, it may not have a definite membrane surrounding it but rather may seem to surround chromosomes that contain the genes encoding ribosomal RNA (rRNA). It is the principal site of rRNA synthesis. Thus, there are hundreds of rRNA genes found on nucleolar-organizing regions

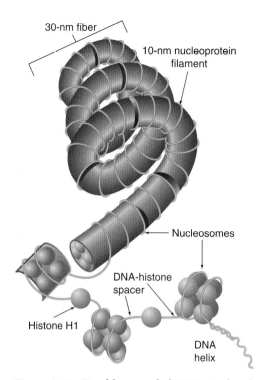

Figure 1-10. Double-stranded DNA in the chromosome. At the bottom of this drawing the DNA is uncoiled to show the attached histone (positively charged) proteins. H1 histone is bound in the linker region between nucleosomes. Each nucleosome contains eight histones.

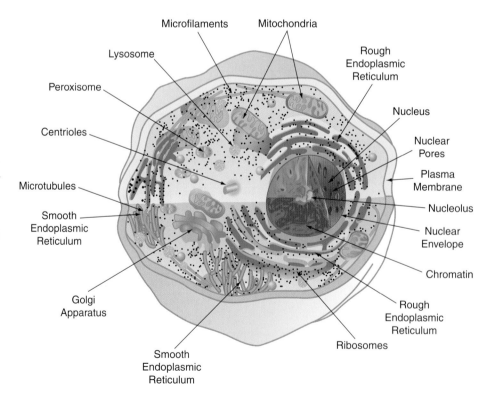

Figure 1-11. Transverse section of a typical cell showing the nucleus *(center right)* with a prominent nucleolus. In the nucleolus, the components of the ribosome are manufactured. Ribosomal genes are transcribed by a resident RNA polymerase. The final products are translocated out of the nucleus through nucleopores in the nuclear membrane and are moved to the cell cytoplasm for assembly into ribosomes. Reproduced from former website http://homepage.emc.edu/wissmann_paul/cell/Default.htm.

12   CHAPTER 1   Introduction and General Considerations

at different positions on chromosomes. Ultimately, rRNA synthesized in the nucleolus is translocated to the nucleoplasm and then to the cell cytoplasm through the nucleopores for assembly into ribosomes, essential for protein synthesis.

The Nucleus in Cell Division

Cell division is subject to the appropriate stage of the **cell cycle,** an outline of which is shown in Figure 1-12. After one round of cell division (mitosis), emerging smaller cells proceed to the Gap 1 (G_1) phase (interphase, the longest phase, usually containing most cells), where they increase their pool of **adenosine 5'-triphosphate (ATP)** and increase in size. Other cells that are no longer dividing move to a G_0 phase, where they await appropriate stimulation to enter the cycle again. Cells that continue into the S phase undergo DNA synthesis, replicating copies of the original DNA, and then move into G_2. Cells in this phase, after having engaged in energy-requiring DNA synthesis in the S phase, recover by acquiring energy (ATP) and grow. DNA synthesis results in the duplication of the chromosomes (Figure 1-13A) so that one set is retained by the parental cell and one is acquired by the daughter cell. Cells in this stage are now ready to divide again and undergo mitosis. In mitosis, the daughter cell splits off (cytokinesis) from the progenitor to form two equal but somewhat smaller cells. The regulation of the cell cycle is complex and is governed by "checkpoints" at which certain growth factors are required to pass the checkpoint and enter the next phase (Figure 1-13B). As shown, there are 3 major checkpoints.

Before cell division, while still in the interphase, nuclear chromatin condenses into chromosomes. These chromosomes can be spread (metaphase spread), stained so that they show characteristic banding patterns, and visualized in the electron microscope (for example, as shown in Figure 1-14A). This technology has become highly developed, and individual genes now can be localized to specific chromosomes. Human cells contain a set of

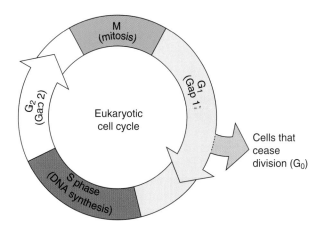

Figure 1-12 Diagram of the cell cycle.

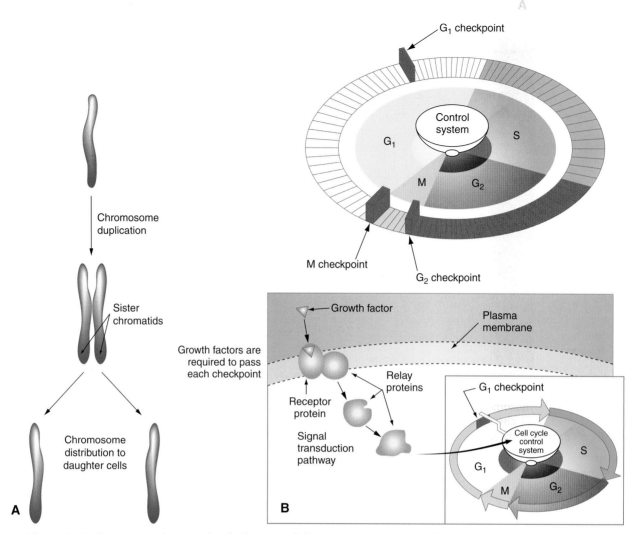

Figure 1-13. **A.** Drawing showing the duplication of chromosomes prior to cell division. **B.** Cell division cycle showing the location of three checkpoints requiring specific growth factors in order to pass on to the next phase.

23 chromosomes inherited from each parent, as shown in Figure 1-14B. After chromatin condensation, the nucleolus disappears. **Centrioles** begin moving to opposite ends of the cell, and fibers appear in the **centromere.** The **kinetochore** is a point of attachment for microtubules of the spindle. The kinetochore is visualized in Figure 1-15. This configuration allows the chromosomes to begin moving to the middle of the cell nucleus in an orderly fashion in preparation for division. When chromatin begins to condense and the nucleolus disappears, this is called **prophase.** When the kinetochores are formed and begin to move to the central latitude of the cell, this is called **prometaphase.** Next, the chromosomes are arrayed around the center of the cell by spindle fibers in **metaphase. Anaphase** ensues when the paired chromosomes dissociate and each member of an identical pair moves to opposite poles of the cell. When all chromatids arrive at each pole, a membrane forms around what will become the daughter nucleus; this is called **telophase.** Finally, cytokinesis occurs when actin at the center of the cell contracts, causing the separation of the two daughter cells. These events are summarized pictorially in Figure 1-16.

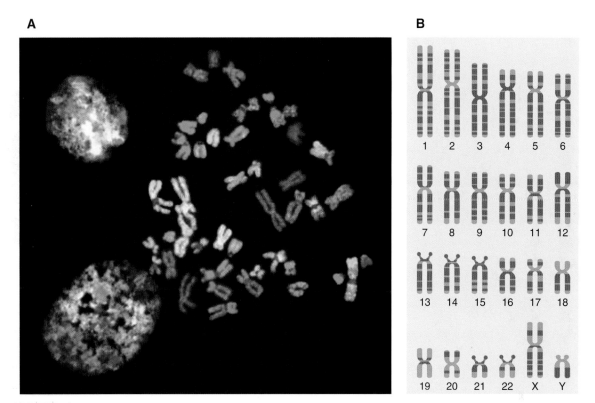

Figure 1-14 **A.** Metaphase chromosomes as viewed under the electron microscope. **B.** Human chromosomes; 23 from each parent. X and Y are the sex determining chromosomes. The remaining 22 chromosomes from each set are autosomes. Part A reproduced wth permission from http://science.cancerresearchuk.org.

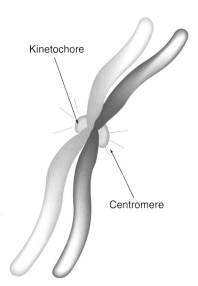

Figure 1-15. A pair of chromosomes attached to the kinetochore. This structure allows chromosomes to move to the center of the cell as shown in Figure 1-14.

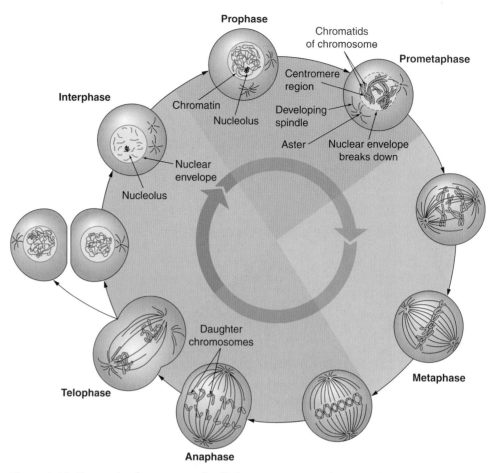

Figure 1-16. Events in the process of cell division, starting from *top/center*. Refer to the text for the actions in each phase.

The Nuclear Membrane

Figures 1-8 and 1-9A show the nucleus and the nuclear membrane. There are two membranes composing the nuclear membrane, not unlike the cell membrane. In these figures, ribosomes can be seen as dots on the cytoplasmic surface of the membrane. The two membranes associate at the nucleopore, many of which dot the surface of the nuclear membrane. Remember that the nucleopores were mentioned earlier as the principal means by which signals, some in the form of proteins, are transferred from the cytoplasm to the interior of the nucleus. The outer membrane of the nuclear membrane is continuous with **rough endoplasmic reticulum,** as seen in Figure 1-8 (lower right). The inner nuclear membrane is stabilized by a layer of thin filaments **(nuclear lamina);** however, these are not present in the location of the nucleopore. A nucleopore occurs at an interruption of the double nuclear membrane. Pores have been visualized in the electron microscope as shown in Figure 1-17. Models of the pore have been developed, and one is shown in Figure 1-18. Molecules up to 44,000 molecular weight can diffuse into the pore, although the larger sizes require more time. Molecules over 60,000 molecular weight and 10 nm in diameter cannot diffuse into the pore and require energized transport. This is because they will have exceeded the diameter of the pore. There are a group of

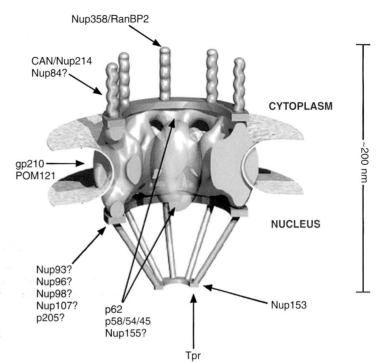

Figure 1-17. Diagram of a nucleopore. Reproduced from http://www.npd.hgu.mrc.ac.uk. G. Dellaire, R. Farrall, and W.A. Bickmore, "The Nuclear Protein Database (NPD): Sub-nuclear localization and functional annotation of the nuclear proteome," *Nucl. Acids Res.*, 31: 328–330, 2003.

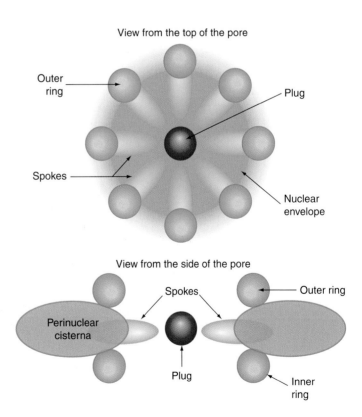

Figure 1-18. A model of the nucleopore, showing eight subunits that attach to the inner and outer nuclear membranes. The diameter of the ring of subunits is about 15 to 20 nm. The subunits project spokes into the center plug.

General Considerations 17

proteins in the cytoplasm called **nucleoporins,** which attach to specific proteins being transported into the nucleus and facilitate the transport through the nucleopore energetically by the hydrolysis of ATP (a protein in this complex has the ability to catalyze this hydrolysis).

Cytoplasm

The cytoplasm fills the volume between the nuclear membrane and the cell membrane and contains a variety of organelles. To visualize this space, refer to Figure 1-4. Sometimes the soluble cytoplasm, exclusive of all its particulate matter and organelles, is referred to as cytosol. Operationally, the cytosol is obtained by centrifuging out all insoluble matter from the cytoplasm; the supernatant would be the cytosol. The cytoplasm is aqueous and exists as a gel. In it are suspended various particulate compartments, such as the mitochondria, ribosomes, rough endoplasmic reticulum, and components of the skeleton (which gives shape, supplies platforms for particles, and provides channels for transport of proteins). The many particulates are visible in Figure 1-4. Despite the aqueous nature of the cytoplasm, its structures are generally not freely floating about the cell but are positioned mainly by components of the cytoskeleton. A discussion of some major components of the cytoplasm follows.

Mitochondria

The mitochondria are structures that appear throughout the cytoplasm as shown in Figure 1-4. Better views are shown in Figure 1-19A and B. A mitochondrion resembles a bacterium in size, and a popular theory is that mitochondria have a bacterial origin emanating from some kind of a symbiotic event. The number of mitochondria in a cell usually corresponds with the need for energy production by the tissue in which the cell is located. Thus, a muscle cell would be expected to have more mitochondria per cell than a skin cell. Mitochondria are able to multiply. When a mitochondrion becomes large, it may divide into two daughter mitochondria (in a manner similar to that of the cell nucleus). Mitochondria have their own complement of DNA, transcriptional, and protein synthetic machinery so that they are able to produce some protein subunits needed within the mitochondrion. Other mitochondrial protein subunits are encoded in genes within the cell nucleus, and their messenger RNAs are translated in the cytoplasm and resulting proteins are transported into the mitochondrion.

The mitochondrion serves the critical function of supplying energy to the cell by the metabolism of fuels to generate energy in the form of ATP molecules (from adenosine diphosphate, or ADP), carbon dioxide, and water. This function is called the **tricarboxylic acid cycle** (TCA cycle). (It is sometimes called the **Krebs cycle,** after its discoverer, or the **citric acid cycle.**) It can function only in the presence of oxygen; thus, this is also referred to as aerobic oxidation. The major substrate for the TCA cycle is pyruvic acid, which is formed from the metabolism of glucose in the cytoplasm by anaerobic glycolysis. This means that the conversion of glucose to pyruvic acid in the cytoplasm can be carried out in the absence of oxygen and possibly anaerobic metabolism (**glycolysis**) was occurring

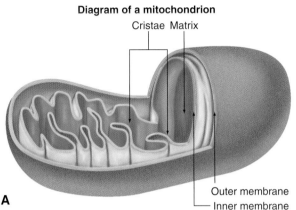
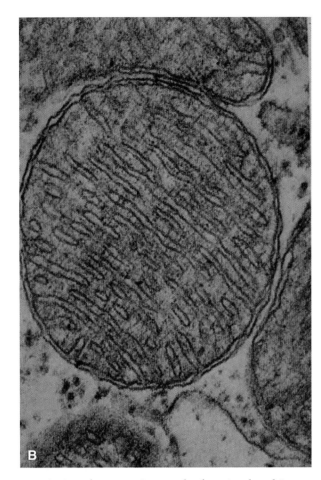

Figure 1-19. **A.** Diagram of a typical mitochondrion. **B.** A transmission electron micrograph of a mitochondrion. Part B reproduced by permission from http://137.222.110.150/calnet/cellbio/page4.htm.

in some organisms before oxygen became plentiful on the planet. There are many enzymatically catalyzed reactions in these two adjoining pathways; these are discussed later, but a simplified overview is shown in Figure 1-20. Energy generated as ATP is formed in the respiratory chain in the mitochondrion, terminating in the action of ATP synthase, an enzyme that catalyzes the conversion of ADP to ATP, thus capturing the energy developed by the TCA cycle and in the associated respiratory chain (consisting of NADH dehydrogenase, cytochrome c reductase, and cytochrome c oxidase, together with ATP synthase) into the high-energy terminal phosphate group of ATP. Four ATPs are formed in glycolysis, converting one molecule of glucose to two molecules of pyruvic acid, but two of these are used in reactions so that the net is two ATPs. In addition, 24 to 28 ATPs generated from the use of two pyruvic acids through the TCA cycle net about 30 ATPs (from ADP) molecules, entrapping energy in the terminal phosphate group, in the complete metabolism of one molecule of glucose.

The genetic component of the mitochondrion was mentioned earlier. It contains 5 to 10 identical circular molecules of DNA encoding information for rRNA, 22 transfer RNAs, and 13 proteins. The proteins are subunits of enzymes located in the inner membrane that include NADH

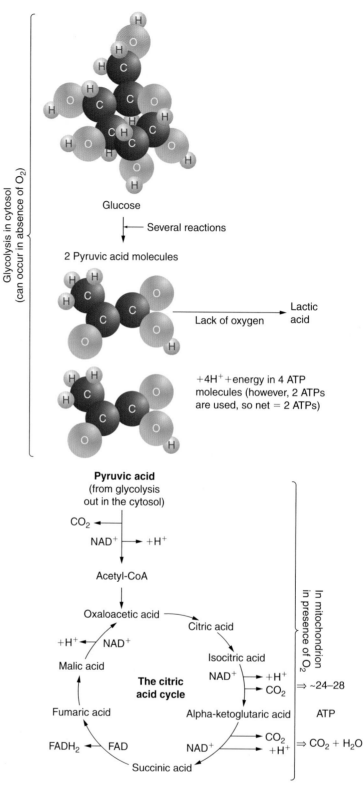

Figure 1-20. Coupling of glycolysis (to produce pyruvic acid from glucose) and the tricarboxylic acid cycle (also citric acid cycle) to produce CO_2, water and energy in the form of 24 to 28 ATPs plus 2 ATP netted from glycolysis to give about 30 ATP molecules.

dehydrogenase, cytochrome c oxidase, and ATP synthase. The subunits synthesized in the mitochondrion, however, are not sufficient to complete the structures of these enzymes. The missing subunits are encoded in the cell nuclear DNA, whose mRNAs are translated in the cytoplasm. The finished protein subunits are translocated into the mitochondrion to complete the structures of the mitochondrial proteins. Mutations in mitochondrial DNA appear to account for several neurological diseases, so this compartment is of great interest in neurobiochemical research. Also, the mitochondrion plays an important role in **apoptosis** (also known as **programmed cell death**), a system critical, among others, to the killing of cancer cells. This general topic is discussed later.

Rough Endoplasmic Reticulum, Smooth Endoplasmic Reticulum, and Ribosomes

The rough endoplasmic reticulum (RER) is pointed out in Figure 1-4. The term refers to a network of interconnected membrane-bound sacs located in the cell cytoplasm. **Smooth endoplasmic reticulum (SER)** refers to the sac structure alone (without attached ribosomes), whereas RER is the SER with ribosomes attached facing out to the cytoplasm (not toward the inner side of the sac). The entire fraction of the cytoplasm that contains the ribosomes and the endoplasmic reticulum is called the **microsomes** when separated by centrifugation. The events occurring on the ribosomes depend upon factors coming from the cytoplasm. Pictures of the endoplasmic reticulum are shown in Figure 1-21. The RER, but not the SER, is involved in protein synthesis. Proteins synthesized in the RER are directed inside the RER by a signal recognition particle that binds to a signal sequence at the N-terminus of the growing polypeptide chain. This sequence of events ensures the deposition of the protein within the RER in preparation for fates discussed later. In the SER, there are activities of cholesterol metabolism, membrane synthesis, detoxification, and calcium ion storage. Details of ribosomes and protein synthesis are presented in Chapter 2.

Golgi Apparatus

The Golgi apparatus is named for Camillo Golgi, an Italian physician and biologist. It is a saclike structure, similar to the SER, and is in intimate relationship to the RER. It modifies lipids and proteins and stores materials that will be exported from the cell in vesicles. These vesicles separate from the Golgi by budding and move to the cell membrane, where the material inside the vesicle is released to the cell exterior. The movement of these vesicles is an energetic process and requires ATP. Some of these vesicles will become lysosomes. A picture of the Golgi apparatus is shown in Figure 1-22. Like the nucleolus, the Golgi disintegrates at the beginning of mitosis but reappears at telophase (Figure 1-16). The intimacy of the Golgi with the RER and the functional aspects of the Golgi are shown in Figure 1-23A and B. Proteins synthesized by the RER are exported by budding to the Golgi, where these vesicles fuse to the Golgi. Ultimately, modifications of the proteins may occur in the Golgi; the finished proteins are packaged into vesicles, separated from the Golgi by budding, and transported to the cell membrane, where they fuse. Eventually their contents are released outside of the cell, as shown in Figure 1-23B. Some

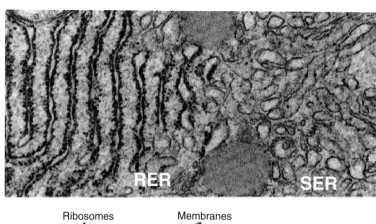
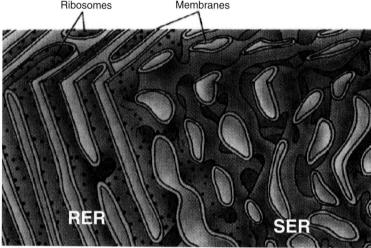

Figure 1-21. Micrograph *(top)* shows the rough endoplasmic reticulum (RER) and smooth endoplasmic reticulum (SER). Ribosome-studded sacs are clearly visible in the RER. At the *bottom* is an idealized picture of these two forms of the endoplasmic reticulum. Reproduced from http://academic.brooklyn.cuny.edu/biology/bio4fv/page/rougher.htm.

vesicles can form at the cell membrane, housing materials from outside the cell, and these **endocytic vesicles** move into the cell cytoplasm. Once in the cytoplasm, they can fuse with organelle membranes or remain as storage sites.

Lysosomes

As mentioned earlier, some vesicles budding off from the Golgi apparatus can form **lysosomes** (Figure 1-23B). Usually, vesicles that contain hydrolytic enzymes (capable of breaking down proteins and other large molecules) are destined to become lysosomes. Moreover, these structures contain an ion pump that, through the hydrolysis of ATP, pumps in hydrogen ions (protons, H^+) to maintain the internal **pH** around 5.0, which is highly acidic and creates an environment that maximizes the activity of its resident 40 or so hydrolytic enzymes for the purpose of digesting macromolecules that enter the cell (Figure 1-24). The products of the enzymatic digestion can diffuse out of the lysosome into the cytoplasm for use by the cell. Organelles that have aged, and even the cell when it

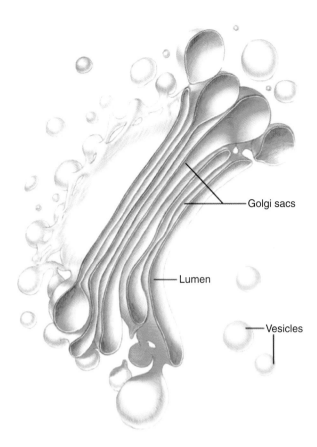

Figure 1-22. An idealized drawing of the Golgi apparatus. Vesicles budding from the ends of the sacs are shown, and individual vesicles are seen moving toward the unseen cell membrane, eventually to export the materials contained within the vesicles. Reproduced from http://ntri.tamuk.edu/cell/golgi.html.

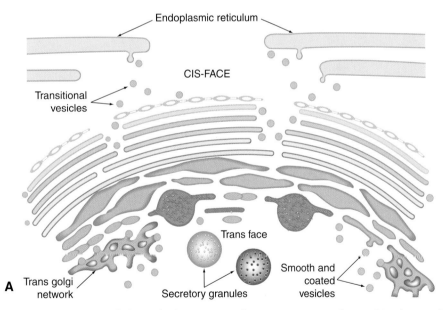

Figure 1-23A. **A.** Interface between the Golgi apparatus and the endoplasmic reticulum. Proteins, synthesized in the rough endoplasmic reticulum are transported to the Golgi as transitional vesicles. In the Golgi, proteins may be modified and packaged as secretory granules destined for export to the cell surface. **B.** The rough endoplasmic reticulum (RER)-Golgi showing proteins in vesicles transported from the RER to the Golgi *(top left)*, fusion with the Golgi and processing in the Golgi to be packaged in secretory vesicles *(bottom right)* and exported through the cell membrane. As seen in the lower center, some vesicles may form into lysosomes. Part A redrawn from http://www.cytochemistry.net/cell-biology/golgi.htm. Part B redrawn from http://www.cytochemistry.net/cell-biology/golgi.htm.

Cont'd

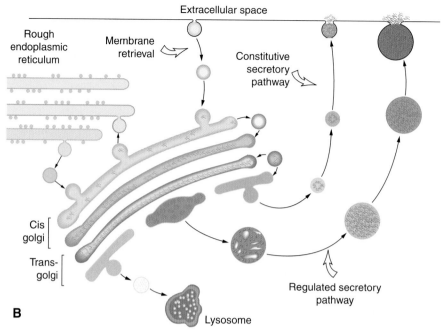

Figure 1-23, cont'd

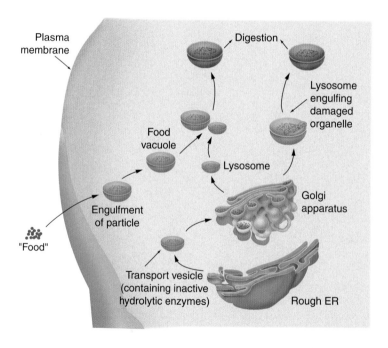

Figure 1-24. Lysosome formation and function. Particles from outside the cell are engulfed in the cytoplasm *(left),* fused with a lysosome and the material is digested by the hydrolytic enzymes in the lysosome. Redrawn from former website http://www.people.virginia.edu/~rjh9u/lysosome.html.

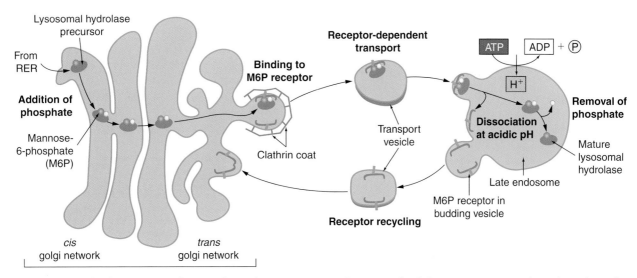

Figure 1-25. The lysosome pathway. The Golgi apparatus is shown on the left. Mannose-6-PO₄ (M6P) is a ligand for the M6P receptor. After binding to the receptor, a vesicle is pinched off from the Golgi containing a lysosomal hydrolase precursor. The receptor-containing vesicle is transported to a lysosome and bound and taken up by the lysosome. Inside the lysosome, phosphate is removed from the hydrolase precursor that activates it and it dissociates into the lysosome. Another vesicle can bud off from the lysosome, containing the empty M6P receptor. This vesicle then travels back to the Golgi and fuses with it to complete the cycle.

dies and turns over, may be aided by the ability of the lysosomes to digest macromolecules and particles. Figure 1-25 shows the lysosome pathway. This pathway is more intricate than what has been shown so far. Figure 1-25 shows the pinching off of a transport vesicle containing a hydrolytic enzyme precursor, from the Golgi apparatus dependent on the binding of mannose-6-phosphate (ligand for the M6P receptor). Once the vesicle fuses with the lysosome, the hydrolase enzyme becomes activated by dephosphorylation, the active enzyme remains in the lysosome, and the empty vesicle is returned to the Golgi apparatus containing the empty receptor, thus demonstrating how the enzyme component of the lysosome can be established.

Peroxisome

A picture of a peroxisome is shown in Figure 1-26, and a micrograph of a peroxisome is shown in Figure 1-27. Peroxisomes are one of a group of microbodies in the cell and measure approximately 0.1 to 1 µm in diameter. Within the peroxisome, organic molecules are degraded oxidatively to produce hydrogen peroxide (H_2O_2), which is converted to oxygen and water by a catalase enzyme. Peroxisomes contain about 50 enzymes, including catalase; urate oxidase and D-amino acid oxidase, which degrade uric acid and amino acids; and other enzymes that degrade long-chain fatty acids. Peroxisomes are present in large numbers in cells that are active in lipid biochemistry, such as in the liver. When peroxisomes increase in size they can divide, but they do not contain DNA and their division therefore differs from that of the nucleus or mitochondria. Proteins that enter the peroxisome contain a signal peptide that directs the protein to that organelle.

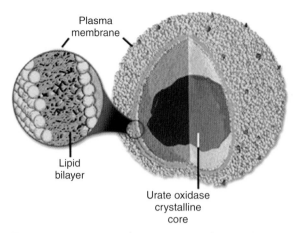

Figure 1-26. Picture of a peroxisome ("microbody"). It is spherical and has a single membrane. The peroxisome functions for detoxification.

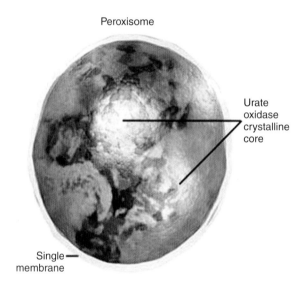

Figure 1-27. Illustration of a peroxisome showing the urate oxidase crystalline core and the single membrane surrounding the particle. Reproduced from http://www.palaeos.com/Eukarya/Lists/EuGlossary/Images/Peroxisome.jpg.

Receptors and Their Cellular Locations

Most receptors receive signals (ligands) that circulate into the extracellular space (from the bloodstream or from other cells in the region). Therefore, most receptors are located in the cell membrane of the target cell with a binding pocket open to the cell exterior. Receptors bind ligands: hormones, neurotransmitters, viruses, antigens, small molecules derived from amino acids, sugars, and many other molecules. An exception to this generality is the receptors for steroid hormones and other members of the nuclear receptor class, which includes thyroid hormone (curiously derived from an amino acid) receptor. These receptors are found in the cell interior, in the cytoplasm, and in the nucleus. In Figure 1-28, this idea is illustrated. In general, receptors either are transported to the nucleus and bind to DNA (promoting gene expression) or are associated with signal

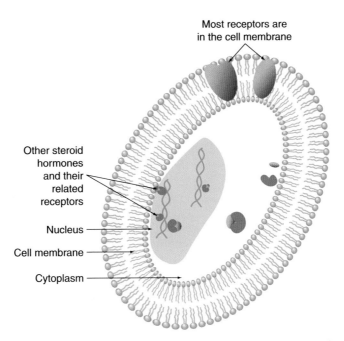

Figure 1-28. Diagram showing that most receptors are in the cell membrane with a ligand binding pocket open to the extracellular space. An exception is the nuclear receptor gene family, which are located in the cytoplasm and cell nucleus. Redrawn from Litwack & Schmidt, *Textbook of Biochemistry*, Wiley, 4th ed.

transduction pathways that culminate in changes mediated by the cell nucleus or in other locations in the cell. Receptors in the cell membrane may have multiple activities, sometimes opening **ion channels** with which they may have an intimate association. Receptors are certainly the key mediators in the process of communication between cells that may be quite distant in the body.

Biological Roles of Water

Water makes up 70% to 80% of our tissues, and life on this planet is impossible without it. The water molecule consists of one oxygen atom and two hydrogen atoms, with the oxygen being more electronegatively charged (the electron density around the oxygen atom is about 10 times that of the hydrogen atoms), resulting in a bond angle of 104.5 degrees, as shown in Figure 1-29. Because of the polar charges of the water molecule and its ability to form hydrogen bonds, it is found in close association with most biological molecules except in highly nonpolar (uncharged) regions. Thus, water plays an important role in the structure of proteins, nucleic acids, and lipids (especially polar groups). In many three-dimensional structures of proteins, for example, shown later, the association of water molecules is evident. Water can form hydrogen bonds with itself and with other molecules, and "hydration shells" form around ions.

Water recently has been found to flow into cells through aquaporin structures, which are channels, specific for water, in the membrane. They are opened by phosphorylation of specific residues and closed by dephosphorylation. Apparently, water molecules flow into the cell through the opened aquaporin channel oxygen first, facing the cytoplasm from the

General Considerations 27

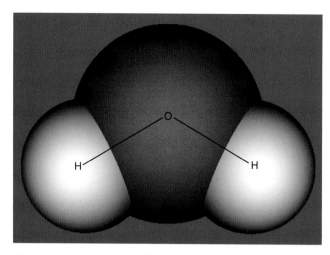

Figure 1-29. A theoretical model of the water molecule. The calculated length of O-H is 0.958 Å (Angstroms). The O-H length is slightly shorter in the gaseous form of water (vapor). The bond angles are likely to change in different hydrogen-bonded environments. Reproduced from http://www.lsbu.ac.uk/water/molecule.html.

outside. As they progress through the channel, water molecules invert so that the hydrogen atoms face the cytoplasmic side and the oxygen atom faces upward (Figure 1-30). When water leaves the cell through this channel, it leaves with the oxygen atom facing upward. The structure of the mammalian aquaporin channel, AQP1, has been solved recently and is shown in Figure 1-31. Humans have 10 different aquaporins located in cells of various tissues, such as kidney, brain, and lens of the eye. It is crucial for these channels to be tight and selective to maintain the electrical potential of cell membranes.

Ion Channels

Because the cell membrane is impermeable to ions, special devices have been developed for the transport of ions into cells in a highly regulated fashion. The ion channels, accomplishing this purpose, are similar in many ways to the aquaporins already described. First, ion channels are transmembrane proteins (Figure 1-32) and they are **glycoproteins** (containing sugar molecules). Ion channels regulate the electrochemical gradient across cell membranes. This gradient is a feature of the different conductance by different ions. Thus, the **cell membrane potential** is a consequence of the flow of these ions and the separation of charges by the nonconducting cell membrane. A more detailed view of a channel is shown in Figure 1-33. Generally, ion channels contain more than one subunit protein, and the subunits aggregate into a cylinder conformation that forms the pore. There are different mechanisms for opening and closing channels. One mechanism is phosphorylation–dephosphorylation. In the example shown in Figure 1-34A and B, the channel is opened by phosphorylation at the gate level and closed by dephosphorylation. Channels also can contain receptor proteins in their structure, and the function of the channel in gating, or in opening and closing, can be regulated by the binding of ligands (one example might be a neurotransmitter) to the resident receptor. An enzyme can be closely associated with a channel, and

Figure 1-30. Simulated movement of water molecules passing through an aquaporin channel single file. At first, from the extracellular side, water molecules are oriented with the oxygen facing downward (toward the cytoplasm). At the middle of the pore, the orientation reverses with the oxygen facing upward (toward the extracellular space). The extracellular space would be at the top of this figure and the intracellular space would be at the bottom.

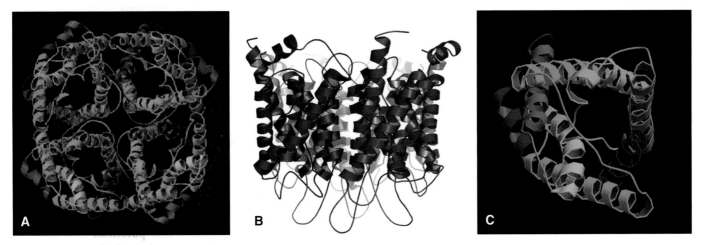

Figure 1-31 A. Three-dimensional structure of the aquaporin AQP1 tetramer looking down the pores from the cytoplasmic side. The monomer on the lower left is represented as a solid space-filling model. B. Side view of aquaporin water channel (AQP1). It selectively conducts water molecules while preventing the passage of ions and other solutes. The channel is composed of four identical subunit proteins. Water molecules pass through the narrowest part of the channel (constriction) in single file. These channels can increase water permeation into the cell by as much as 10-fold. Because of the absorption of water is passive, aquaporin cannot reverse the direction of the osmotic gradient driving the flow of water. There are 12 known types of aquaporins, and 6 of these are located in the kidney. AQP1 picture here, is one of the four most studied channels. Asymmetrical View of a helical ribbon diagram of the AQP1 tetramer atomic model. Part A reproduced from PDB ID: 1J4N. H. Sui, B.G. Han, J.K. Lee, P. Walian, B.K. Jap. Structural Basis of Water-Specific Transport through the AQP1 Water Channel. *Nature* **414** pp. 872 (2001). Part B reproduced from http://wikipedia.org/wiki/Aquaporin. Part C reproduced from PDB ID: 2C32. D.V. Palanivelu, D.E. Kozono, A. Engel, K. Suda, A. Lustig, P. Agre, T. Schirmer. Co-Axial Association of Recombinant Eye Lens Aquaporin-0 Observed in Loosely Packed 3D-Crystals. *J.Mol. Biol.* **355** pp. 605 (2006).

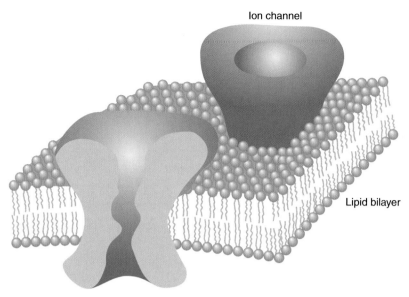

Figure 1-32. Cartoon showing the transmembrane character of an ion channel. Redrawn from http://hebb.mit.edu/courses/8.515/lecture1/sld013.htm.

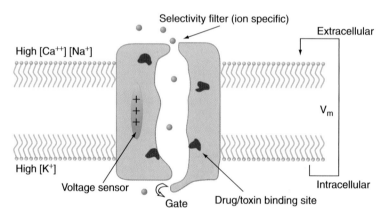

Figure 1-33. Model of a voltage-gated ion channel. The channel separates the extracellular domains, above, containing high concentrations of calcium and sodium ions from the intracellular domain, below, containing a high level of potassium ion. Channels are specific for the ion they conduct. A calcium channel is specific for calcium ion transport into the cell. A sodium channel is specific for sodium ion transport, and so on. The selectivity filter *(top)* selects the ion to be transported. The gate *(bottom)* regulates the flow of ions and is, itself, regulated. Redrawn from http://www.essen-instruments.com/electrophys/tutorial.asp#2.

the activity of the enzyme or its products can influence the action of a channel. This subject comes up later in the text.

pH

Of great importance to the cell is the concentration of hydrogen ions. Water is made up of hydrogen and hydroxyl ions (since $H_2O = H^+ + OH^-$). When the concentration of these two ions is equal, the solution is neutral, or at pH 7.0. When the concentration of H^+ exceeds

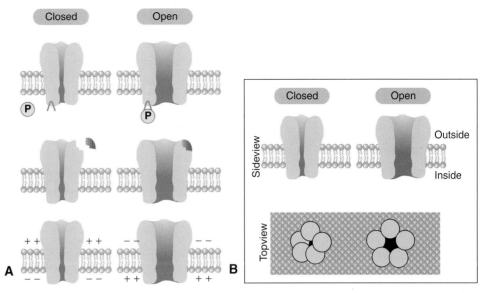

Figure 1-34. **A.** This shows in transverse section, the closed channel *(top left)* in the dephosphorylated form and the open channel in the phosphorylated form *(top right)*. Sometimes the gating can be controlled by the difference between total outer and inner charges. The middle of part A shows a ligand *(green)* binding to a receptor protein that is part of the channel structure. Such a process also could be a factor in control of the gating process. The bottom of part A shows that, as a consequence of channel opening, ions pass from the outside to the inside and the distribution of charges on both sides of the membrane are changed. **B.** This picture shows how the pore would look in the open and closed position as viewed from overhead *(bottom)*. Redrawn from http://www.ifisiol.unam.mx/Brain/ionchan.htm.

that of OH^-, the solution is acidic, having a pH value lower than 7.0. Likewise, when the concentration of H^+, designated $[H^+]$, is lower than the concentration of OH^-, designated $[OH^-]$, the solution is basic, or alkaline, and the pH is higher than 7.0. What is really happening when water dissociates can be written as follows: $H_2O = H^+ + OH^-$; $H^+ + H_2O$ (second water molecule) $= H_3O^+ + OH^-$. But it will be convenient to write these concentrations as $[H^+]$ and $[OH^-]$, anyway. pH is a convenient means to express the $[H^+]$. It is a logarithmic function: $pH = -\log[H^+]$. As stated earlier, when the concentrations of $[H^+]$ and $[OH^-]$ are the same, the pH is 7.0. How this occurs is elaborated as follows. $[H_2O] = 1000$ g/liter (because the density of water is 1.0); then 1000 g/liter $\times$ 1 liter per 18 g/liter = 55.5 M (the molecular weight of water is 18). So, 1 liter of water is 55.5 M. At equilibrium, $K_{eq} = [H^+][OH^-]/[H_2O] = 1.8 \times 10^{-16}$ (K_{eq} can be measured by conductivity); $1.8 \times 10^{-16} \times 55.5$ M $= [H^+][OH^-] = 1 \times 10^{-14}$, because $[H^+] = [OH^-] = 1.0 \times 10^{-7}$ M. $[H^+]$ can range from 100 to 10^{-14}. Biological reactions in cells take place at pH 6.5 to 8.0, with pH 7.4 being the usual physiological range. The "Henderson-Hasselbach" equation (actually Henderson-Hasselbalch) describes how to predict the pH of a solution when an acid or base is present and the concentration of the acid or base is known: $pH = \log_{10} 1/[H^+] = -\log_{10} [H^+]$ (from the discussion earlier). If a weak acid is present, for example, the ionization equilibrium is $HA \rightleftharpoons H^+ + A^-$ (acetic acid, for example). The apparent equilibrium constant, K, would be $K = [H^+][A^-]/[HA]$ (where HA is the amount of acid that is not ionized). Just like pH, the pK of an acid is defined as $pK = -\log K = \log 1/K$ (the pK of an acid is the pH at which the acid is half dissociated). The relationship between pH and the ratio of acid and base is given by $1/[H^+] = 1[A^-]/K[HA]$. Taking the logarithm of both sides, $\log 1/[H^+] = \log 1/K + \log [A^-]/[HA]$. Substituting pH for $\log 1/[H^+]$ and pK for $\log 1/K$, the equation becomes $pH = pK + \log [A^-]/[HA]$. This is the Henderson-Hasselbalch equation. It is important to have an understanding of pH as it applies to the reactions going on inside a cell.

In the foregoing topics, there has been a survey of some of the cell biology to which there are references in later chapters.

Further Reading

Alberts, B., Johnson, A., Lewis, J., Raff, M., Roberts, K., and Walter, P., *Molecular Biology of the Cell*, 4th edition, Garland Publishing, 2002.

Lodish, H.F. (ed.), Berk, A., Matsudaira, P., Kaiser, C.A., Krieger, M., Scott, M.P., Zipursky, S.L., and Darnell, J., *Molecular Cell Biology*, 5th edition, W.H. Freeman and Co., 2003.

A concise discussion of embryonic germ layers may be found in: http://www.training.seer.cancer.gov/module_cancer_disease/unit3_categories1_embryology.html. A general discussion of the Henderson-Hasselbalch equation is given at: http://www.chembuddy.com/?left=pH-calculation&right=pH-buffers-henderson-hasselbalch.

CHAPTER 2

Proteins

Prion Disease: A Fatal Protein Conformation

Prion disease has shocked the scientific community because it turns out not to be a viral or bacterial disease, which most would have expected and a few researchers still believe, but to be caused by the appearance of a protein, with a primary amino acid sequence identical to the normal cellular protein but with an altered conformation (shape in three dimensions). The bovine form of this disease is called bovine spongiform encephalopathy, otherwise known as "mad cow disease." Domestic animals are thought to contract this disease by ingesting feed derived partly from the brains of diseased animals (such as sheep brains with prion disease). Similar diseases occur in other animals, notably scrapie in sheep, and the disease can arise in humans through ingestion of the meat of animals that have the disease. The brains are especially dangerous to consume because this is a major site of pathology and concentration of the lethal protein. It appears that the scrapie form, PrP^{Sc}, ingested can cross the wall of the gut at Peyer's patches (see Figure 2-1A–C), indicating that mucosal-associated lymphoid tissue is involved in the transport. These lymphoid cells may phagocytose PrP^{Sc} and be transported to other lymphoid sites. The prion can replicate at these sites (spleen, nodes, etc.). Because there are neural connections to these sites, PrP^{Sc} may gain access to the nerve, the spinal cord, and the brain.

The fatal prion disease kuru has been known for some time to cause death in tribes of the Fore highlands of Papua New Guinea, whose members ritually ingested the brains of dead relatives. This practice became known in the 1950s, and the tribes were induced to give it up. As a result, the incidence of kuru declined. Also, there has been concern about **blood transfusions** from humans who have the disease. Because the incubation time, before pathology is detected in a human, is usually long, preventive measures pose a challenging problem in terms of detecting the disease in the donor or in the donor's blood. Not much is known about whether other animal products could pose a threat; so far, there is no evidence concerning milk from diseased cows, for example.

Prion diseases in humans are Creutzfeldt-Jacob disease, Gerstmann-Straussler-Scheinker syndrome, fatal familial insomnia, kuru, and Alpers

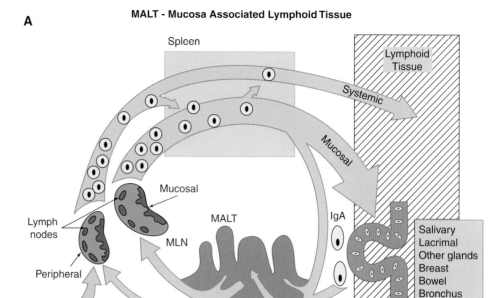

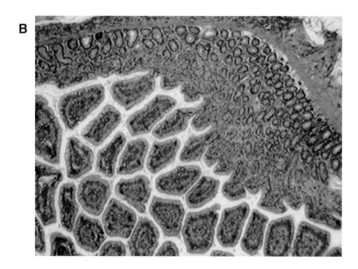

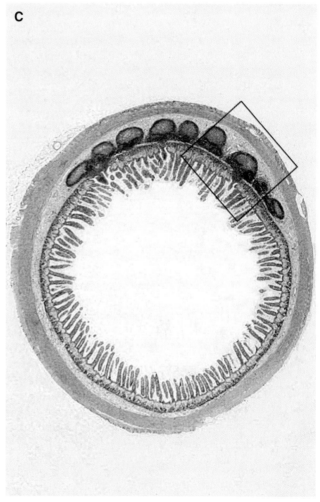

Figure 2-1 **A.** Peyer's patches are part of the mucosa-associated lymphoid tissue (MALT) in the ileum of the small intestine where PrP^{Sc} in ingested meat from diseased animals can be taken up and circulated via the spleen to mucosal and systemic lymphoid tissues and then to neurons in close proximity to the lymphoid tissue and eventually to the brain. **B.** Microscopic image of Peyer's patches, which are large oval lymph tissues in the mucous-secreting lining of the ileum of the human small intestine. The patches become less prominent with age and are more distinctive in younger people. These patches are rich in lymphocytes that play a protective role against infection, since this is a site at which antibodies can be mobilized. **C.** A microscopic view of Peyer's patches in the ileum. Part A redrawn with permission from http://www.chlamydiae.com/images/lycke/Slide1.GIF. Part B reproduced from http://micro.magnet.fsu.edu/optics/olympusmicd/galleries/brightfield/peyerspatchessec.html. Part C reproduced with permission from http://www.bu.edu/histology/p/12001oba.htm.

syndrome. In the human, Creutzfeldt-Jacob disease is of central interest. The incidence of Creutzfeldt-Jacob (pronounced kroits´felt yah´kobb) has been low (about 1 in 1 million), although recently there have been a number of deaths in Britain because of the outbreak of mad cow disease through the ingestion of the meat of diseased animals. Unlike viruses and bacteria, or even the normal cellular **prion protein** (PrP^c), cooking at high temperatures does not affect the fatal prion protein (which is called the scrapie form of the protein, or PrP^{Sc}). Creutzfeldt-Jacob is characterized by loss of motor control leading to paralysis, wasting, dementia, and death. Death often follows a bout of pneumonia. The brain is a highly visible target in this disease, and after death noninflammatory lesions and vacuoles are present and amyloid deposition (similar to Alzheimer's disease) and death of astroglia, the cells that support neurons, occur. There is no cure currently, although considerable ongoing basic research may lead to some therapies.

In addition to infection in humans through ingesting the abnormal form of the prion protein, growth hormone injections may pose a threat. Because this hormone is isolated from brains of animals, if they are diseased, this could be a mode of transmission. Nowadays, human growth hormone is being synthesized through expression of human DNA, which should reduce the possibility of the disease through this route. Corneal transplants with tissue from a diseased human also could be a mode of transmission. There also is evidence of a genetic (autosomal) mode of transmission. As shown in this chapter, the fatal form of the prion protein depends on the unfolding of the innocent native protein to a new conformation that produces the fatal disease. Such a change depends on folding functions of a few amino acids in the protein. Mutations can occur where an amino acid residue changes to a different amino acid so that the resulting prion protein would be more likely to assume the fatal conformation. This situation will become more understandable when protein conformation is discussed later.

Propagation of PrP^{Sc} from PrP^c in the Cell

Knockout experiments in mice, wherein the PrP^c gene (Prnp) becomes nonfunctional and little or no PrP^c is produced in the animal, show that PrP^c is required for the infection with PrP^{Sc} to manifest into disease. In essence, normal prion protein, PrP^c, in the presence of the abnormal prion protein (through ingestion, for example) could result in the conversion of the normal form of the protein to the abnormal form, perhaps by direct interaction of the two forms, although the exact mechanism of the stimulation to the abnormal form is not clear. This leads to an explanation of the propagation of the disease form of the protein from the normal form. The change in conformation seems to occur in endocytic vesicles (see Chapter 1) as shown in Figure 2-2.

The normal prion, PrP^c, is attached to the cell membrane as a modified lipid (one such is a glycosylinositol phospholipid); an example is shown in Figure 2-3. The portion of the protein extending into the extracellular space is capable of binding copper with high affinity, so PrP^c may be involved, in some way, in copper metabolism. A linear model of the protein is shown in Figure 2-4, in which there is a strong copper-binding region (along with some weaker copper-binding regions) in the amino terminal (left of center) part of the molecule. The region of the

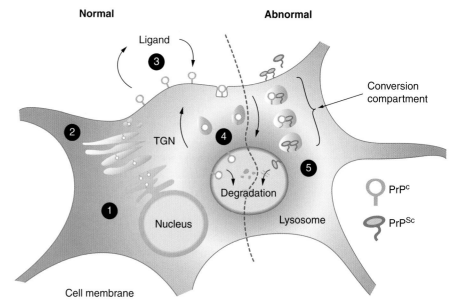

Figure 2-2. Propagation of PrP^{Sc} by conversion of PrP^c after ingestion of PrP^{Sc}. At *1*, PrP^c is in the secretory pathway leading to the cell membrane, *2*, where it is anchored to the membrane through a glycosylinositol anchor, *3*. On the cell membrane, it may act as a receptor, possibly for copper *(3)*. Thereafter, it reenters the cell in an endocytic vesicle *(4)* fusing with a lysosome for degradation *(5)* in the normal process. (See Figure 1-23.) When PrP^{Sc} is present, for example by ingestion *(right side of figure)* it can be endocytized in the presence of PrP^c, and in this endocytic vesicle, the conversion of PrP^c to PrP^{Sc} occurs in the presence of PrP^{Sc}. Presumably, this is the means of propagating the abnormal form. Redrawn with permission from http://www.chemsoc.org/chembytes/ezine/2002/jones_apr02.htm.

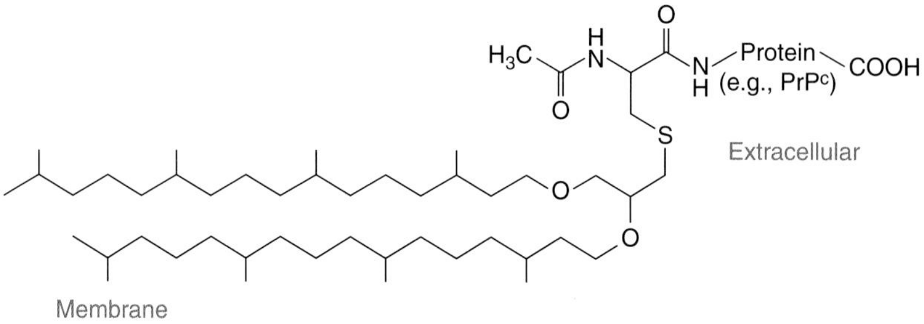

Figure 2-3. Example of a glycosylphosphatidylinositol protein anchor that might anchor PrP^c to the cell membrane.

molecule containing the high-affinity binding site for copper (femptomolar, 10^{-15}M) is apparently necessary for the development of the disease form of prion. Some investigations are focused on aspects of copper functions as a lead to a chemotherapeutic basis to treating the disease.

The three-dimensional structures of the normal, PrP^c, and the abnormal, PrP^{Sc}, forms of the prion protein are shown in Figure 2-5A and B. It is shown that some helical regions of PrP^c have become unwound and present as β-sheets in the structure of PrP^{Sc}. The quantities of exchange of the two forms of peptides are shown in Figure 2-6. In Figure 2-7 is shown the

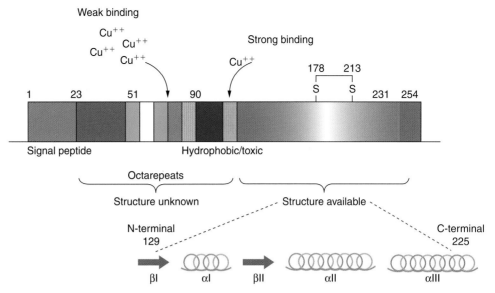

Figure 2-4. Linear cartoon of the prion protein (PrP^c). This model shows weak binding sites for copper and a strong binding site for copper in the N-terminal *(left)* portion of the protein. It also shows portions of the amino acid sequence in the nonhelical β-sheet form *(βI and βII)* as well as helical *(coiled)* forms *(αI, αII and αIII)* at the bottom. Redrawn with permission from http://www.chemsoc.org/chembytes/ezine/2002/jones_apr02.htm.

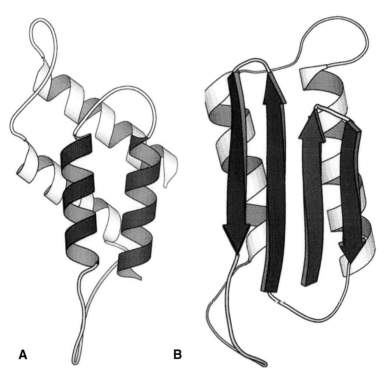

Figure 2-5. Three-dimensional ribbon models of PrP^c *(A)* and PrP^{Sc} *(B)*. The conformations of these two forms are quite different, even though the amino acid sequences (primary structures) are identical. It can be seen that some of the helical structures (αI, αII, and αIII) in Figure 2-4 have become unwound and present in PrP^{Sc} *(B)* as flattened β-sheets. Reproduced with permission from J.G. Black, *Microbiology*, 6th ed., John Wiley, New York.

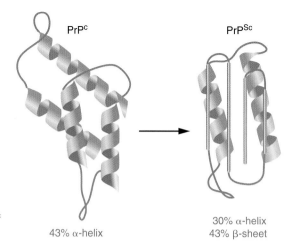

Figure 2-6. Conversion diagram showing amounts of α-helix in PrP^c converted to β-sheet forms in PrP^{Sc}.

43% α-helix

30% α-helix
43% β-sheet

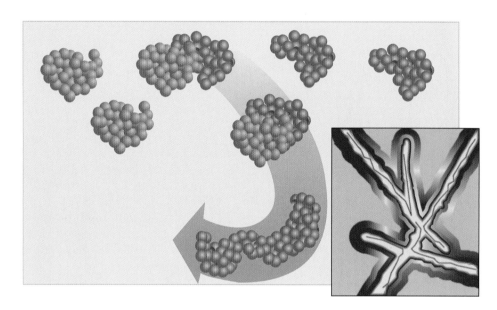

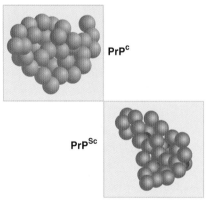

Figure 2-7. Illustration showing how PrP^c in the presence of PrP^{Sc} might be converted to more molecules of PrP^{Sc}. This chain reaction results in fibril formation of the PrP^{Sc} molecules and aggregation of the fibrils that are deposited in the brain and lead to death. Redrawn with permission from http://www.nobel.se/medicine/laureates/1997/illpres/jekyll-hyde.html.

conversion of the pair PrP^c—PrP^{Sc} to PrP^{Sc}—PrP^{Sc}, which comprises the elements of a chain propagation reaction in which increasing amounts of PrP^c are converted to PrP^{Sc}. When enough of the PrP^{Sc} molecules have developed, they form long aggregates of filaments that deposit and damage neural tissue in the brain, accounting for the symptoms described early in this chapter that lead to death. There is some evidence that the conversion just described may involve sulfhydryl (SH) groups so that S–S bonds may be broken, or formed, in the process of the conversion.

Thus, the dreaded prion disease has been described. It becomes clear that to fully understand this or other diseases involving proteins, it is essential to understand protein biochemistry.

Amino Acids

Amino acids are the stuff of which proteins are made. Proteins are the result of linking amino acids through peptide bonds in the order specified by the encoding gene. The protein is folded into its native active state by systems that are reviewed in this book. There are 20 amino acids that commonly comprise the structures of all proteins. They are listed with their structures in Figure 2-8. The anatomy of an amino acid is shown in Figure 2-9. Those starred in Figure 2-8 are **essential amino acids,** as differentiated from the **nonessential amino acids.** The essential amino acids cannot be synthesized in the body; therefore, the human must obtain these from the food source. The essential amino acids are tryptophan, threonine, methionine, valine, leucine, isoleucine, phenylalanine, lysine, arginine, and histidine—10 in all, half the number of the total principal amino acids. In Chapter 13, where nutrition is a topic, there is a discussion about dietary protein and the balance of amino acids in the diet, because some foods are rich in particular amino acids and some are deficient.

The shortened names for amino acids are shown in Figure 2-8 in parentheses. There is an additional abbreviation for each amino acid to facilitate dealing with sequences of peptides and proteins, as shown in Table 2-1.

Many years ago there were attempts to simulate a prebiotic life environment in a **spark chamber** that contained elemental atmospheric gases and initiated with an electrical spark. After sparking, various organic molecules were discovered to be synthesized. Among them were amino acids and nucleic acids, presumably the starting building materials for eventual living organisms. Although this work was of interest, the thrust of these discoveries could not be continued because the prebiotic environment could not be simulated with authority. Recently, however, National Aeronautics and Space Administration (NASA) scientists announced (http://astrobiology.arc.nasa.gov/news/expandnews.cfm?id=1319) that amino acids were formed under conditions mimicking deep space, confirming earlier work with spark chambers but in a more convincing environment. Coupled with this result are observations that recovered meteorites were found to contain amino acids. Both L-forms and D-forms were found, but, as we know, only L-forms are contained in most natural proteins on this planet (interestingly, some D-amino acids are found in the cell walls of

(Text continues on p. 44.)

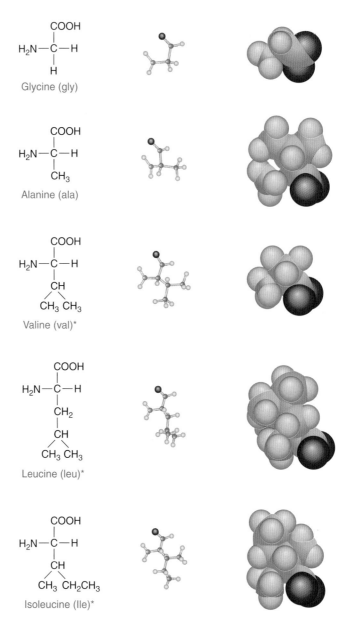

Figure 2-8. The twenty common amino acids. *, essential amino acids.

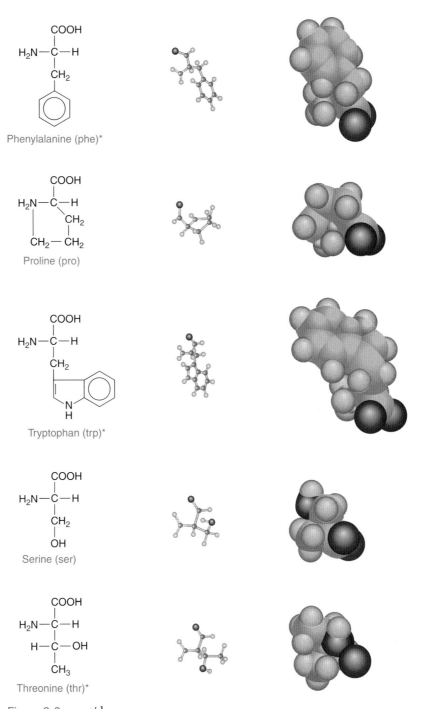

Figure 2-8, cont'd

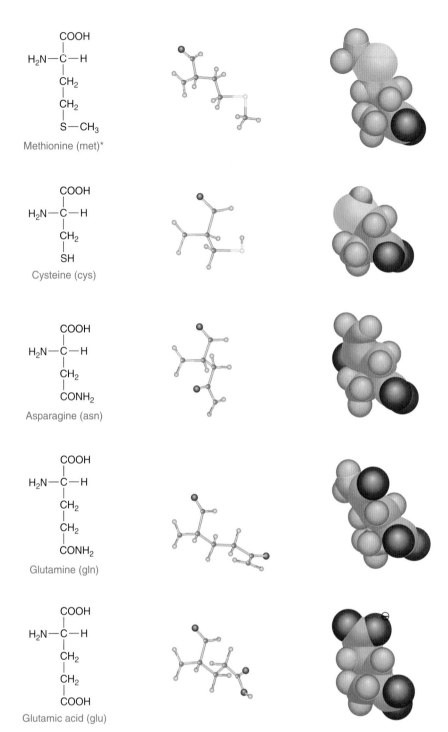

Methionine (met)*

Cysteine (cys)

Asparagine (asn)

Glutamine (gln)

Glutamic acid (glu)

Figure 2-8, cont'd

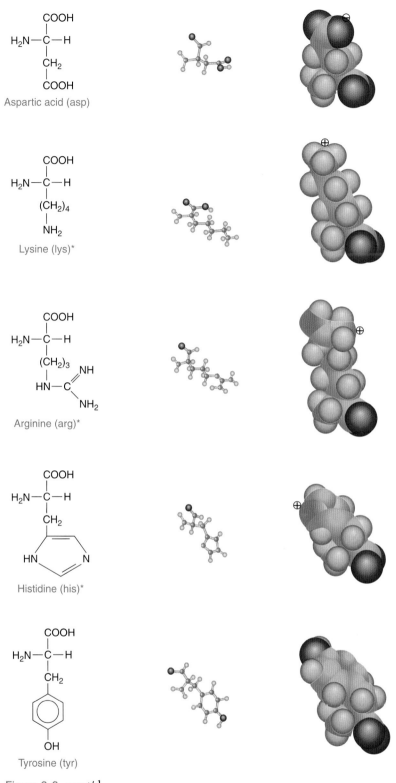

Figure 2-8, cont'd

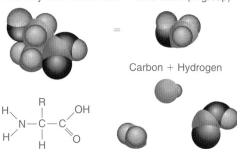

Figure 2-9. Identification of the substituent groups of the asymmetrical carbon of L-valine with space-filling models.

Table 2-1
Names and Abbreviations of Amino Acids

Amino acid	Mol Wt.	Abbreviation	Letter name
Small amino acids			
Glycine	57.05	Gly	G
Alanine	71.09	Ala	A
Serine	87.08	Ser	S
Threonine	101.11	Thr	T
Cysteine	103.15	Cys	C
Hydrophobic amino acids			
Valine	99.14	Val	V
Leucine	113.16	Leu	L
Isoleucine	113.16	Ile	I
Methionine	131.19	Met	M
Proline	97.12	Pro	P
Aromatic amino acids			
Phenylalanine	147.18	Phe	F
Tyrosine	163.18	Tyr	Y
Tryptophan	186.21	Trp	W
Carboxylated amino acids			
Aspartic acid	115.09	Asp	D
Glutamic acid	129.12	Glu	E
Amino/Amide-containing amino acids			
Asparagine	114.11	Asn	N
Glutamine	128.14	Gln	Q
Histidine	137.14	His	H
Lysine	128.17	Lys	K
Arginine	156.19	Arg	R

bacteria). In addition, it has been reported that other organic compounds, as well as membrane-like structures, were formed under deep space–like conditions.

Chirality

L-amino acids are those that, when in solution and a plane-polarized beam of light is passed through, bend light to the left (*levo* in Latin). A solution of a D-amino acid under the same conditions would bend light to the right (*dextro* in Latin). Thus, the *L* and *D* nomenclature was adopted. Specifically, amino acid structures of L- or D-forms depend on the configuration of the amino group in relation to the asymmetrical carbon. An asymmetrical carbon is one in which the four substituents on each carbon bond are different. As a case in point, examine Figure 2-10, where valine is the representative amino acid. The D-form is a mirror image of the L-form, and one form cannot be superimposed upon the other (the definition of **chirality**)—just as looking at the tops of both hands and moving one on top of the other makes the thumb protrude where the little finger is on the bottom hand (Figure 2-11). Sometimes these are referred to as **enantiomers,** a pair of optical

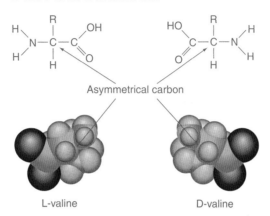

Figure 2-10. L- and D-forms of valine with space-filling models. In the L-form, the amino group is on the left and in the D-form, the amino group is on the right of the asymmetrical carbon.

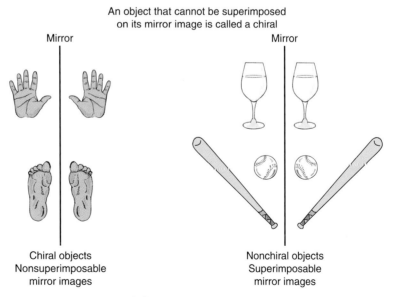

Figure 2-11. Pictorial definition of chirality. Redrawn with permission from http://www.creation-science-prophecy.com/amino/.

Amino Acids

isomers. Compounds with an asymmetrical carbon have more than one form and are, therefore, **stereoisomers.** All amino acids are stereoisomeric except for glycine, which has *two hydrogen atoms* attached to the central carbon in addition to an amino group and a carboxyl group (Figure 2-8). L-amino acids can be converted into D-amino acids as catalyzed by an enzyme called amino acid **racemase.** This process can also occur uncatalyzed but would require much time.

Amino Acids Have Two or More Potential Charges

Amino acids are referred to as **zwitterions** because they have two different charges (or more if they have additional charged substituents). The word *zwitter* comes from German and literally means *hermaphroditic,* in this case referring to charge rather than sex. Thus, amino acids have the property of becoming charged as positive (α-amino group) or negative (α-carboxyl group) depending on the pH of the solution (see Chapter 1). The α-carbon is that carbon with the amino and carboxyl group substituents. In glycine, the α-carbon is shown as $NH_2C\alpha H_2COOH$. In lysine, the α-carbon is the same, followed by other Greek letters, as the carbon skeleton extends from the α-carbon, as shown in Figure 2-12. These charges are expressed by protonation (the proton is H^+) of the NH_2 (to NH_3^+, a +1 charge). The COOH is charged upon deprotonation (to COO^-, a −1 charge). Thus, when glycine, for example, is protonated, the amino group has a +1 charge and the carboxyl group has a 0 charge; when deprotonated, the amino group has a 0 charge and the carboxyl group has a −1 charge. That is to say, $NH_2CH_2COOH + H^+ = NH_3^+CH_2COOH$ and $NH_2CH_2COOH - H^+ = NH_2CH_2COO^-$. Some amino acids have R groups (substituents) that also can be protonated and deprotonated. These are lysine, arginine (extra amine), and histidine (extra secondary amine, guanido group, or histidine as imidazole group); glutamic and aspartic acids have an extra carboxyl group. Thiols and hydroxyls, such as those in cysteine and tyrosine, respectively, can be deprotonated to produce a negative charge (Table 2-2 and Figure 2-13). When amino acids are incorporated into peptides (several amino acids joined through peptide bonds), the internal amino acid α-amino group and the α-carboxyl group cannot be ionized because they are parts of the peptide bond linkage (as shown later);

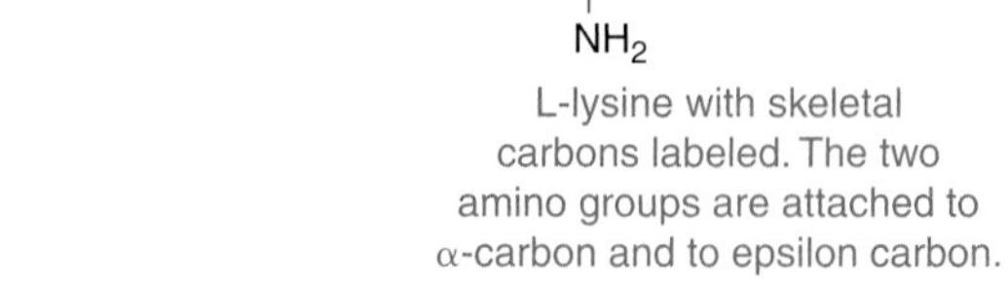

$$H_2N-\underset{\underset{H}{|}}{\overset{\overset{H}{|}}{C_\alpha}}-COOH$$

L-glycine, with α-carbon shown

$$H_2N-\overset{\overset{H}{|}}{C_\alpha}-COOH \quad \text{(alpha)}$$
$$H-\overset{}{C_\beta}-H \quad \text{(beta)}$$
$$H-\overset{}{C_\gamma}-H \quad \text{(gamma)}$$
$$H-\overset{}{C_\delta}-H \quad \text{(delta)}$$
$$H-\overset{}{C_\epsilon}-H \quad \text{(epsilon)}$$
$$NH_2$$

L-lysine with skeletal carbons labeled. The two amino groups are attached to α-carbon and to epsilon carbon.

Figure 2-12. Numbering by Greek letter, the carbons of the amino acid skeleton.

Table 2-2
pKa Values for Amino Acid Side Chain Ionizable Groups and Effects on Charge with Protonation or Deprotonation

Side chain	pKa	Amino acids	Protonated charge	Deprotonated charge
Carboxyl	~4	Glu, Asp	0	−
Amino	10.5	Lys	+	0
Guanido	12.5	Arg	+	0
Thiol	8.5	Cys	0	−
Phenol	10.5	Tyr	0	−
Imidazole	6	His	+	0
Hydroxyl	~13	Ser, Thr	0	−

Reproduced from http://dwb.unl.edu/Teacher/NSF/c10/c10content.html.

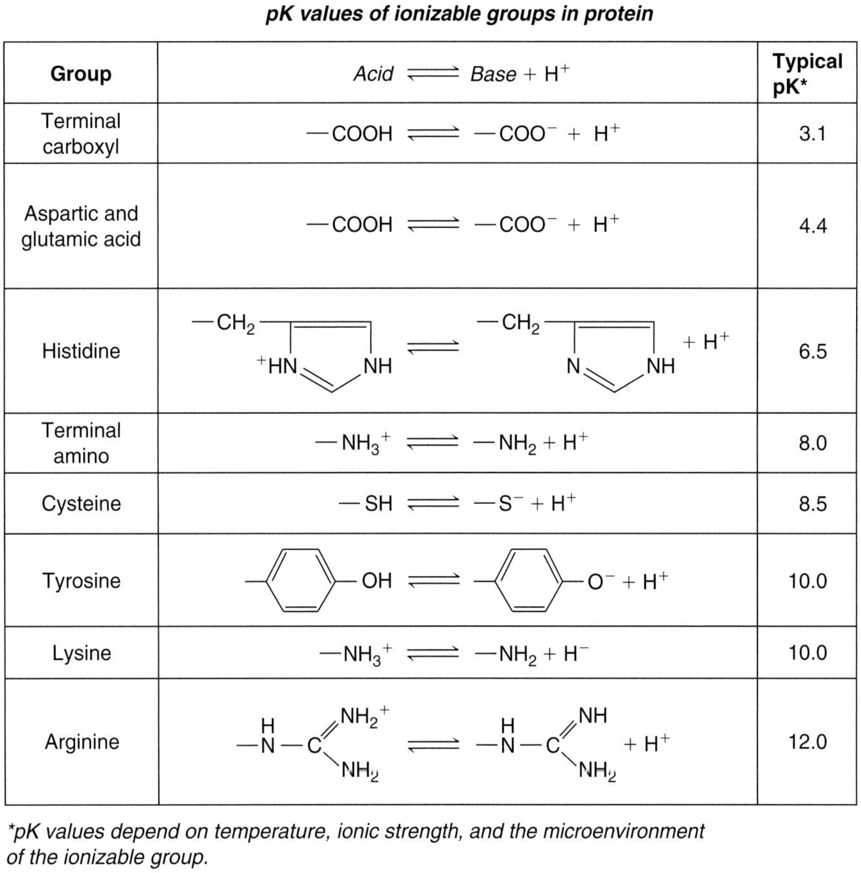

pK values of ionizable groups in protein

Group	Acid ⇌ Base + H⁺	Typical pK*
Terminal carboxyl	—COOH ⇌ —COO⁻ + H⁺	3.1
Aspartic and glutamic acid	—COOH ⇌ —COO⁻ + H⁺	4.4
Histidine	$-CH_2-\text{imidazole}^+ \rightleftharpoons -CH_2-\text{imidazole} + H^+$	6.5
Terminal amino	—NH$_3^+$ ⇌ —NH$_2$ + H⁺	8.0
Cysteine	—SH ⇌ —S⁻ + H⁺	8.5
Tyrosine	—C$_6$H$_4$—OH ⇌ —C$_6$H$_4$—O⁻ + H⁺	10.0
Lysine	—NH$_3^+$ ⇌ —NH$_2$ + H⁻	10.0
Arginine	—NH—C(NH$_2^+$)(NH$_2$) ⇌ —NH—C(NH)(NH$_2$) + H⁺	12.0

*pK values depend on temperature, ionic strength, and the microenvironment of the ionizable group.

Figure 2-13. pK values of terminal ionizable groups in a protein and ionizable groups in the side chains of internal amino acids. Refer to Figure 2-8 for complete structures of amino acids. The equilibrium is shown from the protonated form *(left)* to the deprotonated form *(right)*. Redrawn from http://web.archives.org/web/*/www.agsci.ubc.ca/courses/fnh/410/protein/1_13.htm (December 5, 2004).

however, the *terminal* amino and carboxyl groups are chargeable and are subject to protonation and deprotonation. That is to say, NH_2*peptide*-$COOH + H^+ = NH_3^+$*peptide*$COOH$, and NH_2*peptide*$COOH - H^+ = NH_2$*peptide*COO^-. If the pKa of an ionizable group is known, the charge state can be determined by the Henderson-Hasselbalch equation presented in Chapter 1 (pH = pKa + log [A]/[HA]). If the pH is two units above the pKa, the ionizable group will be nearly completely deprotonated. When the pH is two units below the pKa, the Henderson-Hasselbalch equation indicates that the ionizable group will be nearly completely protonated. When the pH is equal to the pKa, the ionizable group will be half deprotonated. Thus, any compound that has more than one ionizable group will exhibit buffering capacity within two pH units of the pK value, a capacity to resist changes in the pH of the medium. A protein may have, in addition to the terminal amino and carboxyl groups, the side chains of various amino acids within the polypeptide, for example, potential extra NH_3^+ groups in lysine and arginine and potential extra COO^- groups in glutamic and aspartic acids. In these cases, these side chains will contribute to the overall charge on the surface of the protein, and the determination of the isoelectric point (pI), for example, becomes theoretically complex. However, proteins can be titrated in solution similarly to doing so for amino acids, and nowadays this value can be determined by computer (described later). pKa values for amino acid side chain groups are given in Table 2-2. Figure 2-13 shows various ionizable groups of some amino acid side chains in proteins and their pK values. (The groups around the α-carbons of the internal amino acids are participants in peptide bonds and are not ionizable in proteins; only the terminal amino and carboxyl groups and amino acid side chains are.) Of the 20 common amino acids, 13 have two ionizable groups, which are the amino and carboxyl groups substituting the α-carbon. The remaining 7 amino acids have an additional ionizable group in the side chain (see Figure 2-8). The charge on the amino acid is determined by the pH of the medium and the pK of the ionizable group. The pK values (the pH at which the concentration of the protonated form equals that of the unprotonated form) and the pI (the pH at which the charges on a molecule are equal so that the net charge is zero) are determined by titration, as shown for representative amino acids in Figures 2-14 and 2-15. Connecting the values of pKa on either side of the vertical part of the titration and drawing the straight line intersects

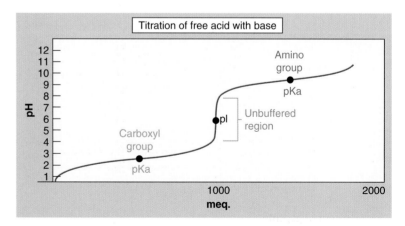

Figure 2-14. Titration of glycine in solution. pH measurement is on the ordinate and the milliequivalents of NaOH added are shown on the abscissa. The pKa values occur at pH levels where the amounts of protonated and unprotonated forms of the group being titrated are equal. The halfway point in the vertical part of the curve (where the net charge is zero) is the isoelectric pH (pI).

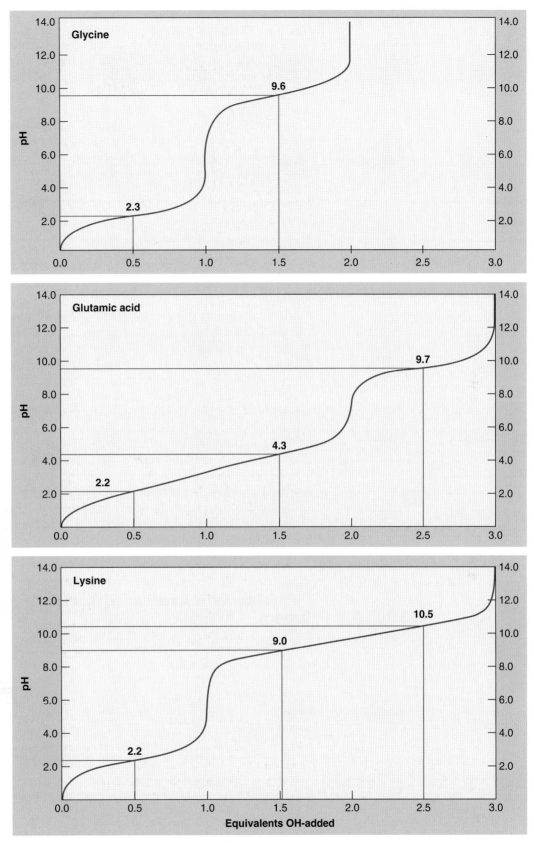

Figure 2-15. Titration curves for glycine *(top)*, glutamic acid *(middle)*, and lysine *(bottom)*. Numbers on the curves are the pK values of each ionizable group. Thus, 2.3 is the pKa of the carboxyl group of glycine and 9.6 is the pKa of the amino group of glycine. The pKa falls on the middle part of the titration for each group giving the pH where the charges of the protonated and unprotonated forms are equal. The pI values at the center of the vertical trace where the net charge on the molecule is zero (nothing to titrate). This value can be estimated by connecting the two pKa values for glycine and the intersection on the vertical line will be the pI, about 5.9. pI values for glutamic acid and lysine are about 7 and 5.6.

the vertical portion approximating the pI value if that point of intersection is extrapolated to the ordinate axis (y-axis). In other words, add the two relevant pKa values and divide by two to calculate the approximate pI. In the case of glutamic acid and lysine, the titrations are more complex because a third ionizable group is involved. This becomes more complicated with a protein in which there may be many ionizable groups in the side chains (Table 2-2) of the composite amino acids. The pI value of a protein may be determined electronically by simply submitting the primary amino acid sequence to the European Molecular Biology Laboratory (EMBL) in Heidelberg, Germany (Figure 2-16). The pI value of a protein is valuable to know because that is the pH at which it is usually precipitable from solution and crystallizable (zero charge, so minimal attraction to the charges on water molecules).

Synthesis and Degradation of Amino Acids

Without going into detail at this point, some important generalizations can be made about the synthesis of nonessential amino acids in the body and the degradation of all amino acids. In principle, the body can provide the carbon skeletons for the nonessential amino acids. In most cases, the required amino group can be added by **transamination** from a preexisting amino acid with the appropriate enzyme (aminotransferase). The carbon skeletons of amino acids are provided by intermediates in the overall metabolism of glucose in the glycolytic pathway (glycolysis) and through the citric acid cycle as shown in Figure 2-17. These cycles are elaborated upon when all participants and products are described. However, in outline, it is clear that *the nonessential amino acids can be formed from 3-phosphoglycerate, pyruvate, oxalacetate, and α-ketoglutarate.* The nonessential amino acids generated are boxed, as are the substrates for these amino acids. An example of this type of conversion can be given in the case of converting a preexisting amino acid that provides an amino group with α-ketoglutarate to generate glutamic acid and a keto acid (corresponding to

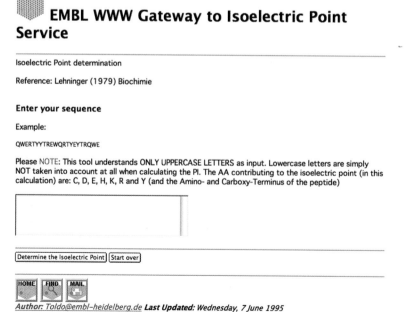

Figure 2-16. A source for the determination of the isoelectric point when the primary sequence of a peptide is provided.

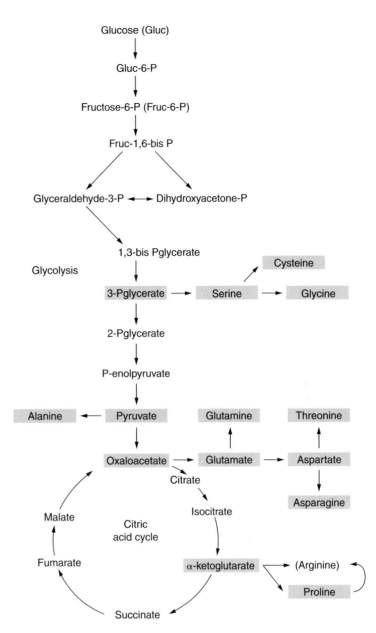

Figure 2-17. Routes of synthesis of the nonessential amino acids in the human. Only the starting intermediates in glycolysis and the citric acid cycle are indicated. The specific pathways, enzymes and cofactors are not shown for simplicity. Clearly the overall metabolism of glucose for energy as adenosine triphosphate (ATP), CO_2 and water provide the intermediates for the synthesis of nonessential amino acids. In some cases a suitable ketoacid can be converted to an amino acid with an appropriate enzyme (an aminotransferase). The ketoacid thus provides the carbon skeleton for the amino acid. The requisite amino group is provided by another amino acid or by NH_4^+. The metabolic scheme shown here is a somewhat further elaboration of Figure 1-20.

the donor amino acid), as shown in Figure 2-18. If the amino acid donor on the left side of figure is aspartate, for example, the enzyme would be called glutamate-aspartate aminotransferase or aspartate aminotransferase. The three-dimensional structure of this enzyme bound to its coenzyme, pyridoxal-P (PLP), is shown in Figure 2-19. Generally, a specific amino acid can be formed from the corresponding α-keto acid skeleton with the amino group generated by transamination with glutamate. Therefore, the nonessential amino acids derive from carbon skeleton molecules generated by the body, whereas the carbon skeletons for essential amino acids cannot be formed in the body. Also, some nonessential amino acids can be formed from essential amino acids ingested in the diet. For example, tyrosine can be formed from phenylalanine by hydroxylation. Additionally, bacteria in the human gut can synthesize amino acids and may even supply small amounts of some essential amino acids (but these cannot satisfy the full requirement for essential amino acids).

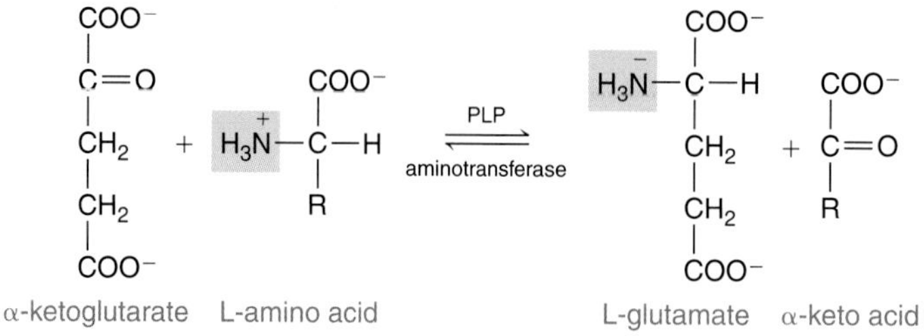

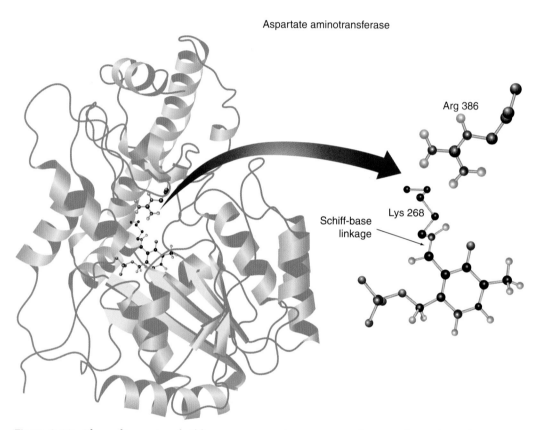

Figure 2-18. A general reaction of transamination by an aminotransferase enzyme in which the carbon skeleton is that of α-ketoglutarate to form glutamate. PLP is the aminotransferase coenzyme, pyridoxal-P. A preexisting amino acid supplies the amino group.

Figure 2-19. Three-dimensional ribbon structure of aspartate aminotransferase bound to its coenzyme, pyridoxal-P. α-Helical protein structure is shown as *coiled ribbons* while β-pleated sheet structure is shown as a *double line*.

Amino acids become degraded when they are not used for protein synthesis. They are first converted to intermediates that can enter the citric acid cycle or be used to form glucose and thus are considered either **ketogenic** or **glucogenic.** If too much protein is ingested, the protein will be degraded to amino acids and some will be converted to glucose and from glucose fat can be synthesized and stored. In addition to transamination, oxidative deamination can occur, in which ammonia as NH_4^+ is generated (Figure 2-20). This reaction of oxidative deamination is catalyzed by glutamate dehydrogenase. Ammonia produced from the degradation of amino

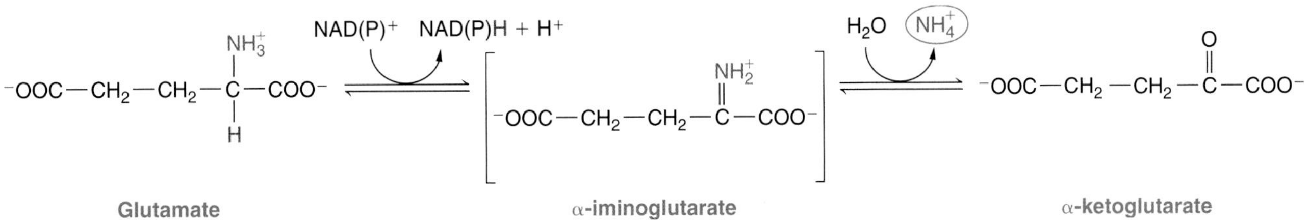

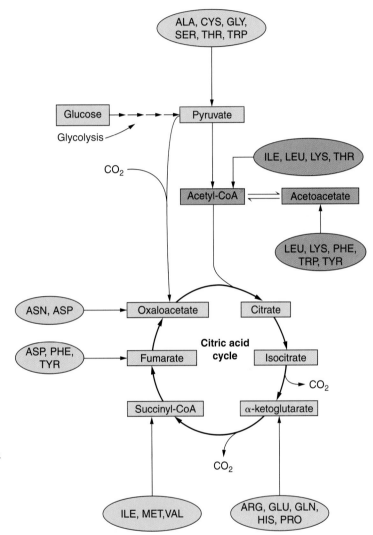

Figure 2-20. Oxidative deamination of glutamate to generate ammonium ion and α-ketoglutarate.

Figure 2-21. Degradation paths of essential and nonessential amino acids. All amino acids can form glucose (glucogenic) except leucine and lysine, which are solely ketogenic and form acetoacetate. Some amino acids can form both glucose and keto acids, such as phenylalanine, tryptophan, tyrosine, isoleucine, and threonine.

acids in this way is excreted as urea through the urea (H_2NCONH_2) cycle. The reaction is as follows:

Ammonia (NH_4^+, ammonium ion, is NH_3, ammonia, in the deprotonated form) + HCO_3^- (bicarbonate) + Aspartate = Urea + Fumarate.

Three ATPs are required and converted to 2 ADP + 2 Pi + AMP + PP. In transamination, ammonia is not generated but the amino group is transferred to another carbon skeleton.

The overall degradation of amino acids is shown in Figure 2-21. The figure shows that the amino acids enter the glycolytic pathway or the

Amino Acids 53

citric acid cycle based on their conversion back to equivalent carbon skeletons. Importantly, all enzymatic steps are reversible, so glucose can be formed from the amino acids—except for leucine and lysine, which are solely ketogenic—and form acetoacetate. Acetoacetate and acetyl-coenzyme A (acetylCoA) are ketone bodies; the structure of acetoacetate is $HOOCCH_2COCH_3$. It is of interest that the Atkins diet reduces the intake of carbohydrates so that the primary energy source becomes keto acids rather than glucose, a condition tolerated well by the heart and the brain. With the relatively high protein and fat intake in this diet, lots of acetoacetate and acetylCoA are being formed and being used directly as fuels, although relatively high quantities of ammonia must be produced from the protein intake, which must be efficiently handled by the kidney.

Proteins

Proteins are composed of amino acids joined through the **peptide bond.** Although the formation of this bond, shown in Figure 2-22, seems to be a simple matter, finding the correct sequence of a large number of different amino acids when constructing a protein requires an apparatus that can translate the sequence encoded first in the gene for the protein by means of mRNA. An interesting example of a dipeptide of commercial importance is **aspartame** (NutraSweet). Aspartame is a noncaloric sweetener used in many kinds of dietary products, particularly as a sweetener for sodas, tea, and coffee. It is a dipeptide of aspartic acid and the methyl ester of phenylalanine. It cannot be consumed by individuals with **phenylketonuria.** The methyl group can be released above 85°F to generate methanol, which is toxic. Methanol is converted to formaldehyde, a carcinogen, in the liver, and it has been claimed that there could be a link between the large consumption of aspartame and an increased incidence of brain tumors. The structure of this dipeptide is shown in Figure 2-23,

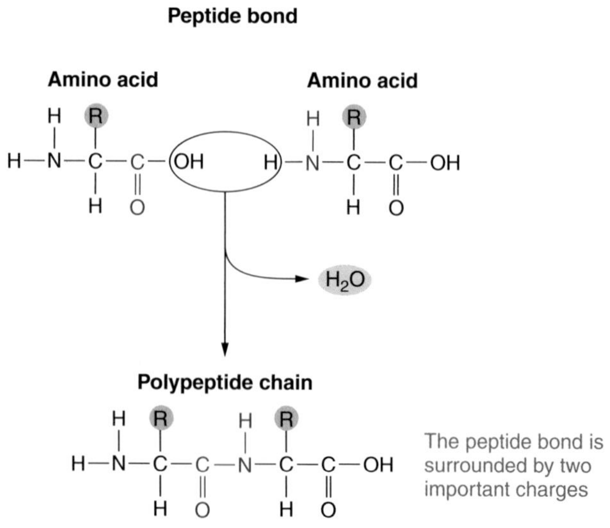

Figure 2-22. Formation of a peptide bond from two amino acids. In the overall process, a carboxyl group on the α-carbon of one amino acid is linked to the amino group of the second amino acid by the exclusion of a water molecule. Thus, the bond is -C-C-N-C-, eliminating the charges on those two groups. In the dipeptide product, the ionizable groups are the amino group on the N-terminal carbon of the left-hand member and the carboxyl group on the C-terminal carbon of the right-hand member. The amino group becomes charged on protonation as NH_3^+ and the carboxyl group becomes charged on deprotonation as COO^-.

The peptide bond is surrounded by two important charges

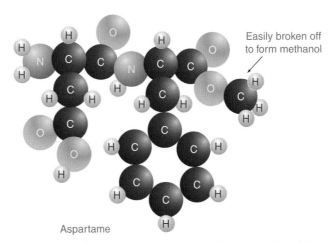

Aspartame

Figure 2-23. Dipeptide of aspartate and methylester of phenylalanine. Aspartyl residue is on the left and the phenylalanine methyl ester is on the right. The peptide bond joining the two utilizes the α-carboxyl group of aspartate and the α-amino group of phenylalanine methyl ester. The residual ionizable groups on the aspartyl residue are the α-amino and the β-carboxyl group of aspartyl residue. These groups are subject to protonation (NH_3^+) and deprotonation (COO^-). The α-carboxyl group of phenylalanine methyl ester ordinarily would be ionizable, except that this deprotonation to COO^- is blocked by the methyl ester.

where aspartyl residue is on the left and phenylalanine methyl ester is on the right. Note that the peptide bond comprises the α-carboxyl group of aspartate and the α-amino group of phenylalanine methyl ester. This leaves the α-amino and β-carboxyl groups of aspartate available for protonation and deprotonation. The α-carboxyl group of phenylalanine is blocked from deprotonation by the methyl ester substitution. An interesting example of a tripeptide that does not require a gene for sequence direction is **glutathione (GSH).** This molecule plays a role in the redox potential of a cell and is therefore important as an antioxidant and a possible protection against cancer. It is γ-glutamylcysteinylglycine (Figure 2-24). It is synthesized in the cell enzymatically (the enzymes are

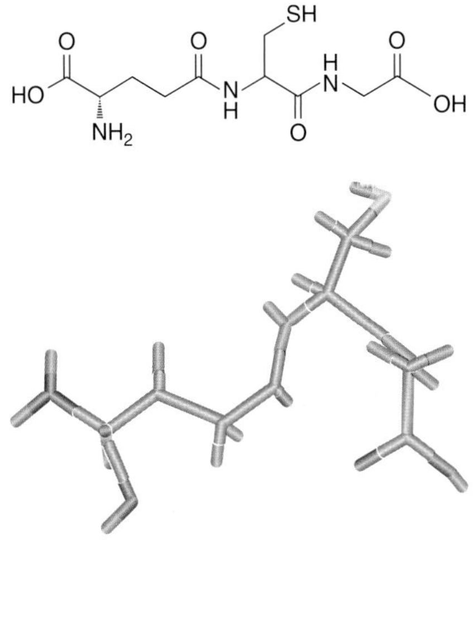

Figure 2-24. Structure of glutathione: γ-glutamylcysteinylglycine. Upper figure.

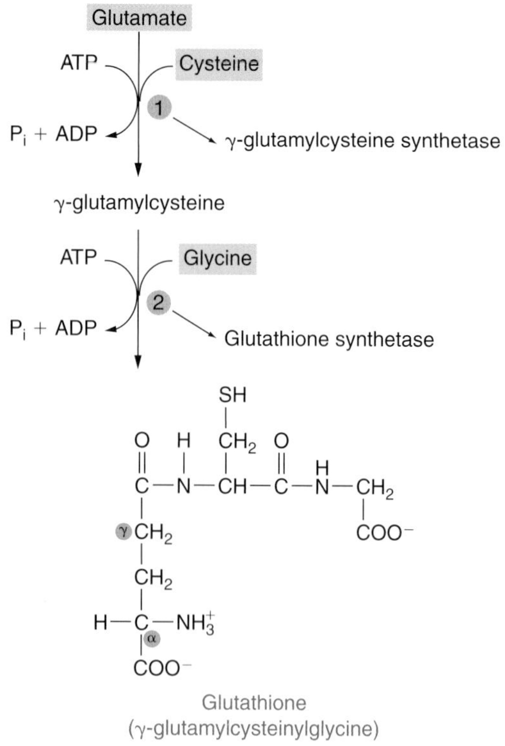

Figure 2-25. Synthetic path of glutathione. The extent of glutathione synthesis is sensitive to the cellular concentration of cysteine. Enzymes that catalyze steps 1 and 2 are gene products. The cysteinyl SH group can form a disulfide bond (S–S) with a second molecule of glutathione and can be converted back to free glutathione by the enzyme, glutathione reductase.

gene products) as shown in Figure 2-25, and the synthetic process is sensitive to the concentration of cysteine in the cell. As can be seen, the first peptide bond linkage involves the γ-carboxyl group of glutamate rather than the α-carboxyl group. In the tripeptide, the ionizable groups subject to protonation and deprotonation are the α-amino and the α-carboxyl groups of γ-glutamyl and the α-carboxyl of glycine. GSH is an important redox agent in the cell, and it can combine with a second GSH to form glutathione disulfide (GSSG), the oxidized form. This form can be reduced back to original form by **GSH reductase.**

There are four levels of complexity in proteins. The first is the primary sequence of amino acids (the specific amino acids and the total number of amino acids are dictated by the gene for the protein). The second is the way in which the primary sequence of amino acids is folded together and involves the interaction of amino acid side chains. The third is the three-dimensional structure of the protein. The fourth is in the subunit structure of the active protein if the protein exists as a multimeric form with subunits, although many exist and function as monomers (single polypeptide chains).

Amino Acid Sequence

The primary structure of a peptide or protein (a protein is a polypeptide) is simply the linear order of amino acids joined by peptide bonds from left (amino terminus) to right (carboxy terminus). An example of a heptapeptide (a seven amino acid containing peptide) is shown in Figure 2-26.

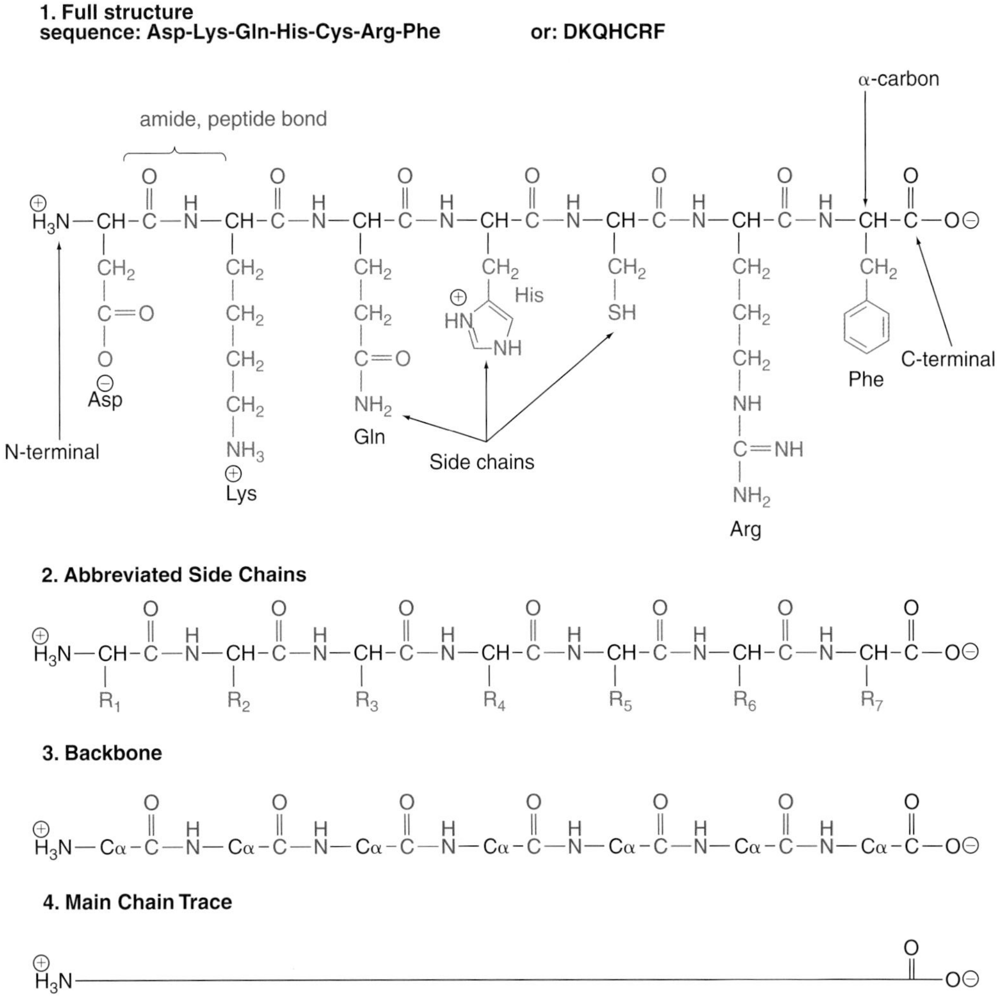

1. Full structure
sequence: Asp-Lys-Gln-His-Cys-Arg-Phe   or: DKQHCRF

Figure 2-26. Representations of the heptapeptide, aspartyl, lysyl, glutaminyl, histidinyl, cysteinyl, arginyl, phenylalanine. The N-terminal amino acid, aspartyl, is on the *left* and the C-terminal amino acid, phenylalanine, is on the *right*. This representation shows only the primary structure (sequence). How this peptide might fold falls under the category of secondary structure. As in 2, amino acids may be abbreviated using R for the side chain or written in even more simple forms in 3 and 4. Formation of peptide bonds using α-carboxyls and α-amino groups is evident in 3.

Secondary Structure

After a protein chain of amino acids has been synthesized on the ribosomal machinery, the protein is folded, usually with the assistance of other proteins, into the appropriate native (active) configuration. The secondary structures involved are much determined by the sequence of the amino acids themselves. Generally, protein sequences fold into two types of structures, α-helical structures typical of globular proteins or β-sheet structures typical of fibrous proteins. Ordinarily, a protein will exhibit both types of folding, although one type could dominate over the other, conferring the major characteristic. This is evident in the discussion of prion protein and the shifts in secondary structure between

Proteins   57

PrP^c and PrP^{Sc} (Figure 2-6). Examples of these structures are shown in Figure 2-27.

The **α-helix** is a characteristic of the globular class of proteins. It is stabilized by hydrogen bonding between amide nitrogens and carbonyl carbons of peptide bonds that are spaced four residues apart (evident in the lower right figure). This structure produces a coil where the R groups (amino acid side chains) lie on the outside of the helix and perpendicular to the axis. Amino acids that favor the formation of the α-helix are alanine, glutamate, aspartate, leucine, isoleucine, and methionine. Glycine and proline (proline tends to fold the chain) residues tend to disrupt this structure. When the chain, even if α-helical, folds to continue in another direction, proline will be found at the turning point. This is because of the pyrrolidine imino group, HN (see Figure 2-13), that restricts movement about the peptide bond and interferes with the further extension of the helix. The β-sheets (also referred to as pleated sheets) are composed of stretches of about 5 to 10 amino acids, and there are usually two to several of these regions. These appear as shown in Figure 2-27 in the upper left. These sheets are stabilized by hydrogen bonding between amide nitrogens and carbonyl carbons to amino acids that are opposite linearly—unlike the α-helix, where the hydrogen bonding occurs in adjacent regions of the helical backbone. Hydrogen bonding in these two structures is shown in Figure 2-28.

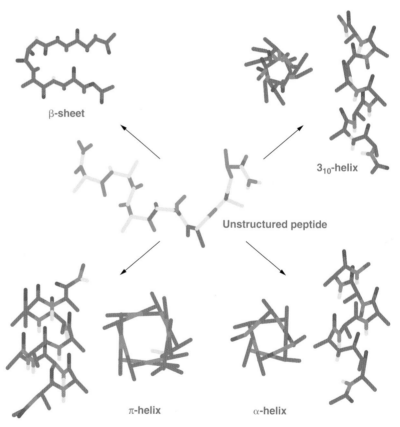

Figure 2-27. Secondary structures of a peptide. The β-sheet form is shown on the *upper left* and the α-helical form is shown on the *lower right*. Other secondary structures are shown, but the α-helix and β-sheet forms are predominant secondary structures.

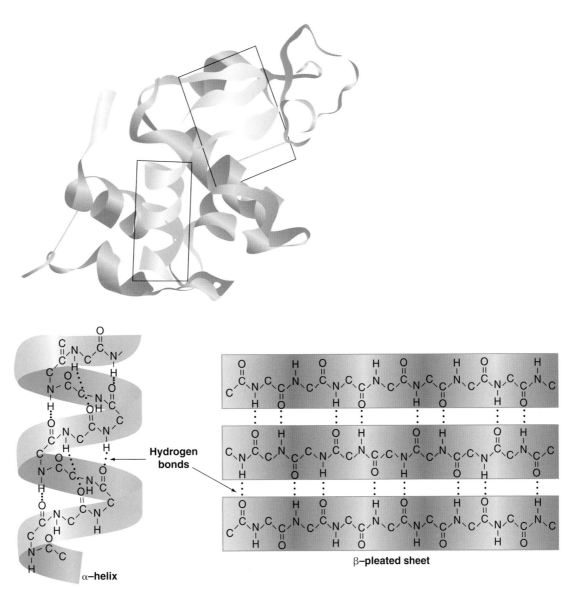

Figure 2-28. Hydrogen bonding between carbonyl carbons and amide mitrogens in the peptide. α-Helix *(bottom left)*. The bonds form between amino acid residues along the axis of the helix *(black dots)*. Hydrogen bonds between amino acid residues in β-sheets form between carboxyl carbons and amide nitrogens of parallel amino acids *(lower right)*. The relevant portions of α-helix and β-sheets in an hypothetical protein are shown at the *top*.

Other types of interactions also stabilize the structures of proteins. These are **electrostatic bonds** between two oppositely charged groups (charge–charge interaction). Ionized side chains of amino acids also interact with water molecules, the opposing charge being attracted to the amino acid–ionized group (charge–dipole interaction). Many of the amino acids, particularly of globular proteins, have side chains that protrude onto the surface of the molecule and are charged and attract water molecules in charge–dipole interactions. When substituent groups of side chains have similar charges and are close together, they exhibit charge repulsion, called **van der Waals forces.** These often play a role in protein folding. Although van der Waals forces may be weaker than electrostatic attractive forces, there are usually a large number of them.

The third category of interactions is **hydrophobic bonding.** This occurs when two nonpolar (uncharged) groups (also called hydrophobic groups) on amino acid side chains are close enough together to attract and create a local environment that excludes water molecules. An important bond for stabilizing the tertiary structure of a protein is the covalent **disulfide bond** that forms when two side chains contain an SH group under oxidizing conditions. Examples of these interactions are shown in Figure 2-29.

Nowadays, it is easy to obtain information on secondary structure knowing only the primary amino acid sequence by using a large database

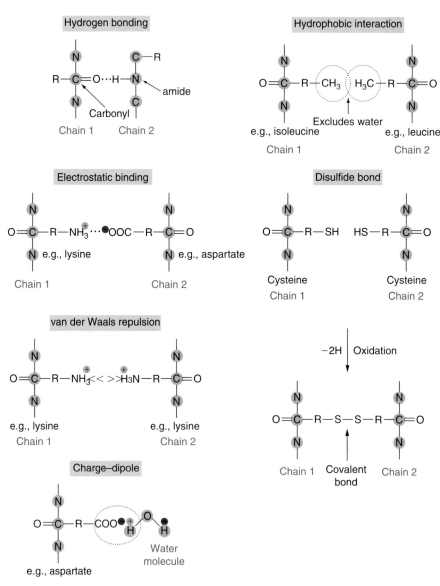

Figure 2-29. Secondary interactions between amino acid residues of adjacent peptide chains of the same molecule that play a role in the tertiary protein structure.

that includes information for many sequences and how they are known to behave. An example is shown in Figure 2-30. The tertiary structure of a protein is the complex folding of the peptide chains involving the secondary structure that produces the protein molecule in the native (active) form. Fully folded active protein conformations are shown in Figures 2-5 and 2-19. **Quaternary structure** applies when a protein in the native form is composed of subunits (each subunit being a separate protein). Whereas the two examples in Figures 2-5 and 2-19 represent proteins in the active state that are monomeric (having a single polypeptide chain), many proteins have to be formed as multimers with subunits. An example is hemoglobin, which is a tetramer, shown in Figure 2-31. The four levels of protein structure are summarized in Figure 2-32. Remembering that most bonds that stabilize the native structure of the protein molecule are noncovalent (excluding covalent disulfide bonds), these active proteins are meant to function under bodily conditions so that proteins, outside of the body, can be disrupted by heat or exposure to extremes of pH and salts. When proteins become disrupted in this way, they are denatured because they cannot function as in the native state. In some cases, proteins can be renatured in the laboratory. Under bodily conditions, proteins are turned over by breaking them down to free amino acids by hydrolytic enzymes (for example, in the lysosome discussed in Chapter 1).

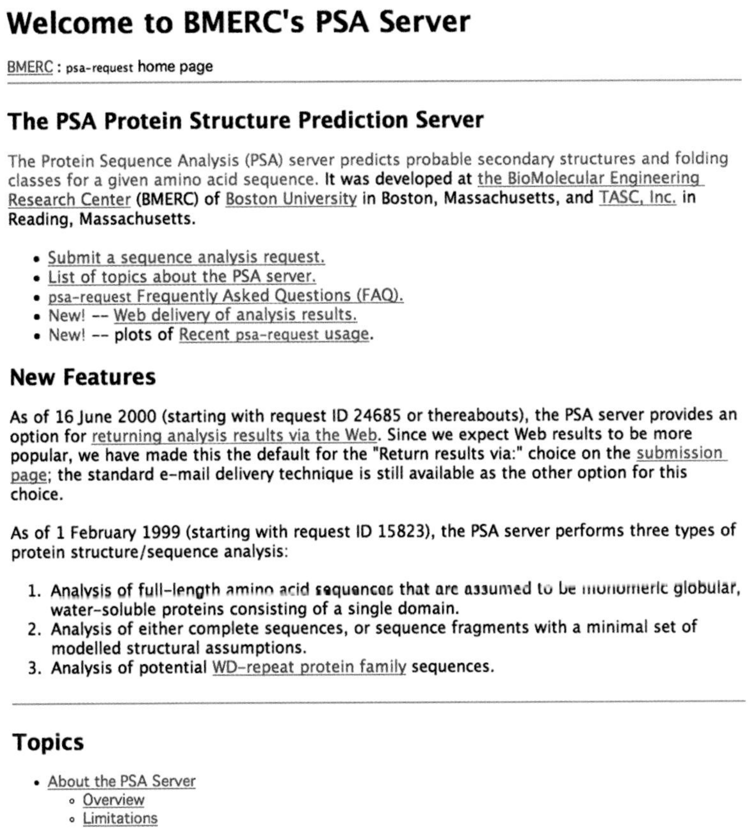

Welcome to BMERC's PSA Server

BMERC : psa-request home page

The PSA Protein Structure Prediction Server

The Protein Sequence Analysis (PSA) server predicts probable secondary structures and folding classes for a given amino acid sequence. It was developed at the BioMolecular Engineering Research Center (BMERC) of Boston University in Boston, Massachusetts, and TASC, Inc. in Reading, Massachusetts.

- Submit a sequence analysis request.
- List of topics about the PSA server.
- psa-request Frequently Asked Questions (FAQ).
- New! -- Web delivery of analysis results.
- New! -- plots of Recent psa-request usage.

New Features

As of 16 June 2000 (starting with request ID 24685 or thereabouts), the PSA server provides an option for returning analysis results via the Web. Since we expect Web results to be more popular, we have made this the default for the "Return results via:" choice on the submission page; the standard e-mail delivery technique is still available as the other option for this choice.

As of 1 February 1999 (starting with request ID 15823), the PSA server performs three types of protein structure/sequence analysis:

1. Analysis of full-length amino acid sequences that are assumed to be monomeric globular, water-soluble proteins consisting of a single domain.
2. Analysis of either complete sequences, or sequence fragments with a minimal set of modelled structural assumptions.
3. Analysis of potential WD-repeat protein family sequences.

Topics

- About the PSA Server
  - Overview
  - Limitations

Figure 2-30. A server can be accessed for determination of protein structure. Reproduced from http://bmerc-www.bu.edu/psa/.

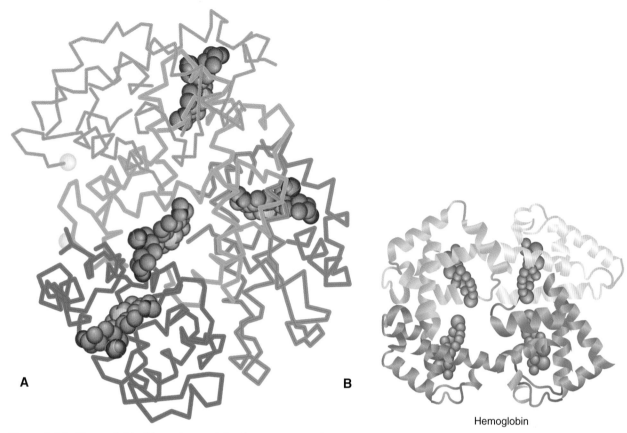

Hemoglobin

Figure 2-31. Hemoglobin (deoxyhemoglobin) contains 4 subunits. On right, the *green* and *blue chains* are α-chains and the *gold* and *aqua chains* are β–chains, four chains in all. The 4 *gray clusters* are non–covalently bonded heme groups (each heme binds one molecule of oxygen). The *gold spheres* at the top are phosphate groups. The *red atoms* are oxygen atoms. Part A reproduced from http://steitzlab1.un.edu/chem-191H-2002/Deame/My%20wels/hemoglobin.htm. *B* shows the extensive α-helical nature of the subunits and the globular nature of the protein. Reproduced from http://www.sci.sdsu.edu/class/bio202/TFrey/chemRev.Images/Hemoglobin.gif.

Protein Synthesis

The synthesis of ribosomal RNA in the nucleolus was discussed in Chapter 1. It was indicated that it was transported out of the nucleus through nuclear pores and was used in the cytoplasm for the synthesis of ribosomes. The ribosomes are the protein synthesis machines. The overall process of protein synthesis directed by a gene is shown in Figure 2-33. Briefly, DNA contains two strands, an antisense strand and a sense strand. The two strands must dissociate so that the transcription machinery responsible for synthesizing mRNA can complex with the start site of transcription. The opening of DNA may be signaled by messages from the cytoplasm in the form of transactivators or by other proteins. Information in the sense strand of DNA is transcribed into mRNA, which carries the information encoded in the gene. DNA is made up of the nucleosides: thymidine, cytidine, guanosine, and adenosine (see Chapter 6). T (thymine) base-pairs with A (adenine), and C (cytosine) base-pairs with G (guanine). Thus, the antisense strand in step 1 of Figure 2-33, in the upper left, shows opened DNA pairs with the complementary sequence below it on the right. Accordingly, ATCGGCT pairs with sense TAGCCGA. This would be only a small section of the gene's code, used alone here just

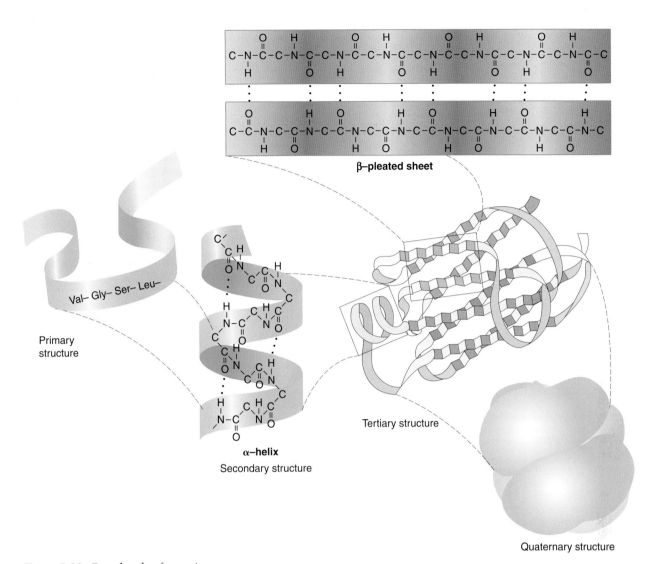

Figure 2-32. Four levels of protein structure.

to make things easier to envision. Then, in step 2, the sense DNA strand transcribes mRNA, where thymine is replaced by uracil in uridine (U). So, after transcription, the sense strand (TAGCCGA) of DNA becomes AUCGGCU in mRNA. mRNA is transported through the nuclear pore to the cytoplasm attaching to a ribosome (step 3). Then a specific transfer RNA (tRNA), each encoded (**anticodon**) with an antisense triplet for a specific amino acid, pairs with a complementary three-base codon in mRNA in step 4. Peptide bonds are formed to produce the growing peptide chain (step 5). The composition of three-letter portions of mRNA (codon), specifying each amino acid, is shown in Figure 2-34.

The Ribosome

The ribosome is a large machine consisting of two subunits. The large subunit has a sedimentation constant (S) of 60S, and the small subunit is 40S. The sedimentation constant is an expression of the rate at which

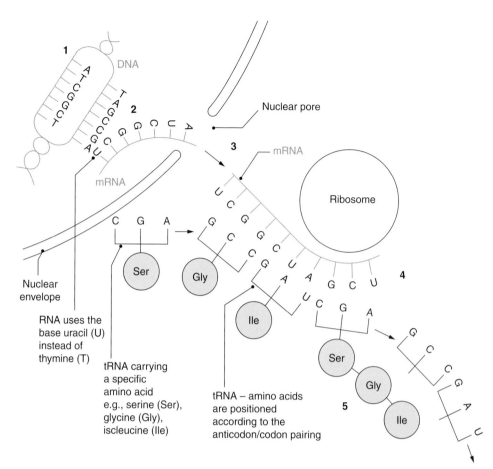

Figure 2-33. Overview of the process of protein synthesis from information encoded in the DNA of the gene to the growing peptide chain. Step 1: The DNA double helix unwinds to expose the transcribable sequence. Step 2: mRNA is produced from the sense strand of DNA. Completed mRNA is transported through the nucleopore to the cytoplasm to the rough endoplasmic reticulum. Step 3: The mRNA complexes with the ribosome. Step 4: Transfer RNAs bearing specific amino acids attach to the mRNA through an anticodon in the tRNA. Step 5: Translation occurs and the polypeptide chain grows and begins to fold.

Figure 2-34. Three bases form a codon for a specific amino acid in messenger RNA. The code is degenerate in that more than one combination will encode one specific amino acid. To the left of each amino acid are given the codes, thus GUU, GUC, GUA, and GUG all code for valine, for example. The codes may be generated by locating the amino acid and finding the first letter *(left)*, then the second letter *(top)*, and then the third letter *(right)* all positioned by the locus of the amino acid residue. The anticodon in a given tRNA will have a 3-letters signal complementary to the code for an amino acid in mRNA. Therefore, Ile, for example, might be AUU in mRNA and the tRNA anticodon would be UAA to attach the tRNA (bearing Ile) to the mRNA.

	Second letter				
First letter	U	C	A	G	Third letter
U	UUU } Phe UUC UUA } Leu UUG	UCU } Ser UCC UCA UCG	UAU } Tyr UAC UAA Stop UAG Stop	UGU } Cys UGC UGA Stop UGG Trp	U C A G
C	CUU } Leu CUC CUA CUG	CCU } Pro CCC CCA CCG	CAU } His CAC CAA } Gln CAG	CGU } Arg CGC CGA CGG	U C A G
A	AUU } Ile AUC AUA AUG Met	ACU } Thr ACC ACA ACG	AAU } Asn AAC AAA } Lys AAG	AGU } Ser AGC AGA } Arg AGG	U C A G
G	GUU } Val GUC GUA GUG	GCU } Ala GCC GCA GCG	GAU } Asp GAC GAA } Glu GAG	GGU } Gly GGC GGA GGG	U C A G

a substance in solution travels from top to bottom in a centrifugal field (in a centrifuge). The ribosome assembled with both subunits is 80S. Ribosomes can constitute as much as 20% of the mass of the cell, and there are about 20,000 ribosomes in a prokaryotic cell. The composition of the subunits is shown in Table 2-3. Some rRNAs and many proteins make up the subunits of the ribosome pictured in Figures 2-35 and 2-36. The ribosome is assembled within the nucleolus, where three of the rRNAs are produced. Apparently, a fourth rRNA is produced outside the nucleolus and transported inside to complete the ribosome. Ribosomal proteins are produced outside the nucleolus then enter and combine with the rRNAs to complete the two subunits of the ribosome, which are transported out individually through the nucleopores. The subunits combine in the cytoplasm. The crystal structures of the prokaryotic ribosomal subunits have been solved and are shown in Figures 2-37 through 2-39. The ribosome is the site of the translation of mRNA into protein. Ribosomes are either free in the cytoplasm or bound as RER. The free ribosomes (individual or grouped as polysomes) are the sites of synthesis of proteins that dissolve in the cell cytoplasm and are used by the cell that synthesizes them. Bound ribosomes occur as RER (Figure 1-21), and the proteins produced at these sites are retained in the endoplasmic reticulum and either are components of membranes or are packaged from this location for storage in the cytoplasm or export outside the cell (Figure 2-40). Ribosomes also are located in the mitochondria. They resemble bacterial ribosomes, which are smaller than eukaryotic ribosomes (the small mitochondrial subunit is 30S, the large subunit is 50S, and the complex is 70S compared to the larger subunits in the human, as shown in Table 2-3). (**Eukaryotic** refers to organisms that have a defined nucleus, including animals,

Table 2-3
Composition of the Subunits of the Ribosome

Subunit	S value	No. proteins	rRNA	No. nucleotides
Large (2.82×10^6D)	60S	45–49	28S, 5.8S 5S	ca 4880
Small (1.4×10^6D)	40S	33	18S	1874
Complex (4.22×10^6D)	80S			

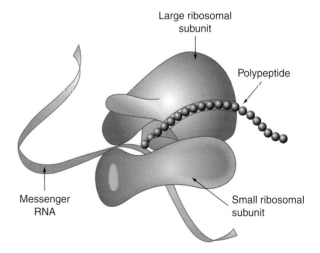

Figure 2-35. Illustration of a ribosome with mRNA attached and a growing polypeptide chain. Redrawn with permission from http://www.bergen.org/ACADEMY/bio/cellbio/cellbio1protsynpg1.html.

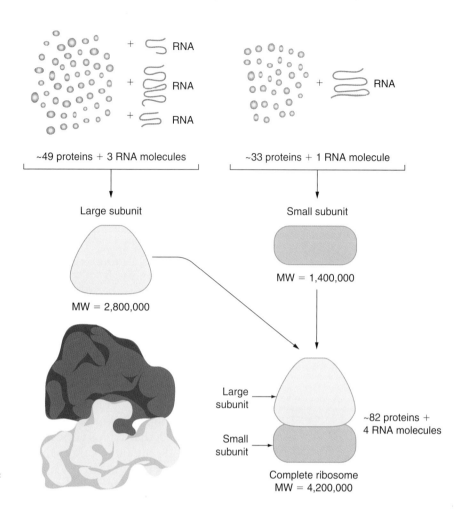

Figure 2-36. The enkaryotic ribosome and the composition of its subunits.

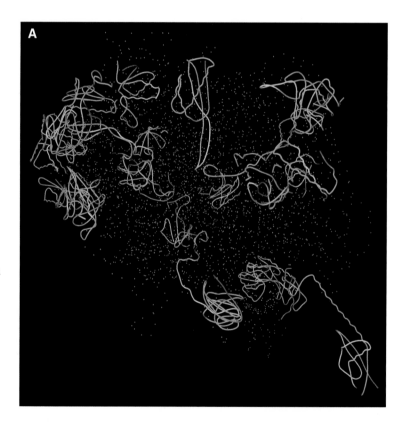

Figure 2-37. **A.** Crystal structure of the ribosome at 5.5 Å resolution. **B.** Small subunit from a prokaryote. Part A from PDB ID: 1GIY. M.M. Yusupov, et al. Crystal Structure of the Ribosome at 5.5 Å Resolution. *Science* **292** pp. 883 (2001). Part B from A. Manuell et al., "Regulation of chloroplast translation: interactions of RNA elements, RNA-binding proteins and the plastid ribosome." *Biochem. Soc. Trans.*, 32: 601–605, 2004.

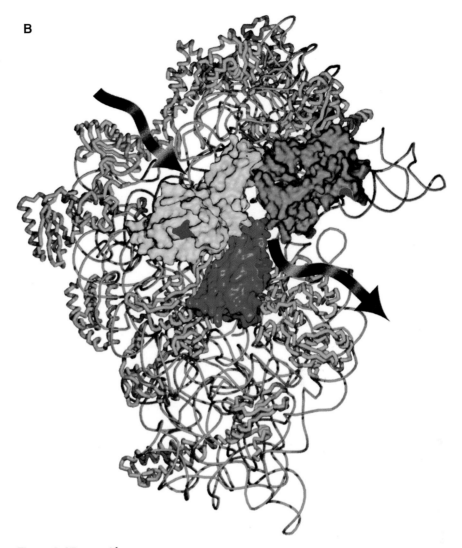

Figure 2-37, cont'd

plants, and fungi; **prokaryotic** refers to organisms with cells that do not have a defined nucleus, which refers mainly to bacteria.) The ribosomes read the mRNA from the 5'-end through to the 3'-end. As the polypeptide chain grows, it is folded and finally released into the endoplasmic reticulum, where it can have one of the three fates shown in Figure 2-40. The reading of mRNA begins with a **start codon,** AUG, and terminates with a **stop codon,** UAG, as shown in Figure 2-41.

Moving Amino Acids to Ribosomes and Forming Peptide Bonds

As shown in Figure 2-33, specific tRNAs bring the correct amino acid to the ribosome, where contact with the mRNA, on the small subunit, is made at the codon by an anticodon in the structure of the tRNA (Figure 2-42). As an example of a specific tRNA, leucyl-tRNA is shown in Figure 2-43A. Structures of two representative amino acid tRNA synthetases are shown in Figure 2-43B. The individual steps taking place on the amino acid tRNA synthetase enzyme are shown in Figure 2-44.

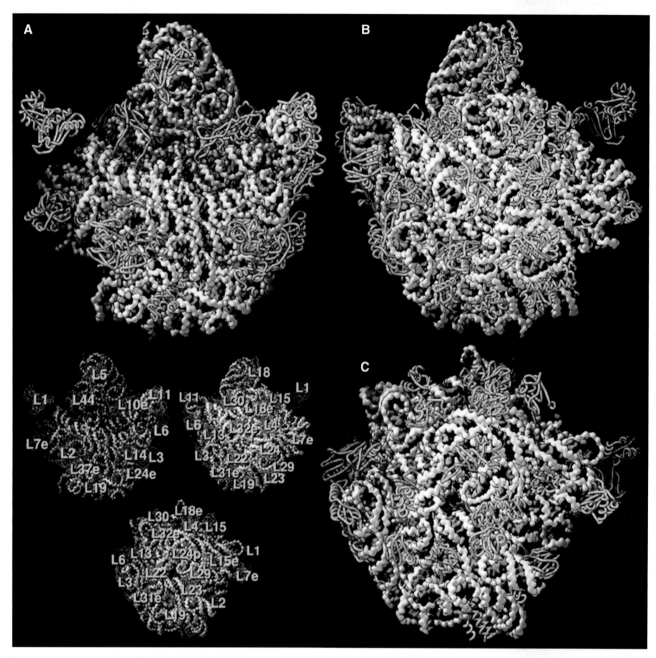

Figure 2-38. Topology of the prokaryotic phage ribosomal subunit. rRNA is in *gray* and ribosomal proteins are in *yellow*, occurring mostly on the surface. A, B, and C show three separate images. Reproduced with permission from the Steitz laboratory (N. Ban et al., *Science*, 289: 905–920, 2000).

Specific sites on the ribosome where these events take place are shown in Figure 2-45. The peptidyl (P) site is the location of the AUG initiation codon and the methionyl-tRNA (Met-tRNA) (AUG also encodes methionine, as shown in Figure 2-34) at the start of translation. The Met-tRNA combines with **guanosine triphosphate (GTP)** using an **initiation factor,** and the Met-tRNA combines with the small ribosomal subunit. The next aminoacyl tRNA is positioned at the **aminoacyl acceptor (A) site.** The covalent bond between the amino acid and the tRNA in the P site is broken, and a

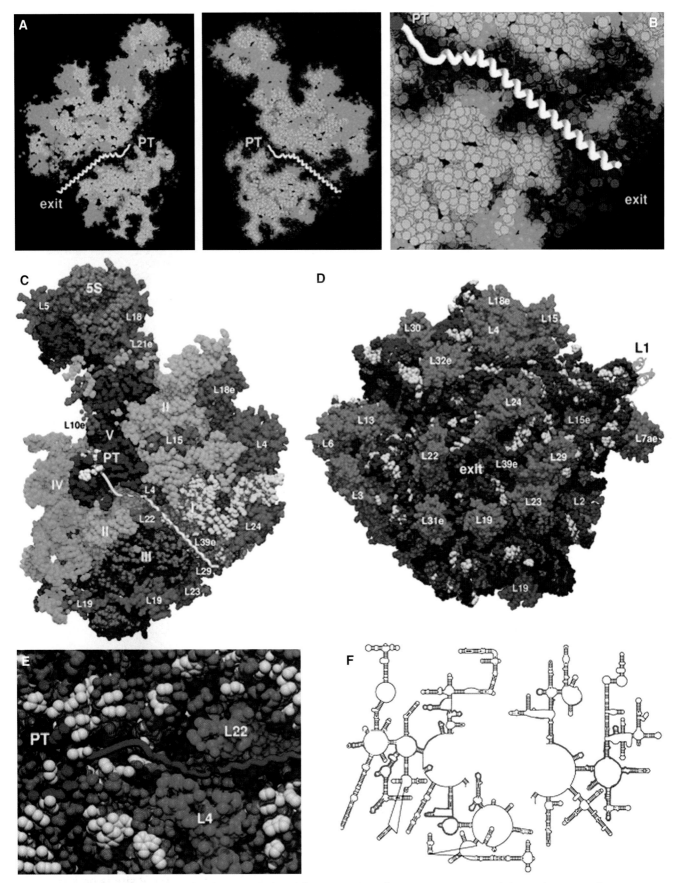

Figure 2-39. A detailed mechanism for peptide bond formation based on x-ray structure of the prokaryotic ribosome. A specific rRNA (A 2486) is the catalytic entity involved in peptide bond formation. These pictures also show the exit canal for the polypeptide chain (A, B, C, and E). The exit is large enough to allow passage of the chain as an α-helix. This is the work of the Steitz group. Reproduced with permission from P. Nissen et al., *Science*, 289: 920–930, 2000.

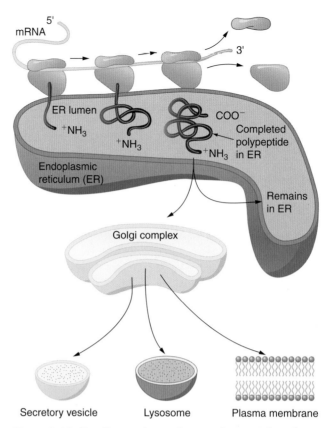

Figure 2-40. Reading and translation of mRNA by ribosomes attached to the ER (rough endoplasmic reticulum [RER]). As the message moves through consecutive ribosomes, the polypeptide chain grows until it is completed. After each protein copy is finished at each ribosomal position, the folded protein is deposited in the ER or transported to the Golgi apparatus, where it can have one of three fates: enclosure in a secretory vesicle and exported to the cell exterior; enclosed in a lysosome for eventual degradation, or translocated to the plasma membrane.

peptide bond is formed between the two amino acids through catalysis by a specific rRNA called **ribozyme,** located at A2486 in the x-ray structure of the ribosome (Figure 2-39). The mechanism of the peptide bond formation catalyzed by the ribozyme is shown in Figure 2-46. The empty tRNA dissociates as shown in the Figure 2-46C. The peptidyl tRNA is translocated to the P site from the A site, and the next aminoacyl tRNA moves into the A site. The E site is the exit site. The polypeptide continues to grow in this way (elongation). Eventually, the stop codon (UAG, UAA, or UGA) at the end of the message is positioned in the A site. No other amino acid will be added because there is no anticodon for a stop codon and therefore no complementary amino acid (Figure 2-34). A termination factor protein binds to the stop codon at the A site, and the termination of translation begins with the release of the newly synthesized protein from the peptidyl tRNA in the P site. The ribosomal subunits dissociate but can reassemble with the mRNA and Met-tRNA to start a new round of translation, as shown in Figure 2-41, to generate copies of the protein encoded in the mRNA. The process should continue until the mRNA breaks down or some required factors become limiting.

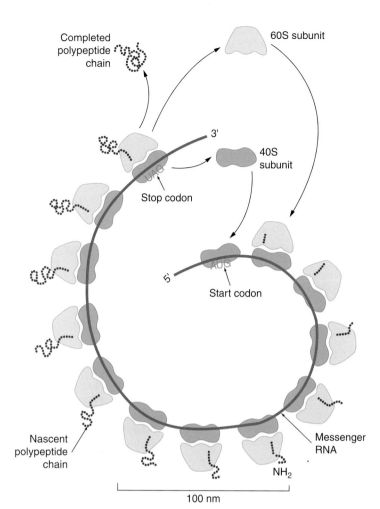

Figure 2-41. Translation is the reading of mRNA from the 5'-end to the 3'-end. The reading starts with the start codon, AUG, and ends with the stop codon, UAG, at which point the protein chain has been completed. Folding occurs while the protein chain is still growing. The completed protein dissociates from the ribosome and the ribosome dissociates into its two subunits that can recombine to start a new cycle of translation. Note that mRNA is in contact with the small subunit of the ribosome.

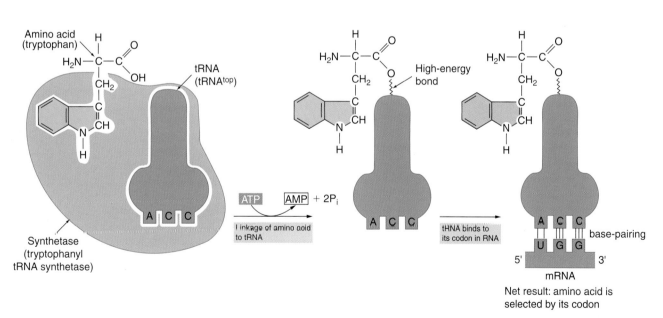

Figure 2-42. Example of tryptophanyl tRNA bringing tryptophan to the elongating polypeptide on the ribosomal subunit. An ATP is required to link the amino acid to the tRNA on the enzyme, tryptophanyl tRNA synthetase. Then tRNA, charged with the amino acid, binds to its codon in mRNA. To do this, the tRNA has a specific anticodon, ACC, in its structure.

Proteins 71

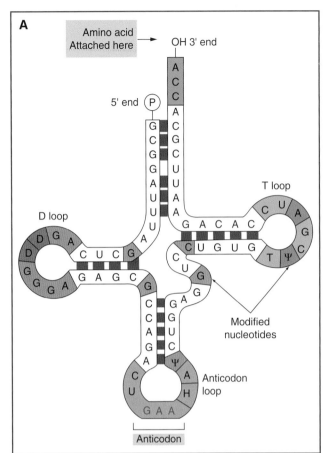
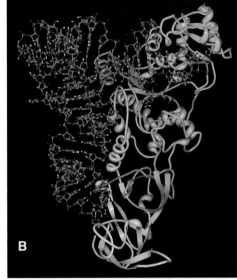
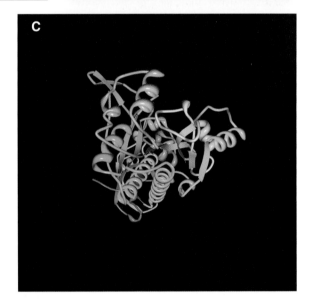

Figure 2-43. **A.** Diagram of a leucine tRNA (anticodon is GAA). The amino acid attaches to the 3′-end to the A residue of the CCA sequence *(green)*. **B.** Glutaminyl tRNA synthetase mutant complexed with glutamine tRNA. The anticodon (GAA) is at the *bottom* as depicted in Figure 2-33. **C.** Asparaginyl tRNA synthetase. α-Helical and β-sheet structures are shown. Figure A is redrawn from http://www.nchi.nim.nih.gov. Figure B is reproduced from PDB ID: 1QRS. J.G. Arnez, T.A. Steitz. Crystal Structure of Three Misacylating Mutants of *Escherichia coli* Glutaminyl-tRNA Synthetase Complexed with tRNA (Gln) and ATP. *Biochem.* **35** pp. 14725 (1996). Figure C reproduced from PDB ID: 1X56. W. Iwasaki et al. *J. Mol. Biol.* **360** pp. 329 (2006).

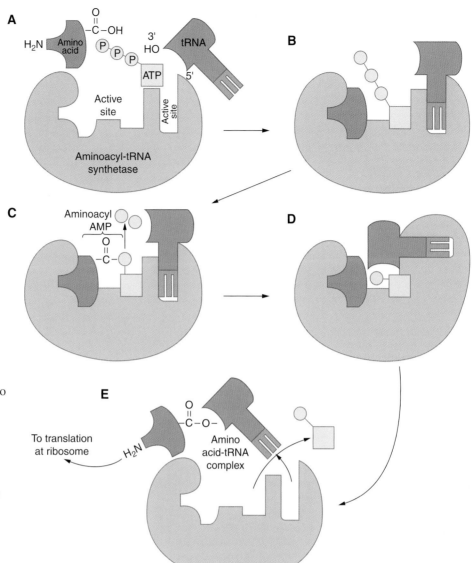

Figure 2-44. Individual steps occurring in the active site of amino acid tRNA synthetase: *(A)* the amino acid, ATP and tRNA bind to the enzyme; *(B)* substrates bound; *(C)* aminoacyl AMP is formed between amino acid and ATP with exclusion of $2P_i$; *(D)* tRNA comes into position to form tRNA amino acyl complex with dissociation of AMP; *(E)* the free anticodon directs the complex to translation at the ribosome.

Initiation

Going back to the beginning of the process, initiation factors (proteins) come into play that have not been discussed. Initiation starts with a complex of **eukaryotic initiation factor-2 (eIF-2)**, GTP, and Met-tRNA^{met}. This complex then binds to the 40S ribosomal subunit, followed by the binding of two additional initiation factors, eIF-3 and eIF-1A. The overall complex with the ribosomal subunit is now at 43S. mRNA combines with other factors to form the 48S initiation complex. The initiation complex finds the first AUG start codon, and the 60S subunit adds on to generate the 80S initiation complex with the hydrolysis of GTP (a function of eIF-2B). Translation starts as described earlier. These events are summarized in Figures 2-47 and 2-48. Figure 2-49 shows the x-ray structure of the largest of three subunits

Proteins   73

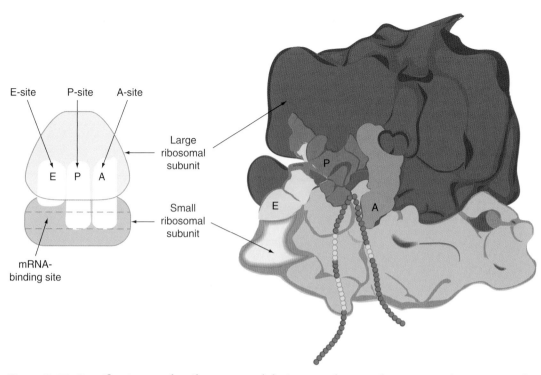

Figure 2-45. Specific sites on the ribosome used during translation. The A-site is the amino acyl acceptor site, the P-site is the peptidyl site, and the E-site is the exit site.

in the prokaryotic initiation factor, archaeal IF-2, similar to eIF-2. Its structure appears to be similar to that of elongation factors. It is likely that the eukaryotic initiation factors resemble the prokaryotic proteins. Initiation is regulated by specific phosphorylation. Ribosomal protein S6 and initiation factor eIF-4F are both activated by phosphorylation. Phosphorylation occurs by the action of various kinases, and the stimulus pathway of phosphorylation starting with the activation of cell surface receptors is shown in Figure 2-50. The phosphorylation of eIF-2A causes it to bind all available eIF-2B, which is needed for the generation of energy through the hydrolysis of GTP (Figure 2-47). 4E-binding protein (4E-BP) that binds to eIF-4E (Figure 2-48) prevents initiation. The structure of eIF-4E is shown in Figure 2-51. **Growth factors** cause the 4E-BP to be phosphorylated, which results in the inactivation of 4E-BP, relieving the inhibition and allowing translation to proceed. The process of initiation is summed up in Figure 2-52.

Elongation

eEF-1 and eEF-2 are the major eukaryotic elongation factors. EF-1α-GTP brings activated tRNA molecules to the A site of the 80S ribosome whose P site is occupied (Figure 2-53A). GTP provides energy for this process when the terminal phosphate group is released. eEF-1α-GDP, the inactive form, is then released by eEF-2-GTP, a translocase (Figure 2-53B). The crystal structure of eEF-2 is shown in Figure 2-54. The cycle is repeated

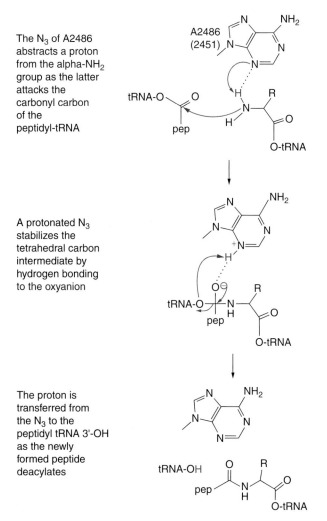

The N$_3$ of A2486 abstracts a proton from the alpha-NH$_2$ group as the latter attacks the carbonyl carbon of the peptidyl-tRNA

A protonated N$_3$ stabilizes the tetrahedral carbon intermediate by hydrogen bonding to the oxyanion

The proton is transferred from the N$_3$ to the peptidyl tRNA 3'-OH as the newly formed peptide deacylates

Figure 2-46. Mechanism by which the ribozyme catalyzes the formation of a peptide bond between two amino acids on the ribosome during translation. Reproduced from http://anx12.bio.uci.edu/~hudel/bs99a/lecture22/lecture3_4b.html. Adapted from T.R. Cech et al., *J. Biol. Chemo.*, 267: 17479–17482, 1992.

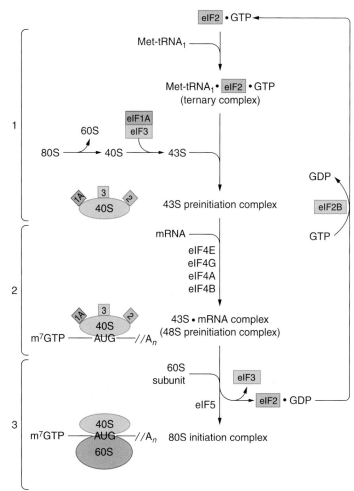

Figure 2-47. Addition of initiation factors eIF2 to form a ternary complex with the 40S ribosomal subunit to generate the 43S preinitiation complex, addition of other initiation factors to form the 48S preinitiation complex, and addition of the 60S subunit to form the 80S initiation complex. eIF2B promotes GTP exchange on eIF2 once the AUG codon is found.

Figure 2-48. Poly-A tail of mRNA and regulation of initiation at the cap. eIF4E is a cap-binding protein; it binds to G and A and is active only with this group = eIF4F. eIFG has binding sites for eIF3 on the small ribosomal subunit and for poly A binding protein Pab1P that bridges the poly A tail and the cap. This is regulated by 4E-BP, which binds to eIF4E to prevent initiation. Growth factors cause 4E-BP to be phosphorylated, resulting in the inactivation of 4E-BP, and allowing translation to proceed.

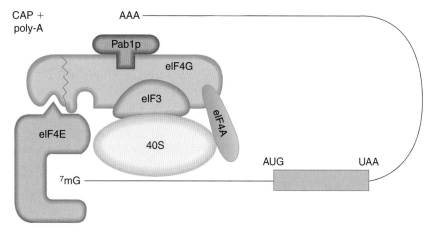

Proteins 75

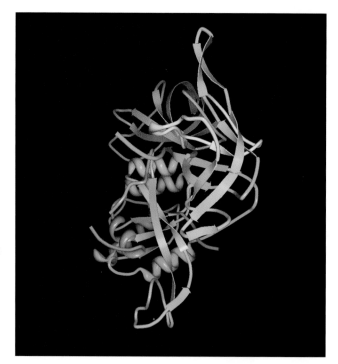

Figure 2-49. Molecular structure of the large gamma subunit of initiation factor eIF2 from *Pyrococcus abyssi*. This factor delivers the initiator methionyl-tRNA to the ribosome. Two of the three subunits including this one, bind zinc tightly. Reproduced from PDB ID: 1KKO. E. Schmitt, S. Blanquet, Y. Mechulam. The Large Subunit of Initiation Factor aIF2 Is a Close Structural Homologue of Elongation Factors. *EMBO J.* **21** pp. 1821 (2002).

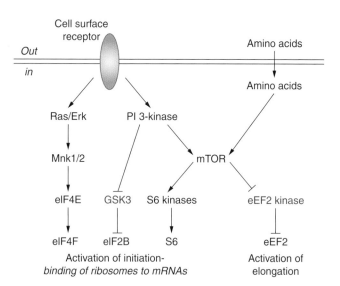

Figure 2-50. Cell surface receptors and their signaling pathways are involved in the activation of mRNA translation. Peptide chain initiation and elongation are stimulated by activating translation factors by altering their phosphorylation states. Two kinases, GSK3 and eEF2, are inactivated as part of this process. Amino acids *(upper right)* positively regulate the signaling events linked to the target of rapamycin, mTOR. eIF4E and eIF2B are eukaryotic initiation factors. S6 is a protein of the small ribosomal subunit. eEF2 is elongation factor 2. Reproduced with permission from http://www.dundee.ac.uk/biocentre/SLSBDIV6egp.htm.

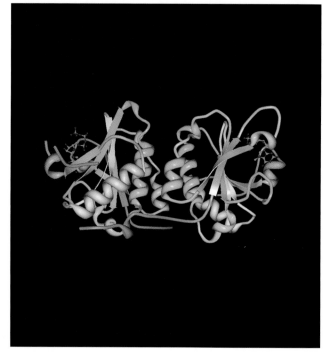

Figure 2-51. Cocrystal structure of the messenger, RNA 5' cap-binding protein (eIF4E) bound to 7methylGPPPG. Reproduced with permission from PDB ID: 1L8B. A. Niedzwiecka, J. Marcotrigiano, J. Stepinski, M. Jankowska-Anyszka, A. Wyslouch-Cieszynska, M. Dadlez, A.C. Gingras, P. Mak, E. Darzynkiewicz, N. Sonenberg, S.K. Burley, R. Stolarski. Biophysical Studies of eIF4E Cap-Binding Protein: Recognition of mRNA 5' Cap Structure and Synthetic Fragments of eIF4G and 4E-BP1 Proteins, *J. Mol. Biol.* **319** pp. 615 (2002).

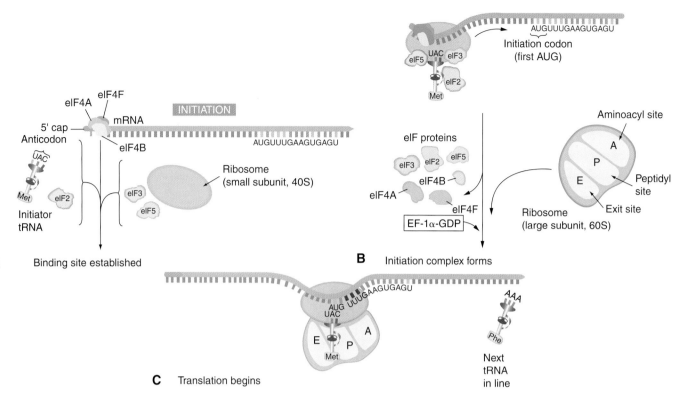

Figure 2-52. Summary of the steps in eukaryotic initiation of translation. Met-tRNAi, the initiator complex, binds to the 40S ribosome. *(A)*. Met is not formylated as it is in prokaryotes. eIF4A and eIF4F bind to the cap at the 5' end of mRNA. eIF4B also binds. This facilitates binding of mRNA and moving it to the translation site. eIF2 plus GTP binds met-tRNA and enters the 40S ribosomal P site in a ternary complex. CAP binding protein (CBP), not shown, binds to 5' cap of mRNA. eIF4A unwinds mRNA secondary structure for about 15 bases, then eIF4B assists further unwinding of mRNA. Energy for this process is provided by ATP. eIF3 helps to bind the met-tRNA 40S ribosome to mRNA, which moves along mRNA to the AUG start codon *(B)* using ATP. eIF-4A and eIF-4B dissociate. eIF-6 moves large ribosomal subunit to met-tRNAi-40S-mRNA complex, and the 60S ribosome combines when eIF-5 causes eIF-2 and eIF-3 to dissociate. eIF-4C also assists in joining 40S and 60S subunits. After joining of subunits is complete, all eIF factors dissociate so that Met-tRNAi-40S-60S-mRNA may engage in translation *(C)*. Redrawn with permission from http://a-s.clayton.edu/hampikian/b4201/HartlOb/chap11GeneExpression.htm.

with the next charged tRNA in line as specified by mRNA (Figure 2-53C). The crystal structure of EF-1α is shown in Figure 2-55.

Termination

Stop codons, which are antisense codons, are used in prokaryotes, as well as in eukaryotes. They are UAG (amber), UAA (ochre), and UGA (opal) (see Figure 2-34). UAG is seen in Figure 2-41. The steps in termination of chain elongation are shown in Figure 2-56. A **release factor** recognizes the stop codon, in this case UGA, and binds in the A site (Figure 2-56B). The polypeptide is released from the last tRNA, and the tRNA is expelled. The ribosome then dissociates into its two component subunits (Figure 2-56C). The message and subunits of the ribosome can be used again in another cycle of protein synthesis.

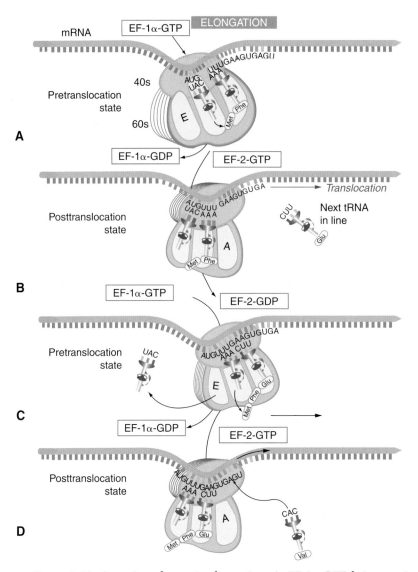

Figure 2-53. Steps in enkaryotic elongation. **A.** EF-1α-GTP brings activated tRNA molecules to the A site of the 80s ribosome, whose P site is occupied. **B.** Conversion of GTP to GDP provides energy. eIF-1d-GDP (inactive) is released by eEF-2-GTP, a translocase. The cycle is then repeated as shown in *C* and *D*. Redrawn with permission from former site: http://a-s.clayton.edu/hampikian/b4201/HartlOb/Chap11GeneExpression.htm.

Transfer of Completed Proteins in the Cytoplasm Targeted to the Mitochondrion

As seen in Chapter 1, the mitochondrion has two membranes with two spaces between the membranes. Proteins synthesized in the cytoplasm but destined for the mitochondria have signal sequences, which are positively charged regions stretching from 10 to 70 amino acids in length. These regions are located at the N-terminal end of the protein. The completed protein forms helices that are charged positively on one side and are hydrophobic on the other. The mitochondrion contains a receptor that binds to the **preprotein** (a completed polypeptide chain

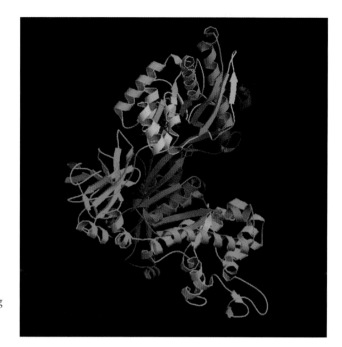

Figure 2-54. Crystal structure of eEF-2 isolated from yeast *(Saccharomyces cerevisiae)* complexed with a small molecule (sordarin) in black dots. From PDB ID: 1NOU. R. Joergensen, P.A. Ortiz, B.L. Mark, D.J. Mahuran, M.M. Cherney, D. Zhao, S. Knapp, M.N.G. James. Crystal Structure of Human Beta-Hexosaminidase B: Understanding the Molecular Basis of Sandhoff and Tay-Sachs Disease. *J. Mol. Biol.* **327** pp. 1093 (2003).

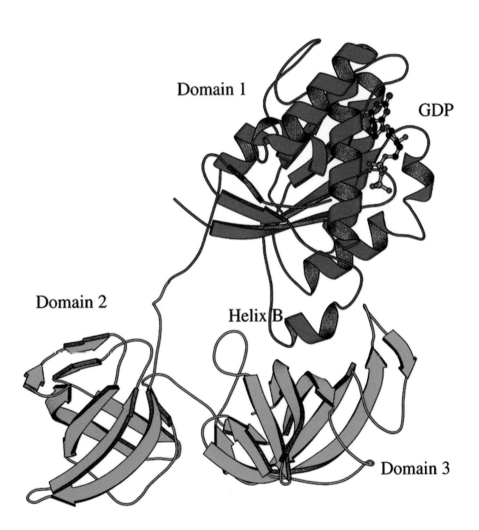

Figure 2-55. Crystal structure of EF-1α-GDP. GDP is shown as a ball and stick structure *(upper right)*. From L. Vitagliano et al., *EMBO J.*, 20: 5305–5311, 2001. Reproduced with permission from http://www.nature.com/emhoj/journal/v20/n19/fig_tab/7594024.fl.html.

Proteins 79

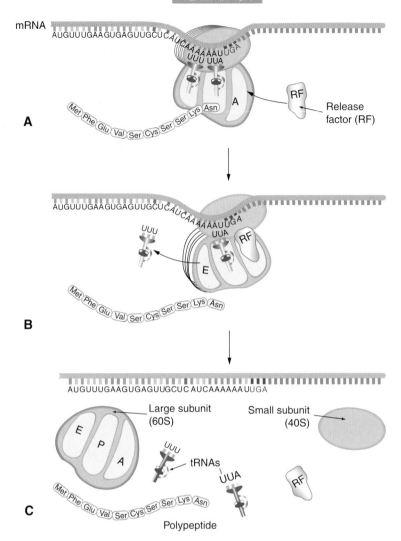

Figure 2-56. Steps in termination of protein synthesis. Release factor (RF) recognizes the stop codon (UGA) and binds to the A site as shown in *A* and *B*. The polypeptide is released from the last tRNA, and the tRNA is expelled. The ribosome then dissociates into its two component subunits as shown in *C*. Redrawn with permission from http://a-s.clayton.edu/hampikian/b4201/HartlOb/Chap11GeneExpression.htm.

with sequences added, usually to the N-terminus, that will be cleaved later), and the protein is transported into the mitochondrion. The outer mitochondrial membrane is called the translocator outer membrane, or **Tom complex.** The translocator inner membrane is called the **Tim complex.** The positive charges on the protein to be moved in must be at the point of entry. Most proteins have to be unwound to pass through the translocators, and this is accomplished with the energy provided by ATP together with **chaperone** proteins, usually 70-kDa heat shock protein (HSP70), as shown in Figure 2-57. The receptor escorts the protein to the translocator involving other proteins. The translocation is stimulated by the membrane potential. Protons (H^+) are pumped from the inner membrane to the intermembrane space so

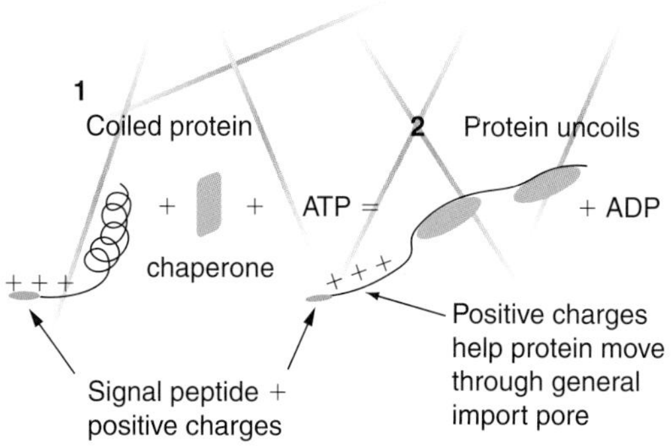

Figure 2-57. Proteins made in the cytoplasm enter the mitochondrion if they contain an appropriate translocation signal at the N-terminus that contains positive changes. The preprotein (containing the signal sequence) is uncoiled with the assistance of chaperones and energy from ATP and the protein enters the mitochondrial translocator N-terminal first.

that the matrix is more electronegative. The protein, having a positively charged signal, is attracted in and moves through the Tim translocators. The internalized protein is folded again with the assistance of HSP60, a **chaperonin,** using the energy of ATP hydrolysis. A more detailed discussion can be found in http://www.cytochemistry.net/cell-biology/mitochondria_lifecycle.htm.

Protein Folding

Protein folding is encoded in the sequence of amino acids, and much of the folding of newly synthesized protein must occur automatically. However, it is known that some HSPs (chaperone proteins) can assist in the folding of polypeptide chains, using their inherent ATPase activity for energy (by hydrolyzing ATP), and they are known to prevent aggregation of completely folded proteins. There are two important classes of chaperonins, HSP70 and HSP60. HSP70 becomes active when there are unfolded regions in newly synthesized protein chains and has an affinity for nonpolar regions. It binds to these nonpolar regions until productive folding of these chains can take place. In eukaryotes, folding of the peptide chain seems to occur spontaneously, resulting from the interactions between neighboring side chains of amino acids as the completed chain folds into the lowest free energy state. The association of subunits of completed proteins also occurs spontaneously when the native protein has a complex multisubunit (for example, quaternary) structure. Some amino acids contribute more to one form of secondary structure than others. Alanine, leucine, and glutamic acid are found in the α-helix configuration more often than other amino acids. Isoleucine, valine, and tyrosine are often found in β-sheet structures. As mentioned before, glycine and proline, as well as asparagine, are found when the polypeptide chain bends. The specific details of protein folding remain a research question and are an area of active study.

Protein Degradation

Proteins can be broken down in the lysosome nonspecifically (see Chapter 1). There is also a specific pathway to degradation called the ubiquitin pathway. **Ubiquitin** is a small, highly conserved protein (in eukaryotes) of molecular weight 8500 D (**Daltons**), which contains 76 amino acid residues. The three-dimensional structure of ubiquitin is shown in Figure 2-58. Proteins that bind ubiquitin are committed to degradation by this pathway. Three proteins are involved in this pathway. There is a ubiquitin-activating enzyme, called E_1; a ubiquitin-carrier or conjugating protein, E_2; and a ligase, E_3. The reactions activating ubiquitin to a form that can combine with the protein substrate are shown in Figure 2-59. Three proteins are involved: two enzymes and a carrier protein. The result is a protein substrate marked by a bound ubiquitin or ubiquitins. The first step produces an adenylate intermediate that resembles tRNA aminoacylation. E_3, the ligase (substrate recognition protein), selects the protein to be marked for degradation, and the enzyme often targets proteins with lysine epsilon amino groups. Proteins selected by the ligase have a free amino terminus, which can be leucine, isoleucine, phenylalanine, proline, tryptophan, asparagine, glutamine, glutamic acid, aspartic acid, lysine, arginine, histidine, or tyrosine but usually not cysteine, glycine, threonine, valine, serine, alanine, or methionine. Some proteins have a

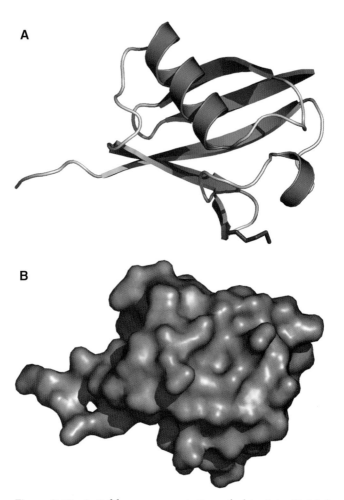

Figure 2-58. **A.** Ribbon representation of ubiquitin. **B.** Molecular surface of ubiquitin. Reproduced from http://en.wikipedia.org/wiki/ubiquitin.

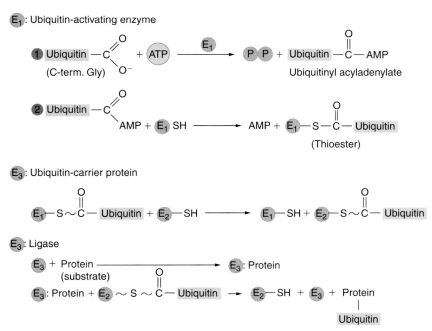

Figure 2-59. Reactions of ubiquitin to generate a form that can bind to a protein substrate with the aid of a ligase, E3. In *1*, the glycine C-terminus of ubiquitin is activated by ATP, catalyzed by ubiquitin-activating enzyme to form ubiquitin acyladenylate. This binds to the activating enzyme through the enzyme's SH group to form the ubiquitin-activating enzyme thioester. This reacts with a carrier protein SH to form a thioester with the carrier protein. Then a ligase *(E₃)* binds the substrate protein and the complex combines with carrier protein thioester ubiquitin to release the carrier protein, the ligase, and the ubiquitinated protein substrate.

PEST sequence (such as Pro-Glu-Ser-Cys), which makes them readily degradable by this pathway. Secreted proteins are good targets, with the N-terminus exposed after transport into the endoplasmic reticulum (the signal sequence is removed). The structure of a ubiquitin-conjugating enzyme is shown in Figure 2-60. Proteins that are ubiquinated are degraded in the 26S proteasome as shown in Figure 2-61. Free amino acids generated from degradation become deaminated and enter glycolytic or citric acid cycles (Figure 2-20) at various intermediates. The amino group, as ammonium, a potentially toxic substance, is excreted through the urea cycle.

Inhibitors of Protein Synthesis

Inhibitors of protein synthesis that affect prokaryotes without affecting or minimally affecting eukaryotes are useful antibiotics because cells that cannot synthesize proteins will die. Antibiotic use in humans is not always without side effects. **Streptomycin,** used successfully to treat

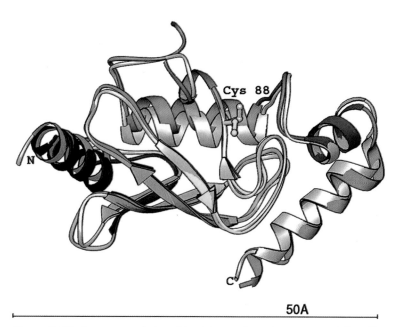

Figure 2-60. Structure of the ubiquitin-conjugating enzyme. The silver ribbon compares the enzyme from *Caenorhabditis elegans*. Reproduced with permission from http://sgce.cbse.uab.edu/sgoc/Structures/1-D6/.

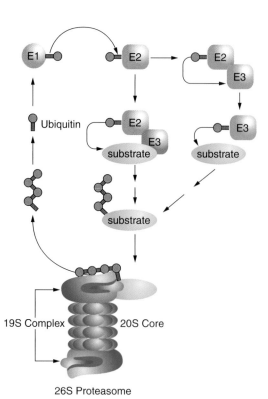

Figure 2-61. The ubiquitin-proteasome system. Ubiquinated proteins are recognized by the 19S regulatory complex (cap complex). Proteins are unfolded by the 19S caps and transported to the proteases of the 20S core complex where degradation occurs. The 20S core is a hollow cylinder of four stacked rings. Each ring is composed of seven subunits. The outer rings are of α-type, whereas inner rings are β-type. Outer cavities are located at the juncture of α- and β-rings. The central cavity is of β-rings and contains the proteolytic activity. E1, E2, and E3 are the enzymes of the ubiquitin pathway (Figure 2-59).

tuberculosis, often caused hearing loss. Streptomycin aminoglycoside causes mRNA misreading, generating mutant proteins that inhibit bacterial growth. **Puromycin** binds the ribosomal A site of prokaryotic and eukaryotic organisms. There it blocks the peptide chain coming from the P site, and protein synthesis is terminated. **Tetracycline** binds to the prokaryotic 30S ribosomal subunit and inhibits subsequent binding of aminoacyl-tRNAs. **Erythromycin** binds to the prokaryotic large (50S) ribosomal subunit, inhibiting translocation. **Chloramphenicol** inhibits the peptidyl transferase enzyme activity of the 50S prokaryotic large ribosomal subunit. **Cycloheximide** also inhibits peptidyl transferase but, in this case, affects the eukaryotic enzyme. Figure 2-62 shows the structures of these inhibitors.

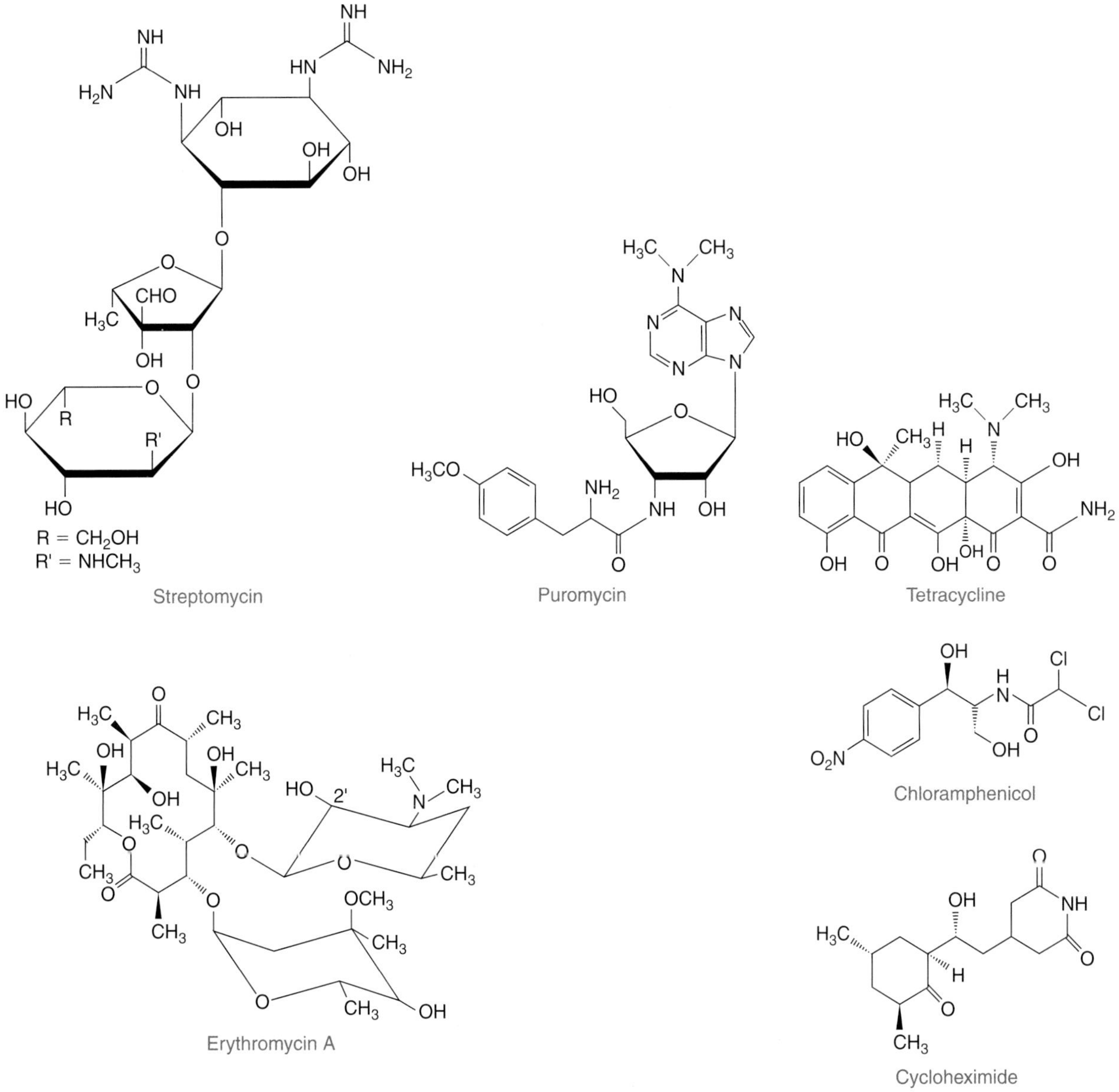

Figure 2-62. Structures of well-known protein synthesis inhibitors.

Classification of Proteins

There are many ways to classify proteins. Size and shape represent one rough classification, as shown in Figure 2-63. Single polypeptide chains vary from 20 to 500,000 amino acid residues. **Globular proteins,** such as albumin and globulin, have high amounts of acidic and basic amino acids, rendering them soluble in dilute salt solutions. **Fibrous proteins,** such as fibrinogen, have high concentrations of a few amino acids such as glycine and proline. Another means to classify proteins is on their solubility in water, salt solutions, and aqueous ethanol, as well as in acidic and alkaline solutions. From solubility characteristics, proteins can be divided into albumins, globulins, prolamins, and glutamins (Figure 2-64). Collagen represents an interesting structure consisting of three α-chains having a Gly-Pro-Pro repeated sequence, clearly a fibrous protein. Collagen structure is shown in Figure 2-65. Half of the Pro side chains are hydroxylated. The helix is formed by every third glycine, because a residue larger than glycine would prohibit the close contact of the three chains. Collagen I makes up 90% of the collagen, and collagen is the most plentiful protein in the body. There are numerous specialized proteins that carry out specific functions, and many of these will be encountered in context. A database exists for protein structure classification called **CATH.** CATH stands for the Class, Architecture, Topology, and Homologous superfamily. There is a further classification based on sequence identity, which indicates highly similar structure and function. Some details of this system are given in Figure 2-66.

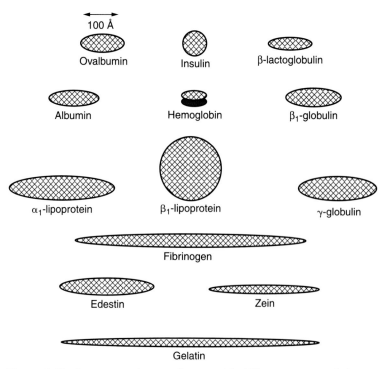

Figure 2-63. Some proteins are shown with different sizes and shapes.

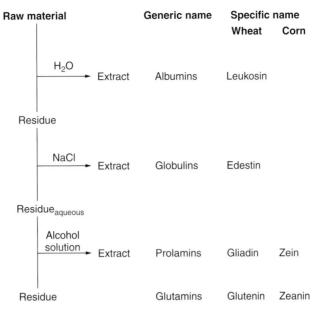

Figure 2-64. A sample classification of certain proteins based on extraction solubility.

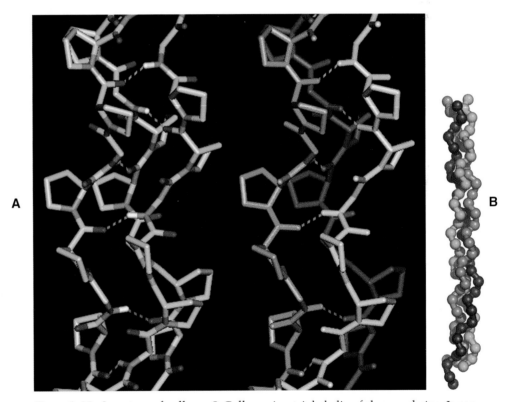

Figure 2-65. Structure of collagen I. Collagen is a triple helix of three α-chains. It contains a repeated sequence of Gly-Pro-Pro in each chain. Hydrogen bonds are shown by dashed lines in A. The three chains are visible by different colors in A (right). Collagen I is a heterotrimer in which two of the α-chains are identical. In B, there is a space-filling model of the same structure. Part A reproduced with permission from http://www.cryst.bbk.ac.uk/PPS2/course/sectionII/II_coll2.gif. Part B reproduced with permission from http://www.cryst.bbk.ac.uk/PPS2/course/sectionII/II_coll3.gif.

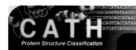

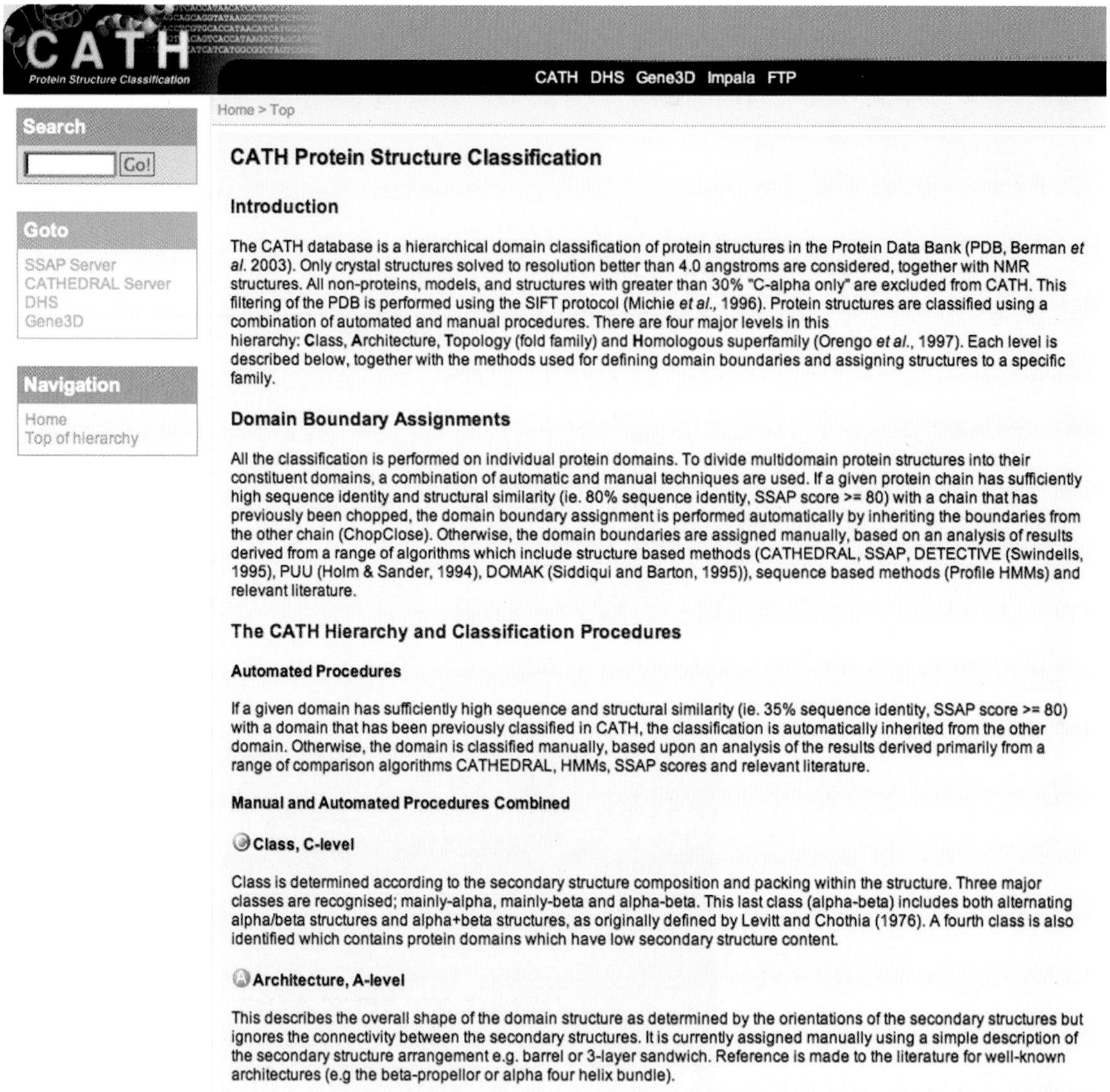

CATH
Protein Structure Classification

CATH DHS Gene3D Impala FTP

Search

Go!

Goto

SSAP Server
CATHEDRAL Server
DHS
Gene3D

Navigation

Home
Top of hierarchy

Home > Top

CATH Protein Structure Classification

Introduction

The CATH database is a hierarchical domain classification of protein structures in the Protein Data Bank (PDB, Berman *et al.* 2003). Only crystal structures solved to resolution better than 4.0 angstroms are considered, together with NMR structures. All non-proteins, models, and structures with greater than 30% "C-alpha only" are excluded from CATH. This filtering of the PDB is performed using the SIFT protocol (Michie *et al.*, 1996). Protein structures are classified using a combination of automated and manual procedures. There are four major levels in this hierarchy: **C**lass, **A**rchitecture, **T**opology (fold family) and **H**omologous superfamily (Orengo *et al.*, 1997). Each level is described below, together with the methods used for defining domain boundaries and assigning structures to a specific family.

Domain Boundary Assignments

All the classification is performed on individual protein domains. To divide multidomain protein structures into their constituent domains, a combination of automatic and manual techniques are used. If a given protein chain has sufficiently high sequence identity and structural similarity (ie. 80% sequence identity, SSAP score >= 80) with a chain that has previously been chopped, the domain boundary assignment is performed automatically by inheriting the boundaries from the other chain (ChopClose). Otherwise, the domain boundaries are assigned manually, based on an analysis of results derived from a range of algorithms which include structure based methods (CATHEDRAL, SSAP, DETECTIVE (Swindells, 1995), PUU (Holm & Sander, 1994), DOMAK (Siddiqui and Barton, 1995)), sequence based methods (Profile HMMs) and relevant literature.

The CATH Hierarchy and Classification Procedures

Automated Procedures

If a given domain has sufficiently high sequence and structural similarity (ie. 35% sequence identity, SSAP score >= 80) with a domain that has been previously classified in CATH, the classification is automatically inherited from the other domain. Otherwise, the domain is classified manually, based upon an analysis of the results derived primarily from a range of comparison algorithms CATHEDRAL, HMMs, SSAP scores and relevant literature.

Manual and Automated Procedures Combined

Class, C-level

Class is determined according to the secondary structure composition and packing within the structure. Three major classes are recognised; mainly-alpha, mainly-beta and alpha-beta. This last class (alpha-beta) includes both alternating alpha/beta structures and alpha+beta structures, as originally defined by Levitt and Chothia (1976). A fourth class is also identified which contains protein domains which have low secondary structure content.

Architecture, A-level

This describes the overall shape of the domain structure as determined by the orientations of the secondary structures but ignores the connectivity between the secondary structures. It is currently assigned manually using a simple description of the secondary structure arrangement e.g. barrel or 3-layer sandwich. Reference is made to the literature for well-known architectures (e.g the beta-propellor or alpha four helix bundle).

Figure 2-66. CATH protein structure classification. Reproduction of a web page describing the CATH base. Reproduced from http://cathwww.biochem.ucl.ac.uk/latest/index.html.

(continued)

Topology (Fold family), T-level

Structures are grouped into fold groups at this level depending on both the overall shape and connectivity of the secondary structures. This is done using the structure comparison algorithm SSAP (Taylor & Orengo, 1989) and CATHEDRAL (Harrison et al. 2002, 2003). Parameters for clustering domains into the same fold family have been determined by empirical trials throughout the databank (Orengo et al. 1992; Orengo et al. 1993; Harrison et al. 2002, 2003). Structures which have a SSAP score of 70 and where at least 60% of the larger protein matches the smaller protein are assigned to the same T level or fold group.

Some fold fgroups are very highly populated (Orengo et al. 1994; Orengo & Thornton, 2005) particularly within the mainly-beta 2-layer sandwich architectures and the alpha-beta 3-layer sandwich architectures.

Homologous Superfamily, H-level

This level groups together protein domains which are thought to share a common ancestor and can therefore be described as homologous. Similarities are identified either by high sequence identity or structure comparison using SSAP. Structures are clustered into the same homologous superfamily if they satisfy one of the following criteria:

- Sequence identity >= 35%, overlap >= 60% of larger structure equivalent to smaller.
- SSAP score >= 80.0, sequence identity >= 20%, 60% of larger structure equivalent to smaller.
- SSAP score >= 70.0, 60% of larger structure equivalent to smaller, and domains which have related functions, which is informed by the literature and Pfam protein family database, (Bateman et al., 2004).
- Significant similarity from HMM-sequence searches and HMM-HMM comparisons using SAM (Hughey &Krogh, 1996), HMMER (http://hmmer.wustl.edu) and PRC (http://supfam.org/PRC).

Sequence Family Levels: (S,O,L,I, D)

Domains within each H-level are subclustered into sequence families using multi-linkage clustering at the following levels:

Level	Name	Sequence Identity Overlap
S	35%	80%
O	60%	80%
L	95%	80%
I	100%	80%

The D-level acts as a counter within each S100 family and is appended to the classification hierarchy to ensure that every domain in CATH has a unique CATHSOLID classification. The sequence identity and overlap used for clustering are obtained from an implementation of the Needleman-Wunsch algorithm (Needleman & Wunsch, 1970) using a gap penalty of 3. The percentage sequence identity is calculated as (100 * Number Of Identical Residues/Length Of The Shortest Sequence) and the percentage overlap is calculated as (100 * Number Of Aligned Residues/Length Of The Longest Sequence).

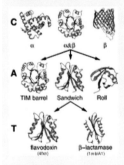

> Back to Index page

Please email any comments or questions to cathteam@biochem.ucl.ac.uk

Figure 2-66, cont'd

Proteome

For many years, there has been an effort to map all proteins in a given cell, that is, the proteins expressed by a genome. The method involves obtaining a homogeneous cell preparation, spotting a sample of the protein content on a polyacrylamide gel plate, and subjecting it to two-dimensional electrophoresis. Proteins are visualized by a protein-specific stain. Individual proteins are separated on the basis of molecular weight and pI. The power of these systems has been increased to thousands of proteins and, in some cases, it is possible to "scoop" the protein spot and subject it to sequencing so that the protein can be identified and its gene can be cloned. A typical gel of hypothalamic cell proteins is shown in Figure 2-67. Cells from many tissues and from many organisms have now been mapped in this fashion, and there are accessible databases available (an example is in Figure 2-68).

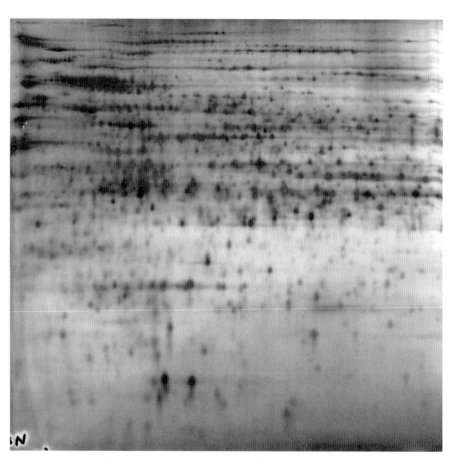

Figure 2-67. Example of a two-dimensional polyacrylamide gel electrophoresis (2DPAGE) of the proteins from a tissue. Reproduced with permission from http://www.pierroton.inra.fr/genetics/2D/pollenb.jpg.

Figure 2-68. Example of a company providing access to proteome databases. Reproduced from http://www.incyte.com/control/researchproducts/insilico/proteome.

Further Reading

Books

Branden, C.I., and Tooze, J., *Introduction to Protein Structure,* Garland, 1991.

Creighton, T., *Proteins: Structure and Molecular Properties,* W.H. Freeman and Co., 1993.

Lesk, A.M., *Introduction to Protein Architecture: The Structural Biology of Proteins,* Oxford University Press, 2001.

Voet, D., and Voet, J.G., *Biochemistry,* Wiley Text Books, 1995.

Reviews

Baskakov, I.V., Legname, G., Gryczinski, Z., and Prusiner, S.B., "The particular nature of unfolding of the human prion protein," *Protein Sci.,* **13:** 586–595, 2004.

Bernhard, S.A., Dahlquist, F.W., and Matthews, B.W. (eds.), *Classic Papers on Protein Structure and Function,* University Science Books, 2001.

Bosshard, H.R., Marti, D.N., and Jelesarov, I., "Protein stabilization by salt bridges: concepts, experimental approaches and clarification of some misunderstandings," *J. Mol. Recognit.,* **17:** 1–16, 2004.

Papers

Regan, L., "What determines where alpha-helices begin and end?" *Proc. Natl. Acad. Sci.,* **90:** 10,907–10,908, 1993.

CHAPTER 3

Enzymes

Clinical Enzymology in Diagnosis of Disease

Clinical enzymology has been in use in diagnostic laboratories for about 50 years. It is based on the premise that when a tissue is damaged, infected, or inflamed, cell membranes become more permeable or are destroyed so that the content of the cell cytoplasm, especially the dissolved substances, are able to pass in solution to the extracellular space and then into the bloodstream. The blood can be drawn, plasma can be prepared, and the enzymatic activity or some other characteristic of the enzyme can be measured. These measurements have been useful, particularly in the diagnosis of damage to the heart, liver, and muscle, and enzyme measurements are useful in following the healing process and the prognosis. The activity of the enzyme must be sufficient to overcome the diluting effects of the total circulation (about 8 quarts). Enzyme activity will increase in the blood proportionately to the extent of tissue damage.

Certain enzymes have been of interest over the years. Those concentrated in the heart and liver are aspartate aminotransferase (also referred to as serum glutamate-oxaloacetate transaminase) and alanine aminotransferase (also called serum glutamate-pyruvate transaminase). Alkaline phosphatase is reflective of bone, intestine, and other tissues. Creatine kinase is reflective of skeletal and cardiac muscle. Lactate dehydrogenase (LDH) is reflective of heart, liver, muscle, and red blood cells. α-Amylase reflects the pancreas, and acid phosphatase, especially tartrate labile acid phosphatase, reflects the prostate gland. By measuring several enzyme activities, a pattern is seen that characterizes an organ. However, the search can be made more specific and defining when the isoenzymes, if a particular enzyme has them, are measured. The case in point is LDH, whose isoenzymes can be determined electrophoretically. LDH consists of four subunits made up of two kinds, a heart-type (H) and a skeletal muscle type (M). Monomers are formed first, then dimerization occurs, and finally dimers, or monomers, interact to form tetramers. This gives rise to five isoenzymes having the following composition of subunits: H_4 (LDH1), H_3M_1 (LDH2), H_2M_2 (LDH3), HM_3 (LDH4) and M_4 (LDH5), as diagrammed in Figure 3-1. The crystal structure of the dogfish M_4 tetramer, shown in Figure 3-2, emphasizes the quaternary structure of the

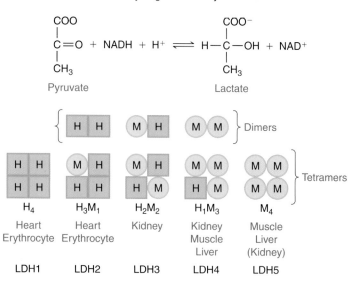

Figure 3-1. Lactate dehydrogenase (LDH). It catalyzes the reaction shown at the top. The bottom figures show the subunit compositions of the dimers and the tetramers. The H_4 (LDH1) enzyme reflects heart myocardium and erythrocyte, and the M_4 (LDH5) reflects muscle and liver primarily.

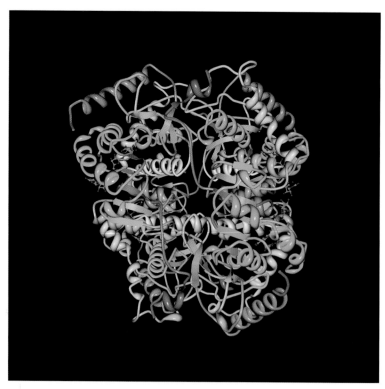

Figure 3-2. Assumed biological molecule of dogfish lactate dehydrogenase M_4 tetramer. Each subunit is shown in a different color. Reproduced from PDB ID: 6LDH. C. Abad-Zapatero, J.P. Griffith, J.L. Sussman, M.G. Rossman. Refined Crystal Structure of Dogfish M4 Apo-lactate Dehydrogenase. *J. Mol. Biol.* **198** pp. 445 (1987).

enzyme. H_4, which is the principal isoform in the heart, also has 3-hydroxybutyrate dehydrogenase activity, which can be assayed simultaneously to confirm the form of the enzyme. In electrophoresis, H_4 is the most negatively charged of the isoforms and therefore is the form closest to the anode. M_4, which predominates in skeletal muscle and liver, is the most positively charged isoform and migrates closest to the cathode in electrophoresis. Therefore, the separation and identification of the isoforms of LDH or the direct biochemical assay of specific isozymes, in concert with the clinical findings, affords a conclusive diagnosis of myocardial infarction, for example.

Creatine kinase is another enzyme of interest in this context. The reaction catalyzed by this enzyme is creatine + ATP = creatine-P + ADP. This is an important enzyme in muscle because creatine-P is a major energy source. This enzyme is a dimer of which one subunit is a muscle type (M) and the other is a brain type (B). Consequently there are three isozymes: MM, MB, and BB. In myocardial infarction, in addition to the LDH isozyme elevations expected from the preceding discussion, there are elevations of the MM and MB isozymes of creatine kinase. In muscular dystrophy and other skeletal muscle diseases, there are elevations of the MM isozyme in serum. Figure 3-3 shows the time course of plasma

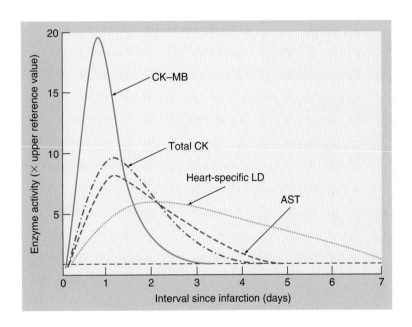

Time course of plasma enzyme activities after myocardial infarction

Enzyme	Onset (hr)	Peak (hr)	Duration (days)
Creatine kinase (MB isoenzyme; CK–MB)	3–10	12–24	$1\frac{1}{2}$–3
Creatine kinase (total; total CK)	5–12	18–30	2–5
Aspartate aminotransferase (AST)	6–12	20–30	2–6
Heart-specific lactate dehydrogenase (LD)	8–16	30–48	5–14

Figure 3-3. Patterns of plasma enzyme activities with time after myocardial infarction. Inspecting the table, it is possible to observe elevations in enzyme activities as much as 48 hours after the attack, and some activities persist for days, even weeks. *AST*, aspartate aminotransferase; *CK-MB*, MB subunit of creatine kinase; *heart-specific LD*, H_4, isozyme of lactate dehydrogenase.

enzyme activities following myocardial infarction. Blood draws are made 18 to 30 hours after the attack and 12 hours and 48 hours after that. From the information given earlier, it is clear that many diseases can be confirmed or diagnosed by measurements of enzymatic activities in plasma. Among these are acute hepatitis, jaundice, cirrhosis, muscular dystrophy, and some cancers.

Enzymes Are Catalytic Proteins

Proteins perform many functions in the body. They are antibodies, structural proteins, transporters, many hormones, receptors, chemokines, and enzymes. Some proteins have more than one function. There already has been reference in this book to heat shock protein 70 (HSP70), which is a transporter and an ATPase enzyme. Enzymes catalyze chemical reactions that otherwise could not occur under bodily conditions. These reactions, carried out in a test tube, require conditions (temperature, pressure, etc.) incompatible with life as we know it. In its simplest form, an enzyme would catalyze the conversion of one substrate to another: for example, $A \leftrightarrows B$. With an enzyme present, the reaction would be

Enzyme (E) + Substrate (S) $\leftrightarrows$ ES (enzyme–substrate complex) = E + P (product)

or simply

$$E + S \leftrightarrows ES \leftrightarrows E + P$$

It has been proven experimentally that a transient enzyme–substrate complex exists, so it is not a theoretical complex. Importantly, the enzyme is regenerated as it was in the beginning and the chemical reaction or reactions occur when the enzyme and substrate are in a complex. Many other nonenzymatic reactions of proteins can occur in much the same way to the level of formation of a complex. Thus, a reaction between antibody (Ab) and antigen (An) can be described as follows:

$$Ab + An \leftrightarrows Ab\text{-}An$$

The interaction of hormone (H) and receptor (R) can be described similarly:

$$H + R \leftrightarrows R\text{-}H$$

So, these can be considered semi-enzymic reactions, only progressing (reversibly) as far as the complex.

For a chemical reaction to proceed, energy is needed. This amount of energy is called the **energy of activation;** it is the energy that must be put into the reaction to make it go on its own. Another way of putting it is the energy that pushes the reactants to the top of the hill so that the reaction can proceed downward. Enzymes are catalysts in the sense that they greatly decrease the energy of activation required, as shown in Figure 3-4.

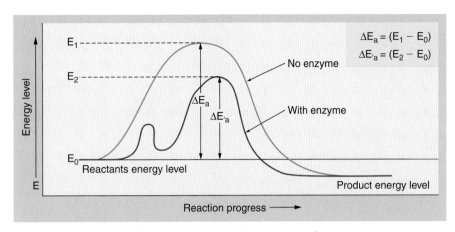

Figure 3-4. Energy of activation. Without the enzyme, the reaction requires the energy level $\Delta E_a = (E_1-E_0)$, E_0 being the energy level of the ground state. The presence of the enzyme lowers the energy level required to $\Delta E'_a = (E_2-E_0)$, clearly a lower energy of activation. Reproduced from http://www.worthington-biochem.com/introbiochem/kinetics.html.

Kinetics

Enzyme kinetics refers to a mathematical expression to describe the progress of a reaction catalyzed by a given enzyme. The simplest case would be a reaction in which there is one substrate. Bearing in mind that the rate of an enzymatic reaction is influenced by the conditions (pH, temperature, salt concentration, etc.), these are specified in a given enzymatic assay for reproducibility and comparisons with other data involving measurements of the same enzymatic activity. If the reaction is first run at several concentrations of the substrate, the rate, or velocity, of the reaction can be determined as shown in Figure 3-5. This assumes that there is a known method to measure the product formation with time or some other means to quantify the reaction. A first order simple curve is generated, which proceeds at initial velocity and then reaches saturation at the maximal velocity of the reaction as a function of substrate (S) concentration, [S], in molarity. Importantly, half of the saturation value on the y-axis gives half-maximal velocity (1/2 V_{max}), and extrapolation to the x-axis gives that concentration of substrate that produces half-maximal velocity. *This value is the* **Michaelis constant** *(Michaelis-Menten constant)* (K_m), *and this concentration of the substrate often approximates the amount of that substance in the cell.* Although the K_m value *does not give the affinity* of the enzyme for the substrate, it is sometimes used to convey an idea of the strength of the interaction between enzyme and substrate. A value of affinity can be achieved when there is only the simple interaction between two substances without release of an altered product, such as in a reaction between hormone and receptor or between antigen and antibody. Another method to view the same set of data is to plot reciprocal values: 1/[S] instead of [S] and 1/v_i instead of v_i (initial velocity). This is a **Lineweaver-Burk plot,** as shown in Figure 3-6. The

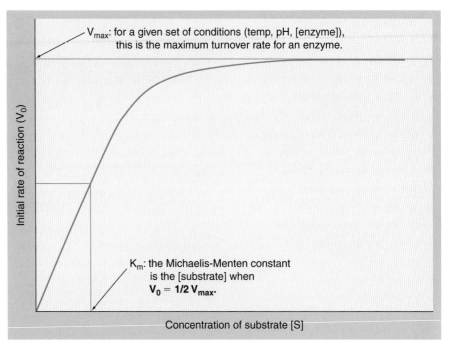

Figure 3-5. Rate or velocity (v) of an enzymatic reaction determined by varying the concentration of the substrate [S]. A first order curve is obtained of the type shown in the figure. Not shown on the curve are the experimentally determined points for each level of [substrate] used. The curve reaches saturation at the maximal obtainable velocity (V_{max}). The half-maximal velocity occurs at a point on the x-axis representing the K_m (Michaelis-Menten constant), which is that concentration of substrate that produces half-maximal velocity. Generally, the K_m reflects the approximate level of the substance, S, in the cell. [S], molar concentration.

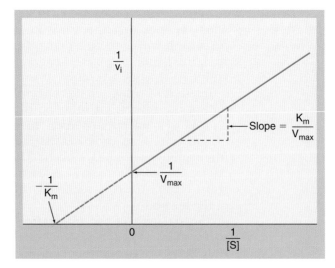

Figure 3-6. Data of Figure 3-5 replotted as reciprocals according to Lineweaver-Burk plot. The value of K_m can be determined from the x-axis intercept in the second quadrant. The value of V_{max} can be determined from the y-axis intercept, or calculated from the slope after obtaining the value of K_m.

Lineweaver-Burk equation is derived from the Michaelis-Menten equation and has the form

$$[1/v] = [K_m(1)/V_{max} [S] + (1)/V_{max}]$$

or

$$1/v_i = (K_m/V_{max})(1/[S]) + 1/V_{max}$$

This equation is in the form of a **straight line equation:**

$$y = mx + b$$

In this case, y is the ordinate (vertical) axis, x is the abscissa (horizontal) axis, m is the slope of the line, and b is the intercept on the y-axis (the intercept on the x-axis would be $-1/K_m$). In the case of the Lineweaver-Burk equation, $y = 1/v$; $m = K_m/V_{max}$; $x = 1/[S]$; and $b = 1/V_{max}$. The value of the Michaelis constant (K_m) can be obtained easily from the intercept on the x-axis and the maximal velocity (V_{max}) can be obtained from the y-axis intercept. Having a straight line to deal with is easier than measuring these values from a curve (Figure 3-5).

The Michaelis-Menten Equation

The Michaelis constant, K_m, discussed earlier is derived from the Michaelis-Menten equation. In the simple enzymatic reaction:

$$E + S \underset{k_{-1}}{\overset{k_1}{\rightleftharpoons}} ES \overset{k_2}{\longrightarrow} P + E$$

The overall reaction is $S \rightleftharpoons P$, because the enzyme (E) is on both sides of the reaction (and therefore drops out). This reinforces the catalytic role of the enzyme because a catalyst only increases the rate of the reaction and does not become changed, itself, in the reaction. The rate or velocity of this reaction is the rate of appearance of the product (P) with time (t), or $v_0 = (d[P]/dt)_0$, v_0 being the initial rate of the reaction. This is also approximately equal to the rate of formation of the enzyme–substrate complex (ES) and its breakdown:

$$v_0 = \frac{d[P]}{dt} = k_2[ES] = \frac{k_2[E]_T[S]}{K_m + [S]}$$

$[E]_T$ is the total amount of the enzyme put into the reaction, and K_m is the Michaelis constant:

$$K_m = \frac{k_{-1} + k_2}{k_1}$$

That is, the Michaelis constant is directly proportional to the rate constants leading from the enzyme–substrate complex and inversely proportional to the rate constant leading to the enzyme–substrate complex. As substrate concentration, [S], becomes large, the initial velocity, v_0, will approach maximal velocity, V_{max}, so that the rate of the reaction will no longer depend on substrate concentration. Thus, V_{max} can be substituted in the equation:

$$v_0 = k_2[E]_T[S]/K_m + [S]$$

or

$$V_{max} = k_2[E]_T = \text{constant}$$

Then the Michaelis-Menten equation becomes the following when V_{max} is substituted for $k_2[E]$:

$$v_0 = \frac{V_{max}[S]}{K_m + [S]}$$

This equation describes the plot in Figure 3-5. The changes in all participants in the reaction are shown in Figure 3-7. Another useful expression is the turnover number (k_{cat}) of an enzyme, which gives the number of enzyme–substrate complexes converted to product per enzyme molecule per unit time. Because k_2 determines the rate of conversion from ES to E + P, it approximates k_{cat}:

$$k_{cat} = k_2 = \frac{k_2[E]_T}{[E]_T} = \frac{V_{max}}{E_T} = \text{turnover number}$$

The turnover number is expressed in reciprocal seconds (1/sec or $\sec^{-1}$).

The Lineweaver-Burk equation is derived from the Michaelis-Menten equation. The reciprocal of both sides of the Michaelis-Menten equation can be taken from the following form of the Michaelis-Menten equation:

$$v_0 = \frac{V_{max}[S]}{K_m + [S]}$$

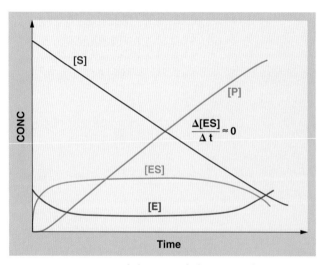

Figure 3-7. Diagram of the rate of change in the concentrations of substrate (S), product (P), enzyme (E), and enzyme–substrate complex (ES). The substrate concentration, [S] is in great excess over the total enzyme concentration [E]. The amount of enzyme-substrate complex is low at the beginning of the reaction and stays constant until the end, when the amount of substrate has become minimal.

Thus, it becomes

$$\frac{1}{v_0} = \frac{K_m}{V_{max}[S]} + \frac{[S]}{V_{max}[S]}$$

Putting this in the form of a straight line equation $(y = mx + b)$, it becomes the following:

$$\frac{1}{v_0} = \frac{K_m}{V_{max}}\frac{1}{[S]} + \frac{1}{V_{max}}$$

$$(y = mx + b)$$

This is the form plotted in Figure 3-6. This is the most convenient plot for the determination of the Michaelis constant and maximal velocity.

Enzyme Inhibition

An inhibitor, when present in an enzymatic reaction, will always decrease the rate of the reaction. There are three types of inhibitors. A **competitive inhibitor** binds in the substrate-binding site and competes with the binding of the substrate (Figure 3-8). This kind of inhibition can be overcome by increasing the amount of substrate relative to the inhibitor. Conversely, as the amount of the inhibitor is increased, the rate of the reaction will decrease more or less proportionately. Figure 3-9 shows a direct plot of enzyme activity with and without the presence of a competitive inhibitor. The two other types of inhibition are **noncompetitive** and **uncompetitive** (usually the rarest type). In Figure 3-9, the presence of a competitive inhibitor decreases the rate of the reaction and increases the apparent Michaelis constant, although it does not change the maximal velocity (if the reaction goes long enough, it will eventually reach the maximal velocity, albeit more slowly than the uninhibited enzyme). The Michaelis constant is therefore modified, in the presence of the inhibitor, by the value $1 + [I]/K_i$, where $[I]$ is the molar concentration of the inhibitor and K_i is the inhibition constant. The inhibition constant

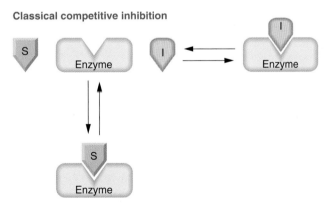

Figure 3-8. Model showing that a competitive inhibitor binds in the substrate-binding site.

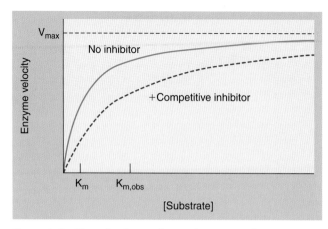

Figure 3-9. Plot of velocity (y-axis) versus substrate concentration (x-axis) in presence or absence of a competitive inhibitor. Values of K_M and v are changed, but V_{max} is unaffected. The K_M observed in the presence of the inhibitor is modified by the quantity $(1 + [I]/K_i)$, where [I] is the molar concentration of inhibitor, and K_i is the inhibition constant of the inhibitor.

can be defined as the dissociation constant of the enzyme and the inhibitor from the reaction:

$$E + I \rightleftharpoons EI$$

Thus,

$$K_i = \frac{[E][I]}{[EI]}$$

The double reciprocal plot of this reaction is shown in Figure 3-10.

A comparison of competitive and noncompetitive inhibition is shown in Figure 3-11, where velocity is plotted against substrate concentration. Figure 3-12 shows a model of noncompetitive inhibition. In this model, it is clear that the inhibitor binds to a site that is different from the substrate-binding site (the case in competitive inhibition). Because of this, excess substrate will not reverse the inhibition, which is the case with competitive inhibition. Figure 3-13 shows the double reciprocal plot for the case of noncompetitive inhibition. Whereas in competitive inhibition the slope of the inhibited curve gives a Michaelis constant, K_m, that is increased by the factor $1 + [I]/K_i$, in the case of noncompetitive inhibition, *both* the slope and the value of the reciprocal velocity, $1/v$, are increased by the same value: $1 + [I]/K_i$. In binding to a site apart from the substrate-binding site on the enzyme, the noncompetitive inhibitor changes the shape of the enzyme, allowing the substrate to bind, but the catalytic event does not occur and product is not released. In the third case of uncompetitive inhibition, the inhibitor binds to the enzyme–substrate complex rather than to the free enzyme: $E + S \rightleftharpoons ES + I \rightleftharpoons ESI$. A model is shown in Figure 3-14. This figure also shows the resulting double reciprocal plot in which a parallel line to the control occurs, indicating that the value of reciprocal maximal velocity, $1/V_{max}$, is modified by the factor $1 + [I]/K_i$ but the value of the slope stays the same as the control. That is, the value of the slope is unchanged. Because the inhibitor causes the enzyme–substrate complex to be blocked (ESI), the reaction cannot proceed to the formation

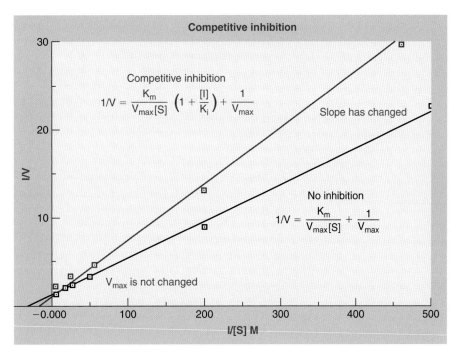

Figure 3-10. Double reciprocal (Lineweaver-Burk) plot of an enzyme reaction +/− a competitive inhibitor. Note that the factor $(1+[I]/K_i)$ now modifies the equation when the inhibitor is present.

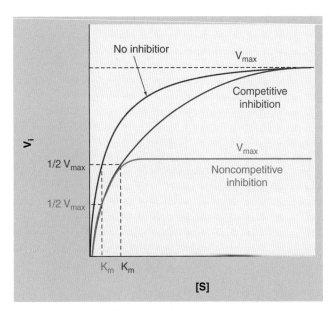

Figure 3-11. Plot of initial velocity, v_i, on the y-axis against substrate concentration [S] in molarity on the x-axis, comparing no inhibitor, competitive inhibitor, and noncompetitive inhibitor. The K_m is increased in the presence of an inhibitor. V_{max} is unchanged with a competitive inhibitor and is decreased with a noncompetitive inhibitor.

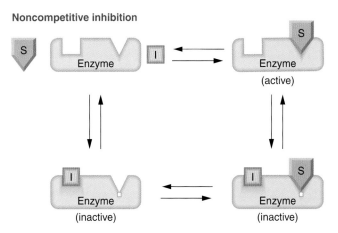

Figure 3-12. A model of noncompetitive inhibition showing that the substrate and the noncompetitive inhibitor bind at different sites on the enzyme.

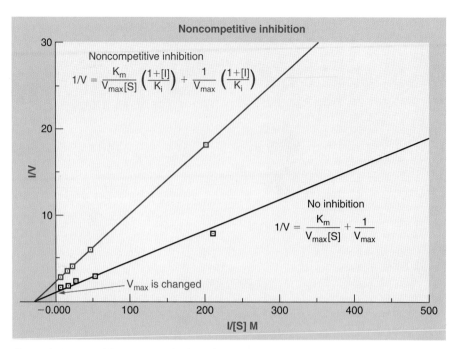

Figure 3-13. Double reciprocal plot of noncompetitive inhibition compared to no inhibitor control. The inhibitor curve is raised over the control. The slope and $1/V_{max}$ values are modified by the value $(1 + [I]/K_i)$. The x-intercept is the same (unchanged), whereas the x-intercept is moved to the right (decreased) in Figure 3-10 for competitive inhibition. The y-intercept is moved upward, whereas in competitive inhibition it is not changed (Figure 3-10).

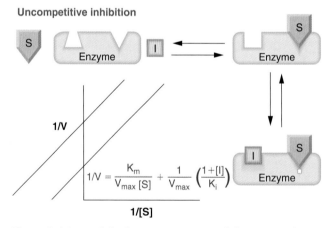

Figure 3-14. Model of uncompetitive inhibition. In this case, the inhibitor binds to ES rather than to E, although the site of inhibitor binding is still distant from the substrate-binding site. The double reciprocal plot also is shown, where only the term $1/V_{max}$ is modified by the quantity $(1 + [I]/K_i)$ and the slope is unchanged.

of products. Formally, uncompetitive inhibition is a special form of noncompetitive inhibition. When this form of inhibition is observed, it usually occurs in a system with more than one substrate.

Allosterism

This phenomenon occurs in complex enzymes, often with multiple subunits, and generates a special form of kinetics. It is important to introduce this topic because the binding of oxygen to hemoglobin, for example, follows much the same set of rules (see Figure 14-23). In short, the sequential binding of substrate to active sites generates changes in the enzyme so that successive binding of a second (and sometimes more than two molecules) to the multi-subunit enzyme produces differing rates of reaction. There is also a special case called half-of-the-sites where two subunits must join before there is an active catalytic center; in this case, each subunit contributes part of the active site.

Allosteric enzymes are those that bind effectors (with positive or negative effects). **Effectors** *are small molecules that bind to sites distinct from the catalytic site and cause changes in the conformation of the enzyme expressed at the catalytic center.* In some cases, the substrate can act allosterically, especially where more than one molecule of substrate binds to an enzyme (where an enzyme might have two catalytic sites, for example). This would be a means for the first molecule of substrate to bind, alter the conformation of the enzyme, and allow the second molecule to bind with greater affinity than the first. Allosteric enzymes are located at branch points of metabolic pathways where controls would be important, especially in prokaryotes but also in humans. These effectors can alter the maximal velocity or the Michaelis constant values. Some allosteric enzymes have separate catalytic and regulatory subunits (the substrate binds to one subunit and the effector binds to another subunit of the enzyme). An example is **protein kinase A,** which has four subunits: two catalytic subunits and two regulatory subunits. When the effector, cyclic adenosine 3',5'-monophosphate (cyclic AMP), binds to the regulatory subunits, the enzyme dissociates into a dimer of the regulatory subunits, with two molecules of cyclic AMP bound to each one and two catalytic subunits, which are dissociated into monomers and are enzymatically active. The catalytic subunits are not active in the tetramer in the absence of cyclic AMP. This mechanism is shown in Figure 3-15. The structure of the catalytic subunit is shown in Figure 3-16.

Classical allosteric kinetics are shown in Figure 3-17. The allosteric enzyme shows an S-shaped curve when initial velocity is plotted as a function of substrate concentration on the x-axis. When an effector substance is present, it binds to a site distinct from the substrate-binding site and changes the conformation of the enzyme so that it processes substrate molecules faster; that is, it speeds up the reaction. As a result, the whole curve is moved to the left so that the Michaelis constant for substrate is reduced. This means that it takes less substrate to achieve a given reaction rate, say, the half-maximal velocity. As a corollary, when an allosteric inhibitor is present, it binds to a site on the enzyme distinct from the substrate-binding site and produces a change in the conformation of the enzyme so that the enzyme requires more substrate to achieve half-maximal velocity, for example (the Michaelis constant is increased), and the whole curve moves to the right (as shown in Figure 3-17).

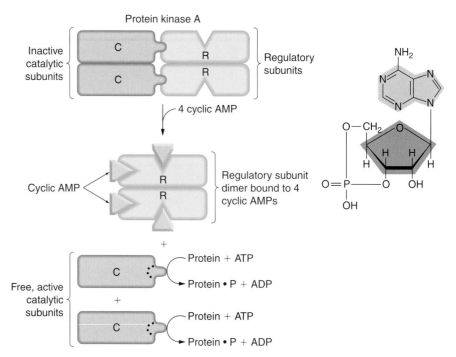

Figure 3-15. Mechanism of protein kinase A and interaction with effector, cyclic AMP, to cause the liberation of active catalytic subunits.

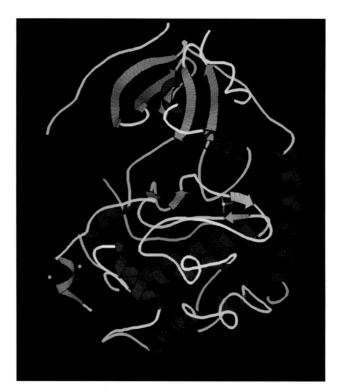

Figure 3-16. Structure of the mammalian catalytic subunit of cAMP-dependent protein kinase and an inhibitor peptide displays an open conformation. Reproduced from PDB ID: 1ctp J. Zheng, D.R. Knighton, N.H. Xuong, S.S. Taylor, J.M. Sowadski, L.F. Ten Eyck. Crystal Structures of the Myristylated Catalytic Subunit of cAMP-Dependent Protein Kinase Reveal Open and Closed Conformations. *Protein Sci.* **2** pp. 1559 (1993).

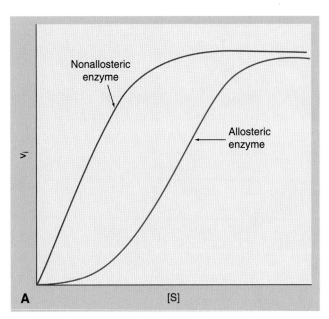 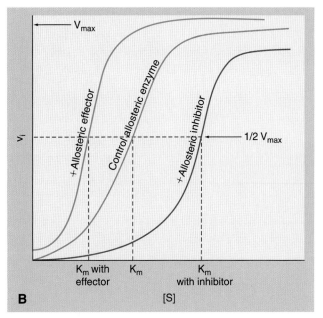

Figure 3-17. Kinetic behavior of allosteric enzymes. **A.** Typical curve of v_i versus [S] for a nonallosteric and allosteric enzyme. The allosteric enzyme shows a sigmoidal or S-shaped curve. **B.** Effects of a positive effector or an allosteric inhibitor on the S-shaped curve. An effector moves the curve to the left, causing a decrease in the K_m value. The presence of an inhibitor causes the curve to move to the right, increasing the K_m value. When the K_m is reduced, the "affinity" of the enzyme for the substrate is increased. When the K_m is increased, the "affinity" of the enzyme for the substrate is reduced. When the K_m is reduced, less substrate is required to achieve $1/2\ V_{max}$.

Classification

Although there is a well-defined and detailed classification of enzymes that has been generated by an international body (the International Union of Biochemistry), only the types of reactions catalyzed by a given class of enzymes are reviewed here. The first are the **oxidoreductases.** *These enzymes either add or remove hydrogen atoms in a given reaction.* The oxidation direction refers to the removal of two hydrogen atoms. The reductase direction refers to the addition of hydrogen atoms. Inspecting Figure 3-1, the LDH in the left-to-right direction is adding two hydrogen atoms to pyruvate to produce lactate, which is, in fact, a reductase activity. In the opposite direction (lactate to pyruvate), it would be an oxidase (removal of hydrogens). So, this enzyme is clearly an oxidoreductase. The direction that this enzyme would take in the cell would depend upon the availability of substrates (pyruvate or lactate) and the concentration of the coenzyme (NADH or NAD^+).

Another class of enzymes is the **transferases.** These enzymes catalyze the transfer of a group from one molecule to another. *Aminotransferases are members of this group, and these enzymes catalyze the transfer of an amino group from one molecule to another.* Aspartate aminotransferase has been mentioned in connection with the diagnosis of myocardial infarction. The enzymatic reaction is shown in Figure 3-18 with the crystal structure of aspartate aminotransferase.

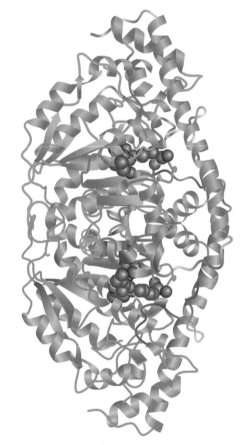

Figure 3-18. Structure of aspartate aminotransferase. Redrawn with permission from http://www.sci.osaka-cu.ac.jp/grad/MOLMS/staff/soukan/hirotsu.html.

Hydrolases are enzymes that *catalyze the hydrolysis of a bond by the addition of water.* An example of this class would be a protease, or a proteolytic enzyme that cleaves the peptide bond. In the case of the viral **human immunodeficiency virus (HIV)** infection, the virus makes long polypeptide chains in the infected cell and these chains contain several proteins that are inactive until they are broken down into the separate functional proteins. This breakdown is achieved by the HIV-1 protease that is encoded in the viral genome and expressed in the human infected cell but is not found in the uninfected cell. The proteins synthesized by following the instructions of the infecting virus (viral RNA + viral reverse transcriptase = complementary DNA + host cell machinery = RNA = transcription = translation + HIV-1 protease = mature functional viral-coded proteins) are used in the production of more viruses inside the cell. Therefore, many researchers are trying to find specific inhibitors of this protease to interfere with the disease process. The HIV-1 enzyme is composed of two symmetrical subunits (Figure 3-19A).

Lyases are another category of enzymes. These *catalyze the addition of water, ammonia, or carbon dioxide to double bonds, or they can remove these substances to create double bonds in their place.* An example of this class is citrate lyase as shown in Figure 3-20. Here, a

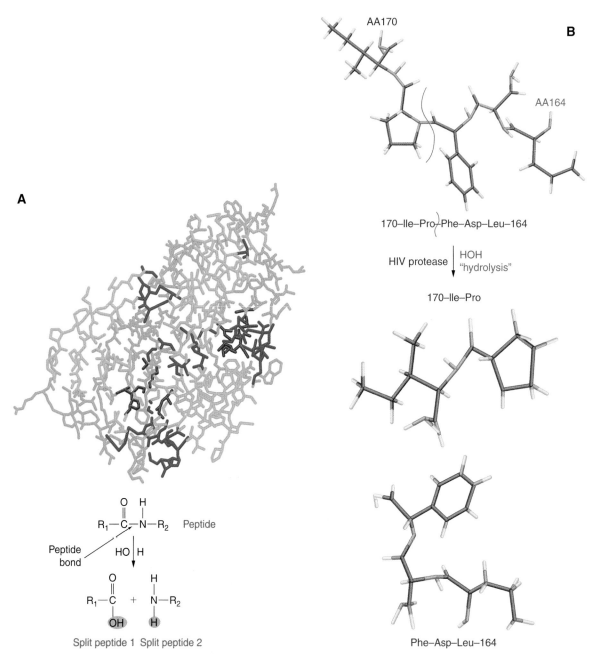

Figure 3-19. HIV-1 protease. **A.** Structure of the enzyme. **B.** The reaction catalyzed by this enzyme. A peptide bond between prolyl and phenylalanyl residues is hydrolyzed by the addition of H_2O. The generic reaction is shown below the structure in *A*.

Classification

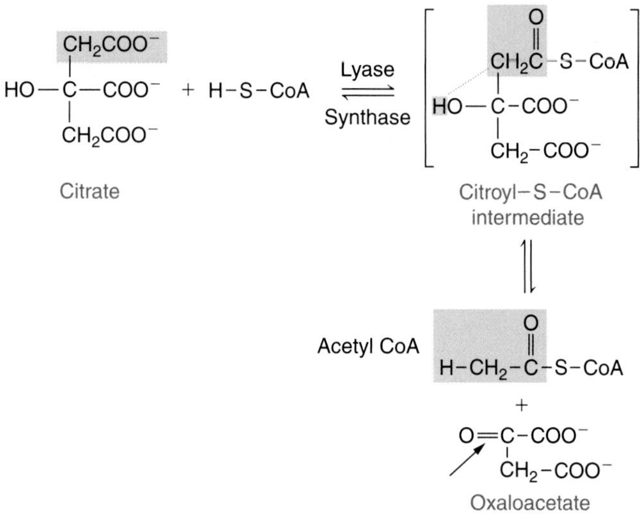

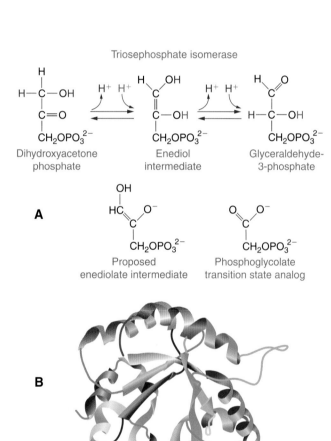

Figure 3-20. The reaction catalyzed by citrate lyase. The double bond of oxaloacetate is created by removal of a hydrogen from a hydroxyl group and carbons 1 and 2 of citrate (acetyl).

Figure 3-21. Triose phosphate isomerase. **A.** The reaction catalyzed by this isomerase. Hydrogen atoms moved in the reaction are marked in *red*. The proposed intermediate and transition state analog are shown below the reaction. **B.** Crystal structure of triose phosphate isomerase. The structure of this enzyme is an αβ barrel.

110 CHAPTER 3 Enzymes

double bond (keto group) is created when a hydrogen atom is removed from carbon 3 of citrate, along with an acetate group (carbons 1 and 2 of citrate). Note that the reverse reaction is the formation of citrate from oxaloacetate or citrate synthase reaction.

Isomerases constitute another class of enzymes. These enzymes *catalyze many kinds of mutase reactions, such as converting L- to D-forms, shifts of chemical groups, and movement of double bonds.* An example of this class is triose phosphate isomerase, which catalyzes the reaction shown in Figure 3-21. This mechanism involves the shifts of protons to create a double bond at carbons 1 and 2 of the enediol intermediate. Glutamate and histidine in the catalytic center of the enzyme extract and donate the protons in the reaction. The crystal structure of the homodimer, shown in Figure 3-21B, showing the $\alpha\beta$ barrel, contains eight parallel β-strands surrounded by eight α-helices. The catalytic center is near the C-terminal ends of the β-strands and the loops connecting the β-strands to the α-helices.

Finally, there are the **ligase** enzymes that *catalyze reactions, using the energy of ATP, that join two chemical groups.* These enzymes are often called synthases or synthetases. One such enzyme, as introduced previously (Figure 2-44), is human aminoacyl-tRNA synthetase. These 20 enzymes, each specific for an amino acid, catalyze the initial step in protein synthesis. A tyrosyl-tRNA synthetase crystal structure and the general mechanism are shown in Figure 3-22.

Some enzymes require no other factors to be active, whereas others require other substances to become active. Those that require other substances, called **coenzymes,** are known as **apoenzymes** in the absence of their coenzyme. In the presence of the coenzyme, the active complex is known as the **holoenzyme:**

$$\text{Apoenzyme} + \text{Coenzyme} \rightleftharpoons \text{Holoenzyme}$$

Some holoenzymes contain a coenzyme that is tightly bound so that when the enzyme is assayed for activity in the test tube little added coenzyme is needed for the reaction to occur at a high rate. On the other hand, many enzymes that require a coenzyme are relatively loosely associated, so the coenzyme becomes a reactant (substrate) in the reaction as measured in the test tube. Thus, the case of LDH mentioned earlier would require the addition of the coenzyme into the reaction being measured in the laboratory. In this case, the reaction as measured has two substrates rather than one: pyruvate and NADH. When kinetics are being determined for an enzyme in this category, it is treated like a two-substrate reaction:

$$A + B = C + D$$

Then the reciprocal initial velocity value, $1/v_i$, would be measured when the substrate, pyruvate, varies in concentration (and plotted as $1/[S]$) and NADH is added at saturating levels. This would give the value of the Michaelis constant for pyruvate. The reaction would be run again with pyruvate at saturating concentration and with a varying concentration of NADH (now as $1/[S]$); the value of the Michaelis constant with respect to NADH would be obtained. Similar approaches would be taken when an inhibitor is present in the system.

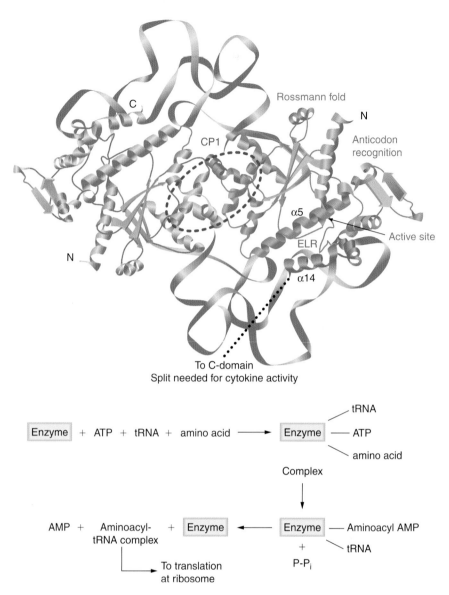

Figure 3-22. The crystal structure of human tyrosyl-tRNA synthetase (a ligase). After a specific cleavage in the structure, the enzyme has cytokine activity. Below the structure is the mechanism catalyzed by the enzyme when tyrosine is the amino acid. The structure is from *PNAS*, 99: 15369–15374, 2002 and is reproduced with permission.

Coenzymes

The role of the coenzyme is to transport chemical groups from one location to another in the enzyme substrate complex, for example, transferring an amino group from an amino acid to a keto acid to generate a new amino acid and a new keto acid. We have already observed this reaction in the case of a transaminase or an aminotransferase (see Figure 3-18):

L-aspartate + α-ketoglutarate ⇌ oxaloacetate + L-glutamate

But while the coenzyme PLP was indicated, its role in the reaction was not defined. The transamination reaction is shown in Figure 3-23A,

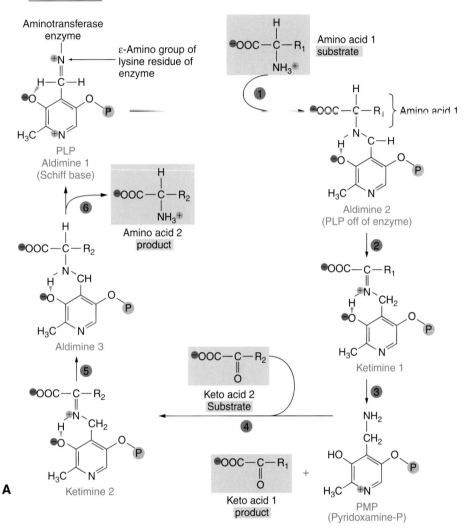

Figure 3-23. **A.** Pyridoxal phosphate (PLP) and the transaminase reaction. **B.** The coenzymes, their structures, and functions. **C.** Structures and possible mechanism of vitamin B_{12} catalyzed reaction. The suggested mechanism is redrawn with permission from http://web.archive.org/web/20060502033641/http://www.chm.bris.ac.uk/motm/vitaminb12/mech.htm.

(continued)

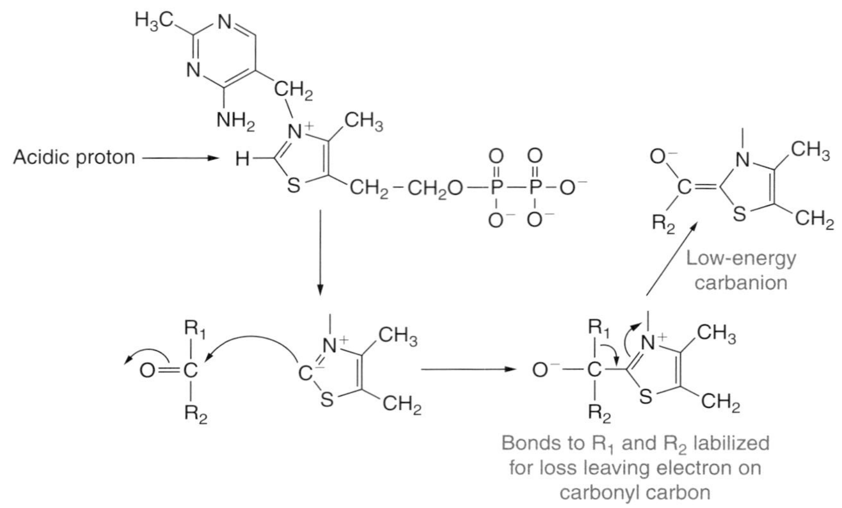
Acidic proton
Low-energy carbanion
Bonds to R1 and R2 labilized for loss leaving electron on carbonyl carbon

The structure of thiamine pyrophosphate showing addition to carbonyl.

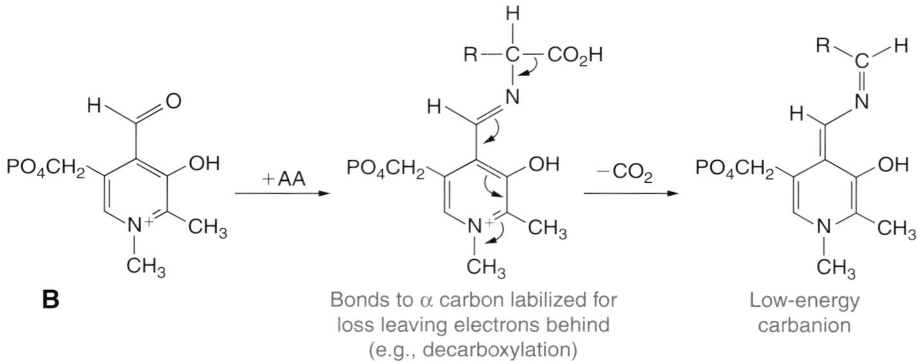
+AA
−CO2
B
Bonds to α carbon labilized for loss leaving electrons behind (e.g., decarboxylation)
Low-energy carbanion

Pyridoxal phosphate showing its activity in decarboxylating an amino acid as a coenzyme with amino acid decarboxylase. These two coenzymes can form a Schiff base or a carbonyl base.

Figure 3-23, cont'd

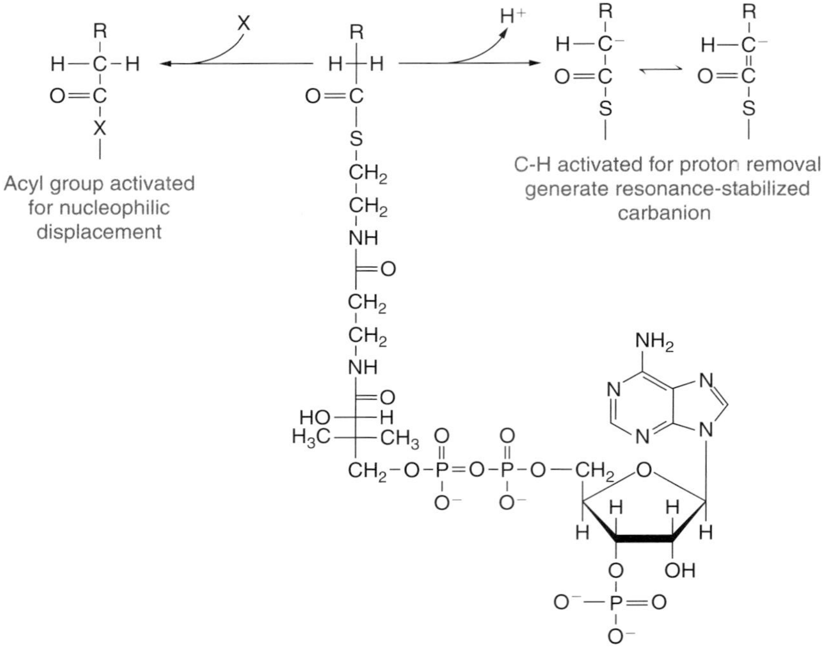
Acyl group activated for nucleophilic displacement
C-H activated for proton removal generate resonance-stabilized carbanion

Coenzyme A carries acyl groups through its –SH in a high energy bond that forms esters by activating acids through a nucleophilic displacement of –SCoA. It can form an enolate-type nucleophilic carbon, which can add to carbonyls. Citrate and acetoacetyl CoA are formed in this way.

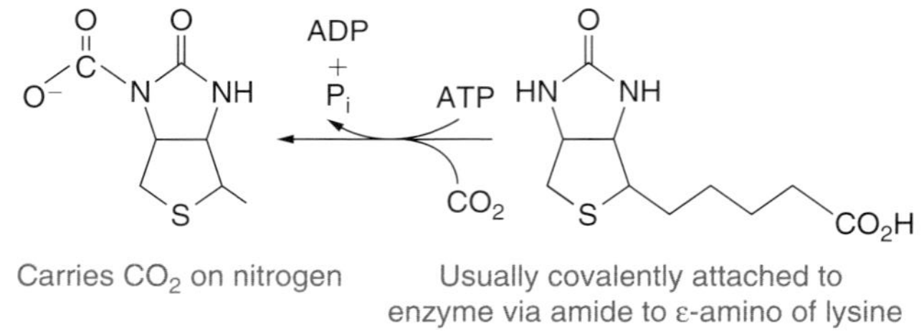
ADP + Pi
ATP
CO2
Carries CO2 on nitrogen
Usually covalently attached to enzyme via amide to ε-amino of lysine

Biotin is a coenzyme that carries CO_2 from one active site to another when CO_2 is attached to the enzyme through a lysine epsilon amino group.

Usually attached covalently to enzyme
by amide with ε-amino group of lysine

Oxidation of aldehyde to acyl group and transfer to thioester.
Removal from TPP takes a couple of steps (note that
adduct is like tetrahedral intermediate in ester hydrolysis).

Lipoic acid is a disulfide that can oxidize aldehydes to acids and carry them as thioesters. It is involved in conversion of pyruvate to acetyl CoA by oxidative α-decarboxylation.

5-methyl   Methylene   5,10-methenyl   5-formyl and other forms

Tetrahydrofolic acid carries one carbon fragments in various oxidation states. The generic structure of tetrahydrofolate (H_4 folate):

Tetrahydrofolate (H_4 folate)

Figure 3-23, cont'd

Coenzymes   115

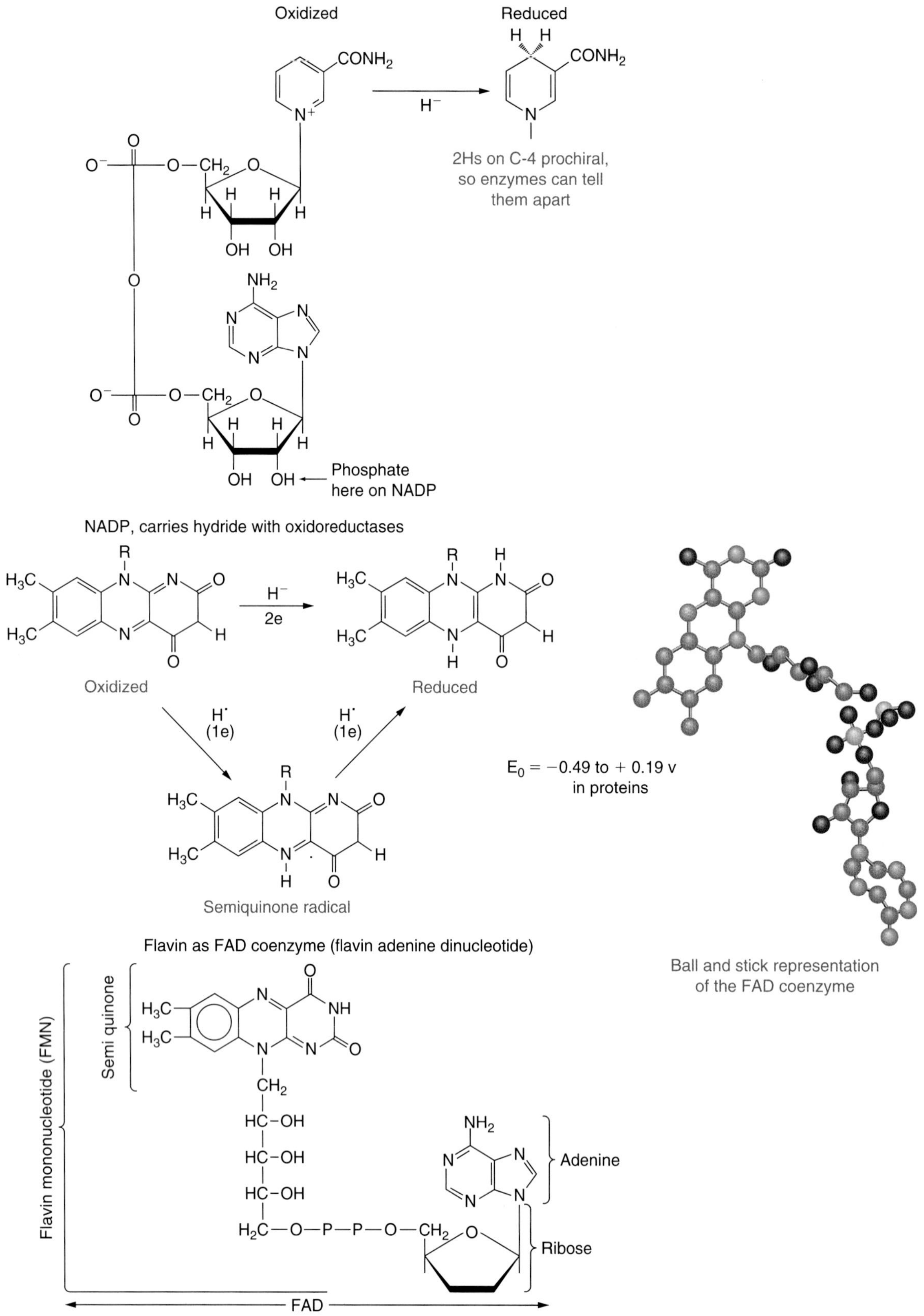

Oxidized

Reduced

2Hs on C-4 prochiral, so enzymes can tell them apart

← Phosphate here on NADP

NADP, carries hydride with oxidoreductases

Oxidized

Reduced

Semiquinone radical

$E_0 = -0.49$ to $+0.19$ v in proteins

Flavin as FAD coenzyme (flavin adenine dinucleotide)

Flavin mononucleotide (FMN)

Semi quinone

Adenine

Ribose

FAD

Ball and stick representation of the FAD coenzyme

Figure 3-23, cont'd

116 CHAPTER 3 Enzymes

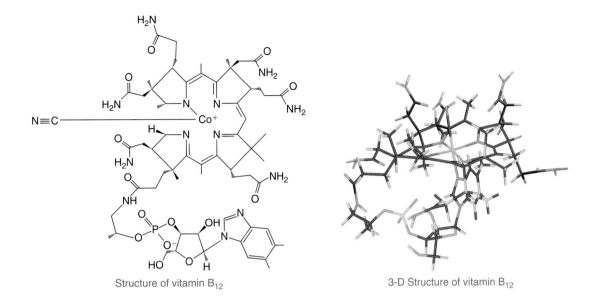

Structure of vitamin B_{12}

3-D Structure of vitamin B_{12}

B_{12} as coenzyme catalyzes a rare homolytic sission reaction whereby a hydrogen and a group on an adjacent carbon atom exchange places.

A possible mechanism:

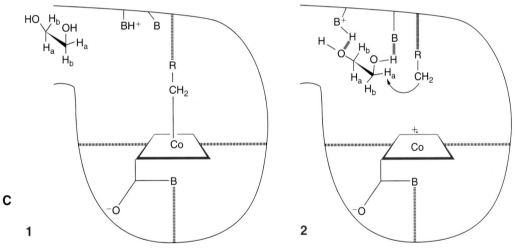

C

Step 1, and the substrate, 1,2-dihydroxyethane, approaches the holoenzyme, and is about to plug the reaction 'bottle.' Then **(Step 2)** the substrate binds and the cobalt–carbon bond breaks homolytically to give the 5'-adenosyl radical and a Co(II) corrin, anchored to the large coenzyme. (Are the amide groups involved?) It has been suggested that the cobalt–carbon bond, already a rather weak bond and readily subject to photolysis if the holoenzyme is subjected to visible light, is further weakened by binding of the metalloenzyme to the large coenzyme.

Figure 3-23, cont'd

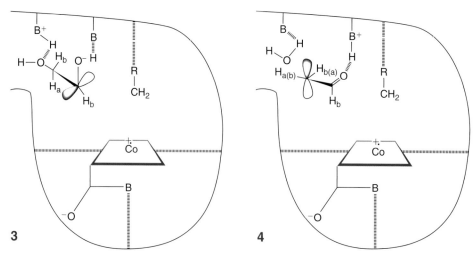

In **Step 3,** a hydrogen atom is abstracted from the substrate by the 5'-deoxyadenosyl radical, converting the dangling CH_2 into a CH_3 Then, **Step 4,** the enzyme base (B) and the acid (BH^+) cause assisted β-hydroxy fragmentation, that is loss of a water molecule.

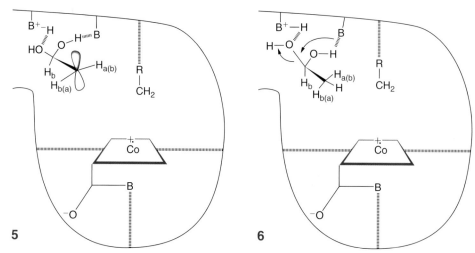

With **Step 5,** comes the readdition of a water molecule, resulting in a substrate reorientation of C_2 toward the 5'-deoxyadenosyl methyl. There must be competing C_1 and C_2 racemisation. A hydrogen atom abstraction occurs, with competing C_2 racemization. The cycle is completed **(Step 6),** by substrate dehydration and release of ethanal and a water molecule, so that the system can pick up another substrate molecule.

Figure 3-23, cont'd

where the first half of the mechanism is shown in detail. Table 3-1 lists the vitamins and their coenzyme forms with the reactions catalyzed. Figure 3-23B shows the structures of the coenzymes and their coenzymatic activities. The coenzymes fall into groups depending on the types of reactions in which they participate. **Thiamine pyrophosphate** and **pyridoxal phosphate** are coenzymes that, with their apoenzymes, add to carbonyls forming imines and **Schiff bases** (aldimines) that permit α-decarboxylations, transaminations, and aldol condensations. Vitamin B_{12} contains a carbon–cobalt bond (Figure 3-23C) can exchange a group for a hydrogen. Coenzymes that carry groups are **coenzyme A, biotin, lipoic acid** and **tetrahydrofolic acid**. Coenzymes participating in oxidore-

Table 3-1
The Vitamins, Their Coenzymes, and Their Chemical Functions

Vitamin	Coenzyme	Reaction catalyzed	Human deficiency disease
Water-Soluble Vitamins			
Niacin (niacinate)	NAD^+, $NADP^+$ NADH, NADPH	Oxidation Reduction	Pellagra
Riboflavin (vitamin B_2)	FAD, FMN $FADH_2$, $FMNH_2$	Oxidation Reduction	Skin inflammation
Thiamine (vitamin B_1)	Thiamine pyrophosphate (TPP)	Two-carbon transfer	Beriberi
Lipoic acid (lipoate)	Lipoate Dihydrolipoate	Oxidation Reduction	—
Pantothenic acid (pantothenate)	Coenzyme A (CoASH)	Acyl transfer	—
Biotin (vitamin H)	Biotin	Carboxylation	—
Pyridoxine (vitamin B_6)	Pyridoxal phosphate (PLP)	Decarboxylation Transamination Racemization $C\alpha$—$C\beta$ bond cleavage α,β-Elimination β-Substitution	Anemia
Vitamin B_{12}	Coenzyme B_{12}	Isomerization	Pernicious anemia
Folic acid (folate)	Tetrahydrofolate (THF)	One-carbon transfer	Megaloblastic anemia
Ascorbic acid (vitamin C)	—	—	Scurvy
Water-Insoluble (lipid-soluble) Vitamins			
Vitamin A	—	—	—
Vitamin D	—	—	Rickets
Vitamin E	—	—	—
Vitamin K	Vitamin KH_2	Carboxylation	—

This table is reproduced from http://wps.prenhall.com/wps/media/objects/724/741576/Instructor_Resources/Chapter_25/Text_Images/FG25_TB01.JPG.

ductase reactions are **nicotinamide adenine dinucleotide (NAD)**, NADH, NADP, NADPH, and flavins (FAD).

Prosthetic Groups

The distinction between coenzymes (sometimes referred to as cofactors) and prosthetic groups is that a coenzyme can leave the enzyme during the reaction and recombine after the reaction is complete. An example is the transamination reaction in Figure 3-23A, in which PLP as an aldimine with the enzyme leaves it to form an aldimine with the amino acid substrate, undergoes the remaining reactions, and is regenerated as PLP so that it can recombine with the free enzyme (apoenzyme) again. A **prosthetic group** is a coenzyme or cofactor that is *tightly bound to the enzyme and is associated with the enzyme throughout the catalytic reaction.* Many prosthetic groups are metals; one such example was covered when the vitamin B_{12} (containing cobalt) mechanism was given in Figure 3-23C. Other examples of prosthetic groups that are tightly bound metals are found in **metalloenzymes.** The most common metals that are tightly bound to enzymes are iron, zinc, copper, and manganese. Iron is found to be bound in heme, which is an important cofactor for enzymes, and in hemoglobin. The chemical structure of heme is shown in Figure 3-24. The crystal structure of heme is shown in Figure 3-25. One example of a metalloenzyme with heme is the **cyclooxygenase (cox)** that catalyzes the reaction shown in Figure 3-26. Cyclooxygenase exists in two forms, **cox-1** and **cox-2.** The crystal structure of cox-1 is shown in Figure 3-27A. Cox-1 is constitutive, is produced at a steady rate, and is responsible for

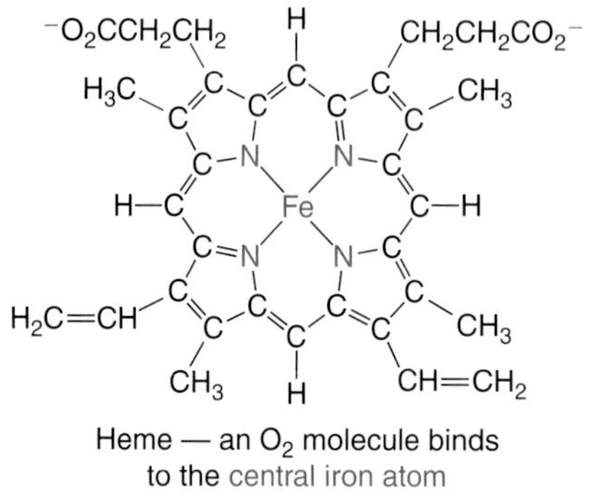

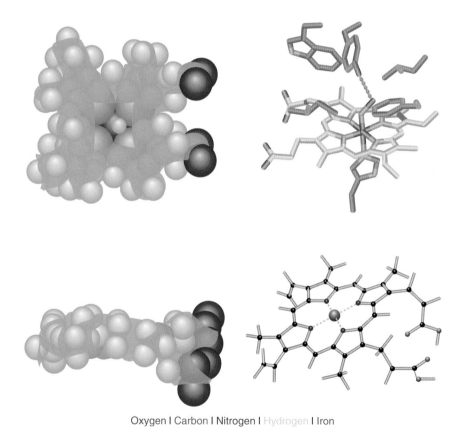

Figure 3-24. The chemical structure of heme.

Oxygen | Carbon | Nitrogen | Hydrogen | Iron

Figure 3-25. Crystal structure of heme.

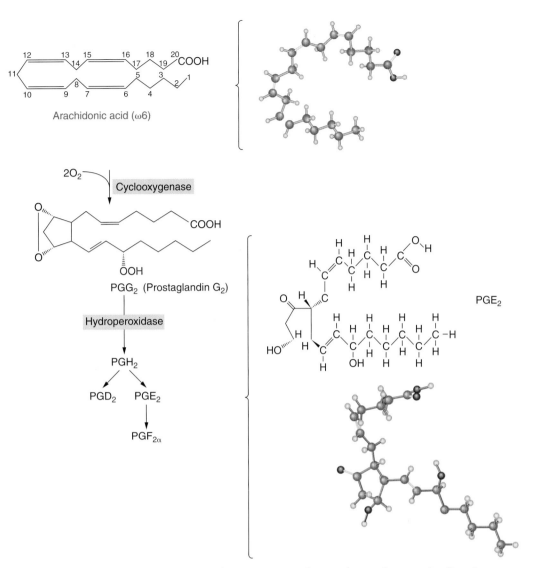

Figure 3-26. Cyclooxygenase (cox) is a key enzyme in the synthesis of prostaglandins from arachidonic acid. The product of the cyclooxygenase reaction (PGG_2) is unstable. Cyclooxygenase exists in 2 forms, cox-1 and cox-2. Cox-1 is a constitutive enzyme (constantly produced in baseline amounts). Cox-2 is inducible. With a stimulus, cox-1 produces thromboxane A_2 (a prostaglandin), prostacyclin, and PGE_2. With an inflammatory stimulus, cox-2 is induced and produces PGE_2, histamine, etc. (Some of these products are indirect.) Aspirin inhibits cox-2 by acetylating a serine residue on the enzyme.

production of the prostaglandin, thromboxane A_2, and prostacyclin (PGI_2), which is involved in anticlotting in the endothelium. In the stomach mucosa, PGI_2 reduces proton secretion and increases bicarbonate ion. Thus when aspirin is present, which inhibits both cox-1 and cox-2 (if it has been induced), the side effect of potential stomach ulceration and bleeding can occur because the output of H^+ from the stomach parietal cells is increased. Cox-1 is also responsible for the production of prostaglandin E2 (PGE_2) in the kidney, leading to dilation of the arterioles and excretion of sodium ion and water. Cox-2 is an inducible enzyme

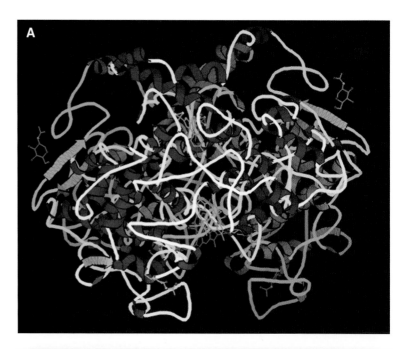

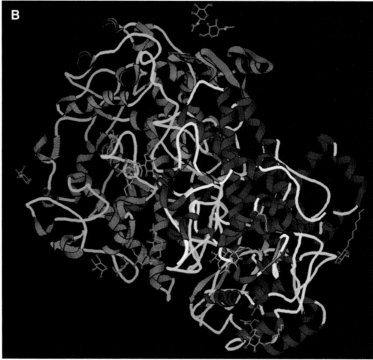

Figure 3-27. **A.** Cycloogenase-2 (prostaglandin synthase-2 complexed with a selective inhibitor SC-558. The enzyme has two different active sites. Together they are called prostaglandin synthase. On one side is the cyclooxygenase active site, and on the other is a peroxidase site. The latter activates the heme groups. Activated hemes participate in the cyclooxygenase reaction. The enzyme is a dimer of identical subunits, giving two cyclooxygenase active sites and two peroxidase sites. There are knobs at the underside that attach to the endoplasmic reticulum membrane. The cyclooxygenase active site is buried in the protein. Arachidonic acid is guided through a tunnel in the knob to reach the catalytic center. **B.** Crystal structure of arachidonic acid bound to the cyclooxygenase active site of COX-2. Part A reproduced from PDB ID: 1pgg. P.J. Loll, D. Picot, O. Ekabo, R.M. Garavito. Synthesis and Use of Iodinated Nonsteroidal Antiinflammatory Drug Analogs as Crystallographic Probes of the Prostaglandin H2 Synthase Cyclooxygenase Active Site. *Biochem.* **35** pp. 7330 (1996). Part B reproduced from PDB ID:1cvu. J.R. Kiefer, J.L. Pawlitz, K.T. Moreland, R.A. Stegeman, W.F. Hood, J.K. Gierse, A.M. Stevens, D.C. Goodwin, S.W. Rowlinson, L.J. Marnett, W.C. Stallings, R.G. Kurumball. Structural Insights into the Stereochemistry of the Cyclooxygenase Reaction. *Nature* **405** pp. 97 (2000).

responding to inflammatory stimuli or tissue injury or chronic arthritis. The enzyme is induced in macrophages and other cells and produces PGE_2 and other inflammatory mediators generating inflammation, swelling, and pain. New drugs are in development that will specifically inhibit cox-2 and diminish the unwanted side effects of the inhibition of cox-1. The structure of cox-2 is shown in Figure 3-27B.

Of interest is the copper-binding activity of prion protein, mentioned in Chapter 2. In Figure 3-28, a theoretical structure for the binding sites of copper in the prion protein (PrP^c) is shown. It is suggested that each atom of copper is complexed by two histidine residues. As yet, it is not known whether PrP has enzymatic activity. Superoxide dismutase (SOD) is an example of another enzyme with a tightly bound metal. The type found in man is bound with iron, but enzymes bound with manganese, copper, or zinc are also found. This enzyme metabolizes the toxic superoxide radical sometimes occurring from respiration:

$$2O_2^- + 2H^+ \xrightarrow{SOD} H_2O_2 + O_2$$

The structure of the human enzyme is shown in Figure 3-29. The copper–zinc SOD is shown in Figure 3-30. The enzyme is a dimer of 32 kDa each. There are eight β-barrels connected by three external loops. The metal-binding sites are located in a cavity formed by two loops containing charged residues, which attract superoxide to the catalytic center. The oxidized copper (Cu^{++}) enzyme contains one copper and one zinc ion. The copper ion is coordinated by three histidine residues and a molecule of water. Zinc is bound by two more histidine residues and an aspartic acid. SOD is present in normal erythrocytes (red blood cells), about 100 μg/ml. In clinical cases, SOD may be administered to reduce the presence of superoxide radicals. SOD has been cloned and is expressed in bacteria, from which it can be purified to homogeneity.

The only RNA enzyme known so far is **ribozyme**, introduced in Chapter 2. This enzyme, which generates the peptide bond during

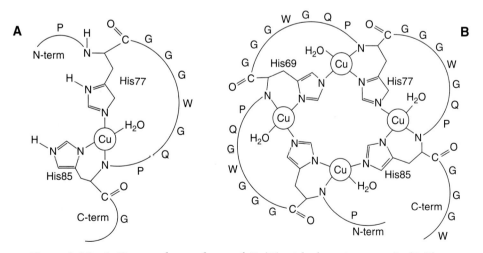

Figure 3-28. **A.** Proposed complexes of Cu(II) with the prion protein (PrP). **B.** Proposed bridge complex of four Cu(II) ions with PrP. Reproduced with permission from http://www.pnas.org/cgi/content-nw/full/96/5/2042/F6

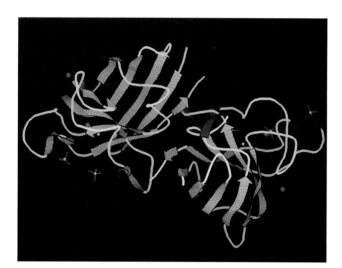

Figure 3-29. Atomic resolution structure of Cu-Zn human superoxide dismutase (SOD). Reproduced from PDB ID: 2c9v. R.W. Strange et al. Variable Metallation of Human Superoxide Dismutase: Atomic Resolution Crystal Structures of Cu-Zn, Zn-Zn and as-Isolated Wild-Type Enzymes. *J. Mol. Biol.* **356** pp. 1152 (2006).

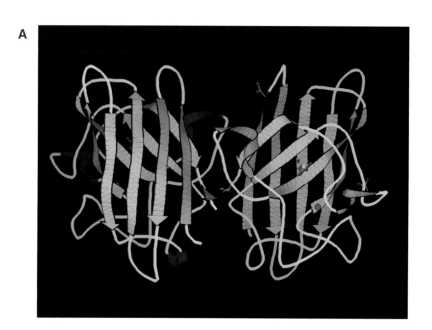

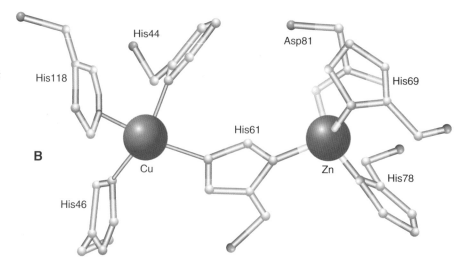

Figure 3-30. **A.** Structure of copper–zinc superoxide dismutase (SOD), top. **B.** The coordination of Cu^{2+} *(left figure, orange central ball)* and that of zinc *(blue ball at right)*. Copper ion is coordinated by four histidine residues and zinc by three histidine residues and an aspartate residue. PDB ID: 2C9U. R.W. Strange et al. Variable Metallation of Human Superoxide Dismutase: Atomic Resolution Crystal Structures of Cu-Zn, Zn-Zn and as-Isolated Wild-Type Enzymes. *J. Mol. Biol.* **356** pp. 1152 (2006).

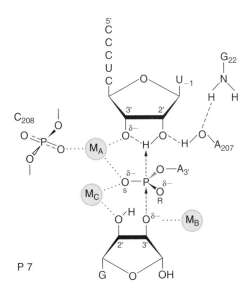

Figure 3-31. Coordination sites for metals (Mg^{2+} or Mn^{2+}) in ribozymes. Electrostatic interactions between the metal ion and the RNA phosphate groups and electrostatic interactions with sugar hydroxyl ions generate the coordination sites.

protein synthesis (Figure 2-46) is a metalloenzyme, and the structure of the RNA is such that it positions the metal for its catalytic function. Most ribozymes are complexed with magnesium and sometimes manganese. The coordination of metals in ribozymes is shown in Figure 3-31. Here it is seen that the phosphate groups of RNA, with sugar hydroxyl groups, form the coordination sites with metals.

Drugs and Enzymes

Enzyme inhibitors are used widely as drugs to treat various diseases. These have been successful in treating certain types of cancer, for example. Cervical cancer has been effectively treated with the drug methotrexate. This drug is an inhibitor of dihydrofolate reductase (DHFR). DHFR catalyzes the conversion of dihydrofolate (H_2 folate) to tetrahydrofolate (H_4 folate), with NADPH as the coenzyme contributor of the hydrogen atom. The reaction is shown in Figure 3-32. H_4 folate is required as a cofactor (donating two hydrogens to generate H_2 folate) for the conversion of deoxyuridine monophosphate to deoxythymidine 5'-monophosphate, catalyzed by thymidylate synthase in the biosynthesis of pyrimidine nucleotides. If H_4 folate is unavailable through the blocking action of methotrexate, then pyrimidine synthesis is greatly reduced, a situation that would compromise growing cells, particularly cancer cells. This strategy has been successful in curing cervical cancer, and it has been used for the treatment of rheumatoid arthritis, multiple sclerosis, and lupus. The structure of methotrexate is given in Figure 3-33 with a view of the enzyme with methotrexate

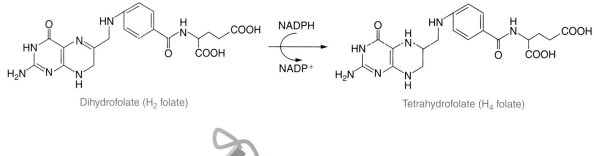

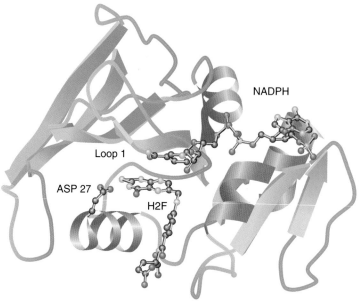

Figure 3-32. Dihydrofolate reductase reaction *(top)*. The crystal structure of DHFR is shown with the substrate (H_2F) attached and the coenzyme (NADPH) attached *(bottom)*. Structure at bottom redrawn or reproduced from http://research.ch4ecm.psu.edu/sjbgroup/images/dhfr.gif.

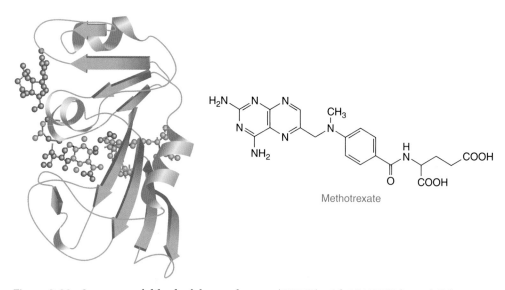

Figure 3-33. Structure of dihydrofolate reductase (DHFR) with NADPH bound *(blue structure)* and methotrexate bound *(pink structure)*. The chemical structure of methotrexate is shown on the right.

bound. Many enzyme inhibitors successfully used as drugs are strong competitive inhibitors.

Recently, new drugs have been developed that are specific inhibitors of the induced cox-2 but not cox-1. As previously discussed, cox-1 is a constitutive enzyme that produces prostaglandins (PGs) in several tissues, and these PGs are required for normal functions. Cox-2, on the other hand, is induced under inflammatory conditions, by cytokines such as interleukins and tissue necrosis factor. The PGs produced by cox-2 produce inflammation and pain. Cox-2 is encoded by a gene separate from that for cox-1. Nonsteroidal anti-inflammatory drugs, like aspirin and relatives, inhibit both cox-1 and cox-2 by acetylating a serine residue in the catalytic center. Although the inhibition of cox-2 is beneficial, the inhibition of cox-1 can result in damage to the stomach. There are now two medications that predominantly inhibit cox-2 with little effect on cox-1. These are rofecoxib and celecoxib. The structures are shown in Figure 3-34. Cox-1 and cox-2 have the same molecular weight of 71 kDa. The major difference between these two enzymes is that in cox-1 there is an amino acid residue, isoleucine, that is replaced by a valine in cox-2. The smaller sized valine, residue 523 in cox-2, leaves an entrance in the enzyme that allows cox-2 inhibitors to bind. The complex prevents arachidonic acid from binding to the catalytic center; however, this does not occur in cox-1 because the larger amino acid residue of isoleucine (compared to valine) prevents the binding of these inhibitors. This specificity is illustrated in Figure 3-35.

Another drug that is an enzymatic competitive inhibitor is **allopurinol**, a drug that has been used in gout as an inhibitor of xanthine oxidase. Gout occurs when uric acid crystallizes as the sodium salt in joints, often in the large toe as an indicator, and it is painful. Uric acid is poorly soluble in aqueous solutions. Allopurinol is a structural analog of hypoxanthine, which is a substrate of xanthine oxidase. The enzyme *catalyzes the conversion of hypoxanthine to xanthine to uric acid* in humans, as well as in

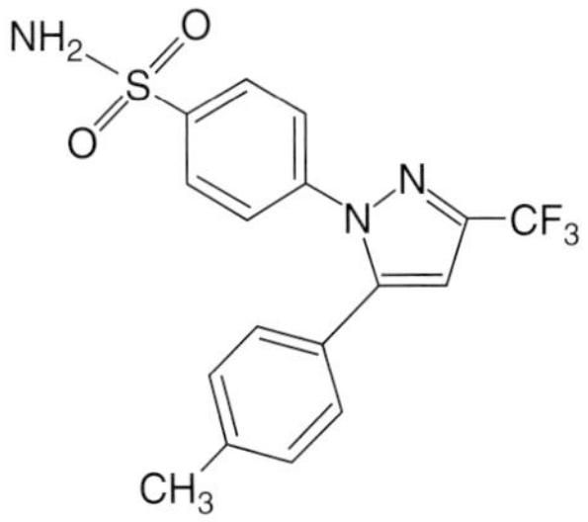

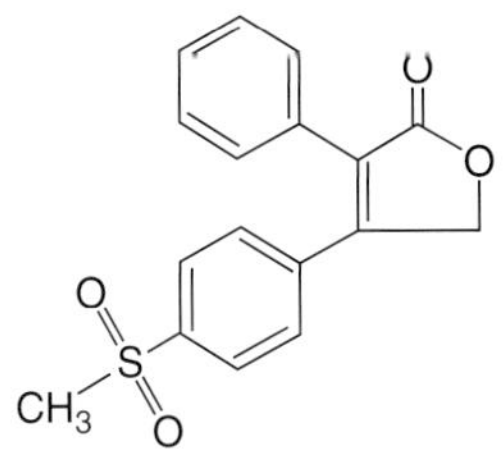

Figure 3-34. Structures of two specific drug inhibitors for COX-2.

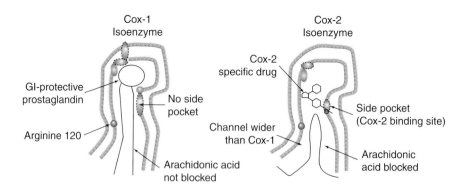

Figure 3-35. Model of specific inhibition of cox-2 that does not affect cox-1. Redrawn from C.J. Hawkey, "Cox-2 inhibitors," *Lancet,* 353: 307–314, 1999, Figure 1, with permission.

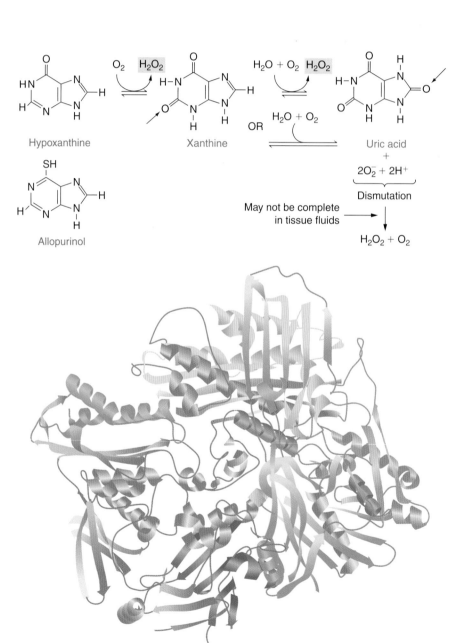

Figure 3-36. Top, reactions catalyzed by xanthine oxidase. The drug allopurinol, a strong competitive inhibitor, is shown. Conversion of xanthine + $2O_2^- + 2H^+$ to products is from F. Lacy, D.A. Gough, G.W. Schmid-Schonbein, *Free Radical Biology and Medicine,* 25: 720–727, 1998. The crystal structure is aldehyde oxidoreductase, a member of the xanthine oxidase family.

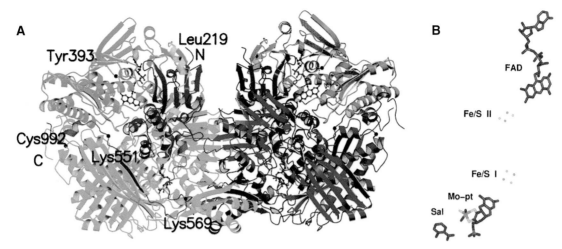

Figure 3-37. Structure of xanthine dehydrogenase dimer showing locations of FAD, Fe/S, and molybdenum. **A.** Molecular structure of the XDH dimer divided into the three major domains and two connecting loops. The two monomers have symmetry-related domains in the same colors, in lighter shades for the monomer on the *left* and in darker shades for the monomer on the *right*. From N to C terminus, the domains are: iron/sulfur-center domain (residues 3–165; *red*), FAD domain (residues 226–531; *green*), and Mo-pt domain (residues 590–1,331; *blue*). The loop connecting the iron/sulfur domain with the FAD domain (residues 192–225) is shown in *yellow*, the one connecting the FAD domain with the Mo-pt domain (residues 537–589) is in *brown*, and the N and C termini are labeled. The FAD cofactor, the two iron/sulfur centers, the molybdopterin cofactor, and the salicylate also are included. **B.** For clarity, the arrangement of the cofactors and salicylate in one subunit of XDH are presented. The Mo ion is in *green*, the iron ions are in *light blue*, and the sulfur atoms in *yellow*. XDH, Xanthine dehydrogenase. Reproduced with permission from C. Enroth et al., *PNAS*, 97: 10723–10728, 2000.

other species, and uric acid is the product of purine degradation in man. The enzyme contains **flavin adenine dinucleotide (FAD)** as coenzyme, along with iron and molybdenum as prosthetic groups. The reactions catalyzed by this enzyme and the crystal structure of a member of the xanthine oxidase family (the enzyme has been crystallized from bovine milk and from other species but not yet from humans) are shown in Figure 3-36. Figure 3-37 shows the structure of xanthine oxidase from milk with the locations of FAD, the coenzyme, molybdenum, and the iron–sulfur center. These examples show that most drugs effective in treating clinical conditions through their effects on enzymes are strong competitive inhibitors.

Further Reading

Copeland, R.A., *Enzymes*, 2nd edition, Wiley-VCH, 2000.

Segel, I.H., *Enzyme Kinetics*, John Wiley & Sons, Inc., 1975.

CHAPTER **4**

Carbohydrates

Diabetes: A Prevalent Disease That Disrupts Glucose Utilization

Diabetes comes from the Greek word for siphon, and *mellitus* derives from the Latin meaning, sweet like honey, referring to the urine of a diabetic. More than 18 million people in the United States have diabetes. The disease is known to have been described as early as 1500 BC. It is a condition in which either insulin is not produced by the pancreas (mainly because the beta cells of the pancreas have been destroyed by viruses or autoimmune disease) **(type 1)** or insulin is produced but is not utilized properly and may be in short supply **(type 2)**. Of diabetics, 5% to 10% are type 1; the rest are type 2. About 4% of pregnant women have gestational diabetes, representing about 135,000 women per year in the United States. Finally, about 20 million in this country have a condition known as **prediabetes,** in which the blood glucose levels are elevated but not high enough to fall into the category of type 2 diabetes.

There may be a genetic disposition to develop diabetes, although the probability of inheriting type 1 diabetes may be small. In any event, the genetics of diabetes is complicated, and if there is a strong genetic component, especially for type 1 diabetes, more than one gene may be involved. It is thought that there may be a viral cause of type 1 coupled with environmental factors. Genetic factors seem to be involved in type 2 diabetes, one of which involves (one or more genes on) the long arm of human chromosome 2 (Figure 4-1). Whereas some forms of diabetes may result from the mutation of a single gene, several mutations may cause other types of diabetes; among these are thought to be mutations in the gene for the glycolytic enzyme, glucokinase, and the hepatocyte nuclear factor (HNF) transcription factors HNF-1α, HNF-4α, and HNF-1β. Some gene mutations have been identified that are involved in type 2 diabetes (maturity onset diabetes in the young, as shown in Table 4-1). Environmental factors play an important role in the development of this disease.

Type 2 diabetes is reaching epidemic proportions in the United States. As an example, the Pima Indians in Mexico, who live in an area without means of refrigeration or transportation, have no diabetes. In contrast, Pimas who live in Arizona have a high incidence, as do other Southwest

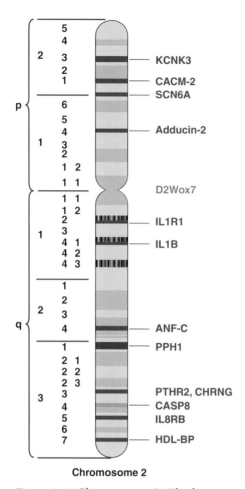

Figure 4-1. Chromosome 2. The long arm is the lower part of the chromosome. Redrawn with permission from Macmillan Publishers Ltd: L.W. Hillier et al., "Generation and annotation of the DNA sequences of human chromosomes 2 and 4," *Nature*, 434: 724–731, 2005.

Table 4-1
Genes Associated with Maturity Onset Diabetes (type 2) of the Young (MODY).

	Gene	Chrom	DM phenotype
MODY 1	HNF-4α	20q12-13	Mild–Severe
MODY 2	Glucokinase	7p15	Mild
	HNF-1α	12q24	Severe
MODY 3	IPF-1	13q12	Mild
MODY 4	HNF-1β	17q	Mild–Severe
MODY 5			

132   CHAPTER 4   Carbohydrates

Indians. Moreover, there has been a fourfold increase in diabetes among Pima children over the past 5 years. Lack of exercise and mixing with American genes of parents may be contributors. A white American with a blood glucose level of 200 has a 20% chance of developing the disease after two decades, and an American Indian with that level has an 80% chance of developing the disease in the same period. It has been theorized that American Indians endured cycles of feast and famine, so a gene may predominate (possibly a mutation) that encodes information to conserve blood sugar. Prolonged stress can contribute to stress-induced diabetes, in which endogenous cortisol can be a factor because its episodic elevation in stresses causes a peripheral (tissue) decrease in cellular uptake of glucose. This produces a rise in serum glucose of about 10%, enough to evoke insulin release from the pancreas. Prolonged periods of stress events are thought to eventually exhaust the beta cells from producing insulin.

Indications of diabetes or prediabetes depend on a glucose tolerance test, as well as on a fasting level of blood glucose. Table 4-2 shows the levels of blood glucose that generate a diagnosis of diabetes or prediabetes and compares them with normal values.

The cellular utilization depends upon the normal functioning of an insulin receptor. In the normal condition, insulin binds and activates the insulin receptor and facilitates the uptake of glucose (thus reducing the blood concentration). In the type 1 diabetes cell where insulin is not present, resulting from the failure of the beta cells of the pancreas to produce insulin, glucose is not taken up by the cell because the insulin receptor is not activated. As a result, the blood level of glucose rises. In the type 2 diabetes where the use of insulin is reduced, less-than-normal cellular uptake of glucose occurs, leading to a rise in blood glucose. The interference with the use of insulin (or, insulin resistance) could be caused by the "masking" of insulin receptors or to some other defect. These situations are reviewed in Figure 4-2.

The beta cells are part of the pancreas, which is located below and behind the stomach (Figure 4-3). The pancreas functions in two ways, separated by the exocrine pancreas and the endocrine pancreas. The **exocrine pancreas** produces digestive enzymes that are secreted into the intestinal tract, whereas the **endocrine pancreas** produces hormones. In the endocrine pancreas there are groups of cells called the **islets of Langerhans** (Figure 4-4). The human pancreas has about 1 million islets, representing about 2% of the total population of cells in the pancreas (Figure 4-5). Cells in the islets produce insulin (beta cells),

Table 4-2
Blood Glucose Levels During Fasting or After Oral Glucose Ingestion That Are Diagnostic of Diabetes or the Prediabetic Condition Compared to Normal Values

	Blood glucose (g/dL)		
	Fasting	30–90'	120'
Normal	<115	<200	<140
Diabetic	>140	>200	>200
Impaired glucose tolerance	<140	>200	140–199

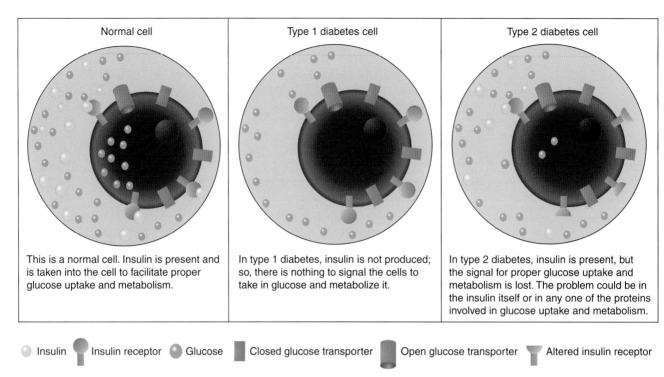

Figure 4-2. Characteristics of tissue cells in type 1 and type 2 diabetes compared to normal. Redrawn with permission from http://darwin.nmsu.edu/~molbio/diabetes/disease.html.

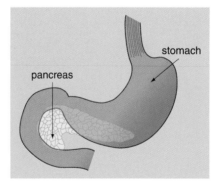

Figure 4-3. The pancreas is located behind the stomach.

glucagon (alpha cells), and somatostatin (delta cells). The endocrine hormones are secreted directly into the bloodstream. Insulin, after it is secreted by signals to the pancreas (elevation of the blood glucose level), circulates in the blood to the tissues. In the plasma membranes of tissue cells are located insulin receptors (Figure 4-6). The insulin receptor consists of two dimers of α- and β-chains. The β-chain courses through the cell membrane and contains **tyrosine kinase** activity. The α-chain binds insulin on the surface. Details explaining conformational changes in the receptor after insulin binding and the transmission of signals are shown in Figure 4-7. Elaboration of the signal transduction emanating from the activated insulin receptor (after tyrosine

(Text continues on p. 138.)

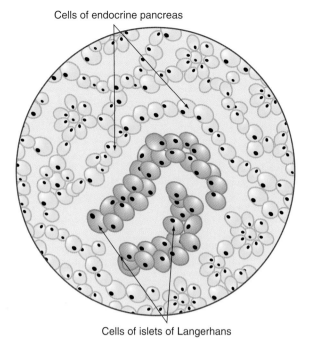

Figure 4-4. Islets of Langerhans in the endocrine pancreas.

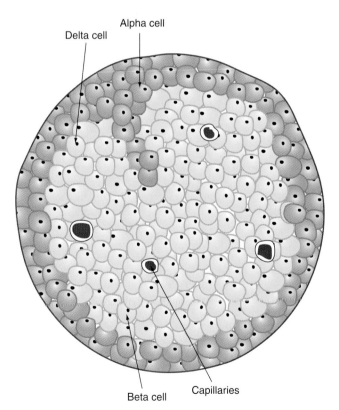

Figure 4-5. Cell types in the islet of Langerhans of the endocrine pancreas.

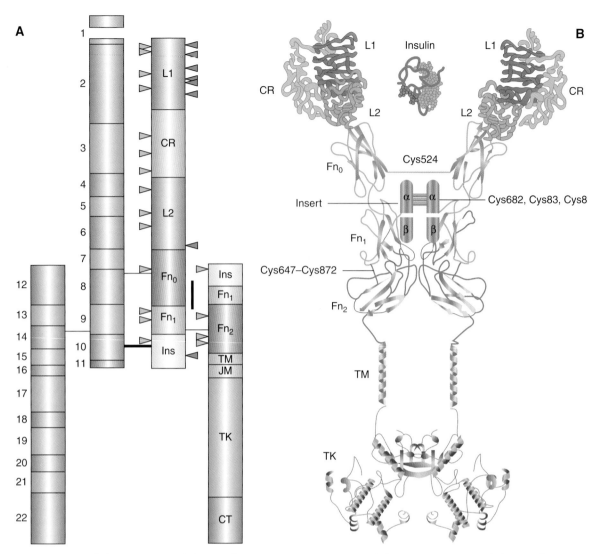

Figure 4-6. Modular structure of the insulin receptor. **A.** Diagram of the $\alpha_2\beta_2$ structure of the insulin receptor. The *left half* of the diagram shows the boundaries of the 22 exons of the gene. The *right half* of the diagram indicates the predicted boundaries of the protein modules, corresponding mainly to exon boundaries L1 and L2, large domains 1 and 2 rich in leucine repeats. *CR*, Cys-rich domain; *CT*, carboxy terminal tail; *Fn$_0$, Fn$_1$, Fn$_2$*, fibronectin type III domain; *Ins*, insert in Fn$_1$; *TM*, transmembrane domain; *JM*, juxamembrane domains; *TK*, tyrosine kinase domains; *the black bar along Fn$_0$*, major immunogenic region; *orange arrowheads*, N-glycosylation sites; *green arrowheads*, ligand (insulin) binding hot spots; *the 2α*, subunits are linked by a disulfide bond in each Fn$_0$ domain. There is a single disulfide bridge between the α- and β-subunits. Exon II is highlighted in *orange*. **B.** Domain organization of the insulin receptor. The binding surface is shown in *yellow*. The insulin backbone is shown in *blue*. Reprinted by permission from Macmillan Publishers Ltd: "Generation and annotation of the DNA sequences of human chromosomes 2 and 4," *Nature Reviews Drug Discovery*, 1: 769, 2002.

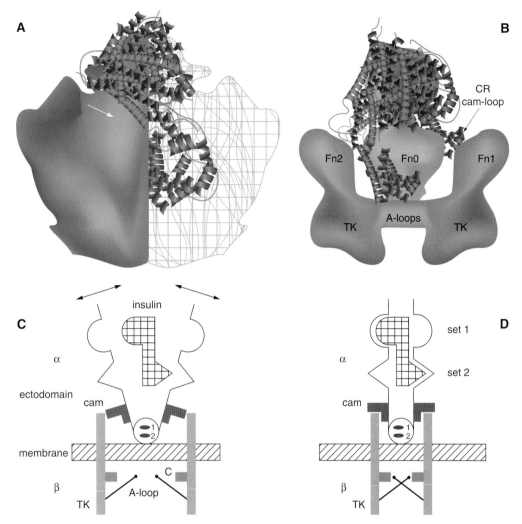

Figure 4-7. Insulin/insulin receptor complex. **A.** End view of the full-mass representation of the IR dimer: *left half,* surface rendering; *right half,* wire mesh representation. Fitted structure of two IR-adapted LCL regions *(green and orange-brown)* showing correspondence of surface features and extent in uppermost regions (L1). *Arrow,* cam-like region on the CR domain. **B.** Higher density solid surface representation of the view rotated slightly from panel A to show CR loop regions (cams) of atomic structure reaching the top portion of Fn2 domains of the 3D reconstruction. **C** and **D.** Simplified schematic of structural changes during activation of the insulin receptor. **C.** Inhibitory state: ectodomain of dimeric subunits each with two differing insulin binding sites and a blocking cam. Unbound bivalent insulin: subunits resting against cams, crossing membrane, with tyrosine kinase (TK) domains separated. *A-loop,* activation loop; *C,* catalytic region. Small on-axis circles (1 and 2) represent the two disulfide bonds. *Arrows* indicate thermally induced motion. **D.** Insulin bound state: blocking cams rotated and subunits resting closer to the center of the ectodomain. TK domains are in position for transphosphorylation via A-loops. Sets 1 and 2 indicate schematically different sets of amino acids from monomers I and II interacting with corresponding different sites on insulin. Redrawn with permission from F.P. Ottensmeyer, et al., "Mechanism of transmembrane signaling: insulin binding and the insulin receptor," *Biochem.,* 39: 12103–12112, 2000. Copyright (2000) American Chemical Society. Reproduced directly from http://lurdwig-sunZ.unil. ch~lantezan/tp-gwup/pathway/diabetes4.htm.

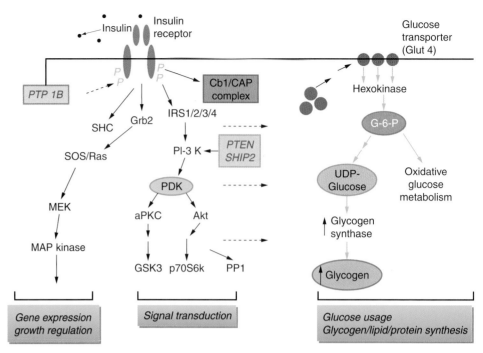

Figure 4-8. Diagram of insulin signal transduction pathways.

kinase phosphorylation) is shown in Figure 4-8. The signaling pathways are complex and involve other kinases. The pathway on the left leads to the function of insulin as a growth stimulator. The pathway on the right shows the insulin-mediated utilization of the glucose pathway. After insulin activation of the insulin receptor, **Glut 4**, a glucose transporter, is moved to the cell membrane (muscle and adipose tissues), where it facilitates the transport of glucose into the cell. Inside the cell, glucose is acted upon by hexokinase to form glucose-6-phosphate on its way to glycogen, the storage form of carbohydrates, if glucose is not entering glycolysis (followed by the citric acid cycle) to be metabolized for energy ("oxidative glucose metabolism" in Figure 4-8).

Insulin

Human insulin now can be produced by recombinant DNA technology. In the past, it was purified from animal sources (pig and cow) for human use (injection). With information about mad cow disease, purification of insulin from bovine pancreas for human use could be problematic.

Insulin is synthesized in the beta cells of the pancreas as a preproprotein called **preproinsulin.** It contains three disulfide bonds in the activated form and is a dimer of two polypeptide chains (A-chain, 21 amino acids; B-chain, 30 amino acids and a molecular weight of 5808 D). It is activated by proteolytic cleavage, removing the connecting peptide (C-peptide) to produce the activated form of insulin. Preproinsulin is shown in Figure 4-9. Once it is activated by removal of the signal peptide

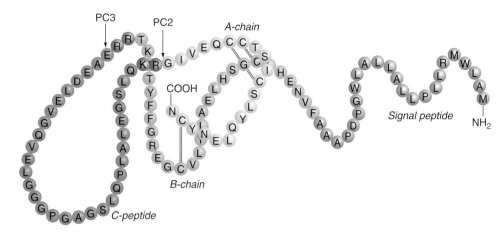

Figure 4-9. The structure of preproinsulin. The A- and B-chains are indicated as well as the three disulfide bonds *(double lines)*. The C-peptide is cleared off proteolytically to activated insulin after the signal peptide (determining secretion) has been removed.

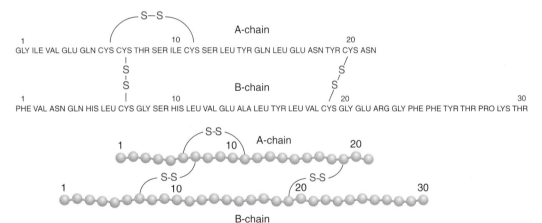

Figure 4-10. Active insulin. At the top is shown the amino acid sequences of the A- and B-chains. At the bottom is a diagram of active insulin.

and the connecting peptide, the active circulating form of insulin develops, as shown in Figure 4-10. Insulin exists in solution (depending on the medium) as monomers or dimers or as crystals of hexamers. The hexamer was crystallized as a zinc–insulin complex and is shown in Figures 4-11 and 4-12. The insulin monomer is shown in Figure 4-13, and the insulin dimer is shown in Figure 4-14. The insulin monomer is the biologically active form of the hormone because it exists in this form at physiological concentrations (about 1 ng/ml). At higher concentrations it

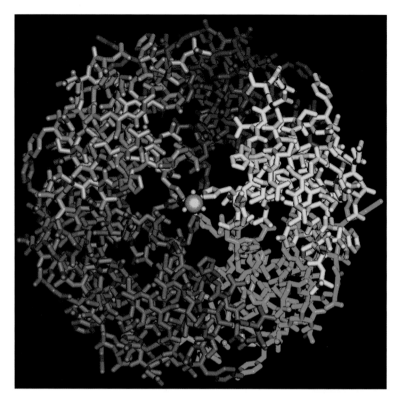

Figure 4-11. Crystal structure of the insulin–zinc hexamer viewed down the threefold axis, indicating the position of the zinc ions and the three water molecules that coordinate with each of them. Each monomer is shown in a different color. Reproduced from http://www.cryst.bbk.ac.uk/PPS2/course/section11/insulin.html.

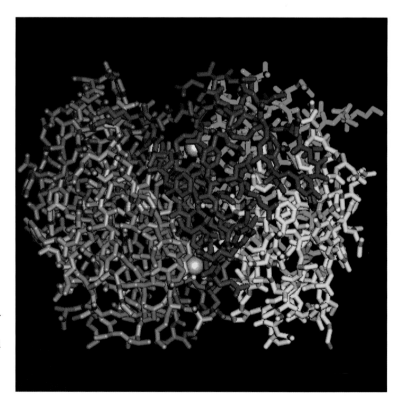

Figure 4-12. Another view of the crystalline hexamer of insulin–zinc complex perpendicular to the threefold axis. Two zinc atoms are present in the hexamer. Reproduced from http://www.cryst.bbk.ac.uk/PPS2/course/section11/ins_sol4.gif.

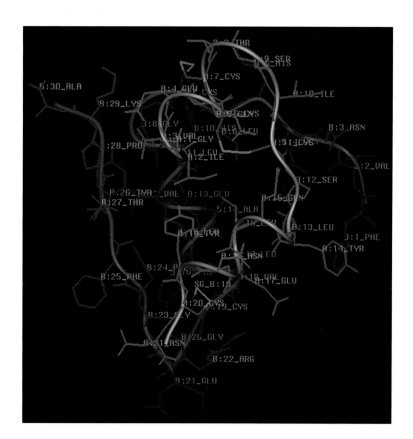

Figure 4-13. Crystal structure of insulin monomer. It is viewed perpendicular to the threefold axis. The disulfide bonds appear in *yellow*. Reproduced from http://www.cryst.bbk.ac.uk/PPS2/course/section11/insulin.html.

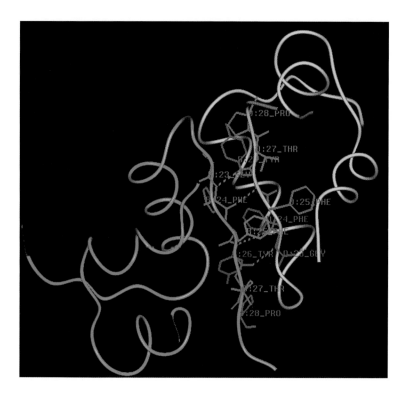

Figure 4-14. Structure of the insulin dimer. Reproduced from http://www.cryst.bbk.ac.uk/PPS2/course/section11/insulin.html.

forms dimers, and in the presence of zinc it forms hexamers (two zinc atoms per hexamer).

The insulin monomer binds to the insulin receptor α-subunits in the target cell membrane. The α-subunit comprises the insulin-binding site with pockets that interact with two regions of the insulin molecule. Insulin binding causes a conformational change in the receptor, in turn, causing regions of the receptor to approach one another. This results in the activation of tyrosine kinase, which autophosphorylates the β-subunits. The receptor β-subunit tyrosine kinase also phosphorylates various substrates in the cell, such as insulin receptor substrate (IRS) 1, IRS-2, and the associated binding protein Grb (Figure 4-8), which are intermediates in the signal transduction pathway leading to biological responses.

There is a motif in the juxtamembrane with the sequence Asn-Pro-Glu-Tyr (the **NPEY motif**). It plays a role in contacting the receptor tyrosine kinase to various cellular signaling molecules. For example, IRS-1 and Shc both interact with the NPEY motif, and mutation of this sequence impairs the phosphorylation of these molecules. Some researchers believe that an aberrant NPEY motif may be a cause of type 2 diabetes. Another view of insulin receptor signal transduction is in Figure 4-15, which shows events at the cell surface, inside the cell, and in the nucleus following insulin binding to the receptor.

The Pancreatic Beta Cell

Insulin is manufactured exclusively in the pancreatic beta cell. Circulating glucose can activate signaling pathways inside the beta cell that set into motion a series of events leading to the synthesis and secretion of insulin into the bloodstream. These signals lead to an increase in cytoplasmic calcium ions that cause the release of insulin (**exocytosis**) from the insulin secretory granule. This event is pictured in Figure 4-16. The islets of Langerhans and a section of the beta cell are shown in Figure 4-17. *When the level of glucose in the blood is increased by 10% or more, this is sufficient to evoke insulin release from the pancreatic beta cell.* The events from the glucose stimulation of the cell to the release of insulin are shown in Figure 4-18. The trigger for insulin secretion is the accumulation of calcium ion in the cell. Calcium comes from internal stores and from the outside (bottom of Figure 4-18). The metabolism of glucose leads to an increase in the cellular ATP and a decrease in [ADP], which affects a closed potassium channel. This causes depolarization that in turn stimulates calcium entry into the beta cell. The exocytosis of insulin may depend on the phosphorylation–dephosphorylation of the membrane proteins of the insulin secretory granules, an area of current investigation. Still, there are many details concerning this mechanism that remain to be described in detail. Transplantation of pancreatic islet cells is being studied as a therapeutic maneuver for type 1 diabetes and efforts are directed toward stimulating the growth of beta cells as an approach to treating type 2 diabetes. Type 2 diabetes is a disease of overweight individuals coupled with lack of exercise and aging and aligned with some genetic background, as discussed earlier. In the United States, it is reaching epidemic proportions.

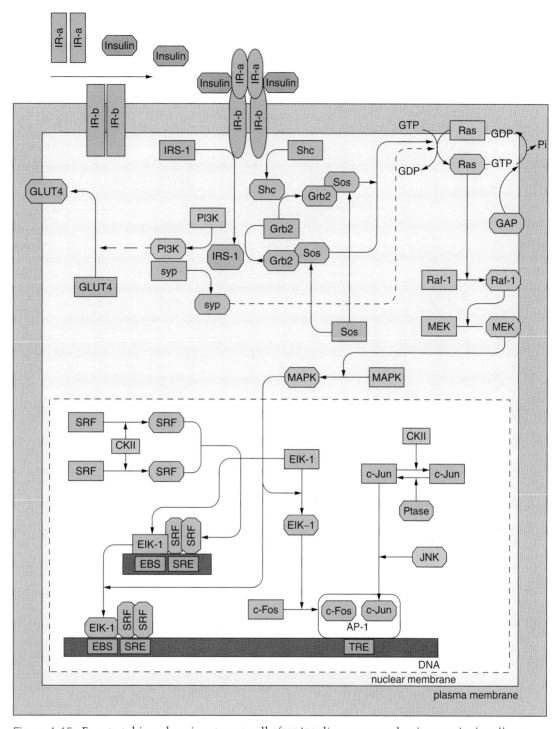

Figure 4-15. Events taking place in a target cell after insulin monomer binding to the insulin receptor. Redrawn with permission from http://www.grt.kyushu-u.ac.jp/spad/images/signalpathway/insulin.gif.

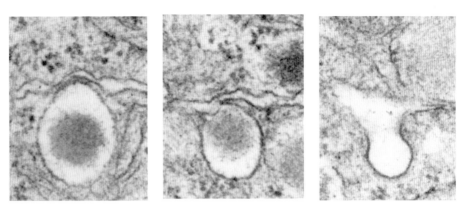

Figure 4-16. Electron micrographs showing secretion of insulin from an insulin-containing granule in the pancreatic beta cell. An insulin secretory granule undergoing an exocytotic event. It first docks with the beta call plasma membrane *(left panel)*. There is then a membrane fusion event between the plasma membrane and the insulin secretory granule membrane (notice these two membranes are continuous in the *center panel*). The insulin secretory contents are then expelled into the extracellular milieu *(right panel)*. Reproduced with permission from http://www.pnri.org/research/rhodes/research.html.

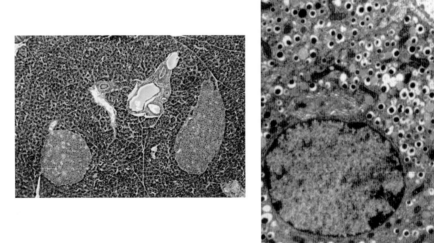

Figure 4-17. Views of the islets of Langerhans *(left)*, and a portion of the beta cell showing an insulin-containing granule. Two islets of Langerhans clearly visible in this stained pancreatic section *(left)*. Electron micrograph section of a pancreatic beta cell *(right)*. Notice the characteristic insulin secretory granules, which contain an insulin crystal consisting of ~250,000 insulin molecules. Reproduced with permission from http://www.pnri.org/research/rhodes/research.html.

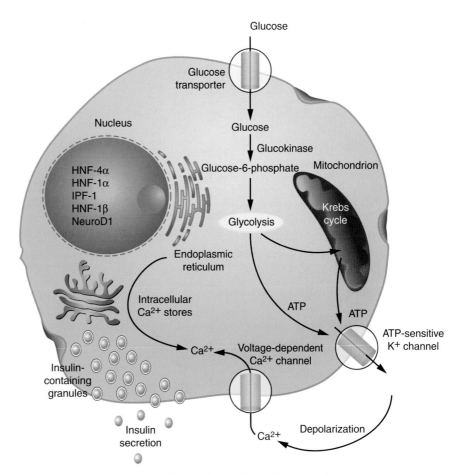

Figure 4-18. Glucose stimulates release of insulin from the pancreatic β-cell.

Effects of Diabetes

Type 1 diabetes is insulin-dependent diabetes, reflecting a failure to produce insulin, often because of the destruction of the beta cells of the pancreas. Some believe that the beta cells are destroyed by a virus. Alternatively, it is considered an autoimmune disease in which antibodies may be produced against the beta cell, causing destruction of these cells. Type 1 diabetes begins at an early age and progresses. Characteristics are hyperglycemia, hyperlipidemia (**chylomicrons** and **very low density lipoproteins,** or **VLDLs**), and ketosis and ketoacidosis. This represents about 10% of the cases of diabetes. Type 2 diabetes is independent of insulin and is primarily a failure to respond to insulin (as if the insulin receptors were blocked or partially blocked). In this condition, there is no increase in circulating ketone bodies. There is hyperinsulinemia, hyperglycemia, and hypertriglyceridemia but no ketosis or ketoacidosis. Of cases of diabetes, 90% to 95% are type 2, and about 6% of people in the United States are prediabetic with polydipsia, polyuria, and glucosuria. Complications of diabetes extend to the cardiovascular system and the nervous system (Figure 4-19).

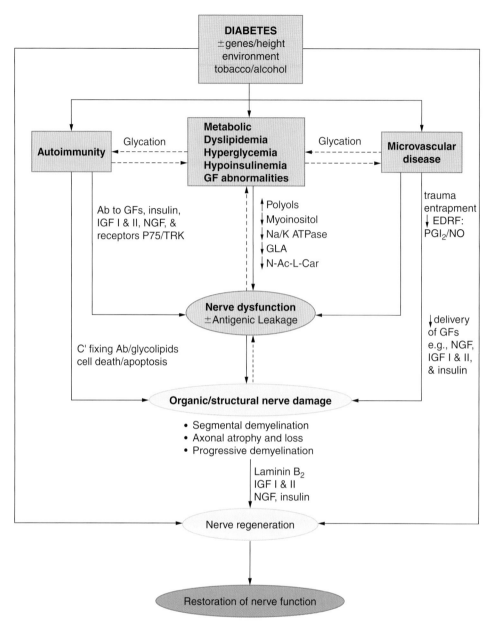

Figure 4-19. Clinical problems caused by diabetes. *Ab,* antibody; *EDRF,* endothelium-derived relaxation factor (related to nitric oxide); *GF,* growth factor; *GLA,* γ-linoleic acid; *IGF,* insulin-like growth factor; *NGF,* nerve growth factor; *NO,* nitric oxide; *PGI$_2$,* prostaglandin I$_2$.

Simple Sugars

The simplest sugars are monosaccharides. Glucose is the prime example and the most important sugar from the point of view of diabetes. Figure 4-20 shows a simple stick structure (Fischer projection) of glucose and how it is closed into a ring structure (Haworth projection) when the carbon-1 (C1) aldehyde (top carbon) becomes proximate to the C5 hydroxyl. The ring is closed (bottom structure) when the C1 aldehyde approaches the C5 hydroxyl. The carbonyl group can be attacked from two

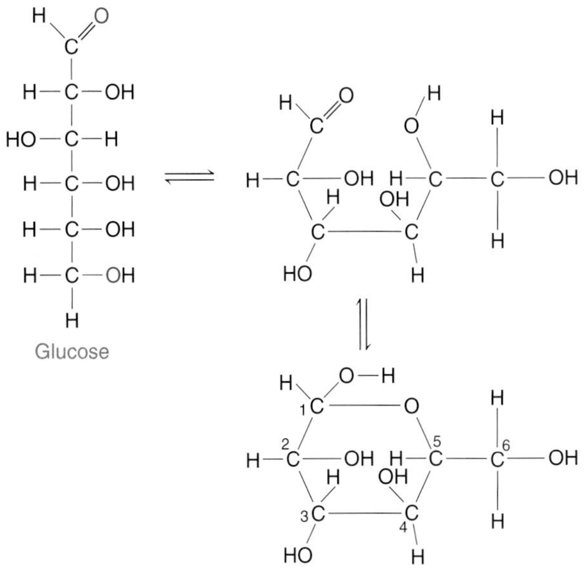

Figure 4-20. Stick model of glucose (carbon 1 is the top carbon and numbering proceeds to the bottom, carbon 6), Carbon 1 aldehyde reacts with carbon 5 hydroxyl to release a molecule of water and close the ring.

sides (thus the possibility of α- or β-forms), and a proton is emitted from the aldehyde group in the process (Figure 4-21). The α- and β-forms of glucose in the ringed form are named based on the position of the hydroxyl on C1, as shown in Figure 4-22. β-D-Glucose has the C1 hydroxyl on the left, and α-D-glucose has the C1 hydroxyl on the right. *D*, as before with the amino acids, refers to *dextro*. When the sugar carbons are in the form of a ring, there are two ways for the carbon bonds to bend: they can assume either a "chair" configuration or a "boat" configuration. One form may be favored over the other depending on the substituents (hydroxyls) of the ring (Figure 4-23). The sugar rings do not form a flat ring like the benzene ring because benzene has three carbon-to-carbon double bonds that are shorter than carbon-to-carbon single bonds. Thus, simple sugars can be represented in five ways, as shown in Figure 4-24 (the fifth way, the boat structure, is not shown). Disaccharides are formed by using the α- (downward bond) or β- (upward bond) hydroxyl on the ring, as shown in the example diagrammed in Figure 4-25. In Figure 4-26, many of the

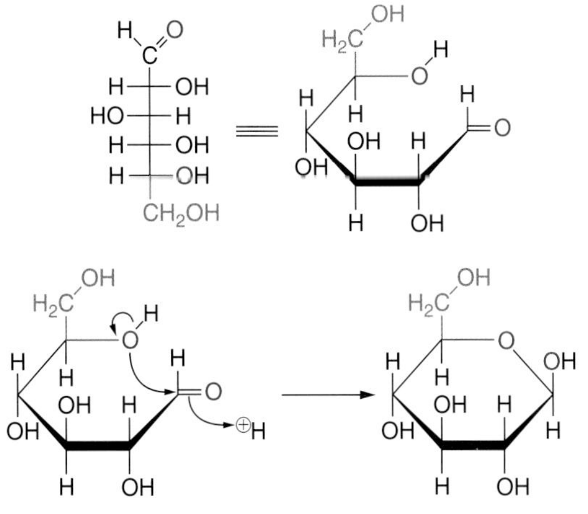

Figure 4-21. Conversion from the Fischer projection of D-glucose to the Haworth projection showing glucose as a pyranose (5-membered) ring structure. The heavy lines at the bottom of the ring indicate that part of the structure extends outward from the page toward the reader.

Simple Sugars 147

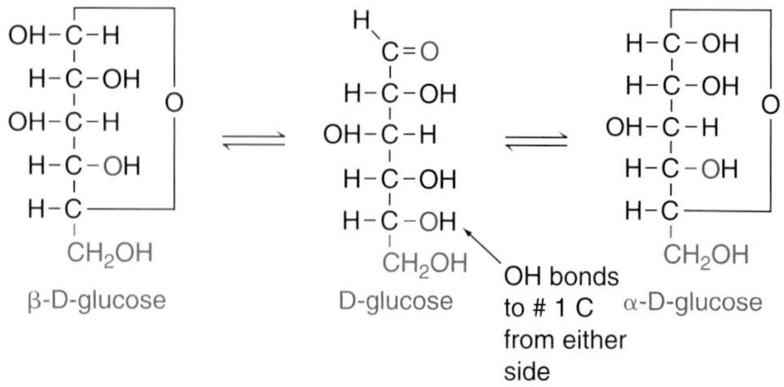

Figure 4-22. β-D-glucose and α-D-glucose ringed forms.

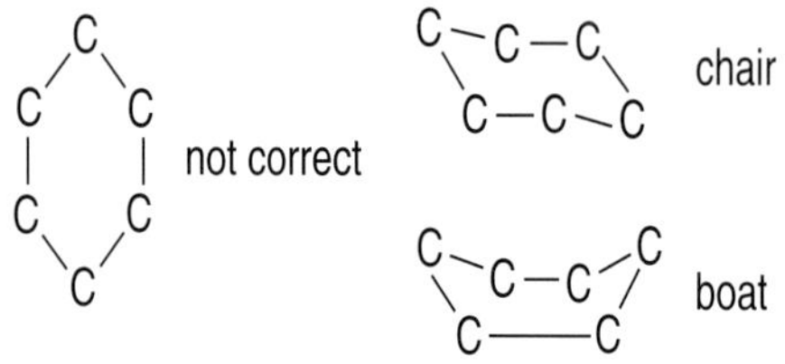

Figure 4-23. Permitted ring structures for sugars.

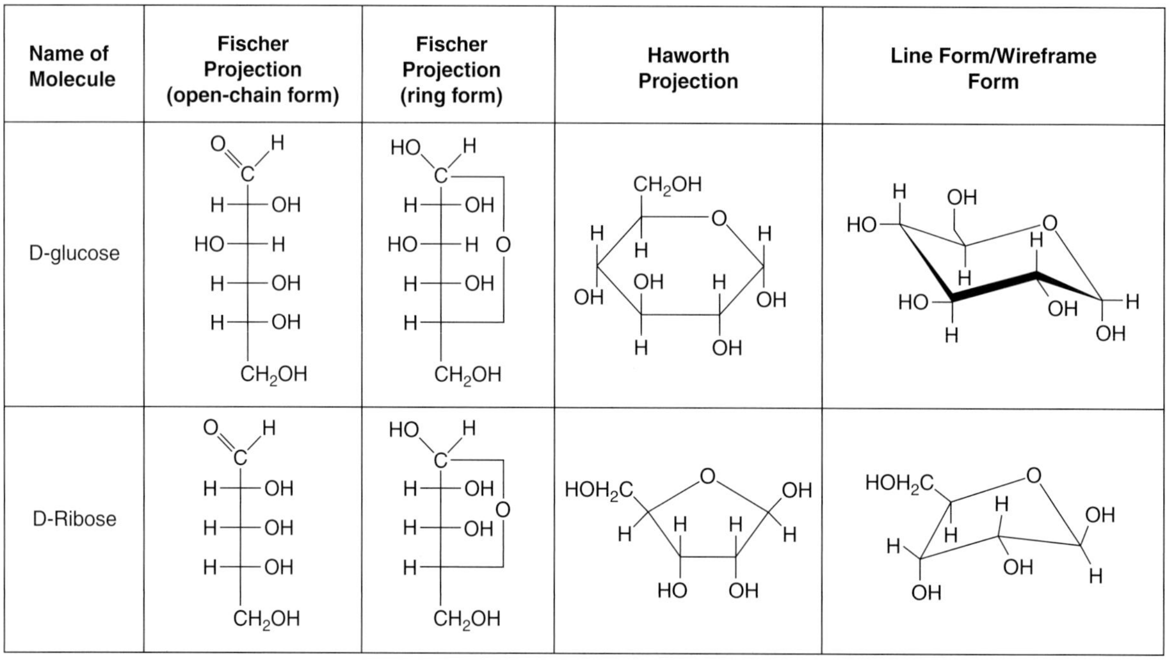

Figure 4-24. Different ways of writing the structures for the hexose, glucose, and for the pentose, ribose. Redrawn from http://chemed.chem.purdue.edu/organic/topicreview/structures/represent/fischer.htm#Fischer%20Projection%20-%20Ring%20Form.

148   CHAPTER 4   Carbohydrates

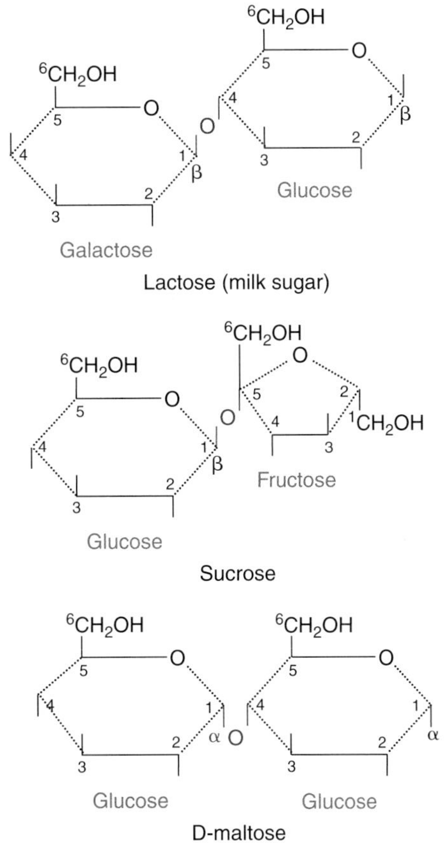

Figure 4-25. Important commonly occurring disaccharides. Sugars interact using α- or β-hydroxyls on the ring. The glucose moiety can open and is a reducing sugar. Both rings are locked in sucrose so it is a non-reducing sugar. The glucose on the left of D-maltose cannot open, whereas the glucose on the right can open, and therefore maltose is a reducing sugar. Maltose, as will be seen, is the repeating unit in starch.

common monosaccharides and disaccharides are shown in chair configurations. Atomic models of some common sugars are shown in Figure 4-27. Sucrose is cane sugar and is the common sweetener.

Recently, emphasis on dieting has spawned a number of synthetic sweeteners, some of which are shown in Figure 4-28. Saccharin has been shown to cause bladder cancer in laboratory animals, although, given the intake in humans, it is not certain how carcinogenic it is for them. It is the oldest artificial sweetener, discovered in 1879, and it is 300 times as sweet as sucrose. It is heat stable and does not affect blood insulin levels. Xylitol and sorbitol are natural sweeteners. Sorbitol is metabolized more slowly than sucrose. Aspartame, the artificial sweetener, is a neurotoxin in high amounts. Splenda, the newest artificial sweetener, contains 2% impurities, including lead, arsenic, and methanol, all of which are toxins. Splenda has been found to increase glycosylated hemoglobin (negative effect), which is a marker for glycemic control in diabetes. In mice, Splenda decreases thymus weight by 40%, and it has many other negative effects.

(Text continues on p. 153.)

Structures of common sugars

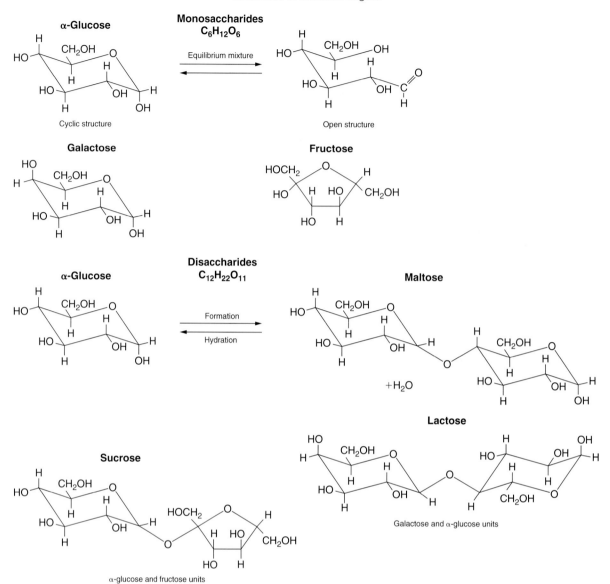

Figure 4-26. Chair structures of common monosaccharides and disaccharides.

CHAPTER 4   Carbohydrates

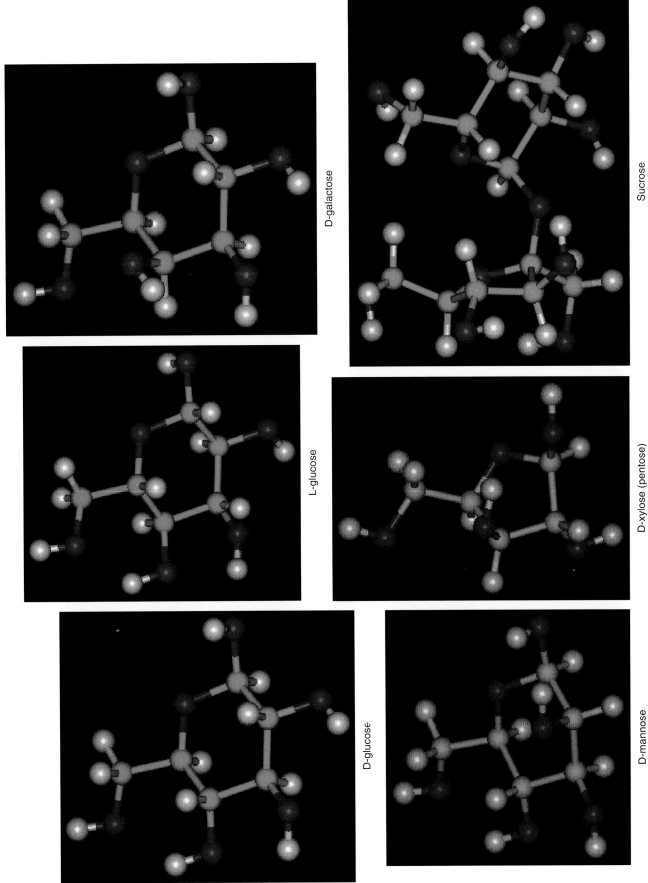

Figure 4-27. Atomic structures of some common sugars. Reproduced with permission from http://www.nyu.edu:80/pages/mathmol/library/sugars/sucrose.gif.

Simple Sugars

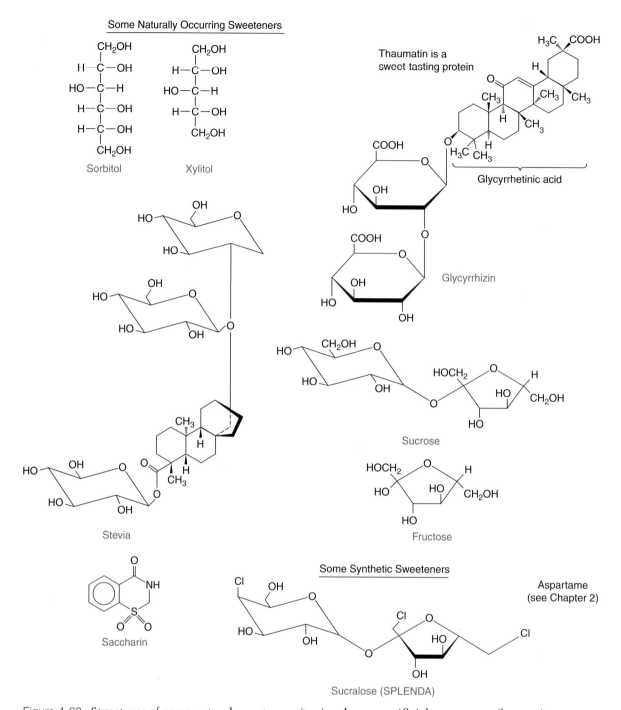

Figure 4-28. Structures of some natural sweeteners *(top)* and some artificial sweeteners *(bottom)*.

CHAPTER 4 Carbohydrates

Starch

Starch is a plant product of photosynthesis and is the major storage form of carbohydrates. Starch is composed of a linear polysaccharide, **amylose,** and a branched polysaccharide, **amylopectin.** Amylose consists of 200 to 2000 anhydroglucoses, as shown in Figure 4-29. The repeating anhydroglucose unit is shown in brackets. Amylopectin is also shown in Figure 4-29 (bottom). Whereas amylose has α-D-1,4 bonds (α-bonds point downward), amylopectin has α-D-1,6 bonds occurring at between 20 and 30 anhydroglucose units. Amylopectin forms a high percentage of waxy starches. Figure 4-30 shows how side-branching chains occur in clusters within the amylopectin molecule.

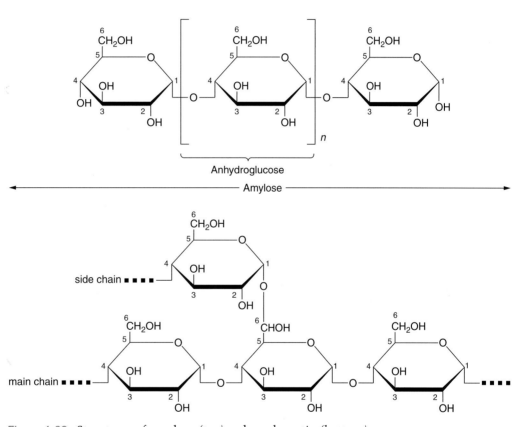

Figure 4-29. Structures of amylose *(top)* and amylopectin *(bottom)*.

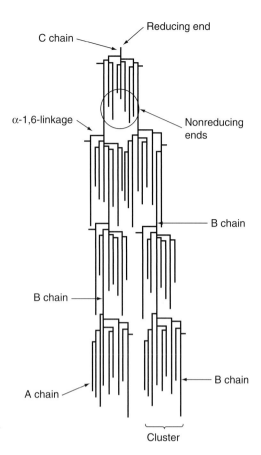

Figure 4-30. Diagram showing how side branching chains are clustered together in the amylopectin molecule.

Glycogen

Glycogen is the carbohydrate storage molecule in humans. When glucose enters a muscle cell, for example, in which it is not needed immediately for energy, it is converted to glycogen. Figure 4-31 shows in outline these relationships. The lower right portion of the figure shows that an enzyme, **glycogenin,** plays the initial role in the formation of glycogen from UDP-glucose in muscle and liver. It catalyzes the addition of linear chains of glucose covalently attached to itself. Tyrosine-194 of the muscle enzyme attaches and adds up to 10 glucose molecules. Glycogenin is a glycosyltransferase. After this addition, glycogen synthase and the branching enzyme complete the synthesis of glycogen. Figure 4-32 shows how a dimer of glycogenin (GN) adds linear chains of glucose molecules, and then glycogen synthase and the branching enzyme complete the structure of glycogen. Figure 4-33 shows the crystal structure of muscle (rabbit) glycogenin dimer (the form active in addition of glucose residues autocatalytically). Figure 4-34 shows the glycogenin monomer with UDP-glucose and manganese (Mn^{2+}), and the bottom of the figure shows how UDP-glucose interacts with manganese and amino acid residues in the active site of glycogenin. Exactly how the glucose chains are added is not completely understood. It is thought that the transfer of the first glucose residues is intermolecular and that subsequent glucose units are

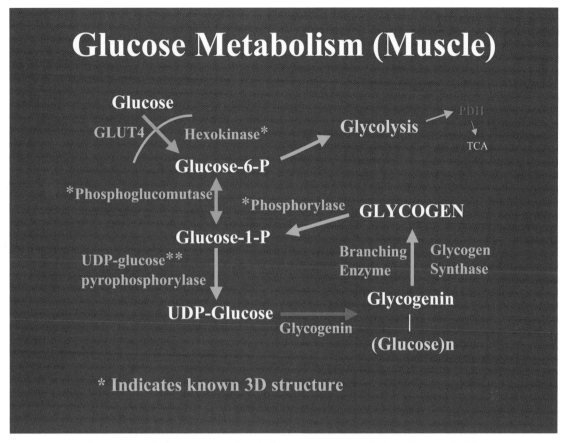

Figure 4-31. Glucose metabolism in muscle. If glucose coming into the cell is needed for energy, it enters glycolysis and then the TCA from which ATPs are generated. If the cell is glucose sufficient, entering glucose *(bottom right)* is converted to the storage form, glycogen. *PDH*, pyruvate dehydrogenase; *TCA*, tricarboxylic acid cycle; *UDP*, uridine diphosphate. Reproduced with permission from Thomas D. Hurley, PhD.

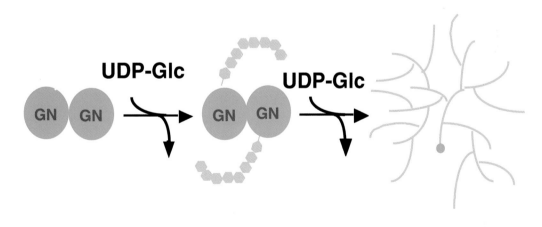

Figure 4-32. The role of the glycogenin (GN) enzyme in initiating glycogen synthesis from glucose and UDP-glucose. Reproduced with permission from Thomas D. Hurley, PhD.

Glycogen   155

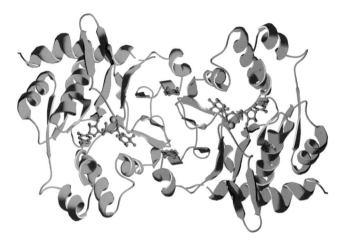

Figure 4-33. Crystal structure of glycogenin dimer. The *gray* balls represent Mn^{2+} ions. UDP-glucose is shown in purple atomic models *(left center and right center)*. Tyrosine-194 residue is shown in *purple* in the center of the structure *(upper and lower)*. Reproduced with permission from http://www.biochemistry.iu.edu/personnel/Hurley/Hurley-web/glycogenin.htm.

attached by intramolecular reaction within the glycogenin dimer. Possible scenarios are conveyed in Figure 4-35.

The partial structure of glycogen, showing straight chains of glucose and branches, is shown in Figure 4-36. A more complete structure of glycogen with glycogenin (G) at the center is displayed in Figure 4-37. In this figure, 5 layers of glycogen structure are shown, whereas there are 12 layers in the completed particle. The completed particle contains as many as 60,000 glucose units. The liver is the primary storage site, with as much as 10% of the body's glycogen, and muscle is second, with about 1% of the body's glycogen. The body has about 40,000 calories of glucose in body fluids (including the blood) immediately available for energy use and about 600,000 calories stored in glycogen. Glycogen is largely made up of glucose units in α-1,4 linkages except for the branches (which are in α-1,6 linkages). Glucose is released from glycogen at the ends (Figures 4-36 and 4-37); because there are many ends, the release can be fast.

Breakdown of Glycogen for Energy Use (Glycogenolysis)

Glucose is released from glycogen by phosphorolysis catalyzed by the enzyme phosphorylase. The reaction is as follows:

$$\text{glycogen} + P_i \rightarrow \text{glycogen (n − 1)} + \text{glucose-1-phosphate}$$

The structure of glycogen phosphorylase (a two-subunit enzyme) is shown in Figure 4-38. Phosphorolysis proceeds to within four residues of a branch point, producing one molecule of glucose-1-phosphate for each glucose released. At this point, the debranching enzyme catalyzes the transfer of a trimer from a branch to the free end of the glycogen molecule (the debranching enzyme is a transglucosylase (α-1,4 to α-1-4). The α-1-6 glucosidase activity of the debranching enzyme then cleaves α-1,6-glucose at the branch point in glycogen (Figure 4-36). About 11 of 12 glucose units in glycogen are released as glucose-1-phosphate by phosphorolysis. These can enter glycolysis at the level of glucose-6-phosphate (catalyzed by phosphoglucomutase) without the addition of ATP. The phosphoglucomutase reaction is shown in Figure 4-39. The structure of phosphoglucomutase is shown in Figure 4-40. The importance of

(Text continues on p. 161.)

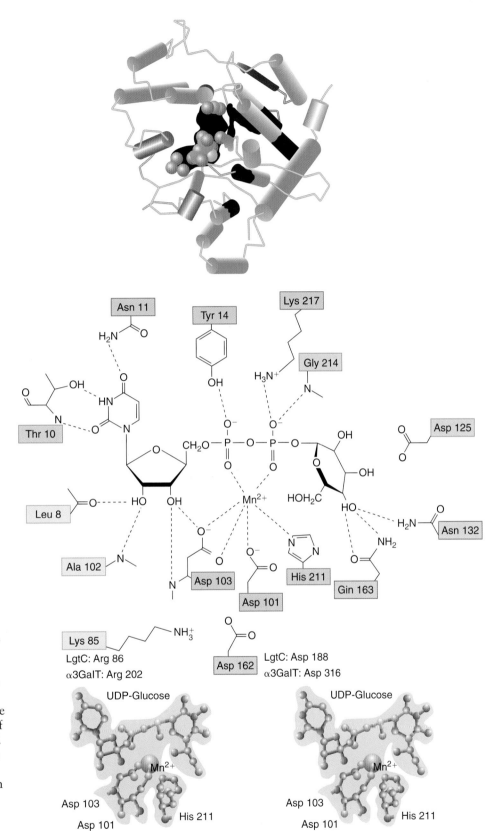

Figure 4-34. Crystal structure of glycogenin monomer *(top)*. Interaction of UDP-glucose with enzyme (glycogenin) amino acid residues and Mn^{2+}. Note that aspartate residues (101 and 103) are essential for the interaction. Redrawn from B.J. Gibbons, P.J.T. Roach, and T.D. Hurley, "Crystal structure of the autocatalytic initiator of glycogen biosynthesis," *J. Mol. Biol.*, 319: 463–477, 2002 with permission. The crystal structure at the top is redrawn from http://www.biochem.ucl.ac.uk/bsm/pdbsum/1ll2/tracel.html.

Glycogen

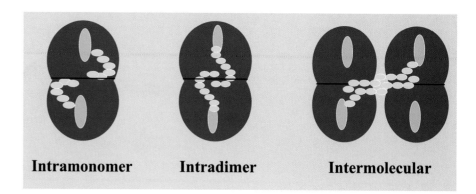

Figure 4-35. Models for glucosylation. Reproduced with permission from Thomas D. Hurley, PhD.

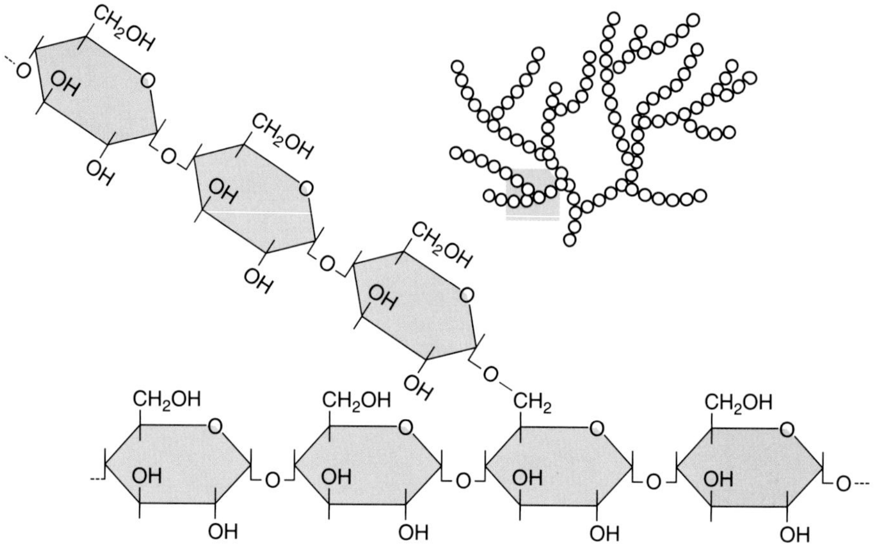

Figure 4-36. Partial structure of glycogen showing straight chain of glucose residues and a branch point. The box in the *upper right hand* figure is amplified at the *bottom*.

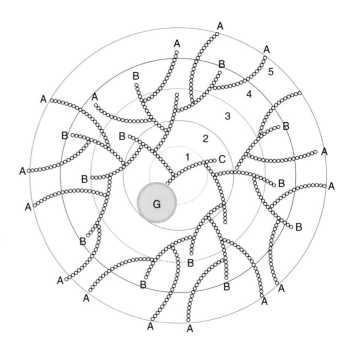

Figure 4-37. Partial structure of the glycogen particle. Five layers are shown here although the completed glycogen particle has 12 layers. Each B chain has 2 branch points, and all chains have the same length. The distribution between the A and B chains is the same. *G*, glycogenin. Redrawn with permission from http://bip.cnrs-mrs.fr/bip10/glycogen.htm.

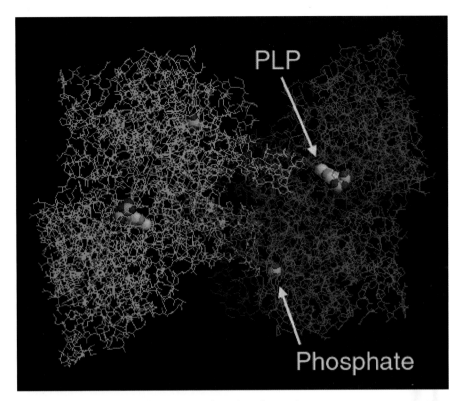

Figure 4-38. Structure of glycogen phosphorylase. There are two protein chains colored in *green* and *blue*. The molecule labeled *PLP* is the coenzyme pyridoxal phosphate. Inorganic phosphate *(Phosphate)* is indicated as bound to each subunit. It is located next to a key threonine residue or a serine residue, depending on the species, which is key to controlling an allosteric change. The subunits work in concert (projections wrap around each other; middle of structure) when responding to small changes in shape that play a role in regulation. Reproduced from http://www.rcsb.org/pdb/molecules/1ygp.gif.

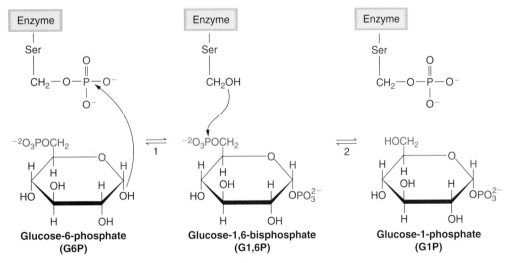

Figure 4-39. The phosphoglucomutase mechanism.

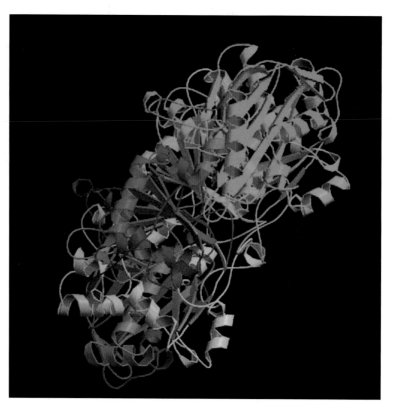

Figure 4-40. Crystalline phosphoglucomutase. From PDB ID: 1c47. S. Baranidharan et al. Binding Driven Structural Changes in Crystalline Phosphoglucomutase Associated with Chemical Reaction. To be published.

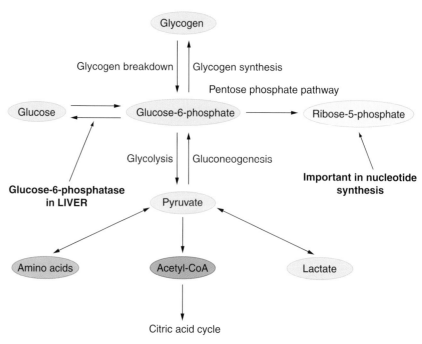

Figure 4-41. Glucose-6-phosphate is a key intermediate in the metabolism of glucose and glycogen.

glucose-6-phosphate as an intermediate in the metabolism of glucose is summarized in Figure 4-41. In the liver, glucose-6-phosphate is converted back to glucose by glucose-6-phosphatase. A mutation in this enzyme can lead to one of the glycogen storage diseases discussed later in this chapter. Glycogen is the preferred storage form of energy in muscle because it can be converted to glucose and burned for energy much faster than storage fat could be used. In addition, blood levels of glucose could not be maintained by fat because fat cannot be converted to glucose. A summary of the breakdown of glycogen by glycogen phosphorylase and the debranching enzyme is shown in Figure 4-42. Glucose is released from every nonreducing end of each branch of glycogen (*black arrows* in Figure 4-43).

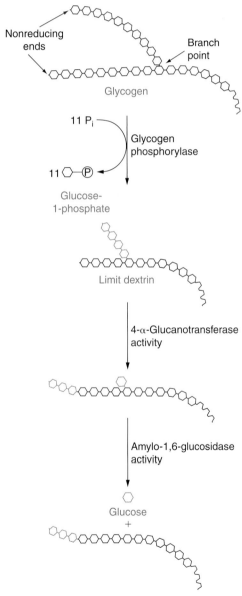

Figure 4-42. Summary of the breakdown of glycogen by glycogen phosphorylase *(first 2 steps)* and by the debraching enzyme *(last step)*.

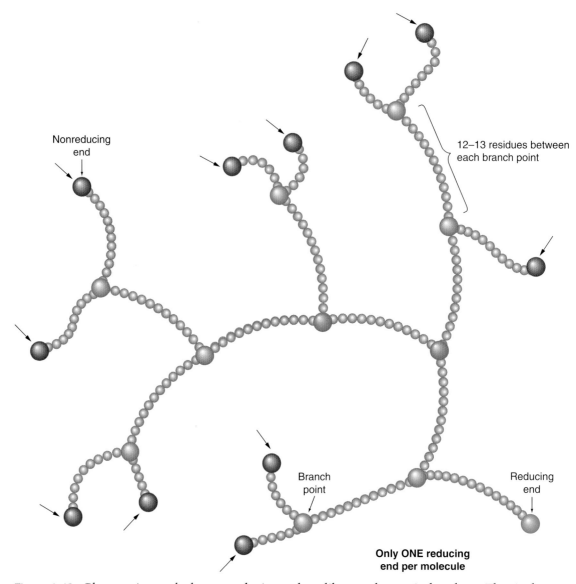

Figure 4-43. Glycogen is attacked at nonreducing ends to liberate glucose-1-phosphate. The single reducing end has glucose with the carbon one free from the ring and able to react.

Glycogen Synthesis

Four enzymes participate in the synthesis of glycogen: glycogenin, UDP-glucose pyrophosphorylase, glycogen synthase, and glycogen branching enzyme (Figure 4-31). When glucose enters the cell, it is converted to glucose-6-phosphate and either enters glycolysis, followed by the citric acid cycle for energy generation, or, if the cell is sufficient in energy, is converted to glucose-1-phosphate by phosphoglucomutase (Figure 4-39). Glucose-1-phosphate is then converted to UDP-glucose by UDP-glucose pyrophosphorylase (Figure 4-44). Glycogenin forms the center of the growing glycogen molecule and attaches UDP-glucose molecules to

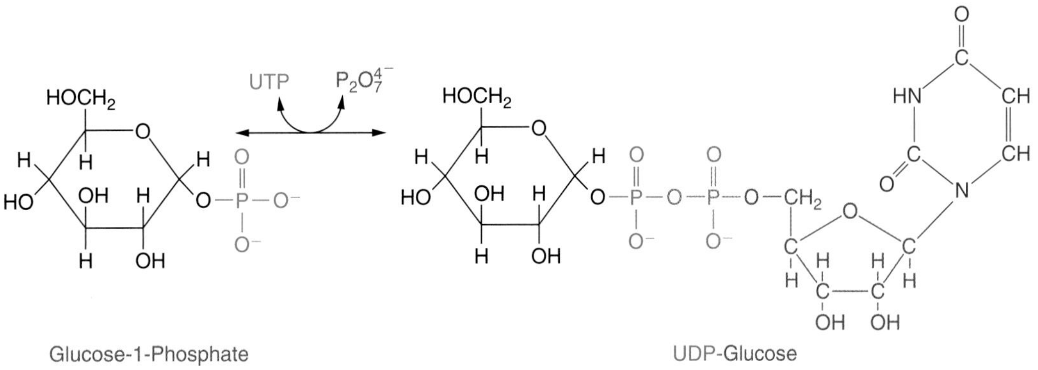

Figure 4-44. Reaction catalyzed by UDP-glucose pyrophosphorylase. $P_2O_7^{4-}$ is pyrophosphate.

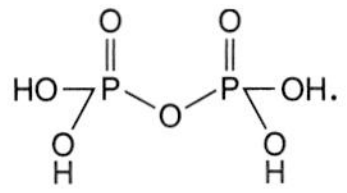

itself, a step that acts as a primer. Then glycogen synthase catalyzes the reaction:

$$\text{UDP-glucose} + \text{glycogen}_{(n \text{ units})} \rightarrow \text{UDP} + \text{glycogen}_{(n+1 \text{ units})}$$

The glycogen particle continues to grow in this way by adding glucose to the nonreducing end. UTP and the hydrolysis of pyrophosphate uses up the equivalent of two ATP high-energy phosphates. The branching enzyme installs the branches by its activity as amylo-(1,4–1,6)-transglycosylase and transfers a terminal fragment of 6 to 7 glucose residues, from a stretch of 11 or more glucose residues, to an internal glucose residue at a C6 position. This process, together with glycogen synthase, continues until the mature particle is complete (Figure 4-37).

Effects of Hormones on Glycogen Breakdown and Synthesis

Glycogen breakdown is stimulated by glucagon and epinephrine. Glucagon is a 29–amino acid protein secreted by the alpha cells of the pancreas, and it consists of a single helix, which is fairly unusual. The alpha cells in the islets of Langerhans occur early in the development of the pancreas. The structure and sequence of glucagon and the structure of epinephrine are shown in Figure 4-45. Here, epinephrine is acting through a β-adrenergic receptor in muscle and in liver, but epinephrine can have a similar effect in liver when it acts through an α_1-adrenergic receptor. This results in the activation of protein kinase C, whose action inactivates the insulin receptor. Both the epinephrine and glucagon hormones operate through the activation of protein kinase A, which results in the phosphorylation of glycogen synthase *a* to glycogen synthase *b* (the phosphorylated form which is *inactive in glycogen synthesis*). The structure of the β-adrenergic receptor is shown in Figure 4-46. The structure of the α-receptor is a seven-membrane–spanning protein. Protein kinase A also phosphorylates phosphorylase kinase to its phosphorylated form, which is active, and the activated phosphorylase kinase phosphorylates glycogen phosphorylase *b*, the inactive form, to glycogen phosphorylase *a*, the

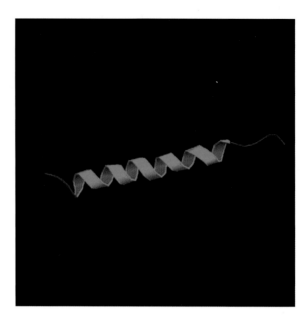

H-His-Ser-Gln-Gly-Thr-Phe-Thr-Ser-Asp-Tyr-Ser-Lys-Tyr-Leu-Asp-Ser-Arg-Arg-Ala-Gln-Asp-Phe-Val-Gln-Trp-Leu-Met-Asn-Thr-OH
  1   2   3   4   5   6   7   8   9 10 11 12 13 14 15 16 17 18 19 20 21 22 23 24 25 26 27 28 29

Figure 4-45. Structure *(top)* and sequence of glucagon. The structure of epinephrine is at the *bottom*. Top structure is reproduced from http://www.cryst.bbk.ac.uk/PPS2/course/section10/1gcn.gif.

active form. This causes the breakdown of glycogen to liberate glucose. These activities are summarized in Figure 4-47.

The effect of insulin (structure in Figure 4-10) is the opposite of that of glucagon and epinephrine. Its action results in the use of blood glucose for synthesis of glycogen, and this occurs through the insulin stimulation of protein phosphatase-1. The structure of Ser/Thr protein phosphatase-1 is shown in Figure 4-48. Figure 4-49 shows the overall effects of insulin on glycogen synthesis and on the inhibition of glycogen breakdown. Insulin acts through the membrane insulin receptor (Figures 4-6 through 4-8) to induce protein phosphatase-1, which has three actions:

- The hydrolysis of the phosphate group from glycogen synthase *b*, the inactive form, converting it to glycogen synthase *a*, the active form

- The hydrolysis of the phosphate group from phosphophosphorylase kinase to dephosphophosphorylase kinase

- The release of the phosphate group from glycogen phosphorylase *a*, the active form, to glycogen phosphorylase *b*, the inactive form

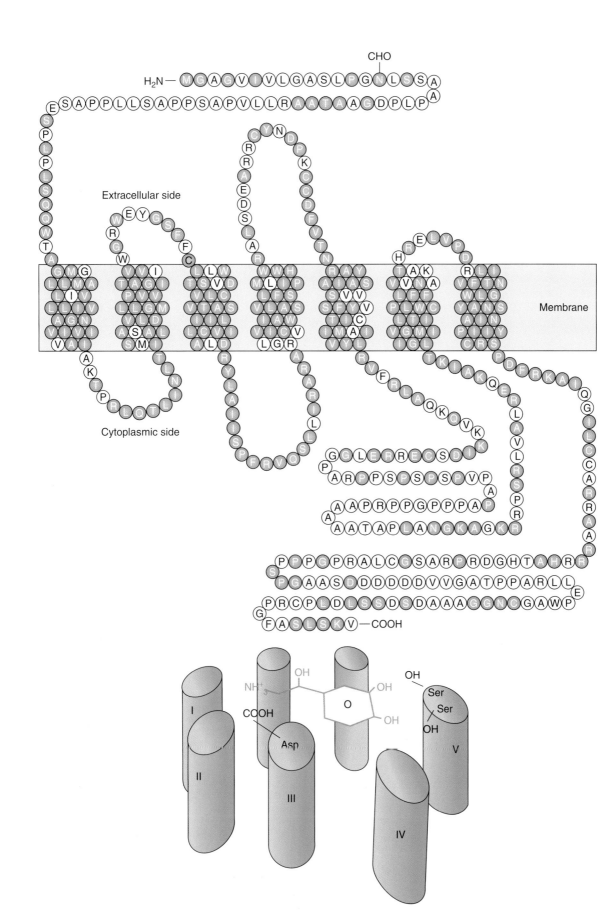

Figure 4-46. Structure of the β-adrenergic receptor. Sequence is shown at the *top*. The receptor is a seven-membrane-spanning protein. The *bottom* diagram shows the binding of norepinephrine to the extracellular part of the receptor.

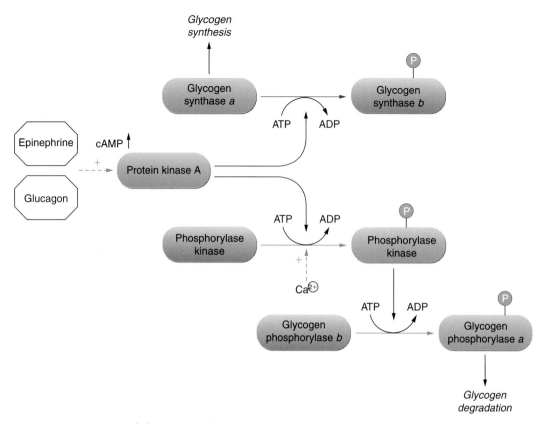

Figure 4-47. Actions of glucagon and epinephrine on glycogen synthesis and breakdown. For this action epinephrine is operating through a β-adrenergic receptor. *Green*, active form; *red*, inactive form.

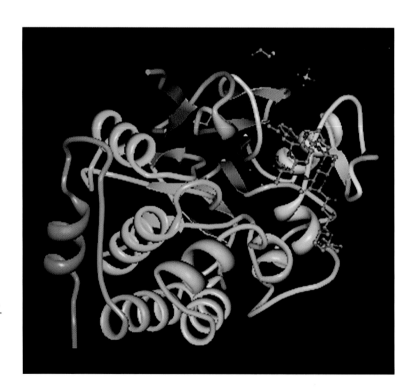

Figure 4-48. Crystal structure of the tumor-promoter okadaic acid bound to protein phosphatase-1. Reproduced from PDB ID: 1jk7. J.T. Maynes, K.S. Bateman, M.M. Cherney, A.K. Das, H.A. Luu, C.F. Holmes, M.N. James. Crystal Structure of the Tumor-Promoter Okadaic Acid Bound to Protein Phosphatase-1. *J. Biol. Chem.*, **276** pp. 44078 (2001).

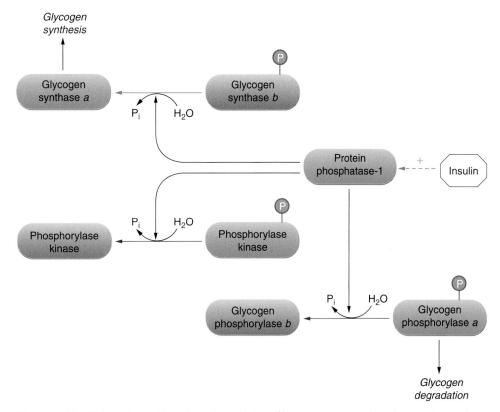

Figure 4-49. The action of insulin, through its effect on protein phosphatase-1, on the activation of glycogen synthesis and the inhibition of glycogen breakdown. *Green*, active; *red*, inactive.

These actions result in the synthesis of glycogen and the inhibition of glycogen breakdown. Figure 4-50 summarizes the actions of glucagon, epinephrine, and insulin on the formation and breakdown of glycogen. Note that in liver, epinephrine can bind to an α_1-adrenergic receptor, which can stimulate protein kinase C. This, in turn, can inactivate the insulin receptor and inhibit glycogen synthesis by blocking insulin action. Epinephrine, therefore, can produce the same effect by acting through two different receptors.

When a cell is in need of stored energy, glucose coming into the cell stimulates glycogen synthesis. In such a cell, protein phosphatase-1 is complexed with glycogen phosphorylase *a*, and in this complex protein phosphatase-1 is inactivated. The incoming glucose binds to glycogen phosphorylase *a*, causing the dissociation of protein phosphatase-1 from the complex. This results in its activation. So activated, protein phosphatase-1 removes the phosphate from glycogen phosphorylase *a*, converting it to glycogen phosphorylase *b* and thus reducing glycogen degradation. The active protein phosphatase-1 also removes the phosphate group from glycogen synthase *b*, converting it to the active glycogen synthase and thus stimulating glycogen synthesis. These activities are summarized in Figure 4-51. The binding site for glucose in the

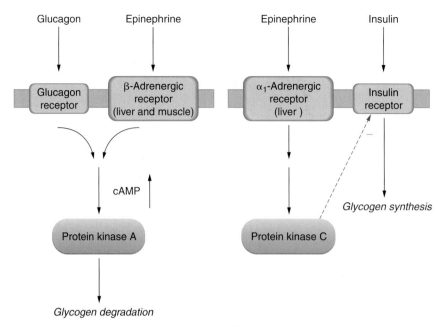

Figure 4-50. Summary of the actions of glucagon, epinephrine, and insulin on the breakdown and synthesis of glycogen. *Green* color indicates the active form of the enzyme.

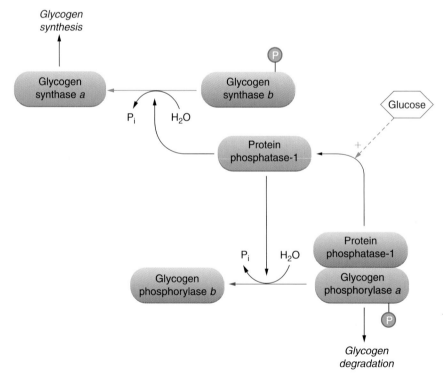

Figure 4-51. Effects of glucose on glycogen synthesis. *Red* indicates inactive form, and *green* indicates active form.

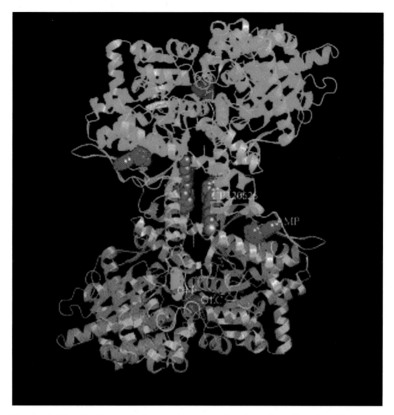

Figure 4-52. Binding of glucose to muscle phosphorylase b. Glucose is a competitive inhibitor of the enzyme, and it is shown in *red* in the center, close to the catalytic center. The molecule in *magenta* is AMP, an activator, which binds in the allosteric site situated at the interface of the two subunits about 30 Å from the catalytic site. The *orange* molecule is a drug bound in a new allosteric site. Reproduced from http://www.sciencedirect.com/science?_ob=MiamiCaptionURL&_method=retrieve&_udi=B6TF8-44K0G17-1&_image=figgr1&_ba=2&_user=10&_coverDate=05% directly and from N.G. Oikonomakos, S.E. Zographos, V.T. Skamnaki, and G. Archontis, *Bioorg. Med. Chem.*, 10: 1313–1319, 2002 with permission. Institute of Biological Research & Biotechnology, The National Hellenic Research Foundation, 48 Vas. Constantinon Avenue, Athens 11635, Greece.

glycogen phosphorylase molecule is shown in Figure 4-52. In addition to glucose, glucose-6-phosphate is an inhibitor of glycogen phosphorylase and an activator (glucose-6-phosphate signals glycogen synthase when glucose is in overabundance) of glycogen synthase (Figure 4-53).

Glycogen Storage Diseases

Although glycogen storage diseases are rare, they deserve mention in a discussion of glycogen. Most depend on specific enzyme deficiencies. Inability to break down glycogen, for example, would be catastrophic, and glycogen buildup would clog the cell. **Von Gierke's disease** is caused by a lack of glucose-6-phosphatase. Glucose-6-phosphate must travel by

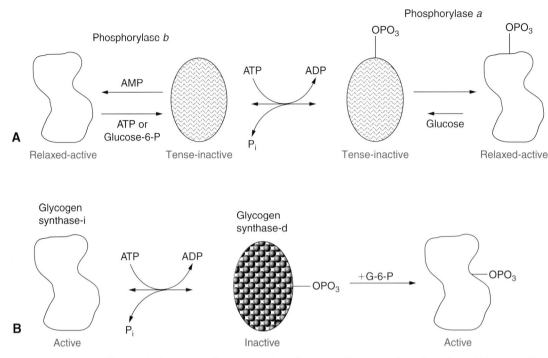

Figure 4-53. Effects of glucose or glucose-6-phosphate on glycogen phosphorylase *(A)* or on glycogen synthase *(B)*. Glucose or glucose-6-phosphate is an inhibitor of glycogen phosphorylase, thus preventing glycogen breakdown. In *B*, glucose-6-phosphate is an activator of glycogen synthase.

way of a translocase in the endoplasmic reticulum membrane to reach the cisternal surface of the endoplasmic reticulum at the location of the enzyme. Thus, the disease could be caused by a mutation of the phosphatase or the translocase. This disease occurs with a frequency of about 1 in 200,000. Table 4-3 summarizes the known glycogen storage diseases.

Is Type 2 Diabetes a Disease of Protein Aggregation?

We have seen in the case of prion diseases that prion scrapie (PrP^{Sc}) has the property of aggregating and is the causative factor. There is also amyloid deposition in Alzheimer's disease, and it may be a causative agent. Interestingly, nearly three quarters of autopsies of people with type 2 diabetes show the accumulation of **amyloid** in the islet (Figure 4-54). What is not clear is whether this is the cause or a result of type 2 diabetes. **Hyaline** (clear membranous material) degeneration was discovered in the islet in patients with diabetes. This material was shown to contain amyloid by staining under the microscope. Later it was discovered that the hyaline staining material was a protein monomer that was 37 amino acids long and was called **islet amyloid polypeptide.** The sequence of this protein, which is named **amylin,** is shown in Figure 4-55. This protein

Table 4-3
Glycogen Storage Diseases

Type: Name	Enzyme Affected	Primary Organ	Manifestations
Type 0	glycogen synthase	liver	hypoglycemia, early death, hyperketonia
Type Ia: von Gierke's	glucose-6-phosphatase	liver	hepatomegaly, kidney failure, thrombocyte dysfunction
Type Ib	microsomal glucose-6-phosphate translocase	liver	like Ia; also neutropenia, bacterial infections
Type Ic	microsomal Pi transporter	liver	like Ia
Type II: Pompe's	lysosomal α-1,4-glucosidase, lysosomal acid α-glucosidase acid maltase	skeletal and cardiac muscle	infantile form = death by 2; juvenile form = myopathy; adult form = muscular dystrophy-like
Type IIIa: Cori's or Forbe's	liver and muscle debranching enzyme	liver, skeletal, and cardiac muscle	infant hepatomegaly, myopathy
Type IIIb	liver debranching enzyme, normal muscle enzyme	liver, skeletal, and cardiac muscle	liver symptoms same as type IIIa
Type IV: Anderson's	branching enzyme	liver, muscle	hepatosplenomegaly, cirrhosis
Type V: McArdle's	muscle phosphorylase	skeletal muscle	excercise-induced cramps and pain, myoglobinuria
Type VI: Her's	liver phosphorylase	liver	hepatomegaly, mild hypoglycemia, hyperlipidemia and ketosis, improvement with age
Type VII: Tarui's	muscle PFK-1	muscle, RBCs	like V; also hemolytic anemia
Type VIb, VIII or Type IX	phosphorylase kinase	liver leukocytes, muscle	like VI
Type XI: Fanconi-Bickel	glucose transporter-2 (Glut 2)	liver	failure to thrive, hepatomegaly, rickets, proximal renal tubular dysfunction

PFK-1, phosphofructo kinase; *RBCs*, red blood cells.

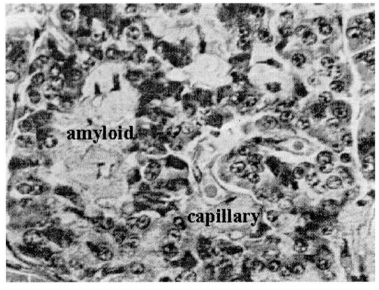

Figure 4-54. Amyloid in the pancreatic islet. Redrawn with permission from Thomas D. Hurley, PhD.

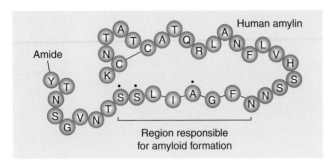

Figure 4-55. The sequence of human amylin. Redrawn with permission from http://www.joplink.net/prev/200107/02.html/.

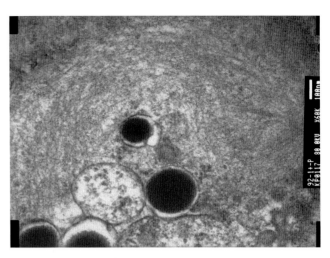

Figure 4-56. Electron micrograph of amyloid. Reproduced with permission from http://www.joplink.net/prev/200107/02.html.

Table 4-4
Various Amyloids. Reproduced with permission from http://www.joplink.net/prev/200107/02.html.

Amyloid designation	Protein (polypeptide)	Associated diseases	Clinical classification
AA	SAA serum amyloid A	[Recurrent inflammation]	Secondary
		R.A., U.C., Crohn's disease	Secondary
		Hodgkins	Secondary
		Cancer	Secondary
		Mediterranean fever	Familial
		Many others	
A beta	Beta amyloid polypeptide	Alzheimer disease	Isolated
Beta amyloid		Down's syndrome	Isolated
AE	Procalcitonin	Medullary Ca thyroid	Isolated
E Endocrine	Amylin	Type 2 diabetes mellitus	Isolated?
AIAPP	Islet amyloid polypeptide	Pancreatic islet cell tumor	Isolated
AF	Serum prealbumin	Familial A. polyneuropathy	Familial
AL	kappa/lambda	Multiple myeloma 20%	Primary, primary idiopathic*
ASc		Cardiac amyloid	Isolated
		Systemic senile amyloidosis	Senile
ATTR	Transthyretin	Senile systemic cardiac	
HA (A beta2M)	Beta 2 microglobulin	Long-standing dialysis	Secondary
IAA (AANF)	Atrial naturetic peptide	Isolated atrial amyloid	Isolated
HCCAA	Cystatin-C	Familial icelandic amyloid	Isolated
Acys	Serum amyloid P	Angiopathy	
AP		Present in all amyloid	
Prion amyloid (AprPSc)	PrPSc	Scrapie/KURU	Isolated
		Creutzfeldt-Jakob disease	Isolated
		Gerstmann-Straussler syndrome	Isolated
		vCJD transmissible bovine SE	Isolated

PrPSc, scrapie form of prion protein; *R.A.*, rheumatoid arthritis; *U.C.*, ulcerative colitis; *vCJD*, variant Creutzfeldt-Jakob disease.

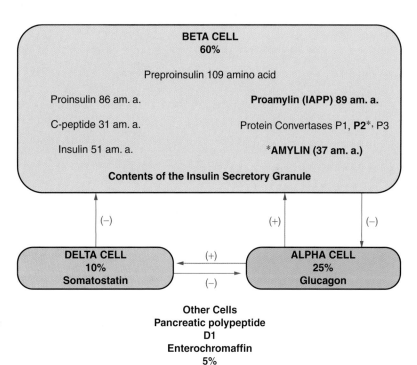

Figure 4-57. Cellular composition of the islet, the proteins contained in each cell type, and effects on other islet cells. Redrawn with permission from http://www.joplink.net/prev/200107/02.html.

undergoes polymerization and aggregation into β-sheets laid down crosswise. The parallel arrays of fibers are 7 to 10 nm in diameter. An electron micrograph of amyloid is shown in Figure 4-56. In type 2 diabetes, there can be a substantial loss of islets through apoptosis that could arise from the pathological effects of aggregation of amylin. There are various amyloids that have different substituents in different diseases, and these are shown with their associated diseases in Table 4-4. The interesting thing about amylin is that it *is contained in the pancreatic islet cell and is cosecreted with insulin in response to increased plasma glucose. It is synthesized with insulin when stimuli other than elevated glucose generate insulin synthesis.* Like insulin, it is secreted as a proprotein of 89 amino acids and is matured by preprotein convertase enzymes to the 37–amino acid amylin. The cellular composition of the islet cells (about 1 million to 1.5 million islets in a human) and their contents are shown in Figure 4-57. Amylin levels are above normal in type 2 diabetes.

Use of Glucose for Energy

Glucose entering a cell in need of energy is metabolized in the cytoplasm by glycolysis, producing two molecules of pyruvate per molecule of glucose. The enzymatic steps are shown in Figure 4-58, where glycolysis proceeds downward to pyruvate. The opposite direction leads to the synthesis of glucose from two molecules of pyruvate. In Figure 4-58, the *red arrows* signify three irreversible reactions; however, when glucose needs to be synthesized from pyruvate, four enzymatic reactions are employed

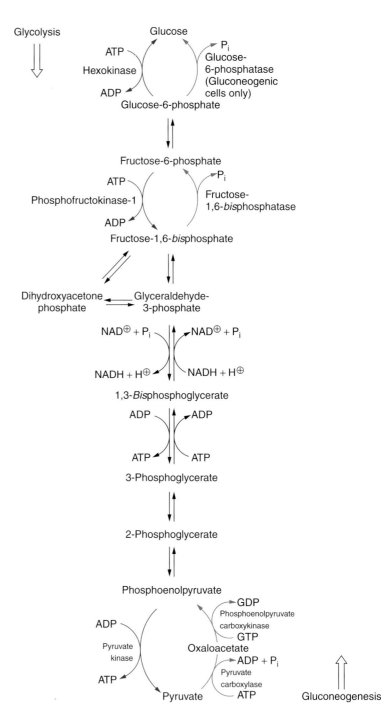

Figure 4-58. Reactions of glycolysis and gluconeogenesis. *Red arrows* signify enzymatic steps that are metabolically irreversible. *Blue arrows* in gluconeogenesis are reactions specifically needed to convert pyruvate to glucose.

that are unique to **gluconeogenesis** (formation of glucose) and these are signified by *blue arrows*. In gluconeogenesis, six ATPs and two NADH + H$^+$ are required, compared to glycolysis from glucose to pyruvate, where two ATPs and two NADH + H$^+$ are needed.

The metabolically irreversible steps of glycolysis consist of reactions catalyzed by hexokinase, phosphofructokinase, and pyruvate kinase. The four unique reactions in the reversal of glycolysis (gluconeogenesis) are catalyzed by pyruvate carboxylase, phosphoenolpyruvate carboxykinase, fructose-1,6-bisphosphatase, and glucose-6-phosphatase. The reactions of glycolysis and the coupling to the citric acid cycle are shown in Figure 4-59

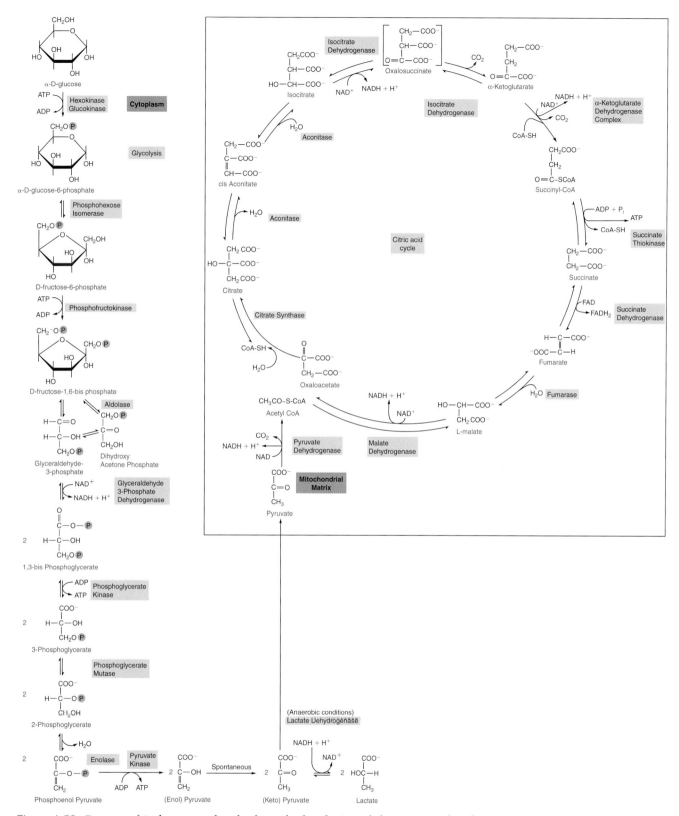

Figure 4-59. Passage of 1 glucose molecule through glycolysis and the citric acid cycle.

Use of Glucose for Energy

in more detail. The six-carbon glucose is converted to three-carbon compounds (so that yields from then on must be multiplied by two during glycolysis) in the form of glyceraldehydes-3-phosphate. Finally, pyruvate enters the mitochondrion from the cytoplasm and is converted to acetyl CoA (irreversibly) by the catalytic action of pyruvate dehydrogenase (Figure 4-60).

For each molecule of acetyl CoA metabolized through one turn of the citric acid cycle, three molecules of NADH and one molecule of $FADH_2$ are produced. Reoxidation of each NADH molecule through the respiratory chain forms three ATP molecules from ADP, and the reoxidation of $FADH_2$ produces two ATPs. One ATP equivalent is produced from GTP by the substrate level phosphorylation catalyzed by succinate thiokinase. Thus, 9 molecules of ATP from $NADH_2$ plus 2 ATP from $FADH_2$ and 1 ATP from the succinate thiokinase reaction gives 12 ATPs generated per turn of the cycle from acetyl CoA, equivalent to one molecule of pyruvate. Because two molecules of pyruvate derive from one molecule of glucose, then one molecule of glucose requires two turns of the citric acid cycle so that 24 ATPs are generated. In glycolysis, two NADH are produced from the glyceraldehyde-3-phosphate dehydrogenase reaction, which when reoxidized through the respiratory chain produces 6 ATPs.

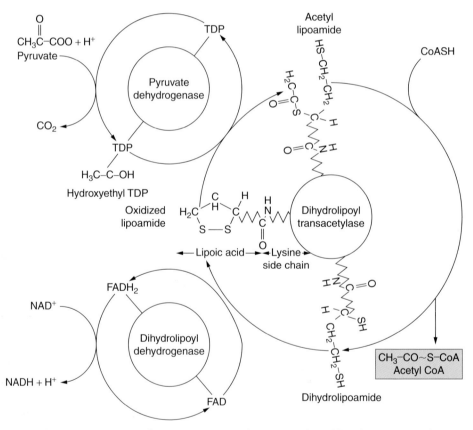

Figure 4-60. Conversion of pyruvate to acetyl-CoA catalyzed by the pyruvate dehydrogenase complex. Lipoic acid is joined by an amide link to a lysine residue of dihydrolipoyl transacetylase forming a long flexible arm that allows the lipoic acid prosthetic group to rotate between the active sites of each enzyme in the complex. *TDP*, thiamine diphosphate.

In the phosphoglycerate kinase reaction, 2 ATPs are produced at the substrate level and in the pyruvate kinase reaction; 2 more ATPs are produced at the substrate level, accounting for 10 ATPs. Subtracting 2 ATPs used in the hexokinase and phosphofructokinase reactions results in a net of 8 ATPs from glycolysis. In the citric acid cycle, 6 ATPs derive from the respiratory-chain oxidation of two NADH molecules, respectively, in the pyruvate dehydrogenase, isocitrate dehydrogenase, α-ketoglutarate dehydrogenase, and malate dehydrogenase reactions, giving a total of 24 ATPs. Then, there is a phosphorylation at the substrate level in the succinate thiokinase reaction, giving 2 ATPs, and a respiratory chain oxidation of 2 $FADH_2$ in the succinate dehydrogenase reaction, giving 4 ATPs. This produces a net of 30 ATPs in the citric acid cycle, giving a total of 38 ATPs under aerobic conditions. About 1 ATP is used in the transport of H^+ into the mitochondrion with pyruvate and malate.

The respiratory chain accounts for the oxidation of reducing equivalents and couples the phosphorylation of ADP to ATP, of which the terminal phosphate is of high energy. In one turn of the citric acid cycle, 11 high-energy phosphates are formed as ATP by reoxidation of reducing equivalents, and an additional high-energy phosphate results from the conversion of succinyl CoA to succinate in the succinate thiokinase reaction at the substrate level, giving 12 ATPs per turn of the cycle. Elements of the respiratory chain are shown in Figure 4-61. In the reoxidation of reducing equivalents from NADH and $FADH_2$, the chain consists of flavoprotein, coenzyme Q, and the cytochromes. The respiratory chain is coupled to proton translocation from the inside of the mitochondrial membrane to the outside. This movement of protons is driven by three respiratory chain complexes (complexes I, III, and IV), and each acts as a proton pump. A sketch of the complexes of the proton-pumping respiratory chain is shown in Figure 4-62, and ideas of the structures of the complexes are shown in Figure 4-63. The respiratory chain is located in the inner mitochondrial membrane. It oxidizes NADH + H^+ and $FADH_2$ using oxygen and water, and ATP is generated from ADP by ATP synthase, which is the terminal protein in the chain. Water is also generated in the process by the oxidation of hydrogen; this is called **metabolic water.** In the respiratory chain, reducing equivalents from NADH + H^+ are passed on to FAD (Figure 4-61), and the resulting $FADH_2$ is oxidized by cytochromes b (several types of b), c, and a. The reduced cytochrome a transfers electrons to oxygen, which reacts with protons to form water. Cytochromes are proteins with heme cofactors, and in the process the iron in heme changes from Fe(III) to Fe(II) (and eventually back again). The cytochromes transport protons only, whereas FAD binds both electrons and protons. These cofactors (cytochromes, FAD, etc.) are bound to proteins, which are organized into components within the inner mitochondrial membrane or closely associated with it. Structures of the various cofactors of the respiratory chain are shown in Figure 4-64. The process terminates with ATP synthase, which catalyzes the formation of ATP from ADP and inorganic phosphate. The enzyme acts as the terminal coupling factor. By this overall process, three ATPs are formed from each NADH + H^+ and two from each $FADH_2$. This concept was originally discovered in 1961 by Peter Mitchell, who hypothesized that protons are transferred through the mitochondrial membrane in a directed fashion, resulting in a pH gradient. This is called the **chemiosmotic theory** of oxidative phosphorylation. As a result of glycolysis and the citric acid cycle from one molecule of glucose, 38 ATPs are generated

(Text continues on p. 182.)

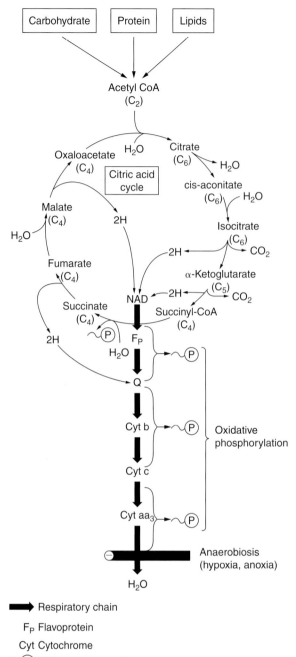

Figure 4-61. Acetyl-CoA is the product of catabolism of carbohydrate, protein, and lipid, and it is taken into the citric acid cycle with H_2O and oxidized to CO_2 with the release of reducing equivalents, 2H. These are oxidized in the respiratory chain leading to the coupled phosphorylation of ADP to ATP. Eleven ~(P) of ATP are generated for one turn of the cycle and one ~(P) derives from the conversion of succinyl-CoA to succinate, giving 12 ATP per turn of the cycle.

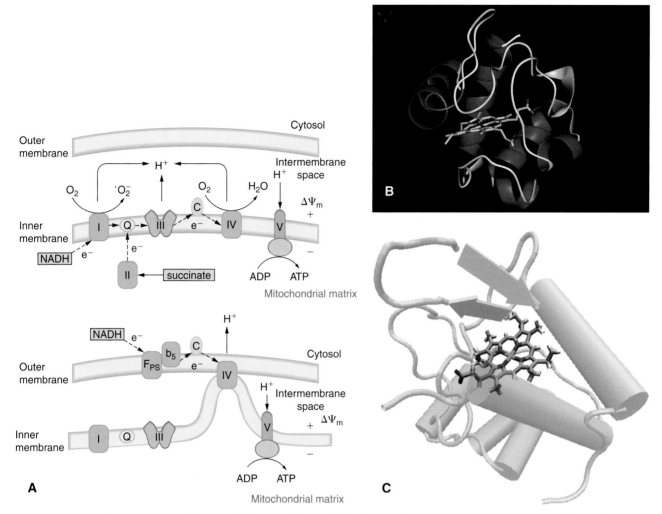

Figure 4-62. **A.** The respiratory chain and the complexes of the chain whose action transports protons from the inner mitochondrial membrane outward during the oxidation of NADH. **B.** Structure of cytochrome c. The heme prosthetic group is in the center of the molecule in stick structure. The iron atom is at the center of the heme. **C.** Structure of cytochrome c. The heme prosthetic group is modeled in sticks at the center and the iron atom is in *green*. Part A redrawn with permission from http://www.mcmaster.ca/inabis98/laher/kristian0829/f3large.jpg. Part B reproduced from the PDB, ID:1YCC. G.V. Louie, G.D. Brayer. High-Resolution Refinement of Yeast ISO-1-cytochrome c and Comparisons with Other Eukaryotic Cytochromes c. *Jour. Mol. Biol.* **214** pp. 527 (1990).

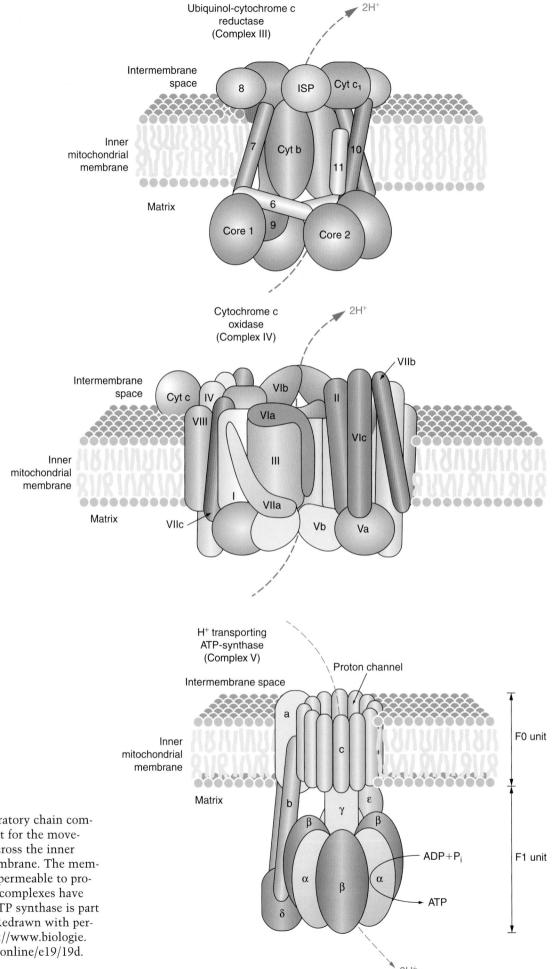

Figure 4-63. Respiratory chain complexes that account for the movement of protons across the inner mitochondrial membrane. The membrane itself is not permeable to protons so that these complexes have been developed. ATP synthase is part of the β-subunit. Redrawn with permission from http://www.biologie.uni-hamburg.de/b-online/e19/19d.htm.

Coenzyme A

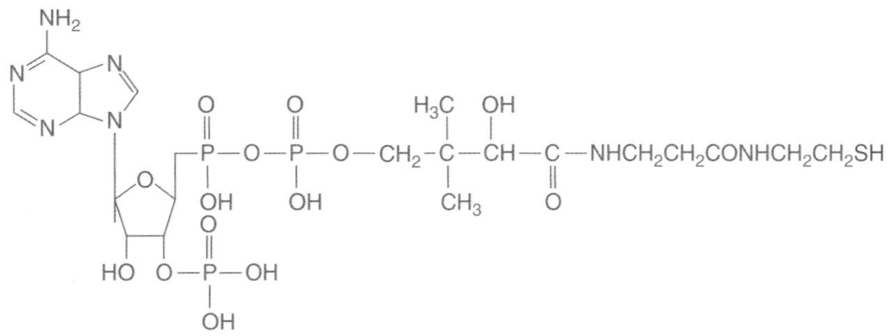

NAD+

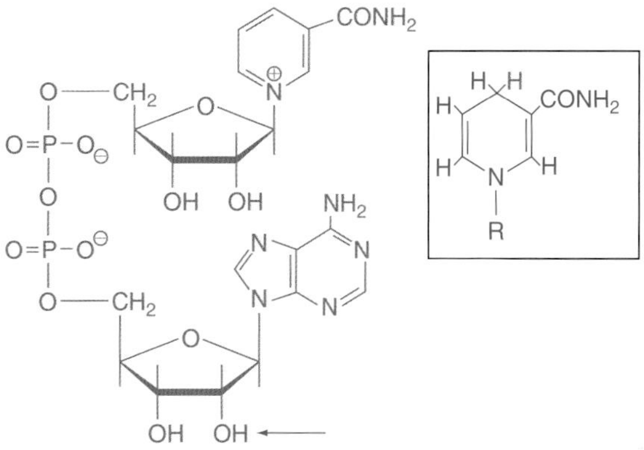

FAD

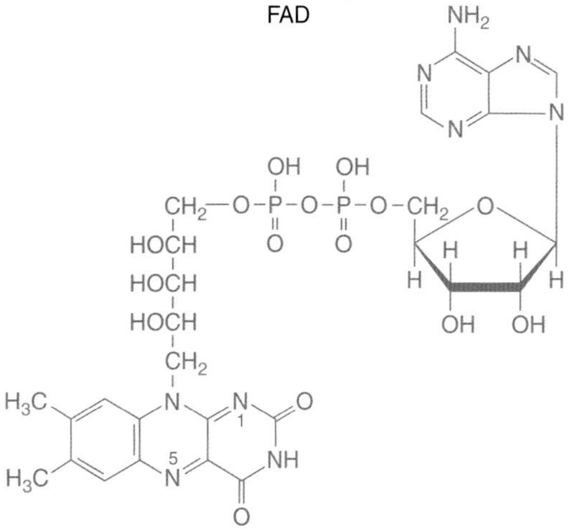

Coenzyme Q_{10}

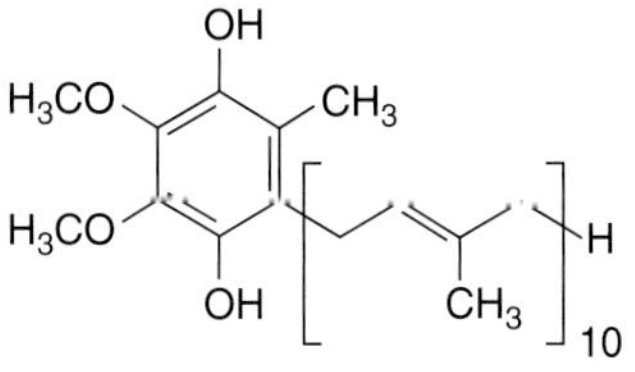

Ubiquinol ($CoQH_2$)

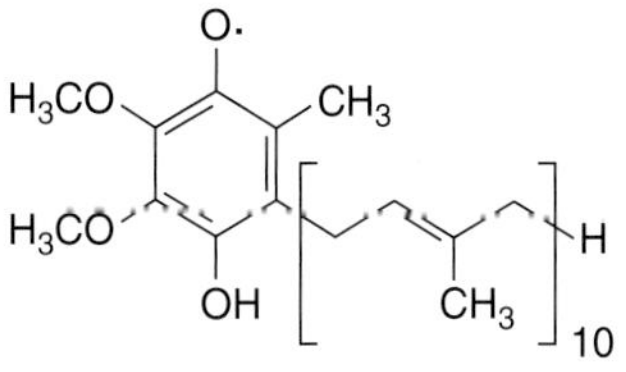

Semiquinone radical (CoQH·)

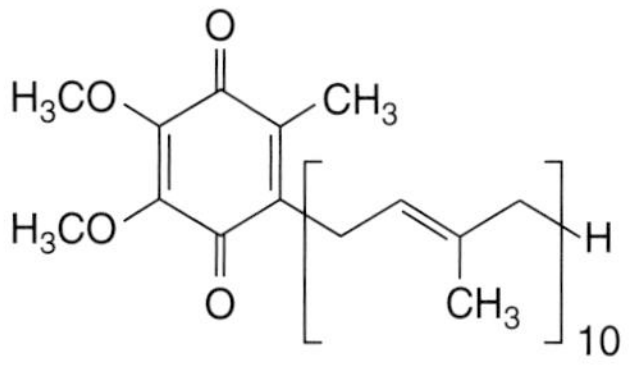

Ubiquinone (CoQ)

Coenzyme Q can exist in three oxidation states: the fully reduced ubiquinol form ($CoQH_2$), the radical semiquinone intermediate (CoQH·), and the fully oxidized ubiquinone form (CoQ).

Figure 4-64. Structures of some of the cofactors in the respiratory chain.

(including 4 ATPs from glycolysis). The high-energy bond of the terminal phosphate of ATP contains 7.3 kcal/mol; the 38 ATP are equivalent to 277 kcal/mol, which calculates to slightly more than 40% efficiency. By comparison, most human-made machines function at 20% efficiency.

Attention has not been given in this chapter to the metabolism of the pentose, ribose. Because ribose and its derivative, deoxyribose, are so critical to the structure of nucleic acids, the metabolism of ribose and its phosphate derivatives (including the pentose phosphate shunt) is included in Chapter 6 on nucleic acids. In addition, the conversion of ribose to deoxyribose, the essential sugar in deoxynucleic acids is described.

Glycerol Can Be Converted to Glucose

During the breakdown of fat (triacylglycerides), glycerol is produced. Glycerol can contribute directly to the formation of carbohydrates, as shown in Figure 4-65.

Glycoproteins

As mentioned on the subject of protein synthesis, carbohydrate moieties are added to the structure of some proteins, mainly to make them more water soluble and for specific recognition. Carbohydrate portions of glycoproteins are often found on the surface of a cell (Figure 4-66). Most sugar

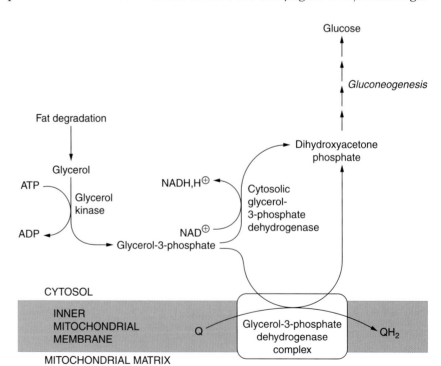

Figure 4-65. Formation of glucose from glycerol. Redrawn with permission from http://cwx.prenhall.com/horton/medialib/mediaportfolio/textimages/FG1313.jpg

182 CHAPTER 4 Carbohydrates

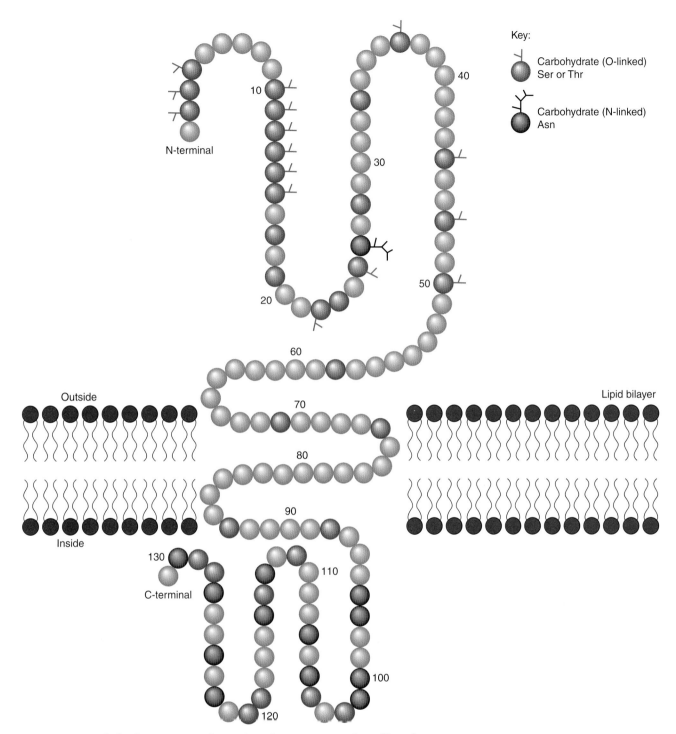

Figure 4-66. Carbohydrate groups of proteins often appear on the cell surface.

substitutions in proteins occur either through asparagine (amide) and are called N-linked sugars. Others are substituted through the hydroxyl group of serine or threonine and are O-linked sugars. The primary substituents at these sites often are *N*-acetylglucosamine and *N*-acetylgalactosamine (Figure 4-67). In addition, some proteins are anchored to the cell membrane through a glycerophosphatidylinositol bridge. Acetylglucosamine is linked through the amide nitrogen of asparagine of the evolving

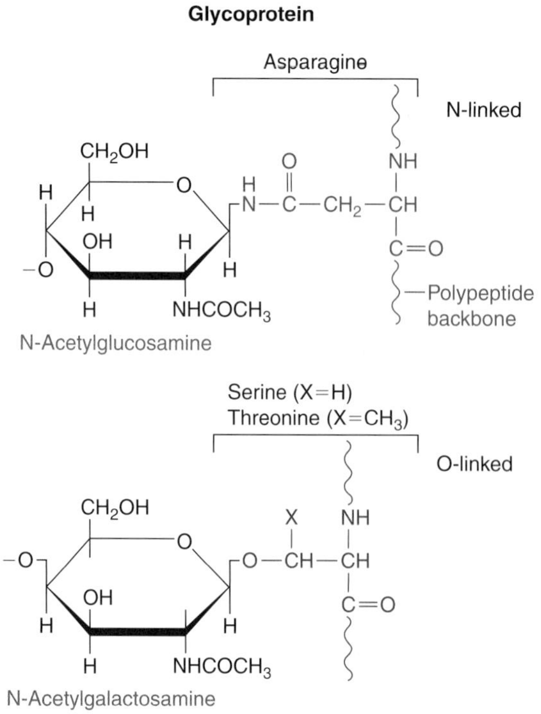

Glycoprotein

Figure 4-67. N-linked and O-linked sugar substituents of proteins.

glycoprotein. The target **sequence for *N*-glycosylation** is Asn-*X*-Ser/Thr, where *X* is any amino acid except Pro or Asp. N-linked carbohydrate substituents of proteins are of three types: a complex type, a hybrid type, or a high mannose type (Figure 4-68). N-linked oligosaccharides have at least five sugar residues. There are many variants of these types. O-linked oligosaccharides are shorter, with one to four sugar residues; however, there are exceptions. An example of a **glycophosphatidylinositol** membrane to protein anchor is shown in Figure 4-69.

In Figure 4-70, the growing protein chain enters the lumen of the endoplasmic reticulum, where oligosaccharide is added to the protein. This is followed by more carbohydrate additions and trimming until the final mannose has been attached. The final carbohydrate modification is carried out in the Golgi apparatus. At this point, the folding of the protein has been completed in the endoplasmic reticulum lumen with the aid of molecular chaperones. If a protein has not been folded correctly, it will be degraded by the **proteasome** in the cytoplasm. It is believed that when improperly folded proteins escape this protective degradative mechanism, they can contribute to neurodegenerative diseases such as Alzheimer's. The carbohydrate portion of glycoproteins in the endoplasmic reticulum can lead to binding to molecular chaperones. Of these, the **calnexin–calreticulin cycle** operates to ensure the correct disulfide bond formation in proteins. Calnexin and calreticulin are lectins (containing carbohydrate moieties themselves), and they are specific for monoglucosylated proteins. Then another protein, ERp57, that contains a disulfide bond, has disulfide isomerase activity, interacts with the protein, and ensures the correct disulfide bond formation in the newly synthesized glycoprotein.

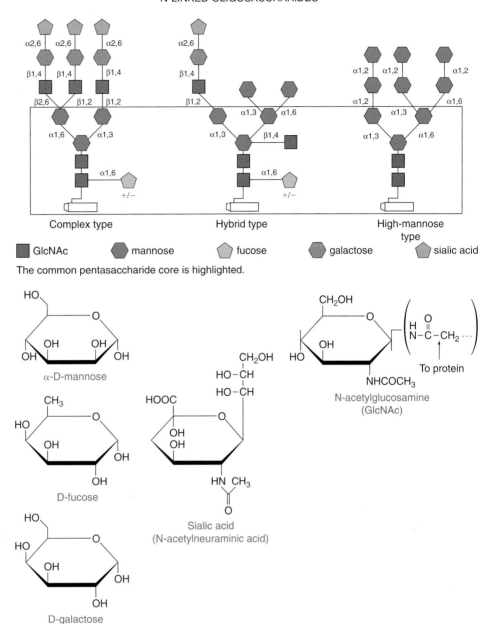

Figure 4-68. Types of carbohydrate substituents in glycoproteins. The linkages are indicated and the protein is represented by the line drawing at the bottom of each type of N-linked of oligosaccharide. *Upper figure* redrawn from http://www.cryst.bbk.ac.uk/pps97/assignments/projects/emilia/Structure.HTM.

Blood Group Proteins

In the human there are three types of blood group antigens. Everyone can synthesize the O antigen (Figure 4-71). A-type blood marks a person with an enzyme that can add N-acetylgalactosamine to the O antigen. The B-type individual has an enzyme that adds galactose to the O antigen. These would be posttranslational modifications, and the starting glycoprotein where the carbohydrate moiety would be added in the endoplasmic reticulum would be considered a cotranslational event. The AB type has both enzymes, and the O type lacks both enzymes. These groupings of carbohydrate substituents are found on glycoproteins and glycolipids.

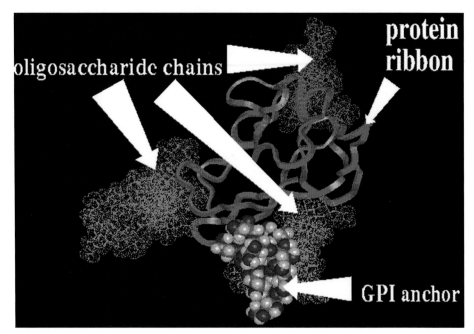

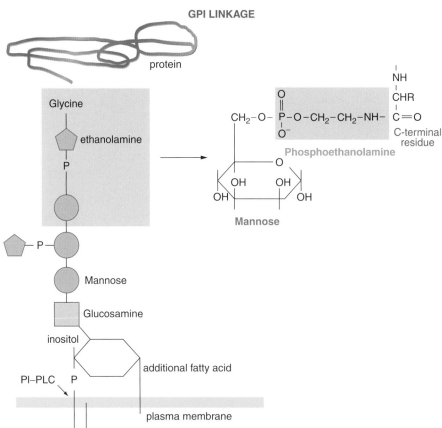

Figure 4-69. Glycophosphatidylinositol (GPI) membrane anchor connecting a protein to the cell membrane. Redrawn from http://www.cryst.bbk.ac.uk/pps97/assignments/projects/emilia/Structure.htm.

The ABO carbohydrates are linked to **sphingolipid** when they are on the cell surface. In the serum, these carbohydrates are complexed to protein as glycoproteins, and they are the so-called secreted forms. Some people produce the glycoprotein forms and some do not (nonsecreters). This distinction can be ascertained and is used in the courts forensically, especially in rape cases.

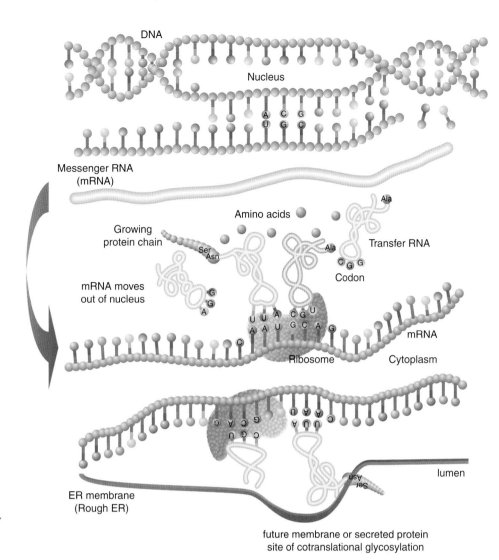

Figure 4-70. Modification of newly synthesized proteins by addition of carbohydrate moieties to form glycoproteins.

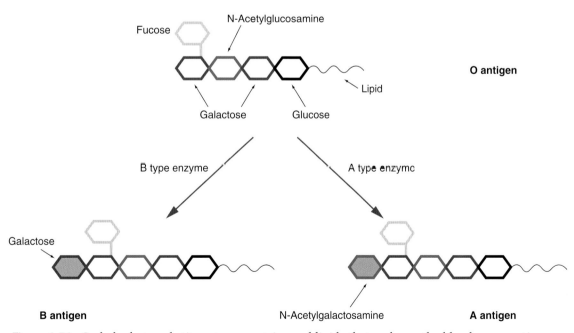

Figure 4-71. Carbohydrate substituents on proteins and lipids that make up the blood group antigens. All individuals synthesize the O antigen. The A type has an enzyme that adds an N-acetyl-galactosame to the O antigen, whereas the B type has an enzyme that adds a galactose to the O antigen. The AB type has both enzymes, and the O type lacks both enzymes.

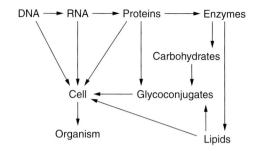

Figure 4-72. The emphasis of glycobiology relating the modifications of proteins and lipids and the roles these macromolecules play in the functions of the cell.

Lactose Intolerance

More than half the people in the world are **lactose intolerant,** which means that they have an allergic reaction to dairy products that contain lactose. Lactose intolerants do not produce the enzyme lactase (β-galactosidase) essential for the digestion of lactose. Lactose is the disaccharide, galactosylglucose joined in a β 1-to-4 linkage, as shown in Figure 4-25. Although a small amount of the enzyme may be produced in the digestive tract of these individuals, it is insufficient for the breakdown of lactose. Consequently, the undigested lactose is metabolized by large intestinal bacteria, which results in diarrhea, dehydration, and bloat. Either dairy products containing lactose are removed from the diet or lactase (Lactaid) is taken orally before ingestion of these foods.

Glycobiology

As carbohydrate biochemistry became more complex, proteins and lipids were found to be modified by the addition of oligosaccharides to their structures. Glycobiology concentrates on the newer aspects of carbohydrate chemistry, namely, the modification of proteins to glycoproteins or proteoglycans and the modification of lipids. Therefore, glycobiology embodies the understanding of the cellular an molecular biology of glycans. The emphasis of this area can be set in context by the overview in Figure 4-72.

Further Reading

Lehmann, J., and Haines, A.H., *Carbohydrates: Structure and Biology,* John Wiley & Sons, 2002.

CHAPTER 5

Lipids

Hypercholesterolemia: A Disease in Which Serum Cholesterol Is Not Properly Imported at the Cellular Level

Cholesterol is an essential component in the structure of the cell membrane, and it comprises about a quarter of the lipids in the membrane. Cholesterol circulating in the blood becomes a problem only when the transporters of cholesterol, the **low-density lipoproteins (LDLs)**, are unable to move their cholesterol content into peripheral tissues (**high-density lipoproteins**, or **HDLs**, transport cholesterol from peripheral tissues to the liver). Movement of LDLs into tissues is accomplished by LDL binding to its receptor on the cell membrane. The structure of the **LDL receptor** is shown in Figure 5-1. The binding site for the LDL is at the N-terminus of the receptor extending outward from the cell membrane. The receptor contains both N-linked and O-linked oligosaccharide domains. The domain extending into the cell cytosol (soluble cytoplasm) is required for aggregation of the receptors during endocytosis. When LDL receptors cannot function because of limited expression or a mutation, LDLs are not properly taken up into tissue cells. This is the cause of familial **hypercholesterolemia.** When LDLs are not transported, the amount of LDLs in the blood increases to a level that can exceed 300 mg/ml (a normal level would be below 200 mg/ml). After a buildup of LDLs in the blood, macrophages ingest LDLs and become sponge cells that are part of the buildup of atherosclerotic plaque.

The normal view of LDL metabolism is shown in Figure 5-2. As shown, LDL receptors in the membrane become part of the surface of coated pits that bind LDL containing cholesteryl esters (CEs). The cholesteryl esters are transported to the interior of the cell inside a coated vesicle that includes the LDL receptors to become an endosome. The endosome can be targeted to a lysosome where free cholesterol (C) can be released to enter the unesterified pool, located mainly in cellular membranes. From this pool, cholesterol can be used as a starting substrate for steroid biosynthesis. Some specialized cells, where steroid hormones are synthesized, contain "droplets" of CE that can be marshaled through the hormonal activation of cholesteryl esterase to

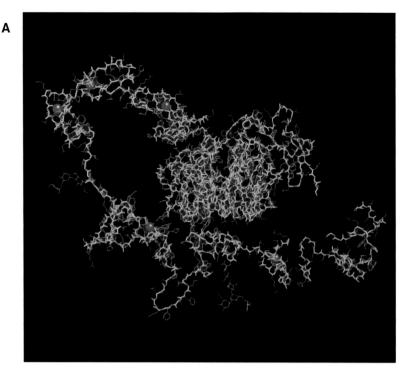

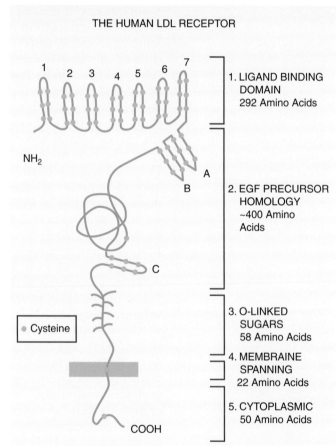

Figure 5-1. Structure of the LDL receptor. **A.** Model of its structure. **B.** Model of the LDL receptor with the N-terminus at the top and the C-terminus (inside the cell) at the bottom. Part B is redrawn with permission from http://www.hhmi.org/biointeractive/museum/exhibit98/images/ldl.gif.

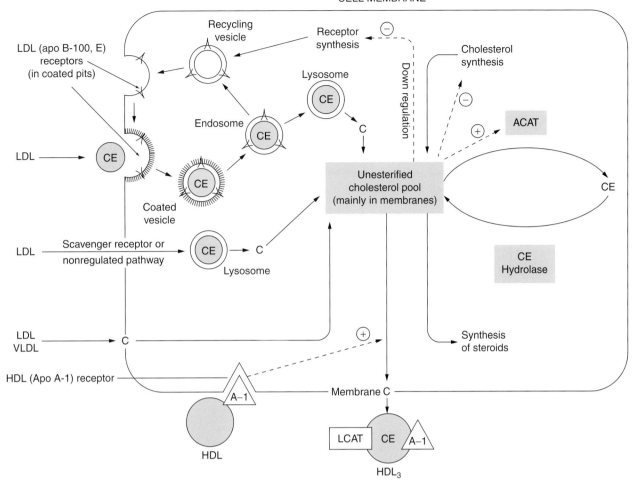

Figure 5-2. Uptake of LDLs from the bloodstream to make cholesterol available to the interior of a tissue cell, such as the liver cell. *ACAT*, Acyl coenzyme A: cholesterol acyltransferase; *apoA-1 receptor*, HDL receptor; *C*, Cholesterol; *CE*, cholesteryl ester; *HDL*, high density lipoprotein; *LCAT*, lecithin-cholesterol acyltransferase; *LDL*, low density lipoprotein; *VLDL*, very low density lipoprotein.

avail free cholesterol to the mitochondria, where steroid hormone biosynthesis begins. The upshot of this process of LDL usage is to avail a store of usable cholesterol to a tissue cell, which reduces the LDL cholesterol circulating in the bloodstream.

Familial hypercholesterolemia is a genetic disease characterized by the absence or a defect of LDL receptors in the liver or in extrahepatic tissues that causes these receptors to function poorly or not at all. Other genetic factors in hypercholesterolemia are **lecithin cholesterol acyltransferase (LCAT)** deficiency, alternative apolipoprotein, and point mutations in the LDL receptor. This type of mutation in the LDL receptor leads to an increase in blood LDLs, as shown in Figure 5-3. In Figure 5-4, a pedigree shows inheritance of familial hypercholesterolemia over three generations. The LDL receptor gene is encoded on chromosome 19, and the inheritance is of the autosomal dominant mode. In this pedigree, there is

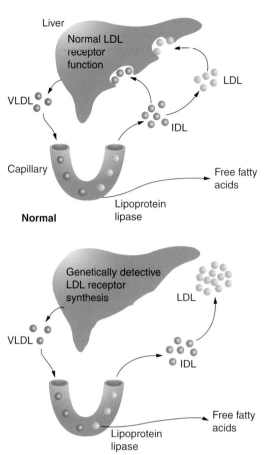

Figure 5-3. Consequence of mutated LDL receptor in familial hypercholesterolemia. Because LDLs are not able to bind to nonfunctional LDL receptors in the liver cell membrane, the LDLs remain in the bloodstream and become concentrated to 300 mg/ml or higher. *IDL,* intermediate density lipoprotein.

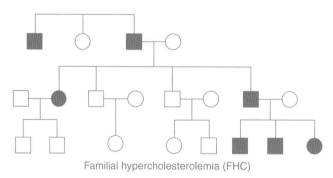

Familial hypercholesterolemia (FHC)

Figure 5-4. Pedigree of three generations showing the incidence of familial hypercholesterolemia. Circles represent females and squares represent males. When the figures are solid, that individual has the disease. In F1, an affected male is married to an unaffected female. Of their four offspring in F2, there are three males and one female. The female and one male have inherited the disease. When the female daughter in F1 marries an unaffected male, their 2 sons in F3 do not have the disease; when the affected son marries an unaffected female, all of their children (2 males and one female) in F3 have the disease.

direct transmittance from the parent carrying the disease to an affected offspring; the transmission is from affected male to affected male. In the progeny from a couple in which one parent has the disease, half of the children will be affected with the disease. These individuals usually die of heart failure as early as their 40s or in the following three decades, depending on the severity of the disease. Today, there are many drugs that successfully lower serum cholesterol and are useful in treating hypercholesterolemia. With a normal person who eats a healthy diet and exercises, cholesterol consumed in the diet added to the cholesterol synthesized in the body should be in equilibrium with the processes that use and remove cholesterol from the body; thus, the concentration in the blood should remain within normal boundaries. When this is not the case and consumption of a high-cholesterol diet increases the circulating level of cholesterol, there is a problem with cholesterol usage.

Biosynthesis of Cholesterol

Cholesterol is synthesized in all cells of the body, but the most active tissues are the liver, the adrenal cortex, and the intestine, as well as the reproductive tissues. The liver accounts for 75% or more of the body's cholesterol synthesis, the remainder coming from the diet, mainly animal tissues. In the adrenal, the ovaries, and the testis, cholesterol is the starting point for the synthesis of adrenal and sex steroidal hormones. The entire cholesterol structure is derived from the acetate of acetyl CoA. The series of complex reactions are outlined in simplified form in Figure 5-5. Three molecules of acetyl CoA are converted in the endoplasmic reticulum to mevalonate (C6 compound). Mevalonate is converted to isopentenyl diphosphate (a C5 fragment, the active isoprene). Six of these isoprenes polymerize to form **squalene** (C30). Squalene, with the removal of three carbons, is then cyclized to form cholesterol.

A key regulatory enzymatic reaction is the conversion of 3-hydroxy-3-methylglutaryl-CoA (HMG-CoA) to mevalonate catalyzed by **HMG-CoA reductase.** The structure of HMG-CoA reductase is shown in Figure 5-6. This enzyme is an important target for the **statins,** drugs used to lower the circulating level of cholesterol substantially (up to 50%). Less than 200 mg/dl (dl = **deciliter** = 100 ml) is defined as a normal level of cholesterol in blood, and elevated cholesterol is considered to be 240 mg/dl or above. Some statins used for this purpose and in the treatment of hypercholesterolemia are shown in Figure 5-7. Cholesterol also forms **bile acids** excreted through the intestine. Because the bile acids are an end product of cholesterol, increasing their excretion has been thought to be a means to reducing cholesterol in the blood, although there are now better drugs for reducing circulating cholesterol levels.

Bile acids function in the intestine in the absorption of fat by facilitating formation of micelles, and their measurement in serum can reflect on the adequate functioning of the liver.

Synthesis of Bile Acids

The synthesis of bile acids from cholesterol is shown in Figure 5-8. The formation of bile acids occurs in the liver, and they are secreted into the gallbladder. Appropriately, usually after a meal, the bile acids enter

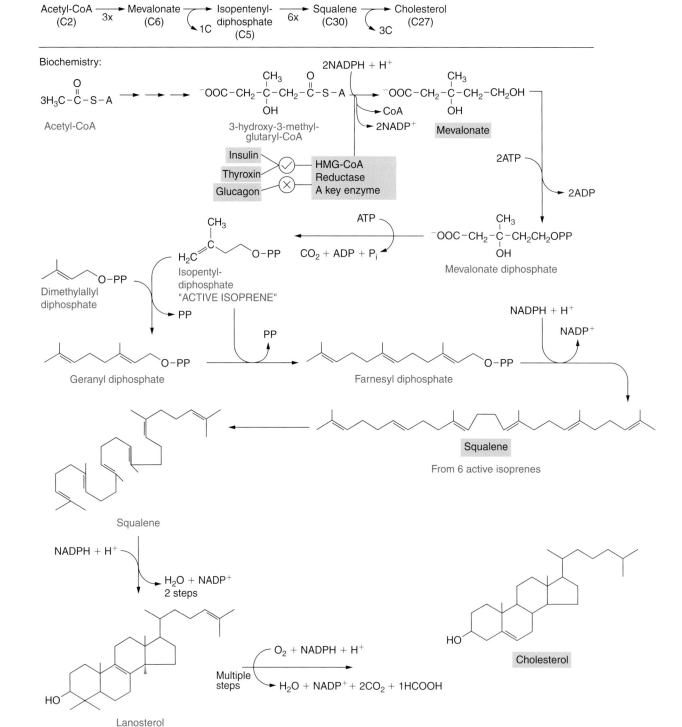

Figure 5-5. Synthesis of cholesterol from acetate (acetyl-CoA).

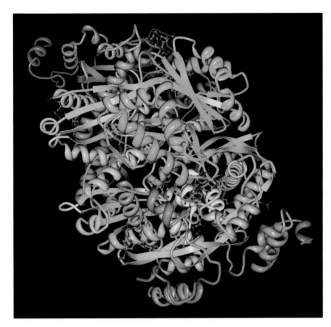

Figure 5-6. Complex of the catalytic portion of human HMG-CoA reductase with HMG (hydroxyl-methyl-glutarate), CoA, and $NADP^+$. From PDB ID: 1dqa. E.S. Istvan, M. Palnitkar, S.K. Buchanan, J. Deisenhofer. Crystal Structure of the Catalytic Portion of Human HMG-CoA Reductase: Insights into Regulation of Activity and Catalysis. *EMBO J.* **19** pp. 819 (2000).

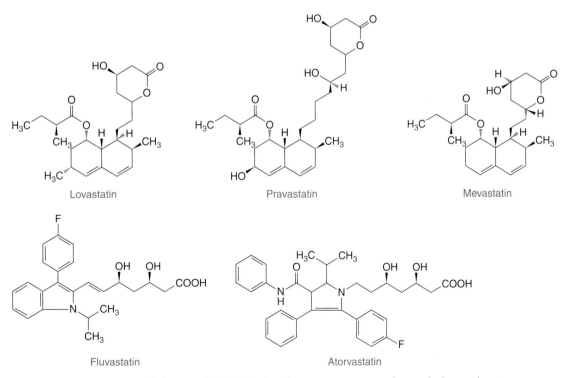

Figure 5-7. Some statin inhibitors of HMG-CoA reductase use to treat hypercholesterolemia.

Hypercholesterolemia

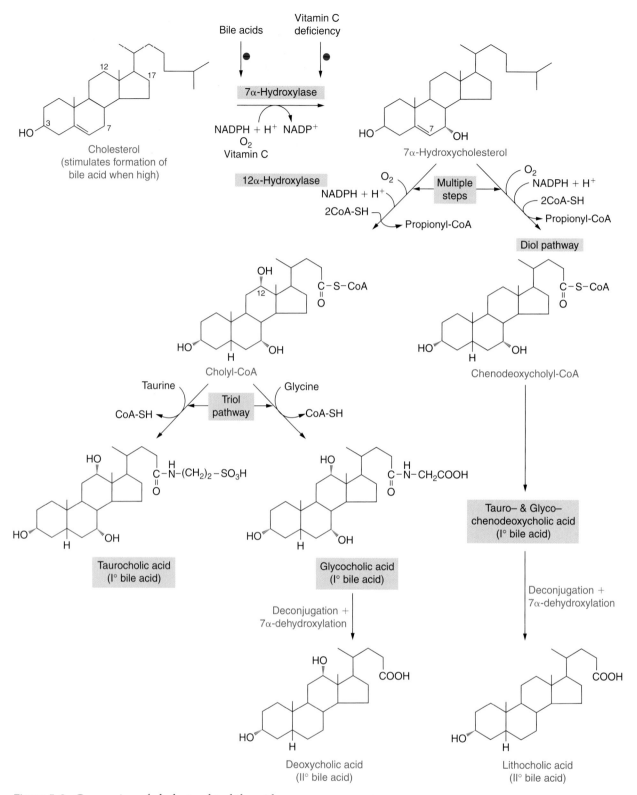

Figure 5-8. Conversion of cholesterol to bile acids.

196   CHAPTER 5   Lipids

the intestine through the bile duct. The rationale behind consumption of fiber in the diet is that fiber increases the elimination of bile acids; bile acids bind to the fiber and are excreted through the intestine. This is also the rationale for the clinical use of ingested resins (for example, cholestyramine), which bind and increase the elimination of bile acids, thus decreasing the recapture of bile acids through the enteric circulation. Figure 5-9 shows that up to 90% of the bile acids are recirculated to the liver and systemic circulation. Thus, the increased fiber and clinical use of resins can remove bile acids through the feces, causing a decreasing return of bile acids to the body. The net result is to decrease the blood content of cholesterol. But the use of cholesterol-lowering drugs (Figure 5-8) through the inhibition of HMG-CoA reductase is far more effective than these maneuvers. When bile acid formation is high,

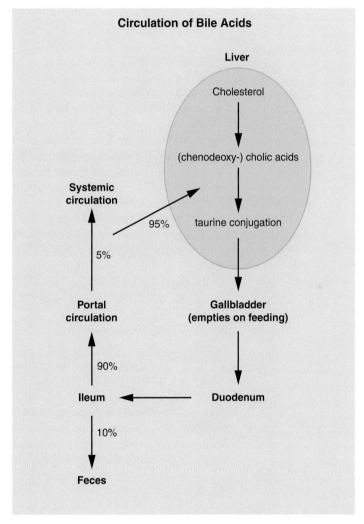

Figure 5-9. Enterohepatic circulation of bile acids. Bile acids are synthesized in the liver and stored in the gallbladder. Upon eating, the gallbladder empties bile acids into the small intestine, and then they move into the large intestine. About 10% are removed from the large intestine into the feces, and the remainder are recirculated back into the liver; some remain in the bloodstream. Reproduced with permission from http://web.vet.cornell.edu/public/popmed/clinpath/CPmodules/chem/bileacid.htm.

there is a negative feedback on further bile acid formation; thus, the level of cholesterol would increase. Conversely, when bile acid production is low, more cholesterol would be converted to bile acids. However, all systems, when operating normally, tend to maintain cholesterol in the blood at a constant concentration. Therefore, it is questionable whether an increased rate of bile acid removal actually affects the endogenous rate of synthesis of cholesterol. Most likely, the major regulator of the endogenous process is HMG-CoA reductase. Recently, it has been found that two nuclear receptors are involved in the regulation of bile acid synthesis. There are two lipid-activated nuclear receptors: the **liver X receptor (LXR)** and the **farnesoid X receptor (FXR).** The LXR binds oxysterols to cause its activation, and the FXR binds bile acids. Both receptors form dimers with the retinoid X receptor, and when activated these receptors bind to gene promoters to produce mRNAs, which, in turn, are translated into proteins that prevent bile salt toxicity (FXR) and overproduction of cholesterol (LXR).

Prognosis

In familial hypercholesterolemia, there is the genetic factor that causes the malfunctioning of the LDL receptor so that LDLs containing cholesterol either cannot be taken up by the tissues or are poorly taken up, principally in the liver. Thus, circulating cholesterol cannot be maintained in equilibrium and the level rises. With this defect or without it, eating diets rich in saturated fat or high in cholesterol contributes to elevated blood levels. There are also many diseases that cause cholesterol to be elevated in the blood, such as obesity, diabetes, hypothyroidism, or kidney or liver disease. High circulating levels of cholesterol are thought to increase the risk of heart disease because cholesterol can deposit in the vessels and atherosclerosis can develop. This can lead to the formation of blood clots, clogging of the coronary arteries, and infarction. In the male population, an initial heart attack among hypercholesterolemics can occur in the 40s or 50s, and by age 60, about 85% of males with this disease have had a heart attack. In women with this disease, the first heart attack comes about 10 years later than it does in men. The incidence of familial hypercholesterolemia is slightly less than 1 in 100 people. With individuals who have a cholesterol level above 300 mg/dl, the risk of a fatal heart attack is about five times that of a person with a cholesterol level 200 mg/dl or below. Today, blood screening occurs frequently enough so that elevations in circulating cholesterol become known early. Drugs can be then used to lower the blood level, diets can be changed, exercise can be increased, and coronary bypass procedures can be used; all are effective in extending the life of a familial hypercholesterolemic.

Fatty Acids and Fat

Fatty acids are hydrocarbon chains with a carboxyl group at one end. When the chain contains no double bonds, it is a saturated fatty acid. When the chain contains one or more double bonds, it is an unsaturated fatty acid. One of each type is shown in Figure 5-10. When double bonds are introduced into the hydrocarbon chain, there is a bend in the structure, as shown in Figures 5-10 and 5-11. This is because the distance between two carbons separated by a double bond is shorter than the distance between two carbons separated by a single bond. Common saturated fatty acids and common unsaturated fatty acids are shown in Figure 5-11. The carboxyl end of the molecule represents a polar, hydrophilic group, and the rest of the molecule (the hydrocarbon chain) is lipophilic (hydrophobic). When one or more double bonds are introduced into the molecule, the hydrogen atoms on either side of the double bond can assume a *cis* or *trans* configuration, as shown in Figure 5-12. When fatty acids are esterified with glycerol, they are known as fats or oils, depending on their physical properties. The physical properties are determined by the degree of unsaturation in the fatty acid constituents and the length of the fatty acid chains. Thus, substitution with saturated fatty acids will produce solid fats, and substitution with unsaturated fatty acids will produce oils (liquid fats, obtained mostly from plants). In the diet, *trans* fatty acids (in fats) can be harmful and may be carcinogenic. They occur particularly when an oil (containing *cis* fatty acids) is continuously reused in cooking, such as in the cooking of french fries, because the configuration of the double bonds can shift from *cis* to *trans* in this process. Some fatty acids are considered essential in that they need to be obtained through the diet; these are unsaturated fatty acids like oleic and linoleic acids. **Arachidonic acid** (20:4 [20 refers to carbon number and 4 refers to number

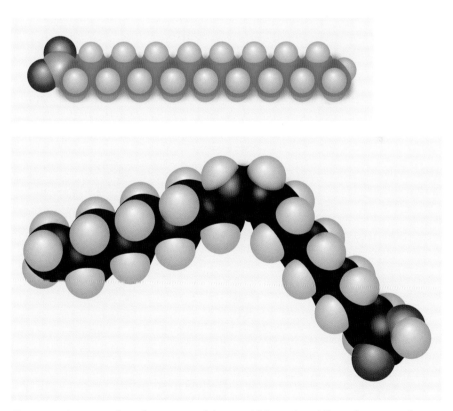

Figure 5-10. Examples of a saturated fatty acid (stearic acid) on the *top* and an unsaturated fatty acid (oleic acid) on the *bottom*.

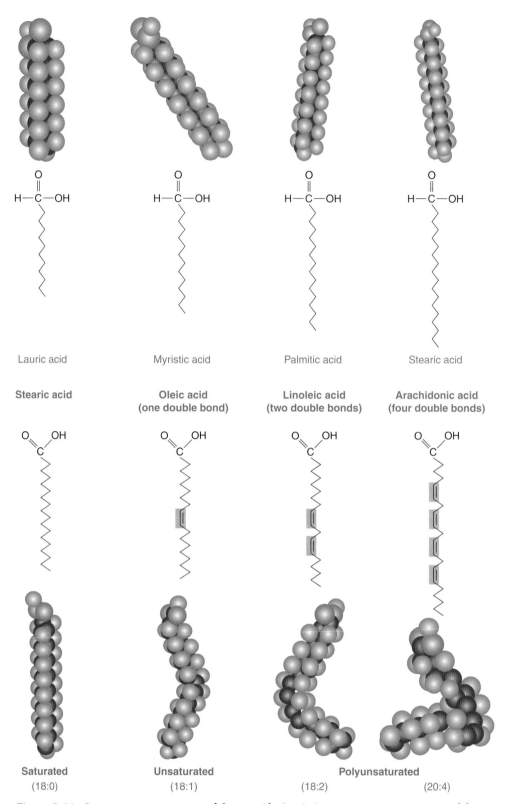

Figure 5-11. Some common saturated fatty acids *(top)*. Some common unsaturated fatty acids in addition to stearic acid (saturated) *(bottom)*.

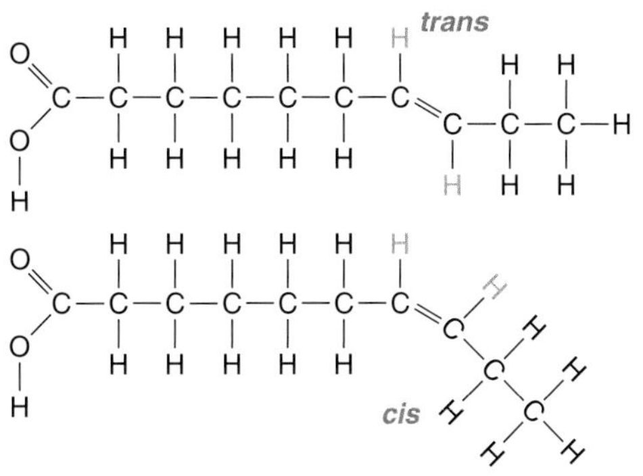

Figure 5-12. *Trans* positions of hydrogen atoms on either side of the double bond in decanoic acid are shown in the top structure, with the hydrogen atoms in yellow. At the bottom, the hydrogen atoms are shown in *cis* configuration.

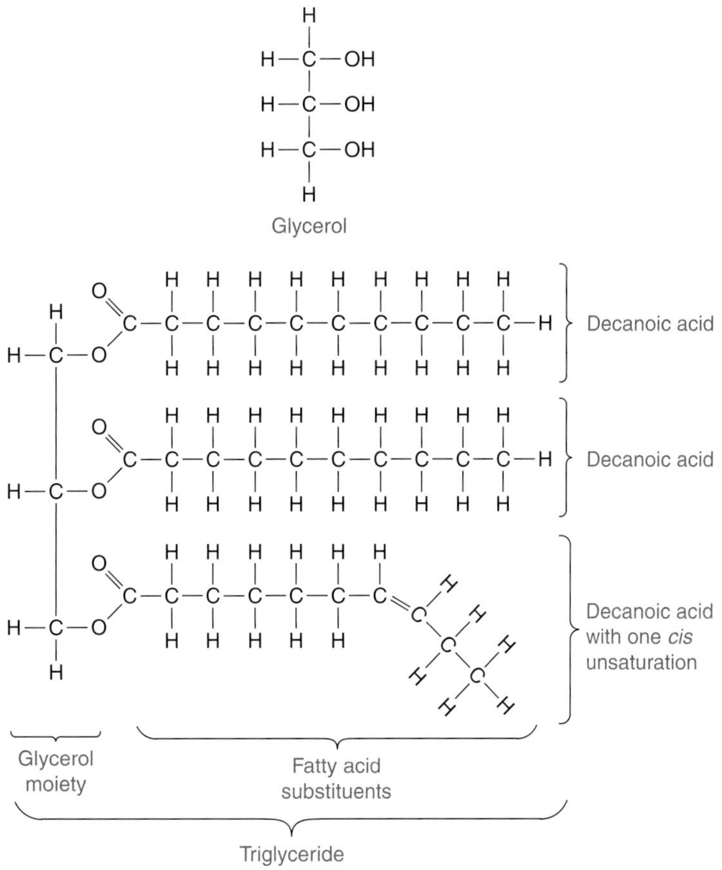

Figure 5-13. Structure of glycerol is at the *top*. Structure of a triglyceride is at the *bottom*.

of double bonds]) is a constituent of the cell membrane and, as will be seen later, is a substrate for the important active substances, prostaglandins, leukotrienes, and lipoxins.

When glycerol has one substituted fatty acid, it is a monoglyceride. When it contains two fatty acids, it is a diglyceride, and it is a triglyceride when all three positions are filled. The structures of glycerol and a triglyceride are shown in Figure 5-13.

Fatty Acid Oxidation

Dietary lipids are absorbed in the intestine, facilitated by the action of bile salts secreted from the gallbladder. Once emulsified, the fats are degraded by pancreatic lipase and phospholipase A_2 in the intestine. The products are free fatty acids and a mixture of mono- and diglycerides. Lipase degrades at the 1 and 3 positions of glycerol, and phospholipids are degraded by pancreatic phospholipase A_2 at the 2 position of the glycerol component, releasing free fatty acid and the lysophospholipid (Figures 5-14 and 5-15). These products are absorbed by intestinal mucosal cells, where the resynthesis of triacylglycerides occurs. These are then solubilized by combination with proteins (lipoprotein complexes) into chylomicrons, which contain lipid droplets, polar lipids, and a layer of proteins (Figure 5-16).

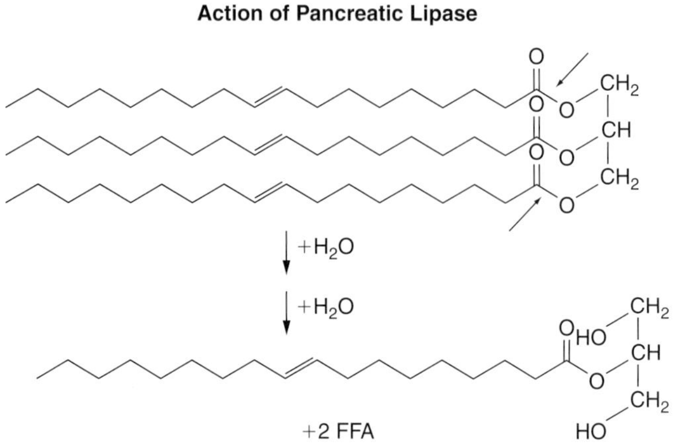

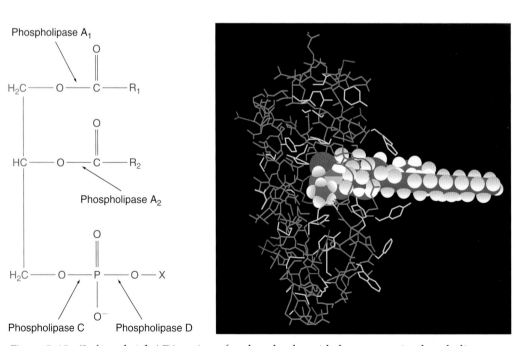

Figure 5-14. Digestion of a triglyceride by pancreatic lipase.

Figure 5-15. *(Left and right)* Digestion of a phosphoglyceride by pancreatic phospholipase. Right side reproduced with permission from Professor E. A. Dennis.

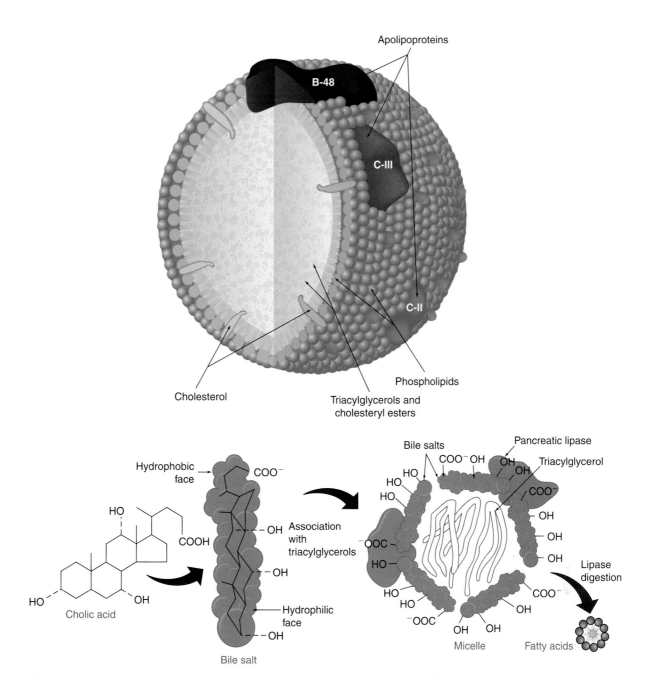

Figure 5-16. Diagram of a chylomicron *(top)*. Formation of a chylomicron *(bottom)*.

The triacylglycerols and phospholipids combine in the endoplasmic reticulum, then apolipoproteins are added to the complex and the chylomicrons are moved by exocytosis into the lymphatic circulation. This circulation funnels into the blood following passage through the liver. The VLDLs and the chylomicrons transport fatty acids to peripheral tissues from the liver, and the LDLs transport cholesterol to the peripheral tissues from the liver. The HDLs carry cholesterol from the peripheral tissues to the liver (recycling). When cholesterol is mobilized from the plasma membrane to the HDL, the reaction is catalyzed by LCAT. The free fatty acids released from fat cells **(adipocytes)** are carried in the circulation while bound to albumin. These transporting mechanisms are summarized in Figure 5-17.

β-Oxidation is the process by which fatty acids are broken down into two carbon units starting from the carboxyl end of the fatty acid. After fat (triacylglycerides) is ingested, it is broken down in the intestine into free fatty acids and monoacylglycerol. Dietary fats are delivered to the extrahepatic tissues as chylomicrons and VLDLs

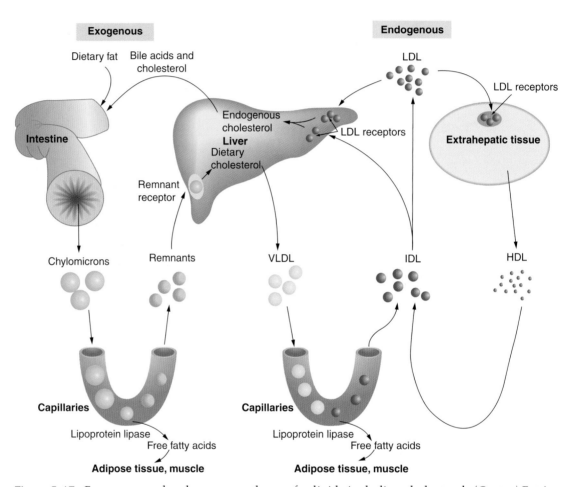

Figure 5-17. Exogenous and endogenous pathways for lipids including cholesterol. *(Center)* Fat is ingested and processed into chylomicrons in the intestine; the chylomicrons are absorbed into the circulation where lipoprotein lipase releases free fatty acids; the free fatty acids are distributed to adipose and muscle tissues (carried by albumin). Partially degraded chylomicrons (remnants) are taken up by the liver by way of a remnant receptor and the cholesterol is packaged in VLDLs. These enter the circulation where lipoprotein lipase releases free fatty acids that are carried to peripheral tissues; the partially degraded VLDLs (IDLs) are carried to the liver via LDL receptors and to extrahepatic tissues. HDLs are produced in the extrahepatic tissues and are carried back to the liver. Release of cholesterol in the liver is carried to the intestine as bile acids; some are excreted but most are recirculated to the liver (enterohepatic circulation). (Figure 5-9.) *HDL,* high density lipoprotein; *IDL,* intermediary density lipoprotein; *LDL,* low density lipoprotein; *VLDL,* very low density lipoprotein.

(Figure 5-17). The oxidation of fatty acids takes place in the matrix of the mitochondrion. This oxidation occurs in four reactions from the carboxyl end of the fatty acyl group until acetyl CoA remains. The series of enzymatic reactions is shown in Figure 5-18. In the first step, two hydrogens are removed from the β- and γ-carbons (second and third carbons starting from the carboxyl group) of the fatty acyl CoA, creating a double bond. In the enoyl CoA hydratase step (second step), a water molecule is added across the double bond so that the β-carbon becomes $-CH_2-$ and the γ-carbon becomes $-HCOH^-$. In the third step, catalyzed by 3-L-hydroxyacyl dehydrogenase, NAD^+ is reduced to $NADH + H^+$ by removal of two hydrogens from the γ-carbon, creating a carbonyl at the γ-carbon. Finally, in the β-ketothiolase catalyzed reaction, the β-carbon is cleaved together with acyl CoA and another CoA is added to the remaining carboxyl (the former γ-carbon). The cycle begins again until the fatty acid is degraded completely to acetyl CoA equivalents. So, for a mole of palmitoyl CoA ($CH_3(CH_2)_{14}COO^-$), as the starting fatty acyl CoA, for example, the products would be $7\ FADH_2 + 7\ NADH + 7H^+ + 8\ CH_3COSCoA$.

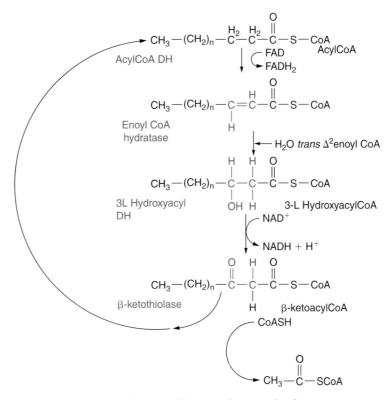

Figure 5-18. Beta-oxidation of fatty acyl CoA. The first reaction is catalyzed by fatty acyl dehydrogenase: $FADH_2$; next is enol CoA hydratase: H_2O; the third enzyme in this pathway is β-ketothiolase + CoASH → Acetyl CoA. The starting fatty acid (as fatty acyl CoA) is continuously broken down in this pathway until the final butyryl CoA is converted to 2 acetyl CoA molecules. The substituents in blue show the positions of hydrogens and hydroxyls at the location of the double bond. *DH*, dehydrogenase.

Activation and Transport of Fatty Acids into Mitochondria

The conversion of fatty acyl CoA molecules into acetyl CoA equivalents by β-oxidation in the mitochondrion has been described. Now the means by which fatty acids enter the mitochondrion from the cytoplasm can be discussed. Short-chain fatty acids can diffuse into the mitochondrion from the cytoplasm, but long-chain fatty acids cannot. The longer fatty acids are converted to acyl CoA molecules in the cytoplasm and carried into the mitochondrion complexed to carnitine. This system is referred to as the **carnitine cycle.** Carnitine is a derivative of lysine, and its structure is shown in Figure 5-19, along with the carnitine cycle. A general scheme of fatty acid metabolism is shown in Figure 5-20.

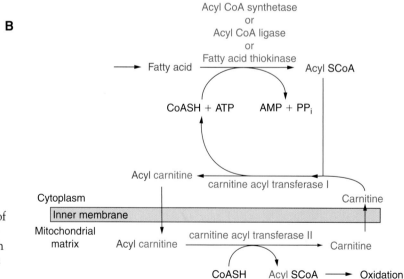

Figure 5-19. **A.** Reactions of the carnitine cycle with the structures of carnitine and acyl carnitine. **B.** Diagram of the carnitine cycle by which larger fatty acids are moved into the mitochondrial matrix.

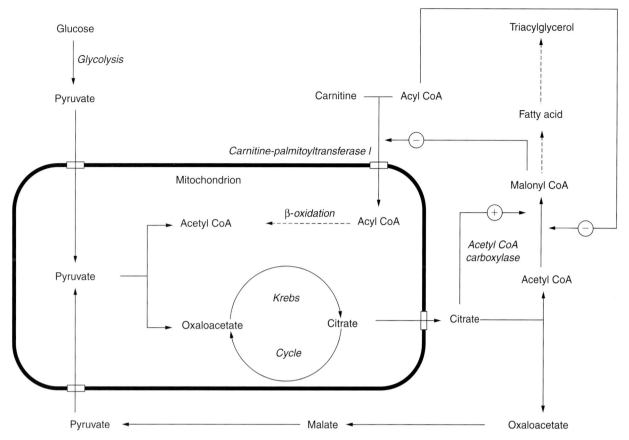

Figure 5-20. Overview of fatty acid metabolism.

Lipid Metabolism and Hormonal Control

Fatty acids are released from triglycerides ingested in the diet by three lipases. The fatty acids and the glycerol released in this process are transported into the bloodstream. In a tissue (liver), glycerol is converted to glyceraldehyde-3-phosphate and enters glycolysis or gluconeogenesis, depending on the state of the individual. If glucose is sufficient, the sugar will enter glycolysis and then be converted further in the citric acid cycle. The fatty acids are converted to acetyl CoA by β-oxidation. Thus, the overall picture is shown in Figure 5-21. Figure 5-22 shows the structures of some enzymes involved in the hydrolysis of triglyceride and the β-oxidation of fatty acids.

In a glucose-sufficient condition, glucose can be converted to fat as shown in Figure 5-23. Like the control of HMG-CoA reductase in the synthesis of cholesterol, the main hormones controlling the breakdown of triglyceride are epinephrine, glucagon, and insulin. **Lipolysis** is activated by epinephrine and glucagon, whereas insulin inhibits lipolysis and stimulates the formation of fat from fatty acids. Three lipases govern the breakdown of triglyceride: triacylglycerol lipase, diacylglycerol lipase, and monoacylglycerol lipase. Of the three, triacylglycerol lipase is the one activated by epinephrine, as shown in Figure 5-24. The mechanism

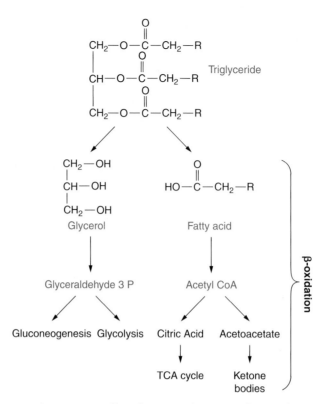

Figure 5-21. Overall utilization of ingested fat (triglyceride).

by which epinephrine activates triacylglycerol lipase is mediated by an adrenergic receptor, which produces cyclic AMP from ATP. The cyclic AMP activates protein kinase, which phosphorylates inactive lipase to produce the active, phosphorylated form (Figure 5-25).

Fat can be formed when the diet is low in fat but high in carbohydrates. Glucose is converted to glyceraldehyde-3-phosphate, which is converted to glycerol that is the precursor of fat when combined with fatty acids. Fatty acids are also produced from glucose through the formation of pyruvate and then acetyl CoA. In starvation, or when protein, carbohydrates, and fat are limiting in the diet, stored fat will be used to form ketone bodies. When fat is broken down to form fatty acids and glycerol, the glycerol is converted to glyceraldehyde-3-phosphate and then forms glucose. The liberated fatty acids form acetyl CoA, which does not enter the citric acid cycle because oxaloacetate will be limiting in a starved person and oxaloacetate is required for acetyl CoA to enter the citric acid cycle. Any oxaloacetate will be used in gluconeogenesis to form glucose. Fatty acid breakdown also will occur to generate needed ATP; as a result, ketone bodies will be formed. Muscle (protein) wasting will occur to supply needed amino acids. Ketone bodies are derived from acetate in the liver, and they are acetone, acetoacetate, and β-hydroxybutyrate. Both acetoacetate and β-hydroxybutyrate can be used as carbon sources by the heart, skeletal muscle, and kidney, and they are well used by the brain in place of glucose during starvation. In the Atkins diet, which consists of very low carbohydrate and normal or elevated intake of protein and fat, endogenous fat (in adipocytes) is broken down quickly (**hormone-sensitive lipase** is activated) and converted to fatty acids (and glycerol), which are further converted to acetyl CoA and then to ketone bodies. The stimulants to activate the breakdown of adipocyte stores of fat are glucagon, epinephrine, and β-corticotropin (insulin is an inhibitor), and they operate through their cognate cell membrane receptors, resulting in the formation of cyclic AMP and the activation of protein kinase A. The kinase phosphorylates hormone-sensitive lipase (active in the phosphorylated form). Hormone-sensitive lipase, together with diacylglycerol lipase, produces monoacylglycerols, which are finally hydrolyzed by monoacylglycerol lipase. The free fatty acid products diffuse out of the adipose cells and combine with blood albumin for transport to other tissues, where they enter cells by free diffusion (passive transport). The process of ketogenesis is shown in Figure 5-26.

Of the ketone bodies formed, acetone is transported to the lungs, which accounts for the characteristic oral odor generated by Atkins dieters who have become ketotic. Acetoacetate and β-hydroxybutyrate are converted back to acetyl CoA (β-hydroxybutyrate is first converted to acetoacetate). Thus, acetyl CoA accumulates during fasting and is converted into ketone bodies because oxaloacetate is not available for its entry into the citric acid cycle. If this process (ketogenesis) is carried to extreme, the high levels of ketone bodies may lead to acidosis (they are acidic); however, excess ketone bodies are excreted in the urine, as well as being used by the tissues under these conditions. Although ketogenesis takes place in the liver, only the tissues peripheral to the liver use them; the liver, itself, cannot. The release of fatty acids from adipose increases ketogenesis in liver occurring by supplying substrate. In untreated insulin-dependent diabetes, diabetic ketoacidosis results from a reduced supply of glucose to the tissues because of reduced circulating insulin. Reduced insulin and increased glucagon (low blood glucose) increase the activity of hormone-sensitive lipase of adipose, generating increases in fatty acids and in fatty acid oxidation. This leads to increased acetyl CoA and ketone body production, often acidifying the blood.

Fatty acid synthesis is controlled by acetyl CoA carboxylase, which is the rate-limiting enzyme. It catalyzes the carboxylation of acetyl CoA to malonyl CoA in two steps, as shown in Figure 5-27. This enzyme contains biotin linked through the terminal carboxyl of the biotin side chain through the ε-amino group of a lysine residue (Figure 5-28). Because this side chain is relatively long, biotin can transit between the two active sites on the enzyme (Figure 5-27). The structure of acetyl CoA carboxylase is shown in Figure 5-29. The enzyme is regulated by phosphorylation (through the activation of protein kinase A stimulated by epinephrine and glucagon through cyclic AMP when blood glucose is low) and by allosteric control of metabolites. For example, citrate stimulates the enzyme and palmitoyl CoA inhibits it (feedback inhibition, as shown in Figure 5-30). When it is phosphorylated, the enzyme dissociates into inactive monomers, acetyl CoA is not metabolized into malonyl CoA, and acetyl CoA is retained for the production of ketone bodies that may be needed in the continued condition of low glucose. A separate AMP-activated kinase also phosphorylates the enzyme, resulting in inhibition. This mechanism is important in the heart, for example, which does not synthesize fatty acids. In tissues of this type, malonyl CoA is an inhibitor

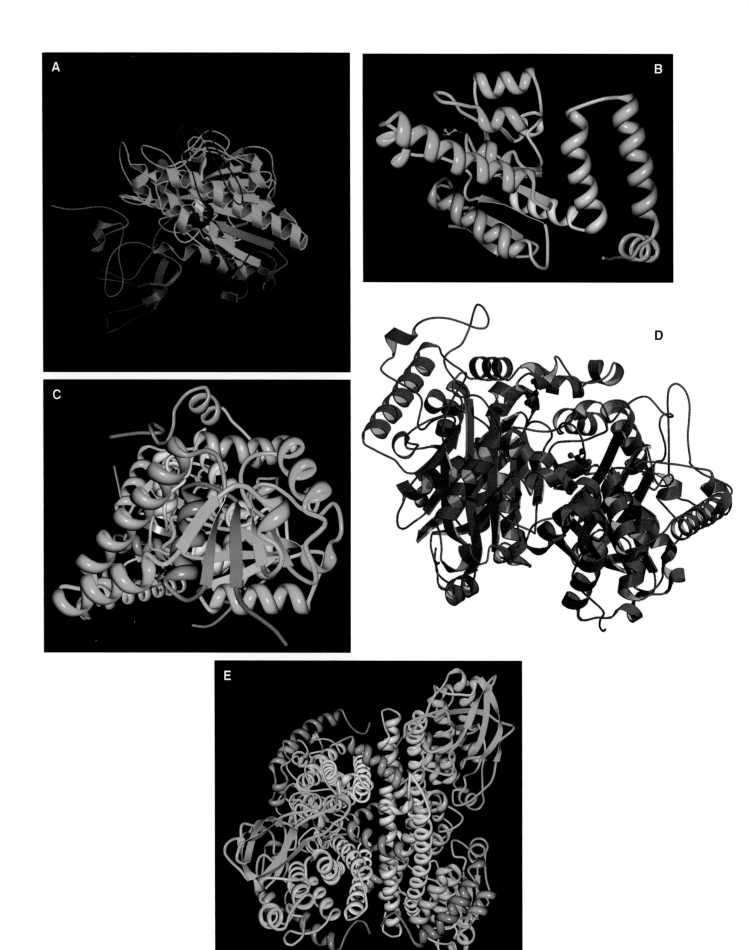

Figure 5-22, cont'd

For legend see opposite page.

Figure 5-22. Some structures of enzymes in the breakdown of triglyceride and in the oxidation of fatty acids. **A.** Structure of the pancreatic lipase-colipase complex. **B.** Crystal structure of enoyl-CoA hydratase from *Thermus thermophilus*. **C.** L-3-hydroxyacyl-CoA dehydrogenase complexed with 3-hydroxybutyryl-CoA. **D.** Dimeric peroxisomal thiolase of *saccharomyces cerevisiae*. **E.** Structure of acyl CoA dehydrogenase, the initial enzyme in the β-oxidation pathway of fatty acid acyl CoA. It is a homotetrameric enzyme. The adenosine portion *(magenta)* protrudes from the active site of each subunit. The adenosine portion of FAD is not visible. Met 165 and trp 166 are probably involved in the electronic path between FAD and the electron transfer flavoprotein. These two amino acid residues are shown in red. Part A reproduced from PDB ID: ln85. H. van Tilbeurgh, L. Sarda, R. Verger, C. Cambillau. Structure of the Pancreatic Lipase-Procolipase Complex. *Nature* **359** pp. 159 (1992). Part B reproduced from PDB ID: 1uiy. B. Bagutdinov, S. Kuramitsu, S. Yokoyama, M. Miyano, T.H. Tahirov. Crystal Structure of enoyl-CoA Hydratase from *Thermus thermophilus* HB8. *To be published*. Part C reproduced from PDB ID: 1f12. J.J. Barycki, L.K. O'Brien, A.W. Strauss, L.J. Banaszak. Sequestration of the Active Site by Interdomain Shifting. Crystallographic and Spectroscopic Evidence for Distinct Conformation of the L-3-hydrovxyacyl-CoA Dehydrogenase. *J. Biol. Chem.* **275** pp. 27186 (2000). Part D reproduced from PDB ID: 1afw. M. Mathieu, Y. Modis, J.P. Zeelen, C.K. Engel, R. A. Abagyan, A. Ahlberg, B. Rasmussen, V.S. Lamzin, W.H. Kunau, R.K. Wierenga. The 1.8A Crystal Structure of the Dimeric Peroxisomal 3-ketoacyl-CoA Thiolase of *Saccharomyces Cerevisiae*: Implications for Substrate Binding and Reaction Mechanism. *J. Mol. Biol.* **273** pp. 714 (1997). Part E reproduced from PDB ID: 2cx9. K. Murayama, T. Kinebuchi, M. Shirouzu, S. Yokoyama. Crystal Structure of Acyl-CoA Dehydrogenase. *To be published*.

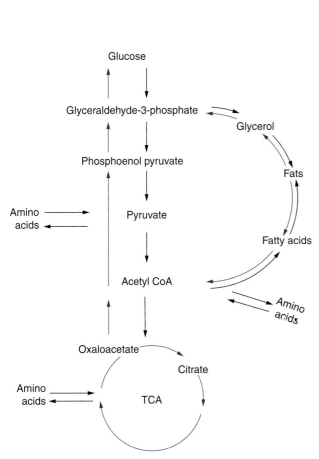

Figure 5-23. Conversion of glucose to fats in a glucose sufficient situation.

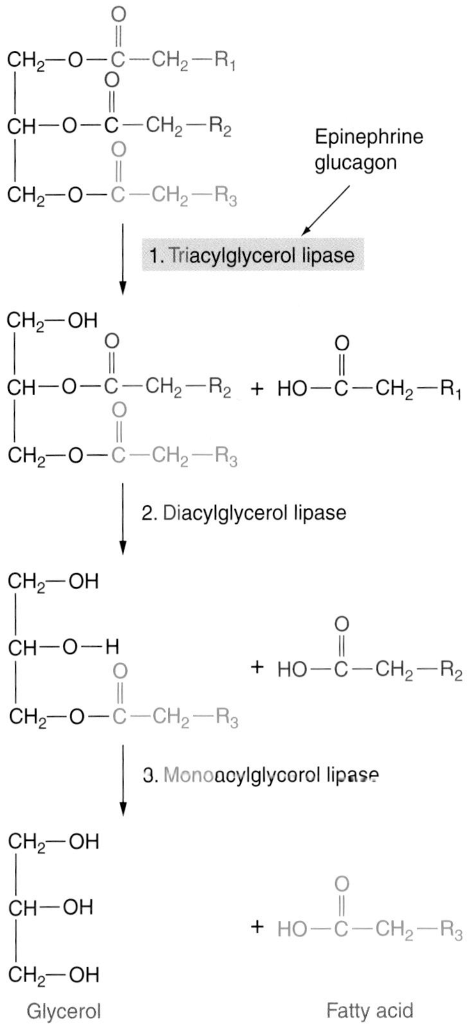

Figure 5-24. Breakdown of triglyceride by three lipases.

Lipid Metabolism and Hormonal Control

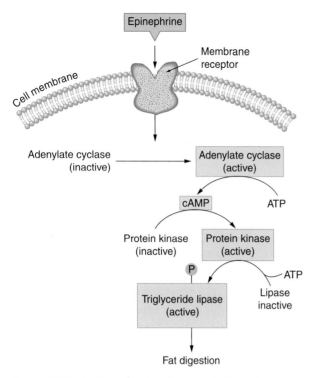

Figure 5-25. Action of epinephrine to induce the active form of triglycerol lipase through phosphorylation by protein kinase.

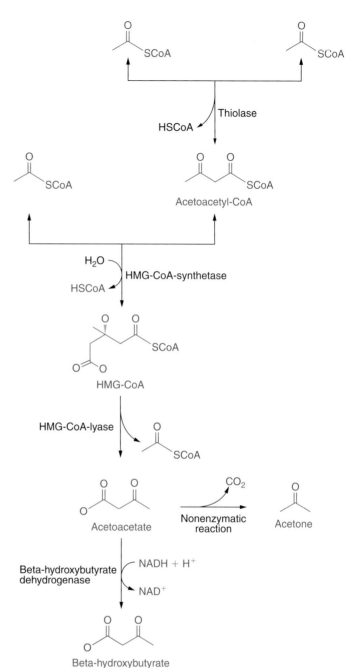

Figure 5-26. Process of ketogenesis. Acetoacetyl CoA is formed from 2 molecules of acetyl CoA. HMG-CoA synthetase catalyzes the formation of HMG-CoA, an enzyme only found in the liver. Some HMG-CoA is converted to cholesterol via HMG-CoA lyase to initially form mevalonate. When glycogen is high in the liver, acetoacetate is converted to β-hydroxybutyrate by β-hydroxybutyrate dehydrogenase. Ketone bodies are utilized by the tissues by conversion to acetoacetate, and then to *acetoacetyl*-CoA by *acetoacetate*: succinyl-CoA transferase that is present in all tissues except the liver, so the liver produces ketone bodies but does not utilize them.

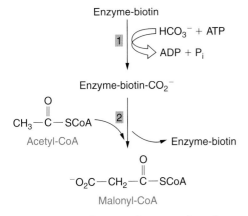

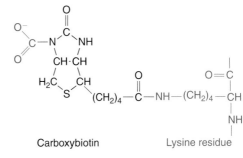

Figure 5-28. The flexible arm of biotin attached to the enzyme acetyl CoA carboxylase.

Figure 5-27. The acetyl CoA carboxylase reaction. The ATP-dependent carboxylation of biotin at one active site (1) is followed by transfer of the carboxyl group to acetyl CoA at a second site (2). The overall reaction is: $HCO_3^- + ATP + acetyl-CoA \longleftrightarrow ADP + P_i + malonyl\ CoA$.

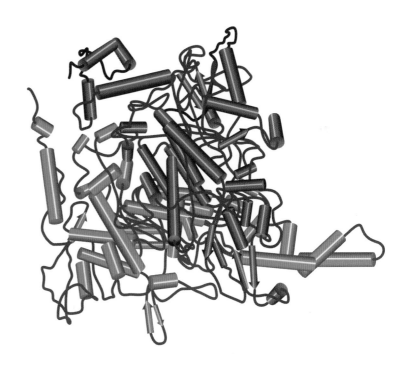

Figure 5-29. Structure of acetyl CoA carboxylase from yeast. Redrawn from PDB ID: 1o12. H. Zhang, Z. Yang, Y. Shen, L. Tong. Crystal Structure of the Carboxyltransferase Domain of Acetyl-Coenzyme A Carboxylase. *Science* **299** pp. 2064 (2003).

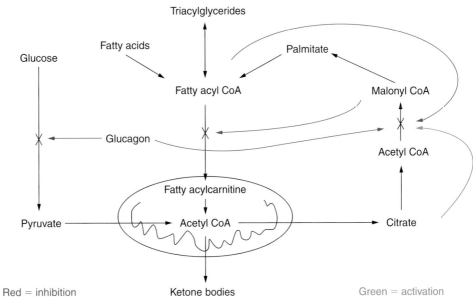

Figure 5-30. Regulatory interactions of fatty acid synthesis and oxidation in the liver.

Red = inhibition

Green = activation

Lipid Metabolism and Hormonal Control

of fatty acid oxidation. When ATP is low and AMP is high, the production of malonyl CoA is reduced, allowing for fatty acid oxidation.

Fatty acid synthesis starting with acetyl CoA and malonyl CoA is catalyzed by fatty acid synthase, a single enzyme that has resulted from gene fusion of several individual enzymes. It is a polypeptide chain with several individual domains and exists as a dimer. Acetyl CoA and malonyl CoA are the substrates (linked as thioester derivatives of coenzyme A through the β-mercaptoethylamine portion of the coenzyme, as shown in Figure 5-31). Fatty acid synthesis occurs in the soluble cytoplasm. The acetyl CoA substrate is produced in the mitochondria and transported to the cytoplasm by a shuttle mechanism with citrate. In the synthase, phosphopantetheine is attached as a phosphate ester of a serine hydroxyl of the acyl carrier protein domain. Like the biotin of acetyl CoA carboxylase, the phosphopantetheine permits the thiol group to transit from one active site to another. The thiols of cysteine and phosphopantetheine attach to the carbonyl groups as thioesters of acetate, malonate, or the growing fatty acid chain (Figure 5-32). Fatty acid synthase is a dimer attached in antiparallel fashion (Figure 5-33). The individual steps of the fatty acid synthase reactions are shown in Figure 5-34, and the reactions can be viewed as a cycle in Figure 5-35. In this set of reactions, the final product is the 16-carbon fatty acid palmitate. Extension of the length of the fatty acid being synthesized beyond 16 carbons is carried out in the mitochondria and the endoplasmic reticulum. Essentially, the fatty acid oxidation system runs in reverse to accomplish elongation, and malonyl CoA is the two-carbon

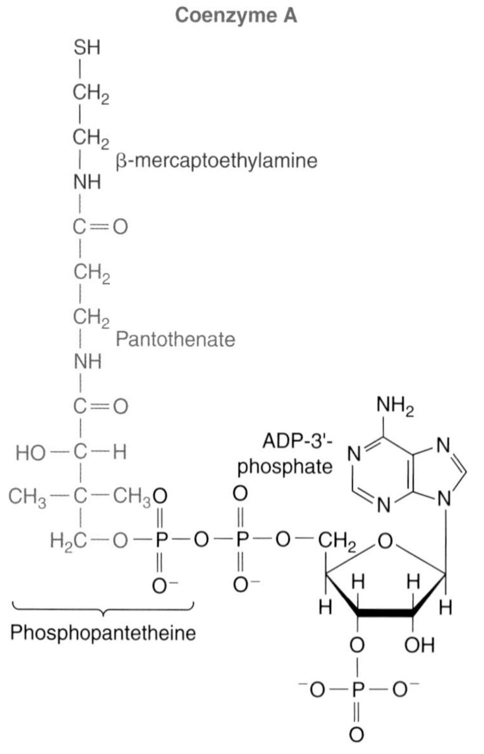

Figure 5-31. Structure of coenzyme A.

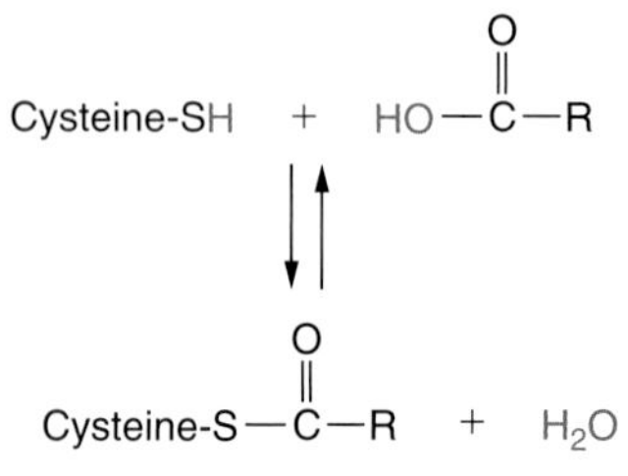

Figure 5-32. Formation of thioester of cysteine with the carboxyl group of acetate, malonate, or a growing fatty acid chain.

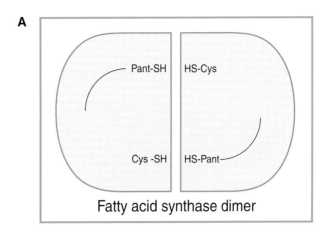

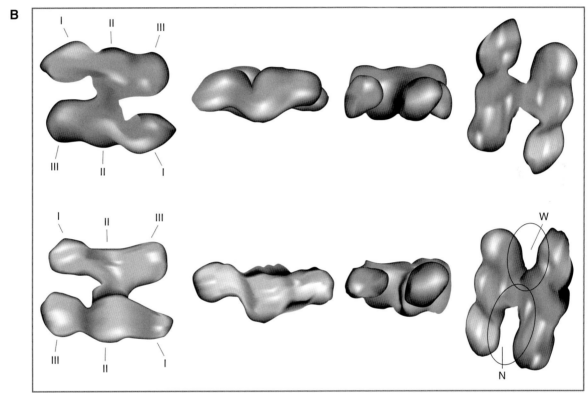

Figure 5-33 **A.** Antiparallel arrangement of the two subunits of fatty acid synthase. At each reaction cycle the growing fatty acid is transferred from the phosphopantetheine (Pant-SH) of one subunit to the cysteine thiol of the other subunit. The distance between Cys and *Pant thiols* is less than the distance between thiols of the same subunit. **B.** 3D reconstructions of fatty acid synthase from electron cryomicroscopy data. Part A redrawn from http://www.dentistry.leeds.ac.uk/biochem/MBWeb/mb2/part1/fasynthesis.htm. Part B redrawn with permission from J. Brink et al., "Quaternary structure of human fatty acid synthase by electron cryomicroscopy," *PNAS*, 99:138–143, 2002.

Steps 1–3 of the **Fatty Acid Synthase** reaction pathway are catalyzed by the catalytic domains listed in the diagram at right.

Shown are the cysteine of one protein subunit and the acyl carrier protein phosphopantetheine (Pant) of the other subunit of the dimeric complex.

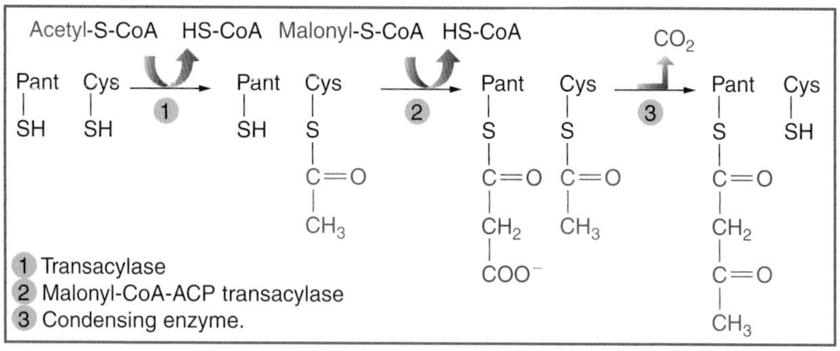

① Transacylase
② Malonyl-CoA-ACP transacylase
③ Condensing enzyme.

In steps 4–6:

The β-ketone is reduced to an alcohol, by electron transfer from NADPH.

Dehydration yields a trans double bond.

Reduction at the double bond by NADPH yields a saturated chain.

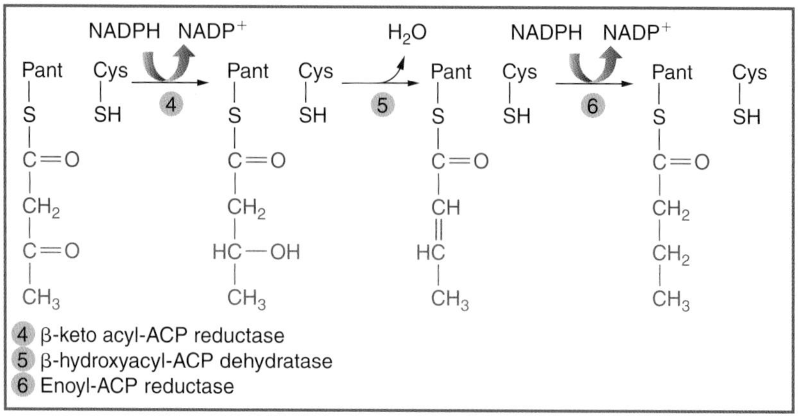

④ β-keto acyl-ACP reductase
⑤ β-hydroxyacyl-ACP dehydratase
⑥ Enoyl-ACP reductase

Following intersubunit transfer of the fatty acid from phosphopantetheine to cysteine sulfhydryl, the cycle begins again, with reaction of another malonyl CoA.

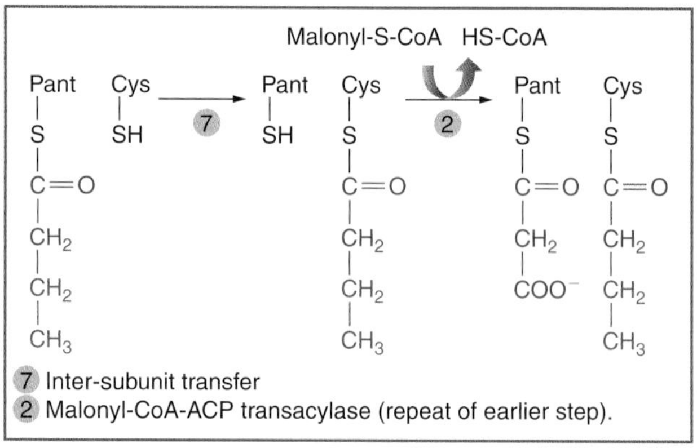

⑦ Inter-subunit transfer
② Malonyl-CoA-ACP transacylase (repeat of earlier step).

Figure 5-34. The individual steps in fatty acid synthesis catalyzed by fatty acid synthase. Redrawn with permission from http://www.dentistry.leeds.ac.uk/biochem/MBWeb/mb2/part1/fasynthesis.

fragment donor. NADPH donates the final electrons. The overall reaction in the synthesis of palmitate, accounting for the ATP synthesis of malonate, is as follows:

$$8 \text{ acetyl CoA} + 14 \text{ NADPH } (+ 14 \text{ H}^+) + 7 \text{ ATP} = \text{palmitate} + 14 \text{ NADP}^+ + 8 \text{ CoA} + 7 \text{ ADP} + 7 \text{ P}_i.$$

Fatty acid double bonds are introduced by desaturases at specific positions within the chain. This requires endoplasmic reticulum enzymes: NADH cytochrome b_5 reductase, cytochrome b_5, and desaturase. Desaturases are mixed-function oxidases, which catalyze a four-electron reduction of oxygen to form two water molecules as a double bond is

Fatty Acids and Triacylglycerols

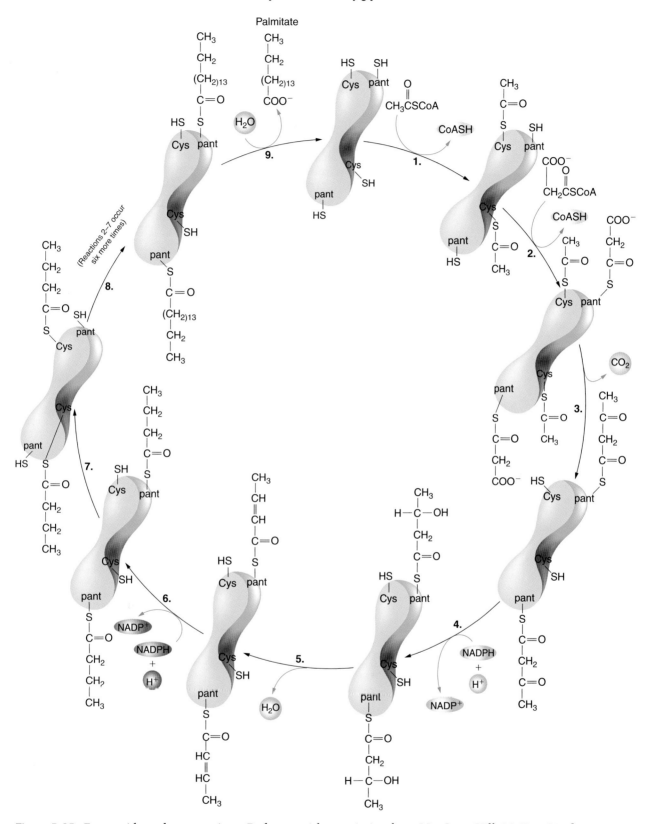

Figure 5-35. Fatty acid synthase reactions. Redrawn with permission from MacGraw-Hill, McKee, *Biochemistry*, 1996, DuBuque, Iowa.

introduced into the fatty acid. The overall reaction for introduction of a double bond in stearate (18:0) to form oleate (18:1 *cis* Δ-9) is the following:

$$\text{stearate} + \text{NADH} + \text{H}^+ + \text{O}_2 = \text{oleate} + \text{NAD}^+ = 2\text{H}_2\text{O}$$

Oleic acid has a double bond at C^{9-10}, as shown in Figure 5-36. Certain fatty acids are required to be ingested in the diet, and they are called *essential fatty acids*. Humans have a limited capacity for synthesizing long-chain fatty acids, and the ingestion of linoleic and α-linolenic acids is essential. Linoleic acid is contained in vegetable oils and is an ω-6 fatty acid (a double bond is located at the sixth carbon from the carboxyl end of the molecule). Linolenic acid is in fish oils and belongs to the ω-3 fatty acids (having a double bond at C3 from the carboxyl end of the fatty acid molecule). Arachidonic acid (20:4) can be formed from linoleic acid (18:2). Structures of important fatty acids are shown in Figure 5-37.

The regulation of fatty acid synthesis is complex. In the liver, under adequate glucose conditions, the synthase is stimulated by insulin evoked by elevated blood glucose, ensuring that incoming glucose is converted and stored as fat. Insulin stimulates fatty acid synthase by increasing certain transcription factors, such as **upstream stimulatory factors (USFs)** and **sterol response element–binding protein (SREBP-1)**. Transcription of fatty acid synthase is reduced by polyunsaturated fatty acids by suppression of SREBP-1.

Leptin is an important hormone that inhibits fatty acid synthase expression and SREBP-1 in adipose cells. It is a 130–amino acid protein containing six helices (Figure 5-38). Leptin is produced in adipose cells,

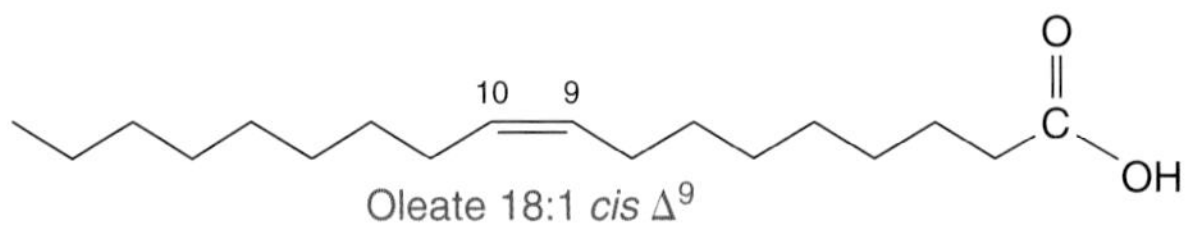

Figure 5-36. Structure of oleic acid (18:1) with a C^{9-10} double bond.

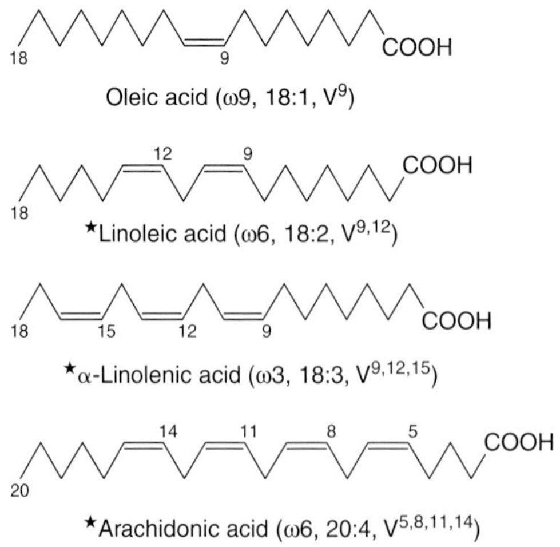

Figure 5-37. Structures of essential fatty acids: linoleic and linolenic and other important fatty acids. ★, indicates fatty acid is essential.

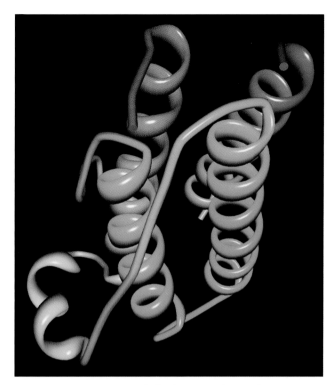

Figure 5-38. Structure of the human obesity hormone, leptin. It contains 167 amino acid residues and has six helices. PDB ID: 1ax8. F. Zhang, M.B. Basinski, J.M. Beals, S.L. Briggs, L.M. Churgay, D.K. Clawson, R.D. DiMarchi, T.C. Furman, J.E. Hale, H.M. Hsiung, B.E. Schoner, D.P. Smith, X.Y. Zhang, J.P. Wery, R.W. Schevitz. Crystal Structure of the Obese Protein leptin-E100. *Nature* **387** pp. 206 (1997).

and the circulating level of leptin is proportional to the total body fat. Leptin is a ligand for brain hypothalamic receptors whose actions promote the synthesis of α-melanocyte stimulating hormone, which is an appetite suppressant. It also counteracts the actions of other feeding stimulants: **neuropeptide Y,** secreted by cells in the hypothalamus and gut, and **anandamide,** a cannabinoid-like substance that binds to receptors and stimulates feeding. The structures of neuropeptide Y and anandamide are shown in Figure 5-39. In contrast to adipose tissue, leptin stimulates the oxidation of fatty acids in liver and muscle mitochondria, which reduces fat storage in those tissues. So far, however, there is not convincing genetic evidence that mutating leptin function does not always occur in obese individuals. There is some evidence that injecting leptin into an obese individual resulted in the loss of weight over a year and that the majority of that weight loss is in stored fat. It is generally believed that leptin will be found to play an important role in the regulation of body weight. Animal experiments have shown evidence for a small protein called **resistin** that causes the liver to reduce its sensitivity to insulin. Resistin is produced in humans in macrophages, and there is a relationship between elevated levels of resistin in the human with obesity and type 2 diabetes.

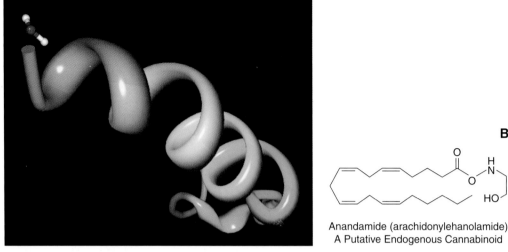

Figure 5-39. **A.** Structure of human neuropeptide Y consists of 1 helix with 13 amino acid residues. **B.** The structure of anandamide, a hormone derived from arachidonic acid. Part A from PDB ID: 1ron. S.A. Monks, G. Karagianis, G.J. Howlett, R.S. Norton Solution Structure of Human Neuropeptide Y. *J. Biomol. NMR* **8** pp. 379 (1996).

Phospholipids

Phospholipids are similar to triglycerides except that one of the glycerol hydroxyls is substituted by a phosphate group instead of a fatty acid. Usually the phosphate group is further substituted with another group; for example, in the case of phosphatidylcholine, the third carbon of glycerol is substituted by a phosphate linked to choline, as shown in Figure 5-40. A variety of glycerophospholipids is shown in Figure 5-41. In all cases, the third carbon of glycerol is substituted by a phosphate group linked to an additional group, such as ethanolamine, serine, choline, inositol, amide, or glycerol. In diphosphatidylglycerol, the 1- and 3-carbons of glycerol are substituted by phosphate groups linked to other groupings. Most glycerophospholipids are found in membranes as shown in Figure 5-42. Phospholipids also act as emulsifiers to solubilize hydrophobic molecules.

Sphingomyelin is another phospholipid that is important in certain tissues, such as brain. It is a member of the sphingolipids, which, like the phospholipids, have a polar head group and two lipophilic tails. Sphingosine, a long-chain amino alcohol is the central molecule of the sphingolipids. The sphingomyelins (only phospholipids of this group) and the glycosphingolipids comprise the sphingolipids, which are cell membrane components that are part of the myelin sheath. The sphingolipids are made up of the cerebrosides, sulfatides, **globosides,** and gangliosides. **Ceramide** is present both in sphingomyelins and in glycosphingolipids. It is synthesized from serine and palmitoyl CoA, as shown in Figure 5-43. Major sphingomyelins contain palmitic acid or stearic acid N-acylated at C2 of sphingosine. Sphingomyelins are formed from

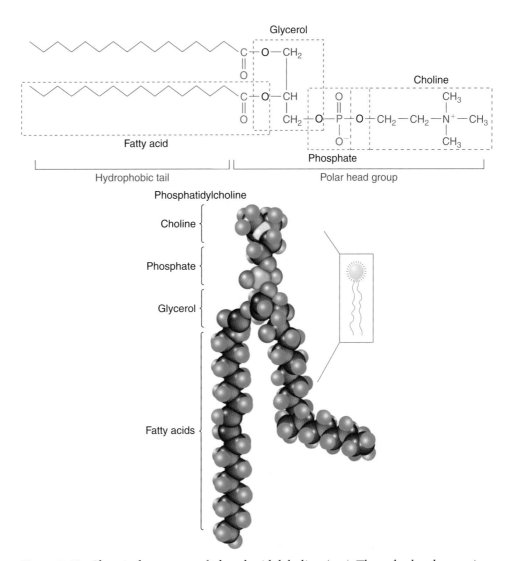

Figure 5-40. Chemical structure of phosphatidylcholine *(top)*. The polar head group is formed by the glycerol portion, the phosphate and choline, and this part of the molecule is hydrophilic. The fatty acid substituents are lipophilic; an atomic model of this structure is shown at the bottom.

ceramide by the donation of phosphorylcholine from phosphatidylcholine catalyzed by sphingomyelin synthase, as shown in Figure 5-44.

Sphingomyelin is broken down by the enzyme **sphingomyelinase** (the structure of the bacterial enzyme is shown in Figure 5-45). The enzyme (called acid sphingomyelinase for the enzyme located in the lysosome) catalyzes the breakdown of sphingomyelin:

sphingomyelin + H_2O = N-acylsphingosine (ceramide) + choline-P

When ceramide is released by the breakdown of cell membrane sphingomyelin, the ceramide, in combination with the proapoptotic factor,

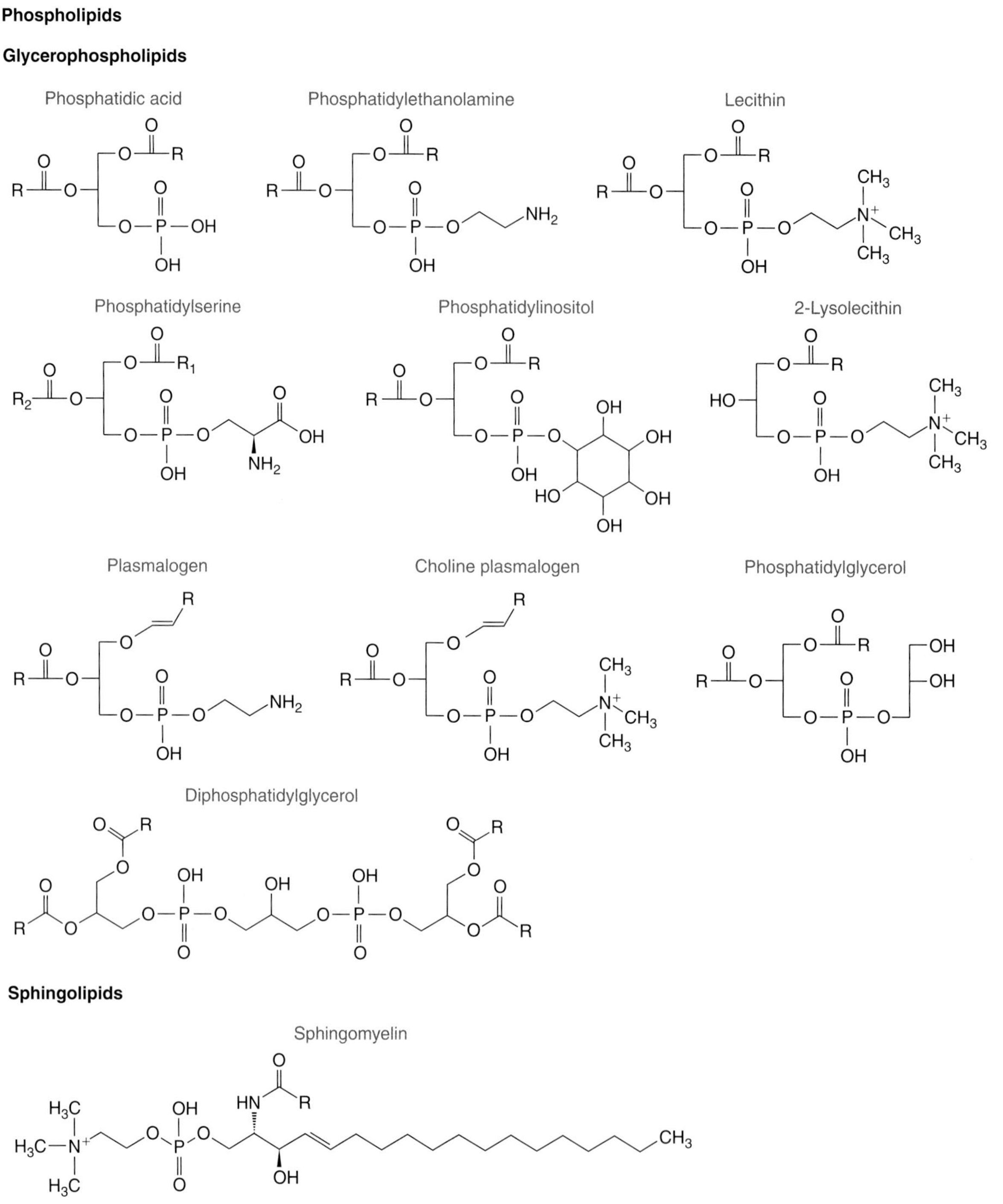

Phospholipids

Glycerophospholipids

Phosphatidic acid

Phosphatidylethanolamine

Lecithin

Phosphatidylserine

Phosphatidylinositol

2-Lysolecithin

Plasmalogen

Choline plasmalogen

Phosphatidylglycerol

Diphosphatidylglycerol

Sphingolipids

Sphingomyelin

Figure 5-41. A variety of phospholipids.

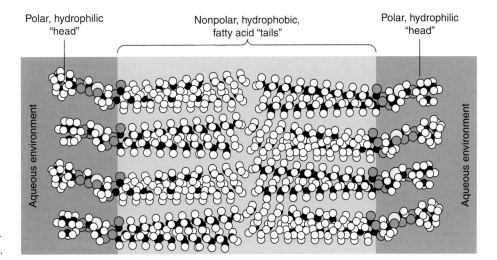

Figure 5-42. Occurrence of glycerophospholipids in membranes.

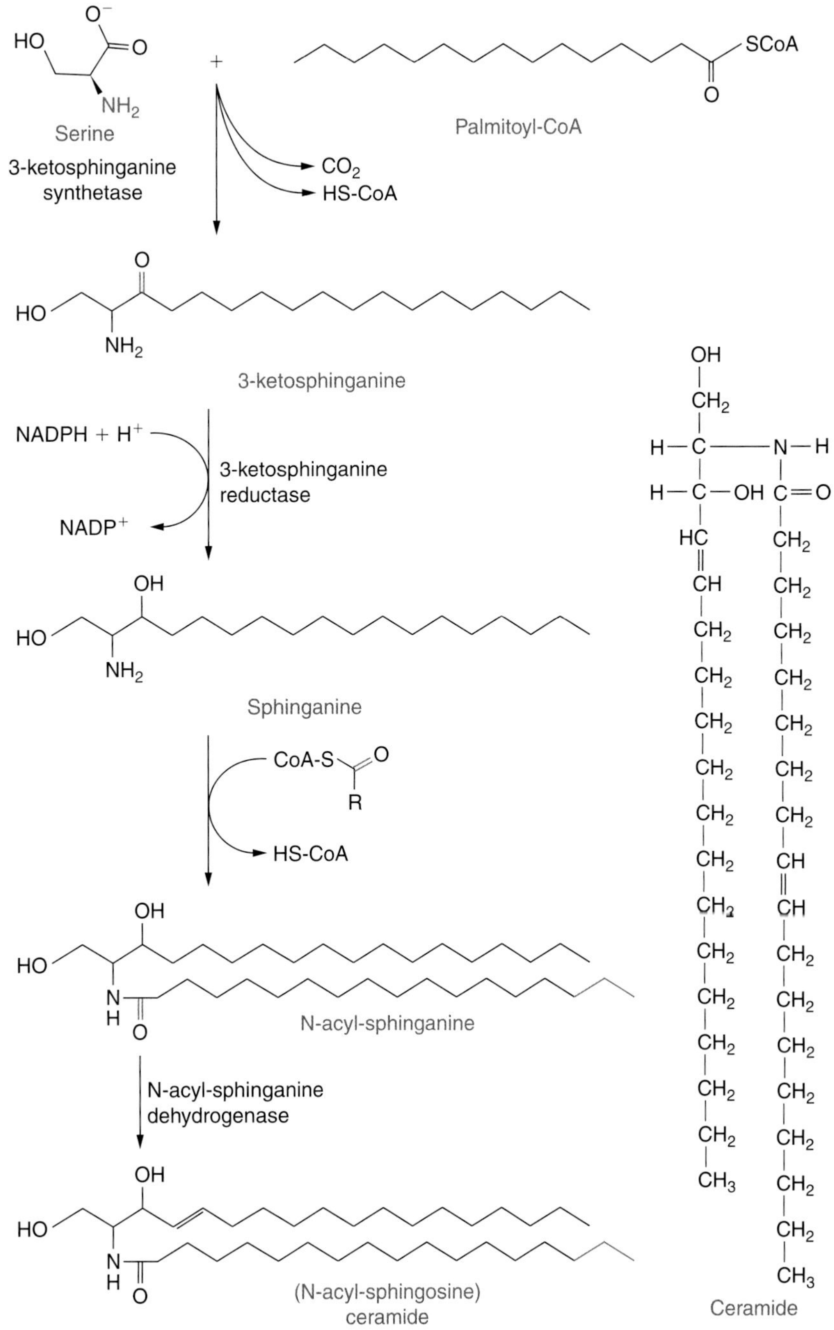

Figure 5-43. Biosynthesis of ceramide (N-acyl-sphingosine) from serine and palmitoyl CoA.

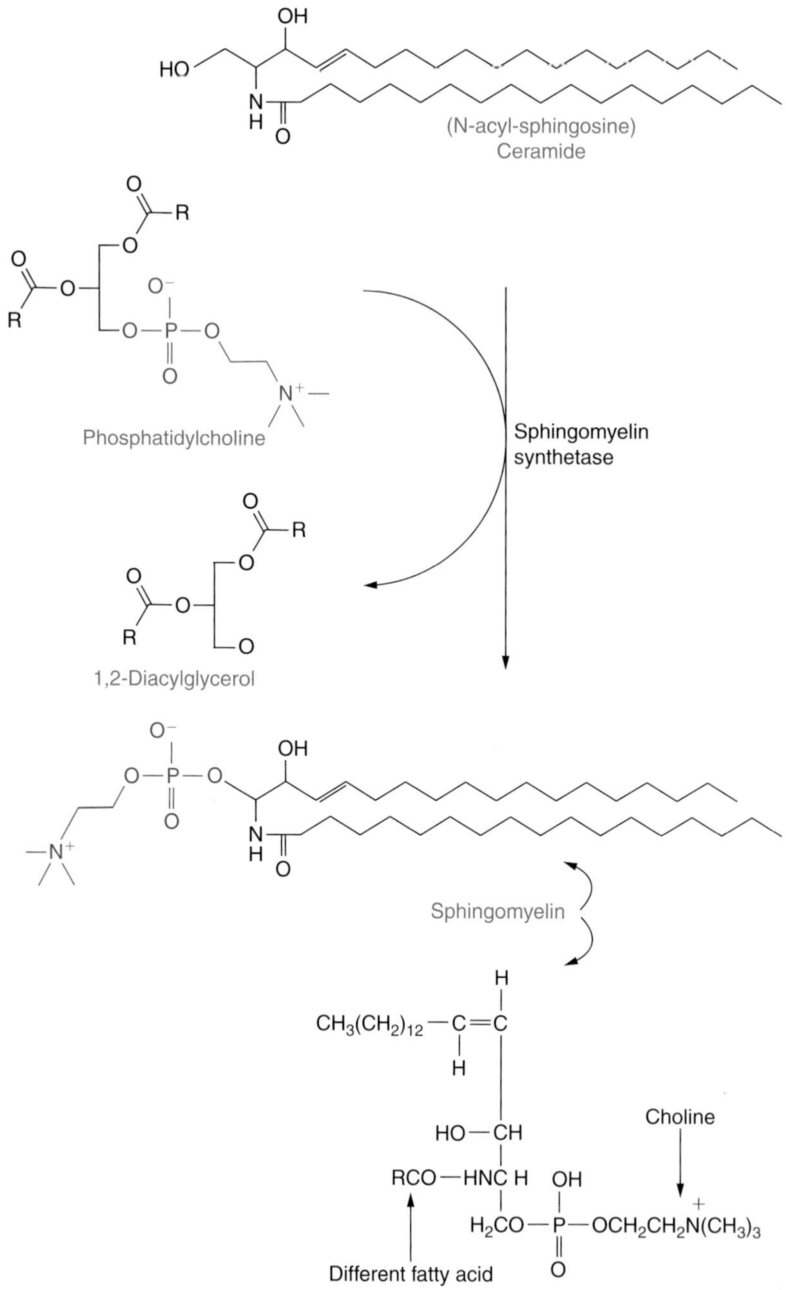

Figure 5-44. Biosynthesis of sphingomyelin from ceramide.

Bax, can lead to programmed cell death (apoptosis) shown in Figure 5-46. The cytokine **tumor necrosis factor-α (TNFα)** activates sphingomyelinase (reaction shown on page 221). Ceramide is further metabolized to sphingosine and then to sphingosine-1-phosphate, which activates the nuclear factor, **nuclear factor kappa B (NF-κB),** which, in endothelial cells, leads to the expression of **cell adhesion molecules (CAMs).** HDL blocks the conversion of sphingosine to sphingosine-1-phosphate catalyzed by sphingosine kinase and is able to reduce the expression of adhesion molecules.

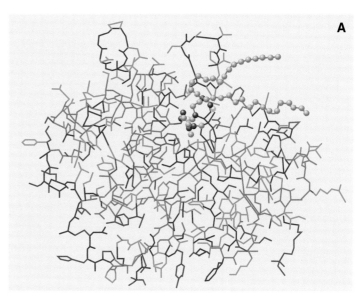
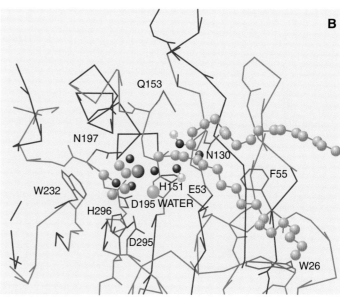

Figure 5-45. Structure of bacterial sphingomyelinase–sphingomyelin complex. **A.** Structural model for the *B. cereus* SMase-SM complex. The backbone of the SMase structure is shown in *pink*, and the side chains in *white*. Side chains of residues that were predicted to be important in the catalysis and the substrate recognition are shown in *navy* and *red*, respectively. The SM structure is shown in ball-and-stick, where carbon atoms are shown in *green*; nitrogens, in *blue*; oxygens, in *red*; hydrogens, in *white*; and a phosphorus, in *pink*. **B.** An enlarged view of the substrate binding region in the model. *SM*, Sphingomyelin; *SMase*, Sphingomyelinase. Reproduced with permission from Y. Matsuo et al., "A distant evolutionary relationship between bacterial sphingomyelinase and mammalian DNase I," *Protein Science*, 5: 2459, 1996.

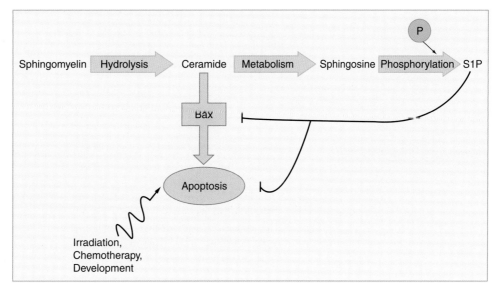

Figure 5-46. Breakdown of membrane sphingomyelin by sphingomyelinase releases ceramide, which, in combination with the proapoptotic Bax, can lead to programmed cell death (apoptosis). *S1P*, Sphingosine-1-phosphate. Redrawn with permission from R.F. Casper and A. Jurisicova, "Protecting the female germ line from cancer therapy," *Nat Med.* **6**: 1100–1101, 2000.

Phospholipids

Table 5-1
Several Diseases Associated with Abnormal Sphingolipid Metabolism

Disorder	Enzyme Deficiency	Accumulating Substance	Symptoms
Tay-Sachs disease	Hexosaminidase A	GM_2 ganglioside	mental retardation, blindness, early mortality
Gaucher's disease	Glucocerebrosidase	Glucocerebroside	hepatosplenomegaly, mental retardation in infantile form, long bone degeneration
Fabry's disease	α-Galactosidase A	Globtriaosylceramide; also called ceramide trihexoside (CTH)	kidney failure, skin rashes
Niemann-Pick disease, more info below Types A and B Type C1 Type C2 Type D	Sphingomyelinase	Sphingomyelin LDL-derived cholesterol LDL-derived cholesterol	all types lead to mental retardation, hepatosplenomegaly, early fatality potential
Krabbe's disease; globoid leukodystrophy	Galactocerebrosidase	Galactocerebroside	mental retardation, myelin deficiency
Sandhoff-Jatzkewitz disease	Hexosaminidase A and B	Globoside, GM_2 ganglioside	same symptoms as Tay-Sachs, progresses more rapidly
GM_1 gangliosidosis	GM_1 ganglioside: β-galactosidase	GM_1 ganglioside	mental retardation, skeletal abnormalities, hepatomegaly
Sulfatide lipodosis; metachromatic leukodystrophy	Arylsulfatase A	Sulfatide	mental retardation, metachromasia of nerves
Fucosidosis	α-L-Fucosidase	Pentahexosylfucoglycolipid	cerebral degeneration, thickened skin, muscle spasticity
Farber's lipogranulomatosis	Acid ceramidase	Ceramide	hepatosplenomegaly, painful swollen joints

Reproduced with permission from http://www.med.unibs.it/~marchesi/lipsynth3.html.

Acid sphingomyelinase, found in the lysosome, is involved in the breakdown of sphingomyelin when it is carried to this organelle. Mutation or poor expression of this enzyme can lead to a lysosomal storage disease known as **Niemann-Pick disease.** There are five types of this disease, only two of which are caused by defects in acid sphingomyelinase. Disorders of sphingolipid metabolism lead to a number of diseases, many of which are shown in Table 5-1.

Glycosphingolipids

Glycosphingolipids have a ceramide constituent, with mono- or oligosaccharides substituting the C1 of sphingosine. The structure of sphingosine is shown in Figure 5-47. There are four main classes of glycosphingolipids: **cerebrosides, sulfatides, globosides,** and **gangliosides.** The cerebrosides, located mainly in neuronal cell membranes, have a single sugar, most commonly galactose, linked to ceramide (galactocerebroside). Glucocerebrosides (Figure 5-48) occur as intermediates in the synthesis or degradation of complex glycosphingolipids. Sulfatides (Figure 5-49) refer to sulfuric acid esters of galactocerebrosides. Excess accumulation of sulfatides is seen in sulfatide lipidosis (Table 5-1). **Globosides** are cerebrosides that contain more than one carbohydrate, such as galactose, glucose, or N-acetylgalactose. Globotriaosylceramide has one glucose and two galactose residues (it accumulates in the kidneys of patients with **Fabry's disease**; see Table 5-1). Gangliosides (Figure 5-50) resemble globosides except for an additional content of N-acetylneuraminic acid shown in Figure 5-51.

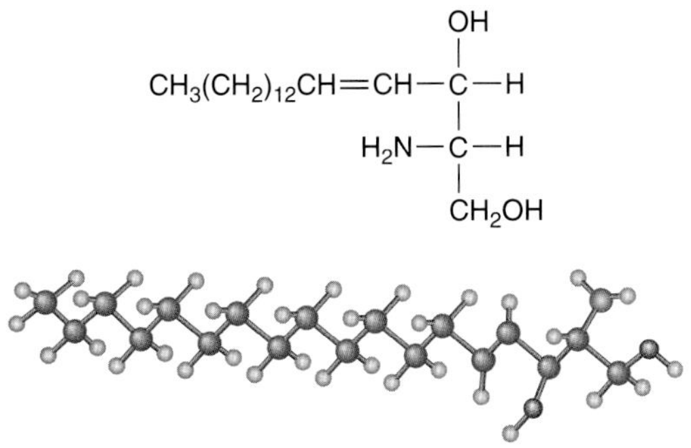

Figure 5-47. The structure of sphingosine.

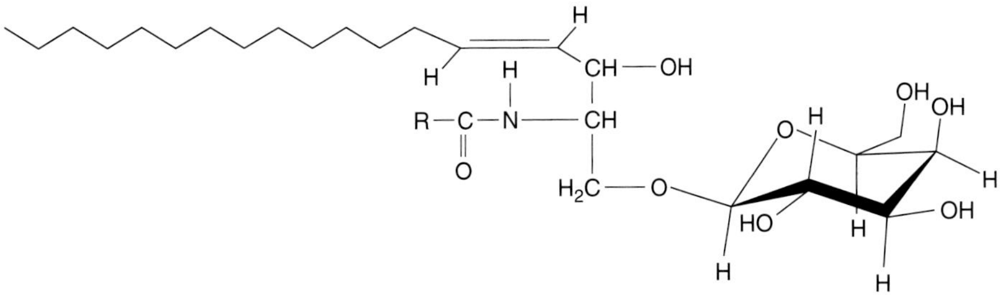

Figure 5-48. Structure of a glucocerebroside.

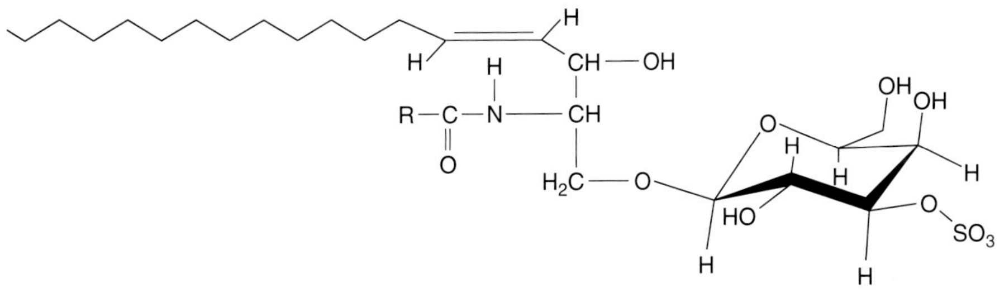

Figure 5-49. Structure of a sulfatide.

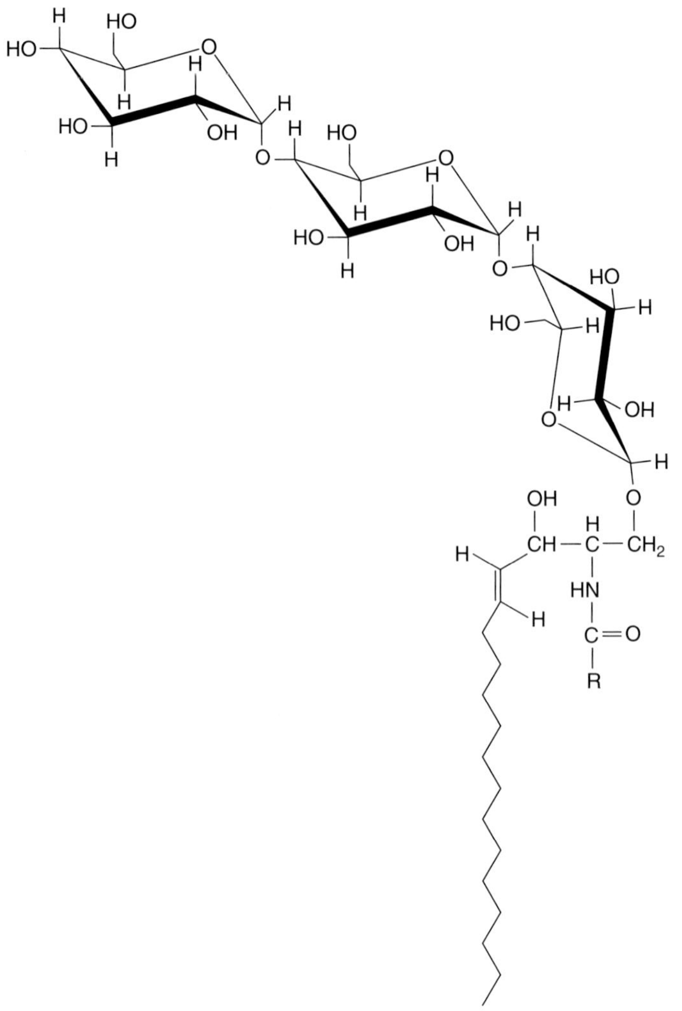

Figure 5-50. Structure of a ganglioside.

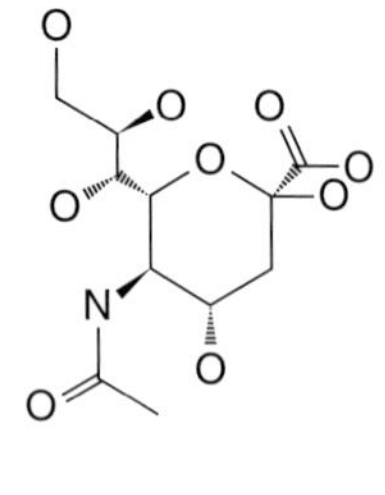

N-acetylneuraminic acid

Figure 5-51. Structure of N-acetylneuraminic acid (NANA).

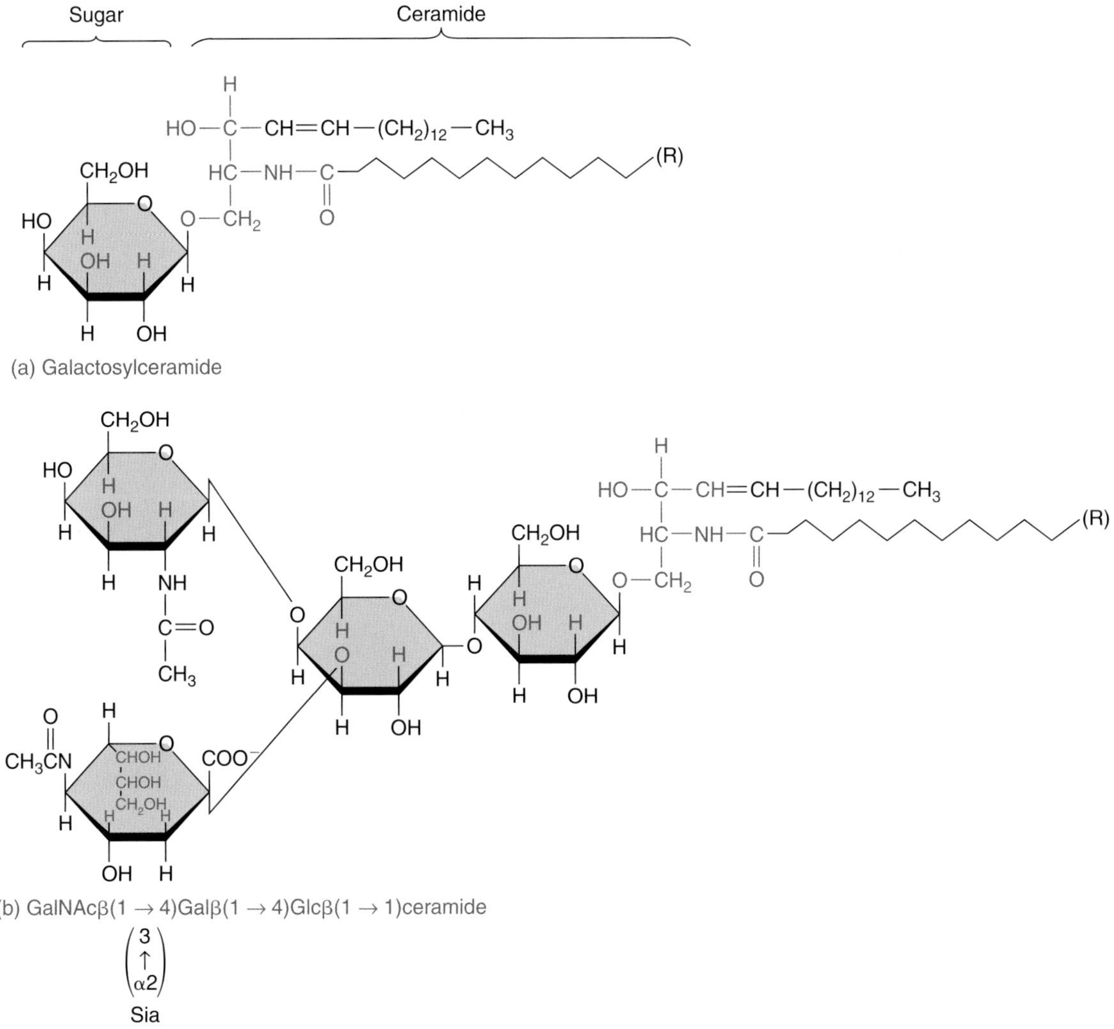

Figure 5-52. Structures of single and complex glycosphingolipids.

Examples of the structures of simple and more complex glycosphingolipids are shown in Figure 5-52. Outlines of the biosynthesis of glycosphingolipids are shown in Figures 5-53A and 5-53B. In Figure 5-54, the degradation of glycosphingolipids is shown. Excess accumulation of glucocerebrosides is a characteristic of **Gaucher's disease** (Table 5-1).

The biosynthesis of gangliosides is shown in Figure 5-55. G refers to ganglioside, and it is often modified by letters or subscripts: M refers to monosialic acid, D to disialic acid, T to trisialic acid, and Q to tetrasialic acid. The structure of **sialic acid** is shown in Figure 5-56. Numbers also may be in a subscript: 1 = Gal-GalNAc-Gal-Glc-ceramide, 2 = GalNAc-Gal-Glc-ceramide, and 3 = Gal-Glc-ceramide (see, for example, Figure 5-53b). Lipid storage diseases are often caused by deficiencies in lysosomal enzymes responsible for the degradation of the carbohydrate portions of gangliosides (Table 5-1).

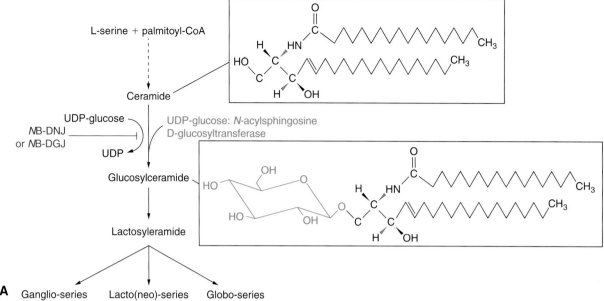

A Biosynthesis of glycosphingolipids (GSLs) in humans

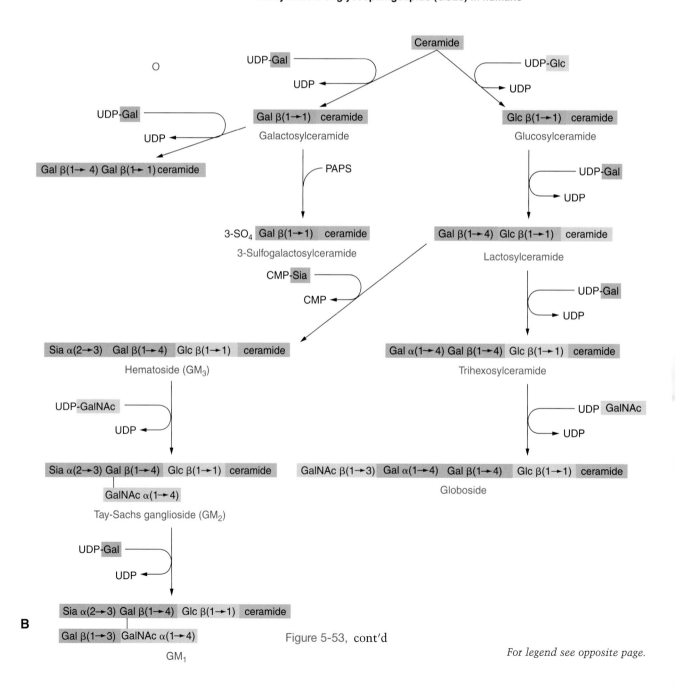

B Figure 5-53, cont'd *For legend see opposite page.*

Figure 5-53 **A.** Biosynthesis of glycosphingolipids in humans. **B.** Biosynthesis of glycosphingolipids. *NB-DNJ*, N-butyldeoxynojirimycin (imino sugar inhibitor); *NB-DGJ*, N-butyldeoxygalactonojirimycin (imino sugar inhibitor).

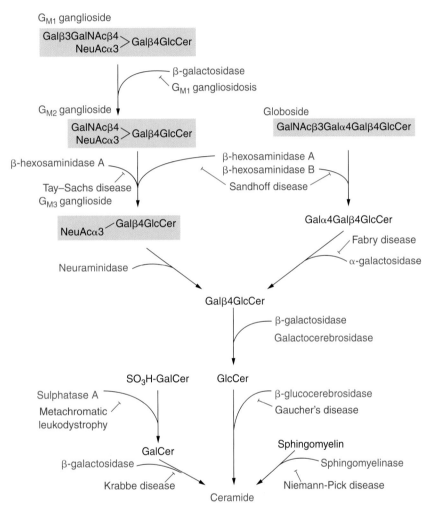

Catabolism of glycosphingolipids in humans: enzymes involved and enzyme-related diseases

Figure 5-54. Catabolism of glycosphingolipids.

Glycosphingolipids

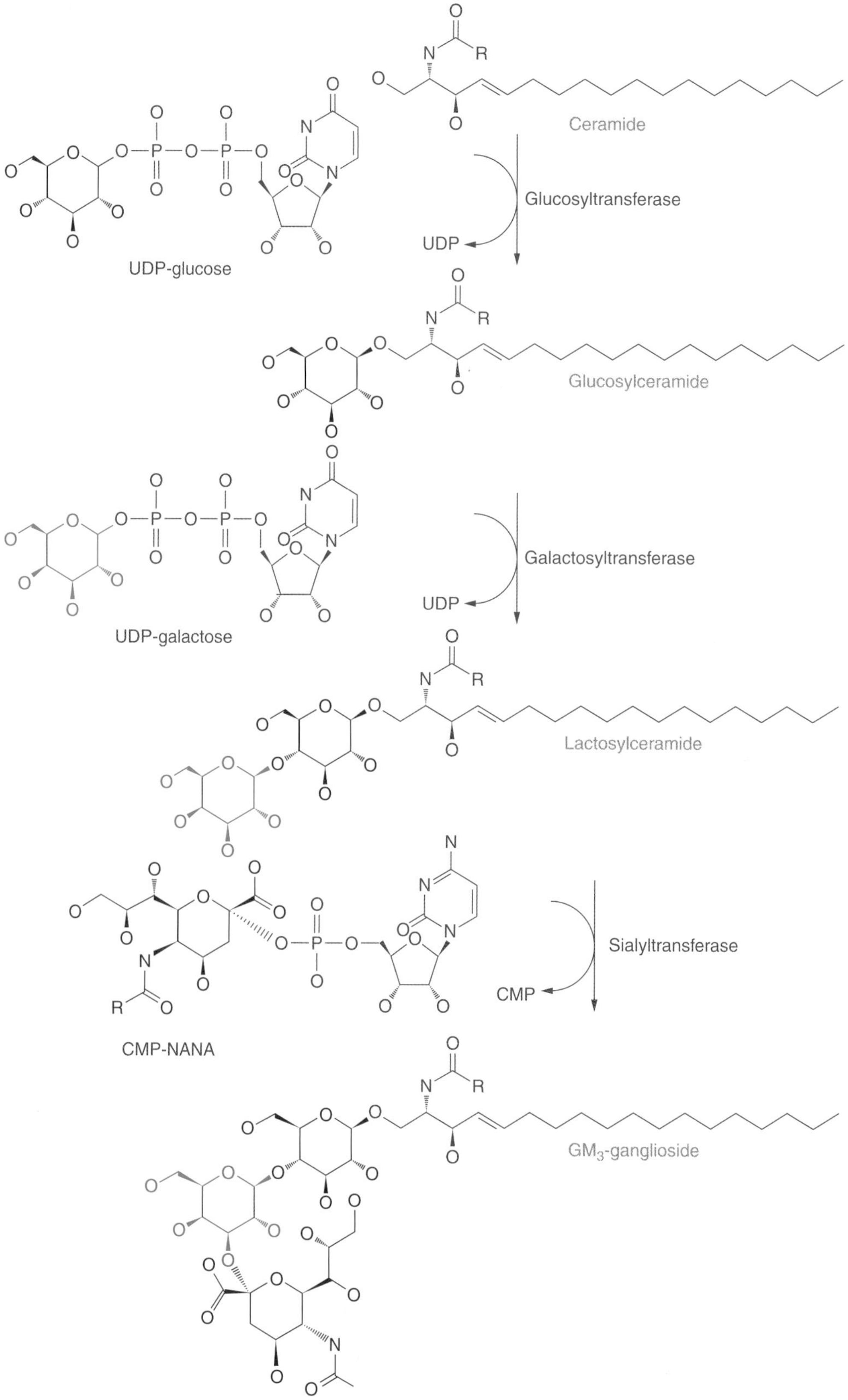

Figure 5-55. Biosynthesis of gangliosides.

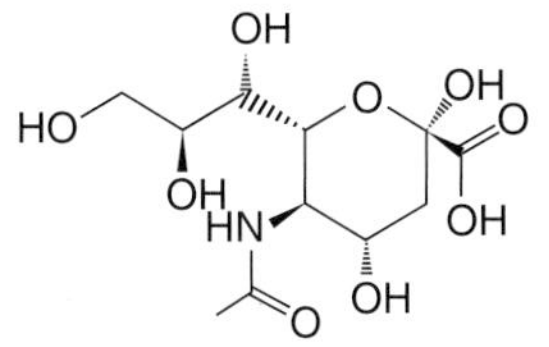

Figure 5-56. Structure of sialic acid.

Lipoproteins

Although fatty acids are transported in the blood by constituent blood proteins, lipids (cholesterol and phospholipids) are transported in lipoprotein particles. Because of the insolubility of lipids, the phospholipids in lipoprotein particles provide the interface; one end is polar and can interact with water. The nonpolar portion of the lipid faces inward in the lipoprotein particle (Figure 5-57). The nonpolar, interior lipids include triglyceride and cholesteryl esters (fatty acid esterified through the 3-hydroxyl position of cholesterol). The classes of lipoproteins are chylomicrons, VLDLs, intermediate-density lipoproteins, LDLs, and HDLs, as shown in Figure 5-58. HDLs are used for the transport of lipid through the circulatory system to the liver, and high levels of HDLs forecast a low risk of heart attack. In contrast, high levels of LDLs (which transport cholesterol to the nonhepatic tissues) are considered cautionary. D stands for "density" in these acronyms because fat has a density of about 0.88 g/ml, whereas protein has a density of about 1.0 g/ml. In consequence, the higher the fat content, the lower the density of the particle. This is shown in Figure 5-59. The functions of these particles are shown in Figure 5-60. High levels of LDL particles in the blood have been associated with susceptibility to atherosclerosis and heart attack. Although the mechanism of initiation of this process is not clear, the formation of fatty plaques in the arteries is a function of **foam cells.** These cells originate from macrophages that have a **scavenger receptor** on their cell surface. The scavenger receptor takes up LDL particles,

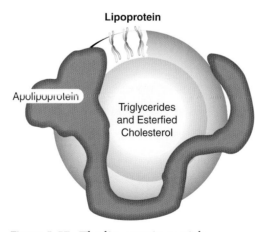

Figure 5-57. The lipoprotein particle.

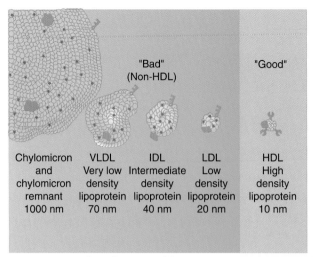

Figure 5-58. Classification of lipoproteins. Redrawn with permission from http://www.medscape.com/pi/editorial/clinupdates/2001/608/art-cu02.fig02.gif.

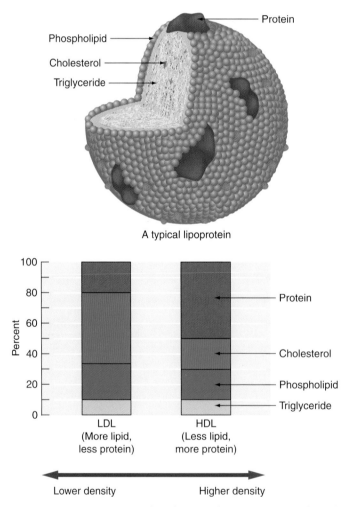

Figure 5-59. Component distribution that gives rise to low density and high density lipoproteins.

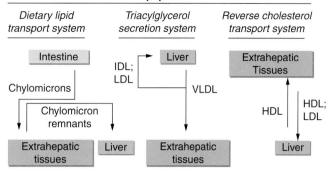

Figure 5-60. Functions of lipoprotein particles. *HDL*, high density lipoprotein; *IDL*, intermediate density lipoprotein; *LDL*, low density lipoprotein; *VLDL*, very low density lipoprotein.

LIPOPROTEINS: LIPID CARRIERS

Type	Formation	Lipid Content	Function
Chylomicrons	Formed from fats in food processed by the intestine	Mostly triglycerides	To transport digested fat (as triglycerides) to muscle and fat cells
Very low density lipoprotein	Formed in the liver	More than $1/2$ triglycerides About $1/4$ cholesterol	To transport triglycerides from the liver to fat cells
Low density lipoprotein	Formed from VLDL after it delivers triglycerides to fat cells	More than $1/2$ cholesterol Less than $1/10$ triglycerides	To transport cholesterol to various cells
High density lipoprotein	Formed in the liver and small intestine	About $1/4$ cholesterol About $1/20$ triglycerides	To remove cholesterol from tissues in the body and transport it to the liver

facilitating the transformation of the macrophage to a foam cell, which becomes rich in cholesterol. Much of the clinical lore connecting cholesterol levels and circulating LDLs remains something of a mystery that will require more research to be understood completely. Nevertheless, the disorders involving lipoproteins circulating in the blood can be described as shown in Table 5-2.

Table 5-2
Susceptibility to Coronary Artery Disease (CAD) Based on Disorders Involving Circulating Lipoproteins

Disorder	Biochemical Disorder	Susceptibility to CAD
Familial combined hyperlipoproteinemia	↑VLDL and LDL particle number, ↑apo B	Increased
Familial hypertrig, with low HDL	↑Triglycerides, ↓HDL	Increased
Familial hypoalphalipoproteinemia	↓HDL	Probably not increased
Familial dyslipidemic hypertension	↑Triglycerides, ↓HDL, High blood pressure	Increased
Atherogenic lipoprotein profile	↑VLDL TG; ↓HDL, small-dense LDL	Increased
Familial Hyperchylomicronemia*	↑chylomicrons	Not increased or slightly increased
Familial Dysbetalipoproteinemia†	↑β-VLDL	Increased

* Caused by lipoprotein lipase deficiency or deficiency of apo CII
† Associated with the apo E 2/2 genotype
CAD, coronary artery disease.
Reproduced with permission from http://www.acclakelouise.com/acc99/htm/gentbl3.jpg.

Lipid Anchoring of Proteins to Membranes

The attachment of nonpolar lipids to proteins is a means to anchor proteins to the cell membrane. A specific amino acid residue in a protein, to be anchored, is either esterified or amidated to a lipid chain, such as myristate or palmitate as shown in Figure 5-61. In this form, the attached lipid can fuse with the membrane lipids, anchoring the protein in a lipid medium. The lipid moiety can be a fatty acid, like the myristate or palmitate shown in Figure 5-61, or it can be an isoprene residue or prenyl residue, such as farnesyl or geranylgeranyl (see Figure 5-5); these are intermediates in cholesterol synthesis. These alterations come under the category of posttranslational modifications of proteins. Other posttranslational modifications include: acetylation, glycosylation, methylation, phosphorylation, hydroxylation, nucleotidylation, and ADP ribosylation. In the case of proteins destined to attach to the membrane, posttranslational modification by myristoylation occurs through the N-terminal of a glycine residue, S-palmitoylation of a cysteine residue, and S-prenylation of cysteine residues near the C-terminus with farnesyl or geranylgeranyl residues. Prenylated proteins form about 2% of the cellular proteins. Prenylated proteins include *ras* proteins or are involved in intracellular transport and ensure the membrane localization required for biological activity. Anchoring of the *ras* protein to the membrane is pictured in Figure 5-62. Membrane insertion is reversible as long as the protein is farnesylated, but when the protein is palmitoylated, the membrane insertion is thought to be irreversible. **Farnesyl transferase** is the enzyme that catalyzes the interaction between the *ras* protein and farnesyl by the mechanism shown in Figure 5-63. This enzyme is a cancer drug target because *ras* is an **oncogene** that plays a role in the generation of certain cancers, so inhibition of the enzyme could prevent an active *ras* and thus interfere with the carcinogenic process. Farnesyl transferase is a barrel-like structure with a central hydrophobic pocket for

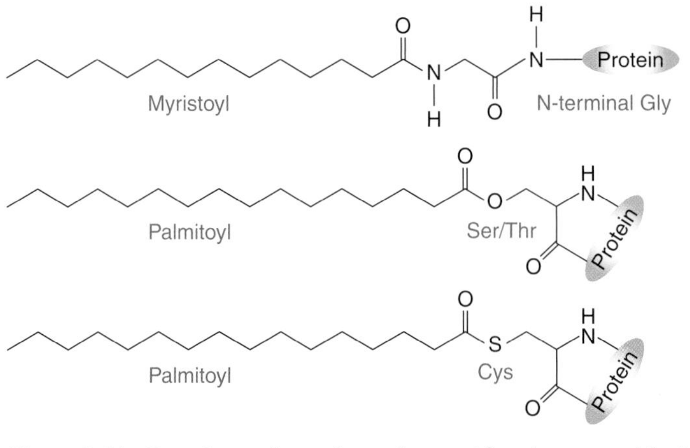

Figure 5-61. Protein anchors through esterification or amidation with fatty acids, such as palmitate or myristate.

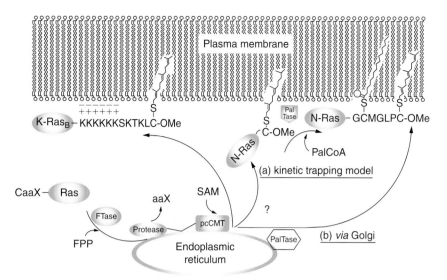

Figure 5-62. Movements of the *ras* protein showing its anchoring to the cell membrane via lipidation through a cysteine residue. Redrawn with permission from the Internet article "Farnesylation of Proteins and Peptides" (v3ss2004 2.pdf): http://www.mpi-dortmund.mpg.de/deutsch/abteilungen/abt4/praktikum/v3ss2004.pdf.

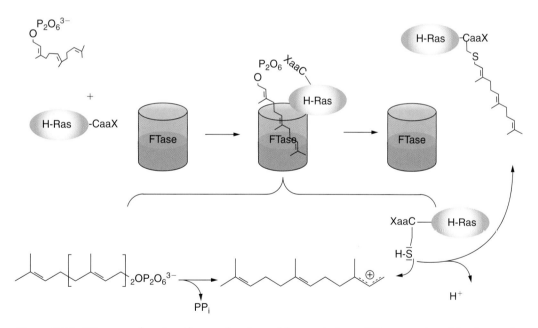

Figure 5-63. Diagram showing the mechanism of farnesylation of H-*ras* catalyzed by farnesyl transferase. Redrawn with permission from http://www.mpi-dortmund.mpg.de/deutsch/abteilungen/abt4/praktikum/v3ss2004.pdf.

Lipid Anchoring of Proteins to Membranes   237

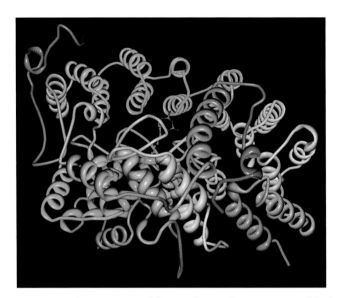

Figure 5-64. Structure of farnesyl transferase responsible for the farnesylation of *ras* protein. PDB ID: 1ft2. S.B. Long, P.J. Casey, L.S. Beese. Cocrystal Structure of Protein Farnesyltransferase Complexed with a Farnesyl Diphosphate Substrate. *Biochemistry* **37** pp. 9612 (1998).

binding farnesyl pyrophosphate (Figure 5-64). The **CaaX motif** of *ras* directly interacts with farnesyl in the active site of the enzyme (Figure 5-63). The cysteine residue of the CaaX motif coordinates a zinc ion in the active site of the enzyme required for the reaction to occur. The glycophosphatidylinositol anchoring is another mechanism for attaching proteins to a membrane that was considered in Chapter 4 (Figure 4-69).

Further Reading

Gurr, M.I., Frayn, K.N., and Harwood, J., *Lipid Biochemistry*, Blackwell Publishing, 2002.

Vance, D.E., and Vance, J.E., *Biochemistry of Lipids, Lipoproteins and Membranes*, Elsevier Science, 2002.

CHAPTER 6

Nucleic Acids and Molecular Genetics

Huntington's Disease: A Trinucleotide Repeat Mutation

There are a great variety of genetic diseases. Many involve a **point mutation** in which a single nucleotide is altered to another so that the resultant protein, after transcription and translation, functions poorly or not at all. Some examples of enzymes that have been mutated, giving rise to disease, have been mentioned in previous chapters. There are other forms of alterations in the genetic material that lead to disease, and some of these will be discussed here. In the example chosen, **Huntington's disease,** first described by George Huntington in 1872, there is a specific mutant gene that causes the disease. The disease is rare, numbering 1 person in 10,000 and causing death in about 1.6 per 1 million. The disease can occur in the second decile, when suicides often happen. The gene locus is on chromosome 4 at p16.3 (Figure 6-1), and the full sequence of the gene is known. The mutant DNA sequence contains an expanded trinucleotide repeat of cytosine, adenine, and guanine (CAG in the mRNA) within the coding region of the gene. The normal gene can have 10 to 26 CAG repeats in the mRNA without causing any abnormality; however, the range of the number of repeats generating the possibility of Huntington's disease varies between 36 and 121 CAG repeats. The intermediate range of 27 to 35 forms a risk of having a child with Huntington's disease. When the repeats number more than 60, juvenile onset is likely, whereas those with 40 to 55 repeats generate adult onset. A study showed that 75 families, all with Huntington's disease, contained extensions of the CAG (CAG in mRNA codes for glutamine; see Figure 2-34). The CAG region is polymorphic (having many forms) because there are many alleles that differ in the number of CAG repeats. The triplet repeat sequence is classified as a short tandem repeat polymorphism, or STRP, because the CAG sequence is repeated many times in tandem. The Huntington's disease gene product in mice is essential for embryonic development, and it has a role in the normal functioning of the basal ganglia. **Huntingtin,** the Huntington's disease gene protein product, is expressed in all types of neurons. The unique protein encoded by the normal Huntington's disease gene (and translated using mRNA) has a molecular weight of about 330 kDa.

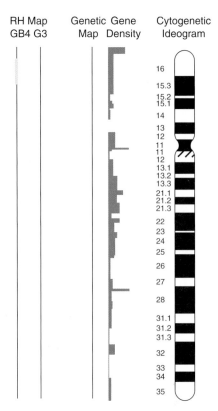

Figure 6-1. The human chromosome 4.

The genetic transmission is autosomal dominant. Each affected person has an affected parent with no generation skipping. Both males and females are affected. About half of the offspring of an affected parent are also affected, but normal siblings of an affected brother or sister have normal children. A typical pedigree showing autosomal dominant transmission (in this case, for hypercholesteremia through three generations) is shown in Figure 6-2A. Figure 6-2B shows a pedigree for Huntington's disease in Venezuela and one in the United States.

There are eight other central nervous system diseases in which proteins are expressed, other than huntingtin, that contain expansions of polyglutamine. Some of these are fragile X syndrome, spinobulbar muscular atrophy, and myotonic dystrophy, in addition to Huntington's (Figure 6-3). Experimentally, it has been shown that a **transgene** (gene placed experimentally in the germline that functions as a normal gene) in mice containing expanded polyglutamine tracts causes the central nervous system disorder; the normal protein with an acceptable number of glutamines does not cause disease. Huntington's disease is a progressive, neurodegenerative disorder. Symptoms are quick, jerky, and purposeless movements called chorea (thus the term **Huntington's chorea** to describe these symptoms). Chorea is the Greek word for *dance.*

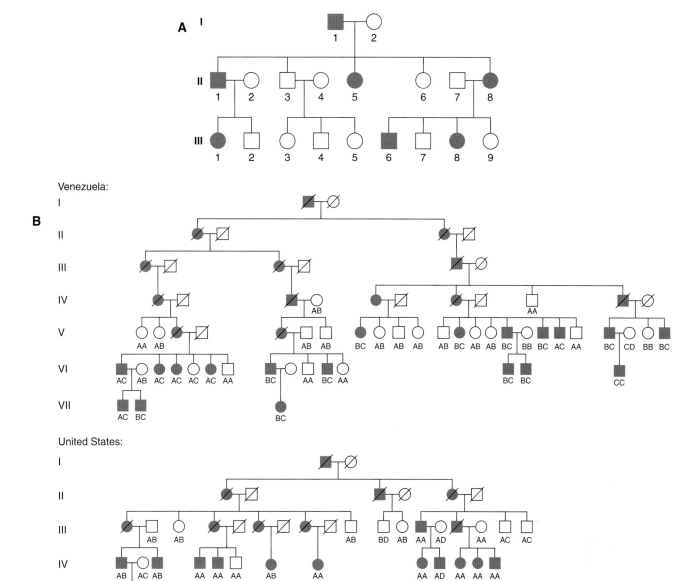

Figure 6-2. **A.** Typical pedigree for an autosomal dominant disease, in this case, for hypercholesterolemia. *Squares*, males; *circles*, females. Filled in characters indicate presence of disease. Approximately half the offspring of an affected parent are ill. Children of a diseased parent will have the disease, because there is no skipping of generations. **B.** Pedigree of Huntington's disease in Venezuela and in the United States. There are 7 generations in the Venezuela pedigree and 5 generations represented in the U. S. pedigree. There is no generation skipping and children of an affected parent have the disease with about 50% incidence. A slash through a square or circle indicates a dead person at the time the pedigrees were constructed.

Huntington's Disease: A Trinucleotide Repeat Mutation 241

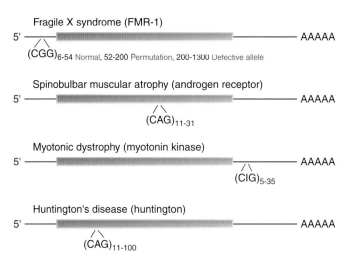

Figure 6-3. Examples of 4 central nervous system diseases that have extended glutamine (CGG or CAG) repeats in the messenger RNAs produced from the abnormal genes. These include fragile X syndrome, spinobulbar muscular atrophy, myotonic dystrophy, and Huntington's disease.

Huntington's disease is a progressive, neurodegenerative disease, which is autosomal non–sex-linked dominant, and it displays a uniform decline of cognitive functions with involuntary movements. Likely, the alteration in the gene to the disease form occurs during sperm development. It could be a familial prion disease. A loss of neurons takes place in the neostriatum and the cortex of the brain, with onset occurring between 35 and 44 years of age. The age of onset could depend on the state of methylation of the Huntington's disease gene locus. The disease can be detected using a blood sample by analyzing the CAG repeat region in the DNA. As a result of neuronal loss, there is decreased metabolic activity (particularly a decrease in the uptake and metabolism of glucose) in certain regions of the brain, characteristic of this disease, as shown in Figure 6-4. Huntington's disease patients usually survive 12 to 18 years before death. As pointed out in Figure 6-4, hypometabolism is followed by cellular degeneration and cell death (programmed cell death or apoptosis) in the thalamus, brain stem, and spinal cord, decreasing the weight of the entire brain by 20% to 30%. The first 10 years of the disease are marked by spasmodic movements of the muscles, which increase with stress. Visual problems also occur. There are substantial changes in the brain, as would be expected. Huntington's disease individuals lose social inhibitions, become aggressive, and later are confined to bed as their disabilities increase, eventually requiring full-time care. The final stages are marked by depression, dementia, and schizophrenia.

It has been found that the **mutant huntingtin protein (Htt)** inhibits normal histone acetylation by blocking histone acetylases or recruiting the histones into aggregates (potentially another aggregation disease). The reduction of histone acetylation by Htt inhibition of the histone acetylases decreases transcription. These effects are diagrammed in Figure 6-5. Htt also is thought to bind abnormally tightly to at least one

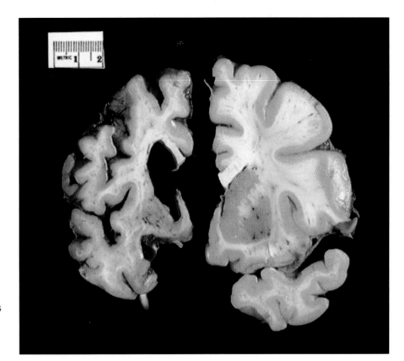

Figure 6-4. Lowered metabolic activity in certain brain regions characteristic of Huntington's disease. Reproduced with permission from http://www.crump.ucla.edu/software/lpp/clinicalcases/neurocases/neuro1/neurdx1.html.

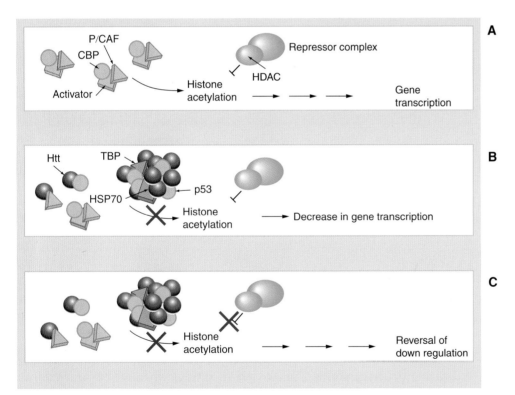

Figure 6-5. Huntington's disease and gene expression. **A.** Gene transcription is switched on by complexes that contain activator proteins and acetyltransferases (such as CBP and P/CAF); acetyltransferases add acetyl groups to the histone proteins that package DNA into a compact form. Transcription is regulated by an interplay between histone acetyltransferases and histone deacetylases (HDACs). **B.** Steffan et al. have found that mutant huntingtin protein (Htt) inhibits histone acetylation by blocking histone acetylases or recruiting them into aggregates. The reduction in histone acetylation leads to decreased transcription. **C.** The balance can be partially redressed by using molecules that prevent histone deacetylation. Redrawn with permission from http://www.nature.com/nature/journal/v413/n6857/fig_tab/413691a0_F1.html.

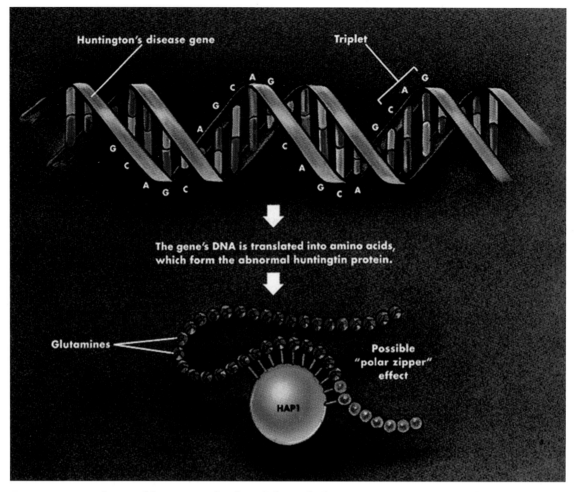

Figure 6-6. Htt, abnormal huntingtin, binds tightly to the brain protein, HAP1. Reproduced with permission from http://web.sfn.org/images/brainbriefings/huntingtons_illus_large.gif.

specific brain protein, **HAP1,** which could contribute to the aggregation phenomenon associated with this disease (Figure 6-6).

Purines and Pyrimidines

There are two common purines, which are adenine and guanine. The purine structure consists of two adjoined rings with five carbon and four nitrogen atoms. The common pyrimidines are cytosine, uracil, and thymine. Pyrimidines have a single ring made up of four carbon and two nitrogen atoms. These structures are shown in Figure 6-7. The numbering of the ring constituents is shown in Figure 6-8 and includes the attachment of deoxyglucose that occurs in DNA (ribose attaches in RNA). Adenine and guanine are present in both DNA and RNA, whereas cytosine and thymine are present in DNA but cytosine and uracil (replacing thymine) are present in RNA. Thymine is exclusive to DNA; it does also

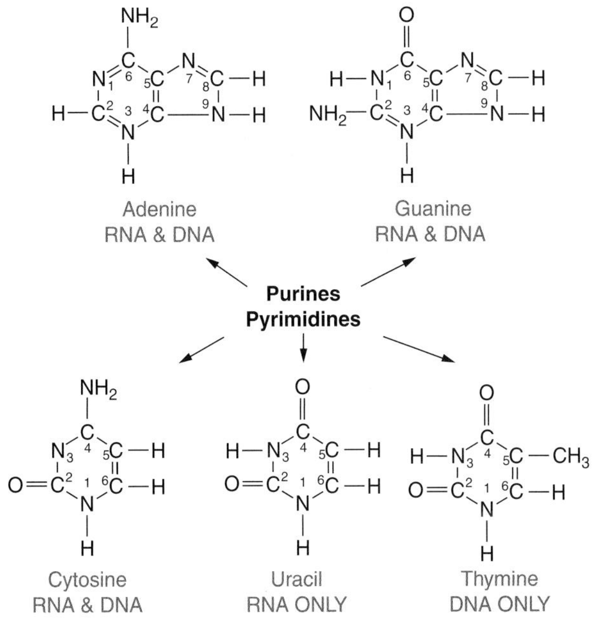

Figure 6-7. Structures of purines and pyrimidines.

occur in tRNA but not in rRNA or in mRNAs. Atomic models of adenine, guanine, thymine, and cytosine are shown in Figure 6-9. Uracil replaces thymine in RNA because, if it occurred in DNA, it would be recognized by the **base excision repair mechanism.** Deamination of cytosine (a mutation) produces uracil, which could be recognized by the DNA repair mechanism because it is not part of DNA. This could be the reason for having thymine in DNA and not having uracil in DNA. Deamination of 5-methylcytosine produces thymine, as shown in Figure 6-10. The sugar moiety in RNA is ribose, and the sugar moiety in DNA is 2'-deoxyribose (Figure 6-11A). The combination of a base, such as adenine, with the pentose sugar produces a nucleoside. When the nucleoside is phosphorylated, it is a nucleotide. With a single phosphate group, it is a mononucleotide (XMP; X standing for the base, which would be G, A, C, T, or U); with two phosphates it is a dinucleotide (XDP), and with three phosphates it is a trinucleotide (XTP) (Figure 6-12). Figure 6-13 shows the structures of the common bases with their name and abbreviation, as well as the naming, with the naming and abbreviation for the respective nucleoside and the respective mononucleotide. In DNA or RNA the bases are present as mononucleotides (Figure 6-11B). There are other pyrimidines and purines in addition to those that occur in nucleic acids, such as xanthine and hypoxanthine, and the pyrimidine **orotic acid,** which occur as metabolites (Figure 6-14). In nucleosides and further phosphorylated forms, the base can exist in two orientations about the N-glycosidic bond. These orientations are identified as either *syn* or *anti* (the predominating form in nucleic acids), as shown in Figure 6-15. Nucleotides occur more often than nucleosides in cells. Phosphorylation of the pentose sugar is at the 5' carbon; the prime is used to distinguish the position on the sugar compared to the position on the purine or

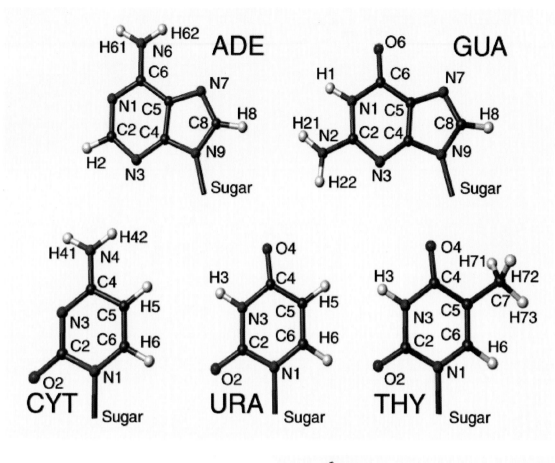

Figure 6-8. Numbering of the atoms of the rings of purines and pyrimidines *(above)*. Attachment of deoxyribose (occurring in DNA) at the bottom. In case of deoxyribose R equals H2″; in case of ribose R equals O2′-H02′. The dashed line shows the nucleotide boundary. The methyl group in thymine is numbered by assuming higher priority for C4 than for C6. So H71 is the hydrogen with the smallest (positive or negative) angle with respect to C4. C4 is chosen as having the higher priority in CIP rules. C6 is closer to the nucleic acid "backbone," and if protein rules assigning priority to backbone and side chain atoms were applied a different outcome would be the result. Reproduced with permission from http://www.bmrb.wisc.edu/referenc/nomenclature/.

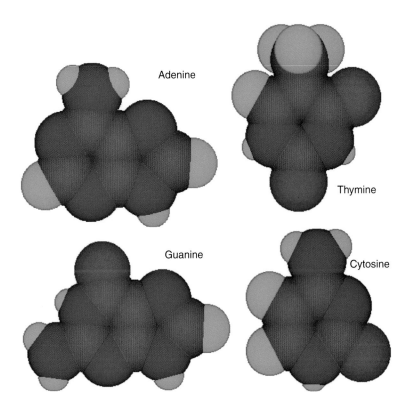

Figure 6-9. Models of the purines and pyrimidines constituting DNA. Reproduced with permission from http://www.historyoftheuniverse.com/nabases.html.

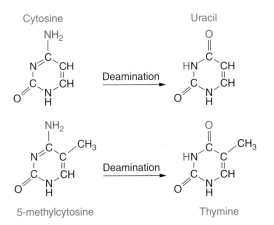

Figure 6-10. Removal of an amino group by deamination. Since uracil is not part of DNA, it can be detected and repaired by a DNA repair mechanism if cytosine is accidentally deaminated.

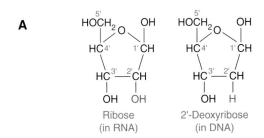

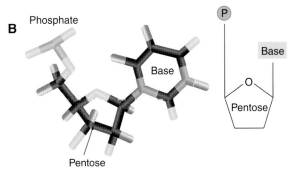

Figure 6-11 **A.** Structures of ribose and 2'-deoxyribose, constituents of RNA and DNA. **B.** Structure of a pyridine mononucleotide.

Purines and Pyrimidines

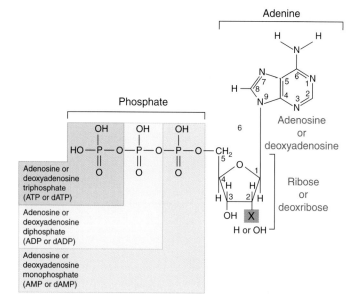

Figure 6-12. Structures of mono-, di- and tri-nucleotides of adenine. Redrawn with permission from http://ntri.tamuk.edu/cell/nucleic.html.

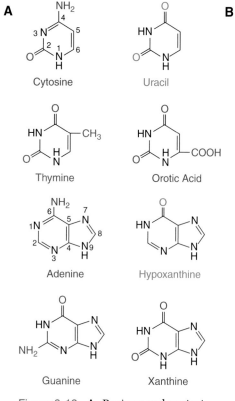

Figure 6-13. **A.** Purines and pyrimidines in addition to the common bases in DNA and RNA. **B.** Structures, names, and abbreviations given for the common bases, nucleosides, and mononucleotides.

Base Formula	Base (X=H)	Nucleoside X=ribose or deoxyribose	Nucleotide X=ribose phosphate
	Cytosine, C	Cytidine, C	Cytidine monophosphate, CMP
	Uracil, U	Uridine, U	Uridine monophosphate, UMP
	Thymine, T	Thymidine, T	Thymidine monophosphate, TMP
	Adenine, A	Adenosine, A	Adenosine monophosphate, AMP
	Guanine, G	Guanosine, G	Guanosine monophosphate, GMP

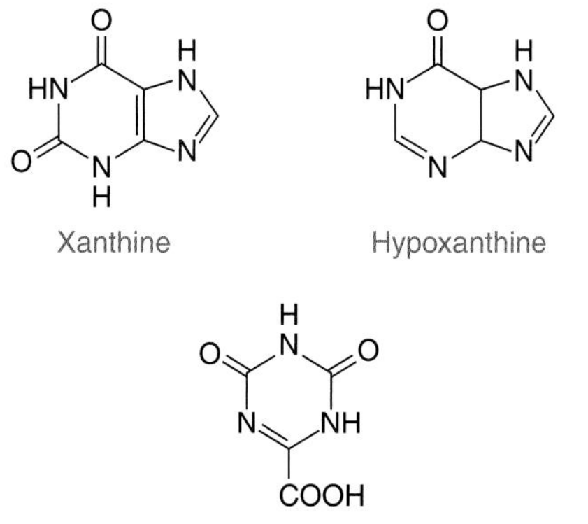

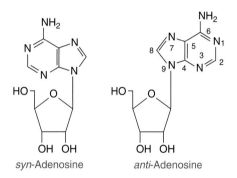

Figure 6-14. Purines and a pyrimidine that do not occur in nucleic acids but occur as metabolites.

Figure 6-15. Conformation of *syn-* or *anti-*adenosine.

pyrimidine base. If a mononucleotide, for example, with adenine, contains ribose, it would be denoted simply as AMP; if it contains deoxyribose, it would be denoted as dAMP. Some modifications of bases can occur; for example, in DNA, 5-methylcytosine is found. Other modified bases are found in tRNAs. Another adenosine derivative is cyclic AMP (Figure 3-16). Guanosine also has a cyclic derivative, cyclic GMP, which sometimes antagonizes cyclic AMP. Whereas adenylate cyclase catalyzes the formation of cyclic AMP, it is guanylate cyclase, as part of a receptor, that catalyzes the formation of cyclic GMP.

Base-Pairing

Natural DNA is double-stranded, and the complementation of the bases in DNA forms part of the force that holds the two strands together. Adenine pairs with thymine (adenine can also form a base pair with uracil), and guanine pairs with cytosine. This is summarized in Figure 6-16. The pairs are held together by hydrogen bonding. Pairing occurs in DNA

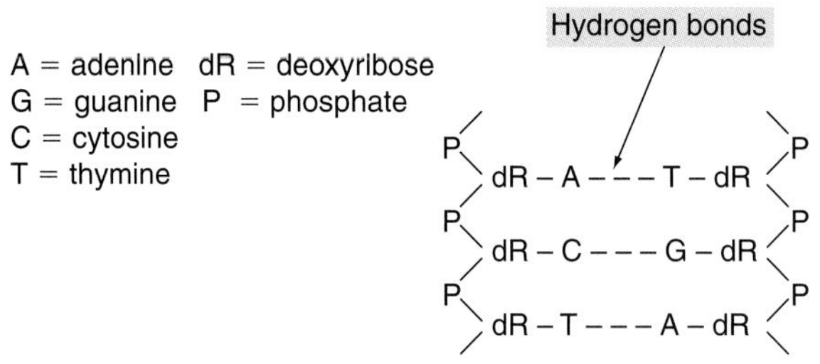

Nucleic Acids

A = adenine   dR = deoxyribose
G = guanine   P = phosphate
C = cytosine
T = thymine

Figure 6-16. Base-pairing in DNA. Adenosine pairs with thymine and guanosine pairs with cytosine.

because two sets of base pairs appear as rungs in a ladder, as shown in Figure 6-17. In double-stranded DNA, the bases are in the interior of the double helix and the pentoses and phosphate groups are on the outside (Figure 6-18). The 3' carbon of one sugar is linked through a phosphodiester bond to the 5' carbon of the next sugar. A single turn of the helix requires 10 base pairs and is 3.4 nm (nanometers; 10^{-9} meters) for each turn. The two DNA strands are antiparallel, forming the double helix with the 5' end at the top in this figure (on the left) and the 3' end at the bottom. Both replication of DNA and transcription of mRNA from the sense strand of DNA occur from the direction of the 5' end to the 3' end. In addition to hydrogen bonding between the internal bases, other forces stabilize the double strand of DNA. There are ionic charges (Van der Waals forces) between the tightly stacked base pairs, and there are hydrophobic attractions between the base pairs in the helix interior as the nonpolar nitrogenous bases are packed tightly enough together to exclude water molecules and form a stable nonpolar interior.

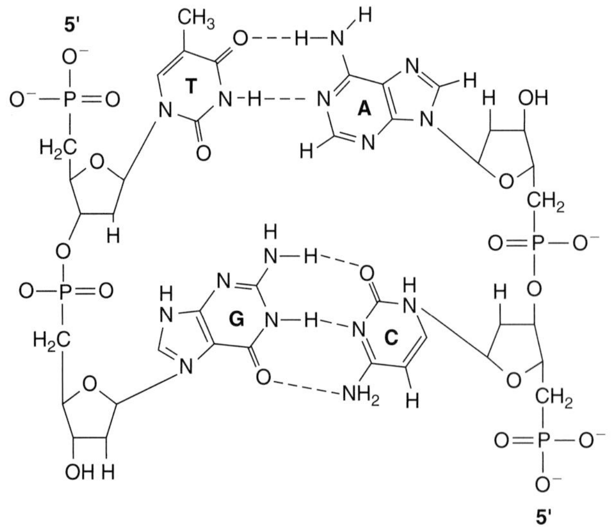

Figure 6-17. Base-pairing as it occurs in DNA. Hydrogen binding holds the bases together and hydrogen bonding is indicated by dashed lines. *T*, Thymine; *A*, adenine; *G*, guanine; and *C*, cytosine.

250   CHAPTER 6   Nucleic Acids and Molecular Genetics

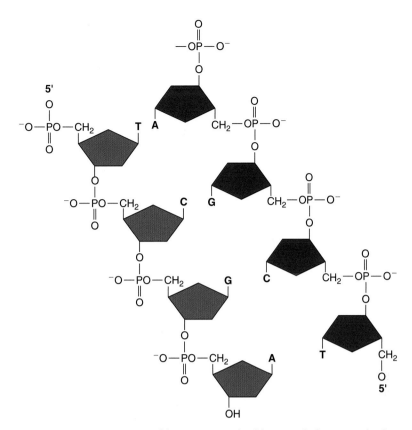

Figure 6-18. Appearance of base pairs is double stranded DNA. The base pairs, bonded by hydrogen bonds are in the interior of the helix while the pentoses and phosphate groups are on the outside.

In the case of RNA, it exists commonly as a single strand, in contrast to DNA, but there are certain secondary structures in RNA forming a double strand. In this case, the base-pairing rule applies; that is, G pairs with C and A pairs with U (Figure 6-19).

DNA double strands can be melted apart in vitro by subjecting them to temperatures in the range of 50° to 60°C. As the temperature increases, hydrogen bonding and other forces holding the strands together are loosened so that the chains come apart. The chains can reassociate when the temperature is returned to normal values (about 37°C). The melting temperature can be an important characteristic of a specific DNA.

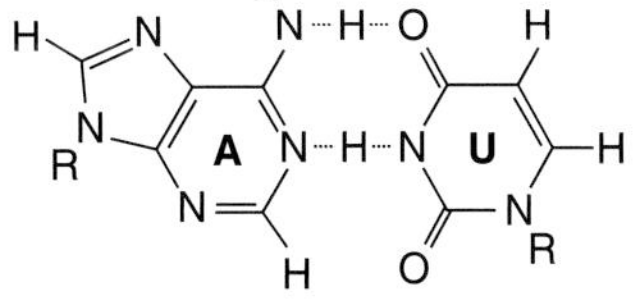

Figure 6-19. Watson-Crick type base-pairing between adenine (A) and uracil (U).

Purines and Pyrimidines

Biosynthesis of Purines and Pyrimidines and Their Catabolism

The biosynthesis of purines occurs in the body, and the structure, in terms of the carbons and nitrogens in the rings, are derived from aspartate, carbon dioxide, glycine, N^{10}-formyl-tetrahydrofolate (N^{10}-formyl-THF), and glutamine, as shown in Figure 6-20. The biosynthetic route starts with ribose-5-phosphate and its conversion to 5-phospho-D-ribosyl-1-pyrophosphate (PRPP), as shown in Figure 6-21. The remaining steps in the synthesis of inosine 5'-monophosphate (IMP), the first fully structured nucleotide, is shown in Figure 6-22. Structures of some of the enzymes in this pathway are shown. Figure 6-23 shows the structure of the enzyme catalyzing the synthesis of PRPP (from ribose phosphate), which is the substrate for purine biosynthesis (Figure 6-22). The structures of the enzymes catalyzing the sequential 10 steps in purine biosynthesis follow in Figures 6-23 through 6-32. The conversion of IMP to the other two purine mononucleotides, AMP and GMP, is shown in Figure 6-33. IMP forms AMP with the intermediate, adenylosuccinate, formed with aspartate, and then fumarate is removed. GMP is formed with the intermediate, xanthosine monophosphate, with the addition of water and then glutamine. A summary of the regulation of purine biosynthesis is shown in Figure 6-34. In essence, the end products inhibit the further formation of ATP and GTP, and the concentration of PRPP is a feed-forward activator, acting much like a substrate rate-limited enzyme reaction.

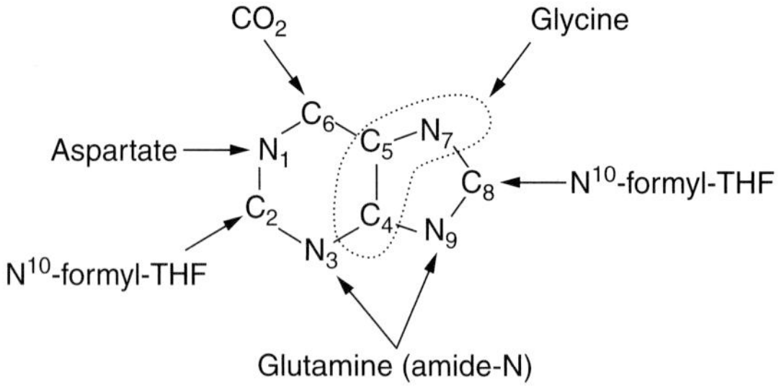

Figure 6-20. Sources of the carbons and nitrogens of the purine ring.

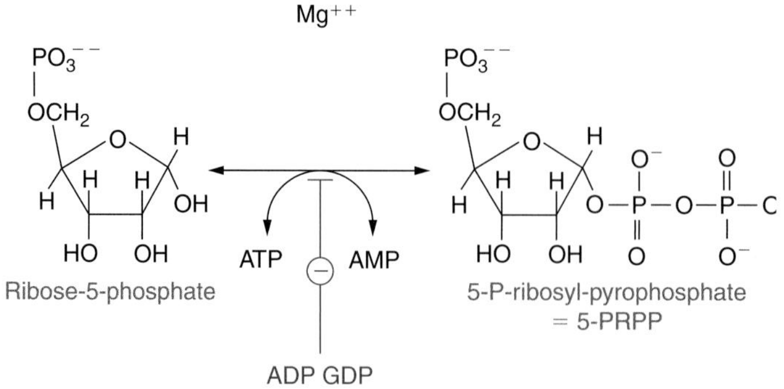

Figure 6-21. Conversion of ribose-5-phosphate to 5-phospho-D-ribosyl-1-pyrophosphate catalyzed by PRPP synthetase.

The carbons and nitrogens of the pyrimidine structure are derived from glutamine amide, aspartate, and bicarbonate (HCO_3^-). This is pictured in Figure 6-35. Pyrimidine biosynthesis starts from **carbamoyl phosphate,** which is formed from glutamine, ATP, and CO_2 (Figure 6-36). The structure of this cytosolic enzyme, carbamoyl phosphate synthetase II (CPSII), is shown in Figure 6-37. This enzyme is distinct from CPSI, which is a component of the urea cycle and forms carbamoyl phosphate from ammonia and bicarbonate in the mitochondrion. CPSII is present in most extrahepatic (outside the liver) tissues. The synthesis of pyrimidine from carbamoyl phosphate to UMP is shown in Figure 6-38. The first step in the use of carbamoyl phosphate is the addition of aspartate with the elimination of inorganic phosphate and a molecule of water and then oxidation, using oxidized quinone to form orotate (orotic acid) and reduced quinone. The first step is catalyzed by aspartate transcarbamylase (ATCase). The structure of an aspartate transcarbamoylase enzyme is shown in Figure 6-39. There are six steps starting with the carbamoyl phosphate synthetase reaction, as shown in Figure 6-40, wherein each

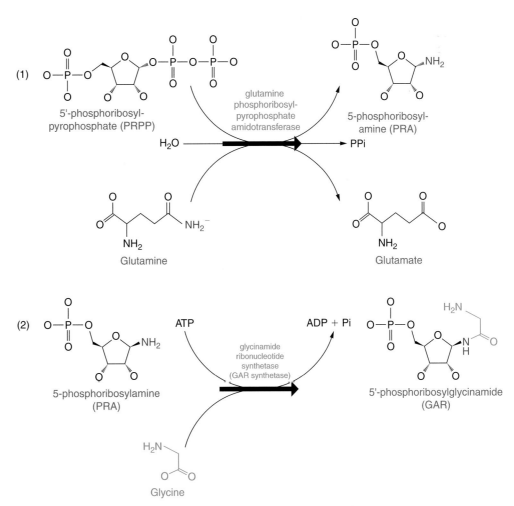

Figure 6-22. Biosynthesis of inosine 5′-monophosphate (IMP) from PRPP in 10 steps. Negative charges in single-bonded oxygens on phosphate groups are omitted.

(continued)

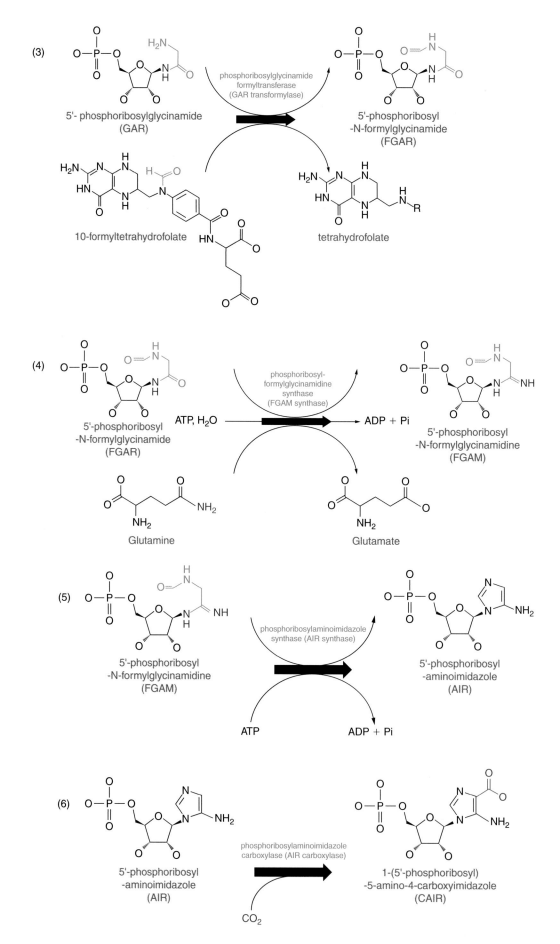

Figure 6-22, cont'd

(7) 1-(5'-phosphoribosyl)
-5-amino-4-carboxyimidazole
(CAIR)

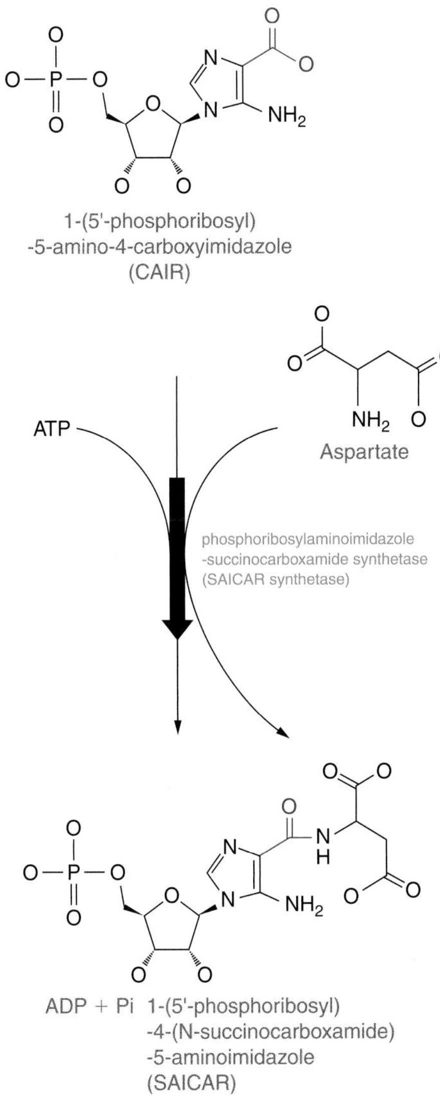

(8) 1-(5'-phosphoribosyl)
-4-(N-succinocarboxamide)
-5-aminoimidazole
(SAICAR)

ATP

Aspartate

phosphoribosylaminoimidazole
-succinocarboxamide synthetase
(SAICAR synthetase)

adenylosuccinate lyase

Fumarate

ADP + Pi 1-(5'-phosphoribosyl)
-4-(N-succinocarboxamide)
-5-aminoimidazole
(SAICAR)

1-(5'-phosphoribosyl)
-5-amino
-4-imidazole carboxamide)
(AICAR)

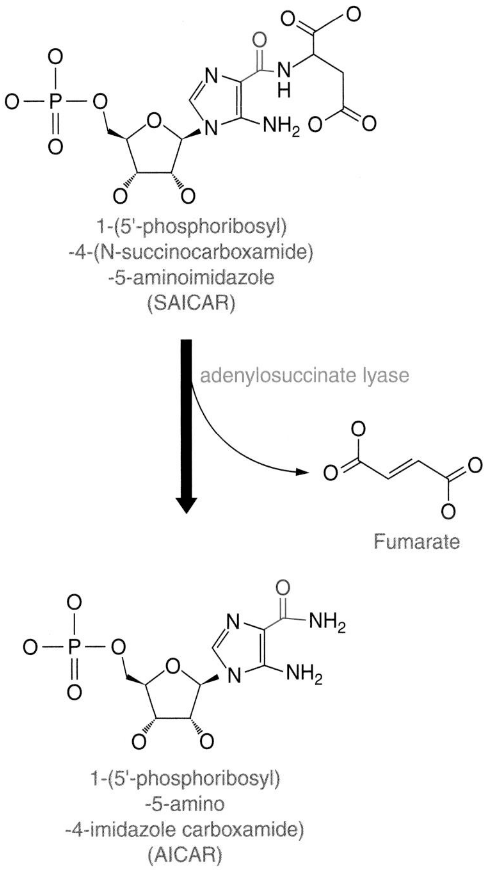

Figure 6-22, cont'd

Purines and Pyrimidines   255

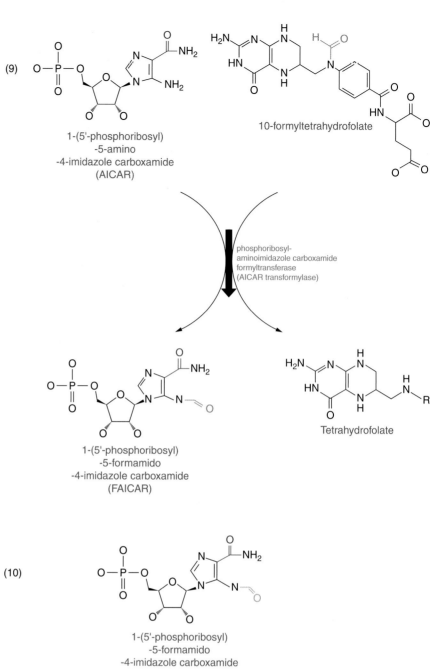

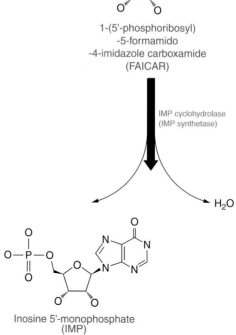

Figure 6-22, cont'd

Figure 6-23. Crystal structure of phosphoribosyl-pyrophosphate synthetase that catalyzes the reaction shown in Figure 6-20. The enzyme was from *Bacillis subtilis*. *Chain A*, blue (298 residues); *Chain B*, red (306 residues). PDB ID: 2c4k. P. Kursula, P. Stenmark, C. Arrowsmith, H. Berglund, A. Edwards, M. Ehn, S. Flodin, S. Graslund, M. Hammarstrom, B.M. Hallberg, L. Holmberg Schiavone, T. Kotenyova, P. Nilsson-Ehle, D. Ogg, C. Persson, J. Sagemark, H. Schuler, M. Sundstrom, A.G. Thorsell, S. Van Den Berg, J. Weigelt, P. Nordlund. Crystal Structure of Human Phosphoribosylpyrophosphate Synthetase-Associated *Protein* 39 (Pap39). *To be published.*

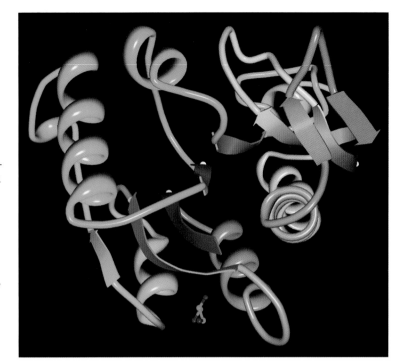

Figure 6-24. Crystal structure of glutamine PRPP amidotransferase from *Basillis subtilis* (expressed in *E. coli*). *Chain A*, yellow; *Chain B*, light blue; *Chain C*, green; *Chain D*, red. This is the first enzyme in the purine biosynthetic pathway. PDB ID: 1AO0. S. Chen, D.R. Tomchick, D. Wolle, P. Hu, J.L. Smith, R.L. Switzer, H. Zalkin. Mechanism of the Synergistic End-product Regulation of *Bacillus subtilis* Glutamine phosphoribosylpyrophosphate Amidotransferase by *Nucleotides, Biochemistry* **36** pp. 10718 (1997).

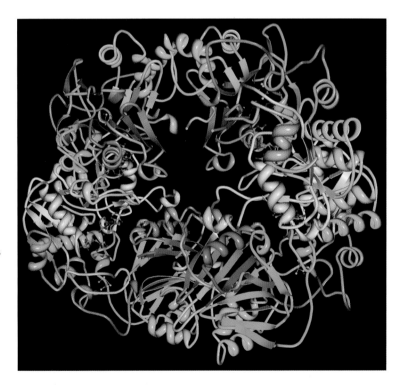

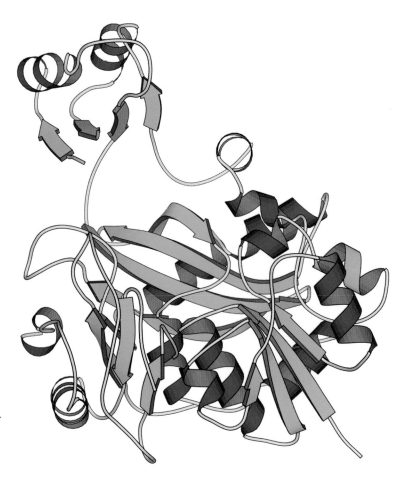

Figure 6-25. Crystal structure of glycinamide ribonucleotide synthetase (GAR synthetase) from *E. coli.* This enzyme catalyzes the second step of purine biosynthesis. Reproduced with permission from http://arginine.chem.cornell.edu/Structures/StructurePictures/1GSO.jpg.

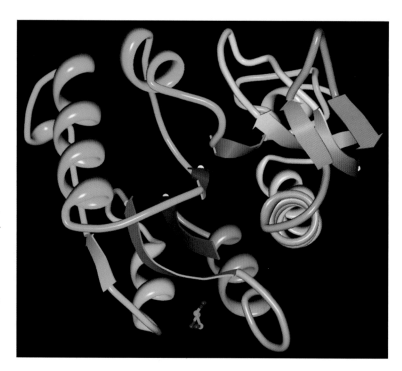

Figure 6-26. The apo structure of human glycinamide ribonucleotide transformlyase (GARtransformlyase). It bears a 40% sequence homology to the human enzyme. This is the third enzyme of the purine biosynthetic pathway. PDB ID: 1zLx. T.E. Dahms, G. Sainz, E.L. Giroux, C.A. Caperelli, J.L. Smith. The apo and Ternary Complex Structures of a Chemotherapeutic Target: Human Glycinamide Ribonucleotide Transformylase. *Biochemistry* **44** pp. 9841 (2005).

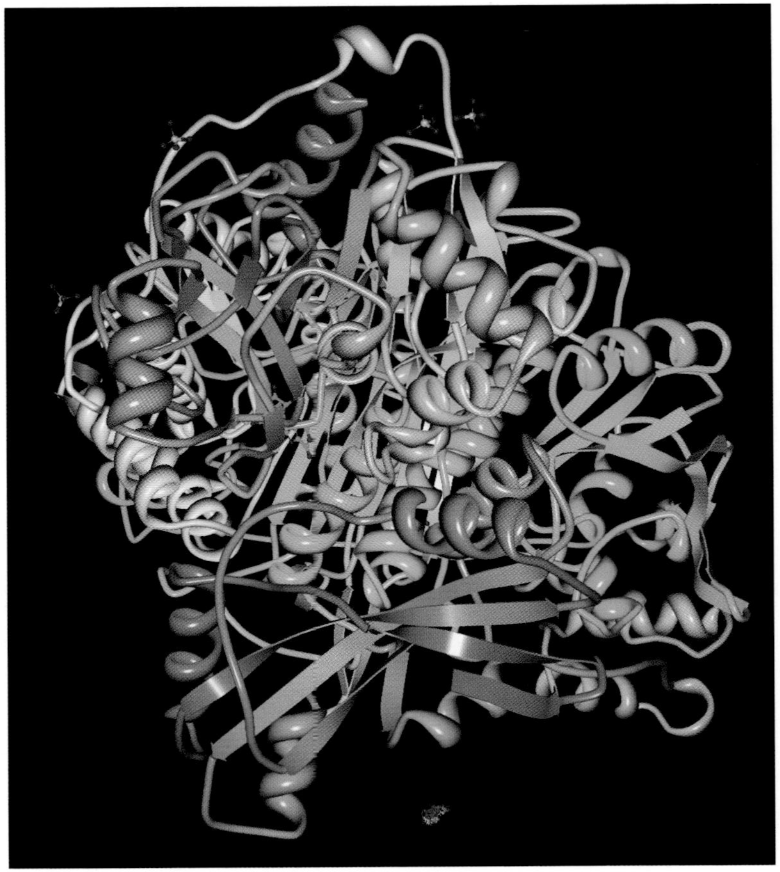

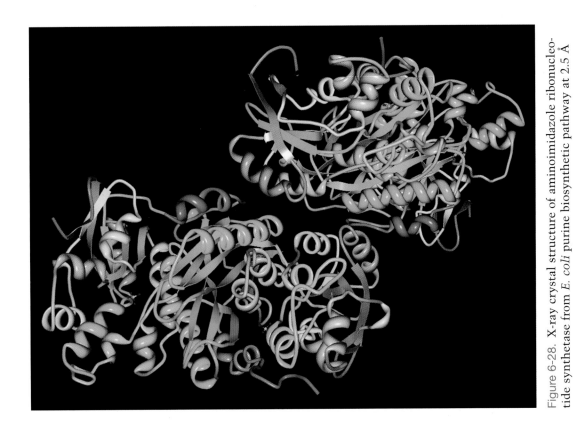

Figure 6-28. X-ray crystal structure of aminoimidazole ribonucleotide synthetase from *E. coli* purine biosynthetic pathway at 2.5 Å resolution. This is the fifth enzyme in the purine biosynthetic pathway. PDB ID: 1cli. C. Li, T.J. Kappock, J. Stubbe, T.M. Weaver, S.E. Ealick. X-ray Crystal Structure of Aminoimidazole Ribonucleotide Synthetase (PurM) from the Escherichia coli Purine Biosynthetic Pathway at 2.5 Å Resolution. *Structure Fold. Des.* **7** pp. 1155 (1999).

Figure 6-27. Structure of formylglycinamide synthase (FGAM synthase). PDB ID: 1t3t. S.E. Ealick, R. Anand, A.A. Hoskin, J. Stubbe. Domain Organization of Salmonella typhimurium Formylglycinamide Ribonucleotide Amidotransferase Revealed by X-ray Crystallography. *Biochemistry* **43** pp. 10328 (2004).

Purines and Pyrimidines   259

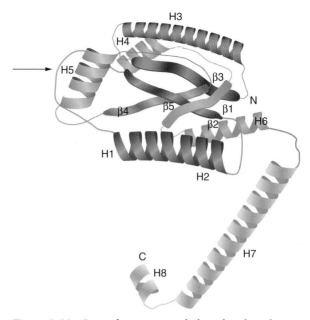

Figure 6-29. Crystal structure of phosphoribosylaminoimidazole mutase or AIR carboxylase involved in the sixth step of purine biosynthesis. The enzyme was obtained from *Thermotoga maritima*. The N-terminus is in *red* and the C terminus is in *yellow*. Pictured is the asymmetric unit of the tetrameric enzyme from PDB ID 1o4v. Redrawn with permission from R. Schwarzenbacher et al., "Crystal structure of a phosphoribosylaminoimidazole mutase PurE (TM0446) from *Thermotoga maritima* at 1.77-Å resolution," *Proteins*, 55: 474–478, 2004.

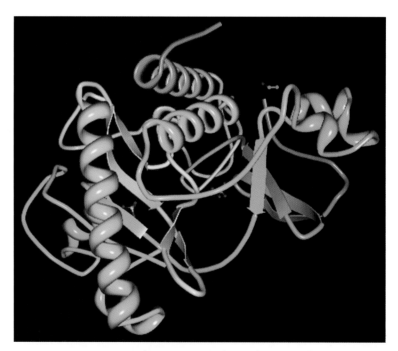

Figure 6-30. Structure of SAICAR synthase, the seventh enzyme in the purine biosynthetic pathway. PDB ID: 1A48. V.M. Lovdikov, V.V. Barynin, A.I. Grabenko, W.R. Melik-Adamyan, W.R. Lamzin, K.S. Wilson. The Structure of SAICAR Synthase: An Enzyme in the *de novo* Pathway of Purine Nucleotide Biosynthesis. *Structure* **6** pp. 363 (1998).

Figure 6-31. Structure of *T. maritime* adenylosuccinate lyase. PDB ID: 1c3u. E.A. Toth, T.O. Yeates. The Structure of Adenylosuccinate Lyase, an Enzyme with Dual Activity in the *de novo* Purine Biosynthetic Pathway. (8th Step). *Structure Fold. Des.* **8** pp. 163 (2000).

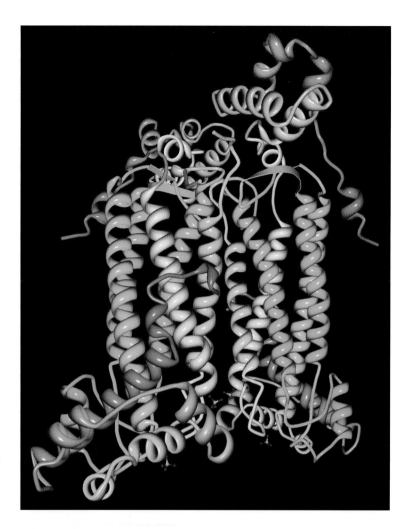

Figure 6-32. Crystal structure of the homodimeric bifunctional transformylase and cyclohydrolase enzyme avian ATIC in complex with AICAR and XMP at 1.93 Å. This enzyme is AICAR tranformylase, the ninth reaction in purine biosynthesis and IMP synthetase, the tenth reaction. PDB ID: 1M9N. D.W. Wolan, S.E. Greasly, G.P. Beardsley, I.A. Wilson Structural Insights into the Avian AICAR Transformylase Mechanism. *Biochemistry* **41** pp. 15505 (2002).

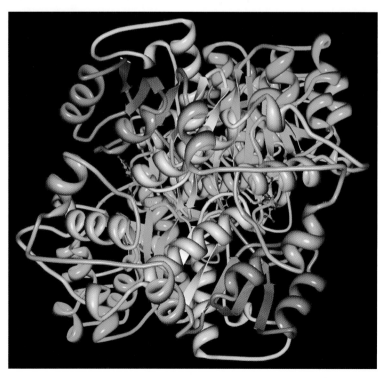

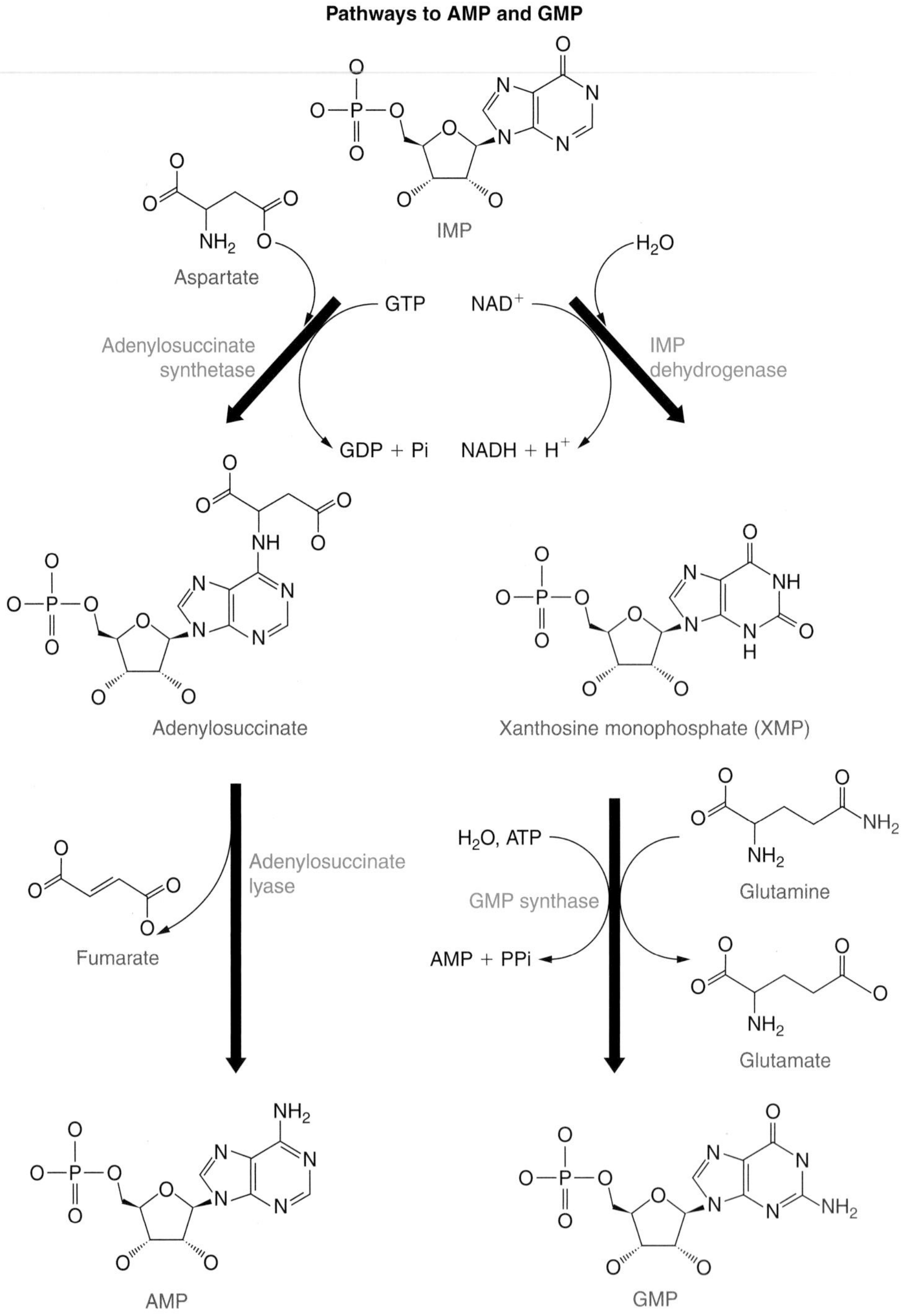

Pathways to AMP and GMP

Figure 6-33. Conversion of IMP to AMP and GMP. Negative charges on oxygens of phosphate or Carboxly groups have been omitted.

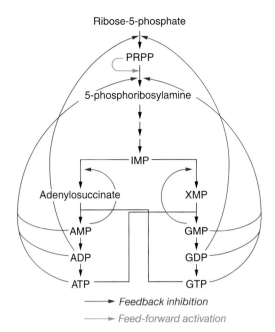

Figure 6-34. Regulation of the purine synthetic pathway. AMP, ADP, ATP, GMP, GDP, and GTP are feedback inhibitors of PRPP utilization, and AMP and GMP also inhibit conversions of IMP to AMP and IMP to XMP. The concentration of PRPP acts as a feed-forward substrate.

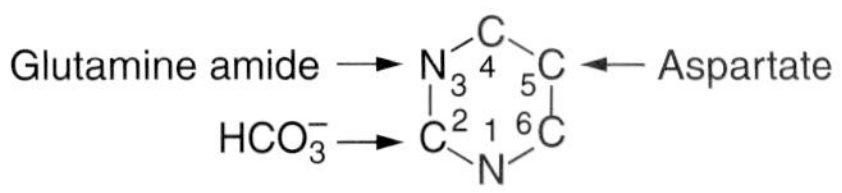

Figure 6-35. Derivation of the ring nitrogens and carbons in the pyrimidine structure.

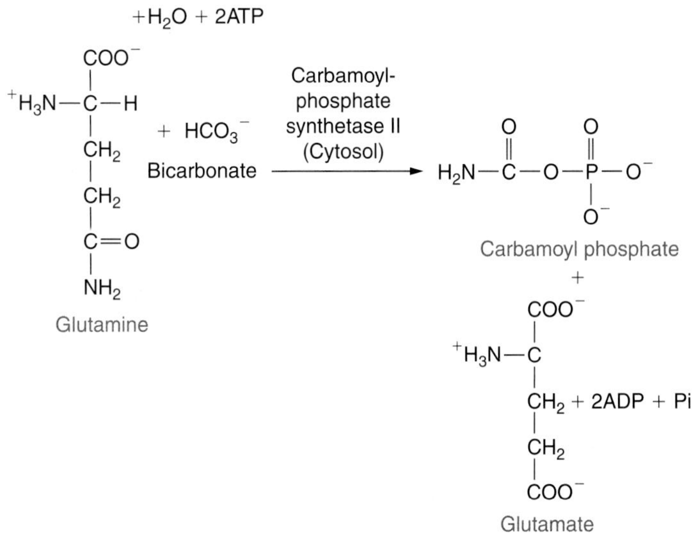

Figure 6-36. Formation of carbamoyl phosphate from glutamine, and bicarbonate in the cytosol by carbamoyl phosphate synthetase II, an enzyme present in extrahepatic tissues, whereas there is another carbamoyl phosphate–forming enzyme, carbamoyl phosphate synthase II, which is part of the urea cycle that catalyzes the formation of carbamoyl phosphate from ammonia and bicarbonate in the mitochondrion.

Purines and Pyrimidines

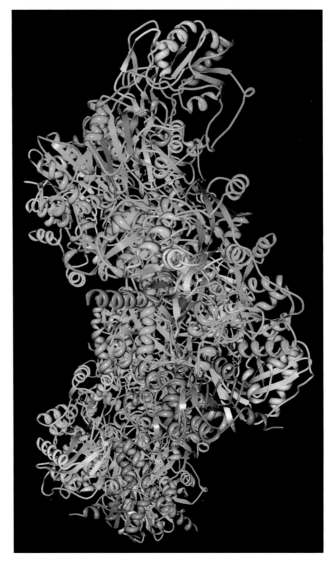

Figure 6-37. Structure of carbamoyl phosphate synthetase complexed with the ATP analog AMPPNP. PDB ID: 1BXR. J.B. Thoden, G. Wesenberg, F.M. Raushel, H.M. Holden. Carbamoyl Phosphate Synthetase: Closure of the B-domain as a Result of Nucleotide Binding. *Biochemistry* **38** pp. 2347 (1999).

step identifies the enzyme that catalyzes the reaction. The structures of three of the enzymes, dihydroorotase, dihydroorotate dehydrogenase, and OMP decarboxylase are shown in Figures 6-41 through 6-43. The structure of orotate phosphoribosyltransferase was not found. The structures of carbamoyl phosphate synthetase II and of aspartate transcarbamoylase are shown in Figures 6-37 and 6-39.

Once UMP is formed, it can be converted to the diphosphate by nucleoside monophosphate kinase:

$$UMP + ATP \rightleftharpoons ADP + UDP$$

It can be further converted to triphosphate by nucleoside diphosphate kinase:

$$UDP + ATP \rightleftharpoons UTP + ADP$$

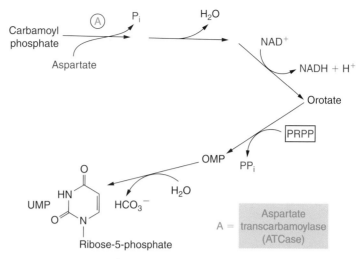

Figure 6-38. Pyrimidine synthesis from carbamoyl phosphate. The first step is catalyzed by aspartate transcarbamoylase (ATCase) with aspartate. In subsequent steps a molecule of water is removed and an oxidation occurs with NAD^+ to form orotate. PRPP is added to form orotate monophosphate (OMP) and then bicarbonate is removed with the addition of a water molecule to form uridine monophosphate (UMP).

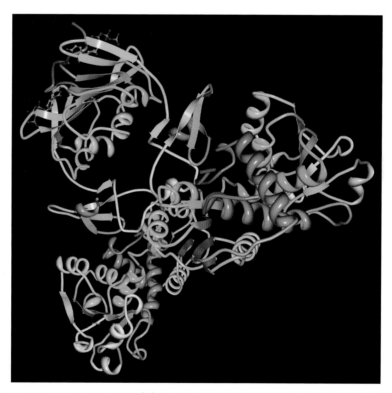

Figure 6-39. Structure of the *E. coli* aspartate transcarbamoylase. Black structures at top are CTP. The structure and mechanisms of the human enzyme are not well understood. PDB ID: 1rab. R.P. Kosman, J.E. Gouaux, W.N. Lipscomb. Crystal Structure of CTP-ligated T State Aspartate Transcarbamoylase at 2.5 Å Resolution: Implications for ATCase Mutants and the Mechanism of Negative Cooperativity. *Proteins* **15** pp. 147 (1993).

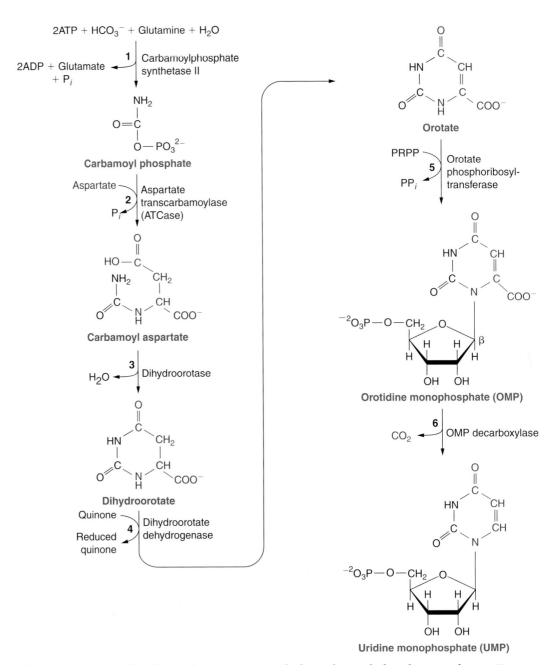

Figure 6-40. Pyrimidine biosynthesis starting with the carbamoyl phosphate synthetase II reaction to generate uridine monophosphate (UMP).

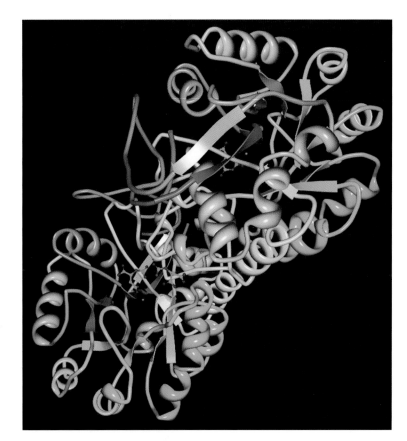

Figure 6-41. Molecular structure of dihydroorotase: a paradigm for catalysis through the use of a binuclear metal center. PDB ID: 1j79. J.B. Thoden, G.N. Phillips Jr., T.M. Neal, F.M. Raushel, H.M. Holden. Molecular Structure of Dihydroorotase: A Paradigm for Catalysis through the Use of a Binuclear Metal Center. *Biochemistry* **40** pp. 6989 (2001).

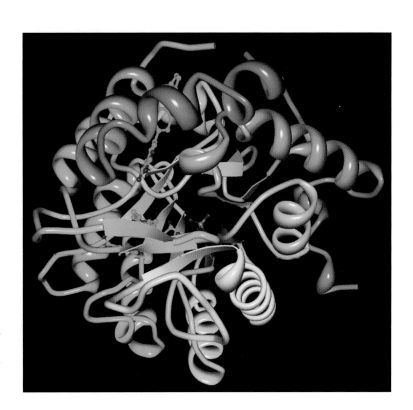

Figure 6-42. Human dihydroorotate dehydrogenase bound to a novel inhibitor. PDB ID: 2b0m. D.E. Hurt, A.E. Sutton, J. Clardy. A Novel Inhibitor of Dihydroorotate Dehydrogenase and Thoughts for Additional Species-Specific Drug Design. *To be published.*

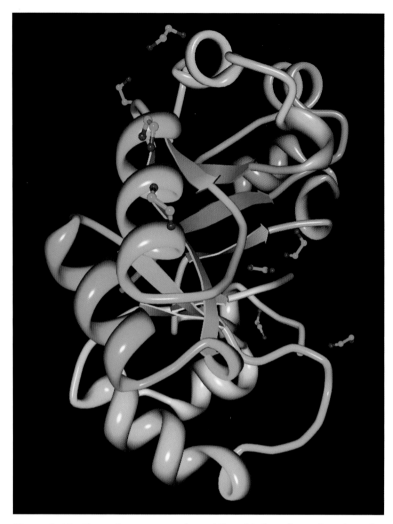

Figure 6-43. Crystal structure of orotidine 5′-phosphate decarboxylase (OMP decarboxylase) from *T. maritima* at 2.00 Å resolution. PDB ID: 1VQT Joint Center for Structural Genomics (JCSG), I.A. Wilson Crystal Structure of Orotidine 5′-Phosphate Decarboxylase (TM0332) from *Thermotoga maritima* at 2.00 Å Resolution. *To be published.*

To form the other pyrimidine, cytosine, the conversion occurs at the nucleotide level by catalysis with the enzyme, CTP synthetase as shown in Figure 6-44. Both UTP and CTP inhibit carbamoyl phosphate synthetase as shown in Figure 6-45, although this enzyme is activated by ATP and PRPP. UMP is a competitive inhibitor of OMP decarboxylase, and CTP synthetase is activated by GTP and inhibited by CTP (end product inhibition).

Purine Interconversions

In purine biosynthesis, IMP is the primary product (Figure 6-22), and it can be converted to adenine and guanine derivatives at the nucleoside monophosphate level, as shown in Figure 6-46. For conversion to AMP, IMP is first converted to adenylosuccinate catalyzed by adenylosuccinate synthetase. In a second step, adenylosuccinate is converted to AMP by adenylosuccinate lyase. For conversion to GMP, IMP is converted to

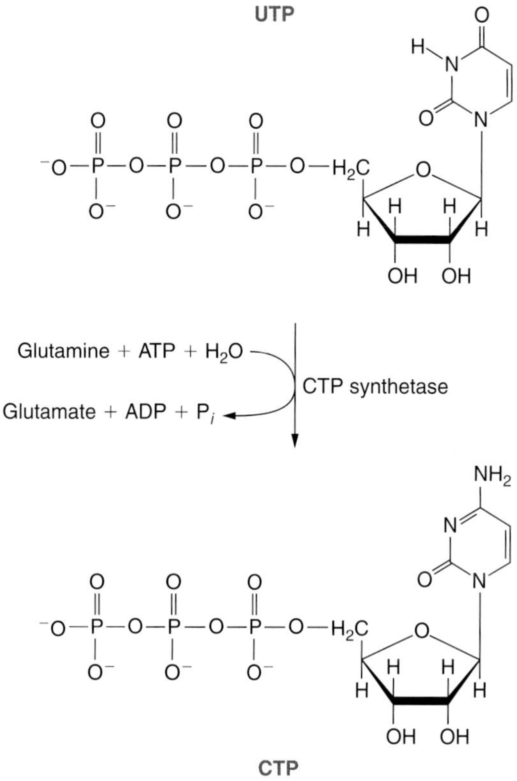

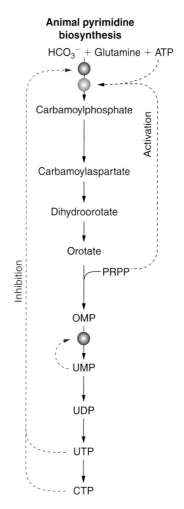

Figure 6-44. Conversion of UTP to CTP catalyzed by CTP synthetase.

Figure 6-45. Synthesis route of UTP and CTP showing negative feedback (inhibition) by UTP, CTP, and UMP. PRPP activates the formation of carbamoyl phosphate, as does ATP.

Purines and Pyrimidines

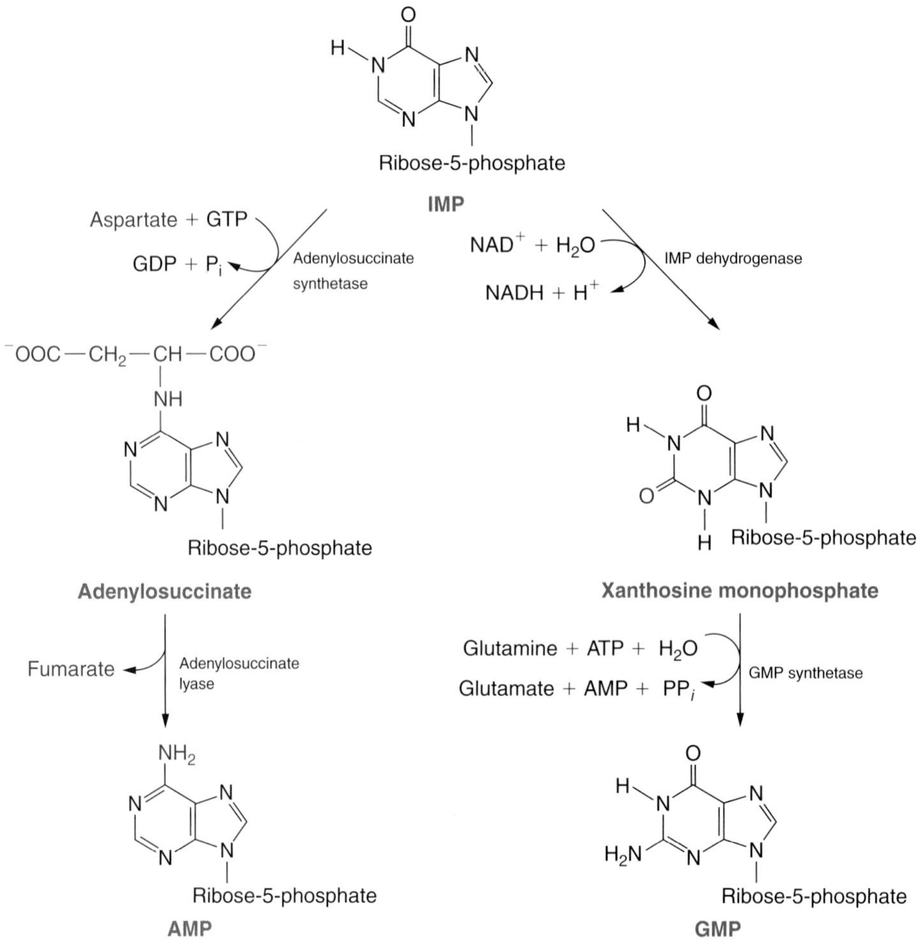

Figure 6-46. Conversion of IMP to AMP and GMP.

xanthosine monophosphate by IMP dehydrogenase, and then xanthosine monophosphate is converted to GMP by GMP synthase (Figure 6-46). The structures of these enzymes are shown in Figure 6-47. The mononucleotides can then be converted to the dinucleotides with the appropriate kinase:

The guanylate kinase reaction:

$$GMP + ATP \rightleftharpoons GDP + ADP$$

The adenylate kinase reaction:

$$AMP + ATP \rightleftharpoons 2\ ADP$$

Then, the nucleoside diphosphates can be converted to the triphosphates catalyzed by nucleoside diphosphate kinase:

$$GDP + ATP \rightleftharpoons GTP + ADP\ \text{(for example)}$$

This is shown in Figure 6-34, also showing the negative feedback regulation by the end products. In *E. coli*, the genes for all of these enzymes of purine synthesis and conversion to the higher phosphates are known.

Purine and Pyrimidine Nucleotide Catabolism

Uric acid is the product of purine nucleotide catabolism, as shown in Figure 6-48. Both AMP and IMP are degraded by nucleotidase, which removes the phosphate group to yield the nucleoside (base plus sugar). Purine nucleotide phosphorylase then removes the sugar moiety to release the free bases, adenine, or hypoxanthine. Adenine can be deaminated to form hypoxanthine, which is oxidized by xanthine oxidase. A similar pathway affects xanthine monophosphate and GMP, which produces either xanthine or guanine. Guanine can be deaminated to form xanthine, which is oxidized to uric acid, also by xanthine oxidase. Xanthine oxidase is an important enzyme in this degradative pathway of purines. The crystal structure of xanthine oxidase is shown in Figure 6-49.

The purine bases generated by catabolism can be recovered through a **salvage pathway.** Thus, adenine, guanine, and hypoxanthine can form nucleotides again by phosphoribosylation. There are two important enzymes involved in this process: **adenine phosphoribosyltransferase (APRT)** and **hypoxanthine-guanine phosphoribosyltransferase (HGPRT).** APRT catalyzes the reaction:

$$\text{adenine} + \text{PRPP} \rightleftharpoons \text{AMP} + \text{PP}_i$$

HGPRT catalyzes the reactions:

$$\text{hypoxanthine} + \text{PRPP} \rightleftharpoons \text{IMP} + \text{PP}_i$$

$$\text{guanine} + \text{PRPP} \rightleftharpoons \text{GMP} + \text{PP}_i$$

Converting IMP to AMP and salvaging IMP from AMP catabolism cause the deamination of aspartate to fumarate through the purine nucleotide cycle (Figure 6-50). The crystal structures of APRT and HGPRT are shown in Figure 6-51.

In the case of thymine nucleotides, because thymine is exclusive to DNA (and uracil exclusive to RNA), the sugar moiety in thymine nucleotides is deoxyribose (ribose in nucleotides comprising RNA). Thymine appears as dTMP, and it arises from dUMP from deoxyuridine. TMP can also arise from thymidine, and these reactions are catalyzed by thymidine kinase:

$$\text{thymidine} + \text{ATP} \rightleftharpoons \text{TMP} + \text{ADP}$$

$$\text{deoxyuridine} + \text{ATP} \rightleftharpoons \text{dUMP} + \text{ADP}$$

The catabolism of pyrimidine nucleotides leads to the formation of β-alanine (CMP or UMP) or to β-aminoisobutyrate + NH_3 + CO_2 (dTMP). The products of catabolism, β-alanine, and β-aminoisobutyrate are trans aminated (with α-ketoglutarate), ultimately to form malonyl-S-CoA and methylmalonyl-S-CoA as shown in Figure 6-52. These products can enter the citric acid cycle.

Uracil is released part way through the catabolism of RNA, and then uridine can be salvaged to form UMP by the action of uridine phosphorylase and uridine kinase in the following reactions:

Uridine phosphorylase:

$$\text{uracil} + \text{ribose-1-phosphate} = \text{uridine} + P_i$$

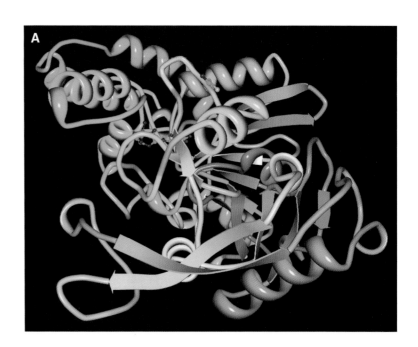

Figure 6-47. Structures of enzymes in the pathways converting IMP to AMP and GMP. A. Adenylosuccinate synthetase in complex with the natural feedback inhibitor AMP. B. *T. maritima* adenylosuccinate lyase. C. Ternary complex of human type-II inosine monophosphate dehydrogenase with 6-CL-IMP and selenazole adenine dinucleotide. D. The crystal structure of GMP synthetase reveals a novel catalytic triad and is a structural paradigm for two enzyme families. Part A PDB ID: 1SON. R. Fonne-Pfister, P. Chemla, E. Ward, M.Girardet, K.E. Kreuz, R.B. Honzatko, H.J. Fromm, H.P. Schar, M.G. Grutter, S.W. Cowan-Jacob. The Mode of Action and the Structure of a Herbicide in Complex with Its Target: Binding of Activated Hydantocidin to the Feedback Regulation Site of Adenylosuccinate Synthetase. *PNAS USA* **93** pp. 9431 (1996). Part B PDB ID: 1c3u. E.A. Toth, T.O. Yeates. The Structure of Adenylosuccinate Lyase, an Enzyme with Dual Activity in the *de novo* Purine Biosynthetic Pathway. *Structure Fold. Des.* **8** pp.163 (2000). Part C PDB ID: 1b3o. T.D. Colby, K. Vanderveen, M.D. Strickler, G.D. Markham, B.M. Goldstein. Crystal Structure of Human Type II Inosine Monophosphate Dehydrogenase: Implications for Ligand Binding and Drug Design. *PNAS USA* **96** pp. 3531 (1999). Part D PDB ID: 1gpm. J.J. Tesmer, T.J. Klem, M.L. Deras, V.J. Davisson, J.L. Smith. The Crystal Structure of GMP Synthetase Reveals a Novel Catalytic Triad and Is a Structural Paradigm for Two Enzyme Families. *Nat. Struct. Biol.* **3** pp. 74 (1996).

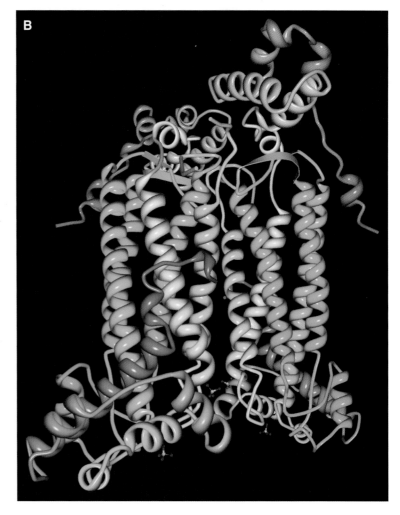

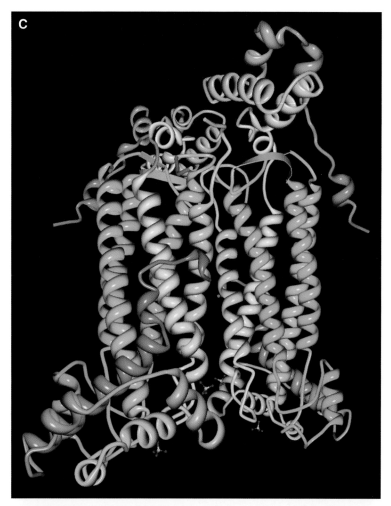
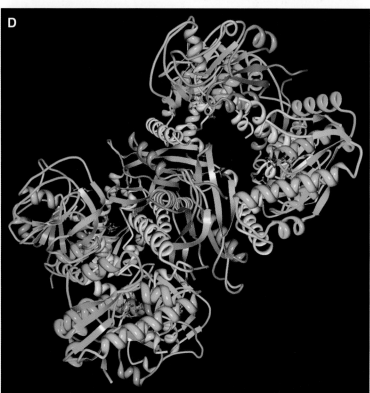

Figure 6-47, cont'd

Purines and Pyrimidines

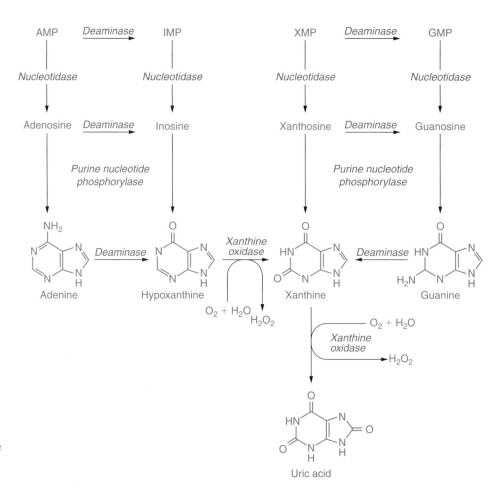

Figure 6-48. Catabolism of purine nucleotides. Note that the penultimate purine is xanthine which is finally converted to uric acid by xanthine oxidase.

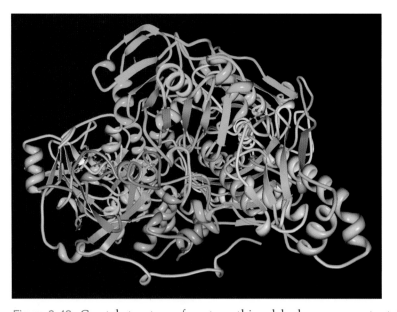

Figure 6-49. Crystal structure of a rat xanthine dehydrogenase mutant. PDB ID: 1wyg. T. Nishino, K. Okamoto, Y. Kawaguchi, H. Hori, T. Matsumura, T., B.T. Eger, E.F. Pai, T. Nishino. Mechanism of the Conversion of Xanthine Dehydrogenase to Xanthine Oxidase: Identification of the Two Cysteine Disulfide Bonds and Crystal Structure of a Non-Convertible Rat Liver Xanthine Dehydrogenase Mutant. *J. Biol. Chem.* **280** pp. 24888 (2005).

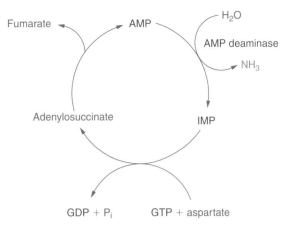

Figure 6-50. The purine nucleotide cycle. Conversion of AMP to IMP and formation of GDP from IMP results in the deamination of aspartate to form fumarate.

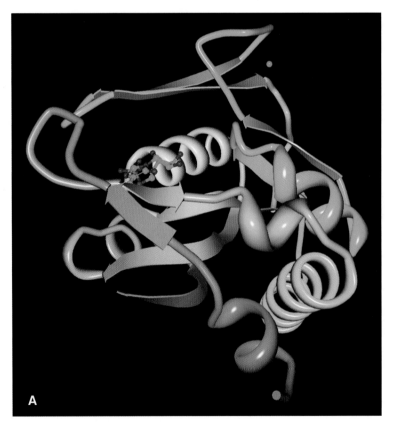

Figure 6-51. **A.** Three-dimensional structure of human adenine phosphoribosyltransferase (APRT) and its relation to DHA-urolithiasis. **B.** Human HGPRTase with transition state inhibitor. Human HGPRT has four chains, each of 214 amino acid residues. Part A reproduced from PDB ID: 1ore. M. Silva, C.H.T.P. Silva, J. Iulek, O.H. Thiemann. Three-Dimensional Structure of Human Adenine Phosphoribosyltransferase and Its Relation to DHA-Irolithiasis. *Biochemistry* **43** pp. 7663 (2004).

(continued)

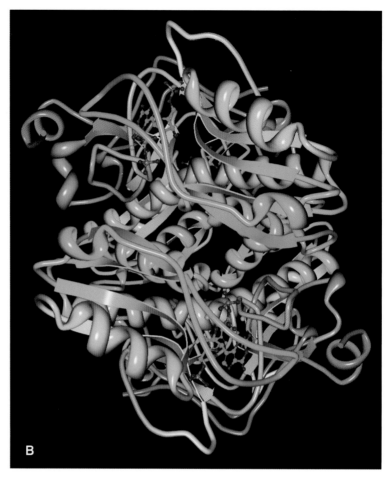

Figure 6-51, cont'd Part B reproduced from PDB ID: 1bzy W. Shi, C.M. Li, P.C. Tyler, R.H. Furneaux, C. Grubmeyer, V.L. Schramm, S.C. Almo. The 2.0 Å Structure of Human Hypoxanthine-Guanine Phosphoribosyltransferase in Complex with a Transition-State Analog Inhibitor. *Nat. Struct. Biol.* **6** pp. 588 (1999).

Uridine kinase:

$$\text{uridine} + \text{ATP} \rightleftharpoons \text{UMP} + \text{ADP}$$

Likewise, thymine can be salvaged by thymine phosphorylase and thymidine kinase:

Thymine phosphorylase:

$$\text{thymine} + \text{deoxyribose-1-phosphate} \rightleftharpoons \text{thymidine} + P_i$$

Thymidine kinase:

$$\text{thymidine} + \text{ATP} \rightleftharpoons \text{dTMP} + \text{ADP}$$

Deoxycytidine is salvaged by deoxycytidine kinase (which also acts on deoxyadenosine and deoxyguanosine, although deoxycytidine is the preferred substrate):

$$\text{deoxycytidine} + \text{ATP} \rightleftharpoons \text{dCMP} + \text{ADP}$$

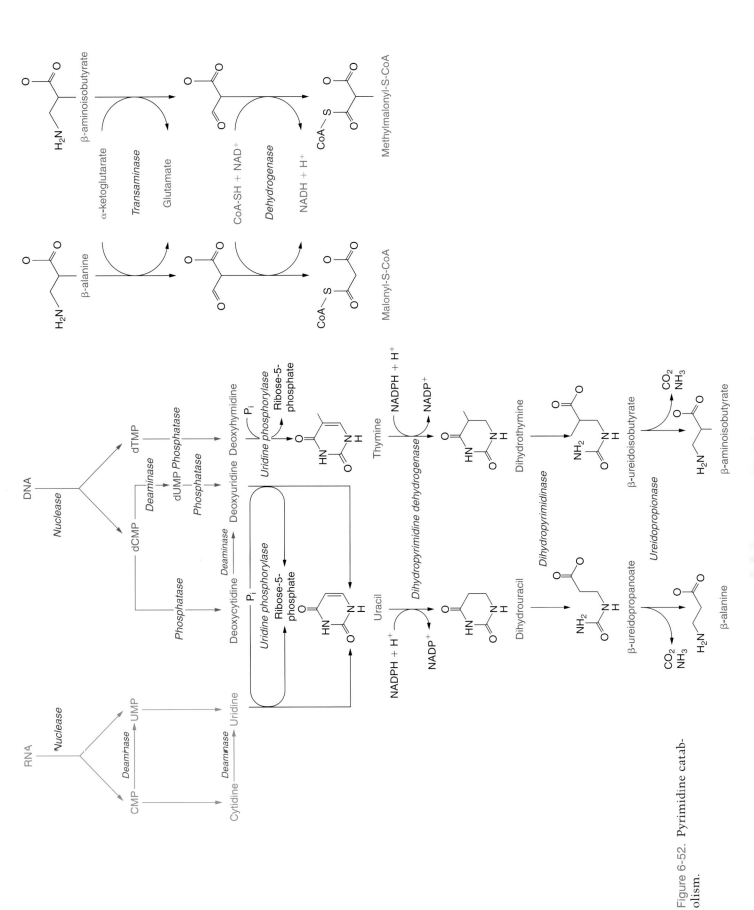

Figure 6-52. Pyrimidine catabolism.

The salvage of pyrimidines by this pathway is considered to be minor in view of the low concentrations of ribose-1-phosphate and the pyrimidine nucleosides.

Thymidine kinase is an important enzyme, and its concentration changes during the cell cycle; its highest concentration appears during DNA synthesis. The conversion of dUMP to dTMP is shown in Figure 6-53. The reaction is catalyzed by thymidylate synthase. Methylene tetrahydrofolate (5,10-methylene THF) is the coenzyme that donates the methyl group to dUMP to form dTMP. The structures of these enzymes are shown in Figure 6-54. An overview of the synthesis of the deoxyribose containing trinucleotides found in DNA is shown in Figure 6-55. Human cellular concentrations of the deoxytrinucleotides are dATP, 0.013 mM (millimolar); dGTP, 0.005 mM; dCTP, 0.022 mM; and dTTP, 0.023 mM. For the corresponding **pentose-containing trinucleotides,** ATP is 2.8 mM, GTP, 0.48 mM; CTP, 0.21 mM; and UTP, 0.48 mM.

The pentose sugars used in the synthesis of nucleic acids arise from the glycolysis pathway and can be shuttled back to the glycolysis pathway when not needed for nucleotide synthesis. The formation of ribose-5-phosphate from glucose-6-phosphate (part of the oxidative pentose phosphate pathway, which is active in tissues forming RNA and DNA) is

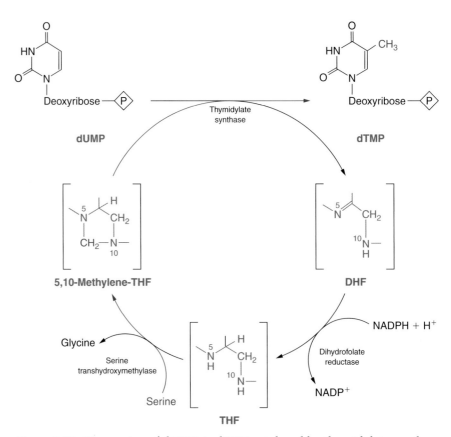

Figure 6-53. Conversion of dUMP to dTMP catalyzed by thymidylate synthase with methylene THF (tetrahydroftolate) serine is converted to glycine in regenerating the methylene form of THF from dihydrofolate (DHF).

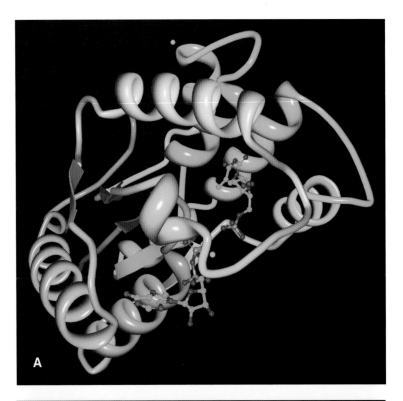

Figure 6-54. **A.** Wild-type human thymidylate kinase complexed with azTMP and ADP. **B.** Thymidylate synthase. Part A reproduced from PDB ID: 1e98. N. Ostermann, A. Lavie, S. Padiyar, R. Brundiers, T. Veit, J. Reinstein, R.S. Goody, M. Konrad, I. Schlichting. Potentiating AZT Activation: Structures of Wild-Type and Mutant Human Thymidylate Kinase Suggest Reasons for the Mutants' Improved Kinetics with the HIV Prodrug Metabolite azTMP. *J. Mol. Biol.* **304** pp. 43 (2000). Part B reproduced from PDB ID: 1njb. J.S. Finer-Moore, L. Liu, C.E. Schafmeister, D.L. Birdsall, T. Mau, D.V. Santi, R.M. Stroud. Partitioning Roles of Side Chains in Affinity, Orientation, and Catalysis with Structures for Mutant Complexes: Asparagine-229 in Thymidylate Synthase. *Biochemistry* **35** pp. 5125 (1996).

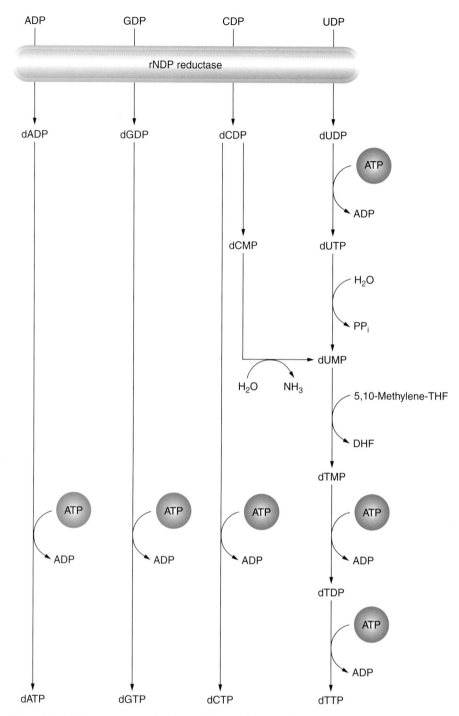

Figure 6-55. An overview of the synthesis of deoxyribose-containing trinucleotides found in DNA.

shown in Figure 6-56. The structure of pentose phosphate isomerase, the enzyme catalyzing the conversion of ribulose-5-phosphate to ribose-5-phosphate (reversibly) is shown in Figure 6-57. The pathway is shown in more detail in Figure 6-58. An overview of the connection between glycolysis and the pentose phosphate pathway is shown in Figure 6-59. Structures of some enzymes in this pathway are shown in Figure 6-60. Again, ribose is the component of RNAs, and deoxyribose is the component of DNAs. These sugars differ only in the carbon-2 position of the pentose, where the deoxy sugar has H-C-H and ribose has H-C-OH (Figure 6-61). *The deoxy sugar is formed while it is a component of the nucleotide* by **ribonucleotide reductase** that is coupled to thioredoxin reductase, which regenerates the oxidized coenzyme as shown in Figure 6-62. The structure of thioredoxin reductase is shown in Figure 6-63.

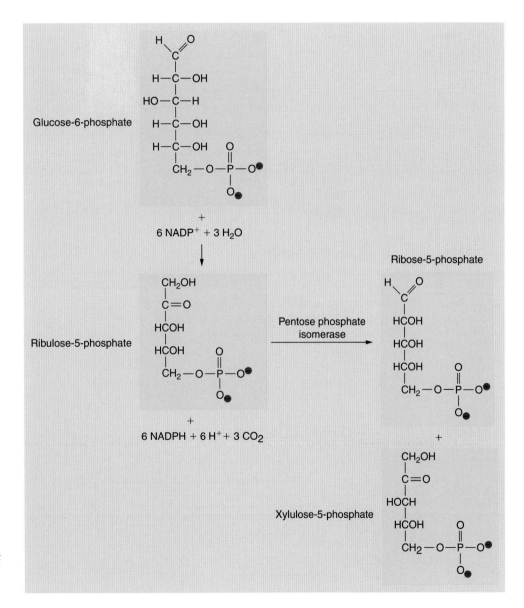

Figure 6-56. Generation of ribose-5-phosphate from glucose-6-phosphate.

Purines and Pyrimidines

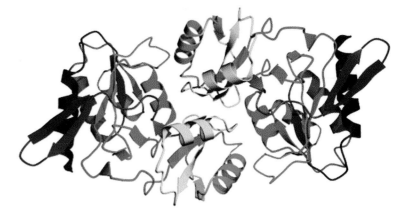

Figure 6-57. Crystal structure of *E. coli* ribose-5-phosphate isomerase. The enzyme consists of two subunits. Reproduced from R-G. Zhang et al., *Structure*, 11: 31–42, 2003 and directly from http://www.sciencedirect.com.

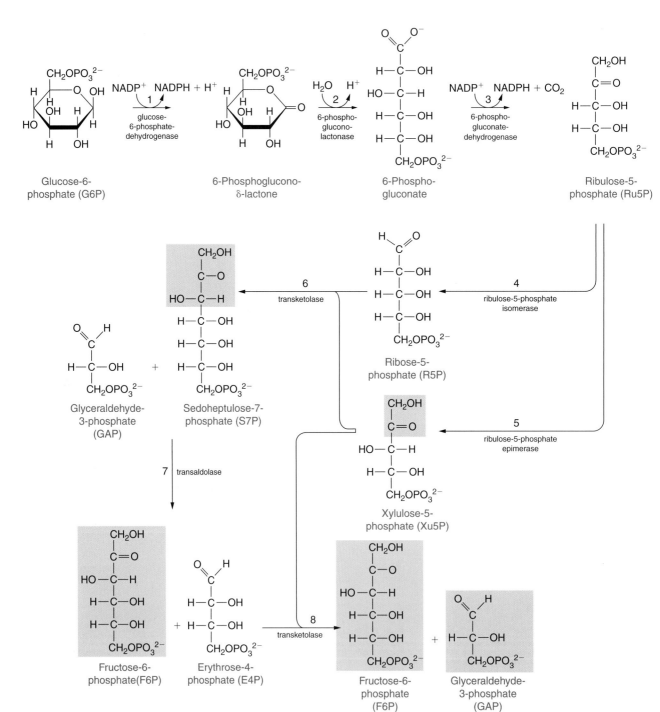

Figure 6-58. The oxidative pentose phosphate pathway with the enzymes involved.

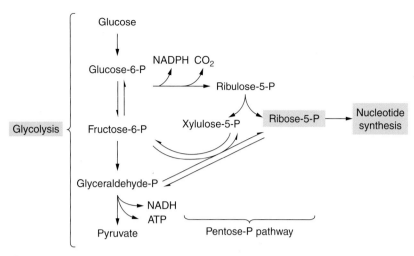

Figure 6-59. Generation of ribose-5-phosphate from glucose-6-phosphate from glycolysis and reversal showing ribose-5-phosphate entry back into the glycolytic cycle when not used for the synthesis of nucleotides.

Disorders of Purine and Pyrimidine Metabolism

One of the major health problems developing from purine metabolism is the overproduction of uric acid, the end product of purine catabolism, because uric acid is poorly soluble. As such, it can crystallize out of solution, and this often occurs in the joints. Initial insults involve the large joint of the large toe, and it is painful. Deposition elsewhere can lead to severe inflammation and arthritis. Drugs are available to treat this condition. One such drug is allopurinol (Figure 6-64), which is a competitive inhibitor of xanthine oxidase in that its structure resembles the normal substrate, hypoxanthine. Anti-inflammatory drugs are useful also. Some health problems associated with purine metabolism are shown in Table 6-1. There can be a genetic loss (sex linked) of the enzyme HGPRT. This gene is found on the X chromosome, and individuals with this trait have severe gout and a central nervous system disorder known as **Lesch-Nyhan syndrome.** This disease is recessive and is relatively rare, with a frequency of 1 in 380,000 births without any geographical or racial concentration. It is particularly rare in females, although they can be carriers of the trait. Self-mutilation is a symptom displayed by individuals with a severe form of this disease. There are no useful drugs for this disease, and often the approach is the wearing of devices to prevent self-mutilation. Stress aggravates the condition. Eventually, **gene therapy** may be able to replace the defective gene product with a normally expressed HGPRT enzyme.

Other disorders are mentioned in Table 6-1. Disorders of the pyrimidine metabolic pathway are less consequential because the pathway is simpler and because the products are more water soluble than uric acid (in the case of the purines). Clinical consequences of disordered pyrimidine metabolism are listed in Table 6-2.

(Text continues on p. 288.)

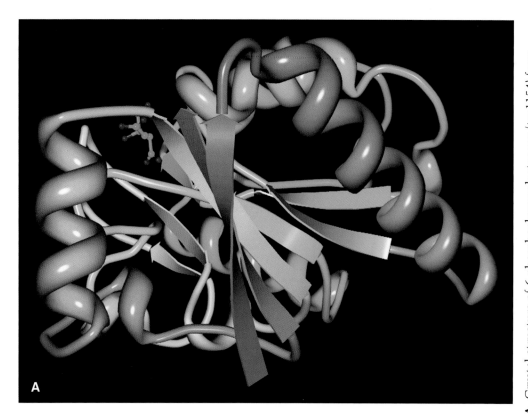

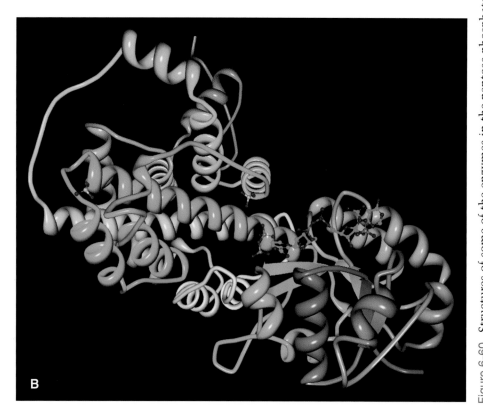

Figure 6-60. Structures of some of the enzymes in the pentose phosphate pathway. **A.** Crystal structure of 6-phosphogluconolactonase (tm1154) from *Thermotoga maritima* at 1.70 Å resolution. **B.** 6-phosphogluconate dehydrogenase. **C.** Yeast transketolase. **D.** Crystal structure of human transaldolase. Part A reproduced from PDB ID: 1vll. Joint Center for Structural Genomics (JCSG) Crystal Structure of 6-Phosphogluconolactonase (TMll54) from *Thermotoga maritima* at 1.70 Å Resolution. *To be published*. Part B reproduced from PDB ID: 1pgn. M.J. Adams, G.H. Ellis, S. Gover, C.E. Naylor, C. Phillips. Crystallographic Study of Coenzyme, Coenzyme Analogue and Substrate Binding in 6-Phosphogluconate Dehydrogenase: Implications for NADP Specificity and the Enzyme Mechanism. *Structure* **2** pp. 651 (1994). Part C reproduced from PDB ID: 1tka. S. Konig, A. Schellenberger, H. Neef, G. Schneider. Specificity of Coenzyme Binding in Thiamin Diphosphate-Dependent Enzymes. Crystal Structures of Yeast Transketolase in Complex with Analogs of Thiamin Diphosphate. *The Journal of Biological Chemistry* **269** pp. 10879 (1994). Part D reproduced from PDB ID: 1f05. S. Thorell, P. Gergely, K. Banki, A. Perl, G. Schneider. The Three-Dimensional Structure of Human Transaldolase. *FEBS Letters* **475** pp. 205 (2000).

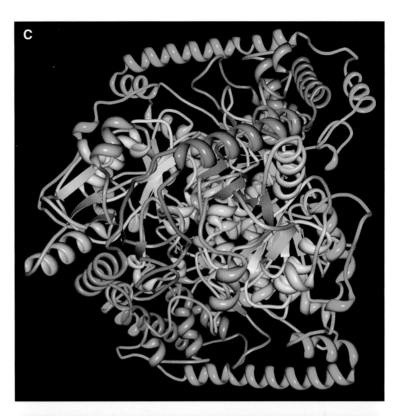
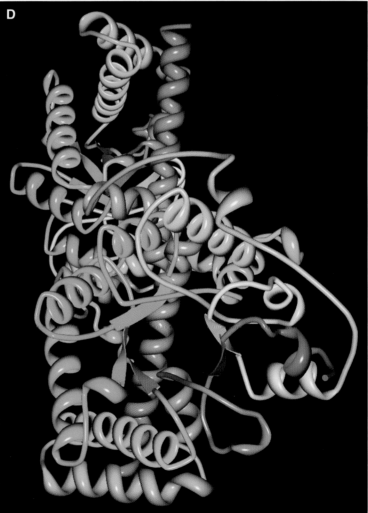

Figure 6-60, cont'd

Purines and Pyrimidines

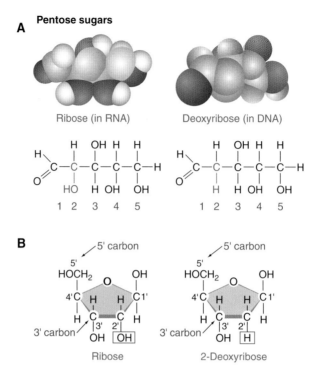

Figure 6-61. **A.** Atomic models of ribose and deoxyribose with open structures beneath. **B.** The ring structures of ribose and deoxyribose.

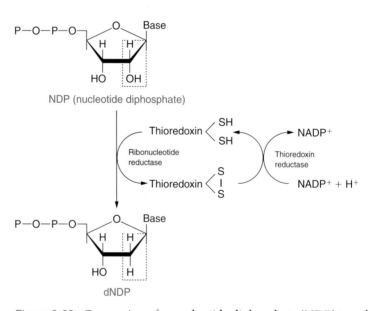

Figure 6-62. Conversion of a nucleotide diphosphate (NDP) to a deoxy nucleotide diphosphate (dNDP) by ribonucleotide reductase in which the enzyme is oxidized (2SH → S-S) in order to reduce the C-2 of ribose. The enzyme is regenerated to its reduced state (2SH) by thioredoxin reductase, which generates oxidized $NADP^+$ from $NADP^+ + H^+$.

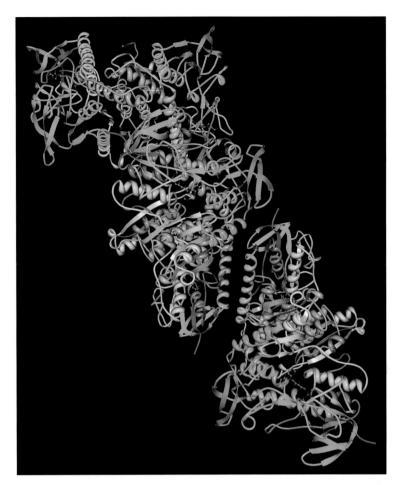

Figure 6-63. Structure of human thioredoxin reductase. PDB ID: 2cfy. J.E. Debreczeni, C. Johansson, K. Kavanagh, P. Savitsky, M. Sundstrom, C. Arrowsmith, J. Weigelt, A. Edwards, F. Von delft, U. Oppermann. Crystal Structure of Human Thioredoxin Reductase 1. *To be published.*

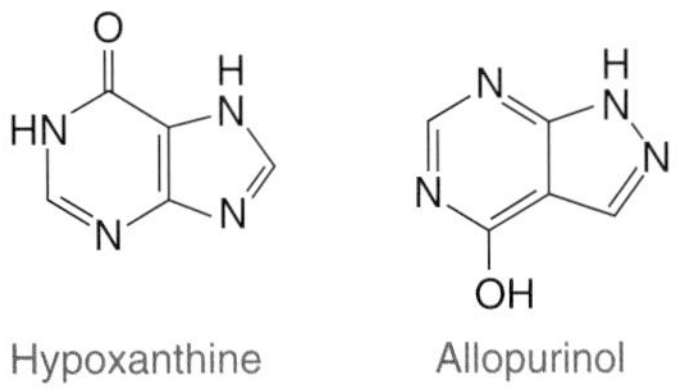

Hypoxanthine   Allopurinol

Figure 6-64. Structures of hypoxanthine and allopurinol.

Table 6-1
Diseases Associated with Disorders of Purine Metabolism

Disorder	Defect	Nature of Defect	Comments
Gout	PRPP[f] synthetase	increased enzyme activity due to elevated V_{max}	hyperuricemia
Gout	PRPP synthetase	enzyme is resistant to feedback inhibition	hyperuricemia
Gout	PRPP synthetase	enzyme has increased affinity for ribose-5-phosphate (lowered K_m)	hyperuricemia
Gout	PRPP amidotransferase	loss of feedback inhibition of enzyme	hyperuricemia
Gout	HGPRT[a]	partially defective enzyme	hyperuricemia
Lesch-Nyhan syndrome	HGPRT	lack of enzyme	
SCID[e]	ADA[b]	lack of enzyme	
Immunodeficiency	PNP[c]	lack of enzyme	
Renal lithiasis	APRT[d]	lack of enzyme	2,8-dihydroxyadenine renal lithiasis
Xanthinuria	Xanthine oxidase	lack of enzyme	hypouricemia and xanthine renal lithiasis
von Gierke's disease	Glucose-6-phosphatase	enzyme deficiency	

a, HGPRT (hypoxanthine-guanine phosphoribosyl transferase); b, adenosine deaminase; c, purine nucleotide phosphorylase; d, adenosine phosphoribosyltransferase; e, severe combined immunodeficiency; and f, 5-phospho-D-ribosyl-1-pyrophosphate.

Table 6-2
Disorders of Pyrimidine Metabolism

Disorder	Defective Enzyme	Comments
Orotic aciduria, type I	Orotate phosphoribosyltransferase and OMP decarboxylase	
Orotic aciduria, type II	OMP decarboxylase	
Orotic aciduria (mild, no hematological component)	The urea cycle enzyme, ornithine transcarbamoylase, is deficient	Increased mitochondrial carbamoyl phosphate exits and augments pyrimidine biosynthesis; hepatic encephalopathy
β-Aminoisobutyric aciduria	Transaminase, affects urea cycle function during deamination of α-amino acids to α-keto acids	Benign, frequent in Asians
Drug-induced orotic aciduria	OMP decarboxylase	Allopurinol and 6-azauridine treatments cause orotic acidurias without a hematological component; their catabolic by-products inhibit OMP decarboxylase

OMP, orotatemonophosphate.

Biosynthesis of Deoxyribonucleic Acids

DNAs are large polynucleotides containing many individual bases in deoxynucleotide form. The ordering of the sequence in which the deoxynucleotides are joined together is determined by a preexisting single strand of DNA that contains the information for the sequence of the bases (and therefore for base-pairing), directing the specific nucleotide base to be added next to the growing molecule. When life was first formed on this planet, the sequence-directing DNA strands had to come from somewhere. Without entering into a philosophical discussion, and for the purposes of understanding how these large molecules containing the genetic information are synthesized in the cell, we can accept that sense strands of DNA were preexistent or that the primitive DNAs were made from preexisting RNA templates.

Simply put, the phosphate group of a nucleotide can react with the 3'-hydroxyl group of a second nucleotide to form a dinucleotide joined through a phosphoric acid ester (Figures 6-65 and 6-66). The first nucleotide will have an unsubstituted 5' phosphate group, and the last nucleotide added will have a free 3'-hydroxyl group. As mentioned earlier, *information in the ultimate polynucleotide will progress from the 5' end to the 3' end.* Thus, writing a theoretical sequence would appear as follows:

$$5'\text{-pGpApTpCpApCpTpG-}3'$$

In the preceding sequence, *p* indicates the nucleotide status and the phosphate group. Discovery of the double-stranded nature of DNA is attributed to James Watson and Francis Crick, whose model predicted that DNA was a helix of two complementary antiparallel strands wound around each other and stabilized by hydrogen bonding between the bases in the two strands. Relative to the axis of the helix, the bases jut inward at about a 90-degree angle to the helical axis. Interaction between the bases has been shown in Figures 6-16 through 6-19. Remember that A forms a base pair with T and G forms a base pair with C. Figure 6-17 showed that when G forms a base pair with C, there are three sets of hydrogen bonds, whereas when A forms a base pair with T, there are two sets of hydrogen bonds, making the interaction between G and C stronger than that between A and T. In the antiparallel double strand of two DNAs, one strand is oriented from 5' to 3' and the opposing strand is oriented from 3' to 5'. Between the ribose phosphate groups along the helices, there is a major groove (larger) and a minor groove (smaller). This is the result of asymmetry of the deoxyribose rings and the structural nature of the upper and lower surfaces of a base pair. The most common form of double-stranded DNA is the B form, resulting from physiological conditions. There exists a novel form of DNA called the **Z-DNA.** This form exists as a 180-degree change in the orientation of the bases relative to the B form. There is also A-DNA, which is similar to the more commonly occurring **B-DNA** but has certain differences in its double helix: one more residue per turn of the helix, a lower degree rotation per residue, a larger tilt of the base relative to the axis of the helix, a narrower major groove, and a shallower minor groove. General appearances of the helices of B-DNA compared to Z-DNA are shown in Figure 6-67. Characteristics of the major DNA helices are shown in Table 6-3.

In the process of cell division, the genetic material has to be copied to provide the daughter cell with the same DNA as the parent. For this to occur, the two strands of the DNA separate and the replication mechanism proceeds such that the DNA polymerase synthesizes a complementary strand. This means that a T occurs for every A and a C occurs for every G, and vice versa, so that the rules of base-pairing are observed in the newly synthesized strand. This is shown in Figure 6-68. The structure of human DNA polymerase is shown in Figure 6-69, and the structure of human DNA ligase (polydeoxyribonucleotide synthase) is shown in Figure 6-70.

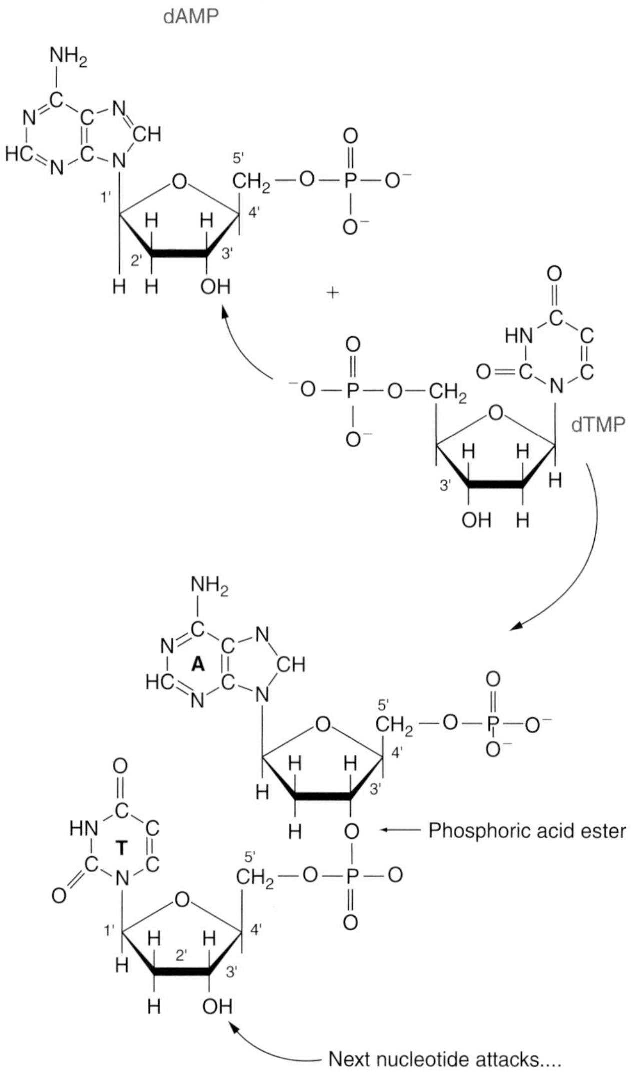

Figure 6-65. Formation of the first dinucleotide with a phosphoric acid ester structure.

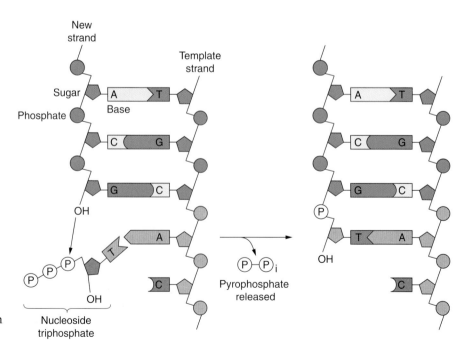

Figure 6-66. The terminal phosphate of the next nucleotide added to the growing chain by DNA polymerase provides the energy for the formation of a new bond.

290   CHAPTER 6   Nucleic Acids and Molecular Genetics

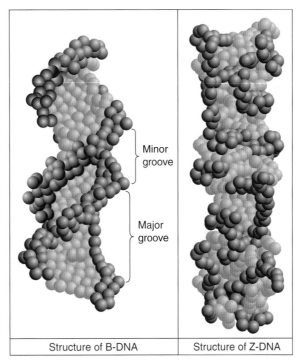

Figure 6-67. General appearances of B-DNA compared to Z-DNA.

Table 6-3
Parameters of the Major DNA Double Helices

Parameters	A Form	B Form	Z Form
Direction of helical rotation	Right	Right	Left
Residues per turn of helix	11	10	12 base pairs
Rotation of helix per residue (in degrees)	33	36	−30
Base tilt relative to helix axis (in degrees)	20	6	7
Major groove	Narrow and deep	Wide and deep	Flat
Minor groove	Wide and shallow	Narrow and deep	Narrow and deep
Orientation of N-glycosidic bond	*anti*	*anti*	*anti* for Py, *syn* for Pu
Comments	—	Most prevalent within cells	Occurs in stretches of alternating purine–pyrimidine base pairs

The DNA polymerases have grown to a large-sized family of enzymes. In humans there are as many as 13 enzymes, and the list may not be complete. For the purposes here, the enzyme will be referred to only as DNA polymerase.

Many enzymes are needed in the process of DNA replication. Topoisomerase I begins the unwinding of the double-stranded DNA to be copied. It does this by breaking the supercoiled DNA with a nick in a single strand, which releases the twisting tension on the double helix. This will

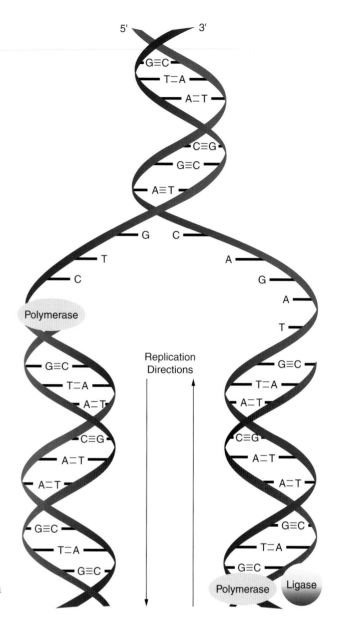

Figure 6-68. Opening of the parental double helix and synthesis of the complements of each opened strand by DNA polymerase and ligase. Newly synthesized strands are shown in *brown*.

cause the coils to untwist. Topoisomerase II nicks both strands, and these enzymes remain bound to the DNA. The double strand is unwound by the enzyme helicase after the supercoiling has been relaxed by topoisomerase. Helicase uses the energy of ATP to break the strands apart and overcome the forces of hydrogen bonding by the base pairs. DNA polymerase travels along the single-stranded DNA (in the direction from 5' to 3'), using free deoxynucleotide triphosphates (dNTPs) to form a hydrogen bond (base pair) with the next dNTP on the single strand. This occurs so that A will complement T and G will complement C, forming a phosphodiester bond with the previous nucleotide of the same strand. The triphosphate is the source of energy for the binding of the subsequent nucleotide (Figure 6-66).

The DNA polymerase responsible for the synthesis of new strands is DNA polymerase III (DNA pol III). About 100 nucleotides are added to the growing chain per second in human cells. However, the polymerase

Figure 6-69. DNA polymerase β (pol β) complexed with seven base pairs of DNA; soaked in the presence of BaCl2 and NaCl. PDB ID: 1zqn. H. Pelletier, M.R. Sawaya. Characterization of the Metal Ion Binding Helix-Hairpin-Helix Motifs in Human DNA Polymerase Beta by X-Ray Structural Analysis. *Biochemistry* **35** pp. 12778 (1996).

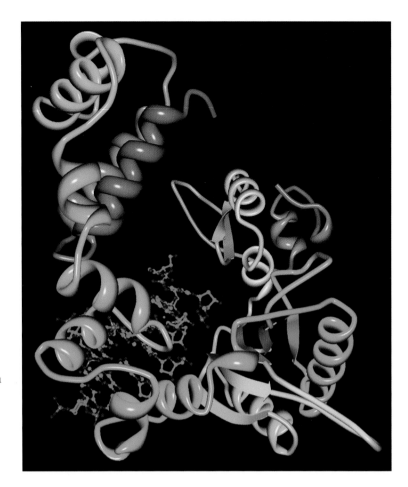

Figure 6-70. NMR structure of human DNA ligase IIIα BRCT domain. PDB ID: 1imo. V.V. Krishnan, K.H. Thornton, M.P. Thelen, M. Cosman. Solution Structure and Backbone Dynamics of the Human DNA Ligase III alpha BRCT Domain. *Biochemistry* **40** pp. 13158 (2001).

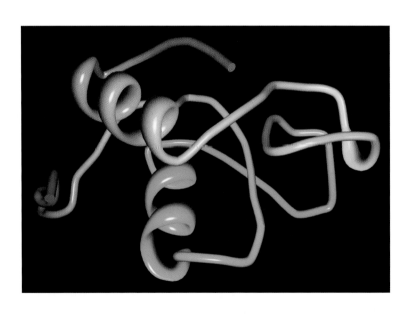

needs to have a primer with a free 3' hydroxyl for attaching a dNTP. The DNA polymerase actually has a proofreading capacity to ensure the correct base is added, and it can remove (through 3',5' exonuclease activity) any mistakes. One of the subunits of this enzyme acts like a clamp to hold the DNA polymerase to the DNA template.

A **primeosome** is a group of proteins containing a primase, which attaches a small RNA primer to the single-stranded DNA. This primer provides a substitute 3' hydroxyl group for DNA polymerase to start its synthetic activity. Eventually, the RNA primer is deleted by RNaseH and the gap is filled by DNA pol I. The unattached gap left when the RNA primer is removed is filled in by ligase. Ligase promotes the formation of a phosphodiester bond at the 3' end. The stability of the replication fork is maintained by single-stranded binding (SSB) proteins. These proteins assist helicase in availing the single-stranded template for DNA pol III. Because DNA polymerase travels only in the direction of 5' to 3', which is a smooth process for the leading strand (it starts with the 5' end), it has a more difficult time with the lagging strand, which has the 3' end to start with on the uncoiled second strand. To continue its work in the 5' to 3' direction, it synthesizes stretches of the lagging strand in the 5' to 3' direction. These stretches are called **Okazaki fragments.** These stretches are closed up by DNA pol I and ligase to complete the lagging strand.

Gyrase is one of the class of topoisomerases that controls topological transitions of DNA and, in this case, induces the unwinding of the double strand. These processes are pictured in Figure 6-71 and in somewhat more detail in Figure 6-72. If this were the totality of the process, it is estimated that there would be one mistake in about every 1000 base pairs. Therefore, a DNA proofreading and repair mechanism was developed that can recognize and repair an improper base pair. The proofreading mechanism of DNA polymerase is pictured in Figure 6-73. DNA pol II is more involved in proofreading than the other polymerases, whereas DNA pol I completes the chain synthesis between the Okazaki fragments on the lagging strand. Given the existence of this protective system, the actual error rate in DNA synthesis is about 1 in 1 million or 1 billion. A lesion in one of the two DNA strands is excised by a repair enzyme, and the correction is made referring to the undamaged strand. Corrections are accomplished by two methods: DNA glyoxylase removes an altered base after the removal of the resulting sugar phosphate in a process known as **base excision repair;** alternatively, a small region of the DNA strand surrounding the damage can be removed as an oligonucleotide through a process called **nucleotide excision.** As described earlier, the gap left, in both cases, would be filled by DNA polymerase and DNA ligase. One example of a mismatched base pair can occur when cytosine is deaminated to form uracil (an unacceptable event in the sequence of DNA). Cytosine and uracil are alike except that cytosine has an amino group in the 4-position and uracil has a keto group in the 4-position (Figure 6-74). One method to repair this unacceptable transition would be to remove the uracil and replace it with a cytosine. Alternatively, the erroneous uracil might go on to form a base pair with adenine (continuing replication process) and give rise to an unacceptable base pair. The repair mechanism erroneously could excise the uracil and replace it with adenine (an A–U base pair) to give rise to a mutant DNA. This would be an infrequent phenomenon.

Originally, a number of models of DNA replication were considered. First, there was the *conservative model* in which the parental double helix

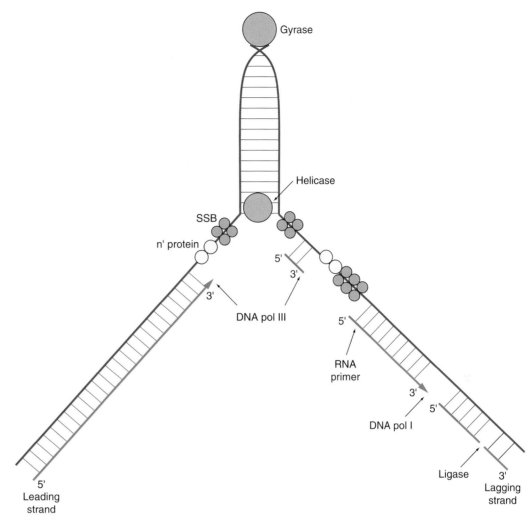

Figure 6-71. The process of DNA replication.

would remain intact and an all-new copy would be made. Then a *semiconservative model* predicted that the two strands of the parental molecule would separate, each functioning as a template for synthesis of a new complementary strand so that the first generation of molecules would be a hybrid, with one strand from the parent and one new strand. The third model was the *dispersive model* in which each strand of the daughter molecules contained a mixture of the old and newly synthesized DNAs. The semiconservative model was proven experimentally and is shown in Figure 6-75. Double-stranded DNA in the B configuration is shown as an atomic model in Figure 6-76. The base pairs can be seen in the center of the molecule lying flat.

In mitochondria, which contain only 40 genes in a circular DNA, the genetic material is not replicated in concert with nuclear DNA; the inheritance of these genes is strictly maternal. Mitochondrial DNA replication begins when the size of a mitochondrion has grown to the point at which the mitochondrion begins to divide by fission. The molecular

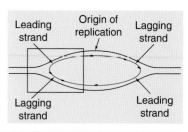
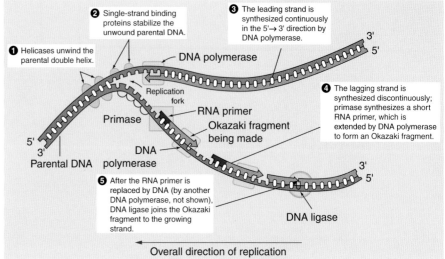
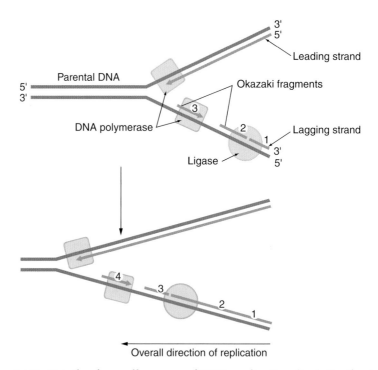

Figure 6-72. Details of overall process of DNA replication *(top)*. Synthesis of the lagging strand on the DNA template *(bottom)*. Redrawn from http://fajerpc.magnet.fsu.edu/Education/2010/Lectures/25_DNA_Replication.htm.

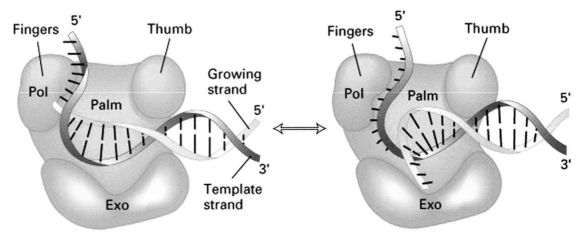

Figure 6-73. Proofreading mechanism of DNA polymerase. *Pol*, Polymerase; *EXO*, excision site. Reproduced with permission from H. Lodish et al., *Molecular Cell Biology*, 4th ed., W.H. Freeman and Company, New York, NY, 2000.

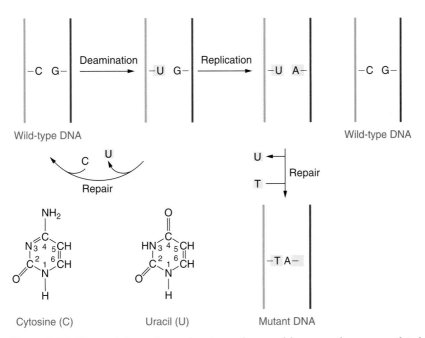

Figure 6-74. Potential repair mechanisms that could occur when a uracil is base-paired with guanine instead of a cytosine (in DNA). This can happen if external influences cause cytosine in DNA to be deaminated at the 4-position to form uracil. In one case the correct base repair is made after uracil is excised. In the other rare case uracil would be removed and replaced by thymine (to pair with A) to produce or mutation. Redrawn with permission from H. Lodish et al., *Molecular Cell Biology*, 4th ed., W.H. Freeman and Company, New York, NY, 2000.

mechanism of mitochondrial DNA replication is similar to bacterial DNA replication. In mitochondrial DNA replication, both strands of mitochondrial DNA are synthesized continuously as leading strands (5′ to 3′ direction). The first strand begins to be synthesized and then pauses, waiting for a specific signal for completion. This pausing causes the formation of D-loops, as shown in Figure 6-77. After the fork of the first strand synthesis passes the origin of the second strand, the synthesis of the second strand is started. Thus, the origins for left- and right-strand replication are separate for mitochondrial DNA but coincident for

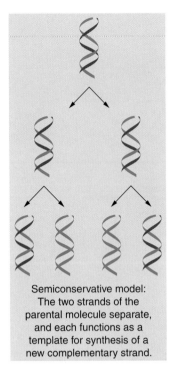

Figure 6-75. Semiconservative model of DNA synthesis which has been confirmed experimentally.

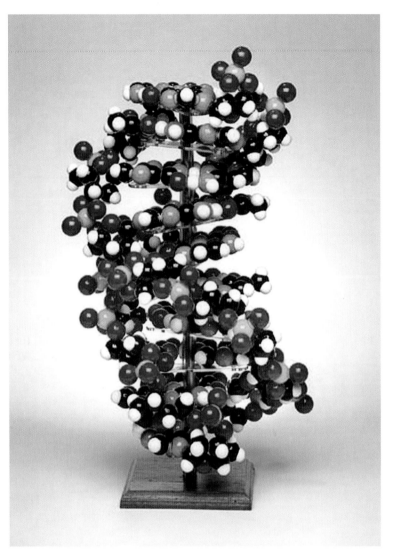

Figure 6-76. Space-filling atomic model of BDNA in a double strand. Reproduced with permission from http://www.edvotek.com/1509.html.

nuclear DNA. After completion, there are two daughter circular double-stranded mitochondrial DNAs for distribution to the two daughter mitochondria; some of these will be smooth circles, and others will have D-loops.

Mutations and Damage to DNA

Mutations sometimes occur in nuclear DNA. These mutations can be passed on to future generations, although some alterations of DNA can be corrected by repair mechanisms. As indicated, DNA polymerase has a proofreading mechanism that greatly reduces any mistakes in replication. Even with this mechanism in place, mistakes can occur at a frequency of about 1 in 1 billion (10^9) nucleotides. There are other mechanisms for the generation of mutations. Ultraviolet light, chemicals (carcinogens), radioactivity, and even heat can produce changes in DNA.

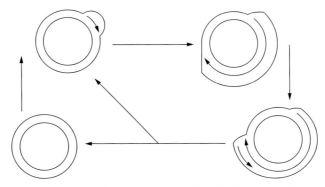

Figure 6-77. Mechanism of mitochondrial DNA replication. The mitochondrial DNA (containing about 40 genes) is a double-stranded circle *(lower left)*. One strand is replicated (as a leading strand), pauses, and waits for a signal to complete the synthesis. When the replication fork passes through the origin of the second strand (also a leading strand) replication of the second strand begins. The result is the generation of 2 double strands available for each daughter mitochondrion.

It is thought that, occasionally, cosmic radiation (gamma rays) makes a direct hit on cellular DNA, causing a mutation or breakage. Heat can destroy the linkage between purines and deoxyribose, and ultraviolet light can cause the dimerization of adjacent thymines in the same strand to form a thymine dimer. Chemicals can cause cytosine to be methylated, and radioactivity can break the strands of double-stranded DNA. Some repair of damage by heat can be made by repair enzymes, but other changes are irreversible.

The appearance of a thymine dimer, induced by ultraviolet light, between two adjacent thymine bases in the same DNA strand is shown in Figure 6-78. This lesion can be repaired by a nuclease and DNA polymerase (Figure 6-79). Figure 6-80 shows some of the direct-acting and

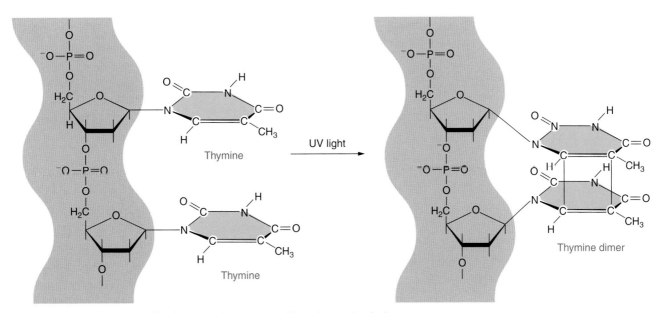

Figure 6-78. Appearance of a thymine dimer caused by ultraviolet light.

Biosynthesis of Deoxyribonucleic Acids 299

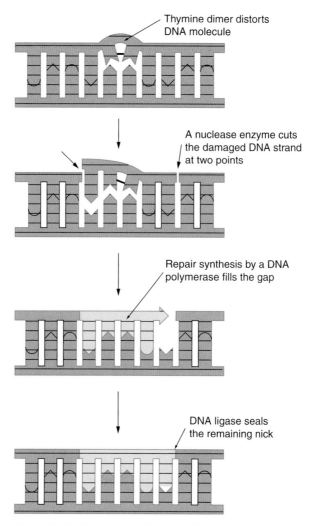

Figure 6-79. Repair of a thymine dimer in one DNA strand by a nuclease and DNA polymerase. Redrawn with permission from http://fajerpc.magnet.fsu.edu/Education/2010/Lectures/25_DNA_Replication.htm.

indirect-acting carcinogens. And Table 6-4 summarizes the various types of damage to DNA.

Specific Nucleases: Restriction Enzymes

The discovery of nucleases that could cut DNA at specific sequences revolutionized nucleic acid research and permitted enormous progress in DNA technology. These enzymes were discovered in bacteria, and they are called **restriction enzymes** or **restriction endonucleases.** Exonucleases are enzymes that cut DNA from the ends of the molecule, and endonucleases cut in the interior of the molecule. After the enzyme binds to the double strand, it travels along the molecule until it encounters its target signal sequence. Then it stops and performs its catalytic function. These

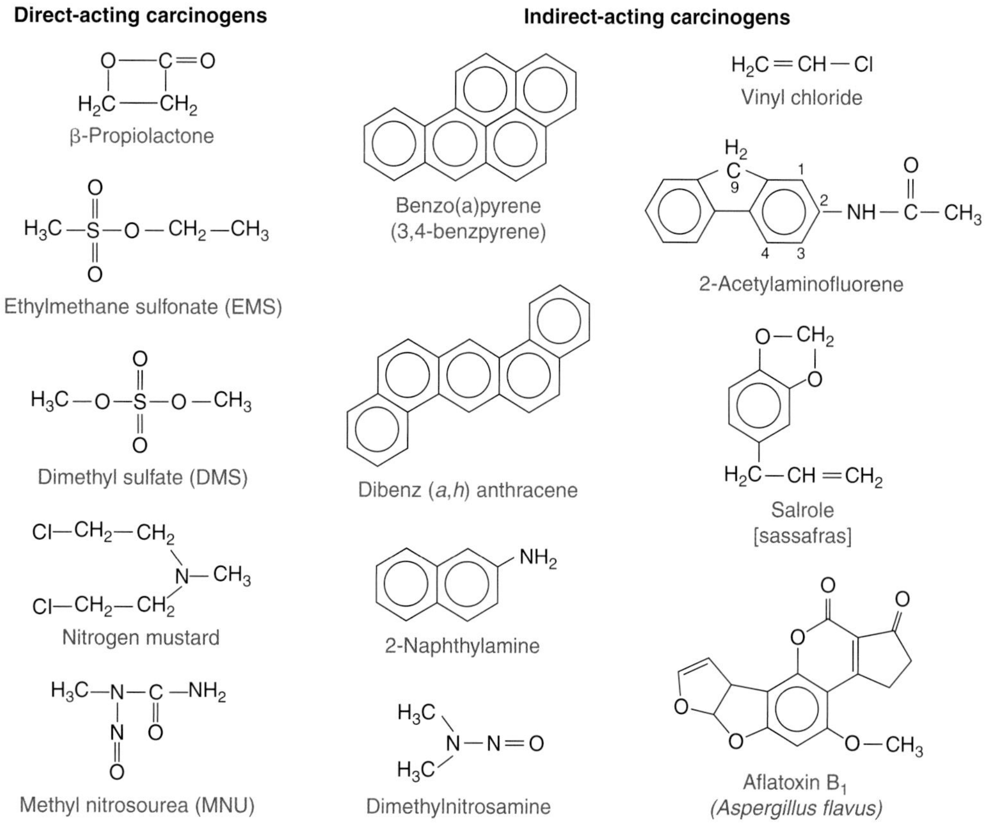

Figure 6-80. Direct-acting and indirect-acting chemical carcinogens that cause lesions in DNA and that can produce cancer.

Table 6-4
Types of Damage to DNA

I. Single base alteration
   A. Depurination
   B. Deamination of cytosine to uracil
   C. Deamination of adenine to hypoxanthine
   D. Alkylation of base
   E. Insertion or deletion of nucleotide
   F. Base analog–incorporation
II. Two base alteration
   A. UV light–induced thymine-thymine (pyrimidine) dimer
   B. Bifunctional alkylating agent cross-linkage
III. Chain breaks
   A. Ionizing radiation
   B. Radioactive disintegration of backbone element
   C. Oxidative free radical formation
IV. Cross-linkage
   A. Between bases in same or opposite strands
   B. Between DNA and protein molecules (e.g., histones)

Data from *Harper's Illustrated Biochemistry*, R.K. Murray et al., 26th ed. (2000), p. 335.

Biosynthesis of Deoxyribonucleic Acids

enzymes catalyze the cleavage of double-stranded DNA at specific sites, usually reading the target site of four, six, or eight bases in sequence from 5' to 3' direction. The specific enzymes are named for the organisms from which they were derived, in abbreviated form.

To visualize how these enzymes work, the example of the *Eco* RI restriction endonuclease is taken. As shown in Figure 6-81, this enzyme catalyzes the cleavage of DNA between the G and the A of the sequence: 5'-GAATTC-3' to give the products 5'... G and AATTC ... 3'. Examples of four restriction enzymes with their cleavage sites and products are shown in Figure 6-82. This figure shows that *Hae* III and *Sma* I each produce blunt-ended products; that is, the terminal bases emerging from either strand after cutting align with each other in space. On the other hand, *Eco* RI and *Hind* III produce "sticky" ends after cutting because one set of bases overlaps the other. The structure of *Eco* RI bound to double-stranded DNA, and the cleavage products of the reaction are shown in

Figure 6-81. The action of *E. coli* restriction endonuclease on double-stranded DNA. Redrawn from http://www.bioteach.ubc.ca/MolecularBiology/RestrictionEndonucleases/endonuclease%202.gif.

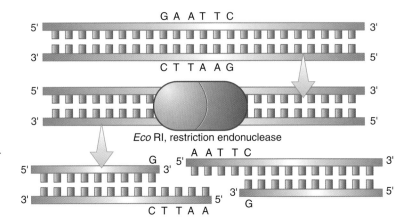

Four Restriction Enzymes

Enzyme	Recognition site	Cleavage products	
Eco RI	— GAATTC —	— G	AATTC —
	— CTTAAG —	— CTTAA	G —
Hae III	— GGCC —	— GG	CC —
	— CCGG —	— CC	GG —
Hind III	— AAGCTT —	— A	AGCTT —
	— TTCGAA —	— TTCGA	A —
Sma I	— CCCGGG —	— CCC	GGG —
	— GGGCCC —	— GGG	CCC —

Figure 6-82. Examples of four restriction enzymes with their recognition sites and cleavage products, which are either blunt-ended or overhanging "sticky" ends.

Figure 6-83. The top strand sequence is the same as the bottom strand sequence, except backward because both strands are read in the 5′ to 3′ direction and the two strands have opposite orientations to one another: one is in the 5′ to 3′ direction, and the other strand is in the 3′ to 5′ direction. The overhanging ends produced in the products of cleavage are called "sticky" ends because the base pairs formed between the two overhanging portions will glue the two pieces together even though the backbone is cut. *Importantly, these sticky ends allow for the insertion of foreign DNAs, which can be designed to have the appropriate sticky end sequences.* Thus, in the case of *Eco* RI, the overhanging end is TTAA, which could glue together with the overhang of the other strand, AATT (which would form a base pair with TTAA; AA forms base pairs with TT, and TT forms a base pair with AA). Molecular structures of *Eco* RV bound to DNA before and after cutting the two strands (double arrow on the bottom figure) are shown in Figure 6-84.

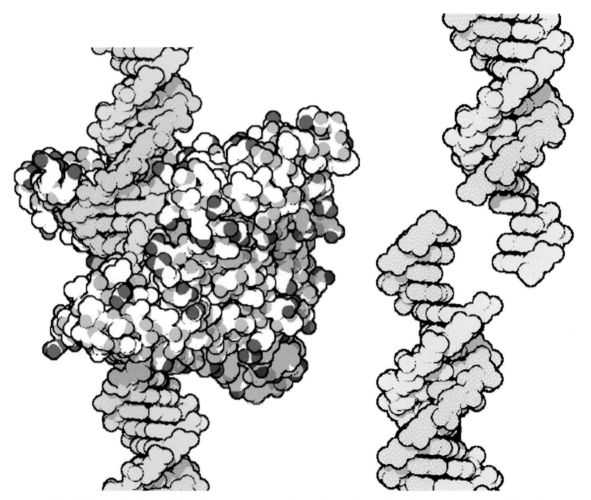

Figure 6-83. *Left*, atomic models of *Eco* RI restriction endonuclease bound to double-stranded DNA. *Right*, release of the specifically cleaved DNA strands. The intact sequence of the *Eco* RI target sequence is shown in the upper left, and the sequence of the cleared products of the two strands is in the lower right. Reproduced from the Protein Data Bank: http://www.rcsb.org/pdb/molecules/pdb8_2.html.

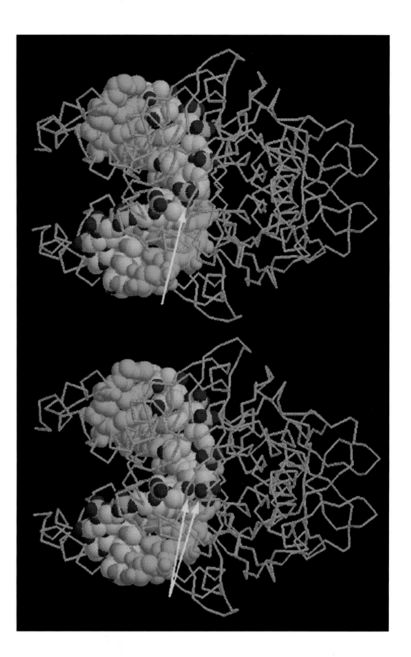

Figure 6-84. Structures of *Eco* RV bound to DNA. Above, before the cleavage of DNA. Bottom, after the cleavage of the double strand *(double arrow)*. Reproduced with permission from http://www.rcsb.org/pdb/molecules/pdb8_3.html.

For a given length of DNA, there will be a number of restriction sites for a specific restriction enzyme. An example of a number of *Eco* RI target sites can be seen in the case of a large DNA of 50 kilobases (a cosmid), where six specific sites exist that would create a signature of products. The *Eco* RI enzyme molecules bound to their specific target sites (enzymatic action has not taken place) are shown in Figure 6-85.

The roster of restriction enzymes and their specificities is quite large, and an extensive list is presented in Table 6-5. Use of restriction enzymes for mapping is one way to characterize a DNA molecule. It is a means to isolate a specific fragment from linear or circular DNA for further manipulations. The DNA fragments are separated out on gel electrophoresis on the basis of size. The gel can be stained with ethidium bromide, which binds to the fragments and is visible under ultraviolet light. The bands are recorded on film (ethidium bromide intercalates between the base pairs of DNA and fluoresces yellow–green under ultraviolet light).

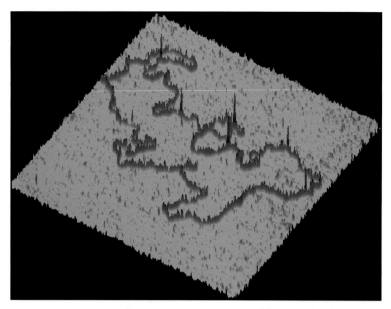

Figure 6-85. A map of six *Eco* RI sites on a 50-kb cosmid. The bound enzymes are visualized as vertical red bars at various points on the double-stranded DNA *(blue molecule)*. If the catalytic action took place, a number of fragments of different sizes would characterize the cleavage pattern. Reproduced from http://homer.hsr.ornl.gov/cbps/mappage.htm.

A hypothetical digest is shown in Figure 6-86. Using this methodology, detailed restriction maps can be developed especially for vectors used as the vehicles to carry a particular DNA fragment of interest. An example of the map of pBR322, an important vector, is shown in Figure 6-87.

The introduction of a foreign DNA into a plasmid using an *Eco* RI restriction site, as an example, is shown in Figure 6-88, and the details of the cutting and attachment to the sticky ends are shown in Figure 6-89. This procedure (recombination and the production of recombinant DNA) makes it possible to express substances that can be used for therapy when these are proteins. The expressed protein then contains the *human sequence* so that little or no immunological response is expected when treating a human as opposed to proteins isolated from animals (porcine insulin, for example) that would produce an immunological response, partially neutralizing the hormone. Also, this process can be used to avert the danger of prion disease that could accompany the use of proteins derived from animal tissues, especially brain (certain hormones like growth hormone, for example). An example of the cloning of the human insulin gene for the purpose of producing human insulin for therapeutic use is shown in Figure 6-90.

Natural Genomic DNA

Mammalian genes have introns and exons. Introns are sequences that are not used in the production of the gene-encoded protein. The exons are joined into the meaningful message, *excluding the introns* (but it is still possible that mutations may occur by incorrect splicing that include parts of introns in the mRNA and the expressed protein). To make an

Table 6-5
A List of Many Restriction Endonucleases with Their Recognition Sites in DNA

Enzyme	Recognition Site	Enzyme	Recognition Site	Enzyme	Recognition Site
Aat II	GACGI▼C	Cla I	AT▼ CGAT	Nde I	CA▼ TATG
AccI	GT▼ (A/T)(T/G)AC	Csp I	CG▼ G(A/T)CCG	NgoM I	G▼ CCGGC
AccIll	T▼CCGGA	Csp 45 I	TT▼ CGAA	Nhe I	G▼ CTAGC
Acc65 I	G▼GTACC	Dde I	C▼ TNAG	Not I	GC▼ GGCCGC
AccB7 I	CCANNNN▼NTGG	Dpn I	G^meA▼ TC	Nru I	TCG▼ CGA
AcyI	G(A/G)▼ CG(T/C)C	Dra I	TTT▼ AAA	Nsi I	ATGCA▼ T
Age I	A▼ CCGGT	EclHK I	GACNNN▼ NNGTC	Pst I	CTGCA▼ G
Alu I	AG▼ CT	Eco47 III	ACG▼ GCT	Pvu I	CGAT▼ CG
A/w26 I	G▼ TCTC(1/5)	Eco52 I	C▼ GGCCG	Rvu II	CAG▼ CTG
A/w441	G▼ TGCAC	Eco72 I	CAC▼ GTG	Rsa I	GT▼ AC
Apa I	GGGCC▼ C	EcoI CR I	GAG▼ CJC	Sac I	GAGGCT▼ C
Ava I	C▼ (T/C)CG(A/G)G	Eco RI	G▼ AATTC	Sac II	CCGC▼ GG
Ava II	G▼ G(A/T)CC	Eco RV	GAT▼ ATC	Sal I	G▼TCGAC
Ba/ I	TGG▼ CCA	Fok I	GGATG(9/13)	Sau3A I	▼ GATC
BamH I	G▼ GATCC	Hae II	(A/G)GCGC▼ (T/C)	Sau96 I	G▼ GNCC
Ban I	G▼ G(T/C)(A/G)CC	Hae III	GG▼ CC	Sca I	AGT▼ ACT
Ban II	G(A/G)GC(T/C)▼ C	Hha I	GCG▼ C	Sfi I	GGCCNNNN▼ NGGCC
Bbu I	GCATG▼ C	Hinc II	GT(T/C)▼ (A/G)AC	Sgf I	GCGAT▼ CGC
Bc/ I	T▼ GATCA	Hind III	A▼ AGCTT	Sin I	G▼ G(A/T)CC
Bgl I	GCCNNNN▼ NGGC	Hinf I	G▼ ANTC	Sma I	CCC▼ GGG
Bg/ II	A▼ GATCT	Hpa I	GTT▼ AAC	SnaB I	TAC▼ GTA
BsaM I	GATTGCN▼	Hpa II	C▼ CGG	Spe I	A▼ CTAGT
BsaO I	CG(A/G)(T/C)▼ CG	Hsp92 I	G(A/G)▼ CG(T/C)C	Sph I	GCATG▼ C
Bsp1286 I	G(G/A/T)GC(C/A/T)▼ C	Hsp92 II	CATG▼	Ssp I	AAT▼ ATT
BsrBR I	GATNN▼ NNATC	I-Ppo I	CTCTCTTAA▼ GGTAGC	Stu I	AGG▼ CCT
BsrS I	ACTGGN▼	Kpn I	GGTAC▼ C	Sty I	C▼ C(A/T)(T/A)GG
BssH II	G▼ CGCGC	Mbo I	▼ GATC	Taq I	T▼ CGA
Bst71 I	GCAGC(8/12)	Mbo II	GAAGA(8/7)	Tru9 I	T▼ TAA
Bst98 I	C▼ TTAAG	Mlu I	A▼ CGCGT	Tthlll I	GACN▼ NNGTC
Bst E II	G▼ GTNACC	Msp I	C▼ CGG	Vsp I	A▼ TAAT
Bst O I	CC▼ (A/T)GG	MspA I	C(A/C)G▼ C(G/T)G	Xba I	T▼ CTAGA
Bst XI	CCANNNNN▼ NTGG	Nae I	GCC▼ GGC	Xho I	C▼ TCGAG
Bst ZI	C▼ GGCCG	Nar	GG▼ CGCC	Xho II	(A/G)▼ GATC(T/C)
Bsu36 I	CC▼ TNAGG	Nci I	CC▼ (G/C)GG	Xma I	C▼ CCGGG
Cfo I	GCG▼ C	Nco I	C▼ CATGG	Xmn I	GAANN▼ NNTTC

Data from http://www.bioscience.org

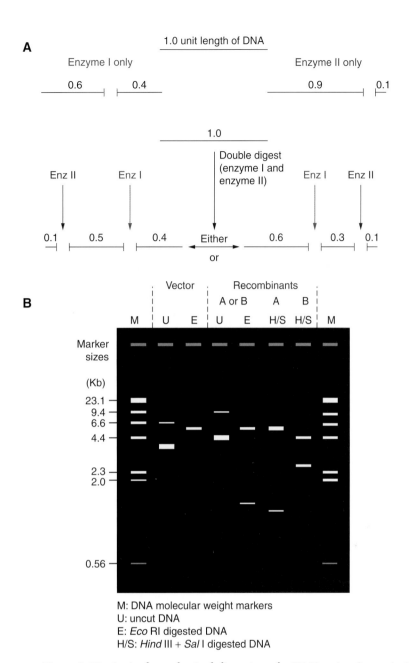

Figure 6-86. **A.** An hypothetical digestion of a DNA using 2 restriction enzymes. **B.** Appearance of a restriction enzyme digest as visualized by ethidium bromide after gel electrophoresis. Reference is made to the positions of molecular weight markers (M).

artificial DNA complementary to a human (or other species) gene, the mRNA, which contains exons spliced together (after the removal of the introns, called **editing**), must be reverse transcribed to form a complementary DNA (cDNA). This process is pictured in Figure 6-91. The structure of the enzyme, reverse transcriptase, is shown in Figure 6-92. By preparing the cDNA from the mRNA (intron information excluded), the cDNA can be used to encode a message directly and then translate to a protein that will be functional.

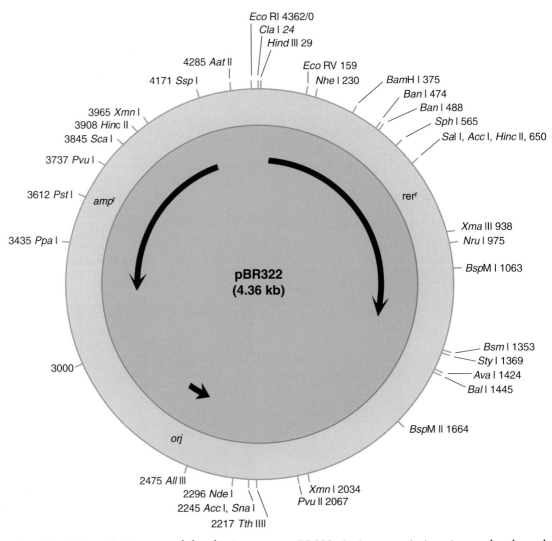

Figure 6-87. A restriction map of the cloning vector, pBR322. A given restriction site can be cleaved and a foreign DNA can be inserted when the sticky ends of the cleavage site are hybridized with the sticky ends (that are complementary) of the foreign DNA. About 10% of the total number of restriction sites are shown in this figure.

Sequencing DNA

To learn the sequence (of bases) in a single strand of DNA, the complementary strand will be synthesized in the laboratory. The synthesis will be catalyzed by DNA polymerase with the addition, first, of dideoxyATP (ddATP) in the place of dATP. The dideoxy nucleotide does not have a hydroxyl group (only a hydrogen) in the 3' position, which is required for the further addition of nucleotides to a growing chain (Figure 6-65). In consequence, the chain will add further nucleotides until an A is to be added and then the reaction will stop because a further nucleotide cannot be added without an available hydroxyl at the 3' position. The same procedure will be carried out except for the addition, individually, of ddCTP in place of dCTP, ddGTP in place of dGTP, and ddTTP in place of dTTP. In each case, a series of fragments will be synthesized until the final

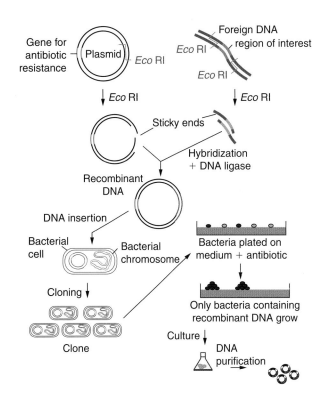

Figure 6-88. Introduction of a foreign DNA into a plasmid, known as gene cloning, using an *EcoR*1 restriction site.

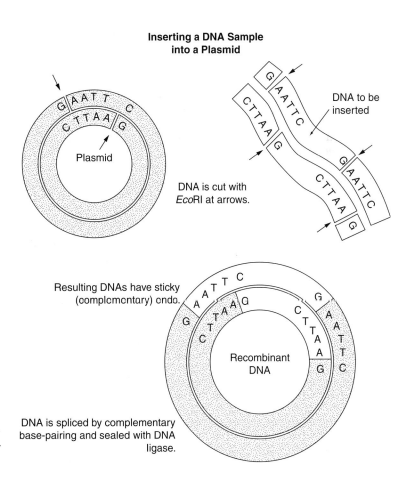

Figure 6-89. Details of insertion of a foreign DNA into a plasmid using an *EcoR*1 restriction site. Redrawn with permission from http://web.mit.edu/esgbio/www/rdna/cloning/html.

Biosynthesis of Deoxyribonucleic Acids   309

How the Insulin Gene Is Transferred

Plasmids are small circles of DNA found in bacterial cells, separate from the bacterial chromosome.

Restriction enzymes cut across the two strands, leaving loose ends to which cDNA can be attached.

Special linker sequences are added to the human cDNA so that it will fit precisely into the loose ends of the opened plasmid DNA ring.

The plasmid cotaining the human gene is now ready to be inserted into a living organism.

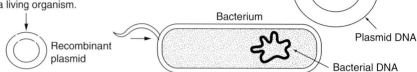

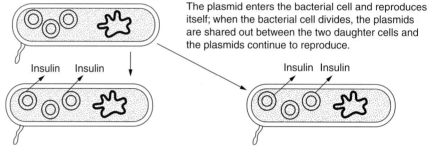

Cloning the Human Insulin Gene

The plasmid enters the bacterial cell and reproduces itself; when the bacterial cell divides, the plasmids are shared out between the two daughter cells and the plasmids continue to reproduce.

In this way, a clone of identical cells is formed. If the human gene incorporated encodes for the hormone insulin, then such a clone can provide a reliable insulin source.

Figure 6-90. Cloning of the human insulin gene. Redrawn with permission from http://web.mit.edu/esgbio/www/rdna/cloning.htm.

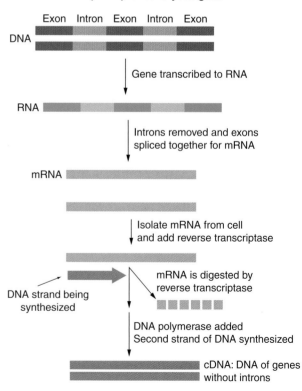

Figure 6-91. Producing a complementary DNA (cDNA) from an isolated eukaryotic gene.

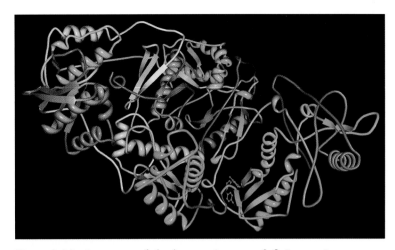

Figure 6-92. Structure of the human immunodeficiency virus reverse transcriptase (PDB 1hmi).

dideoxy nucleotide is added, representing the last base. The fragments are then separated by size on an electrophoretic gel and exposed—if the dideoxy compounds were radiolabeled or each was conjugated to a different dye so that their positions on a gel could be determined by exposure to x-ray film or ordinary film. Then, the size of the fragments would determine their order and the number of bases until an A is reached, until a G is reached, until a C is reached, and until a T is reached would be known. In this way, the sequence can be constructed. Because this would yield the sequence of the complementary strand, the sequence of the original template strand could then be written (using the rules for base-pairing. Where an A was found, the original strand would have a T; where a G was found, the original strand would have a C, and vice versa).

Inhibition of DNA Synthesis

Until recently, one of the key methods to "knock out" a gene has been to introduce an antisense DNA that has a sequence complementary to the sense strand (like the naturally occurring double strand). Because DNA transcribes single-stranded mRNA, the antisense DNA can form a duplex with the single-stranded mRNA, thus preventing translation of the protein it encodes. Likewise, antisense RNA has been used to achieve the same result because single-stranded mRNA can form a duplex with an antisense strand. Although these methods have been useful, they have drawbacks. In some cases, a given antisense molecule, after being introduced into a cell, will function better than others.

Lately, a new exciting technique has been discovered. This involves **posttranscriptional gene silencing (PTGS)** using double-stranded RNA (dsRNA) to knock out expression of specific genes. Hopefully, this new work may lead to usefulness in gene therapy. This method is also referred to as RNA interference (RNAi). The double-stranded RNA consists of sense and antisense strands (where the sense strand, like mRNA, carries the information for the translation of a specific protein). The use of dsRNA is more efficacious than the introduction of only a single strand. In mammalian cells, the dsRNAs are broken down into short inhibitory RNAs (siRNAs) when the dsRNA is introduced into the cell by a transgene (introduced into an organism by artificial means) or a virus. Interestingly, when dsRNA of more than 30 nucleotides is used, the organism may respond nonspecifically in an antiviral response that leads to nonspecific gene suppression. However when dsRNAs of about 21 nucleotides in length with 2 nucleotide 3' overhangs (like sticky ends) are used, the nonspecific response is bypassed. Transient transfection of these siRNAs could lead to more than 90% reduction in the target RNA and specific protein levels. The sequence of the sense dsRNA has to match the target mRNA closely because any mismatch will significantly reduce the ability to silence the gene expression. This work is still being developed and should form the basis for eventual specific knockouts of defective genes that could be replaced with functional ones.

Functional Genomics

The full sequencing of the human genome (the genes contained on all chromosomes) opened a huge field of exploration that will increase the understanding of our species and find new and specific drugs to cure diseases. Another objective is to map all human genes (some 80,000 to 100,000 genes) to their chromosomes and determine their precise position on a given chromosome, an area that has been progressing rapidly. One way to use this information is to compare two cells, perhaps a tumor cell with its normal parent, to determine the genetic changes that have taken place in the oncogenic process, as well as to discover new oncogenes (genes that cause cancer). Any two situations in a given cell can be compared as to the nature of the genetic alterations that have occurred, as result of exposing the cell to some agent, such as a drug or a hormone. Although comparing a cancer cell to its parent may involve a relatively huge number of changes in gene expression, comparing the early effects of a given chemical in one cell could give discrete information on how that chemical acts on the cell. For example, a cell might be under the control of a given hormone. The two samples for investigation using microarray analysis might be the cell at 0 time of exposure and the cell after 30 minutes exposure to the hormone.

Changes in the levels of mRNAs are correlated to the differences in the state or the type of the cell. For each sample, the total RNA is prepared (usually from cells in tissue culture), then the single-stranded cDNAs are prepared to the total RNA. With appropriate enzymes (DNA ligase, DNA pol I, and RNaseH), the double-stranded cDNAs are formed. From these, the cRNAs are made. These are labeled with fluorescent tags; green for one sample and red for the other. The two sets of cRNAs from the two samples are mixed together and washed over the microarray (probably a glass surface). Each probe is then excited with a laser beam, and its fluorescence at each element is detected with a scanning confocal microscope. A pattern of dots is generated, one dot for each gene equivalent. When the dot is green, the amount of mRNA in the first sample exceeds the amount in the second sample. When the spot is red, the amount of mRNA in the second sample exceeds the amount in the first sample. When the dot is yellow, the amounts of mRNA in each sample are about equal.

The intensity of the red or green fluorescence is then determined, and the fold induction or suppression, as the cases may be, is determined. Because of some baseline noise in the system, it is useful to inspect genes that are induced, or repressed, perhaps by a factor of three in the first measurement. When genes of interest are established on this basis, induction (for example) can be confirmed by other methods, such as Northern blotting, with time after treatment with the hormone to confirm that the gene in question has indeed been induced. This is an extremely valuable approach for learning about the patterns of genes induced and/or suppressed by a specific hormone or drug, for classifying tumors, and for inspecting the genetic changes during development. In Figure 6-93, the process of the gene chip array is summarized without a lot of detail.

An extension of this technique is the proteome, whereby large two-dimensional electrophoretic gels the proteins in the cell are separated and

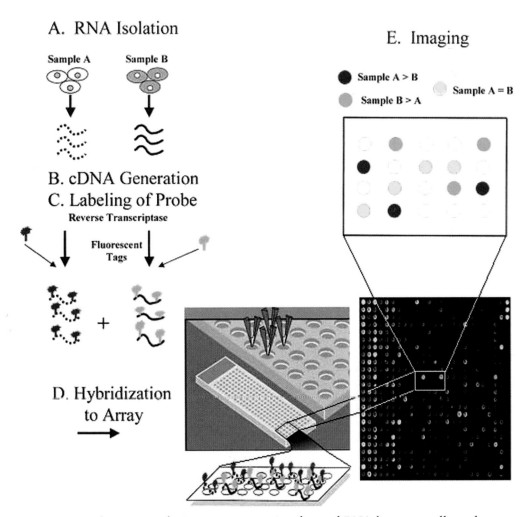

Figure 6-93. The process of microarray comparing the total RNA from two cells or the same cell, one bearing a treatment with an agent. Reproduced with permission from http://www.fao.org/DOCREP/003/X6884E/x6884e00.jpg.

identified. Each gel can examine thousands of proteins, although a human cell may contain 30,000 proteins or more. The proteome analysis can be used to compare different cells or the same cell treated differently and establish patterns of proteins whose expression is induced or suppressed.

Gene Therapy

The hope offered by gene therapy is the ability to replace malfunctioning (or nonfunctioning) genes with normal genes. In this way, a normal protein could be produced that is either absent or nonfunctional and causing disease. Although some experiments have succeeded in animals and a few have worked in humans, the field is in its infancy. Nevertheless, the future utility of the approach should pay dividends. The provision of genes that, when activated, would cause the death of a cancer cell would be another objective, as well as knocking out a gene by some methods mentioned earlier. Some diseases are obvious targets for gene therapy. Cystic fibrosis affects 1 in 2500 individuals, and about 1 in 31 people are carriers of this recessive disease. A major difficulty in this disease is a mutation in a chloride channel gene (**cystic fibrosis transmembrane regulator, CFTR**) affecting chloride ion absorption. This produces a thick mucus that clogs the airways and leads to dangerous lung infections. The mucus also prevents pancreatic enzymes from reaching the intestines, so food is not digested properly. Duchenne muscular dystrophy features a huge DMD gene that encodes dystrophin. There are mutations, deletions, and duplications in the gene that cause progressive proximal muscular deterioration and death at an early age. Another prospect is sickle cell disease, which affects 1 in 10 African Americans, meaning that about 3 million people in the United States are carriers of this disease. About 80,000 individuals have the disease actively. There is a substitution of valine for a glutamate in β-globin, producing abnormally sticky hemoglobin molecules that cause a deformation of the red blood cell and reduce the oxygen-carrying capacity of the blood. Familial hypercholesterylemia has been discussed, and the gene therapy target would be to supply an appropriate LDL transporter to the liver and possibly other tissues. Other diseases of no lesser importance that would be targets for gene therapy are hemophilia and SCID with a deficiency of adenosine deaminase.

Up to this time, the main problem seems to be in the delivery of the normal gene to replace a malfunctioning gene. Even when the gene is successfully replaced in a human, often the effect lasts only for several months rather than indefinitely. A number of vectors to carry the replacement gene have been proposed, and many have been tried. Table 6-6 lists some of these vectors with their advantages and disadvantages. There are two popular approaches. The first consists of **ex vivo therapies,** in which cells are removed from the patient's body, treated with a functional gene, and then readministered back to the patient; these often involve blood cells. The other is *in vivo* therapy, in which the therapeutic gene is administered to the patient directly using one of the gene delivery systems. Promising experiments may lead to a large-scale clinical trial; however, there are many roadblocks to be hurdled, especially related to the safety of the patient, before a gene therapy clinical trial can take place. Institutional and federal committees are involved in the review process.

Table 6-6
Some Viral and Nonviral Vectors Used or Tested in Gene Therapy Experiments

Vectors	Advantages	Disadvantages	Notes
Retroviruses (e.g., MMLV)	1. Long-lasting gene expression due to stable integration 2. Enters cells efficiently	1. Only infects dividing cells 2. Potential insertional mutagenesis 3. Hard to produce (low yield)	Well suited for HSC GT Note: 37% of all GT trials use retroviruses
Lentiviruses Retroviral subclass (i.e., HIV)	1. Long-lasting gene expression due to stable integration 2. Will infect both dividing and nondividing cells	1. Potential insertional mutagenesis	10% of at GT trials
Adenoviruses (dsDNA virus)	1. Enters cells efficiently 2. High rate of delivery and expression of theraputic gene 3. Does not integrate into chromosome 4. Infects nondividing cells	1. Immunogenic—cleared rapidly from body 2. Can cause inflammation and tissue damage	Short expression time (~14 days) good for one-time or infrequent treatments 20% of all GT trials
Adenoassociated viruses (AAV)	1. Much less immunogenic and toxic than adenovirus 2. Long-term expression possible 3. Wide host cell range	1. Difficult to produce in high quantities	Stable in nucleus as a nonreplicating extrachromosomal form 1% of all GT trials
Herpes simplex virus	1. Produced at high levels 2. Targets nondividing nerve cells 3. Can carry a great deal of DNA 4. Maintained as a nonreplicating extrachromosomal form in neuronal and bone marrow target tissue	1. Immunogenic 2. Potentially toxic—can cause encephalitis	6% of all GT trials
Liposomes	1. Not immunogenic 2. Can deliver large quantities of DNA 3. High affinity for uptake into cytoplasm by endocytosis	1. Low rate of delivery 2. Transient expression	Binds to and spontaneously condenses DNA
Plasmid therapy	1. No viral component	1. Transient gene expression 2. Difficult to target to specific tissues	3% of all GT trials

DNA, deoxyribonucleic acid; *GT*, gene therapy; *HIV*, human immunodeficiency virus; *HSC*, hematopoietic stem cell; and *MMLV*, Moloney murine leukemia virus.
Reproduced from http://www.biology.iupui.edu/biocourses/Bio1540/15genetherapy2k1.html.

The first successful gene therapy experiment was done in 1990 when a 4-year-old girl with SCID (adenosine deaminase deficiency) was treated by the insertion of a normal adenosine deaminase gene into the girl's cells. As a result, her immune system could function at 25% of normal, allowing her to lead a normal life, although the process involved regular injections of T cells.

Ribonucleic Acids

RNAs resemble DNAs, but in general they are single-stranded. As mentioned earlier, thymine in DNA is always replaced by uracil in RNA. The sugar moiety in RNA is ribose, compared to deoxyribose in DNA. The biosynthesis of purines and pyrimidines, including UMP (Figures 6-38 through 6-40), has been presented already.

There are three major types of RNA, along with some others that will be noted. The major RNAs (rRNA, mRNA, and tRNA) are transcribed from their cognate genes (Figure 6-94). Insofar as mRNA is concerned, it is transcribed from its gene to form nuclear RNA, which contains information from introns in DNA, as well as exons. About 1% of the total DNA is transcribed into mRNA. After splicing, the introns are removed and the exons are attached in sequence to form mRNA (Figure 6-95). As described previously, mRNA is exported from the nucleus to the cytoplasm, where it is translated into protein by the ribosome.

tRNA is single-stranded with a cloverleaf appearance because it folds back upon itself, generating regions of base-pairing as shown in Figure 6-96. tRNAs deliver amino acids from the cytoplasm to the ribosome where a protein is being synthesized. Some detail of both mRNAs and tRNAs has been given in the chapter on proteins, specifically on protein synthesis. The three-letter anticodon at the end of tRNA complements the amino acid code in mRNA, directing the sequence of attaching amino acids to the growing polypeptide chain (see Figures 2-41 through 2-44A, Figures 2-45 through 2-48, Figure 2-52, Figure 2-53, and Figure 2-56).

In tRNA, it has been seen that the single strand can fold back on itself to form a double strand for an interim. This double-strandedness is stabilized by base-pairing. RNA can have other types of secondary structures, including duplexes, single-stranded regions, **hairpins,**

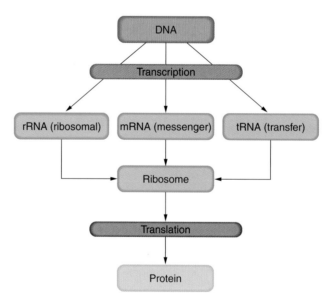

Figure 6-94. Transfer of information from the gene (DNA) through RNA to expression of protein. The types of RNAs are transcribed from their respective genes.

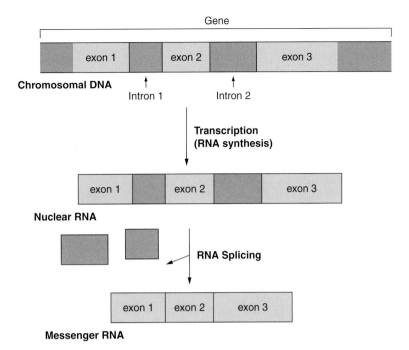

Figure 6-95. Pathway of messenger RNA synthesis from its cognate gene.

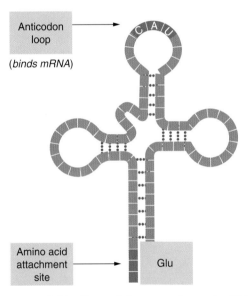

Figure 6-96. General features of transfer RNA (tRNA). Where the single strand folds back upon itself, base-pairing occurs.

internal loops, bulge loops or bulges and junctions. These structures are summarized in Figure 6-97. Some of these structures are identifiable in tRNA. When a secondary structure pairs with a complementary sequence outside the loop, a pseudoknot can be formed, representing a tertiary structure. As shown in Figure 6-98, this can occur when the loop at the top crosses the major groove and the loop at the bottom crosses the minor groove (Figure 6-98A). It also can occur when one loop crosses the major groove of DNA and the other group bridges the

Ribonucleic Acids 317

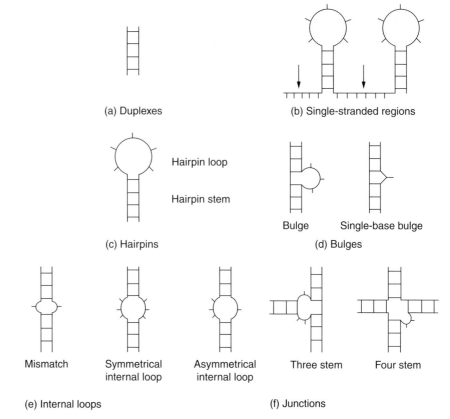

Figure 6-97. Secondary structures of RNA.

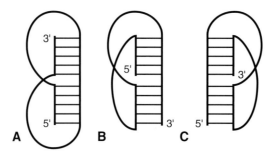

Figure 6-98. Formation of three types of pseudoknots in RNA when it interacts with DNA.

whole helix, as shown in Figure 6-98B. A pseudoknot also can be formed when one loop of RNA crosses the minor groove and the other loop bridges the entire helix (Figure 6-98C). There are other possibilities for the formation of pseudoknots.

In addition to mRNA and tRNA, there is rRNA, small nuclear RNA (snRNA), heterogeneous nuclear RNA (hnRNA), small nucleolar RNA (snoRNA), and RNA that serves as the enzyme, ribozyme.

As discussed previously, rRNA forms ribosomes in complexes with proteins (Figure 2-38). snRNA, of 100 to 300 base pairs, forms small nuclear ribonucleoprotein particles, which are involved in pre-mRNA splicing. In Figure 6-99, a diagram of an snRNA is shown with some

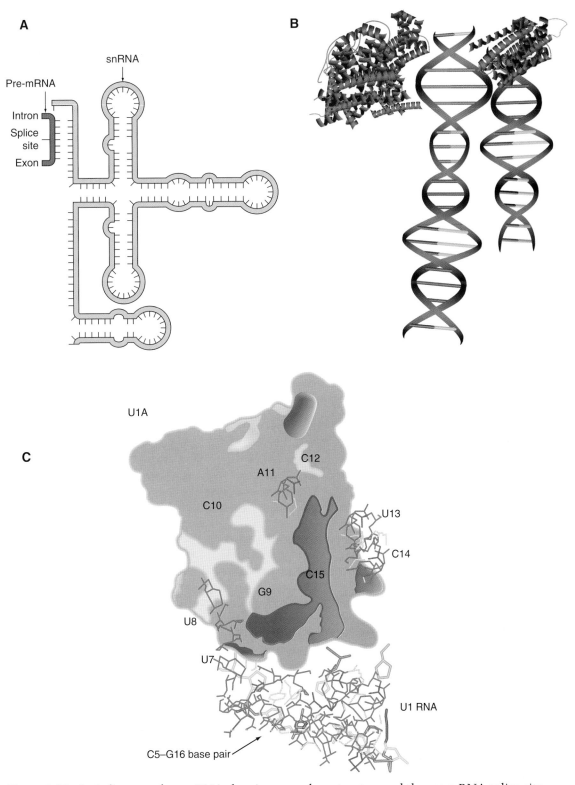

Figure 6-99. **A.** A diagram of an snRNA showing secondary structure and the pre-mRNA splice site. **B.** The crystal structure of a human SRP ternary complex consisting of the M domain of SRP54, SRP19, and 7SL RNA. **C.** Crystal structure at 1.92 Å resolution of the RNA binding domain of the U1A protein complexed with an RNA hairpin. Part B reproduced with permission from A. Kuglstatter, C. Oubridge, and K. Nagai. *Nat. Struct. Biol.* 9: 740–744, 2002, from http://www2.mrc-lmb.cam.ac.uk/personal/kn/NewFiles/Mrc_ach.html. Part C reproduced with permission from C. Oubridge et al., "Crystal structure at 1.92 Å resolution of the RNA-binding domain of the U1A Spliceosomal protein complexed with an RNA hairpin," *Nature,* 372: 432–438, 1994, from http://www2.mrc-lmb.cam.ac.uk/personal/kn/NewFiles/Mrc_ach.html.

structures of a 7SL RNA bound to two proteins, SRP54 and SRP19, and another structure of the RNA-binding domain of the U1A protein complexed with an RNA hairpin. hnRNA is the primary transcript from RNA polymerase II (RNA pol II). It is called pre-RNA because it is the precursor of all of the mRNAs and it contains the information from introns, as well as exons. To obtain meaningful mRNA, the introns are spliced out with the help of snRNAs. hnRNAs are about 5000 to 20,000 nucleotides long compared to about 2000 bases found in mRNA, and the hnRNA has a half-life of minutes compared to hours for mRNA. This suggests that 3000 to 17,000 bases can be attributed to intron sequences. Figure 6-100 shows the addition of the polyA tail to mRNA

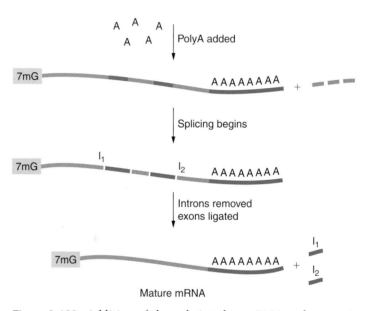

Figure 6-100. Addition of the polyA tail to mRNA and processing to mature mRNA.

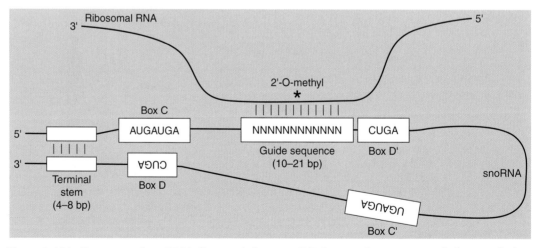

Figure 6-101. Structure of snoRNA *(bottom)* showing C/D box guide to 2'-O-methylation of ribosomal RNA. *bp*, Base pairs. Redrawn from http://lowelab.ucsc.edu/~lowe/thesis/node14.html.

and the further processing to mature mRNA. snoRNA is located in the nucleolus within the nucleus, where it functions in the processing and ribose methylation of rRNA. snoRNA contains one or two 10- to 20-base-pair sequences that complement rRNA to position that RNA for 2′-O-methylation. In addition, there are other stretches of base sequences (boxes) for localization of snoRNA to the nucleolus and association with ribonucleoprotein particles. The C′ and D′ boxes function as guides for the methylation function, as shown in Figure 6-101. A more detailed structure of a snoRNA is shown in Figure 6-102. The ribozyme has been discussed in the context of its enzymatic activity in the formation of the peptide bond during protein synthesis (Figure 2-46).

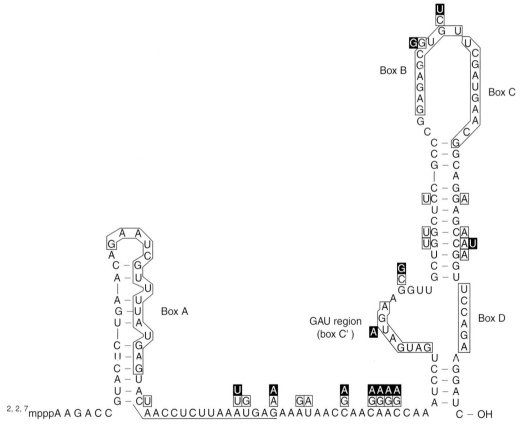

Figure 6-102. A detailed structure of a small nucleolar RNA (trypanosomal U3sno RNA). Redrawn with permission from T. Hartshorne and N. Agabian, "A common core structure for U3 small nucleolar RNAs," and redrawn directly from http://www.pubmedcentral.nih.gov/pagerender/fcgi?artid=331778&pageindex=2.

Further Reading

Bloomfield, V.A., Crothers, D.M., Tinoco, I., Hearst, J.E., Wemmer, D.E., Killman, P.A., and Turner, D.H., *Nucleic Acids: Structures, Properties, and Functions*, University Science Books, 2000.

CHAPTER 7

Transcription

Asbestosis: A Disease of Aberrant Transcription

Asbestos is a general term for a number of naturally occurring mineral fibers. Among these are actinolite, amosite, anthophyllite, crocidolite, chrysotile, and tremolite. The appearance of tremolite fibers is shown in Figure 7-1. As early as the fifth century B.C., the Greeks used asbestos fibers woven into lamp wicks. In the late nineteenth century, asbestos was widely used in building products, especially in pipe construction and acoustic plaster. By the late 1970s, this usage was banned in the United States, but the use of asbestos is still common in vinyl floor tile, cement pipe, and asphalt roofing. Canada and Brazil still export asbestos products. Asbestos fibers constitute a hazard when they become airborne and are inhaled and deposited in the lungs. This can result in asbestosis, which is a debilitating lung disease, and this can lead to **mesothelioma,** which is a rare cancer of the lung or stomach lining. Asbestosis is the scarring of the lung tissue that can cause mild impairment to disability to death. Mesothelioma is a cancer of the pleural or peritoneal cavity. It may spread to tissues surrounding the lungs or other organs. Most mesotheliomas are attributable to exposure to asbestos. The probability of developing lung cancer from asbestos is greatly increased by smoking. The period between asbestos exposure and the onset of disease ranges from 10 to 40 years. An overview of asbestos exposure is shown in Figure 7-2. Figure 7-3 shows the results of asbestosis in the lungs with pulmonary fibrosis and pleural thickening.

NFκB is a protein DNA-binding factor required for the transcription of proinflammatory molecules and has been implicated in lung diseases, including occupational and environmental lung disease. These proinflammatory molecules include **intercellular adhesion molecule (ICAM-1)**, inducible nitric oxide synthase, cyclooxygenase 2, **interleukin-1-β (IL-1β)**, TNFα, IL-6, and chemokines like IL-8. NFκB is the critical mediator of the inflammatory cascade. The overall activity of NFκB as it is activated by TNFα is shown in Figure 7-4. RelA, the larger subunit, is 65,000 in molecular weight; thus, it is also referred to as p65 (the Rel is a family of proteins of which p65 is one). NFκB can bind to DNA either as a homodimer of p50 or as the heterodimer of p50

Figure 7-1. The appearance of tremolite fibers. Reproduced with permission from Daniel Friedman http://www.inspect-ny.com.

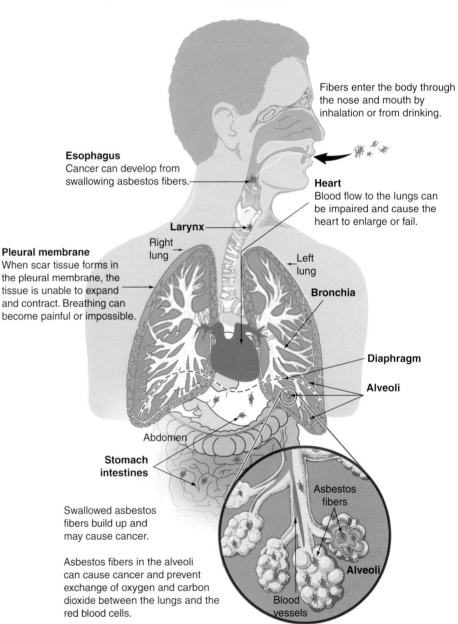

Figure 7-2. An overview of asbestos exposure.

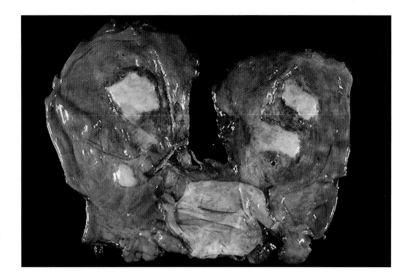

Figure 7-3. Asbestosis lungs with pulmonary fibrosis and plural thickening. Reproduced with permission from http://www.mesotheleoma.com/_diagnosis.htm.

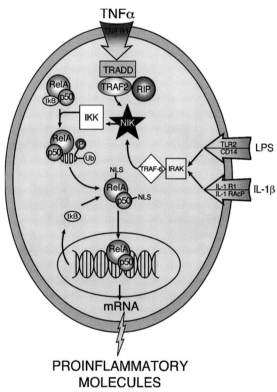

Figure 7-4. Overview of the activation of NFκB by TNF-α. NFκB is a cytoplasmic protein consisting of RelA and p50 subunits and the subunit IκB. Upon activation, IκB is dissociated from the complex after its phosphorylation by IκB kinase and the active NFκB (RelA and p50) enter the nucleus and binds to promoters activating the transcription of proinflammatory molecules. TNF-α binds to the type 1 TNF receptor (TNFR1) which results in an association with TNFR1-associated death domain protein (TRADD), the receptor-interacting protein (RIP) and the TNF receptor-associated factor-2 (TRAF-2). These cytoplasmic proteins form an active signaling complex that interacts with NF-κB–inducing kinase (NIK). Activation of NIK results in phosphorylation of IκB kinases (IKK) which cause phosphorylation of IκB. Phosphorylated IκB is targeted for destruction by the ubiquitinization/proteasome degradation pathway, allowing the translocation of NF-κB to the nucleus. IL-1β binds to the type 1 IL-1 receptor (IL-1R1) and the IL-1 receptor accessory protein (IL-1RAcP), which facilitates and interaction between IL-1 receptor–associated kinase (IRAK) and TNF receptor–associated factor-6 (TRAF-6). The interaction between IRAK and TRAF-6 can also be triggered by endotoxin (LPS). LPS binds with high-affinity to CD14 and to toll-like receptor 2 (TLR2). These proteins form an active signaling complex that also results in activation of NIK and IKK, leading to the sequence of events that results in activation of NF-κB. Activation of NF-κB results in expression of messenger RNA for a variety of proinflammatory mediators that are involved in the pathogenesis of lung inflammation. IκB is also induced by NF-κB activation and contributes to the down-regulation of this intracellular signaling cascade. Reproduced with permission from J.W. Christman, R.T. Sadikot, and T.S. Blackwell, "Impact of basic research on tomorrow's medicine," *Chest*, 117: 1482–1487, 2000. The Role of Nuclear Factor-κ B in Pulmonary Diseases.

and p65. It is likely that the transcriptional effects are more pronounced with the binding of the heterodimer to DNA. *In vitro* work suggests that acetylation of the DNA-binding domain of NFκB aids in its nuclear translocation and enhances its binding to DNA (for example, to the long terminal repeat of HIV-1) and enhances its transcriptional effect. p50 has been shown to be acetylated by the coactivators CBP and p300 dependent on the **histone acetyltransferase (HAT)** domain and by TAT proteins. This results in the acetylation of three lysine residues: K431, K440, and K441. The consensus sequence to which NFκB binds is GGGAATTCCC. The binding of the homodimer of p50 to DNA is shown in Figure 7-5, and the binding of the heterodimer of NFκB to DNA is shown in Figure 7-6.

It has been suggested that iron present in asbestos fibers induces cellular oxidation–reduction changes by producing intracellular reactive oxygen species that result in the activation of NFκB (phosphorylation of IκB and its dissociation from the p65 or p50 dimer). But activation of NFκB also occurs with inhalation of non–iron-containing fibers. In addition to NFκB, activator protein-1 (AP-1; another transcription factor) is activated

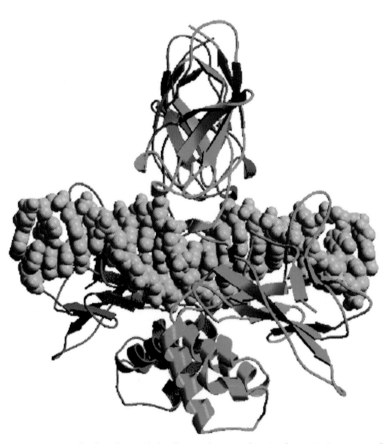

Figure 7-5. The binding of the homodimer of p50 of NFκB *(green and purple)* to DNA *(pink)*. Reproduced from http://www.biochem.ucl.ac.uk/bsm/xtal/teach/trans/nfkb2.gif.

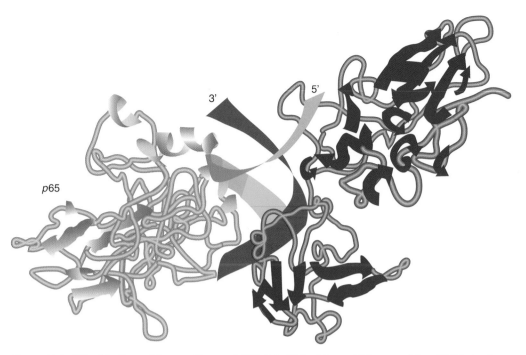

Figure 7-6. The binding of heterodimer of NFκB (p50 in *blue*; p65 in *purple*) to DNA *(red and yellow strands)*. Reproduced with permission from http://www.biochem.umd.edu/biochem/kahn/molmachines/enhancesomes/NFkB.html.

by reactive oxygen species. In Figure 7-7, AP-1 appears in diagram form as the Jun–Fos dimer. Its structure is shown also when it binds to DNA with the **nuclear factor of activated T cells (NFAT).** AP-1 is a dimer of C-Jun and of c-Fos or c-Jun with another member of the Jun family. Dimerization is accomplished through leucine zippers on either molecule, consisting of four leucine residues on each molecule arranged closely so that the two molecules can interact through these zippers with strong hydrophobic bonding (Figure 7-8). NFκB binding to DNA leads to promotion of the transcription of IL-8 and IL-6 in bronchial epithelial cells and in alveolar macrophages. This transcription factor binds to DNA in airway epithelial and pleural mesothelial cells increasingly after inhalation of asbestos particles.

DNA microarrays have been used to determine what other genes are activated in murine pulmonary cells after exposure to asbestos. This study revealed that the genes responding to asbestos exposure involved transcription factors (already discussed). A number of genes involved in inflammatory pathways and a number of kinases (a kinase is required to phosphorylate IκB in the activation of NFκB).

Presently, there is little, clinically, to rescue sufferers of severe asbestosis, and many of these cases result in death.

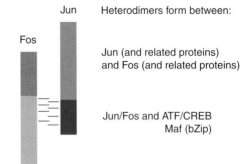

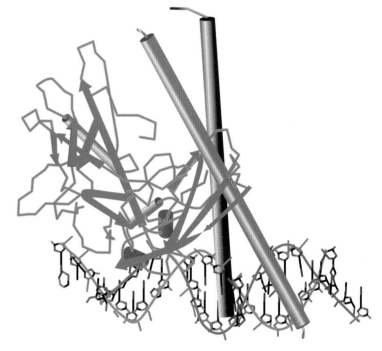

Figure 7-7. *Top*, diagram of AP-1 as a dimer of Jun and Fos proteins joined through leucine zippers (a sequence of four leucine residues on either protein) through a hydrophobic interaction. AP-1 also may be composed of heterodimers of different species of Jun. *Bottom*, structure of C-Fos *(green single helix)* and C-Jun *(red single helix)* together with N-fat transcriptional factor in green (280 residues) on left. DNA is composed of 20 bases in each strand. Bottom redrawn from http://www.biochem.ucl.ac.uk/bsm/pdbsum/1a02/tracel_b.html.

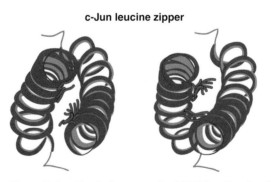

Figure 7-8. c-Jun belongs to the bZIP family of eukaryotic transcription factors. In order to transactivate target genes it must either homodimerize or heterodimerize with c-Fos via its leucine zipper (LZ) domain. The c-Jun LZ homodimer is a classical coiled coil in which extended helices from each monomer wrap around one another with a left-handed superhelical twist. The supercoiling is obvious in the adjacent view down the long axis of the dimer. The hydrophobic dimer interface is composed of alternating "rungs" of leucine residues (shown in red). Redrawn from F.K. Junius et al., *J. Biol. Chem.*, 271: 13663–13667, 1996.

Transcription Factors and Transcription Complex

Transcription is the process of synthesizing an RNA from a DNA template. The process involves three phases: **initiation, elongation,** and **termination.** Initiation is the binding of RNA polymerase to a double-stranded DNA molecule. The double strand of DNA must be opened (Figures 6-71 and 6-72) so that the RNA polymerase can bind to the DNA (gene) promoter. For this to take place, various transcription factors (proteins) bind to other proteins, DNA, or RNA polymerase to construct the transcription apparatus. Eukaryotic transcription factors are summarized in Figure 7-9. The core transcriptional elements are the **TFIIB** recognition element (BRE), TATA box, the initiator element (INR), and the downstream promoter element (DPE). These elements are shown in Figure 7-10, where the sequences constituting each box are defined. The

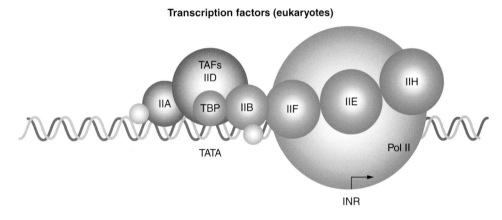

Figure 7-9. A summary of eukaryotic transcription factors. *TAFs*, transcription activating factors; *TBP*, TATA box binding protein; *INR*, initiator element ($P_yP_yAN_A^TP_yP_y$); *TATA*, TATA box (TATA A_T^A A A ^G_A).

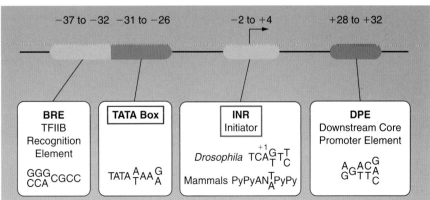

Figure 7-10. Elements of the basal eukaryotic promoter. Redrawn from S.T. Smale and J.T. Kadonga, *Ann. Rev. Biochem.*, 72: 449–479, 2003.

BRE is the site for the binding of TFIIB and has a consensus sequence of G/C,G/C,G/A,CGCC. The 3' end of BRE is where the 5' of the TATA begins (Figure 7-10). A CCAAT box, not shown in Figure 7-10, is sometimes found upstream at -50 to -100, but in promoters lacking a TATA box, the CCAAT box can be close to the initiator (INR). A CCAAT box is found in half of vertebrate promoters, and it binds nuclear factor-1 (NF-1).

The **TATA-binding protein** is 27 kDa and is required by all three RNA polymerases (pol I, II, and III) to initiate transcription. It binds to the TATA box (TATAAA consensus sequence). The structure of the TATA-binding protein bound to double-stranded DNA containing the TATA box sequence is shown in Figure 7-11. The TATA box is located about 25 to 30 base pairs upstream (toward the 5' end of DNA) from the start site near the INR. In humans, about 32% or 1031 potential promoter regions have a TATA box, but transcription can occur when the TATA box is absent in the promoter. The preinitiation complex (PIC) consists of the transcriptional factors, the TATA-binding protein, and the RNA pol II bound to the promoter (Figure 7-12). The constituent proteins of the PIC are TFIID of 750 kDa; the TATA-binding protein and the TATA-binding protein **transcriptional activator factors (TAFs)**; the general transcription factors TATA-binding protein, TFIIB, TFIIE, TFIIF, and TFIIH; and RNA pol II (Figure 7-13), consisting of 12 subunits (Table 7-1). Structures of these factors are shown in Figures 7-14 and 7-15.

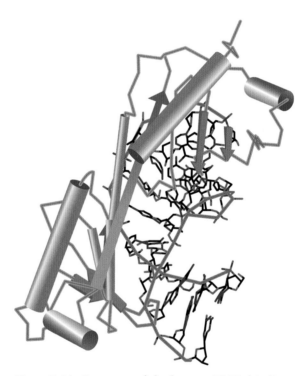

Figure 7-11. Structure of the human TATA binding protein bound to DNA duplex that contains the TATA box consensus sequence. The TATA binding is a single chain of 180 amino acid residues *(in green)* with 4 helices and 9 strands. 12 bases are contained in each strand of DNA.

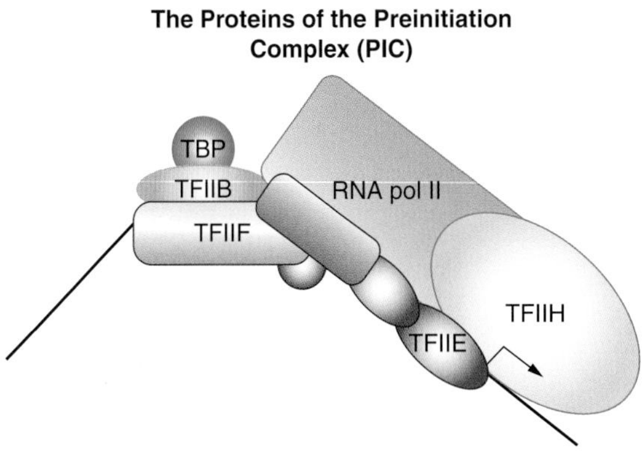

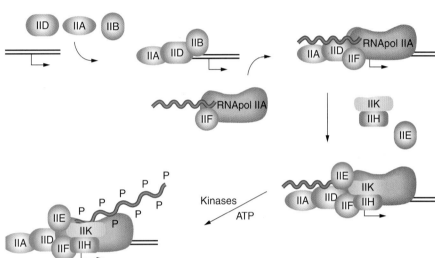

Figure 7-12. Constituents of the preinitiation complex (PIC).

Figure 7-13. RNA pol II in the process of initiation.

Table 7-1
Subunits of RNA Polymerase II

Subunit	Mass(KDa)
Rpb1	191.6
Rpb2	138.8
Rpb3	35.3
Rpb4	25.4
Rpb5	25.1
Rpb6	17.9
Rpb7	19.1
Rpb8	16.5
Rpb9	14.3
Rpb10	8.3
Rpb11	13.6
Rpb12	7.7
Total	513.6

Subunits 1, 2, 3, 5, 6, 8, 10, 11, and 12 are conserved in the RNA polymerases I, II, and III. The subunits 4, 7 and 9 are unique to RNA Pol II.
Reproduced with permission from P. Cramer et al., "Architecture of RNA polymerase II and implications for transcription mechanism," *Science*, 288: 640-649, 2000.

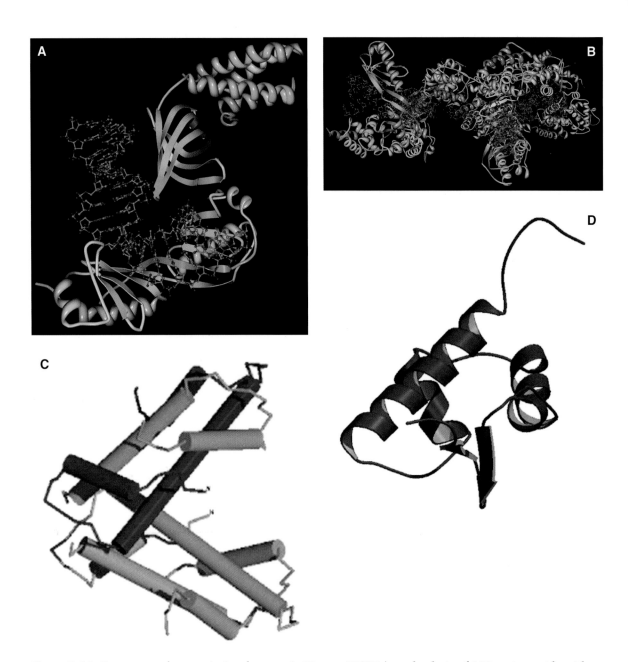

Figure 7-14. Structures of transcription factors. **A.** Human TFIID (*purple chain* of 180 amino acid residues TATA binding protein), TIIAA (two chains *red* and *brown* of 43 and 47 residues) and TIIAG (one chain, *orange*, of 97 residues). **B.** Human TFIIB consisting of five chains *(purple, blue, light purple, blue, dark blue)*, TATA box binding protein, and DNA containing TFIIB recognition element. **C.** Human TFIID 135-kDa subunit. **D.** Human TFIIEβ, a single chain of 81 amino acid residues. Part A reproduced from PDB ID: 1nvp. M. Bleichenbacher, S. Tan, T.J. Richmond. Novel Interactions between the Components of Human and Yeast TFIIA/TBP/DNA complexes. *J. Mol. Biol.* **332** pp. 783 (2003). Part B reproduced from PDB ID: 1c9b F.T. Tsai, P.B. Sigler. Structural Basis of Preinitiation Complex Assembly on Human Pol II Promoters. *EMBO J.* **19** pp. 25 (2000). Part C reproduced from PDB ID: 1h3o. S. Werten, A. Mitschler, C. Romier, Y.G. Gangloff, S. Thuault, I. Davidson, D. Moras. Crystal Structure of a Subcomplex of Human Transcription Factor TFIID Formed by TATA Binding Protein-Associated Factors hTAF4 (hTAF(II)135) and hTAF12 (hTAF(II)20). *J. Biol. Chem.* **277** pp. 45502 (2002). Part D reproduced from PDB ID: 1d8j. M. Okuda, Y. Watanabe, H. Okamura, F. Hanaoka, Y. Ohkuma, Y. Nishimura. Structure of the Central Core Domain of TFIIEβ with a Novel Double-Stranded DNA-Binding Surface. *EMBO J.* **19** pp. 1346 (2000).

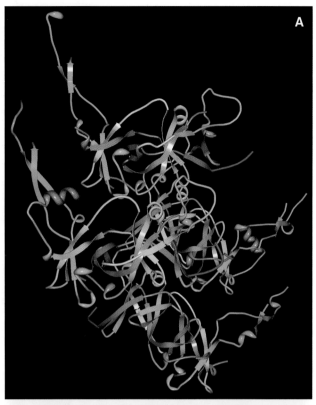

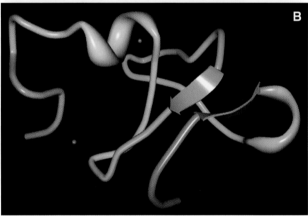

Figure 7-15. Structures of transcription factors. **A.** Human TFIIFβ subunit (consisting of four chains in *purple, gold, and light blue*). The α subunit consists of four chains *(red, pink, green, and yellow)*. **B.** Human TFIIH 44 kDa subunit consisting of 59 amino acid residues with two helices and two strands. The atoms in *blue* are Zn. Reproduced from http://www.biochem.ucl.ac.uk/bsm/pdbsum/1e53/tracel.html.

Binding of the TATA-binding protein to the TATA box has the effect of bending DNA, providing a saddle structure for the other transcription factors (Figure 7-12). In most cases, the TATA-binding protein associates with the promoter before the other transcription factors bind. The TAFs involve HATs, protein kinases, coactivators, functions of some transcription factors, and other activities. The INR marks the start site of transcription, which interacts with the TATA-binding protein and the coactivator, SP1 (spacing relative to TATA-binding protein and SP1 on DNA is important for the INR), in addition to the largest

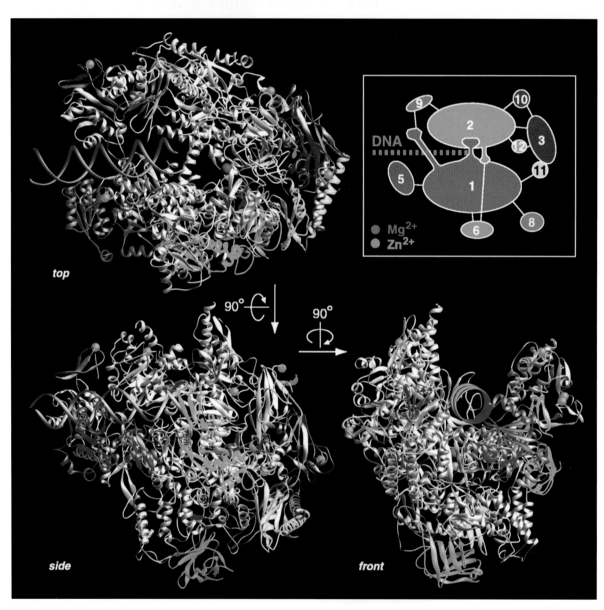

Figure 7-16. RNA pol II structure. The enzyme has an overall negative charge on its surface with a positively charged saddle structure that helps to bind DNA and RNA. Reproduced with permission from P. Cramer, D.A. Bushnell, and R.D. Kornberg, "Structural basis of transcription: RNA polymerase II at 2.8 A resolution," *Science*, 292: 1863–1876, 2001.

Figure 7-17. Architecture of RNA polymerase II. Note the entry of double-stranded nucleic acid (DNA template and RNA transcript) into the saddle structure *(upper left)*. Reproduced with permission from P. Cramer et al., "Architecture of RNA polymerase II and implications for transcription mechanism," *Science*, 288: 640–649, 2000.

subunit of RNA pol II, the TAFs, and various transcription factors (TFIIB, TFIID, and TFII-I). In the sequence of the INR, $YYA_{+1}NT/AYY$, Y is a pyrimidine and the A is the transcriptional start site. The INR-binding protein, TFIID (its subunits), and TFII-I stimulate transcription and may act synergistically with another factor, upstream stimulating factor 1, to stimulate transcription. The INR box binds RNA polymerase directly. Initiation of transcription occurs when RNA pol II dissociates from the TATA-binding protein and the TAFs and begins a forward movement over the opened strand of DNA (Figures 6-71 and 6-72) requiring helicase and an ATPase. Some transcription factors dissociate, although TFIIB remains at the initiation site, and the remaining transcription factors are important to this process. RNA pol II progresses by use of a clamp mechanism during elongation. RNA pol II, in addition to the clamp structure, has a positively charged saddle structure that facilitates the binding of DNA and RNA (Figures 7-16 and Figure 7-17). A detailed structure of the clamp is shown in Figure 7-18. The fold of the clamp is stabilized by three zinc ions, and the clamp closes on DNA and RNA to trap the DNA template and the RNA transcript. The strands of DNA template and transcribed RNA in the structure of RNA pol II are shown in Figure 7-19. Thus, the clamp is a major element stabilizing RNA pol II during transcription as it closes on the active site to retain the nucleic acids. The active site formed by the hybrid of DNA and the transcribed RNA prefers the entrance of nucleotide triphosphates instead of deoxynucleotide triphosphates, a selectivity expected for the synthesis of RNA. The growing RNA and the addition of subsequent nucleotides to the RNA are shown in Figure 7-20. A summary of the NA polymerase transcription process (elongation) is shown in Figure 7-21.

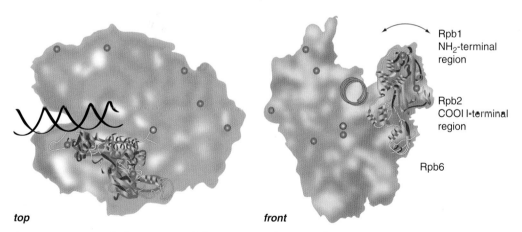

Figure 7-18. Detailed structure of the CLAMP of RNA polymerase II. It is comprised of the N-terminal regions of Rpb1 and Rpb6 and the C-terminal regions of Rpb2. Redrawn with permission from P. Cramer et al., "Architecture of RNA polymerase II and implications for transcription mechanism," *Science*, 288: 640–649, 2000.

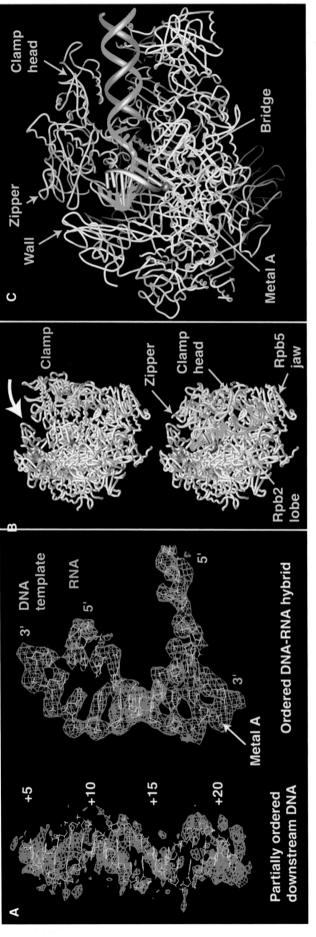

Figure 7-19. Structure of RNA pol II showing the DNA template strand and the RNA transcribed strand (A) and the role of various structures on the enzyme, such as the clamp head and others (B and C). Reproduced with permission from A.L. Gnatt et al., "Structural basis of transcription: an RNA elongation complex at 3.3 Å resolution," *Science*, 292: 1876–1882, 2001.

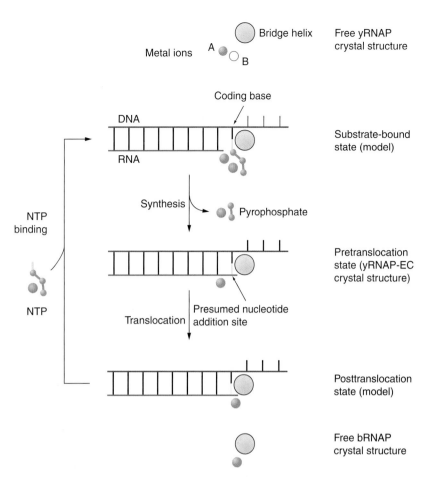

Figure 7-20. RNA polymerase (RNAP) is bound to DNA and it is shown in the complex where the next nucleotide will be added to the growing RNA. Redrawn with permission from P. Cramer, "Multisubunit RNA polymerase protein–nucleic acid interactions," *Curr. Opin. Struct. Biol.*, 12: 89–97, 2002.

Enhancers are factors involved in transcription that bind further from the distal promoter (DPE; see Figure 7-10). The consensus sequence of the DPE is A/G$_{+28}$,G,A/T,G/A/T. The DPE helps position TFIID with INR at +28. It is clear that *transcription factors come in groups in which more than one factor can recognize the same binding site.* Enhancers facilitate recognition over long distances, and their actions involve protein–protein contacts, alterations of proteins, phosphorylation, changes in chromatin structure, nuclear localization, etc.

Coactivators and Corepressors

Coactivators and corepressors are well known for the gene superfamily of nuclear receptors. The family is summarized in Figure 7-22. There are three major groups in this superfamily: the endocrine receptors, which

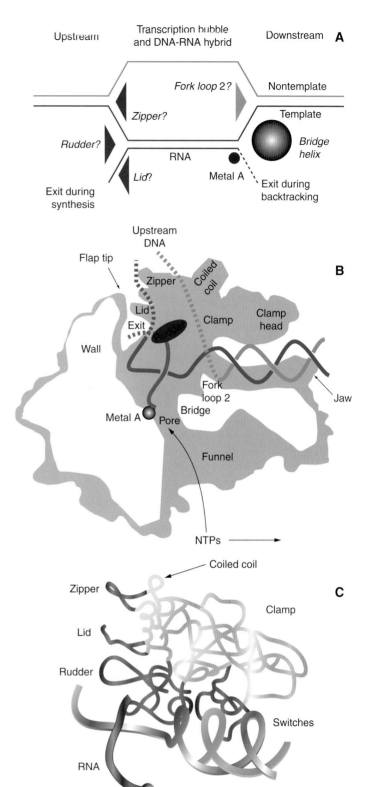

Figure 7-21. The RNAP elongation complex. **A.** Schematic presentation of the arrangement of nucleic acids during RNA chain elongation. The DNA template and nontemplate strands are in blue and green, respectively, and the RNA is in red. The active site metal ion A is indicated by a red sphere. Protein elements that are proposed to be involved in the maintenance of the arrangement of nucleic acids are indicated. **B.** Cutaway view of the yRNAP elongation complex. Cut surfaces are lightly shaded. During transcription, DNA enters the enzyme from the right (the polymerase moves to the right). Structural features that appear to be important for function are labeled. Coloring of nucleic acids is as in *A*. Exiting RNA and DNA strands are not revealed in the electron density map, but their anticipated locations are indicated by dashed lines. In this view, the clamp swings over the active center from back to front. Only one of the jaws (the lower jaw) is visible in the cutaway view. **C.** Ribbon diagram of the clamp and nucleic acid backbones in the yRNAP-EC structure. The view is related to that in *B* by a 40 degrees rotation around the vertical axis. Coloring of nucleic acids is as in *A*. The switch regions at the base of the clamp are in *pink*, and the three loops emanating from the edge of the clamp are in *red*. Redrawn with permission from P. Cramer, "Multisubunit RNA polymerase protein-nucleic acid interactions," *Curr. Opin. Struct. Biol.*, 12: 89–92, 2002.

Nuclear receptor superfamily

Endocrine receptors		Adopted orphan receptors		Orphan receptors	
High-affinity, Hormonal lipids		Low-affinity, Dietary lipids		Unknown	
ERα,β	Estrogen	RXRα,β,γ	9-cis RA, DHA	SF-1	?
PR	Progesterone	PPARα,β,γ	Prostanoids, FA	LRH-1	?
AR	Androgen	LXRα,β,	Oxysterols	SHP	?
GR	Glucocorticoid	FXR	Bile acids	TLX	?
MR	Mineralocorticoid	PXR/SXR	Xenobiotics	PNR	?
		CAR	Xenobiotics	NGFI-Bα,β,γ	?
RARα,β	Retinoic acid			RORα,β,γ	?
TRα,β,γ	Thyroid hormone			ERRα,β,γ	?
VDR	Vitamin D			RVRα,β,γ	?
				GCNF	?
				TR2,4	?
				HNF-4	?
				COUP-TFα,β,γ	?

Figure 7-22. The nuclear receptor superfamily.

can be subdivided between the steroid receptor subclass; receptors that form heterodimers with the **retinoic acid receptor (RAR)**; and the large group of **orphan receptors** for whom either the ligand is unknown or the function is obscure. For this discussion, the endocrine receptors will be used as models. These receptors are located either in the cytoplasm or in the nucleus. Those receptors located in the cytoplasm of the hormone target cell exist in the "repressed" state in the absence of the hormone. The hormone will derive from a cell that is different from the target cell. When it arrives in the target cell, it binds to the inactive receptor complex, alters the conformation of the receptor protein by binding to the ligand-binding domain, and causes the active monomeric receptor to dissociate from the other proteins in the inactive "repressed" complex. This enables it to translocate into the nucleus, homodimerize with another liganded receptor molecule, and bind to the hormone-responsive element (HRE) on a gene promoter. This is visualized in Figure 7-23. The overall mechanisms for monomeric nuclear receptors located in the nucleus (not in the cytoplasm) and the heterodimeric receptors that form dimers with the RAR or the **retinoic acid X receptor (RXR)** are shown in Figures 7-24 and 7-25.

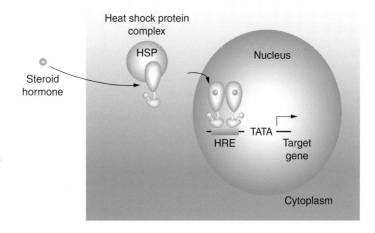

Figure 7-23. The overall mechanism of a cytoplasmic steroid receptor, such as the glucocorticoid or mineralocorticoid receptor. *HRE*, hormone-responsive element on a target gene promoter; *HSP*, heat shock protein.

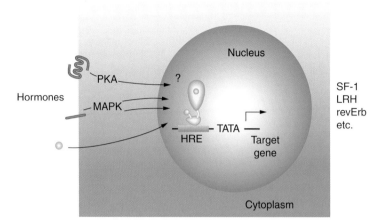

Figure 7-24. Overall mechanism for a monomeric nuclear receptor. Hormonal targets operate on cell membrane receptors that have one or more possible signal transduction pathways, such as leading to activation of protein kinase A or MAPkinase. These signals lead to activation of a monomeric receptor located on DNA at an HRE.

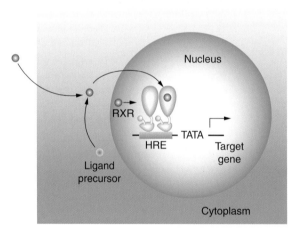

Figure 7-25. Overall mechanism of heterodimeric nuclear receptors.

The cytoplasmic steroid hormone receptors are in the repressed state in the absence of the hormonal ligand, and these receptors are bound in a complex with proteins, mostly heat shock proteins that occlude the DNA-binding domain of the steroid receptor. Thus, the receptor in this form is not able to bind to DNA or translocate to the nucleus through the nucleopore. When the hormone enters the cell and binds to the ligand-binding domain of the receptor, the receptor undergoes a conformational change and the accessory proteins in the complex are dissociated, generating the active monomeric receptor now capable of translocating to the nucleus, homodimerizing, and binding to the cognate hormonal response element on a gene promoter. Once bound to the HRE, the receptor forms part of a structure with other proteins that generate the pretranscriptional complex, leading to transcription and expression. One of the proteins that combines with the receptor while it is complexed to DNA is a coactivator protein, which binds to the receptor through a **LXXLL** motif (Figure 7-26) and facilitates transcription. Some coactivators and corepressors are summarized in Figure 7-27. A more complete list of coactivators is given in Table 7-2.

In the case of the cytoplasmic receptors, their repressed form is the complex in the absence of the hormonal ligand; therefore, these receptors do not complex with corepressors because they are corepressed in the absence of ligand in the cytoplasm. However, once they are activated and enter the nucleus and bind to the HRE, they will complex with a coactivator. On the other hand, receptors found initially in the nucleus bound to DNA (at the HRE site) are complexed with a corepressor, and such molecules (for example, thyroid hormone receptor and RAR) are activated while attached to DNA by the entry of hormonal ligand and the displacement of the corepressor by a coactivator (Figure 7-27). Crystal structure of a coactivator, SRC-1, bound to the ligand-binding domain of the heterodimer, RXRα and PPAR is shown in Figure 7-28, and the

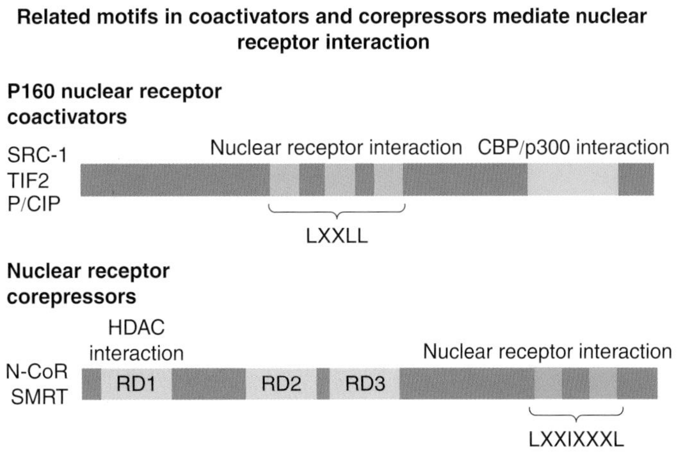

Related motifs in coactivators and corepressors mediate nuclear receptor interaction

P160 nuclear receptor coactivators

SRC-1
TIF2
P/CIP

Nuclear receptor interaction   CBP/p300 interaction

LXXLL

Nuclear receptor corepressors

N-CoR
SMRT

HDAC interaction
RD1   RD2   RD3

Nuclear receptor interaction

LXXIXXXL

Figure 7-26. Cartoons of typical coactivators and corepressors with the LXXLL and related motif required for interaction with the nuclear receptor class.

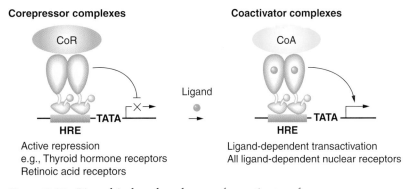

Figure 7-27. Ligand-induced exchange of coactivators for corepressors.

Table 7-2
Classes of Coactivators That Interact with Nuclear Receptors

Class	Coactivator	Sequence[a]
Class I		SRLXXLL
	GRIP1	TKLLQLL
	SRC-1	HKLVKLL
	AIB-1	KKLLQLL
Class II		PΦLXXLL
	TRAP220	PILTSLL
		PMLMNLL
	RIP140	PILYYML
Class III		(S/T)ΦLXXLL
	RIP140	TYLEGLL
		TLLASLL
		SLLLHLL
		TLLQLLL
		TVLQLLL
	PGC-1	SLLKKLL
	DAX-1	SILYNLL
		SILYSML
		SILYSLL
	SHP	TILYALL
		SILKKIL

[a]X, any amino acid; Φ, hydrophobic amino acid. Conserved amino acids in each class are in boldface type.
Reproduced with permission from P. McDonnell, "Dissection of the LXXLL nuclear receptor-coactivator interaction motif using combinatorial peptide libraries: a discovery of peptide antagonists of estrogen receptors α and β," *Mole. Cell. Biol.*, 19: 8226–8239, 1999 and directly from http://mcb.asm.org/cgi/content-nw/full/19/12/8226/T 1..

crystal structure of the corepressor, NCoR, is shown bound to the ligand-binding domain of the human PPARα in Figure 7-29. In Figure 7-30, the binding of the corepressor, SMRT, to PPARα is shown. The binding of the coactivator, **thyroid receptor–binding protein (TRBP),** to the **estrogen receptor-α (ERα)** is shown in Figure 7-31.

The p160 coactivators are a family of proteins that includes SRC-1, already described. In this group are TF2/**GRIP1** and ACTR/AIB1, in addition to SRC. *They act as platform proteins for the binding of other proteins that can modify chromatin,* such as p300, **CREB-binding protein (CBP)** and **CARM1,** an arginine methyltransferase. The p160s interact with signal transducer and activator of transcription (STAT) 3, MEF2C, and NFκB, which are also activators. Models of the p160 group of coactivators are shown in Figure 7-32.

The CBP and p300 are large nuclear proteins involved in many cellular processes. They are coactivators for nuclear receptors, as well as other transcription factors. CBP and p300 have an acetyltransferase activity, known to acetylate histones, transcription factors, and nuclear transport proteins. Acetylation of p160s themselves by CBP results in their dissociation from nuclear receptors. Activities of CBP are summarized in Figure 7-33. Sites are indicated on the cartoon structure that show which other molecules interact with CBP. Coactivators are located in the cellular nuclei. The binding of these coactivators to various nuclear recep-

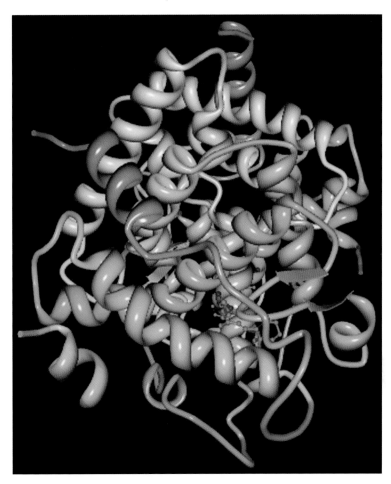

Figure 7-28. Crystal structures of human RXRα and PPAR heterodimer with RXRα ligand-binding domain bound with 9-*cis* retinoic acid and PPAR bound with the agonist ligand gi262570 and the SRC-1 peptide. Reproduced from http://www.biochem.ucl.ac.uk/bsm/pdbsum/1fm9/tracel.html/.

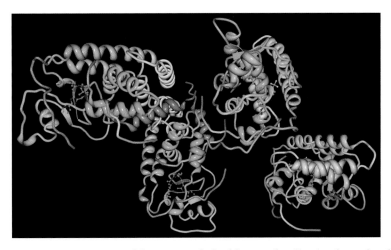

Figure 7-29. NCoR-2 *(blue, green, light blue, and yellow)* is bound to human PPARα *(purple, red, brown, and magenta)*. Reproduced from PDBsum ID 1kkg, http://www.biochem.ucl.ac.uk/bsm/pdbsum/1kkq/tracel.html.

SMRT-RID/PPARα LBD/Antagonist complex

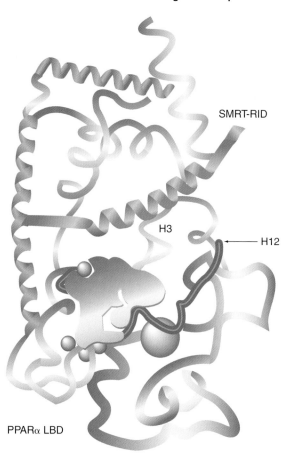

Figure 7-30. Structures showing the binding of SMRT corepressor through its nuclear receptor interaction domain (RID) to the coactivator binding side of PPARα ligand-binding domain (LBD) bound with an antagonist. The relevant structure of SMRT is shown in red. Redrawn from Internet from unknown source.

tors all depend on the LXXLL motif. A summary of these motifs in the sequences of several coactivators is shown in Figure 7-34.

Some activities associated with coactivators are nucleosome remodeling, histone acylation, histone methylation, and recruitment of RNA pol II. Activities associated with corepressors are DNA methylation, histone methylation, and histone deacetylation. These activities are summarized in Figure 7-35.

In the case of the nuclear receptors located in the cytoplasm in the absence of ligand, the entrance of the ligand into the cell and its binding to the ligand-binding domain of the cytoplasmic receptor generates a conformational change in the receptor and the dissociation of the other proteins in the inactive complex. These events allow the ligand–receptor complex to enter the nucleus, dimerize, and bind to the cognate HRE on DNA, where the coactivator binds to the ligand-binding domain. The elements of these activities are shown in Figure 7-36. For those unliganded nuclear receptors located in the nucleus and bound to the HRE, a corepressor is bound to the ligand-binding domain of the receptor. When the ligand becomes available to the cell and enters the nucleus, it binds to the receptor, evoking a conformational change, and the corepressor is replaced by a coactivator. These actions are seen in simplified form in Figure 7-37.

(Text continues on p. 349.)

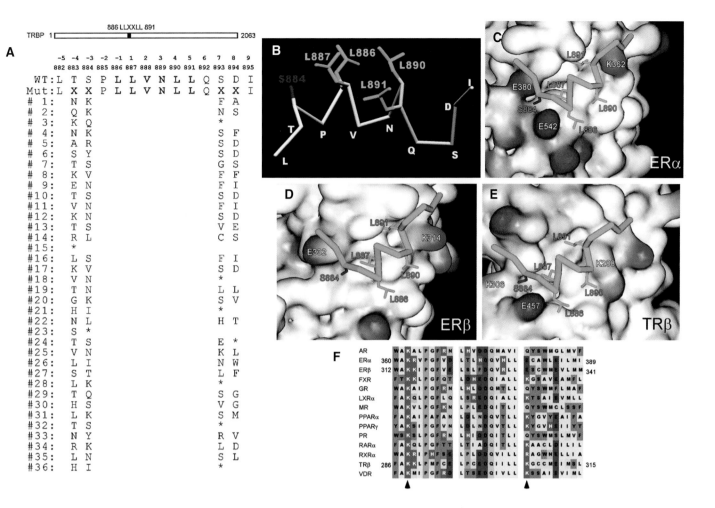

Figure 7-31. Binding of the coactivater, TRBP, to ERα and TRβ. The S884 residue of TRBP displays an important role in receptor selectivity. **A.** Diagram of a TRBP LXXLL motif (882–895) is shown. Full-length TRBP with the LXXLL motif is depicted on top. The amino acids of the LXXLL motif depicted in *bold* are wild type *(WT)*. **B.** A computationally engineered TRBP peptide (882–895) shows that the side chain of S884 *(red)* has the same orientation as the leucines *(green)* in the LXXLL motif. The side chain of S884 is near that of L887. **C.** The engineered coactivator TRPB peptide bound to ERα-LBD is shown in a molecular surface view. Coordinates of DES/ERα-LBD are from 3ERD. The side chains of Leu885, Leu887, Leu890, and Leu891 of TRBP are shown in *green*, and the side chain of S884 is shown in *red*. The negatively charged *(red)* E380 in helix 4, E542 in helix 12, and positively charged *(blue)* K363 in helix 3 of ERα are also indicated. **D.** The surface view of ERβ-LBD (1QKM) with TRBP peptide is shown. Side chains of residues in TRBP are labeled in an identical manner as above. The negatively charged *(red)* E332 in helix 4 and positively charged *(blue)* K314 in helix 3 of ERβ are indicated. Helix 12 of ERβ is not shown. **E.** A similar surface view with TRβ-LBD is shown. TRβ (1BSX) structure was obtained from the Protein Data Bank. The K306 *(blue)* in helix 4, conserved E457 *(red)* in helix 12, and K288 *(blue)* in helix 3 of TRβ are indicated. **F.** Amino acid alignment of nuclear receptors within LBD helices 4 and 5. The *numbers* indicate the amino acids of ERα, ERβ, and TRβ. The aromatic amino acids are shown in *brown*, prolines in *green*, threonines and serines in *magenta*, positively charged residues in *dark blue*, negatively charged residues in *red*, cystines and methionines in *yellow*, glutamines and asparagine in *light blue*, and hydrophobic or small residues in *gray*. The conserved lysines in helix 3 (ERα/K362, ERβ/K314, and TRβ/K288) and aligned ERα/E380, ERβ/E332, and TRβ/K306 in helix 4 are indicated by *arrows*. Redrawn with permission from L. Ko, G.R. Cardona, T. Iwasaki, K.S. Bramlett, T.P. Burris, and W.W. Chin, "Ser-884 adjacent to the LXXLL motif of coactivator TRBP defines selectivity for ERs and TRs," *Mol. Endocrinol.*, 16: 128–140, 2002.

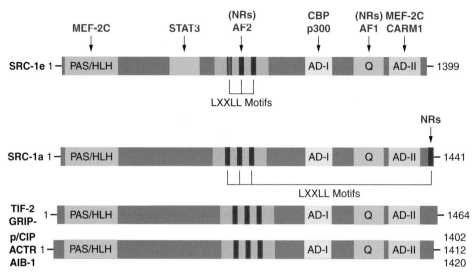

Figure 7-32. The p160 coactivators.

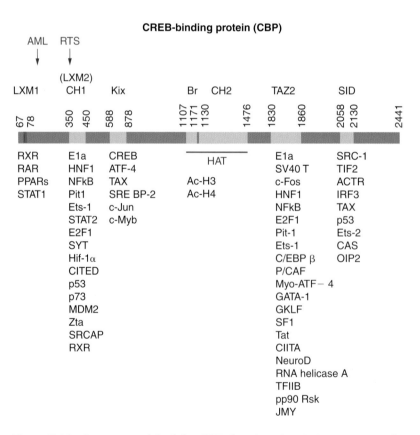

Figure 7-33. Cartoon model of the CBP showing sites in its structure that interact with other proteins.

Coactivator	Partial sequence	Amino acid residues in sequence
SRC-1	Q T S H K L V Q L L T T T A E	628–642
	E R H K I L H R L L Q E G S P	685–699
	K D H Q L L R Y L L D K D E K	744–758
	Q Q K S L L Q Q L L T E	1430–1441
TIF2	K G Q T K L L Q L L T T K S D	636–650
	E K H K I L H R L L Q D S S S	685–699
	K E N A L L R Y L L D K D D T	740–754
ACTR	K G H K K L L Q L L T C S S D	626–640
	E K H R I L H K L L Q N G N S	690–704
	E N N A L L R Y L L D R D D P	743–757
ARA70	Q Q A Q Q L Y S L L G Q F N C	87–101
TIP60	H E R A M L K R L L R I D S K	488–502
PGC1	E E P S L L K K L L L A P A N	137–151
ASC2	L T S P L L V N L L Q S D I S	882–896
dFTZ	E R P S T L R A L L T N P V K	107–122
NRIF3	K R S L K L D G L L E E N S F	4–18
TIF1	Y P R S I L T S L L N S S Q	718–732
TRAP220	S Q N P I L T S L L Q I T G N	599–613
	K N H P M L M N L L K D N P A	640–654
ARA267	E L S A A L P G L L S D K R D	721–735
	Q N C E K L G E L L L C E A Q	1278–1292
RIP140	I V L T Y L E G L L M H Q A A	16–30
	Q D S T L L A S L L Q S E S S	128–142
	Y A S S H L K T L L K K S K V	180–194
	V A C S Q L A L L L S S E A H	262–276
	A N N S L L L H L L K S Q T I	375–389
	Q K V T L L Q L L L G H K N E	495–509
	E R R T V L Q L L L G N P K G	708–722
	S K N G L L S R L L R Q N Q D	814–828
	K S F N V L K Q L L L S E N C	931–945
CBP	S K H K Q L S E L L R G G S G	64–78
	L I Q Q Q L V L L H A H K C	352–366
	I S P S A L Q D L L R T L K S	2063–2077

Figure 7-34. A summary of LXXLL domains in several coactivators. Redrawn with permission from D.M. Heery et al., *Nature*, 387: 733–736, 1997.

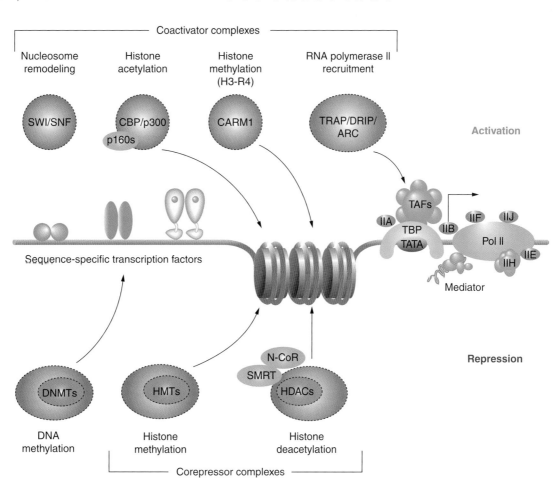

Figure 7-35. Activities associated with coactivators and corepressors. Redrawn from an unknown source on the Internet.

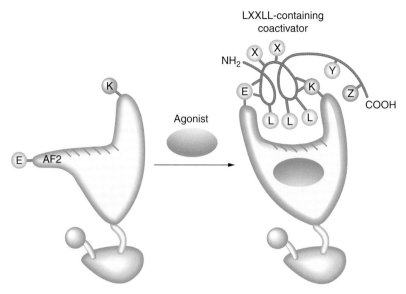

Figure 7-36. Agonist binding to a cytoplasmic (nuclear) receptor causes conformational change in the receptor structure and allows for interaction with a coactivator once the receptor has dimerized and is bound to its HRE. Redrawn from an unknown source on the Internet.

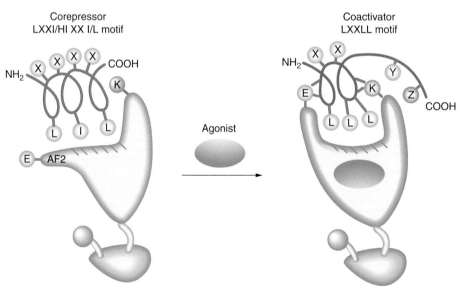

Figure 7-37. Nuclear receptors bound with a corepressor to the ligand-binding domain (LBD). Upon entry of the ligand (agonist) there is a conformational change in the receptor LBD and a substitution of the corepressor by a coactivator. Redrawn from an unidentified source on the Internet.

The Glucocorticoid Receptor as a Model Transcription Factor

The glucocorticoid receptor is one of the nuclear receptors, located in the cell cytoplasm, that was among the first transcription factors to be described, although it was not known to be a transcription factor for some time. These receptors activate gene transcription in response to ligand binding, in this case, a steroid hormone. The general properties of a steroid hormone receptor are shown in the model in Figure 7-38. The variable N-terminal domain (Figure 7-38A) contains a major epitope (antigen site), as well as transactivation factor activity. The central portion contains the highly conserved DNA-binding domain (Figure 7-38A, B, and C) consisting of two **zinc finger** structures, followed by a nuclear translocation signal (consisting of basic amino acids), variable "hinge region" (D), and then the conserved ligand-binding domain (D–H). Transcriptional activating activity also resides in the ligand-binding domain, along with the binding sites for heat shock proteins (to form the inactive complex in the absence of ligand) and sites for dimerization, which also include the DNA-binding domain.

The nuclear receptor superfamily falls into four classes of receptors in terms of their dimerization and DNA-binding properties (Figure 7-39). The glucocorticoid receptor is located in the cytoplasm bound to a **heat shock protein 90 (HSP90)** dimer and other proteins in the absence of ligand. Upon entrance of the ligand (cortisol) into the cell, the hormone binds to the steroid-binding domain of the receptor, producing a conformational change and a shedding of the HSP90 and other proteins from the complex. The receptor in this form passes into the nucleus and forms a homodimer with another glucocorticoid receptor molecule. Then it binds to the HRE of DNA, allowing the pretranscriptional complex to develop, including the binding of coactivator. Next, RNA pol II binds and transcription of the relevant gene occurs. The estrogen receptor is found in the nucleoplasm bound with HSP90 in the absence of ligand (estradiol). When estradiol binds to the receptor, HSP90 is released and the receptor homodimerizes and binds to the estrogen-responsive site on specific target gene promoters. Similarly, this allows the buildup of the pretranscriptional complex and transcription. The retinoid receptors (RAR–RXR) are also found in the nucleus attached to DNA in the absence of ligand, but transcription is blocked by a corepressor. Binding of the ligand allows for the exchange of the repressor for a coactivator, and transcription is facilitated.

Class I receptors bind to inverted repeats on DNA. In the case of the ligand-bound homodimerized glucocorticoid receptor, the **glucocorticoid response element** in double-stranded DNA is as follows (where n is any nucleotide):

$$5'\ AGAACAnnnTGTTCT\ 3'$$

$$3'\ TCTTGTnnnACAAGA\ 5'$$

The natural ligand for the human glucocorticoid receptor is cortisol (also called hydrocortisone), although there exist several synthetic glucocorticoids that are higher-affinity ligands than cortisol for the receptor (many are used as drugs, such as **triamcinolone** and **dexamethasone;** see the chapter on steroid hormones). The structures of these steroids are shown in Figure 7-40. The crystal structures of the receptor's ligand-binding domain and the DNA-binding domain have been solved. The ligand-binding domain of the glucocorticoid receptor

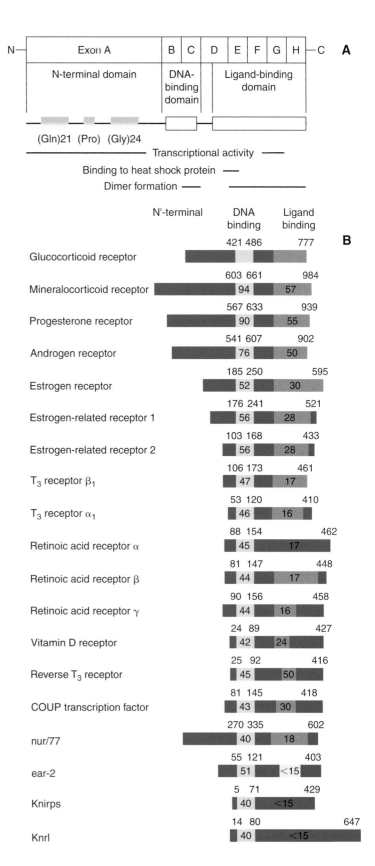

Figure 7-38. **A.** A model of the activities ascribed to a steroid receptor. This is general for all the steroid receptors except that the information concerning amino acid residues is specific for the androgen receptor. **B.** Several members of the steroid receptor superfamily using a similar model.

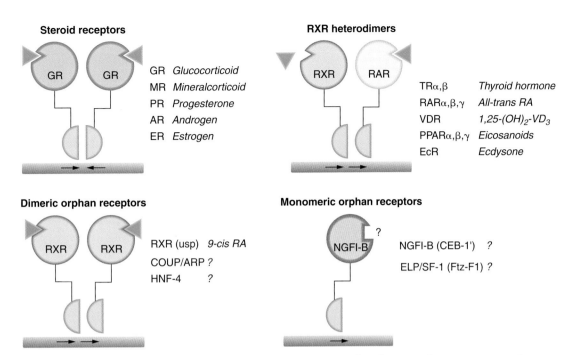

Figure 7-39. Classes of receptors in the nuclear receptor superfamily: steroid receptors, RXR heterodimers, dimeric orphan receptors, and monomeric orphan receptors. The glucocorticoid receptor (GR) is shown in the steroid receptor class.

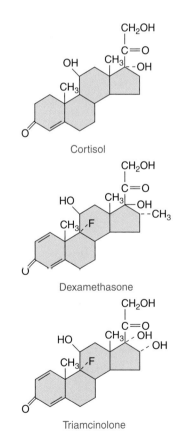

Figure 7-40. Structures of cortisol, dexamethasone, and triamcinolone.

complexed with cortisol is shown in Figure 7-41, and the DNA-binding domain bound to the double-stranded DNA glucocorticoid response element is shown in Figure 7-42. The DNA-binding domain of the receptor contains two zinc fingers that form a complex with the HRE of DNA, and there are specific amino acids in the fingers that account for the sequence recognition in the DNA-binding region and the spacing between the two half-sites of the HRE, as shown in Figure 7-43.

Transcription of the gene encoding the glucocorticoid receptor allows for alternative splicing, giving rise to two transcripts, an mRNA for the natural receptor (GRα) and the isoform, GRβ (Figure 7-44). GRβ is truncated by 35 amino acid residues in the C-terminal region. The two forms are identical through amino acid 727, but GRα adds 50 amino acids toward the C-terminal and GRβ adds 15 nonhomologous amino acids. GRα is the classical receptor occurring in nearly all human cells (except individual *pars intermedia* cells of pituitary and hepatobiliary cells), and GRβ occurs at lower concentrations in many cell types. GRβ is located mainly inside the nucleus, does not bind the ligand (cortisol), and is transcriptionally inactive but inhibits the transcriptionally active GRα. Its activity increases with its concentration. Consequently, it may be a factor in glucocorticoid resistance. GRα resides in the cell cytoplasm in the absence of the ligand. The ligand, cortisol, is produced in the adrenal cortex in response to stress or serotonin (endogenous rhythm) and is secreted after each round of stress or other stimulus. It is carried in the blood, with 90% bound to **transcortin** (or corticosteroid-binding globulin) and the remaining 10% in free form. It is the unbound hormone that diffuses through the target cell membrane to enter its cytoplasm. In the target cell, such as the liver cell, GRα is present as a monomer in a large complex with an HSP90 dimer and other proteins. The HSP90 proteins cover up the zinc fingers of the DNA-binding domain of the receptor and mask the nuclear translocation signal (rich in basic amino acids) downstream (toward the C-terminus) from the zinc fingers. When the cortisol enters the cell, it binds to the receptor (binding constant in the 50-nM range; about 1 nM for synthetic glucocorticoids). This causes a conformational change in the receptor and a shedding of the external proteins to generate a form capable of translocating to the nucleus and of binding to DNA. In addition, five serine residues in the receptor become phosphorylated. In this form, the receptor is translocated through the nucleopore (see Chapter 1) into the nucleus, where it heterodimerizes and binds to the glucocorticoid responsive element (GRE). Here, the receptor can enhance the binding of other proteins leading to transcription, or it can associate with other transcription factors, such as AP-1 and NFκB (Figure 7-45). Once the receptor dimer is bound to the GRE, other coactivators are needed, which are directed by the activator factor regions (AF-1 and AF-2) in the receptor structure in the N-terminal and C-terminal domains, as shown in Figure 7-46. Proteins are added, such as the p160 coactivators, including SRC-1 and the glucocorticoid receptor–interacting protein 1 (GRIP1; Figure 7-47), p300/CBP cointegrators, p300/CBP-associated protein (p/CAF), the switching/sucrose nonfermenting (SWI/SNF) complex, and the vitamin D receptor–interacting protein/thyroid hormone–associated protein (DRIP/TRAP) complex (see Figures 7-32 through 7-34). The NRB site on GRIP1 in the amino terminus contains three LXXLL motifs for interaction with the liganded receptor at AF-2 and another site at its carboxy terminus that interacts with AF-1 of the receptor (independent of ligand). HAT activity is a

(Text continues on p. 356.)

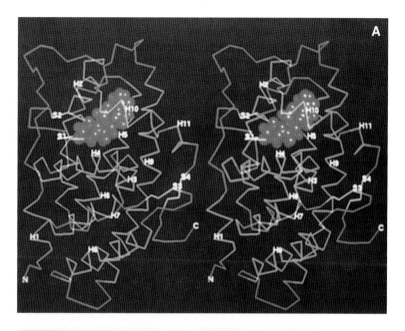
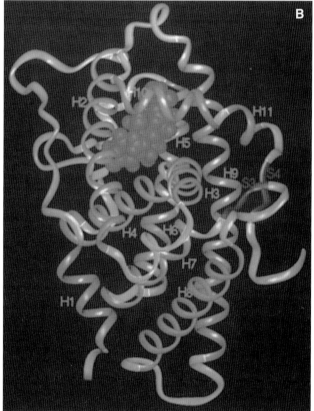

Figure 7-41. The ligand-binding domain of the glucocorticoid receptor.
A. Stereoscopic view of a trace of the mouse glucocorticoid receptor ligand-binding domain (LBD) complexed with cortisol. H and S correspond to α-helix and β-sheet. Bound cortisol in the space-filling model. **B.** Ribbon diagram of the mouse glucocorticoid LBD complexed to cortisol (space-filling model). Reproduced with permission from R. Dey, P. Roychowdhury, and C. Mukherjee, "Homology modeling of the ligand-binding domain of glucocorticoid receptor: binding site interactions with cortisol and corticosterone," *Protein Eng.*, 14: 565–571, August 2001.

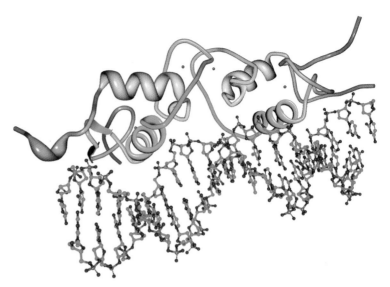

Figure 7-42. DNA-binding domain of the rat liver glucocorticoid receptor homodimer *(red and purple)* bound to the glucocorticoid response element in double-stranded DNA (major groove). *Blue* atoms are zinc (two per zinc finger). Reproduced from PDB ID: 1glu. B.F. Luisi, W.X. Xu, Z. Otwinowski, L.P. Freedman, K.R. Yamamoto, P.B. Sigler. Crystallographic Analysis of the Interaction of the Glucocorticoid Receptor with DNA. *Nature* **352** pp. 497 (1991).

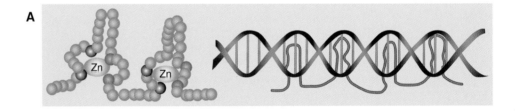

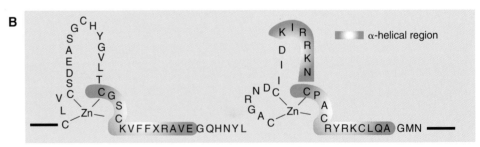

Figure 7-43. **A.** On the left are shown the 2 zinc fingers of the glucocorticoid receptor. The figure on the right shows how the zinc fingers fit into the major groove of DNA. **B.** The amino acid sequence of the 2 zinc fingers of the glucocorticoid receptor. The finger on the left is upstream (toward the N-terminal) to finger on the right. The amino acids responsible for recognition of the bases in DNA and the spacing between the half sites are in *blue*. The finger on the left is referred to as the proximal finger and the finger on the right is the distal finger.

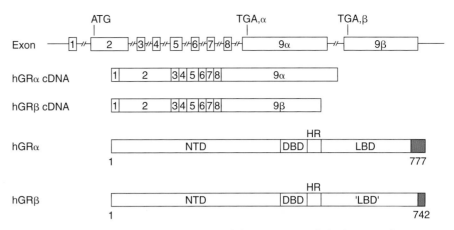

Figure 7-44. Schematic representation of the structure of the human glucocorticoid receptor (hGR) gene. Alternative splicing of the primary transcript gives rise to the two mRNA and protein isoforms, hGRα and hGRβ. The functional domains and subdomains are indicated beneath the linearized protein structures. *AF*, activation function; *DBD*, DNA-binding domain; *LBD*, Ligand-binding domain; *NLS*, nuclear localization signal. Redrawn with permission from E. Charmandari, T. Kino, and G.P. Chrousos, "Molecular mechanisms of glucocorticoid action," February 2004. Edited by Sebastiano Filetti.

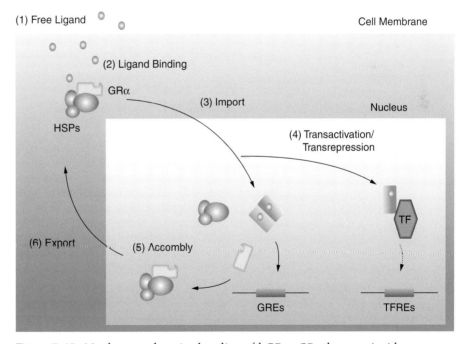

Figure 7-45. Nucleocytoplasmic shuttling of hGRα. *GR*, glucocorticoid receptor; *GREs*, glucocorticoid-response elements; *HSP*, heat shock proteins; *TF*, transcription factor; *TFREs*, transcription factor–response elements. Redrawn with permission from E. Charmandari, T. Kino, and G.P. Chrousos, "Molecular mechanisms of glucocorticoid action," *Orphanet* encyclopedia, February 2004: http://www.orpha.net/data/patho/GB/uk-glucocorticoidaction.pdf.

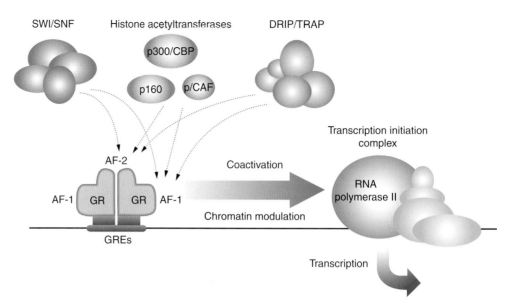

Figure 7-46. Schematic representation of the interaction of coactivators with the AF-1 and AF-2 domains of the glucocorticoid receptor and their role in transcriptional regulation. *AF,* activation function; *GR,* glucocorticoid receptor; *GREs,* glucocorticoid-response elements. Redrawn with permission from E. Charmandari, T. Kino, and G.P. Chrousos, "Molecular mechanisms of glucocorticoid action," *Orphanet* encyclopedia, February 2004: http://www.orpha.net/data/patho/GB/uk-glucocorticoidaction.pdf.

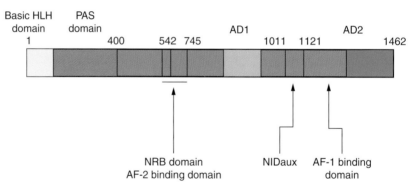

Figure 7-47. Linearized GRIP1 molecule and distribution of its functional domains. *AD1,* activation domain 1; *AD2,* activation domain 2; *HLH,* helix-loop-helix; *NIDaux,* auxiliary nuclear receptor interacting domain; *NRB,* nuclear receptor binding; *PAS,* period arylhydrocarbon receptor. Redrawn with permission from A. Vottero et al., *J. Clin. Endocrinol. Metab.,* 87, 6: 2658–2667, 2002. Redrawn with permission from E. Charmandari, T. Kino, and G.P. Chrousos, "Molecular mechanisms of glucocorticoid action," *Orphanet* encyclopedia, February 2004: http://www.orpha.net/data/patho/GB/uk-glucocorticoidaction.pdf.

feature of p160, CBP/p300, and p/CAF for the acetylation of promoter histones, the net effect of which is to enhance transcription. (These activities are summarized generically in Figure 7-48.) As a result, mRNAs of glucocorticoid target genes are translocated to the cytoplasm and translated into new protein molecules that alter the character of the cell, reflecting the actions of the hormone on that cell.

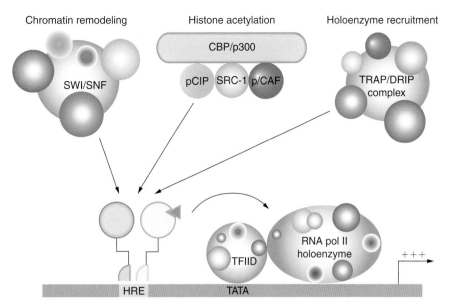

Figure 7-48. Ligand-dependent recruitment of multiple coactivator complexes. Upon ligand binding, the receptors recruit different coactivator complexes. The complex CBP/p160/PCAF possesses histone acetyltransferase activity, the SWI/SNF complex possesses ATP-dependent chromatin remodeling activity, and the TRAP/DRIP complex may recruit the RNA polymerase ll (RNAP ll) holoenzyme. Recruitment of the complexes may be sequential or combinatorial. It is conceivable that chromatin remodeling complexes are initially recruited to the promoter. These factors may relieve the repression imposed by high-order chromatin structure and allow a second acetylation-dependent step on gene activation. Activation would require the combinatorial of subsequent action of additional complexes that include the TRAP/DRIP complex. Redrawn from *EMBO Rep.* 4: 1122–1126, 2003. Récepteurs à la Provençale EMBO Workshop on the Biology of Nuclear Receptors.

Chromatin

A discussion of chromatin was presented in Chapter 1; some further aspects will be developed here with respect to gene expression. Chromatin is tightly packaged double-stranded DNA with units of histones contained within loops of DNA and uniform stretches of DNA between the histone units. Recent ideas suggest that DNA itself determines the locations of nucleosomes on DNA. A visualization of a chromatin fiber in condensed form and then stretched out to show the repeating histone-containing units is presented in Figure 7-49. The histone repeating groups are octamers consisting of pairs of histones: H2A, H2B, H3, and H4. The minimal unit of the chromatin structure is the nucleosome, consisting of 146 base pairs of DNA and the eight histones with the DNA wrapped (nearly two turns) around the histone core (Figure 7-50). H1 and H5 are linker histones and are important to the solenoid structure, where the "beads-on-a-string" stretched-out chromatin itself is turned into a solenoid (Figure 7-51). The chromatin has to be opened to be transcriptionally active. The role of many coactivators and other related molecules is

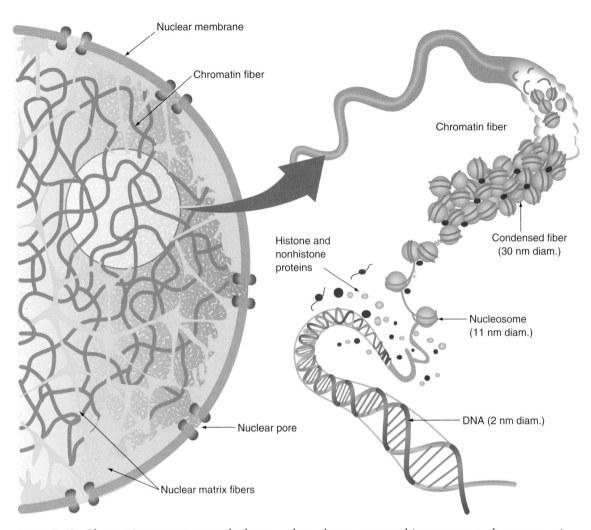

Figure 7-49. Chromatin structure stretched out to show the component histone groups that occur at intervals of 200 base pairs. Between the histone groupings are individual proteins complexed to DNA that include histone H1 and nonhistone proteins. The DNA connecting histone groups (octamers) is called linker DNA.

to open the chromatin for this purpose, accomplishing this through the associated enzymatic activities, causing acetylation of histones. This results in the relaxation of chromatin while repressors deacetylate histones and other factors (Figure 7-52), causing chromatin to tighten up. The acetylation of histone by HATs occurs on the lysine residue:

$$\text{Histone-lys} + \text{acetyl CoA} \rightleftharpoons \text{Histone acetyl-lys} + \text{CoA-SH}$$

A number of coactivator proteins have HAT activity, and those that are known are listed in Table 7-3. The crystal structure of human pCAF histone acetyltransferase bound to CoA is shown in Figure 7-53. Acetylated histones H3 and H4 are deposited into newly formed DNA and are soon deacetylated. The two enzymes involved in these modifications each have the subunit p48. The histone chaperone, CAF1, forms a complex with p48, H3, and H4 (Figure 7-54). As shown in this figure, acetylation takes place on the lysine residues of histone tails, and this substitution eliminates the positive charge of the lysine terminus. Because the histone tails lie outside the core of the nucleosome and interact with nearby

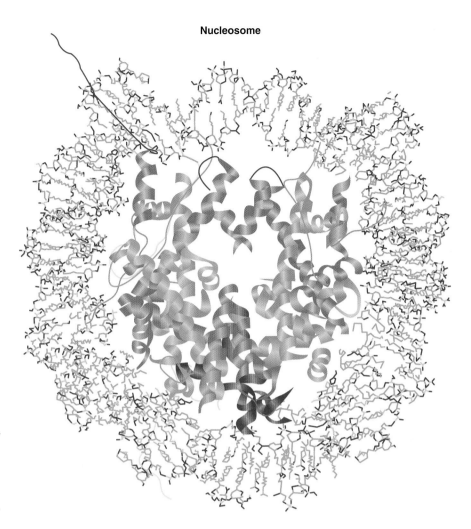

Figure 7-50. Structure of the nucleosome showing the doublestrand of DNA surrounding the octamer of histones: H2A, H2B, H3, and H4. Reproduced from http://www.webbooks.com/MoBio/Free/Ch3D1.htm.

nucleosomes, acetylation opens up this highly ordered structure, making the chromatin accessible to large complexes of proteins to facilitate transcription (Figure 7-54). With the energy provided by ATP, the nucleosome is remodeled to make the individual strands of DNA available to RNA polymerase. Certain factors, such as TFIII, are held in place on the template through the actions of **bromodomains.** *These domains (for example, in hTAF250) are spaced to bind consecutive acetyllysines in the tail of histone H4, favoring propagation of acetylation to immediate neighbors.* Other HAT complexes are recruited by coactivators, including pCAF (p300/CBP-associated factor), which contains a GCN5 subunit, the HAT bromodomain, and a histone-like TAF complex similar to TFIID. Repression of chromatin occurs through the activity of histone deacetylase associated with corepressors, returning chromatin to its original, tightly packaged state that excludes the transcriptional machinery.

Nucleosome remodeling is mediated by the human 2 mating type SWI/SNF complex (Figure 7-48), where an important component is an ATP-dependent helicase. The tail of H2A may be affected in this process, which also decreases the length of DNA per nucleosome, renders the DNA more susceptible to nucleases, and increases the ability of the histone octamer to slide along DNA, enabling a bulge to slip off the octamer as the polymerase moves through.

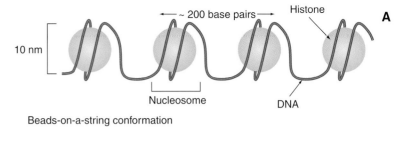

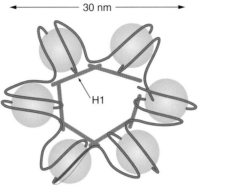

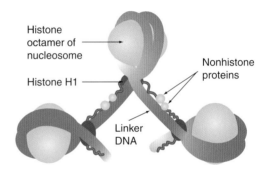

Figure 7-51. Conformations of chromatin. **A.** Beads-on-a-string where nucleosome units are stretched out. **B.** One turn of solenoid showing multiple nucleosomes twisted into a solenoid and stabilized by H1 histone. **C.** Another view of beads-on-a-string conformation showing the location of histone H1 and other proteins.

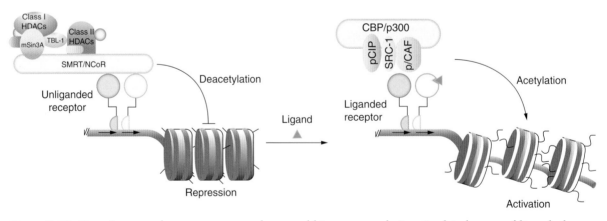

Figure 7-52. Coactivator and corepressor complexes and histone acetylation. In the absence of ligand, the nuclear hormone receptor heterodimer is associated with corepressor complexes. The corepressors (SMRT/NCoR) recruit histone deacetylases (HDACs) either directly or through their interaction with Sin3. Many other proteins must belong to these complexes, whose exact composition is still unknown. Deacetylation of histone tails leads to chromatin compaction and transcriptional repression. Ligand binding causes the release of the corepressor complex and the AF-2-dependent recruitment of a coactivator complex that contains at least p160 coactivators (such as p/CIP or SRC-1), CBP/p300, and PCAF. All of these proteins possess histone acetyltransferase (HAT) activity that allows chromatin decompaction and gene activation. Multiple protein–protein interactions exist among the different components: CBP/p300 contacts the receptor, the p160 coactivators, and PCAF through independent domains. Similarly, the receptor binds CBP/p300, p160 coactivators, and PCAF, and PCAF can also bind directly to CBP/p300, p160 coactivators, and the receptor.

Table 7-3
Summary of Known and Putative Human HATs

HAT	Organisms known to contain the HAT	Known transcription-related functions/effects	HAT activity demonstrated in vitro[b]	Histone specificity of recombinant enzyme in vitro[a,b]	Known native HAT complexes and nucleosomal histone specificities in vitro
GNAT superfamily					
Hat1	Various (yeast to humans)	None (histone deposition-related B-type HAT)	Yes	**H4**	Yeast HAT-B, HAT-A3 (no nucleosome acetylation)
Gcn5	Various (yeast to humans)	Coactivator (adaptor)	Yes	**H3/H4**	Yeast ADA, SAGA (**H3**/H2B); human GCN5 complex, STAGA, TFTC (**H3**)
PCAF	Humans, mice	Coactivator	Yes	**H3/H4**	Human PCAF complex (**H3**/weak H4)
Tip60	Humans	HIV TAT interaction	Yes	**H4/H3/H2A**	Tip60 complex
MOZ	Humans	Leukemogenesis, upon chromosomal translocation	ND		
MORF	Humans	Unknown (strong homology to MOZ)	Yes	**H4/H3/H2A**	
HBO1	Humans	ORC interaction	Yes*	ND*	HBO1 complex
p300/CBP	Various multicellular	Global coactivator	Yes	**H2A/H2B/H3/H4**	
Nuclear receptor coactivators		Nuclear receptor coactivators (transcriptional response to hormone signals)			
SRC-1	Humans, mice		Yes	**H3/H4**	
ACTR	Humans, mice		Yes	**H3/H4**	
TIF 2	Humans, mice		ND		
TAF$_{II}$250	Various (yeast to humans)	TBP-associated factor	Yes	**H3/H4**	TFIID
TFIIIC		RNA polymerase III transcription initiation			TFIIIC (**H2A/H3/H4**)
TFIIIC220	Humans		Yes*	ND	
TFIIIC110	Humans		Yes	ND	
TFIIIC90	Humans		Yes	**H3**	

[a]Histones that are the primary in vitro substrates for a given HAT are bold; other histones listed are acetylated weakly or in a secondary manner.
[b]Asterisks indicate proteins for which HAT activity has been suggested indirectly or demonstrated in an incomplete manner. Elp3 can acetylate all four histones but has only been tested with them individually in in-gel assays. The HAT function of HBO1 has primarily been shown by the in vitro free histone **H3/H4**-acetylating activity of a purified human complex containing it, although recombinant GST-HBO1 (and the complex) did weakly acetylate nucleosomes. Finally, TFIIIC220 was identified as a HAT only in in-gel assays, and its activity has not yet been confirmed by recombinant protein studies. ND, not determined. ACTR, activator of thyroid and retinoic acid receptor; ADA, member of the SAGA complex; GCN5, histone acetyltransferase acetylating N-terminal lysines on histones H2B and H3; GNAT, GCN5-related N-acetyltransferase; HAT, histone acetyltransferase; Hat1, histone acetyltransferase1 (human); HBO1, part of a multi-subunit complex possessing histone H3 and histone H4 acetyltransferase activities; MORF, monocytic leukemia zinc finger protein-related factor; MOZ, monocyte leukemia zinc finger HAT; p300/CBP, CREB-binding protein; PCAF, p300/CBP-associated factor; SAGA, Spt-ADA-GCN5-acetyltransferase (complex); TAF$_{II}$250, complex consisting of TBP and associated factors (TAFs) that bind to DNA; TFIII90, subunit of TFCIII2; TFIIIC110, a subunit of human TFCIII2; TFIIIC, multi-subunit transcription factor; TFIIIC220, a subunit of human TFCIII2; TFCIID, horseshoe-shaped structure that includes TFIIA and TFIIB, the tope of which interacts with DNA, TIF2, translation initiation factor eIF4A; Tip60, substrate for HAT activity of p300/CBP; SRC-1, steroid receptor coactivator-1 with HAT activity (also known as NCOA-1);
[c]S. John and J.L. Workman, unpublished result.
Reproduced in part from http://mmbr.asm.org/cgi/content/full/64/2/435/T1, from Table 1 of D.E. Sterner and S.L. Berger, *Microbiol. & Molec. Biolo. Revs.*, 64: 435–459, 2000.

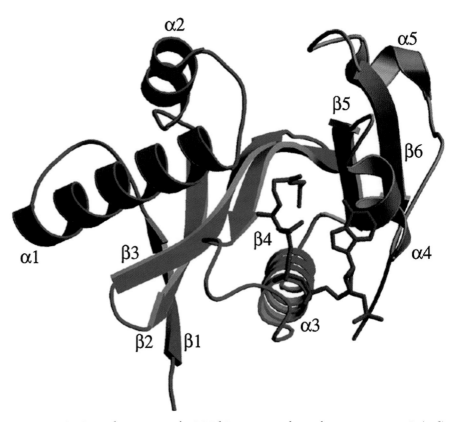

Figure 7-53. Crystal structure of pCAF histone acetyltransferase coenzyme A *(red)* complex. N- and C-terminal protein segments flanking the core are in magenta and gold, respectively. Reproduced with permission from A. Clements et al., "Crystal structure of the histone acetyltransferase domain of the human PCAF transcriptional regulator bound to coenzyme A," *The EMBO J.*, 18: 3521–3532.

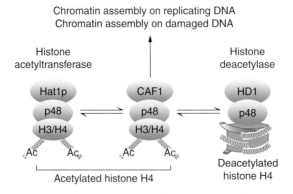

Figure 7-54. Composition of histone acetyltransferase and histone deacetylase that act on H3 and H4.

Further Reading

Books

Alberts, B., Roberts, K., Lewis, J., Raff, M., and Hopkin, K., *Essential Cell Biology*, Taylor and Francis, 2003.

Lodish, H., Matsudaira, P., Berk, A., Zipursky, S.L., and Scott, M.P., *Molecular Cell Biology*, W.H. Freeman & Co., 2003.

Reviews

Furia, B., Deng, L., Wu, K., Baylor, S., Kehn, K., Li, H., Donnelly, R., Coleman, T., and Kashanchi, F., "Enhancement of Nuclear Factor-kappa B Acetylation by Coactivator p300 and HIV-1 Tat Proteins," *J. Biol. Chem.* **277**: 4973–4980, 2002.

CHAPTER 8

Polypeptide Hormones

Panhypopituitarism: A Malfunction of the Hypothalamus–Pituitary–End Organ Axis

Organisms as complicated as a human must have an elaborate set of communication systems because of organ and tissue specialization and the spatial separation of the organs. A major system for altering functions of many organs is a hormonal system whose signals start in the brain. As part of this function, incoming signals from the outside environment (or from within the body) are processed by the brain and result in electrical and chemical signals to the hypothalamus (in the brain). The signal causes the secretion of a specific **releasing hormone** from a hypothalamic neuronal nerve ending, which travels a brief distance through a closed portal circulation to reach the environment of the pituitary. In the pituitary, the releasing factor binds to a receptor in the membrane of a target cell. The target cell responds through a second messenger and a signal transduction mechanism (a set of biochemical changes) that lead to changes in transcription and more immediate cellular changes, causing the release (and often synthesis) of a specific anterior pituitary hormone into the general circulation. *The narrow closed portal system carrying the releasing hormone from the hypothalamus to the anterior pituitary is enveloped by a delicate narrow stalk.* This stalk can be severed by trauma to the head (traumatic brain injury), sometimes as a result of an automobile, or other, accident. When the stalk is damaged, there is disruption of the blood supply between the hypothalamus and the anterior pituitary, leading to infarction of the anterior pituitary. The blood supply also can be damaged by edema. If this occurs, the releasing factors do not reach the pituitary and the pituitary hormones are not signaled to be released into the bloodstream, causing a severe malfunctioning of the end organs in the system and, without treatment, resulting in death. The hypothalamus–pituitary system connected by the pituitary stalk is shown in Figure 8-1

The frequency of traumatic brain injury is about 2 million people per year in the United States. Of these, about 80,000 have associated neuroendocrine disorder. Another 50% or so of people with traumatic brain injury are discovered postmortem to have had neuroendocrine disorder

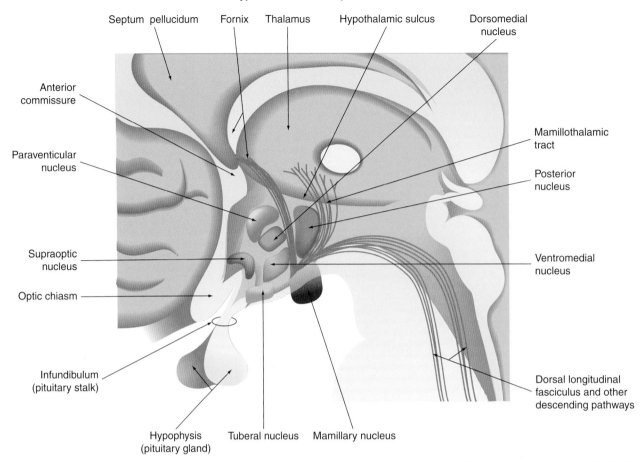

Figure 8-1. Appearance of the human hypothalamus, pituitary stalk and anterior and posterior pituitary. Individual cells of the *pars intermedia* are located (without a discrete structure) between the anterior and posterior pituitary lobes.

(they were undiagnosed). There are about 52,000 deaths each year from traumatic brain injury, and there may be as many as 6.5 million people alive with the consequences of this injury. The symptoms of **panhypopituitarism** may not show up until weeks or months following the brain injury (until the effective supply or hormones is used) and may be signaled by appearance of malaise and decreased vital signs with progressive lethargy, anorexia, hypothermia, slow heart rate, or hypotension with hyponatremia (low blood sodium). Blood levels of end organ hormones—cortisol, testosterone, **triiodothyronine (T_3), tetraiodothyronine (T_4, thyroxine)**, and the anterior pituitary hormone, **thyroid-stimulating hormone (TSH)**—and sodium levels are confirmatory.

Panhypopituitarism can be a childhood disease when it is congenital (at birth) because of poor development, especially in the midline brain structure *(septum pellucidum)*. Sometimes, in children, tumors arise in the hypothalamic–pituitary region, especially in childhood craniopharyngioma. In childhood hypopituitarism, **growth hormone,** which determines stature and growth rate, is the most affected. Central hypothyroidism can occur when there is a deficiency of TSH from the anterior pituitary. Deficiency in the gonadotropic hormones, **luteinizing hormone (LH),** and **follicle-stimulating hormone (FSH),** does not show up until puberty, when breast development and menstrual cycles in females are affected and an enlargement of penis and testicles occurs in males. In females, the situation can be detected by ultrasound, which indicates pubertal-sized ovaries and uterus that can be diagnostic. Interference with elaboration of **vasopressin (antidiuretic hormone, ADH)** from the posterior pituitary leads to *diabetes insipidus* and excessive urination.

Hypopituitarism can also occur in pregnancy. It may occur as a result of profound blood loss during and after childbirth, known as **Sheehan's syndrome.** Blood loss can lead to death of the cells of the anterior pituitary. The chronic form of Sheehan's syndrome can become known months or years after childbirth, and the rarer acute form appears shortly after delivery. In some cases of Sheehan's syndrome, pituitary autoimmunity has been reported. Treatment involves supplying the end organ hormone in most cases, and surgery may be needed to remove a tumor.

Humoral Mechanism

The relevant parts of the brain focusing on the hypothalamus and pituitary are shown in Figure 8-1, and the vasculature is shown in Figure 8-2. In Figure 8-2, it is clear that the blood vessels within

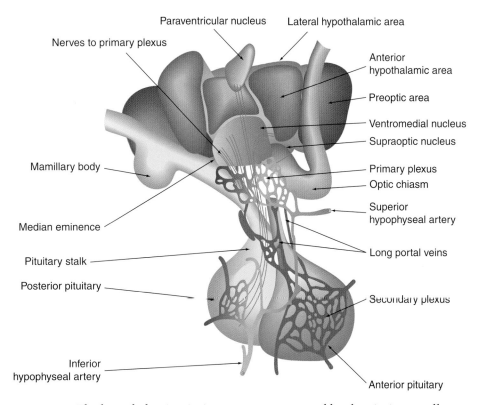

Figure 8-2. The hypothalamic–pituitary system connected by the pituitary stalk. Releasing hormones from neurons in the hypothalamus course down the neuronal axons in the pituitary stalk to the anterior or posterior pituitary, where they find their receptors on the membranes of target cells.

the stalk are delicate. It is important that the releasing hormones are contained within the small volume of the blood vessels in the stalk and do not enter the general circulation. Thus, the small amounts (nanogram quantities) of the releasing factors, released in each secretory event, are undiluted so that they are concentrated enough to reach the binding constant of their cognate (related to the structure of the hormone) receptors. The transmission of hormonal signals, by way of the blood (the humoral mechanism), from the hypothalamus to the pituitary to the end organs is shown in Figure 8-3. The releasing hormones of the hypothalamus are summarized in Table 8-1. Releasing hormones of known structure and function are truly hormones, whereas substances that act like releasing hormones but whose structures and physiological functions are not understood are called factors. The well-understood releasing hormones are corticotropic-releasing hormone (CRH), growth hormone–releasing hormone (GHRH), somatostatin, gonadotropin-releasing hormone (GnRH) and thyrotropin-releasing hormone (TRH). Prolactin-releasing factor (PRF) and **prolactin release–inhibiting factor (PIF)** have functions to release **prolactin (PRL)** and inhibit its release, but it is not clear what the identities of the physiological hormones really are. Nevertheless, there are known substances that are active agonists and antagonists (Table 8-1).

The functions of the releasing hormones are to release the relevant hormones of the anterior pituitary (in some cases, to stimulate their syntheses). The anterior pituitary hormones are listed in Table 8-2. Some of the hormones listed in Table 8-2 arise from a single gene product known as **preproopiomelanocortin.** To understand better where the anterior pituitary hormones arise from, as well as hormones from the *pars intermedia*–like cells (PILC) located between the anterior pituitary and the posterior pituitary, a model of the preproopiomelanocortin precursor protein is shown in Figure 8-4. Hormones are split out of the precursor by proteases that are specific for the Arg–Lys or Lys–Arg linkages, the locations of which are indicated at the top of the figure. Some hormones arise preferentially in the anterior pituitary cells, and others arise in the PILCs. Thus, the release of γ-melanocyte–stimulating hormone (γ-MSH), adrenocorticotropic hormone (ACTH) and β-lipotropin (β-LPH) occurs mainly in the anterior pituitary cells. In the PILCs, ACTH is broken down further so that α-MSH and corticotropic-like inhibitory peptide (CLIP) are released and β-LPH is broken down to γ-LPH and β-endorphin. γ-LPH and β-endorphin also serve as precursors for β-MSH and Met-enkephalin. The anterior pituitary (also known as the **adenohypophysis**) is composed of essentially five cell types. The largest cell population is made up of **somatotrophs,** which produce growth hormone and occupy 50% of the cells. The **lactotrophs** (producing PRL) make up 20% of the cells, as do the **corticotrophs** that produce ACTH. The thyrotrophs (producing TSH) and the **gonadotrophs** (producing LH and FSH) each occupy 5% of the cells. Sometimes these cells are referred to as "tropes" rather than "trophs." The "troph" ending refers to the ability of the hormone to produce growth of the target cells, and "trope" refers to the ability to induce changes in the target cells. Often the hormone in question accomplishes both functions. For example, ACTH not only causes the synthesis and release of cortisol from the adrenal cortex cells (primarily the *fasciculata* layer of cells) but also is an important growth factor for

(Text continues on p. 378.)

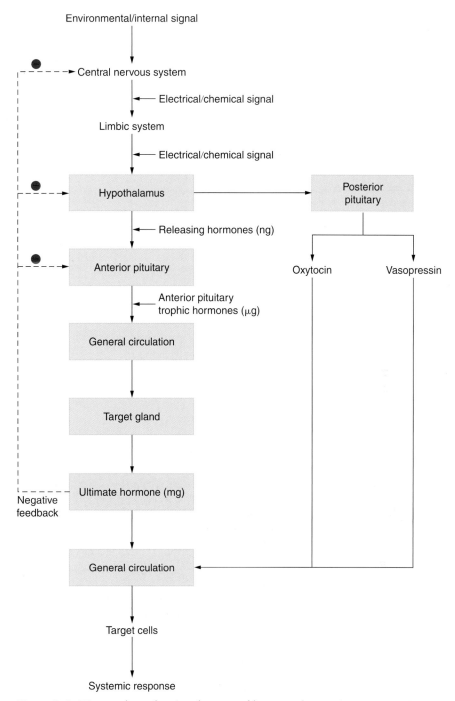

Figure 8-3. Humoral mechanism by way of hormonal secretions connecting the hypothalamus to the anterior pituitary to the end organ or the hypothalamus to the posterior pituitary. The vessels carrying hormonal signals are shown in Figure 8-2. The target gland is the last organ in the pathway: an example would be the corticotrophic releasing hormone from the hypothalamus, causing the release of ACTH from the anterior pituitary, which causes the release of cortisol from the adrenal cortex (target gland) in this case. In parentheses are shown the masses of each type of hormone released.

Table 8-1
Releasing Hormones of the Hypothalamus

Releasing Hormone	Number of Amino Acid Residues	Action	Structure/Sequence*
Corticotropin-Releasing Hormone (CRH)	41	Binds receptor on corticotropic cell of anterior pituitary to release ACTH and β-endorphin and lipotropin	SQEPPISLDLTFHLLREVLEMTKADQLAQQ-AHSNRKLLDI-Ala-NH$_2$
Growth Hormone–Releasing Hormone (GHRH) (Somatocrinin)	(40)-44	Binds to receptor on sommatotroh of anterior pituitary to stimulate growth hormone (GH) secretion	YADAIFTNSYRKVLGQLSARKLLQDIMSR-QQGESNQERGARAR-Leu-NH$_2$
Somatostatin	14	Binds to receptor on sommatotroph to inhibit release of GH	

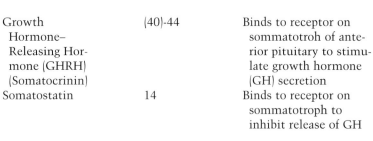

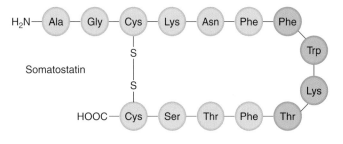

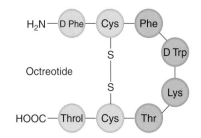

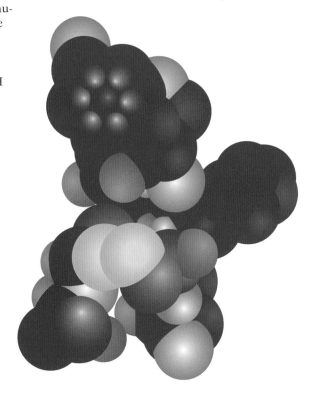

*The one-letter abbreviation for amino acids is generally used.

Releasing Hormone	Number of Amino Acid Residues	Action	Structure/Sequence
		"Octreotide" (8 amino acids with active portion of somatostatin)	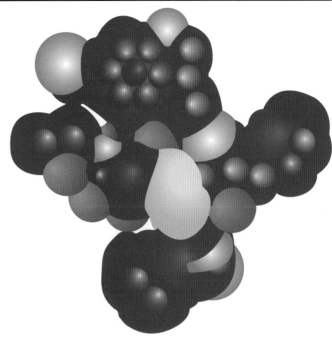
Gonadotropin Releasing Hormone (GMRH)	10	Binds to receptor on lactotrope to release luteinizing hormone (LH) and to receptor on folliculotrope to release follicle-stimulating hormone (FSH). LH and FSH can be released from the same cell (gonadotrope)	PGLU-HWSYGLRP-GLY-NH$_2$

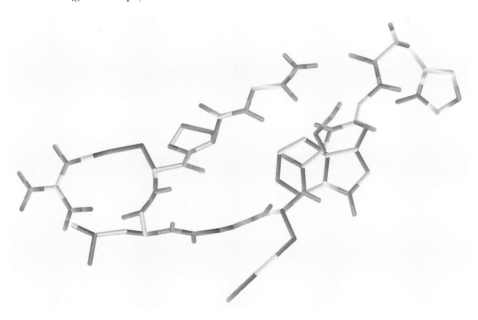

Pyroglutamate (N-Terminus) is in upper right and glycine-amide (N-Terminus) is at end of fold in right center.

Continued

Table 8-1
Releasing Hormones of the Hypothalamus—cont'd

Releasing Hormone	Number of Amino Acid Residues	Action	Structure/Sequence
Thyrotropin-Releasing Hormone (TRH)	3	Binds to receptor on thyrotrope of anterior pituitary to release thyrotropic stimulating hormone (TSH)	PGlu-H-Pro-NH$_2$ 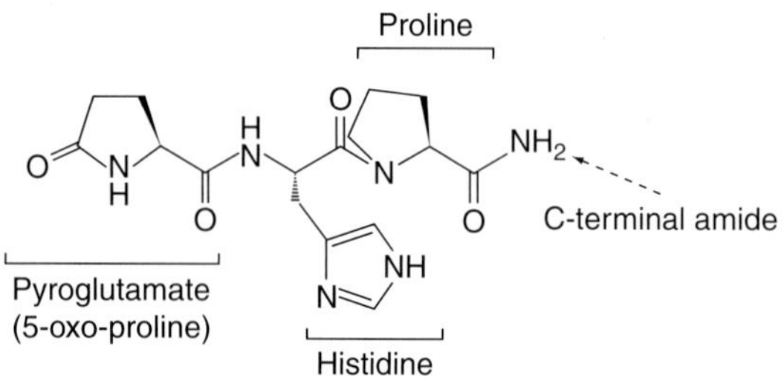 (Pyroglutamate (5-oxo-proline) – Histidine – Proline with C-terminal amide)
Prolactin Releasing Factor (PRF)	?	This factor could be TRH. PRF binds to a receptor on the lactotrope to release prolactin (PRL). PRF may also be a discrete factor different from TRH or vasopressin. There are 20- and 31- amino acid peptides known to release PRL in human pituitary. The sauvagine peptide (40 amino acids) also releases PRL.	Prolactin-releasing peptides (P20, P31) and sauvagine (SAU) P20(b) TPDINPAWYAGRGIRPVGRF P20(r) TPDINPAWYTGRGIRPVGRF P20(h) TPDINPAWYASRGIRPVGRF P31(b) SRAHQHSMEIRTPDINPAWYAGRGIRPVGRF P31(r) SRAHQHSMETRTPDINPAWYAGRGIRPVGRF P31(h) SRTHRHSMEIRTPDINPAWYASRGIRPVGRF SAU ZGPPISIDLSLELLRKMIEIEKQEKEKQQAANNRLLLDTI
Prolactin–Release Inhibiting Factor (PIF)	56?	May be derived from for GNRH. Acts on lactotrope to inhibit release of prolactin (PRL). One precursor peptide active as a PIF is identical to residues 27–52 of N-terminal region of proopiomelanocortin precursor (A). A second PIF is residues 109–147 of the vasopressin-neurophysin precursor (β).	WCLESSECQDLSTESNLLACIRACKP (a) ASDRSNATLLDGPSGALLLRLVQLAG–APEPAEPAQPGVY (b)

Structure reproduced from http://www.bigochem.ucl.ac.uk/bsm/pdbsum/1go9tracel.html. Structures reproduced from http://www.biochem.ucl.ac.uk/bsm/pdbsum/2soc/trace1.html. Reproduced from http://fulcrum.physbio.mssm.edu/~hwlab/online/fg1-fig1.gif.

Table 8-2
Hormones of the Anterior Pituitary

Anterior Pituitary Hormone	Number of Amino Acid Residues or Molecular Weight	Action	Structure/Sequence
Adrenocorticotropic Hormone (ACTH)	39 Amino acids	Signals adrenal cortex to release cortisol (and aldosterone secondary stimulus).	
Growth hormone (GH, Somatotropin)	191 Amino acids 22,124 Daltons	Promotes growth; stimulates lipid and carbohydrate metabolism, especially in liver and adipose tissue. Releases IGF-1. Stimulates bone sulfation.	

```
        1                          16
        SYSMEHFRWGKPVGKK          0.1*
                                R
        α-MSH (may be acetylated) R 18   6
                                  P
           β Cell Tropin          V 20
                                  K 21   111
                                  V
                                  Y
        ─FELPFAEASEDEAGNP 24      103
        39      35 33
        Common to   Immunological
        All Species (species vary)

                   CLIP
```
*Numbers represent extent of full activity as ACTH for the given length of the partial peptide.

Follicle-Stimulating Hormone (FSH)	82 Amino acids in α-subunit 118 Amino acids in β-subunit	Signals ovaries and testes. Acts to stimulate growth and maturation of ovarian follicles. Together with estrogens, stimulates formation of LH receptors on granulosa cells in late follicular phase. With testosterone, supports the process of spermatogenesis in males.	

Human follicle-stimulating hormone FSH, LH, and TSH all share the same α-subunit; differences in the β-subunit allow for specific receptor recognition. Structure from PDB ID: 1Fl7. K.M. Fox, J.A. Dias, and P. Van Roey. Human Follicle Stimulating Hormone. **15** pp. 378 (2001).

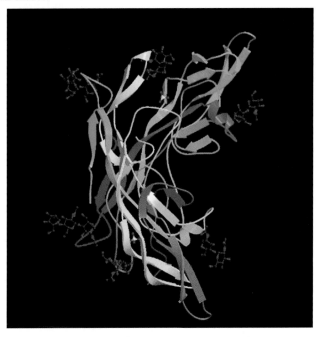

Continued

Table 8-2
Hormones of the Anterior Pituitary—cont'd

Anterior Pituitary Hormone	Number of Amino Acid Residues or Molecular Weight	Action	Structure/Sequence
			Human growth hormone (Protein Data Bank 1HGU; King Appelet Structure) 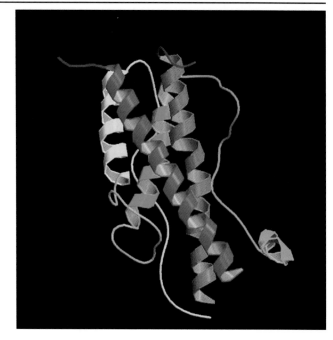
Luteinizing Hormone (LH)	~22,000 Daltons 2 subunits: α-subunit 96; β-subunit 121 amino acids	Stimulates secretion of sex steroids from gonads of males and females. In males, LH acts on leydig cells to stimulate synthesis and secretion of testosterone; in females, LH stimulates theca cells to secrete testosterone, which is then converted to estrogen in nearby granulosa cells.	
			Solution structure of the α-subunit of human chorionic gonadotropin (similar to LH) [Modeled with diantennary glycan at Asn78] (1HD4, Protein Data Bank. 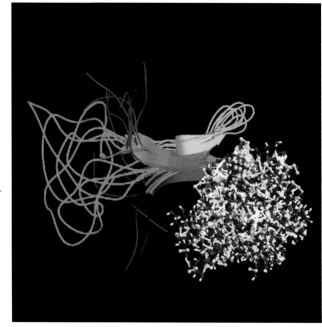

Anterior Pituitary Hormone	Number of Amino Acid Residues or Molecular Weight	Action	Structure/Sequence
Prolactin (PRL)	199 Amino acids 23,000 Daltons	Growth and development of mammary gland, synthesis of milk, and maintenance of milk secretion; enhances progesterone secretion; some role in immune response.	Solution structure of prolactin (1RWS Protein Data Bank; King Appelet Structure) 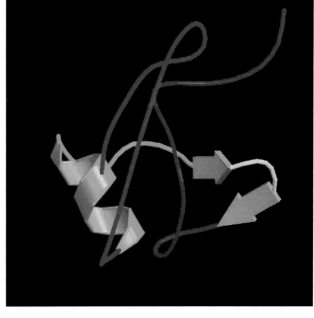
Thyroid-stimulating hormone (TSH)	211 Amino acids 28,300 Daltons 2 subunits	Stimulates secretion and synthesis of thyroid hormone (T_4 and T_3) from thyroid gland. The α-subunit is identical to that for FSH and LH, above. Grossman et al., *Endocrine Review*, 18: 476–501, 1997.	

Continued

Table 8-2
Hormones of the Anterior Pituitary—cont'd

Anterior Pituitary Hormone	Number of Amino Acid Residues or Molecular Weight	Action	Structure/Sequence

The schematic drawing of hTSH showing domains important for bioactivity. For clarity, the carbohydrate chains are not shown. The α-subunit backbone is shown as gray line, and the β-subunit chain is shown as a black line. The functionally critical domains are marked directly within the line drawings. The peripheral β-hairpin loops are marked as follows; αL1, αL3 in the α-subunit; βL1, βL3 in the β-subunit. Two long loops are αL2 with α-helical structure and βL2, a loop analogous to the "Keutmann loop" in the human chorionic gonadotropin β-subunit.

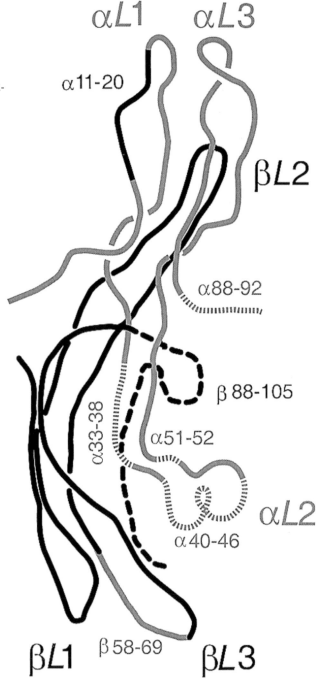

Anterior Pituitary Hormone	Number of Amino Acid Residues or Molecular Weight	Action	Structure/Sequence
Melanocyte-stimulating hormone (MSH)	α Polypeptide = 13 β Polypeptide = 18 γ Polypeptide = 12	Skin darkening; CNS functions	α-MSH & Clip (corticotropin-like-inhibitory peptide) arise from the breakdown of ACTH in the PARS intermedia-like cells between anterior and posterior pituitary. γ-MSH arises in anterior pituitary

$$\text{H}_2\text{N—S—YSMEHFRWCKPV—COOH} \atop \alpha\text{-MSH}$$
(positions 1, 10, 13)

$$\left[\text{H}_2\text{N—KVYPNGAEDESAEAFPLEF—COOH} \atop \text{CLIP}\right]$$
(positions 10, 19)

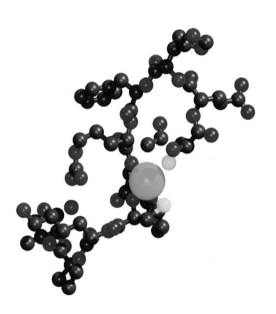

Trithiol α MSH cyclized through rhenium coordination (2 molecules of MSH)

Reproduced from Figure 3 of M.W. Szkudlinski et al. "Thyroid-stimulating hormone and thyroid-stimulating hormone receptor structure-function relationships." *Physiol. Rev.*, 82: 473–502, 2002.

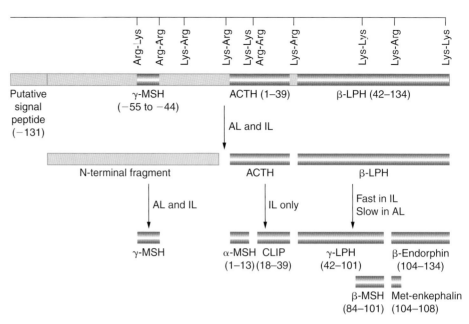

Figure 8-4. Propreproopiomelanocortin is a precursor of the anterior pituitary hormones γ MSH, ACTH and β-lipotropin as well as hormones specific to *pars intermedia-like cells* (PILCs), α-MSH and CLIP. Schematic representation of the preproopiomelanocortin molecule formed in pituitary cells, neurons, and other tissues. The numbers in parentheses identify the amino acid sequences in each of the polypeptide fragments. For convenience, the amino acid sequences are numbered from the N-terminus of ACTH and read toward the C-terminal portion of the parent molecule. The locations of Lys-Arg and other pairs of basic amino acid residues are also indicated; these are the sites of proteolytic cleavage in the formation of the smaller fragments of the parent molecule. *AL*, anterior lobe; *IL*, intermediate lobe.

those cells. In the absence of ACTH, cells of the adrenal cortex will die off. From Figure 8-3, it is seen that the pituitary hormones enter the general blood circulation and bind to receptors in the cell membrane of the target cell. In general, *all polypeptide hormones and neurotransmitters, including those derived from amino acids, are ligands for receptors in the cell membranes of the target cell, whereas steroidal hormones pass through the cell membrane of the target cell and bind to receptors either in the cell cytoplasm or in the cell nucleus.* An apparent exception is the thyroid hormone receptor whose ligand, T_3, is derived from the amino acid tyrosine. The thyroid hormone receptor is considered to be a member of the steroid receptor gene superfamily and is located in the nucleus. The target end organs for the anterior pituitary hormones are given in Table 8-2 and are more clearly summarized in Figure 8-5.

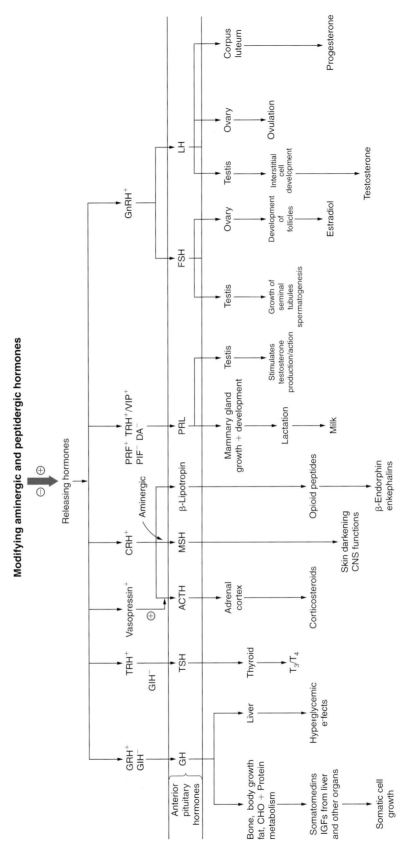

Figure 8-5. Overview of the anterior pituitary hormones showing the connections between the aminergic hormones and neurotransmitters of the CNS, the releasing hormones from the hypothalamus, and the anterior pituitary hormones together with the organs upon which they act and their general effects. *ACTH*, adrenocorticotropic hormone; *CHO*, carbohydrate; *CRH*, corticotropic-releasing hormone; *FSH*, follicle-stimulating hormone; *GH*, growth hormone; *GIH*, growth hormone release–inhibiting hormone, or somatostatin; *GnRH*, gonadotropic releasing hormone; *GRH*, Growth hormone–releasing hormone, or somatocrinin; *IGFs*, insulin-like growth factors; *LH*, luteinizing hormone; *MSH*, melanocyte-stimulating hormone; *PIF*, prolactin release-inhibiting factor; *PRF*, prolactin releasing factor; *PRL*, prolactin; T_3, triiodothyronine; T_4, thyroxine; *TRH*, thyroid-stimulating hormone releasing hormone; *TSH*, thyrotropic-stimulating hormone. Superscript plus or minus signs or encircled plus or minus signs refer to positive or negative actions.

Humoral Mechanism

Posterior Pituitary

The posterior pituitary stores and releases two important hormones, **oxytocin** and **vasopressin** (sometimes abbreviated VP, but usually referred to as ADH). These hormones are synthesized in hypothalamic neuronal cell bodies, together with a **neurophysin** that accompanies the hormone from the cell body through the axon to the nerve ending awaiting a signal to release the complex into the bloodstream. The neuronal cell bodies synthesizing oxytocin are located primarily in the paraventricular nucleus, and the neuronal cell bodies synthesizing vasopressin are located primarily in the supraoptic nucleus (Figure 8-1); both hormones are made in other locations as well, but in lesser amounts. They are released into the general blood circulation after the appropriate signal. Oxytocin and vasopressin are under separate controls and are released separately. Both hormones are nine amino acid–containing peptides (nonapeptides), and their structures are shown in Figure 8-6. At the nerve ending in the posterior pituitary, either oxytocin or vasopressin is stored as a complex with a neurophysin, a 79 amino acid–containing protein. Estrogen can release oxytocin, but not vasopressin, from the posterior pituitary; nicotine can release vasopressin, but not oxytocin, from the posterior pituitary. The biological controls for releasing vasopressin, for example, are shown in Figure 8-7. Signals for the release of vasopressin come from **interneurons** that sense blood pressure (baroreceptor), respond to pain or fright, and so on, by releasing norepinephrine (adrenergic neuron) or sense the sodium ion concentration in the blood (osmoreceptor).

Release of vasopressin from the nerve endings of the vasopressinergic neuron causes uptake (reabsorption) of water in the distal kidney (Figure 8-7), expanding the fluid content of the blood and thus raising blood pressure. Oxytocinergic neurons are excited to release oxytocin by estrogen or by the suckling response, as shown in Figure 8-8. The suckling stimulus is mediated by neural pathways in the spinal cord, and within milliseconds the signal appears in the paraventricular nucleus to culminate in the release of oxytocin from the posterior pituitary into the general circulation. It binds to its receptor in mammary gland muscle cells to cause their contraction, which forces the milk in the gland out through the nipple. In this process, the secretion of PRL also is stimulated, which results in the secretion of milk from the milk-secreting cells of the mammary gland.

The receptors for vasopressin and oxytocin are similar. Both contain seven membrane-spanning subunits, as shown in Figures 8-9 and 8-10. How vasopressin, as the receptor ligand, makes contact with amino acid residues in four of the seven transmembrane-spanning domains is shown as a surface view in Figure 8-11. The amino acid sequence of vasopressin receptor and oxytocin receptor are similar, and Figure 8-12 shows the amino acid residues conserved within the oxytocin or vasopressin receptor subfamily. The organization of the gene expressing oxytocin and vasopressin is shown in Figure 8-13.

(Text continues on p. 387.)

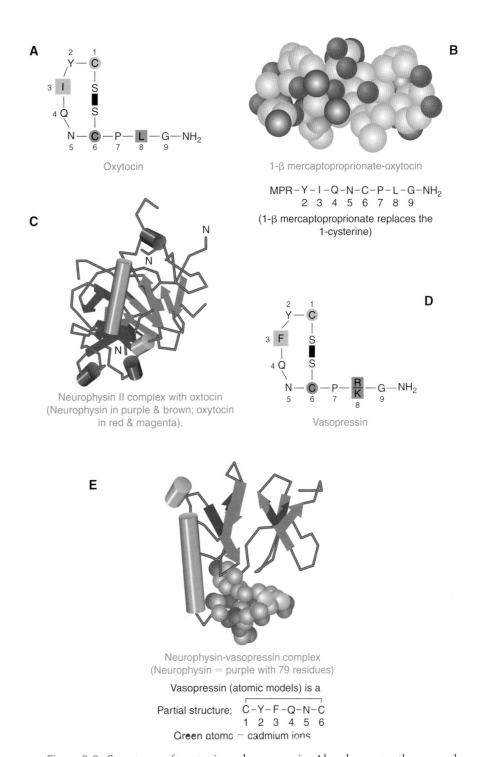

Figure 8-6. Structures of oxytocin and vasopressin. Also shown are the neurophysins, proteins to which the hormones bind in their storage forms. At some point after release into the circulation the complex dissociates allowing the free hormones to circulate (with a shorter half-life than in the complex with neurophysin). Oxytocin and neurophysin differ in the amino acid residues 3 and 8 (boxed in structures). The one letter abbreviations for amino acids are used: c, cysteine; y, tyrosine; f, phenylalanine; i, isoleucine; q, glutamine; n, asparagine; p, proline; l, leucine; r, arginine; k, lysine; g, glycine. Crystal structures are redrawn from the Protein Data Bank.

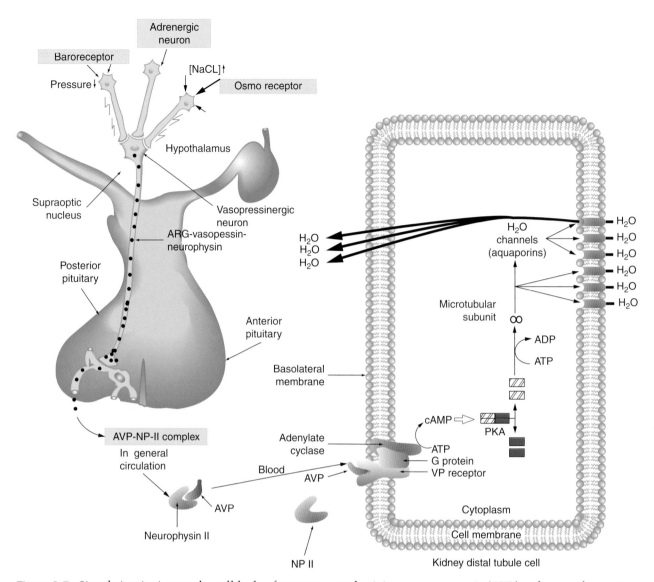

Figure 8-7. Signals impinging on the cell body of a neuron synthesizing arg-vasopressin (AVP) and neurophysin II (NPII, vasopressinergic neuron). The complex of AVP-NPII is transported down the long axon to the nerve ending in the posterior pituitary. AVP-NPII is released into the general circulation and later the complex dissociates. Free AVP binds to its receptor in the cell membrane of a distal kidney tubule cell. The activated receptor is coupled to a G protein and to adenylate cyclase, which converts ATP to cyclic AMP (cAMP). cAMP activates protein kinase A, which phosphorylates microtubular subunits to form aquaporins (water channels) in the apical membrane. Water is taken up and transported across the cell into the basolateral space outside the cell. This water dilutes the blood and increases its partial pressure; therefore blood pressure rises.

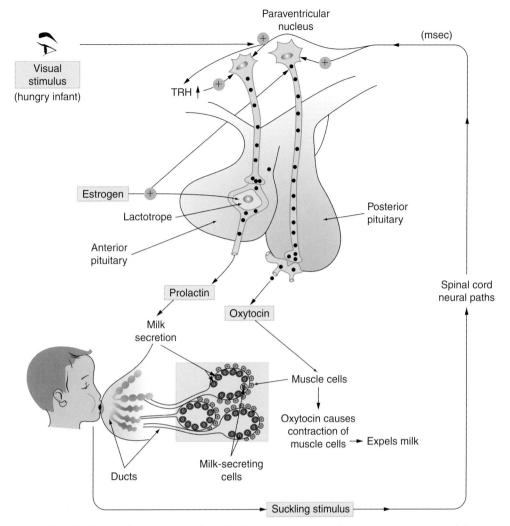

Figure 8-8. Release of oxytocin and prolactin in nursing mother in response to suckling or visual stimuli. Neuronal cells in the hypothalamus are stimulated affecting release of OT from posterior pituitary and PRL from anterior pituitary. PRL binds to receptor in milk-secreting cells in mammary gland to cause secretion of milk. OT binds to receptor in muscle cells surrounding milk-producing cells, causing contraction and expulsion of milk into the mammary ductal system and out through the nipple.

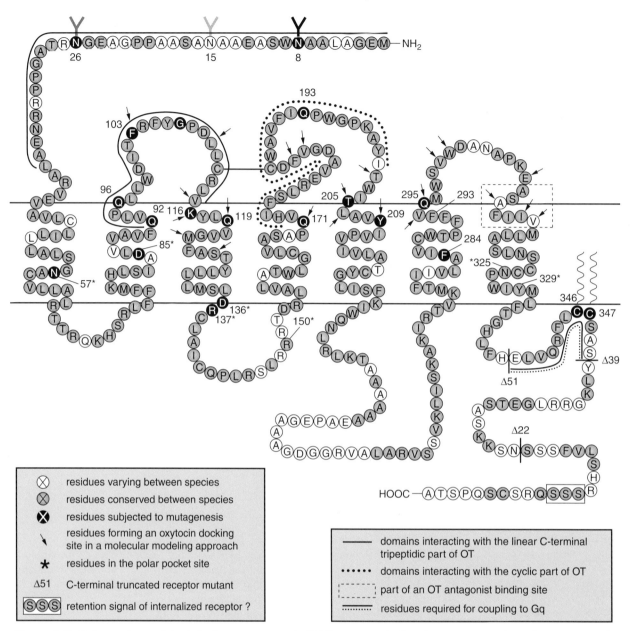

Figure 8-9. Schematic model of the human oxytocin (OT) receptor indicating amino acid residues putatively involved in ligand-binding and associated signal transduction events. The amino acid residues in *orange solid circles* are conserved (identical) between the OT receptors from different mammalian species (human, rhesus monkey, pig, bovine, sheep, rat, and mouse). Residues in open circles show interspecies variation. At these positions, an amino acid substitution may be tolerated through mammalian evolution without influencing the functional properties of the receptor. Residues in *black solid circles* have been subjected to mutagenesis. The glutamine and lysine residues highly conserved within the vasopressin/OT receptor family may partly define an agonist-binding pocket common to all different subtypes of this receptor family. According to a molecular modeling approach, an OT docking site has been proposed (corresponding residues are marked by arrows). In the inactive receptor conformation, the highly conserved arginine (R137) may be constrained in a pocket formed by polar residues (indicated by asterisks). After agonist binding, this arginine side chain may be shifted out of the "polar pocket," thereby unmasking a G protein–binding site. Receptor domains putatively interacting with OT, a peptide OT antagonist, and $G_q\alpha$ are marked by lines. Redrawn from G. Gimpl and F. Fahrenholz, "The oxytocin receptor system: structure, function, and regulation," *Physiol. Revs.*, 81: 629–683, 2001.

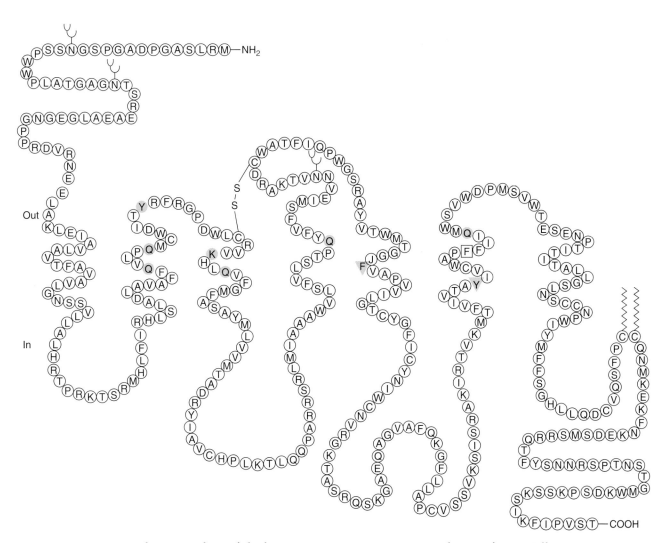

Figure 8-10. Transmembrane topology of the human vasopressin V1a receptor showing functionally important residues. Amino acids highlighted in orange circles are critically involved in agonist binding that in the square is possibly involved in antagonist binding; those in orange triangles modulate the process of receptor activation in oxytocin receptor. Potential glycosylation (on Asn 14, 27 and 196) and palmitoylation (on Cys 365 and 366) sites are also indicated.

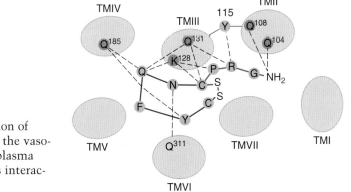

Figure 8-11. Schematic representation of the interaction of vasopressin with the membrane-spanning domains of the vasopressin V1a receptor, viewed from the surface of the plasma membrane. Amino acids shown to be involved in this interaction are indicated. *TM*, transmembrane.

Posterior Pituitary

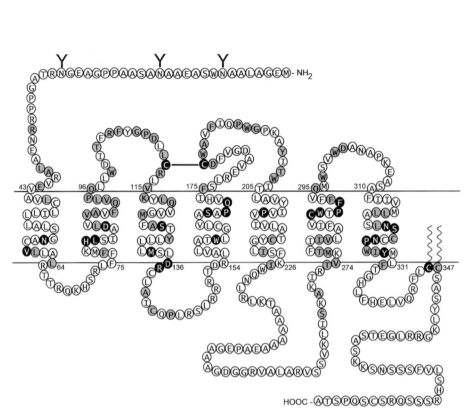

Figure 8-12. Schematic structure of the human oxytocin (OT) receptor with amino acid residues shown in one-letter code. Residues conservative within the OT/vasopressin receptor subfamily are outlined in *gray*, and residues conservative for the whole G protein–coupled receptor superfamily are outlined in *black*. The putative *N*-glycosylation ("Y") and palmitoylation (at C346/C347) sites are marked. Reproduced from http://physrev.physiology.org/cgi/content-nw/full/81/2/629/F4.

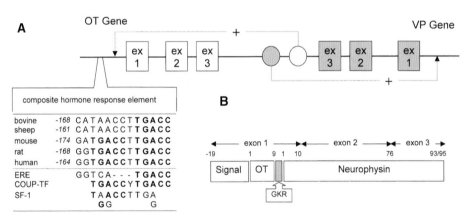

Figure 8-13. Organization of the oxytocin (OT) and vasopressin (VP) gene structure, including schematic depiction of the putative cell-specific enhancers (*open circle*, enhancer of OT gene; *shaded circle*, enhancer of VP gene). **A.** Details of the approximately 160 base-pair region (composite hormone response element) of the upstream OT gene promoter conserved across five species, including the sequences of the response elements estrogen response element (ERE), chicken ovalbumin upstream promoter transcription factor I (COUP-TF), and steroidogenic factor-1 (SF-1) are indicated. **B.** Domain organization of preprooxytocin, including the processing sites. The precursor is split into the indicated fragments by enzymatic cleavages, one involving a glycyl-lysyl-arginine (GKR) sequence and leaving a carboxamide group at the COOH-terminal end of OT. Reproduced from http://physrev.physiology.org/cgi/content-nw/full/81/2/629/F1.

Actions of Releasing Hormones and Anterior Pituitary Hormones

CRH–ACTH–Cortisol Pathway

The human releasing factor, CRH, has the sequence shown in Figure 8-14. The sequences of human urocortin and frog sauvagine, which are capable of binding to the CRH receptor in the membrane of the corticotroph, are given as well. It is apparent that there are some conserved amino acid residues, such as 4-proline, 7-serine, 9-aspartate, 10-leucine, 11-threonine (not in sauvagine), 15-leucine, 16-arginine, 20-glutamate, 31-alanine, 34-asparagine, and 35-arginine. In addition to the major human CRH receptor, hCRH(F)$_1$, there are two subtypes: hCRH(F)$_{2\alpha}$ and hCRH(F)$_{2\beta}$. Properties of these subtypes and their locations are shown in Table 8-3. The structures of the CRH receptors and their subtypes are shown in Figure 8-15. Location of the binding of the ligand, CRH, to its receptor is shown in Figure 8-16. Here, it is seen that CRH interacts with the amino terminal tail and transmembrane domains 1, 2, and 3, counting from the left to the right (Figure 8-16B). The promoter of the gene for CRH is complex because it is regulated by many factors, including glucocorticoids (cortisol), the end product of the hypothalamic–pituitary–adrenal cortex axis, CREB (via cyclic AMP), and Fos, as shown in Figure 8-17. CRH binds to its receptor in the anterior pituitary corticotropic cell, setting off signal transduction events that lead to the release of ACTH, shown in Figure 8-18. As seen, of the ACTH containing 24 amino acids, amino acids 5 through 24 can give full agonist activity, as well as the full 24–amino acid peptide.

Binding of CRH to its receptor causes stimulation of adenylate cyclase activity through an associated G protein. The generation of cyclic AMP causes the activation of protein kinase and subsequent phosphorylation events required to release ACTH. ACTH is released into the general circulation and finds its receptor in the cell membranes of cells of the adrenal cortex; the highest concentration of these receptors is in the middle layer of cells *(zona fasciculata)*, which synthesizes and releases cortisol. The other two layers are the inner layer *(zona reticulosa)*, which releases the weak androgen, **dehydroepiandrosterone (DHEA),** and the outer layer *(zona granulosa)*, which releases the steroid, aldosterone. The ACTH receptor, when activated by ACTH, also opens a calcium channel so that

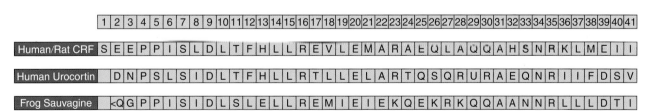

Notes: Uppercase denotes natural "L" form of the amino acids.
< Q = Pyroglutamic acid residue.

Figure 8-14. Sequences of human CRH (CRF used here), human urocortin, and frog sauvagine, all of which are ligands for the CRH receptor in the anterior pituitary corticotrope. Redrawn from http://www.bentham.org/cmccnsa1/sample/cmccnsa1-1/saunders/saundersms.htm.

Table 8-3
Properties and Locations of Human CRH (CRF) Receptor Subtypes

Receptor Subtype	Amino Acids	% Homology to hCRF$_1$	% Homology to rCRF$_1$	Binding Affinity (K$_i$, nM)			Efficacy (EC$_{50}$, nM)			Receptor Expression (highest density only)
				r/hCRF	urocortin	sauvagine	r/hCRF	urocortin	sauvagine	
hCRF$_1$	415	100	97.1	10	0.6	1.6	1.1	0.8	1.7	Widespread in brain esp. cortex and cerebellum (low in hypothalamus), high levels in anterior pituitary.
hCRF$_{2\alpha}$	411	68	67.5	416	2.0	5.2	10	0.2	0.2	Brain and periphery (esp. lateral septum, amygdala), low in pituitary
hCRF$_{2\beta}$	431	66.5	64.9	50.1a	1.6a	3.2a	3.0	0.1	0.3b	Cerebral arterioles, choroid plexus, heart, and skeletal muscle
hCRF$_{2\gamma}$	397	67.3	68	25b	1.4b	—	32b	3.0b	4.0b	Septum, hippocampus, and amygdala

Reproduced from http://www.bentham.org/cmcnsal/sample/cmccnsa1-1/saunders/saundersms.htm.

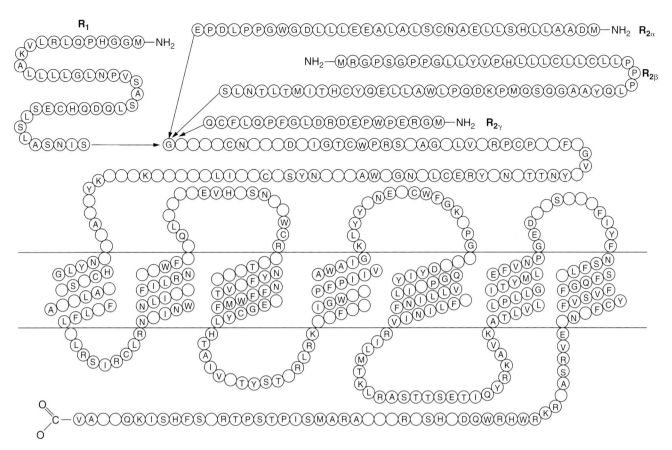

Figure 8-15. Structures of human CRH receptor and subtypes. Redrawn from http://www.bentham.org/cmccnsa/sample/cmccnsa1-1/saunders/saundersms.htm.

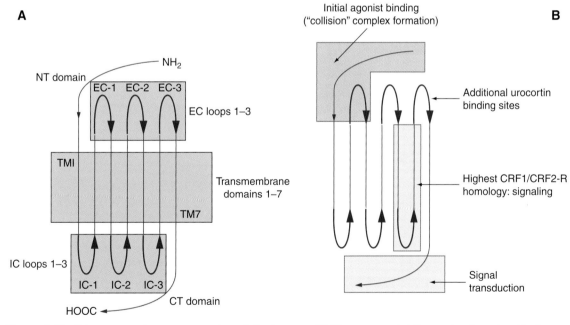

Figure 8-16. Diagrammatic representation of the human CRH receptor. **A.** The major domains. **B.** Area of interaction with CRH and areas involved in signal transduction. Reproduced from D.E. Grigoriadis et al., "The CRF receptor: structure, function and potential for therapeutic intervention," *Curr. Med. Chem.*, 1: 63–97, 2001 and directly from http://www.bentham.org/cmccnsa/sample/cmccnsa1-1/saunders/saundersms.htm.

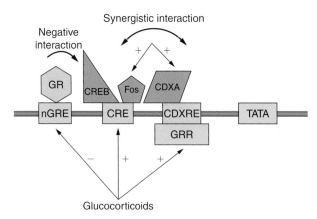

Figure 8-17. Schematic model of CRH promoter regulation in the hypothalamus. *nGRE*, negative glucocorticoid regulatory element; *CRE*, cAMP regulatory element; *GRR*, 213 to 99 bps region stimulated by glucocorticords; *CDXRE*, caudal type homeobox respsonse element; *TATA*, TATA box. +, stimulatory; −, inhibitory; *thin arrows*, regulatory effects by cAMP and glucocorticoids through the different elements. *Thick arrows*, negative and synergistic stimulatory ± double-headed arrow interactions between sites. Redrawn from R.C. Nicholson, B.R. King, and R. Smith, "Complex regulatory interactions control CRH gene expression," *Frontiers in Bioscience*, 9: 32–39, 2004.

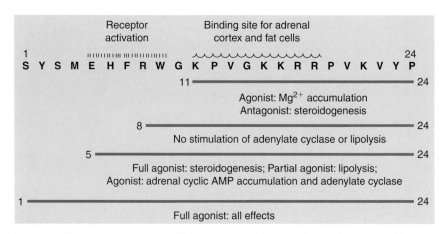

Figure 8-18. Sequence of ACTH and its activities. Redrawn from http://www.neurosci.pharm.utoledo.edu/MBC3320/ACTH.htm.

calcium ions from the outside can enter the cell. Protein kinase A is activated by the stimulation of adenylate cyclase, mediated by a G protein associated with the activated receptor. Protein kinase A phosphorylates cholesteryl esterase, which becomes activated (it is not active in the unphosphorylated form). Cholesteryl esterase converts cholesterol esters, stored in the lipid droplet, to free cholesterol. Cholesterol enters the mitochondrion and undergoes a series of conversions within the mitochondrion and outside in the microsomes (for hydroxylations), culminating in the synthesis of cortisol (see Chapter 9), which is released into the general blood circulation. The overall system from ACTH binding to its adrenal cortex receptor to the release of cortisol is shown in Figure 8-19. The entry of free cholesterol into the mitochondrion overcomes the substrate-limited steroid synthetic system, the side chain of cholesterol is cleaved by the enzymatic activity of the **steroid acute response (StAR)**

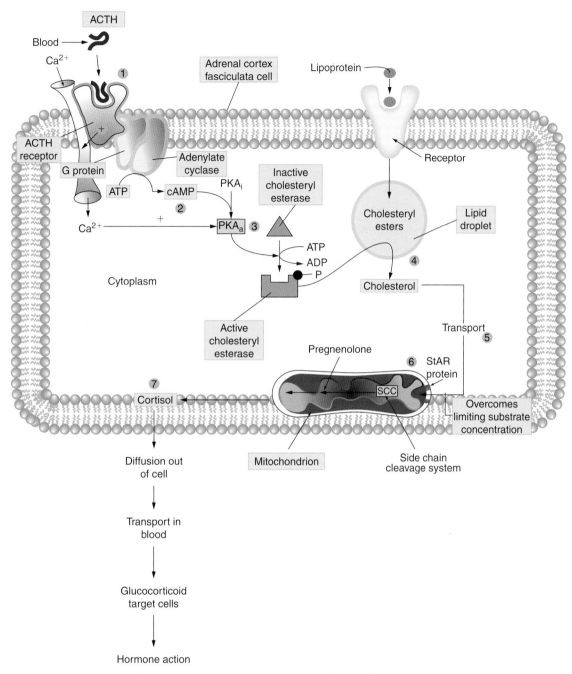

Figure 8-19. Overview of ACTH action on the *zona fasciculata* cell producing cortisol. **1.** binding of ACTH to ACTH receptor; **2.** activated receptor activates adenylate cyclase through G protein; **3.** cAMP from adenylate cyclase reaction activates protein kinase A and PKA phosphorylates inactive cholesteryl esterase to the active form; **4.** cholesteryl esterase hydrolyzes cholesteryl esters in the lipid droplet (derived from lipoproteins in the blood) to form free cholesterol; **5.** cholesterol is transported into the mitochondrion and activates substrate-limited cortisol synthesis in a first side chain cleavage step catalyzed by the STAR protein (steroid acute response protein); **6.** cortisol synthesis proceeds; and finally in **7.** cortisol is released into the blood circulation.

protein, and enzymatic steps follow that culminate in the synthesis of cortisol. Cortisol diffuses outwardly through the cell membrane and gains access to the general circulation. Unbound cortisol in the blood reaches a high enough level to exceed the binding constant for the cortisol receptor, and as cortisol diffuses into cells it binds to the cytoplasmic receptor proportionately to the number of receptor molecules in a given cell. When the number of receptors is small and the binding sites are filled, free excess cortisol will diffuse back out of the cell into the general circulation. The major targets for cortisol, which have high levels of the receptor (for example, 50,000–70,000 molecules per cell), will act as sink for the hormone, but all of the sites need not be filled to obtain a hormonal effect on the cell. The liganded receptor then translocates in the active form to the nucleus and carries out its action as a transcriptional activator or repressor, as the case may be (Chapter 7).

The overall pathway of secretions, starting with CRH, is shown in Figure 8-20.

Growth Hormone–Releasing Hormone— Growth Hormone—Bodily Growth Path

The system for growth hormone begins with GHRH. This hormone is sometimes called somatocrinin, sermorelin, or **somatoliberin.** Various signals impinge upon the neuron that produces GHRH, including aminergic and peptidergic neurons. In essence, the secretion of the anterior pituitary growth hormone is controlled positively by GHRH and negatively by somatostatin. Growth hormone in blood feeds back negatively on the further secretion of GHRH, and the end organ product, **insulin-like growth factor (IGF-I),** feeds back positively on the release of somatostatin. Thus, when growth hormone is being synthesized and released from the **somatotrope** and the end product of the axis (hormone from the end organ) is produced, the net effect is to reduce the higher signals, which lead to more growth hormone release. In addition to these controls on growth hormone, a newly discovered hormone from the stomach, **ghrelin,** stimulates the secretion of growth hormone from the anterior pituitary. The action of ghrelin may cause a greater release of growth hormone than GHRH. In addition to releasing growth hormone, GHRH stimulates the synthesis of growth hormone, whereas ghrelin releases preformed growth hormone but does not stimulate growth hormone synthesis. These relationships are shown in Figure 8-21. The controls on GHRH secretion derive from interneurons (neurons that interface with the cell body producing, in this case, GHRH) that produce a variety of neurotransmitters and peptides, as shown at the top of Figure 8-21.

GHRH positively affects the release of growth hormone, and the opposing peptide, somatostatin, inhibits the release of growth hormone. Release of growth hormone into the circulation feeds back positively on the somatostatin-producing neuron (somatostatinergic neuron; the ending "ergic" means *work*, and the cell that produces the substance in question can be stimulated) to reduce the further release of growth hormone. The release of growth hormone, like many of the anterior pituitary hormones, is pulsatile (several events, or episodes, of secretion during the day and therefore several spikes in the blood levels of growth hormone; Figure 8-22). The pulsatile nature of the release of growth hormone is caused by the several inputs at the level of GHRH. GHRH is transported through the

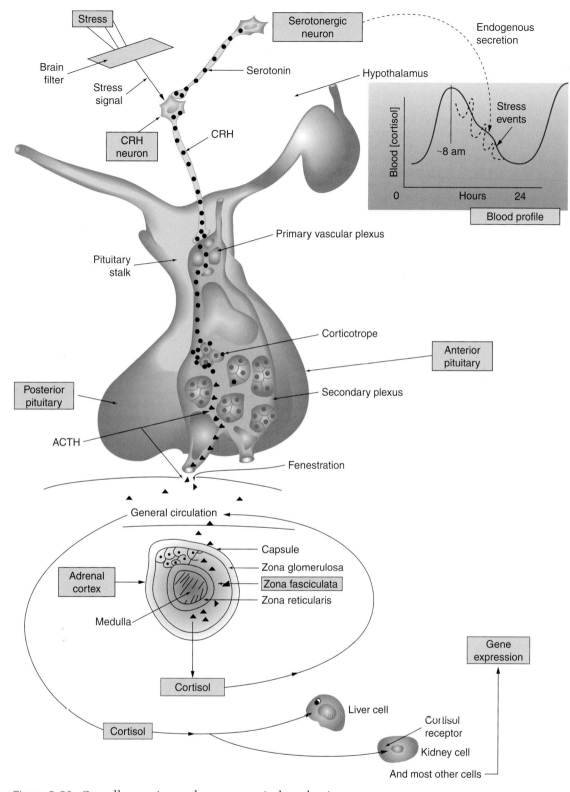

Figure 8-20. Overall secretion pathway to cortisol production.

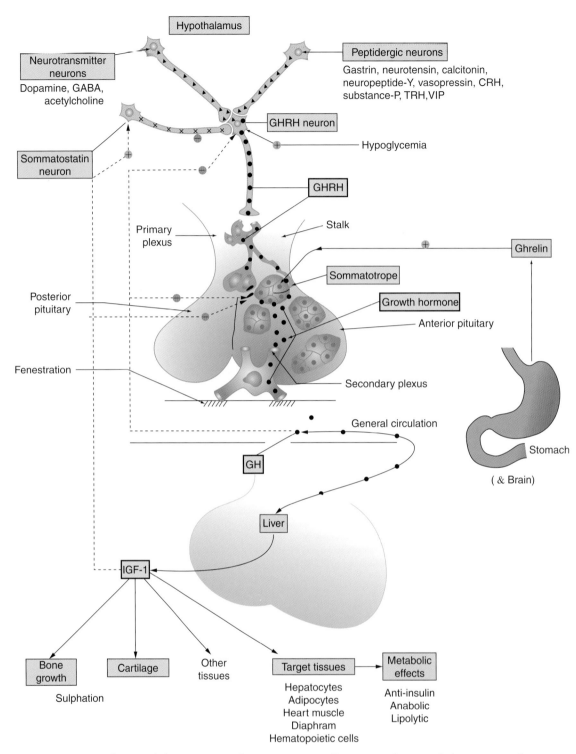

Figure 8-21. Regulation of the secretion of GHRH, GH, and IGF-1 in the growth hormone pathway. *CRH*, corticotrophin-releasing hormone; *GABA*, gamma aminobrutyric acid; *GH*, growth hormone; *GHRH*, growth hormone releasing hormone; IGF-1, insulin-like growth factor-1; *TRH*, thyrotropin releasing hormone; *VIP*, vasoactive intestinal peptide.

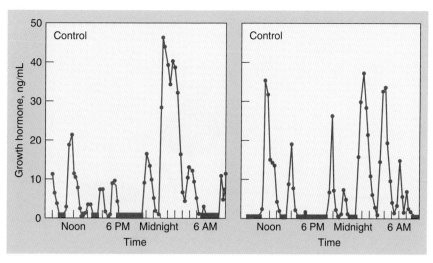

Figure 8-22. The characteristic pulsatile pattern of GH secretion in normal children. Note the maximal GH secretion during the night. Redrawn from http://www.endotext.org/pediatrics/pediatrics1/pediatrics1a_2.htm.

axon of the GHRHergic neuron to the primary vascular plexus, where GHRH is released from the nerve ending. GHRH courses through the plexus to the secondary vascular plexus (of the closed vascular system) and gains access to the cells of the somatotrope, where it binds to a receptor in the cell membrane. This causes the production of cyclic AMP, activation of protein kinase A, and calcium ion influx within the cell, which triggers the release of growth hormone through biochemical events (signaling). Growth hormone enters the general circulation through a thin section of the major blood vessels (fenestration), which allows entry into the general bloodstream. From the general circulation, growth hormone finds its receptors in the cell membranes of the various targets listed in Figure 8-21. Primarily in the liver, growth hormone causes the release of IGF-I, which has stimulatory effects on bone and other tissues, including sulfation, that result in growth.

Ghrelin acts positively on the somatotropic cells (along with GHRH) to release growth hormone. The biochemical events following ghrelin binding to its membrane receptor are shown in Figure 8-23. Ghrelin levels fluctuate in the blood; it is lowest following a meal and rises during the fasting period just before the next meal (Figure 8-24).

The gene for GHRH consists of five exons, shown in Figure 8-25, encoding a mature protein (exons 2–5) of 44 amino acids (Table 8-1), although the amino terminus segment contains most of the activity (amino acid residues 1–29) and synthetic peptides are based on the shorter version. The GHRH receptor is a seven-transmembrane protein (Figure 8-26), like many other membrane receptors, and it is encoded by 7 exons from a 10-exon gene (Figure 8-27).

Ghrelin, in addition to GHRH (which is the major positive regulator of growth hormone synthesis and release), is a secretagogue that stimulates primarily the secretion of preformed growth hormone. The ghrelin gene and the final protein product are shown in Figure 8-28. Note that Ser-3 of ghrelin protein is acetylated. The ghrelin receptor, also in the cell membrane of the somatotropic cell, is a seven-membrane protein

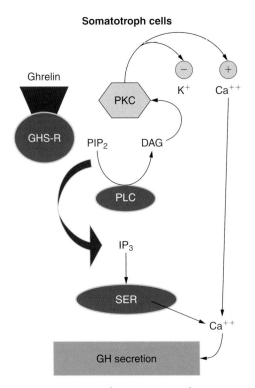

Figure 8-23. Signaling events in the somatotropic cell following binding of ghrelin to its cell membrane receptor (GHS-R). The phosphatidylinositol pathway is stimulated so that phosphatidylinositol bisphosphate (PIP2) is converted to diacylglycerol (DAG) and inositol triphosphate (IP3). IP3 binds to a receptor in the smooth endoplasmic reticulum (SER), which releases Ca^{2+} from a calcium ion store. DAG activates protein kinase C (PKC) to increase Ca^{2+} influx and the elevation of intracellular Ca^{2+} causes the release of GH from storage sites in the cell. PLC, phospholypase C. Redrawn from F. Lago et al., *Vitamins & Hormones*, 71: 406–432, 2005.

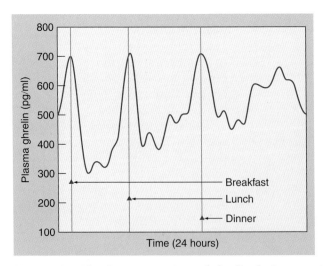

Figure 8-24. Blood concentrations of ghrelin during a typical day. Data are from 10 persons. Adapted with permission from D.E. Cummings et al., "A prepandial rise in plasma ghrelin levels suggests a role in meal initiation in humans," *Diabetes*, 50: 1714, 2001.

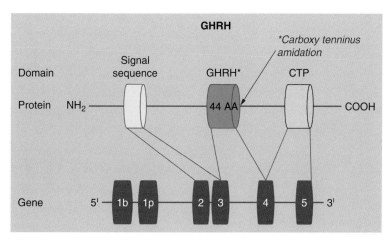

Figure 8-25. Diagram showing structure of the GHRH gene and proprotein. Note that exon 1 is noncoding. CTP refers to C-terminal protein. Redrawn from http://www.endotext.org/pediatrics/pediatrics1/pediatrics1a_2.htm.

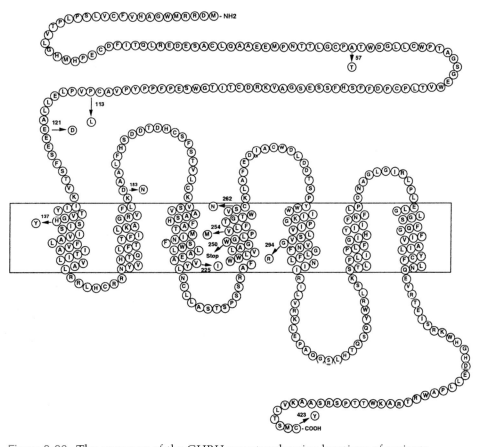

Figure 8-26. The sequence of the GHRH receptor showing locations of variants *(arrows)*. Taken from E.J. Lee et al., "Absence of constitutively activating mutations in the GHRH receptor in GH-producing pituitary tumors," *J. Clin. Endocrinol. & Metabol.* 86: 3989–3995, 2001 and directly from http://jcem.endojournals.org/cgi/content/vol86/issue8/images/large/eg0187732002.jpg.

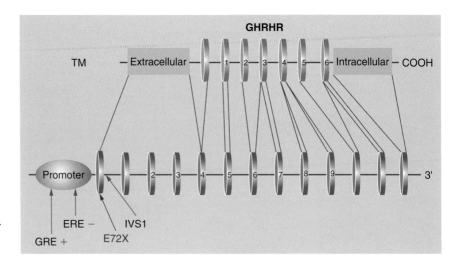

Figure 8-27. Gene encoding the GHRH receptor. Exons 2–4 are noncoding. *ERE*, estrogen response element; *GRE*, glucocorticoid response element. Redrawn from http://www.endotext.org/pediatrics/pediatrics1/pediatrics1a_2.htm.

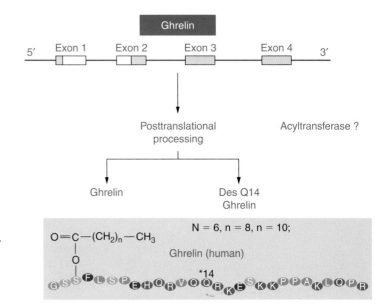

Figure 8-28. Gene processing and human ghrelin structure. Ghrelin is a 28 amino acid peptide in which the serine in position 3 is modified by an addition of an acyl group, primarily n-octanoic acid. This acyl modification is essential for ghrelin's biological activity. The Ghrelin gene and final protein product taken from F. Lago et al., *Vitamins & Hormones*, 71: 406–432, 2005.

(Figure 8-29), and it is encoded (by way of mRNA) by two exons in its gene (Figure 8-30).

Somatostatin is an important inhibitor of growth hormone secretion (Figure 8-21). The structure of somatostatin and the synthetic drug, **octreotide,** are shown in Figure 8-31 with the structure of one of the somatostatin receptors (human somatostatin 5 receptor). The synthetic octreotide is based on the somatostatin structure. Octreotide produces the same effects as somatostatin, but they last longer. The gene for somatostatin contains two exons (Figure 8-32) and through its mRNA encodes a preprosomatostatin protein of 116 amino acid residues. This is matured to the 14 amino acid molecule (Figure 8-31B), as well as a 28–amino acid precursor isoform. Although the hypothalamic somatostatin inhibits the release of GHRH (and is known as GHRIH, the growth hormone release–inhibiting hormone), the pancreatic somatostatin inhibits the release of both insulin and glucagon (insulin lowers blood glucose levels, and glucagon raises them). There are five different receptors for somatostatin

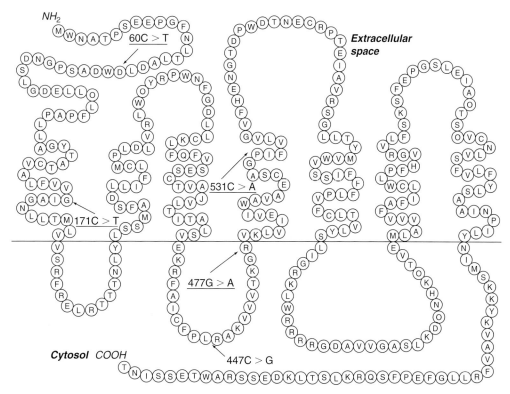

Figure 8-29. The ghrelin receptor. Genomic screening for sequence variations in the type 1a growth hormone secretagogue receptor (GHS-R1a) gene revealed five single-nucleotide polymorphisms (SNPs). The SNPs were found in the codons of amino acids, which are indicated by arrows in this presentation of the predicted structure of the receptor. Reproduced with permission from J. Vartiainen et al., *Eur. J. Endocrinol.*, 150: 457–463, 2004.

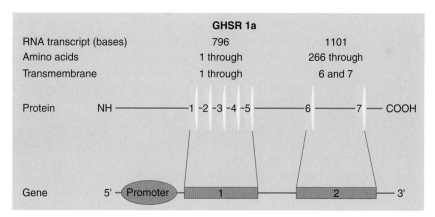

Figure 8-30. Gene encoding information for translation of ghrelin receptor (GHSR1a). *GHSR*, growth hormone secretogogue receptor. Redrawn from http://www.endotext.org/pediatrics/pediatrics1/pediatrics1a_2.htm.

encoded by different genes. The effects of somatostatin on the pituitary are mediated by SSTR2 and SSTR5, receptors of the G protein–coupled class (Figure 8-31A), where it inhibits the formation of cyclic AMP, the mediator of GHRH effects. The gene and somatostatin product are diagrammed in Figure 8-32. The genes for somatostatin receptors 2 and 5 are shown in Figure 8-33.

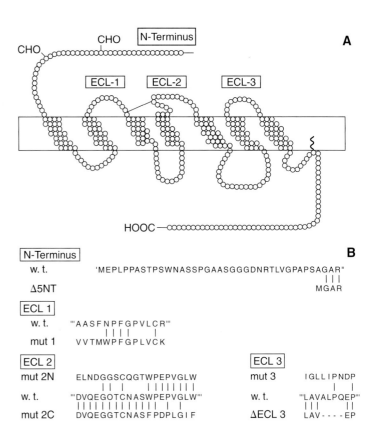
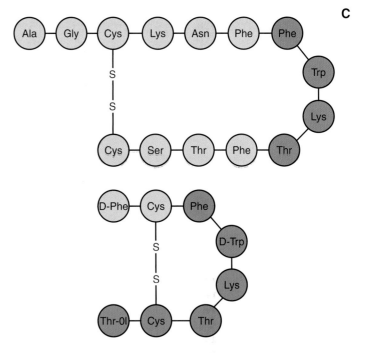

Figure 8-31. A general structure for the human somatostatin 5 receptor and comparison of the sequences of the hsst$_5$R mutants with their wild-type counterparts. **A.** Locations and borders of extracellular regions. NH_2, amino terminus; *HOOC*, carboxyl terminus; *CHO*, putative glycosylation site; also depicted is the potential palmitoylation site in the intracellular carboxyl tail, which serves as a membrane anchor. **B.** Below the structure is alignment of the ECL, NH$_2$-terminal deletion, and ECL3 deletion mutants with their wild-type (w.t.) counterparts reveals residues that were conserved as well as the conservative exchanges. Bold wild-type sequences are numbered according to their location in the receptor sequence. Thirty-five residues of the amino acid terminus of hsst$_5$R were removed to create the mutant Δ5NT del. An A36M mutation was introduced to create a new translational initiation codon of the mutant; therefore, only 3 of 39 amino-terminal hsst$_5$R residues remain in the mutant. **C.** Structure of somatostatin. Structure of octreotide *(below)*. Parts A and B reproduced with permission from M.T. Greenwood et al., *Mold. Pharmacol.*, 52: 807–814, 1997.

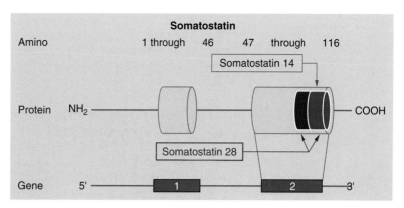

Figure 8-32. Gene and protein product (preprosomatostatin) for somatostatin. Redrawn with permission from http://www.endotext.org/pediatrics/pediatrics1/pediatrics1a_2.htm.

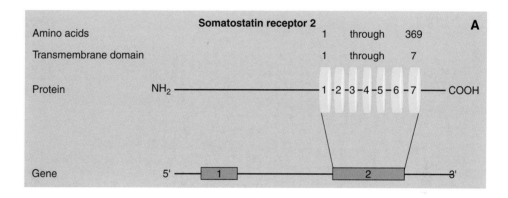

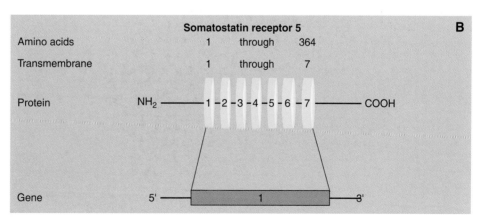

Figure 8-33. **A.** Diagram representing the gene and protein product for somatostatin receptor 2 (SSTR2). **B.** Diagram representing the gene and protein product for somatostatin receptor 5 (SSTR5). Redrawn with permission from http://www.endotext.org/pediatrics/pediatrics1/pediatrics1a_2.htm.

Growth hormone, the product of the somatotropic cell of the anterior pituitary, is a protein consisting of 191 amino acid residues with two intramolecular disulfide bonds (Figure 8-34) and a molecular weight of 22,128 Daltons. Growth hormone is required for childhood growth and for the optimal functioning of most organ systems throughout most of the normal lifetime. Growth hormone, like the other hormones discussed so far, acts through growth hormone receptors in cell membranes. Its product, in most cases, is the production and release of IGF-I (sometimes referred to as **somatomedin C**), which is a growth factor for many tissues, especially bone (Figure 8-21). IGF-I is released from the target cell (for example, liver) and acts as a hormone by affecting distant cells, as well as acting locally in the same cell that produces it **(autocrine)** or in nearby cells **(paracrine).** IGF-I stimulates bone growth of the fetus and epiphyseal growth (ends of bone growth) of children, as well as increasing lean muscle and bone mineral density and decreasing fat.

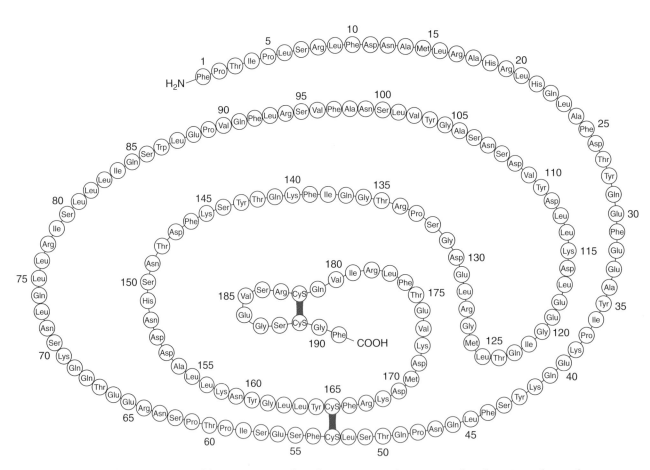

Figure 8-34. Covalent structure of hGH. Reprinted with permission from R.K. Chawla, J.S. Parks, and D. Rudman, "Structural variants of human growth hormone: biochemical, genetic and clinical aspects," *Ann. Rev. Med.*, 34: 519–547, 1983. Reprinted with permission from W.H. Daughaday, "Growth hormone, insulin-like growth factors, and acromegaly," in *Endocrinology*, 3rd ed., Vol. 1, L.J. De Groot, Editor, W.B. Saunders, Philadelphia, 1995.

The gene for growth hormone is located on chromosome 17q22-q24 with five genes, of which four are for growth hormone or a variant and the fifth is for a pseudogene. The gene encoding growth hormone through mRNA is diagrammed in Figure 8-35. The growth hormone receptor is located on chromosome 5 (5p13-12), and its gene contains 10 exons encompassing about 300 kilobases of genomic DNA (Figure 8-36). The receptor, like many other membrane receptors that bind peptides or

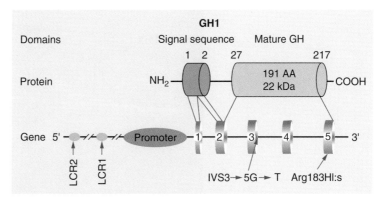

Figure 8-35. Gene for human growth hormone. Redrawn from http://www.endotext.org/pediatrics/pediatrics1/pediatrics1a.htm.

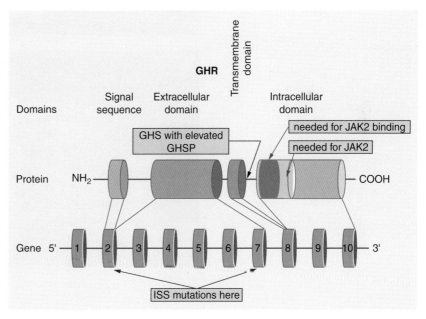

Figure 8-36. Diagram for the gene encoding information for the growth hormone receptor and the mature protein. Mutations can occur in the GH receptor (red arrows) that can lead to growth hormone resistance and other conditions. GHR, growth hormone receptor; GHS, growth hormone secretagogue; GHSP, GHS protein; JAK2, janus kinase 2. Redrawn from previous site http://www.endotext.org/pediatrics/pediatrics1/pediatrics1a_3.htm.

proteins, has an extracellular ligand-binding domain, a transmembrane domain (anchoring the protein in the membrane), and an intracellular domain that has tyrosine kinase activity. The structure of the growth hormone receptor with and without bound growth hormone is shown in Figure 8-37. Growth hormone, like other growth factors, binds to its receptor in the cell membrane of a target cell, and this binding causes the

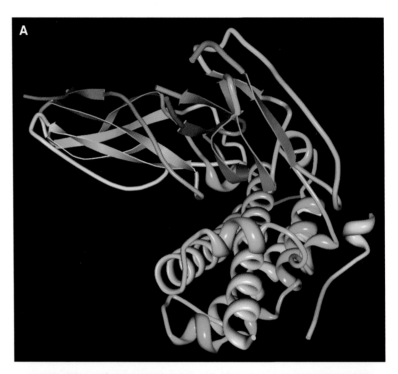

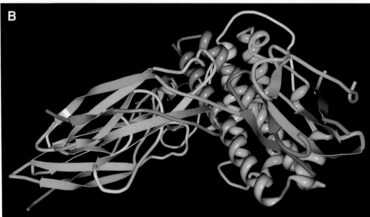

Figure 8-37. **A.** Structure of the dimerized growth hormone (GH) receptor. **B.** Structure of the dimerized GH receptor (human) complexed with GH *(yellow green)*. Binding of monomer GH receptor to GH causes the receptor to dimerize and activates the signal transduction process (initial step is tyrosine autophosphorylation). Part A reproduced from PDB ID: la22. T. Clackson et al. Human Growth Hormone Bound to Single Receptor. *J. Mol. Biol.* **277** pp. 1111 (1998). Part B reproduced from PDB ID: 3hhr. A.M. DeVos et al. Human Growth Hormone and Extracellular Domain of Its Receptor: Crystal Structure of the Complex. *Science* **255** pp. 306 (1992).

receptor to dimerize. The dimerization activates tyrosine kinase activity of the receptor, which autophosphorylates the receptor. The receptor associates with Janus tyrosine kinase (JAK) 2, which also becomes phosphorylated (and activated). A number of signaling cascades become activated, including STATs, **mitogen-activated protein kinase (MAPK),** and **phosphatidylinositol-3-kinase (PI3K),** whose activities are directly involved in regulating growth hormone responsive genes (Figure 8-38). A more detailed picture of possible growth hormone action is shown in Figure 8-39. A number of signaling pathways (for example, STAT, MAPK, and PI3K) are involved in cell proliferation.

Again, somatostatins oppose the actions of growth hormone. Five somatostatins have been found; three of them cause cell cycle arrest, and the other two contribute to apoptosis (programmed cell death), as shown in Figure 8-40.

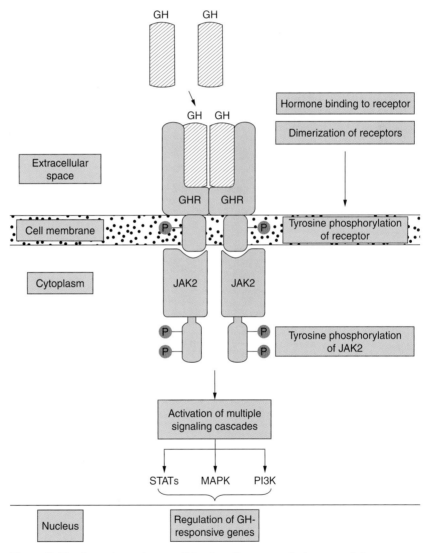

Figure 8-38. Overview of a possible signaling scenario for growth hormone action. *GH*, growth hormone; *GHR*, growth hormone receptor; *JAK2*, Janus kinase 2; *MAPK*, mitogen-activated protein kinase; *P*, phosphate; *PI3K*, phosphatidylinositol 3−; *STAT*, signal transducers and activators of transcription protein.

Actions of Releasing Hormones and Anterior Pituitary Hormones

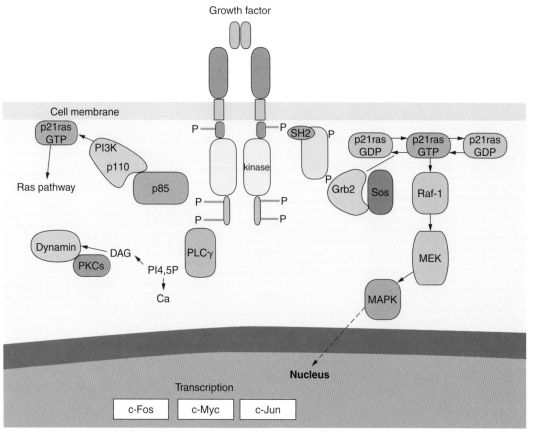

Figure 8-39. Growth factor signal transduction.

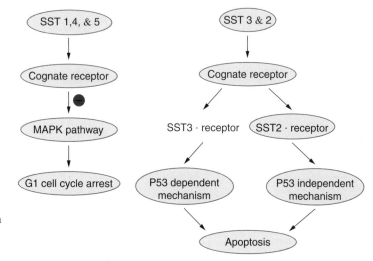

Figure 8-40. Five somatostatins have been found and they negatively control cell proliferation by the general pathways shown in this figure. Information from G. Fer Joux et al., *J. Physiol. Paris*, 94: 205–210, 2000.

Gonadotropins

The gonadotropins are two anterior pituitary hormones: LH and FSH. The sequences and structures of these hormones are given in Table 8-2. These hormones drive the ovarian cycle in the female and the testosterone formation and spermatogenesis in the male. The LH receptor sequence is shown in Figure 8-41 with a diagram of the receptor

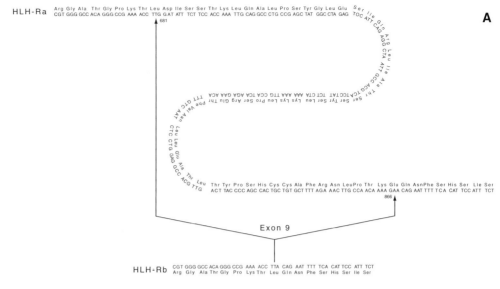
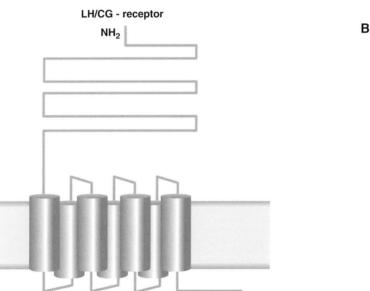

Figure 8-41. **A.** DNA and protein sequence of the LH/hCG receptor. Human luteinizing hormone/human chorionic gonadotrophin (LH/HCG) receptor variant sequence. Comparison of HLH-Ra and HLH-Rb cDNA nucleotide and amino acid sequences. The 62 amino acids encoded by exon 9 are indicated. **B.** Putative secondary structure of the LH/hCG receptor (hCGR). N-terminus is extracellular, C terminus intracellular. The receptor can be divided into the following domains: a 331 residues long extracellular domain representing also the major ligand binding domain; a transmembrane domain consisting or 7 membrane spanning alpha helices linked together by hydrophilic extracellular and intracellular loops; an intracellular domain. Portions of the intracellularly located subdomains are likely to be involved in G Protein-coupling and desensitization by phosphorylation. The receptors for FSH and TSH are of similar overall architecture. Part A redrawn from T. Minegishi et al., "Expression of luteinizing hormone/human chorionic gonadotrophin (LH/HCG) receptor mRNA in the human ovary," *Molecular Human Reproduction*, 3: 101–107, 1997. Part B redrawn from http://info.uibk.ac.at/c/c5/c511/hcgr.html.

structure. The structures of the hormones, LH and **human chorionic gonadotropin (hCG),** secreted from accessory tissues during pregnancy, are so similar that they bind to the same receptor and promote the same signal cascade. LH and FSH receptors, as well as the TSH receptor, belong to a superfamily of seven-membrane G protein–coupled receptors. These receptors have unusually long extracellular domains with leucine-rich repeats forming a structure capable of surrounding the ligand. In the male, LH promotes the synthesis of testosterone in the Leydig cells (interstitial cells) of the testis. LH binds to its membrane receptor on the Leydig cell and activates the receptor, which associates with a G protein to, in turn, activate adenylate cyclase and form cyclic AMP from ATP (Figure 8-42). Cyclic AMP then causes the activation of protein kinase A, which phosphorylates cholesterol ester hydrolase to an active form that converts cholesteryl esters in the lipid droplet to free cholesterol. The free cholesterol can now enter the mitochondrion to serve as substrate for the synthesis of testosterone

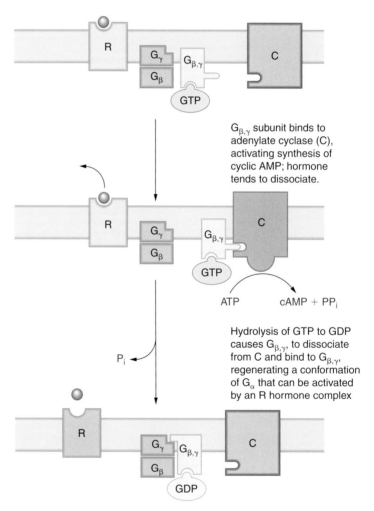

Figure 8-42. Liganded LH receptor (R) associates with a G protein ($G_\beta G\gamma$) that, in turn, activates adenylate cyclase (C), which catalyzes the formation of cAMP from ATP.

(Figure 8-43). The synthetic system depends on the availability of cholesterol to the mitochondrion, because the concentration of cholesterol in the mitochondrion is rate limiting. Free cholesterol is moved into the mitochondrion by the StAR protein (Figure 8-44). The StAR protein appears with each new episode of LH activity. Although it is not completely clear how the StAR protein transports the cholesterol molecule into the mitochondrion, the transfer may involve a pore or the generation of a pore through which the cholesterol molecule could move.

The steroid synthetic mechanism is similar for all steroid hormones, and the target cell (in this case, the Leydig cell under the influence of LH) contains the specific enzymes that catalyze the substitutions to the steroid molecule to generate the specific hormone. The testosterone generated in this process is used by the Sertoli cell to amplify spermatogenesis in the seminiferous tubules, as shown when the actions of FSH are discussed. An overview of LH and FSH actions in the male is shown in Figure 8-45. Similar to the situation with ACTH stimulation of cortisol synthesis, LH stimulates the synthesis of testosterone, again by providing

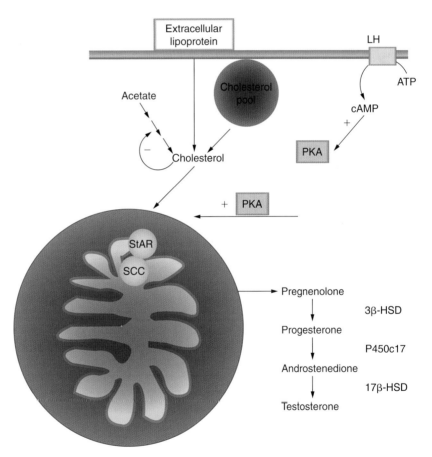

Figure 8-43. LH binding to its receptor in the Leydig cell stimulates the synthesis and release of testosterone. Hydroxylation takes place in the microsome (P450c17). *3β-HSD*, 3β-hydroxysteroid dehydrogenase; *17β-HSD*, 17β-hydroxysteroid dehydrogenase; *cAMP*, cyclic AMP; *LH*, luteinizing hormone; *P450c17*, a microsormal hydroxylase; *PKA*, protein kinase A; *SCC*, side chain cleavage (of cholesterol); *STAR*, steroid-activated response protein.

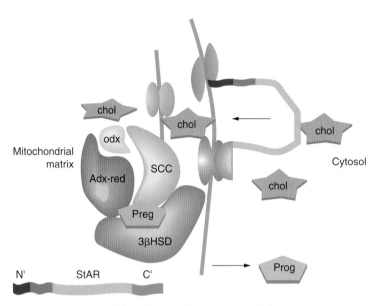

Figure 8-44. A model of the mode of action of the StAR protein. StAR may associate (N'- and C'-terminals of StAR) with sites on the mitochondrial membrane to allow cholesterol (chol, in *blue*) to pass through from the outside to the inside. It is converted to Δ^5-pregnenolone in the matrix by the side chain cleavage complex (SCC) and then converted to progesterone by catalysis by 3β-hydroxysteroid dehydrogenase (3βHSD). *Adx*, adrenochrome; *adx-red*, reduced adrenochrome; *Preg*, pregenolone; *Prog*, progesterone. Redrawn from http://tigger.uic.edu/~dbhale/research.htm.

free cholesterol to the mitochondrion. The exact steps in the synthesis of these steroid hormones are given in Chapter 9. However, the androgen receptor, although it will bind testosterone as an agonist, prefers to bind a metabolite of testosterone, 5α-dihydrotestosterone. The structures of testosterone and 5α-dihydrotestosterone are shown in Figure 8-46. The testosterone formed in the Leydig cell is used, after binding to the androgen receptor, in the Sertoli cell of the seminiferous tubule to induce proteins for the formation of sperm (Figure 8-45). It is also indicated that there is a negative feedback by testosterone on the release of LH and interference with the action of GnRH in the anterior pituitary. Some aspects of the negative feedback mechanism have been established and are complex, as shown in Figure 8-47.

FSH (Table 8-2) binds to the FSH receptor in the cell membrane of FSH target cells. The FSH receptor is a seven-membrane protein with a large extracellular domain (Figure 8-48), similar to the LH receptor. The mode of action of the FSH receptor is similar to that of the LH receptor in that it is coupled to a G protein that, when activated following ligand binding to the receptor, activates adenylate cyclase to produce cyclic AMP from ATP. Cyclic AMP activates protein kinase A, which opens

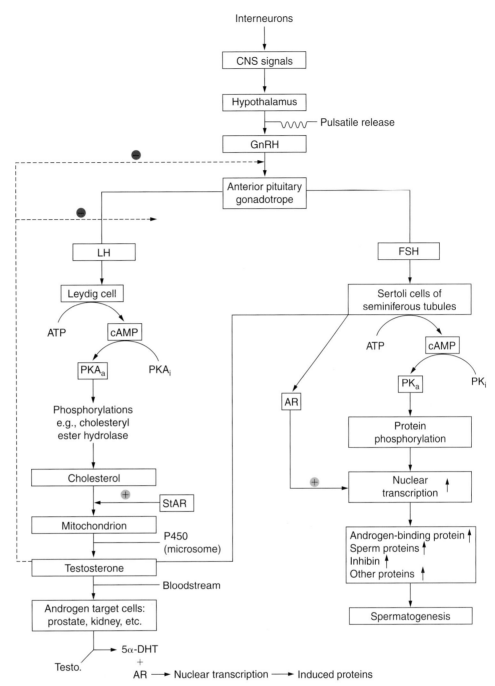

Figure 8-45. Overview of the actions of LH and FSH in the male. *AR*, androgen receptor; *cAMP*, cyclic *AMP*; *FSH*, follicle-stimulating hormone; *GuRH*, gonadotropic releasing hormone; *LH*, luteinizing hormone; PKA_a, activated protein kinase A; PKA_i, inhibited protein kinase A; *StAR*, steroid acute response protein; *Testo*, testosterone; *5α-DHT*, 5α-dihydrotestosterone.

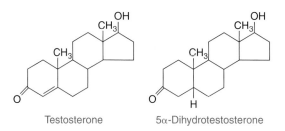

Figure 8-46. Structures of testosterone and 5α-dihydrotestosterone.

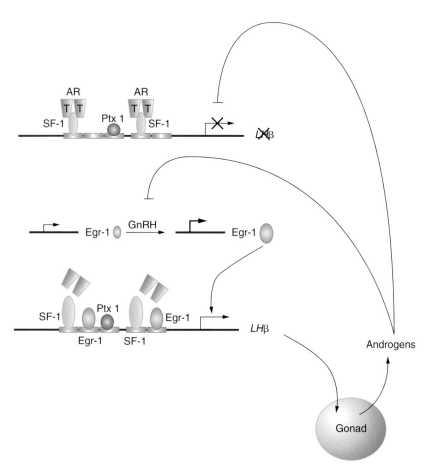

Figure 8-47. A model for LHβ promoter repression by a physical interaction between AR and SF-1. In the presence of elevated androgen concentrations, androgen receptor (AR) interacts with steroidogenic factor 1 (SF-1) on LHβ promoter, preventing interaction of early growth response factor 1 (Egr-1) with its response elements. LHβ is then turned off. A GnRH pulse (resulting from a reduction in circulating androgen concentrations), favors Egr-1 transcription, increasing its accumulation. In turn, Egr-1 displaces AR from SF-1 and favors an active setting of transcription factors on LHβ promoter. LHβ transcription is then triggered. The resulting LH protein production stimulates androgens production by the gonads. Reproduced from http://www.nuclear–receptor.com/content/1/1/8/figure/F6.

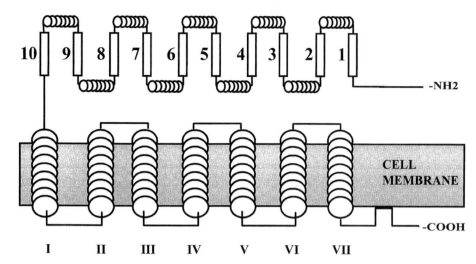

Figure 8-48. Schematic model of the FSH receptor. The extracellular domain consists of several leucine-rich repeats made up of alternating β-sheets *(indicated by rectangles numbered consecutively in Arabic numerals)* and α-helices *(indicated by the coils)*. The transmembrane domain consists of seven hydrophobic segments *(Roman numerals)* spanning the cell membrane and connected by intra- and extracellular loops. Within the intracellular domain a putative fourth intracellular loop is depicted. From M. Simoni, J. Gromoll, and E. Nieschlag, *Endocrine Revs.* 18: 739–773, 1997.

calcium channels to admit calcium ion into the cell and phosphorylates proteins (transcription factors). These, in turn, stimulate transcription of specific genes (Figure 8-49).

FSH and its receptor are critical to the ovarian cycle. Its level increases at the beginning of the follicular phase and through ovulation and then falls in the luteal phase (Figure 8-50). An overall view of the ovarian cycle showing the hormonal changes is shown in Figure 8-51. On day 1 of the 28-day cycle, LH and progesterone levels in the blood are low; estrogen level is low but slowly rising to drive the developing follicle (follicular phase). As this phase continues, FSH levels slowly rise and estrogen level peaks about day 12 (LH also increases markedly in the **LH spike**), driving ovulation. The residual follicle, after ovulation **(Graafian follicle)**, now secretes progesterone. This secretion drops off by day 28, when menstruation occurs. If fertilization should occur about the time of ovulation, hCG is elaborated from the placenta to maintain a luteal phase (corpus luteum of pregnancy) and progesterone is elaborated from the placenta, reaching high levels by termination of pregnancy. This high level of progesterone competes with endogenous cortisol for the glucocorticoid receptor (progesterone is a glucocorticoid antagonist), so lactation is held in abeyance, as well as other functions that depend on the glucocorticoid receptor. The overall secretion and feedback system for the gonadotropins are diagrammed in Figure 8-52.

(Text continues on p. 417.)

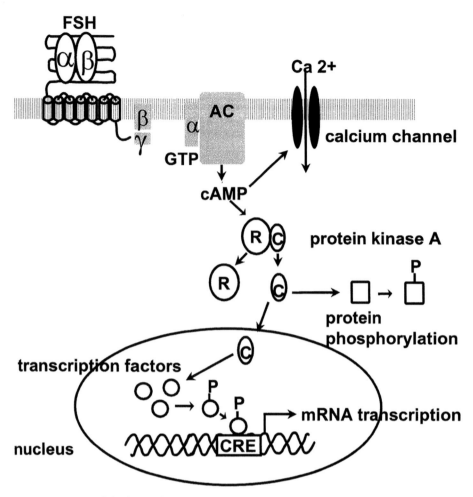

Figure 8-49. Model of signal transductional pathways of the FSH receptor. Upon binding of FSH to the FSH receptor, the $G_{s\alpha}$ subunit dissociates. Together with GTP, this complex directly activates adenylyl cyclase, thereby leading to cAMP synthesis. PKA is activated by cAMP, which causes the dissociation of the catalytic subunit (C) from the regulatory subunit (R). The active catalytic site can activate proteins by phosphorylation. Conversely, the production of cAMP leads to an intracellular rise of Ca^{2+}, presumably due to the gating of calcium channels. In the nucleus the catalytic subunit of the PKA can phosphorylate transcription factors such as CREB and CREM, which then bind to CREs preceding certain genes. Finally, mRNA synthesis of primary response genes of FSH action starts. Reproduced from http://edrv.endojournals.org/cgi/content-nw/full/18/6/739/F7.

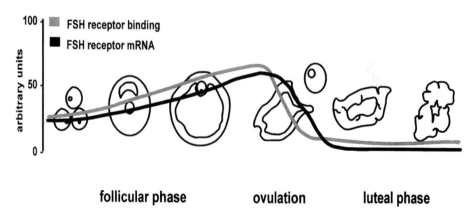

Figure 8-50. Expression of the FSH receptor during estrus. Follicle maturation is indicated by the development of primary follicles into a Graafian follicle. After ovulation, the corpus luteum and the corpus albicans are shown. Expression levels of the FSH receptor are given in an arbitrary scale ranging from 0 to 100 U. Reproduced from http://edrv.endojournals.org/cgi/content-nw/full/18/6/739/F6.

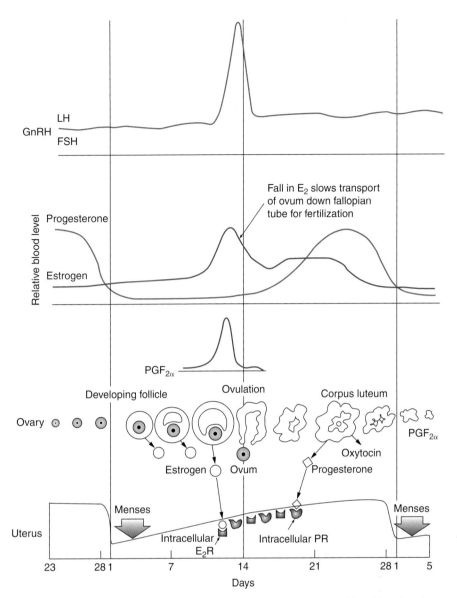

Figure 8-51. The ovarian cycle. In the upper diagram, relative blood levels of GnRH LH, FSH, progesterone, estrogen, and $PGF_{2\alpha}$ are shown. In the lower diagram, events in the ovarian follicle, corpus luteum, and uterine endometrium are diagrammed. E_2, estradiol; E_2R, intracellular estrogen receptor; *FSH*, follicle-stimulating hormone; *GnRH*, gonadotropin-releasing hormone; *LH*, luteinizing hormone; $PGF_{2\alpha}$, prostaglandin $F_{2\alpha}$; *PR*, intracellular progesterone receptor. Reproduced from G. Litwack and T.J. Schmidt, "Peptide hormones" in *Textbook of Biochemistry*, 5th ed., T. Devlin, Editor, Wiley-Liss, 2002.

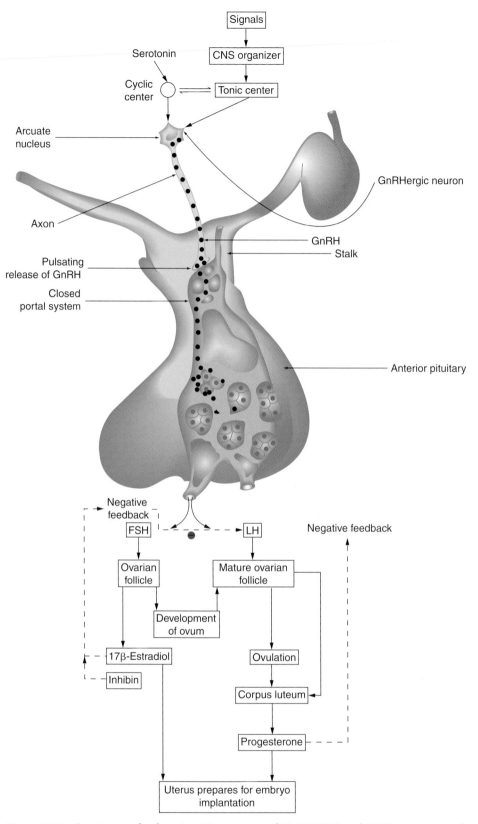

Figure 8-52. Ovarian cycle showing. Generation of GnRH, LH and FSH, estrogen and progesterone. The hormone, inhibin, inhibits FSH secretion. 17β-estradiol inhibits LH release.

Thyrotropin

Thyrotropin is the anterior pituitary hormone whose release is stimulated by TRH (Table 8-1). Structures of TRH and TSH are given in Tables 8-1 and 8-2. This system governs the release of thyroid hormone from the thyroid follicle cell. The overall pathway governing the secretion of thyroid hormones is shown in Figure 8-53. Different signals impinge on the hypothalamus to cause the release of TRH, in particular α-adrenergic stimulation. TRH, from the nerve ending of the TRHergic neuron, courses through the closed circulation connecting the hypothalamus to the anterior pituitary and binds to its receptor on the surface of the

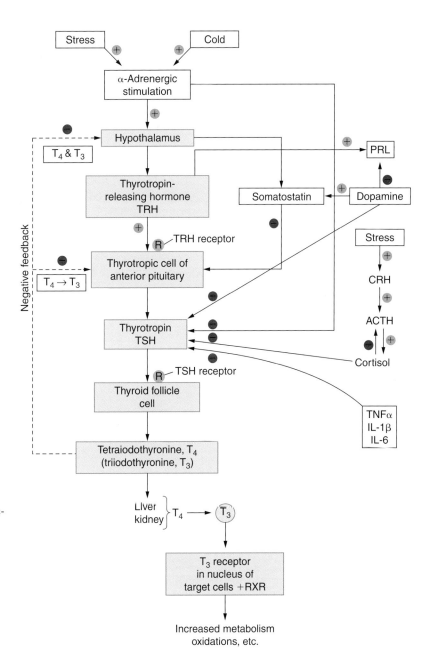

Figure 8-53. Pathway of thyroid hormone release from TRH, TSH and other regulators. *ACTH*, adrenocorticotropic hormone; *CRH*, corticotropin-releasing hormone; *IL-1β*, interleukin-1β; *IL-6*, interleukin-6; *PRL*, prolactin; *SST*, somatostatin; T_3, triiodothyronine; T_4, tetraiodothyronine; *TNFα*, tumor necrosis factorα. Negative feedback by thyroid hormones on TSH and TRH cause a decrease in their syntheses.

thyrotrophic cell. This causes biochemical changes through signal transduction, culminating in the release of thyrotropin (TSH). The signal transduction of TRH in the pituitary is complex and involves hydrolysis of phosphatidylinositol. This causes the activation of phospholipase C, which increases the concentration of **inositol 1,4,5-triphosphate (IP3)** and **diacylglycerol (DAG).** Elevated IP$_3$ increases the cellular concentration of Ca^{2+}, stimulating calcium–calmodulin-dependent protein kinase (**Ca–CAM kinase**). Elevated DAG activates **protein kinase C** and subsequent phosphorylations of several proteins in the nucleus, plasma membrane, and cytosol resulting in hormone secretion (of TSH) and gene expression. TSH then binds to its receptor in the membrane of the thyroid follicle cell. From the colloid (containing thyroglobulin) of the thyroid follicle cell (Figure 8-54), two hormones are split out and secreted: T$_4$ and T$_3$. T$_4$ is the preponderant circulating form (100 nmol/liter total with 0.02% in the free form) with only a small amount of T$_3$ (2 nmol/liter total with 0.4% in the free form), although T$_3$ is the most active agonist for the thyroid hormone receptor. Nearly all thyroid hormones (99%) circulate in the bound form; 70% is bound to **thyroid-binding globulin (TBG)**, 10–15% is bound to transthyretin, and 15 to 20% is bound to albumin. In some target tissues, T$_4$, taken up from the blood, is converted to T$_3$, which then enters the nucleus and binds to the thyroid hormone receptor.

TRH is a tripeptide of L-pyroglutamyl–L-histidinyl–L-prolinamide (Table 8-1). It is transcribed from its gene and translated to a precursor, proTRH, from which the tripeptide is excised enzymatically (carboxypeptidases), amidated (to form prolinamide by peptidyl glycine α-amidating monooxygenase), and finally cyclized (to form pyroglutamyl). The regulation of the secretion of TRH is positively controlled by α-MSH and negatively by the **Agouti receptor protein.** The formation of α-MSH through its precursor, proopiomelanocortin (POMC) is stimulated by the hormone **leptin,** and the Agouti receptor protein is inhibited by leptin. MSH causes

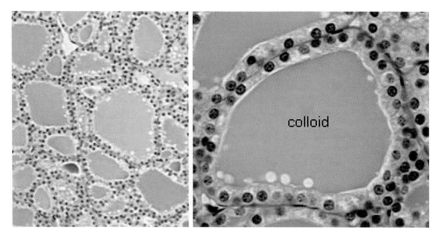

Figure 8-54. Thyroid epithelial cells are arranged in spheres called thyroid follicles *(left)*. These follicles are filled with colloid, a storage form of the thyroid hormone precursor *(right)*. In spaces between the thyroid follicles are parafollicular or C cells, which secrete the hormone calcitonin. Reproduced from http://arbl.cvmbs.colostate.edu/hbooks/pathphys/endocrine/thyroid/anatomy.html.

the release of TRH by agonizing the melanocortin 4 receptor on the surface of the TRHergic neuron. Leptin also directly stimulates the release of TRH from the TRHergic neuron. In turn, TRH stimulates the release of TSH from the thyrotrophic cell of the anterior pituitary and somatostatin inhibits the release of TSH (somatostatin arises during fasting). These relationships are shown in Figure 8-55.

The TRH receptor is a typical seven-membrane protein (Figure 8-56). TRH is able to increase (up-regulate) the TRH receptor concentration through signal transduction that does not involve receptor endocytosis. There are two forms of the TRH receptor, TRHR1 and TRHR2. These are G protein–coupled forms whose signal transduction process culminates in stimulating the release and synthesis of both TSH and PRL. TRHR1 mRNA levels are up-regulated in the pituitary and hypothalamus after suckling, positing a physiological role for the secretion of PRL by TRH. The TRHR1 gene contains several promoter elements, including two glucocorticoid response elements.

The TSH receptor in the membrane of the thyroid follicular cell (also called thyrocyte) has three states: a closed conformation (does not bind TSH), an open conformation, and an active conformation stabilized by the binding of TSH (containing two subunits, α- and β-subunits), as diagrammed in Figure 8-57. The amino acid sequences of the subunits are given in Figure 8-58. The α-subunit is virtually identical to the α-subunits in FSH, LH, and hCG, and the β-subunit is the specific subunit (also for FSH and LH) that is unique for each hormone, identifies the ligand, and is recognized specifically by the TSH receptor. The TSH–TSH receptor complex with the carbohydrate substitutions in the TSH molecule are shown in Figure 8-59. The action of the TSH–TSH receptor complex is to initiate signal transduction involving a G protein (Gs), which activates adenylate cyclase and the formation of cyclic AMP. This, in turn, activates protein kinase A. An associated Gq protein phosphorylates

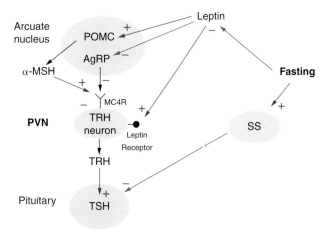

Figure 8-55. Role of fasting, somatostatin (SS) pathways and leptin on TRH and TSH secretion. α-*MSH*, α-melanocyte-stimulating hormone; *AgRP*, Agouti receptor protein; *MC4R*, melanocortin 4 receptor; *POMC*, proopiomelanocortin. Red arrows, stimulation; blue arrows, inhibition.

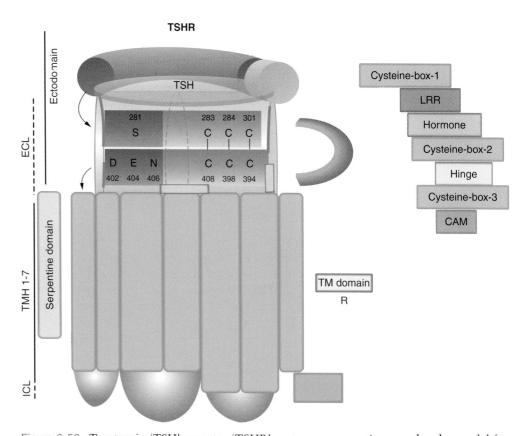

Figure 8-56. Tyrotropin (TSH) receptor (TSHR) cartoon representing a molecular model for the tight packing of the ectodomain (LPR, cysteine-box-2, cystein-box-3) closely located or in between the extracellular loops of the serpentine domain. A new LRR template was introduced whose "scythe blade" shape allows an interaction of the hormone parallel to the LRR structure. New template for Cystein-boxes-2 and -3 was identified based on the structure of IL8 and a portion of the N-terminal tail of the IL8 receptor. Also hypothesized is a disulfide bridge between Cys398/Cys408 (C-b-3) either to Cys283/Cys284 (C-b-2) or in a reverse manner with Cys408/Cys398. The hydrophilic amino acids Asp403, glu404, and Asn406 of C-b-3, spatially located in proximity to Ser281, may be involved in intransmolecular signal transduction from the ECD towards the serpentine domain. Redrawn from http://www.hotthyroidology.com/editorial_145.htm.

and activates phospholipase C and the production of diacylglycerol and IP_3. These activate protein kinase C and a cellular increase in calcium ion from the endoplasmic reticulum. In the human, the protein kinase A pathway predominates with normal concentrations of TSH. These activities are shown in Figure 8-60.

There are two genes encoding thyroid hormone (T_3) receptors, one on chromosome 17 for the α receptor and one on chromosome 3 for the β receptor. There are alternative mRNA splice products for each gene, so there are four T_3 receptors: α1, α2, β1, and β2. The α2 receptor does not bind T_3. These receptors and heterodimer formation with the RXR are shown in Figure 8-61. The crystal structure of the DNA-binding domains of thyroid hormone β receptor with RXR in the heterodimer binding to the thyroid response element of DNA is shown in Figure 8-62. The

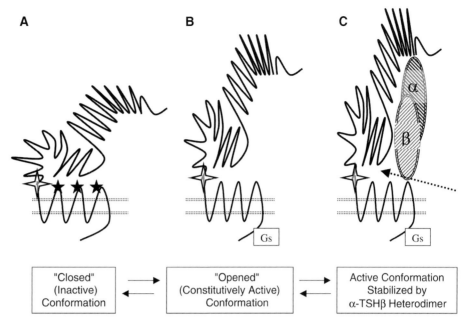

Figure 8-57. The two-state model of TSH receptor (TSHR) activation. **A.** "Closed" (inactive) TSHR conformation; three five-point stars in the closed (inactive) conformation illustrate putative interactions between the extracellular domain and the extracellular loops of transmembrane domain; these interactions are critical for an inactive state and may include electrostatic interactions. Four-point star depicts the location of TSHR 409 418 fragment, which may contribute to receptor activation, when the constraints of extracellular domain–transmembrane domain interaction are removed. **B.** "Opened" (unliganded) TSHR conformation. **C.** "Opened" (hormone-activated) TSHR conformation. Reproduced from M.W. Szkudlinski et al., "Thyroid-stimulating hormone and thyroid-stimulating hormone receptor structure-function relationships," *Physiol. Revs.*, 82: 473–502, 2002.

influence of TSH on the follicular cells (thyrocytes) culminates in the release of T_3 and T_4, which are hydrolyzed from thyroglobulin. Thyroglobulin, stored in the colloid, is a huge dimeric molecule of 660,000 molecular weight (each monomer contains 2769 amino acid residues with a molecular weight of 303,218). It also contains carbohydrate moieties. If thyroglobulin contained 1.6% iodine, it would not be saturated with iodine but, at that level, it would contain six T_4 molecules and two T_3 molecules. Although crystal structures could not be found, an HLA class II histocompatibility antigen contains a thyroglobulin-like structure, which is shown in Figure 8-63.

Iodide ion (I^-) is taken up by the thyroid follicle cell (thyrocyte) and is transported inward by a **cotransporter** (a membrane exchanger that admits one ion in exchange for emitting another different ion) in exchange for Na^+. The uptake of iodide is stimulated by TSH. Iodide,

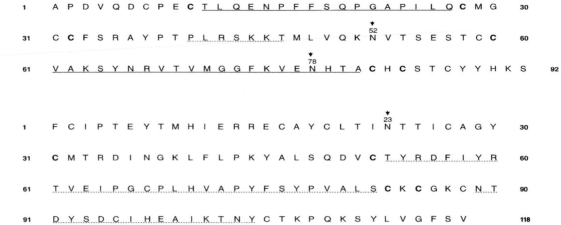

Figure 8-58. The primary structure of the human α-subunit (A) and the human TSH (hTSH) β-subunit (B). Functionally important residues are numbered and highlighted in boldface in the wild-type sequence. Specific mutations are shown either above the corresponding wild-type residue (if the mutation resulted in an increase of hTSH activity) or below (if the mutation caused a decrease of activity). Mutations αN^{78}Q and βN^{23}Q, which disrupt a glycosylation recognition sequence (highlighted by the outlined font) decrease *in vivo* but not *in vitro* activity, whereas disruption of the glycosylation recognition sequence at αN^{52} (αN^{52}Q, αN^{52}D) increases *in vitro* but slightly decreases *in vivo* activity. Multiple residue mutations are underlined. Also depicted is the location of selected structural features. The β-hairpin loops correspond to the continuous lines above the primary structure flanked by arrows. The bold part of these lines indicates β-sheet, and the thin part of the line indicates the actual loops. α-Helix (between α40 and 46) is marked with an interrupted line below the sequence (–). The seat belt between βC88 and βC105 is marked by a double line (=). Chimeric substitutions in this not only decreased hTSH activity but also changed receptor specificity. Reproduced from http://physrev.physiology.org/cgi/content-nw/full/82/2/473/F2.

within the cell, is oxidized by H_2O_2, catalyzed by thyroid peroxidase, to form active iodine. Iodine is then transported across the apical surface of the thyrocyte, and it iodinates tyrosine residues of thyroglobulin to form mono- and di-iodotyrosines (MIT and DIT). MIT and DIT are taken up in the colloid, where about 1% or more of the colloid is removed each day for normal usage (Figure 8-64). TSH also causes colloid droplets to be taken up from the colloidal space back into the thyrocyte, where colloid droplets are fused with lysosomes and thyroglobulin is hydrolyzed to release T_4 and T_3. Released T_4 and T_3 are secreted from the cell to enter the bloodstream (Figure 8-65). The structures of T_4 and T_3 are shown in Figure 8-66. When T_4 enters a target cell, it is deiodinated to T_3 (which is 1000 times more active than T_4) by one of three selenium-containing deiodinases (D1, D2, D3) (Figure 8-67). The locations of these three deiodinases in various tissues are shown in Figure 8-68. The K_m of T_4 for D2 is 1000 times lower (meaning a much higher approximate affinity; *affinity* can be applied to

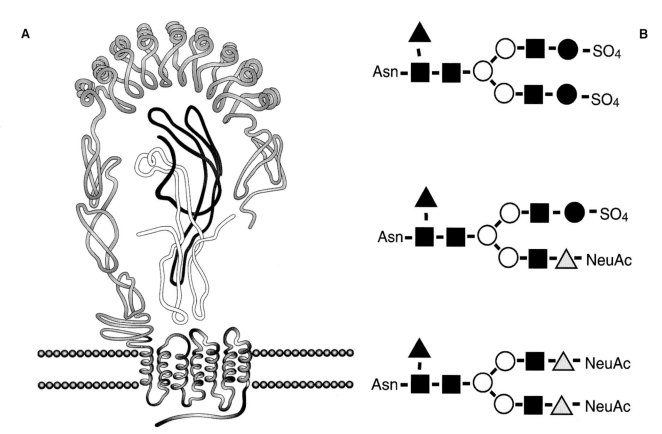

Figure 8-59. **A.** Schematic configuration of TSH-TSHR complex. Two parallel β-hairpin loops of the α-subunit (αL1, αL3) are located in the bottom part of the model and may participate in the interaction with the extracellular loops of the receptor transmembrane domain. Two analogous loops of the β-subunit (βL1, βL3) are shown in the top part with proposed binding site within the concave of leucine-rich repeats (LRR).
B. N-linked oligosaccharides of TSH. The sulfated biantennary structure *(a)* represents that of bovine TSH and bovine luteininzing hormone (LH) The sulfated and sialylated oligosaccharide *(b)* is more typical of pituitary-derived hTSH as well as LH. The sialylated nonsulfated structure *(c)* represents that of recombinant hTSH expressed in Chinese hamster ovary cells, pituitary human follicle-stimulating hormone, as well as placental human chorionic gonadotropin. Carbohydrate residues are marked as follows: mannose (O), *N*-acetylglucosamine (■), *N*-acetylgalactosamine (●), fucose (▲), galactose (∆), and sialic acid (NeuAc). Reproduced from http://physrev.physiology.org/cgi/conent-nw/full/82/2/473/F7.

reactions forming a reversible complex but not strictly to reactions that progress beyond the complex, like enzymatic reactions that form products in a direction different than back to the ligand) than for D1. The tissue targets for the thyroid hormone are skin and appendages, eyes, cardiovascular system, respiratory system, developing central nervous system, skeleton, musculature, gastrointestinal tract, renal system, blood protein components, pituitary especially for growth hormone and ACTH, adrenal medulla (primary catechols), female reproductive system, and developing tissues. Thyroid actions are upon

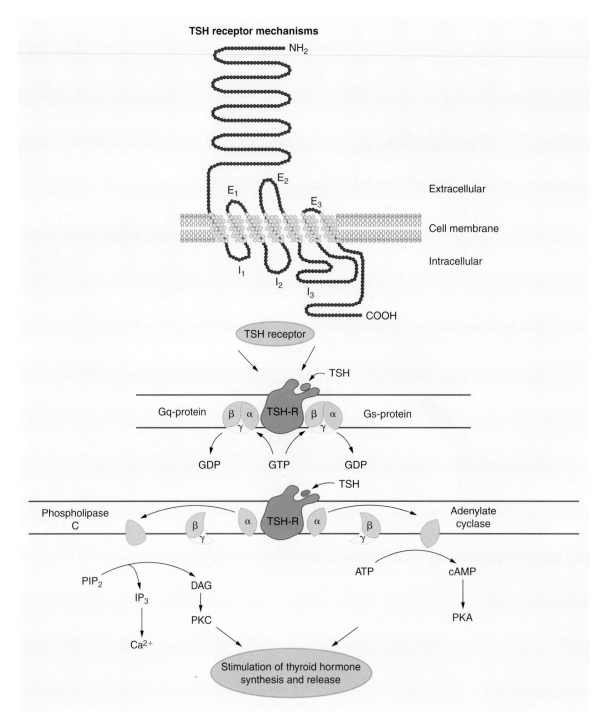

Figure 8-60. The TSH receptor signaling pathway. Redrawn from http://www.ncbi.nlm.nih.gov/books/bv.fcgi?rid=endocrin.box.291.

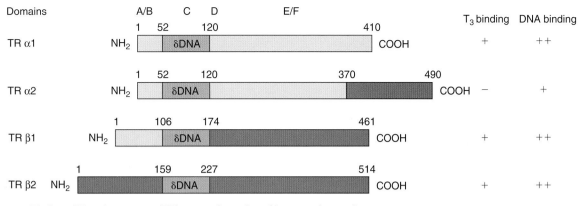
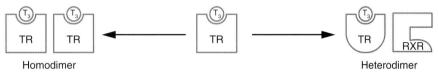
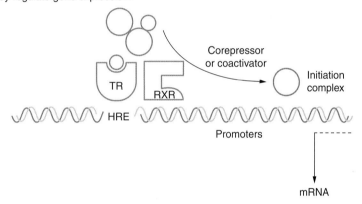

Figure 8-61. Thyroid hormone (T₃) receptors, heterodimer formation, and gene activation. Redrawn from http://www.ncbi.nlm.nih.gov/books/bv.fcgi?rid=endocrin.box.283.

energy metabolism, oxygen consumption, DNA, RNA, cell number, cell size in all cells up to 2 years of age, and thermoregulation to maintain the core body temperature. Adult brain, spleen, retina, and testes are not affected by thyroid hormone. The actions of T_3 are mediated by the thyroid hormone receptor and the promoters of genes that bear the thyroid response element.

(Text continues on p. 430.)

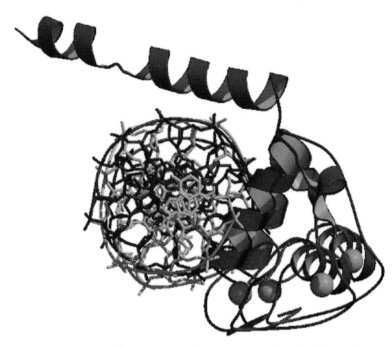

Figure 8-62. Crystal structures of T$_3$ receptor β *(red)* and RXR *(purple)* DNA-binding domains bound to the thyroid response element of DNA. Green atoms represent zinc ions. Reproduced from PDB ID: 2NLL. F. Rastinejad et al. Retinoid X Receptor-Thyroid Hormone Receptor DNA-Binding Domain Heterodimer Bound to Thyroid Response Element DNA. *Nature* **375** pp. 203 (2005).

Figure 8-63. Thyroglobulin-like structure in HLA class II histocompatibility antigen (gamma chain). Reproduced from C. Chiva, P. Barthe, A. Codina, M. Gairi, F. Molina, C. Granier, M. Pugniere, T. Inui, H. Nishio, Y. Nishiuchi, T. Kimura, S. Sakakibara, F. Albericio, E. Giralt. Synthesis and NMR Structure of P41ICF, a Potent Inhibitor of Human Cathepsin L. *J. Am. Chem. Soc.* **125** pp. 1508 (2003) and http://www.ebi.ac.uk/Thornton-srv/databases/cgi-bin/pdbsum/getpage.pl.

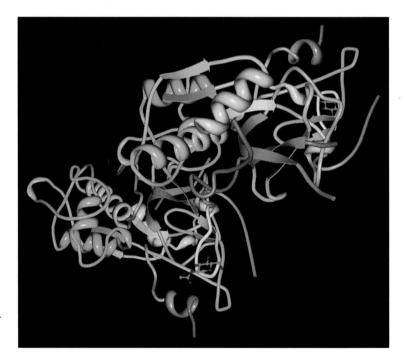

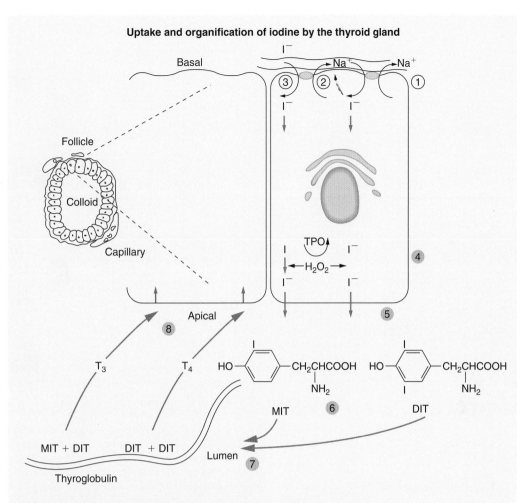

Figure 8-64. Uptake and organification of iodine by the thyroid gland. Redrawn from http://www.ncbi.nlm.nih.gov/books/bv.fcgi?rid=endocrin.box.250.

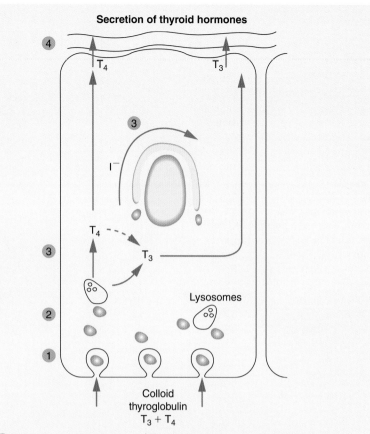

Figure 8-65. Secretion of thyroid hormones. Redrawn from http://www.ncbi.nlm.nih.gov/books/bv.fcgi?rid=endocrin.box.279.

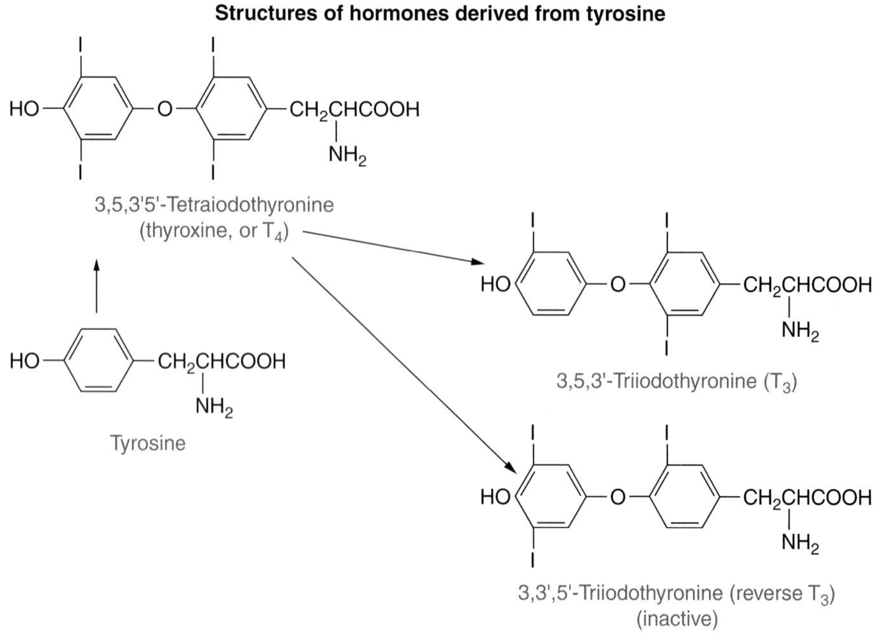

① Under the influence of TSH, colloid droplets consisting of thyroid hormones within the thyroglobulin molecules are taken back up into the follicular cells by pinocytosis.
② Fusion of colloid droplets with lysosomes causes hydrolysis of thyroglobulin and release of T_3 and T_4.
③ About 10% of T_4 undergoes mono-deiodination to T_3 before it is secreted. The released iodide is reused. Several-fold more iodide is reused than is taken from the blood each day, but in states of iodide excess there is loss from the thyroid.
④ On average, 100 μg of T_4 and 10 μg of T_3 are secreted per day.

Figure 8-66. Structures of thyroid hormones T_4, T_3, and reverse T_3 derived from tyrosine.

428   CHAPTER 8   Polypeptide Hormones

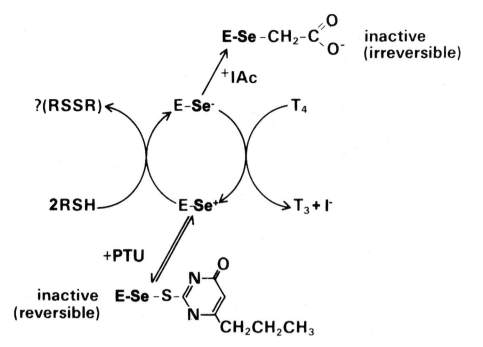

Figure 8-67. Deiodination mechanism for D1-catalyzed T_4-to-T_3 conversion. The steps in the enzymatic reaction cycle at which iodoacetic acid and PTU are thought to act to inhibit catalysis are indicated. *RSH* is a reduced sulfhydryl compound and *R-SS-R* is the oxidized counterpart; *PTU* is propylthiouracil inhibitor; the structure at the bottom of the figure is the inactive selenium-containing enzyme. Reproduced with permission from A.C. Bianco et al., "Biochemistry, cellular and molecular biology, and physiological roles of iodothyronine selenodeiodinases," *Endocr. Rev.*, 23: 38–89, 2002.

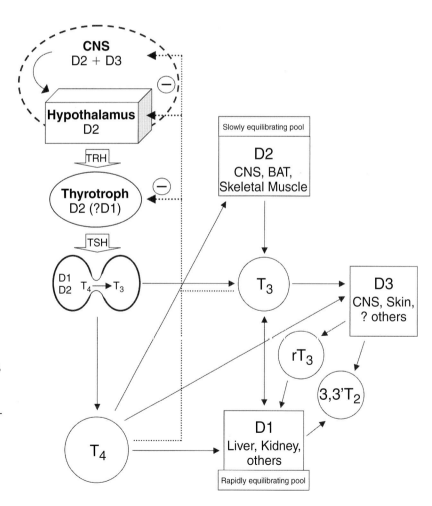

Figure 8-68. Schematic diagram of the human thyroid axis depicting the role and probable tissue location of D1, D2, and D3 in the production and inactivation of plasma T_3 and in feedback regulation of thyroid function. Reproduced with permission from A.C. Bianco et al., "Biochemistry, cellular and molecular biology, and physiological roles of iodothyronine selenodeiodinases," *Endocr. Rev.*, 23: 38–89, 2002.

Prolactin

Information about PRL-releasing hormone and PRL are given in Tables 8-1 and 8-2. Some information about the pathway of releasing PRL from the anterior pituitary in response to the suckling stimulus is given in Figure 8-8. In addition to recently discovered PRL-releasing peptides, TRH causes the release and synthesis of PRL by operating through the TRH receptor on PRLergic neurons of the anterior pituitary (although this conclusion is still controversial). The PRL-releasing peptides may turn out to be the physiologically important ligands for the release of PRL. The TRH signal transduction process mediated by its receptor initiates phospholipid turnover (hydrolysis of phosphatidylinositol bisphosphate, PIP_2) through a G protein, resulting in the formation of IP_3, which increases the cytosolic concentration of Ca^{2+} by mobilizing the calcium store from the endoplasmic reticulum. Diacylglycerol also is formed, which activates protein kinase C and stimulates (opens) calcium influx channels. Like the other hypothalamic-releasing hormones, the half-life of this tripeptide is short (minutes). TRH is inactivated by **thyroliberinase** (also known as pyroglutamyl peptidase II, a zinc-containing enzyme that cleaves the pyroglutamyl–histidine bond in TRH). The enhanced calcium ion concentration in the lactotrophic (mammotrophic) cell causes the release of preformed PRL (or TSH from the thyrotrophic cell).

PRL acts on the mammary gland (Figure 8-8) through its receptor to stimulate breast development and the syntheses of milk proteins during lactation. Primarily, the signal transduction mechanism from the PRL receptor involves the JAK2/STAT5 pathway (Figure 8-69). Enhancement of milk protein synthesis (β-lactalbumin, casein, and whey acidic protein) also can arise from other membrane receptor pathways (not fully understood), which act through the MAPK system. Although other hormones can activate STAT5 (tyrosine phosphorylation of STAT5), the transcriptional activation of casein is specific to PRL. In addition, the glucocorticoid receptor pathway is important because glucocorticoids play a positive role in milk synthesis. In pregnancy, when the level of progesterone is high, the agonist, cortisol, is replaced by progesterone (its high concentration becomes competitive) and milk synthesis is braked until near term. Then the progesterone level falls, allowing the normal levels of cortisol to activate the glucocorticoid receptor and initiate milk synthesis. Although its action is complicated in the male, PRL may increase the level of LH receptors in the Leydig cell and enhance the formation and release of testosterone. At higher levels (for example, from a tumor secreting superphysiological amounts of PRL), PRL may have inhibitory effects on testosterone production.

The structure of the PRL receptor is shown in Figure 8-70. Release of PRL from the lactotroph can be inhibited by PIF. This potential hormone may be derived from other hormones (Table 8-1), but clearly dopamine is an important inhibitor of PRL release. Release of PRL from the anterior pituitary is complex, because a number of factors influence the release, both positively and negatively (Figure 8-71). The positive influences are mating, suckling, the hormones estradiol and progesterone, glucocorticoids, and plasma osmolarity (Na^+ concentration). The negative influences are light, odor, sound, and stress. Negative influences lead to the release of PIF, and positive influences lead to the release of PRF from the

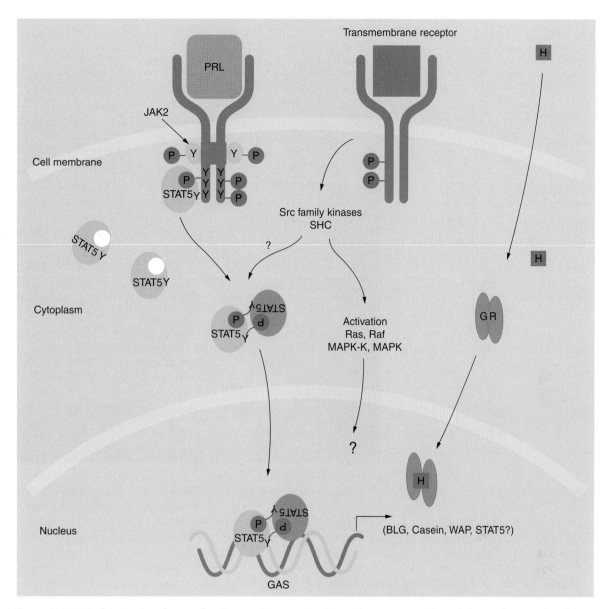

Figure 8-69. Prolactin signal transduction pathways. A schematic representation of the pathways that have been implicated in prolactin (PRL) signal transduction. The Jak (Janus kinase)/Stat [signal transducer and activator of transcription]) pathway has an essential role in mediating the response to prolactin in mammary epithelial cells. Engagement of the prolactin receptor induces homodimerization resulting in activation of the associated Jak 2 kinase, which then tyrosine phosphorylates the receptor and leads to activation of Stat5 through specific SH2 (Src-homology 2) domain-phosphotyrosyl interactions. This results in the dimerization of Stat5, which translocates to the nucleus and binds to its recognition site, where it activates transcription. Activated Jak2 may also associate with the SH2 domain of SHC, which then interacts with Grb2 (an adaptor protein) following tyrosine phosphorylation, thereby activating the MAP (mitogen activated protein) kinase pathway. The mechanism of interaction between prolactin signal transduction and glucocorticoid and other signaling events has not been established. Y and H represent tyrosine and glucocorticoid, respectively. *GR*, glucocorticoid receptor; *GAS*; TTCCNGGAA. Redrawn from C.J. Watson and T.G. Burdon, "Prolactin signal transduction mechanisms in the mammary gland: the role of the Jak/Stat pathway," *Rev. Reprod.*, 1: 1–5, 1996.

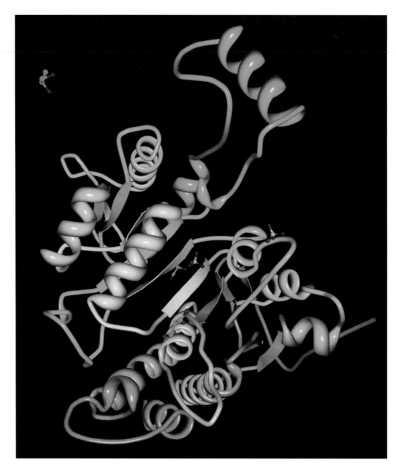

Figure 8-70. The x-ray structure of a growth hormone-prolactin receptor complex. Reproduced from PDB ID: 1BP3. W. Somers, M. Ultsch, A.M. DeVos, A.A. Korssiakoff. The X-Ray Structure of a Growth Hormone-Prolactin Receptor Complex. *Nature* **372** pp. 478 (1994).

neurons in the hypothalamus. Then, from the median eminence, elaborated in response to PRF, are TRH, oxytocin, and neurotensin. PIF release causes the release of dopamine, somatostatin, and γ-aminobutyric acid (GABA). At the level of the anterior pituitary, PRL is released from the lactotroph to the extent dictated by the positive and negative influences. Evidence that dopamine may be an important regulator (PIF) of PRL secretion is in PRL's negative feedback effect on the release of dopamine. The neurotransmitter, dopamine, is elaborated by the **tuberoinfundibular dopaminergic system (TIDA),** and the release of dopamine is under the control of many factors (Figure 8-72). Dopamine is derived from L-tyrosine, as shown in Figure 8-73. Dopamine operates through a membrane receptor (Figure 8-74) located in the lactotrophic cell membrane. The dopamine D2 receptor operates through a G protein to stimulate the formation of cyclic AMP, followed by the activation of protein kinase A. Protein phosphorylation is the mechanism by which dopamine inhibits the release of PRL.

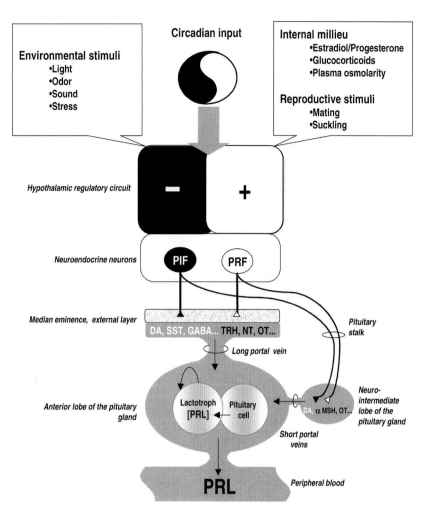

Figure 8-71. An overview of the regulation of prolactin (PRL) secretion. PRL secretion is paced by a light-entrained circadian rhythm, which is modified by environmental input, with the internal milieu and reproductive stimuli affecting the inhibitory or stimulatory elements of the hypothalamic regulatory circuit. The final common pathways of the central stimulatory and inhibitory control of PRL secretion are the neuroendocrine neurons producing prolactin-inhibiting factors (PIF), such as dopamine (DA), somatostatin (SST), and γ-aminobutyric acid (GABA), or prolactin-releasing factors (PRF), such as thyrotropin-releasing hormone (TRH), oxytocin (OT), and neurotensin (NT). PIF and PRF from the neuroendocrine neurons can be released either at the median eminence into the long portal veins or at the neurointermediate lobe, which is connected to the anterior lobe of the pituitary gland by the short portal vessels. Thus, lactotrophs are regulated by blood-borne agents of central nervous system or pituitary origin (α-melanocyte–stimulating hormone) delivered to the anterior lobe by the long or short portal veins. Lactotrophs are also influenced by PRF and PIF released from neighboring cells (paracrine regulation) or from the lactotrophs themselves (autocrine regulation). Reproduced with permission from http://physrev.physiology.org/cgi/content-nw/full/80/4/1523/F3.

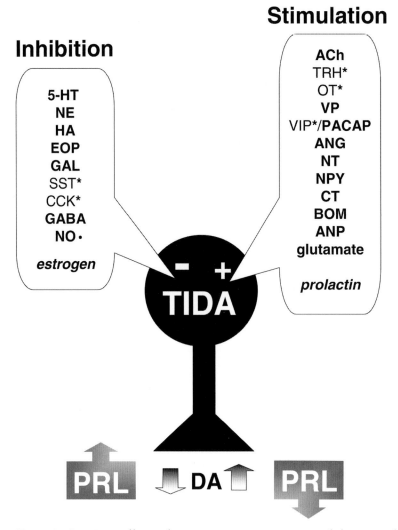

Figure 8-72. Direct effects of neurotransmitters, neuromodulators, and peripheral hormones on the activity of tuberoinfundibular dopaminergic system (TIDA). The inhibitory agents *(left)* will promote an increase of prolactin secretion as a result of diminishing TIDA activity. On the other hand, the stimulatory neurotransmitters and progesterone *(right)* will tend to decrease prolactin secretion as a result of increasing output of TIDA neurons. It should be noted, however, that many of these agents have multiple levels of action, often with opposing biological effect. Therefore, in some cases, effects on PRF and/or directly at the lactotrophs will prevail over the influence on TIDA activity. *Ach*, acetylcholine; *ANG II*, angiotensin II; *ANP*, atrial natriuretic peptides; *BOM*, bombesin-like peptides (gastrin-releasing peptide, neuromedin B, neuromedin C); CCK_8, cholecystokinin-8; *CT*, calcitonin; *EOP*, endogenous opioid peptides (endorphin, enkephalin, dynorphin, nociceptin/orphanin); *GABA*, γ-aminobutyric acid; *GAL*, galanin; *HA*, histamine; *NE*, norepinephrine; *NO*, nitric oxide; *NPY*, neuropeptide Y; *NT*, neurotensin; *OT*, oxytocin; *PACAP*, pituitary adenylate cyclase–activating peptide; *SST*, somatostatin; *TRH*, thyrotropin-releasing hormone; *VIP*, vasoactive intestinal polypeptide; *VP*, vasopressin; *5-HT*, serotonin. Reproduced from M.E. Freeman et al., *Physiol. Revs.*, 80: 1523–1631, 2000 and directly from http://physrev.physiology.org/cgi/content–nw/full/80/4/1523/F5.

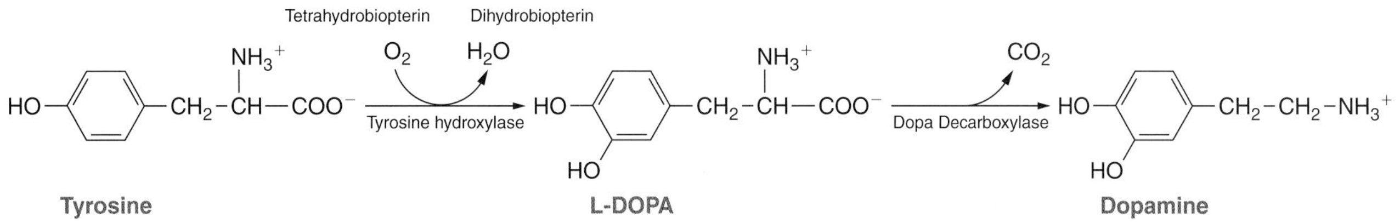

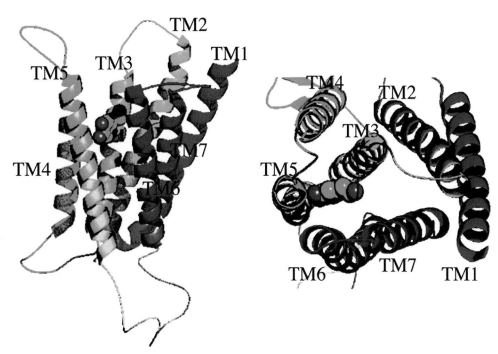

Figure 8-73. Synthesis of dopamine from tyrosine.

Figure 8-74. Predicted binding site of dopamine (shown in spheres) in the predicted structure of human dopamine D2 receptor. *TM*, transmembrane helix. Reproduced from M. Yashar et al., "The predicted 3D structure of the human D2 dopamine receptor and the binding site and binding affinities for agonists and antagonists," *PNAS USA*, 101: 3815–3820, 2004.

Gastrointestinal Hormones

The major gastrointestinal hormones are listed in Table 8-4 with the activity of each hormone and the stimulus for its release. Many gastrointestinal hormones are synthesized in the brain ("brain–gut" hormones). Gastrin stimulates gastric acid secretion and proliferation of the gastric epithelium. Gastrin is released from the **G cells** in the antrum of the stomach in response to the presence of peptides, amino acids, and other foodstuffs in the gastric lumen. Should the pH of the gastric lumen drop below 3, the secretion of gastrin would be inhibited. In response to gastrin, the **enterochromaffin-like (ECL) cells** (endocrine cells in the

Table 8-4
Major Gastrointestinal Hormones, Their Functions, and Stimuli Causing Their Release

Hormone	Major Activities	Stimuli for Release
Gastrin	Stimulates gastric acid secretion and proliferation of gastric epithelium	Presence of peptides and amino acids in gastric lumen
Cholecystokinin	Stimulates secretion of pancreatic enzymes and contraction and emptying of the gallbladder	Presence of fatty acids and amino acids in the small intestine
Secretin	Stimulates secretion of water and bicarbonate from the pancreas and bile ducts	Acidic pH in the lumen of the small intestine
Ghrelin	Appears to be a strong stimulant for appetite and feeding; also a potent stimulator of growth hormone secretion (also releases growth hormone; see Fig. 8-23)	Not clear, but secretion peaks before feeding and diminishes with gastric filling
Motilin	Apparently involved in stimulating housekeeping patterns of motility in the stomach and small intestine	Not clear, but secretion is associated with fasting
Gastric inhibitory Polypeptide	Inhibits gastric secretion and motility and potentiates release of insulin from beta cells in response to elevated blood glucose	Presence of fat and glucose in the small intestine

Reproduced in large part from "Pathophysiology of the Endocrine System," Colorado State University, rbowen@lamar.colostate.edu.

acid-producing part of the stomach) secrete histamine, which stimulates the **parietal cells** to secrete acid. Gastrin binds to its receptor (CCK2 receptor) to stimulate calcium ion entry through Ca^{2+} channels (L- and N-types). Adenylate cyclase–activating peptide stimulates Ca^{2+} uptake through an L-type channel. Factors such as somatostatin, misoprostol (prostaglandin E1), and galanin bind to receptors that block L-type calcium channels through inhibitory G proteins. Gastrin is a growth factor for ECL cells and for the acid-producing part of the stomach (stem cells).

Gastrin induces histidine decarboxylase, the enzyme that forms histamine (Figure 8-75). Histamine binds to the histamine H_2 receptor (also in vascular smooth muscle, suppressor T cells, neutrophils, heart, and central nervous system) in the gastric parietal cells to stimulate gastric acid secretion. There are two other histamine receptors, histamine H_1 and H_3 (Table 8-5). Gastrin is a small protein of 2216 Daltons containing 17 amino acids, as shown in Figure 8-76. Gastrin, as shown, is a linear peptide formed from a prepropeptide and cleaved to a family of peptides, all

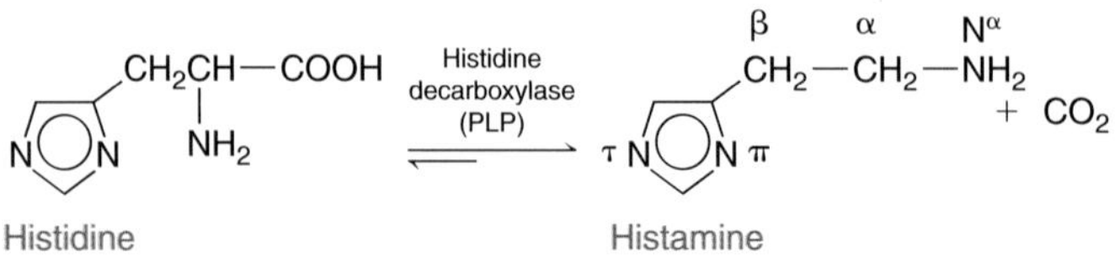

Figure 8-75. L-Histidine decarboxylase reaction forming histamine from L-histidine.

Table 8-5
Operational Characteristics of Histamine Receptors

Receptor	Location	Response	Agonists	Antagonists
Histamine H_1	Most smooth muscle, endothelial cells, adrenal medulla, heart, CNS	Smooth muscle contraction, stimulation of NO formation, endothelial cell contraction, increased vascular permeability, stimulation of hormone release, negative inotropism, depolarization (block of leak potassium current) and increased neuronal firing, inositol phospholipid hydrolysis and calcium mobilization, hyperpolarization by Ca^{2+}- dependent potassium current	Histamine[a] 2-[3-(trifluoromethyl)-phenyl]histamine 2-Thiazolylethylamine 2-Pyridylethylamine 2-Methylhistamine	Mepyramine (+) and (−) chlorpheniramine Triprolidine Temelastine Diphenhydramine Promethazine
Histamine H_2	Gastric parietal cells, vascular smooth muscle, suppressor T cells, neutrophils, CNS, heart, uterus (rat)	Stimulation of gastric acid secretion, smooth muscle relaxation, stimulation of adenylate cyclase, positive chronotropic and inotropic effects on cardiac muscle, decreased firing rate, hyperpolarization or facilitation of signal transduction in CNS, block of Ca^{2+}-dependent potassium conductance (I AHP, accommodation of firing, AHP), increase of hyoperpolarization-activated current, inhibition of lymphocyte function	Histamine[a] Amthamine Dimaprit Impromidine[b] Arpromidine[b]	Cimetidine Ranitidine Tiotidine Zolantidine Famotidine
Histamine H_3	CNS, peripheral nerves (heart, lung, gastrointestinal tract), endothelium, enterochromaffin cells	Inhibition of neurotransmitter release, endothelium-dependent relaxation of rabbit middle cerebral artery, inhibition of gastric acid secretion (dog), increase in smooth muscle voltage–dependent Ca^{2+} current, inhibition of firing of tuberomammilary (histaminergic) neurons	Histamine[a] R-α-methylhistamine Imetit immepipN$^\alpha$-methylhistamine[a]	Thioperamide Clobenpropit Iodophenpropit Iodoproxyfan

AHP, after hyperpolarization; CNS, central nervous system; NO, nitric oxide.
[a]Nonselective.
[b]H_3- antagonist.
Reproduced from Endocrine Index, rbowen@colostate.edu.

```
1                    10                      17
P—G—P—W—L—E—E—E—E—E—A—Y—G—W—M—D—F
```

Figure 8-76. Sequence of gastrin. One letter abbreviations for amino acids are used.

of which have the identical carboxy terminus (Figure 8-77). "Big gastrin," a peptide of 34 amino acids, is the major circulating form. The form with the highest biological activity is "mini gastrin," made up of 14 amino acids. Interestingly, the five C-terminal amino acids of gastrin and **cholecystokinin (CCK)** are identical, explaining overlapping biological effects. Gastrin binds to the gastrin/CCK_B receptor, which also binds CCK. CCK_B is a G protein–coupled receptor whose action results in the activation of protein kinase C, PIP_2 hydrolysis, and increased cytosolic content of Ca^{2+}. A fragment of the extracellular domain of the CCK_B receptor is shown in Figure 8-78.

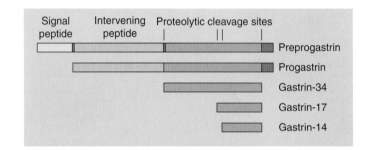

Figure 8-77. Preprogastrin and cleavage products. Redrawn from Endocrine Index, rbowen@colostate.edu.

Figure 8-78. Solution NMR structure of the CCK2E3. PDB ID: 1L4T. C. Giragossian, D.F. Mierke. Intermolecular Interactions between Cholecystokinin-8 and the Third Extracellular Loop of the Cholecystokinin-2 Receptor. *Biochemistry* **41** pp. 4560 (2002).

When a meal is consumed, the response to eating can be divided into four phases: cephalic (or head), gastric, intestinal, and interdigestive. The interdigestive phase is concerned with the events ongoing between meals. The cephalic phase refers to the senses (smell, taste, swallowing, etc.). The parasympathetic nerves activate cholinergic reflexes that cause salivation. Vagus nerve reflexes relax the proximal stomach and stimulate acid secretion by the parietal cells of the stomach. Secretion of acid, as well as histamine, by the ECL cells occurs. Histamine also increases acid (HCl) secretion from the parietal cell. Vagal reflexes to the antrum stimulate secretion of gastrin from G cells. Gastrin then further increases acid secretion from parietal cells and stimulates histamine secretion from ECL cells. These events are capitulated in Figure 8-79, and the cellular composition of the gastric mucosa is shown in Figure 8-80.

The cellular events leading to the secretion of HCl by the parietal cell are shown in Figure 8-81. The gastric phase is marked by distension of the stomach, which alters a number of activities. As acid is secreted, the pH of the lumen lowers, and somatostatin will be released from the D cells of the stomach. Somatostatin prevents excessive acid production by inhibiting gastrin secretion from G cells and histamine secretion from ECL cells. Cholinergic reflexes stimulate release of pepsinogen from **chief cells** (Figure 8-80). Some proteins are digested in the stomach by the HCl and pepsin. The acid cleaves an inhibitory peptide in pepsinogen, leading to active pepsin. **Chyme** (partially digested food) enters the small intestine, marking the intestinal phase. This stimulates the exocrine (discharging a secretion by means of a duct) pancreas to secrete digestive enzymes. Entrance of the chyme into the small intestine stimulates duodenal mucosal S cells to secrete the hormone, secretin. Secretin

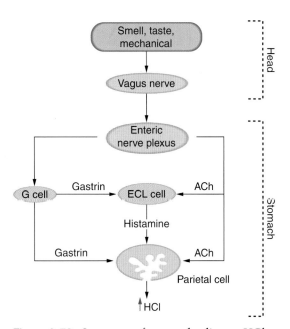

Figure 8-79. Sequence of events leading to HCl secretion in the stomach.

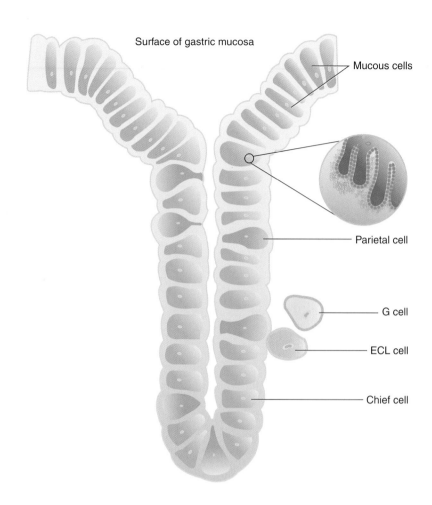

Figure 8-80. Cellular composition of the gastric gland.

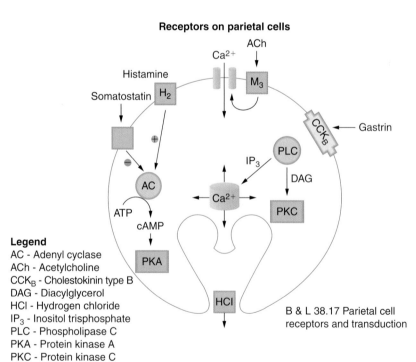

Figure 8-81. Cellular events in the parietal cell leading to the secretion of HCl. *ACh,* acetylcholine. Redrawn from http://www.bris.ac.uk/Depts/Physiology/ugteach/ugindex/s1_index/gi_lect/lectures/wS1gi4/sld007.htm.

440   CHAPTER 8   Polypeptide Hormones

stimulates pancreatic ductal cells to secrete water and bicarbonate and I cells in the duodenum and upper jejunum to secrete CCK. CCK (and gastrin) stimulates enzyme secretion from pancreatic acinar cells. Bicarbonate and water are secreted from the epithelial cells in the pancreatic ducts. The mechanism for bicarbonate secretion depends on carbonic anhydrase and is similar to the secretion of acid from the parietal cells of the stomach. Carbonic anhydrase catalyzes the following reaction:

$$CO_2 + H_2O \rightleftharpoons HCO_3^- + H^+$$

The enzyme has a molecular weight of 29,000 and is complexed to an atom of zinc (Figure 8-82).

The enzymes derived from the pancreas into the duodenum are proteases (trypsin and chymotrypsin), lipase, amylase, and other enzymes, such as ribonuclease, deoxyribonuclease, gelatinase, and elastase. Trypsin and chymotrypsin appear as the precursors trypsinogen and chymotrypsinogen. Trypsinogen is activated by enterokinase, an enzyme in the intestinal mucosa. Enterokinase recognizes the sequence, D-D-D-K-X, and cleaves the bond between lys (K) and Ile (in this case, X, at positions 6–7) in trypsinogen. Active trypsin then activates chymotrypsinogen to form chymotrypsin. These two enzymes digest proteins into small

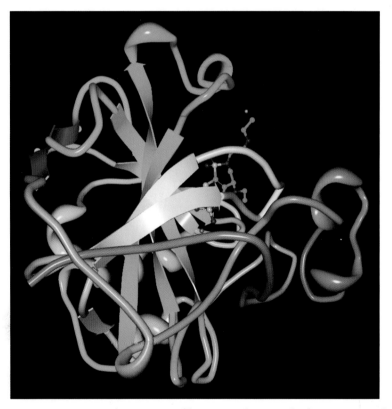

Figure 8-82. Crystal structure of human carbonic anhydrase II. Complexed with an inhibitor, brinzolamide. PDB ID: 1A42. T. Stams, Y. Chen, P.A. Boriack-Sjodin, J.D. Hurt, J. Liao, J.A. May, T. Dean, P. Laipis, D.N. Silverman, D.W. Christianson. Structures of Murine Carbonic Anhydrase IV and Human Carbonic Anhydrase II Complexed with Brinzolamide: Molecular Basis of Isozyme-drug Discrimination. *Protein Sci.* **7** pp. 556 (1998).

peptides. Other pancreatic proteases, like carboxypeptidase (from procarboxypeptidase activated by trypsin) and peptidases in the intestinal epithelial cells, digest the small peptides into amino acids, which are absorbed through transporters. These transporters move amino acids and sodium ion into the cell cytoplasm of the cells of the intestinal lumen. There are four types of **sodium-dependent amino acid transporters (symports)** and one type each for neutral, acidic, basic, and branched amino acids. Di- and tripeptides are also absorbed independently of sodium. Inside the enterocyte, di- and tripeptides are digested to amino acids by cytoplasmic peptidases. Absorption of amino acids at the apical side (closest to the lumen) of the enterocyte generates an osmotic gradient calling for absorption of water (similar to Na$^+$ uptake). On the basolateral side (closest to the interior) of the enterocyte amino acids are exported to the extracellular space (and then to the bloodstream) through transporters that are sodium independent. Structures of a branched chain amino acid transporter and a glutamate transporter are shown in Figures 8-83 and 8-84. Other pancreatic enzymes are amylase and lipase. Amylase

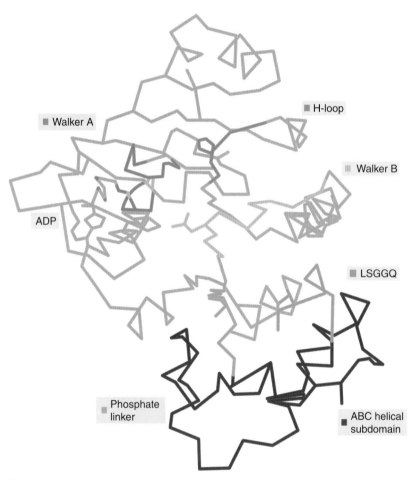

Figure 8-83. Crystal structure of a branched chain amino acid transporter from *Methanococcus jannaschii*. ADP is found at the nucleotide site. Colors show conserved motifs and key amino acids. ADP binding causes Walker B segment to move toward Walker A segment and other conformational changes to transport substrates. Redrawn from http://web.uct.ac.za/depts/chempath/images/mjstruct.gif and from http://www.uct.ac.za/depts/chempath/pumps.htm.

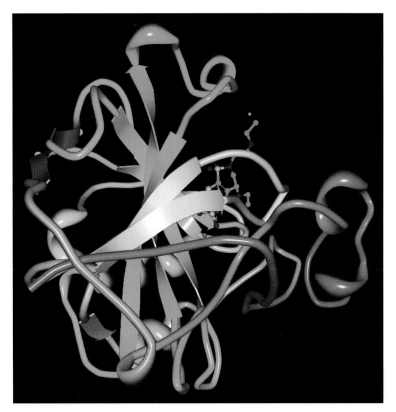

Figure 8-84. Structure of a glutamate transporter from *Pyrococcus horikoshii*. PDB ID: 1XFH. D. Yernool, O. Boudker, Y. Jin, E. Gouaux. Structure of a Glutamate Transporter Homologue from *Pyrococcus horikoshii*. *Nature* **431** pp. 811 (2004).

digests starch (Chapter 4), and lipase digests triglycerides (Chapter 5). Pharmaceutical attempts are ongoing to inhibit intestinal lipase to reduce the absorption of fats and reduce body weight.

Other gastrointestinal hormones are ghrelin, **motilin,** and **gastric inhibitory peptide (GIP).** Ghrelin and its action were covered earlier with respect to growth hormone secretion (Figure 8-23). Ghrelin secreted by the stomach induces fasted motor activity in the stomach and duodenum. It acts on ghrelin receptors on the vagal afferent nerve terminals and then activates neuropeptide Y (NPY) neurons in the brain; however, NPY is not involved in the ability of ghrelin to induce fasted motor activity. Motilin is a peptide consisting of 22 amino acids and has no sequence homology to other hormones. The sequence of human motilin is F-V-P-I-F-T-Y-G-E-L-Q-R-M-Q-E-K-E-R-N-K-G-Q. The structure of motilin is a single helix (Figure 8-85). Motilin is secreted from the endocrine cells of the duodenal epithelium into the circulation in the fasted state (between meals) at intervals of 100 minutes, causing contractions that clear the stomach and small intestine of undigested foodstuffs. GIP (previously called enterogastrone) is a peptide containing 42 amino acids with the sequence Y-A-E-G-T-F-I-S-D-Y-S-I-A-M-D-K-I-H-Q-Q-D-F-V-N-W-L-L-A-Q-K-G-K-K-N-D-W-K-H-N-I-T-Q. It is derived from a preproGIP of 144

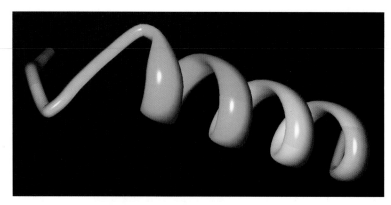

Figure 8-85. Structure of human motilin. Reproduced from PDB ID: 1lbj. A. Andersson, L. Maler. NMR Solution Structure of Motilin in Phospholipid Bicellar Solution. *J. Biol. Mol. NMR* **24** pp. 103 (2002).

amino acids containing a signal peptide. The bond cleaved in the prepro-GIP to release the 42–amino acid GIP is arg–arg. It is a member of the secretin family, secreted from the epithelial cells of the intestinal mucosa, and inhibits gastric motility and secretion of gastric acid. In addition to its direct action in the gut, GIP enhances the release of insulin from the pancreas in response to elevations of circulating glucose (also called glucose-dependent insulinotropic peptide).

There are many other topics related to peptide hormones, and these will be covered in other chapters. Included in this group are metabolic hormones, like insulin and glucagon, and neuropeptide hormones.

Further Reading

Norman, A.W., and Litwack, G., *Hormones*, 2nd edition, Academic Press, 1997.

CHAPTER 9

Steroid Hormones

Stress: A State That Can Have Serious Pathological Consequences

Stress can arise from internal or external environmental events or a combination of the two. There can be healthy stress and damaging stress. The latter will be addressed primarily. The human has relatively narrow conditions for survival: limits of temperature, clean air, uncontaminated food, positive interactions with other humans, and so on. Ability to adapt to large changes in these and wother parameters is limited. Approaching the limits of temperature toleration, for example, creates a stress and sets into motion bodily reactions. Although all bodily responses are not completely understood, the mechanisms involving the adrenal cortex and the nervous system are well known. In general, these mechanisms allow for adaptation unless the stress is too extreme, in which case there can be serious deleterious consequences.

There are both acute and chronic stresses. Acute stresses happen over a short period and invoke the "flight or fight" response. Walking in the woods and suddenly seeing a huge bear appear 30 yards directly ahead would constitute an acute stress. All systems would start pumping, and the net result would be that the person would turn and run as quickly as humanly possible. When the danger passed, all systems would normalize (relaxation response). Epinephrine and norepinephrine, as well as cortisol (the steroidal glucocorticoid hormone), would have been poured into the bloodstream to allow the bodily changes to martial the intense reaction needed to reach safety. Other acute stresses could be a loud noise, crowding, danger (as just described), infection, and even remembering a dangerous situation.

Chronic stress calls for the suppression of the "flight or fight" response and lasts over a long period. Long-term relationship problems, loneliness, financial worries, long recoveries from surgery, and a highly pressured work environment qualify, among others. These long-term stresses can lead to high blood pressure, heart disease, suppression of the immune system, and even diabetes (stress-induced diabetes) and other pathological conditions. Stress over long periods causes the blood level of cortisol to rise. In addition to the outcomes previously mentioned, depression can set in, caused by the negative feedback of cortisol on the **limbic system** (mainly hippocampus and amygdala). The continued negative feedback of elevated cortisol causes the hippocampus to shrink. Because this organ is the center for memory, this can account for memory losses. Also, in short-term stress there can be brief periods of short-term memory loss, which are usually reversed during recovery. Because serotonin accounts for the endogenous stimulation of the hypothalamic–pituitary–adrenal cortex axis (promoting a normal background of cortisol secretion that is not stress related) and serotonin plays a role in generating feelings of well-being, the negative effect of elevated cortisol on suppressing serotonin release may contribute to depression. This can be a problem in aged people, for whom this form of depression is not uncommon. Long-term stress causes a depression of the immune system, opening the subject to more infection.

In contrast to all of these negative effects of stress, there is an appropriate positive stress that motivates a person to higher levels of achievement and provides excitement in good circumstances. A person experiencing positive stress will have an effective relaxation response.

A 17–amino acid neuropeptide called **orphanin** (FQ or nociceptin) is a newer anxiolytic substance that modulates anxiety generated by acute stress in experimental animals. It could be a factor in the recovery stage from stress. Orphanin does not bind to the classical **opioid** receptors (κ, μ, and δ receptors) but binds to a nociceptin receptor (Figure 9-1) in neurons, where it reduces the activation of adenylate cyclase (that forms cyclic AMP) and reduces the activity of calcium channels. However, it opens potassium channels, similarly, to the action of opioids. Further research on orphanin could lead to new drugs (agonists) for treating stress disorders.

An overview of the effects of stress in promoting the release of cortisol is shown in Figure 9-2. A stress event is filtered in the brain and an electrochemical signal goes to the hypothalamus, which sends impulses through the spinal cord to preganglionic fibers innervating the adrenal medulla. This results in the secretion of epinephrine and norepinephrine into the general circulation. Elevated levels of these catecholamines in the blood

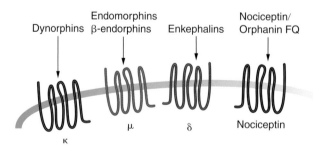

Figure 9-1. Orphanin (FQ/nociceptin) binds to a membrane receptor in neurons that is different from the classical opioid receptors.

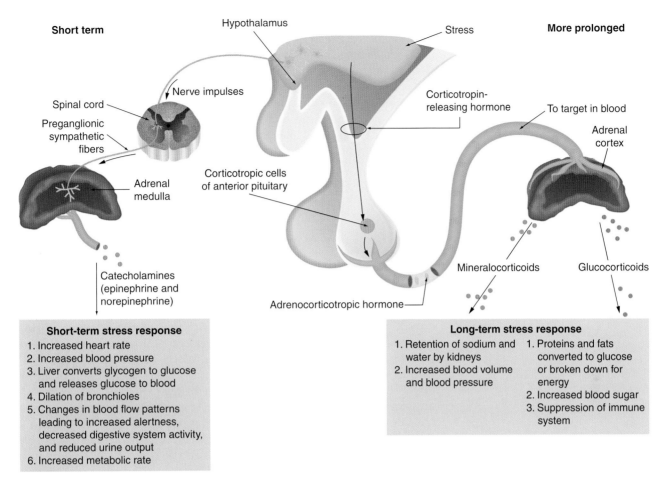

Figure 9-2. Responses to stress. Sympathetic nervous system responses *(left)* and hormonal responses *(right)*.

cause increases in blood pressure, heart rate, and metabolic rate, in addition to stimulating the conversion of glycogen to glucose in the liver and dilating the bronchioles. This is a rapid response. The hypothalamus also releases corticotropin-releasing hormone (CRH) into the closed portal system connecting the hypothalamus and the anterior pituitary, resulting in the release of **adrenocorticotropic hormone (ACTH)** from the corticotropic cells. ACTH in the general circulation finds its receptor in the cell membrane of the cells in the adrenal cortex, resulting in the release of cortisol and **aldosterone** (the main mineralocorticoid). Aldosterone is also released by acetylcholine (from stress) and by angiotensin II.

Cortisol in the bloodstream circulates to the tissues, where proteins and fats are broken down by lipolysis and converted to glucose; blood sugar is increased (the peripheral uptake of glucose from the blood is inhibited) and, in cases of prolonged stress, there is suppression of the immune system, breakdown of protein (muscle), and a decrease of protein synthesis. In an acute stress event, cortisol can cause blood glucose to increase by about 10%, enough to evoke insulin secretion. This is not a harmful event by itself, but if this process is continued indefinitely, the small increases in blood glucose eventually could exhaust the beta cells of the pancreas, leading to stress-related diabetes.

Aldosterone, a true stress hormone, causes the retention of sodium by the kidneys, which is followed by increased blood volume (reabsorption of water) and pressure. Although cortisol is secreted in stress, it is also secreted endogenously under the control of serotonin (circadian rhythm), whereas aldosterone is secreted primarily as a result of stress. This accounts for the much higher circulating level of cortisol (up to 25 μg/dl of plasma) than of aldosterone (up to about 30 ng/dl), about a thousandfold difference. The level of aldosterone can be increased substantially with higher-than-normal salt intake.

Adrenal Medulla

The adrenal gland is pictured in Figure 9-3. The adrenal cortex (derived from mesoderm) occupies 80 to 90% of the gland, and the medulla (derived from neuroectoderm) occupies 10 to 20%. The medulla, through the autonomic nervous system, responds rapidly to stress; the cortex produces longer-lasting effects. The medulla contains **chromaffin cells** that are specialized postganglionic neurons (these cells do not have axons). The chromaffin cell is directly stimulated by acetylcholine in the stress response, culminating in the release of epinephrine (85%) and norepinephrine (15%) from storage vesicles. These catecholamines sometimes are called adrenaline and noradrenaline. The acetylcholine receptor is located in the membrane of the chromaffin cell (Figure 9-4). This receptor opens a calcium channel, causing inflow of calcium ions into the cell that results in the release of catecholamines (and **enkephalin,** an opioid peptide) from chromaffin cell secretory vesicles into the bloodstream.

Catecholamines in the chromaffin cell are synthesized from tyrosine in four steps. The first step is the rate-limiting **tyrosine hydroxylase** step, converting tyrosine to L-dihydroxyphenylalanine (L-DOPA). This is acted on by **amino acid-α decarboxylase (AAAD)** to form dopamine. Then dopamine-β hydroxylase (DBH) catalyzes the conversion of dopamine to norepinephrine. Finally, norepinephrine is converted to epinephrine by **phenylethanolamine N-methyl transferase (PMNT),** as shown in Figure 9-5. *PMNT is induced by cortisol from the adrenal cortex during the stress response.* When cortisol is low in concentration, the medulla synthesizes more norepinephrine than epinephrine. These catecholamines (much of the norepinephrine diffuses back into the cytoplasm for further synthesis of epinephrine) are stored in secretory vesicles, awaiting a stress signal (through acetylcholine) for release into the bloodstream. In the bloodstream, approximately half of the epinephrine and half of the norepinephrine are bound to albumin. Epinephrine, entirely from the adrenal medulla, reaches blood levels of 20 to 50 ng/ml and norepinephrine reaches levels of 100 to 350 ng/ml (a small fraction of norepinephrine is derived from the adrenal medulla, but most is derived from nerve endings of the sympathetic nervous system). The half-life of the catecholamines is 10 to 15 seconds. The catecholamines act on adrenergic receptors, of which there are four types, in target tissues (heart, liver, skeletal muscle, adipocytes, vascular and bronchial smooth muscle, and others). The receptor types are α_1, α_2, β_1, and β_2, and they interact either with the phosphatidyl inositol system or the formation and inhibition of cyclic AMP (Table 9-1).

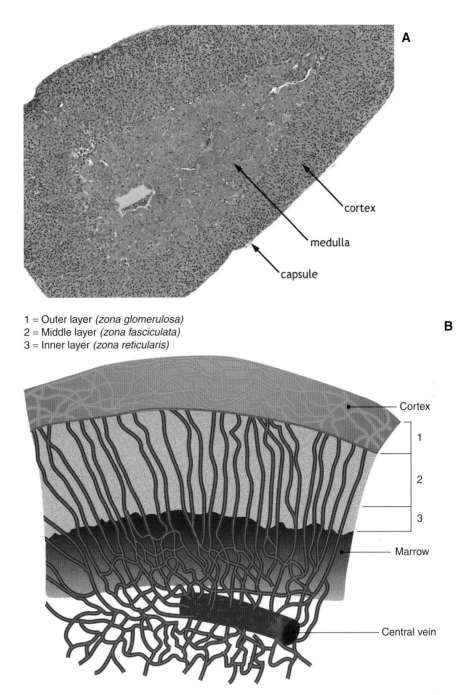

Figure 9-3. The adrenal gland. **A.** Cross-section of the adrenal showing the three layers of the cortex and the adrenal medulla in the center, made clear in the diagram. **B.** Blood supply of the adrenal cortex. Note that blood flows through the cortex and then bathes the medulla before entering the general circulation. In this way, the medulla is exposed to increased secretion of cortisol.

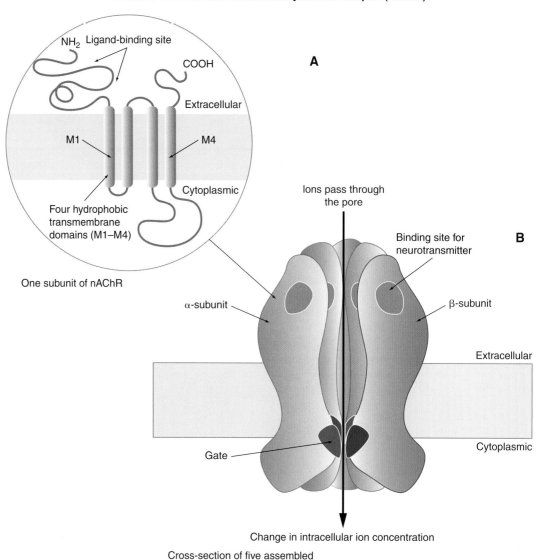

Figure 9-4. **A.** Diagram of a single subunit of the acetylcholine receptor. **B.** Cross-section of five assembled subunits of the acetylcholine receptor. Binding site for neurotransmitter is site of acetylcholine binding. Ca^{2+} ions pass through the center *(opened)* channel. Redrawn with permission from Expert Reviews in Molecular Medicine, Cambridge University Press, 1999.

Thus, the adrenal medulla and the cortex act together to help the body respond to stress. The catecholamines generate glucose in the blood, increase blood flow to the skeletal muscles, and break down triglycerides to free fatty acids, all to power the "fight or flight" response. Cortisol also increases fat breakdown to free fatty acids, shuts down the uptake of blood glucose by peripheral tissues, increases the conversion of amino acids to glucose and then to glycogen, and generally protects from stress. Cortisol is a key hormone, without which survival in the face of stress or adaptational requirements would not occur. These activities are summarized in Figure 9-6.

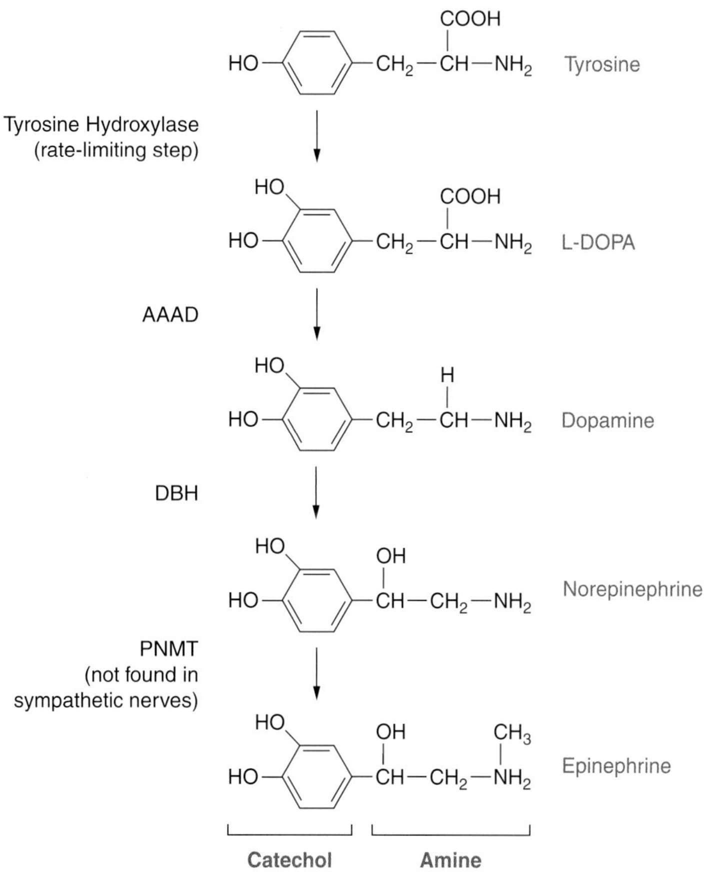

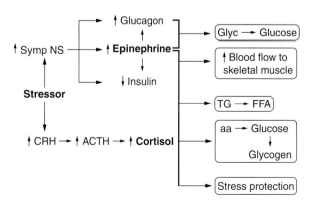

Figure 9-5. Conversion of L-tyrosine to norepinephrine and epinephrine in chromaffin cells of the adrenal medulla. *AAAD*, aromatic amino acid decarboxylase; *DBH*, dopamine β-hydroxylase; *PNMT*, phenylethanolamine N-methyltransferase.

Table 9-1
Adrenergic Receptor Subtypes

Receptor Type	Action	Relative Affinity	Response
α_1	↑PI(IP$_3$)Gp	E > NE	Vasoconstriction, uterine contraction, pupil dilation
α_2	↓cAMP Gi	NE > E	Presynaptic ACh-mediated NE release, smooth muscle contraction
β_1	↑cAMP Gs	E = NE	Cardiac output stimulation, lipolysis
β_2	↑cAMP Gs	E >> NE ca 10X	Bronchodilation, vasodilation, uterine relaxation, glycogenolysis, smooth muscle relaxation

ACh, acetylcholine; *cAMP*, cyclic adenosine monophosphate; *E*, epinephrine; *NE*, norepinephrine.

Figure 9-6. Summary of effects of catecholamines and cortisol in the stress reaction. *aa*, amino acid; *FFA*, free fatty acid; *Glyc*, glycogen; *Symp NS*, sympathetic nervous system; *TG*, triglyceride.

Stress: A State That Can Have Serious Pathological Consequences

Adrenal Cortex

The adrenal cortex consists of three layers of cells. These are called the *zona glomerulosa* (starting from the outside of the gland just below the capsule), and this layer of cells secretes **mineralocorticoids** (named for salt-retaining action), mainly aldosterone, in response to stress signals. The middle layer is the *zona fasciculata*, which secretes most of the cortisol from the gland, and the inner layer is the *zona reticularis*, which secretes primarily weak androgens, in particular, dehydroepiandrosterone (DHEA, and some androstenedione); most DHEA circulates in the form of the sulfate derivative (DHEA-S). These layers of cells are depicted in Figure 9-3. The reactions leading to the individual steroid hormones are shown in Figure 9-7. The mineralocorticoid pathway shows the series of steps leading to the synthesis of aldosterone that takes place in the *zona glomerulosa*. The glucocorticoid pathway depicts the events occurring primarily in the *zona fasciculata* cells, leading to the formation of cortisol. The androgen pathway depicts the events leading to the synthesis of DHEA (the major product of the *zona reticularis*). The fetal adrenal gland consists mainly of *zona reticulosa* type cells. The other layers develop near birth. Other more potent androgens shown in Figure 9-7 are chiefly the products of the testes. *Importantly, all of the steroid hormones are synthesized with Δ5-pregnenolone as the required intermediate.*

Aldosterone

Like cortisol, aldosterone levels fluctuate during the day, with the highest concentration at 8 a.m. and the lowest concentration about 11 p.m. These fluctuations align with cortisol (because of the levels of ACTH); however, the circulating levels of aldosterone are extremely low compared to cortisol. *Like cortisol, ACTH will stimulate the synthesis of the aldosterone but, unlike cortisol, the stimulation is short term.* Sodium depletion, in the absence of ACTH, activates the renin–angiotensin system, which stimulates aldosterone synthesis. The renin–angiotensin system is depicted in Figure 9-8. Hemorrhage, for example, would greatly reduce the extracellular volume and the sodium ion content of blood, which would stimulate the juxtaglomerular cells of the kidney to release the proteolytic enzyme, renin (Figure 9-9), into the bloodstream. This enzyme cleaves angiotensinogen (N-DRVYIHPFH*LL*VS-*continued peptide;* which is an α_2-globulin, a proprotein synthesized in the liver and secreted) at the -*L*-*L*- bond to form angiotensin I (DRVYIHPF*HL;* not a hormone). Next, the **converting enzyme** (located in vascular epithelium, lung, liver, adrenal cortex, pancreas, kidney, spleen, and neurohypophysis) cleaves the C-terminal dipeptide, *H-L,* to form the octapeptide hormone, angiotensin II (*D*RVYIHPF), which has a short half-life of about 1 minute. Angiotensin III (RVYIHPF) also can be formed by aminopeptidase, which cleaves the N-terminal aspartate (D) of angiotensin II. Angiotensin III is a second hormone that binds to the angiotensin II receptor (AT$_1$). Angiotensin II or angiotensin III binds to the angiotensin receptor in the membrane of the *zona glomerulosa* cell to stimulate the synthesis and secretion of aldosterone. Angiotensin II activates the receptor to stimulate the level of DAG, which activates protein kinase C. Protein kinase A is also activated, and these enzymes phosphorylate proteins,

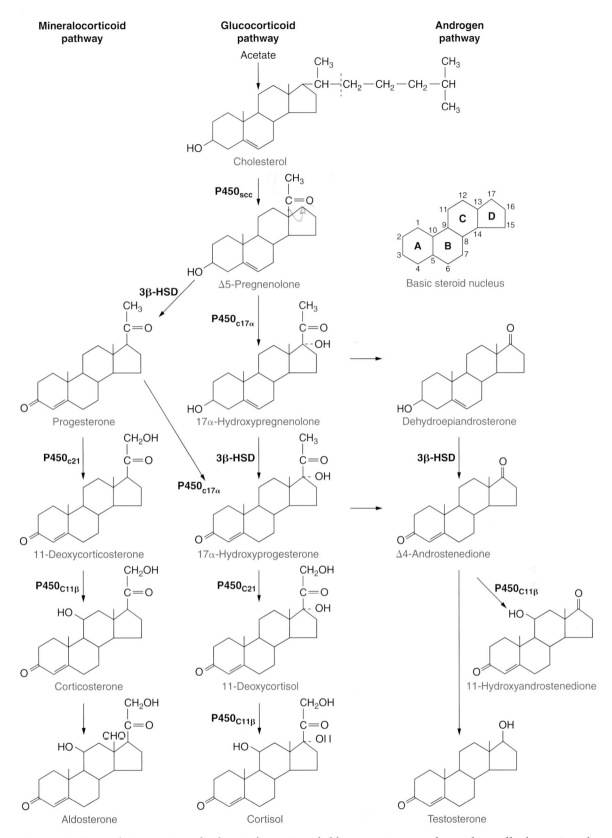

Figure 9-7. Biosynthetic reactions leading to formation of aldosterone in *zona glomeralosa* cells: formation of cortisol, primarily in *zona fasciculata* cells and formation of dehydroepiandrosterone, the major product along with its sulfate, in *zona reticularis* cells. The basic steroid nucleus and the numbering system are shown in the *upper right*. *3β-HSD*, 3beta-hydroxysteroid dehydrogenase; $P450_{C17\alpha}$, enzyme hydroxylating the 17α position; $P450_{C11\beta}$, enzyme hydroxylating the C11 position; $P450_{C21}$, microsomal hydroxylase hydroxylating the C21 position; $P450_{scc}$, side chain cleavage enzyme system.

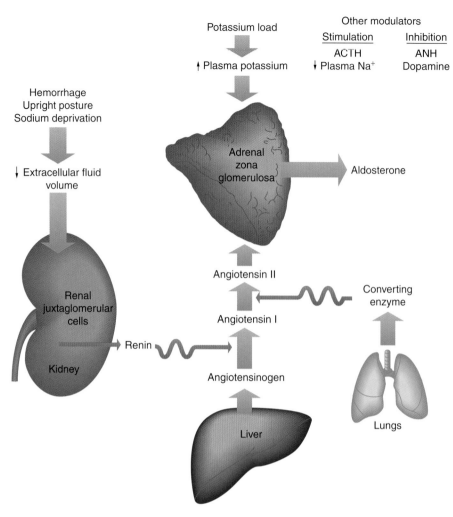

Regulation of aldosterone secretion: Activation of renin–angiotensin system in response to hypovolemia is predominant stimulus for aldosterone synthesis.

Figure 9-8. Renin-angiotensin system, which is stimulated by a decrease in blood volume or sodium ion concentration, and which stimulates the synthesis and release of aldosterone from the *zona glomerulosa*. ANH, atrionatriuretic hormone.

including **cholesterol esterase,** which stimulates the synthesis of aldosterone by cleaving cholesterol esters to free cholesterol (Figure 9-7).

When blood volume increases, the heart (atrial myocytes) secretes an **atrionatriuretic peptide** (**ANP,** also called atriopeptin) containing 32 amino acids, which binds to its receptor (Figure 9-10) in the membrane of the *zona glomerulosa* cell. The activation of this receptor, which contains guanylate cyclase activity, stimulates the synthesis of cyclic GMP (Figure 9-11) from GTP. *Cyclic GMP opposes cyclic AMP and inhibits synthesis of aldosterone* (ANP is secreted from the heart atrial myocytes when blood volume is high [hypervolemia]). Angiotensin II binds to its receptor, AT_1, resulting in the stimulation of aldosterone synthesis. There is a second aldosterone receptor called AT_2, although it is less well understood. A diagram of the AT_1 angiotensin II receptor is shown in Figure 9-12. Stress produces ACTH and acetylcholine, for which there are

454   CHAPTER 9   Steroid Hormones

Figure 9-9. Human renin complexed with a substituted piperidine. PDB ID: 1pr8. C. Oefner et al. Renin Inhibition by Substituted Piperidines: A Novel Paradigm for the Inhibition of Monomeric Aspartic Proteinases? *Chem. Biol.* **6** p. 127 (1999). H.M. Berman, J. Westbrook, Z. Feng, G. Gilliland, T.N. Bhat, I.N. Shindyalov and P.E. Bourne. The Protein Data Bank. *Nucleic Acids Research* **28** p. 235 (2000), http://www.pdb.org.

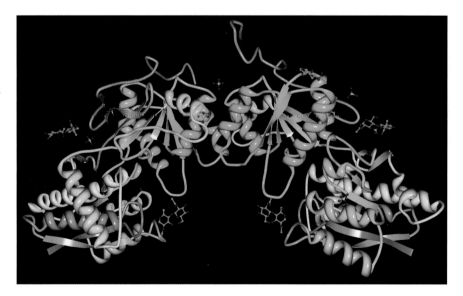

Figure 9-10. Dimierzed hormone-binding domain of the atrionatriuretic receptor. *Red* and *purple* represent the receptor. *Green* atoms are Cl ions. *Black* atoms are 4 NAG-NAG; 1 NAG; 9 Sulfate ions. PDB ID: 1dp4 F. Van Den Akker, et al. Structure of the Dimerized Hormone-Binding Domain of a Guanylyl-Cyclase-Coupled Receptor. *Nature* **406** pp. 101 (2000). H.M. Berman, J. Westbrook, Z. Feng, G. Gilliland, T.N. Bhat, I.N. Shindyalov and P.E. Bourne. The Protein Data Bank. *Nucleic Acids Research* **28** p. 235 (2000), http://www.pdb.org.

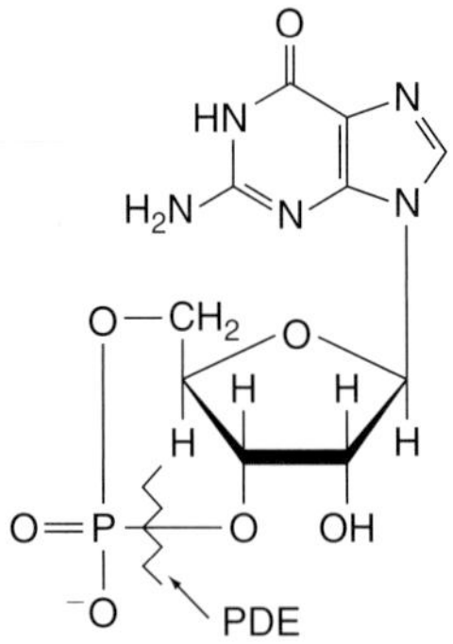

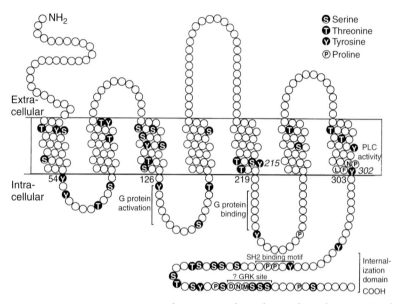

Figure 9-11. Structure of cyclic GMP. Arrow points to bond attacked by phosphodiesterase (PDE).

Figure 9-12. Representation of amino acids and signaling domains in the rat AT_{1a} receptor. The helixes were positioned on the basis of modeling of G protein–coupled receptors. Depicted are Ser, Thr, and Tyr residues, as well as residues thought to be involved in specific signal transduction events. Highlighted residues in transmembrane domains may be important in ligand binding even if they are not phosphorylated. Reproduced with permission from B.C. Berk and M.A. Corson, *Circ. Res.*, 80: 607–616, 1997.

receptors in the *zona glomerulosa* cell. ACTH plays a more transient role than it does on the *zona fasciculata* cell. These activities are summarized in Figure 9-13.

In various tissues other than the *zona glomerulosa* of the adrenal cortex, angiotensin II is being discovered to have complex effects. Angiotensin II has been found to stimulate tyrosine kinases ($pp60^{c-src}$, focal adhesion, and Janus [JAK2, TYK2] kinases). These kinases appear to be involved in vasoconstriction (vascular smooth muscle), protooncogene expression, and protein synthesis.

Aldosterone, released from the *zona glomerulosa* cell, circulates in the bloodstream and reaches its cellular targets, principally, the distal kidney tubular cell. As a result of binding to its receptor in the cytoplasm of this cell (not in the cell membrane), the action of aldosterone is to cause

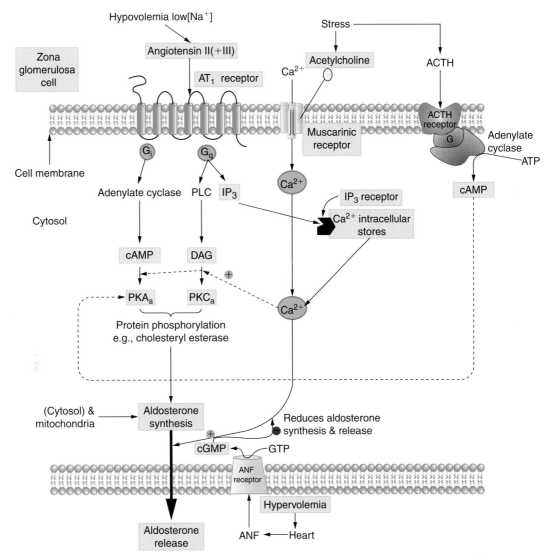

Figure 9-13. Activity of angiotensin II on the *zona glomerulosa* cell. Also shown are activities of acetylcholine, ACTH, and ANF, all of which influence aldosterone synthesis and secretion, as well as *mobilization* of Ca^{2+} that stimulates release of aldosterone.

sodium ion to be reabsorbed from the tubular urine while secreting potassium ion into the urine. Sodium ion is reabsorbed through the action of aldosterone signal transduction, which increases **epithelial sodium ion channels (ENaC)** at the apical side of the cell and Na^+/K^+-ATPase at the basolateral side of the cell. These are the late responses of aldosterone action. The early responses elevate **serum and glucocorticoid-inducible kinase (Sgk)**, the **corticosteroid hormone-induced factor (CHIF)**, and Kirsten *ras* (Ki-*ras*).

Aldosterone mechanism of action on the renal tubular cell is shown in Figure 9-14. The early inducible proteins (Sgk, CHIF, and Ki-*ras*) increase the number of transport proteins so that sodium ion is reabsorbed rapidly from the tubular urine. Water is then passively reabsorbed to maintain sodium ion at a constant concentration, expanding the extracellular volume and increasing blood pressure. When plasma volume is

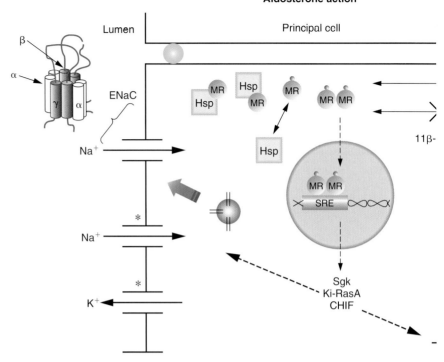

Figure 9-14. Mechanism of aldosterone action on the renal tubular cell. *Dashed red lines* and *arrows* point to Na$^+$ ion channels *(ENaC)* on the apical surface and Na$^+$/K$^+$-ATPase on the basolateral side. *11β-HSD*, 11β-hydroxysteroid dehydrogenase 2; *CHIF*, corticosteroid hormone-inducible factor; *ENaC*, sodium conductance channel; *HSP*, heat shock protein; *Ki-RasA*, Kirsten-RasA; *MR*, mineralocorticoid receptor; *Sgk*, serum-glucocorticoid inducible kinase; *SRE*, steroid receptor enhancer.

decreased (hypovolemia), renin secretion takes place and the renin–angiotensin system is called into play, resulting in increased aldosterone release from the adrenal and increased tubular reabsorption of Na$^+$, which expands plasma volume. Aldosterone induces Na$^+$/K$^+$-ATPase in the tubular cell, which moves potassium ion from the extracellular space (and the blood) into the urine (Figure 9-14). *Potassium is a stimulator for aldosterone synthesis (depolarizes the zona glomerulosa cell); therefore, the removal of K$^+$ is a negative feedback on aldosterone synthesis* (in addition to ANF). The time course of aldosterone action on the renal tubular cell in terms of Na$^+$ transport from the lumen to the extracellular fluid is shown in Figure 9-15. A structure of the ATP-binding fragment of Na$^+$/K$^+$-ATPase is shown in Figure 9-16. A model of the ENaC is shown in Figure 9-17. ENaC is composed of multiples of three types of subunits: α, β, and γ. Aldosterone is reported to act, in addition to gene regulation, directly on the cell membrane to promote ion transport into the cell. This has been reported for certain cell types, including the lymphocyte, and could explain rapid effects of the hormone that take place long before transcriptional and translational events occur, as summarized in Figure 9-18.

The structure of the mineralocorticoid receptor (MR) is similar to the glucocorticoid receptor (GR) except for the N-terminal domain. The receptors of the nuclear receptor class are diagrammed in Figure 9-19 and show

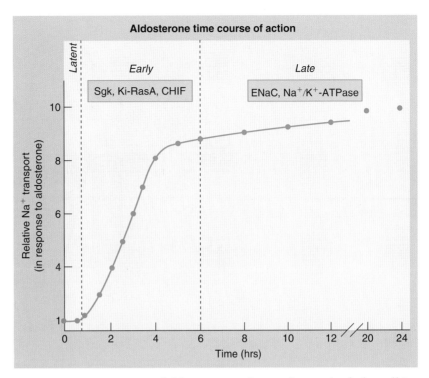

Figure 9-15. Time course of aldosterone action on the renal tubular cell in terms of sodium ion transport into the cell.

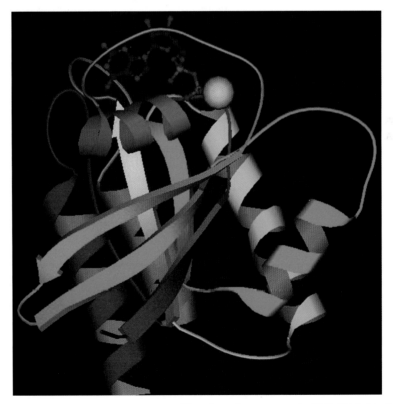

Figure 9-16. Na$^+$K$^+$ ATPase. Ball and stick structure is ATP. PDB ID: 1mo8. M. Hilge et al. ATP-Induced Conformational Changes of the Nucleotide-Binding Domain of Na, K-ATPase. *Nat. Struct. Mol. Biol.* **10** p. 468 (2003).

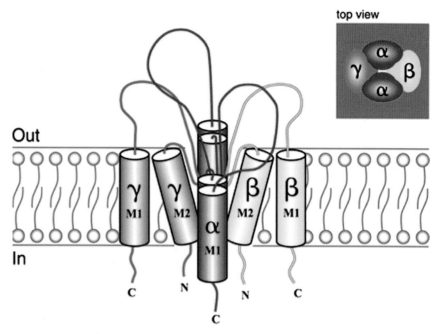

Figure 9-17. Structure of ENaC and membrane topology. ENaC is composed of three partly homologous subunits, α, β, and γ, which are inserted into the membrane with a proposed stoichiometry of 2α:1β: 1γ. Small amiloride-sensitive sodium currents can be produced by the α-subunit alone, but coexpression with β and γ ENaC subunits greatly augments the current to a level similar to that of the native channel. Reproduced with permission from B.C. Rossier et al., *Ann. Rev. Physiol.*, 64: 877–897, 2002 and reproduced directly from K. Gormley, Y. Dong, and G.A. Sagnella, *Biochem. J.*, 371: 1–14, 2003.

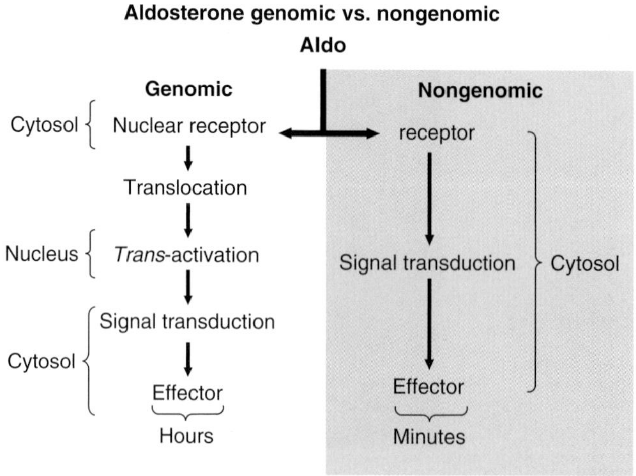

Figure 9-18. Genomic and reported nongenomic rapid effects of aldosterone. The nongenomic activity of aldosterone (ALDO) may involve a membrane receptor.

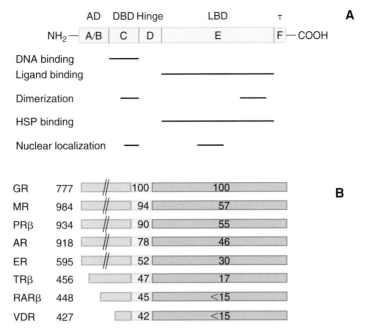

Figure 9-19. **A.** Locations of various activities of steroid hormone receptors. **B.** The genomic organization for most steroid/thyroid hormone receptors have been cloned. For PR, GR, AR, ER, MR, TR, and RAR, the exon-intron organization is quite complex. The genes usually encompass about 60 kb and are interrupted by numerous *introns*. The promoter region resembles that of a housekeeping gene and is often embedded in a GC-rich island. Multiple sites of transcription initiation are the rule. Not shown are a variety of orphan receptors that are in the class of nuclear hormone receptors. Numbers to the left of each diagram refer to the total number of amino acid residues in the protein. The AD region is highly variant between the receptors. The DNA binding domains of the GR and MR are similar and both bind to the same promoters in many cases. The LBDs are overlapping for GR, MR, PRβ, and AR, to a certain extent; all four bind some of the same ligands to different degrees, although a ligand for one receptor (e.g., PRβ) may be an antagonist for another (e.g., GR). Numbers in DBDs and LBDs are % homogeneity to GR. Redrawn with permission from M.J. Tsai and B.W. O'Malley, "Steroid hormone receptors," *Ann. Rev. Biochem.*, 63: 451–486, 1994. *AD*, antigenic domain; *AR*, androgen receptor; *DBD*, DNA binding domain; *ER*, estrogen receptor; *GR*, glucocorticoid receptor; *LBD*, ligand-binding domain; *MR*, mineralocorticoid receptor; *PRβ*, progesterone receptor β; *RARβ*, retinoic acid receptorβ; τ, tau; *TRβ*, thyroid hormone receptorβ; *VDR*, vitamin D receptor.

the percentage homologies among the DNA-binding and the **ligand-binding domains** in the various proteins. The ligand-binding domains of the glucocorticoid and mineralocorticoid receptors are shown in Figures 9-20 and 9-21. The DNA-binding domain consists of two zinc fingers in the steroid hormone receptors, as shown in Figure 9-22. The left-hand zinc finger is closer to the N-terminus, and the right-hand zinc finger is downstream (further from the N-terminus toward the C-terminus). The zinc fingers associate with DNA in the major groove. The proximal box (P-box) determines binding to the specific **hormone-responsive element (HRE),** and the distal box (D-box) is a dimerization site that stabilizes the receptor dimer in its binding to DNA. Figure 9-23A shows the glucocorticoid receptor monomeric DNA-binding domain (73 amino acid residues).

Adrenal Cortex 461

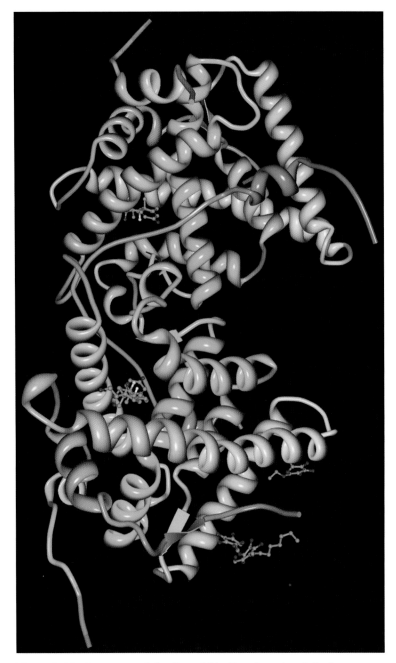

Figure 9-20. Structure of the ligand-binding domain of the human GR with dexamethasone. PDB ID: 1m2z. R.B. Bledsoe et al. Crystal Structure of the Glucocorticoid Receptor Ligand-Binding Domain Reveals a Novel Mode of Receptor Dimerization and Coactivator Recognition. *Cell* **110** pp. 93 (2002).

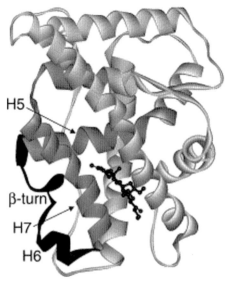

Figure 9-21. Model of the mineralocorticoid receptor ligand-binding domain, based on the crystal structure of the progesterone receptor ligand-binding domain. The region (amino acids 804–874) found to be critical for aldosterone binding from analysis of the first set of chimeras is shown in dark gray. The region (amino acids 820–844) found to be critical for aldosterone binding from analysis of the second series of chimeras is shown in black. The atomic model in black is aldosterone. Reproduced from F.M. Rogerson, F.E. Brennan, and P.J. Fuller, *J. Steroid Biochem. & Mol. Biol.* 85: 389–39b, 2003, and directly from http://www.sciencedirect.com/science?_ob=ARTICLEURL&_udi=B6T8X-4&_user=10&_coverDate=06%2F30%2F2003&_alid=2.

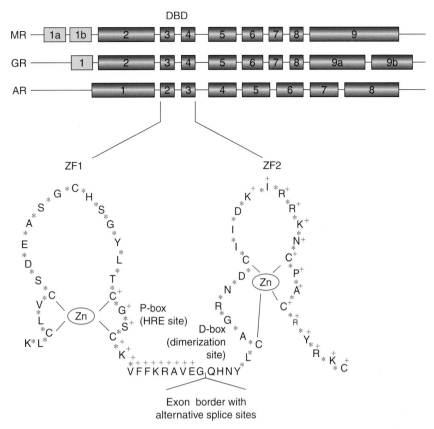

Figure 9-22. Organization and structure of the DNA-binding domain of the MR, GR, and AR genes. *Beige boxes,* untranslated exons. *Blue boxes,* translated exons. *Below,* amino acid sequence is shown in one letter code. Amino acids marked with a plus sign form α helices. *DBD,* DNA-binding domain; *D-box,* distal box; *black letters,* DNA identification sequence; *HRE,* hormone responsive enhancer; *P-box,* proximal box; *ZF,* zinc finger. Redrawn from L. Wickert and J. Selbig, *J. Endocrin.,* 173: 429–436, 2002.

The proximal zinc finger is upper right (blue atoms are zinc), and the distal zinc finger is lower left. Figure 9-23B shows the DNA-binding domain of the glucocorticoid receptor dimer in association with the HRE site. One glucocorticoid receptor monomer (81 amino acid residues) is in purple, and the other is in red (81 amino acid residues). The full-length protein is 777 amino acid residues. The zinc fingers are binding in the major groove of DNA. Note that the distal zinc fingers (top, center) are dimerized.

Cortisol binds to the mineralocorticoid receptor, as well as aldosterone, but cortisol circulates at a concentration about 1000 times that of aldosterone. *Aldosterone target tissues are protected from the excess of cortisol by the enzyme 11β-**hydroxysteroid dehydrogenase 2** (11β-**HSD2**) that catalyzes the conversion of cortisol to cortisone. Cortisone is inactive as a ligand for either glucocorticoid receptor or mineralocorticoid receptor.* Aldosterone is not a substrate for 11β-HSD2. The reactions catalyzed by HSD1 and HSD2 (separate gene products) are shown in Figure 9-24. If a patient has a mutation in the 11β-HSD2 enzyme rendering it less active or inactive, the result is apparent

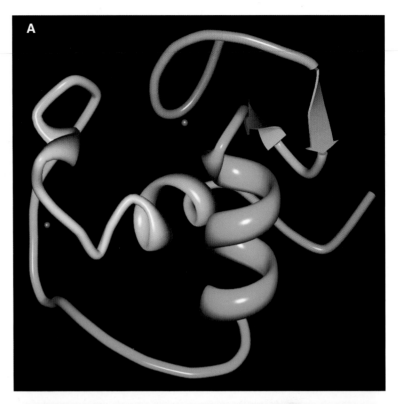

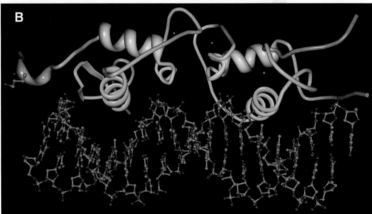

Figure 9-23. **A.** DNA-binding domain of glucocorticoid receptor. Part A reproduced from PDB ID: 1gdc. H. Baumann et al. Refined Solution Structure of the Glucocorticoid Receptor DNA-Binding Domain. *Biochemistry* **32** pp. 13463 (1993). Part A Reproduced from http://www.ebi.ac.uk/thornton-srv/databases/cgi_bin/pdbsum/GetPage.pl?pdbcode=1gdc. **B.** Interaction of the glucocorticoid homodimeric receptor with DNA. Part B reproduced from PDB ID: 1glu. B.F. Luisi, W.X. Xu, Z. Otwinowki, L.P. Freedman, P.B. Sigler. Crystallographic Analysis of the Interaction of the Glucocorticoid Receptor with DNA. *Nature* **352** pp. 497 (1991). H.M. Berman, J. Westbrook, Z. Feng, G. Gilliland, T.N. Bhat, I.N. Shindyalov and P.E. Bourne. The Protein Data Bank. *Nucleic Acids Research* **28** pp. 235 (2000), http://www.pdb.org.

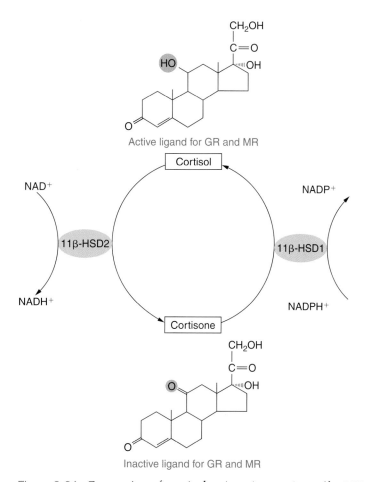

Figure 9-24. Conversion of cortisol to inactive cortisone (for MR and GR) catalyzed by 11β-HSD2. 11β-HSD1 enzyme catalyzes the reverse reaction. 11β-HSD2 is in the kidney, and its action prevents hypertension by converting cortisol to cortisone (an inactive ligand for MR). 11β-HSD1 is found in the liver, for example. Mutations in 11β-HSD2 that reduce or obliterate its activity can lead to hypertension, because excess cortisol in the kidney would bind to the MR to cause reabsorption of Na$^+$ and increased vascular volume and pressure. Such a condition is known as apparent mineralocorticoid excess and can be treated with spironolactone, an MR inhibitor.

mineralocorticoid excess (excess cortisol binding to the kidney mineralocorticoid receptor), which causes Na$^+$ reabsorption from the urine followed by water reabsorption from the urine to increase vascular volume and blood pressure. This effect of excess kidney cortisol would be greater than could be countered by atrionatriuretic peptide (ANP; same as ANF), which would stimulate Na$^+$ and water excretion. Hyperaldosteronism is the overproduction of aldosterone (**Conn's syndrome**). It can be caused by a tumor or by hyperplasia of the adrenal gland. Surgery may be indicated. Reduced cardiac output, in addition, can stimulate the synthesis of aldosterone in cardiac and vascular cells, which acts on the mineralocorticoid receptor of those cells, resulting in sustained hypertension. Aldosterone can act on mineralocorticoid receptor at sites in addition to the kidney to cause hypertension. Both apparent mineralocorticoid excess and hyperaldosteronism can be

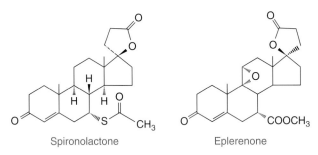

Figure 9-25. Structures of spironolactone and eplerenone.

treated by the mineralocorticoid receptor inhibitor, **spironolactone,** or by a newer drug, **eplerenone** (Figure 9-25), related to spironolactone (which appears to be safer in patients with acute myocardial infarction with left ventricular dysfunction, although hyperkalemia may be a problem).

Recently, a point mutation (ser810leu in the ligand-binding domain) of the mineralocorticoid receptor generates a form of the receptor (change in the ligand-binding domain) so that progesterone becomes an agonist rather than its traditional role as an antagonist. This is of importance in pregnancy, when the progesterone level is high enough to inhibit the glucocorticoid receptor in mammary gland and prevent the formation of milk proteins until term. Then the progesterone level drops dramatically, and the normal levels of cortisol can again compete to activate the glucocorticoid receptor. This mutation in the mineralocorticoid receptor causes the high level of progesterone to activate the mineralocorticoid receptor, permanently activating the receptor (as long as high levels of progesterone are present) and leading to hypertension (preeclampsia), which can be fatal to mother, fetus, or both. Apparently, as Yale University scientists have observed, this condition occurs in about 6% of pregnancies. Also, less severe hypertension increasing during pregnancy could be related to this mutation. This may represent a new cause of hypertension, which needs to be evaluated in the male.

Cortisol

The generation of the glucocorticoid cortisol, in reaction to a stress event is shown in Figure 9-2. The glucocorticoid receptor mechanism is reviewed in Figures 7-23, 7-25, 7-26, 7-36, and 7-38 through 7-48, and the synthesis of cortisol in the adrenal gland is shown in Figure 9-7. Cortisol circulates in the blood with 75 to 80% bound to transcortin (or corticosteroid-binding globulin), 15% to albumin, and 5 to 10% in free form. It is the free form that enters cells by free diffusion to bind to receptors in the cytoplasm, which are activated subsequently and translocate to the nucleus to stimulate (or depress) transcription. Nearly all cells of the body contain glucocorticoid receptors in various amounts, ranging from about 2000 molecules per cell to about 60,000 molecules per cell. The highest concentration is in liver. Two cell types are devoid of the glucocorticoid receptor: *pars intermedia*–type cells in the pituitary and hepatobiliary cells.

The glucocorticoid response in a given cell will be roughly proportional to the number of receptors in the cell up to a point, although some cells have an excess of receptor to induce a response. Cortisol induces surfactant protein (Figure 9-26) in the developing lung that allows the newborn to breathe on its own. Cortisol is a major defense against inflammation, as well as stress. In many tissues subject to inflammation, cortisol, through the glucocorticoid receptor, induces a protein called lipocortin I (also **annexin I,** calpactin II, and p35). This protein interacts directly or indirectly to inhibit cytoplasmic phospholipase A_2 (PLA_2). PLA_2 releases (by hydrolysis) arachidonic acid from the cell membrane to the interior of the cell, where it is a substrate for many inflammatory substances, such as prostaglandins (especially thromboxane), leukotrienes, and lipoxins. These pathways are summarized in Figure 9-27. Lipocortin I either binds PLA_2 directly to cause its inhibition or acts indirectly by activating a phosphatase that dephosphorylates PLA_2 to an inactive form. Recent evidence suggests that there is a direct interaction in which the binding activity of lipocortin I is attributed to amino acid residues 175 to C-terminal

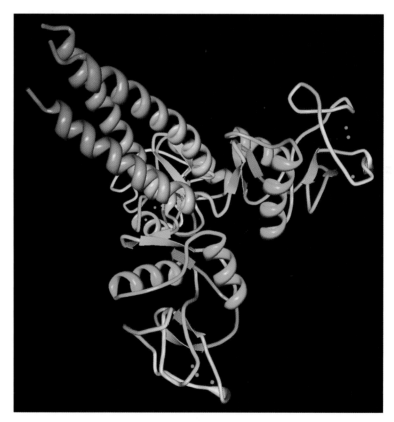

Figure 9-26. Structure of human lung surfactant protein D fragment. Reproduced from PDB ID: 1b08. K. Hakansson et al. Crystal Structure of the Trimeric Alpha-Helical Coiled-Coil and the Three Lectin Domains of Human Lung Surfactant Protein D. *Sturcture Fold. Des.* **7** pp. 255 (1999).

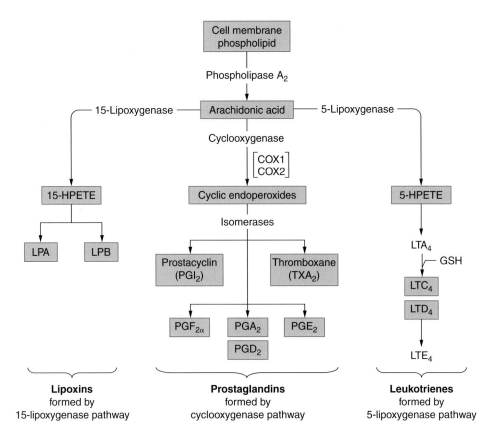

Figure 9-27. Prostaglandius (PGs), leukotrienes (LTs), and lipoxins (LPs) derived from arachidonic acid. Overview of enzymatic pathways giving rise to major prostaglandin relatives. Subscript number refers to the number of double bonds in the structure (most have equivalent structures with one double bond not shown here). COX1 is constitutive cyclooxygenase; the conjugated leukotrienes and LTC_4, LTD_4. α in $PGF_{2\alpha}$ refers to the steric position of the C-9 substituent. HPETE, Hydroperoxlicosatetraenoic acid. Redrawn from Figure 16-2 of A.W. Norman and G. Litwack, *Hormones*, 2nd ed., Academic Press, San Diego, 1997.

(346). Lipocortin I interacts with the calcium-dependent lipid-binding domain of cytoplasmic PLA_2 (the lipocortin N-terminus did not react). By either mechanism, lipocortin I inactivates PLA_2, which reduces the availability of arachidonic acid to the cell cytoplasm for conversion to inflammatory substances (Figure 9-28). Figure 9-29 shows a model of the interaction between the PLA_2 and the inner cell membrane. It is suggested that when PLA_2 is associated with the membrane, two to three lipid molecules are desolvated (removal of electrostatically bound water) to facilitate the binding of substrate (phospholipid) and increasing the reaction rate.

Other effects of cortisol involve the inhibition of the synthesis of cell adhesion molecules (CAMs) and intercellular adhesion molecules (ICAMs), which are inflammatory. For example, through the action of inflammatory mediators, such as $TNF\alpha$ and IL-1, vascular endothelial cells increase their expression of adhesion molecules, such as ICAMs and vascular cell adhesion molecules (VCAMs). A proinflammatory cytokine, IL-8, is produced by vascular epithelial cells, and the action

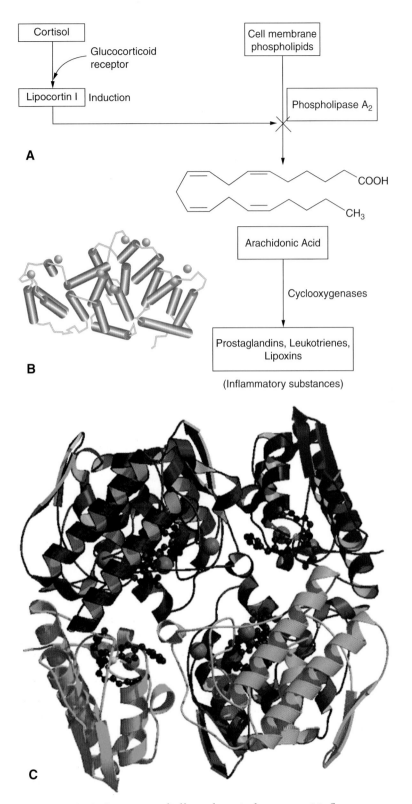

Figure 9-28. **A.** Summary of effect of cortisol, as an anti-inflammatory agent, on the release of arachidonic acid from the cell membrane phospholipids by phospholipase A2 (PLA$_2$). **B.** Structure of lipocortin I (annexin I). **C.** Structure of PLA$_2$ complexed with a highly potent substrate analog. *Blue atoms* in Part B *Pink atoms in part* C are Ca^{2+} ions. Part B is redrawn from PBD ID: 1mcx. H. Luecke, A. Rosengarth. A Calcium-Driven Conformational Switch of the N-terminal and Core Domains of Annexin A1. *Journal of Molecular Biology* **326** pp. 1317 (2003).

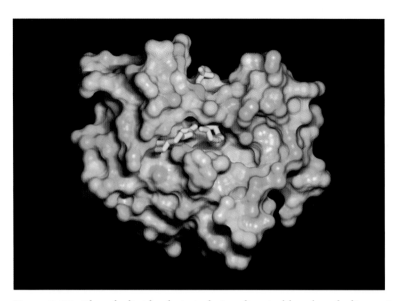

Figure 9-29. Phospholipid substrate being digested by phospholipase A_2. Repulsive interactions with the membrane are shown in *blue; greenish colors* represent the attraction interactions. Reproduced courtesy of Pieter Meulenhoff and Frans van Hoesel.

of IL-8 triggers the activation of **integrins** on the leukocyte surface. The rolling leukocytes bind to adhesion molecules (ICAMs and VCAMs) on the inner surface of the vascular endothelial cells. This enables the leukocytes to flatten and squeeze between the endothelial cells and leave the blood vessel. Leukocytes leaving the blood vessel can infiltrate tissues to reach an infected site, for example, and this process generates inflammation. Chronic inflammation can cause tissue damage and scarring. The role of adhesion molecules in the leakage of leukocytes from the blood vessels is shown in Figure 9-30.

Cortisol also inhibits the synthesis of induced cyclooxygenase 2 (COX2), an enzyme that responds to the inflammatory process and, at higher levels, increases the rates of formation of inflammatory prostaglandins and leukotrienes (Figure 9-31A). The endogenous COX1 maintains a level of prostaglandins and leukotrienes needed for the normal organism. COX1 has a slight structural difference from COX2 (Figure 9-31B), allowing for the design of drugs that specifically inhibit COX2.

Many cortisol-related compounds are used as drugs. These are shown in Table 9-2 with their potencies as anti-inflammatories (glucocorticoid action) and salt-retaining activity (mineralocorticoid action) measured against the activity of cortisol, arbitrarily taken as one for both activities (cortisol binds equally well to the glucocorticoid receptor and to the mineralocorticoid receptor). A double bond at C4, a keto group at C3, and hydroxyls at C11 and C17 are needed for anti-inflammatory activity. Halogen substituents greatly increase anti-inflammatory activity in the order F > Cl > Br > I.

Overproduction of cortisol by the adrenal gland produces a syndrome called **Cushing's disease.** In this case, most often, there is a

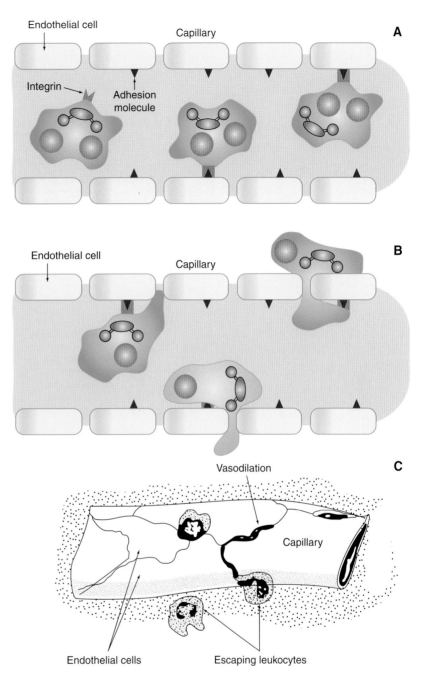

Figure 9-30. Role of adhesion molecules in the escape of leukocytes from blood vessels to invade tissue sites of inflammation. **A.** Integrins on the leukocyte surface bind to adhesion molecules on the inner surface of vascular endothelial cells. **B.** Leukocytes flatten and squeeze out between endothelial cells, increasing vascular permeability. **C.** External view of leukocytes squeezing out of a blood vessel. Redrawn with permission from http://www.cat.cc.md.us/courses/bio141/lecguide/unit1/prostrucI/inflam.htm. Doc Kaiser's Microbiology home page.

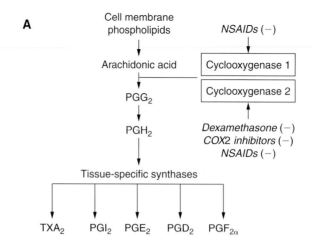

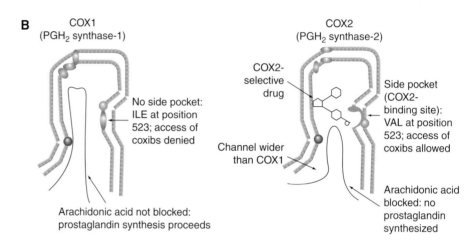

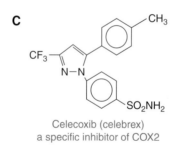

Celecoxib (celebrex)
a specific inhibitor of COX2

Figure 9-31. **A.** Cortisol inhibits the synthesis of COX2 (dexamethasone is a potent glucocorticoid). Nonsteroidal anti-inflammatories such as aspirin directly acetylate the active center of COX1 and COX2. **B.** Because of a point difference (in one amino acid residue) in the ligand binding site, specific drugs such as celecoxib *(C)* can specifically inhibit COX2 but not COX1. Redrawn from B.N. Cronstein, *Cleveland Cli. J. Med.*, 1: 13–19, 2002.

Table 9-2
Compounds Related to Cortisol That Have Glucocorticoid Activity (Anti-inflammatory) and Mineralo-corticoid Activity (Salt Retaining)

		Mineralocorticoid	Anti-inflammatory
	cortisol	1	1
	aldosterone	800	1
	fludrocortisone	800	5–40
	11-deoxycorticosterone	40	0
	prednisolone	0.6	4
	6a-methyl prednisolone	0	5
	betamethasone	0	5–100

Continued

Table 9-2
Compounds Related to Cortisol That Have Glucocorticoid Activity (Anti-inflammatory) and Mineralo-corticoid Activity (Salt Retaining)—cont'd

		Mineralocorticoid	Anti-inflammatory
	dexamethasone	0	10–35
	flumethasone	0	>100

tumor of the adrenal gland responsible for the overproduction. The converse situation is called **Addison's disease,** where there is an underproduction of cortisol. Some adrenal cortical tissue can be destroyed by certain diseases, such as tuberculosis, which accounts for the underproduction of cortisol. Surgery can be effective in Cushing's disease, with or without the use of glucocorticoid receptor inhibitors. Cortisol, or a related drug (Table 9-2), can be taken orally in the case of Addison's disease.

Dehydroepiandrosterone

DHEA is produced in the *zona reticularis* cells, the innermost layer of cells of the adrenal cortex (Figure 9-3). It circulates in the blood as the sulfate derivative (DHEA-S). A tissue receptor for DHEA or DHEA-S (Figure 9-32) has not been found, although it does bind to a subunit of **glucose-6-phosphate dehydrogenase (G6PD)** and is a noncompetitive inhibitor of the enzyme. The vascular endothelium increases its activity of G6PD, which is an antioxidant enzyme. G6PD is the first enzyme in the **pentose phosphate pathway** (Figure 6-57) and a major source of cellular NADPH. In addition to its reducing equivalent activity, NADPH is a cofactor for **nitric oxide synthase (NOS):**

L-arginine + NADPH + O_2 ⇌ citrulline + nitric oxide + $NADP^+$

There are three forms of NOS: a neuronal form (nNOS), an epithelial form (eNOS), and an induced form (iNOS). NOS is located in the

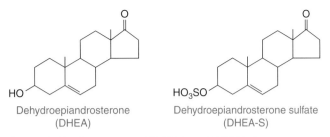

Figure 9-32. Structures of DHEA and DHEA-S.

cell attached to the cytoplasmic side of the endoplasmic reticulum, Golgi, or plasma membrane through myristoylation or palmitoylation (Figure 5-61). Endotoxins or cytotoxins that stimulate **cytokine** secretion cause the induction of iNOS. nNOS and eNOS are expressed constitutively and are regulated by calmodulin interaction with Ca^{2+}, reflecting the concentration of calcium. The monomer of NOS contains a reductase and an oxygenase with FAD, NADPH, and flavin mononucleotide or heme and **tetrahydrobiopterin (BH4)** cofactors. BH4 is required for the dimerization of NOS. The structure of iNOS synthase is shown in Figure 9-33.

This inhibitory activity of DHEA toward G6PDH and consequently NADPH plays a role in promoting endothelial cell oxidant stress

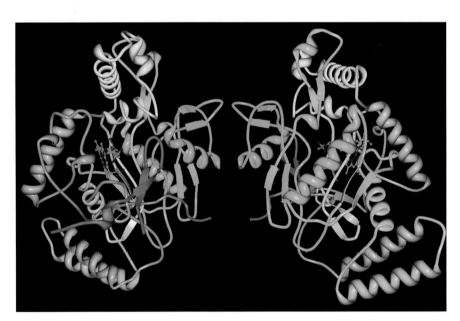

Figure 9-33. Crystal structure of induced nitric oxide synthase with 7-nitroindazole bound. Reproduced from PDB ID: 1m8e. R.J. Rosenfeld, E.D. Garcin, K. Panda, G. Andersson, A. Aberg, A.V. Wallace, D.J. Stuehr, J.A. Tainer, E.D. Getzoff. Conformational Changes in Nitric Oxide Synthases Induced by Chlorzoxazone and Nitroindazoles: Crystallographic and Computational Analyses of Inhibitor Potency. *Biochemistry* **41** pp. 13915 (2002).

Table 9-3
Plasma Concentrations of Adrenal Cortical Steroids
Average 8 a.m. Plasma Concentration and Secretion Rates of Adrenocortical Steroids in Adult Humans

	Plasma concentration (μg/dl)	Secretion rate (mg/dl)
Cortisol	13	15
Corticosterone	1	3
11-Deoxycortisol	0.16	0.40
Deoxycorticosterone	0.07	0.20
Aldosterone	0.009	0.15
18-OH Corticosterone	0.009	0.10
Dehydroepiandrosterone sulfate	115	15

and may decrease bioavailability of nitric oxide in endothelial cells. DHEA-S levels in blood decrease markedly during aging, and many regard this substance as a protective agent against cancer and other pathological conditions. There have been many controversial researches on its potential protective activity. DHEA-S is the most concentrated steroid derivative in the blood and circulates at a level about 100-fold that of cortisol (Table 9-3). DHEA is the major secretory product of the fetal adrenal gland and may have a role in development.

Steroid Hormone Structures

X-ray structures for steroid hormones that are ligands for the steroid receptor class are shown in Figure 9-34. The structures shown to this point, for example, in Figures 9-7 and 9-25 and in Table 9-2, are drawn as if they were standing on end on a surface. If the steroid is laid flat on this surface, the structures appear as in Figure 9-34. For the most part, the four rings appear to be planar (they practically lie flat on the surface) except for estradiol, the female sex hormone. These steroids enter the binding pocket of their respective receptors A-ring first (left end of structures). Testosterone, progesterone, aldosterone, and cortisol appear to be nearly identical in terms of the A-ring and C19 methyl group. The four rings of these structures all lie nearly flat, so that the other rings appear similar except for substitutions, particularly in the D-ring. Predictably, on this basis, these hormones interact (cross-react) with one another's receptors to various degrees. In the case of estradiol, its A-ring is unsaturated (three double bonds); the A-rings of the other steroids contain one double bond. The distance between two carbon atoms connected by a double bond is shorter than the distance between two carbon atoms connected by a single bond. Therefore, estradiol has a constricted A-ring that pulls it out of the plane of the other three rings (top structure). Because of this difference and because the steroids

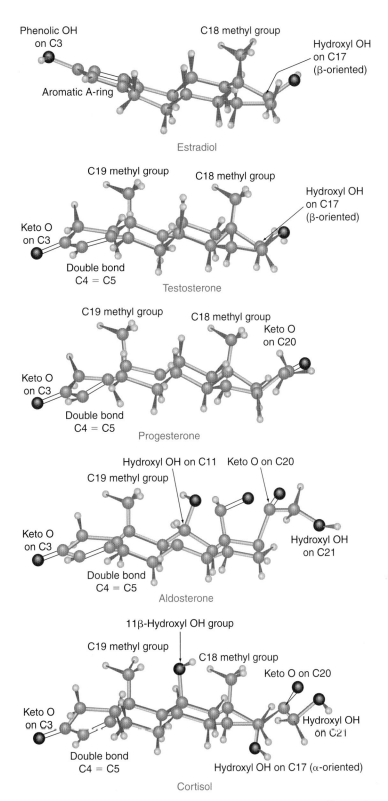

Figure 9-34. Ball-and-stick representations of some steroid hormones determined by X-ray crystallographic methods. Details of each structure are labeled. In aldosterone, the acetal grouping is $R-\underset{\underset{OR_2}{|}}{\overset{\overset{OR_1}{|}}{C}}H$ and the hemiketal grouping is $\underset{\underset{R_2}{}\diagdown\diagup\underset{OH}{}}{\overset{\overset{R_1}{}\diagdown\diagup\overset{OR_3}{}}{C}}$, where R_1, R_2, and R_3 refer to different substituents. Redrawn from J. Glusker in *Biochemical Actions of Hormones*, G. Litwack, Editor, Academic Press, New York, 1979, pp. 121–204.

enter the binding pocket A-ring first, estradiol binds to the estrogen receptor exclusively and not to the others, particularly, the testosterone receptor. Also, testosterone (and dihydrotestosterone) does not bind to the estrogen receptor. By comparing the structures of cortisol and aldosterone, it is obvious why the two ligands bind equally to the mineralocorticoid receptor. On the other hand, aldosterone does not bind as well as cortisol to the glucocorticoid receptor. The architecture around the C- and D-rings may explain that, suggesting the binding pocket may be deep.

The location of these receptors in the cell in the unliganded form differs. Well-known receptors that are members of the steroid hormone receptor group are mostly located within the nucleus in the unliganded (absence of hormonal steroid) form. These include receptors for thyroid hormone, retinoic acid, estrogen, and androgen. Some receptors are divided between the nucleus and cytoplasm: vitamin D receptor and mineralocorticoid receptor. Unliganded glucocorticoid receptor is almost exclusively in the cytoplasm in the unliganded form, and the preponderance of the unliganded mineralocorticoid receptor is located in the cytoplasm.

Unliganded Forms of Receptors and Mechanism of Activation

In the inactive, unliganded form, cytoplasmic receptors are complexed with other proteins, such as heat shock proteins and other proteins (HSP90 dimer, HSP70, immunophilin, and other smaller proteins). The greatest affinity for glucocorticoid by the cytoplasmic unactivated glucocorticoid receptor complex is the form to which a dimer of HSP90, a p23-protein, and an immunophilin other than FKBP51 are bound. With the immunophilin, **FKBP51,** the unactivated complex has somewhat lower affinity for the ligand (Figure 9-35). The partial structure of human HSP90 is shown in Figure 9-36. A cyclophilin 40 immunophilin structure is shown in Figure 9-37. This unactivated complex, in the absence of ligand, is typical of the steroid receptor class, whether the receptor resides in the cytoplasm or in the nucleus. The general mechanism of receptors for steroid hormones, thyroid hormone, and vitamins D and A is similar and corresponds to the mechanism shown in Figure 9-38. The ligand diffuses into the cell from the bloodstream (free form not bound to plasma proteins) by free diffusion, and the receptor, upon binding its ligand, undergoes a conformational change, causing the proteins associated with the receptor (HSP90 dimer, HSP70, p23, and p59) to dissociate, releasing the ligand-bound receptor monomer followed by dimerization of the liganded receptor. The dimer is transported through the nucleopore by transporting proteins. It associates with its specific response element (SRE) on a gene promoter and, with adapter proteins and coactivators, facilitates the transcription of the gene by RNA pol II (Figures 7-9, 7-10, 7-42, and 7-45). This mechanism leads to positive or negative regulation of gene expression. The activation mechanism is seen in more detail for the different hormone receptors in Figure 9-39. The activation process involves multiple phosphorylations of the receptor. In Figure 9-40 is shown the sites of phosphorylation on the progesterone receptor and a potential role for one of the phosphorylation sites in inducing an interaction with a protein that

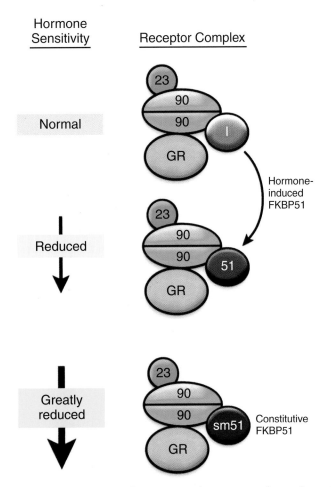

Figure 9-35. FKBP51 influences on glucocorticoid signaling. In the absence of hormone, GR exists in a complex with a dimer of HSP90, a subunit of p23, and any one of several immunophilin-related proteins (I). In the presence of immunophilins other than FKBP51, GR has a high affinity for hormone *(top panel)*. Glucocorticoids stimulate expression of FKBP51, enhancing the likelihood for some time after hormone withdrawal that GR complexes will contain FKBP51 and have a lowered affinity for subsequent hormone exposures *(middle panel)*. In squirrel monkeys, FKBP51 is constitutively expressed at high levels and more greatly depresses GR's affinity for hormone *(bottom panel)*. Reproduced from J. Cheung and D.F. Smith, *Mol. Endocrinol.*, 14: 939, 2000.

forms a bridge between the general transcription factors and the receptor. Dopamine, through a dopamine receptor, modulates the transcriptional activity of progesterone receptor, estrogen receptor, thyroid hormone receptor, and chicken ovalbumin upstream promoter transcription factor (COUP-TF) but not glucocorticoid receptor and other steroid hormones receptors. In uterine cells, some polypeptide growth factors, such as insulin-like growth factor IGF-I, TGFα, and **epidermal growth factor (EGF)** are capable of activating endoplasmic reticulum–responsive genes, in the absence of estrogen. These effects can be inhibited by the endoplasmic reticulumantagonist (or ER) ICI164, suggesting crosstalk between receptor systems.

(Text continues on p. 483.)

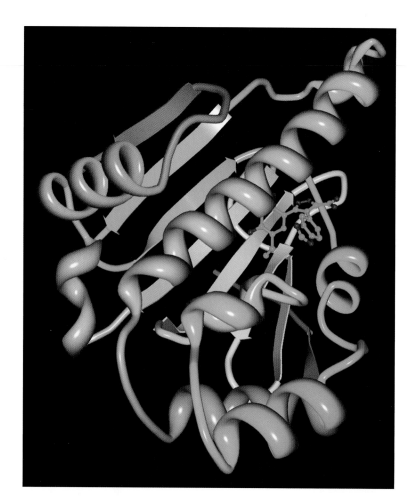

Figure 9-36. Crystal structure of human HSP90α complexed with dihydroxyphenylpyrazoles. Geldamycin is a specific inhibitor of HSP90. PDB ID: 1yc3. A. Kreusch, et al. Crystal Structures of Human HSP90α Complexed with Dihydroxyphenylpyrazoles *Bioorg. Med. Chem. Lett.* **15** pp. 1475 (2005). H.M. Berman, J. Westbrook, Z. Feng, G. Gilliland, T.N. Bhat, I.N. Shindyalov and P.E. Bourne. The Protein Data Bank. *Nucleic Acids Research* **28** pp. 235 (2000) http://www.pdb.org.

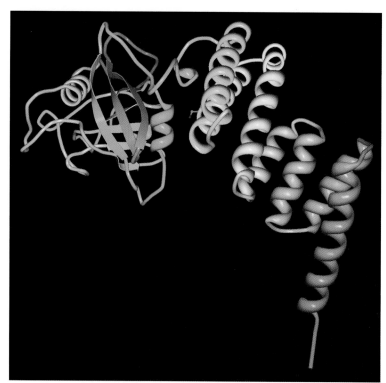

Figure 9-37. An immunophilin similar to one that can be found in the inactivated glucocorticoid receptor complex. This is cyclophilin 40, monoclinic form. Immunophilins have enzymatic activity that is peptidyl-prolyl Cis-trans isomerase. PDB ID: 1ihg P. Taylor, et al. Two Structures of Cyclophilin 40: Folding and Fidelity in the TPR Domains *Structure* **9** pp. 431 (2001). H.M. Berman, J. Westbrook, Z. Feng, G. Gilliland, T.N. Bhat, I.N. Shindyalov and P.E. Bourne. The Protein Data Bank. *Nucleic Acids Research* **28** pp. 235 (2000) http://www.pdb.org.

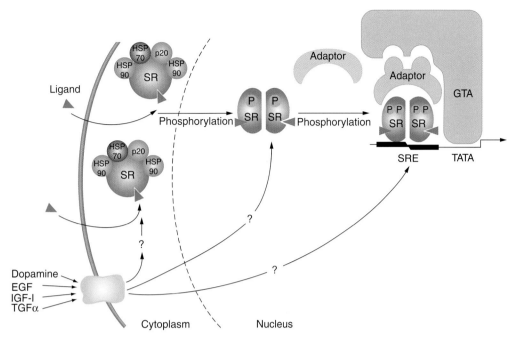

Figure 9-38. Activation of nuclear hormone receptors. *GTA*, general transcription apparatus; *P*, phosphate; *SR*, steroid receptor; *SRE*, steroid response element in DNA. Redrawn with permission from M.J. Tsai and B.W. O'Malley, *Ann. Rev. Biochem.*, 63: 451–486, 1994.

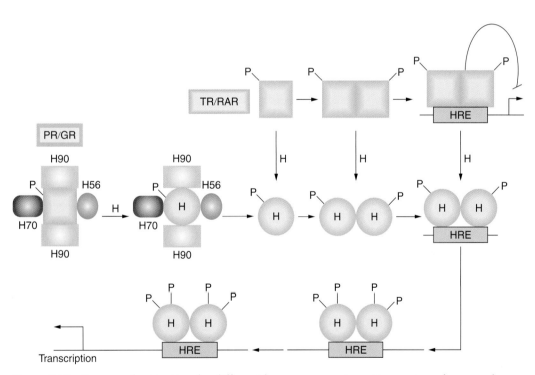

Figure 9-39. Process of activation for different hormone receptors. Conversion of square shape to circular shape indicates conformational change. *GR*, glucocorticoid receptor; *H90*, HSP90; *H56*, HSP56; *H70*, HSP70; *H*, hormone; *HRE*, hormone responsive element; *PR*, progesterone receptor; *RAR*, retinoic acid receptor; *TR*, thyroid hormone receptor. Redrawn with permission from M.J. Tsai and B.W. O'Malley, *Ann. Rev. Biochem.*, 63: 451–486, 1994.

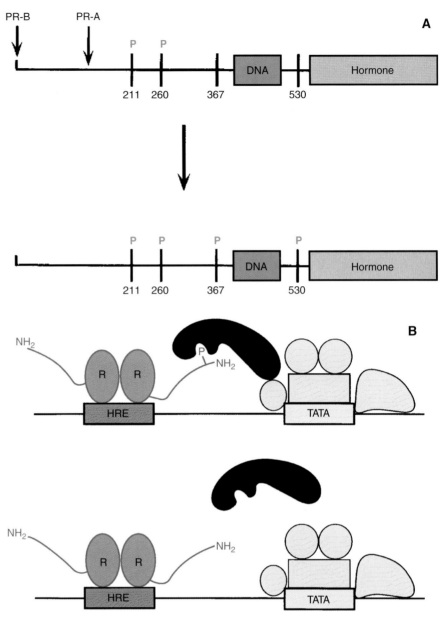

Figure 9-40. Phosphorylation of the progesterone receptor. **A.** Location of the four phosphorylation sites on the cytoplasmic progesterone receptor (PR). PR-A and PR-B are two forms of the progesterone receptor. The presence of progesterone allows for phosphorylation of amino acid residues 367 and 530, and for more extensive phosphorylation on amino acid residues 211 and 260 (involving more receptor molecules). **B.** Suggests a role for amino acid 211 (serine), which may induce an interaction with a protein, forming a bridge between the receptor and the general transcription factors. Reproduced with permission from N.L. Weigel, *Biochem. J.*, 319: 657–667, 1996.

Ligands and Receptor Conformation: The Sex Hormones

The sex hormones refer to estrogen (17β-estradiol), progesterone, testosterone, and dihydrotestosterone. The synthesis of these steroidal hormones is shown in Figure 9-7, and the conversion of testosterone to dihydrotestosterone, the preferred ligand for the androgen receptor, is shown in Figure 8-46. As discussed earlier, the binding of ligand (for example, cortisol) to the inactive receptor complex results in an altered conformation of the receptor that leads to the dissociation of other proteins complexed with the receptor. Agonists and antagonists, exemplified in the case of the progesterone receptor, bind to different regions of the ligand-binding domain by first contacting a low-affinity site and then making higher-affinity contacts with different binding sites (Figure 9-41). The ligand-binding domain of the estradiol receptor is shown in Figure 9-42, and the

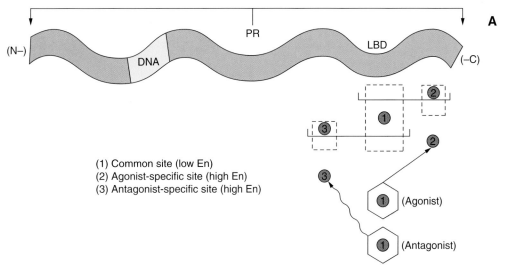

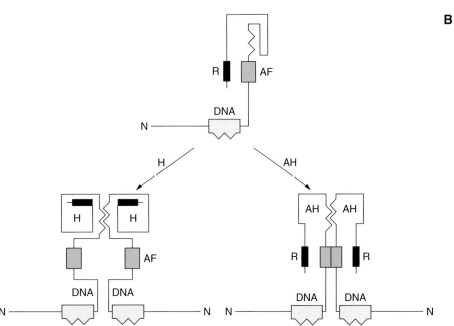

Figure 9-41. **A.** Binding of agonist or antagonist to different sites in the ligand-binding domain of the progesterone receptor (PR). **B.** Antagonist produces a conformation of the receptor dimer *(bottom right)* that is incapable of removing a repressor function. In the presence of agonist, a conformation of the dimeric receptor aligns with the activation function (AF), allowing transcription to occur. *AH*, antihormone; *H*, hormone; *R*, repressor. Redrawn with permission from M.J. Tsai and B.W. O'Malley, *Ann. Rev. Biochem.*, 63: 451–456, 1994.

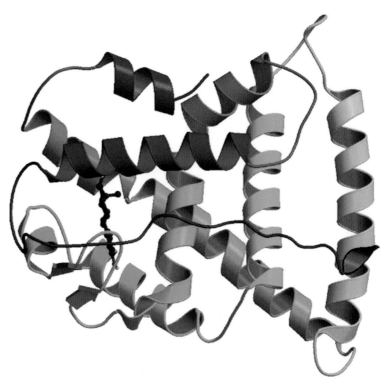

Figure 9-42. Homologous–extension-based model of human estrogen receptor with bound estradiol (theoretical models, purple structure). From PDB ID: 1akf. G.J. Malouf, W. Xu, T.F. Smith, S.C. Mohr. Homology Model for the Ligand-Binding Domain of the Human Estragen Receptor. *J. Biomol. Struct. Dyn.* **15** pp. 841 (1998).

DNA-binding domain of this receptor is similar to that of the other steroid hormone receptors (Figure 9-43). Because the A-ring of the steroidal hormone enters the binding pocket first and the A-ring of estradiol is unsaturated (three double bonds), there are many other natural and synthetic compounds in the environment with structures that resemble the A-ring of estradiol, bind to the estrogen receptor, and act as either agonists or antagonists. Some of these compounds are shown in Figure 9-44. Many of these compounds are organochlorine pesticides, such as DDT, toxaphene, dieldrine, and chlorodecone. Other compounds, such as polychlorinated biphenyls and polycyclic aromatic hydrocarbons, also bind. Many of these are carcinogens, producing tumors that can result from overstimulation of the estradiol receptor pathway. Antagonists of the estrogen receptor can have deleterious effects on female agricultural workers in a field sprayed routinely with pesticides (for example, in the past, DDT), which can produce masculinization (for example, **hirsutism;** unwanted hair in females) and even cancers.

In general, the role of estrogens and androgens is to develop and mature the secondary sex characteristics. In the female, uterine weight increases with uterine glycogen and cell proliferation in the uterus and breast. In the male, development of the prostate, penis, gonads, bodily hair, low voice, and other changes, including some in the kidney, occurs. There is crosstalk between the estrogen receptors and the

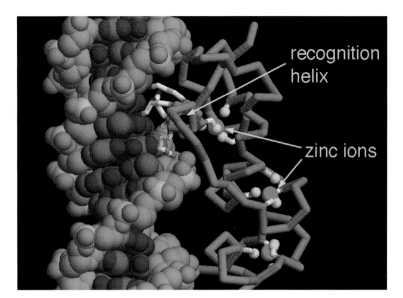

Figure 9-43. The DNA-binding domain of the estrogen receptor in contact with the estrogen responsive element in DNA. The zinc atoms are indicated in the figure. The HRE for the estrogen receptor is AGGTCnnnTGACCT. A half-site would be AGGTC or TGACCT. The structure of the partial ER is in blue. Reproduced originally from PDB ID 1hc9 http://www.rcsb.org/static.do?p=education_discussion/molecule_of_the_month/pdb45_3.html.

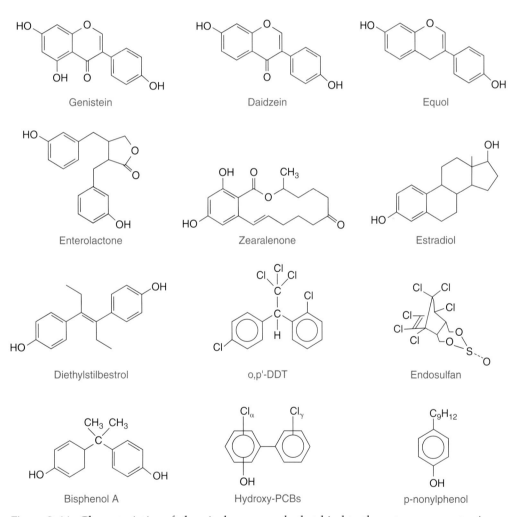

Figure 9-44. Characteristics of chemical compounds that bind to the estrogen receptor include an unsaturated ring with a hydroxyl group (phenolic ring). Hydrophobic substituents para to the phenolic hydroxyl are necessary.

Ligands and Receptor Conformation: The Sex Hormones

polypeptide hormone receptors. IGF-I, TGFα, and EGF are able to activate the same genes that are activated by the estrogen receptor in uterine cells in the absence of estrogen. Because the polypeptide hormone receptors activate protein kinases, the status of phosphorylation in the cell plays an important role in determining the extent of expression of estrogen responsive genes. One effect that can be accomplished by a peptide growth factor, like EGF, is the phosphorylation (and activation) of the estrogen receptor in the absence of estrogen. Phosphorylation sites on the estrogen receptor are shown in Figure 9-45.

Specific roles of estrogens, progestins, and androgens have been discussed in terms of the ovarian cycle and spermatogenesis. The roles of estradiol and progesterone in the ovarian cycle are shown in Figure 8-51. The role of testosterone (and dihydrotestosterone) in spermatogenesis is shown in Figure 8-45. Blood levels of estradiol in the premenopausal adult are 23 to 361 pg/ml (pg = 10^{-12}g)/; for postmenopausal adult, <30 pg/ml; and for prepubertal female, <20 pg/ml. Following puberty, androgen levels in the blood increase dramatically during adolescence, facilitating muscle growth and lean body mass. In the male adult, the blood level of testosterone is 300 to 1100 ng/dl (ng = 10^{-9}g); for the female, 20 to 90 ng/dl. As males age (70s to 80s), the androgen level in blood falls to a range of 450 to 500 ng/dl with a decreased lean body mass and tendency to falls and fractures. Prostate cancer is thought to be related to the stimulatory effects of androgens after the age of 40. The structure of the human androgen receptor ligand-binding domain with a bound ligand is shown in Figure 9-46, and the DNA-binding domain in contact with the HRE is shown in Figure 9-47. Nonsteroidal androgen receptor ligands have been discovered, and some are shown in Figure 9-48.

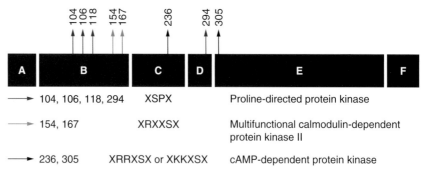

Figure 9-45. Model indicating the six potential ER phosphorylation sites studied and their locations in different domains (*lettered A–F*) of the receptor protein. The locations of two additional, potential cAMP-dependent protein kinase sites are also indicated. Below the receptor schematic is a listing of these serines and their positions in known protein kinase consensus sequences. Redrawn with permission from M. Tsai and B.W. O'Malley, *Ann. Rev. Biochem.*, 63: 451–456, 1994.

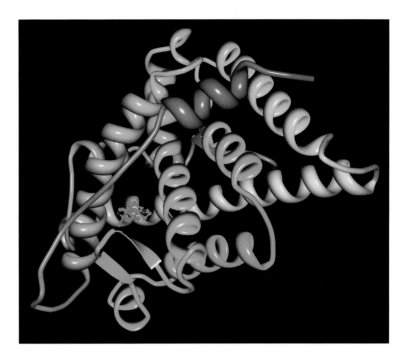

Figure 9-46. The androgen receptor. Human receptor ligand-binding domain (amino acid residues 447–709) complexed with the agonist, metribolone (R1881, *green structure and red dots*). Reproduced from PDB ID: 1e3g (right view). P.M. Matias et al. Structural Evidence for Ligand Specificity in the Binding Domain of the Human Androgen Receptor. Implications for Pathogenic Gene Mutations. *J. Biol. Chem.* **275** pp. 26164 (2000).

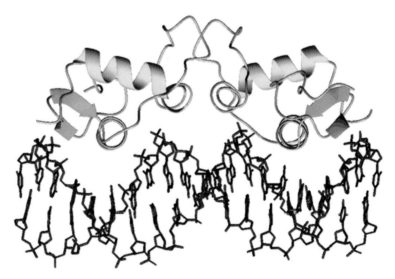

Figure 9-47. The androgen receptor DNA-binding domain *(purple)* of androgen receptor chimera bound to the androgen hormone-sensitive element of DNA *(bottom)*. Reproduced from P.L. Shafter, A. Jervain, and D.E. Doleris, The Nucleic Acid Database Project, Rutgers, The State University of New Jersey, 1996–2004 and http://www.molfunction.com/images/gallery2/dbd15.gif.

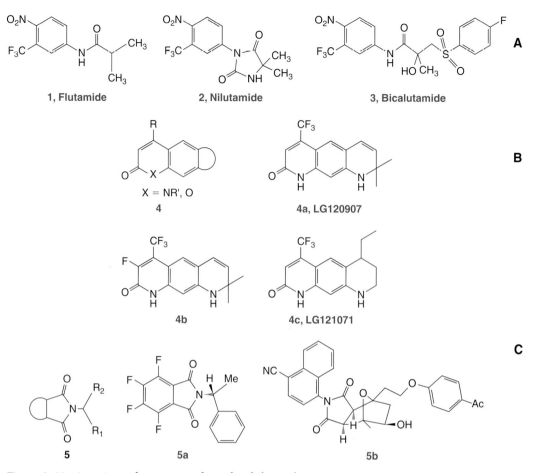

Figure 9-48. Agonist and antagonist ligands of the androgen receptor. **A.** Antagonists that are nonsteroidal for the androgen receptor. **B.** Compounds 4a and 4b are antagonists, and 4c is an agonist. All of these compounds are nonsteroidal. **C.** Antagonists.

Peroxisome Proliferators and Orphan Receptors

Peroxisome proliferators are composed of industrial and pharmaceutical chemicals, such as plasticizers, herbicides, and hypolipidemic drugs. The **peroxisome proliferator–activated receptor (PPAR)** is relegated to the class of orphan receptors (for which either ligands or functions (or both) are unknown. More information is becoming available about PPARs. PPARs, as well as many orphan receptors, are considered members of the nuclear receptor superfamily. Peroxisomes are subcellular organelles found in animal and plant cells (Figures 1-26 and 1-27), and they function to metabolize H_2O_2, perform β-oxidation of fatty acids, and metabolize cholesterol. Peroxisome proliferators lead to an increase in the size and number of peroxisomes and their enzymes involved in the β-oxidation of fatty acids and cause hypertrophy of hepatocytes (liver cells). They are liver carcinogens for experimental rodents, but it is unclear whether they are carcinogens for humans. Some peroxisome proliferators are shown in Figure 9-49. Similarities exist in these compounds. All of these contain at least one unsaturated ring. The crystal structure of the heterodimer ligand-binding domains of PPARγ and RXRα, together with ligands, is shown in Figure 9-50. Endogenous activators that are ligands of PPAR are prostaglandins, leukotrienes, and

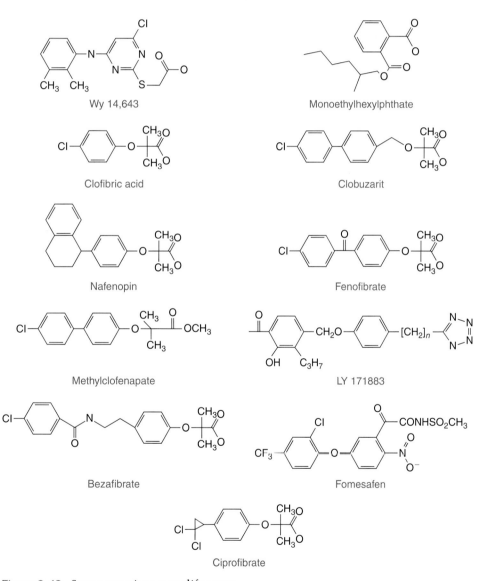

Figure 9-49. Some peroxisome proliferators.

fatty acids. As shown in Figure 9-50, PPAR, with rosiglitazone bound to PPAR and 9-*cis*-retinoic acid bound to RXR, in the heterodimer associates with DNA through two zinc fingers. The responsive element for this heterodimer is TGACCTXTGTCCT, and the transcriptional complex directs the synthesis of mRNA. Thus, PPAR is a genuine member of the steroid hormone receptor superfamily. The basic structure of PPAR with its various activities is shown in Figure 9-51, where the

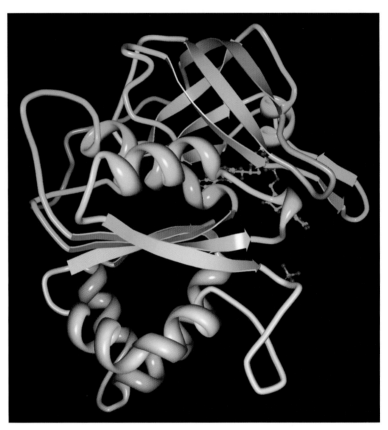

Figure 9-50. Crystal structure of the human RXRα and PPARγ ligand-binding domains (heterodimer) complexed with 9-*cis* retinoic acid (RXRα ligand) and rosiglitazone (PPARγ ligand) and coactivator peptides. Reproduced from PDB ID 1fmb, http://www.rcsb.org/pdb/explore.do?structureId=1FM6. R.T. Gampe et al. *Mol. Cell.* **5** pp. 545 (2000).

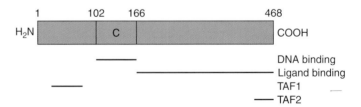

Figure 9-51. Model of the functions and their locations of the peroxisome proliferator activated receptor (PPAR). Redrawn with permission from M.J. Tsai and B.W. O'Malley, *Ann. Rev. Biochem.*, 63: 451–456, 1994.

Table 9-4
Some Orphan Receptors

Name or human gene	Ligand	Comments
Orphan A1; GPR3		There is a proposed ligand
Orphan A12; GPR23		There is a proposed ligand
Orphan A13; P2RY10		
Orphan A14; GPR142		
Orphan A15; GPR159		Also known as RDC1
Orphan A6; GPR63	Dioleophospholiolic acid (low affinity)	
Orphan SREB; GPR85		
LNB7TM; EMR1	Mucin-like receptor	
COUP-TF		
HNF-4		

COUP-TF, chicken ovalbumin upstream promoter-transcription-factor; *EMR1*, EGF-like containing mucin-like hormone receptor, an orphan *BGPR*; *GPR*, G-protein coupled receptor; *HNF-4*, hepatocyte nuclear factor-4; LNB7TM, oligo sense EGF and mucin-like receptor (7 transmembrane domains); P2RY, purinergic receptor; *RDC1*, Chemokine orphan receptor 1; *SREB*, receptor expressed in brain.

similarity to the steroid hormone receptors is obvious in terms of the functions of the receptor and their locations in the molecule. There are three separate gene products (isoforms) that are versions of PPAR (α, β/δ, and γ) generated from separate genes. PPAR isoforms differ in their ligand preferences. The α-form (mainly in liver and kidney) binds saturated fatty acids and peroxisome proliferators (Figure 9-49) better than the other forms. The β-form (highest in kidney, brain, muscle, spleen, lung, and adrenal and immune systems) has its highest affinity for prostaglandin J2 (PGJ2), and the γ-form (highest in white adipose and the immune system) has its highest affinities for PGA1 and other compounds. All receptor isoforms form heterodimers with RXR and bind to the same PPAR element on DNA, although the regions surrounding the PPAR element may determine the extent of transcriptional activity. Table 9-4 lists a few orphan receptors.

Programmed Cell Death (Apoptosis) Induced by Glucocorticoids

Programmed cell death, or apoptosis, is a cell suicide mechanism produced by various agents that cause the transcription of genes that sponsor this process. Apoptosis is important in normal development and tissue homeostasis; it is the balance to cell proliferation. If the apoptotic process is disrupted, it can lead to formation of cancer. Conversely, cancers can be destroyed if the cancer cells can be induced to undergo programmed cell death. Most anticancer agents operate by killing cancer cells in this way. Glucocorticoids, such as cortisol, can induce the apoptotic program in certain cell types, usually those of the hematopoietic system and in cancers derived from hemopoietic cells. Apoptosis can occur in response to a variety of signals in different cell types, but the cell death program in all cases is mediated by the activation of specific proteases called **caspases.** These enzymes are aspartate-specific cysteine proteases. In thriving cells, the caspases are present as inactive proenzymes composed of three domains: an N-terminal prodomain, a large subunit, and a small subunit (Figure 9-52). The actions of caspases, once activated, can lead to the death of the cell by hydrolyzing various substrate proteins and generating the cleavage of DNA. The activation of caspases involves the proteolytic processing of the prodomain and the subunits, followed by assembly of the large and small subunit (heterodimer) and dimerization of two heterodimers into the catalytically active form of the enzyme (Figure 9-53). The proteolysis occurs at sites specific for caspase cleavage (Asp-X), and the initial activation of a caspase (or other enzyme) that can initiate activation can be brought about by the stimulation of a factor at the cell membrane that causes caspase aggregation, which leads to an activated enzyme. This pathway is called the extrinsic pathway. On the other hand, glucocorticoids, for example, could transcriptionally activate a gene encoding the molecule, Bax, which causes the mitochondrion to leak cytochrome c into

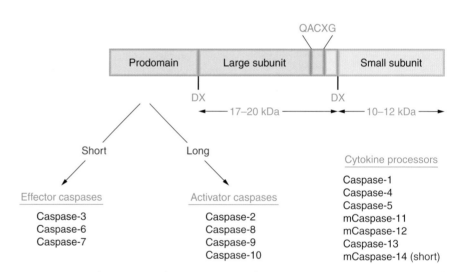

Figure 9-52. The structural composition of caspases. Caspases are comprised of three domains: an amino-terminal prodomain, a large subunit, and a small subunit. Each caspase contains a conserved pentapeptide motif QACXG, which includes the active site cysteine residue. Activation of caspases involves proteolytic processing between domains at critical aspartic acid residues (DX), resulting in the removal of the prodomain and self-association of large and small subunit heterodimers to form an active tetramer. Caspases are characterized based on the length of their prodomains. Caspases 2, 8, 9, and 10 are activators, whereas caspases 3, 6, and 7 are considered effectors. The remaining caspases are characterized as cytokine processors with the majority having long prodomains. Of these, caspases 11, 12, and 14 have been found only in mice. Reproduced from S.L. Planey and G. Litwack, *Biochem. Biophys. Res. Commun.*, 279: 307–312, 2000.

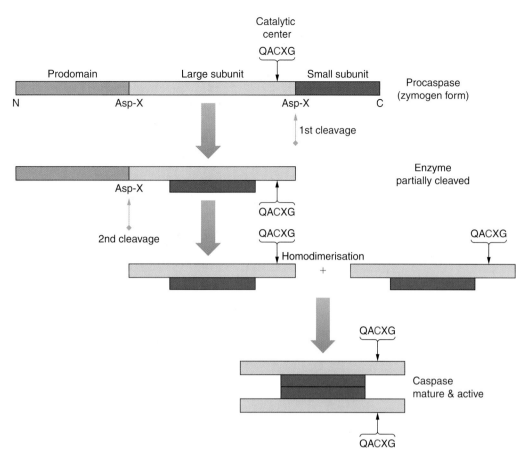

Figure 9-53. Activation of a caspase by proteolytic cleavage of the procaspase precursor. In the initial cleavage of the proenzyme, the small subunit is split from the prodomain and large subunit. In the second cleavage, the large subunit is split from the prodomain. The cleavages occur at a caspase-specific site, Asp-X. The released large and small subunits form a heterodimer; then two heterodimers dimerize to form the catalytically active enzyme. It is possible for this process to occur by autocatalysis, through an external, initial activation of a proenzyme or through aggregation of caspase induced by an external factor. In the latter case, the aggregated form of the enzyme can autocatalyze its activation.

the cytoplasm. With other factors (**Apaf-1**), cytochrome c can bring about the activation of caspase-9. This caspase, in turn, activates other caspases, leading to proteolysis of key proteins and the activation of an endonuclease, which cleaves DNA, all combining to cause the death of the cell. The two general pathways, extrinsic and intrinsic, are shown in Figure 9-54. Glucocorticoid-induced apoptosis in hematopoietic cells (blood cells) involves the intrinsic pathway of apoptosis. Glucocorticoids induce the Bax protein to initiate the intrinsic pathway, as shown in Figure 9-55. Proteins involved in the mitochondrial response in the intrinsic pathway are shown in Figure 9-56. The structure of caspase-9 is shown in Figure 9-57.

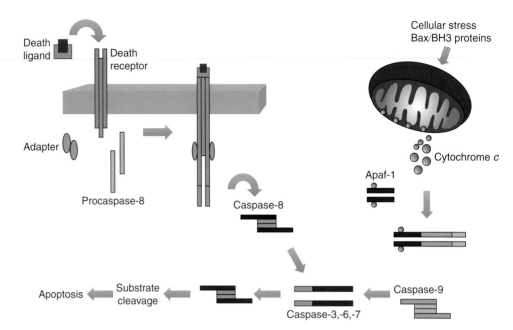

Figure 9-54. Pathways of caspase activation and apoptosis. Two distinct apoptotic pathways have been described that function to initiate the activation of caspases and other pro-apoptotic activities. Ligation of death receptors after ligand binding recruits caspase-8 or -10, which activate downstream caspases such as caspase 3, 6, or 7. In the second pathway, cellular stress causes the release of cytochrome c from the mitochondria, which binds Apaf-1. This complex activates caspase 9, which subsequently activates downstream caspases. Redrawn from S.L. Planey and G. Litwack, *Biochem. Biophys. Res. Commun.*, 279: 307–312, 2000.

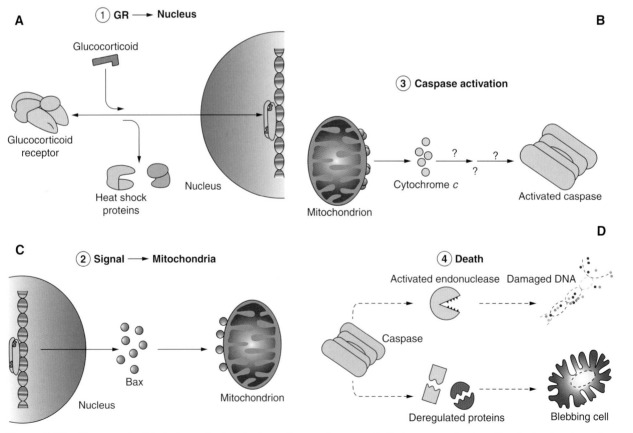

Figure 9-55. Glucocorticoids induce the intrinsic pathway of apoptosis in hematopoietic cells. **A.** Activation of the glucocorticoid receptor by a glucocorticoid molecule. **B.** Expression of Bax protein by the glucocorticoid receptor through transcription-translation of the Bax gene. **C.** Caspase activation by cytochrome c released from the mitochondrion by the action of Bax. **D.** Cell death following proteolysis of key proteins in the cell and the activation of a nuclease that cleaves DNA.

Figure 9-56. Proteins involved in mitochondria as part of the intrinsic pathway of apoptosis.

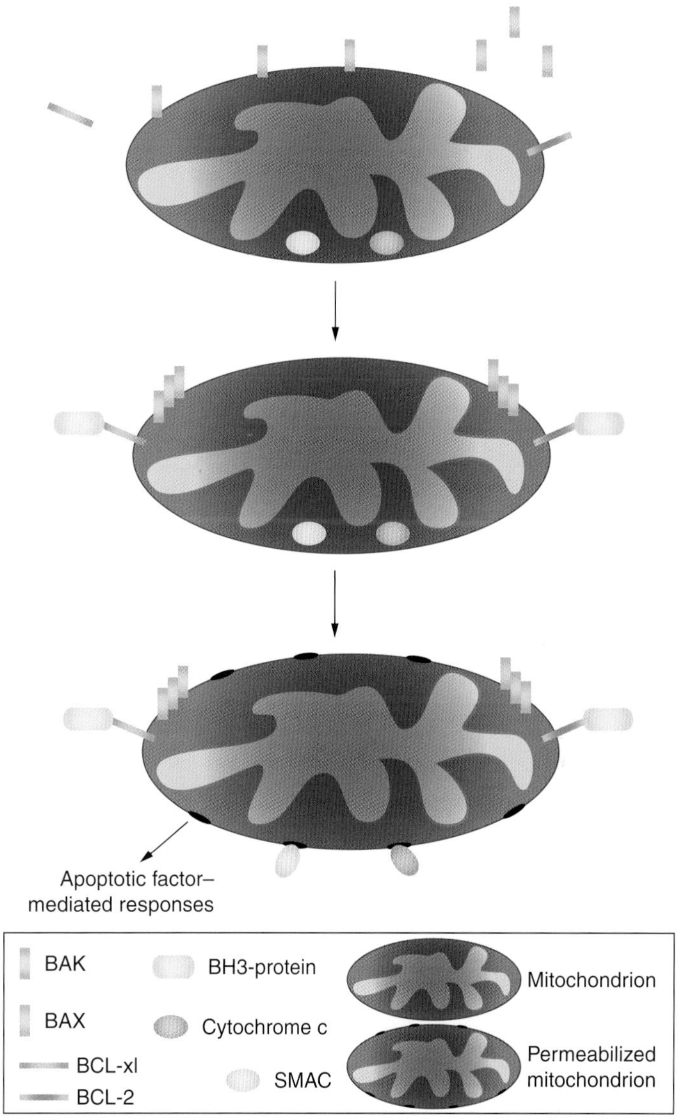

Apoptotic factor–
mediated responses
BAK
BH3-protein
Mitochondrion
BAX
Cytochrome c
BCL-xl
BCL-2
SMAC
Permeabilized
mitochondrion

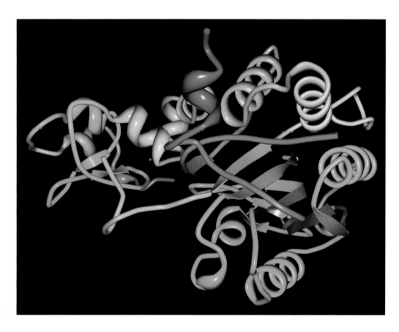

Figure 9-57. Crystal structure of dimeric caspase-9. PDB ID: 2ar9. Y. Chao et al. Engineering a Dimeric Caspase-9: A Re-Evaluation of the Induced Proximity Model for Caspase Activation *Plos. Biol.* **3** pp. 1079 (2005). H.M. Berman, J. Westbrook, Z. Feng, G. Gilliland, T.N. Bhat, I.N. Shindyalov and P.E. Bourne. The Protein Data Bank. *Nucleic Acids Research,* **28** pp. 235 (2000), http://www.pdb.org.

Further Reading

Norman, A.W., and Litwack, G., *Hormones*, 2nd edition, Academic Press, 1997.

CHAPTER **10**

Metabolism

Hyperammonemia and Disruptions of the Urea Cycle

Excess Ammonium Ion and Urea in the Blood Can Be Lethal

Ammonium ion is the end product of protein breakdown and amino acid metabolism. Ammonia is toxic and is converted to urea by the urea cycle for excretion. **Hyperammonemia** occurs if there is a defect in the urea cycle. Defects are genetic and involve deficiencies in enzymes of the urea cycle, such as carbamoyl phosphate synthetase and ornithine transcarbamylase. Other deficiencies are important in that the enzymes involved, when deficient, lead to an accumulation of the substrate: argininosuccinic acid synthase deficiency leads to citrullinuria, and argininosuccinate lyase deficiency leads to argininosuccinic aciduria. The normal blood level of ammonia is in the range of 10–40 mmol/liter, and blood urea nitrogen is in the range of 6–20 mg/dl. In the newborn, urea cycle disorders occur in about 1 in 30,000. In a normal adult with 5 liters of blood, the total soluble ammonia is about 150 mcg (150 × 10^{-5} grams), whereas urea nitrogen is about 1000 mg (about 800 times the amount of ammonia). Most ammonia derived from amino acids circulates in the form of urea, indicating that the conversion of ammonia to urea is highly efficient. Efficient conversion of ammonia to urea protects the central nervous system from toxicity. Elevated blood ammonia occurs when an infant has a genetic defect (for example, missing enzymes or low enzyme activity) or when an adult has a diseased liver. Recently, it has been observed that an increasing incidence of underlying genetic disease is cropping up in the previously normal adult. Increased blood levels of ammonia, usually the result of urea cycle abnormalities, produces a variety of symptoms, such as poor growth, irritability, lethargy, vomiting, anorexia, rapid breathing, disorientation, combativeness, coma, cerebral edema, and death if treatment is ineffective. Respiratory failure can occur in late-stage disease. There are clear signs of central nervous system involvement: poor coordination, hyper- or hypotonia (stiff or flaccid muscles), ataxia, tremor, seizures, decerebrate posturing (rigid extension of arms and legs), and coma. In the brain (cerebellum and striatum), ammonia overactivates the excitatory **N-methyl-D-aspartate (NMDA) receptors** at its glycine site, resulting in accumulation of cyclic guanosine monophosphate (cyclic GMP). NH_4^+ in the brain directs glutamate dehydrogenase to convert α-ketoglutarate and NH_4^+ to glutamate, depressing the citric acid cycle by lowering the level of α-ketoglutarate and subsequently oxaloacetate (Figure 10-1). Depression of aerobic oxidation causes cell damage and cell death. If hyperammonemia is suspected, the plasma

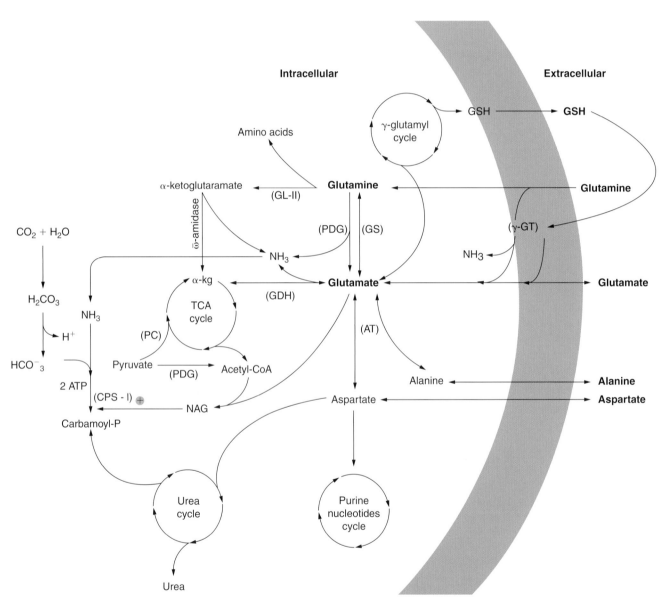

Figure 10-1. Metabolism of amino acids produces ammonia that enters the urea cycle after combining with bicarbonate to form carbamoyl-phosphate. Schematic illustration of glutamine and glutamate metabolism and their interaction with the tricarboxylic acid (TCA) cycle, urea cycle, γ-glutamyl cycle (formation of GSH), and transamination reactions. *AT*, aminotransferase reactions; *CPS-I*, carbamoyl-phosphate synthetase-I; *γ-GT*, γ-glutamyl transpeptidase; *GL-II*, glutamine aminotransferase pathway; *GS*, glutamine synthetase; *GDH*, glutamate dehydrogenase; *NAG*, N-acetylglutamate; *PC*, pyruvate carboxylase; *PDG*, phosphate-dependent glutaminase; *PDH*, pyruvate dehydrogenase; "+" indicates activation. Redrawn from http://isu.indstate.edu/mwking/nitrogen-metabolism.html.

level of ammonia must be tested. The functionality of the liver should be tested in terms of measurement of serum transaminases, prothrombin time, alkaline phosphatase activity (see clinical enzymology in Chapter 3), and bilirubin. Primary genetic disease may manifest as increased plasma levels of citrulline or argininosuccinic acid. Liver disease is suspected when there is a general increase in plasma amino acid levels. Dysfunctions of amino acid catabolism can generate partial inhibition of the urea cycle, resulting indirectly as excess ammonia in the blood. In this case, intermediates of amino acid catabolism may be increased in the blood, such as propionic acid, methylmalonic acid, and isovaleric acid. Urinary amino acid profiles characterize argininosuccinic aciduria, hyperornithinemia, hyperammonemia, homocitrullinuria, or intolerance to dibasic amino acids in ingested protein (lysinuric protein intolerance). The concentration of lactic acid in the blood can rule out mitochondrial diseases (if pyruvate is not well used by the mitochondrial citric acid cycle, lactate will increase in the blood). Increased blood ammonia stimulates the respiratory system so that blood pH may be higher than normal. Blood urea nitrogen will be less than 3 mg/dl, compared to the normal range of 8–20 mg/dl, in disorders of the urea cycle. Immediate treatment of hyperammonemia is reducing protein intake and replacing the protein with calories from other sources; sometimes hemodialysis is required, or sodium, phenylacetate, or benzoate (Ucephan) may be given intravenously (Ucephan stimulates the excretion of nitrogen as phenylacetylglutamine and hippuric acid). Molecular genetics is used to confirm the diagnosis of genetic disease of the urea cycle. The longer high levels of ammonia in the blood go untreated, the more damage to the central nervous system will occur in the form of cell death, culminating in cerebral edema, increased intracranial pressure, and death.

The Urea Cycle

Of excreted nitrogen in the form of urea, about 80% is synthesized (from ammonia) in the liver. Conversion of excess glutamine from the liver to ammonia is catalyzed by a phosphate-dependent kidney glutaminase:

$$\text{HOOC-CH(NH}_2\text{)-CH}_2\text{-CH}_2\text{-CH}_2\text{-CH}_2\text{NH}_2 + H_2O \rightleftharpoons$$
$$\text{glutamine}$$

$$\text{HOOC-CH(NH}_2\text{)-CH}_2\text{-CH}_2\text{-CH}_2\text{-COOH} + NH_4^+$$
$$\text{glutamate}$$

The reaction producing urea from ammonia involves five enzymes, of which the initial two are located in the mitochondrial matrix. The other enzymes of the urea cycle are located in the soluble cytoplasm (cytosol). The urea cycle was discovered by Hans Krebs and Kurt Henseleit and is often referred to as the Krebs-Henseleit cycle.

Bacteria and plants are able to convert nitrogen in the atmosphere, by nitrogen fixation, into amino acids and proteins ingested in the diet. After digestion of food proteins, the resulting amino acids (and very short peptides) are absorbed by the intestinal tract to form human proteins (Figure 10-2). Amino acids not needed for protein synthesis are deaminated and enter the tricarboxylic acid or Krebs cycle to be metabolized (Figure 2-21). Ammonia released from amino acid metabolism is combined with bicarbonate to form carbamoyl phosphate in the first step of the urea cycle (Figure 10-1) that will convert the ammonia into urea. The urea cycle is shown in Figure 10-3. Ammonium ion and bicarbonate enter the mitochondrial matrix (upper left of Figure 10-3) and are converted to carbamoyl phosphate by the action of carbamoyl phosphate synthetase-I and ATP. Carbamoyl phosphate is converted to citrulline by the action of ornithine transcarbamylase with ornithine (generated from arginine in the final reaction of a previous turn of the cycle). Citrulline is transported out of the mitochondrion to the cytosol where, with aspartate, argininosuccinate synthetase catalyzes its conversion to argininosuccinate. Argininosuccinate lyase then converts argininosuccinate to fumarate and arginine (arginine also is available from dietary protein or from protein degradation), and arginine is converted by the terminal enzyme in the cycle, arginase, to

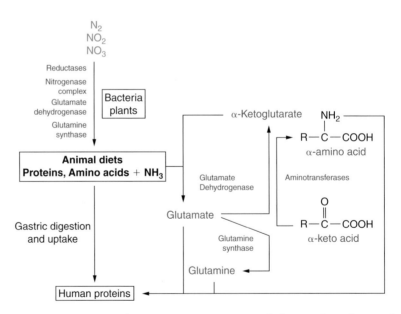

Figure 10-2. Atmospheric nitrogen is converted ultimately to human bodily proteins. Redrawn from I. Nissim, "Newer aspects of glutamine/glutamate metabolism: the role of acute pH changes," *Am. J. Physiol. Renal Physiol.*, 277: F493–F497, 1999.

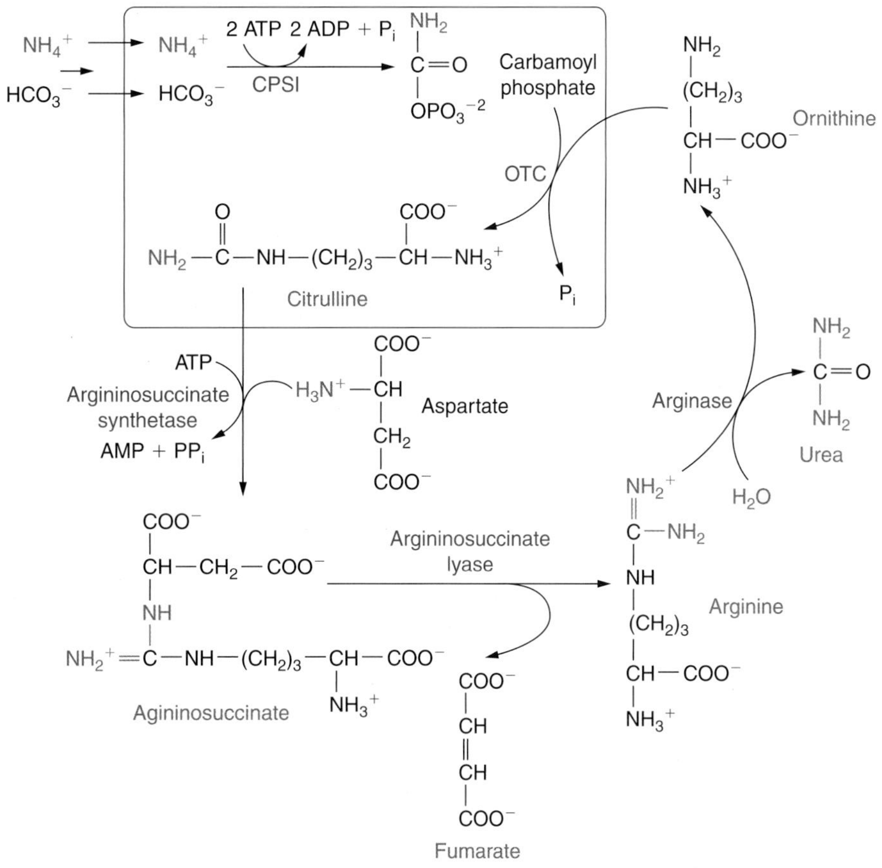

Figure 10-3. Diagram of the urea cycle. The reactions of the urea cycle that occur in the mitochondrion are contained in the red rectangle. All enzymes are in *red*. CPS-I, carbamoyl phosphate synthetase-I; OTC, ornithine transcarbamoylase. Redrawn from http://isu.indstate.edu/nitrogen-metabolism.html.

one molecule of urea and one molecule of ornithine, which can again be used in the mitochondrial reactions. Urea is excreted into the urine by the kidney proximal convoluted tubule. Structures of some enzymes of the urea cycle are shown in Figure 10-4. Deficiencies of any enzymes in the urea cycle pathway (genetic defect) will allow the substrate of the deleted enzyme to accumulate, thereby increasing the concentration of ammonia in the blood. Defects in argininosuccinate synthetase or argininosuccinate lyase will cause the accumulation of citrulline (citrullinuria) or argininosuccinate (argininosuccinic aciduria). A summary of urea cycle defects is given in Table 10-1.

The Urea Cycle   501

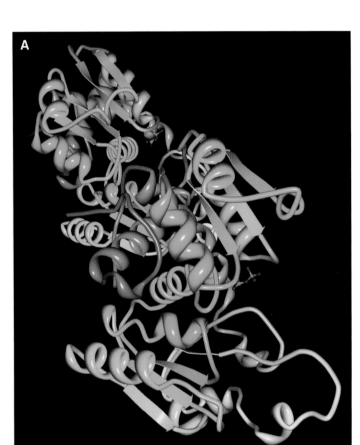
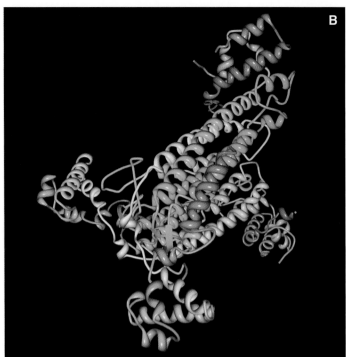
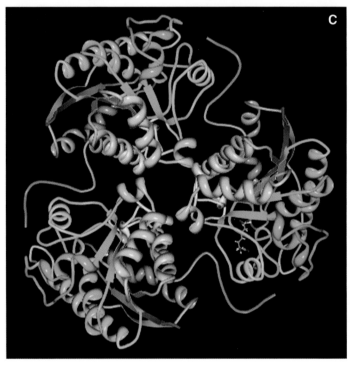

Figure 10-4. Structures of some of the enzymes in the urea cycle. **A.** Crystal structure of human ornithine transcarbamoylase complexed with carbamoyl phosphate (atomic model). Each chain contains 321 amino acid residues. **B.** Human argininosuccinate lyase. Each chain contains 434 amino acid residues. **C.** Crystal structure of human arginase. Part A reproduced from PDB ID: 1FVO, D. Shi, H. Morizono, X. Yu, L. Tong, N.M. Allewell, M. Tuchman. Human Ornithine Transcarbamylase: Crystallographic Insights into Substrate Recognition and Conformational Changes. *The Biochemical Journal* **354** pp. 501 (2001) and from http://www.rcsb.org/pdb/cgi/explore.cgi?pdbId=1fvo. Part B reproduced from PDB ID: 1AOS, M.A. Turner, A. Simpson, R.R. McInnes, P.L. Howell. Human Argininosuccinate Lyase: A Structural Basis for Intragenic Complementation. *Proceedings of the National Academy of Sciences of the United States of America* **94** pp. 9063 (1997) and from http://www.rcsb.org/pdb/cgi/explore.cgi?pdbId=1aos. Part C reproduced from PDB ID: 2AEB, L. Di Costanzo, G. Sabio, A. Mora, P.C. Rodriguez, A.C. Ochoa, F. Centeno, D.W. Christianson. Crystal Structure of Human Arginase I at 1.29 a Resolution and Exploration of Inhibition In the Immune Response. *Proceedings of the National Academy of Sciences of the United States of America* **102** pp. 13058 (2005) and from http://www.rcsb.org/pdb/cgi/explore.cgi?pdbId=2aeb.

Table 10-1
Summary of Urea Cycle Defects (UCDs)

UCD	Enzyme Deficiency	Symptoms/Comments
Type 1 hyperammonemia	Carbamoyl phosphate synthetase 1	24–72 hours after birth infant becomes lethargic, needs stimulation to feed, and experiences vomiting, increasing lethargy, hypothermia, and hyperventilation; without measurement of serum ammonia levels and appropriate intervention infant will die. Treatment with arginine, which activates N-acetylglutamate synthetase.
N-Acetylglutamate synthetase deficiency	N-Acetylglutamate synthetase	Severe hyperammonemia, mild hyperammonemia associated with deep coma, acidosis, recurrent diarrhea, ataxia, hypoglycemia, hyperornithinemia. Treatment includes administration of carbamoyl glutamate to activate CPS1.
Type 2 hyperammonemia	Ornithine transcarbamoylase	Most commonly occurring UCD. Only X-linked UCD, ammonia and amino acids elevated in serum, increased serum orotic acid due to mitochondrial carbamoyl phosphate entering cytosol and being incorporated into pyrimidine nucleotides, which leads to excess production and consequently excess catabolic products. Treat with high-carbohydrate, low-protein diet, ammonia detoxification with sodium phenylacetate or sodium benzoate.
Classic citrullinemia	Argininosuccinate synthetase	Episodic hyperammonemia, vomiting, lethargy, ataxia, seizures, eventual coma. Treat with arginine administration to enhance citrulline excretion, also with sodium benzoate for ammonia detoxification.
Argininosuccinic aciduria	Argininosuccinate lyase (argininosuccinase)	Episodic symptoms similar to those of classic citrullinemia, elevated plasma and cerebral spinal fluid argininosuccinate. Treat with arginine and sodium benzoate.
Hyperargininemia	Arginase	Rare UCD, progressive spastic quadriplegia and mental retardation, ammonia and arginine high in cerebrol spinal fluid and serum, arginine, lysine, and ornithine high in urine. Treatment includes diet of essential amino acids excluding arginine, low-protein diet.

CPS1, carbamoyl phosphate synthetase 1.
Reproduced from http://web.indstate.edu/thcme/mwking/nitrogen-metabolism.html.

Nitrogen Flow, Amino, and Amide Group Transfers in Amino Acid Metabolism

Glutamate dehydrogenase uses ammonia for amino acid formation, and glutamine synthetase converts glutamic acid into glutamine. Glutaminase breaks down glutamine to glutamate and ammonium ion (discussed earlier). Amino groups and amide groups from these amino acids are transferred to other carbon skeletons by reactions of **transamination** and **transamidation.** The glutamate dehydrogenase and glutamine synthase (synthetase) reactions are shown in Figures 10-5 and 10-6. Glutamate dehydrogenase converts free ammonia and α-ketoglutarate into glutamate, and the reverse reaction provides an entry (α-ketoglutarate) for amino acids into the citric acid cycle for energy and production of reduced NADH or NADPH + H$^+$ (glutamate dehydrogenase can use either cofactor). This enzyme, representing a branch point linking amino acids to

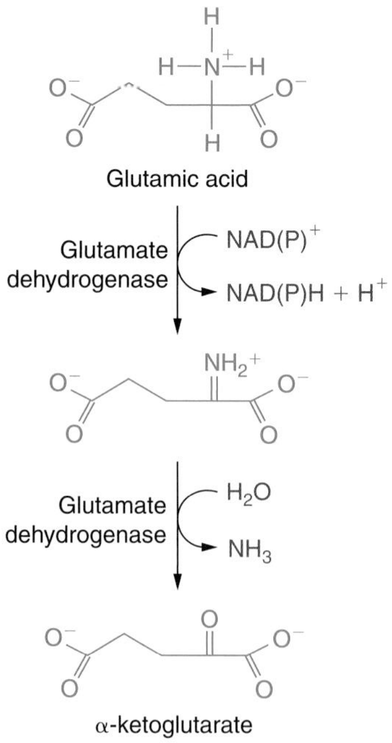

Overall reaction: Glutamate + NADP$^+$ + H$_2$O $\rightleftharpoons$ α-Ketoglutarate + NADPH + H$^+$ + NH$_4^+$

Figure 10-5. Glutamate dehydrogenase catalyzed reaction.

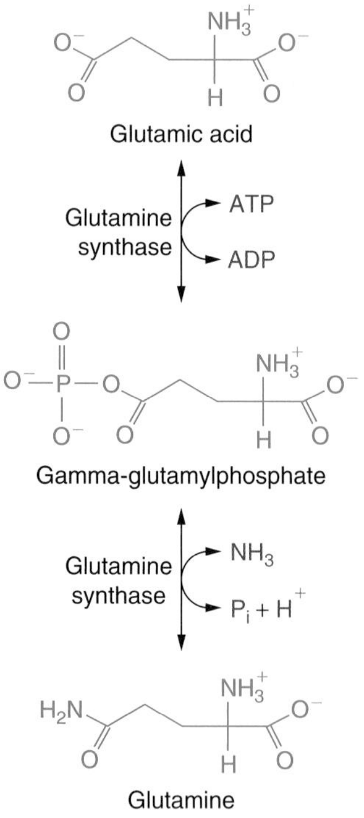

Overall reaction: Glutamate + NH$_4^+$ + ATP $\longrightarrow$ Glutamine + ADP + P$_i$ + H$^+$

Figure 10-6. Glutamine synthase reaction.

504 CHAPTER 10   Metabolism

energy metabolism, is an allosteric enzyme (see Chapter 3) regulated by the effectors, ATP and GTP, positively for the formation of glutamate and by the positive allosteric effectors, ADP and guanosine diphosphate(GDP), in the reverse direction, forming NH_4^+ and α-ketoglutarate. The direction of the reaction depends on the ATP concentration in the cell. When [ATP] is high, conversion of glutamate to α-ketoglutarate is limited, because the need for additional energy is small. When [ATP] is low, glutamate is converted to α-ketoglutarate for oxidation in the citric acid cycle and production of energy through ATP formation (through the electron transport chain). Glutamate is a key amino group donor to other amino acids through the actions of transaminases.

Glutamine synthetase converts glutamate with ammonium ion to glutamine, an important circulating amino acid that carries ammonia (NH_3) from various tissues to the kidney, where glutaminase releases ammonium ion for excretion. Ammonium ion can be used to balance slight acidity in urine. Ammonium ion is more toxic that ammonia. Ammonium ion is used in the brain by glutamate dehydrogenase with α-ketoglutarate to form glutamate, a reaction that can run down the level of α-ketoglutarate and then oxaloacetate and depress the citric acid cycle, leading to cell damage and possibly cell death. The structures of glutamate dehydrogenase and glutamine synthetase are shown in Figure 10-7.

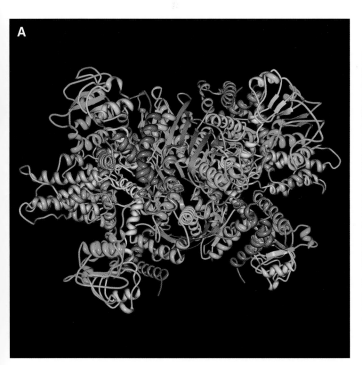
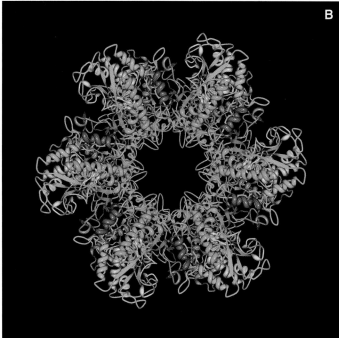

Figure 10-7. **A.** Structure of human glutamate dehydrogenase apo-form. There are six polypeptide chains, each of which is composed of 496 amino acid residues. **B.** Refined atomic model of glutamine synthetase (bacterial). Each of the 12 polypeptide chains is shown in a different color and each consists of 468 amino acid residues. Twenty-four green atoms are manganese ions. Part A reproduced from PDB ID: 1L1F, T.J. Smith, T. Schmidt, J. Fang, J. Wu, G. Siuzdak, C.A. Stanley. The Structure of Apo Human Glutamate Dehydrogenase Details Subunit Communication and Allostery. *Journal of Molecular Biology* **318** pp. 765 (2002) and from http://www.rcsb.org/pdb/cgi/explore.cgi?pdbId=1L1F. Part B reproduced from PDB ID: 2GLS. M.M. Yamashita, R.J. Almassy, C.A. Janson, D. Cascio, D. Eisenberg. Refined Atomic Model of Glutamine Synthetase at 3.5 Å Resolution. *The Journal of Biological Chemistry* **264** pp. 17681 (1989) and from http://www.ebi.ac.uk/thornton-srv/databases/cgi-bin/pdbsum/GetPage.pl?pdbcode=2gls.

In the liver, glutamine synthetase is located in cells (hepatocytes) that are **perivenous** (surrounding a vein, but not the portal vein), whereas glutaminase is located in cells surrounding the portal vein **(periportal cells).** Figure 10-8 shows hepatic cells as they are organized into lobules with respect to their positions about the portal vein (right) and about the central hepatic vein. *Glutamine is the most important nitrogen transporter between tissues, but glutamine conversion through glutamate occurs inside the cells, with minimal exchange between tissues.* In periportal hepatocytes, glutamine is used for the synthesis of glucose by way of glutaminase, transamination to α-ketoglutarate, and

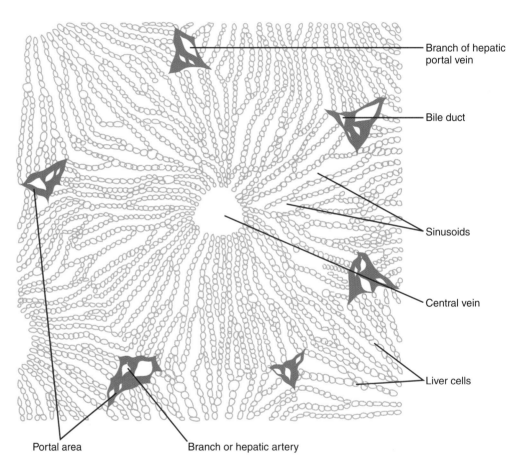

Figure 10-8. Anatomy of a traditional hepatic lobule. Cords (plates) of hepatocytes, 1–2 cells thick, with intervening sinusoids are arranged radially around a central hepatic venule. At the periphery of the lobule are multiple portal triads, each containing a portal venule, a hepatic arteriole, and 1–2 intralobular bile ducts. Bile is secreted by the hepatocytes into an anastomosing network of bile canaliculi, then drains peripherally into ductules at the margins of the portal triads, and from there empties into the interlobular bile ducts. Redrawn from http://www.ariess.com.

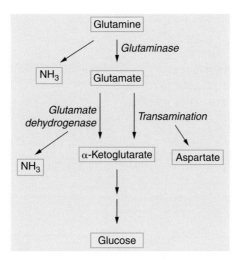

Figure 10-9. Pathways of glutamine utilization in periportal hepatocytes. Redrawn from M. Watford, "Glutamine and glutamate metabolism across the liver sinusoid," *J. Nutr.*, 130: 983S–987S, 2000.

entry into the citric acid cycle and up the glycolytic pathway in reverse to form glucose (Figure 10-9). Ammonia, as NH_4^+, from the glutamate dehydrogenase reaction enters the urea cycle (Figure 10-3) to form urea. In perivenous hepatocytes, glutamine is formed from lactate and arginine (Figure 10-10). An overall summary of these events is shown in Figure 10-11 in terms of what forms of nitrogen derive from the two hepatic cell locations and movement into and out of these cells.

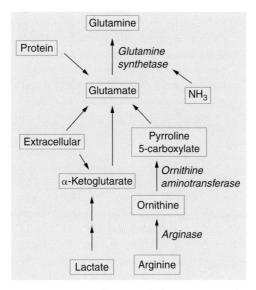

Figure 10-10. Pathways of glutamine synthesis in perivenous hepatocytes. Redrawn from M. Watford, "Glutamine and glutamate metabolism across the liver sinusoid," *J. Nutr.*, 130: 983S–987S, 2000.

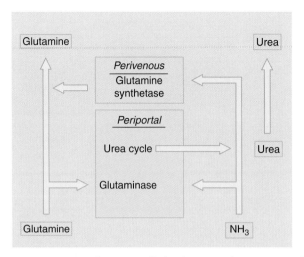

Figure 10-11. The intercellular hepatic glutamine cycle. M. Watford, "Glutamine and glutamate metabolism across the liver sinusoid," *J. Nutr.*, 130: 983S–987S, 2000.

Transamination

In the metabolism of free amino acids, transaminations are a principal means to remove nitrogen from their structures. Transaminations generate other amino acids that can be oxidatively deaminated, producing ammonia, or the amine groups are converted to urea in the urea cycle. Transaminations are reversible reactions removing the α-amino group from one amino acid to the keto carbon of an acceptor, and this class of enzymes generally contains pyridoxal phosphate as a coenzyme, although there are transaminases (aminotransferases) that use pyruvate as a cofactor. A key transaminase is aspartate aminotransferase, whose structure is similar across many species (Figure 10-12A). It catalyzes the reversible reaction shown in Figure 10-12B:

$$\text{L-aspartate} + \alpha\text{-ketoglutarate} \rightleftharpoons \text{oxaloacetate} + \text{L-glutamate}$$
$$\text{(2-oxoglutarate)}$$

There are two forms of this enzyme differing in primary structure, one in the cytosol and one in the mitochondrion. Because the enzyme often is called GOT (glutamate-oxaloacetate transaminase), they may be abbreviated s-GOT and m-GOT. The catalytic reactions of both enzymes are the same but with minute variations. S-GOT is dimeric (Figure 10-12A). The reaction mechanism showing the interaction between the pyridoxal phosphate of aspartate aminotransferase and the ensuing reactions of the coenzyme with aspartate to the ternary complex, which yields oxaloacetate and pyridoxamine phosphate, are shown in Figure 10-13. Note that the bond between the coenzyme and the enzyme involving lysine 258 (pig heart aspartate aminotransferase) is shifted to the amino group of the substrate, aspartate, and forms a Schiff base. A hydrogen is removed from aspartate β-carbon to form a double bond between the aspartate amino

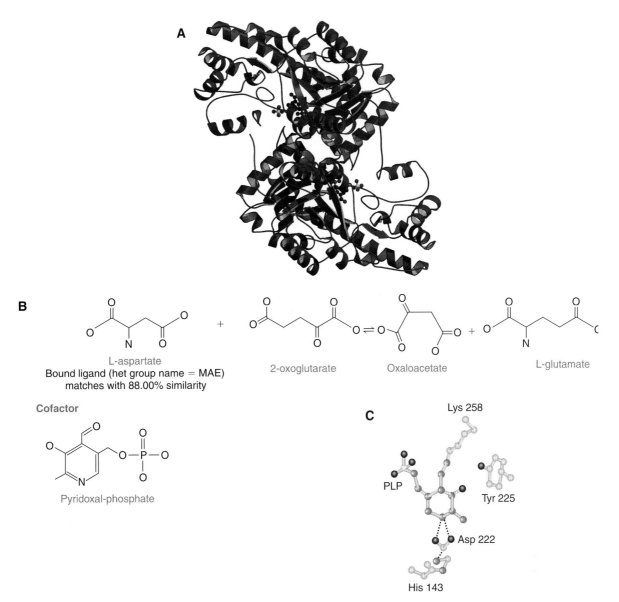

Figure 10-12. **A.** Crystal structure of bacterial aspartate aminotransferase dimer. Atomic structures are pyridoxal phosphate coenzyme and substrate maleate. **B.** Aspartate aminotransferase catalyzed reaction showing the structure of the coenzyme, pyridoxal phosphate (below). **C.** Pyridoxal phosphate and its interactions with amino acid residues of aspartate aminotransferase. Parts A and B redrawn from PDB ID: 1AIC. V.N. Malashkevich, J. Jager, M. Ziak, U. Sauder, H. Gehring, P. Christen, J.N. Jansonius. Structural Basis for the Catalytic Activity of Aspartate Aminotransferase K258H Lacking the Pyridoxal 5′-Phosphate-binding Lysine Residue. *Biochemistry* **34** pp. 405 (1995) and from http://www.ebi.ac.uk/thornton-srv/databases/cgi-bin/pdbsum/GetPage.pl?pdbcode=1aic.

group and the former aldehyde of pyridoxal phosphate, and the hydrogen is shifted to the same pyridoxal phosphate carbon to form the intermediate complex. The double bond then can be broken to form the corresponding keto acid in the case of transamination. This intermediate also can proceed to decarboxylation of the amino acid substrate in the case of a decarboxylase, which also uses pyridoxal phosphate as coenzyme, or to racemization of the amino acid to the D-form by a racemase, which uses

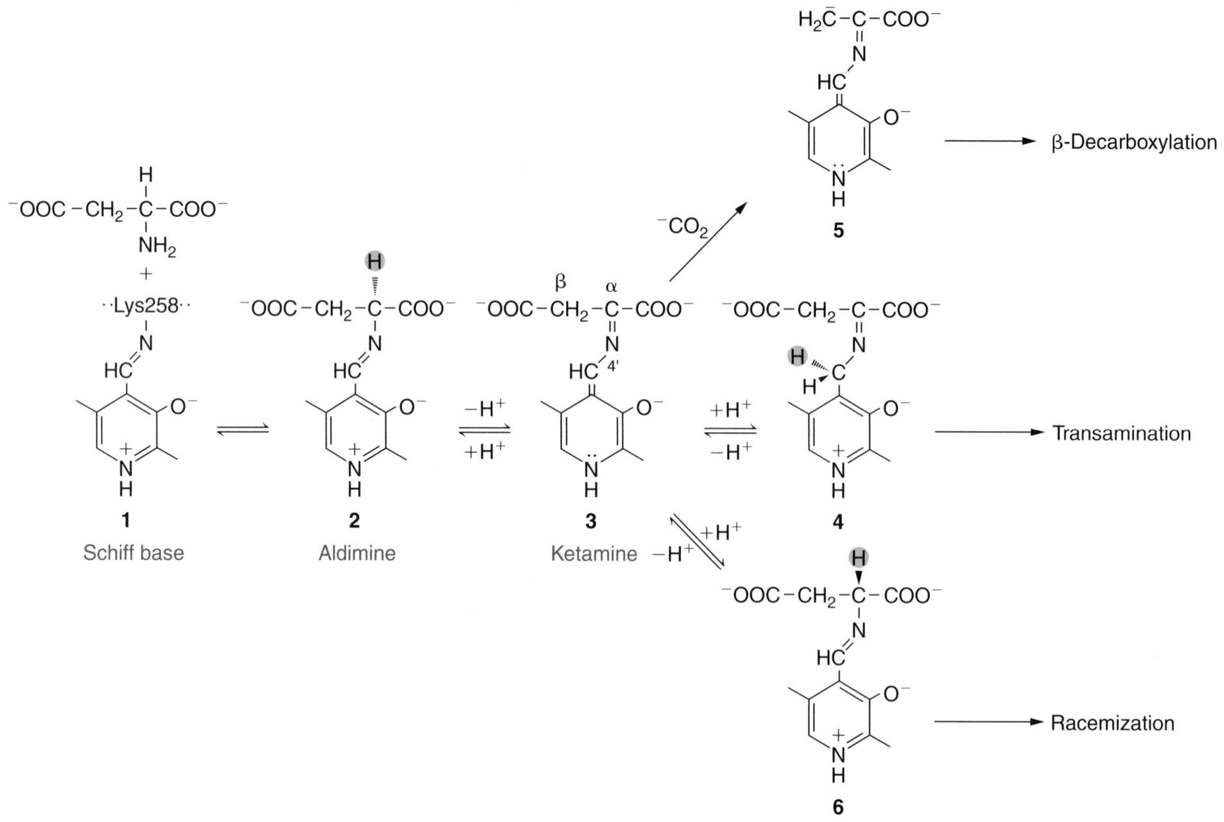

Figure 10-13. Reaction mechanism of transamination (and other pyridoxal phosphate catalyzed reactions) in the pig heart aspartate aminotransferase reaction. This shows an intermolecular transfer between the enzyme lysine residue (lys258) and the amino group of the aspartate substrate to form in second schiff base in the first step of the reaction.

pyridoxal phosphate as a coenzyme. The transamination reaction mechanism is shown in Figure 10-13, and the overall reaction is as follows:

Enz-lysN-PLP + asp ⇌ Enz-PLP=Nasp−H$^+$ ⇌ Enz-PLP-N=asp ⇌ oxaloacetate (OAA) + Enz-PMP + αKG ⇌ Enz-PLP-N-αKG ⇌ Enz-PLP=N-αKG ⇌ Enz-lysN-PLP + L-glutamate

Thus, the original form of the enzyme (Enz) bound to the coenzyme (PLP or PMP) is rejuvenated, and the overall reaction (Figure 10-12B) is as follows:

L-asp + αKG = OAA + L-glutamate

Thus, the amino group of aspartate (asp) is transferred to α-ketoglutarate (αKG). Pyridoxal phosphate (PLP) bound to the aminotransferase through the ε-carbon of a specific lysine (lys) residue in a Schiff base (Enz-lysN-PLP) is subject to a shift of that bond to the incoming amino acid substrate to form another Schiff base (Enz-PLP=Nasp). The coenzyme, pyridoxal phosphate, is still bonded to the enzyme by interactions other than through the aldehyde carbon of pyridoxal phosphate (Figure 10-12B), for example, tyrosine and histidine residues (Figure 10-12C). The structures of the coenzymes, pyridoxal phosphate and pyridoxamine phosphate (PMP), as well as the vitamin forms, pyridoxine and pyridoxal, and a Schiff base (aldimine) linkage are shown in Figure 10-14. Other transaminases are as follows:

Glutamate-pyruvate aminotransferase:

$$\text{L-glutamate} + \text{pyruvate} \rightleftharpoons \alpha KG + \text{L-alanine}$$

Glutamate-α-ketoisovalerate (αKIV) aminotransferase:

$$\text{L-glutamate} + \alpha KIV \rightleftharpoons \alpha KG + \text{L-valine}$$

or

HOOC-HC(NH$_2$)-(CH$_2$)$_2$-COOH (L-glutamate) + (CH$_3$)$_2$CH-C(O)-COOH (αKIV) $\rightleftharpoons$ HOOC-C(O)-(CH$_2$)$_2$-COOH (αKG) + (CH$_3$)$_2$CH-CH(NH$_2$)-COOH (L-valine)

In addition, aspartate aminotransferase can be coupled with glutamate-pyruvate aminotransferase:

$$\overset{E1}{\text{L-asp} + \alpha KG \rightleftharpoons \text{OAA} +} \overset{E2}{\textbf{L-glutamate} + \text{pyruvate} \rightleftharpoons \alpha KG + \text{L-alanine}}$$

E1 is aspartate aminotransferase and E2 is glutamate–pyruvate aminotransferase; the aminated product of the first reaction is glutamate, which is a cosubstrate for the coupled reaction. Thus, in effect, the amino group

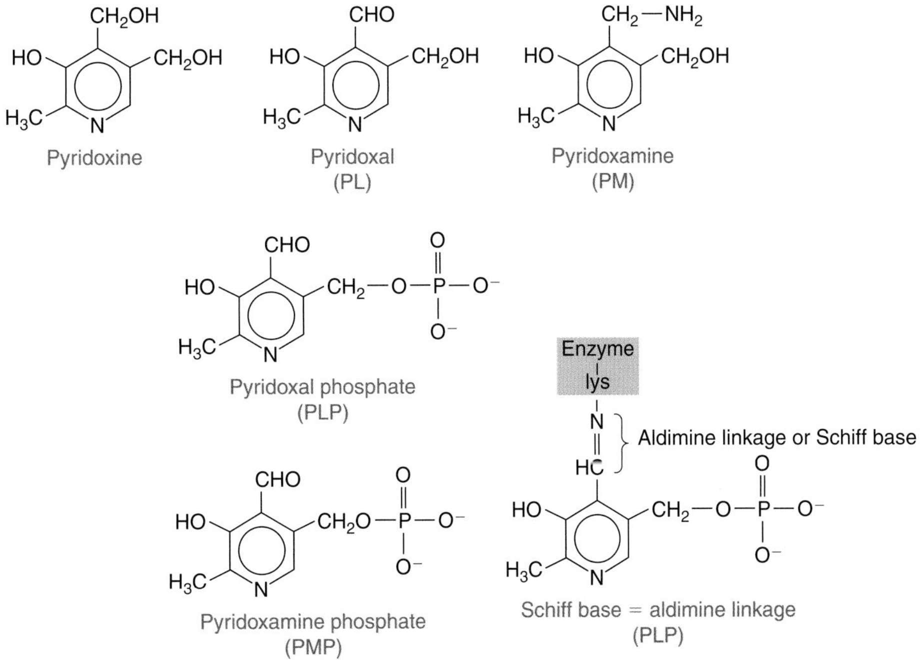

Figure 10-14. Structures of the coenzymes, pyridoxal phosphate (PLP) and pyridoxamine phosphate (PMP), and the vitamin forms, pyridoxine (PN) and pyridoxal (PL). The Schiff base, or aldimine linkage, is also shown.

of aspartate is transferred to become the amino group of alanine. Transamination reactions allow for the transfer of amino acid amino groups to aminate carbon skeletons to generate other amino acids.

Transamidation

In the synthesis of aminosugars, for example, a special case of transamination takes place whereby the ε (C5) amide of glutamine contributes its amide group to fructose-6-phosphate to form glutamate and glucosamine-6-phosphate. Glucosamine-6-phosphate with acetyl CoA is the starting point for the synthesis of some acetylated sugars, such as acetylglucosamine phosphate and uridine diphosphate (UDP)-acetylglucosamine (Figure 10-15).

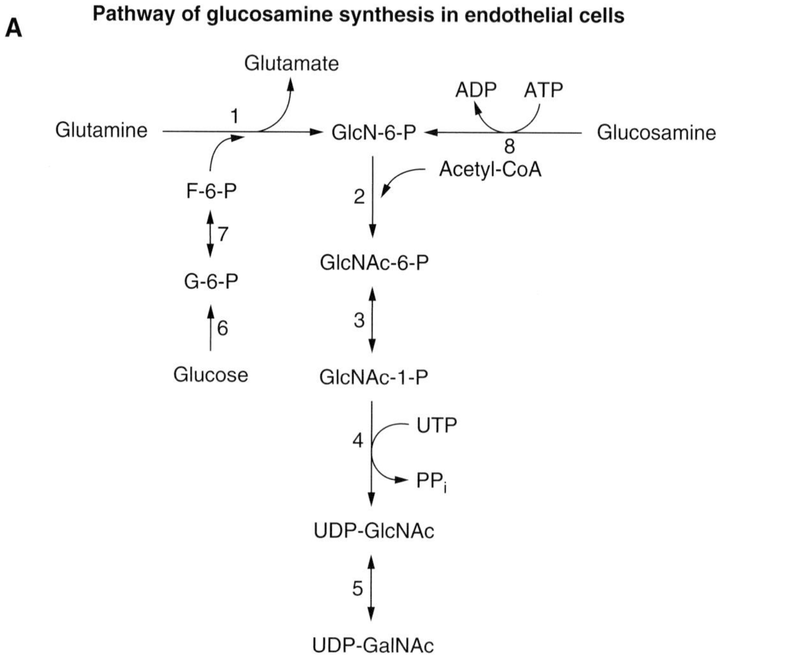

A Pathway of glucosamine synthesis in endothelial cells

Enzymes that catalyse the indicated are: 1, GFAT; 2, glucosamine-phosphate N-acetyltransferase; 3, phosphoacetylglucosamine mutase; 4, UDP-GlcNAc pyrophosphorylase; 5,UDP-GlcNAc 4-epimerase; 6, hexokinase; 7, phosphohexose isomerase; 8, glucosamine kinase. Abbreviations: G-6-P, glucose 6-phosphate; F-6-P, fructose 6-phosphate.

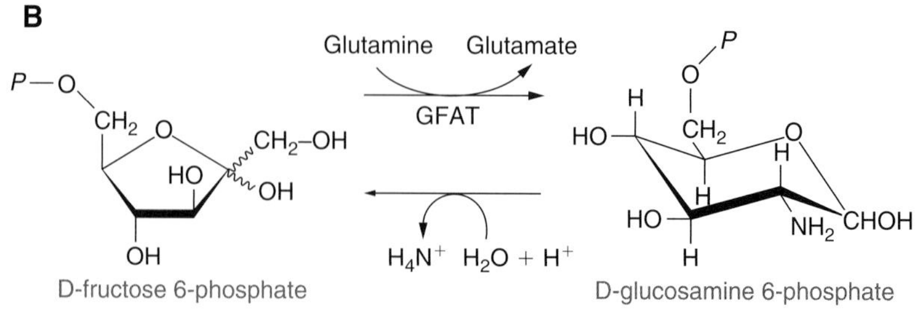

Figure 10-15. **A.** Scheme showing the synthesis of D-glucosamine 6-phosphate from D-fructose 6-phosphate that is transamidated by GFAT (L-glutamate: D-fructose 6-phosphate transaminase) from glutamine, moving the distal amide group and producing glutamate. D-glucosamine 6-phosphate is used for the synthesis of other related sugar derivatives. **B.** The forward reaction of GFAT.

Deamination

Deamination of serine (and threonine) is catalyzed by serine dehydratase (and threonine dehydratase), a pyridoxal phosphate–requiring enzyme (Figure 10-16). Deamination by the hydrolysis of glutamine and asparagine amide nitrogens also occurs by pyridoxal phosphate–catalyzed enzymatic reactions. Histidase deaminates histidine to generate **urocanate** and ammonia (Figure 10-17).

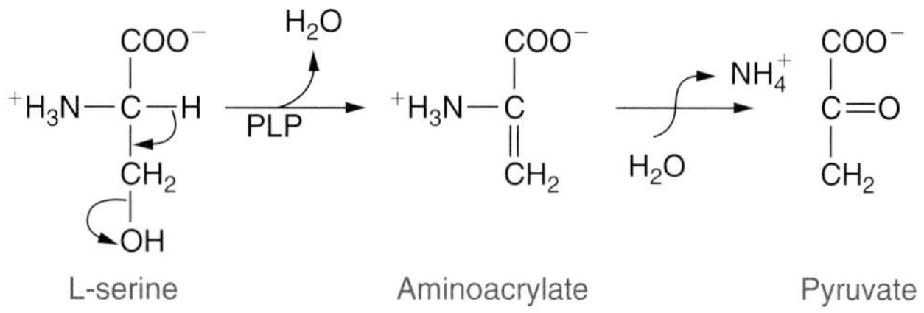

Figure 10-16. Deamination of serine catalyzed by the pyridoxal phosphate enzyme, serine dehydratase. Water is removed from the amino acid (arrows) and then the amine is removed to produce ammonium ion (NH_4^+) and pyruvate by the hydration of aminoacrylate. The analogous reaction occurs with L-threonine, catalyzed by threonine dehydrase where the product of the reaction is 2-ketobutyrate.

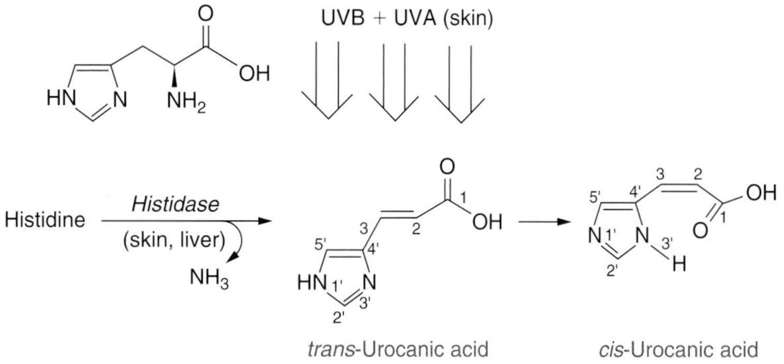

Figure 10-17. Histidase reaction (liver and skin) produces urocanic acid and ammonia. In skin, ultraviolet light isomerizes urocanic acid to *trans*-urocanic acid and *cis*-urocanic acid.

Amino Acid Oxidation

There are both L- and D-amino acid oxidases, which catalyze the direct oxidation of amino acids—especially when amino acids are in excess or are at toxic levels. They are located in liver and kidney **peroxisomes** (Figures 1-26 and 1-27). These enzymes are flavoproteins and exhibit wide specificities (Figures 10-18 and 10-19).

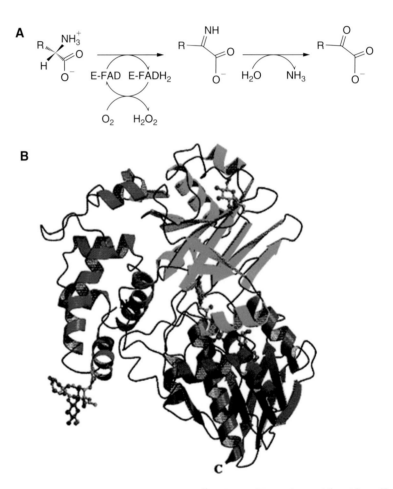

Figure 10-18. **A.** Reaction mechanism of L-amino acid oxidase (from *Calloselasma rhodastoma*). E-FAD, enzyme bound flavin adenine nucleotide (coenzyme). Shown are generic structures for L-amino acid, intermediate and corresponding keto acid. **B.** Structure of L-amino acid oxidase. FAD is the atomic structure *(ball and stick)*. Part B reproduced from P.D. Pawelek, et al., "The structure of L-amino acid oxidase reveals the substrate trajectory into an enantiomerically conserved active site," *The EMBO J.*, 19: 4204–4215, 2000.

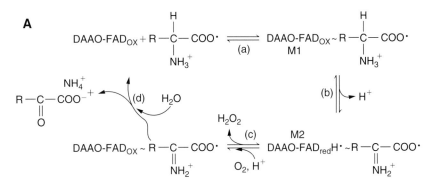

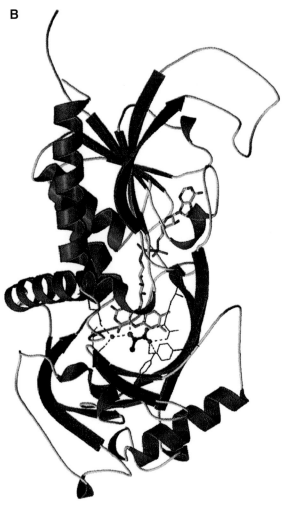

Figure 10-19. **A.** Reaction mechanism of D-amino acid oxidase *(Rhodotorula gracilis)* at pH <8. *(a)* Binding of the D-amino acid to form the Michealis complex (M1). *(b)* Reversible conversion into the reduced enzyme-produce complex (M2) involving release of H^+. *(c)* (Re)oxidation by dioxygen, and *(d)* release of products α-keto acid and NH_4^+. *DAAO*, D-amino acid oxidase; *FAD*, flavin adenine dinucleotide (coenzyme). **B.** Structure of D-amino acid oxidase. The substrate D-alanine is located above the reduced flavin *Re*-side. Dashed lines (---) denote hydrogen bonds involved in substrate fixation. Parts A and B reproduced from S. Umhau et al., "The x-ray structure of D-amino acid oxidase at very high resolution identifies the chemical mechanism of flavin-dependent substrate dehydrogenation," *PNAS USA*, 97: 12463–12468, 2000, and from http://www.pnas.org/cgi/content/full/97/23/12463.

Amino Acid Decarboxylation

Amino acid decarboxylases are enzymes that contain pyridoxal phosphate as coenzyme. There is a decarboxylase for most amino acids. An example is aromatic amino acid decarboxylase (broader spectrum of substrates than most), which catalyzes the decarboxylation of dihydroxyphenylalanine (DOPA), tryptophan, or hydroxytryptophan. The general reaction is as follows:

(A) L-aromatic amino acid $\rightleftharpoons$ aromatic amine + CO_2

This enzyme is responsible for the conversion of L-DOPA to dihydroxyphenylethylamine (dopamine) and L-5-hydroxytryptophan to serotonin. DOPA decarboxylase catalyses the following reactions:

(B) L-aromatic amino acid $\rightleftharpoons$ aromatic amine + CO_2

(C) aromatic amine + $\frac{1}{2} O_2 \rightleftharpoons$ aromatic aldehyde + NH_3

The reaction in (C), consuming oxygen, occurs with dopamine but is not characteristic of pyridoxal phosphate enzymes using other substrates. The more typical reaction is shown in (A) and defined in the scheme shown in Figure 10-13. A number of amino acid decarboxylases are listed in Table 10-2. The crystal structure of DOPA decarboxylase (aromatic amino acid decarboxylase) is shown in Figure 10-20.

Table 10-2
A List of Some Amino Acid Decarboxylases with Pyridoxal Phosphate as Coenzyme

Enzyme	Substrate(s)	Products
Aspartate α-decarboxylase	L-aspartate	β-alanine + CO_2
Valine decarboxylase	L-valine (or L-leucine)	2-methylpropanamine + CO_2
Glutamic acid decarboxylase	L-glutamate (in brain: L-cysteate,* 3-sulfino-L-alanine, L-aspartate)	4-aminobutanoate + CO_2
Lysine decarboxylase	L-lysine (hydroxy-L-lysine)	cadavarine + CO_2
Arginine decarboxylase	L-arginine	agmatine + CO_2
Histidine decarboxylase	L-histidine (PLP or pyruvate as coenzyme)	histamine + CO_2
Aromatic L-amino acid decarboxylase	tryptophan (DOPA, hydroxy-tryptophan)	tryptamine + CO_2
Phenylalanine decarboxylase	L-phenylalanine (tyrosine + other aromatic amino acids)	phenylethylamine + CO_2
Methionine decarboxylase	L-methionine	3-methylthiopropanamine + CO_2

*2-amino-3-sulfoproprionate

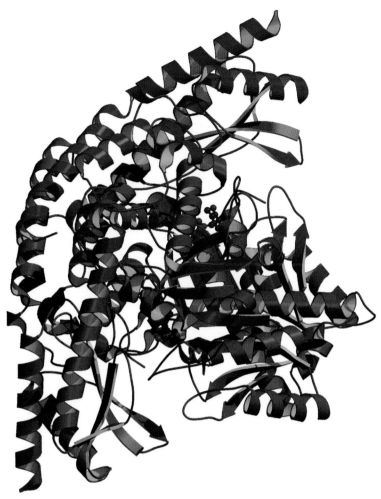

Figure 10-20. Crystal structure of pig kidney DOPA decarboxylase or aromatic amino acid decarboxylase as a dimer with two molecules of pyridoxal phosphate bound (atomic structures). Each polypeptide chain consists of 464 amino acid residues with 22 helices and 11 strands. Reproduced from PDB ID: 1JS6. P. Burkhard, P. Dominici, C. Borri-Voltattorni, J.N. Jansonius, V.N. Malashkevich. Structural Insight into Parkinson's Disease Treatment from Drug-inhibited DOPA Decarboxylase. *Nature Structural Biology* **8** pp. 963 (2001) and from http://www.ebi.ac.uk/thornton-srv/databases/cgi-bin/pdbsum/GetPage.pl?pdbcode=1js6.

Metabolism of Specific Amino Acids to Key Substances

Methionine

Methionine is one of eight essential amino acids (along with tryptophan, lysine, phenylalanine, threonine, valine, leucine, and isoleucine; histidine and arginine are essential in children). The nonessential amino acids are glutamate, glutamine, aspartate, asparagine, alanine, cysteine, tyrosine, proline, serine, and glycine (and ornithine). Methionine and cysteine are encoded in the genetic code, but homocysteine and cystine are not. Methionine has only one codon, AUG, whereas other amino acids, except for tryptophan (UGG), have two or three codons (Figure 2-34). Homocysteine is a product of the methyl donor, S-adenosyl methionine (SAM). Recently, measurement of homocysteine level in blood has been used as an indicator of possible coronary artery disease (15–30 mmol/liter for moderate, >30–100 mmol/liter for intermediate, and >100 mmol/liter for severe). **Hyperhomocysteinemia (HHCE)** occurs in 5% to 7% of the population, and patients with mild HHCE often experience premature coronary artery disease in their 30s and 40s, as well as recurrent venous and arterial thrombosis. HHCE can be caused by defects in the enzymes (cystathionine β-synthase deficiency), as well as deficiencies in B vitamins, especially folate, vitamin B_{12}, or vitamin B_6. Some drugs and chronic disease states can lead to moderate HHCE. High levels of homocysteine in the blood damage arterial endothelial cells and promote smooth muscle growth, leading to the formation of plaque and possibly to the disruption of the blood clotting mechanism, increasing the risk of heart attack or stroke. It is, as yet, unknown whether normalizing the level of homocysteine in the blood will reverse these risks. Methyltetrahydrofolate reductase polymorphisms produce moderate HHCE.

Methionine occurs as the N-terminal amino acid of all proteins in eukaryotes (although it can be removed during posttranslational modification of the protein) and in other positions in proteins. Methionine and homocysteine are interconvertible, and cysteine can be derived from homocysteine. These are the sulfur-containing amino acids. Methionine is involved in cysteine, carnitine, and taurine syntheses by the **transsulfuration pathway.** It is also involved in the production of **lecithin** (significant in brain, lung, and spleen) and the synthesis of **phosphatidylcholine,** as well as other phospholipids. Dipalmitoyl lecithin has the following structure:

$$CH_2OC(O)(CH_2)_{14}CH_3$$
$$|$$
$$CHOC(O)(CH_2)_{14}CH_3$$
$$|$$
$$CH_2OP(O^-)OCH_2CH_2N^+(CH_3)_3$$

The sn3 (carbon 3 of glycerol) position, which contains a phosphoric acid ester of choline and lecithin, is phosphatidyl choline. SAM is important in the synthesis of phosphatidyl choline from phosphatidylethanolamine, as shown in Figure 10-21. Betaine, derived from choline, is the immediate methyl donor for the conversion of homocysteine to methionine. The structure of betaine is as follows:

$$(CH_3)_3N^+CH_2COO^-$$

Some aspects of methionine metabolism are enlarged with structures in Figure 10-22. The synthesis of SAM from methionine is shown in Figure 10-23. SAM is an important methyl group donor. All phosphate groups of ATP are lost as P_i and PP_i in the synthesis of SAM. Homocysteine is condensed with serine in the production of cystathionine (Figure 10-22). Cystathionase cleaves cystathionine to cysteine and α-ketobutyrate, and these two reactions summate as transsulfuration. α-Ketobutyrate is converted to proprionyl CoA after decarboxylation. Cysteine negatively regulates cystathionase allosterically, and cysteine inhibits the expression of the gene for cystathionine synthase. Cystathionine synthase and

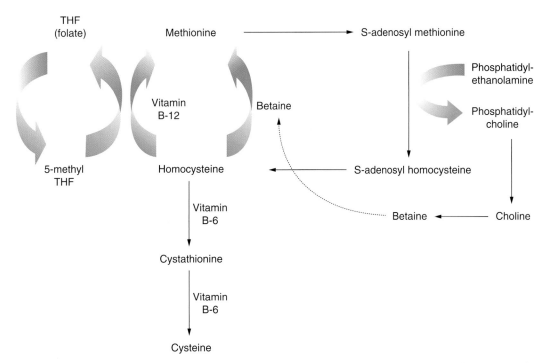

Figure 10-21. Role of S-adenosylmethionine in formation of phosphatidyl-choline (lecithin) and the donation of betaine methyl group (from choline) for the conversion of homocysteine to methionine.

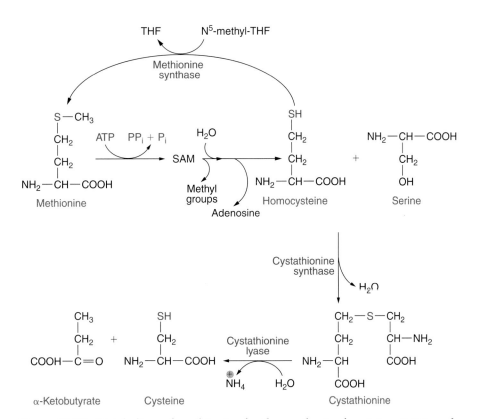

Figure 10-22. Metabolism of methionine for the synthesis of cysteine. *SAM*, s-adenosylmethionine; *THF*, tetrahydrofolate.

Metabolism of Specific Amino Acids to Key Substances

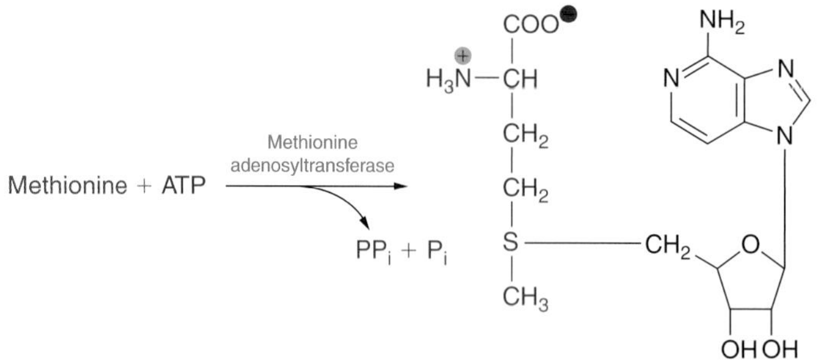

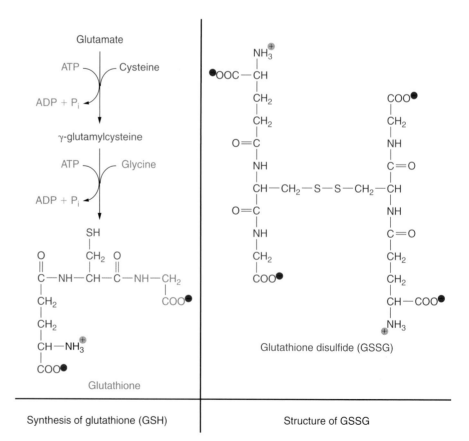

Figure 10-23. Synthesis of 5-adenosylmethionine from methionine.

cystathionase contain pyridoxal phosphate as their cofactor. Homocystinuria, sometimes reflecting mental retardation, occurs with impaired function of cystathionine synthase, which often can be improved by administration of pyridoxine (vitamin B_6), possibly reflecting poor binding to its coenzyme, pyridoxal phosphate, to the synthase. Diminished activity of cystathionase causes benign cystathionine excretion in the urine.

Glutathione (GSH), formed from cysteine, reduces cystine (Cys-S-S-Cys) nonenzymatically (cystine is produced from the oxidation of two cysteines). GSH is a subsequent product of cysteine generated in the transsulfuration pathway. The synthesis of GSH is shown in Figure 10-24.

Figure 10-24. Synthesis of glutathione (GSH) and its oxidation product, glutathione disulfide (GSSG). Redrawn from http://isu.indstate.edu/mwking/nitrogen-metabolism.html.

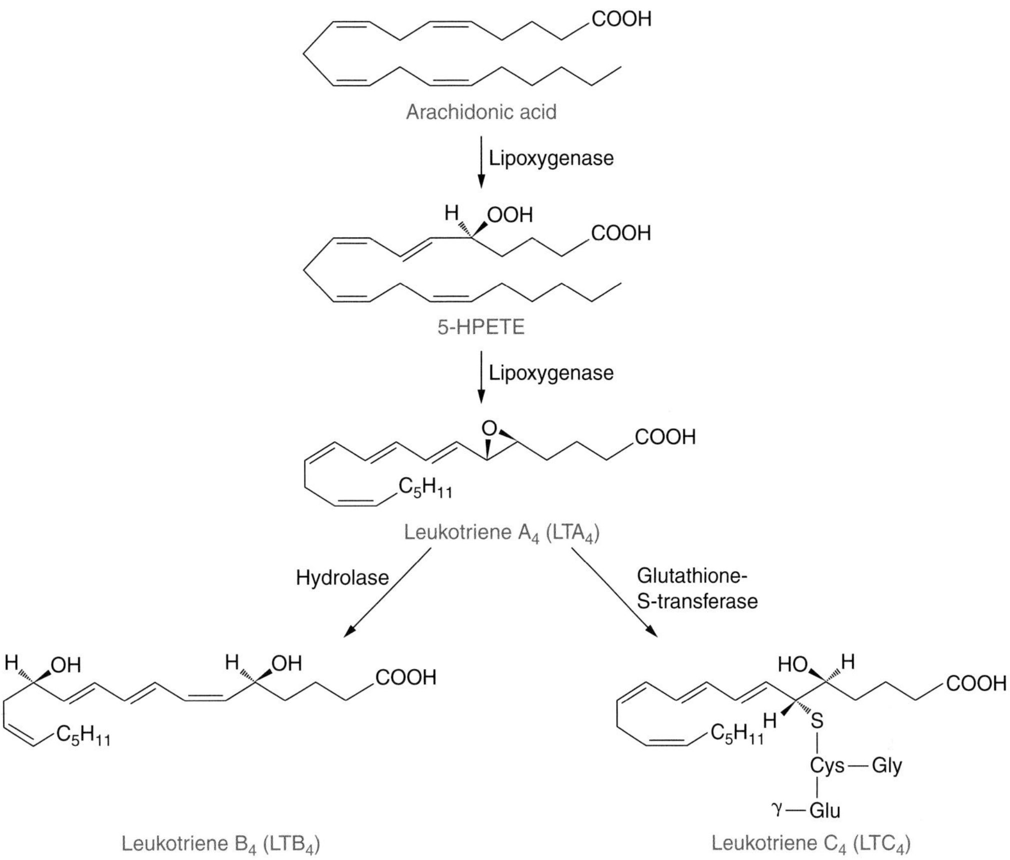

Figure 10-25. Synthesis of leukotriene C_4 (LTC$_4$) containing glutathione (glu-cys-gly) from arachidonic acid (from the cell membrane released by phospholipase A_2.). *HPETE,* hydroperoxyeicosatetraenoic acid. The amino acids of glutathione are released upon degradation.

GSH is a cellular reducing agent and is conjugated to a variety of drugs and **xenobiotics** (not natural components of the body) by way of the GSH S-transferase family of enzymes. It is involved in the γ-glutamyl cycle in the transport of amino acids across cell membranes and is part of the structure of peptide-containing leukotrienes. The synthesis and structure of a GSH-containing leukotriene (LTC$_4$) is shown in Figure 10-25. The γ-glutamyl cycle is shown in Figure 10-26.

Phenylalanine and Tyrosine

Phenylalanine, an essential amino acid, is taken in through the diet and is needed in large amounts to form tyrosine, a nonessential amino acid, when tyrosine is insufficient in the diet. Tyrosine is produced in the body by the hydroxylation of phenylalanine in the liver. In a diet low in tyrosine, about half of the phenylalanine eaten will be converted to tyrosine. If tyrosine is supplied in the diet in large amount, the need for phenylalanine in the diet is reduced by half. Tyrosine, either through ingestion or formed from phenylalanine, serves as the precursor of catecholamines, **melanin,** and other

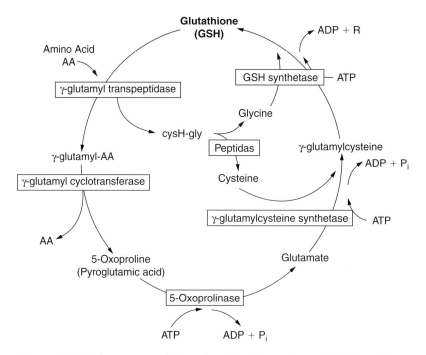

Figure 10-26. The gamma-glutamyl cycle. *AA*, amino acid. This cycle can aid the transport of an amino acid (AA) from outside the cell into the cell interior. Redrawn from http://www.ojrd.com/content/2/1/1b/figure/f1.

metabolites. Phenylalanine hydroxylase (PAH), expressed only in the liver, is a tetrameric enzyme that converts phenylalanine to tyrosine with the incorporation of one atom of oxygen into the needed hydroxyl group and another atom of oxygen into water. Consequently, it is a mixed function oxygenase. Its coenzyme is tetrahydrobiopterin (BH4). BH4 donates two hydrogen atoms in the PAH reaction, and the reduction state of dihydrobiopterin (after donating its hydrogens) is replenished by dihydrobiopterin reductase, using NADH + H$^+$ as coenzyme (Figure 10-27). Figure 10-27 shows an oversimplified reaction. In particular, two hydrogen atoms are transferred to one oxygen atom to form water and another hydrogen atom is transferred to the hydroxyl group of tyrosine, whereas the mechanism shown here allows for the transfer of only two atoms of hydrogen between BH4 and dihydrobiopterin. The mechanism is complex and not completely understood. The enzyme, PAH, requires erythro-BH4, which contains an atom of **nonheme iron** (Figure 10-28), and it is this iron atom with which oxygen forms a superoxide. The iron atom Fe(III) is first reduced to Fe(II) (the mechanism is not known), followed by a decrease in affinity for two water molecules near the iron. BH4 then binds to the enzyme, followed by the binding of phenylalanine. Dioxygen (O:O) binds where one of the two water molecules was previously located, and a peroxy-BH4 is formed. The oxygen–oxygen bond breaks down followed by the breakdown of peroxyhydrobiopterin into hydroxybiopterin plus an activated oxygen intermediate (peroxypterin iron). The products, L-tyrosine and hydroxytetrabiopterin, are released in the final step. The partial structure of the tetrameric PAH is shown in Figure 10-29. A deficiency of PAH or its mutation to a less active or inactive form (there are more than 400 known

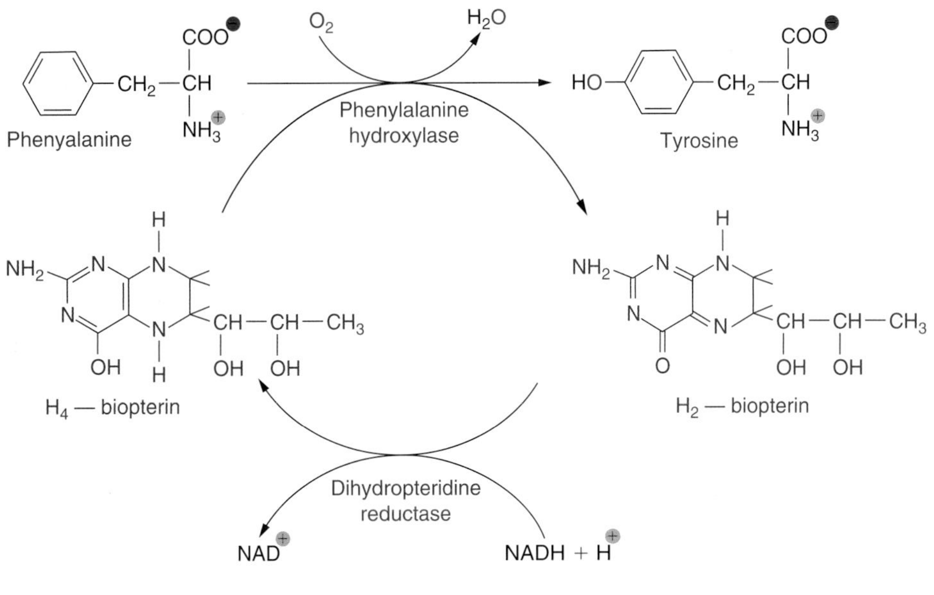

Figure 10-27. Conversion of L-phenylalanine to L-tyrosine by the action of phenylalanine hydroxylase, a mixed function oxygenase. Hydrogens are donated to atoms of oxygen for the formation of the hydroxyl group on tyrosine and for the formation of water.

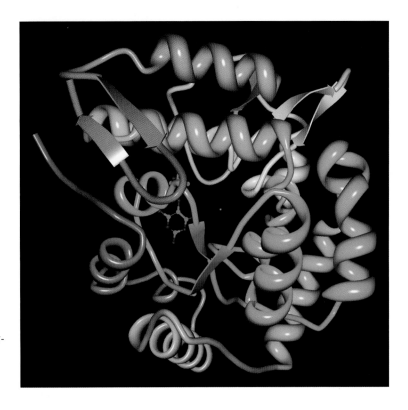

Figure 10-28. Crystal structure of double truncated human phenylalanine hydroxylase with bound 7,8-dihydro-L-biopterin. Iron atom is in center. Reproduced from PDB ID: 1DMW. H. Erlandsen, E. Bjorgo, T. Flatmark, R.C. Stevens. Crystal Structure and Site-specific Mutagenesis of Pterin-bound Human Phenylalanine Hydroxylase. *Biochemistry* **39** pp. 2208 (2000) and from http://www.rcsb.org/pdb/cgi/explore.cgi?pdbId=1dmw.

mutations) leads to phenylketonuria in children. When phenylalanine cannot be converted to tyrosine, phenylalanine levels in the blood rise and the serum concentration of phenylalanine indicates the severity of the disease (normal, 1 mg/dl or 0.061 mM; benign hyperphenylalaninemia, or HPA, 4–10 mg/dl or 0.24 to 0.605 mM; variant HPA, 10–20 mg/dl or 0.605–1.21 mM; classic phenylketonuria, >20 mg/dl or >1.21 mM). When PAH is defective or when there is a defect in generating adequate amounts of the cofactor, BH4, HPA occurs only in a small percentage of cases, which cannot be controlled by limiting phenylalanine in the diet. Phenylalanine

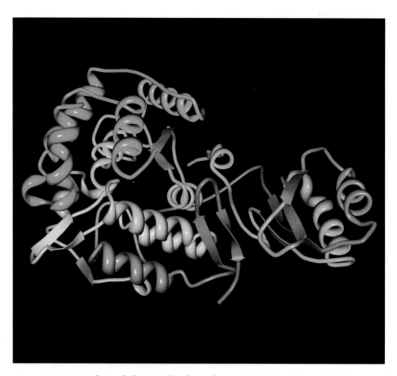

Figure 10-29. Phenylalanine hydroxylase. PDB ID: 1phz. B.Kobe, et al. Structural Basis of Autoregulation of Phenylalanine Hydroxglase. *Nature Structural Biology* **6** pp. 442 (1999).

is not converted to tyrosine; instead, phenylalanine is transaminated to phenylpyruvate (Figure 10-30). Two enzyme defects in the BH4 synthetic pathway have been identified in patients with HPA; these are defects in GTP cyclohydrolase, the first enzyme in the pathway, and pyruvoyl tetrahydropterin synthase, the second enzyme in the pathway.

Under normal conditions, when phenylalanine is present in excess over the need for tyrosine formation, it will be transaminated with α-ketoglutarate to phenylpyruvate and glutamate. The phenylpyruvate formed will not be enough to accumulate excessively and cause problems. Phenylalanine, in excess of its normal conversion to tyrosine, is also converted to phenylethylamine and to other metabolites shown in Figure 10-31. The accumulation of phenylpyruvate *when PAH is poorly active or inactive* can lead to mental retardation in infants. Dietary phenylalanine must be restricted. A future plan, where PAH is clearly defective, might be the introduction of the normal PAH gene to a phenylketonuria child (gene therapy). Whether this would halt further damage or cause partial or complete reversal would be of great interest.

Formation of Catecholamines

Tyrosine is the precursor of the catecholamines: dopamine, norepinephrine (noradrenalin) and epinephrine (adrenaline). Both norepinephrine and epinephrine are secreted from nerve endings, and norepinephrine is the

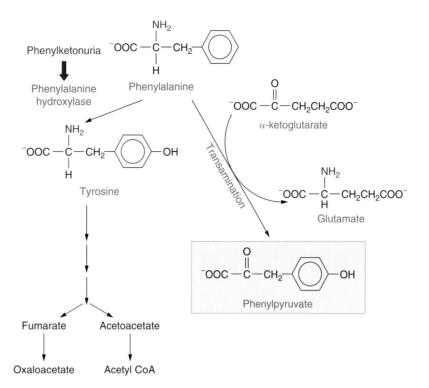

Figure 10-30. A deficiency of phenylalanine hydroxylase (PAH) prevents conversion of phenylalanine to tyrosine. Instead, phenylalanine is transaminated to phenylpyruvate. Accumulation of phenylpyruvate can lead to mental retardation in infants, which is called phenylketonuria (PKU).

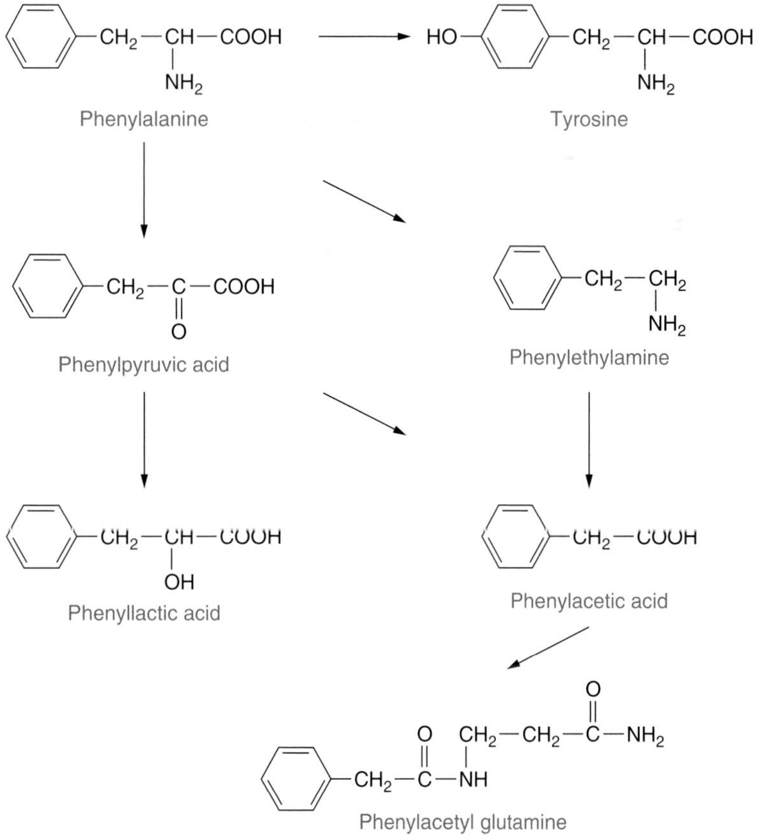

Figure 10-31. Conversion of phenylalanine to metabolites in addition to tyrosine under normal conditions.

Metabolism of Specific Amino Acids to Key Substances   525

principal mediator at postganglionic sympathetic endings. Catecholamine synthesis takes place in the adrenal medulla and catecholamine-secreting neurons, where tyrosine is imported as the precursor. The first enzymatic step is catalyzed by tyrosine hydroxylase, whose mechanism resembles PAH, and it has BH4 as cofactor. As with PAH, dihydropteridine reductase reduces the cofactor to its original state as in the PAH reaction. The product of the tyrosine hydroxylase reaction is DOPA (3,4-dihydroxyphenylalanine). DOPA is converted to dopamine by the catalytic action of DOPA decarboxylase. Dopamine is converted to norepinephrine, followed by the action of **phenylethanolamine N-methyltransferase (PNMT),** producing epinephrine from norepinephrine and using SAM to donate the methyl group. PNMT is induced by cortisol secreted from the adrenal cortex, and it courses through vascular connections to the adrenal medulla before reaching the general circulation (see Chapter 9). Consequently, when a stress event occurs, signaling the adrenal cortex to secrete cortisol, the levels of epinephrine in the adrenal medulla increase and enter the circulation. Although this overall reaction occurs in the adrenal medulla, other regions in the brain promote these reactions but terminate at the formation of dopamine. The reactions occurring in the adrenal medulla are shown in Figure 10-32.

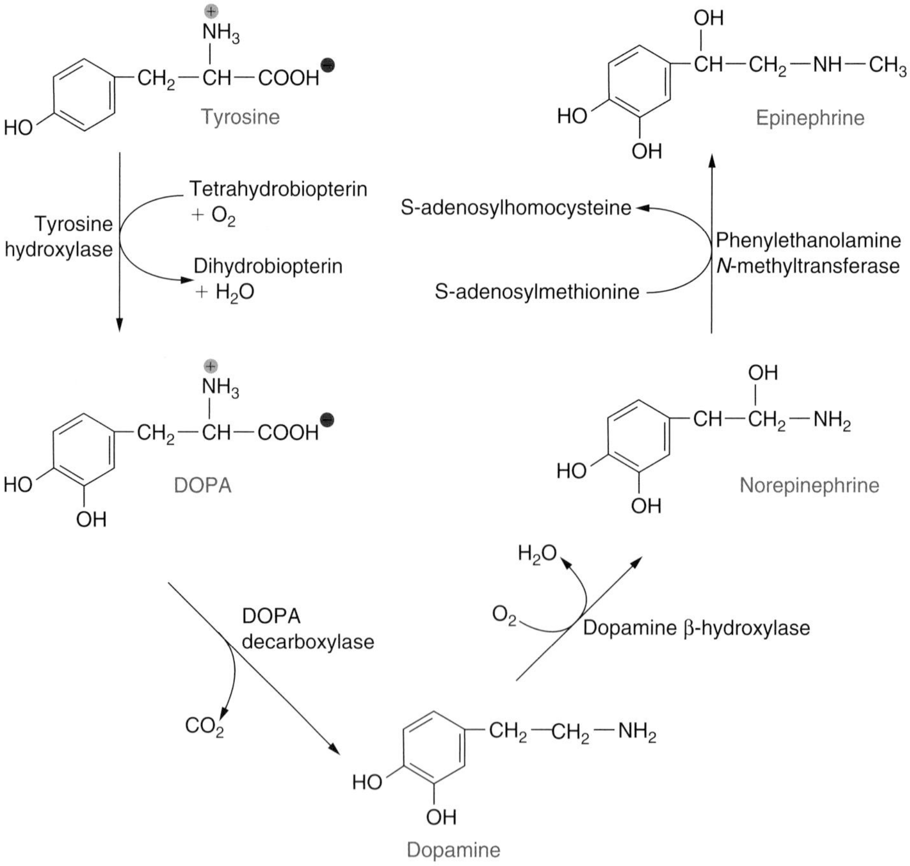

Figure 10-32. Conversion of L-tyrosine to epinephrine and other catecholamines in the adrenal medulla. Some areas of the brain have this set of reactions, which end at the level of dopamine. *DOPA,* dihydroxyphenylalanine.

Formation of Melanin

Melanin is a family of pigments that are polymeric, having differing colors. The synthesis of melanin from tyrosine is complex and is derived from DOPA, the tyrosine oxidation product catalyzed by tyrosinase. Tyrosinase is a 62,610 molecular weight enzyme, also known as diphenol oxidase. It contains glycosylation sites and histidine-rich sites that serve to bind copper, its cofactor. A theoretical scheme for the synthesis of melanin is shown in Figure 10-33. Tyrosinase catalyzes the conversion of tyrosine to DOPA and the conversion of DOPA to dopaquinone in the first two steps of the pathway. A number of intermediates follow that derive from DOPA, finally producing indolequinone and polymerization to form melanin. These events take place in the melanocyte (Figure 10-34). The melanocytes are located in the inner layers of the skin, where they produce melanin and carotene that blend to produce the color of the skin (and produce color in other organs, such as eyes and hair). Red color in hair is the product of phaeomelanin with spherical melanosomes (melanin granules), compared to the oblong melanosomes containing the usual black-colored eumelanin.

Melanin granules are uniformly distributed in the cell (Figure 10-35) to absorb ultraviolet rays from the sun and protect the body from injurious rays, up to a point. Enough light is admitted to allow the conversion of the precursor 7-dehydrocholesterol to vitamin D with 10 minutes of sunlight daily sufficing to supply the human requirement for vitamin D. The formation of vitamin D is shown in Figure 10-36. The receptor for vitamin D is a member of the steroid nuclear receptor gene family, and its action is similar to that of other steroid receptor transactivation factors (Figure 9-19). It is located mainly (about 95%) in the target cell

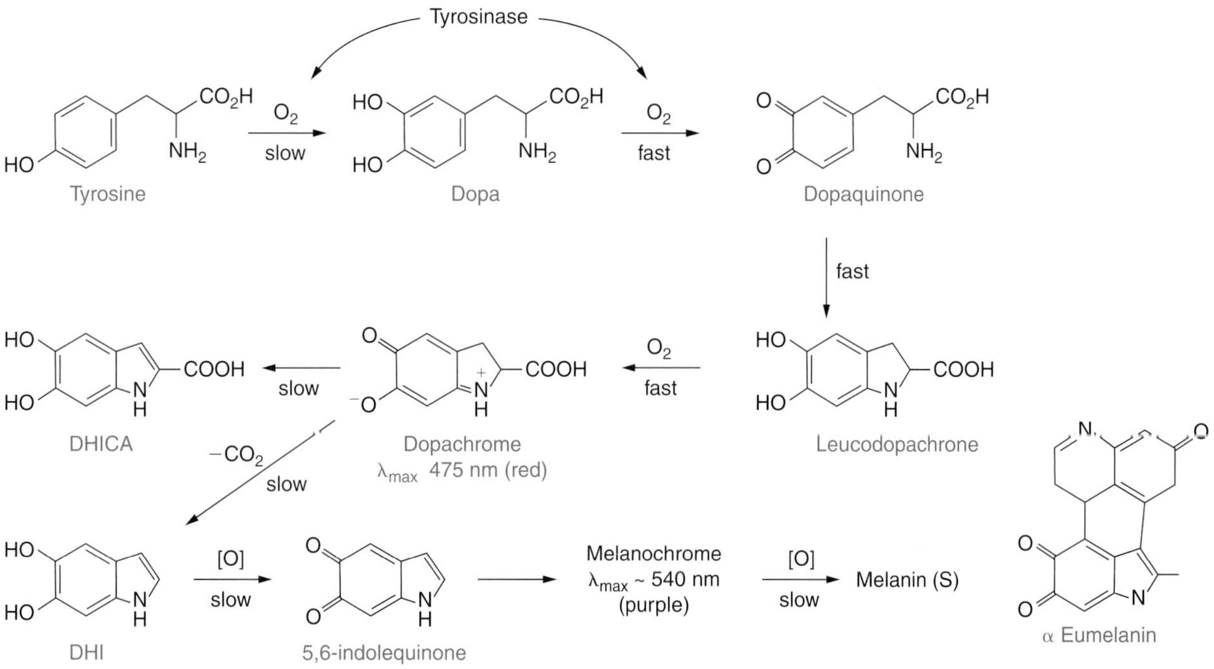

Figure 10-33. Synthesis of melanin from tyrosine. *DHI*, dihydroxyindole; *DHICA*, dihydroxyindole catecholamine.

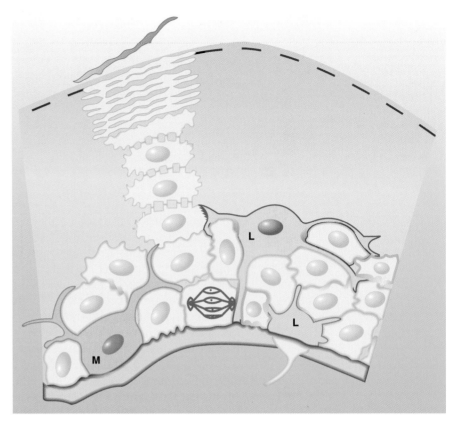

Figure 10-34. Location of a melanocyte in the skin. One of the melanocytes is in the basal layer and labeled M. Reproduced with permission from http://www.nurseminerva.co.uk/images/skin3.gif.

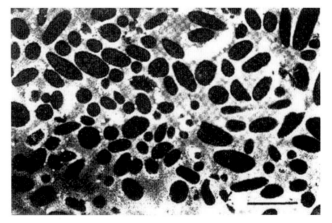

Figure 10-35. Electron micrograph of melanosome fraction.

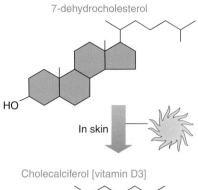
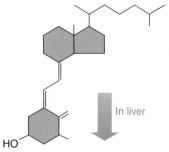
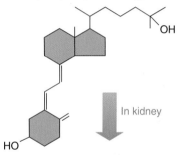
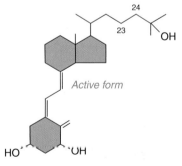

Figure 10-36. Conversion of 7-dehydrocholesterol to vitamin D with the aid of sunlight, in skin. Double bonds within rings are not shown.

Metabolism of Specific Amino Acids to Key Substances

nucleus in the absence of its ligand, **1,25-dihydroxyvitamin D3 (1,25-dihydroxycholecalciferol).** The structure of the ligand-binding domain of the vitamin D receptor bound to dihydroxyvitamin D3 is shown in Figure 10-37. In its interaction with DNA, it resembles the other steroid hormone receptors (Figure 9-23), although the vitamin D receptor forms a heterodimer with the retinoid X receptor, or RXR (similarly, the thyroid hormone receptor complexes with retinoic acid receptor, or RAR). Natural diets do not contain sufficient vitamin D, and sunlight is essential to fulfill this requirement, although certain foods are now supplemented with vitamin D. The liver hydroxylates cholecalciferol to 25-hydroxycholecalciferol mediated by the enzyme, 25-hydroxylase. The kidney contains 1-α-hydroxylase, which produces 1,25-dihydroxycholecalciferol from the 25-hydroxycholecalciferol from the liver. Vitamin D is transported in the blood by the vitamin D–binding protein. The function of vitamin D, a lipid soluble substance, is mediated by the nuclear receptor. This leads to the induction of calcium-transporting proteins, which carry calcium from the lumen of the intestine, across the epithelial cells to the basolateral side of the cell, and eventually into

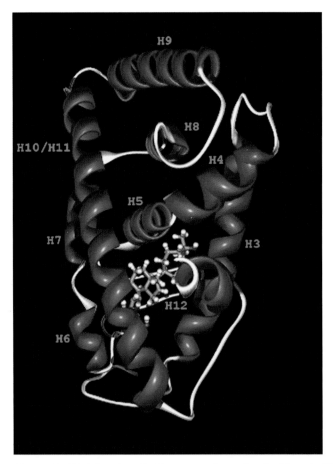

Figure 10-37. The vitamin D receptor bound to its ligand, dihydroxyvitamin D3. The ligand is shown as a ball-and-stick model. *H*, helix. Reproduced with permission from P. Rotkiewicz et al., "Model of three-dimensional structure of vitamin D receptor and its binding mechanism with 1-alpha,25-hihydroxyvitamin D3," *Proteins*, 44: 188–199, 2001.

the blood. Thus, the role of the vitamin is to ensure the proper balance between calcium and phosphorous for bone mineralization. Because the receptor is located in many of the cells of the body, the role of the vitamin is not limited to its effects on bone mineralization.

A number of other diseases are associated with blocks in the metabolic conversions of phenylalanine and tyrosine, as shown in Figure 10-38. When the conversion of phenylalanine to tyrosine is blocked (PAH nonfunctional), phenylalanine is converted to phenylpyruvate, resulting in phenylketonuria. When the conversion of tyrosine to melanin is blocked, a deficit of pigmentation occurs, known as **albinism.** When conversion of

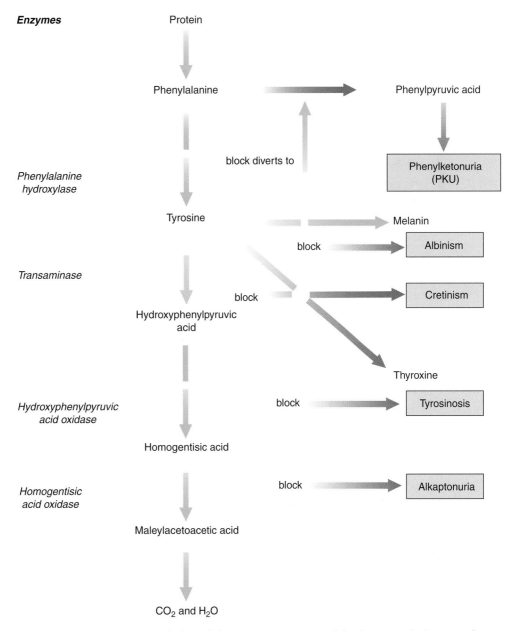

Figure 10-38. Conversion of phenylalanine to tyrosine and further metabolism. Deficits in the functioning of enzymes catalyzing individual steps lead to clinical conditions cited.

tyrosine to iodinated derivatives and then to thyroxine is blocked, cretinism is caused during development. When the conversion of hydroxyphenylpyruvate to homogentisic acid is blocked, tyrosinosis occurs. Finally, when the conversion of homogentisic acid to maleylacetoacetate is blocked, alkaptonuria follows. Besides specific normal fates of phenylalanine and tyrosine, these amino acids are incorporated into proteins, along with the other 18 amino acids.

Tryptophan

Tryptophan is an essential amino acid and is incorporated into proteins, along with the other amino acids. In addition to this fate, tryptophan is converted to the important neurotransmitters, serotonin and melatonin. The biosynthesis of these compounds occurring in the pineal gland is shown in Figure 10-39. These reactions are sensitive to lightness and

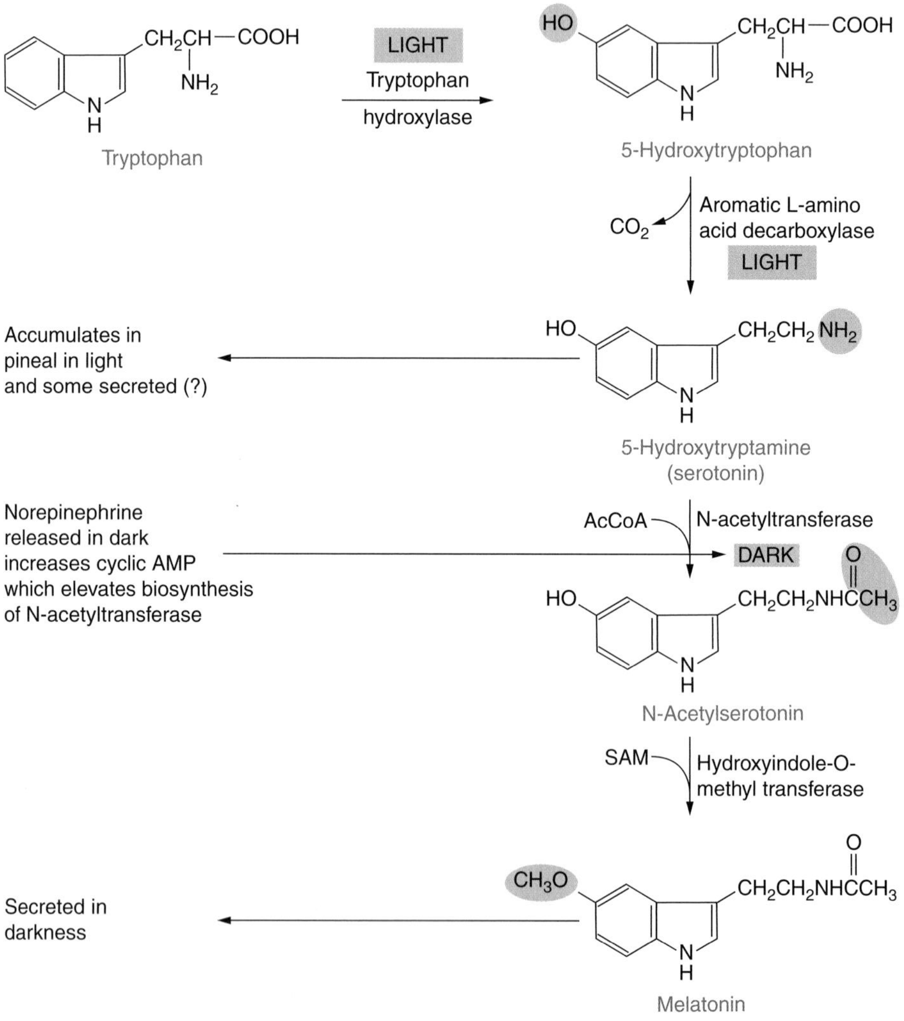

Figure 10-39. Synthesis of serotonin and melatonin from tryptophan in pinealocytes. Circles indicate the group altered in the reaction. Redrawn from A.W. Norman, and G. Litwack, *Hormones*, 2nd ed, Academic Press, San Diego, 1997.

darkness. Serotonin accumulates in the pineal gland during daylight, and it is converted to N-acetylserotonin and then to melatonin in the dark. Melatonin is secreted from the gland in the dark. Figure 10-40 shows the neural connections between the eyes and the pineal gland. Norepinephrine, released during darkness, binds to α- and β-receptors in the pinealocyte to induce the synthesis of N-acetyltransferase, which converts serotonin (accumulated in the light) to N-acetylserotonin and then to melatonin through hydroxyindole-O-methyltransferase.

The structures of two of the enzymes in the synthetic pathway, tryptophan hydroxylase and N-acetyltransferase, are shown in Figure 10-41. The pineal gland (about 172 mg in the human adult) exerts a role in human reproductive physiology and is involved in sexual maturation; melatonin receptors have been found in the gonads. It is claimed that autistic individuals have an increased level of blood serotonin. Once secreted from the serotonergic neuron, most serotonin is reabsorbed by a reuptake mechanism, thus limiting the amounts of effective serotonin. The basis of action of certain antidepressant drugs, such as **Prozac,** is to inhibit the reuptake process, resulting in prolonged serotonin effects in the synaptic cleft.

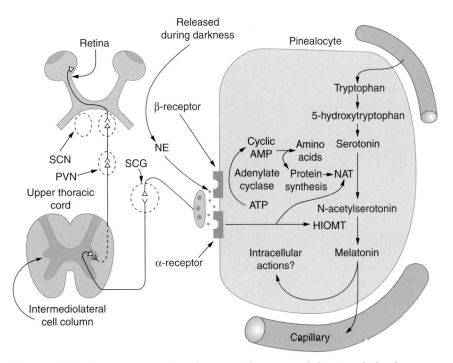

Figure 10-40. Neural connections between the eyes and the pineal gland as demonstrated in various mammals and synthesis of melatonin within the pineal gland. *HIOMT,* hydroxyindole-O-methyltransferase; *NAT,* N-acetyltransferase; *NE,* norepinephrine, *PVN,* paraventricular nucleus; *SCG,* superior cervical ganglia, *SCN,* suprachiasmatic nucleii. Redrawn with permission from M. Shafii and S.L. Shafii (eds.), *Biological Rhythms, Mood Disorders, Light Therapy, and the Pineal Gland,* American Psychiatric Press, Washington DC, 1990. Redrawn directly from A.W. Norman and G. Litwack, *Hormones,* 2nd ed., Academic Press, San Diego, p. 490, 1997.

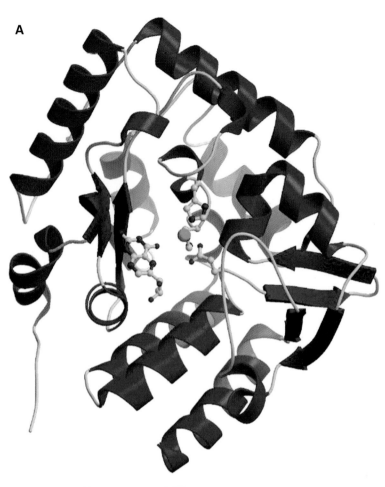

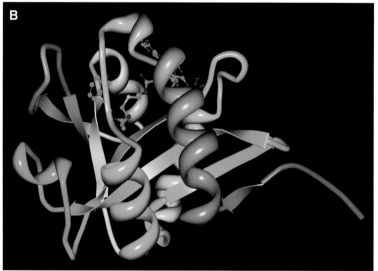

Figure 10-41. **A.** Human tryptophan hydroxylase catalytic domain monomer. Dihydrobiopterin cofactor is ball-and-stick model; ferric iron is green. **B.** Structure of serotonin N-acetyltransferase. Part A reproduced with permission from L. Wang, "Three-dimensional structure of human tryptophan hydroxylase and its implications for the biosynthesis of the neurotransmitters serotonin and melatonin," *Biochemistry*, 41: 12569-12574, 2002. Part B reproduced with permission from PDB ID: 1KUX. E. Wolf, J. De Angelis, E.M. Khalil, P.A. Cole, S.K. Burley. X-ray Crystallographic Studies of Serotonin N-acetyltransferase Catalysis and Inhibition. *Journal of Molecular Biology* **317** pp. 215 (2002) and from http://www.rcsb.org/pdb/cgi/explore.cgi?pdbId=1kux.

Melatonin increases serum prolactin concentrations, and there may be a role for melatonin in the cyclicity of menstruation. The amount of melatonin produced by the pineal is linked to the sleep pattern, and it plays a key role in setting the biological clock, especially the sleep **circadian rhythm.** Recently, melatonin has been credited with non–receptor-mediated activity in scavenging hydroxy radicals and superoxide anion, thus protecting, particularly at night (where melatonin reaches concentrations 10 times that during daylight), nuclear DNA, proteins, and membrane lipids in the brain and elsewhere. The pineal gland plays many regulatory roles, as shown in Figure 10-42. A model of the MT_2 melatonin receptor is shown in Figure 10-43. In addition to the syntheses of serotonin and melatonin (Figure 10-39), serotonin may be further metabolized to other compounds, as shown in Figure 10-44.

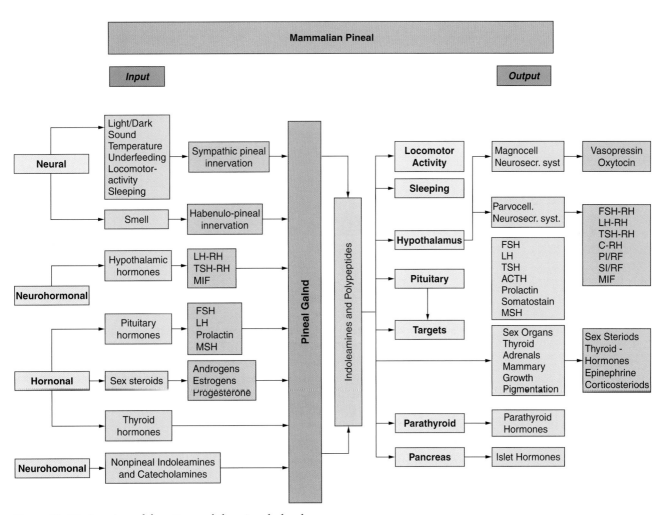

Figure 10-42. Inputs and functions of the pineal gland.

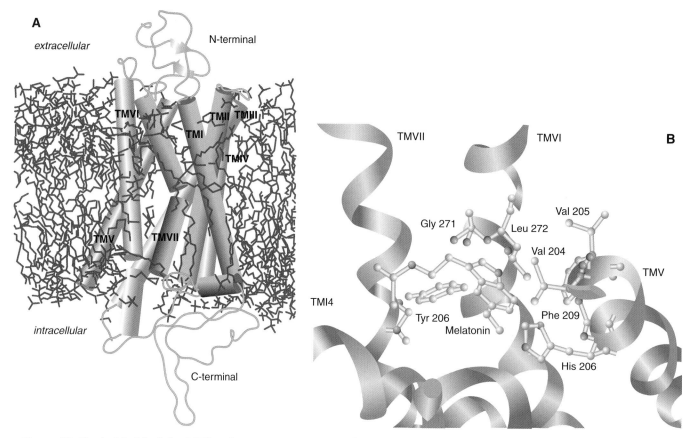

Figure 10-43. **A.** Model of the MT^2 melatonin receptor. **B.** Melatonin binding site of the melatonin receptor. Reproduced from Z. Sovova et al., "A structural model of human MT^2 melatonin receptor and its melatonin recognition site," Laboratory of High Performance Computing, Institute of Physical Biology USB, and Institute of Landscape Ecology AS CR, University of South Bohemia, Czech Republic, and from http://www.xray.cz/setkani/abst/2004/sovova.htm.

Arginine

Arginine, an essential amino acid in childhood (along with histidine), in addition to being incorporated into proteins, is the precursor of creatine and creatinine. Arginine gives rise to citrulline (urea cycle), agmatine, putrescine (by way of ornithine), **nitric oxide (NO),** proline, and glutamate (Figure 10-45). In the kidney, on the pathway to creatine and creatinine, arginine and glycine are converted to guanidoacetate and ornithine, catalyzed by the action of arginine-glycine transamidinase (Figure 10-46). The next step is catalyzed by guanidoacetate methyltransferase of liver. This converts guanidoacetate to phosphocreatine with SAM, which donates a methyl group to the amido group, and ATP, which donates a phosphate to the amino group. Phosphocreatine is acted upon by creatine kinase in muscle to generate creatine, regaining the high-energy phosphate in the process by conversion of ADP to ATP. Phosphocreatine, in muscle, is converted partly to creatinine, nonenzymatically, which is excreted in the urine. The structure of creatine kinase is shown in Figure 10-47.

(Text continues on p. 540.)

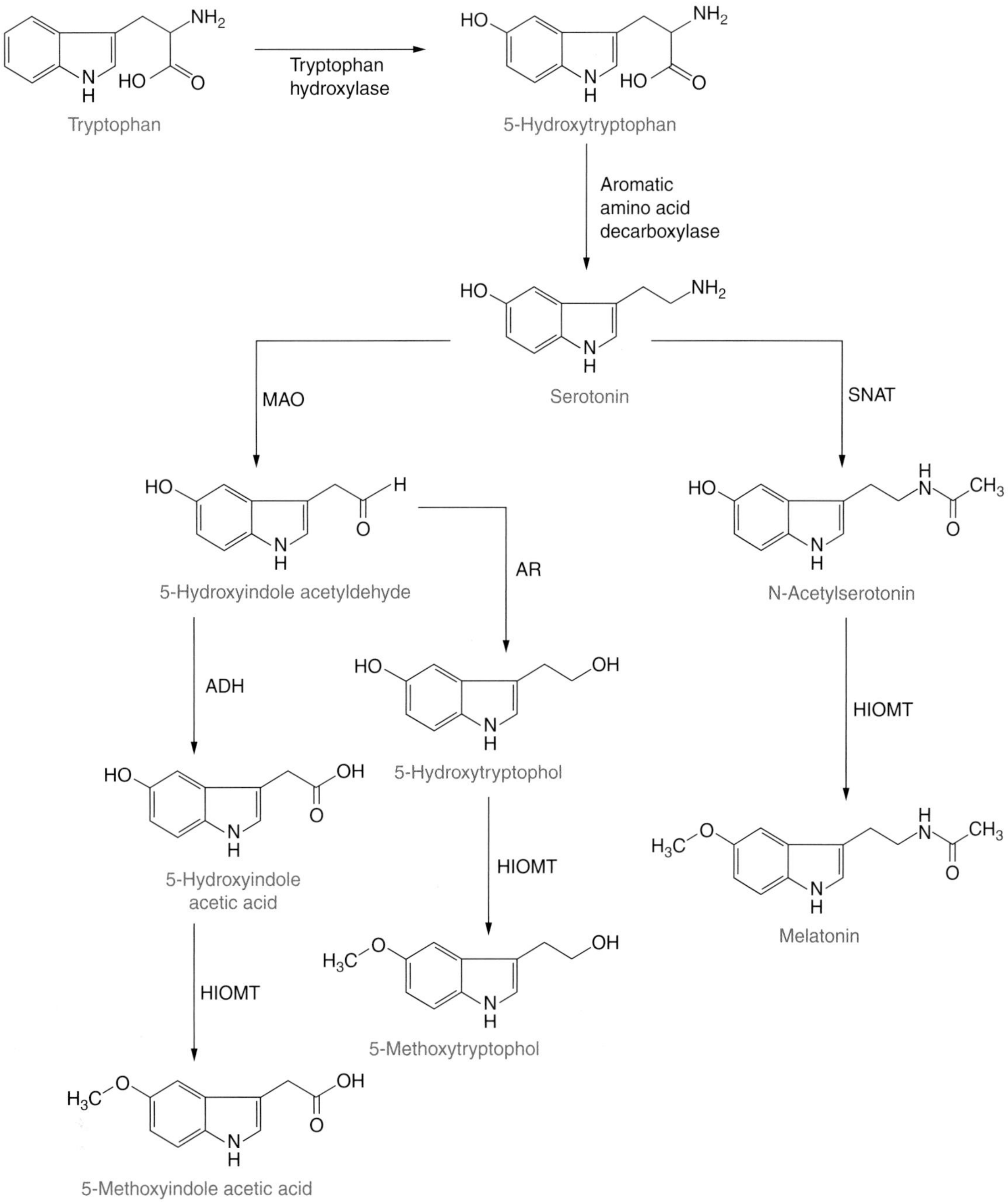

Figure 10-44. Biosynthesis of pineal methoxyindoles from serotonin. Serotonin may be either acetylated to form N-acetylserotonin through the action of the enzyme serotonin-N-acetyltransferase (SNAT) or oxidatively deaminated by monoamine oxidase (MAO) to yield an unstable aldehyde. This compound is then either oxidized to 5-hydroxyindole acetic acid by the enzyme aldehyde dehydrogenase (ADH) or reduced to form 5-hydroxytryptophol by aldehyde reductase (AR). Each of these 5-hydroxyindole derivatives of serotonin is a substrate for hydroxyindole-O-methyltrasferase (HIOMT). The enzymatic transfer of a methyl group from S-adenosylmethionine to these hydroxyindoles yields melatonin (5-hydroxy-N-acetyltryptamine), 5-methoxyindole acetic acid, and 5-methoxytryptophol, respectively. Pineal serotonin is synthesized from the essential amino acid tryptophan by 5-hydroxylation, followed by decarboxylation. The first step in this enzymic sequence is catalyzed by tryptophan hydroxylase. The second step is catalyzed by aromatic L-amino acid decarboxylase.

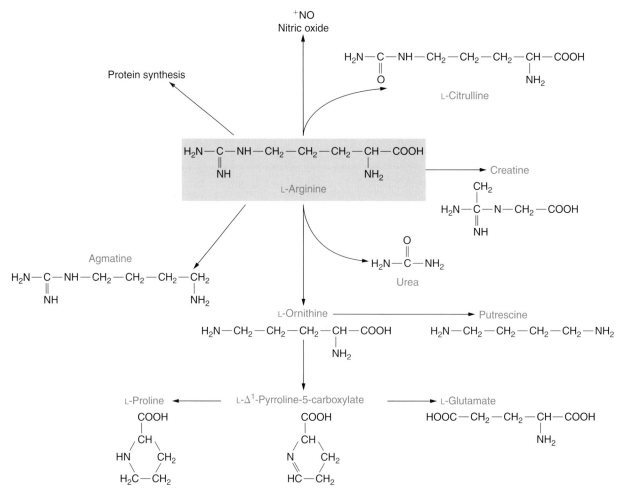

Figure 10-45. Metabolic fates of arginine in mammalian cells. The five enzymes on which the central limbs of the pathways are based include *(clockwise from the top):* nitric oxide synthase (NOS), arginine:glycine amidinotransferase, arginase, arginine decarboxylase, and arginyl-tRNA synthetase. Redrawn from G. Wu and S.M. Morris, Jr., "Arginine metabolism: nitric oxide and beyond," *Biochem. J.,* 336: 1–17, 1998.

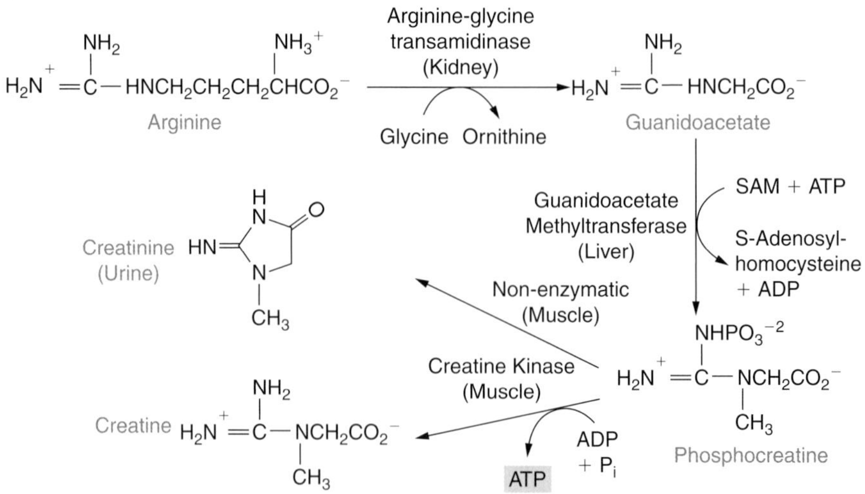

Figure 10-46. Formation of creatine and creatinine from L-arginine. *SAM,* S-adenosylmethionine.

538 CHAPTER 10 Metabolism

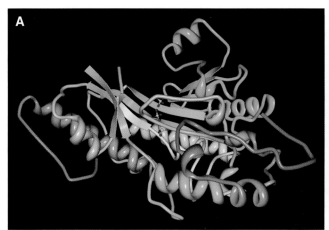

Figure 10-47. **A.** Muscle creatine kinase. **B.** Crystal structure of chicken brain-type creatine kinase. Part A reproduced from PDB ID: 2CRK. J.K. Rao, B. Bujacz, A. Wlodawer. Crystal Structure of Rabbit Muscle Creatine Kinase. *FEBS Letters* **439** pp. 133 (1998) and from http://www.rcsb.org/pdb/cgi/explore.cgi?pdbId=2crk. Part B reproduced from PDB ID: 1QH4. M. Eder, U. Schlattner, A. Becker, T. Wallimann, W. Kabsch, K. Fritz-Wolf. Crystal Structure of Brain-type Creatine Kinase at 1.41 A Resolution. *Protein Science* **8:** pp. 2258 (1999) and from http://www.rcsb.org/pdb/cgi/explore.cgi?pdbId=1qh4.

Ninety-five percent of the creatine is contained in skeletal muscle, and the remainder is in heart, brain, and testes. Creatine supplementation improves muscular performance when muscle creatine levels are increased by 20%. Phosphocreatine provides an energy source in terms of ATP.

Another important product of arginine is NO, which functions in the body at low concentrations as a biological signal involved in the control of blood pressure, neurotransmission, learning, and memory. NO acts as a cellular messenger in a range of events, both normal and pathological. It brings about vasodilation by activating guanylate cyclase, which increases the cellular level of cyclic GMP. Normally, a phosphodiesterase opens the cyclic structure to form GMP and to mitigate the vasodilation effect of cyclic GMP. **Sildenafil (Viagra)** blocks this enzyme and so prolongs the vasodilation caused by NO through cyclic GMP. NO is formed from arginine through the action of nitric oxide synthase (NOS). This enzyme catalyzes a five-electron oxidation of a guanidino nitrogen of L-arginine. Two successive monooxygenation reactions take place to generate N^{ω}-hydroxy-L-arginine as intermediate, as shown in Figure 10-48. The reaction uses 2 mol of O_2 and 1.5 mol of NADPH per mol of NO formed. This is a complex enzyme, requiring five bound cofactors or prosthetic groups: FAD, flavin mononucleotide (FMN), heme, BH4, and Ca^{2+}-calmodulin (CaM). There are NOS isozymes derived from three different genes: neuronal (nNOS), cytokine-inducible (iNOS), and endothelial (eNOS); their characteristics are shown in Table 10-3. The crystal structure of inducible dimeric nitric oxide synthase ($iNOS_{ox}$) in the oxidized form is shown in Figure 10-49 in the swapped conformation. Domain swapping affects the N-terminal hook conformation. iNOS and nNOS are cytosolic, whereas eNOS is associated with the endothelial membrane

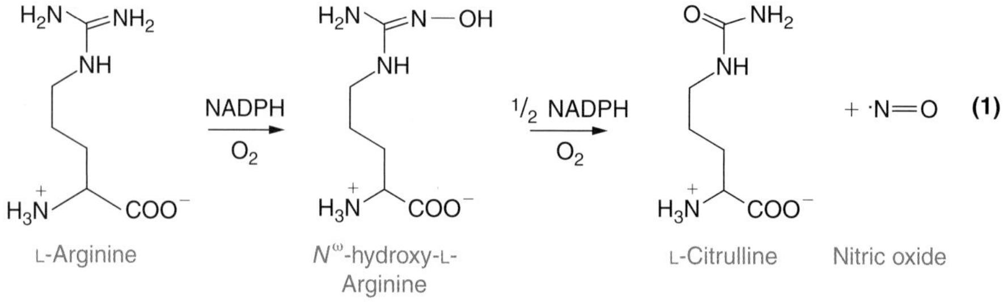

Figure 10-48. Nitric oxide synthase (NOS) catalytic reaction generating nitric oxide (NO) and L-citrulline from L-arginine.

Table 10-3
Properties of Mammalian Nitric Oxide Synthase Isozymes

Enzyme	Gene	No. of exons	No. of residues	Subcellular location	Regulation
nNos	*NOS1*	29	1429–1433	Mainly soluble (brain); mainly particulate (skeletal muscle)	Ca^{2+}/CaM
iNos	*NOS2*	27	1144–1153	Mainly soluble	Cytokine-inducible; Ca^{2+}-independent
eNOS	*NOS3*	26	1203–1205	Mainly particulate	Ca^{2+}/CaM

Reproduced from http://metallo.scripps.edu/PROMISE/NOS.html.

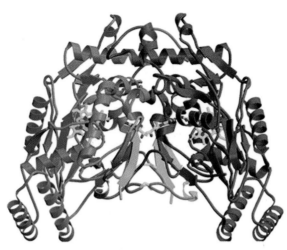

Figure 10-49. Crystal structure of inducible nitric oxide synthase in the oxidized form. In this representation, the dimer is in a swapped form in which the N-terminal hook regions *(cyan and orange)* interact primarily with the opposite subunit. The heme *(yellow bonds)* is cupped in the inward-facing palm of the central webbed β-sheet of the "catcher's mitt" subunit fold. A self-symmetrical disulfide bond *(yellow, bottom center)* links the two subunits. Two molecules of BH4 *(yellow, center, on edge)* are also bound at the interface and line the active-center channels leading to the hemes. Reproduced with permission from Figure 1a of B.R. Crane et al., "N-terminal domain swapping and metal ion binding in nitric oxide synthase dimerization," *EMBO J.*, 18: 6271–6281, 1999.

by N-terminal myristoylation and posttranslational palmitoylation. As shown in Figure 10-49, the enzymes are homodimers, with each monomer having an N-terminal oxygenase domain and a C-terminal reductase domain, homologous to NADPH:P450 reductase. The linker between the two domains contains a CaM-binding sequence. CaM triggers electron flux from FMN to heme to couple the oxygenase and reductase domains, and CaM facilitates NADPH-dependent reduction of cytochrome c and ferricyanide in BH4- and heme-depleted nNOS.

Arginine also gives rise to ornithine through the action of arginase in the urea cycle (Figure 10-3 and Figure 10-45). About 10% of the ornithine produced is converted to putrescine by ornithine decarboxylase in the polyamine pathway (Figure 10-50). Putrescine is converted further to spermidine and spermine. In the mitochondria, ornithine is converted to the amino acids proline and glutamate through the intermediate, glutamate semialdehyde. The structures of the polyamines are shown in Figure 10-51. Agmatine is decarboxylated arginine (Figure 10-45). Some arginine can be synthesized from citrulline by argininosuccinate synthase, which produces L-argininosuccinate and then L-arginine by the action of argininosuccinate lyase. Little, if any, arginine would be expected to be produced in the growing child by this route.

Histidine

Histidine is required by the growing child and must be consumed in the diet, but histidine is not essential in the adult. An important fate of histidine, besides its incorporation into proteins, is the formation of

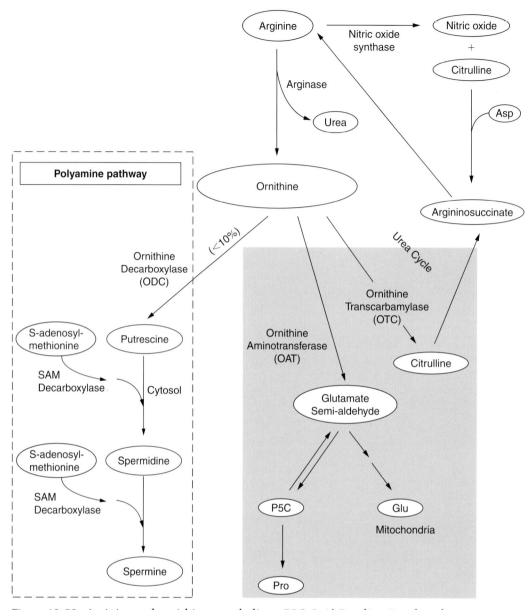

Figure 10-50. Arginine and ornithine metabolism. *P5C*, L-Δ^1-Pyroline-5-carboxylate.

histamine by histidine decarboxylase (Figure 8-75). Histamine stimulates the secretion of gastric acid mediated by histamine H_2 receptors. Drugs used to block histamine action at H_2 receptors are **cimetidine (Tagamet)** and **ranitidine (Zantac).** Histamine is synthesized and released by the mast cells, where the amine mediates the allergic response by causing vasodilation and bronchoconstriction through the action of histamine H_1 receptors. Drugs that inhibit the H_1 receptor are **diphenhydramine (Benadryl)** and **loratadine (Claritin).** Structures of some drugs affecting H_1 and H_2 receptors are shown in Figure 10-52. The degradation of histidine is shown in Figure 10-53. Note that glutamate is formed in the fourth

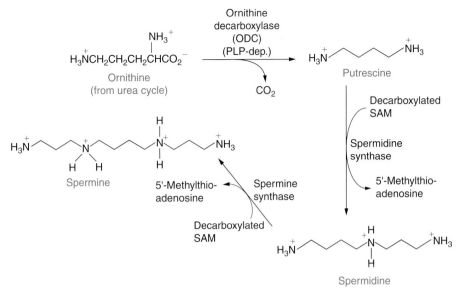

Figure 10-51. Polyamine biosynthesis.

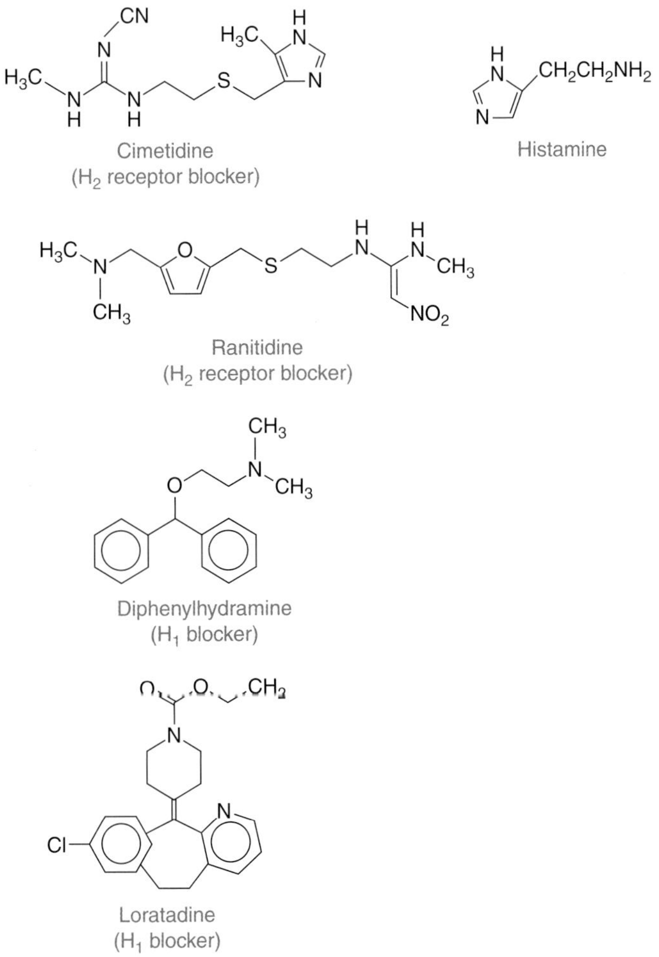

Figure 10-52. Some blockers of H_1 and H_2 histamine receptors.

Metabolism of Specific Amino Acids to Key Substances 543

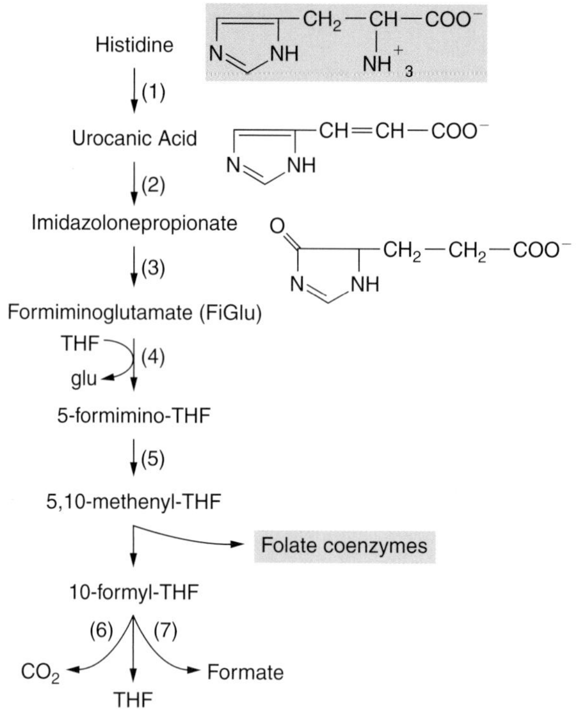

Figure 10-53. The catabolism of histidine. In the first three steps, histidine is sequentially degraded to formiminoglutamate (FiGlu) by (1) histidase, (2) urocanase, and (3) imidazolepropionate amino hydrolase. The formimino group of FiGlu, originating from the ring-2-carbon of histidine, enters the one-carbon pool as 5-formiminotetra-hydrofolate (5-formimino-THF) in a reaction catalyzed by (4) FiGlu formiminotransferase. The formimino group attached to 5-formimino-THF is deaminated by (5) 5-formimino-THF deaminase to yield 5,10-methenyl-THF. The one-carbon moiety of 5,10-methenyl-THF can proceed reductively through the one-carbon pool (i.e., to 5,10-methylene-THF, 5-methyl-THF, and methionine), or oxidatively to the folate coenzyme 10-formyl-THF. As the formyl group of 10-formyl-THF, the ring-2-carbon of histidine can be released as either CO_2 or formate, reactions catalyzed by (6) 10-formyl-THF dehydrogenase and (7) 10-formyl-THF hydrolase, respectively, thereby regenerating the THF molecule. Redrawn from K.L. Schalinske and R.D. Steele, "Quantification of the carbon flow through the folate-dependent one carbon pool using radiolabeled histidine: effect of altered thyroid and folate status," Arch. Biochem. Biophys., 328: 93–100, 1996.

degradative step. The final products are CO_2, formate, and regeneration of tetrahydrofolate cofactor used in reaction 4.

Glutamate

Glutamate has been discussed in the context of generating glutamine, dehydrogenase reactions, incorporation into GSH, and transamination reactions involving glutamate (Figures 10-1, 10-2, 10-5, 10-6, 10-9 through 10-11, and 10-24). An additional fate of glutamate is the inhibitory neurotransmitter in the brain, which is γ-**aminobutyric acid (GABA)**. GABA is formed by the action of glutamate decarboxylase, as shown in Figure 10-54. This neurotransmitter exerts a presynaptic inhibitory

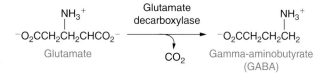

Figure 10-54. Formation of GABA from L-glutamate by the action of glutamate decarboxylase.

action on excitatory but not inhibitory innervation of certain muscles. $GABA_A$ receptors have multiple subunits, and the β-subunit may be required for the functioning of the receptors. These receptors contain ion channels, but the molecular basis of the spontaneous opening of these channels is not understood.

α-Ketoglutarate formed from glutamate by transamination (Figure 10-1) or by glutamate dehydrogenase (Figure 10-5) is ultimately broken down in four steps to form proline and in five steps to form ornithine and then arginine in the urea cycle.

Serine

Serine is a nonessential amino acid that can be synthesized from glucose. Glucose forms 3-phosphoglycerate in the glycolytic pathway, which is converted to 3-phosphohydroxypyruvate. The latter can be transaminated to 3-phosphoserine and then converted to serine by the action of a phosphatase (Figure 10-55). Serine can also be converted to the nonessential amino acid, glycine by serine hydroxymethyltransferase, as shown in Figure 10-56.

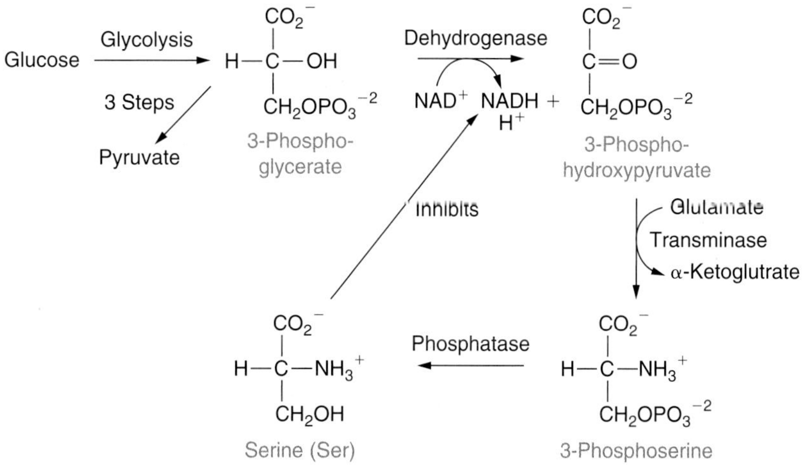

Figure 10-55. Conversion of glucose to the nonessential amino acid, serine.

Metabolism of Specific Amino Acids to Key Substances   545

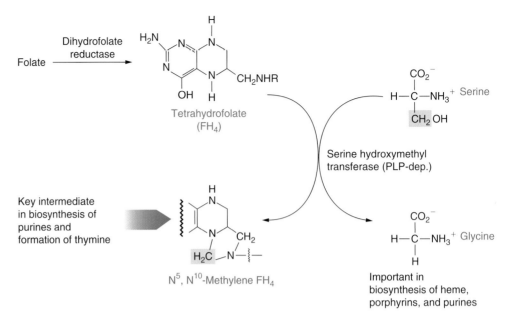

Figure 10-56. Conversion of serine to glycine.

Catabolism of Amino Acids

When amino acids are not needed for the synthesis of proteins, they can be metabolized to compounds that enter the citric acid cycle (Figure 10-57). They are metabolized for the production of energy in the form of ATP and account for 10% or more of the energy production in the body. Although each amino acid is catabolized by a separate pathway, all amino acid derivatives are reduced to five intermediates that enter the citric acid cycle, where they are converted to water, CO_2, and energy, or that form fatty acids (ketogenic) or glucose (glucogenic). Metabolites of glucogenic amino acids are pyruvate, α-ketoglutarate, succinyl CoA, fumarate, or oxaloacetate. Ketogenic amino acids form acetoacetate or acetyl CoA. Some amino acids are both glucogenic and ketogenic: tryptophan, phenylalanine, tyrosine, isoleucine, and threonine. Lysine and leucine are ketogenic only. The rest of the amino acids are glucogenic (arginine, glutamic acid, glutamine, histidine, proline, valine, methionine, aspartic acid, asparagine, alanine, serine, cysteine, and glycine).

Alanine, serine, and cysteine are metabolized to pyruvate. Serine is converted to pyruvate by the catalytic action of serine dehydratase (Figure 10-58), which removes the hydroxyl group through an intermediate that is then hydrolyzed to produce pyruvate and ammonia. Glycine is also convertible to pyruvate by the action of serine hydroxymethyltransferase (N^5,N^{10}-methylene-tetrahydrofolate cofactor), which converts glycine to serine. Serine is then converted to pyruvate by serine dehydratase. Cysteine is also convertible to pyruvate, and the three alkyl carbons of tryptophan can be converted to alanine and then to pyruvate by alanine aminotransferase.

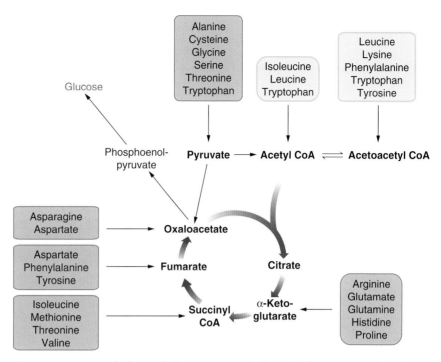

Figure 10-57. Catabolism of the amino acids through the citric acid cycle. Amino acids in green boxes are ketogenic; amino acids in blue boxes are glucogenic. Some amino acids can form both glucose or fatty acids: isoleucine, tryptophan, phenylalanine, and tyrosine.

Threonine is converted to glycine and acetyl CoA by threonine dehydrogenase (NAD+) and α-amino-β-ketobutyrate lyase (incorporating CoA), as shown in Figure 10-59. The intermediate is α-amino-β-ketobutyrate. Glycine is converted to serine and serine to pyruvate by serine dehydratase (as described earlier). The crystal structure of serine hydroxymethyltransferase is shown in Figure 10-60.

Aspartate and asparagine are both convertible to oxaloacetate. Asparaginase converts asparagine into aspartate and ammonia, and then aspartate is converted into oxaloacetate by aspartate aminotransferase (Figure 10-61). The structure of aspartate aminotransferase is shown in Figure 10-12.

A number of amino acids are convertible to α-ketoglutarate. Glutaminase converts glutamine into glutamate, and then aspartate aminotransferase converts it to α-ketoglutarate. Proline forms pyrroline 5-carboxylate, catalyzed by proline oxidase, and then pyrroline 5-carboxylate hydrolyzes spontaneously to form glutamate γ-semi-aldehyde and then glutamate. The conversion of glutamate γ-semi-aldehyde to glutamate is catalyzed by glutamate 5-semi-aldehyde dehydrogenase. Glutamate γ-semi-aldehyde is also formed from ornithine, and ornithine is formed from arginine by arginase (Figure 10-3). These reactions are shown in Figure 10-62.

Again, glutamate is converted to α-ketoglutarate by aspartate aminotransferase. Histidine also can form α-ketoglutarate because it can be converted to glutamate. Histidine ammonia lyase deaminates histidine

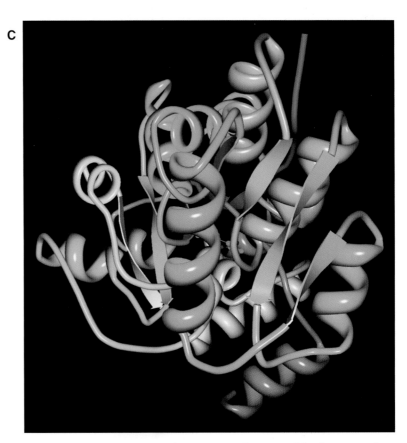

Figure 10-58. **A.** Reaction of L-serine with serine dehydratase to form an amino acrylate intermediate that tautomerizes to the imine, which is hydrolyzed to produce water and ammonia. The hydroxyl group of serine is removed by β-elimination. **B.** Structure of the coenzyme, pyridoxal phosphate. **C.** Structure of human liver serine dehydratase containing 319 amino acid residues and pyridoxal phosphate coenzyme (atomic structures). Reproduced from PDB ID: 1P5J. L. Sun, X. Li, Y. Dong, M. Yang, Y. Liu, X. Han, X. Zhang, H. Pang, Z. Rao. Crystallization and Preliminary Crystallographic Analysis of Human Serine Dehydratase. *Acta Crystallographica* **59** pp. 2297 (2003) and from http://www.rcsb.org/pdb/cig/explore.cgi?pdbId=1p5j.

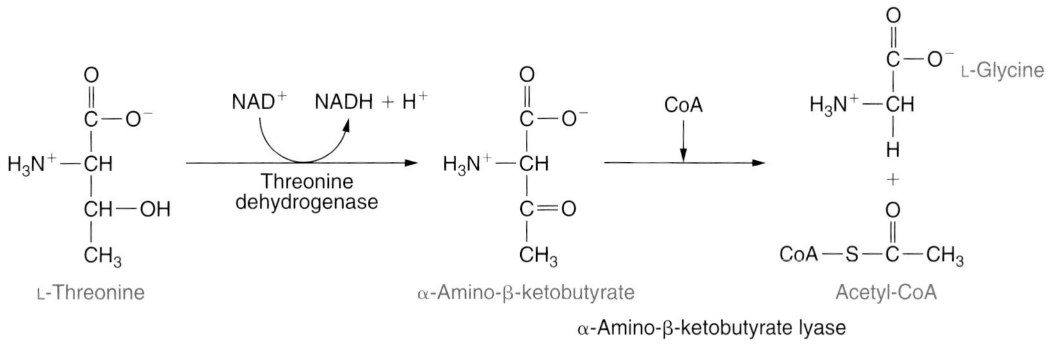

$$\text{L-Threonine} \xrightarrow[\text{dehydrogenase}]{\text{Threonine} \atop \text{NAD}^+ \to \text{NADH} + \text{H}^+} \alpha\text{-Amino-}\beta\text{-ketobutyrate} \xrightarrow[\alpha\text{-Amino-}\beta\text{-ketobutyrate lyase}]{\text{CoA}} \text{L-Glycine} + \text{Acetyl-CoA}$$

Figure 10-59. Conversion of L-threonine to acetyl-CoA and L-glycine by the sequential action of threonine dehydrogenase and α-amino-β-ketobutyrate lyase.

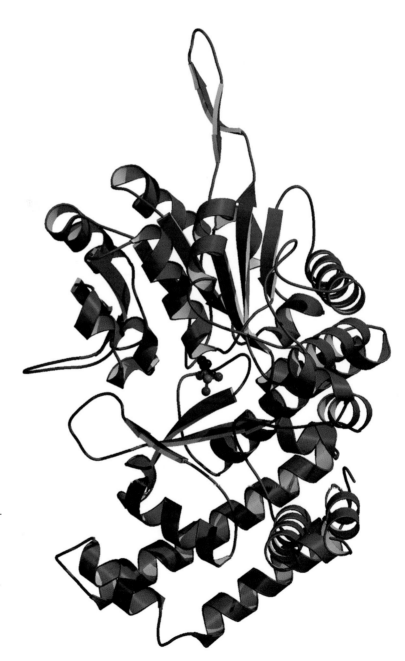

Figure 10-60. Crystal structure of human (breast) serine hydroxymethyl-transferase, a single polypeptide chain consisting of 470 amino acid residues. Reproduced from PDB ID: 1BJ4. S.B. Renwick, K. Snell, U. Baumann. The Crystal Structure of Human Cytosolic Serine Hydroxymethyltransferase: A Target for Cancer Chemotherapy. *Structure* **6** pp. 1105 (1998) and from http://www.ebi.ac.uk/thornton-srv/databases/cgi-bin/pdbsum/GetPage.pl?pdbcode=1bj4.

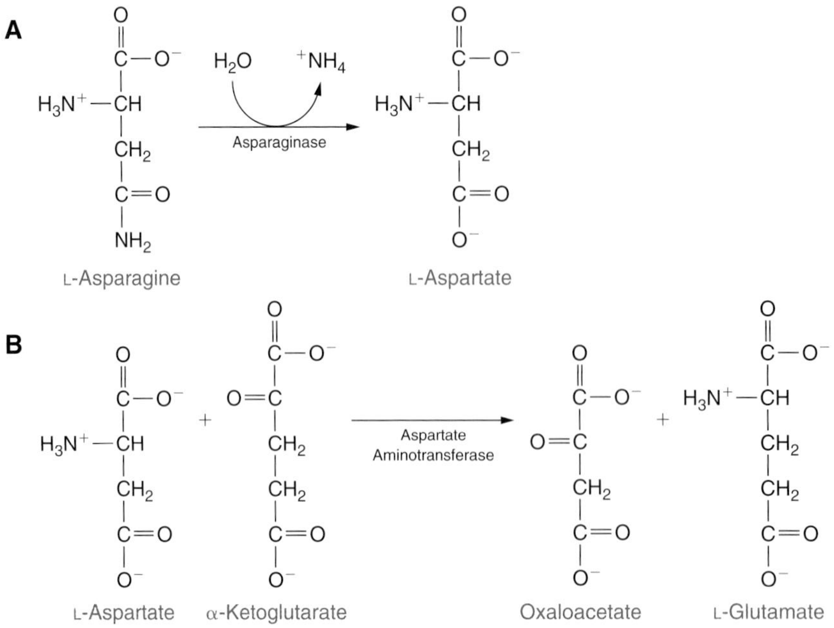

Figure 10-61. **A.** Conversion of L-asparagine to L-aspartate and ammonia by asparaginase. **B.** Conversion of L-aspartate and α-ketoglutarate to oxaloacetate and L-glutamate by aspartate aminotransferase.

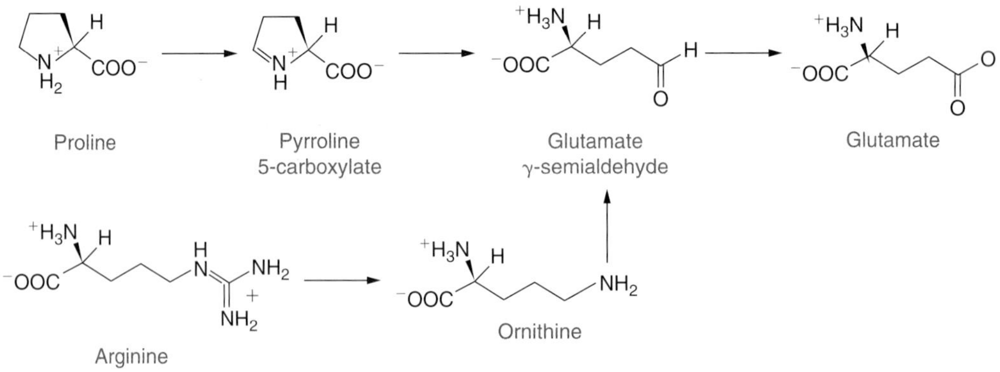

Figure 10-62. Reactions of proline and arginine that lead to the formation of glutamate.

to urocanate. Then urocanate hydratase forms 4-imidazole-5-propionate by the addition of water to urocanate. 4-Imidazole-5-propionate is then hydrolyzed to N-formiminoglutamate by imidazole propionase. The formimino group is transferred to tetrahydrofolate by glutamate formiminotransferase to form N^5-formimino-tetrahydrofolate and glutamate (Figure 10-63).

Methionine, valine, and isoleucine are broken down ultimately to succinyl CoA (Figure 10-57). They are converted first to propionyl CoA, which is converted to D-methylmalonyl CoA by propionyl CoA carboxylase. D-Methylmalonyl CoA is acted on by methylmalonyl CoA racemase to form L-methylmalonyl CoA. This is converted to succinyl CoA by methylmalonyl CoA mutase. The reactions involving methionine are not simple; they require nine steps involving the synthesis of SAM (Figure 10-64).

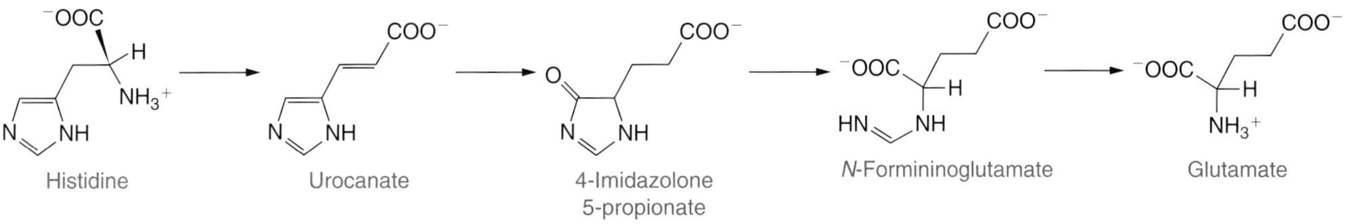

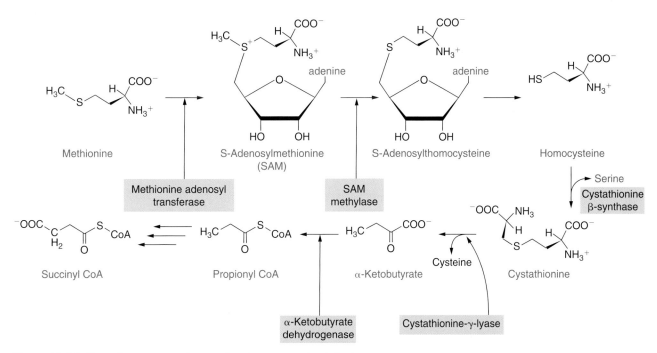

Figure 10-63. Reactions of histidine leading to glutamate.

Figure 10-64. Reactions converting methonine to succinyl CoA.

Branched-chain amino acids are catabolized in muscle, adipose, kidney, and brain, but not in the liver, as occurs with other types of amino acids. They are acted on first by branched-chain amino acid aminotransferase to produce the corresponding keto acids. Then branched-chain α-keto acid dehydrogenase, with CoA-SH and NAD$^+$, converts the keto acids to corresponding CoA derivatives. α-Keto acid dehydrogenase is a large multienzyme complex homologous to pyruvate dehydrogenase, including the same cofactors; it is described in Figure 4-60. The overall reactions of branched-chain amino acids are shown in Figure 10-65. Ingestion of branched-chain amino acids leads to activation of a phosphatase, which activates the phosphorylated, inactive branched-chain α-keto acid dehydrogenase. This enzyme, if genetically altered and not expressed, results in an accumulation of branched-chain amino acids in the blood and urine, imparting the odor of maple syrup in the urine. The disease, consequently, is called **maple syrup disease,** and it progresses to

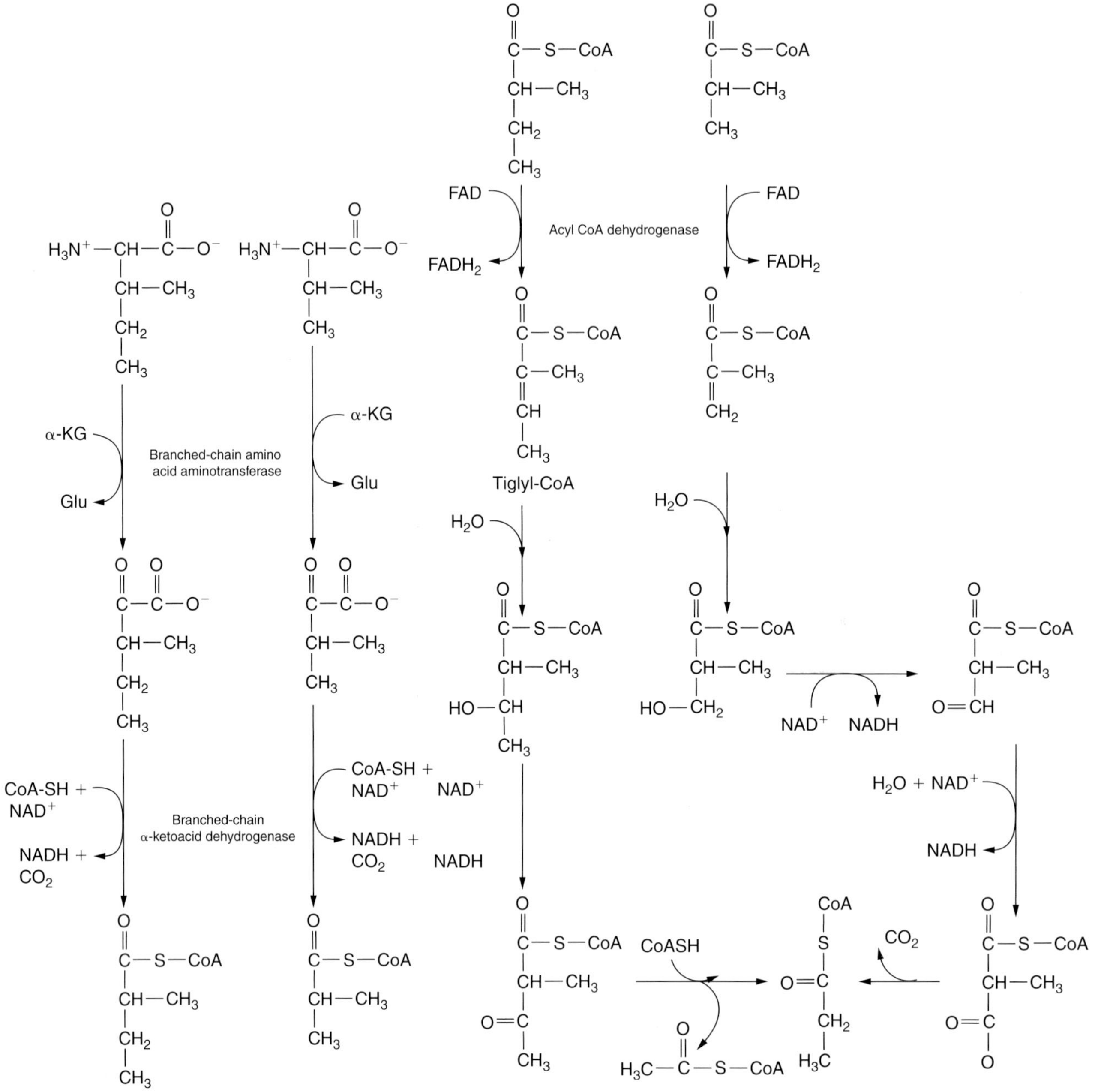

Figure 10-65. Catabolic pathways of branched-chain amino acids. Tiglyl-CoA refers to tiglic acid.

mental retardation unless the diet is restricted by omitting intake of valine, leucine, and isoleucine in early life.

Lysine and leucine are purely ketogenic amino acids. Leucine is transaminated to α-ketoisocaproate and then oxidatively decarboxylated by branched-chain α-keto acid dehydrogenase to isovaleryl CoA, like the other branched-chain amino acids (Figure 10-65). Isovaleryl CoA is dehydrogenated by isovaleryl CoA dehydrogenase to form β-methylcrotonyl CoA.

Methylcrotonyl CoA carboxylase, a biotin-containing enzyme, converts isovaleryl CoA to β-methylglutaconyl CoA. A hydratase then converts it to β-hydroxy-β-methylglutaryl CoA, which is then hydrolyzed by HMG-CoA lyase to acetyl CoA and acetoacetate. These reactions are summarized in Figure 10-66.

The catabolism of aromatic amino acids is shown in Figure 10-38 for phenylalanine and tyrosine. Tryptophan is converted to acetoacetate in a series of reactions shown in Figure 10-67.

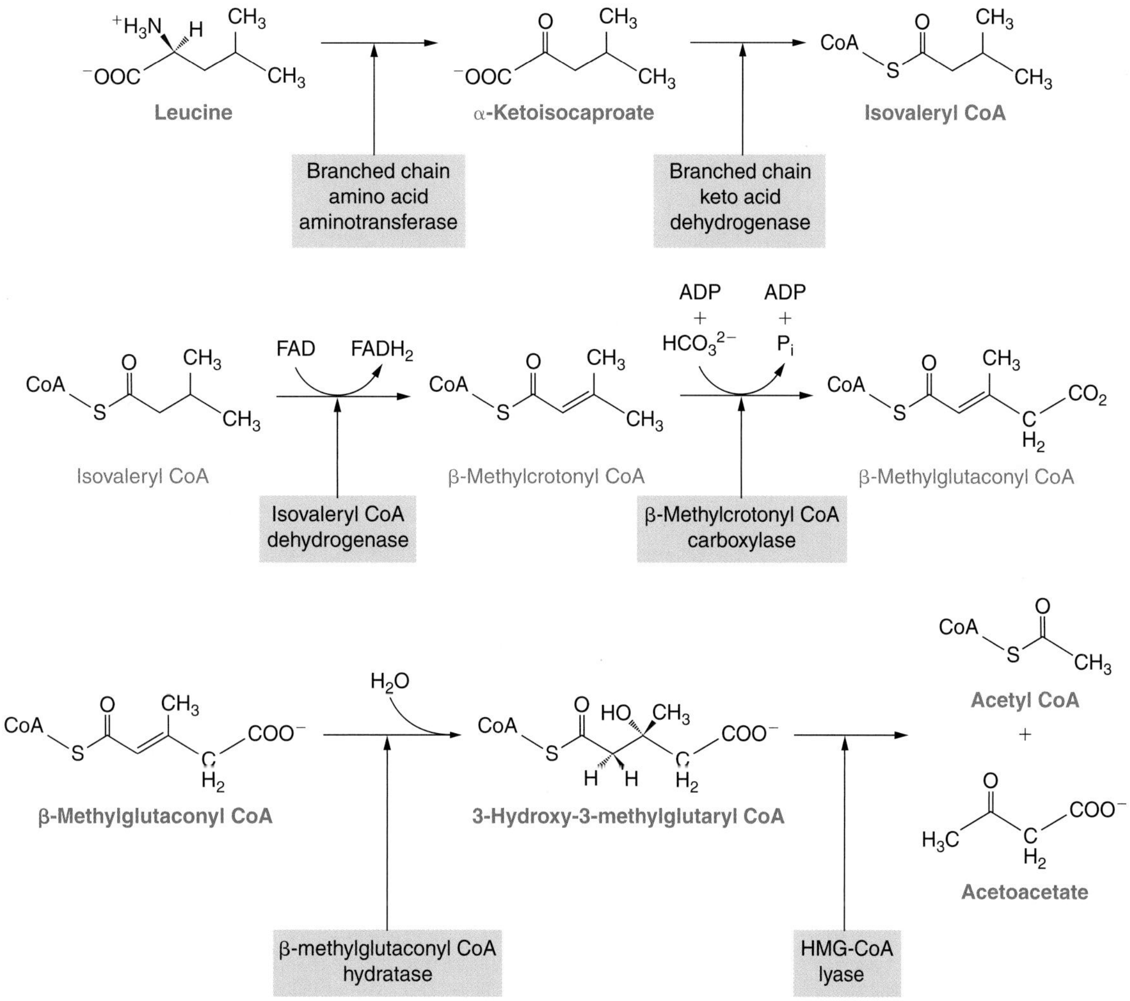

Figure 10-66. Conversion of leucine, a purely ketogenic branched-chain amino acid, to acetoacetate and acetyl CoA.

Catabolism of Amino Acids 553

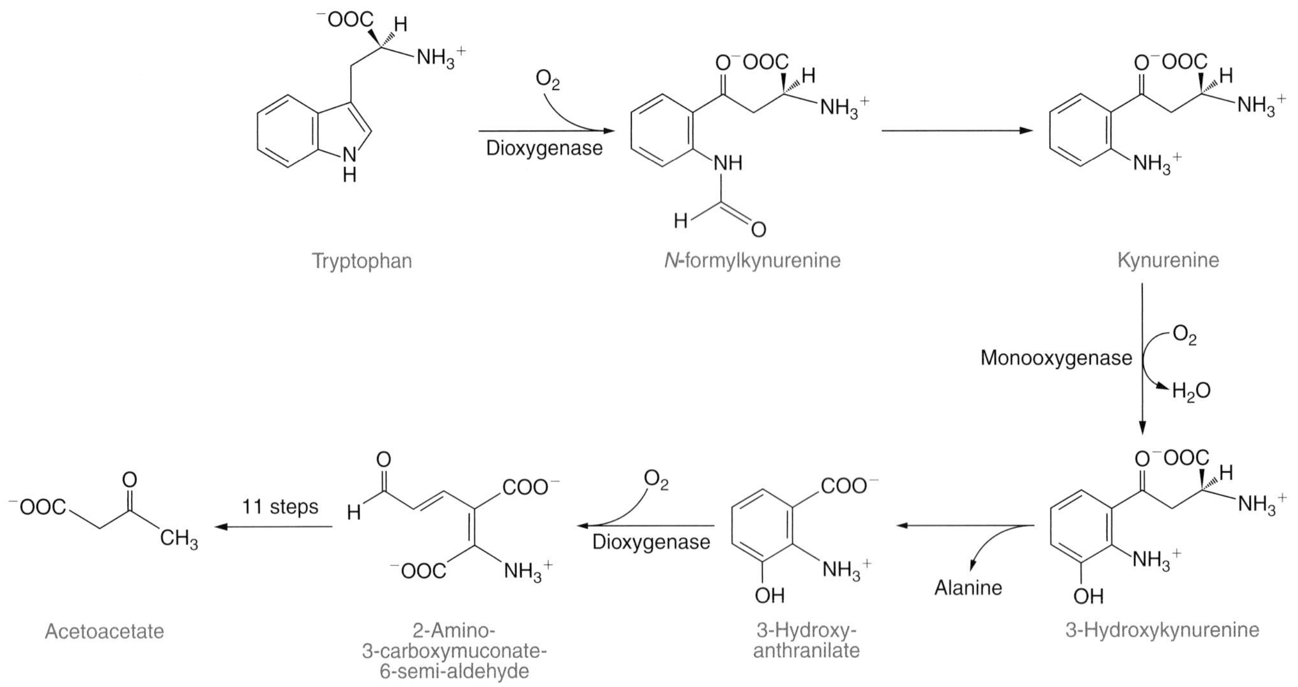

Figure 10-67. Catabolism of tryptophan.

Metabolism of Lipids

All of lipid metabolism is encompassed by fatty acid biosynthesis, covered in Chapter 5; sterol and bile acid biosyntheses, covered in Chapter 5; the synthesis and degradation of ketone bodies; steroid hormone metabolism, discussed in Chapter 9; breakdown of fats; and the metabolism of androgens and estrogens. Fatty acid biosynthesis is shown in Figure 5-35. Fatty acid synthesis (the overall summary reaction is 8 acetyl CoA + 7 ATP + 14 NADPH + 14 H$^+$ → CH$_3$(CH$_2$)$_{14}$CO$_2^-$ + 7 ADP + 7 P$_i$ + 14 NADP$^+$ + 7 H$_2$O) takes place in the cytoplasm. This is separated from the oxidation of fatty acids, which takes place in the mitochondria (Figure 5-18). Both the synthetic route and the oxidative pathway use acetyl CoA. CoA exists transiently as a substrate in the synthesis of fatty acids to form malonyl CoA, and the synthesis of malonyl CoA by acetyl CoA carboxylase (ACC; Figure 10-68) initiates the synthetic pathway (Figure 5-5). *This enzyme represents the principal site of regulation of fatty acid synthesis.* ACC becomes active when the monomer is polymerized, and this alteration is stimulated by citrate and end product inhibited by long-chain fatty acids. ACC is inhibited by phosphorylation by protein kinase A and by AMP kinase, which respond to the action of glucagon or epinephrine. This is the mechanism by which these hormones inhibit fatty acid synthesis. Because *insulin acts oppositely to glucagon and epinephrine,* its action leads to increased glycogen and triacylglyceride synthesis (Figure 10-69A). Figure 10-69B shows the actions of various phospholipases on a triglyceride. Insulin decreases the level of cyclic AMP, depressing activities of protein kinase A and AMP kinase and activating protein phosphatases that dephosphorylate and activate ACC to stimulate fatty acid synthesis. Insulin also

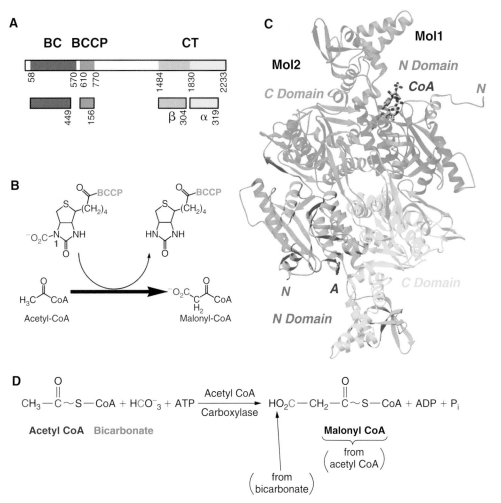

Figure 10-68. Structures of ACCs. **A.** Schematic drawing of the primary structures of eukaryotic multidomain ACC and bacterial multisubunit ACC. **B.** The chemical reaction catalyzed by CT. The N1 atom of biotin is labeled. **C.** Schematic drawing of the structure of the CT domain dimer of yeast ACC. The N and C domains of one monomer are colored cyan and yellow, whereas those of the other monomer are colored purple and green. The CoA molecule bound to one monomer is shown as a stick model. Only the adenine base was observed in the other monomer *(labeled A)*. Part C was produced with ribbons. **D.** Overall ACC catalyzed reaction showing reactants and products. *ACC,* Acetyl-CoA carboxylase; *CT,* carboxyltransferase domain. Parts A, B, and C reproduced with permission from H. Zhang et al., "Crystal structure of the carboxyltransferase domain of acety-coenzyme A carboxlase," *Science,* 299: 2064–2067, 2003, and from http://www.sciencemag.org/cgi/content/full/299/5615/2064/F1. Part D reproduced from http://www_medlib.med.utah.edu/NetBiochem/fattyAcids/5_2a.html-7k.

Metabolism of Lipids

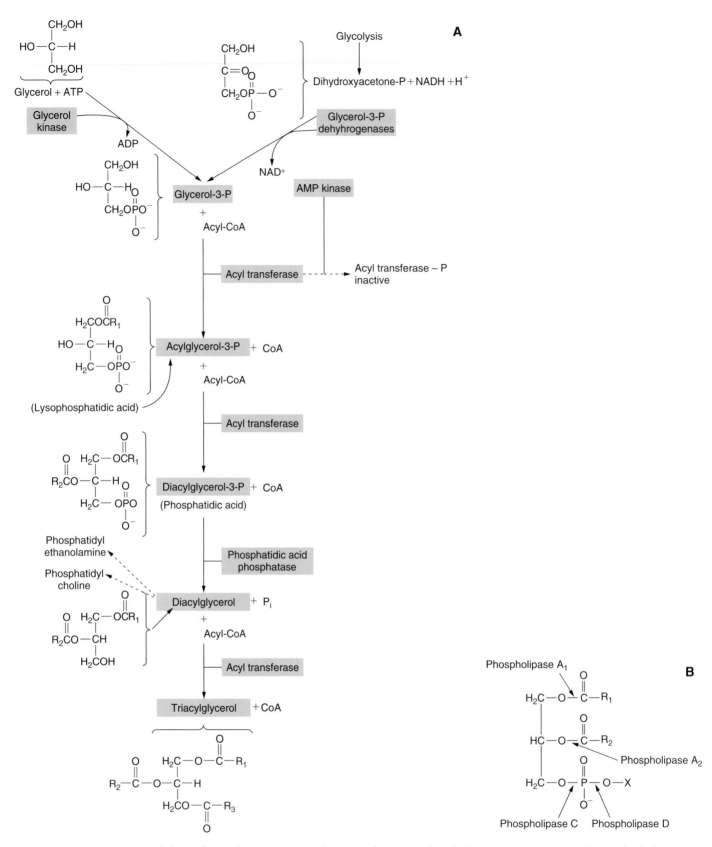

Figure 10-69. **A.** Triacylglyceride synthesis. **B.** Sites of action of various phospholypases (A_1, A_2, C, and D) in hydrolyzing portions of a triglyceride. Products of phospholipase action are lysophospholipids. Phospholipase A_2 releases arachidonic acid from cell membrane C-2 position of phospholipids and arachidonic acid is a precursor of prostaglandins, leukotrienes, and other inflammatory products.

can stimulate several cyclic AMP–independent kinases, which can activate ACC by phosphorylating different sites than those that inhibit ACC. Once a fatty acid is synthesized, double bonds can be introduced in its structure by fatty acid desaturase. Fatty acid desaturase is a metalloenzyme, containing two iron atoms in the active site, located in endoplasmic reticulum membranes (Figure 10-70A). There are four major human fatty acid desaturases (Δ-9-desaturase, Δ-6-desaturase, and Δ-5(4) desaturase and Δ-6 desaturase). They are named according to the position in the fatty acid being desaturated. The enzyme associates with cytochrome b5 and cytochrome b5 reductase, which uses NADH and O_2 for the introduction of the double bond into fatty acids. Actions of desaturases on fatty acids are shown in Figure 10-70B. Oleic and linoleic acids are essential unsaturated fatty acids in the diet.

The breakdown of fats (triglycerides) is accomplished by hormone-sensitive lipase (triacylglycerol lipase) as described later in this chapter (in the "Fat as Storage Energy" section).

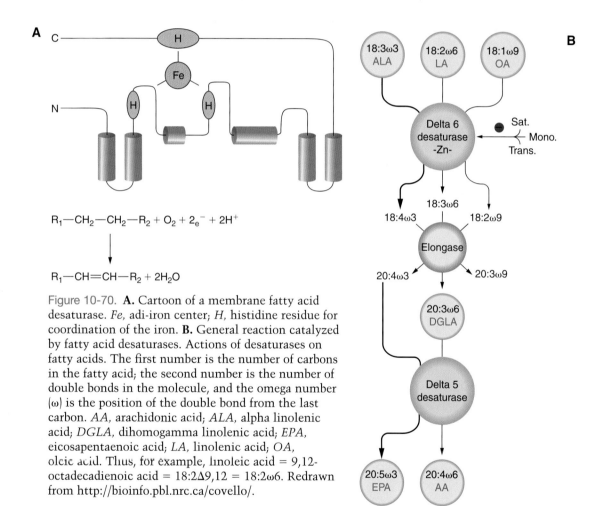

Figure 10-70. **A.** Cartoon of a membrane fatty acid desaturase. *Fe*, a di-iron center; *H*, histidine residue for coordination of the iron. **B.** General reaction catalyzed by fatty acid desaturases. Actions of desaturases on fatty acids. The first number is the number of carbons in the fatty acid; the second number is the number of double bonds in the molecule, and the omega number (ω) is the position of the double bond from the last carbon. *AA*, arachidonic acid; *ALA*, alpha linolenic acid; *DGLA*, dihomogamma linolenic acid; *EPA*, eicosapentaenoic acid; *LA*, linolenic acid; *OA*, oleic acid. Thus, for example, linoleic acid = 9,12-octadecadienoic acid = 18:2Δ9,12 = 18:2ω6. Redrawn from http://bioinfo.pbl.nrc.ca/covello/.

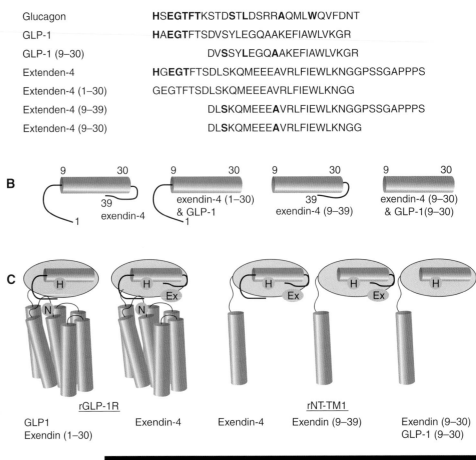

Figure 10-71. **A.** Amino acid sequence of glucagon and glucagon-like peptides, GLP-1 and extenden-4. **B.** Cartoons of portions of extender-4 as in A. **C.** Cartoons of the seven-membrane spanning receptor for GLP-1. Redrawn from http://www.astbury.leeds.ac.uk/Report/2002/Report/8Donnelly.pdf. **D.** Structure of glucagon. Reproduced from http://upload.wikimedia.org/wikipedia/en/6/67/Glucagon_stereo_animation.gif.

Glucagon

Glucagon is a polypeptide hormone synthesized in the alpha cells of the pancreatic islets. It causes the level of blood glucose to rise and the synthesis of fatty acids to decrease. It is signaled to be secreted when the blood level of glucose falls, just as insulin is secreted from the beta cells when the blood level of glucose is elevated. Glucagon is a 29–amino acid peptide that binds to a seven-membrane–spanning receptor, like the receptor for a related peptide, as shown in Figure 10-71. There is a **glucagon-like peptide (GLP-1)** that binds to a receptor in the intestine and in the pancreatic beta cell and is released into the bloodstream following food intake. GLP-1 stimulates glucose-dependent secretion of insulin and thus works differently than glucagon. Extendin-4, shown in Figure 10-71, is found in the venom of the Gila monster and can agonize the GLP-1 receptor. The N-terminal region of these peptides has the greatest binding activity. The interrelationships among glucagon, insulin, and GLP and their effects on blood glucose levels during feeding or fasting are shown in Figure 10-72. During feeding, GLP-1 is released from the intestine, binds to a receptor on the pancreatic beta cell membrane, and causes the release of insulin to be stimulated in the presence of glucose, similarly to the action of gastric inhibitory peptide (GIP). GIP is a 42–amino acid peptide that inhibits gastric acid secretion in the stomach and stimulates insulin secretion from the pancreatic beta cell in the presence of glucose. In turn, insulin released from the beta cell, under these conditions, inhibits the release of glucagon from the alpha cell. During fasting, when blood glucose will be low, the alpha cell is stimulated by low blood glucose to release glucagon, which binds to the **glucagon receptor (GLU-R)** to produce cyclic AMP; glucagon binds to receptors on the beta and delta cells. Somatostatin is released from the delta cell and binds to a receptor on the beta cell, causing the release of insulin to be inhibited during fasting, just as it

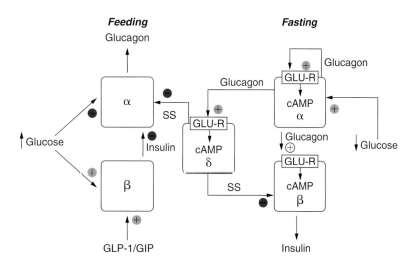

Figure 10-72. A model depicting the roles of glucagon in insulin-secreting islet cells (β), glucagon-secreting cells (α), and somatostatin (ss)-secreting cells (δ). Bold arrows emphasize major actions that are either stimulatory (+) or inhibitory (−). *GIP*, gastric acid inhibitory peptide; *GLP*, glucagon-like peptide; *GLU-R*, glucagon receptor.

Fatty Acid Degradation

Fatty acids are degraded in the mitochondria by the β-**oxidation pathway** to produce acetyl CoA. First, the fatty acid is converted to a fatty acetyl CoA derivative. This is followed by an oxidation step by an FAD dehydrogenase. The next steps involve hydration, oxidation by NAD$^+$-dehydrogenase, and cleavage of the chain to release acetyl CoA and a fatty acetyl CoA reduced in length by two carbons. The same set of reactions is repeated until the fatty acid is fully degraded to acetyl CoA. This system is referred to as the fatty acid spiral (Figure 10-73). The acetyl

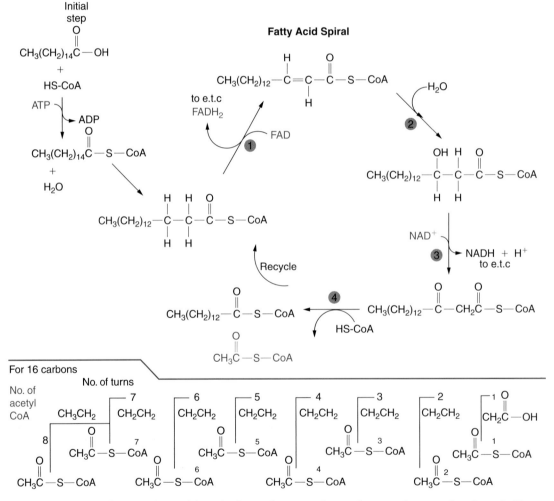

Figure 10-73. The fatty acid spiral by which mechanism of β-oxidation a fatty acid is degraded by two carbons at a time to form acetyl CoA until the fatty acid is completely broken down. This takes place in mitochondria and the acetyl-CoA produced can enter the citric acid cycle for conversion to ATP energy (Figure 4-59). *e.t.c.*, electron transport chain.

CoA produced can enter the citric acid cycle for conversion to ATP (Figure 5-23). When fatty acid degradation takes place at a rate faster than glycolysis, an excess of acetyl CoA is produced while pyruvate from glycolysis is not formed to a similar extent. Because pyruvate is relatively low in concentration, little oxaloacetate is produced, limiting the use of acetyl CoA by the citric acid cycle (Figure 4-59; Figure 5-23). When this condition occurs, excess acetyl CoA is converted to ketone bodies rather than entering the citric acid cycle. The ketone bodies are acetone, acetoacetate, and β-hydroxybutyrate (Figure 5-25).

Fat as Storage Energy

Fatty acids that can be used for energy by way of acetyl CoA and the citric acid cycle can come from fat absorption in the form of triglycerides in chylomicrons, where the fatty acids are released by extracellular lipase. There are free fatty acids complexed with albumin circulating in the bloodstream that can be used. But a major source is found in adipocytes (fat cells) of adipose tissue, where **hormone-sensitive lipase (HSL)** converts stored fat into free fatty acids in response to lipolytic hormones, such as norepinephrine and glucagon. This enzyme is depressed by insulin. The substrates of this enzyme and its structural organization are shown in Figure 10-74. The three-dimensional structure of HSL is shown in Figure 10-75. This enzyme is regulated by reversible phosphorylation so that lipolytic agents, such as norepinephrine or glucagon, increase phosphorylation in the regulatory domain of HSL. Elevated insulin levels produce an inhibition of adipocyte HSL by decreasing the phosphorylation on the regulatory domain of HSL. However, the phosphorylation level probably facilitates a translocation of HSL involving an accessory protein, perilipin. **Perilipin** is a protein (two forms, 57 kDa and 46 kDa, derived from RNA splicing of a single gene where the 46-kDa form is the result of skipping exon 6) that coats lipid storage droplets in adipocytes, protecting the adipocyte until breakdown by HSL.

Perilipin, the principal substrate of protein kinase A in adipocytes, becomes phosphorylated (and activated) by protein kinase A in response to lipolytic stimulation (glucagon, norepinephrine) and probably plays a transport role in effecting HSL action. Perilipin is localized to the surface of intracellular neutral lipid droplets, and it is confined to adipocytes, whereas other members of this "PAT" protein family, such as ADRP and TIP47, have broad tissue distributions. Underphosphorylated adipocyte HSL resides in the soluble cytoplasm, apart from its triacylglycerol substrate in the lipid droplet. When the adipocyte becomes stimulated by glucagon or catecholamine, cyclic AMP is formed and activates protein kinase A, which phosphorylates both perilipin and HSL. Phosphorylated perilipin translocates HSL to the lipid droplet surface to engage its substrate, increasing lipolysis some 30-fold over unstimulated cells. A similar mechanism takes place in steroidogenic cells, where hormones, such as adrenocorticotropic hormone acting at the cell membrane, stimulate the level of cyclic AMP and activate a mechanism culminating in the release of free cholesterol from lipid droplets, where it is stored as

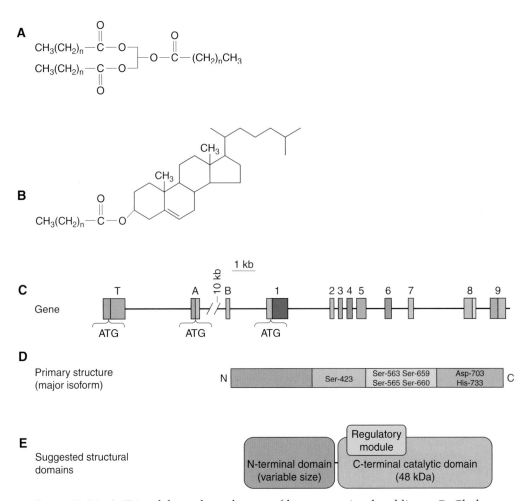

Figure 10-74. **A.** Triacylglycerol, a substrate of hormone-stimulated lipase. **B.** Cholesterol ester, a substrate of hormone-sensitive lipase in steroidogenic cells. **C.** Gene for human hormone-sensitive lipase showing intron *(heavy line)* and exon *(boxes)* structure. **D.** Linear representation of amino acid sequence divided into three functional regions. Exons encoding each of these regions bear the same shading (thus, the N-terminal domain derives from exons 1-4, etc.). **E.** Domain structure of hormone-sensitive lipase with two major structural domains, an N-terminal binding domain, and a C-terminal catalytic domain containing a regulatory module having multiple phosphorylation sites. Redrawn with permission from S.J. Yeaman, "Hormone-sensitive lipase—new roles for an old enzyme," *Biochem. J.*, 379: 11–22, 2004, and directly from http://www.biochemj.org/bi/319/bj3190411.htm-17k.

cholesterol esters. Perilipin may emerge as a new drug target in the treatment of obesity.

The domain structures of perilipin and its family members, TIP47 and ADRP, are shown in Figure 10-76 with the overall protein structure. In addition, residues 206–431 of TIP47 bear a structural similarity to residues 23–166 of apolipoprotein E shown in this figure. The lipid components of droplets and lipoprotein particles (e.g., apolipoprotein E) consist of a triglyceride and cholesterol core surrounded by a phospholipid monolayer. Thus, similar structures may provide stability to lipoproteins, as well as docking sites for receptors. A model summarizing HSL regulation is shown in Figure 10-77.

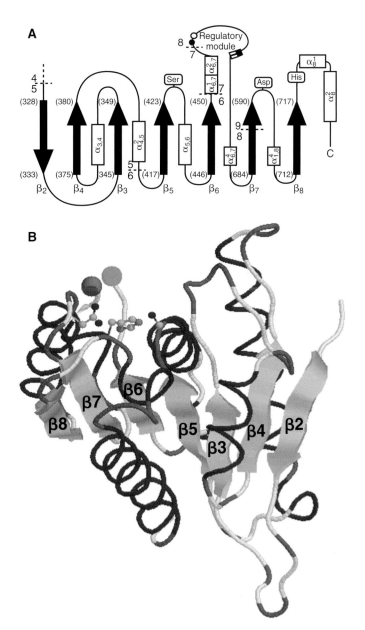

Figure 10-75. **A.** Schematic representation of the catalytic domain of hormone-stimulated lipase (HSL). Exon limits are indicated with dashed lines and the corresponding exon numbers. Also residue positions are indicated. The three catalytic triad residues are indicated with Ser, Asp, and His. The region containing the phosphorylation sites protrudes from the protein. The regulatory site and basal site are shown with circles. The estimated position of the newly discovered phosphorylation site is indicated with squares. **B.** Ribbon representation of the model for the catalytic domain of HSL. The strands of the central β-sheet are numbered accordingly to the enzymes of the carboxylesterase B family. The residues of the catalytic triad are shown in ball-and-stick representation. The model does not include a vast area located immediately behind the catalytic triad, inserted in the primary sequence between β-strands 6 and 7, that constitutes a regulatory module. The N- and C-terminal residues of this regulatory module are indicated by a light green and dark green sphere, respectively (kindly provided by J.A. Contreras and C. Holm). Reproduced from http://demeijer.com/biology/scriptie.pdf. J. De Meijer, a thesis written at the Biochemical Physiology Research Group, Department of Experimental Zoology, University of Utrecht, The Netherlands, 1998.

Fat as Storage Energy   563

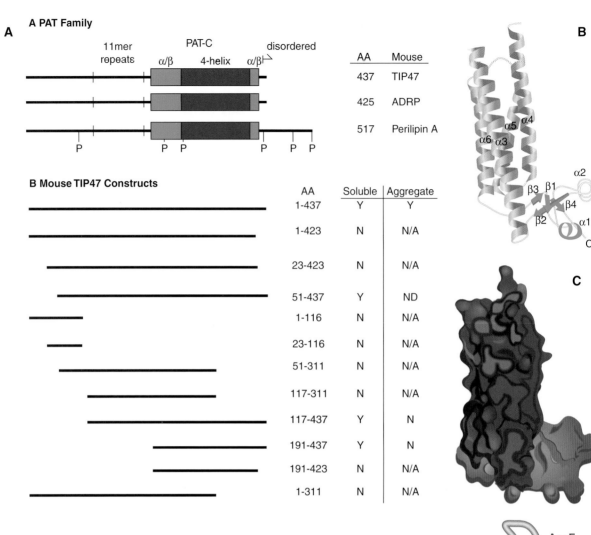

Figure 10-76. **A.** Domain Structure of PAT Proteins. The PAT-C region is shown in filled rectangles, green for the α/β domain, magenta for the four-helix bundle domain. The 11-mer repeat regions are delineated by vertical bars at the N- and C terminus of the repeat region. PKA phosphorylation sites are denoted by the letter "P." **B** and **C.** Overall structure of TIP47 PAT-C. The domains are colored as in A. **D.** Structural similarity of TIP47, residues 206-431, and apolipoprotein E, residues 23-166. Parts A, B, C, and D reproduced from S.J. Hickenbottom et al., "Structure of a lipid droplet protein: the PAT family member TIP47," *Structure*, 12: 1199–1207, 2004.

564   CHAPTER 10   Metabolism

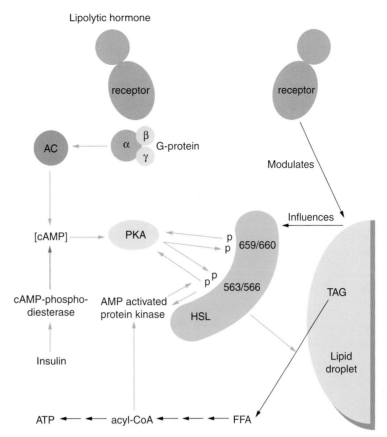

Figure 10-77. A model for HSL regulation. *Red arrows,* activation; *blue arrows,* inhibition; *black arrows,* another kind of relationship; *green arrows,* protein kinase activity. Lipolytic hormone *(top),* glucagon or catecholamine. *AC,* adenylate cyclase; *cAMP,* cyclic AMP; *FFA,* free fatty acid; *HSL,* hormone stimulated lipase (numbers on the molecule are phosphorylated amino acid residues); *PKA,* protein kinase A; *TAG,* triacylglycerol. Note that insulin stimulates cAMP phosphodiesterase, which degrades cAMP and suppresses PKA activity.

Lipid and Carbohydrate Metabolism Are Jointly Regulated

The metabolism of lipids and carbohydrates (carbohydrate metabolism is covered in a separate section later in this chapter) differs from one organ to the next. The rate at which metabolites enter the brain, muscle, and adipose tissue is controlled by the liver, which also controls the level of blood glucose (with additional controls from the pancreas in the form of insulin and glucagon). Fatty acid metabolism in adipose tissue is coordinated with glycolysis, accounting for the conversion of glucose, from the liver, to glycerol in adipose to which fatty acids are esterified to form triglycerides. Glucose, fatty acids, and ketone bodies can be oxidized in muscle, and muscle forms lactate, which is transported to the liver for conversion to glucose (gluconeogenesis). Glucose is the main source of energy for the brain—except under conditions of fasting or starvation, when it can use ketone bodies effectively.

β-Oxidation of fatty acids, which is coupled to oxidative phosphorylation, is dependent on the **energy charge** in the cell (ADP/ATP ratio). When the energy charge is low (ADP/ATP is high and [ADP] is high), degradation of fatty acids is stimulated to produce acetyl CoA, which enters the citric acid cycle to regenerate ATP. When the level of ATP is high in the cell, both fatty acid synthesis and phosphatidic acid synthesis are stimulated.

The transport of fatty acids across the mitochondrial inner membranes is another control mechanism mediated by the carnitine acyltransferase system (Figures 5-19 and 5-20). When the level of glucose is high, malonyl CoA levels are high (malonyl CoA is an inhibitor of mitochondrial carnitine-acyltransferase I in the mitochondrial outer membrane), fatty acid transport into the mitochondrion is inhibited, and acetyl CoA is used for fatty acid synthesis.

Triacylglycerides and cholesterol are carried by lipoproteins in the blood (Figures 5-2, 5-58, 5-59, and 5-60). Lipids in the diet are digested and emulsified by bile acids (Figure 5-9) and are transported by chylomicrons (Figure 5-16). Fatty acids released from adipose tissue are transported by serum albumin. The liver synthesizes triacylglycerols, which are transported by VLDLs (Figure 5-17). Cholesterol is also synthesized in the liver, and it is transported mainly by LDLs. When cells die, their components are released; cholesterol from dead cells is transported by HDLs.

Steroid Hormone Metabolism

The formation of bile acids from cholesterol has been discussed in Chapter 5 (Figure 5-8). However, the metabolites of steroid hormones have not been discussed. The formation of adrenal steroidal hormones is shown in Figure 9-7. This figure shows the hormonally active steroids but not the metabolites. Δ^4-Androstenedione and testosterone are subject to the action of the enzyme **aromatase,** which occurs in female sexual tissues, such as the mammary gland, and leads to the formation of 17β-estradiol or estrone and subsequently to estriol:

Cholesterol ⇌ pregnenolone ⇌ 17α-hydroxypregnenolone ⇌ dehydroepiandrosterone ⇌ androstenedione ⇌ testosterone ⇌ 17β-estradiol

Pregnenolone ⇌ progesterone ⇌ 17α-hydroxyprogesterone ⇌ androstenedione ⇌ estrone ⇌ estriol

17β-Estradiol is the principal female sex hormone, although estrone can agonize the estradiol receptor. The conversion of testosterone to 17β-estradiol is catalyzed by aromatase, as is the conversion of androstenedione to estrone. The aromatase reaction is shown in Figure 10-78. Because aromatase is contained in mammary glands and 17β-estradiol is a growth factor for mammary cancer (aromatase can convert dehydroepiandrosterone from the adrenal cortex to 17β-estradiol), sometimes aromatase inhibitors are used to decrease the endogenous production of the female sex hormone. Some of these inhibitors are shown in Figure 10-79. 17β-Estradiol can be inactivated in a number of ways: oxidation of the

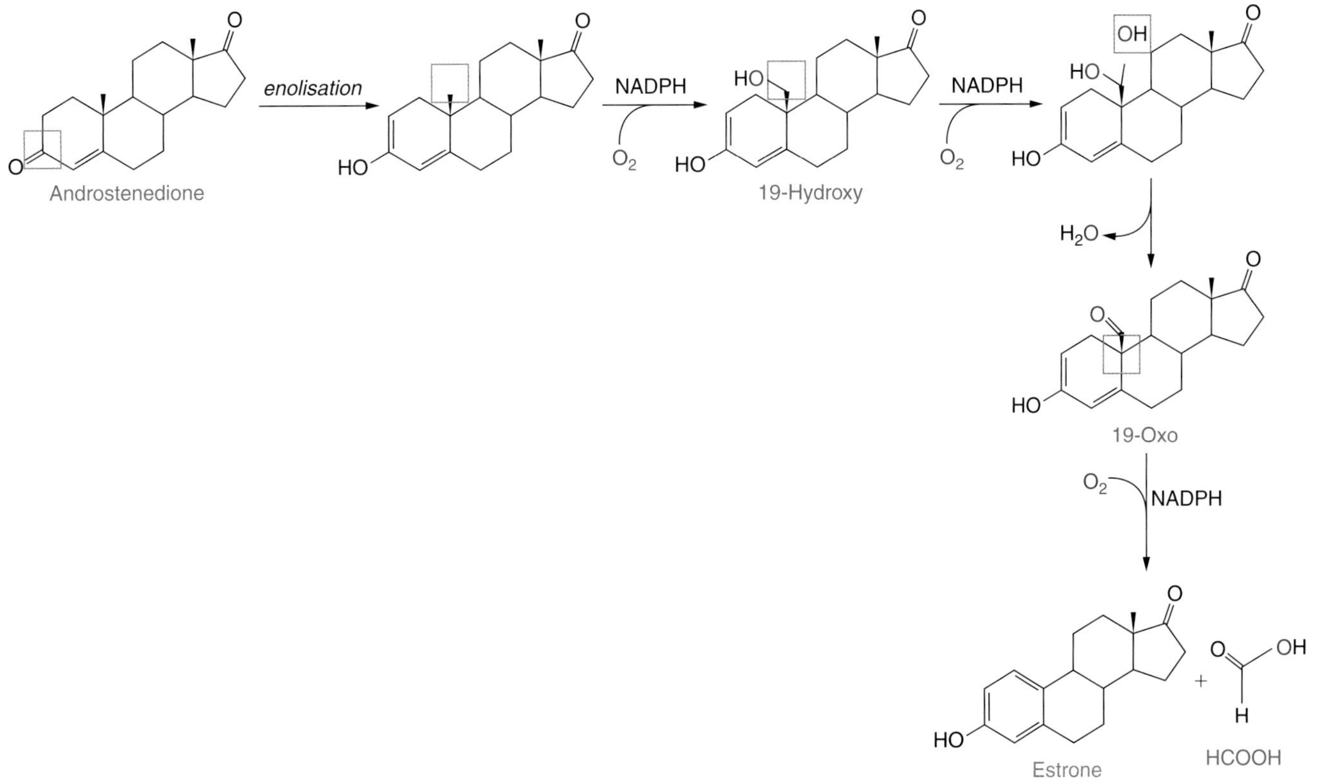

Figure 10-78. Aromatase reaction converting androstenedione to estrone. The reaction is mediated by cytochrome P-450.

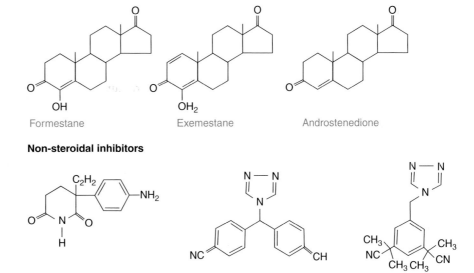

Figure 10-79. Some inhibitors of aromatase.

Steroid Hormone Metabolism

17β-hydroxyl to the ketone; hydroxylation at C2, followed by methylation; further hydroxylation or ketone formation at a number of positions (C6, C7, C14, C15, C16, or C18, for example); and glucuronidation of the 3-hydroxyl. Many similar derivatives of estrone occur. Glucuronidation of positions 3 and 17 of 17β-estradiol are shown in Figure 10-80.

The formation of testosterone is shown in Figure 9-7. Testosterone can be inactivated by reduction of the 4-ene-3-one and oxidation of the C17 hydroxyl to a ketone. Some urinary excretion products are androsterone and etiocholanolone, as shown in Figure 10-81. These compounds are excreted also as the glucuronide or sulfate derivatives of the C3 hydroxyl group.

Progesterone is inactivated by reduction of the C20 ketone (Figure 9-7) to the hydroxyl and reduction of the 4-ene-3-one, saturating the double bond and converting the C3 ketone to a hydroxyl. Pregnanediol and its C3 glucuronide are the excreted products of progesterone (Figure 10-82).

The glucocorticoid, cortisol, is inactivated by reduction of the 4-ene-3-one (Figure 9-7) saturating the double bond and converting the C3 ketone to a hydroxyl group, reducing the C20 ketone to an alcohol, and cleaving the side chain. The excretion products are 11β-hydroxyandrosterone and allo tetrahydrocortisone (Figure 10-82), as well as their C3 glucuronides.

The mineralocorticoid, aldosterone, is inactivated by reduction of the 4-ene-3-one to saturate the double bond (Figure 9-7), converting the C3

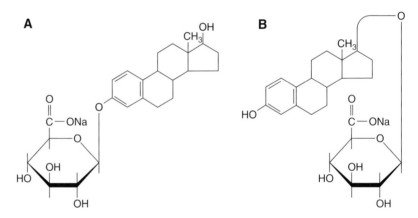

Figure 10-80. **A.** Structure of 3-glucuronide of 17β-estradiol. **B.** Estradiol-17β-D-glucuroride.

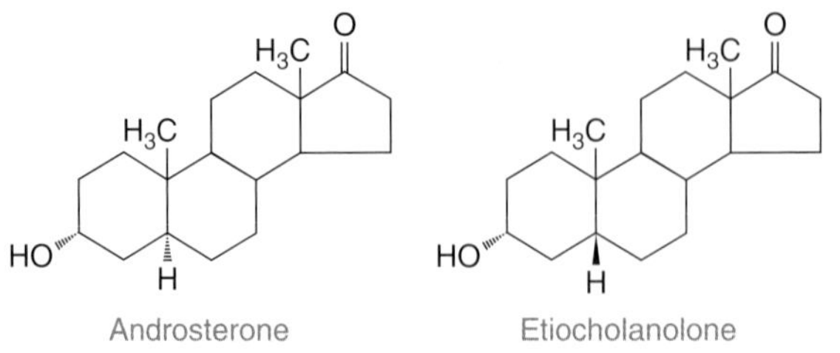

Figure 10-81. Structures of androsterone and etiocholanolone. Both are excreted as glucuronides or sulfate derivatives of the C-3 hydroxyl group.

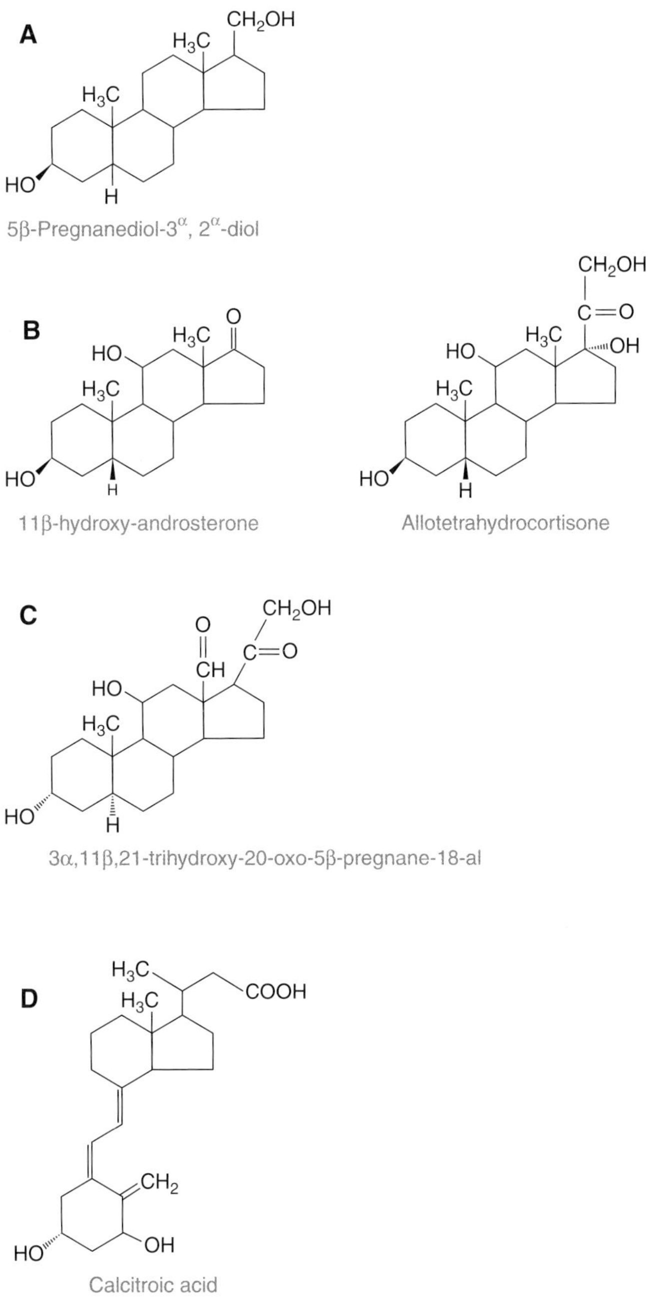

Figure 10-82. Excretion products of various steroid hormones. **A.** Excretion product of progesterone as pregnanediol and its glucuronide derivative. **B.** Excretion products of cortisol that also occur as the glucuronide derivatives. **C.** Major aldosterone excretion product, also occurring as the glucuronide derivative. **D.** Excretion product of vitamin D_3.

ketone into a hydroxyl group. The primary excretion product is 3α, 11β, 21 trihydroxy-20-oxo-5β-pregnane-18-al (Figure 10-82), and the glucuronide derivative of the C3 hydroxyl.

Vitamin D is inactivated by cleavage of the side chain between C23 and C24 (Figure 10-36) to form calcitroic acid (Figure 10-82), the main excretion product.

The normal level of pregnanediol in female urine is 0.2–6 mg/day; in the follicular phase it is 0.1–1.3 mg/day, and in the luteal phase it is 1.2–9.5 mg/day. In the male, it is 0.2–1.2 mg/day in the urine. The levels of 17-hydroxycorticosteroids in the urine are 2.0–6.0 mg/day in the female and 3.0–10.0 mg/day in the male. Aldosterone derivatives in the urine are 6–25 µg/day on a normal salt diet.

Nucleic Acid Metabolism

Nucleic acid metabolism consists of the synthesis of purine and pyrimidine mono-, di-, and triphosphates, which was covered in Chapter 6 (Figures 6-21, 6-38, 6-40, and 6-44 through 6-46). The degradation of purines to uric acid is shown in Figure 6-48, and the catabolism of pyrimidines to malonyl CoA and methylmalonyl CoA, which can be used

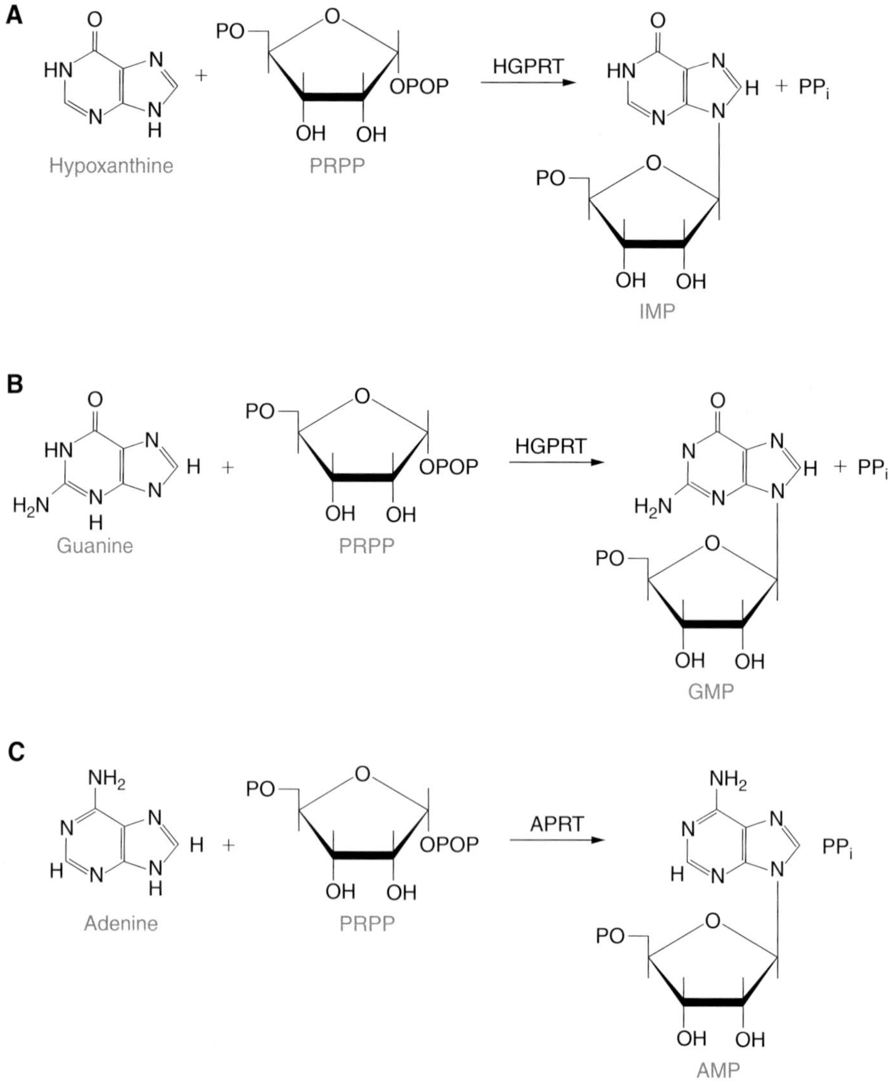

Figure 10-83. **A.** Salvage of hypoxanthine by hypoxanthine-guanine phosphoribosyl-transferase (HGPRT) to form IMP (inosine monophosphate). **B.** Salvage of guanine by HGPRT to form guanosine monophosphate (GMP). **C.** Salvage of adenine by adenine phosphoribosyltransferase (APGT) to form adenosine monophosphate (AMP). *PRPP*, phosphoribosyl pyrophosphate; *PPi*, pyrophosphate. PPi is formed as the magnesium salt.

subsequently to generate energy, is shown in Figure 6-52. Other aspects of metabolism involve purine salvage and pyrimidine salvage. Free purine bases may be reused through salvage as an alternative to complete catabolism to uric acid. In this case, hypoxanthine-guanine phosphoribosyltransferase (HGPRT) converts hypoxanthine to inosine monophosphate (IMP) with phosphoribosylpyrophosphate (PRPP) shown in Figure 10-83A. It can salvage guanine to form GMP as well (Figure 10-83B). When there is a genetic deficiency of HGPRT, a severe neurological disease occurs, called the Lesch-Nyhan syndrome, in which there is uncontrollable self-mutilation. Such patients have high blood uric acid levels because of uncontrolled synthesis and inability to control the levels of IMP and GMP in salvage pathways. Adenine can be salvaged by adenine phosphoribosyltransferase (APRT), as shown in Figure 10-83C. A crystal structure of APRT is shown in Figure 6-51A. In pyrimidine salvage, pyrimidine phosphorylase, the enzyme that cleaves the sugar from the base, catalyzes a reversible reaction and is able to convert thymine to thymidine. Thymidine can be converted further by thymidine kinase to form thymidine monophosphate, shown in Figure 10-84, and a similar set of reactions takes place to salvage uracil. A crystal structure of thymidine kinase is shown in Figure 6-54, and structures for HGPRT are in Figure 10-85.

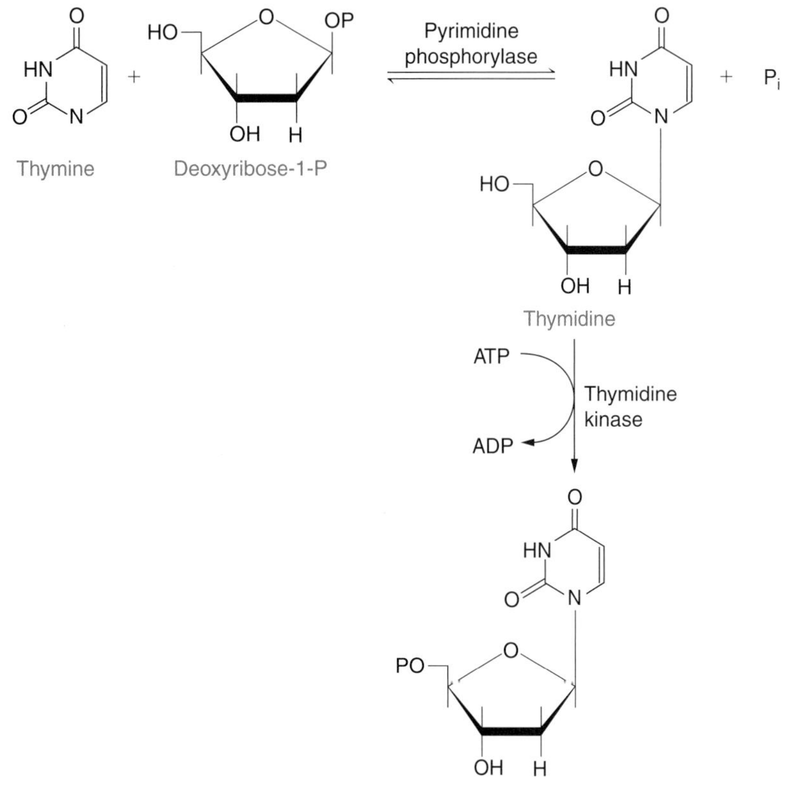

Figure 10-84. Salvage of the pyrimidine, thymine, by pyrimidine phosphorylase to form thymidine (with deoxyribose-1-phosphate). Thymidine can be converted to thymidine monophosphate (TMP) by thymidine kinase and ATP. In addition to salvage of thymine, there is a uridine phosphorylase and a uridine kinase that can salvage uracil in a similar set of reactions.

Nucleic Acid Metabolism   571

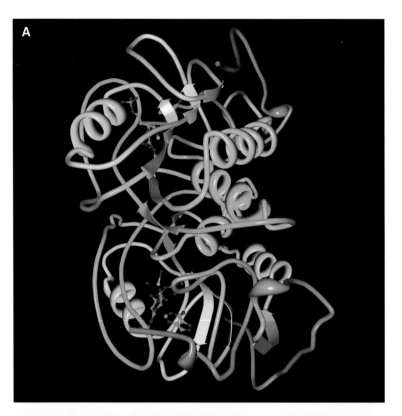
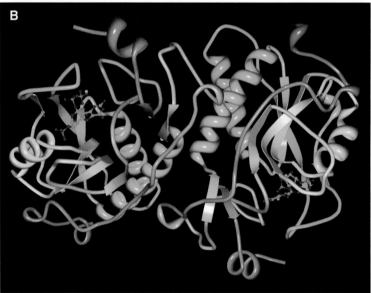

Figure 10-85. **A.** Ternary complex structure of human HGPRT (hypoxanthine guanine phosphoribosyltransferase), PPRP, Mg^{2+}, and the inhibitor HPP reveals the involvement of the flexible loop in substrate binding. **B.** HGPRT from *Toxoplasma gondii* showing four polypeptide chains of 228, 231, 213, and 227 residues. Parts A and B reproduced from PDB ID: 1D6N. G.K. Balendiran, J.A. Molina, Y. Xu, J. Torres-Martinez, R. Stevens, J.P. Focia, A.E. Eakin, J.C. Sacchettini, S.P. Craig III. Ternary Complex Structure of Human HGPRTase, PRPP, Mg^{2+}, and the Inhibitor HPP Reveals the Involvement of the Flexible Loop in Substrate Binding. *Protein Science* **8** pp. 1023 (1999) and from http://www.rcsb.org/pdb/cgi/explore.cgi?pdbId=1d6n.

Although much attention was given to endonucleases (restriction enzymes) in Chapter 6, little has been said about the breakdown of DNA and RNA in more general terms. In the adult, DNA in the cell is stable, although it may be able to turn over individual bases without breaking down the DNA molecule. DNA breaks down in the process of cell death, and this involves DNase activity. DNA in the cell can be damaged (see Table 6-4 and Figures 6-78 through 6-79) and repaired by various DNA repair enzymes (for example, see Figure 6-74).

DNA Damage and Repair

Chemical agents and radiation can injure DNA, which can dysregulate cellular processes and cause cancer. As a protection mechanism, the human genome has more than 130 DNA repair genes encoding proteins that screen the genome to eliminate damage. DNA repair consists of three major mechanisms: **base excision repair, nucleotide excision repair,** and **mismatch** repair. The repair mechanisms in the human are similar to those in bacteria (for example, *Escherichia coli*). Bases in DNA can be changed by deamination or alkylation. Deamination can change cytosine to uracil and 5-methylcytosine to thymine (Figure 10-86). Nitrous acid (HNO_2) is a common deaminating reagent that can convert cytosine to uracil, adenine to hypoxanthine, and guanine to xanthine. This alters the hydrogen bonding of the modified base, which results in mispairing.

Alkylation is the addition of an alkyl group (C_nH_{2n+1}) to another molecule, which can alter base pairing. An alkylating agent can alter guanine, which normally pairs with cytosine, to 7-ethylguanine, which pairs with thymine. This alteration in pairing can lead to a mutation. An agent that can accomplish this alkylation is ethylmethane sulfonate:

$$CH_3\text{-}CH_2\text{-}O\text{-}SO_2\text{-}CH_3$$

Other well-known alkylating agents are sulfur mustard (di-(2-chloroethyl)sulfide)

$$Cl\text{-}CH_2\text{-}CH_2\text{-}S\text{-}CH_2\text{-}CH_2\text{-}Cl$$

and nitrogen mustard (di-(2-chloroethyl)methylamine)

$$Cl\text{-}CH_2\text{-}CH_2\text{-}N(CH_3)\text{-}CH_2\text{-}CH_2\text{-}Cl$$

Free radicals can change the structure of a base, as can NH_2OH (Figure 10-87). The base modified by deamination or alkylation is denoted the abasic site (AP site). DNA glycosylase can recognize this site and remove the altered base. AP endonuclease removes the AP site and the neighboring nucleotides, and the gap is filled by DNA polymerase I and DNA ligase, as shown in Figure 10-88. An example of nucleotide excision involves the removal of the nucleotides damaged by ultraviolet light, forming a dimer. After enzymatic removal, the gap is filled by DNA polymerase I and DNA ligase. Nucleotide excision in *E. coli* (similar to the human situation except the names of the genes are different) is shown in Figure 10-89. Mismatch repair, which is incompletely

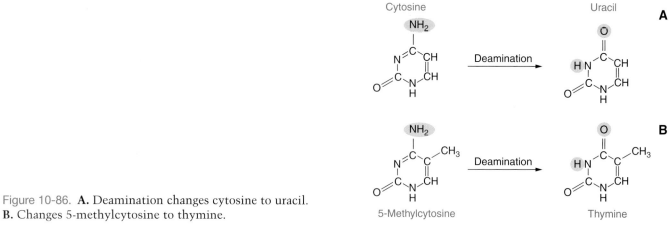

Figure 10-86. **A.** Deamination changes cytosine to uracil. **B.** Changes 5-methylcytosine to thymine.

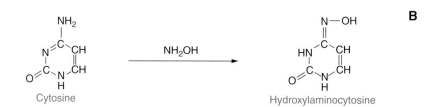

Figure 10-87. **A.** Base structures induced by free radicals. **B.** The base change induced by NH₂OH.

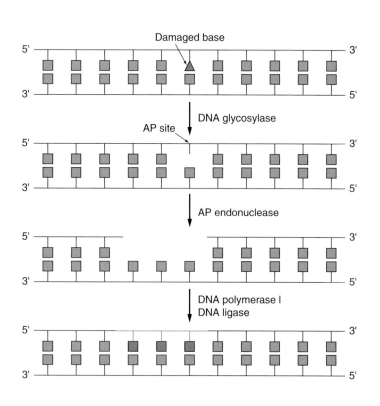

Figure 10-88. DNA repair by base excision. Redrawn from http://www.web-books.com/MoBio/Free/Ch7G.htm#Base.

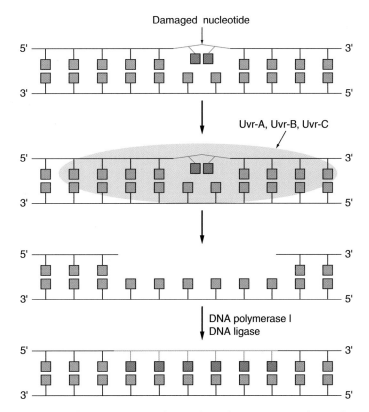

Figure 10-89. DNA repair by nucleotide excision. Redrawn from http://www.web-books.com/MoBio/Free/Ch7G.htm#Base.

understood in eukaryotes, is known for *E. coli.* The protein MutS binds to mismatched base pairs, and then MutL joins the complex and activates MutH, which cleaves the unmethylated strand at the guanine-adenine-thymine-cytosine (GATC) site. Exonuclease removes the segment from the cleavage site to the mismatch assisted by other proteins. For cleavage occurring on the 3' side of the mismatch, exonuclease I degrades a single strand in the 3' to 5' direction. For cleavage on the 5' side of the mismatch, degradation of the single-stranded DNA is accomplished by exonuclease VII or RecJ. The gap is filled by DNA polymerase III and DNA ligase (Figure 10-90). Mismatch repair is similar in eukaryotes with homologs of MutS (MSH1 and MSH5) and of MutL (MLH1, PMS1, and PMS2).

Cell Death

Because metabolism includes catabolism, cell death can be discussed in this context. There are two types of cell death: necrosis and programmed cell death (apoptosis). Necrosis is the death of cells in a tissue, usually

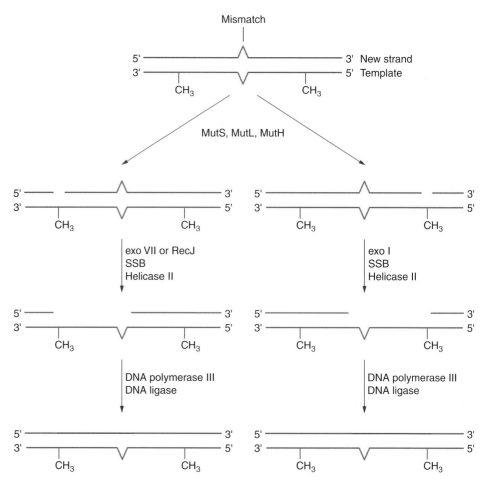

Figure 10-90. Mismatch repair. Redrawn from http://www.web-books.com/MoBio/Free/Ch7G.htm#Base

as a consequence of some toxic factor, whereas programmed cell death is the result of a genetic program that induces the cell to undergo an orderly suicide in a process often beneficial to the organism. For example, during development, some tissues die off while new tissues are formed and the precursor tissues are removed by apoptosis. In the ovarian cycle, when the thickened uterine wall collapses during menses so that another cycle of tissue proliferation can occur, those cells are killed by apoptosis. Apoptosis is the balance to cell proliferation so that organs can maintain their appropriate size. DNA is broken down in both cases by DNases. In the case of necrosis, the breakdown is more random than in the case for apoptosis, in which DNA is broken down to internucleosomal stretches. Contents of dead cells are phagocytized by macrophages, and macromolecules are generally broken down by lysosomal enzymes (Chapter 1). Major DNases are DNase I and

DNase II. DNase I is found in the nuclei of many cell types, whereas DNase II is located in the lysosomes. Although some regions of DNA are more sensitive to hydrolysis by DNase I than others, the macromolecule is usually protected by proteins bound to the nucleic acid. When unprotected, DNA is broken down by DNase I; the cleavage process continues to the individual deoxyribonucleotides. DNase I has a more alkaline pH optimum than DNase II, which would be expected to have an acidic pH optimum in the context of the lysosome. RNA is broken down by ribonuclease, and the resulting pyrimidines are degraded to common metabolites, which enter the citric acid cycle for energy production; the amino groups of pyrimidines enter the urea cycle. Purines derived from nucleic acids are converted to uric acid for excretion, and amino groups from purines can enter the urea cycle.

Carbohydrate Metabolism

An overview of glucose metabolism is shown in Figure 10-91. Dietary carbohydrate is broken down in the intestine into monosaccharides, which enter the blood and are taken up by various tissues, especially the liver. If not needed as glucose per se, they can be synthesized into glycogen, the storage form. Glucose enters the liver cell cytoplasm through the transporter, Glut 2 (in the brain, the transporter for glucose is Glut 3). The specific glucose transporters in various tissues are shown in Figure 10-92. The structure of Glut 1, a transporter present in all human cells, is shown in Figure 10-93. Inside the liver cell, glucose enters the glycolysis pathway to form pyruvate. Pyruvate can enter the mitochondrion and, with oxaloacetate, forms citrate and proceeds through the citric acid cycle to generate energy in the form of ATP. In the glycolytic pathway, dihydroxyacetone phosphate is formed, which can be converted to glycerol-3-phosphate, a precursor for membrane phospholipid. Pyruvate can be converted to lactate or to alanine catalyzed by alanine aminotransferase. The amino group of alanine can be excreted by way of the urea cycle; the carbon skeleton of alanine is converted back to pyruvate. This is the case in the context of the glucose–alanine cycle, where alanine is exported to the liver from muscle (muscle protein breakdown).

Glucose, upon entering the liver cell, encounters the first enzyme of the glycolysis pathway, hexokinase. Hexokinase is a family of four enzymes, including glucokinase (hexokinase IV in liver cells and pancreatic islet cells), which catalyzes the phosphorylation of glucose to form glucose 6 phosphate. Glucose-stimulated insulin release from the beta cells of the pancreas is tightly regulated by glucokinase, which acts as a sensor to couple glucose metabolism to the release of insulin. Figure 10-94 shows the crystal structure of yeast hexokinase and the conformational change induced in human brain hexokinase upon binding a molecule of glucose. This conformational change in the hexokinase molecule results in the positioning of the 6-hydroxyl group of glucose in the vicinity of ATP. Fructose is a commonly ingested monosaccharide in sucrose, honey, corn syrup, fruits, and vegetables—it can

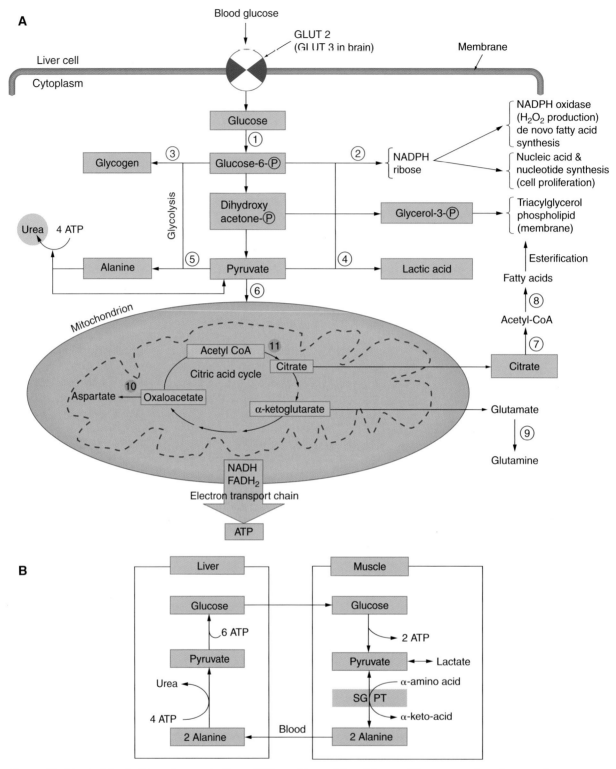

Figure 10-91. **A.** Metabolism glucose in liver. (1) hexokinase/glucokinase; (2) pentose-phosphate pathway; (3) glycogen synthesis; (4) lactate dehydrogenase; (5) alanine aminotransferase = soluble glutamate-pyruvate transaminase (SGPT); (6) pyruvate dehydrogenase; (7) ATP-citrate lyase; (8) fatty acid synthesis; (9) glutamine synthetase; (10) aspartate aminotransferase; (11) citrate synthase. **B.** Glucose-alanine cycle by means of which muscle can eliminate nitrogen as alanine, especially during muscle breakdown. Alanine is transported to the liver, where it is converted back to pyruvate and then to glucose by way of gluconeogenesis. Glucose can then go back to muscle for energy. Not shown are multiple steps between glucose-6-phosphate and dihydroxyacetone phosphate, and between dihydroxyacetone phosphate and pyruvate.

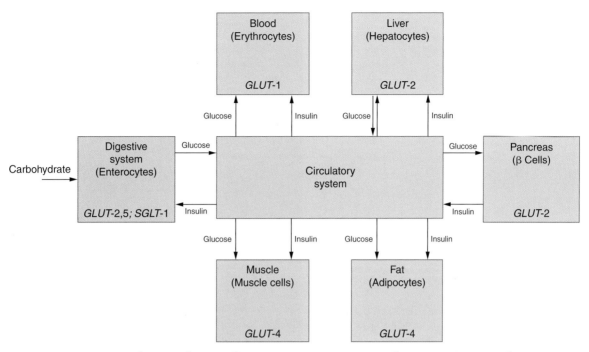

Figure 10-92. Diagram showing the specific GLUT transporter in specific tissues. Arrows indicate movement of a substance (glucose or insulin) from one compartment to another. GLUT-3 is not shown in this diagram. It is located in brain and nerves, placenta, kidney, fibroblasts, liver, and cardiac muscle. Redrawn from *Quantitative Physiology; Cells and Tissues*, Lecture 11, 1: 6.6-6.7.4, 2004.

Figure 10-93. Ribbon representation of Glut1 (human facultative transporter), the major glucose transporter present in all human cells. It shows a space-filling representation of the main channel *(yellow)*, helices *(colored)*, and loops *(white)*. Residues in space-filling rendering correspond to several conserved motifs around the channel; Gln^{279}, Leu^{280}, and Ser^{281} (QLS motif) *(red)*; Tyr^{292} and Tyr^{293} *(purple)*; Gln^{282} *(green)*; and Trp^{412} *(cyan)*. Residues 388–412 implicated in the putative binding site for cytochalasin B are colored by atom; cysteines are shown as sticks *(red)*. Reproduced from F.A. Zuniga et al., "A three-dimensional model of the human facilitative glucose transporter Glut1," *J. Biol. Chem.*, 276: 44970–44975, 2001 and from http://www.jbc.org/cgi/content/full/276/48/44970/F3.

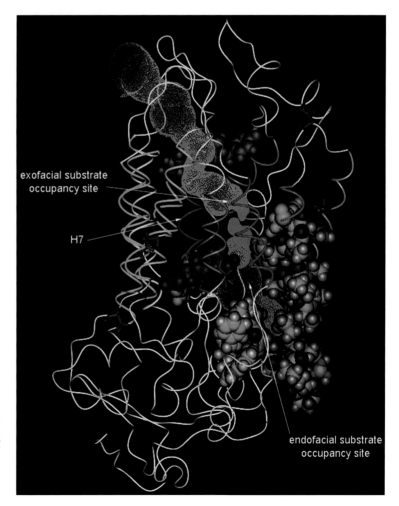

Carbohydrate Metabolism

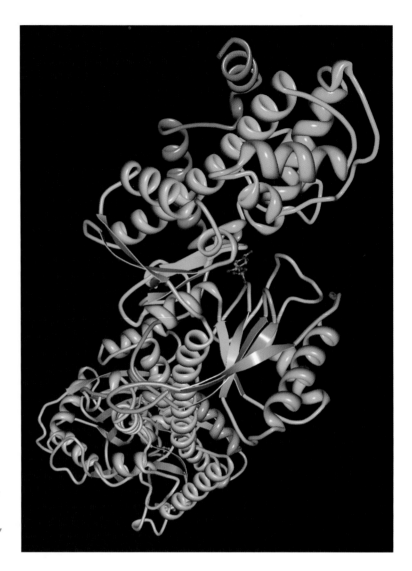

Figure 10-94. Human hexokinase. A.E. Aleshin, et al. "Regulation of hexokinase 1: Crystal structure of recombinant human brain hexokinase complexed with glucose and phosphate," *J. Mol. Biol.*, 282: 345–357, 1998.

enter the glycolytic pathway through the action of either fructokinase or hexokinase:

Fructokinase: fructose + ATP ⇌ fructose-1-phosphate + ADP

Fructose-1-phosphate ⇌ dihydroxyacetone phosphate + glyceraldehyde

(Dihydroxyacetone phosphate is converted to glyceraldehyde-3-phosphate by triose phosphate isomerase; glyceraldehyde is converted to glyceraldehyde-3-phosphate by triose kinase.)

Hexokinase: fructose + ATP ⇌ fructose-6-phosphate + ADP

Phosphofructokinase: fructose-6-phosphate + ATP ⇌ fructose-1,6-bisphosphate

Fructose-1-phosphate formed by fructokinase bypasses phosphofructokinase, which is the rate-limiting step of glycolysis. Too much dietary fructose leads to the accumulation of fructose-1-phosphate, which can elevate levels of acetyl CoA and lipogenesis and can even lead to gout (excess dietary fructose can raise levels of uric acid). On the other hand, *ingestion of fructose, compared to glucose, does not raise the blood level of glucose much, because fructose is cleared rapidly by the liver and does not evoke the beta cells of the pancreas to produce as much insulin.* Other importantly consumed monosaccharides are mannose and galactose. They are introduced to glycolysis by phosphomannose isomerase, which converts mannose into glucose, and by phosphorylation to galactose-1-phosphate and then to UDP-galactose, which becomes epimerized to the glucose-1-phosphate that can be converted to glucose-6-phosphate. Galactose comes mainly from lactose, a disaccharide of glucose and galactose, which is hydrolyzed in the intestine by lactase, an enzyme that can be missing or poorly functional in subjects with lactose intolerance. In lactose intolerance, Lactaid (containing lactase) can be consumed before eating dairy products.

Regulation of Blood Glucose Level

This is an important feature of carbohydrate metabolism. Dietary carbohydrates are ingested and first attacked in the oral cavity by α-amylase, which cleaves 1–4 linkages of polysaccharides. Some further digestion occurs in the acid medium of the stomach. In the small intestine, there are many disaccharidases and oligosaccharidases that break down carbohydrates further to simple sugars, which are absorbed. Glucose is absorbed through cells of the small intestinal epithelium using a glucose–sodium symport (transporter of two substances in the same direction) at the apical side (facing the lumen) of the cell and moved out of the cell to the extracellular space and eventually to the bloodstream through a glucose uniport (transporter of a single substance) at the basolateral side of the cell (facing the extracellular space on the way to the bloodstream) (Figure 10-95). The symport system links sodium with glucose transport, wherein sodium ion travels down its own gradient and glucose travels against its gradient (the carrier or transporter releases glucose inside the cell into a glucose concentration that is higher than that on the outside of the cell where glucose was initially recognized and bound by the carrier or transporter). The energy to transport glucose *against* its gradient comes from the movement of sodium ion *down* its concentration gradient. At the basolateral side of the cell, sodium ion is pumped out to the extracellular space and potassium ion is moved to the cell interior from the extracellular space by Na^+/K^+-ATPase, an **antiport** (transporter moving two molecules in opposite directions), powered by ATP. As much as 25% of the cell's ATP may be used to maintain the sodium ion gradient. In muscle and adipose cells, glucose (from the bloodstream) is transported inwardly by Glut 4 (a uniport), as shown in Figures 10-92 and 10-96.

Glucagon, cortisol, epinephrine, and growth hormone (also prolactin) raise the level of glucose in the blood, and insulin lowers the level of blood glucose. Insulin is released from the beta cells of the pancreas

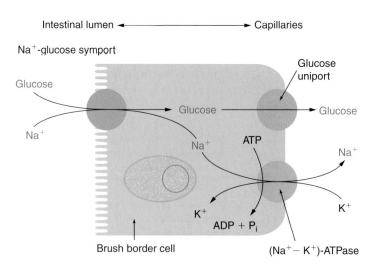

Figure 10-95. Glucose transport in the small intestinal epithelium.

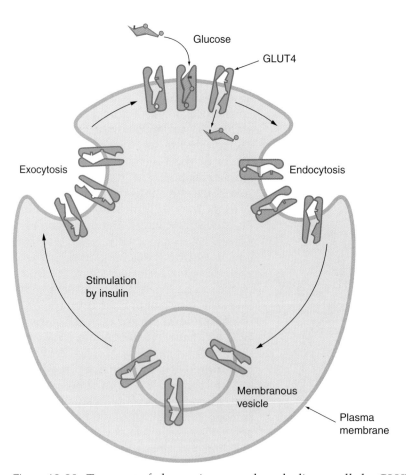

Figure 10-96. Transport of glucose into muscle and adipose cells by GLUT4.

following stimulation by glucose, mannose, leucine, and certain other amino acids and following vagus nerve stimulation. Essentially, eating a meal releases these substances. Glucose is a major nutrient of the brain, and it forms glycogen in liver and muscle, representing the storage form. Glucose can be converted to fatty acids and triglycerides. These fates of glucose are reversible. The anabolic effects of released insulin, after glucose is sensed by glucoreceptors of beta cells, are to cause the uptake of glucose and amino acids from the blood into muscle and the uptake of glucose from the blood to form triglycerides in fat cells. In the liver, glucose is converted to glycogen and/or metabolized to pyruvate. In glycolysis, insulin *stimulates:* the uptake of glucose into the liver cell; the activities of hexokinase and phosphofructokinase, which increase the production of glyceraldehyde-3-phosphate; and the activities of pyruvate dehydrogenase, which increases the rate of formation of acetyl CoA from pyruvate. Acetyl CoA enters the citric acid cycle to produce ATP, carbon dioxide and water (metabolic water). The effects of insulin also include a *reduction* in the conversion of fructose-1,6-bisphosphate to fructose, pyruvate to oxaloacetate, and oxaloacetate to phosphoenolpyruvate by **phosphoenolpyruvate carboxykinase (PEPCK).** The effect of insulin on lipid metabolism primarily in the liver and fat cells is to stimulate ACC (acetyl-CoA carboxylase) to increase the rate of formation of malonyl CoA from acetyl CoA. This increases the rate at which triglycerides are formed from fatty acids and glycerol. At the same time, insulin action reduces the rate at which triglycerides are broken down to form fatty acids and glycerol. Thus, fat storage is increased.

The opposing regulator to insulin is glucagon. It is released from the alpha cells of the islets of Langerhans in the pancreas when the blood level of glucose falls. In the liver, glucagon stimulates the breakdown of glycogen and the conversion of pyruvate to glucose, culminating in the release of glucose into the bloodstream. In adipose tissue, glucagon affects the hydrolysis of triglycerides to fatty acids, which are taken up by the liver for the formation of keto acids. Amino acids are taken up by the liver and converted to pyruvate and then to glucose. Glucagon, acting through its membrane receptor, interacts with a G protein, which activates adenylate kinase to form cyclic AMP from ATP. Cyclic AMP activates protein kinase A, and this kinase phosphorylates inactive phosphorylase *b* kinase to its active form. Phosphorylase *b* kinase phosphorylates inactive phosphorylase *b* to its active form, phosphorylase *a*. Phosphorylase *a* breaks down glycogen to glucose-1-phosphate, which is then converted to glucose. Inactive glycogen synthase kinase is phosphorylated by protein kinase A to its active form glycogen synthase kinase, which converts active glycogen synthase D to its inactive form, glycogen synthase I. Thus, the immediate action of glucagon on the liver is to cause the breakdown of glycogen to make glucose available to the bloodstream and to inhibit the machinery responsible for the synthesis of glycogen. Glucagon actions are summarized in Figure 10-97. Glucagon is released from the alpha cells of the pancreas when the fall in blood glucose is sensed by a glucose sensor. The glucose sensor consists of Glut 1 (Figure 10-93) and glucokinase. Glucose entry into the alpha cell would occur at levels lower than in the beta cell because the Glut 1 Km (Michaelis-Menten constant) (glucose) is about 1 mM, whereas the Km (glucose) of glucokinase is about 5.5 mM. Thus, the alpha cell sensor is active in the lower physiological range of

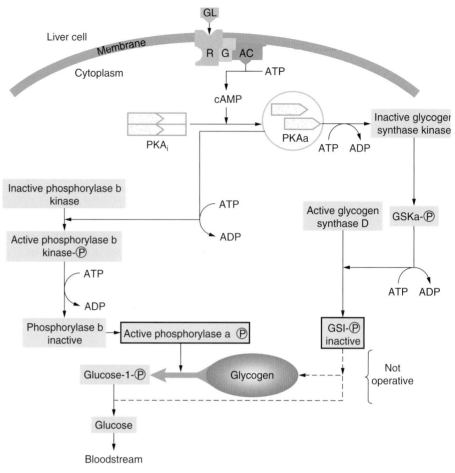

Figure 10-97. Effects of glucagon *(GL)* on the liver cell. Glucagon binds to its receptor *(R)* in the cell membrane and the activated receptor interacts with a G protein *(G)* and it activates adenylate cyclase *(AC)* to form cyclic AMP *(cAMP)* from ATP. cAMP activates protein kinase A *(PKA)* from its inactive form *(PKKi)* to the active form by release of the catalytic subunit *(PKAa)*. PKAa phosphorylates inactive glycogen synthase kinase, which phosphorylates active glycogen synthase D to its inactive form *(GSI)*, so that more glycogen is not synthesized. PKAa also phosphorylates inactive phosphorylase b kinase and the active form phosphorylates inactive phosphorylase b to active phosphorylase a, which breaks down glycogen to form glucose-1-phosphate. G-1-P is converted to glucose, which enters the bloodstream.

glucose concentrations. The membrane potential of the alpha cell is determined by a potassium channel of the K_A type and a tetrodotoxin-sensitive K^+ channel. The secretion of glucagon is coupled to the influx of Na^+ and Ca^{2+}, which generate action potentials, whereas the action potential is terminated by a voltage-dependent K^+ channel. Glucagon is released from the readily releasable pool by the entry into the cell of Ca^{2+}. Increased blood glucose concentrations reduce the activity of the alpha cell as a result of the K^+_A, Na_{TTX}, and Ca^{2+} channels becoming inactive, and the secretion of glucagon falls to basal levels.

In contrast to the liver, glycogen breakdown to glucose in muscle is stimulated by cortisol but not by glucagon. Glucose released from glycogen enters glycolysis and generates ATP for muscle work, and pyruvate is converted to lactate. Lactate enters the blood and is transported to the liver, where it is converted back to pyruvate by pyruvate carboxylase (stimulated by cortisol) and is further converted back to glucose, which circulates through the blood to other tissues. PEPCK in liver is elevated by cortisol to stimulate the rate of conversion of oxaloacetate (formed from pyruvate) to phosphoenolpyruvate and then to glucose through the remainder of gluconeogenesis. Glucose is further derived from triglycerides in adipose tissue, where glucagon stimulates the breakdown of triglycerides to glycerol and fatty acids. Glycerol kinase converts glycerol to glycerol-3-phosphate, and glycerol-3-phosphate is converted to dihydroxyacetone phosphate by glycerol phosphate dehydrogenase (stimulated by glucagon). The process of gluconeogenesis continues culminating in the formation of glucose, which is transported in the bloodstream to other tissues.

Overview

Metabolism deals with the dietary intake of food constituents—proteins, nucleic acids, polysaccharides, and lipids—and their utilization, catabolism, and resynthesis in the body. Proteins are digested to amino acids in the gut, absorbed, and circulated to tissues for the synthesis of specific proteins. Proteins are also broken down into amino acids again and either reused or converted to keto acids and ammonia (converted to urea for excretion). Ingested nucleic acids can be broken down into pentoses and bases. The bases can be reused for the synthesis of nucleotides and nucleic acids. Polysaccharides are broken down in the gut to monosaccharides, especially glucose, which can enter glycolysis for the formation of pyruvate and ATP. Pyruvate can be converted to acetyl CoA for entry into the citric acid cycle to produce NADH and coenzyme QH_2 that can be used later by the cell, ATP for the energy pool, CO_2, and H_2O. Also, glucose can be used for the formation of pentose sugars using the hexose-monophosphate pathway (Figures 6-55, 6-57, and 6-58), which generates NADPH that can be used for other synthetic processes. During glycolysis, NADH is produced for the cell's pool. All of these processes can lead to gluconeogenesis, in which the products of glycolysis can reused to form glucose when needed. Fats are hydrolyzed to glycerol and fatty acids. Glycerol can enter glycolysis, and fatty acids can be oxidized in mitochondria (β-oxidation) to form acetyl CoA for the citric acid cycle. Alternatively, acetyl CoA can be used for the synthesis of isoprenoids and cholesterol. An overview of these processes is shown in Figure 10-98.

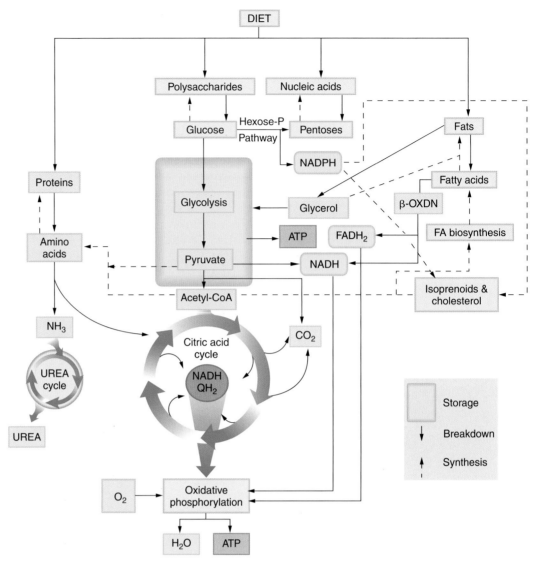

Figure 10-98. Overview of metabolism.

Further Reading

Salway, J.G., *Metabolism at a Glance,* Blackwell Publishing, 2004.

CHAPTER **11**

Growth Factors and Cytokines

New Approaches to Ovarian Cancer, Such as the Action of TRAIL (TNF-Related Apoptosis-Inducing Ligand) Might Form the Basis of a Treatment

Ovarian cancer is probably the most difficult female cancer to treat. It is often not discovered until it has developed into a late stage. Consequently, most of the phenomenology and scientific information comes from studies of advanced tumors, and there is little genetic information about early developing cancer cells. Statistically, 1 female in 70 will develop ovarian cancer during her lifetime, and there are more than 14,000 deaths from ovarian cancer annually in the United States. Ovarian cancer represents about 4% of all new cases of cancer. About 5% to 10% of ovarian cancers are considered to be hereditary. The 5-year survival for advanced cancer (stages 3 and 4) is from 5% to 20%. One-half of deaths from ovarian cancer occur later than age 55, and one-fourth of deaths from this cancer occur in women age 34 or younger. If early diagnosis is made, the cure rate can be as high as 95%.

Although the cause of ovarian cancer is unknown, increased risk is associated with a family history of this disease (or of breast cancer), an early menopause, and no pregnancies (the more children a woman has, the lower the risk for this cancer). Mutations of oncogenes, if present, increase the risk of this disease (as well as of breast cancer). These oncogenes are breast cancer (**BRCA**) 1 and 2, although the presence of BRCA1, especially hypermethylated BRCA1, is related more to ovarian cancer than the expression of mutated BRCA2. BRCA1 occurs on chromosome 17, and BRCA2 on chromosome 13. Loss of heterozygosity (LOH) may be indicative (when cells divide, the daughter cell should contain a heterozygous marker if that was the case in the parent cell; if, instead, the daughter allele has become homozygous or hemizygous through mitotic recombination, deletion, or gene conversion, this LOH is an event important in tumor progression).

Ovarian tumors result from a complex pathway that involves a number of oncogenes and tumor suppressor genes. These include HER-2/neu, K-*ras*, p53, BRCA1, and some tumor suppressor genes on chromosome 17. Grade 2 and grade 3 tumors show induced expression of genes associated with the cell cycle, signal transducer and activator of transcription (STAT) 1 or 3 or Janus tyrosine kinase (JAK) 1 or 2. Inherited ovarian cancer involves mutation of the BRCA1 gene, a tumor suppressor. Use of talc, exposure to asbestos, high dietary fat intake, and childhood mumps are also suspected as factors, although none of these has been proven. Symptoms are few and unspecific: vaginal bleeding, abnormal intestinal gas, nausea, or lack of appetite. The first sign could be an enlarged ovary. Because the ovaries are located deep in the pelvic cavity, ovarian enlargement may not be detected for some time, and enlargement only may be caused by benign fibrosis. Symptoms for the advanced disease are swollen abdomen, lower abdominal and leg pain, sudden change in body weight in either direction, change in bowel or bladder function, nausea, and swelling of the legs.

A physician will usually detect an enlarged ovary during a pelvic examination. Pelvic ultrasound may confirm an enlarged ovary. To confirm the disease, a **laparoscope** (a tube with a camera lens for viewing) may be inserted through a small incision in the abdomen, allowing sampling of the fluid and tissue. These samples are tested for the presence of cancer. If warranted, a **laparotomy** (incision through the abdominal wall) may be performed, in which a surgeon removes one or both ovaries (**oophorectomy**) and may remove other tissues (hysterectomy), which include fallopian tubes, uterus, and neighboring lymph glands, depending on whether there is a spread of the cancer. The ovarian tumor can shed cells that implant and grow in other tissues: uterus, bladder, bowel, and lining of the bowel wall (omentum). These secondary tumors can develop even before the primary cancer is suspected. Recently, however, a biochemical test has been developed using proteomics (Figures 2-67 and 2-66) that analyzes the blood proteins of women suspected of this disease at an early stage. A pattern of proteins shed from the cancer cells into the bloodstream (not seen in healthy women) is seen in the blood of a woman with early stage ovarian cancer, and this technique should make it possible soon to detect, without surgery for the diagnosis, the cancer at a curable stage. The current therapy for advanced disease is the combination of two drugs, Taxol and carboplatin, together with surgery to reduce the bulk of the tumor. Unfortunately, most patients die during treatment because of the development of drug resistance. In ovarian cancer cell culture, increased expression of STAT1 leads to a three- to fivefold resistance to cisplatin (similar to carboplatin) compared to parental cells that have lower levels of STAT1. cDNA microarray techniques (Figure 6-92) are being used to determine means to overcome resistance to therapeutic drugs.

The location of the ovaries is shown in Figure 11-1. The appearance of ovarian carcinoma is shown in Figure 11-2. There are four stages of ovarian cancer. In stage 1, the cancer is limited to one or both ovaries (Figure 11-3A). Of individuals in this stage, 90% have a 5-year survival rate. Stage 2 is characterized by the extension of the tumor into the pelvic region without spread to the abdomen. It is found in the uterus, fallopian tubes, bladder, sigmoid colon, or rectum. Unfortunately, few patients are diagnosed at this stage. Of the people with stage 2 cancer, 80% will survive for 5 years. In stage 3, the spread has progressed beyond the pelvis to the abdomen, abdominal wall, small bowel, lymph nodes, or surface of the liver. Of women with stage 3 cancer, 20% to 50% will survive for 5 years. Stage 4 is the most advanced form of the cancer, in which it has metastasized to the liver, spleen, or lung. Of people with stage 4 cancer, 10% to 20% will survive for 5 years. The stage of ovarian cancer is determined when surgery takes place to

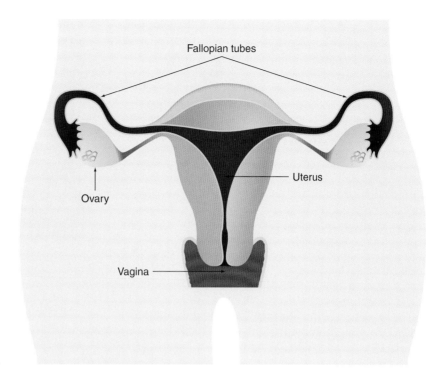

Figure 11-1. Location of the ovaries.

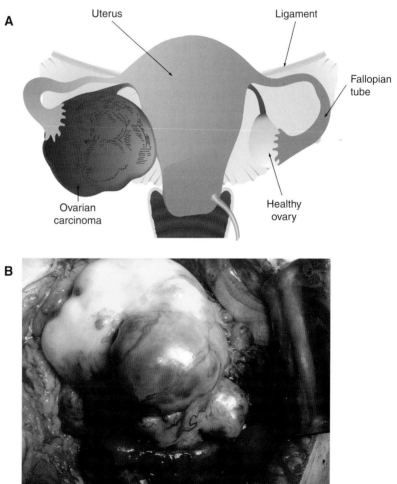

Figure 11-2. **A.** Appearance of ovarian carcinoma compared to the normal ovary. **B.** Appearance of the tumor. Part B reproduced from http://medicalimages.allrefer.com/large/ovarian-cancer-metastasis.jpg.

New Approaches to Ovarian Cancer 589

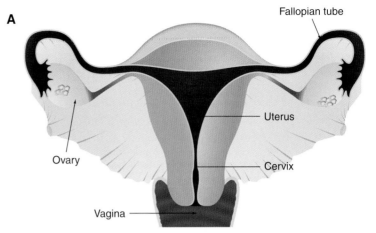

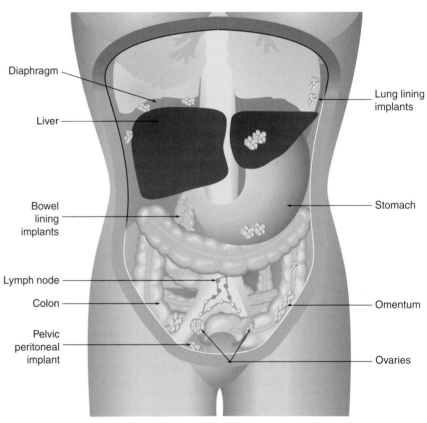

Figure 11-3. **A.** Stage 1 of ovarian cancer where the cancer is limited to one or both ovaries. **B.** Pattern of metastases of ovarian cancer. Reproduced from http://www.cancerfacts.com.

remove as much of the cancer as possible, and treatment is adjusted to the stage. The grade of a tumor corresponds to the stage. A grade 1 tumor is still well differentiated and appears like normal tissue. Grade 2 is somewhat differentiated, but grade 3 is poorly differentiated and is clearly abnormal. Sometimes ovarian cancer reoccurs after presumably successful therapy has been completed, and it usually retains its original stage. Unfortunately, up to now, the outcomes of ovarian cancer have been grim. The current biochemical focus is to find new methodology to detect ovarian cancer in its early stages when the disease can be cured. Proteomics (see Chapter 2) may offer this methodology when applied to blood samples.

The **tumor necrosis factor–related apoptosis-inducing ligand (TRAIL)** is a protein related to the TNF family of **growth factors** that has the capacity to stimulate the growth of some cells and kills most tumor cells without affecting normal cells. In ovarian cancer cell cultures, TRAIL kills the cells. But in subjects bearing an ascites (ovarian cell) tumor, the cancer cells secrete a substance, **interleukin (IL) 8,** that reduces the killing action of TRAIL. Consequently, treatment of the cancer with TRAIL will require another substance that will inactivate IL-8. However, the use of TRAIL in fighting cancers is a hopeful development because TRAIL is a natural substance, produced in the body, and has little or no toxicity. Treatment of ovarian cancer with TRAIL is enhanced when combined with chemotherapeutic drugs, such as paclitaxel (a Taxol relative).

TRAIL binds to cell membrane receptors R1, R2, R3, R4, and R5. R1 (death receptor 4, or DR4) and R2 (DR5) are structurally complete receptors that have transmembrane segments and cytoplasmic segments. These receptors affect a signaling mechanism into the target cell cytoplasm that culminates in apoptosis of the cell. R3 (decoy receptor 1, or DcR1) and R4 (DcR2) are incomplete receptors that bind TRAIL; R3 has a glycophospholipid anchor to the membrane but does not have a transmembrane domain, so the death signal does not reach the interior of the cell when TRAIL binds to R3. R4 does have a cytoplasmic domain but it does not have the **death domain (DD)** in its structure. Therefore R3 and R4 can bind TRAIL, but the complexes are unproductive; these are referred to as "decoy" receptors. R5, **osteoprotegerin (OPG)** is a secreted receptor that lacks a membrane anchor, a transmembrane domain, and a cytoplasmic domain. Thus, binding of TRAIL to R3, R4, or R5 represents a decoy function in which TRAIL resistance could occur if the level of these receptors is higher than the productive receptors, R1 and R2. These ideas are conveyed in Figure 11-4. Thus, a tumor cell will be sensitive to the killing action of TRAIL if the number of R1 and/or R2 receptors is substantially greater than the numbers of R3, R4, and R5 decoy receptors. Conversely, cellular resistance to TRAIL can occur when one or more of the decoy receptors exceeds the numbers of R1 and R2 receptors.

The killing activity of TRAIL occurs through the signal transduction process following the binding of TRAIL to a productive membrane receptor. The cytoplasmic DDs interact to form an aggregate (disc), which further interacts with the DD of another intracellular protein, **fas-associated death domain protein (FADD).** This causes the proenzyme of a caspase, caspase-8, to become activated, and activated caspase-8 (a protease) cleaves proapoptotic Bid to tBID, which translocates to the mitochondria and promotes the release from the

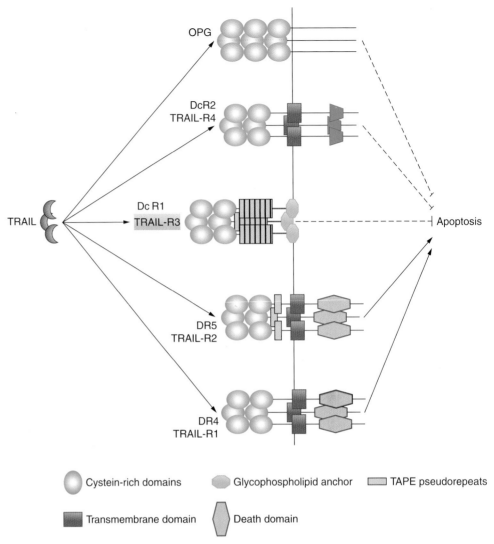

Figure 11-4. TRAIL system. Homotrimers of TRAIL interact with homotrimers or heterotrimers of TRAIL-R1 and -R2, inducing apoptosis through their cytoplasmic death domains. Trimers of TRAIL-R3 and TRAIL-R4 are also capable of binding TRAIL but do not trigger the apoptotic signal. Osteoprotegerin (OPG) is a soluble receptor that binds TRAIL and inhibits its activity. Redrawn from C.R. de Almodovar et al., "Transcripturial regulation of the TRAIL R3 gene," *Vitam. Horm.*, 67: 51–63, 2004.

mitochondria of cytochrome c and **Smac/DIABLO.** Cytochrome c triggers the activation of caspase-9 through the apoptosis factor Apaf-1. Smac is released from the mitochondrial intramembrane space when the outer membrane of the mitochondrion is disrupted. Smac binds to the endogenous **inhibitor of apoptosis (IAP)**, relieving inhibition and promoting the cleavage of procaspase-9 to the active form and the activation of procaspase-3 to its active form. Caspase-3 and other downstream caspases cleave a number of cellular proteins that lead to fragmentation of DNA and cell death. Smac relies on the N-terminal peptide in its structure to promote caspase-3 activation. These events are summarized in Figure 11-5. This is known as the intrinsic pathway

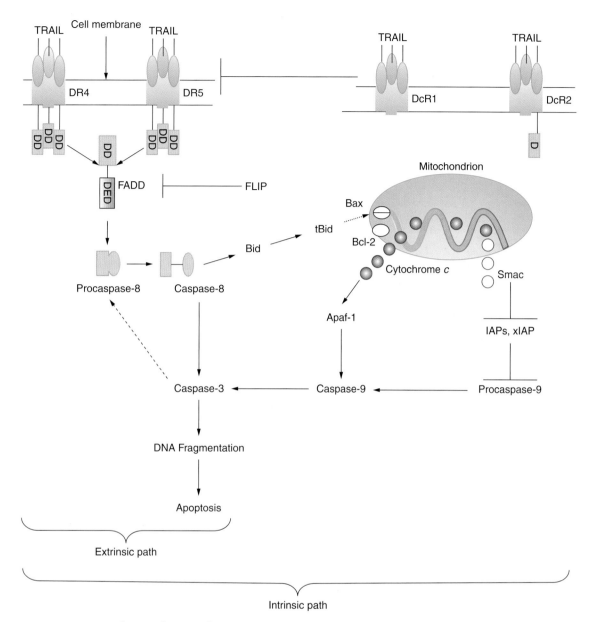

Figure 11-5. Signaling pathways of TRAIL receptors. Stimulation of DR4 and DR5 causes recruitment of procaspase-8 through an interaction with FADD. Decoy receptors, DcR1 and DcR2 can inhibit the signal by competitively and nonproductively binding TRAIL. After procaspase-8 is recruited, it autocleaves and becomes activated. Activated caspase-8 activates downstream caspases such as caspase-3. Caspase-8 also cleaves proapoptotic Bid to form tBid, which translocates to mitochondria to promote the release of cytochrome c and Smac. Cytochrome c triggers the activation of caspase-9 through Apaf-1, and Smac facilitates caspase-9 by blocking IAPs (inhibitors of apoptosis). Activated caspase-9 activates caspase-3 and leads to cleavage of cellular proteins and fragmentation of DNA followed by cell death. Caspase-3 also induces caspase-8 activation. Redrawn from X-X. Wu, O. Ogawa, and Y. Kakehi, "TRAIL and chemotherapeutic drugs in cancer therapy," *Vitam. Horm.*, 67: 365–383, 2004.

of apoptosis. There is also an extrinsic pathway, which does not involve the mitochondrion (left side of Figure 11-5, excluding the mitochondrion).

TNF Superfamily

TRAIL is a member of an extensive tumor necrosis factor superfamily (TNFSF), and there are a number of receptors for the TNF superfamily of ligands called the **tumor necrosis factor receptor superfamily (TNFRSF).** The overall group of molecules, consisting of ligands and receptors, is involved with regulation of the immune system and inflammation and, in some cases, antitumor activity. They are factors that may be responsible for the spontaneous regression of tumors that sometimes occurs in cancer patients. A factor isolated in 1975 that was shown to have strong antitumor activity against implanted skin tumors of mice was designated as the TNF. The TNF family of ligands (and coreceptors) and receptors is shown in Table 11-1. The family is larger, with many more ligands and receptors, but this table lists a number of important members. TRAIL's gene (THF-like-2, or TL2) maps to human chromosome 3 (3q36), and TRAIL is sometimes referred to as the 10th member of the TNF ligand superfamily (TNFSF10). TRAIL occurs as both membrane-bound and soluble forms and induces DNA fragmentation and apoptosis in a number of cancer and transformed cell lines. TRAIL is expressed in many tissues including spleen, thymus, prostate, lung, kidney, and intestine. Alternative spliced variants occur in both neoplastic and normal tissues and are designated TRAILβ (lacks exon 3 of the TRAIL gene) and TRAILγ (lacks exons 2 and 3 of the TRAIL gene). Although both tumor cells and normal cells express the TRAIL receptor, DR4, TRAIL induces apoptosis in the tumor cells (transformed cells) but not in normal, untransformed cells. In some cases, resistance to TRAIL is a function of the overriding expression of decoy receptors (those that bind TRAIL but do not signal in the death program). TRAIL can induce apoptosis in certain T cell cultures that have been stimulated by IL-2. The structure of TRAIL in its complex with the extracellular domain of the DR5 receptor is shown in Figure 11-6. The structure shows that TRAIL contains a uniquely elongated loop resulting from an insertion of 12–16 amino acids, which other members of the TNF family do not have. There are extensive interactions between the extracellular region of the receptor and the extended loop of TRAIL.

There are some similarities among the members of the TNF family. For comparison with the structure of TRAIL, Figure 11-7 shows the structure of a mutated trimer of human TNFα. Trimers of TNFα bind to the TNF receptor (TNFR1) and form an aggregate on the membrane, which may amplify the signal, facilitate clearance of the aggregate from the membrane after its function is over, or both. A diagram of the aggregation is shown in Figure 11-8.

(Text continues on p. 600.)

Table 11-1
The TNF Superfamily (TNFSF and TNFRSF)

Ligands/Coreceptors*	Amino Acid Residues (AAs); Molecular Weight	Comments and Characteristics
NGF (Nerve Growth Factor)	120AA; 12.5 kDa	From a propeptide with signal sequence; homodimer 1525 kDa; binds to LNGFR member of the TNFRSF
CD40L	261AA; 39 kDa; 22AA cytoplasmic domain; 215AA extracellular domain	Membranes of B cells, CD4+ & CD8+ T cells, mast cells, basophils, eosinophils, dendritic cells, monocytes, NK cells, and gd cells. Also as proteolytically cleaved, cytoplasmic form of 15–18 kDa with biological activity; forms trimeric structures like TNFα
CD 137L/4-1BBL	309AA; 50 kDa; 82AA and 34-kDa cytoplasmic region; 21AA transmembrane segment; 206AA extracellular domain	Expressed in B cells, dendritic cells, and macrophages
TNFα	233AA; 26 kDa in membrane with 29AA cytoplasmic domain; 28AA transmembrane domain; 176AA extracellular domain	Either transmembrane or soluble protein is biologically active; expressed in many cell types, including macrophages, CD4+ and CD8+ T cells, adipocytes, keratinocytes, mammary and colon epithelia, osteoblasts, mast cells, dendritic cells, pancreatic β-cells, astrocytes, neurons, monocytes, and steroid-producing cells of *zona reticularis*
CD134/OX40L	183AA; 21AA cytoplasmic domain; 23AA transmembrane segment; 139AA extracellular domain	OX40L exists as a trimer; limited expression: activated CD4+ and CD8+ T cells, B cells, and vascular epithelial cells
CD27L/CD70	50 kDa; 193AA transmembrane glycoprotein; 20AA cytoplasmic segment; 18AA transmembrane segment; 155AA extracellular domain	Expressed by NK cells, B cells, CD45RO+, CD4+ and CD8+ T cells, gd T cells, and some leukemic cells; may be involved in antibody production in B cells
FasL (Fas Ligand)	40-kDa transmembrane protein 281AA; 80AA cytoplasmic domain; 179AA extracellular domain	Can occur as circulating trimer; can be cleaved by a protease to give an active 70-kDa trimer of 26-kDa monomers; F273L mutation results in gld/gld generalized lymphoproliferative disease; expressed by: type II pneumocytes and bronchial epithelium, monocytes, LAK cells, NK cells, dendritic cells, B cells, macrophages, CD4+ and CD8+ T cells, colon, and lung carcinoma cells
CD30L	40-kDa; 234AA transmembrane glycoprotein: 46AA cytoplasmic domain; 21AA transmembrane segment; 172AA extracellular domain	Expressed by monocytes and macrophages, B cells, activated CD4+ and CD8+ T cells, neutrophils, megakaryocytes, resting CD2+ T cells, erythroid precursors, and eosinophils
TNFβ/LT-α	Circulates as 171AA, 25-kDa glycosylated polypeptide; a larger form (205AA) exists, suggesting proteolytic processing; no transmembrane form, but it can be membrane associated since it can bind to membrane anchored LTβ, forming a heterotrimer	Circulating TNFβ is ~150 pg/ml heterotrimer binds to LTβR and TNFRI receptor, but TNFRI activation will not occur.

*TNFSF usually forms trimeric structures, ligands and receptors of the TNSF and TNFRSF undergo clustering during signal transduction; monomers of TNFSF are two-sheet structures composed of β-strands.

Continued

Table 11-1
The TNF Superfamily (TNFSF and TNFRSF)—cont'd

Ligands/Coreceptors	Amino Acid Residues (AAs); Molecular Weight	Comments and Characteristics
LTβ	33-kDa type II transmembrane glycoprotein; 244AA; 16AA cytoplasmic segment; 31AA transmembrane domain; 197AA extracellular region.	LTβ forms a heterotrimer with TNFβ on membrane; LTβ is not secreted
TRAIL	32 kDa, 281AA; 17AA cytoplasmic domain; 21AA transmembrane segment; 243AA extracellular domain	Homotrimer in membrane; many tissues express TRAIL, including lymphocytes; may have anticancer cell activity

RECEPTORS (TNFRSF)**

Ligands/Coreceptors	Amino Acid Residues (AAs); Molecular Weight	Comments and Characteristics
(LNGFR/p75 Human Low-affinity Nerve Growth Factor Receptor)	75 kDa; 427AA with extracellular N-terminus; 25AA signal sequence; 225AA extracellular domain; 23AA transmembrane segment; 154AA cytoplasmic domain	Transmembrane-glycoprotein; can appear as a 200 kDa disulfide-linked homodimer; neurotrophins bind to LNGFR with K_D ~1–3 mM, no inherent tyrosine kinase activity; death domain in cytoplasmic domain; protease cleavage-35 to 45-kDa LNGFR; cells expressing LNGFR: oligodendocytes, B cells, bone marrow fibroblasts, autonomic and sensory neurons, Schwann cells, follicular, dendritic cells, select astrocytes, and mesenchymal cells.
CD40	50 kDa; 277AA transmembrane glycoprotein (B cell proliferation and differentiation); 20AA signal sequence; 173AA extracellular domain; 22AA transmembrane segment: 62AA cytoplasmic domain	4 Cys-rich motifs in extracellular region with juxtamembrane sequence rich in SER and THR: CD40 up-regulates FAS to prime cells for subsequent FAS-mediated apoptosis; CD40 pathway involves NFκβ and protein kinase (LYN) activation: cells expressing CD40: monocytes, basophils (not mast cells), eosinophils, endothelial cells, interdigitating dendritic cells, Langerhans cells, blood dendritic cells, fibroblasts, keratinocytes, and Reed-Sternberg cells of Hodgkin's disease and Kaposi's sarcoma cells.
CD137/4-1BB/ILA	30-35 kDa; monomer and dimer on cell surface 255AA; 17AA signal sequence, 169AA extracellular region, 27AA transmembrane segment; 42AA cytoplasmic domain	Cys-rich motif in extracellular domain CD137 binds its ligand, CD137L, at K_D of ~30 pM; alternative splicing event can give rise to soluble form; CD137 ligation can interrupt cell apoptotic program associated with activation-induced cell death; cell expressing CD137: fibroblasts, thymocytes, monocytes, and CD4$^+$ and CD8$^+$ T Cells
TNFR1/p55/CD120a	55 kDa; 455AA transmembrane glycoprotein; 190AA extracellular domain; 25AA transmembrane segment; 220AA cytoplasmic domain	Expressed in all nucleated mammalian cells; four Cys-rich motifs in extracellular region; first Cys-rich motif required for binding; 80AA death domain in cytoplasmic region; NFκB is activated by TNFR1; TNFR1 binds both TNFα and TNFβ; K_D ~20-60 pM; for soluble TNFα; K_D for TNFβ = 650 pM; TNFR1 most important for circulating TNFα; membrane bound TNFα associates with TNFR2; soluble TNFR1 blocks TNFα activity (decoy) and occurs in blood and urine at 1–3 ng/mL; protease activity gives soluble forms of 32 kDa and 48 kDa; cells expressing TNFR1: hepatocytes, monocytes, and neutrophils; cardiac muscle cells; endothelial cells; and CD34$^+$ hematopoietic progenitors

** TNFRSF is usually trimeric or multimeric, stabilized by intracysteine disulfide bonds; it exists in both membrane-bound and soluble forms; many forms transduce apoptotic signals in a variety of cells.

Ligands/Coreceptors	Amino Acid Residues (AAs); Molecular Weight	Comments and Characteristics
TNFR2/p75/CD120b	75 kDa; 461AA transmembrane glycoprotein; 240AA extracellular region; 27AA transmembrane segment; 173AA cytoplasmic domain	TNFR2 binds TNFα and transfers it to TNFR1, which becomes activated; TNFα binding to TNFR2 induces apoptosis in rhabdomyosarcoma cells and cell migration in Langerhans cells; soluble TNFα binds TNFR2 with a K_D of 300 pM; TNFα levels are usually at 100 pM so that it should normally bind to TNFR1; therefore, TNFR2 acts as a decoy; cells expressing TNFR2: monocytes, endothelial cells, Langerhans cells, and macrophages
CD134/OX40/ACT35	48 kDa; 250AA; 188AA in extracellular region; 26AA transmembrane segment; 36AA cytoplasmic domain	Expressed in $CD4^+$ and $CD8^+$ T cells only
CD27	50–55 kDa; mature CD27 is 27kDa, 242AA; 175AA in extracellular domain; 21AA transmembrane segment; 46AA cytoplasmic domain	Expressed as homodimer on cell surface; it has no death domain but induces apoptosis by associating with Siva cytoplasmic protein, which has a death domain; blood and urine contain a soluble 32-kDa CD27 (probably from proteolysis) cells expressing CD27: NK cells, B cells, $CD4^+$ and $CD8^+$ T cells, and thymocytes
FAS/CD95/APO-1	43 kDa; 335AA; 156AA extracellular region; 20AA transmembrane segment; 144AA cytoplasmic domain	On fibroblasts FAS ligation can lead to either proliferation or apoptosis depending on number of expressed FAS molecules; 3 Cys-rich motifs; 68AA death domain in cytoplasmic region identical to one found in TNFR1 cytoplasmic domain; death domain associates FADD protein (with FAS) or TRADD protein with TFNR1; both transmit apoptotic signals; alternative gene splicing produces soluble forms of FAS
		Soluble blood FAS circulates as dimer and trimer at low mg/ml concentrations; cells expressing FAS: $CD34^+$ stem cells, fibroblasts, NK cells, keratinocytes, hepatocytes, B cells and B cell precursors, monocytes, $CD4^+$ and $CD8^+$ T cells, $CD45RO^+$ gd T cells, eosinophils and thymocytes
CD30/Ki-1	105–120-kDa transmembrane glycoprotein; mature CD30 is 577AA; 18AA signal sequence, 365 extracellular region, 24AA transmembrane segment; 188AA cytoplasmic domain	6 Cys-rich motifs in extracellular region; patients with $CD30^+$ lymphomas have 85-kDa soluble CD30 in blood; cells expressing CD30 Reed-Sternberg cells, $CD8^+$ T cells and $CD4^+$ T cells.
LT-βR	75-kDa transmembrane glycoprotein; 201AA extracellular domain; 26AA transmembrane segment; 187AA cytoplasmic domain	4 Cys-rich motifs in extracellular domain; LT-βR binds heterotrimers (1 TNFβ + 2 LT-β) over LT-β homotrimers: first 2 Cys-rich motifs resemble TNFR1, and third & fourth Cys-rich motifs resemble TNFR2; LT-βR activates NF–kB and induces cell death via TRAF-3; LT-βR activates genes for IL-8 and Rantes; LTβR expressed by monocytes, fibroblasts, smooth muscle, and skeletal muscle cells

Continued

Table 11-1
The TNF Superfamily (TNFSF and TNFRSF)—cont'd

Ligands/Coreceptors	Amino Acid Residues (AAs); Molecular Weight	Comments and Characteristics
DR3/WSL-1/TRAMP/APO-3/LARD (Death Receptor 3)	54 kDa; 417AA; 24AA signal sequence; 178AA extracellular domain; 23AA transmembrane segment; 192AA cytoplasmic domain	Can activate both NFκB and induce apoptosis like TNFR1; four Cys-rich motifs in extracellular region; many alternate splice forms of DR3, some of which may be soluble; cells expressing DR3: T and B cells and human umbilical vein endothelial cells
DR4 (Death Receptor 4)	468AA; 23AA signal sequence; 226AA extracellular domain; 19AA transmembrane segment; 220AA cytoplasmic domain	One of three known receptors for TRAIL; two Cys-rich motifs in extracellular domain; expressed by activated T cells
DR5 (Death Receptor 5)	411AA; 51AA signal sequence; 132AA extracellular domain; 22AA transmembrane segment; 206AA cytoplasmic domain	Second receptor for TRAIL; triggers apoptotic program like DR4 without FADD participation; two Cys-rich motifs in extracellular domain
DcR1/TRID (decoy receptor 1) (TRAIL receptor without an intracellular domain)	259AA; 23AA signal sequence; 217AA extracellular domain; 19AA transmembrane domain	Membrane receptor for TRAIL with no intracellular domain (Figure 11-4 = TRAIL Receptor 3), two Cys-rich motifs in extracellular region 50% to 60% identical to AA sequences in same regions of DR4 & DR5; inhibits responsiveness to TRAIL (decoy receptor)
TR2	32 kDa; 283AA; 36AA signal sequence; 165AA extracellular region; 23AA transmembrane segment; 59AA cytoplasmic domain	No known ligand as yet; found on T cells, B cells, monocytes, and endothelium; four Cys-rich motifs in extracellular domain.
GITR (Glucocorticoid-Induced TNFR Family-Related)	228AA; 19AA signal sequence; 134AA extracellular domain; 23AA transmembrane segment; 52AA cytoplasmic domain	Inducible during T cell activation; three Cys-rich motifs in extracellular region; ligation interrupts TCR-DC3-induced apoptosis in T cells
OPG (Osteoprotegerin)	55 kDa; 380AA	Inhibits osteoclasts and protects bone from breakdown; secreted member of TNFRSF; similar to TNFR2 and CD40; no transmembrane segment (see Figure 11-4); circulates as a disulfide-linked homodimer; ligand unknown

Information in this table is taken from http://www.rndsystems.com/asp/g_sitebuilder.asp?bodyid=227#top.

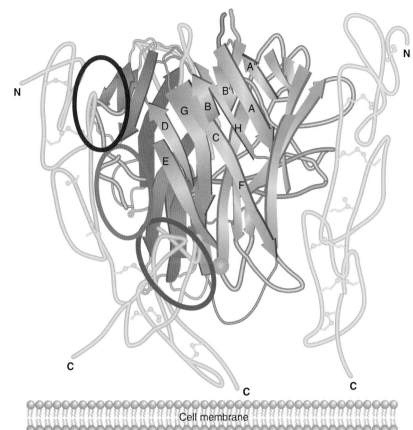

Figure 11-6. Crystal structure of TRAIL-DR5 complex. Each β strand of a TRAIL monomer is labeled with capital letters. The loop connecting strands A and A″ is named AA″ loop. 3 sDR5 molecules are shown in *cyan*. TRAIL subunits are represented by different colors and the AA″ loop is shown in *green* and *blue*. The *yellow lines* in sDR5, where *s* means *soluble*, indicate disulfute bonds. The *green sphere* is bound zinc. Reproduced from S.-S. Cha et al., "Crystal structure of TRAIL-DR5 complex identifies a critical role of the unique frame insertion in conferring recognition specificity," *J. Biol. Chem.*, 275: 31171–31177, 2000.

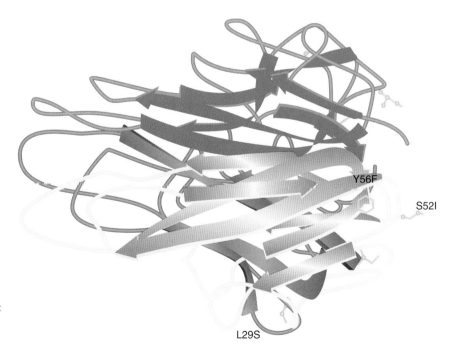

Figure 11-7. Ribbon drawing of TNFα mutant, M3S trimer. Each subunit is represented by *yellow, red,* and *magenta*, respectively. The three mutated residues are in *green ball-and-stick models*. Ser^{29} is located on a loop. Ile^{52} is at the start of a loop, and Phe^{56} is at the beginning of a β-strand. Redrawn from S.- Cha, J.-S. Kim, H.-S. Cho, N.-K. Shin, W. Jeong, H.-C. Shin, Y.J. Kim, J.H. Hahn, and B.-H. Oh, "High resolution crystal structure of a human tumor necrosis factor-α mutant with low systemic toxicity," *J. Biol. Chem.*, 278: 2153–2160, 1998.

Figure 11-8. A schematic representation of an aggregate of TNF trimers bound to parallel dimers of sTNF-R1. TNF is shown as triangles and the receptor as semicircles. Redrawn from J.H. Naismith et al., "Seeing double: crystal structures of type 1 TNF receptor," *J. Mol. Recog.*, 9: 113–117, 1996.

Growth Factors

Growth factors are proteins that bind to receptors on the target cell surface and promote signal transduction that leads to cell division (proliferation) and often differentiation. Some growth factors have this effect on many different cell types, and others are specific in the cell type they stimulate. Some of the more common growth factors are listed in Table 11-2, along with characteristics of each one.

The **platelet-derived growth factor (PDGF)** is an important **mitogen** (stimulator of cell division) for certain cell types, especially connective tissue cells. It is a dimer made up of either A or B polypeptide chains so that it has the forms of a homodimer (AA or BB) or a heterodimer (AB). However, the A and B chains are similar, and the dimer is held together by disulfide bonds. There are two structurally related receptors for PDGFs, which are protein tyrosine kinase receptors called PDGFRα and PDGFRβ. Activation of these receptors by binding of PDGF generates cell growth, and their effects can alter cell shape and motility. PDGF can cause reorganization of the actin filament system in the cytoplasm and lead to **chemotaxis** (movement of a cell toward a chemical gradient of PDGF). PDGF is stored in the α-granules of platelets and was purified from platelets, but it is also synthesized by other cell types.

Table-11-2
Common Growth Factors and Their Properties

Factor	Principal Source	Primary Activity	Comments
PDGF	Platelets, endothelial cells, placenta	Promotes proliferation of connective tissue, glial and other cells	Two different protein chains form three distinct dimer forms; AA, AB, and BB
EGF	Submaxillary gland, Brunners gland	Promotes proliferation of mesenchymal, glial, and epithelial cells	
TGF-α	Common in transformed cells	May be important for normal wound healing	Related to EGF
FGF	Wide range of cells; protein is associated with the extracellular matrix	Promotes proliferation of many cells; inhibits some stem cells; induces mesoderm to form in early embryos	At least 19 family members, 4 distinct receptors
NGF		Promotes neurite outgrowth and neural cell survival	Several related proteins first identified as protooncogenes; trkA (*trackA*), trkB, trkC.
Erythropoietin	Kidney	Promotes proliferation and differentiation of erythrocytes	
TGF-β	Activated TH$_1$ cells (T-helper) and natural killer (NK) cells	Anti-inflammatory (suppresses cytokine production and class II MHC expression), promotes wound healing, inhibits macrophage and lymphocyte proliferation	At least 100 different family members
IGF-I	Primarily liver	Promotes proliferation of many cell types	Related to IGF-II and proinsulin, also called somatomedin C
IGF-II	Cells	Promotes proliferation of many cell types primarily of fetal origin	Related to IGF-I and proinsulin

EGF, epidermal growth factor; *FGF*, fibroblast growth factor; *IGF-I*, insulin-like growth factor I; *IGF-II*, insulin-like growth factor II; *MHC*, major histocompatibility complex; *NGF*, nerve growth factor; *PDGF*, platelet-derived growth factor; *IGF-α*, tumor growth factor-α; *TGF-β*, tumor growth factor-β. Reproduced from http://web.indstate.edu/theme/mwking/growth-factors.html.

The generation of the active dimeric forms of PDGF is shown in Figure 11-9, and the crystal structure of PDGF-BB is shown in Figure 11-10. The structure of the PDGF β-receptor (170–180 kDa) is shown in Figure 11-11. The cytoplasmic domain of the receptor contains a tyrosine kinase catalytic center, which catalyzes the phosphorylation of protein tyrosine residues, as shown in Figure 11-12. The gene for the human PDGF α-receptor is located on chromosome 4 (4q12) in proximity to the genes for the stem cell factor (SCF) receptor and the vascular endothelial growth factor (VEGF) receptor-2. The gene for the PDGF β-receptor is located on chromosome 5 near the gene for the **colony-stimulating factor (CSF)** 1 receptor.

Not unlike the case for TRAIL, PDGF also has soluble proteins that bind it. Primarily, α$_2$-macroglobulin binds PDGF-BB but not PDGF-AA, which is a regulatory mechanism for controlling the amount of PDGF available for binding to its receptor. Another protein has been found in a neural retinal cell (PDGF-associated protein) that binds PDGF with low affinity and enhances the activity of PDGF-AA but lowers the activity of

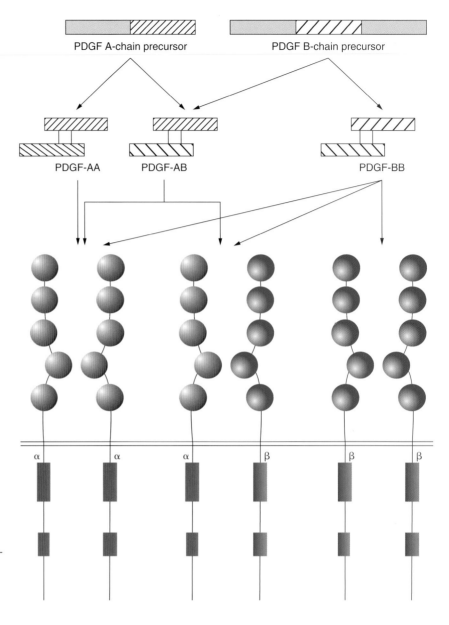

Figure 11-9. Three dimeric forms of PDGF. Redrawn with permission from C-H. Heldin and B. Westermark, "Mechanism of action and *in vivo* role of platelet-derived growth factor," *Physiol. Rev.*, 79: 1283–1316, 1999.

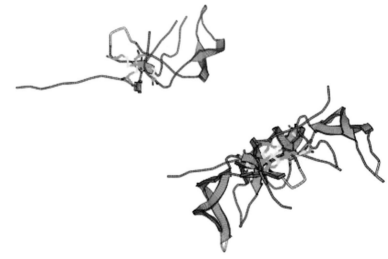

Figure 11-10. Crystal structure of platelet-derived growth factor-BB (PDGF-BB). Reproduced from PDB ID: 1PDG. C. Oetner, A. D'Arcy, F.K. Winkler, B. Eggimann, M. Hosing. *EMBO J.* **11** pp. 3921 (1992).

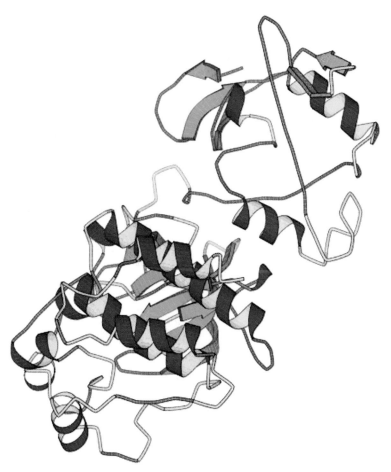

Figure 11-11. Crystal structure of platelet-derived growth factor receptor-β (PDGFR-β). Reproduced from PDB ID: 1LWP. K. Gayathri, A. Muthuvel. To be published.

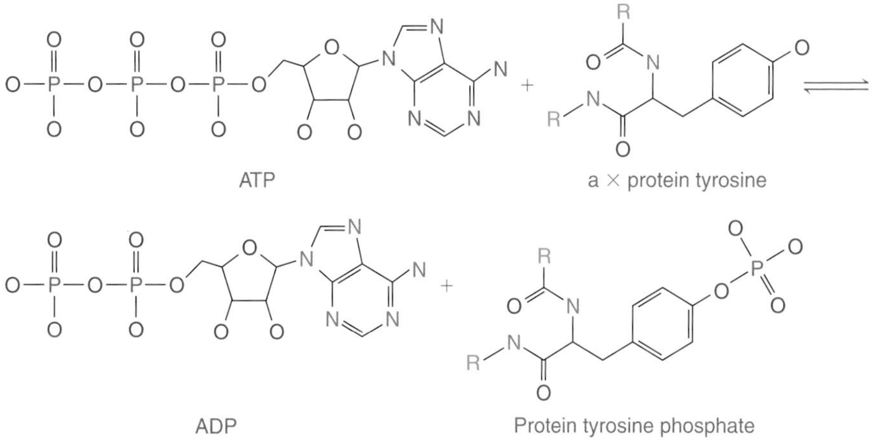

Figure 11-12. Phosphorylation of protein tyrosine residues catalyzed by protein tyrosine kinase.

Growth Factors 603

Table 11-3
Effects of PDGFRα and PDGFRβ on Cells

Effect	α-Receptor	β-Receptor
Cell growth	Stimulation	Stimulation
Actin reorganization	Stimulation of edge ruffling and loss of stress fibers	Stimulation of edge ruffling, loss of stress fibers, and stimulation of circular ruffles
Chemotaxis	Stimulation or inhibition depending on cell type	Stimulation
Ca^{2+} mobilization	Weak stimulation	Stimulation
GAP junctional communication	?	Inhibition
Apoptosis	?	Inhibition

GAP, GTPase activating protein.
Reproduced from Table 2 of C.H. Heldin, and B. Westermar, "Mechanism of action and *in vivo* role of platelet-derived growth factor," *Physiol. Rev.*, 79: 1283–1316, 1999.

PDGF-BB. In Table 11-3, the cellular effects of PDGFRα and PDGFRβ are listed.

As for most tyrosine kinase receptors, the PDGF receptors are activated in dimerized form (Figure 11-8), which allows for the autophosphorylation in *trans* of each monomer. PDGF itself is a dimer that can bind two receptors at once, forming a bridge between them, like the VEGF receptor. The dimeric receptor complex is stabilized further by receptor–receptor interaction mediated by one of the Ig domains (Ig domain 4) of the receptors. The structure of the growth factor domain of PDGF-C, similarly to PDGF-B, is shown in Figure 11-13, and the crystal structure of the dimeric growth factor domains of PDGF-CC is shown in Figure 11-14. A general view of the signal transduction mechanism for PDGF for either homodimer is shown in Figure 11-15. The dimeric growth factors bind to their receptors, resulting in activation of the receptors, which autophosphorylate their cytoplasmic domains. PDGF-BB activates protein kinase C, which leads to the activation of mitogen-activated protein kinase (MAPK). The phosphorylated forms of the cytoplasmic receptor domains interact with Src homology domains, or SH2 domains (as shown in Figure 11-16), of *Ras*-GTPase activating protein (*Ras*-GAP), which signals *Ras* (an oncogene product that binds GTP and has GTPase activity), then Raf-1 (a leukemia oncogene product), and then **MAPK/extracellular signal-regulated kinase (ERK) kinase (MEK),** which stimulates MAPK. Activated MAPK translocates to the nucleus and stimulates genes that result in cell division. Components of the MAPK pathway associate on scaffolds and scaffolding proteins, in this case, MP1, and the scaffold in mitogenesis appears as shown in Figure 11-15, inset. This complex of reactions, incompletely understood, leads to cell division and proliferation.

Epidermal Growth Factor

Epidermal growth factor (EGF) binds to specific high-affinity receptors on the target cell surface. This receptor also has a tyrosine kinase catalytic center in its cytoplasmic domain. When EGF binds to the receptor, the

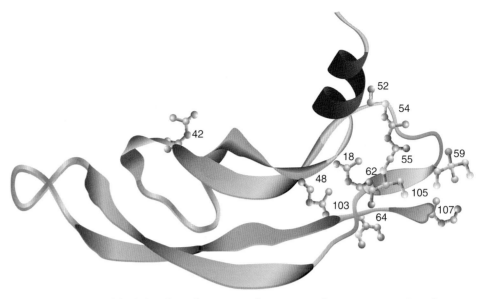

Figure 11-13. Model of the three-dimensional structure of GFD-PDGF-C based on sequence alignment with PDGF-B and VEGF. **A.** A-helixes are shown in *red*, and β-sheets are shown in *blue*. Cystine residues are numbered and shown in a *stick representation*, sulfur is colored in *yellow*, and nitrogen and oxygen are colored in *blue* and *red*, respectively. The first intramonomeric cystine bond is Cys^{18}-Cys^{62}, the second is Cys^{48}-Cys^{103}, and the third is Cys^{55}-Cys^{105}. The two intermonomeric cysteines making the dimer are Cys^{42} and Cys^{54}. *GFD*, growth factor domain. Reproduced from Figure 7 of L.J. Reistad et al., *J. Biol. Chem.*, 278: 17114–17120, 2003.

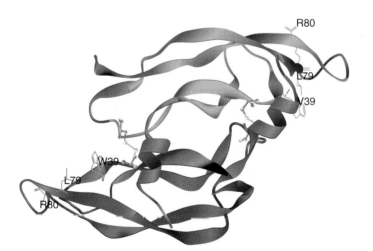

Figure 11-14. Model structure of the dimeric GFD-PDGF-CC based on the structural alignment with dimeric VEGF-AA. The two monomers, colored *red* and *green*, are bound together by the two interdisulfide bonds *(yellow)*. The *blue color* indicates putative areas for binding to the receptor. Reproduced from Figure 8 of L.J. Reigstad et al., "Platelet-derived growth factor (PDGF)-c, a PDGF family member with a vascular endothelial growth factor-like structure," *J. Biol. Chem.*, 278: 17114–17120, 2003.

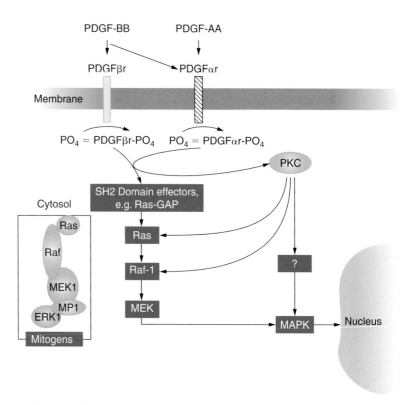

Figure 11-15. Schematic of signal transduction pathways of MAPK activation initiated by PDGF ligands. PDGF-AA induces only the tyrosine phosphorylation of PDGFαr, whereas PDGF-BB activates either receptor subunit. Receptor phosphorylation causes the activation of downstream effector molecules that contain an SH2 domain (e.g., Ras-GAP) and the sequential activation of Ras, Raf-1, MAPK kinase (MEK), and eventually MAPK. Alternatively, receptor phosphorylation can activate PKC, which in turn can stimulate MAPK by either a direct effect on Ras-Raf-1 or through an unknown intermediate molecule. Ethanol can affect receptor phosphorylation and PKC activity in the MAPK cascade. Inset shows a diagram of the scaffolding of factors in the PDGF signaling pathway. The final product can be *erk* or MAPK. MP1 is the adaptor which facilitates scaffold formation. Redrawn with permission from J. Luo and M.W. Miller, "Platelet-derived growth factor-mediated signal transduction underlying astrocyte proliferation: site of ethanol action," *J. Neurosci.*, 19: 10014–10025, 1999. Inset redrawn from H.J. Schaeffer and M.J. Weber, "Mitogen-activated protein kinases: specific messages from ubiquitous messengers," *Mol. Cell. Biol.*, 19: 2435–2444, 1999.

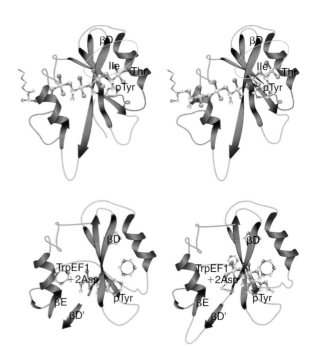

Figure 11-16. Structure of SH2 domain of various proteins. The SH2 domain is in ribbon representation with α-helices in *blue* and β strands in *red*. Various peptide ligands are shown in stick representation. Taken from Figure 4 from G. Waksman, S. Kumaran, and O. Lubman, "Structure of the SAP and Grb2 homology 2 (SH2) domains," *Expert Reviews in Molecular Medicine*, 6, January 30, 2004, http://www.expertreviews.org.

tyrosine kinase domain becomes activated and autophosphorylates the receptor. Then it phosphorylates other proteins in the signaling pathway. EGF causes proliferation of mesodermally and ectodermally derived cells. The structure of human EGF is shown in Figure 11-17, and the structure of EGF bound to the EGF receptor is shown in Figure 11-18.

Unlike the PDGF, which is cleared by the cell on the surface, EGF is absorbed by the cell by the process of endocytosis. The uptake

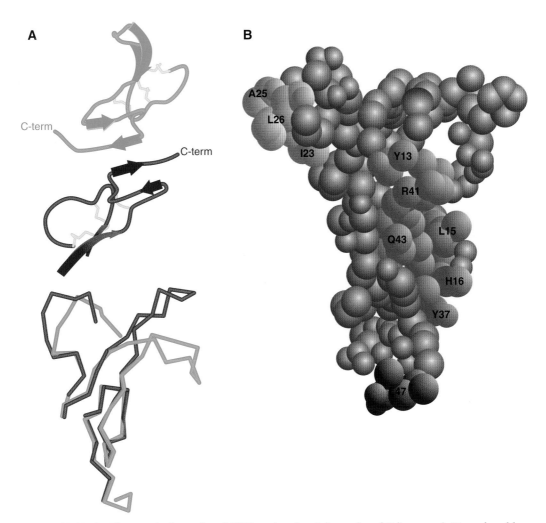

Figure 11-17 **A.** The two independent hEGF molecules A (in *red*) and B (in *green*). Top related by a noncrystallographic two-fold axis and form a potential dimer in the crystals. The three disulfide bridges, Cys^6–Cys^{20}, Cys^{14}–Cys^{31}, and Cys^{33}–Cys^{42} are shown in *yellow*. Bottom structural superposition of hEGF molecules A and B based on Cα atoms of rigid segments 13–21 and 30–47. The *N*-terminal segment (residues 1–12) and the residues 22–29 are adjacent to each other in the *upper* part of the figure. **B.** Space-filling model of the EGF molecule. The figure shows the distribution of some surface residues known to be important for EGF binding to its receptor: Tyr^{13}, Leu^{15}, His^{16}, Tyr^{37}, Arg^{41}, Gln^{43} (in *green*), Ile^{23}, Ala^{25}, Leu^{26} (in *light green*), and Leu^{47} (in *red*). Part A redrawn from Figure 2 and Part B reproduced from Figure 3 of H.-S. Lu, J.-J. Chai, M. Li, B.-R. Huang, C.-H. He, and R.-C. Bi, "Crystal structure of human epidermal growth factor and its dimerization," *J. Biol. Chem.*, 276: 34913–34917, 2001.

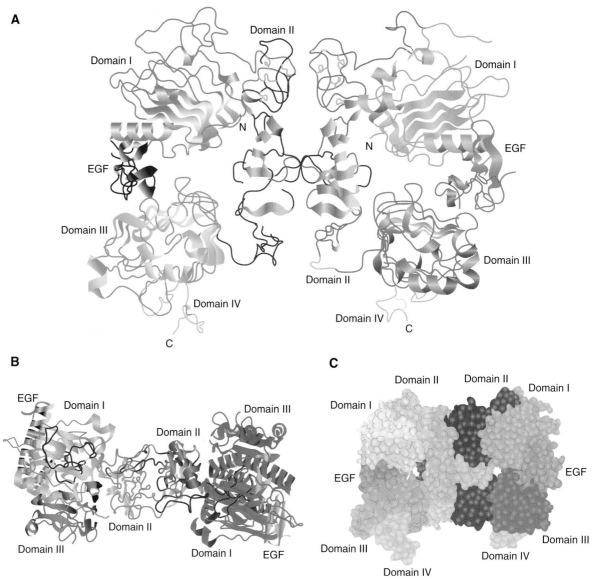

Figure 11-18. Crystal structure of the 2:2 EGF-EGFR complexes. **A.** Ribbon diagram with the approximate two-fold axis oriented vertically. One EGF chain in the 2:2 EGF-EGFR complex is *pale green*, and the other EGF chain is *pink*. Domains I, II, III, and IV in one receptor in the dimer are colored *yellow, orange, red*, and *gray*, respectively. Domains I, II, III, and IV in the other receptor are colored *cyan, dark blue, pale blue*, and *gray*, respectively. Most of domain IV is disordered. The disulfide bonds are shown in *yellow*. The intervening parts that were not assigned are transparent. **B.** The top view of *A*. **C.** A surface model corresponding to *A*. Redrawn from Figure 1 of H. Ogiso et al., "Crystal structure of the complex of human epidermal growth factor and receptor extracellular domains," *Cell*, 110: 775–787, 2002.

process allows for the down-regulation of the ligand–receptor complexes after the signal for mitosis has been transmitted. The uptake process is mediated by a protein, **clathrin,** which is a major internalization process for cell membrane proteins that become transformed into the early endosome. The formation of clathrin-coated vesicles is a complex process in which different proteins interact with one another and interact with the membrane itself (Figure 11-19). The

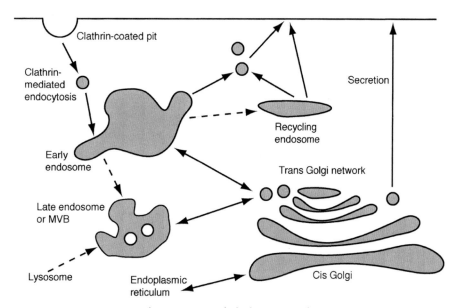

Figure 11-19. Diagram of movement of clathrin-coated pit, containing a ligand-receptor complex, from cell surface to interior structures, detailing the cellular trafficking. *Solid arrows* indicate trafficking events mediated by vesicle transport, whereas *dashed lines* show events mediated by direct fusion or fission of organelles. *MVB,* multivesicular body. Reproduced from Figure 1 of D.J. Owen, "Linking endocytic cargo to clathrin: structural and functional insights into coated vesicle formation," *Biochem. Soc. Trans.,* 32: 1–14, 2004.

appearance of a clathrin-coated pit on the cell membrane and details of the clathrin molecule as it forms a triskelion and a membrane vesicle are shown in Figure 11-20. Further details on the triskelion structure and how the triskelia incorporate into the vesicle are shown in Figure 11-21. Appearance of the clathrin-coated vesicle is shown in Figure 11-22A and B. The clathrin network is first built up as an outline structure, which is subsequently filled in to form a coat surrounding the entire vesicle. The network contains 36 triskelia in a structure of hexagons and pentagons. The clathrin-removing process is catalyzed by the enzyme, "uncoating ATPase." Three such ATPases will bind to one triskelion in the absence of ATP but not in its presence. ATP binds to the enzymes, initiating the catalysis, and the clathrin molecules disassemble.

The signal transduction pathway initiated by EGF binding to the EGF receptor is complex. Two EGF molecules bind to receptors, which dimerize on the cell surface, and the receptor tyrosine kinase autophosphorylates on the cytoplasmic domains and phosphorylates other proteins. The EGF receptor activates *ras* and the MAPK pathway, which leads to the phosphorylation of various transcription factors, such as c-fos, creating AP-1 and elk-1 that cause proliferation. EGF also activates JAKs, which activate STAT1 and STAT3 transcription factors. These also contribute to proliferation. In addition, EGF produces phosphatidylinositol signaling and calcium release and the activation of protein kinase C. EGF signaling also intersects with other signaling pathways. These activities of EGF are summarized in Figure 11-23.

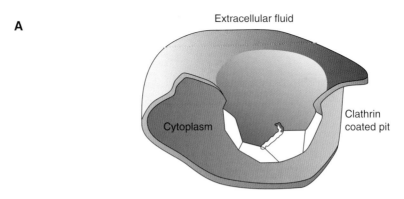
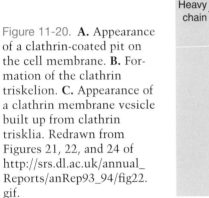
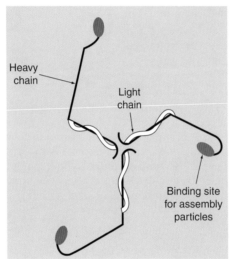
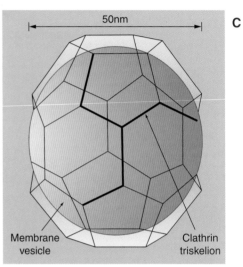

Figure 11-20. **A.** Appearance of a clathrin-coated pit on the cell membrane. **B.** Formation of the clathrin triskelion. **C.** Appearance of a clathrin membrane vesicle built up from clathrin trisklia. Redrawn from Figures 21, 22, and 24 of http://srs.dl.ac.uk/annual_Reports/anRep93_94/fig22.gif.

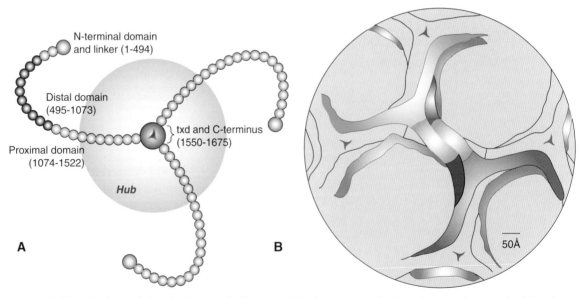

Figure 11-21. The legs of the clathrin triskelion are joined at a central trimerization domain (txd) in *A*. Formation of the polyhedral lattice *(B)* involves association of neighbor proximal domains in *yellow* and neighbor distal domains in *purple*, beneath. Redrawn from http://www.bio.indiana.edu/~ybelab/media/clathrindomains.jpg.

610   CHAPTER 11   Growth Factors and Cytokines

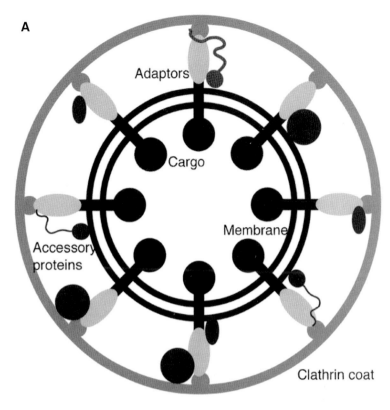

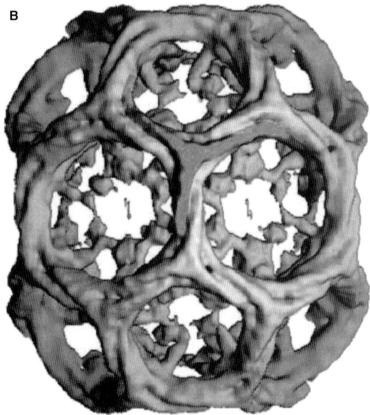

Figure 11-22. **A.** An electron micrographic reconstruction of a clathrin-coated vesicle. **B.** A schematic representation of a clathrin-coated vesicle. The central membrane layer with its embedded transmembrane cargo (in this case, ligand-receptor complexes) is coupled to the outer mechanical clathrin-scaffold layer by a middle layer of adapters. Proteins that play other roles in clathrin-coated vesicle formation may also be found in the middle layer. Reproduced from Figure 1 of D.J. Owen, "Linking endocytic cargo to clathrin: structural and functional insights into coated vesicle formation," *Biochem. Soc. Trans.*, 32: 114, 2004.

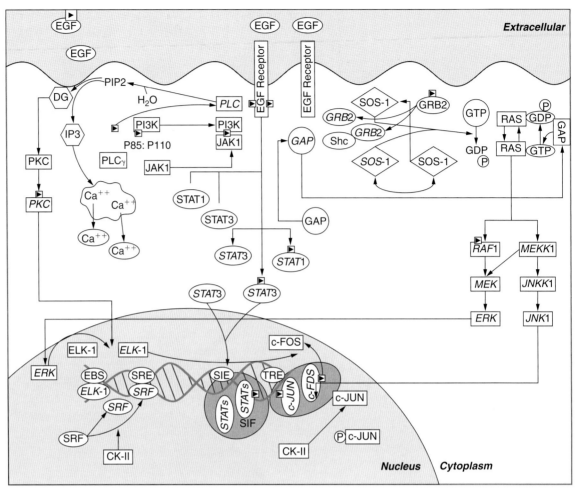

Figure 11-23. Overview of EGF-receptor signaling pathways leading to cell proliferation. Redrawn from http://www.biocarta.com/pathfiles/h_egfPathway.asp.

Transforming Growth Factor

There are two proteins of interest with transforming growth factor, TGFα and TGFβ.

Transforming Growth Factorβ

TGFβ is a protein secreted from tumor lines in culture. It is capable of reversibly inducing (by removal of TGFβ from the culture medium) a transformed phenotype in untransformed cells in culture. There are many relatives of TGF, four of which have extensive similarities in amino acid sequence. The TGFβ family includes the proteins **activins** (A, B, and AB proteins) and **inhibins** (A and B proteins), **Mullerian-inhibiting substance,** and **bone morphogenic protein** regulatory factors. There are as many as 100 proteins that contain a region of sequence homology to TGFβ. The TGFβ receptors have serine/threonine kinase activities associated with their cytoplasmic domains, a situation that differs from receptors for EGF,

PDGF, and **fibroblast growth factor (FGF)**, which contain protein tyrosine kinase activities. TGFβ proteins affect mesenchymal and epithelial cell types to induce proliferation and differentiation of mesodermal cells in early embryos. TGFβ also can have negative growth effects on endothelial cells, macrophages, and T and B lymphocytes.

The TGFβ gene family evolved from a common ancestral gene, and the gene family consists of four members: TGFβ1 (same as TGFβ4), TGFβ2, TGFβ3, and TGFβ5. The amino acid sequences of human TGFβ1 and TGFβ2 are shown in Figure 11-24. TGFβ2 is a homodimer with monomers linked by disulfides (cystines). Each monomer has 112 amino acid residues with a molecular weight of 12,720. TGFβ2 is 71% identical in amino acid sequence to TGFβ1 and is 30% to 40% identical to inhibins and activins and to the C-terminal region of Mullerian-inhibiting substance. The crystal structure of TGFβ1 and the TGFβ type II receptor complexed with TGFβ3 is shown in Figure 11-25. TGFβ binds to the type II receptor, and this binding can be amplified by the presence of the type III receptor. The type II receptor recruits the type I receptor and phosphorylates it after the type II receptor is activated by the binding of TGFβ. The complex activates Smad2 and Smad3 by phosphorylation (a process inhibited by Smad7), and these Smads form heterodimers with Smad4 that translocate to the nucleus. The Smad complex regulates gene expression. Heterodimers of Smad2 and Smad4 and of Smad3 and Smad4 are shown in Figure 11-26. The signal transduction pathway of TGFβ is shown in Figure 11-27. Through the microarray method (Figure 6-92) surveying 15,000 genes, TGFβ was found to quickly induce cyclin D3 while down-regulating cyclin D2, cMyc, Cdk4, E2F1, E2F5, and cyclin A. After 6 hours, Cdc25A

```
              1     5        10        15        20
hTGF-β1      A L D T N Y C F S S T E K N C C V R Q L
hTGF-β2      A L D A A Y C F R N V Q D N C C L R P L

                      25        30        35        40
hTGF-β1      Y I D F R K D L G W K W I H E P K G Y H
hTGF-β2      Y I D F K R D L G W K W I H E P K G Y N

                      45        50        55        60
hTGF-β1      A N F C L G P C P Y I W S L D T Q Y S K
hTGF-β2      A N F C A G A C P Y L W S S D T Q H S R

                      65        70        75        80
hTGF-β1      V L A L Y N Q H N P G A S A A P C C V P
hTGF-β2      V L S L Y N T I N P E A S A S P C C V S

                      85        90        95       100
hTGF-β1      Q A L E P L P I V Y Y V G R K P K V E Q
hTGF-β2      Q D L E P L T I L Y Y I G K T P K I E Q

                     105       110
hTGF-β1      L S N M I V R S C K C S
hTGF-β2      L S N M I V K S C K C S
```

Figure 11-24. Amino acid (one letter abbreviation) sequences of human TGF-β1 and TGF-β2 (hTGF-β1, hTGF-β2).

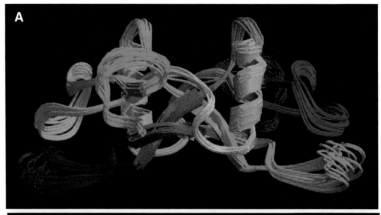

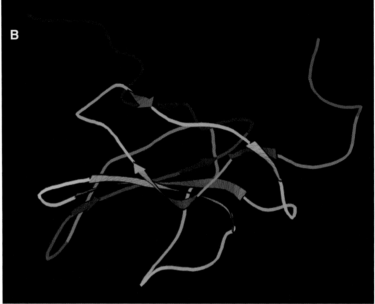

Figure 11-25. **A.** Dimer of TGF-β1. **B.** TGFβ type II receptor extracellular domain. Part A reproduced from PDB ID: 1KLA. A.P. Hinck, S.J. Archer, S.W. Qian, A.B. Roberts, M.B. Sporn, J.A. Weatherbee, M.L. Tsang, R. Lucas, B.L. Zhang, J. Wenker, D.A. Torchia. Transforming Growth Factor Beta 1: Three-Dimensional Structure In Solution and Comparison with the X-Ray Structure of Transforming Growth Factor Beta 2. *Biochemistry* **35** pp. 8517 (1996) and from http://www.rcsb.org/pdf/cgi/explore.cgi?job=graphics;pdbId=1KLA;page=&.opt=show&size=500. Part B reproduced from PDB ID: 1plo. S. Deep et al., Solution Structure and Backbone Dynamics of the TGFβ Type II Receptor Extracellular Domain. *Biochemistry* **42** pp. 10126 (2003).

was induced and cMyc, Id2, Cdk2, p107, E2F1 and E2F5 were down-regulated. After 24 hours, several proteins were down regulated: cyclin D2, p107, Id2, E2F5, cyclin A, cyclin B, and cyclin H. These effects of TGFβ lead to S-phase progression in the cell cycle.

Transforming Growth Factorα

TGFα contains 50 amino acid residues with three disulfide bridges. This mitogenic factor is similar to EGF, sharing 30% sequence identity, and TGFα competes with EGF for the same cell membrane receptor

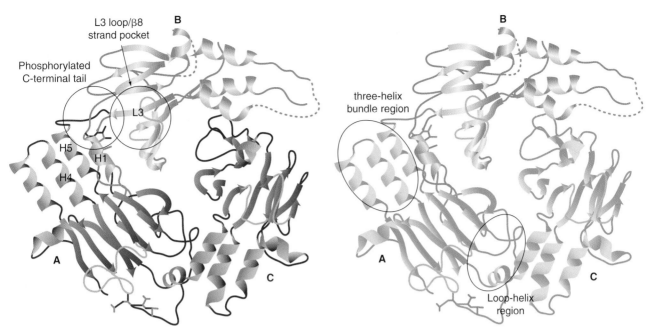

Figure 11-26. Stoichiometry of R-Smad/Smad4 complex. **A.** Crystal structures of the Smad2/Smad4 *(left)* and Smad3/Smad4 *(right)* complexes. The Smad2 subunits are in *red*. The Smad3 subunits are in *green*. The Smad4 subunits in both complexes are in *blue*. The L3 loops are in *yellow*. The phosphoserine side chains are in *stick presentation*. Redrawn from Figure 1 of B.M. Chacko et al., "Structural basis of heteromeric Smad protein assembly in TGF-β signaling," *Mol. Cell.*, 15: 813–823, 2004.

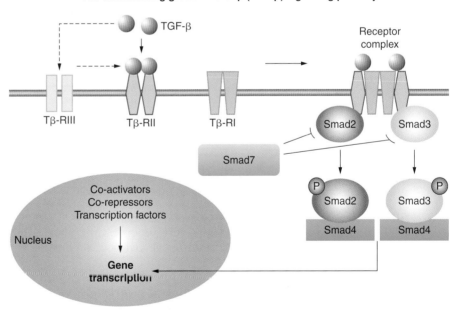

Figure 11-27. The transforming growth factor β *(TGF-β)* signalling pathway. TGF-β binds to the receptor Tβ-RII; this binding might be enhanced by the presence of Tβ-RIII. After binding to TGF-β, Tβ-RII recruits and phosphorylates Tβ-RI, leading to activation of Smad2 and Smad3 by phosphorylation *(P)*. This process is inhibited by Smad7. Activated Smad2 and Smad3 form heterodimers with Smad4 and translocate to the nucleus. Together with co-activators, co-repressors, and other transcription factors, the Smad complex regulates gene expression. Redrawn from Figure 4 from A.Y. Hui and S.L. Friedman, *Expert Reviews in Molecular Medicine*, 5: 14, 2003 and http://www-ermm.cbcu.cam.ac.uk/03005726a.pdf.

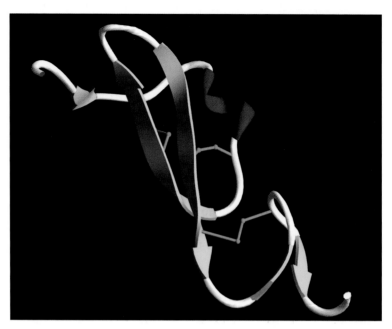

Figure 11-28. Structure of transforming growth factor-alpha (TGFα). The three disulfide bonds are Cys8-Cys21, Cys16-Cys32, and Cys34-Cys43. The side chains play a role in the functionality of TGFα and are Phe 15, Arg 42, and Leu48. Reproduced from PDB ID: 1yud, http://www.rcsb.org/pdb/explore.do?structureId=1YUD.

(EGFR). Some cancer cells express high levels of TGFα or EGF receptor complexes, and TGFα is often secreted by human cancers. TGFα may act, along with TGFβ and similar growth factors, to produce cellular changes in certain cells. With EGF and its relatives, TGFα plays a role in wound healing and formation of tumors. Tumors may be generated when too much TGFα binds to EGF receptors and dysregulates the EGF signaling pathway. The structure of TGFα is shown in Figure 11-28. The structures of TGFα proteins bound to dimeric EGF receptor molecules are shown in Figure 11-29. Because TGFα binds to the EGFR, it either agonizes the EGF pathway or dysregulates the EGF pathway, depending on the number of TGFα–EGFR complexes formed compared to EGF–EGFR complexes. There is a TGFα receptor distinct from the EGF receptor, but TGFα is expressed only in a few cell types, such as activated macrophages, keratinocytes, and some other epithelial cells, and by a variety of carcinomas. TGFα is known to be a potent keratinocyte growth factor.

Fibroblast Growth Factor

FGF represents a superfamily of growth factors now numbering 22 in the human, all of which contain a 120–amino acid core region with 6 identical amino acids distributed within the core. Some FGFs induce cell proliferation, including fibroblasts and basic FGF (FGF-2), which stimulates cell division of endothelial cells, chondrocytes, smooth muscle cells, and melanocytes. It also causes differentiation of adipocytes, and in some cell types it induces the production of specific ILs and prolongs the survival

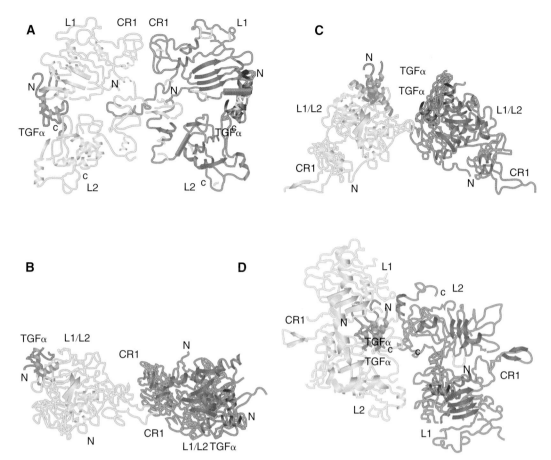

Figure 11-29. Polypeptide trace for the 2:2 TGFα:sEGFR501 complex. **A.** Side view of the back-to-back dimer. The sEGFR501 molecules are shown in *orange* (molecule A) and *magenta* (molecule B). The two TGFα molecules are colored *green* and *lilac*. Disulfide bonds are drawn in *yellow*. **B.** The back-to-back dimer viewed down the dimer axis. **C.** Side view of the head-to-head dimer. **D.** The head-to-head dimer viewed down the dimer axis. Redrawn from Figure 1 of T.P.J. Garrett et al., "Crystal Structure of a Truncated Epidermal Growth Factor Receptor Extracellular Domain Bound To Transforming Growth Factor α," *Cell*, 110: 763–773, 2002.

of neuronal cells. Members of this superfamily act through four FGF receptors with tyrosine kinase activity. In addition to induction of cell division by FGFs, they are involved in development, angiogenesis, hematopoiesis, and tumorigenesis. FGFs are about 18 kDa, although higher molecular weight forms of about 21.5 kDa and 22 kDa also occur. The higher molecular weight forms have some activities in addition to the cell proliferative activity demonstrated by both forms. Native molecular weights of these factors actually vary from the 7 kDa FGF-1 to the 38 kDa of FGF-5. The higher molecular weight forms may cause cells to become binucleate. A characteristic shared among the 23 human family members is a structural fold consisting of 12 antiparallel β-strands (β-trefoil motif), and this motif may be involved in FGF receptor recognition. Table 11-4 lists the human FGFs and the FGF receptors (FGFRs). The table shows which ligands bind to which receptors, although some

Table 11-4
The 23 Human FGFs That Bind to Receptors FGFR1, FGFR2, FGFR3, and FGFR4. FGF-1 Binds to All of the Receptors, Including the Subtypes of FGFR1, FGFR2, and FGFR3

Ligands	Receptors						
	FGFR1		FGFR2		FGFR3		FGFR4
	IIIb	IIIc	IIIb	IIIc	IIIb	IIIc	
FGF-1	●	●	●	●	●	●	●
FGF-2	●	●		●		●	●
FGF-3	●		●				
FGF-4		●		●			●
FGF-5		●					
FGF-6		●		●			●
FGF-7			●				
FGF-8a	—	—	—	—	—	—	—
FGF-8b				●		●	●
FGF-8e						●	●
FGF-8f				●		●	●
FGF-9				●	●	●	●
FGF-10	●		●				
FGF-11							
FGF-12							
FGF-13							
FGF-14							
FGF-16							●
FGF-17b				●		●	●
FGF-18							
FGF-19							●
FGF-20							
FGF-21							

●, Binding
Reproduced from Table 3 of "Fibroblast Growth Factors," R & D Systems and from prior site http://www.rndsystems.com/asp/g_sitebuilder.asp?bodyid=308.

information is lacking for some of the human FGFs. Figure 11-30 shows the structure of FGF-7 based on the structure of FGF-2. FGF-2 (basic FGF, HBGF-2, EDGF) is an 18-kDa peptide with intracellular and extracellular activities. A 24-kDa form of FGF-2 increases IL-6 expression, and the 18-kDa form decreases IL-6 production. The 18-kDa monomeric form is secreted and targets a cell membrane receptor, and the 24-kDa form targets the nucleus. The secreted monomer is sequestered on the cell surface heparan sulfate or matrix glycosaminoglycans, where FGF-2 may dimerize, although FGF-2 target cells can block this dimerization by ribosylating 18 kDa in the receptor-binding domain. After FGF-2 is internalized, it can be degraded to bioactive fragments (4–10 kDa) or translocated to the nucleus. The high molecular weight forms of FGF-2 are translocated to the nucleus through a nuclear translocation motif in the N-terminus. A variety of cells express FGF-2.

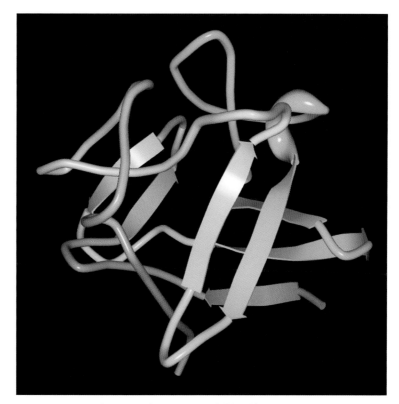

Figure 11-30. The crystal structure of fibroblast growth factor 7. Reproduced from PDB ID: 1991. S. Ye et al. Structural Basis for Interaction of FGF-1, FGF-2, and FGF-7 with Different Heparin Sulfate Motifs. *Biochemistry* **40** pp. 14429 (2001).

FGF-1 (acidic FGF, FGFa, ECGF, HBGF) is 18 kDa and is expressed by cells derived from the three germ layers (ectoderm, mesoderm, and endoderm). FGF-1 is released as a disulfide-linked dimer (differently from FGF-2) that may be reduced to monomeric form by a reducing agent, such as glutathione. The monomeric form binds to a cell membrane receptor or to heparan sulfate (glycosaminoglycan component) of the extracellular matrix, which acts as a storage depot. FGF-1 is released from this depot as needed and can interact with any of the four FGFRs (Table 11-4). FGF-1 also functions within the cell after binding to a cell membrane receptor, being internalized and translocating to the nucleus signaled by a **nuclear localization motif** (NYKKPLK) between amino acids 22 and 28. FGF-1 is expressed by intestinal enterochromaffin cells, renal proximal tubule cells, smooth muscle cells, hepatocytes, neurons, endothelial cells, skeletal muscle cells, macrophages, keratinocytes, and fibroblasts.

Mutated FGFRs are responsible for some human diseases, and these are summarized in Table 11-5. The functional FGF receptor consists of three immunoglobulin-like domains in the extracellular region, an acidic box at the amino terminus, and a tyrosine kinase catalytic center in the cytoplasmic domain (Figure 11-31). Alternative splicing can give rise to up to 14 isoforms, as shown in Figure 11-32. The structure of the FGFR3c in a complex with FGF-1 (without its N-terminal region) is shown in Figure 11-33. In Figure 11-34, cartoon structures of the FGFR1, FGFR2, and FGFR3 are shown, indicating the locations of mutations that give rise to various human diseases (Table 11-5). This figure also shows the signaling pathway initiated by the FGF-complexed FGFR dimer ending in the activation of the MAPK pathway, which leads to cell proliferation after key nuclear proteins have been phosphorylated by MAPK. The

Table 11-5
Mutations of Human FGF Receptors Lead to Various Disease Syndromes

Affected Receptor	Syndrome	Phenotypes
FGFR1	Pfeiffer	Broad first digits, hypertelorism
FGFR2	Apert	Midface hypoplasia, fusion of digits
FGFR2	Beare-Stevenson	Midface hypoplasia, corrugated skin
FGFR2	Crouzon	Midface hypoplasia, ocular proptosis
FGFR2	Jackson-Weiss	Midface hypoplasia, foot anamolies
FGFR2	Pfeiffer	Same as for FGFR1, mutations
FGFR3	Crouzon	Midface hypoplasia, *acanthosis nigricans*, ocular proptosis
FGFR3	Nonsyndromatic craniosynostosis	Digit defects, hearing loss

Reproduced from http://web.indstate.edu/thcme/mwking/growth-factors.html.

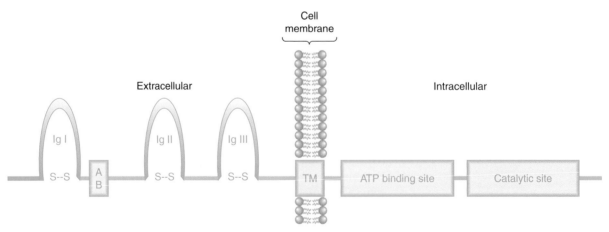

Figure 11-31. Domain structure of the FGF receptor. From http://jorde-lab.genetics.utah.edu/people/reha/Reha.html. *AB*, acidic box; *Ig*, immunoglobulin; *TM*, transmembrane sequence.

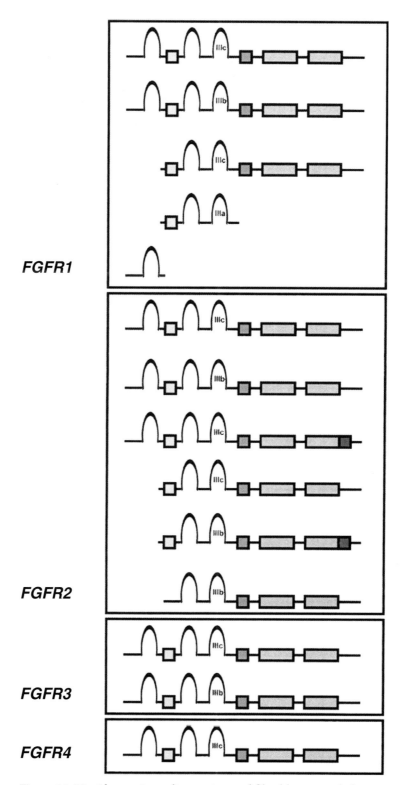

Figure 11-32. Alternative splice variants of fibroblast growth factor receptors. The *green box* shows alternative C-terminal end. Reproduced from Figure 2 of R. Toydemir, "Transmembrane signaling by fibroblast growth factor receptors," and http://ijorde-lab.genetics.utah.edu/people/reha/Reha.html.

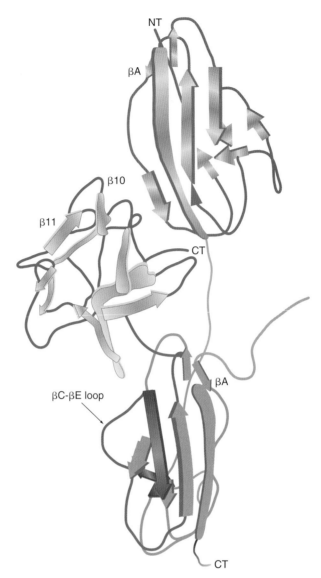

Figure 11-33. The FGFR3c-FGF1 structure. A ribbon representation of the FGFR3c-FGF1 complex. FGF1 is *orange*, FGFR3c D2 is *green*, D3 is *cyan*, and the D2-D3 linker is *black*. The alternatively spliced C-terminal half of D3 is *purple*. The N and C termini of FGF1 are labeled NT and CT, respectively. The FGF1 N-terminal region not included in the truncated FGF1 construct used in the FGFR1c and FGFR2c structures is *gray*. Reproduced from Figure 1 of S.K. Olsen et al., "Insights into the molecular basis for fibroblast growth factor receptor autoinhibition and ligand-binding promiscuity," *PNAS US*, 101: 935–940, 2004.

activated FGFR complex also can lead to the activation of protein kinase C through stimulation of phospholipase C, as shown in Figure 11-35. Tyrosine kinase–containing receptors can be divided into nine subfamilies, of which the FGFRs are one member. These receptors include EGFR, insulin receptor, PDGFR, FGFR, VEGFR, human growth factor receptor, Trk, Elk/ERK, and Axl, as shown in Figure 11-36.

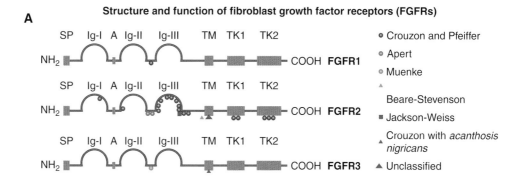

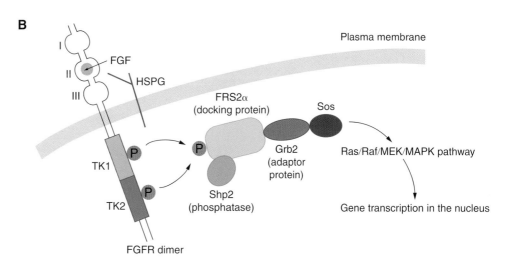

Figure 11-34. Structure and function of fibroblast growth factor receptors (FGFRs). **A.** Schematic representation of three of the four FGFRs and distribution of mutations causing craniosynostoses. The extracellular region of the FGFRs contains three immunoglobulin (Ig)-like loops (I, II, and III), a signal peptide (SP), and an acidic box (A). The tyrosine kinase (TK) domain is split into two subdomains, TK1 and TK2. The receptor is anchored to the cell membrane through a transmembrane (TM) domain. Positions of mutations accounting for seven different craniosynostosis syndromes are shown. **B.** Binding of FGF to FGFRs in association with heparan sulphate proteoglycan (HSPG) induces receptor dimerization and autophosphorylation. This triggers phosphorylation (P) of downstream target proteins, examples of which are shown here. FRS2a, Grb2, Shp2, and Sos are involved in the mitogen-activated protein kinase (MAPK) signaling pathway. Reproduced from Figure 3 of J. Bonaventure and V. El Ghouzzi, "Structure and function of fibroblast growth factor receptors (FGFRs)," *Expert Reviews in Molecular Medicine*, http://www.expertreviews.org/. Accession information: DOI: Vol. 5; 29 January 2003.

Nerve Growth Factor

Nerve growth factor (NGF) is a member of the neurotrophin family. It plays a role in the survival of neurons (Table 11-2) and promotes eurite extension during development. NGF is a survival factor for **nociceptive** (pain-transmitting) neurons. In addition, NGF is up-regulated in adults in response to nociceptive stimuli and is required for extreme sensitivity to pain (hyperalgesia) associated with inflammation. *Inflammatory hyperalgesia is reduced by the binding of NGF, which implies that an*

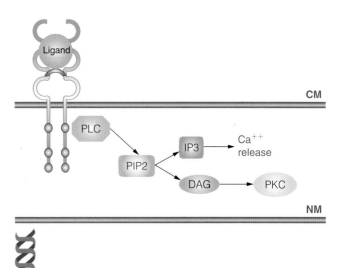

Figure 11-35. Transmembrane signal transduction through PLC gamma pathway. The *red arch* on the receptors shows heparin sulfate. *CM*, cellular membrane; *NM*, nuclear membrane. Reproduced from Figure 5 of R. Toydemir, "Transmembrane signaling by fibroblast growth factor receptors," *Endocrine Related Canon*, 3: 165, 2000 at http://jorde-lab.genetics.utah.edu/people/reha/Reha.html.

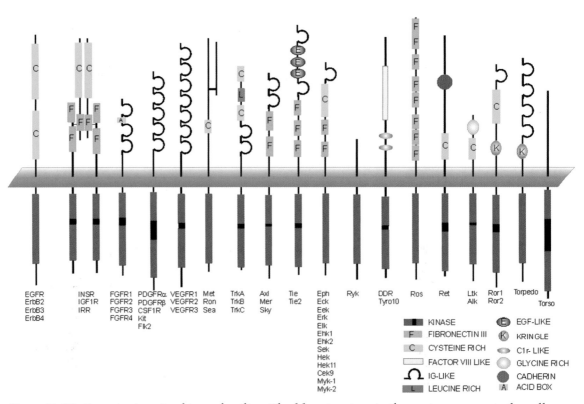

Figure 11-36. Receptor tyrosine kinase families. The *blue* structure in the center represents the cell plasma membrane. The structural features are indicated in the key *(lower right)*. Reproduced from http://www.ludwig.edu.au/angiogenesis/image35.gif.

624 CHAPTER 11 Growth Factors and Cytokines

elevated level of NGF is required to generate the full hyperalgesic response. This activity is specific to NGF among the neurotrophins. Interestingly, snake venom is a rich source of NGF, as well as male mouse submaxillary gland. High concentrations are also found in guinea pig prostate gland and in bovine seminal plasma, although the roles of NGF in these tissues are not understood. NGF forms a homodimer whose structure is shown in Figure 11-37. NGF is a member of the "cystine knot" family consisting of NGF, TGFβ, PDGF, and human chorionic gonadotropin. In the case of NGF, the dimer is linked not by disulfide bonds (although the dimers are connected by disulfide bonds in the other members of the family) but rather chiefly by hydrophobic contacts.

There are two receptors for NGF on the target cell surface. One is called the p75 receptor, also known as the **low-affinity neurotrophin receptor (LANR),** which binds all the ligand members of the neurotrophin family. The structure of the p75 receptor dimer in complex with the NGF dimer is shown in Figure 11-38. The other receptor for NGF is the trkA NGF receptor, which is a member of the tyrosine kinase–containing receptors. This receptor has different specificities for different members of the neurotrophin family. The structure of the dimeric trkA receptor bound with a dimeric NGF is shown in Figure 11-39.

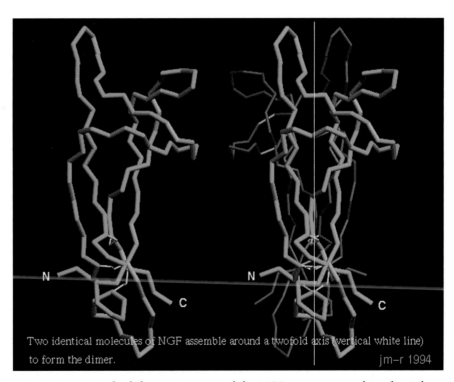

Figure 11-37. On the left is a structure of the NGF monomer, and on the right is a model of the homodimer in a head-to-head arrangement around a two-fold axis *(vertical white line)*. Reproduced from http://people.cryst.bbk.ac.uk/~ubcg09j/neurotrophins/gifs/www_dfor.gif.

Growth Factors

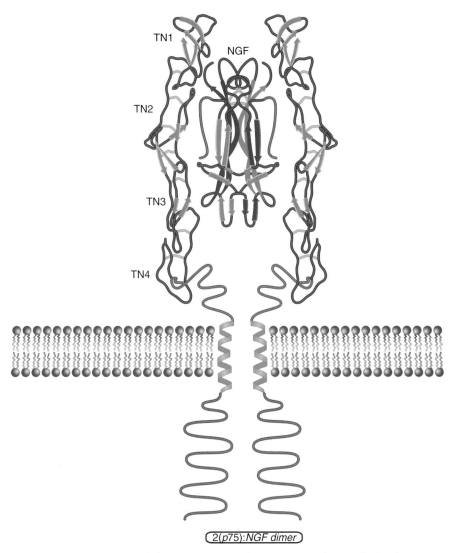

Figure 11-38. Structure of the p75 receptor dimer in complex with the dimeric NGF. Redrawn from former site http://people.cryst.bbk.ac.uk/~ubcg09j/neurotrophins/gifs/www_dfor.gif

Signal transduction mediated by the p75 receptor for NGF is shown in Figure 11-40. The dimeric receptor bound with the dimeric NGF links to phosphatidylinositol-3-kinase (PI3K) and activates the Shc–Grb2–Sos pathway that couples to *Ras* signaling through Raf-1 and MEK and results in the activation of the MAPK pathway. MAPK, once in the nucleus, stimulates Elk-1, which results in the activation of cFos binding to c-Jun to form the AP-1 complex that regulates gene expression. The signal transduction pathway starting with the trkA receptor dimer complexed with dimeric NGF is shown in Figure 11-41. This pathway also ends in the activation of the MAPK pathway and leads to the formation of AP-1, with some slight differences from the p75 pathway. The tyrosine kinase activity of the trkA receptor phosphorylates the phospholipase C enzyme, causing its activation, which also leads to the activation of protein kinase C.

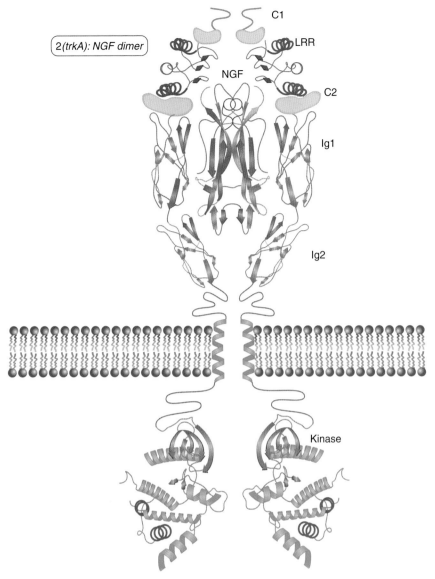

Figure 11-39. Dimeric trkA receptor bound to NGF dimer. Redrawn from http://people.cryst.bbk.ac.uk/~ubcg09j/neurotrophins/gifs/www_dfor.gif

Colony-Stimulating Factor

CSFs are required for survival, proliferation, and differentiation of progenitors (parental or ancestral cells) of hemopoietic cells (which promote the formation of blood cells). CSFs are named for the cells that each factor stimulates. Macrophage CSF (M-CSF, CSF-1) stimulates that cell type, granulocyte CSF (G-CSF) stimulates granulocytes, and granulocyte–macrophage CSF (GM-CSF, CSF-2) stimulates both cell types. There is a multi-CSF, which is IL-3. This is where the definition of growth factors and cytokines becomes cloudy; in general, growth

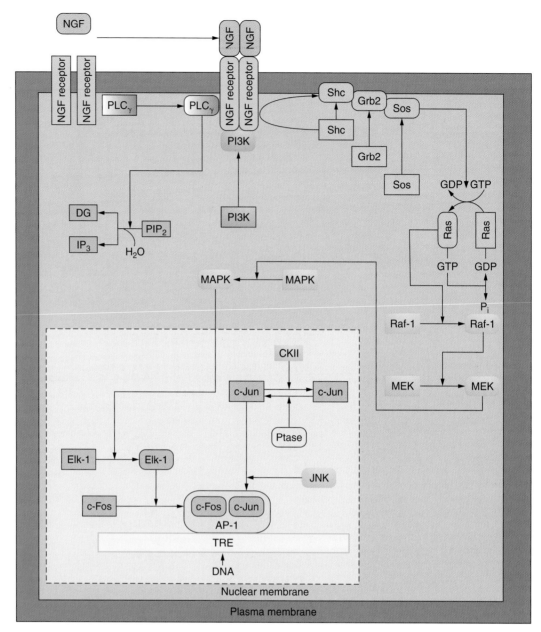

Figure 11-40. Signal transduction pathway of the p75 receptor complexed with its ligand, NGF dimer. Redrawn from former site http://www.grt.kyush-u.ac.jp/spad/images/signalpathway/NGF.GIF

factors are referred to here as agents that cause proliferation (cell division), whereas cytokines produce changes in target cells but can also, but not necessarily, produce cell growth. Strictly speaking, *growth factors stimulate cell division, differentiation, and proliferation, whereas cytokines facilitate communication among cells of the immune system.* Cytokines are produced by immune system cells, and generally they affect the immune response. However, many authors use the terms interchangeably.

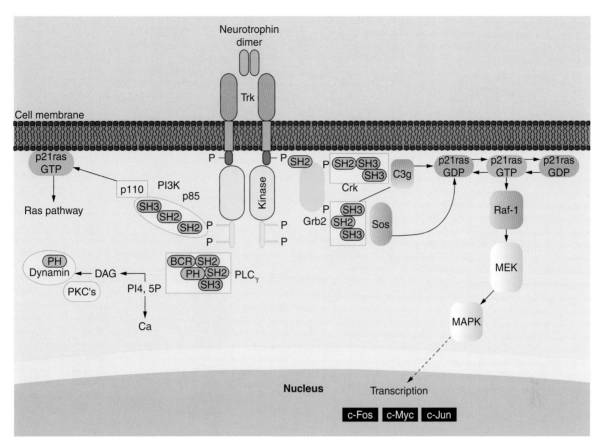

Figure 11-41. Signal transduction pathway of dimeric NGF complexed with trKA dimeric receptor. Redrawn from former site http://people.cryst.bbk.ac.uk/~ubcg09/sigtrans/gifs/sigtemp.gif.

GM-CSF was first purified from human urine, and it was demonstrated to cause the proliferation of hemopoietic cells in the mouse. It is a glycoprotein of 45 kDa and forms a homodimer. The chromosomal locus of GM-CSF is on chromosome 5 (5q21-q32) within 9 kilobases of the IL-3 gene. The formation of hemopoietic cells from stem cell progenitors is shown in Figure 11-42, with the participation of various CSFs in cellular differentiation steps.

The structures of M-CSF, G-CSF, and GM-CSF are shown in Figure 11-43. The receptors for CSFs are members of the IL-1 receptor family (CSF-1R; Figure 11-44A) or of the cytokine receptor superfamily (G-CSF-R and GM-CSF-Rb; Figure 11-44B). The structure of the receptor for M-CSF is CSF-1R, which is shown in Figure 11-45A. It has a marked structural similarity to another receptor, Flt3, a type III tyrosine kinase receptor, that plays a key role in hematopoiesis. Flt3 is closely related to PDGF receptors. Structures for Flt3 are shown in Figure 11-45B. G-CSF-R is shown in Figure 11-46 with a dimer of G-CSF complexed with the homodimeric receptor. Figure 11-47 shows the crystal structure of the complete extracellular domain (ligand-binding portion of the receptor) of the β–common receptor (GM-CSF-R) that binds IL-3, IL-5, and GM-CSF.

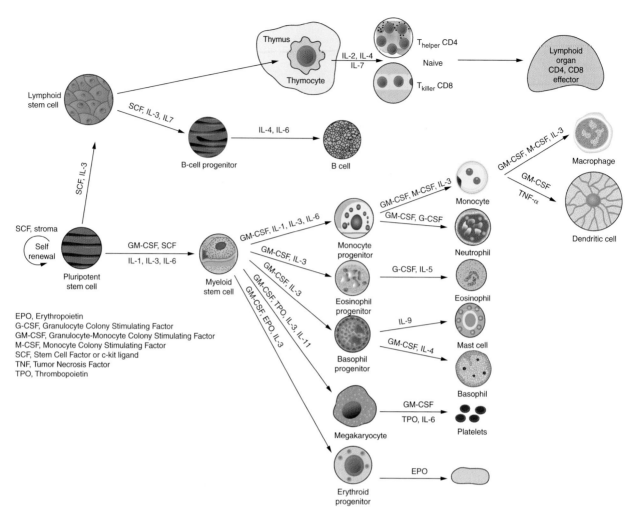

Figure 11-42. Formation of hemopoietic cells from stem cell precursors showing the roles of growth factors and cytokines in the process of differentiation. Note that GM-CSF is involved in the transition; pleuripotent stem cell to myeloid stem cell; myeloid stem cell to the following: monocyte progenitor; eosinophil progenitor; basophil progenitor; megakaryocyte and erythroid progenitor. It is further utilized in the transition: monocyte progenitor to monocyte; basophil progenitor to basophil; megakaryocyte to platelets; monocyte to macrophage; and monocyte to dendritic cell. G-CFS is involved in the conversion of monocyte progenitor to neutrophil, eosinophil progenitor to eosinophil, and M-CSF is involved in the conversion of monocyte to macrophage. Redrawn from former site http://edu.med.image.ch/apprentissage/module4/dif-imm/apprentissage/problemes/hemoporse-csf.jpg.

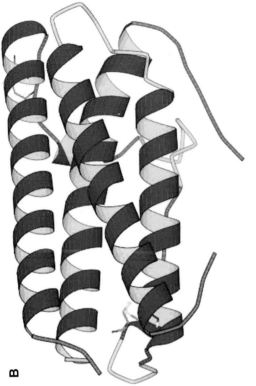
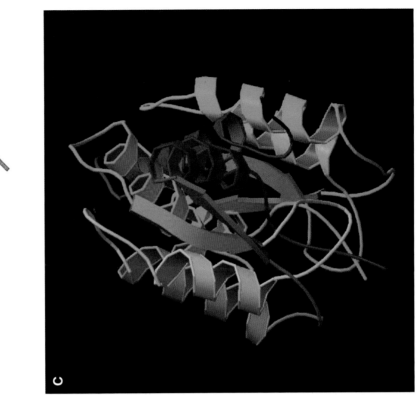
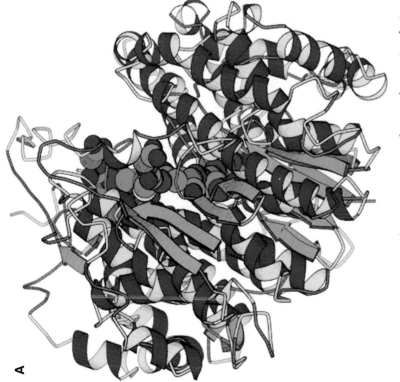

Figure 11-43. Structures of CSFs. **A.** Macrophage colony-stimulating factor (M-CSF). **B.** Granulocyte Colony-Stimulating Factor (G-CSF). **C.** Granulocyte-macrophage Colony-Stimulating Factor (GM-CSF). Part A reproduced from http://bip.weizmann.ac.il/oca-bin/ccpeek?id=1HMC. Part B reproduced from http://bip.weizmann.ac.il/ora-bin/ccpeek?id=1BGC. Part C reproduced from PDB ID 1csg, http://www.imb-jena.de/cgi-bin/ImgLib.pl?CODE=1csg.

Growth Factors 631

M-CSF (along with RANKL) plays a role in the formation and activity of osteoclasts (multinucleate cells that absorb bone matrix and are involved in bone remodeling to form canals or cavities; they are associated with bone breakdown), as shown in the diagram in Figure 11-48. M-CSF is an important osteoclast differentiation factor, which increases osteoclast motility and consequently reduces osteoclastic bone resorption. The regulation of macrophage and osteoclast development by M-CSF (CSF-1) is shown in Figure 11-49A. Figure 11-49B shows the signaling pathways regulated by CSF-1R in myeloid cells (cells derived from bone marrow that distinguish a separate erythroid lineage, often specifically referring to mononuclear phagocytes and granulocytes).

G-CSF signals by activating the JAK/STAT pathway. By activating this pathway in cardiomyocytes, G-CSF prevents cardiac remodeling after myocardial infarction and plays a key role in cardiac recovery. Figure 11-50 shows a typical pathway wherein JAK phosphorylates STAT1, leading to transcriptional activation.

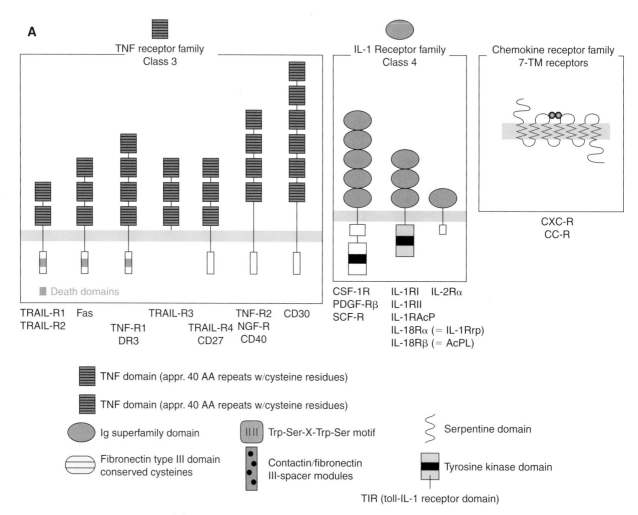

Figure 11-44. **A.** Cartoons of the TNF receptor family, the IL-1 receptor family, and a model of the chemokine receptor family. CSF1R is a member of the IL-1 receptor family.

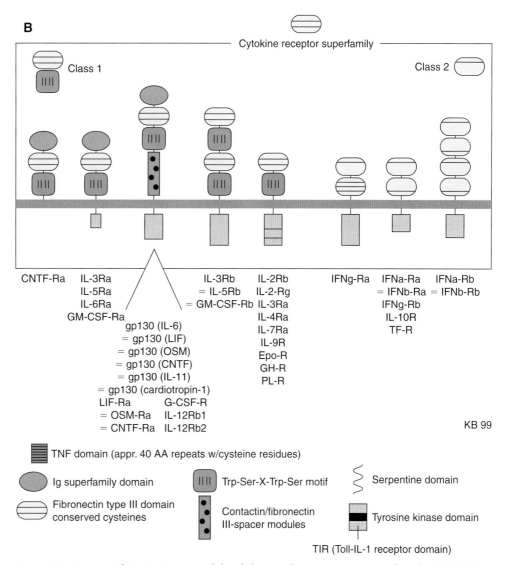

Figure 11-44, cont'd **B.** Cartoon models of the cytokine receptor superfamily. G-CSF-R and GM-CSF-Rb are members of class 1 cytokine receptor subfamily. Parts A and B redrawn from http://www.iir.suite.dk/IIR/10CK/tCKRoth.htm.

GM-CSF has proliferative effects on both granulocytes and macrophages. It operates through the β–common receptor (Figure 11-47), for which agonists are IL-3, IL-5, and GM-CSF. This receptor signals JAK2, which phosphorylates STAT5 to bring about transcriptional effects (Figure 11-51) leading to proliferation.

Erythropoietin

Erythropoietin regulates erythropoiesis (formation of red blood cells) by stimulating proliferation and differentiation of immature erythrocytes. In addition, it induces growth of erythroid progenitor cells and differentiation of erythrocyte colony-forming units into erythroblasts. The two organ systems that synergize to produce red blood cells are the bone marrow, which produces red cells, and the kidney, which produces erythropoietin.

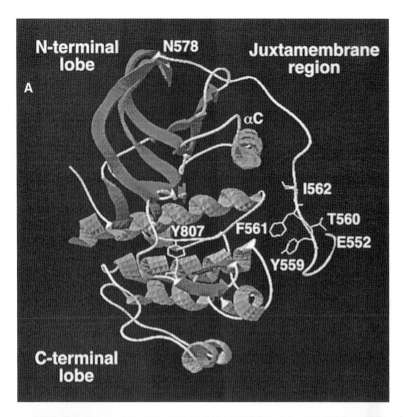

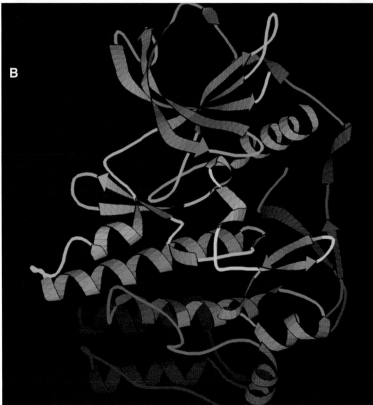

Figure 11-45. **A.** Structure of CSF-1R, the receptor for macrophage colony-stimulating factor. **B.** *(Top)* Structure of Flt3 and *(bottom)* crystal structure of flt3. Part A reproduced with permission from C.M. Rohde, J. Schrum, and A.W.-M. Lee. "A juxtamembrane tyrosine in the colony stimulating factor-1 receptor regulates ligand-induced Src Association, receptor kinase function and down-regulation," *J. Biol. Chem.*, 279: 43448–43461, 2004. Part B reproduced from PDB ID: Flt3. J. Griffim et al. The Structural Basis for Autoinhibition of FLT3 by the Juxtamembrane Domain. *Mole. Cell* **13** pp. 169 (2004).

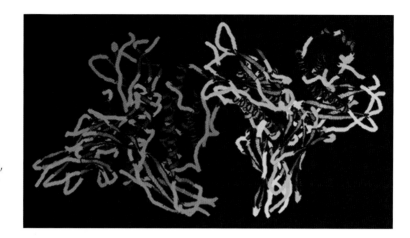

Figure 11-46. Two-to-two complex of G-CSF with its receptor. Reproduced from PDB ID: 1PGR. M. Aritomi, N. Kunishima, T. Okamoto, R. Kuroki, Y. Ota, K. Morikawa. Atomic Structure of the Gcsf-Receptor Complex Showing a New Cytokine-Receptor Recognition Scheme. *Nature* **401** pp. 713 (1999).

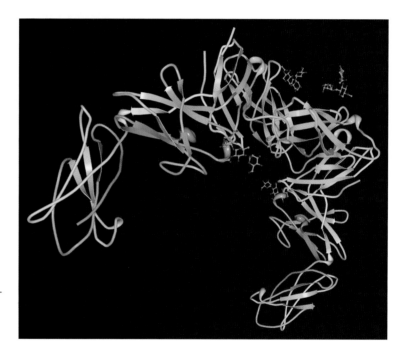

Figure 11-47. Crystal structure of the complete extracellular domain of the dimeric β-common receptor of IL-3, IL-5, and GM-CSF. PDB ID:1 gh7. P.D. Carr, S.E. Gustin, A.P. Church, J.M. Murphy, S.C. Ford, D.A. Mann, D.M. Woltring, I. Walker, D.L. Ollis, I.G. Young. Structure of the Complete Extracellular Domain of the Common Beta Subunit of the Human Gm-Csf, Il-3, and Il-5 Receptors Reveals a Novel Dimer Configuration. *Cell* **104** pp. 291 (2001).

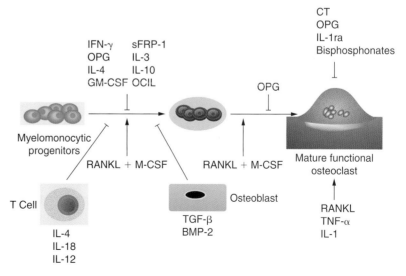

Figure 11-48. Osteoclast formation and activity—direct and indirect action of inhibitors. Shown centrally is a scheme of osteoclast differentiation in response to M-CSF and RANKL. Above the proposed differentiation pathway are examples of direct inhibitors, whereas below are indirect inhibitors signalling through intermediate cell types. Redrawn from Figure 1 of J.M.W. Quinn and M.T. Gillespie, "Modulation of osteoclast formation," *Vertebrate Skeletal Biology*, 328: 739–745, 2005.

Growth Factors

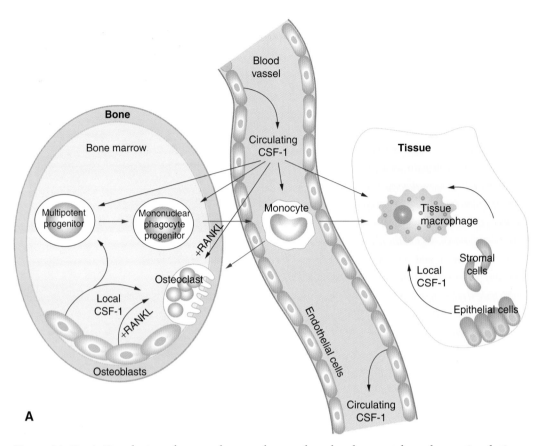

Figure 11-49. **A.** Regulation of macrophage and osteoclast development by colony-stimulating factor-1 (CSF-1). Circulating CSF-1, produced by endothelial cells in blood vessels, together with locally produced CSF-1, regulates the survival, proliferation, and differentiation of mononuclear phagocytes and osteoclasts. CSF-1 synergizes with hematopoietic growth factors (HGFs) to generate mononuclear progenitor cells from multipotent progenitors, and with receptor activator of NF-kB ligand (RANKL) to generate osteoclasts from mononuclear phagocytes. *Red arrows* indicate cell differentiation steps; *blue arrows* indicate cytokine regulation.

Synthesis and release of erythropoietin is signaled when oxygen in the kidney becomes low (hypoxia). For example, anemia is characterized by a low red blood cell count or a low level of hemoglobin, which would limit the oxygen-carrying capacity of the blood. An iron-containing protein in the kidney senses inadequate delivery of oxygen to the tissue, and erythropoietin production is increased in the kidney. This protein complex is the **hypoxia-inducible factor (HIF-1)**. HIF-1β, a component of HIF-1, is expressed constitutively; however, HIF-1α *only appears in hypoxic cells.* The level of oxygen binding to the iron component of HIF-1α elicits a cascade that results in the production and release of erythropoietin. HIF-1 is bound to another protein called factor-inhibiting HIF-1 (FIH-1) under normal oxygen tension. FIH-1 suppresses the transcriptional activity of HIF-1 by preventing its interaction with coactivators of transcription. When hypoxia occurs, FIH-1α, previously bound in a complex (and inactive), is released to affect transcriptional events leading to the formation and release of erythropoietin. The structure of the protein, FIH-1, is shown in Figure 11-52, along with a model of the complex that attaches

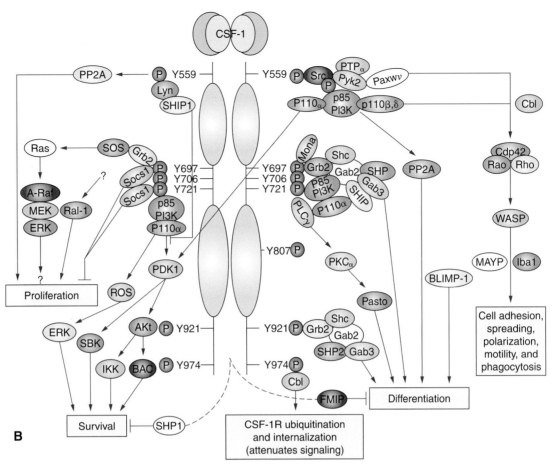

Figure 11-49, cont'd **B.** Signaling pathways regulated by colony-stimulating factor-1 receptor (CSF-1R) in myeloid cells. Binding of colony-stimulating factor-1 (CSF-1) stabilizes a dimeric form of the CSF-1R and leads to activation of the CSF-1R kinase, its tyrosine phosphorylation, and the direct association of signaling molecules with the receptor through their phosphotyrosine-binding domains. Molecules shown touching specific phosphotyrosines are those that associate directly with a particular phosphotyrosyl sequence motif. For clarity, the plasma membrane associations of several molecules are not shown. The precise involvement of the Ras-MEK-MAPK pathway in the CSF-1-regulated proliferation and differentiation of myeloid cells is not clear, but Raf-1 seems to signal independently of this pathway. Differences in signaling pathways are also expected to exist between macrophage progenitor cells and macrophages 25 and 26. *BLIMP*, B lymphocyte–induced maturation protein; *Cb1*, Casitas B lineage; *FMIP*, FMS-interacting protein; *Iba*, ionized Ca^{2+}-binding adaptor protein; *IKK*, IKB kinase; *MAYP*, macrophage actin–associated and tyrosine-phosphorylated protein; *MAPK*, mitogen-activated protein kinase; *MEK*, MAPK kinase; *Mona*, monocyte adaptor; *P*, phosphate; *PDK1*, 3'-phosphoinositide-dependent kinase-1; *PI3K*, phosphatidylinositol 3-kinase; *PK*, protein kinase; *Pkare*, PKA-related gene; *PLC*, phospholipase C; *PP*, protein phosphatase; *Pyk*, proline-rich and Ca^{2+}-activated tyrosine kinase; *ROS*, reactive oxygen species; *SH*, Src homology domain; *SHIP*, SH2-domain-containing polyinositol phosphatase, *SHP*, SH2 domain containing phosphatase, *SOS*, Son of sevenless. Part A reproduced from Figure 1 and Part B redrawn from Figure 3 of F.J. Pixley and E.R. Stanley, "CSF-1 regulation of the wandering macrophage: complexity in action," *Trends Cell Biol.*, 14: 628–638, 2004.

HIF-1α until hypoxia occurs, the complex is broken down or fails to form, and HIF-1α is released to perform its function. The blood, carrying newly released erythropoietin, circulates to the bone marrow, where new red blood cell production is activated. The increase in red blood cells enlarges the capacity of the blood to carry oxygen (hemoglobin, Figure 2-31, in red blood cells) so that hypoxia can be overcome. These relationships are

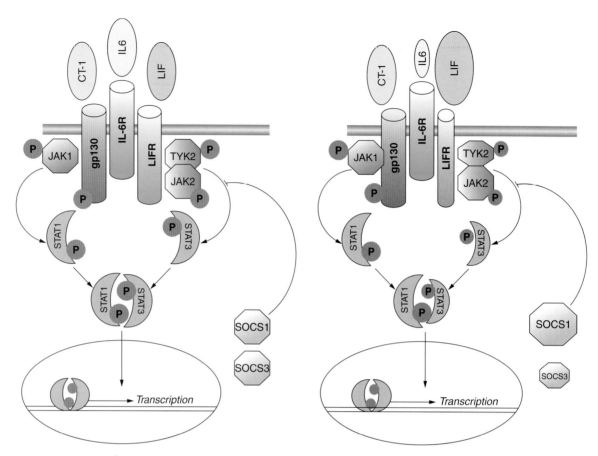

Figure 11-50. A schematic representation of gp130-JAK-STAT signaling alterations in end-stage dilated cardiomyopathy. LIF, leukemia inhibitory factor. Redrawn from Figure 2B of "Alterations in janus kinase (JAK)-signal transducers and activators of transcription (STAT) signaling in patients with end-stage dilated dardiomyopathy," *Brief Rapid Communications*, 2003.

shown in Figure 11-53. Erythropoietin is a 166–amino acid-containing polypeptide with two disulfide bonds, as shown in Figure 11-54, which also shows the crystal structure. The crystal structure of erythropoietin bound to the erythropoietin receptor (EPOR) is shown in Figure 11-55. Erythropoietin activates cells by binding to two cell surface EPORs, which signal the intracellular cascade of phosphorylations through JAK2 and STAT5 (Figure 11-56) reminiscent of the early actions of IL-3 (Figure 11-51).

Interferon-γ

Interferons (IFNs) are small protein cytokines released by tissue cells, macrophages, and lymphocytes in response to viral infection. Released IFN binds to receptors on the membranes of adjacent cells, inducing a specific protein that inhibits the synthesis of viral proteins and thus prevents the spread of the virus. Specifically, IFN-γ is a homodimeric protein participating in the immune response to viral infection. Of the type I IFNs, IFN-α is produced by leukocytes and IFN-β is produced by fibroblasts. These IFNs

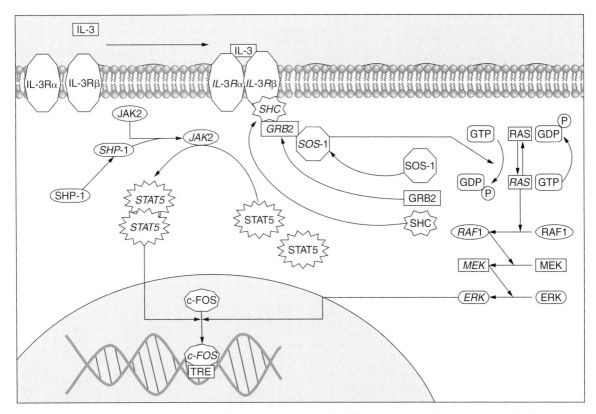

Figure 11-51. Interleukin-3 promotes the proliferation and differentiation of hematopoietic cells through binding to its receptor. The receptor for IL-3 is a heterodimer with a ligand-specific alpha chain (70 kDa, CD123) and a common beta chain (shared with IL-5 and GM-CSF). Signaling is believed to be primarily through Stat5 and the MAPK pathways. Redrawn from http://www.biocarta.com/pathfiles/il3Pathway.asp.

have genes located on chromosome 9, and they bind to type I IFN receptors. Both types of IFNs have similar activities but differ in their binding affinities to receptor. IFN-α and IFN-β bind to one of two receptors, IFNa-R1 and IFNa-R2, either of which is a heterodimeric receptor on the cell surface. Both IFNa-R1 and IFNa-R2 participate in signal transduction. IFN-γ, produced by B cells, natural killer T cells, and antigen-presenting cells, binds to its receptor (IFNg-R2) when the IFNg-R1 chain is present. IFN-γ, encoded on chromosome 12, produced in response to immune inflammatory stimuli rather than viral infection, is a type II IFN that binds to type II receptors. After ligand binding, receptor oligomerization occurs, which initiates signal transduction by multiple phosphorylations of transcriptionally active proteins. These translocate to the nucleus as a trimeric complex, which then activates transcription of the IFN-stimulated genes. The overall action of type I IFNs is shown in Figure 11-57. The structure of IFN-γ is shown in Figure 11-58. The complex of IFN-γ with the IFN-γ receptor containing a high-affinity receptor-binding subunit (IFN-γRa) and a species specific accessory factor (IFN-γRb; also AF-1) is shown in Figure 11-59. The receptor is composed of two ligand-binding IFNg-R1 chains associated with two IFNg-R2 chains, the latter of which is a concentration-limiting factor in IFN-γ responsiveness because IFNg-R1 is usually present in much higher concentration. The IFNg-R1 cytoplasmic domain contains

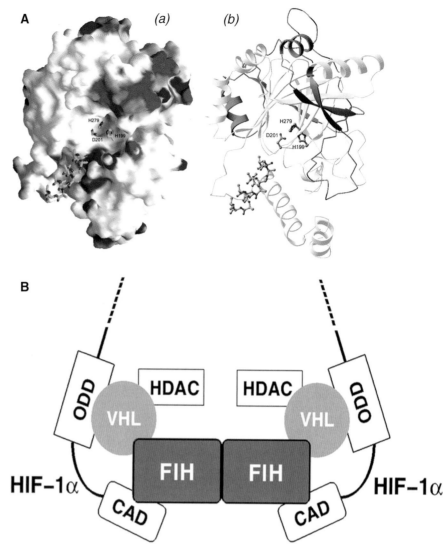

Figure 11-52. **A.** The putative binding sites for HIF-1α and VHL. *(a),* The electrostatic potential surface of FIH-1 is presented with the docked peptide representing the region near the hydroxylation site of HIF-1α CAD. Positive and negative potentials are colored *blue* and *red,* respectively. In the figure, an ideal α-helical polyalanine model (13 residues) was manually docked on the prominent groove near the active site of FIH-1. The facial triad residues (His-199, Asp-201, and His-279) in the active site also are shown in the figure. *(b),* The putative HIF-1α CAD- and VHL-binding sites are represented on a ribbon diagram of FIH-1 with the same orientation as *a.* Two parts of the N-terminal 126 residues were colored differently (residues 12–88, *pink;* residues 89–126, *orange*). The rest of the molecule is colored *gray.* The α-helical polyalanine model docked in the putative substrate-binding groove and the facial triad residues are shown as in model at left. *CAD,* C-terminal activation domain; *VHL,* von Hippel-Lindau ubiquitin-protein ligase complex (proteasome targeting). **B.** A hypothetical model of the complex formation by the FIH-1 dimer, VHL, and HIF-1α. The model based on the structural and functional implications of FIH-1 in the hypoxia regulation is presented in the figure. The model was constructed to represent that FIH-1 acts as a bridge for the association between VHL and HIF-1α CAD, whereas VHL interacts also with the HIF-1α ODD domain and histone deacetylases. In normoxia, VHL binds to the HIF-1α ODD domain through hydroxyproline in the domain, which brings the VHL-associated FIH-1 around HIF-1α CAD. FIH-1 then hydroxylates Asn-803 of HIF-1α CAD, which prevents transcription coactivators from binding to CAD. The dimerization of FIH-1 leads to a large complex consisting of two molecules of FIH-1, two molecules of HIF-1α, two molecules of VHL, and several pairs of associated histone deacetylases. In hypoxia, the complex is not likely to form due to the lack of the proline hydroxylation in the HIF-1α ODD domain. The N-terminal region of HIF-1α, including the DNA-binding domain and Per-Arnt-Sim homology domain, is not shown in the figure (indicated as *broken lines*). *ODD,* Oxygen-dependent degradation domain; *HDAC,* histone deacetylase. Reproduced from Figures 4 and 5 of C. Lee et al., "Structure of human FIH-1 reveals a unique active site pocket and interaction sites for HIF-1 and von Hippel-Lindau," *J. Biol. Chem.,* 278: 7558–7563, 2003.

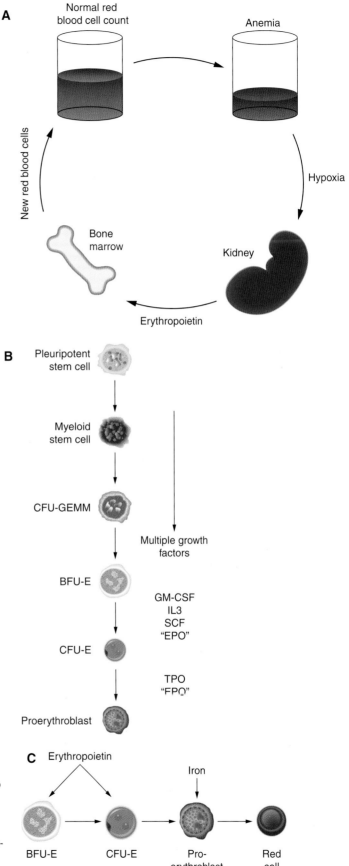

Figure 11-53. **A.** System that normalizes the red blood cell count in blood. **B.** Production of proerythroblasts from red cell precursors. EPO is required in the last two steps. **C.** Maturation of stem cells under the influence of EPO and iron (iron is part of the hemoglobin structure, Figure 2-31). *EPO,* erythropoietin. Redrawn from Figures 1, 2, and 3 of "The interaction of iron and erythropoietin," http://sickle.harvard.edu/epovionmodel. gif&imagrefurl.

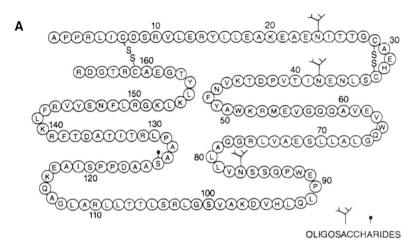

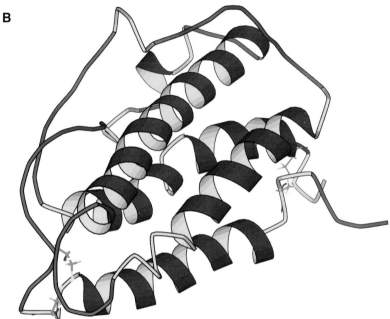

Figure 11-54. **A.** Primary structure of erythropoietin. **B.** NMR structure of human erythropoietin. Part A reprinted with permission from the Athens Medical Society. Part B reproduced from J.C. Cheethan et al., "NMR structure of human erythropoietin and a comparison with its receptor bound conformation," *Nat. Struct. Biol.*, 5: 861–866, 1998 and directly from http://bip.weizmann.ac.il/oca-bin/ccpeek?id=1BUY.

the binding sites for JAK1 and STAT1, as shown in the signal transduction pathway in Figure 11-60. The phosphorylations on the receptor subunits and those activating STAT1 binding are also shown. The α chains bind the ligand (IFN-γ), dimerize, and associate with two signal-transducing β chains. Phosphorylation of a tyrosine residue on the cytoplasmic domain occurs, and JAK1 and JAK2 become activated. STAT1 binds and is phosphorylated, and the complex is translocated to the nucleus to activate a range of IFN-γ-responsive genes. One protein produced from this type of gene expression would be an antiviral protein. An example of this is the antiviral guanylate-binding protein induced by IFN-γ during macrophage induction. The guanylate-binding protein can bind GTP, GDP, and GMP and has GTPase activity. Likely, other types of antiviral proteins are

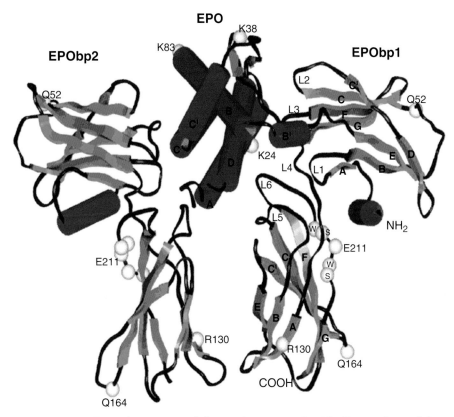

Figure 11-55. Crystal structure of the erythropoietin-(EPObp)2 complex. *Alpha*-Helices are shown as *red cylinders* and β-sheets are shown as *green ribbons*. Erythropoietin: *alpha* A(8-26) up, *alpha* B'(47-52), *alpha* B(55-83) up, *alpha* C(90-112) down, *alpha* C' (114–121), *alpha* D(138-161) down, β1(39-41), β2(133-135). Disulphide bridges: Cys 7-Cys 161, Cys 29-Cys 33. D1 domain of EPObp: *alpha*1 (9-22), bA(26-30), βB(36-42), bC(52-59), bC' (65-68), βD(69-74), βE(78-85), βF(95-103), βG(106-113). Disuphide bridges: Cys 28-Cys 38, Cys 67-Cys 83. D2 domains of EPObp: βA(119-132), βB(137-143), βC(154-162), bC'(170-174), βE(179-183), βF(190-201), and βG(216-220). EPObp loops that interact with erythropoietin are labelled L1(31-35), L2(60-64), L3(86-94), L4(114-118), L5(144-153), and L6(202-205). The locations of residues in the WSXWX box, dimerizing mutation site É2;É3; R130C, two Asn ->. Gln mutation sites (52, 164) on EPObp, and three Asn -> Lys mutations (24, 38, 83) on erythropoietin are depicted as white spheres. Reproduced with permission from R.S. Syed et al., "Efficiency of signaling through cytokine receptors depends critically on receptor orientation," *Nature*, 395: 511–516, 1998.

induced in other cell types after viral infection. When the signaling is complete, the ligand-binding chains dissociate and are recycled to the cell surface.

Insulin-like Growth Factors

Insulin-like growth factors (IGFs) are part of the insulin family, although they are antigenically distinct from insulin. IGF-I is structurally related to insulin, sharing 48% of its amino acid sequences, with 50% identity in amino acid sequences to IGF-II. These two growth factors are induced

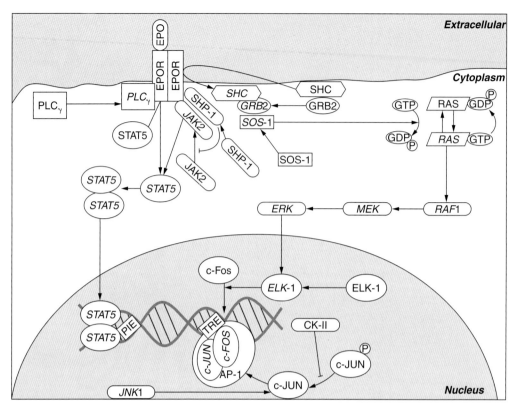

Figure 11-56. Erythropoietin functions to increase the number of red blood cells. Thus, it has found utility as a drug for those needing to replenish erythrocytes for a number of reasons. The signaling mechanism includes multimerization of the receptor upon ligand binding, activation of MAPK cascade, and phosphorylation and activation of Stat5. *TRE*, thyroid hormone response element. Redrawn from www.biocarta.com/pathfiles/epoPathway. gif&imgrefurl=http://www.biocarta.com/pathfiles/epoPathway.asp&h=475m=646&s2

by growth hormone, and they stimulate the growth of skeletal tissues. In particular, IGF-I stimulates the growth of bone (and other tissues) and has autocrine (stimulating growth in the cell producing it) and paracrine (stimulating the growth of nearby cells) activities. IGF-I is identical to the previously known growth factor, somatomedin C, and IGF-II is identical to a factor known as multiplication stimulation factor. IGF-I can bind to the insulin receptor with much lower avidity than insulin. The concentration of insulin is widely variable in the bloodstream, responding to the circulating level of glucose, whereas IGFs are relatively constant in concentration in the blood because they are bound to carrier proteins. Humans have five **IGF-binding proteins (IGFBPs).** The single polypeptide chain structures of IGF-I and IGF-II are similar to proinsulin (Figure 11-61).

In addition to their effects on growth and development, these factors are powerful antiapoptotic factors, commensurate with their positive effects on growth. The structure of the IGF-I receptor tyrosine kinase domain is shown in Figure 11-62A, and comparison with the kinase domain of the insulin receptor is shown in Figure 11-62B. The kinase domains of

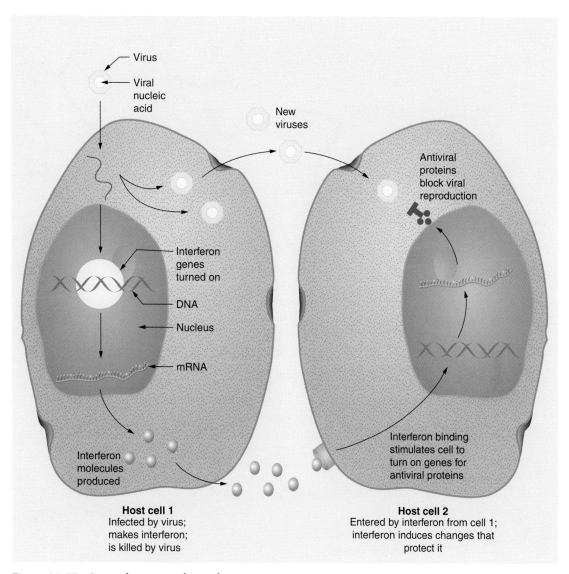

Figure 11-57. General actions of interferons.

the IGF-I receptor and the insulin receptor are similar. The IGF-II receptor is the mannose-6-phosphate receptor, which functions as a transport protein almost exclusively in embryonic and neonatal tissues. Its level falls significantly after birth. It plays a role in integrating lysosomal enzymes containing mannose-6-phosphate residues to the lysosome. Recently, however, IGF-II has been found to induce the expression of COX2 in human keratinocytes, and this expression is mediated through tyrosine kinase–Src–ERK and tyrosine kinase–PI3K pathways but not through the MAPK pathway, suggesting that IGF-II may play a role in the inflammatory process in some cells. The structure of the **IGF-II receptor** is shown in Figure 11-63.

Growth Factors

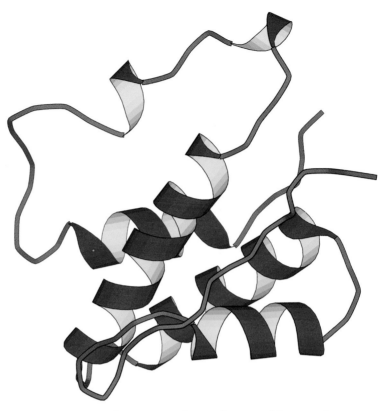

Figure 11-58. Crystal structure of recombinant rabbit interferon-γ at 2.7-Å resolution. Reproduced from http://imb.jena=2RIGPDBsum, ID=2RIG.

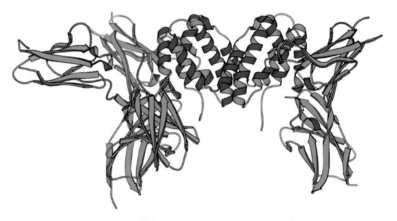

Figure 11-59. Diagram of the 3:1 sIFN-γ complex. The complex is formed by combining 2:1 complex with receptor R3. Reproduced from D.J. Thiel et al., "Observation of an unexpected third receptor molecule in the crystal structure of human interferon-g receptor complex," *Structure* 8: 927–936, 2000, and directly from http://www.rcsb.org/pdb/explore.do?structureid=1fgg; PDB ID 1fgg.

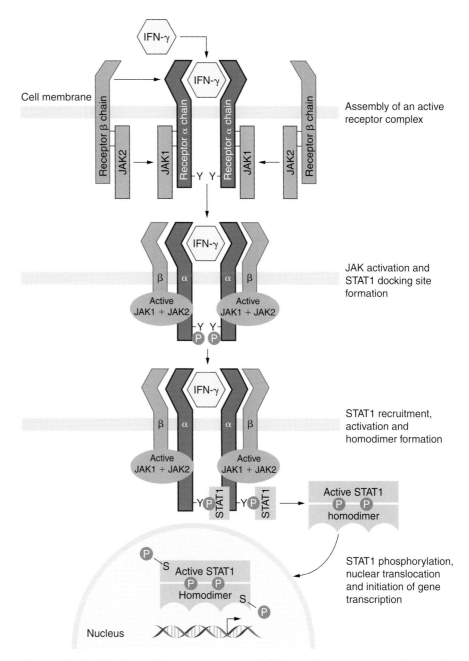

Figure 11-60. A schematic representation of the interferon-γ receptor (IFNγ-R) and its signaling pathway. The receptor for IFN-γ has two subunits: IFNγ-R1, the ligand-binding chain (also known as the "a chain") and IFNγ-R2, the signal-transducing chain (also known as the "b chain," or accessory factor 1) These proteins are encoded by separate genes (*IFNGR1* and *IFNGR2*, respectively), which are located on different chromosomes. Reproduced with permission from Figure 2 of M. Newport, "A schematic representation of the interferon-γ receptor (IFN-γR) and its signaling pathway," *Expert Reviews in Molecular Medicine*, 5: 1–13, 2003.

Figure 11-64 shows a summary of the ligands that bind to cognate (recognizing ligand) receptors in the IGF system. Both IGF-I and IGF-II bind to their respective receptors and to the IGFBPs. IGF-I can bind to the insulin receptor with low avidity and may have an insulin-like function when the concentration of insulin is very low and the level of IGF-I is sufficient. Interestingly, insulin can bind to the IGF-II receptor. The **insulin receptor-like receptor (IRR)** is also indicated without a specific ligand, and the IGF-II receptor is the mannose-6-phosphate receptor. The structures of some IGF serum–binding proteins are shown in Figure 11-65. The half-lives of free IGF-I in the blood, IGF-I bound to IGFBP-1, and IGF-I bound to IGFBP-2 are all 10 minutes. When IGF-I is bound to IGFBP-3, the half-life is greater than 6 hours. Because the molecular weight of IGFBP-3 is about 40,000 and it binds to an acid-labile protein of molecular weight about 100,000, the

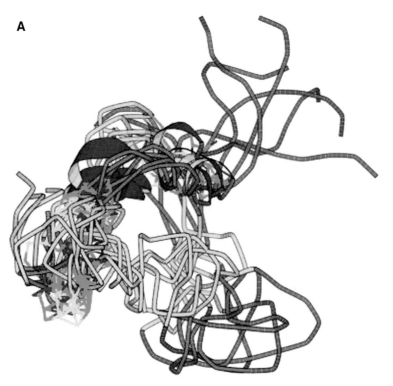

Figure 11-61. **A.** Three-dimensional structure of insulin-like growth factor-I (IGF-I) 1h-NMR and distance geometry. **B.** Human IGF-II. **C.** Mini-proinsulin-two chain analog mutant. Part A reproduced from PDBID: 1but. A. Sato, S. Nishimura, T. Ohkubo, Y. Kyogoku, S. Koyama, M. Kobayashi, T. Yasuda, Y. Kobayashi. Three-Dimensional Structure of Human Insulin-Like Growth Factor-I (Igf-I) Determined By 1h-Nmr and Distance Geometry. *Int J Pept Protein Res* **41** pp. 4340 (1993) and http://bip.weizmann.ac.il/oca-bin/ccpeek?id-1BQT.

Continued

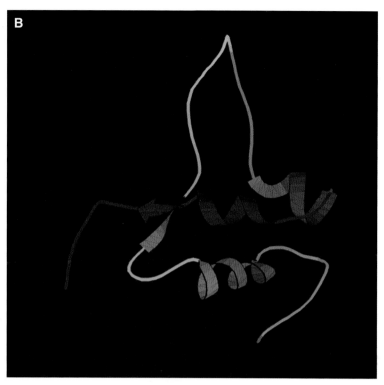

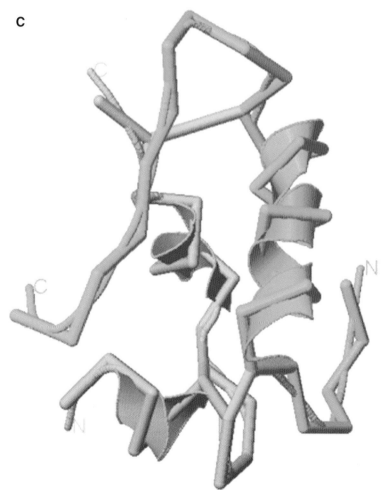

Figure 11-61, cont'd Part B reproduced from PDB ID: 1tgl. A.M. Torres, B.E. Forbes, S.E. Aplin, J.C. Wallace, G.L. Francis, R.S. Norton. Solution Structure of Human Insulin-Like Growth Factor ii. Relationship to Receptor and Binding Protein Interactions. *J. Mol. Biol.* **248** pp. 385 (1995). Part C reproduced from PDB ID:1sjt. Q.X. Hua, S.Q. Hu, W. Jia, Y.C Chu, G.T Burke, S.H. Wang, R.Y. Wang, P.G. Katsoyannis, M.A Weiss. Mini-Proinsulin and Mini-IGF-I: Homologous Protein Sequences Encoding Non-Homologous Structures. *J. Mol. Biol.* **277** pp. 103 (1998).

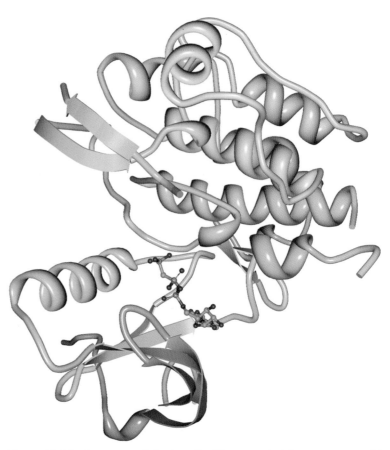

Figure 11-62. Structure of the insulin-like growth 1 factor receptor kinase. Reproduced from PDB ID: 1K3A. S. Favelyukis et al. Structure and Autoregulation of the Insulin-Like Growth Factor I Receptor Kinase. *Nat. Struct. Biol.* **8** pp. 1058 (2001).

complex has a molecular weight of about 150,000, which is large enough that it cannot be excreted as such; bound IGF-I in this complex will not escape from the vascular system until it is broken down.

The signaling pathway of IGF-I mediated by the IGF-IR is shown in Figure 11-66. The receptor can signal to the MAPK pathway through SHC (also through Ork II) and the *Ras* system for the generation of growth and differentiation. The receptor also can signal *Ras* through molecules of **insulin receptor substrates I and II (IRSI and IRSII)** and through PI3K and Akt to generate BAD, a regulator of apoptosis and leading to cell survival.

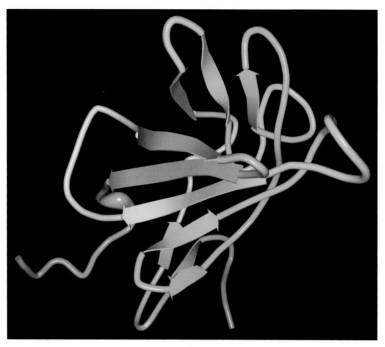

Figure 11-63. Insulin growth factor II interacting with IGF II receptor domain. Reproduced from PDB ID: 2cnj. C. Williams et al. Studies on the Interaction of Insulin Growth Factor II (IGFII) with GFIIR Domain 11. *To be published.*

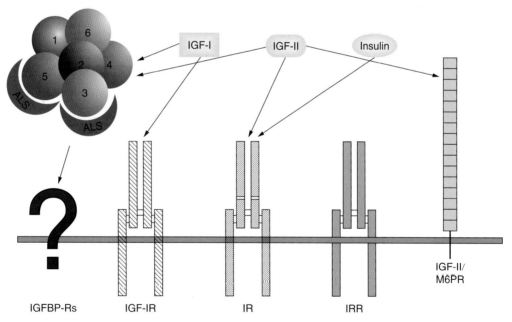

Figure 11-64. The IGF system. Shown are the ligands, cell-surface receptors, and the IGF-binding proteins that constitute the IGF system. As indicated, IGF-I interacts with the IGF-IR and the IGFBPs; IGF-II interacts with the IGF-IR, the IGF-IIR, the exon 11-lacking (A) form of the IR, and the IGFBPs; and insulin interacts with the IR. Some of the IGFBPs exert effects that are independent of their modulation of IGF signaling through the IGF-IR and the IR, and these may involve novel IGFBP receptors. Redrawn from Figure 1 of D. LeRoith and C.T. Roberts, Jr., "The insulin-like growth factor system and cancer," *Cancer Let.* 195: 127–137, 2003.

Growth Factors

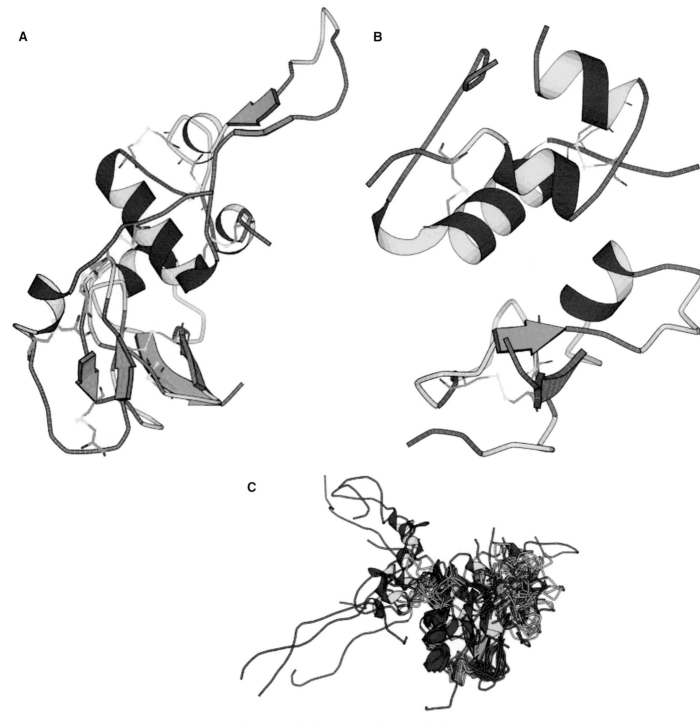

Figure 11-65. Structures of some IGF-binding proteins. **A.** IGFBP-4 with IGF-Ib. **B.** IGFBP-5 with IGF-I. **C.** C-terminal domain of IGFBP-6 with IGF-II. Part A reproduced from I. Siwanowicz, G.M. Popowicz, M. Wisniewska, R. Huber, K.P. Kuenkele, K. Lang, R.A. Engh, and T.A. Holak on http://www.imb-jena.de/cgi-bin/ImgLib/pl?CODE=1wgj. Part B reproduced from W. Zeslawski, H. G. Beisel, M. Kamionka, W. Kalus, R. A. Engh, R. Huber, and T. A. Holak on http://www.inb-jena.de/cgi-bin/ImgLib.pl?CODE-1h59. Part C reproduced from S. J. Headey, D. W. Keizer, S. Yao, G. Brasier, P. Kantharidis, L. A. Bach, and R. S. Norton on http://www.imb-jena.de/cgi-bin/Imglib.pl?CODE-1rmj.

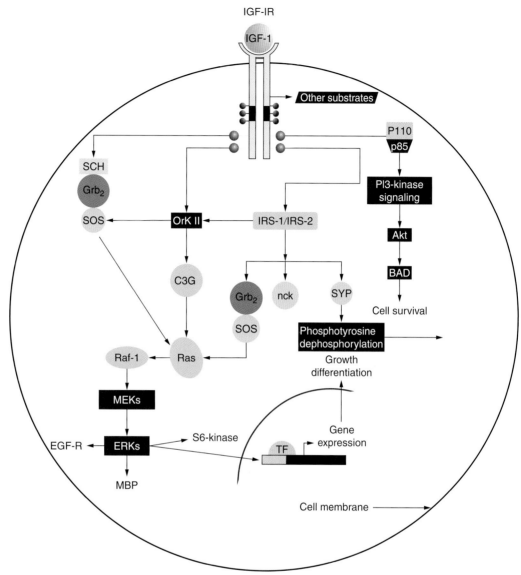

Figure 11-66. Diagram of intracellular signaling pathways mediated by the insulin-like growth factor I receptor (IGF-IR). IGF-IR, when bound to IGF-I, undergoes autophosphorylation on its tyrosine residues. This enhances its intrinsic tyrosine kinase activity and phosphorylates multiple substrates, including insulin receptor substrate I (IRS-I), IRS-2, and Src homology/collagen (SHC). IRS-1, upon phosphorylation, associates with the p85 subunit of the PI3-kinase (PI3K) and phosphorylates PI3-kinase. PI3K, upon phosphorylation, converts phosphoinositide-3 phosphate (PI-3P) into PI-3,4-P2, which in turn activates a serine-threonine kinase Akt (protein kinase B). Activated Akt kinase phosphorylates the proapoptotic factor Bad on a serine residue, resulting in its dissociation from B-cell lymphoma-X (Bcl-X_L). The released Bcl-X_L is then capable of suppressing cell death pathways that involve the activity of apoptosis protease activating factor (Apaf-1), cytochrome C, and caspases. A number of growth factors, including platelet-derived growth factor (PDGF) and IGF 1 promote cell survival. Activation of the PI3K cascade is one of the mechanisms by which growth factors mediate cell survival. Phosphorylated IRS-1 also associates with growth factor receptor bound protein 2 (Grb2), which bind son of sevenless (SOS) and activates the Ras-Raf-mitogen activated protein (ras/raf-MAP) kinase cascade. SHC also binds Grb2/SOS and activates the Ras/raf-MAP kinase cascade. Other substrates for IGF-I are phosphotyrosine phosphatases and SH_2 domain containing tyrosine phosphatase (Syp). *MBP*, myelin basic protein; *nck*, −an adaptor protein composed of SH2 and SH3 domains; *TF*, − transcription factor. Reproduced from Figure 17-10 of former site http://creserverO.nkf.med.ualberta.ca/Schreier/Volume1/chap17/ADK1_17_7-9pdf.

Interleukins

The ILs are an enlarging group of proteins under the heading of cytokines, which are secreted by lymphocytes (sometimes called lymphokines because of their cellular origin). ILs can affect cellular processes in leukocytes and other cells of hemopoietic lineage. Of the 29 or so ILs, 25 are listed in Table 11-6, which indicates their cellular sources and their primary activities. IL-1 is an important modifier of the immune response. In response to an antigen, IL-1 stimulates the activation of T cells. This leads to production of IL-2 and its receptor by the T cells. IL-1 also induces IFN-γ (TNF-α from activated macrophages also induces IFN-γ). There are two IL-1 proteins, α and β, with 26% similarity in primary sequence. IL-1s are secreted by macrophages and from neutrophils, endothelial cells, smooth muscle cells, glial cells, astrocytes, fibroblasts, keratinocytes, hepatocytes, and B and T cells when these cells have been stimulated. IL-1 also can induce proliferation of nonlymphoid cells. The cellular targets for IL-1 are hypothalamus, T cells, macrophages, and hepatocytes. IL-1 is a proinflammatory cytokine but also plays noninflammatory roles in hematopoiesis, fever, appetite control, and bone metabolism. There are three types of IL-1: IL-1α, IL-1β, and IL-1 receptor antagonist (IL-1ra). IL-1α is cytosolic (located in the soluble cytoplasm), IL-1β is cleaved to become active, and IL-1ra is an antagonist in that it can bind to the IL-1 receptor but signal transduction does not follow from the complex. Structures of IL-1α and IL-1β are shown in Figure 11-67.

There is now an expanding list of different IL-1 ligands, and there is a complicated list of receptors that bind various IL-1 ligands. Table 11-7 lists 10 IL-1-type ligands with their cognate receptors and chromosomal location of the ligand. These ligands are expressed by different cell types and are engaged in various activities. There are several receptors for IL-1s, although the main ones are IL-1 receptor I (IL-1RI) and IL-1 receptor II (IL-1RII). IL-1RI has an extracellular ligand-binding domain; IL-1RII is a decoy receptor in that it can bind ligand but does not transmit a signal (like the TRAIL decoy receptors, shown in Figure 11-4). Both IL-1RI and IL-1RII bind IL-1α and IL-1β. They are also referred to as Toll-like receptors (from the similar insect receptor), and the so-called TIR domain is the Toll-like/IL-1R domain. The IL-1 receptor is involved in the activation of NFκB and AP-1 transcription factors. The crystal structure of the TIR domain of human Toll-like receptor 1 (which binds IL-1α and IL-1β) and the IL-1RI complexed with IL-1β are shown in Figure 11-68. A list of receptors that bind IL-1s is given in Table 11-8.

IL-1 also synergizes with IL-6 and TNFα in effecting the acute phase response (increase in acute phase proteins after infection or inflammation), neutrophil mobilization, and T and B cell activation. The signal transduction pathway characteristic of the actions of IL-1 is shown in Figure 11-69. Binding of IL-1 to its receptor activates the receptor-associated protein tyrosine kinase, **Interleukin-1 receptor associated kinase (IRAK).** It then phosphorylates and activates the cytosolic kinase that activates NFκB by phosphorylating IκB, the inhibitor of NFκB. This causes IκB to dissociate from the inactive complex and to be broken down in the proteasome. NFκB translocates to the nucleus and activates responsive genes. Among the proinflammatory genes activated by NFκB are adhesion

(Text continues on p. 658.)

Table 11-6
Partial List of 25 of a Growing Number of Interleukins Indicating Their Cellular Sources and Primary Activities

Interleukins	Principal Source	Primary Activity
IL1-α and -β	Macrophages and other antigen-presenting cells (APCs)	Costimulation of APCs and T cells, inflammation and fever, acute phase response, hematopoiesis
IL-2	Activated TH_1 cells, NK cells	Proliferation of B cells and activated T cells, NK functions
IL-3	Activated T cells	Growth of hematopoietic progenitor cells
IL-4	TH_2 and mast cells	B cell proliferation, eosinophil and mast cell growth and function, IgE and class II MHC expression on B cells, inhibition of monokine production
IL-5	TH_2 and mast cells	Eosinophil growth and function
IL-6	Activated TH_2 cells, APCs, other somatic cells	Acute phase response, B cell proliferation, thrombopoiesis, synergistic with IL-1 and TNF on T cells
IL-7	Thymic and marrow stromal cells	T and B lymphopoiesis
IL-8	Macrophages, other somatic cells	Chemoattractant for neutrophils and T cells
IL-9	T cells	Hematopoietic and thymopoietic effects
IL-10	Activated TH_2 cells, $CD8^+$ T and B cells, macrophages	Inhibits cytokine production, promotes B cell proliferation and antibody production, suppresses cellular immunity, mast cell growth
IL-11	Stromal cells	Synergistic hematopoietic and thrombopoietic effects
IL-12	B cells, macrophages	Proliferation of NK cells, IFN-γ production, promotes cell-mediated immune functions
IL-13	TH_2 cells	IL-4-like activities
IL-15	Endothelial cells and monocytes	Effects are similar to IL-2
IL-16	CD8 T cells	Chemo attracts CD4 T cells
IL-17	Activated memory T cells	Promotes T cell proliferation
IL-18	Macrophages	Induces IFN-γ production
IL-19	Belongs to IL-10 family	Produces IL-6 and TNFα; TNFα leads to apoptosis
IL-20	Belongs to IL-10 family	Activates STAT pathway; epidermal function
IL-21	Activated $CD4^+$ T cells	Affects NK and T cell responses; proliferation of activated T cells; maturation of NK cells; with IL-15 or IL-18, enhances INF-γ production in NK and T cells
IL-22	Belongs to IL-10 family	
IL-23		Critical cytokine for autoimmune inflammation of the brain; recruits inflammatory cells
IL-25	Related to IL-17 family; produced by mast cells and helper T cells	Induces production of IL-4, IL-5, and IL-13
IL-26	Member of IL-10 family; transformed T cells	Activates STAT 3
IL-27	Spleen cells	Proliferation of main $CD4^+$ T cells; enhances IFN-γ production by activated T cells and NK cells; induces STAT 1 and STAT 3 phosphorylation

Reproduced from http://web.indstate.edu/thcme/mwking/growth-factors.html.

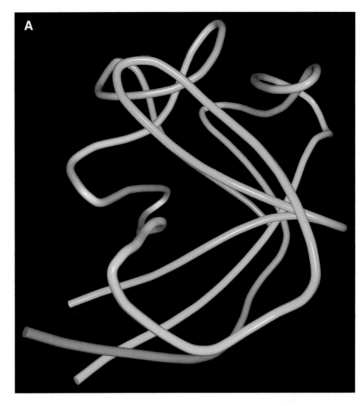

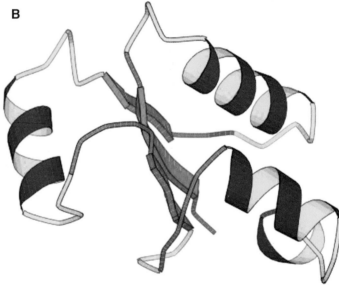

Figure 11-67. Structures of IL-1α *(A)* and IL-1β *(B)*. Part A reproduced from http://www.rcsb.org/pdb/explore/images.do?structureId=2ILA. Part B reproduced from B.C. Finzel, K.D. Watenpaugh, and H.M. Einspahr, http://www.rcsb.org/pdb/explore.do?structureid=1i1b.

Table 11-7
Family of IL-1-Type Ligands, Their Chromosomal Locus, and Their Cognate Receptor

Systematic Name	Alternative Name	Receptor	Chromosome
IL-1F1	IL-1α	IL-1 R1, IL-1 R2	2
IL-1F2	IL-1β	IL-1 R1, IL-1 R2	2
IL-1F3	IL-1ra	IL-1 R1, IL-1 R2	2
IL-1F4	IL-18/IGIF/IL-1γ	IL-1 R5	11
IL-1F5	IL-1Hy1/FIL1d/IL-1H3/IL-1RP3/IL-1L1/IL-1d	IL-1 R6	2
IL-1F6	FIL1e	?	2
IL-1F7	FIL1Z/IL-1H4/IL-1RP1/IL-1H	IL-1 R5	2
IL-1F8	FIL1h/IL-1H2	?	2
IL-1F9	IL-1H1/IL-1RP2/IL-1e	IL-1 R6	2
IL-1F10	IL-1Hy2/FKSG75	IL-1 R1	2

Reproduced from former site http://www.rndsystems.com/asp/g_sitebuilder.asp?bodyid=469.

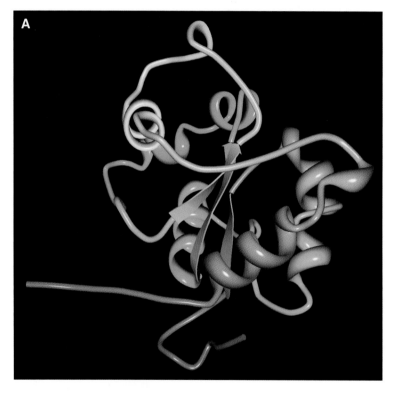

Figure 11-68. **A.** Crystal structure of the TIR domain of human Toll-like receptor 1. **B.** IL-1RI complexed with IL=1β. Part A reproduced from Y. Xu, X. Tao, B. Shen, T. Horng, R. Medzhitov, J.L. Manley, and L. Tong, http://www.imb-jena.de/cgi-bin/ImgLib.pl?CODE=1fyv. Part B reproduced from G.P. Vigers et al., "Crystal Structure of the type-1 interleukin-1 receptor complexed with interleukin-1beta," *Nature*, 386: 190–194, 1997, http://www.ebi.ac.uk/thornton-srv/databases/cgi-bin/pdbsum/GetPage.plCODE=1itb.

Table 11-8
Receptors for Various IL-1 Ligands, Their Chromosomal Location, and the Ligands They Bind

Systematic Name	Alternative Name	Ligands	Chromosome
IL-1 R1	IL-1RI	IL-1α, IL-1β, IL-1ra	2
IL-1 R2	IL-1RII	IL-1α, IL-1β, IL-1ra	2
IL-1 R3	IL-1RAcP	Signaling component	3
IL-1 R4	ST-2/T1/DER4/Fit-1	?	2
IL-1 R5	IL-18 Ra/IL-1Rrp	IL-1F8, IL-IF7	2
IL-1 R6	IL-1 Rrp2/IL-1RL2	IL-1F9 and F5	2
IL-1 R7	IL-18 Rb/AcPL	Signaling component	2
IL-1 R8	TIGIRR-2/IL-1RAPL	?	X
IL-1 R9	TIGIRR-1/IL-1RAPL2	?	X

Reproduced from http://www.rndsystems.com/asp/g_sitebuilder.asp?bodyid=469.

molecules, cytokines and chemokines, matrix metalloproteinases, and tissue factor (Figure 11-70). The induction of proinflammatory genes by NFκB is inhibited by **glucocorticoids** (e.g., cortisol) through its receptor, accounting, in part, for their anti-inflammatory effects. In cells in which NFκB operates with the transcription factor ERK, glucocorticoids can induce MKP-1, the endogenous inhibitor of ERK, as a means to inhibit the action of NFκB. In cells that respond to glucocorticoids with an antiproliferative effect or death, the glucocorticoid receptor can induce IκBα, which binds to NFκB to form an inactive complex.

IL-2 has a complex receptor, and IL-2 is required for T cell activation, proliferation, and survival. The IL-2 receptor is made up of three separate polypeptide chains named α, β, and γ. These chains are represented as cartoons in the left inset of Figure 11-71. IL-2Rγ is also a component of receptors: IL-4R, IL-7R, and IL-15R. IL-2Rβ is required for IL-2 signaling and is a component of IL-15R. *The polypeptide chains of the IL-2R do not generate signaling by themselves but interact with other signaling proteins in the cytoplasm.* The IL-2 signaling pathway operates through the multiple pathways to generate cell proliferation, among which are the activation of MAPK signaling and the inhibition of apoptosis. The details are given in the legend to Figure 11-71. IL-2 also can generate cellular activation and proliferation of B cells and T cells by signaling through the JAK/STAT pathway, like so many of the other cytokines, as shown in Figure 11-72. The IL-2 complexed to its receptor is shown in Figure 11-73.

IL-3 also promotes cell proliferation and differentiation by activation of the JAK/STAT and MAPK pathways (Figure 11-51). IL-3 has a heterodimeric receptor with a ligand-specific α-chain and a β-chain shared with IL-5 and GM-CSF.

IL-4 is another IL that operates through the JAK/STAT and MAPK pathways. The structures of IL-4 and the complex with its receptor are shown in Figure 11-74. The receptor is made up of two chains: the IL-4Rα and the IL-2Rγ chain. IL-4 also operates through the JAK/STAT and MAPK pathways to affect the development of Th2 cells (a subset of helper-inducer T lymphocytes that synthesize and secrete IL-4, IL-5, IL-6,

(Text continues on p. 664.)

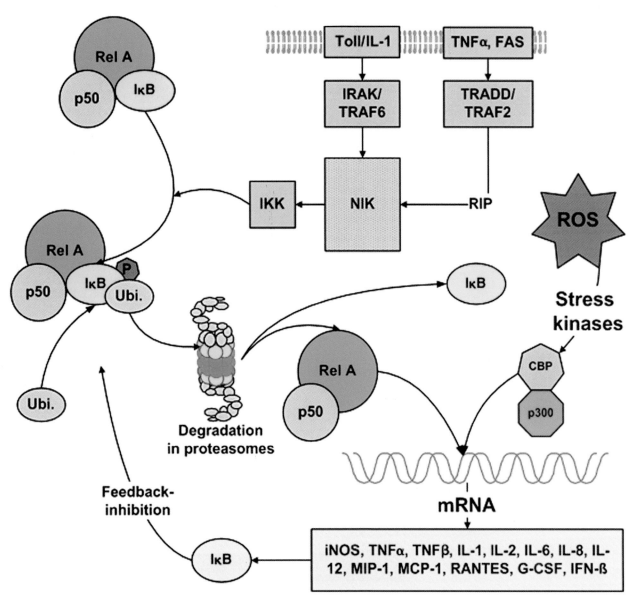

Figure 11-69. Signal transduction pathway of IL-1. Ligand binding to different cell membrane receptors (Toll-like/IL-1, TNF) activates intracellular signal transduction molecules such as the receptor associated protein kinases *Interleukin 1 receptor-associated kinase* (IRAK) and *TNF receptor–associated factor* (TRAF) and then the cytosolic protein kinase *NFκB-inducing kinase* (NIK). This kinase phophorlylates *1κB kinase* (IKK), which in turn activates IκB. Phosphorylated IκB binds to ubiquitin, which transfers IκB to cellular proteasomes for degradation. Following degradation of IκB within the proteasome, NFκB is translocated to the nucleus and promotes the transcription of genes with appropriate responsive elements. Reactive oxygen species also activate stress-activated-protein kinases, which, in turn, activate auxiliary proteins such as CBP and p300. These enhance the NFκB binding to its responsive elements. Reproduced from Figure 11 of http://www.egms.de/figures/journals/cto/2004-3/cto000002.f11.png

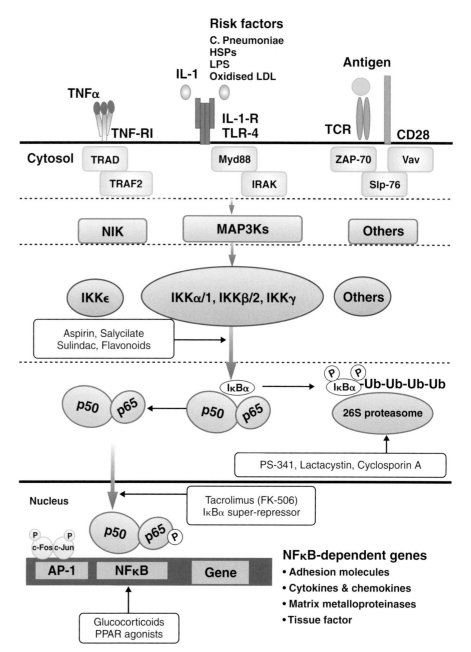

Figure 11-70. NFκB signaling in atherosclerotic lesions—effect of available NFκB-blocking agents. The NFκB pathway is a convergence point for many proatherogenic stimuli. In the model illustrated, oxidized LDL, LPS, and HSPs from infectious agents, as well as cytokines and activated T cells, interact with different receptors at the cell surface. These, in turn, activate intracellular signaling proteins via a set of adaptor proteins and kinases such as NIK, initiating a cascade of phosphorylations. Central to the NFκB cascade is the large multi-subunit IKK complex, which is the point of convergence of multiple signals. Genes regulated by NFκB include adhesion molecules, chemokine, MMP, and cytokines, which play pivotal roles in atherosclerosis. The diagram also shows the NFκB inhibitors currently available. Other agents that might inhibit NFκB, such as statins, have not been included in this diagram as their mechanisms of action on the pathway have not been clarified. Redrawn from Figure 1 of C. Monaco and E. Paleolog, "Nuclear factor κ B: A potential therapeutic target in atherosclerosis and thrombosis," *Cardiovas. Res.*, 61: 671–682, 2004.

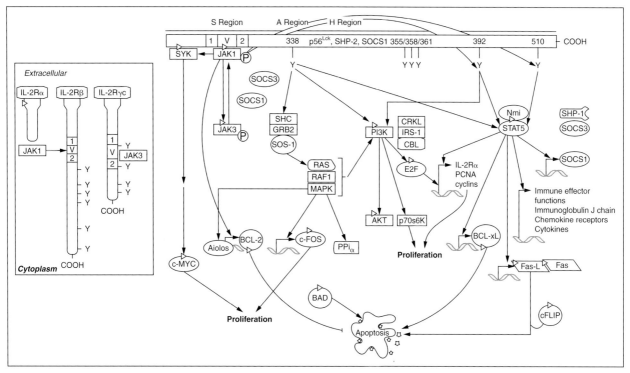

Figure 11-71. IL-2 signaling and the IL-2 receptor components *(left inset)*. The IL-2 receptor is a key component of immune signaling and is required for the activation, proliferation, and survival of T cells. This receptor is composed of three polypeptide chains: the alpha, beta, and gamma chains. The IL-2 receptor gamma chain is a common component for several other cytokine receptors, including IL-4, IL-7, IL-9, and IL-15. The IL-2 receptor beta chain is essential for IL-2 signaling and is also a component of the IL-15 receptor complex. The polypeptides of the IL-2 receptor do not themselves have intrinsic catalytic activity, but interact with cytoplasmic signaling proteins to transduce signals. Different regions of the cytoplasmic domain of the IL-2 receptor beta chain interact and couple with distinct signaling pathways and cellular responses. JAK1 associates with the beta chain, and JAK3 with the gamma chain. Binding of IL-2 induces heterodimerization of receptor subunits, and activation of JAK kinase activity. Tyrosine residues in the beta chain cytoplasmic domain are phosphorylated during activation, recruiting other factors to the phosphorylated tyrosine residues through src homology 2 (SH2) domains. The adaptor protein Shc binds to phosphorylated tyrosine 338 of the beta chain. When bound, Shc is phosphorylated and couples through Grb2 and SOS-1 to activate Ras and stimulate T cell proliferation. Another key proliferative pathway activated by IL-2 is phosphorylation of STAT-5 by JAK kinases. STAT-5 is recruited to IL-2 beta phosphorylated tyrosines at multiple positions, including Y338, Y392, and Y510. Once phosphorylated, STAT-5 enters the nucleus to regulate the transcription of several genes, some proliferative such as cyclin genes and others that are involved in T cell immune function such as cytokine genes. The suppressors of cytokine activation, SOCS-3 and SOCS-1, oppose phosphorylation and activation of STAT-5 and JAK1 caused by IL-2. PI3 kinase is another protein recruited to IL-2 receptor beta chain tyrosines when phosphorylated. Activation of PI3 kinase also contributes to the proliferative activity of IL-2 in T cells. The role of other tyrosines in the IL-2 receptor beta chain—Y355, Y358, and Y361—is not yet clear, but may be involved in signaling by the protein kinase p56lck. In addition to stimulating T cell activation and proliferation, IL-2 activation blocks T cell apoptosis through multiple pathways. Among the genes activated by STAT-5 are BCL-xL, an inhibitor of apoptosis, and Fas-ligand, an activator of apoptosis in cells expressing the Fas receptor. PI3 kinase also contributes to anti-apoptotic activity of IL-2 through AKT activation. T cell responses to IL-2 must be coordinated in part in the complex protein-protein interactions with the IL-2 receptor beta chain. Redrawn from http://www.biocarta.com/pathfiles/h_il2Pathway.asp.

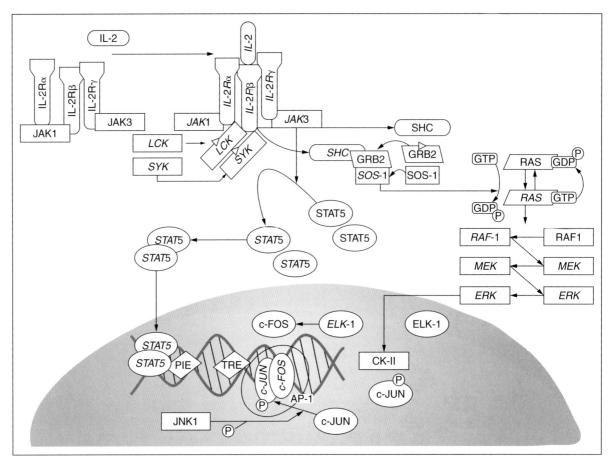

Figure 11-72. Interleukin 2 (IL-2) is a potent cytokine that can lead to cellular activation and proliferation. IL-2 receptors are found on activated B-Cells, LPS-treated monocytes, and many T cells. The receptor is formed from three chains: alpha (CD25), beta (CD122), and gamma (CD132). Primary signaling is through the JAK/Stat pathway and MAPKs. Reproduced from http://www.biocarta.com/pathfiles/h_il2Pathway.asp.

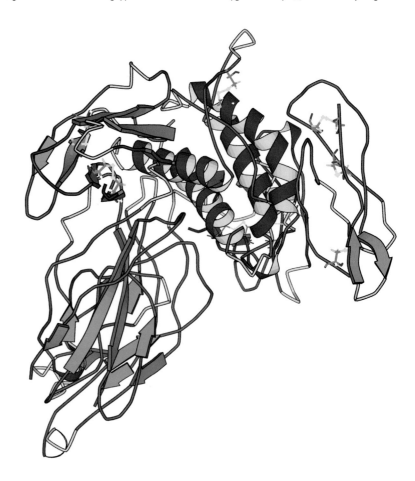

Figure 11-73. IL-2–IL-2R complex. Reproduced from http://www.imb-jena.de.

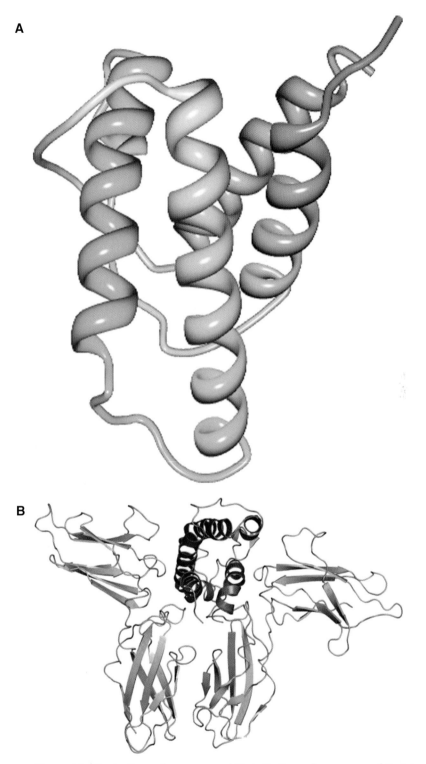

Figure 11-74. **A.** Crystal structure of IL-4. **B.** Crystal structure of IL-4 in complex with IL-4 receptor. Part A reproduced from PDB ID: 1ITM. C. Redfield, L.J. Smith, J. Boyd, G.M. Lawrence, R.G. Edwards, C.J. Gershater, R.A. Smith, C.M. Dobson. Analysis of the Solution Structure of Human Interleukin-4 Determined by Heteronuclear Three-Dimensional Nuclear Magnetic Resonance Techniques. *J. Mol. Chem.* **238** pp. 23 (1994). Part B is reproduced from http://www.imb-jena.de/cgi-bin/ImgLib.pl?CODE=1III.

and IL-10, which influence B cell development, antibody production, and augment humoral responses), as shown in Figure 11-75.

IL-5 stimulates proliferation, maturation, and activation of **eosinophils** (cytotoxic white blood cells, granulocytes, that increase in allergy, inflammation, and infection) as part of the inflammatory response shown in Figure 11-76.

IL-6 has a broad range of activities. It is involved in inflammation, hematopoiesis, neuronal differentiation, and bone loss. Its receptor comprises an α-subunit that contains the IL-6 binding site and a gp130 subunit that is also shared with other cytokines in the IL-6 family of receptors. IL-6 signals through the JAK/STAT pathway and the *Ras* pathway to activate MAPK. The crystal structure of IL-6 is shown in Figure 11-77. A structure showing the components of the receptor complex: IL-6, IL-6R, and gp130 is shown in Figure 11-78A, and the proposed dimerized receptor structure is shown in Figure 11-78B. The signaling pathways of IL-6 are shown in Figure 11-79.

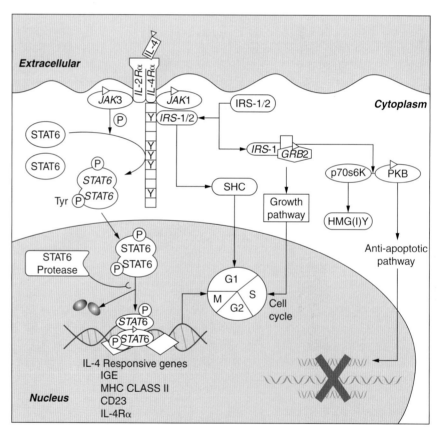

Figure 11-75. Interleukin 4 (IL-4) is a cytokine that can lead to the development of Th2 cells. The 140-kDa IL-4 receptor (CD124) is found on many cell types, even those of nonhematopoietic origin. The receptor is formed from two chains: IL-4R (alpha) and the IL-2R gamma chain (CD132). Primary signaling is through the JAK/Stat6 pathway and MAPKs. Redrawn from http://www.biocarta.com/pathfiles/h_il4Pathway.asp.

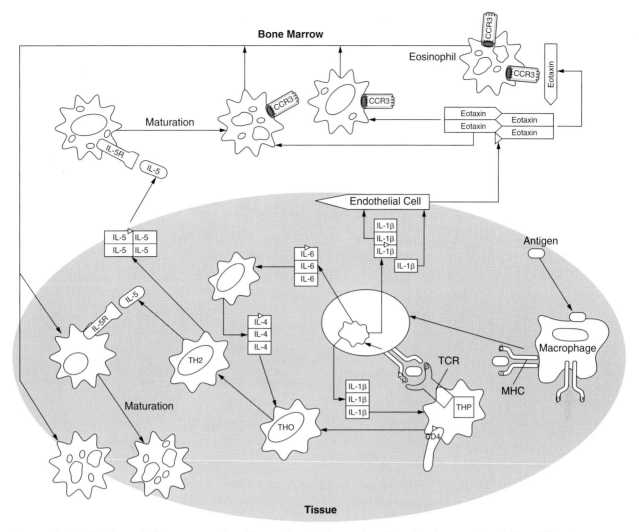

Figure 11-76. IL-5 is an inflammatory signaling molecule that primarily stimulates eosinophil proliferation, maturation, and activation. Eosinophils are leukocytes involved in inflammatory responses that defend against parasites and cause some aspects of asthma, allergic reactions, and perhaps autoimmune disorders. The action of IL-5 begins with an immune response in tissues, such as activation of macrophages and T cells that secrete IL-1, IL-4, and IL-6. The immune response can lead to IL-5 secretion by T cells, eosinophils, and mast cells. Secreted IL-5 stimulates production and maturation of eosinophils in bone marrow that migrates to tissues in response to eotaxin and release factors that damage tissues, causing some of the undesirable consequences of inflammation. The receptor for IL-5 is a heterodimer of an alpha subunit that is required for IL-5 selective binding and a beta subunit that is also part of the IL-3 and GM-CSF receptors. Binding of IL-5 to the IL-5 receptor at the cell surface activates JAK/STAT signaling pathways that regulate transcription, proliferation, and differentiation. Redrawn from http://www.biocarta.com/pathfiles/h_il5Pathway.asp.

IL-7 is essential for the normal development of B cells and T cells. Like IL-6, the IL-7 receptor is a heterodimer of a γ-chain (which is also part of receptors for IL-2, IL-4, IL-9, IL-15, and IL-21) and an α-chain (which is unique for the IL-7 receptor and contains the IL-7 binding domain). Ligand binding to the α-domain results in dimerization with the γ-subunit. Signaling from this receptor complex leads to proliferation, cell survival, and B and T cell development. Familiar pathways are used. A theoretical

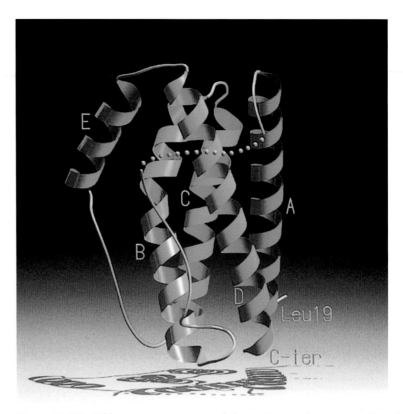

Figure 11-77. Ribbon representation of the IL-6 crystal structure. The four main helices are labeled *A, B, C,* and *D*. The extra helix in the final long loop is labeled *E*. The missing part of the first cross over connection is indicated by a *dashed line*. Reproduced from Figure 1 of W. Somers, M. Stahl, and J.S. Seehra, "A crystal structure of interleukin 6: implications for a novel mode of receptor dimerization and signaling," *EMBO J.* 16: 198–997, 1997, and from http://www.nature.com/cgi-taf/DynaPage.taf?file=emboj/Journal/v16/n5/full/7590092a.html.

model of the structure of IL-7 is shown in Figure 11-80. The signaling pathways for IL-7 are shown in Figure 11-81.

IL-8 binds to glycosaminoglycans, and it is likely that it binds to heparan sulfate on the endothelial cell surface to retain the chemokine for presentation to leukocytes. The dimeric form of IL-8 is the form that binds to heparan sulfate because the monomeric form binds weakly. Each binding domain is an N-sulfated group of six monosaccharide units contained within an about 23-mer sequence, and these sites are separated by 14 acetylated monosaccharide residues. The heparan sulfate binding region looks like a horseshoe over two antiparallel helical regions of the IL-8 dimer, as shown in Figure 11-82. Also shown in this figure is the structure of IL-8.

IL-10 is an anti-inflammatory cytokine affecting activated macrophages. IL-10 represses the expression, by activated macrophages, of TNF-α, IL-6, and IL-1. Although IL-10 uses the JAK/STAT pathway, the anti-inflammatory action of this cytokine involves the induction of heme oxygenase-1 (HO-1) by way of the MAPK pathway with iron and

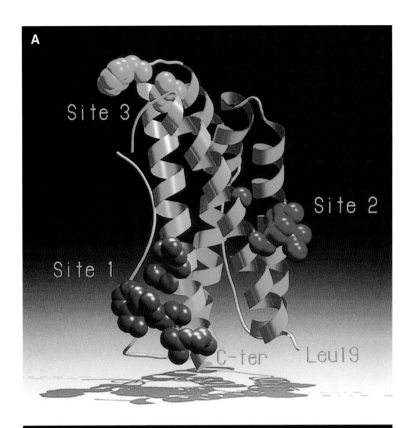

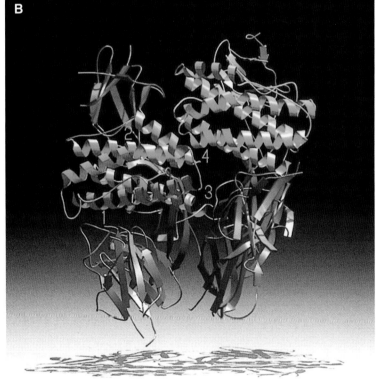

Figure 11-78. **A.** A ribbon representation of the IL-6, IL-6R, and gp130 hexamer signaling model. The IL-6 crystal structure is shown in *green*, IL-6R in *blue*, and gp130 in *red*. The proposed binding sites are labeled. Site 1 is the site of IL-6–IL-6R Interactions. Site 2 is the region where IL-6 interacts with gp130 in the trimer. Site 3 is the site of IL6–gp130 interactions between trimers. Site 4 is the location of IL-6–IL-6 interactions between trimers. **B.** A ribbon representation of IL-6 with space-filling atoms of exposed side chains found to alter IL-6R binding (site 1) or gp130 binding (sites 2 or 3) when mutated. Parts A and B reproduced from Figure 4 of W. Somers, M. Stahl, and J.S. Seehra, "A crystal structure of interleukin 6: implications for a novel mode of receptor dimerization and signaling," *EMBO J.* 16: 198–997, 1997.

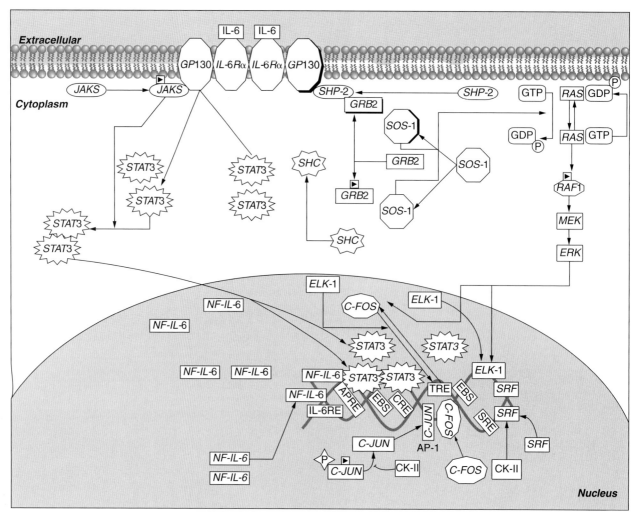

Figure 11-79. Interleukin-6 (IL-6) is a cytokine that provokes a broad range of cellular and physiological responses. In addition to playing a role in inflammation and hematopoiesis, IL-6 is involved in other processes such as neuronal differentiation and bone loss. To produce these effects IL-6 signals through a receptor composed of two different subunits, an alpha subunit that produces ligand specificity and gp130, a receptor subunit shared in common with other cytokines in the IL-6 family. Binding of IL-6 to its receptor initiates cellular events including activation of JAK kinases and activation of ras-mediated signaling. Activated JAK kinases phosphorylate and activate STAT transcription factors, particularly STAT3, that move into the nucleus to activate transcription of genes containing STAT3 response elements. The ras-mediated pathway, acting through Shc, Grb-2, and SOS-1 upstream and activating MAP kinases downstream, activates transcription factors such as ELK-1 and NF-IL-6 (C/EBP-beta) that can act through their own cognate response elements in the genome. These factors and other transcription factors like AP-1 and SRF (serum response factor) that respond to many different signaling pathways come together to regulate a variety of complex promoters and enhancers that respond to IL-6 and other signaling factors. Redrawn from http://www.biocarta.com/pathfiles/h_il6Pathway.asp

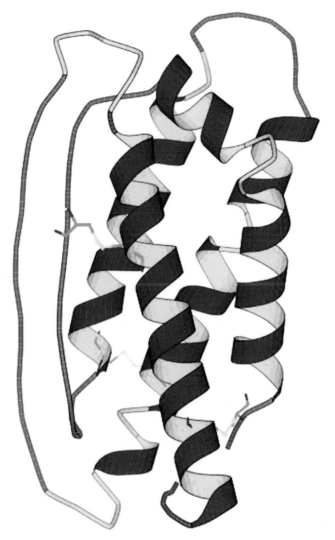

Figure 11-80. A theoretical model of IL-7. Reproduced from http://www.imb-jena.de/cgi-bin/imgLib.pl?CODE=1il7 and http://www.fli_leibniz.de/cgi-bin/imglib.pl?CODE=1il7.

the p38 kinases. HO-1 is involved in the biosynthesis of heme. The structure of IL-10 and the structure of IL-10 complexed to the IL-10 receptor (sIL-10R1) are shown in Figure 11-83. Signaling pathways of IL-10 are shown in Figure 11-84.

IL-12 induces Th1 cell differentiation (Th1 cell is a helper T cell that produces IFN-γ and IL-12) and activates T cells and NK cells. IL-12 operates through the JAK/STAT pathway to induce the ERM gene (ERM is a member of the *Ets* family and codes for an *ets*-related protein, referred to as ETV5, *ets* variant gene 5; it also codes for erythromycin-resistant methylases). ERM and cJUN activate genes for IFN-γ, IL-18 receptor, and

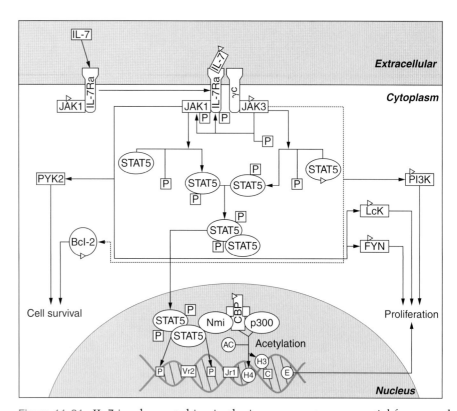

Figure 11-81. IL-7 is a key cytokine in the immune system, essential for normal development of B cells and T cells. Mice with the IL-7 receptor deleted lack B and T cells. Some humans with SCID (severe combined immunodeficiency disease) also have a mutation of their IL-7 receptor gene leading to an absence of T cells and greatly impaired B cell production. The IL-7 receptor includes two polypeptides, a gamma chain and an alpha chain. The alpha-chain is unique to the IL-7 receptor, whereas several other cytokines use the same gamma receptor chain as IL-7, including IL-2, IL-4, IL-9, IL-15, and IL-21. Binding of IL-7 to the alpha chain leads to dimerization of the alpha and gamma chains. JAK3 associated with the gamma chain tyrosine phosphorylates the alpha chain after dimerization. The importance of JAK3 in IL-7 signaling is supported by the similarity of the immune defects in JAK3 knockout mice and IL-7 knockout mice. The phosphorylated alpha chain serves as the site for recruiting other signaling molecules to the complex to be phosphorylated and activated, including STAT5, src kinases, PI3 kinase, Pyk2, and Bcl2 proteins. Some targets of IL-7 signaling contribute to cellular survival, including Bcl2 and Pyk2. Other targets contribute to cellular proliferation, including PI3 kinase, src family kinases (LcK and FYN), and STAT5. The transcription factor STAT5 contributes to activation of multiple different downstream genes in B and T cells and may contribute to VDJ recombination through alteration of chromatin structure. The cell survival and cell proliferation signals induced by IL-7 combine to induce normal B and T cell development. Redrawn from http://www.biocarta.com/pathfiles/h_il7Pathway.asp.

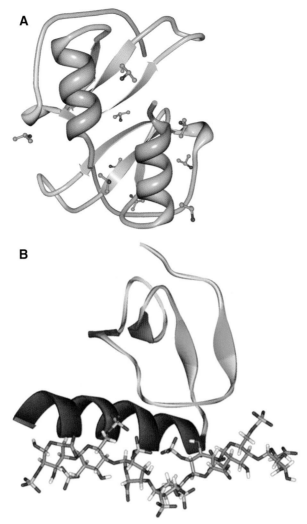

Figure 11-82. **A.** Structure of IL-8 dimer. **B.** Model of heparin/heparin sulfate binding to the IL-8 dimer. Two identical binding motifs within N-sulfated stretches of HS are shown to interact with the α-helical heparin/HS-binding domains of each IL-8 monomer. The N-sulfated regions (≤6 sugar units; *closed circles*) are bridged by a sequence (≤12–14 monosaccharide units; *open circles*) that may be either N-acetylated or N-sulfated. Note the polarity of the sugar chain *(arrow)*. Part A reproduced from PDB ID: 1m8a. D.M. Hoover et al., The Structure of Human Macrophage Inflammatory Protein-3 Alpha/CCL20 Linking Antimicrobial and CC Chemokine Receptor-6-Binding Activities with Human Beta-Defensins. *J. Biol. Chem.* **277** pp. 37647 (2002). Part B reproduced with permission from H. Potzinger et al., "Developing Chemokine mutants with improved proteoglycan affinity and knocked-out GPCR activity as anti-inflammatory recombinant drugs," *Biochem. Soc. Trans.*, 34: 435–437, 2006 and directly from http://www.biochemsoctrans.org/bst/034/0435/bst0340435f02.htm.

CCR5, a chemokine receptor. The structure of human dimeric IL-12 is shown in Figure 11-85. The signaling pathways employed by IL-12 are shown in Figure 11-86.

IL-13 is a recently discovered anti-inflammatory cytokine produced by activated Th0, Th2, and CD8 T cells. It promotes the growth of human B cells and directs naïve B cells to switch to the synthesis of IgE and IgG4 antibodies and induces expression of CD23 and class II **major**

(Text continues on p. 675.)

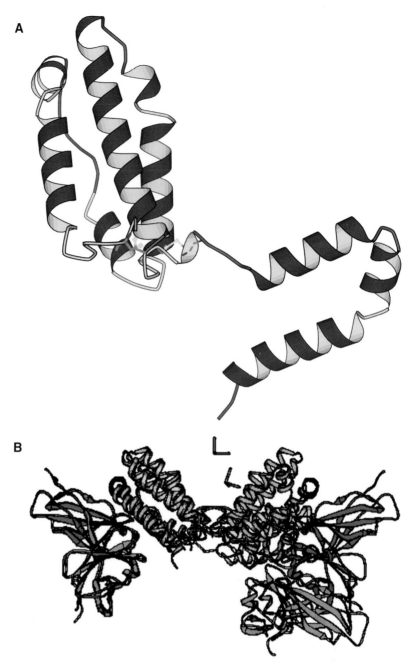

Figure 11-83. **A.** Structure of IL-10. **B.** Superposition of cmvIL-10 and HIL-10 1:2 receptor complexes. *(A)* Superposition of sIL-10R1 bound to cmvIL-10 *(green)* and hIL-10 *(magenta)*. CmvIL-10 and hIL-10 are colored *yellow* and *blue*, respectively. Axes, in the same color as their respective cytokines, are shown representing the location of the two-fold axes of cmvIL-10 and hIL-10 dimers. Part B reproduced from Figure 2 of B.C. Jones et al., "Crystal structure of human cytomegalovirus IL-10 bound to soluble human IL-10R1," *Immunology*, 99: 9404–9409, 2002 and from http://www.pnas.org/cgi/content-nw/full/99/14/9404/F2.

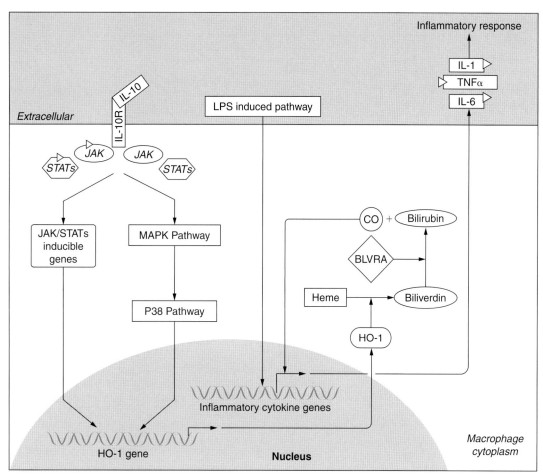

Figure 11-84. Signaling pathways of IL-10. IL-10 is a cytokine with potent anti-inflammatory properties, repressing the expression of inflammatory cytokines such as TNF-alpha, IL-6, and IL-1 by activated macrophages. The IL-10 receptor is in the JAK/STAT class of receptors but activation of the JAK/STAT pathways by IL-10 does not appear on its own to be responsible for the anti-inflammatory properties of this cytokine. The anti-inflammatory actions of IL-10 appear to require induction of the enzyme heme oxygenase-1 (HO-1) through a MAP kinase pathway involving the p38 kinases. HO-1 is involved in the biosynthesis of heme, and catalyzes a reaction producing carbon monoxide, free iron, and the heme precursor biliverdin. HO-1 is induced by IL-10 and is also induced by oxidative stress. Blocking HO-1 with inhibitors or antisense blocks the anti-inflammatory actions of IL-10. The anti-inflammatory actions of HO-1 appear to be the result of signaling by carbon monoxide it produces because removal of CO blocks the anti-inflammatory action of IL-10 and HO-1. The anti-inflammatory actions of IL-10 may be therapeutically useful either directly or through modulation of HO-1 activity. Redrawn from http://www.biocarta.com/pathfiles/h_il10Pathway.asp

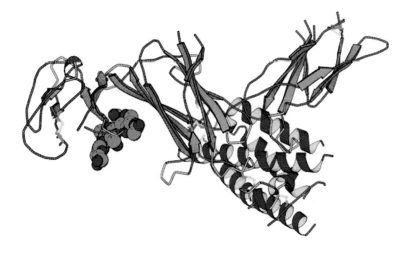

Figure 11-85. Human IL-12. Reproduced from PDB ID: 1f45. C. Yoon, S.C. Johnston, J. Tang, M. Stahl, J. F. Tobin, W. S. Somers. Charged Residues Dominate A Unique Interlocking Topography in the Heterodimeric Cytokine Interleukin-12. *EMBO J* **19** pp. 3530 (2000) Reproduced from http://imb-jena.de/cgi-bin/imgLib..pl?CODE=1f45.

Interleukins

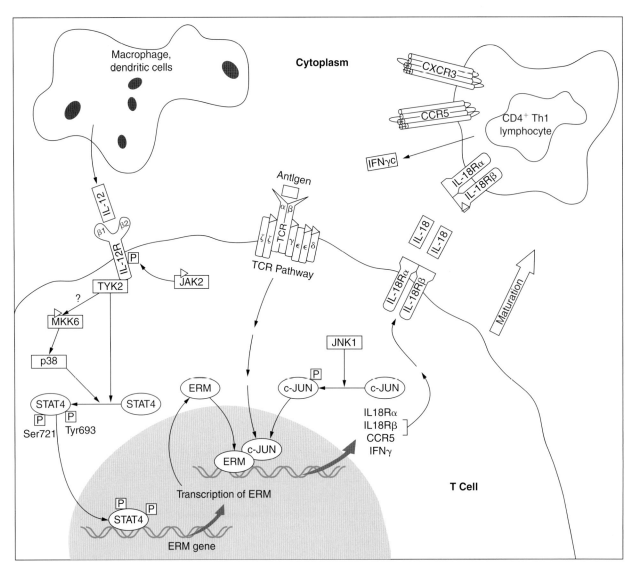

Figure 11-86. Signaling pathway of IL-12. Interleukin-12 (IL-12) promotes cell-mediated immunity by inducing Th1 cell differentiation and activation of both T cells and NK cells. Dendritic cells and macrophages in peripheral tissues act as antigen-presenting cells and secrete IL-12 as one component of the antigen response, Th1 differentiation. The role of IL-12 in cellular immunity is largely mediated by the STAT-4 transcription factor. STAT-4 is essential for IL-12 activity and the phenotype of mice lacking STAT-4 is very similar to the phenotype of mice lacking the IL-12 receptor or IL-12. The role of IL-12 in Th1 differentiation may not be to induce the Th1 cell fate, but to stimulate growth of cells determined for the Th1 cell fate by the T-bet transcription factor. Several signaling pathways contribute to IL-12 activation of STAT-4 to regulate cell-mediated immune responses. The JAK kinases such as JAK2 and TYK2 interact with the activated IL-12 receptor and tyrosine phosphorylate the IL-12 receptor and STAT-4. IL-12 also activates a MAP kinase pathway activating the MAP kinase kinase MKK6 and p38. Phosphorylation of STAT-4 on serine 721 by p38 contributes to the full transcriptional activation of genes by STAT-4. Some of the events downstream of IL-12 appear to include genes activated indirectly by STAT-4, such as genes activated by the transcription factor ERM. ERM is in the Ets family of transcription factors, is activated by IL-12, and activates IL-12-inducible genes such as interferon-gamma that are not activated by STAT-4 itself. Interferon-gamma transcription in T cells is also activated by other signals such as from the T cell receptor. Other proteins activated transcriptionally downstream of IL-12 and STAT-4 include the chemokine receptor CCR5 and IL-18 and its receptor. Some viruses, including HIV, repress cell-mediated immunity by blocking IL-12 signaling. Redrawn from http://www.biocarta.com/pathfiles/h_IL12Pathway.asp.

histocompatibility complex (MHC). Both class I and class II MHC proteins are composed of two polypeptide chains. There are three polypeptide folds in class I molecules: α-1, α-2, and α-3. α-3 associates with a 12-kDa β2 microglobulin. Class I molecules are found on nearly every cell in the human body, whereas class II molecules are found only on the surface of B cells. Class II molecules are also made up of two polypeptide chains, α and β. Class I molecules present antigen to cytotoxic T cells, and class II molecules present antigen to helper T cells. Class I proteins bind endogenous antigens, and class II proteins bind exogenous antigens. An endogenous antigen might be a fragment of a tumor protein, and an exogenous antigen might be a fragment of a bacterium or virus. Figure 11-87 presents views of class I and class II molecules of the MHC.

Among other activities, IL-13 inhibits the synthesis of proinflammatory cytokines, such as TNFα, IL-1β, IL-6, and IL-8. It also inhibits the synthesis of IL-12, NOS, and COX2. IL-13 precursor is comprised of 132 amino acid residues, including a signal peptide that is 20 amino acids long. IL-13 binds to a heterodimeric receptor of IL-13Rα and IL-4Rβ. As would be expected, IL-13 and IL-4 have similar activities. The structure of IL-13 and the heterodimeric receptor of IL-13 (IL-13Rα1 with IL-4Rα) are shown in Figure 11-88.

IL-16 is also referred to as the **lymphocyte chemoattractant factor (LCF).** It comprises 130 amino acid residues and has a molecular weight

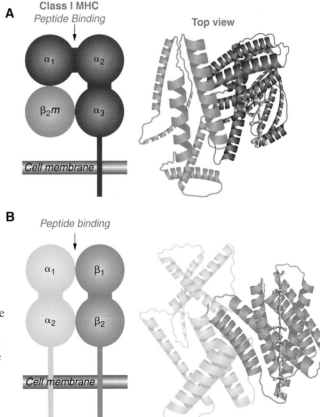

Figure 11-87. **A.** Structure of a class I MHC protein *(left)* and *(right)* crystal structure *(top view)* MHC encoded polypeptide is shown blue; β2 macroglobulin is in *green* and the peptide antigen is in *red (right figure)*. **B.** Structure of class II MHC protein. Left shows two polypeptide chains folding into two separate domains, α1 and α2 for the α-polypeptide and β1 and β2 for the β-polypeptide. The crystal structure is shown on the right where α is in *yellow*, b is in *green*, and the antigen is in *red*. Redrawn from http://www.cehs.siu.edu/fix/medmicro/mhc.htm.

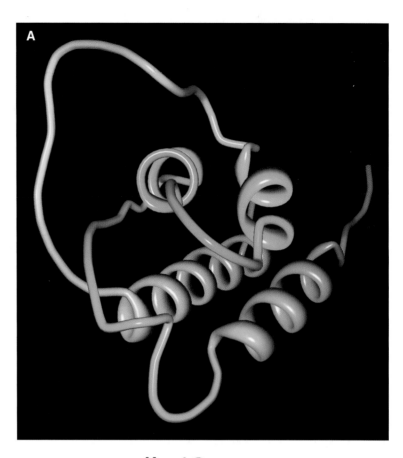
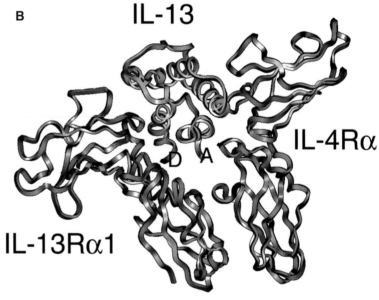

Figure 11-88. **A.** NMR structure of IL-13. **B.** Predicted model for IL-13 interaction with its receptors. Homology model of the CRH domain of IL-13Rα1, IL-4Rα subunits, and IL-13. The three-dimensional model is shown as a *ribbon diagram*. A and D indicate α-helix A and D of the IL-13 molecule, respectively. Part A reproduced from E.Z. Eisenmesser et al., "Solution structure of interleukin-13 and insights into receptor engagement," *J. Mol. Biol*, 310: 231, 2001, http://www.imb-jena.de/cgi-bin/ImgLib.pl?CODE=1ga3. Part B reproduced from Figure 8 of Y. Oshima and R. K. Puri, "Characterization of a powerful high affinity antagonist that inhibits biological activities of human interleukin-13," *J. Biol. Chem.*, 276: 15185–15191, 2001 and directly from http://www.jbc.org/cgi/content/full/276/18/15185.

of 13.5 kDa. IL-16 stimulates a migratory response in CD4⁺ T cells, monocytes, and eosinophils. It also induces the expression of the IL-2 receptor by T cells. From eosinophils, IL-16 stimulates the release of **leukotriene** C4 and IL-4 by way of the CD4⁻ and autocrine CCR3-chemokine–mediated signaling. CCR3 is a G-protein-coupled receptor present on basophils, eosinophils, and Th2 cells. Activation of PI3K, extracellular signal-regulated kinases 1 and 2, and p38 MAPK are involved in the signaling mechanism. The structure of IL-13 is shown in Figure 11-89.

IL-17 is a proinflammatory cytokine and is secreted by activated T cells. Although the secretion of IL-17 is restricted to certain types of T cells, its receptor is expressed on many cell types in the body, including fibroblasts and epithelial cells. IL-17 can induce the release of IL-6, IL-8, G-CSF, and SCF from fibroblasts. CSF stimulates mast cell adhesion to tissues, such as the connective tissue matrix. The signaling pathway of IL-17 is shown in Figure 11-90.

Macrophage IL-18 is a proinflammatory cytokine similar in activity to IL-1β. The inactive precursor to IL-18 is cleaved to the active form by caspase-1 (interleukin-1 converting enzyme, or ICE) as shown in Figure 11-91. The crystal structure of IL-18 is shown in Figure 11-92. The signaling pathway of IL-18 is shown in Figure 11-93.

IL-19 shares a receptor complex with IL-20, and the activities of these two ILs play a role in skin development and function. The crystal

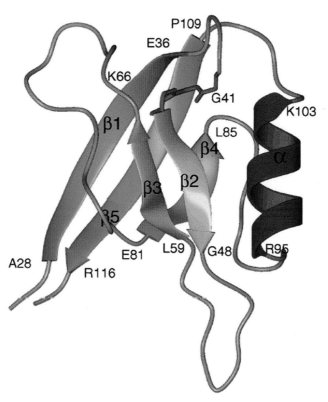

Figure 11-89. Structure of IL-16. Letters refer to amino acids. Reproduced from http://www.biochem.mpg.de/moroder/nmr/il16.gif.

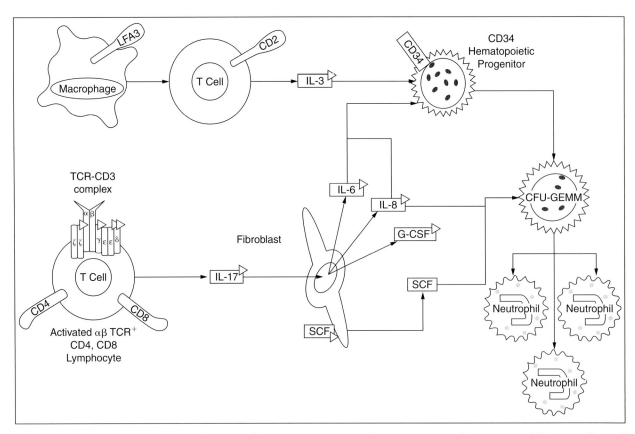

Figure 11-90. Signaling pathway of IL-17. Inflammation is a complex response involving many different cells and signaling molecules, including the secretion of the cytokine IL-17 by activated T cells. IL-17 secretion is restricted to specific subsets of T cells but the receptor for IL-17 is widely expressed throughout the body, including fibroblasts and epithelial cells. Inflammatory responses involving IL-17 probably contribute to arthritis, asthma, skin immune reactions, and autoimmune disorders. Fibroblasts and other cells stimulated by IL-17 are induced themselves to secrete inflammatory and hematopoietic cytokines, including IL-6, IL-8, G-CSF, and stem cell factor (SCF). These cytokines in turn provoke a range of activities, including the stimulation of neutrophil proliferation and differentiation. Redrawn from http://www.biocarta.com/pathfiles/h_il17Pathway.asp.

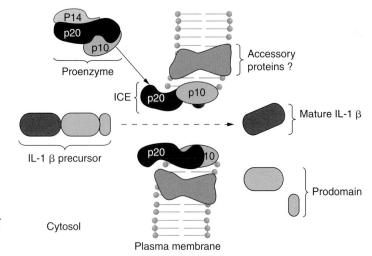

Figure 11-91. Activation of ICE from the proenzyme form. Active ICE (caspase-1) is cleaved from the proenzyme *(top)*, which associates with the cell membrane and cleaves the IL-1β precursor to its active form in the secretion process. Redrawn from M.J. Tocci, "Structure and function of interleukin-1β converting enzyme," *Cell Death Proteins, Vitamins and Hormones*, 53: 27, 1997.

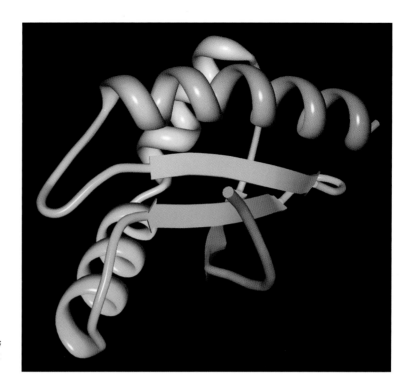

Figure 11-92. Crystal structure of human interleukin-18 (1L-18). Reproduced from http://www.rcsb.org/pdb/cgi/explore.cgi?job=graphics;pdbId=1JOS;page=;pid=251661071124103&opt=show&size=500&cyl=1.

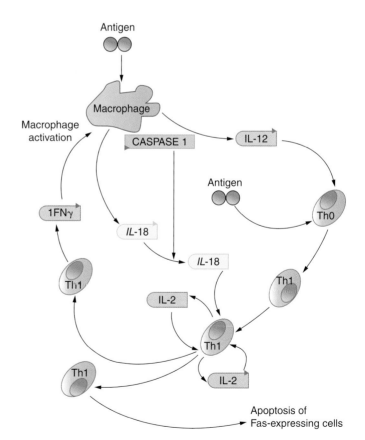

Figure 11-93. Signaling pathway of IL-18. IL-18 is a pro-inflammatory cytokine similar in structure and mechanism of action to IL-1 β. Formation of active IL-18 by macrophages requires cleavage of an inactive precursor by caspase-1 protease, also termed the IL-1-converting enzyme (ICE). One of the key biological responses induced by IL-18 is that, in combination with IL-12, it stimulates Th1 cell differentiation and involvement in immune responses. IL-18 was originally known as interferon-γ-inducing factor, named after its induction of interferon-gamma production. Up-regulation of Fas ligand in Th1 cells by IL-18 may increase apoptosis of Fas receptor expressing cells that interact with the activated Th1 cells. This induction of apoptosis may allow IL-18 to have anti-tumor activity and may also play a role in chronic inflammatory and autoimmune conditions. Reproduced from http://www.biocarta.com/pathfiles/h_il18Pathway.asp.

structure of IL-19 is shown in Figure 11-94. The receptor for IL-19 may be a heterodimeric complex of IL-20R1 and IL-21R2 subunits, as conjectured in Figure 11-95.

IL-22 is a proinflammatory cytokine produced by T cells and is related to IL-10. In addition to a complex receptor, IL-22 can be bound by a decoy receptor that lacks both cytoplasmic and transmembrane domains. The crystal structure of IL-22 is shown in Figure 11-96, and the productive

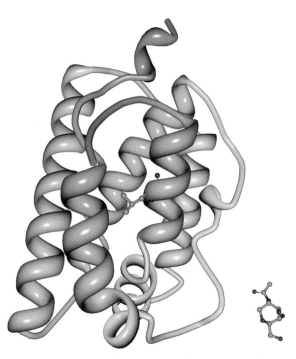

Figure 11-94. Crystal structure of human interleukin-19. PDB sum ID: 1n1f. C. Chang et al. Crystal Structure of Interleukin-19 Defines a New Subfamily of Helical Cytokines. *J. Biol. Chem.* **278** pp. 3308 (2003) and directly from http://www.ebi.ac.uk/msd-srv/msdlite/images/1n1f600.jpg.

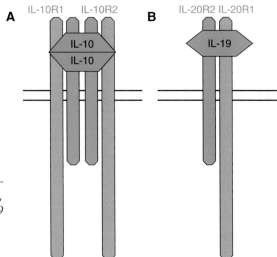

Figure 11-95. Schematic diagram of the IL-10 and IL-19 receptor complexes. **A.** An IL-10 homodimer interacts with four receptor subunits, two molecules of IL-10R1, and two molecules of IL-10R2. **B.** An IL-19 monomer is likely to signal through heterodimeric receptor complexes consisting of IL-20R1 and IL-20R2 subunits.

receptor structure is a heterodimer of IL-22R1 and IL-20R2 and indicated in Table 11-9. The signaling pathway of IL-22 is shown in Figure 11-97.

Not all ILs have been discussed here. There is a growing list; however, some generalizations can be drawn. They are either anti-inflammatory or proinflammatory in their actions. Many use the JAK/STAT pathway and the MAPK pathway of signaling, although there are many variations on this theme. Growth factors and cytokines are increasingly important in biochemistry, and many of these are becoming pharmaceutical targets for chemotherapy of disease.

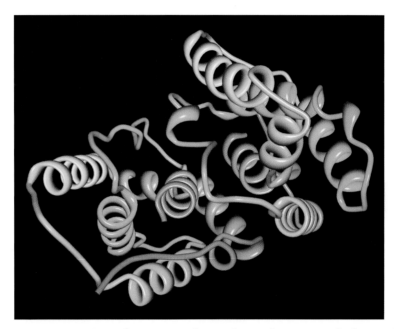

Figure 11-96. Crystal structure of recombinant human interleukin-22 (IL-22). Reproduced from and http://www.imb-jena.de/cgi-bin/htmlit.pl?color=white& id=GIF&src=1m4r.gif&name=Image%20Library%20Thumb%20Nail%201M4 R. R.A.P. Nagem, D. Colau, L. Dumoutier, J.-C. Renauld, C. Ogata, and I. Polikarpov, "Crystal structure of recombinant human interleukin-22," *Structure* 10: 1051, 2002.

Table 11-9
Heterodimeric Receptors of Ligands IL-10, 1L-20, 1L-22 and IL-24

R1/R2	IL-10R2	IL-20R2
IL-10R1	IL-10	?
IL-20R1	?	IL-20, IL-24
IL-22R1	IL-22	IL-24

IL-10, family of cytokines and receptors.
Reproduced from Table I of M. Wang et al., "Interleukin 24 (MDA-7/MOB-5) signals through two heterodimeric receptors, IL-22R1/IL-20R2 and IL-20R1/IL-20R2," *J. Biol. Chem.*, 277: 7341–7347, 2002.

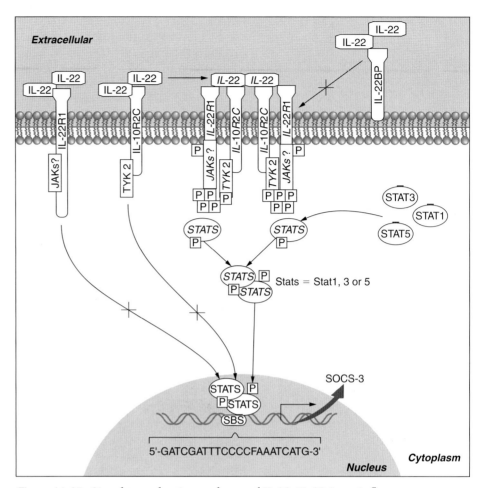

Figure 11-97. Signal transduction pathway of IL-22. IL-22 is an inflammatory cytokine related to IL-10 that is produced by T cells and that induces a response in cells through a heterodimeric cell surface receptor composed of IL-22R1 and IL-10R2C. One of the actions of IL-22 appears to be the induction of the acute phase inflammatory response in hetapocytes, acting through activation of STATs and transcriptional regulation. A gene with homology to the extracellular domain of the cell surface IL-22R1 receptor component was identified that lacked transmembrane and cytoplasmic domains. The protein derived from this gene was found to bind IL-22 and block its interaction with the cell surface receptor. The IL-22 soluble receptor (IL-22BP, IL-22 binding protein) also blocks some of the downstream effects of IL-22 such as STAT activation and the transcriptional induction of genes involved in the immune and inflammatory responses. The IL-22 soluble receptor may act as an anti-inflammatory agent. Redrawn from http://www.biocarta.com/pathfiles/h_il22bpPathway.asp.

Further Reading

Leroith, D., and Bondy, C. (eds.), *Growth Factors and Cytokines in Health and Disease,* John Wiley & Sons, 1997.

CHAPTER **12**

Membrane Transport

Cystic Fibrosis: A Genetic Disease Involving Aberrant Ion Transport

Cystic fibrosis (CF) is a disease generated by a mutation of a chloride ion transport protein called the **cystic fibrosis transmembrane regulator (CFTR).** Because of the many possible mutations in the gene for CFTR, any one mutation produces the disease characterized by a nonfunctional CFTR. Absence of a functioning CFTR impairs chloride ion transport in and out of the cell. This defect occurs in many tissues, including sweat glands, lung, intestines, and pancreas. Because the transport of chloride is followed by water and water transport would be deficient, a thick mucus develops in the lung, usually leading to infection; sweat becomes concentrated in sodium chloride; and pancreatic (digestive) enzyme secretion is deficient. Pancreatic insufficiency is detected by measuring immunoreactive trypsinogen (IRT), a secretory product of the pancreas, in children and in adults and is part of newborn screening for CF, especially when there are no stools in the first 24 to 48 hours of life (*meconium ileus* occurs; the abnormal pancreas is low on secretions and produces a mass in the intestines called meconium that obstructs the intestine, as shown in Figure 12-1). The sweat is well above normal in concentrations of sodium and chloride because of the lower water content, and this forms the basis of a common diagnostic test. The abnormality in the sweat is established at birth and continues during development, making diagnosis possible in the newborn. After a weak electrical current, together with a chemical stimulus (for example, the cholinergic agent pilocarpine, Figure 12-2) of sweating, is applied to an area of the skin on an arm or leg, sweat is collected on a filter paper and is sent to the laboratory. A normal concentration of chloride is less than 40 mmol/liter, values between 40 mmol/liter and 60 mmol/liter are considered borderline, and values greater than 60 mmol/liter are diagnostic for CF. Although the CFTR protein is encoded by a gene on chromosome 7, a gene on chromosome 1 encodes a protein called the **CF antigen,** which occurs at high levels in the sera of patients with overt CF and with carriers. There is some evidence that CFTR may interact with CF antigen, because the structure of the CF antigen is reminiscent of an ion transport regulator.

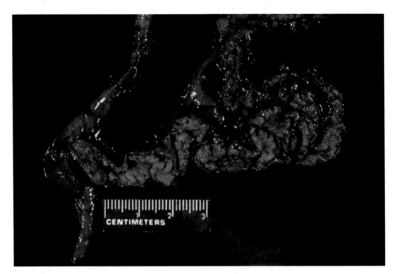

Figure 12-1. Intestines showing *meconium ileus*. *Meconium ileus* is most often seen in the first few days of life in neonates with cystic fibrosis (CF) but can occur rarely in infants with a normal pancreas. In CF, the abnormal pancreatic secretions lead to inspissated meconium that produces intestinal obstruction. The dilated coils of ileum are opened here to reveal the inspissated green meconium (which may also be tarry or gritty), while the unopened colon at the upper left and the appendix at the lower left beyond the ileocecal valve are not dilated, and little or no meconium is passed per rectum. Reproduced with permission from http://www-medlib.med.utah.edu/WebPath/PEDHTML/PED046.html. Spencer S. Eccles Health Sciences Library, University of Utah Health Sciences Center, Salt Lake City, UT.

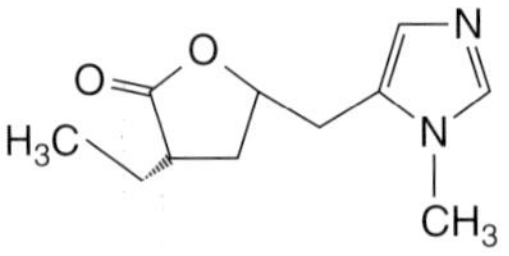

Figure 12-2. Structure of pilocarpine, a water-soluble cholinergic agent.

Males and females are affected equally by CF. There is one CF child in every 2000 births, and currently there are about 30,000 children and young adults with overt CF. CF is the most common lethal hereditary disease of Caucasians (1 in 2000). The incidence in some Native Americans also is strong (Pueblo, 1 in 4000; Zuni, 1 in 1500) and less frequent in Hispanics (1 in 8000); in Africans the frequency is 1 in 15,000 and in Asians, 1 in 32,000. In the United States, 1 in 20–25 Caucasians is a carrier (unaffected recessive), and 1 in 29 Ashkenazi Jews is a carrier, making this group a lower risk among Caucasians than any other. There are about 10 million carriers in all. Symptoms include persistent diarrhea (probable implication of chloride ion transport defect in the intestine), smelly and greasy stool, frequent pneumonia, chronic coughing, salty skin, and poor growth. Therapeutically, digestive enzyme supplements of pancreatic enzymes are fed, chests are percussed routinely to clear mucous buildup, drugs are used to break up mucus, and antibiotics are given. When the lungs begin to fail, a lung transplant may extend life. There are many CF treatment centers in the United States.

CF is an autosomal recessive disorder in which the development of overt CF in a child comes from both parents, who may be normal (unaffected) but who carry a recessive gene. Inheritance of an autosomal recessive disease is shown in Figure 12-3 with an example of a family

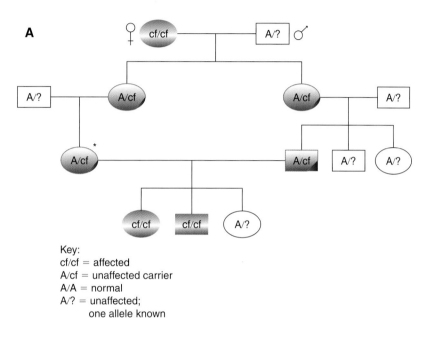

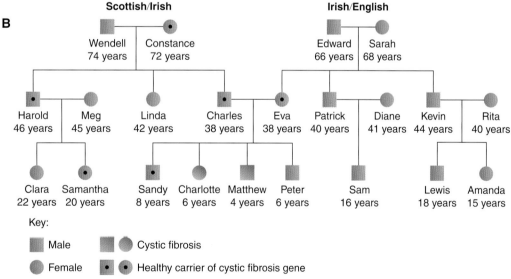

Figure 12-3. **A.** Generic pedigree of autosomal recessive disorder in which affected offspring can have unaffected parents. Heterozygotes (A/cf) have a normal phenotype. Two affected parents (who can be unaffected carriers) will always have affected children. Affected parents with homozygous dominant mates will have unaffected children. Close unaffected relatives who reproduce are more likely to have affected children if they have joint affected relatives. Males and females are affected with the same frequency. **B.** Example of a family pedigree for CF.

pedigree. In Figure 12-3A, the dominant gene, *A*, would represent the normal CFTR protein and the recessive, *cf*, would carry information for a mutation, which would not actually occur unless the mate were also recessive, carrying another *cf*. In the family history in Figure 12-3B, lower part, both parents recessive for CF (carry *cf*) have offspring with overt diseases, as well as an unaffected carrier and one normal child. Thus, the outcomes of matings will be as follows: if one parent is unaffected and the other is an unaffected carrier, there will be a 50% chance of an unaffected child, a 50% chance of a child carrier, and no chance of a child with overt CF; if the parents are both carriers, there will be a 25% chance of an unaffected child, a 50% chance of a child carrier, and a 25% chance of a child with overt CF; if one parent is unaffected and the other has overt CF, there is no chance of having an unaffected child, there is 100% chance of having child carriers, and there is no chance of a child with overt CF; and if one parent is a carrier and the other has overt CF, there is no chance of having an unaffected child, a 50% chance of having a child carrier, and a 50% chance of having a child with overt CF. Sometimes death follows shortly after birth of a child with overt CF, but in manageable cases, with modern treatment, life expectancy can extend to about 30 years of age.

The CFTR protein is complex and is made up of five domains: two membrane-spanning domains (MSDs), two nucleotide-binding domains (NBDs), and a regulatory domain (R). The MSDs form a transmembrane pore, which allows chloride ions (Cl^-) to pass in both directions across the cell membrane. The CFTR resembles the class of **ABC transporters,** which carry substances back and forth across inner cell membranes. Another member of this class, in addition to CFTR, is the multidrug-resistance protein (MDR), whose cellular induction forms the basis of the development of resistance to many chemotherapeutic agents (MDR can pump small molecules, like drugs, out of the cell). Models of the ABC transporter class and the CFTR are shown in Figure 12-4. The MSDs form the Cl^- permeating pore. To open the pore to allow the ions to pass, the R domain is phosphorylated by cyclic AMP–dependent protein kinase A. ATP binding to the NBDs and subsequent hydrolysis of the ATP regulate the opening and closing of the channel. The NBDs operate together because when they are dissociated experimentally their ability to hydrolyze ATP is greatly decreased; they, in fact, form a heterodimer. One ATP-binding site binds ATP tightly, and the other facilitates hydrolysis of ATP to ADP (Figure 12-4C). CFTR is returned to an inactive state when protein phosphatase dephosphorylates the R domain. The structure of NBD1 of the CFTR with bound ATP is shown in Figure 12-5.

A graphic depicting the formation of the CFTR protein from the gene on the long arm of chromosome 7 to the folded protein transporter is shown in Figure 12-6. A large variety of mutations in the CFTR gene is possible, as shown in the figure, the most frequent of which is the single amino acid deletion of F508 (phenylalanine 508) in exon 10, which codes for part of the nucleotide-binding function in NBD1. Table 12-1 lists the most common CFTR mutations. In the case of δ-F508, the nonfunctional mutant protein is synthesized but not modified in the **endoplasmic reticulum** and undergoes degradation. CFTR is also involved in the regulation of bicarbonate ion (HCO_3^-) in the pancreatic duct, where CFTR increases a sodium–proton exchanger

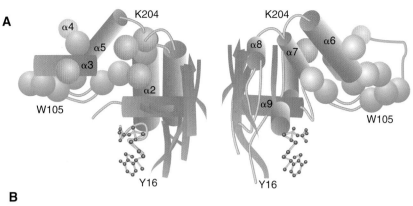
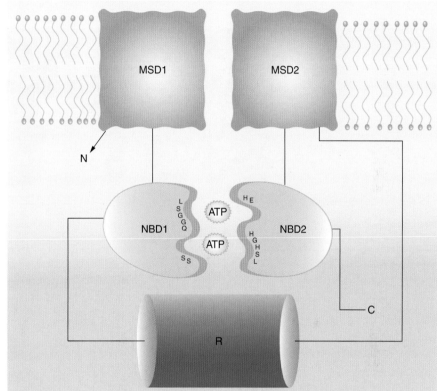
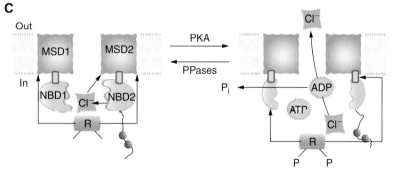

Figure 12-4. **A.** Model of the ABC class of transporters of which CFTR is a member. **B.** Domain organization of CFTR. The five domains of CFTR are shown. Also indicated is a putative nucleotide-binding domain. **C.** Regulation of the CFTR Cl⁻ channel. This simplified model shows the regulation of CFTR by cAMP-dependent phosphorylation at the R domain and cycles of ATP binding and hydrolysis at the NBDs. *MSD*, membrane-spanning domain; *NBD*, nucleotide-binding domain; *P*, phosphorylation of the R domain; P_i, inorganic phosphate; *PKA*, protein kinase A; *PPase*, protein phosphatase; *R*, regulatory domain. *In* and *out* denote the intra- and extracellular sides of the cell membrane. Part A redrawn with permission from http://www.lbl.gov/Science-Articles/Archive/abc-cystic-fibrosis.html. Part B redrawn from http://www.nature.com/emboj/journal/v23/n2/full/7600040a.html. H.A. Lewis et al., "Structure of nucleotide-binding domain 1 of the cystic fibrosis transmembrane conductance regulator," *EMBO J.*, 23: 282–293, 2004. Published online December 18, 2003.

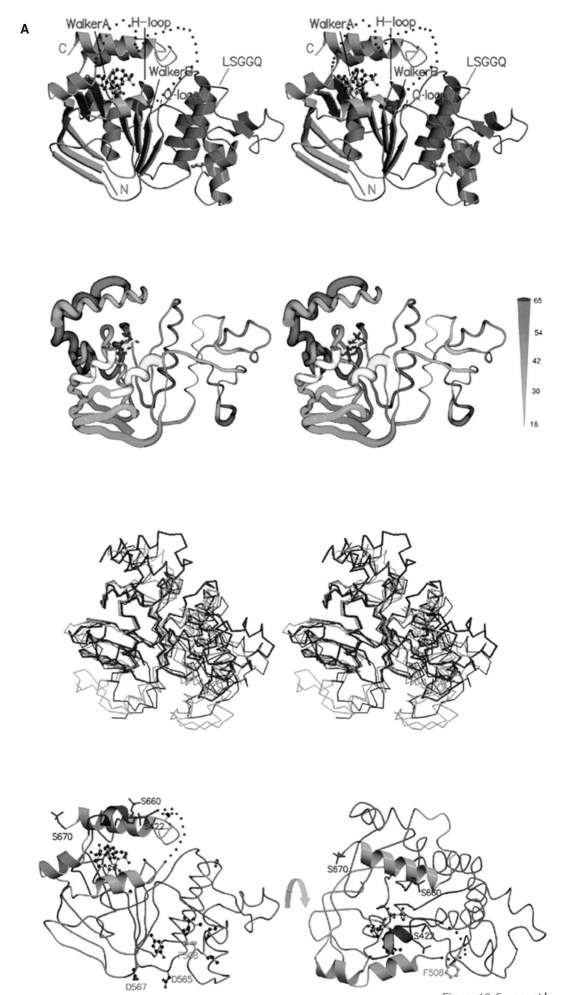

Figure 12-5, cont'd
For legend see opposite page

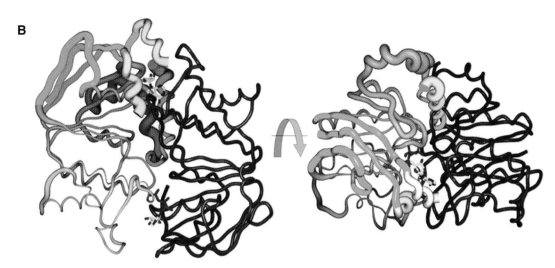

Figure 12-5. **A.** Structure of the nucleotide-binding domain 1 (NBD1) (see *Figure 12-4B*) in stereo ribbon diagram. ATP is shown as a ball-and-stick representation. The subdomains are color-coded. The *dotted red line* indicates residues missing from the structure. **B.** Worm figure of the NBD1–NBD2 interaction. NBD2 is in red and white. NBD1 is color-coded as in A. The figure at the right is rotated 90 degrees toward the viewer relative to the left figure. Part A reproduced with permission from Figure 3 and Part B reproduced from Figure 6 of http://www.nature.com/emboj/journal/v23/n2/full/7600040a.html, H.A. Lewis et al., "Structure of nucleotide-binding domain 1 of the cystic fibrosis transmembrane conductance regulator," *EMBO J.*, 23: 282–293, 2004.

(isoform 3); some of these relationships are viewed in Figure 12-7. During absorption, epithelial cells of the pancreas and lung absorb sodium through the epithelial sodium channel (ENaC), whereas chloride enters through the tight junctions between cells and water is absorbed from the lumen to the basolateral side (adjoining the extracellular space toward the tissues) of the cell. Elevated cyclic AMP activates the CFTR channels, allowing chloride ions to pass into the lumen from the cell interior, and sodium ions follow by passing outwardly through the tight junctions (Figure 12-8). In this process, the activation of the CFTR channels causes inhibition of the ENaC channel, inhibiting the cellular uptake of sodium ions (normally followed by water uptake) from the lumen so that an osmotic gradient results, causing secretion of water into the lumen. CFTR, when stimulated by cyclic AMP, may inhibit ENaC function by an interaction of the two channels binding through **ezrin-radixin-moesin binding phosphoprotein (EBP50)** (containing PDZ domains involved in the linkage of integral membrane proteins to the cytoskeleton), as shown in Figure 12-9. A PDZ domain structure is shown in Figure 12-10. Also, it is theorized that in the lung that maintains high levels of GSH the CFTR modulates GSH efflux in response to stress occurring with infections. There may be other, as yet, undiscovered functions of CFTR.

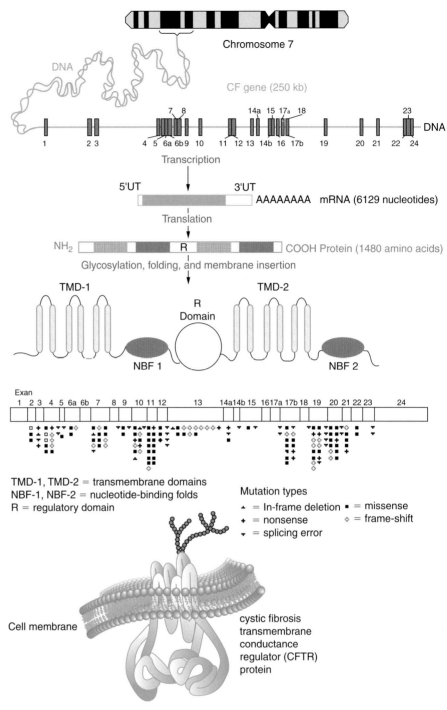

Figure 12-6. Formation of the CFTR protein from gene to fold protein in form of ion channel. Many mutations are possible as shown. Redrawn from http://wsrv.clas.virginia.edu/~rjh9u/cfmap.html. The graphic was taken, in part from R. Lewis, *Human Genetic: Concepts and Applications*, McGraw-Hill, Wm. C. Brown, 1994.

Table 12-1
Common *CFTR* Mutations[1]

Mutation	Type	Frequency (%)
DeltaF508	deletion	28,948 (66.0)
G542ter	nonsense	1062 (2.4)
G551D	missense	717 (1.6)
N1303K	missense	589 (1.3)
W1282ter	nonsense	536 (1.2)
R553ter	nonsense	322 (0.7)
621+1 G->T	splice junction	315 (0.7)
1717-1 G->A	splice junction	284 (0.6)
R117H	missense	133 (0.3)
3849+10kb C->T	alternative splice	104 (0.2)

[1]Data are from the Cystic Fibrosis Genetic Analysis Consortium based on 43,000 CF chromosomes examined through 1994. Mutations are indicated by amino acid residue or nucleotide position; ter, termination codon. Reproduced from Table 1 of http://www.bmb.leeds.ac.uk/teaching/icu3/mdcases/ws4/.

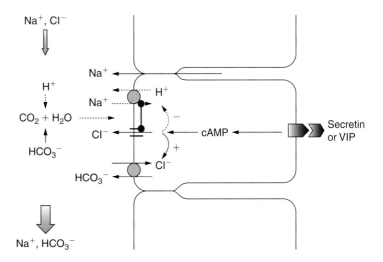

Figure 12-7. A model of HCO_3^- homeostasis in the resting and stimulated pancreatic duct. In the resting state, CFTR (and possibly EBP50) does not interact with the HCO_3^- salvage mechanisms, which operates at maximal capacity. Upon cell stimulation, CFTR, EBP50, and NHE3 are assembled into a complex to inhibit HCO_3^- salvage and at the same time stimulate HCO_3^- secretion. Redrawn from http://www.jbc.org/content/vol276/issue20/images/large/bc1711928009.jpeg and Figure 9 from W. Ahn et al., Regulatory interaction between the cystic fibrosis transmembrane conductance regulator and HCO_3^- salvage mechanisms in model systems and the mouse pancreatic duct. *J. Biol. Chem.*, 20: 17236–17243, 2001.

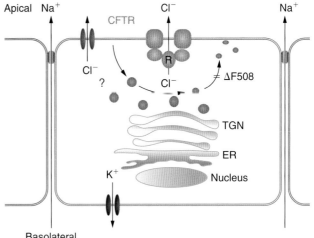

Figure 12-8. Activation of CFTR involves movement of Cl out of the cell into the lumen and movement of Na^+ outward through tight junctions. It is thought that vesicles containing CFTR are also moved to the apical membrane for insertion of additional CFTR channels. *TGN*, trans-Golgi network; *ER*, endoplasmic reticulum. Redrawn with permission from http://www.cellscience.com/reviews2/vesicular.jpg, R.J. Walters, "Cystic fibrosis: search for the North West Passage," *Cell Science Reviews*, 1: 13–23, 2004.

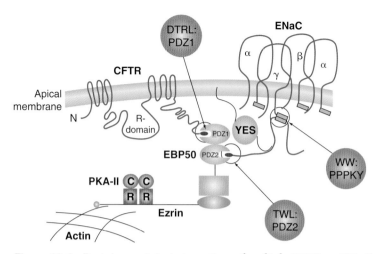

Figure 12-9. Protein–protein interactions that link CFTR to ENaC.

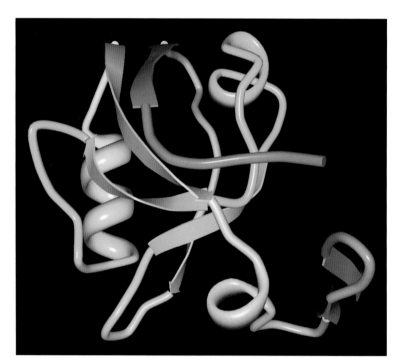

Figure 12-10. The third PDZ domain from the synaptic protein PSD-95. Reproduced from PDBsum, MSD, ID: 1bfe. http://www.ebi.ac.uk/msd-srv/msdlite/images/1bfe600.jpg.

Types of Membrane Transport

Absorption of Large Molecules

Macromolecules that bind to cell surface receptors are absorbed either by the process of receptor-mediated endocytosis as described for EGF (Figure 11-18) or by **pinocytosis**. Pinocytosis is the process of ingesting fluid and macromolecules in small vesicles of less than 150 nm in diameter. Invaginations of the cell membrane form vesicles enclosing the fluid or macromolecules to be ingested. *This is a process carried out by all cells.* For engulfment of particles, the process of phagocytosis is used by phagocytes specialized for this function. In phagocytosis, the cell membrane is extended (pseudopodia) to enclose the particle. The particle size is greater than 250 nm in diameter. Comparison of the processes of pinocytosis and phagocytosis is made in Table 12-2. Pinocytosis and phagocytosis are compared further in Figure 12-11.

Exocytosis

When macromolecules are to be moved out of the cell, this process occurs by exocytosis (movement out of the cell of macromolecules stored in vesicles awaiting signals for release to the outside). An example of this would be the secretion of an anterior pituitary hormone in response to a signal for its release (regulated exocytosis, as in Figure 8-21) or constitutive secretion, as shown in Figure 12-12. Constitutive secretion refers to particles continually secreted without a specialized extracellular signal; the signal for secretion is built into the structure of the protein (specific amino acid sequence). The amino acid sequences signaling secretion can be in the N-terminus; for example, the secretion signal (in single letter abbreviations of amino acids) for human growth hormone is MATGSRTSLLLCLPWLQEGSASARNRQKR, and the secretion signal for human fibrillin-1 is MRRGRLLEIALGFTVLLASYTSHGADA.

Table 12-2
Comparison of Pinocytosis and Phagocytosis

	Pinocytosis	Phagocytosis
Substrate	Fluids, macromolecules	Particles
Mechanism	Membrane invagination	Pseudopodia formation
Endosome	Pinocytic vesicle (~150 nm)	Phagosome (>250 nm)—determined by particle
Energy requirement	Energy independent	Energy dependent
Performed by	All cells	Phagocytes

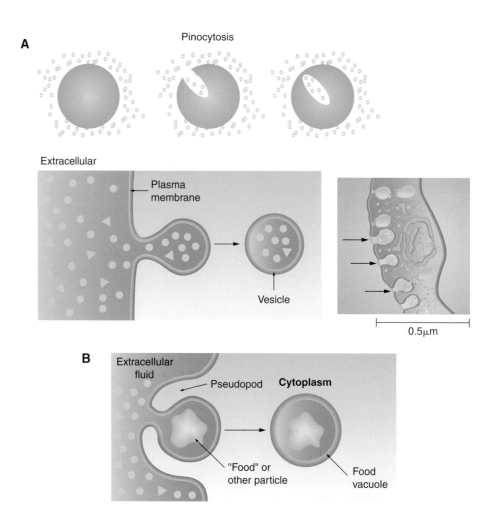

Figure 12-11. Processes of pinocytosis and phagocytosis. **A.** *Upper,* diagram of pinocytosis. *Lower left,* pinocytosis showing the invagination at the cell membrane to engulf outside particles and form vesicle within the cell. *Lower right,* picture of several pinocytic invaginations. **B.** Diagram of the process of phagocytosis in which pseudopodia are formed from the cell membrane, which surround particles to be engulfed. **C.** Photograph of a phagocyte ingesting two red blood cells. Part C reproduced with permission from Alberts, *Essentials of Cell Biology,* Garland Publishing Company.

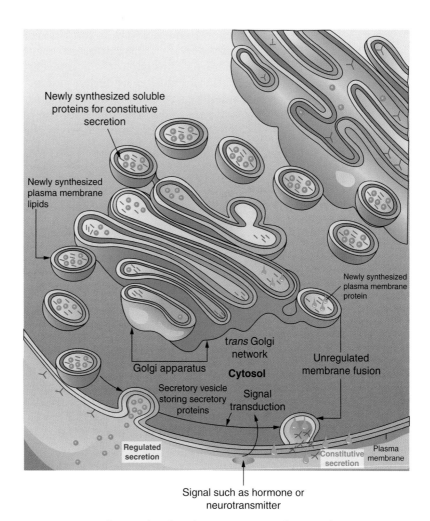

Figure 12-12. The regulated and constitutive pathways of exocytosis. The two pathways diverge in the trans-Golgi network. Many soluble proteins are continually secreted from the cell by the constitutive secretory pathway, which operates in all cells. This pathway also supplies the plasma membrane with newly synthesized lipids and proteins. Specialized secretory cells also have a regulated exocytosis pathway by which selected proteins in the trans-Golgi network are diverted into secretory vesicles, where the proteins are concentrated and stored until an extracellular signal stimulates their secretion.

Passive Diffusion or Osmosis

Osmosis

Osmosis is passive membrane transport; that is, no energy is required to transport some substances, such as water, carbon dioxide, and oxygen, across a semipermeable membrane that is otherwise selective for the solutes that will be transported. When molecules other than water pass through the membrane without energy expenditure and specific proteins are not required to make the passage occur, this process (similar to water transport) is called passive transport. This differs from facilitated diffusion, which requires specific proteinaceous channels or permeases. In general, charged ions do not pass freely through membranes, and in most cases specific transporters are used. The energy-free movement of water (osmosis) through a semipermeable membrane is dictated by the concentration of solutes on either side of the membrane. Thus, when, on one side, the solute concentration is high compared to the other side, water will move from the side with the lower solute concentration to the side with the higher solute concentration. In a solution containing two solutes, one in which the solute concentration is higher on one side of the membrane than on the other and one solute permeates the membrane while the other does not, water will move toward the solution that is more concentrated in the solute that does not permeate the membrane. In other words, the permeable solute will reach an equilibrium whose concentration will end up the same on either side of the membrane, and water moves toward the nonpermeable solute that remains more concentrated on one side of the membrane (water moves *down* its concentration gradient, from where there is more water to where there is less water). The net result will be as equal a distribution of solutes on either side of the membrane as possible and the movement of water to the more concentrated solution to equalize the tonicity of the two solutions (tonicity is the effective particle concentration, **osmolality,** which exerts an osmotic force across a membrane). Thus, there will be no further movement of water when both compartments are equally concentrated, or isotonic, with respect to each other. A hypotonic solution has more water and less solute, so a cell in hypotonic solution will take up water (swell). A hypertonic solution is characterized by a high solute concentration, so a cell put in a hypertonic solution will lose water. Bear in mind that some of the preceding comments refer to an artificial system with a semipermeable membrane, whereas the cell membrane is selective in addition to its inherent permeability. In a cell, there are proteinaceous channels for the passive uptake of dissolved molecules (except for water). Normally, the intracellular solute concentration is balanced by the extracellular sodium chloride concentration. When cells need to take up water from the outside, in addition to the process of osmosis, they can resort to energized transport using the aquaporin mechanism (Figure 1-30 and Figure 1-31) in response to a specific signal, such as vasopressin (if cells express the appropriate cell membrane receptor), which generates cyclic AMP and activated protein kinase A (Figure 12-13).

Facilitated Diffusion

This is one step beyond osmosis or simple diffusion, where there is a proteinaceous channel for a solute to enter the cell but the transport process does not require energy beyond that of free diffusion. In this case, molecules will move into the cell based on the outside concentration. If the outside concentration is higher than the inside concentration, the molecules in solution can be expected to move to the cell interior (down an electrochemical gradient). Facilitated diffusion occurs with dissolved particles that diffuse through channels, constructed of proteins. Transport must be facilitated when a molecule, to be transported to the cell interior, is impermeable to the lipid bilayer of the membrane (for example, glucose and ions), and this process consists of protein channels that span the membrane, thus overcoming the need for the solute to contact (for this type of molecule) the impermeable bilayer (Figure 12-14). There

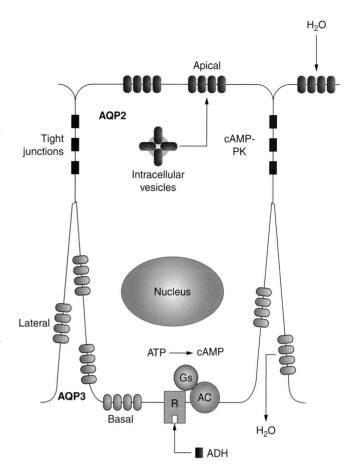

Figure 12-13. Vasopressin (ADH) mechanism for signaling aquaporins (AQs) to take up water. Diagram representing vasopressin-regulated transcellular water permeation of apical membrane of collecting duct principal cell via AQP2 and constitutive water permeation of basolateral membrane via AQP3. *ADH,* antidiuretic hormone, vasopressin; *R,* vasopressin receptor; *AC* , adenylate cyclase; AQP_2, aquaporins on the apical cell membrane, which take up water and transport it to the basolateral side of the cell (AQP3). Aquaporins are activated into vesicles by phosphorylation mediated by protein kinase A (cAMP-PK).

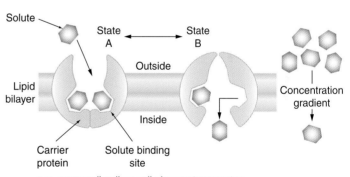

Figure 12-14. Facilitated diffusion in the case of glucose transport down its concentration gradient (more concentrated on the outside than inside).

are three high-affinity transporters for glucose, Glut 1, Glut 3, and Glut 4 and one low-affinity glucose transporter, Glut 2. These operate by facilitated diffusion. High-affinity transporters are found in intestine, kidney, and liver, where there is higher glycolytic activity than in other tissues. Glut 2 also will be present in these tissues because these tissues carry high glucose fluxes. The expression of these transporters is regulated by glucose, as well as different hormones. The crystal structure of Glut 1 is shown in Figure 12-15. Some small molecules, on the other hand, are lipophilic, such as neutral steroids (cortisol, for one), which

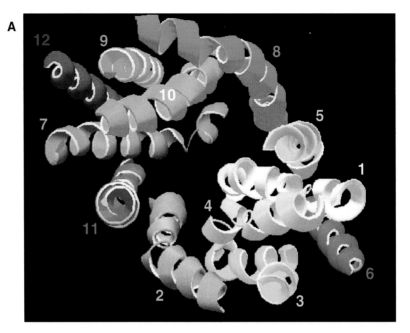

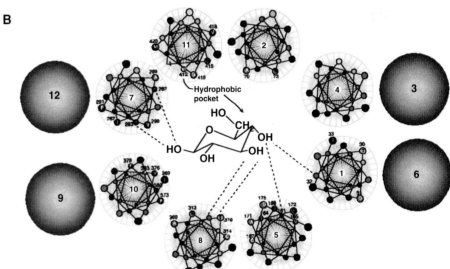

Figure 12-15. Two-dimensional models for the arrangement of the 12 transmembrane helices of GLUT-1 and for the exofacial sugar binding site. **A.** Theoretical arrangement of the 12 putative transmembrane helices of GLUT-1 in the cytoplasmic facing conformation based on homology modeling using the *Escherichia coli* lac permease as the template. **B.** Proposed low-resolution model of the exofacial glucose-binding site. The model is consistent with numerous experimental observations. Glucose is not drawn to scale. The *dotted lines* represent possible hydrogen bonds formed between glucose hydroxyl groups and various side chains on GLUT-1. *Numbered residues* are accessible to pCMBS from the external solvent. Reproduced from http://www.jbc.org/cgi/content/full/279/11/10494/FIG5. M. Mueckler and C. Makepeace, "Analysis of transmembrane segment 8 of the GLUT1 glucose transporter by cysteine-scanning mutagenesis and substituted cysteine accessibility," *J. Biol. Chem.*, 279: 10494–10499, 2004.

dissolve in the lipid bilayer and diffuse (simple diffusion) into the cell without facilitation. Passive and facilitated diffusion are illustrated by Figure 12-16, and the form of kinetics of solute uptake in these two cases is shown in Figure 12-17. Although the passive process obtains a rate directly proportional to the (outside) solute concentration, the facilitated process reflects an enhanced rate (greater than proportional to the outside solute concentration) as the concentration of the solute increases.

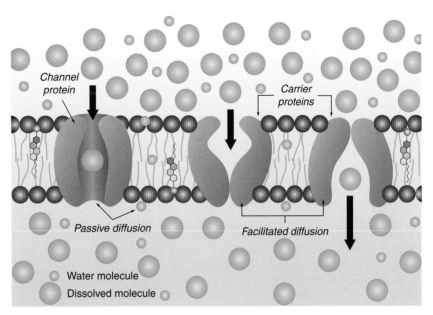

Figure 12-16. Examples of passive *(left)* and facilitated diffusion *(right)*. Openings are formed by channel proteins through which small, dissolved particles, especially ions, diffuse by passive transport.

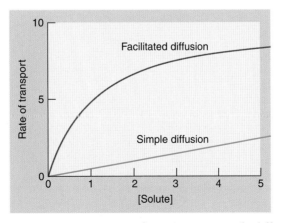

Figure 12-17. Kinetics of transport in simple diffusion and in facilitated diffusion as a function of the concentration of the solute (x-axis).

Energy-Requiring Transport: Active Transport

In this case, a solute is transported across a cell membrane from one side of a membrane to another, usually against a concentration gradient (where the solute concentration is greater on the opposite side), expending energy from the terminal phosphate in the ATP molecule characteristic of ATPase action. The system, again, is a construct of a protein channel that opens with the expenditure of energy and then closes after the transport is complete. In many cases, the solute or solutes are being pumped inward or outward against a concentration gradient. An important example is sodium/potassium ATPase (Na^+/K^+ ATPase), as shown in Figure 12-18, and the 10-msec pumping cycle of the Na^+/K^+ ATPase is

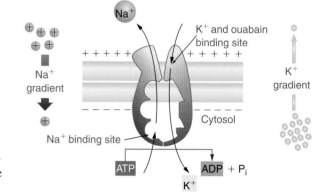

Figure 12-18. Diagram showing functionalities of Na^+/K^+ ATPase. This is an antiport system in which Na^+ is pumped outward against a concentration gradient and K^+ is pumped inward against a concentration gradient. The enzyme utilizes the energy of ATP to achieve the transport of these ions.

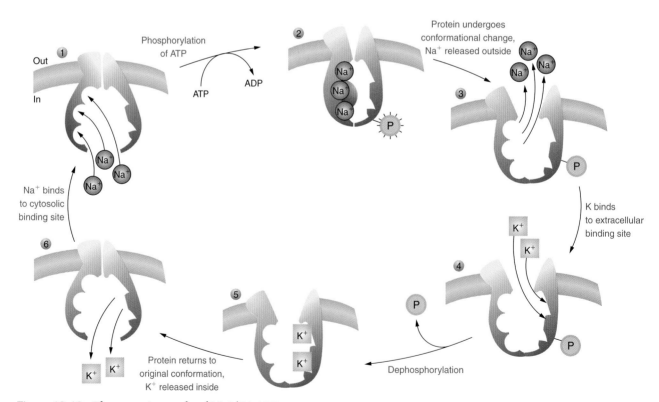

Figure 12-19. The pumping cycle of Na^+/K^+ ATPase.

shown in Figure 12-19. This system is found in all human cells, particularly nerve and muscle cells. In each cycle, three sodium ions are transported out of the cell and two potassium ions are transported into the cell (Figure 12-20). Often found on the basolateral membrane, Na$^+$/K$^+$ ATPase may account for as much as 30% of the total ATP consumption. Operating the Na$^+$/K$^+$ ATPase pump consumes about one-third of the body's

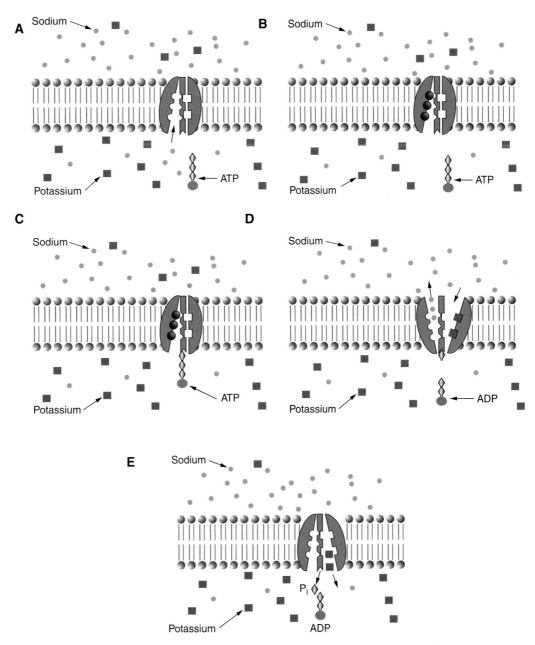

Figure 12-20. Diagrammatic mechanism of Na$^+$/K$^+$ ATPase pump. **A.** Membrane pump open at interior. **B.** Three Na$^+$ ions from interior bind to the pump. **C.** ATP binds to the pump. **D.** The terminal phosphate group of ATP is cleaved (to form ADP, Pi, and energy) and the three Na$^+$ ions are released to the outside and the two potassium ion binding sites are exposed to the outside, allowing two K$^+$ to enter the pump. **E.** When the phosphate group detaches from the pump, the pump returns to its original shape, the two K$^+$ ions enter and the three Na$^+$ ions leave.

energy expenditure. This system is responsible for maintaining a 10- to 30-fold higher Na^+ concentration outside the cell than inside (creating an inwardly directed electrochemical gradient) and a greater K^+ potassium concentration inside the cell compared to the outside. This pump is also an example of one pump transporting more than one ion.

Simple and Coupled Transporters

A protein channel involved in the transport of a single solute in one direction is called a **uniport.** When a solute is transported in one direction and an ion is transported with it in the same direction, this is called a **symport.** When a solute is transported in one direction and, at the same time, an ion is transported in the opposite direction by the same coupled transporter, this is called an **antiport.** Thus, when a solute gradient is established, for example, a sodium gradient where the concentration of sodium ion is greater on the outside than on the inside of a cell, the tendency for inward movement (osmotic pressure) of the sodium ion can be used to drive the transport of a second molecule; in this case, the sodium ion is the cotransported ion moving through a symport or a coupled transporter (Figure 12-21). Likewise, when an ion in the interior of the cell is in higher concentration than on the outside, its transport to the exterior (down its electrochemical gradient) can be used to drive another molecule into the cell by an antiport. These relationships are diagrammed in Figure 12-22.

Another example is the transport of glucose across intestinal epithelial cells. This is different from the situation in which glucose is more concentrated outside the cell, creating a concentration gradient favoring direct transport into a cell (Figure 12-19). In the case of the intestinal epithelial cell, the concentration of glucose may be higher inside the cell than on the outside. To overcome the opposite glucose gradient, glucose transport can be coupled to the uptake of Na^+ ions, where $[Na^+]$ is much higher outside of the cell; thus, the positive electrochemical gradient for

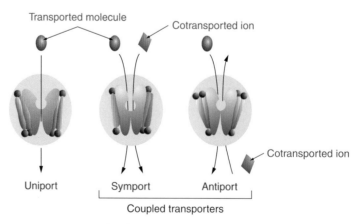

Figure 12-21. Kinds of transporters.

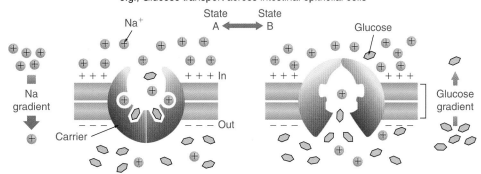

Figure 12-22. Transporting glucose against its gradient, in intestinal epithelia, by using the positive sodium ion gradient in coupled transport; an example of a symporter.

sodium can be used (coupled) to transport glucose against its gradient by a coupled transporter, as shown in Figure 12-21. After glucose is pumped into and across the intestinal epithelial cell, it is further transported on the basolateral side of the cell by a passive glucose transporter (uniport) (Figure 12-23). From the basolateral side of the cell, glucose is transported down its concentration gradient because the concentration of glucose is higher inside the cell than in the extracellular space; thus, passive transport can be used.

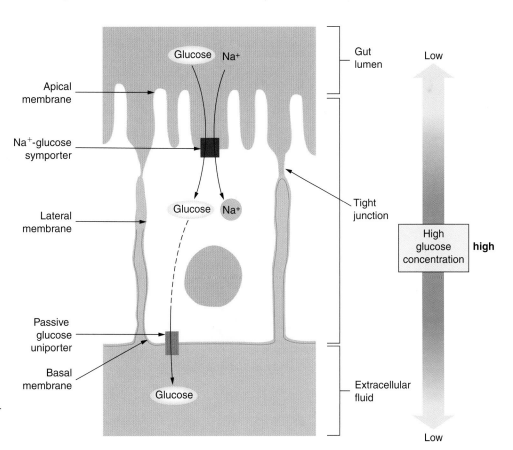

Figure 12-23. Overall transport of glucose from the intestinal lumen across the epithelial cell to the basolateral membrane and to the internal extracellular space.

Simple and Coupled Transporters

Ions and Gradients

Important ions that need to be controlled with respect to inside and outside concentrations are Na^+, K^+, Mg^{2+}, Ca^{2+}, H^+, and Cl^- (Table 12-3). Osmotic balance is maintained in several ways, including the movement of water from low to high solute concentration and extracellular sodium chloride that balances the concentration of intracellular solute. The balance is further maintained by the action of NA^+/K^+ ATPase, described earlier. In addition to these considerations, most cell membranes have a membrane potential derived from a voltage across the membrane, and this potential exerts an influence on the movement of ions. Membranes have different charges on their inner and outer surfaces. Relative to the outside of the membrane, the cytoplasmic side of the membrane is usually at a negative potential, so the electrostatic force drives anions (negatively charged) out and draws cations (positively charged) into the cell. Thus, when charged solutes are moving across a membrane, the concentration gradient must be considered, as already discussed, and the transmembrane potential difference plays a role (Figure 12-24).

Table 12-3
Concentrations of Ions Inside and Outside Cells

Ion	Intracellular	Extracellular
Na^+	5–15 mM	145 mM
K^+	140 mM	5 mM
Mg^{2+}	0.5 mM	1–2 mM
Ca^{2+}	10^{-7} mM	1–2 mM
H^+	$10^{-7.2}$ M (pH 7.2)	$10^{-7.4}$ M (pH 7.4)
Cl^-	5–15 mM	110 mM
Fixed anions	High	0 mM

Intracellular vs extracellular ion concentrations
[intracellular] very different from [extracellular] cations (+ve charged species) balanced by anions (-ve)

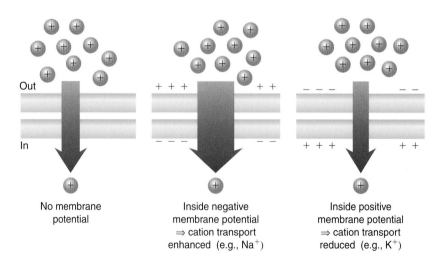

Figure 12-24. Electrochemical gradient established by charges on both sides of a cell membrane.

Calcium ion is a second messenger and is involved in many cellular processes. The potential to move calcium ion into the cell is great because the concentration of calcium ion is vastly greater outside the cell than its intracellular concentration (Table 12-3). Ca^{2+} is actively pumped out of the cell with the energy of ATP by pumps in the plasma membrane and in the endoplasmic reticulum (intracellular calcium store). A hormonal signal can evoke the release of calcium ion from the calcium store when this ion is needed for depolarization to produce exocytosis of another protein or hormone to be released from the cell or when calcium is needed for many other processes, such as muscular contraction, cell proliferation, neuronal activity, or cell death (Figure 12-25). A hormonal signal acting through a receptor

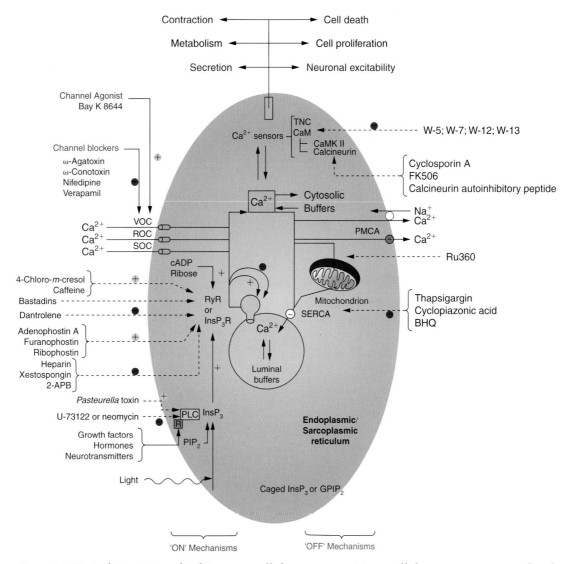

Figure 12-25. Calcium is involved in many cellular processes. Many cellular processes are regulated by the second messenger Ca^{2+}, which is derived from two separate sources. The ON mechanisms depend on Ca^{2+} entry through channels in the plasma membrane or Ca^{2+} release through ryanodine receptors (RYRs) or inositol trisphosphate receptors (InsP₃Rs). The OFF mechanisms remove Ca^{2+} from the cytoplasm using pumps. Also illustrated are the sites of action of some of the products capable of affecting these ON and OFF mechanisms. Redrawn from http://www.emdbiosciences.com/html/cbc/calcium_signaling.htm.

Ions and Gradients

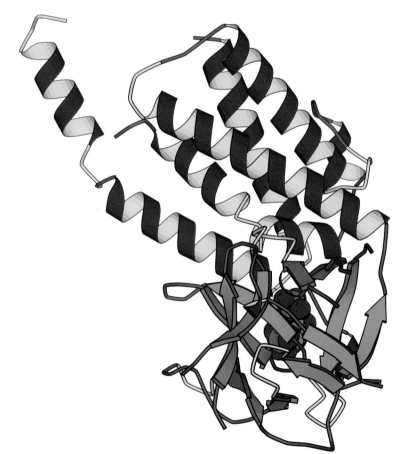

Figure 12-26. External signaling pathway leading to the release of endoplasmic reticulum stores of Ca^{2+} into the cytoplasm, as well as uptake of Ca^{2+} from the extracellular space through a Na^+/Ca^{2+} cotransporter (TRP channel). Redrawn from http://dir.niehs.nih.gov/dir1st/groups/birnbaumer/images/fig-pathway.jpg. Dr. Mariel Birnbaumer, National Institute of Environmental Health Sciences, U.S. Institutes of Health, Laboratory of Signal Transduction.

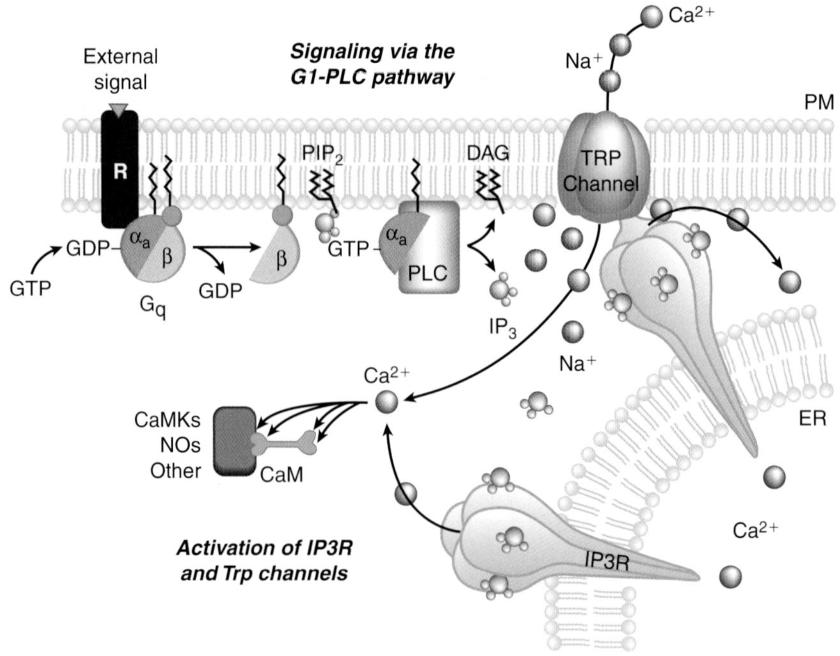

Figure 12-27. Crystal structure of the IP_3 receptor binding core in complex with IP_3. Reproduced from PDBsum ID: 1n4k. I. Bosanac et al. Structure of the Inositol 1, 4, 5-Trisphosphate Receptor Binding Core in Complex with Its Ligand. *Nature* **420** pp. 696 (2002), http://www.fli-leibniz.de/cgi-bin/imgLib.pl?code-1n4k.

at the cell membrane can operate through the phospholipase C pathway to produce **diacylglycerol (DAG)** and **inositol triphosphate (IP$_3$)** from phosphoinositol-*bis* phosphate (PIP$_2$). IP$_3$ is an intracellular signal for the release of Ca^{2+} from internal stores; it binds to the IP$_3$ receptor (IP$_3$R) on the endoplasmic reticulum membrane, generating the release of calcium ions from the store through the IP$_3$ receptor channel (Figures 12-26 and 12-27). Recent studies suggest that IP$_3$ induces calcium ion pumping indirectly, and the channel may respond to the emptying of calcium ions rather than a direct activation of the channel. Calcium ions are admitted into the cell from the outside through a Na$^+$/Ca^{2+} cotransporter, the TRP channel, as shown in Figures 12-26 and 12-28.

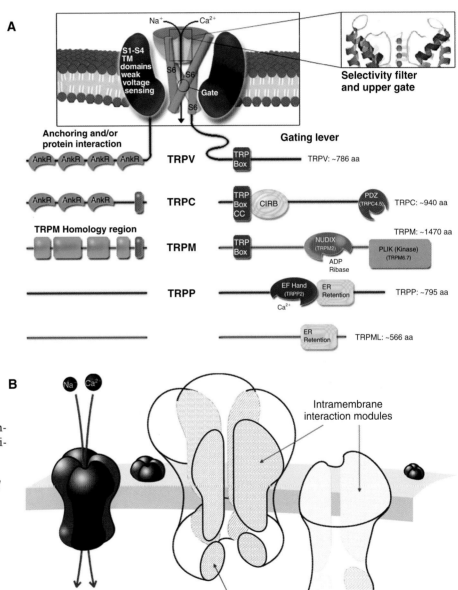

Figure 12-28. **A.** One of 5 TRP channels (Na$^+$/Ca^{2+} cotransporter). **B.** Diagram of the TRP channel. Part A reproduced with permission from http://www.stefanweb.com/figures/clapham2.jpg. J. Stefan Kaczmarek, Harvard Medical School. Part B redrawn from http://www.ukbf.fu-berlin.de/pharma/agSchaefer-Multi.html. Prof. Dr. med. Walter Rosenthalemail, Institut für Pharmakologie, Campus Benjamin Franklin, Charité-Universitätsmedizin, Berlin, Germany.

A second channel for the release of calcium ions from the endoplasmic reticulum to the cytoplasm, in addition to the IP$_3$R, is the ryanodine receptor (RyR), which operates with cyclic adenosine diphosphate ribose (cyclic ADPr) (Figure 12-29). Cyclic ADPr is a new second messenger for the intracellular release of Ca^{2+}, as also appears to be a metabolite of NADP, which is nicotinic acid adenine dinucleotide phosphate (NAADP). This structure is also shown in Figure 12-29. Presumably specific receptors will be characterized for these molecules. The structure of the outward-pumping RyR channel is shown in Figure 12-30. Calcium ion movements in and out of the cell are summarized in Figure 12-31. As shown in the figure, calcium ions leave the cell by way of Ca^{2+}/Na$^+$ antiports (exchangers) or by energized pumps, calcium ATPases. The structure of a calcium ATPase (outward pumping) is shown in Figure 12-32. Figure 12-33 provides a summary of the ligands that interface with G-protein-linked receptors on the cell surface that operate through IP$_3$ and DAG. DAG is an activator of protein kinase C, which plays a role in cell responses. At the bottom of the figure is shown the tyrosine kinase–linked receptors that operate through PIP$_3$ and the MAPK pathway to produce cellular responses, some of which lead to mitogenesis.

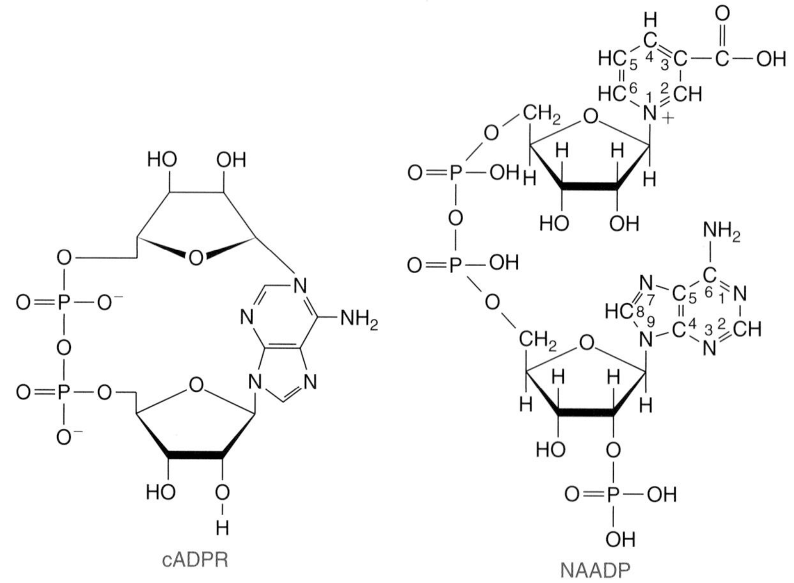

Figure 12-29. Structure of cyclic ADPR. The structure of NAADP (nicotinic acid adenine dinucleotide phosphate).

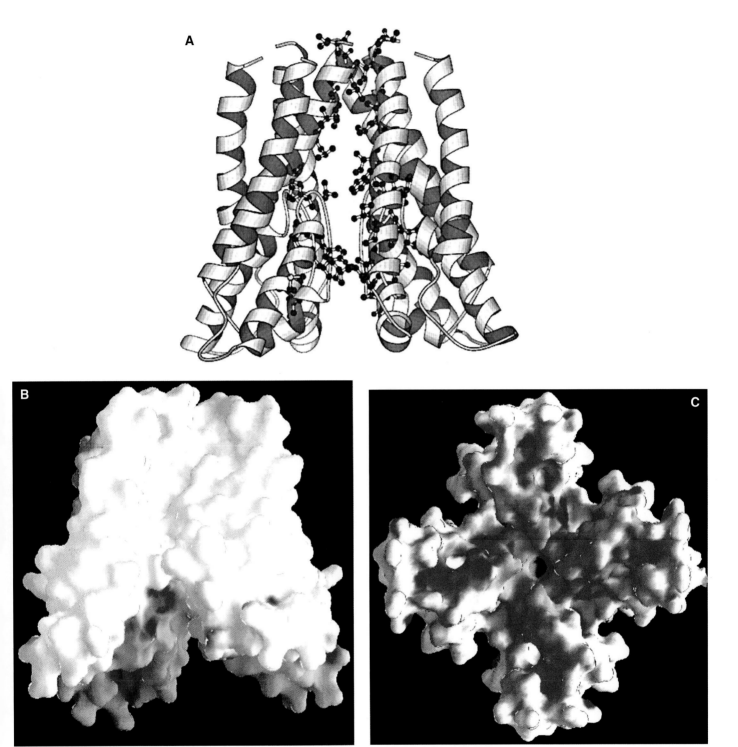

Figure 12-30. Three-dimensional model of the last two predicted TM helices of human ryanodine receptor (RyR). This corresponds to the highest conserved region among most (RyRs) and inositol triphosphate receptors (Insp₃Rs). Because of the similarity between Ca^{2+} channels and K^+ channels, the model was built by extrapolating from the K^+ channel structure. Ca^{2+} ions pass through the pore helix and the selectivity filter. Two transmembrane helices are shown to be important for channeling activity. **A.** Ribbon diagram of the Ca^{2+} channel tetramer shown using MOLSCRIPT. Several hydrophobic residues (shown for two adjacent protomers) line the protomer interface along the TM helices and stabilize the tetramer. **B.** Electrostatic potential representation of the Ca^{2+} channel tetramer. GRASP was employed for this representation. Acidic residues are indicated by red patches, and blue patches indicate basic residues. **C.** Same as *B* but down the tetramer pore helix axis. A broad red patch at the mouth of the channel shown in this three-dimensional model of the tetramer might explain how Ca^{2+} ions are attracted toward the channel. Reproduced with permission from Figure 4 of http://peds.oxfordjournals.org/cgi/content/full/14/11/867/F4. P.K. Shah and R. Sowdhamini, "Structural understanding of the transmembrane domains of inositol triphosphate receptors and ryanodine receptors towards calcium channeling," *PEDS*, 14: 867, 2001.

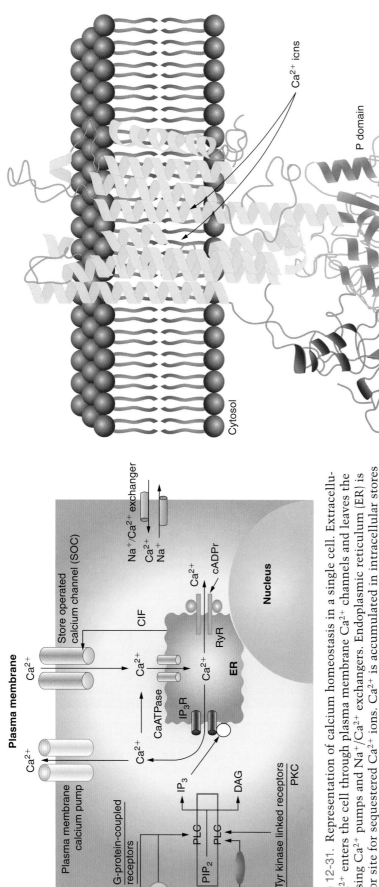

Figure 12-31. Representation of calcium homeostasis in a single cell. Extracellular Ca^{2+} enters the cell through plasma membrane Ca^{2+} channels and leaves the cell using Ca^{2+} pumps and Na^+/Ca^{2+} exchangers. Endoplasmic reticulum (ER) is a major site for sequestered Ca^{2+} ions. Ca^{2+} is accumulated in intracellular stores by means of Ca^{2+} pumps and released by inositol 1,4,5-trisphosphate (IP_3) via IP_3 receptors (IP_3R) and by cyclic adenosine diphosphate ribose (cADPr) via ryanodine receptors (RyR). Store-operated calcium channels (SOCs) open in response to depletion of the (ER) Ca^{2+} stores. Calcium influx factor (CIF) has been postulated to mediate the signal from IP_3R to the plasma membrane store-operated calcium channels (SOCs).

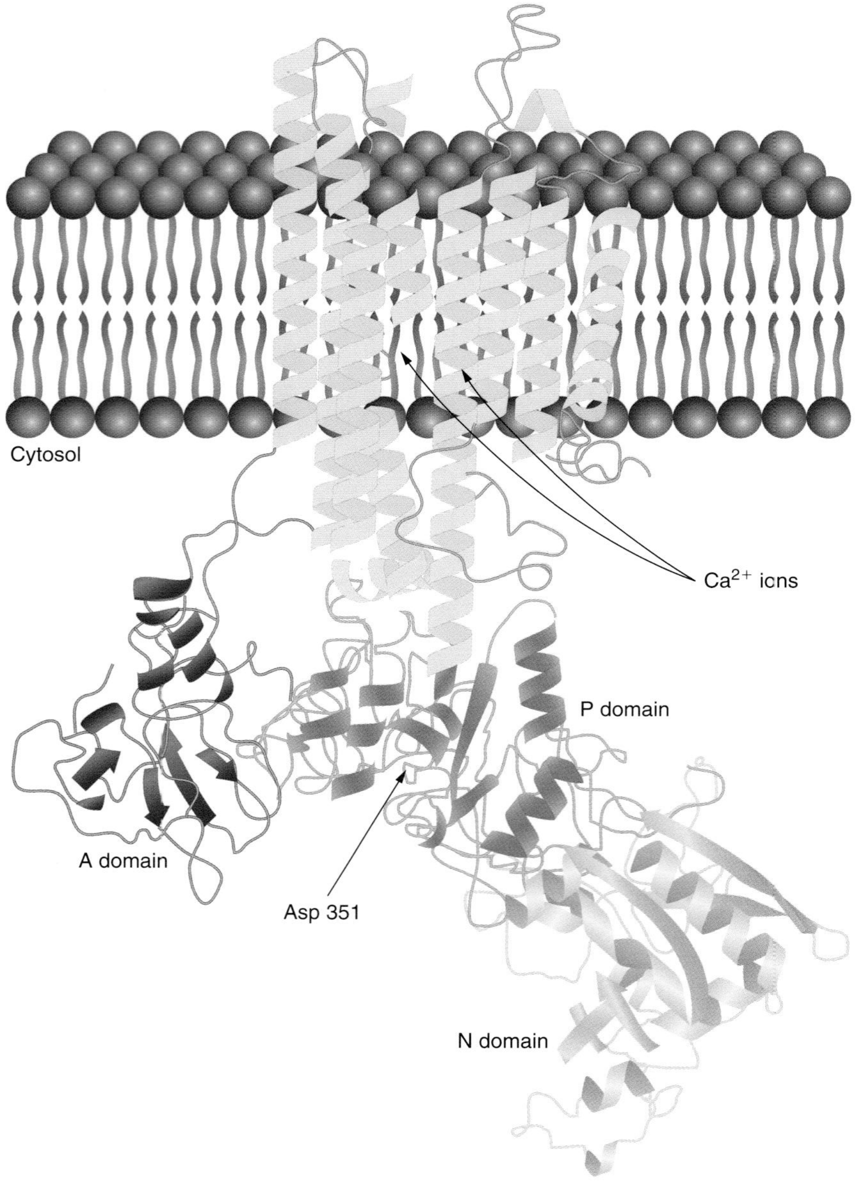

Figure 12-32. Structure of a sarcoplasmic reticulum Ca^{2+} ATPase.

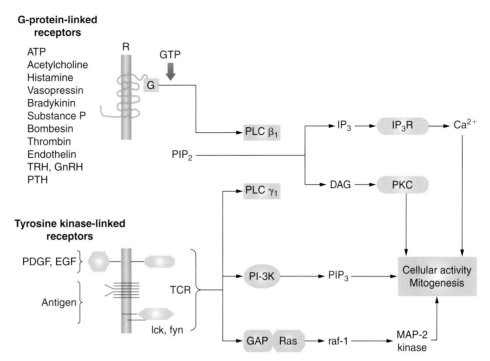

Figure 12-33. Summary of the two major receptor-mediated pathways for stimulating the formation of inositol triphosphate (IP_3) and diacylglycerol (DAG). Many agonists bind to 7-membrane spanning receptors (R), which use a GTP-binding protein (G) to activate phospholipase C-1 (PLC-1), whereas PLC-1 is stimulated by the tyrosine kinase-linked receptors. The latter activate other effectors, such as the phosphatidylinositol-3 kinase (PI-3K), which generates the putative lipid messenger phosphatidylinositol (3,4,5)-triphosphate (PIP_3) and the GTPase-activating protein (GAP) that regulates Ras. *GnRH*, gonadotrophin-releasing hormone; *IP_3R*, IP_3 receptor; *PKC*, protein kinase C; *PTH*, parathyroid hormone; *TRH*, thyrotropin-releasing hormone.

How Do Magnesium and Other Divalent Ions Enter Cells?

Only recently has evidence begun to emerge to characterize transporting molecules for magnesium ions (Mg^{2+}) and other divalent ions (Fe^{2+}, Co^{2+}, Ni^{2+}, Cu^{2+}, Zn^{2+}, Sr^{2+}, Cd^{2+}, Ni^{2+}, and Ba^{2+}). Magnesium, iron, and zinc are particularly important components of biological macromolecules, although some other trace metals occur in the structures of enzymes (cofactors). Aside from new evidence supporting claims for specific transporters, to be discussed, there are suggestions that organic components complex Mg^{2+} and form magnesium salts. The salt enters the outer layer of the cell membrane, the organic portion is metabolized, and the magnesium is released as the ion in the membrane. This substance is 2-aminoethylphosphoric acid. L-aspartate can form a magnesium salt, which enters the inner layer of the cell membrane, and its metabolism releases magnesium as the ion. Orotic acid can form a magnesium salt, which penetrates the outer cell membrane and is carried into the cell as such. It becomes metabolized on the membranes of the mitochondria and in the cytoplasm where the ion is released. Although these ideas are plausible, they are not specific and newer studies support the existence of specific transporters.

MagT1 is a unique magnesium ion transporter of 335 amino acids reported from the laboratory of Goytain and Quamme. It has five transmembrane regions and a number of phosphorylation sites. It has a wide tissue distribution. MagT1 mediates saturable Mg^{2+} voltage-dependent uptake with a Michaelis constant of 0.23 mM. The transporter is specific for Mg^{2+}. Another transporter reported from the laboratory of Goytain and Quamme, SLC41A2, is a magnesium ion transporter regulated by extracellular magnesium. It has a Michaelis constant for $[Mg^{2+}]$ of 0.34 mM and mediates voltage-dependent, saturable Mg^{2+} uptake. Unlike MagT1, it transports a range of divalent ions in addition to Mg^{2+}—Ba^{2+}, Ni^{2+}, Co^{2+}, Fe^{2+}, and Mn^{2+}—but not Ca^{2+}, Zn^{2+}, or Cu^{2+}. Mg^{2+} transport by this carrier is inhibited by large concentrations of Ca^{2+}. It is suggested that this transporter is involved in Mg^{2+} homeostasis in epithelial cells.

Another transporter, **ancient conserved domain protein (ACDP2)** transports Mg^{2+} with saturable uptake and a Michaelis constant of 0.56 mM in a voltage-dependent manner that is not coupled to Na^+ or Cl^- ions. Present in kidney, brain, and heart, with smaller amounts in liver, small intestine, and colon, this transporter transports, in addition to Mg^{2+}, Co^{2+}, Mn^{2+}, Sr^{2+}, Ba^{2+}, Cu^{2+}, and Fe^{2+} but not Ca^{2+}, Cd^{2+}, Zn^{2+} or Ni^{2+}. Magnesium deficiency up-regulates this transporter, particularly in the distal convoluted tubule cells, kidney, heart, and brain. In the brush border membrane of the upper small intestine, the **divalent ion transporter (DMT) 1** transports Mn^{2+}, Fe^{2+}, Co^{2+}, Ni^{2+}, and Cu^{2+}. This is an ATP-driven high-affinity copper transport system, which may be the main route for copper entry in the intestine. It is probable that *specific transporters are used by human cells to transport biologically important ions into cells.*

Proton (H$^+$) Transport

Movement of protons into a medium results in acidification, so transporters are designed to move protons wherever needed to maintain pH homeostasis. One such transporter is the sodium–proton antiport in which sodium and a proton move in opposite directions. Exchanging a proton for a sodium ion across a lumenal membrane will result in the acidification of the lumen, as shown in Figure 12-34. One effect of this transport is to drive bicarbonate ion into the cell against its concentration gradient. The Na^+/H^+ antiport is sensitive to **amiloride** (Figure 12-35), which classifies this transporter as amiloride-sensitive. Amiloride is a potent inhibitor of this channel and can be used to reduce Na^+ uptake into cells. Consequently, it is effective in the treatment of hypertension (a potassium-sparing diuretic) and edema (fluid retention), because the cellular uptake of Na^+ is followed by water uptake and blocking Na^+ uptake will block the consequent water uptake. The Na^+/H^+ antiporter and pH regulator occurs in most cells. There are seven known isoforms that are products of

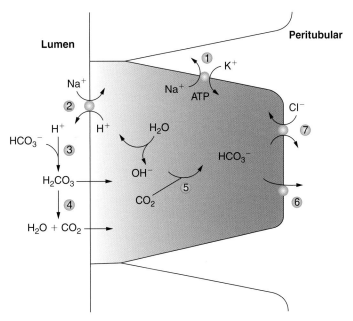

Figure 12-34. Cellular model of acidification involving the Na$^+$/H$^+$ antiport.

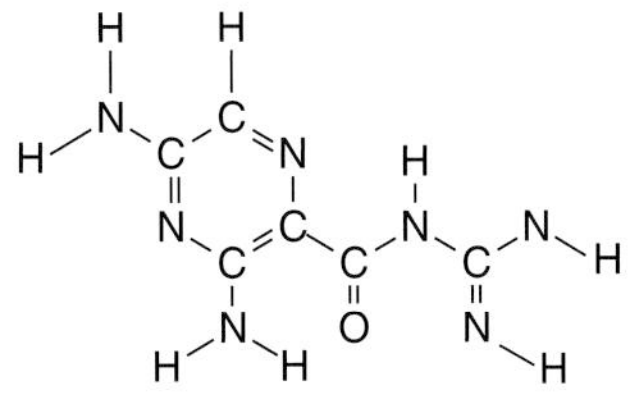

Figure 12-35. Structure of amiloride.

separate genes. NHE1 is the first one discovered that is present in the cell membranes of all tissues. It contains 2 transmembrane helices, like the monocarboxylate transporter family (shown later in this chapter in Figure 12-38). The C-terminal regulatory domain, containing about 315 amino acid residues, mediates reactions with cytoskeletal components. Transmembrane segments 4, 7, and 9 are important to the antiporter function. A schematic representation of the Na$^+$/H$^+$ antiporter and its connections to the cytoskeletal network is shown in Figure 12-36.

An example of a proton-linked channel is the monocarboxylate transporter (MCT), in which a monocarboxylic acid, such as lactate or pyruvate, is transported in tandem with the transport of a proton in the same direction. This type of transporter is critical to cellular metabolism at the cell membrane and the mitochondrial membrane, as can be seen in Figure 12-37. The MCT family of transporters has 12 transmembrane domains, as shown in Figure 12-38. The N-terminal region of this transporter family is important for energy coupling, membrane insertion, or

Proton (H$^+$) Transport

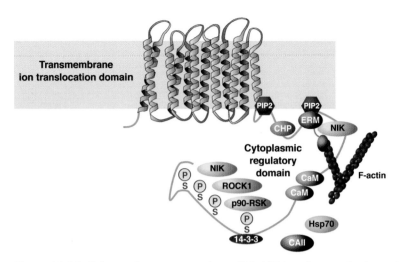

Figure 12-36. Schematic representation of Na$^+$/H$^+$ exchanger isoform 1 (NHE1) and interacting signaling molecules. Kinases are depicted in *yellow*, and regions of serine (S) phosphorylation (P) at the COOH terminus are indicated. Binding regions are, from NH$_2$ terminus to COOH terminus, phosphatidylinositol 4,5-bisphosphate (PIP$_2$) and ezrin, radixin, and moesin (ERM): 512–520 and 550–565; calcineurin homologous protein (CHP), 520–550; Nck-interacting kinase (NIK): 538–638; calmodulin (CaM): 640–686. Kinases that phosphorylate the COOH terminus include the p90-ribosomal protein S6 kinase (p90RSK), NIK, and Rho kinase 1 (ROCK1). Carbonic anhydrase II (CAII) and heat shock protein 70 (HSP70) bind to NHE1, but the binding regions have not yet been determined. Filamentous actin (F-actin) is shown in *red*. Reproduced with permission from http://ajpcell.physiology.org/cgi/content/full/287/4/C844/F1 and from Figure 1 of M. Baumgartner, H. Patel, and D.L. Barber, "Na$^+$/H$^+$ exchanger NHE1 as plasma membrane scaffold in the assembly of signaling complexes," *Am. J. Physiol. Cell Physiol.*, 287: C844-C850, 2004.

the correct structural maintenance, whereas the C-terminal portion is important for recognizing the substrate (that is, lactate or pyruvate) and its binding. There is a conserved aspartate residue (asp^{15} of human MCT1) in the N-terminal half of the protein that may be involved in proton binding, because this is a cotransporter of H$^+$ and a monocarboxylate molecule. This amino acid residue is the second residue on the cytoplasmic side before the beginning of the first transmembrane domain (starting at the N-terminus on the left of Figure 12-38). An arginine residue (arg^{313} in human MCT1) is a putative residue involved in binding the carboxylate anion (negative charge) in the C-terminal region.

In skeletal muscle, the lactate–proton cotransporter contains at least three isoforms, which have different properties and functional roles. The distribution of this transporter is fiber dependent in that the slow-twitch fibers have a higher capacity than the fast-twitch fibers. Lactate and proton effluxes from intense muscle activity and recovery are mediated principally by the lactate–proton transporter, accounting for the reduction in lactate concentration and the drop in internal pH that may be involved in muscle fatigue. So, the MCT is involved in the pH regulation associated with muscle activity. The MCT is also important for the uptake of lactate into resting muscle and other tissues. This transporter can

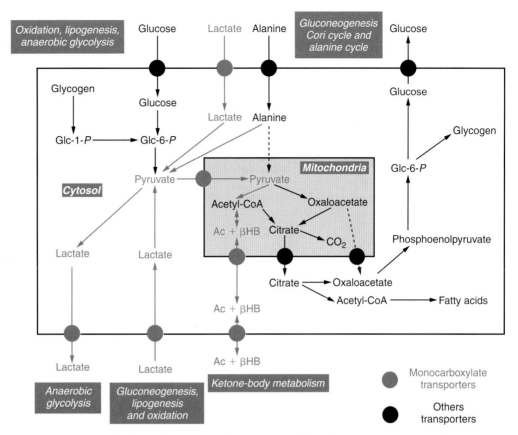

Figure 12-37. Metabolic pathways involving monocarboxylate transport across the mitochondrial and plasma membranes. *Glc-1-P*, glucose 1-phosphate; *Glc-6-P*, glucose 6-phosphate; *AC+βHB*, acetoacetate plus β-hydroxybutyrate. Redrawn from Figure 1 from http://www.biochemj.org/bj/343/0281/bj3430281.htm. A.P. Halestrap and N.T. Price, "The proton-linked monocarboxylate transporter (MCT) family: structure, function and regulation," *Biochem. J.*, 343: 281–299, 1999.

undergo adaptive changes because the capacity for transport of lactate and proton is increased by intense training and reduced by inactivity.

The citrate–proton transporter (symport) moves a proton and citrate from the mitochondrion to the cytoplasmic compartment (Figure 12-37) as a carbon source for fatty acid synthesis and gluconeogenesis. The gene for the human citrate transporter maps to chromosome 22 (22q11.21) in a region associated with allelic losses that produce the clinical diseases: **DiGeorge syndrome**, velo-cardio-facial syndrome, and a subtype of schizophrenia. The DiGeorge syndrome results from a chromosome 22q11 microdeletion. It is classified with velo-cardio-facial syndrome and is characterized by several lesions of the heart. Most patients have abnormalities of the palate, including cleft palate, and facial irregularities including small mouth, wide-set and down-slanting eyes, long face, and other structural defects. Learning disorders are not uncommon. Parents with the 22q11 microdeletion will pass the deletion to half of their offspring. Alcohol consumption during fetal development may be a causative factor in the microdeletion.

In addition to the citrate–proton transporter, another citrate transporter is coupled to sodium transport (Figure 12-22), the citrate–sodium transporter (NaCT). It is expressed predominantly in human liver but has

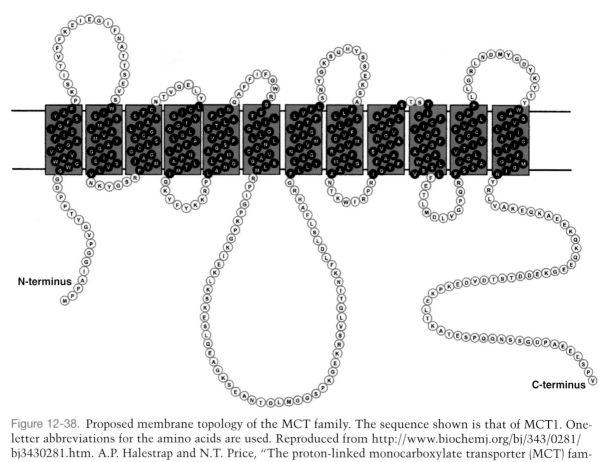

Figure 12-38. Proposed membrane topology of the MCT family. The sequence shown is that of MCT1. One-letter abbreviations for the amino acids are used. Reproduced from http://www.biochemj.org/bj/343/0281/bj3430281.htm. A.P. Halestrap and N.T. Price, "The proton-linked monocarboxylate transporter (MCT) family: structure, function and regulation," *Biochem. J.*, 343: 281–299, 1999.

been found also in the liver, brain, and testes of experimental animals. The transporter interacts with other di- and tricarboxylates but with lower affinity than with citrate. NaCT contains 572 amino acid residues and has structural similarities to members of the Na^+-dicarboxylate cotransporter/Na^+-sulfate cotransporter (NaDC/NaSi) gene family. In the brains of experimental animals, NaCT is expressed in cerebral cortex, cerebellum, hippocampus, and olfactory bulb. NaCT is important in the cellular use of blood citrate for the synthesis of fatty acids and cholesterol in the liver and for the generation of energy in the brain. Interestingly, human NaCT can become occupied by Li^+, which causes a stimulation of NaCT activity, but this stimulation does not occur with animal NaCT. Lithium replaces two of the four Na^+-binding sites in the transporter, and this effect is abolished when a single base mutation in the gene for NaCT is made that replaces a phenylalanine residue with leucine. Stimulation of NaCT may be an important consequence of the use of lithium in the treatment of bipolar disease, because lithium concentrations that produce the stimulation of NaCT are equivalent to concentrations found in humans treated for the disease.

Previously, it was mentioned that proteins are digested in the stomach and intestine and are absorbed in the intestine as single amino acids or as di- or tripeptides. The transport of di- and tripeptides across the luminal border of the intestine is mediated by the human **proton-dependent**

dipeptide transporter (PEPT1). Free amino acids and peptides containing four or more amino acid residues are not transported by PEPT1. PEPT1 is located in the apical membrane of enterocytes in the intestine, where it operates as an electrogenic proton–peptide transporter of di- and tripeptides. There is a variable proton to peptide stoichiometry, allowing for the uptake of neutral and singly charged or multiply charged peptides. PEPT1 has a large capacity and accounts for the efficient uptake of peptidic drugs and amino acid–conjugated drugs. Single nucleotide polymorphisms in the gene for PEPT1 have been discovered, and these conditions lead to a reduced transport capacity. There is also a second transporter called PEPT2. These are members of the **proton-dependent oligopeptide transporters (POT)** gene family. Two human members of this family have been identified, hPHT1 and hPHT2. hPHT1 is expressed primarily in skeletal muscle and spleen, and hPHT2 is found in spleen, placenta, lung, leukocytes, and heart. It appears that the human gene family will be quite complex. This family of transporters is responsible for the absorption and distribution of cephalosporins and other proteoid drugs. The predicted structures of hPHT and hPHT2, which have 11 or 12 transmembrane domains, are shown in Figure 12-39.

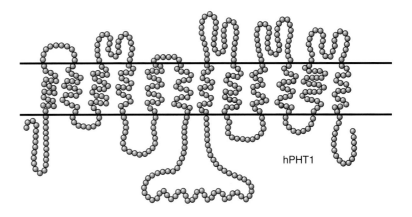

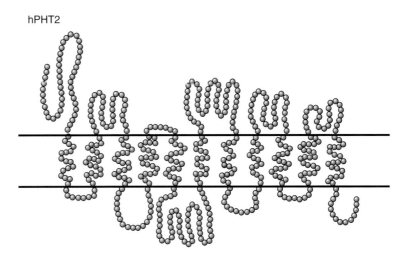

Figure 12-39. Predicted structures of hPHT1 and hPHT2. Redrawn from Figure 4 of http://www.aapsj.org/view.asp?art=ps020216. C.W. Botka et al., "Human proton/oligopeptide transporter (POT) genes: identification of putative human genes using bioinformatics," *AAPS Pharm. Sci.*, 2: 1208, 2000.

Amino Acid Transporters

The amino acid transporter family consists of a variety of transporters grouped according to the specificity of the amino acid molecule being transported and the dependence on Na^+ for the activity. The L-amino acid transporter (LAT1) has 12 membrane-spanning domains (MSDs) and is different from the other 12 MSDs transporters in that it uses an additional single membrane–spanning protein called 4F2 heavy chain (4F2hc:CD98) for its function. This transporter is independent of Na^+ and transports neutral amino acids, such as leucine, isoleucine, and valine, through the plasma membrane, as well as aromatic and branched chain amino acids and amino acid–related drugs, such as L-DOPA. It is one of the major transport systems in the **blood-brain barrier (BBB)**. The LAT1 transporter is pictured schematically in Figure 12-40. The heavy subunit is involved in movement of the heterodimer to the plasma membrane, and the light subunit confers transport function and specificity. Similar transporters are part of the exchange between mother and fetus, where 15 transport systems for amino acids have been identified in the human placenta. In addition to the L-amino acid transporters system, the main transporter for cationic amino acids is system y^+, whose transport

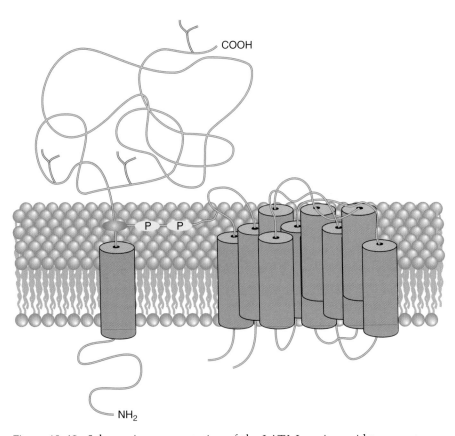

Figure 12-40. Schematic representation of the LAT1 L-amino acid transporter (4F2hc) in a heterodimeric complex with a single transmembrane-spanning protein required for its transporter function.

function is independent of Na⁺. This system allows lysine to be transported from the maternal side to the fetal side.

Another cell membrane system, the human amino acid transporter **hNAT3,** consists of 547 amino acid residues, which has its highest specificity for the transport of L-alanine. Its gene is located on human chromosome 12 (12q12-q13). hNAT3 is a low-affinity transporter that is Na⁺ and pH dependent. Some of its Na⁺-binding sites can be replaced by Li⁺, as might occur in patients being treated for bipolar disease. Its chief location is the liver, with lower concentrations in muscle, kidney, and pancreas.

Human ATB^{0+} (hATB^{0+}) is a novel member of the Na⁺/Cl⁻-dependent neurotransmitter transporter family, with some sequence similarity to glycine and proline transporters. It is expressed in trachea, salivary gland, mammary gland, stomach, and pituitary gland. It transports both neutral and cationic (positively charged) amino acids, with highest affinity for hydrophobic amino acids and lowest for proline. Transport depends on Na⁺ and Cl⁻ and is inhibited by an inhibitor of the B^{0+} system, characterizing it as a member of this transport system.

Cystine–glutamate antiporter is another system having a heavy (4F2hc; 12 transmembrane domains) and a light (xCT; single transmembrane domain) subunit dedicated to the exchange of anionic L-cystine and L-glutamate by a 1:1 exchange (system x_c^-). This system takes up cystine in exchange for the efflux of glutamate because of the low intracellular concentration of cystine (cystine is rapidly reduced in the cell to cysteine for synthesis of GSH) and the high intracellular concentration of glutamate. xCT provides an antioxidant defense mechanism, especially in regions of inflammation. The location of this system in brain suggests that it contributes to the maintenance of the redox state in cerebrospinal fluid. The predicted structure of the heavy subunit, 4F2hc, having 12 transmembrane domains, is shown in Figure 12-41.

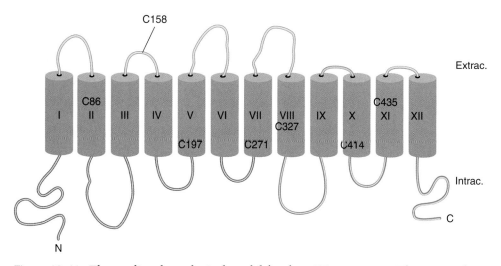

Figure 12-41. The predicted topological model for the xCT transporter. The protein has 12 transmembrane domains with intracellular N and C termini. Location of the seven endogenous cysteines is shown.

Table 12-4
Excitatory Amino Acid Transporters (EAATs)

	EAAT1	EAAT2	EAAT3	EAAT4	EAAT5
Currently Accepted Name					
Alternate Name	GLAST	GLT-1	EAAC1	None	None
Structural Information	542 aa (human)	574 aa (human)	525 aa (human)	564 aa (human)	561 aa (human)
Uptake Inhibitors	trans-2,4-PDC (P 7575)[a]	trans-2,4-PDC (P 7575)[a] Dihydrokainate (D 1064)[b] Kainate (K 0250)[b] TBOA[c] trans-2,3-PDC[d]	trans-2,4-PDC (P 7575)[a]	trans-2,4-PDC (P 7575)[a]	trans-2,4-PDC (P 7575)[a]
Radiolabeled Substrates[e]	L-[^{3}H]-Glutamate D-[^{3}H]-Aspartate	L-[^{3}H]-Glutamate D-[^{3}H]-Aspartate	L-[^{3}H]-Glutamate D-[^{3}H]-Aspartate	L-[^{3}H]-Glutamate D-[^{3}H]-Aspartate	L-[^{3}H]-Glutamate D-[^{3}H]-Aspartate

Abbreviations
trans-2,3-PDC: L-trans-Pyrrolidine-2,3-dicarboxylic acid
trans-2,4-PDC: L-trans-Pyrrolidine-2,4-dicarboxylic acid
TBOA: threo-β-Benzyloxyaspartate

[a] trans-2,4-PDC is a nonselective EAAT inhibitor. K_i values for EAATs 2, 4, and 5 are between 5 and 10 μM and for EAAT1 and EAAT3 are between 50 and 100 μM.
[b] Selective inhibitors of the EAAT2 subtype with K_i values between 15 and 60 μM.
[c] TBOA is a nonsubstrate inhibitor of EAAT2 and has seven-fold selectivity over EAAT1. Not fully characterized with respect to other EAATs. Also inhibits EAAT3. Nonselective.
[d] trans-2,3-PDC is a selective nonsubstrate inhibitor of EAAT2 (compared with EAAT1 and EAAT3).
[e] L-Glutamate as the endogenous substrate and D-aspartate, a nonmetabolizable substrate analogue are most frequently employed in uptake studies.
Reproduced from page 27 of http://www.sigmaaldrich.com/area_of_interest/Life_Science/Cell_Signaling/Key_Resources/eHandbook/eHandbook.html.

The glutamate synapse contains vesicles that import glutamate from the cytosol in exchange with protons (glutamate–proton antiport). The vesicular antiporter has a high specificity for L-glutamate and uses a favorable proton gradient, exchanging a proton for a glutamate to accumulate glutamate against its concentration gradient (the concentration of glutamate in the cytosol is high). ATP drives the protons inside the vesicle, and the high proton concentration inside the vesicle is used by the transporter. The **vesicular monoamine transporter (VMAT)** is coupled to proton transport as well.

Excitatory amino acid transporters (EAATs) are important in the central nervous system. Glutaminergic neurotransmission at excitatory synapses is terminated by sodium-dependent L-glutamate transporters present on the plasma membranes of neurons and astroglia. These transporters produce an efficient clearance of extracellular L-glutamate driven by the sodium gradient across the membranes. In addition to L-glutamate transport, the other excitatory amino acids, L-aspartate, and L-cysteine are substrates of this transporter. There are five human EAATs, EAAT1–5, of which EAAT1 and EAAT2 are the two most abundant. These two transporters are located in astroglia throughout the central nervous system except for the retina, where EAAT2 is neuronal. Before partitioning into astroglia of the adult, the developing nervous system contains EAAT2. EAAT4 is located predominantly in the Purkinje cells of the cerebellum. EAAT3 is located in both neurons and astroglia, as well as in other tissues. EAAT5 is located in the retina. The EAATs remove L-glutamate from the extracellular space, preventing L-glutamate toxicity. Experimentally, a deficiency of EAAT1 results in visual and movement abnormalities, and dysfunction of these transporters is likely to be involved in neurodegenerative diseases, central nervous system disorders, and psychiatric disorders. Information about EAATs is summarized in Table 12-4.

Fatty Acid Uptake

A six-member family of membrane-associated proteins is called **fatty acid transport proteins (FATP)**, which are found in all tissues that use fatty acids. The transporters that have been characterized are designated FATP1-6. FATP1, FATP2, and FATP6 each have a highly conserved AMP-binding region that takes part in the activation of very long–chain fatty acid (VLCFA) formation of acyl CoA derivatives. The uptake into the cell of VLCFAs is closely related to acyl CoA synthase activity. The CoA esters are required for the structural integrity of lipid rafts in cellular membranes. These rafts are required for the cellular uptake of LCFAs. Lipid rafts are constructed in the membrane around cholesterol molecules, and raft formation is stimulated by palmitoylation. The structure of a raft onto which specific proteins may be glycosylphosphatidylinositol-anchored is shown in Figure 12-42. Figure 12-43 shows that resting rafts preformed in the plasma membrane can be aggregated. In this example, the CD45 protein (a receptor-like protein with a glycosylated external domain and a large cytoplasmic domain containing two protein tyrosine phosphatases) has access to its substrates. In the aggregated state, CD45 is excluded, allowing the accumulation of protein phosphotyrosines on raft-associated proteins. Thus, aggregation of rafts is a signaling mechanism. The enzyme of the plasma membrane, fatty acid translocase (FAT/CD36), *is located exclusively on lipid rafts, and this enzyme is required for the uptake of LCFAs*, such as oleic acid in adipocytes. The transport of LCFAs into mitochondria from the soluble cell cytoplasm (cytosol) is a function of the mitochondrial carnitine system (Figure 5-30). The transport system consists of malonyl-CoA–sensitive carnitine palmitoyltransferase (CPT-1), which is localized in the outer mitochondrial membrane; carnitine acylcarnitine translocase, which is localized in the mitochondrial inner membrane; and carnitine palmitoyltransferase II, which is located on the matrix side of the inner mitochondrial membrane. Two proposed models for the LCFA uptake into mitochondria from the cell cytosol are shown in Figure 12-44. The overall transport of fatty acids from the extracellular space into the skeletal muscle cell and into particulates of the cell is shown in Figure 12-45. This shows that the fatty acid is bound to a fatty acid–binding protein (FABP), which carries it to the plasma membrane FAT/CD36 and allows entry into the cytoplasm, where it can undergo further metabolism. Entry of fatty acids into the other subcellular particles also is mediated by fatty acid translocase

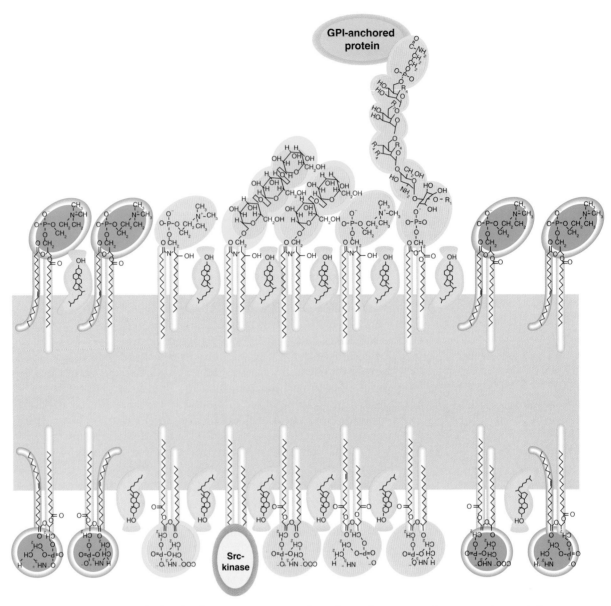

Figure 12-42. Structure of a lipid raft in cross-section. The raft on the inner and outer surfaces of the membrane are in yellow and cholesterol molecules are in orange. Redrawn from http://www.emblheidelberg.de/nmr/sattler/private/filipp/1sl/practicals/03_biocomputing/lipid_raft.jpg. Fabian Volker Filipp, EMBL Heidelberg, Structural and Computational Biology, NMR Sattler Group, Heidelberg, Germany.

(FAT/CD36) as shown. In cardiac myocytes, contraction is the major physiological stimulus that can cause the movement of FAT/CD36 from intracellular stores to the **sarcolemma** (the plasma membrane covering the outer surface of a muscle fiber), or the phosphorylation of the translocase already at the sarcolemma is a process that could enhance the uptake of fatty acids. The crystal structure (unavailable for mammals) of a fatty acid transporter from *Escherichia coli* is shown in Figure 12-46.

(Text continues on p. 729.)

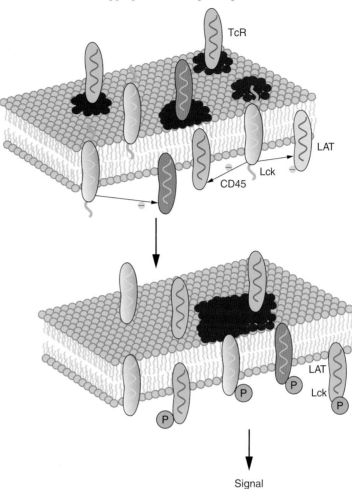

Figure 12-43. Upper figure shows lipid rafts in the membrane in orange with proteins attached on the inner and out sides of the membrane. The lower figure shows aggregation of the rafts in which certain inner bound proteins are excluded (proteins shown in *blue* and *light green*). CD45 is shown in light green and it is excluded by raft aggregation so that remaining proteins can be phosphorylated on tyrosine residues in this example. Phosphatidylmositol lipid metabolism occurs in lipid rafts in response to T cell antigen (TCR in *green*) triggering.

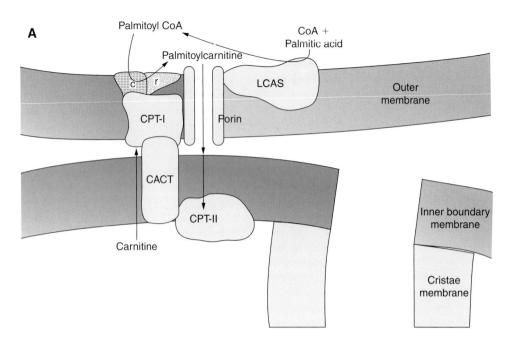

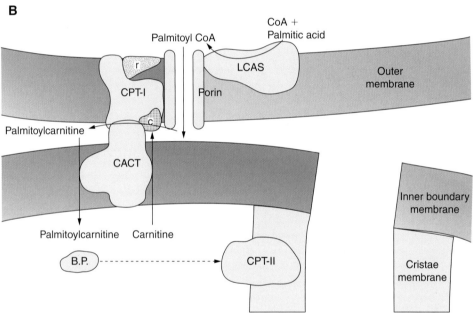

Figure 12-44. **A.** Proposed model of mitochondrial fatty acid uptake with the malonyl-CoA binding site (r) and catalytic site (c) of CPT-I exposed to the cytosol. *B.P.*, postulated matrix binding protein for palmitoylcarnitine; *CACT*, carnitine/acylcarnitine translocase; *CPT-I*, malonyl-CoA–sensitive carnitine palmitoyltransferase; *CPT-II*, malonyl-CoA insensitive carnitine palmitoyltransferase; *LCAS*, long-chain acyl-CoA synthetase. **B.** Proposed model of mitochondrial fatty acid uptake with the malonyl-CoA–binding site *(r)* of CPT-I exposed to the cytosol and the catalytic site *(c)* exposed to the intermembrane face of the outer membrane. Parts A and B were redrawn from figures 1 and 2 of J. Kerner and C. Hoppel, "Fatty acid import into mitochondria," *Biochem. Biophys. Acta. Mol. Cell Biol. Lipids*, 1486: 1–17, 2000.

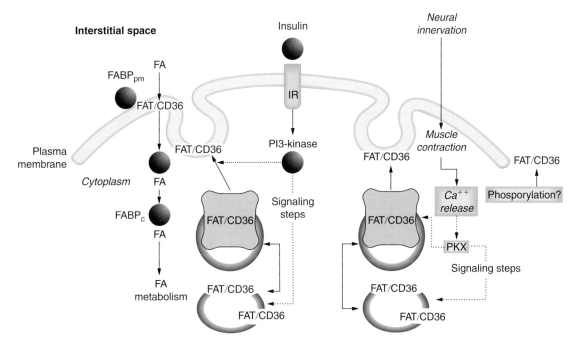

Figure 12-45. Roles of fatty acid translocase in movement across the muscle plasma membrane and to internal particles. Fatty acids (FA) cross the plasma membrane by FA transport or binding proteins. FAs are trapped by the peripheral membrane protein FABPpm to be donated to FAT/CD36 at the plasma membrane. Inside the cell, the FA is complexed to FABPc and then metabolized. FAT/CD36 can be recruited to the plasma membrane from an intracellular pool by muscle contraction or insulin involving the PI 3 kinase signaling system. There may be separate insulin- and contraction-sensitive intracellular depots of FAT/CD36, since insulin and contraction stimulate FA uptake additively in isolated muscles. Contraction-induced translocation may involve a Ca^{2+}-activated signaling system and the phosphorylation surface of FAT/CD36 may be a mechanism to increase transport capacity of plasma membrane associated FAT/CD36. Redrawn with permission from Figure 8 of A. Bonen et al., "Regulation of fatty acid transport and membrane transporters in health and disease," *Mol. Cel. Biochem.*, 239: 181–192, 2002.

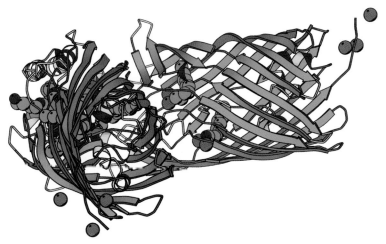

Figure 12-46. Crystal structure of the bacterial fatty acid transporter (FADL) from *E. coli*. Reproduced with permission from Leibniz Institute for Age Research-Fritz Lipmann Institute (formerly IMB Jena), Germany. B. Van Den Berg et al., "Crystal structure of the long-chain fatty acid transporter FADL," *Science*, 304: 1506, 2004.

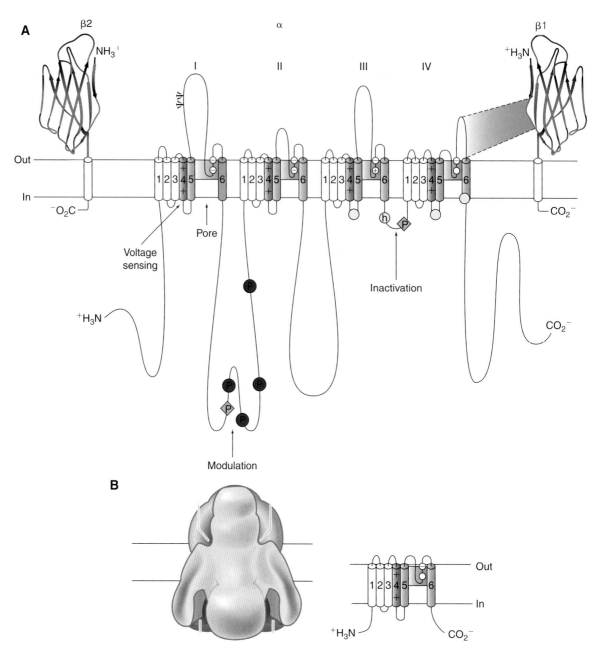

Figure 12-47. Structure of voltage-gated sodium channels. **A.** Schematic representation of the sodium-channel subunits. The α subunit of the $Na_v1.2$ channel is illustrated together with the β1 and β2 subunits; the extracellular domains of the β subunits are shown as immunoglobulin-like folds, which interact with the loops in the α subunits as shown. Roman numerals indicate the domains of the α subunit; segments 5 and 6 (shown in *orange*) are the pore-lining segments and the S4 helices *(yellow)* make up the voltage sensors. *Yellow circles* in the intracellular loops of domains III and IV indicate the inactivation gate IFM motif and its receptor (h, inactivation gate); P, phosphorylation sites (in *red circles*, sites for protein kinase A; in *green* diamonds, sites for protein kinase C); Ψ, probable *N*-linked glycosylation site. The circles in the re-entrant loops in each domain represent the amino acids that form the ion selectivity filter (the outer rings have the sequence EEDD and inner rings DEKA). **B.** The three-dimensional structure of the Na_v channel α-subunit at 20Å resolution, compiled from electron micrograph reconstructions. **C.** Schematic representation of NaChBac, the bacterial voltage-gated sodium channel. Redrawn with permission from G.G. Matthews, *Cellular Physiology of Nerve and Muscle*, 2nd ed., Blackwell Scientific, Boston, 1991 and http://genombiology.com/2003/4/3/207/figure/F1?highres=y.

Sodium Conductance and Voltage-Gated Sodium Channels

The sodium–proton antiport was discussed earlier under proton transport. Sodium channels are important in the initiation and propagation of **action potentials** in neurons. They are also important in endocrine cells and in myocytes. As discussed previously, the hormone aldosterone stimulates the synthesis of **ENaC** (Figure 9-17), the sodium conductance channel in the renal tubular cell (Figure 9-14), but it is not completely related to the voltage-gated sodium channels.

Only a few millivolts are required to depolarize a cell membrane, which results in the activation of sodium channels. Then, just as rapidly, they are inactivated again. Sodium ions enter through the activated channel and further depolarize the membrane, which raises the action potential. The channel is composed of an α-subunit and one or more β-subunits. The β-subunits are required for the opening and closing of the channel, and the α-subunit is the ion-conducting aqueous pore. The α-subunit is 260 kDa, and the β-subunits are about 35 kDa each. The α-subunit folds into four domains (I through IV) and contains six α-helical transmembrane segments (S1–S6). The voltage sensor is in the S4 segments containing basic amino acid residues (positively charged) every three amino acids. Amino and carboxyl termini are part of the face of the sodium channel. Although the three-dimensional structure is not complete, the evidence points to a structure like that shown in Figure 12-47. Another view of the structure of this channel is shown in Figure 12-48. Cell skeletal proteins, **ankyrin** and **neurofascin** (Figure 12-49), are associated with the sodium

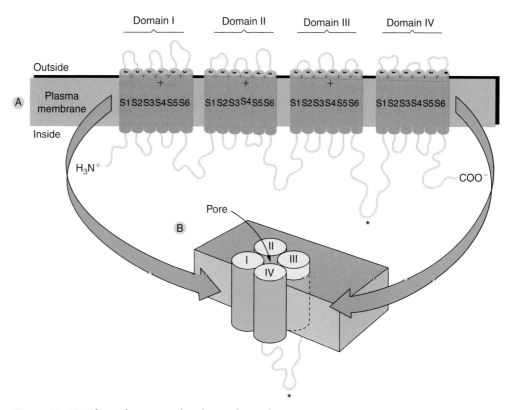

Figure 12-48. The voltage-gated sodium channel.

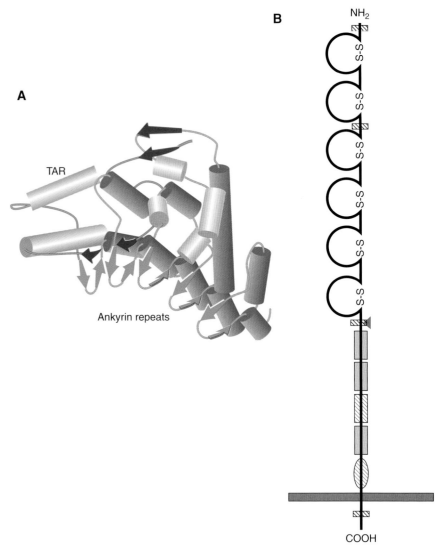

Figure 12-49. **A.** Structure of the ankyrin repeat domain. Ankyrin repeats for β-β-α-α motifs that self-associate. TAR (transcriptional activation region) interacts extensively with the ankyrin repeat stack and is inactive in this conformation. **B.** Basic domain organization of the neurofascin polypeptide. Ig-like domains of the C2 subcategory are shown as loops and are closed by putative disulphide bridges. FNIII-related repeats are represented as rectangles and the segment rich in proline, alanine, and threonine is indicated by an ellipse. The cytoplasmic domain is indicated by a short line at the COOH terminus. The third FNIII-like repeat and the PAT domain are hatched to indicate that they might be alternatively spliced; other potential pre-mRNA splice variants in the Ig-, between the Ig- and the FNIII-like region and in the cytoplasmic domain, are indicated by small hatched boxes. The black arrowhead indicates a major proteolytic cleavage site.

channel. These channels are located at initial segments of axons, postsynaptic folds, and nodes of Ranvier (gaps between segments of the myelin sheath representing the junction between adjacent neuroglial cells; the gaps are about 1 mm apart along the length of the axon, as shown in Figure 12-50), where ankyrin is required for the clustering. Tetrodotoxin and saxitoxin (Figure 12-51) are blockers of these channels. There are high levels of these types of channels in brain, skeletal muscle, heart, uterus, and spinal cord. A number of diseases are associated with mutations of this channel. The superfamily of the voltage-gated sodium channels also includes voltage-gated potassium ion channels and voltage-gated calcium ion channels.

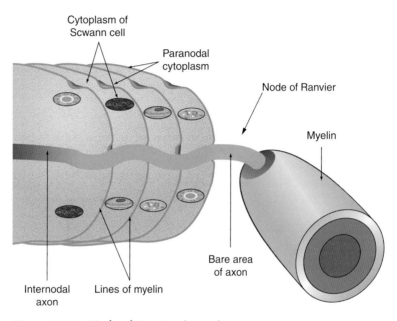

Figure 12-50. Node of Ranvier *(arrow)*.

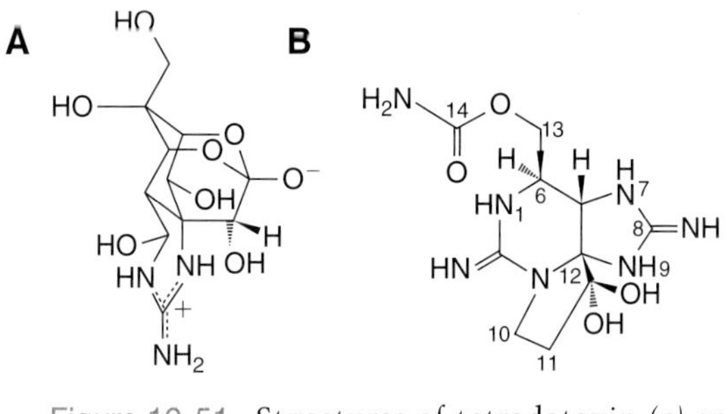

Figure 12-51. Structures of tetrodotoxin *(a)* and saxitoxin *(b)*.

Multidrug Resistance Channel (MDR) of the ABC Transporter Superfamily

The ABC superfamily is composed of seven known transporters, including the MDR channel (also P-glycoprotein of 170 kDa) and the CFTR discussed at the beginning of this chapter. The two channels are compared diagrammatically in Figure 12-52, showing strong similarity. MDR splits ATP and pumps chemotherapeutic drugs out of cancer cells, but recently, prompted by its similarity to the CFTR chloride ion channel, it has been found to be linked to volume-activated chloride ion channel activity. CFTR is a cyclic AMP–activated Cl^- channel, and it may be, in addition, a Cl^-/HCO_3^- exchanger (Figure 12-7). Models of the human MDR-related protein (190 kDa) MRP1 transmembrane domains are shown in Figure 12-53. Although there are many other transporters, the few covered in this chapter should serve as an overall introduction.

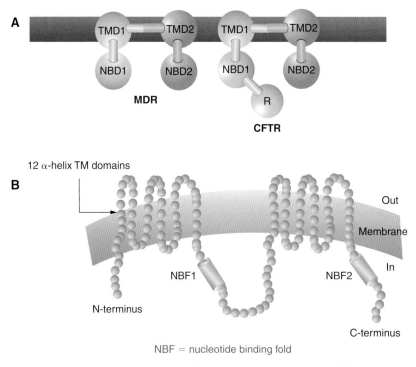

Figure 12-52. **A.** Comparison of MDR with CFTR. **B.** Model of MDR1 (P-glycoprotein). TM, transmembrane. Part A redrawn with permission from Figure 1 in part from D.B. Luckie et al., "CFTR and MDR: ABC transporters with homologous structure but divergent function," Curr. Genomics, 4: 109–121, 2003.

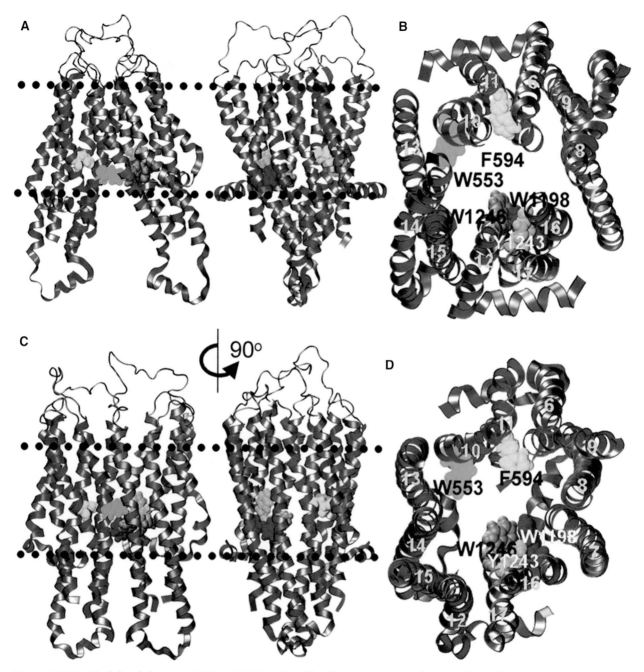

Figure 12-53. Models of the core TMDs (TM6 to TM17) of human MRP1. The models are based on VC-MsbA and closed EC-MsbA TMDs and are shown in *panels A* and *B* and *panels C* and *D*, respectively. Models are shown as viewed from the plane of the bilayer in *A* and *C* and from the extracellular surface in *B* and *D* (extracellular loops are shown as *gray tubes* and are removed in *B* and *D* for clarity). The Trp[553] *(W553)*, Phe[594] *(F594)*, Trp[1198] *(W1198)*, Tyr[1243] *(Y1243)*, and Trp[1246] *(W1246)* residues are shown in *green, yellow, red, blue,* and *orange*, respectively. *TMD*, transmembrane domain. Reproduced with permission from Figure 1 of http://www.jbc.org/cgi/content/full/279/1/463/FIG1. J.D. Campbell et al., "Molecular modeling correctly predicts the functional importance of Phe594 in transmembrane helix 11 of the multidrug resistance protein, MRP1 (ABCC1)," *J. Biol. Chem.*, 279: 463-468, 2004.

Blood-Brain Barrier

Only certain substances in the blood can enter the brain because of the mechanism of the BBB. Endothelial cells lining the blood vessels in the brain form a barrier between the blood circulation and the brain cells. The BBB cannot be penetrated by lymphocytes, monocytes, and neutrophils, so immune responses in the brain are limited. This protects the neuronal network from damage that could be caused by a full immune response. There is the capacity for a local immune response in the rare event that viral or fungal infections or prions occur in the brain. The endothelial cells are surrounded by a basement membrane made up of lamin, fibronectin (Figure 12-54), and other proteins acting as a barrier and mechanical supporter. Figure 12-55 shows the structure of the BBB

Figure 12-54. Structure of lamin A protein. **A.** Diagram of lamin A amino acid sequence showing the domains of the protein, and the position of the four laminopathy-associated missense mutations in DFPLD and AD-EDMD. **B.** Structure of the C-terminal globular domain of lamin A showing the relative positions of the FPLD associated R482W missense mutation and the AD-EDMD associated L530P mutation. Reproduced from http://www.ocms.ox.ac.uk/idc/structures/.

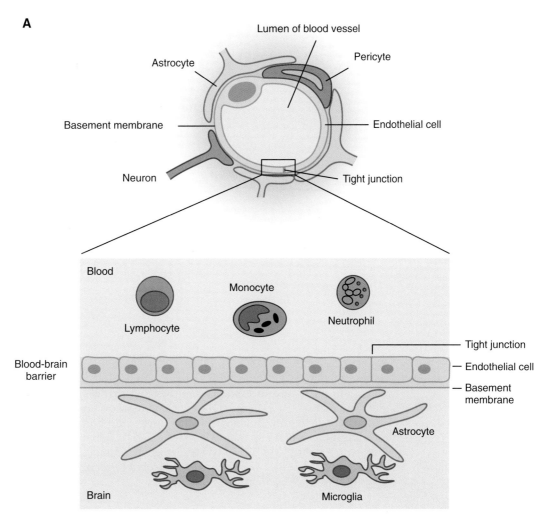

Figure 12-55. **A.** The blood-brain barrier (BBB), is created by the tight apposition of endothelial cells lining blood vessels in the brain, forming a barrier between the circulation and the brain parenchyma (astrocytes, microglia). Blood-borne immune cells such as lymphocytes, monocytes, and neutrophils cannot penetrate this barrier. A thin basement membrane, comprising lamin, fibronectin, and other proteins, surrounds the endothelial cells and associated pericytes, and provides mechanical support and a barrier function. Thus, the BBB is crucial for preventing infiltration of pathogens and restricting antibody-mediated immune responses in the central nervous system, as well as for preventing disorganisation of the fragile neural network. This, together with a generally muted immune environment within the brain itself, protects the fragile neuronal network from the risk of damage that could ensue from a full-blown immune response.

On rare occasions, pathogens (e.g., viruses, fungi, and prions) and autoreactive T cells breach the endothelial barrier and enter the brain. A local innate immune response is mounted in order to limit the infections challenge, and pathogens are destroyed and cell debris is removed, a vital process that must precede tissue repair.

Continued

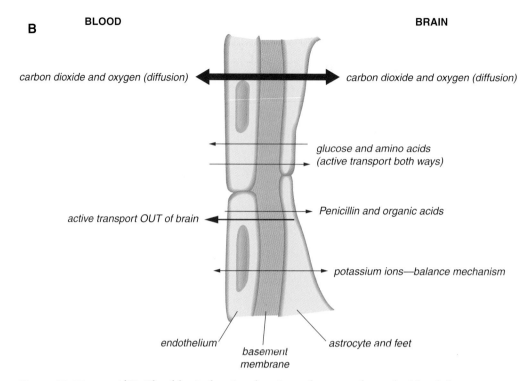

Figure 12-55, cont'd **B.** Blood-brain barrier showing substances from the blood that can pass through to the brain and substances in the brain that can pass through to the blood circulation. Part A reproduced with permission from http://www-ermm.cbcu.cam.ac.uk/03006264h.htm. K. Francis et al., "Innate immunity and brain inflammation: the key role of complement," *Exp. Rev. Mol. Med.*, 5: 1-19, 2003. Part B reproduced with permission from page 5 of http://137.222.110.150/calnet/Blood/image/blood-brain%20barrier-schematic%20diag.jpg. Dept of Anatomy, Univ of Bristol, CALnet Programmes for Veterinary Science Students.

and some components in blood that can enter through the barrier and leave the brain into the circulation. Figure 12-56 summarizes the transport mechanisms of the blood-brain barrier. For specialized secretions, for example, the anterior pituitary hormones, which are of a large size, enter the circulation through thin "windows" in the local capillaries called fenestrations, where the vessel is thinner and will admit molecules of higher molecular weight.

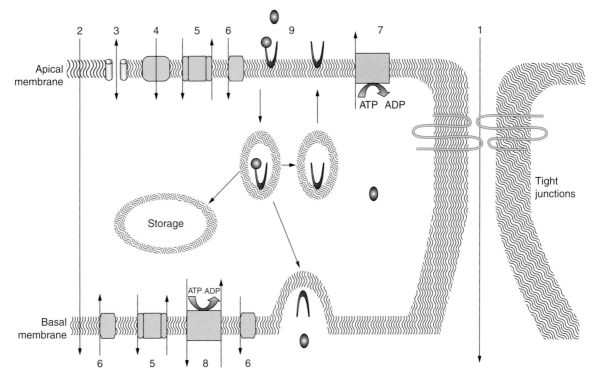

Figure 12-56. Transport mechanisms at the blood-brain barrier (BBB). *1*, paracellular diffusion (sucrose); *2*, transcellular diffusion (ethanol); *3*, ion channel (K^+ gated); *4*, ion-symport channel ($Na^+/K^+/Cl^-$ cotransporter); *5*, ion-antiport channel (Na^+/H^+ exchange); *6*, facilitated diffusion (glucose via GLUT-1); *7*, active efflux pump (P-glycoprotein); *8*, active-antiport transport (Na^+/K^+ ATPase); *9*, receptor-mediated endocytosis (transferrin and insulin). Redrawn from Figure 3 of http://users.ahsc.arizona.edu/davis/bbbfra1.gif.

Further Reading

Books

Layton, H.E., and Weinstein, A.M., *Membrane Transport and Renal Physiology*, Springer, 2002.

Schultz, S.G., *Basic Principles of Membrane Transport*, Cambridge University Press, 1980.

Stein, W.D., *Channels, Carriers and Pumps: an Introduction to Membrane Transport*, Academic Press, 1990.

Reviews

Goytain, A., and Quamme, G.A., "Functional characterization of the human solute carrier, SLC41A2," *Biochem. Biophys. Res. Commun.* **330:** 701–705, 2005.

Goytain, A., and Quamme, G.A., "Identification and characterization of a novel mammalian $Mg2^+$ transporter with channel-like properties," *BMC Genomics*, **1:** 48, 2005.

Inoue, K., Zhuang, L., Maddox, D.M., Smith, S.B., and Ganapathy, V., "Human sodium-coupled citrate transporter, the orthologue of *Drosophila indy*, as a novel target for lithium action," *Biochem. J.*, **374**: 21–26, 2003.

Inoue, K., Zhuang, L., Maddox, D.M., Smith, S.B., and Ganapathy, V., "Structure, function, and expression pattern of a novel sodium-coupled citrate transporter (NaCT) cloned from mammalian brain," *J. Biol. Chem.*, **277**: 39,469–39,476, 2002.

Kim, M.H., Billiar, T.R., and Seol, D.W., "The secretable form of trimeric TRAIL, a potent inducer of apoptosis," *Biochem. Biophys. Res. Commun.*, **321**: 930–935, 2004.

Knopfel, M., Smith, C., and Solioz, M., "ATP-driven copper transport across the intestinal brush border membrane," *Biochem. Biophys. Res. Commun.* **330**: 645–652, 2005.

Juel, C., "Lactate–proton cotransport in skeletal muscle," *Physiol. Rev.* **77**: 321–358, 1997.

CHAPTER 13

Dietary Metals, Iron, Micronutrients, and Nutrition

Iron-Deficiency Anemia

Iron-deficiency anemia occurs when inadequate amounts of iron are taken in the diet. It is the most prevalent deficiency in the world. It has economic consequences because of the diminished capacity of affected individuals to perform labor. In children, it has a blunting effect on growth and learning. Among women, 3.3 million are anemic and 7.8 million are iron deficient (4–8% of premenopausal women are iron deficient but not necessarily anemic). Disorders of iron deficiency affect up to 1 billion people worldwide. During a menstrual period, loss of iron ranges from 4 to 100 mg. With each pregnancy, a woman loses about 500 mg of iron. A newborn consumes iron-deficient milk, and infants consuming cow's milk have a greater risk of iron deficiency because cow's milk has a high level of calcium that competes with iron for absorption. A deficiency of dietary iron leads to a reduction in the number of red blood cells that depend on hemoglobin (iron component of heme) to carry oxygen to the tissues. Blood loss through trauma (external and internal bleeding), heavy menstruation, or internal bleeding from ulcers; tumors, particularly colon cancer; or other causes lead to anemia, as well as deficient dietary iron intake. Small blood losses from gastrointestinal bleeding can be causative of anemia yet may not be detected because the regular test for hemoglobin in the stool may not be sensitive enough. Aspirin and **nonsteroidal anti-inflammatory drugs (NSAIDs)** can stimulate internal bleeding, especially with ulcers. Dietary iron-deficiency anemia is most common in women (20%; 50% of pregnant women and 3% of men).

Males have larger stores of iron than females. The normal bodily concentration of iron is 60 parts per million, and the uptake of iron is regulated by absorptive cells in the small intestine, which correct the bodily losses of iron. There are about 3.7 g of iron in a 154-pound male adult. The internal need for iron is about 20 mg per day, but only 1 mg (10% of dietary iron) is absorbed per day from the typical daily diet that contains 10–20 mg of iron. Because a red blood cell has a 120-day life span,

0.8% of red blood cells are destroyed and replaced per day; 2.5 g of iron is incorporated into hemoglobin, with a daily turnover of 20 mg for synthesis of hemoglobin and degradation and another 5 mg for other requirements. A few tenths of a gram are bound to myoglobin, and about 0.02 g is distributed in many proteins involved in electron transfer, such as the proteins of the electron transport chain that generate most of the body's ATP supplies. About 1 g is stored inside **ferritin** for future use. The majority of this iron is reused through the plasma, but amounts of iron above these needs are deposited in ferritin or **hemosiderin,** iron storage forms. Some forms of iron are not absorbed in the small intestine, especially nonheme iron, usually in the ferric state (Fe^{3+}) from plant sources. Dietary myoglobin and hemoglobin account for two thirds of the body's iron.

Because iron plays a role in temperature regulation, anemic people usually feel cold. Iron is a component of heme, so its deficiency causes lowered levels of myoglobin in muscle cells, as well as a reduced hemoglobin content of red blood cells, limiting the amount of oxygen that can be delivered to the mitochondria for electron transport and energy production in the form of ATP. Consequently, the production of lactic acid is increased. Some enzymes involved in mitochondrial metabolism, such as NADH dehydrogenase and succinate dehydrogenase, contain nonheme iron, and iron is required for DNA synthesis because ribonucleotide reductase (Figure 6-61) is iron dependent. Other nonmitochondrial enzymes, such as **catalase,** and peroxidases contain heme and protect cells against reactive oxygen species derived from hydrogen peroxide. The ability of tissues to form ATP energy is related to body temperature, and it is believed that the connection of iron to copper may affect functioning of the thyroid, which is an important factor in body temperature regulation. Other causes of anemia relate to folic acid and vitamin B_{12} (cobalamin). Vitamin C enhances the absorption of nonheme iron by reducing dietary Fe^{3+} to Fe^{2+} and forming an ascorbic acid–iron complex that is absorbable. Fe^{2+} is absorbed, and Fe^{3+} needs to be reduced to be absorbed. Sufficient stomach acid is needed to release Fe^{3+} from ingested food, so the condition of achlorhydria (a lack of hydrochloric acid in digestive juices) can lead to insufficient absorption of iron. Vegetarians are prone to iron deficiency because they refrain from eating red meat that contains iron and may ingest suboptimal levels of vitamin B_{12}. Some plant sources of nonheme iron, like cereals and soybeans (tofu derivation), also contain **phytic acid,** which binds iron, as well as other metals, and renders it nonabsorbable in the intestine (Figure 13-1). Table 13-1 lists some common food sources and their contents of iron. Egg yolks are also a good source of iron, as are broccoli and spinach. Excellent sources of dietary iron are meats, clams, and oysters. Recommended daily allowances for dietary iron are given in Table 13-2.

Symptoms will not be noticeable with mild anemia; however, frank anemia may cause pale skin color, fatigue, shortness of breath, leg cramps, palpitations on climbing stairs, irritability, dizziness, weakness, sore tongue, atrophy of taste buds, sores at the corners of the mouth, brittle nails, frontal headache, difficulty in sleeping, and decreased appetite. Advanced iron-deficiency anemia can lead to problems in swallowing because webs of tissue form in the throat and esophagus (Plummer-Vinson syndrome). Iron-deficient individuals may consume nonfood items (like starch or clay), a condition known as **Pica,** which is characterized as a behavioral disturbance of iron deficiency. Anemia is easily diagnosed by measurement of blood

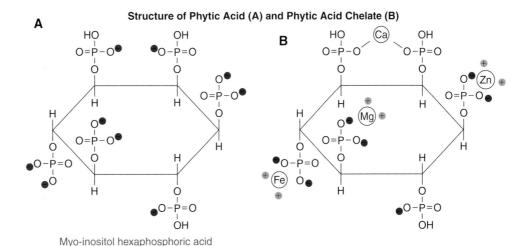

Figure 13-1. **A.** structure of phytic acid (myo-inositol hexaphosphoric acid) and its complexes with metals *(B)*.

Table 13-1
Content of Iron in Various Foods

Food	Serving	Iron Content (mg)
Beef	3 ounces,* cooked	2.31
Chicken, dark meat	3 ounces, cooked	1.13
Oysters	6 medium	5.04
Shrimp	8 large, cooked	1.36
Tuna, light	3 ounces, canned	1.30
Black-strap molasses	1 tablespoon	3.50
Raisin bran cereal†	1 cup, dry	5.00
Raisins, seedless	1 small box (1.5 ounces)	0.89
Prune juice	6 fluid ounces	2.27
Prunes, dried	~5 prunes (1.5 ounces)	1.06
Potato, with skin	1 medium potato, baked	2.75
Kidney beans	1/2 cup, cooked	2.60
Lentils	1/2 cup, cooked	3.30
Tofu, firm‡	1/4 block (~1/2 cup)	6.22
Cashew nuts	1 ounce	1.70

*3 oz of meat = size of a deck of cards
†Cereals contain much nonheme iron, which is not absorbed as well as heme iron.
‡Tofu also contains phytic acid, which blocks absorption of nonheme iron.
Reproduced from http://lpi.oregonstate.edu/infocenter/minerals/iron/.

hemoglobin level (Table 13-3); also given are **hematocrit** values that are measurements of the volume of red blood cells as a percentage of the total blood volume (43–49% in males; 37–43% in females). Other diagnostic values are the size of red blood cells, the serum iron level, and the iron-binding capacity of the blood.

Measurements of serum ferritin and stainable iron in tissue stores are useful for differentiating iron deficiency from anemia caused by chronic disease. Ferritin saturation with iron does not become abnormal until the

Table 13-2
Recommended Daily Dietary Allowances for Iron Intake in Various Individuals

Life Stage	Age	Males (mg/day)	Females (mg/d)
Infants	0–6 months	0.27 (AI)	0.27 (AI)
Infants	7–12 months	11	11
Children	1–3 years	7	7
Children	4–8 years	10	10
Children	9–13 years	8	8
Adolescents	14–18 years	11	15
Adults	19–50 years	8	18
Adults	51 years and older	8	8
Pregnancy	all ages	—	27
Breastfeeding	18 years and younger	—	10
Breastfeeding	19 years and older	—	9

AI, adequate intake, which is established when an RDA cannot be determined; d, day.
Reproduced from page 5 from http://lpi.oregonstate.edu/infocenter/minerals/iron/.

Table 13-3
Hemoglobin and Hematocrit Values in Blood in Various Levels of Anemia

	Normal Values	Values in Anemia
Hemoglobin (measured in grams (g) per deciliter (dl)		Mild: 9.5–10.9 g/dl (grade 1)
		Moderate: 8–9.4 g/dl (grade 2)
		Severe: 6.5–7.9 g/dl (grade 3)
		Life-threatening: less than 6.5 g/dl (grade 4)
Men	14–18 g/dL	Less than 14 g/dl
Women	12–16 g/dL	Less than 12 g/dl
Hematocrit		Mild: 30–36%
		Moderate: 25–30%
		Severe: less than 25%
Men	42%–52%	Less than 42%
Women	36%–48%	Less than 36%

Reproduced from Table 1 on page 2 from http://ohiohealth.cancersource.com/LearnAboutCancer/detail_frame.cfm?DiseaseID=1&ContentID=22234-1&Page=33&subjectID=3&TypeID=2.

tissue stores of iron have been depleted. When this happens, the hemoglobin concentration decreases for lack of iron, but values of red blood cells may not become abnormal until several months after the tissues stores of iron have been depleted (red blood cells turn over every 120 days). Periods of rapid growth require high levels of iron, so infants and children, as well as adolescents and pregnant women, are especially vulnerable to iron deficiency. Oral administration of ferrous iron (Fe^{2+}) as the sulfate is commonly used and will usually correct the deficiency. Blood hemoglobin levels can be followed to ascertain that normalization has been achieved. In serious clinical conditions where oral iron cannot be absorbed or acute bleeding has occurred, intravenous transfusions may be required. Where colon cancer (or other neoplasm) is discovered as a cause of blood loss, surgery to remove the cancer usually is indicated.

Ingestion and Uptake of Iron

Food contains heme iron and nonheme iron. Iron is released from heme inside of the intestinal cells by endoplasmic reticulum heme oxygenase (Figure 13-2) that converts heme into biliverdin, and the released iron is transferred into the body as nonheme iron. Any hemin, containing Fe^{3+}, is reduced to heme (Fe^{2+}) before it is oxidized by heme oxygenase. Biliverdin is reduced subsequently by biliverdin reductase, with NADPH as hydrogen donor to form bilirubin (Figure 13-3). Ultimately, in hepatocytes, bilirubin is made more water soluble by the addition of two glucuronic acids to form bilirubin diglucuronide (Figure 13-4) catalyzed by UDP glucuronyl transferase. It is conjugated to cholesterol and excreted as a bile acid pigment through the intestine.

The main site of absorption of dietary iron is the duodenum and upper jejunum, where iron transport across intestinal epithelial cells is accomplished by the divalent ion transporter (DMT1; Figure 13-5). Figure 13-5B shows the structure of the transmembrane domain 4 (TM4) of the DMT1 because the TM4 is crucial for the function of the transporter. Dietary iron is transported from the intestinal lumen across the villus cell and moved out of the cell at the basolateral side mediated by **transferrin** (see Figure 13-11), the carrier protein that regulates the transport of iron from the site of absorption (villus and crypt cells, as shown in the figure) to all tissues of the body. Transferrin binds two iron atoms, and the normal condition is that transferrin is 20–45% saturated with iron. When there is an iron deficiency, the crypt cells tend to move toward the villi to increase the efficiency of transfer to the internal tissues. Heme iron is ingested mainly as hemoglobin (over 10^{20} molecules in the body), and myoglobin and these structures are shown in Figure 13-6. As indicated in the legend of Figure 13-7, the uptake and regulation of dietary iron is complex, and some believe that it will involve a cascade mechanism similar to the blood-clotting mechanism. An overview of the normal iron absorption process (the transferrin cycle) is shown in Figure 13-8. Of the absorbed iron, 75% is bound to proteins involved in oxygen transport (for example, hemoglobin). In addition, 10–20% goes to a storage pool that can be tapped for formation of red blood cells (erythropoiesis). A single ferritin molecule can store as many as 4000 iron atoms in its mineral core, and more ferritin is produced when excess iron is taken in the diet for added iron storage. *There is no physiological mechanism for excretion of excess iron,* so large excesses of stored iron can lead to disease conditions (to be discussed later). The structures of ferritin, mitochondrial ferritin, and transferrin are shown in Figures 13-9 through 13-11. The crystal structure of the human transferrin receptor with transferrin bound to it is shown in Figure 13-12. Transferrin is the means by which iron is moved about, from the intestine to other tissues involved in hemoglobin production (in the immature red blood cell), and to the liver cell for storage. Signals in the overall process may be developed from the bone marrow to the intestinal cell and possibly from macrophages that

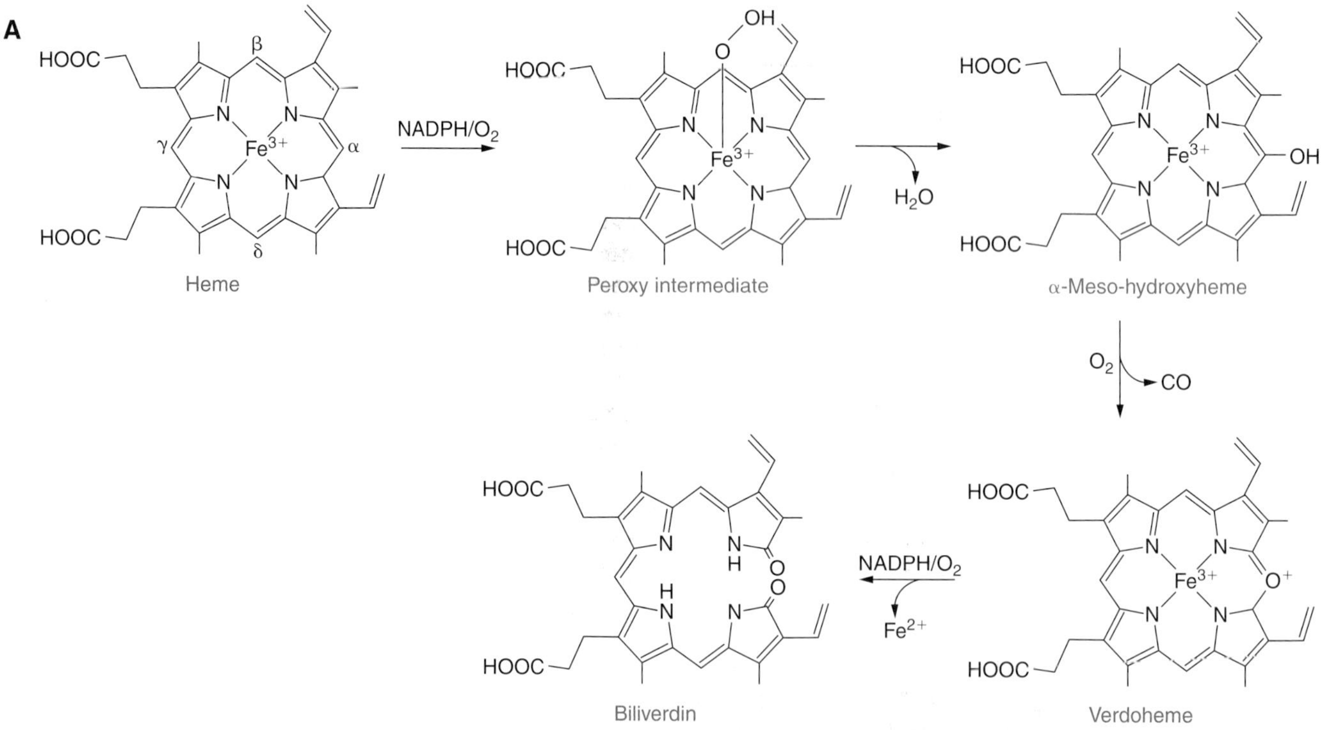

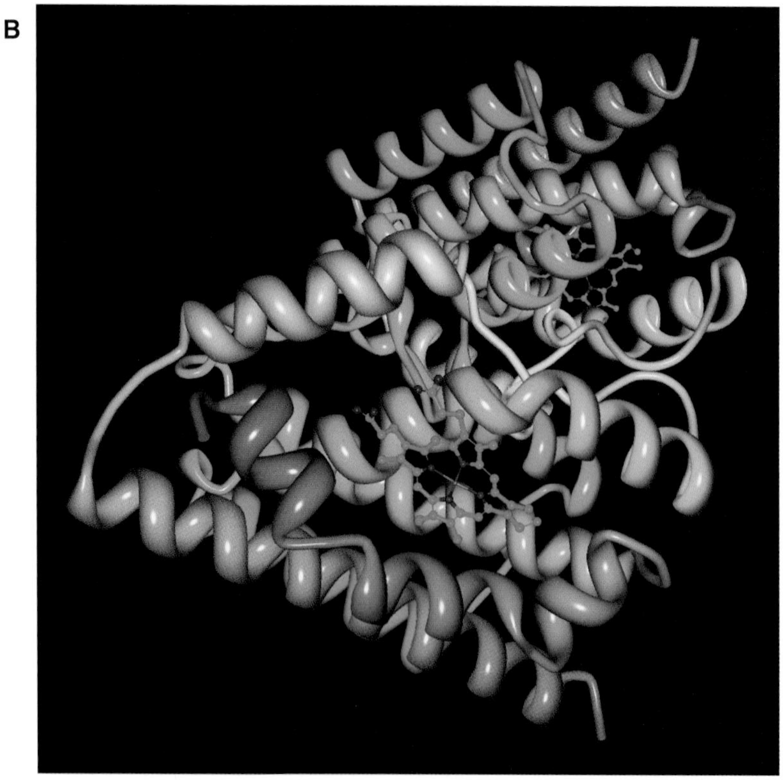

Figure 13-2. **A.** Conversion of heme to biliverdin by human heme oxygenase-1. **B.** Crystal structure of human heme oxygenase-1 in complex with its substrate heme, crystal form B. Atomic structure is of protoporphyrin IX containing Fe. Free atoms *(light green)* are chloride ions. Reproduced from http://www.ebi.ac.uk/msd-srv/msdlite/images/1n3u600.jpg. PDB ID: 1n3u. L. Lad, D.J. Schuller, H. Shimizu, J. Friedman, H. Li, P.R. Ortiz de Montellano, T.L. Poulos. Comparison of the Heme-Free and -Bound Crystal Structures of Human Heme Oxygenase-1. *Journal of Biological Chemistry* **278** pp. 7834 (2003).

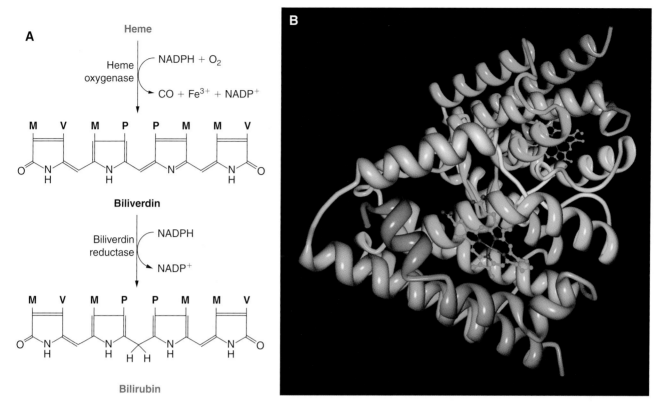

Figure 13-3. **A.** Pathway for the degradation of heme to bilirubin. Conversion of biliverdin to bilirubin catalyzed by biliverdin reductase. This is the only enzymatic reaction in the body that produces carbon monoxide (CO) expired by the lungs. **B.** Crystal structure of rat biliverdin reductase. *M*, Methyl; *P*, propionic; *V*, vinyl. Part A reproduced from http://isu.indstate.edu/mwking/nitrogen-metabolism.html. Part B reproduced from http://www.ebi.ac.uk/msd-srv/msdlite/images/1gcu600.jpg. PDB ID: 1gcu. D. Sun, M. Sato, T. Yoshida, H. Shimizu, H. Miyatake, S. Adachi, Y. Shiro, A. Kikuchi. Crystallization and Preliminary X-ray Diffraction Analysis of a Rat Biliverdin Reductase. *Acta Crystallographica*. **56** pp. 1180 (2000).

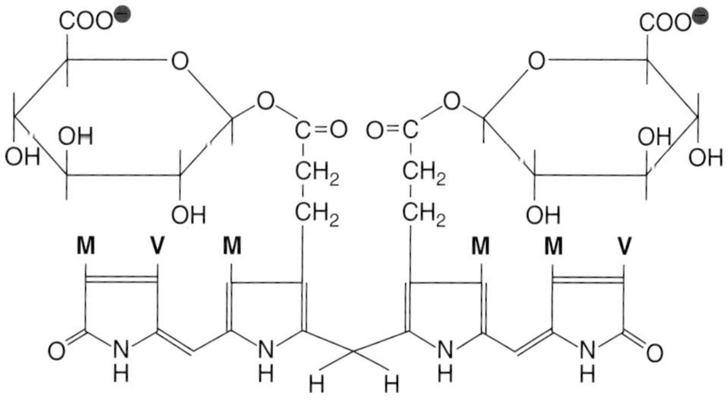

Figure 13-4. Structure of bilirubin diglucuronide. *M*, methyl; *V*, vinyl. Redrawn from http://isu.indstate.edu/mwking/nitrogen-metabolism.html.

Ingestion and Uptake of Iron

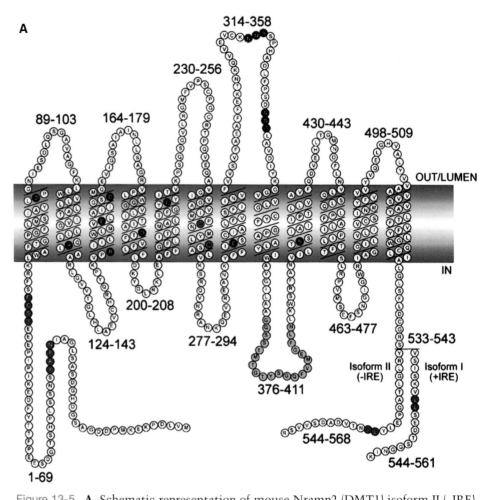

Figure 13-5. **A.** Schematic representation of mouse Nramp2 (DMT1) isoform II (–IRE) and isoform I (+IRE). The 12 transmembrane domains are predicted from hydropathy profiling, calculations of hydrophobic moment, and other computer-assisted analyses and from direct epitope mapping studies. Individual predicted intracellular and extracellular segments are identified, and their position within the primary sequence is shown. Amino acid residues defining sequence landmarks and signature motifs are depicted in different colors, including negatively and positively charged residues within predicted transmembrane (TM) domains *(dark blue)*, conserved histidine residues in TM6 *(red)*, glycine residues in TM4 altered in anemic *mk/Belgrade* mutants (Gly185Arg), and mutated (in *Nramp1*) in mice susceptible to infections (Gly184Asp) *(yellow)*. Also identified are Asn-linked glycosylation signals in the TM7-TM8 extracytoplasmic loop *(black)*, predicted membrane targeting/sorting motifs (tyrosine-based and dileucine) *(green)*, and consensus transport signature common to Nramp orthologs and present in the cytoplasmic face of membrane anchors of bacterial periplasmic permeases *(orange)*. The two different C-termini of the protein generated by alternative mRNA splicing containing or not an iron-response element (isoform I, +IRE; isoform II, –IRE) in the 3′ untranslated region are identified, with corresponding numbering. Finally, the polarity of the protein and membrane domains with respect to the membrane *(light blue)* is indicated *(in, out/lumen)*. TM4, transmembrane domain 4.

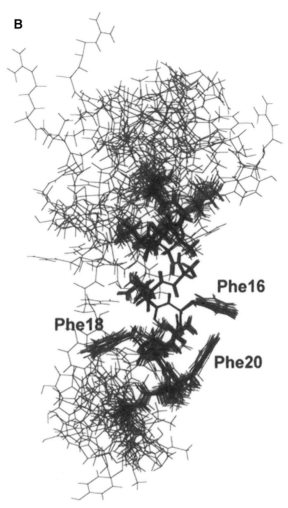

Figure 13-5, cont'd **B.** Stereo view of NMR structures of DMT1-TM4 in sodium dodecyl sulfate micelles. Part A reproduced from Figure 1 of S. Lam-Yuk-Tseung et al., "Iron transport by Nramp2/DMT1: pH regulation of transport by 2 histidines in transmembrane domain 6," *Blood* 101: 3699–3707, 2003. Part B reproduced from H. Li et al., "Structure and topology of the transmembrane domain 4 of the divalent metal transporter in membrane-mimetic environments," *Eur. J. Biochem.*, 271: 1938–1951, 2004 and http://content.febsjournal.org/content/vol271/issue10/images/large/ejb_4104_f3.jpeg.

have engulfed iron atoms for transport using transferrin to the bone marrow (Figure 13-13).

Diseases of iron overloading also occur; the most prevalent disease is hemochromatosis (increased uptake of iron, which is deposited in various tissues, where it is toxic in high amounts). The incidence of this disorder in Caucasian Americans may be 1 in 200. If untreated, it can lead to impotence, arrhythmia, diabetes, and liver failure, and it often eludes diagnosis. *Hfe* is the gene causing genetic hemochromatosis when the gene is mutated. The normal HFE protein (Figure 13-7 and Figure 13-13) binds to the transferrin receptor, causing inhibition of the receptor (acts like a regulator). When the gene is mutated, presumably to interfere with the normal activity of the HFE protein, the activity of the transferrin receptor

(Text continues on p. 753.)

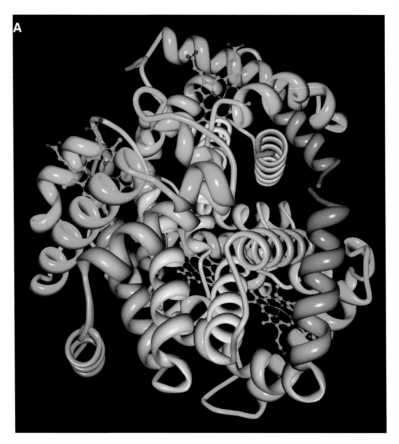

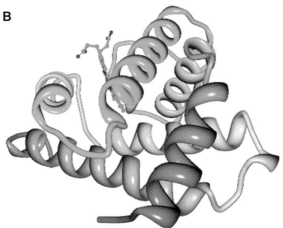

Figure 13-6. **A.** Crystal structure of human deoxyhemoglobin in absence of any anions. **B.** Human myoglobin. Heme group is shown in *green structure*. Part A reproduced from http://www.ebi.ac.uk/msd-srv/msdlite/images/1kd2600.jpg. PBD ID: 1kd2. F.A. Seixas, W.F. de Azevedo, M.F. Colombo. Crystallization and X-ray Diffraction Data Analysis of Human Deoxyhaemoglobin A(0) Fully Stripped of Any Anions. *Acta Crystallographica.* Section D, Biological crystallography. **55** pp. 1914 (1999) Part B reproduced from PDB ID: 2mml. S.R. Hubbard et al. X-Ray Crystal Structure of Recombinant Human Myoglobin Mutant at 2.8 Angstroms Resolution. *J. Mol. Biol.* **213** pp. 215 (1990).

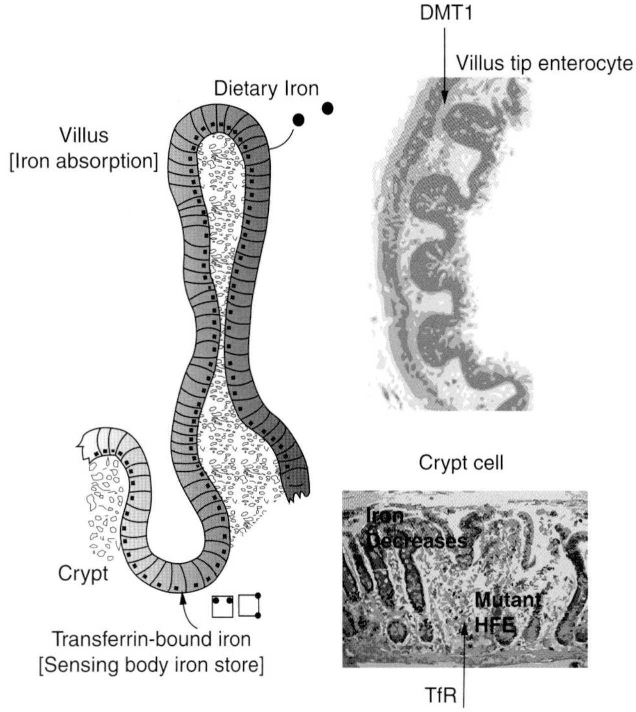

Figure 13-7. Dietary iron is transported into the intestinal villus cell by the divalent metal ion transporter (DMT1). Iron in the crypt cells is transported to the blood by the carrier, transferrin (Tf), which interacts with a transferrin receptor (TFR). A hemochromatosis gene (HFE) encodes a protein (HFE protein) that is highly expressed in duodenal crypt cells (and in other cells of the body). When part of the hemochromatosis (HFE) is deleted experimentally, depressing the function of HFE protein, iron overload occurs. In this situation, DMT1 is increased, suggesting a relationship between DMT1 expression and HFE protein. The HFE protein interacts with and inhibits the transferrin receptor (TFR) on the basolateral membrane of crypt cells. Decreased intracellular iron content leads to increased expression of DMT1, which increases the uptake of dietary iron.

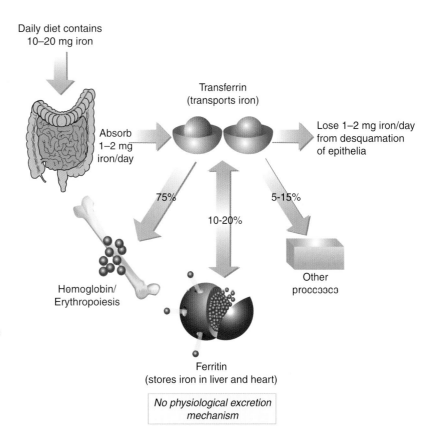

Figure 13-8. Overview of normal iron absorption. Iron is bound and transported in the body via transferrin and stored in ferritin molecules. Once iron is absorbed, there is no physiological mechanism for excretion of excess iron from the body other than blood loss, (i.e., pregnancy, menstruation, or bleeding). Redrawn from http://www.cdc.gov/hemochromatosis/training/images/iron_cycle.jpg.

Ingestion and Uptake of Iron

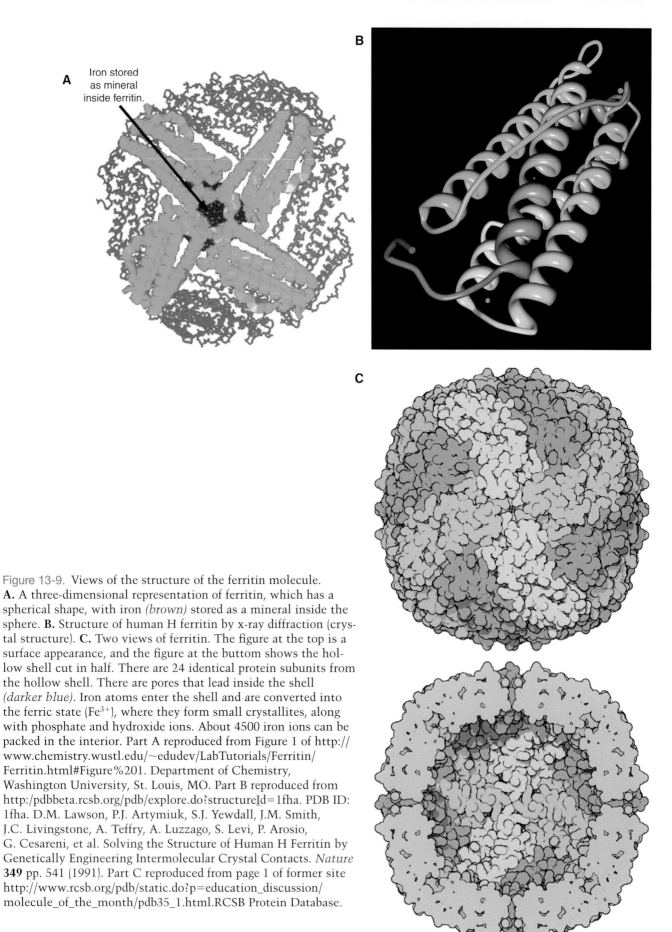

Figure 13-9. Views of the structure of the ferritin molecule. **A.** A three-dimensional representation of ferritin, which has a spherical shape, with iron *(brown)* stored as a mineral inside the sphere. **B.** Structure of human H ferritin by x-ray diffraction (crystal structure). **C.** Two views of ferritin. The figure at the top is a surface appearance, and the figure at the buttom shows the hollow shell cut in half. There are 24 identical protein subunits from the hollow shell. There are pores that lead inside the shell *(darker blue)*. Iron atoms enter the shell and are converted into the ferric state (Fe^{3+}), where they form small crystallites, along with phosphate and hydroxide ions. About 4500 iron ions can be packed in the interior. Part A reproduced from Figure 1 of http://www.chemistry.wustl.edu/~edudev/LabTutorials/Ferritin/Ferritin.html#Figure%201. Department of Chemistry, Washington University, St. Louis, MO. Part B reproduced from http:/pdbbeta.rcsb.org/pdb/explore.do?structureId=1fha. PDB ID: 1fha. D.M. Lawson, P.J. Artymiuk, S.J. Yewdall, J.M. Smith, J.C. Livingstone, A. Teffry, A. Luzzago, S. Levi, P. Arosio, G. Cesareni, et al. Solving the Structure of Human H Ferritin by Genetically Engineering Intermolecular Crystal Contacts. *Nature* **349** pp. 541 (1991). Part C reproduced from page 1 of former site http://www.rcsb.org/pdb/static.do?p=education_discussion/molecule_of_the_month/pdb35_1.html.RCSB Protein Database.

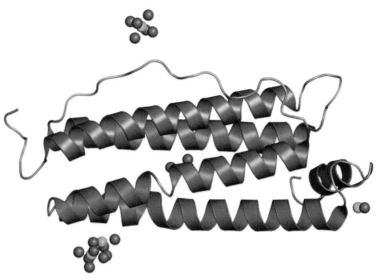

Figure 13-10. Crystal structure of human mitochondrial ferritin. Green atoms are Mg^{2+} ions. Reproduced from http://www.ebi.ac.uk/msd-srv/msdlite/images/1r03600.jpg. PBD ID: 1r03. B. Langlois d'Estaintot, P. Santambrogio, T. Granier, B. Gallois, J.M. Chevalier, G. Précigoux, S. Levi, P. Arosio. Crystal Structure and Biochemical Properties of the Human Mitochondrial Ferritin and Its Mutant Ser 144A1a. *Journal of Molecular Biology* **340** pp. 277 (2004).

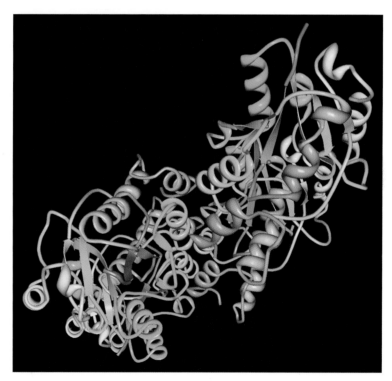

Figure 13-11. Human serum transferrin. Reproduced from PDB ID: 1d3k. A.H. Yang, R.T. MacGillivray, J. Chen, Y. Luo, Y. Wang, G.D. Brayer, A.B. Mason, R.C. Woodworth, M.E. Murphy. Crystal Structures of Two Mutants (K206Q, H207E) of the N-lobe of Human Transferrin with Increased Affinity for Iron. *Protein Science* **9** pp. 49 (2000).

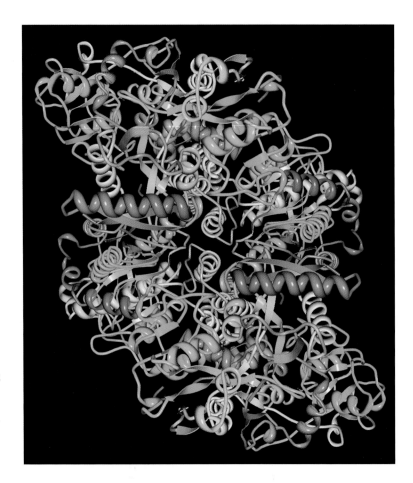

Figure 13-12. Structure of human transferrin receptor–transferrin complex. Reproduced from http://www.rcsb.org/pdb/explore/images.do?structureId=1suv. PDB ID: 1suv. Y. Cheng, O. Zak, P. Aisen, S.C. Harrison, T. Walz. Structure of the Human Transferrin Receptor-Transferrin Complex. *Cell* **116** pp.565–576 (2004).

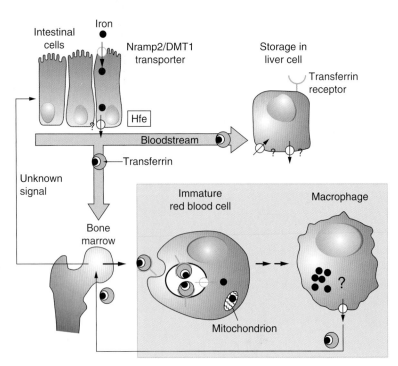

Figure 13-13. Overview of iron transport to various cell types via transferrin. This model suggests that signals may develop from the bone marrow to the intestinal cell to take up more or less iron as the level of hemoglobin is balanced. Redrawn from http://focus.hms.harvard.edu/1999/Apr2_1999/images/Focus1.gif.

is unchecked and excessive amounts of iron may be deposited in various tissues, causing hemochromatosis. In this situation, it appears that as much as three times the normal amount of dietary iron may be absorbed. The treatment for hemochromatosis is bloodletting, which will continue to be the therapy until a better molecular understanding of the mechanism is known so that specific drugs can be used.

Heme Synthesis

The synthesis of heme begins with succinyl-CoA and glycine in the mirochondria, which forms δ-aminolevulinic acid (ALA) through the action of ALA synthase. ALA is exported to the cytoplasm, where ALA dehydrase converts it to porphobilinogen that is converted to uroporphyrinogen III (or uroporphyrinogen I) by uroporphyrinogen I synthase and uroporphyrinogen II cosynthase. Uroporphyrinogen III is converted to coporphyrinogen III by uroporphyrinogen decarboxylase, and coporphyrinogen III enters the mitochondrion, where it is converted to protoporphyrinogen IX by the action of coproporphyrinogen III oxidase. It is then converted to protoporphyrin IX by protoporphyrinogen IX oxidase. Finally, protoporphyrin IX is converted to heme by the action of **ferrochetalase** that incorporates iron into the molecule. The outline of these events is shown in Figure 13-14, and the structures involved are shown in Figure 13-15. The final event in heme

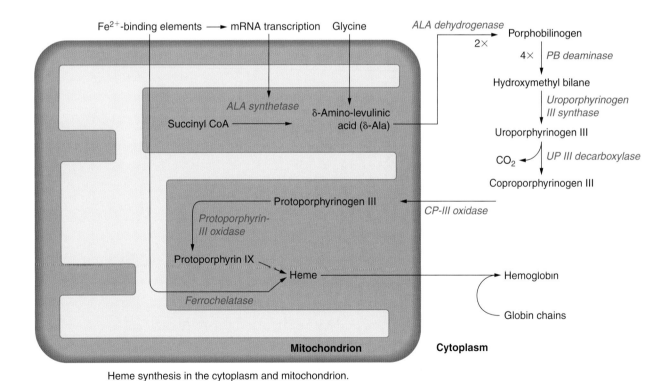

Heme synthesis in the cytoplasm and mitochondrion.

Figure 13-14. Overview of the synthesis of heme showing the events occurring inside the mitochondrion and in the cytoplasm. The process begins in the mitochondrion with succinyl-CoA, from the Kreb's cycle (TCA cycle), and glycine to form delta-aminolevulinic acid. Redrawn from page 2 of http://en.wikipedia.org/wiki/Heme.

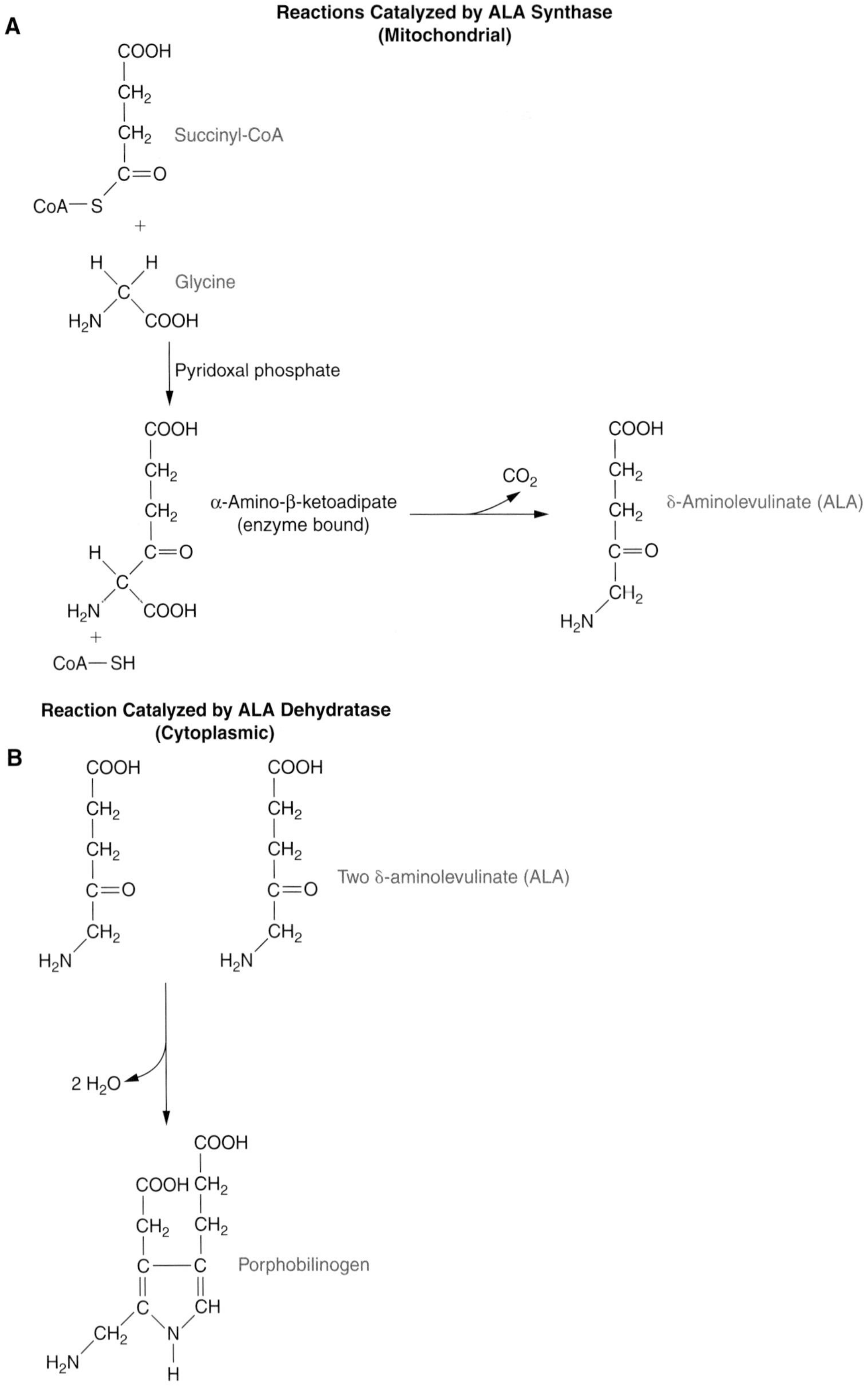

Figure 13-15. Chemical reactions in the synthesis of heme.

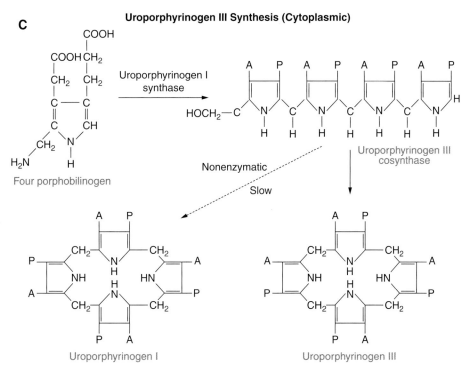

Figure 13-15, cont'd

Heme Synthesis 755

E Reaction Catalyzed by Coproporphyrinogen III Oxidase (Mitochondrial)

Coproporphyrinogen III → Protoporphyrinogen IX

F Reaction Catalyzed by Protoporphyrinogen IX Oxidase (Mitochondrial)

Protoporphyrinogen IX → Protoporphyrin IX

G Reaction Catalyzed by Ferrochelatase (Mitochondrial)

Protoporphyrin IX + Fe^{++} → **Heme**

Figure 13-15, cont'd

synthesis is the insertion of iron into the heme molecule (protoheme IX) catalyzed by ferrochelatase (Figure 13-16). Because all of the intermediates in heme synthesis are tetrapyrroles classified as porphyrins, the overall reaction can be referred to as porphyrin synthesis. A similar process is used by other species for the synthesis of vitamin B_{12} (cobalamin). The rate-limiting enzyme for heme synthesis is the enzyme catalyzing the initial step, ALA synthase, and it is regulated by intracellular concentrations of heme and iron. When cellular (chiefly the liver and bone marrow cells) iron is low as in dietary iron deficiency, there is a decrease in the synthesis of porphyrin, so toxic intermediates do not accumulate. A mutation in the gene for ALA synthase, or any enzyme in the heme synthetic pathway, leads to porphyria (inability to form mature heme); in this case, hematin or heme arginate can be infused to stem a severe attack. Ferrochelatase

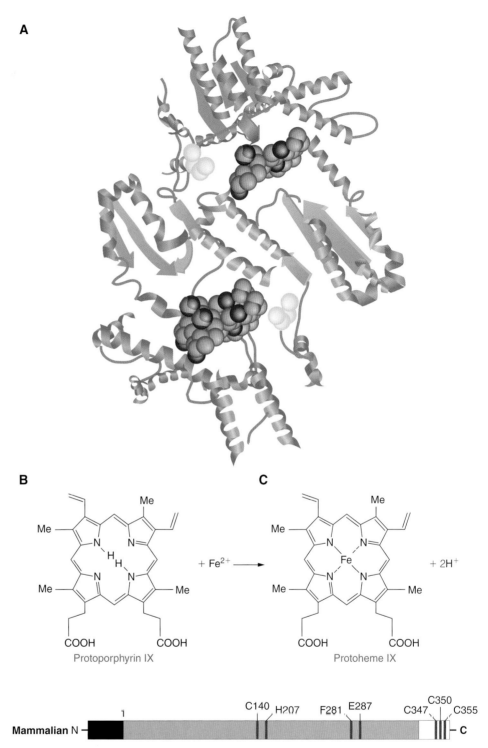

Figure 13-16. **A.** Crystal structure of human ferrochelatase. **B.** Ferrochelatase reaction. **C.** Schematic representation of the primary structure of mammalian ferrochelatase. Catalytic core is in *gray*, presequence is in *black*, and the C-terminal extension is in *white*. F207 is proposed to be involved in metal substrate binding; E287 is proposed to be involved in catalysis; F281 is proposed to interact with the protoporphyrin substrate; and C140, C347, C350, and C355 are proposed to be the [2Fe-2s] cluster ligands. Amino acid numbering is based on the murine mature ferrochelatase. Part A reproduced from C.K. Wu et al., "The 2.0 A structure of human ferrochelatase, the terminal enzyme of heme biosynthesis," *Nat. Struct. Biol.*, 8: 156, 2001. pl?CODE=1hrk. Part B redrawn from http://www.sciencedirect.com. Part C redrawn from G.C. Ferreira, "Ferrochelatase," *Intl. J. Biochem. Cell Biol.*, 31: 995–1000, 1999.

(*ferro* = iron and *chelate* = combination of a metal ion with a compound containing different atoms; therefore, ferrochelatase is an enzyme that catalyzes the formation of such a complex) is encoded by a single gene, and mutations in this gene also lead to porphyria (erythropoietic protoporphyria). Various porphyries and their causes are outlined in Table 13-4. The prosthetic group of human ferrochelatase is an iron–sulfur cluster, (2Fe-2S), presumably in the form of a chelate of iron with cystines. There is an essential histidine residue of ferrochelatase, H207, that is involved in the binding of iron and a conserved glutamate residue, E287, that is functional in the catalytic center. The precise molecular and chemical events of the catalytic mechanism are not yet known. The ferrochelatase enzyme is associated with the inner mitochondrial membrane, with its active site facing the mitochondrial matrix space. The ferrochelatase reaction at the inner mitochondrial membrane is thought to involve another protein known as **frataxin** that acts as the iron donor for the iron–sulfur cluster and for the heme pathway. Holofrataxin (not yet bound with the iron–sulfur cluster) binds to ferrochelatase on the inner mitochondrial membrane. It also binds to the iron–sulfur cluster synthetic unit (ISU). Ferrochelatase dimer has a binding site for frataxin on its matrix side. Holofrataxin is a high-affinity iron-binding partner for holoferrochelatase. It can deliver iron to ferrochelatase and mediates the final step in heme synthesis. Frataxin is involved in both heme sand iron–sulfur biosyntheses; the overall mechanism is pictured in Figure 13-17. The crystal structure of frataxin is shown in Figure 13-18, which also shows how the iron atom is coordinated to alanine, aspartate, and histidine residues in this molecule. As might be expected, heme synthesis responds to a deficiency of oxygen (hypoxia) by cellular increases in the mRNA encoding ferrochelatase. There exists a hypoxia-inducible factor, HIF-1, and two binding motifs for HIF-1 have been identified in the ferrochelatase gene promoter.

Table 13-4
Various Porphyrias, Their Causes, and Their Symptoms

Porphyria	Enzyme Defect	Primary Symptom
Erythropoietic Class		
Congenital erythropoietic porphyria, CEP	Uroporphyrinogen III cosynthase	Photosensitivity
Erythropoietic protoporphyria, EPP	Ferrochelatase	Photosensitivity
Hepatic Class		
ALA dehydratase deficiency porphyria, ADP	ALA dehydratase	Neurovisceral
Acute intermittent porphyria, AIP	Prophobilinogen deaminase	Neurovisceral
Hereditary coproporphyria, HCP	Coproporphyrinogen oxidase	Neurovisceral, some photosensitivity
Variegate porphyria, VP	Protoporphyrinogen oxidase	Neurovisceral, some photosensitivity
Porphyria cutanea tarda, PCT	Uroporphyrinogen decarboxylase	Photosensitivity
Hepatoerythropoietic porphyria, HEP	Uroporphyrinogen decarboxylase	Photosensitivity, some neurovisceral

ALA, δ-aminolevulinic acid.
Reproduced from pages 12–13 of http://www.med.unibs.it/~marchesi/heme.html.

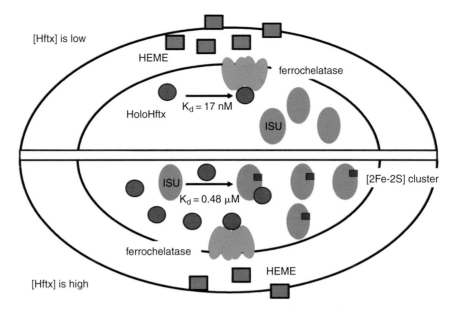

Figure 13-17. Cellular model for the regulation of frataxin chemistry in iron–sulfur cluster and heme biosynthesis. Holofrataxin *(Hftx)* is used as an iron donor for both heme and iron–sulfur cluster biosynthetic pathways Under normal cell growth conditions, the frataxin concentration is sufficient for both heme and iron–sulfur cluster syntheses. The level of frataxin is down-regulated in erythroid differentiation, as is the iron–sulfur cluster biosynthesis pathway. However, heme biosynthesis remains essentially normal as a consequence of the distinct binding affinities of frataxin to ISU and ferrochelatase. Reproduced from Figure 3 of T. Yoon and J.A. Cowan, "Frataxin-mediated iron delivery to ferrochelatase in the final step of heme biosynthesis," *J. Biol. Chem.*, 279: 25943–25946, 2004 and http://www.jbc.org/content/vol279/issue25/images/large/zbc0250428030003.jpeg.

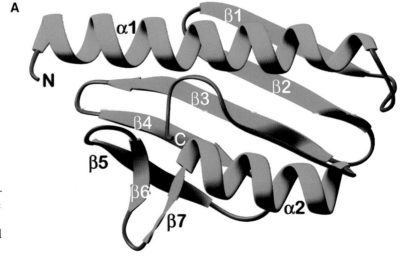

Figure 13-18. **A.** Structure of frataxin. Ribbon diagrams showing the fold of frataxin, a compact αβ sandwich, with α helices colored *turquoise* and β strands in *green*. Strands β1–β5 form a flat antiparallel β-sheet that interacts with the two helices, α1 and α2. The two helices are nearly parallel to each other and to the plane of the large β-sheet. A second, smaller β-sheet is formed by the C-terminus of β5 and strands β6 and β7.

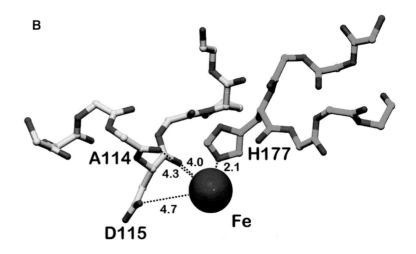

Figure 13-18, cont'd **B.** Iron binding. Adjacent frataxin molecules are colored *green* and *yellow*. Iron, depicted as a *red sphere*, is coordinated to His-177. Distances in Å *(dotted lines)* are between the nucleus of iron and its closest neighbors. Part A reproduced from Figure 2 of S. Dhe-Paganon et al., "Crystal structure of human fataxin," *J. Biol. Chem.*, 275: 30753–30756, 2000 and http://www.jbc.org/content/vol275/issue40/images/large/bc3608148002.jpeg. Part B reproduced from Figure 5 of S. Dhe-Paganon et al., "Crystal structure of human fataxin," *J. Biol. Chem.*, 275: 30753–30756, 2000 and http://www.jbc.org/content/vol275/issue40/images/large/bc3608148005.jpeg.

Formation of Hemoglobin

Hemoglobin synthesis occurs when erythroid cells are differentiating (immature red blood cells) into mature red blood cells. *When the red cells mature, the synthesis of hemoglobin ceases and the hemoglobin already formed has to survive for 120 days, the lifetime of the erythrocyte.* Heme interacts directly with globin to form a specific heme–globin complex, and when this occurs the heme pocket collapses around the porphyrin and a bond is formed between the proximal histidine residue and the heme iron atom. This interaction does not require enzymatic activity but is governed by the concentrations of the reactants, heme and globin. Hemoglobin is a heterotetramer containing two α-globin molecules and two β-globin molecules, each of which has a heme associated with it (four hemes altogether; Figure 13-6A). The structures of these two globin molecules are similar (Figure 13-19). When insufficient β-globin is formed, there is a significant excess of α-globin over β-globin and a group of diseases called β-thalassemias is produced. In this condition, α-globin is unstable by itself and forms aggregates on the membranes of red blood cells that damage the red blood cells, resulting in anemia.

Normally, there is a slight excess of α-globin over β-globin that does not cause thalassemia. The normal excess of α-globin is bound by a chaperone-like protein called **α-hemoglobin stabilizing protein (AHSP)** whose interaction with α-globin is less avid than the interaction between α-globin and β-globin. When a molecule of β-globin appears, it can displace the AHSP from the AHSP–α-globin complex and interact with the released α-globin. The association constant for the interaction of AHSP with α-globin is $10^7 M^{-1}$. Compared to the interaction of α-globin with β-globin, which is $10^{10} M^{-1}$, this indicates that the two globins interact with each other about 1000 times more avidly than the interaction between α-globin and AHSP. AHSP forms a three-helix bundle, and there is a specific surface on the α-globin for the binding of AHSP, as shown in Figure 13-20; these two components form a dimer. When β-globin appears (β-globin does not contain an AHSP-binding surface), it will cause the dissociation of the AHSP–α-globin dimer by competition and form the hemoglobin tetramer (two dimers each of α-globin–β-globin). In this tetramer, the AHSP-binding surface on the α-globin component, but not on the β-globin component, is buried (unavailable) in the interface between α-globin and β-globin in the hemoglobin

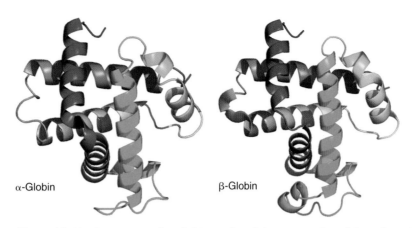

Figure 13-19. Structures of α-globin and β-globin. Reproduced from http://www.science.org.au/sats2004/images/mackay9.jpg. Australian Academy of Science.

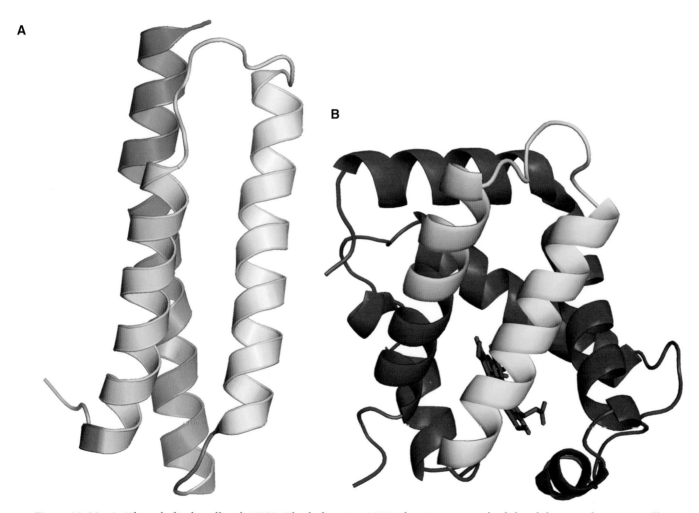

Figure 13-20. **A.** Three-helix bundle of AHSP. The helices in ASHP that interact with alpha globin are shown in *yellow*. **B.** AHSP-binding surface of α-globin. The AHSP interaction surface of alpha globin is shown in *yellow*. The heme is in *blue*. Part A reproduced from http://www.science.org.au/sats2004/images/mackay17.jpg. Australian Academy of Science. Part B reproduced from http://www.science.org.au/sats2004/images/mackay18.jpg. Australian Academy of Science.

molecule (Figure 13-21). Normal red blood cells contain a concentration of hemoglobin that is 4 millimolar, and the normal excess of α-globin is 10–20%, which makes its concentration (of the excess α-globin) about the same as cellular AHSP—that is, about 0.4 millimole.

In β-thalassemia, the larger excess α-globin breaks down, causing the release of the iron atom from heme. Free iron can be toxic and generate reactive oxygen free radicals, and the liberated Fe^{2+} (normally in heme) can interact with peroxide to form Fe^{3+} and hydroxyl ions (.OH + OH^-). Fe^{3+} can interact with H_2O_2 to form Fe^{2+} and another radical, [.OOH + H^+]. Experimentally, the addition of AHSP can reduce the free radical formation generated by the excess of α-globin. In this case, AHSP traps the iron associated with the α-globin in the Fe^{3+} state so that it does not interact with peroxide and cycle back to Fe^{2+}. When the α-globin–AHSP complex is dissociated by β-globin, the resulting tetramer contains four hemes, two of which (previously bound to α-globins associated with AHSP) contain iron in the Fe^{3+} form. After the tetrameric hemoglobin is formed, a reductase reduces the iron atoms associated with the α-globin to the Fe^{2+} state that is normal for hemoglobin. Thus, AHSP could be a candidate for the gene therapy of thalassemia.

α-Thalassemias also can occur when there is a defect in the synthesis of α-globin. β-Thalassemia is characterized by decreased production of β-peptide chains, and this condition is autosomal recessive. Heterozygotes are carriers with mild to moderate microcytic anemia (called thalassemia minor). α-Thalassemia is characterized by decreased production of α-chains. Whereas a single gene is involved in β-thalassemia, two genes are involved in α-thalassemia. Heterozygotes with a single gene defect have α-thalassemia-2 that does not express symptoms. Homozygotes with defects in two genes have α-thalassemia-1 that expresses anemia, presenting a clinical picture similar to β-thalassemia. Characteristics of the thalassemias are summarized in Table 13-5.

The overall system for the formation of hemoglobin in the red blood cell is shown in Figure 13-22. In the normal red blood cell, compound B

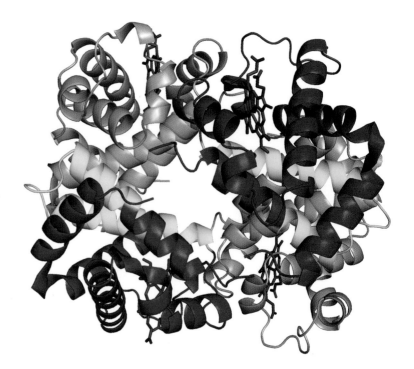

Figure 13-21. The tetrameric hemoglobin (2α-globins/2β-globins) containing four hemes (atomic structures in *blue*). The AHSP-binding surfaces of α-globins are shown in *yellow* and cannot bind AHSP in the tetrameric form. Reproduced from http://www.science.org.au/sats2004/images/mackay19.jpg. Australian Academy of Science.

Table 13-5
Characteristics of the Thalassemias

Category	Anemia	MCV	% Hb A_2	% Hb F
β-Thalassemia				
Heterozygous	Mild	↓	↑	Variable
Homozygous	Severe	↓	Variable	↑ up to 90%
β-δ-Thalassemia				
Heterozygous	Mild	↓	N or ↓	>5%
Homozygous	Moderate to severe	↓	Absent	100%
α-Thalassemia				
Single-gene defect	None	N to ↓	N	N
Double-gene defect	Mild	↓	N to ↓	<5%
Triple-gene defect	Moderate	↓	N to ↓ (Hb H or Bart's present)	Variable

MCV, mean corpuscular volume: ↓, decreased: ↑, increased; N, normal.
Reproduced from former site http://images.google.com/imgres?imgurl=http://a248.e.akamai.net/7/...ssemias%26hl%3Den%26lr%3D%26client%3Dsafari%26rls%Den-us%26sa%3DN.

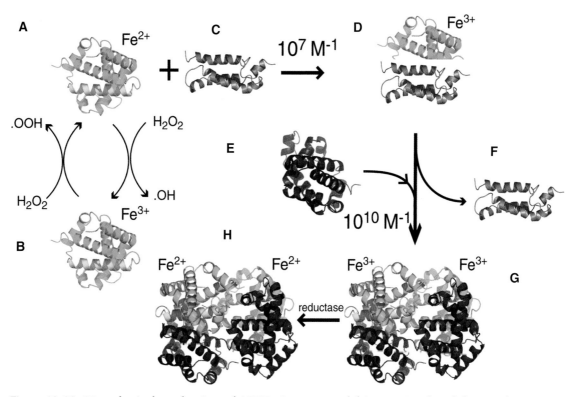

Figure 13-22. Hypothetical mechanism of AHSP. *A*, excess α-globin associated with heme whose iron is in the ferrous form; it can generate Fe^{3+} by interaction with reactive oxygen species, producing α-globin Fe^{3+} *(B)*. *C*, AHSP that interacts with α-globin *(A)* to form an AHSP–α-globin complex *(D)*. This reacts with a molecule of β-globin (containing heme with Fe^{2+}) *(E)*, which dissociates AHSP *(F)* and forms the tetrameric hemoglobin (from two dimers of α-globin–Fe^{3+} and β-globin–Fe^{2+}), *G*. The two Fe^{3+} iron atoms of hemoglobin are reduced to Fe^{2+} by a reductase to form the functional hemoglobin tetramer with four hemes, all of which contain Fe^{2+} *(H)*. Reproduced from http://www.science.org.au/sats2004/images/mackay23d.jpg. Australian Academy of Science. *AHSP*, alpha-hemoglobin stabilizing protein.

Formation of Hemoglobin

will not form, nor will the reactive oxygen radicals appear. Thus, the reaction will involve molecule A and C to form D, and so on, for the slight excess of α-globin over β-globin. As soon as β-globin is formed, it binds heme and forms hemoglobin with the slight excess of α-globin. Thus, β-globin synthesis seems to be the rate-limiting event in the normal formation of hemoglobin in immature red blood cells.

Dietary Metals

Some trace metals important dietary factors, besides calcium and iron, are copper, selenium, zinc, magnesium, and molybdenum; the nonmetal halogen, iodine, is also key. Many of these trace metals are cofactors for enzymes. Zinc, for example, is a cofactor in more than 100 enzymatic reactions. Some examples of enzymes that contain trace metals are discussed here, in addition to other activities of these nutrient metals. Like most essential nutrients, including iron, an excess intake can cause problems (hemochromatosis), either as a consequence of a given diet or as a genetic problem.

Copper

Copper is present in virtually every tissue (stored primarily in the liver), but the total amount of copper in the body is less than 100 mg. Copper is an essential component of many enzymes. Of copper in the blood, 90% is incorporated into the protein, **ceruloplasmin.** Ceruloplasmin is a transporter, carrying copper to various tissues. In addition to its transport function, ceruloplasmin has enzymatic activity and catalyzes the oxidation of minerals, iron in particular. Thus, *copper is a necessary adjunct to the use of iron, and a deficiency of copper will impair the functioning of iron (producing iron-deficiency anemia) even when iron is ingested in normal amounts.* The crystal structure of human ceruloplasmin is shown in Figure 13-23; the enzymatic reaction it catalyzes to convert Fe^{2+} to Fe^{3+} also is shown. *The oxidation of iron to the Fe^{3+} is necessary because Fe^{3+} is the form bound to transferrin (Figure 13-9), its transport protein (Fe^{2+} is the form absorbed in the intestine).*

Superoxide dismutase (SOD) is an important copper–zinc enzyme that catalyzes the removal of superoxide radicals generated during normal metabolic processes, as well as during the phagocytosis of infecting viruses and bacteria in white cells. A deficiency of copper diminishes the activity of SOD, and *superoxides that are not removed efficiently damage cell membranes.* In SOD, both copper and zinc are needed. These metals function together, as seen in the structure of the enzyme. In the case of SOD, a chaperone that transports the copper into the zinc SOD molecule forms a heterodimer with the enzyme and generates conformational rearrangements in both chaperone and SOD (Figure 13-24). As can be seen in the model of the human enzyme (Figure 13-24C), copper becomes complexed by the enzyme amino acid residues, histidine 46 (H46), H120, and H48, and zinc is complexed by aspartate 83 (D83), H80, H71,

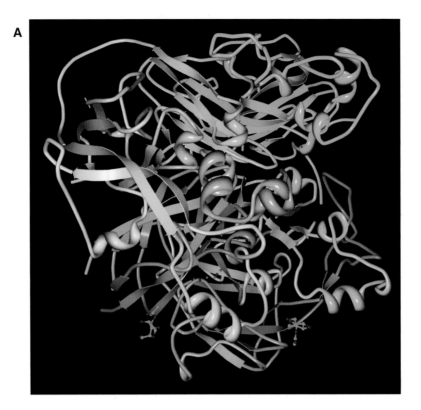

B $4 \times Fe(2+) + 4 \times H(+) + \underset{O(2)}{O=O} = 4 \times Fe(3+) + 2 \times H(2)O$

Figure 13-23. **A.** Crystal structure of human ceruloplasmin. Eight copper ions are depicted in *green* bound to the protein. **B.** Oxidation of ferrous iron (Fe^{2+}) to ferric iron (Fe^{3+}) catalyzed by ceruloplasmin. Reproduced from http://www.ebi.ac.uk./msd-srv/msdlite/images/1kcw600.jpg. PBD ID: 1kcw. I. Zaitseva, V. Zaitseva, G. Card, K. Moshkov, B. Bax, A. Ralph, P. Lindley. The X-ray Structure of Human Serum Ceruloplasmin at 3.1 Angstrom: Nature of the Copper Centures. *Journal of Biological Inorganic Chemistry* **1** pp. 15 (1996).

and H63. This enzyme catalyzes a number of reactions, two of which are shown in Figure 13-25.

The enzyme **lysyl oxidase** contains copper and plays an important role in the synthesis of collagen and elastin, which are major structural components of bone and connective tissue. Lysyl oxidase catalyzes the deamination of peptidyllysine residues to form aldehyde cross-links in the protein. The structure of lysyl oxidase and its reaction are shown in Figure 13-26. The copper-binding sequence in the enzyme is W–x–W–II–x–C–II–x–H–[YN]–H–S–[MI]–[DE] (using single-letter abbreviations of amino acids; x = nonspecific amino acid). The histidines are important in the binding of copper. Lysyl oxidases deaminate some lysine and hydroxylysine residues in collagen and elastin to form aldehydes referred to as **allysine (aminoadipic-δ-semialdehyde)** and hydroxyallysine. The cross-links formed in collagen (Figure 2-66) provide tensile strength and stability of the fibrils. Production of the aldehyde is followed by condensation reactions that generate bi-, tri-, and tetrafunctional cross-links. Aldehyde derivatives of lysine and hydroxylysine can react with similar aldehydes on adjacent polypeptide chains,

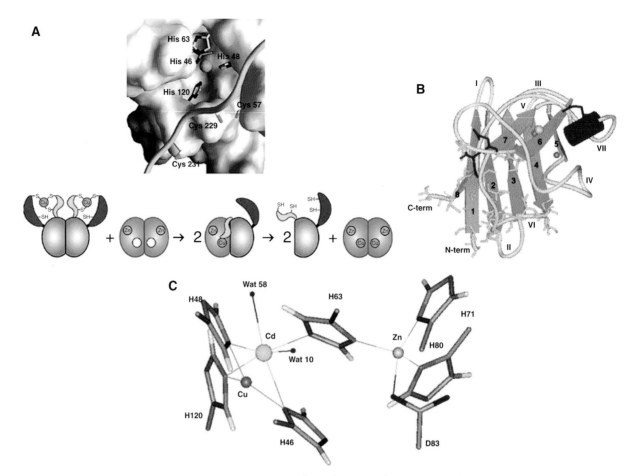

Figure 13-24. Proposed mechanism of copper transfer between chaperone yCCS and superoxide dismutase (SOD1). **A.** Structure-based model of the SOD1 copper site in a heterodimer formed between yCCS and wild-type SOD1. Domain III of yCCS is shown as a *yellow* coil, and SOD1 is shown as a *green* molecular surface. Cys 229 from yCCS domain III is within 10 Å of the copper of the copper ion and SODI His 120. A combination of conformational flexibility and reduction of the Cys 229–Cys 57 disulfide bond could bring both yCCS domain III Cys residues within 5 Å of the His ligands. Schematic illustrating formation of the heterodimeric complex from the yCCS and SOD1 homodimers and delivery of copper by yCCS domain III. **B.** Schematic view of the human Q133M2SOD structure displaying the secondary structure elements. The side chains of the residues homologous to those involved in the subunit–subunit interface in the wild-type dimer are shown as yellow sticks. The mutated residues E50, E51, and Q133 are shown as *red sticks*. *Orange, yellow,* and *gray spheres* of arbitrary radius represent copper, cadmium, and zinc ions, respectively. Cadmium is used to crystallize SOD and is not a physiological component of the enzyme. **C.** Final model of the active site showing the metal coordination. Large spheres of arbitrary radius represent copper *(orange)*, cadmium *(yellow)*, and zinc *(gray)*. The cadmium-bound water molecules are shown as small *red* spheres. Part A reproduced from A.L. Lamb et al., "Heterodimeric structure of superoxide dismutase in complex with its metallochaperone," *Nat. Struct. Biol.*, 8: 751–755, 2001 and http://www.nature.com/nsmb/journal/v8/n9/full/nsb0901-751.html. Parts B and C reproduced from Figures 1 and 5 of M. Ferraroni et al., "The crystal structure of the monomeric human SOD mutant F50E/G51E/E133Q at atomic resolution: The enzyme mechanism revisited," *J. Mol. Biol.*, 288: 413–426, 1999 and http://www.sciencedirect.com.

A $Cu^{+2}-SOD + O_2^- \longrightarrow Cu^{+1}-SOD + O_2$
   $Cu^{+1}-SOD + O_2^- + 2H^+ \longrightarrow Cu^{+2}-SOD + H_2O_2$

B $Cu(II)-SOD1 + H_2O_2 \rightleftharpoons Cu(I)-SOD1 + O_2^- + 2H^+$, (1)

   $Cu(I)-SOD1 + H_2O_2 \rightleftharpoons [Cu(I)O \leftrightarrow Cu(II)/{}^*OH \leftrightarrow Cu(III)]-SOD1 + OH^-$, (2)

Figure 13-25. Two sets of reactions catalyzed by superoxide dismutase (SOD). **A.** In this reaction, the oxidation state of copper changes between +1 and +2. **B.** Interactions of SOD with copper in three states of oxidation. Part A reproduced from http://www.nationmaster.com/encyclopedia/Superoxide-dismutase. Part B reproduced from D.C. Ramirez, S.E. Gomez Mejiba and R.P. Mason, "Mechanism of hydrogen peroxide–induced Cu,Zn–superoxide dismutase–centered radical formation as explored by immuno-spin trapping: The role of copper– and carbonate radical anion–mediated oxidations." *Free Radical Biol. Med.* 38: 201–214, 2005.

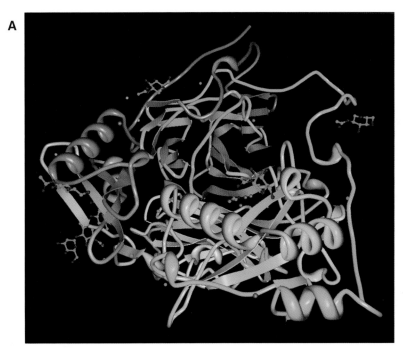

Figure 13-26. **A.** Crystal structure of lysyl oxidase (yeast). Many different ions are bound in this preparation, but there is only one copper atom in the molecule. **B.** Reaction catalyzed by lysyl oxidase. Part A reproduced from http://www.rcsb.org/pdb/cgi/explore.cgi?pdbId=1rky. PDB ID: 1rky. A.P. Duff, D.M. Trambaiolo, A.E. Cohen, P.J. Ellis, G.A. Juda, E.M. Shepard, D.B. Langley, D.M. Dooley, H.C. Freeman, J.M. Guss. Using Xenon as a Probe for Dioxygen-Binding Sites in Copper Amine Oxidases. *J. Mol. Biol.* **344** pp. 599 (2004).

forming bifunctional cross-links (for example, lysinonorleucine and hydroxylysinonorleucine). Aldol condensation products also can react with a histidine residue to form an aldol histidine, which, in turn, can react with another lysine residue to form a tetrafunctional cross-link (histidinohydroxymerodesmosine). Figure 13-27 shows the tropocollagen fiber triplex, the maturation process from procollagen to the fully formed cross-linked collagen, and the appearance of collagen fibers in tissues.

Cytochrome c oxidase is an oxidoreductase containing copper that catalyzes the oxidation of heme Fe^{2+} to heme Fe^{3+} by oxygen to produce water. The structure of the enzyme and the reaction catalyzed are shown in Figure 13-28. Cytochrome c oxidase resides on the inner mitochondrial membrane as one of five enzyme complexes, NADH dehydrogenase, succinate dehydrogenase, cytochrome c reductase, cytochrome c oxidase, and ATP synthase, shown in Figure 13-29. According to the reaction catalyzed by cytochrome c oxidase in Figure 13-28B,

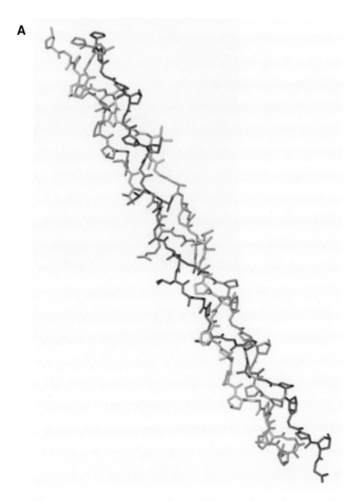

Figure 13-27. **A.** Tropocollagen fiber triplex. The strands contain the repeating sequence (gly-pro-hydroxypro)$_n$ with occasional appearance of other amino acids, such as His.

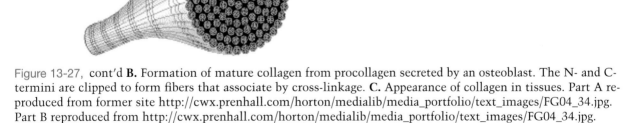

Figure 13-27, cont'd **B.** Formation of mature collagen from procollagen secreted by an osteoblast. The N- and C-termini are clipped to form fibers that associate by cross-linkage. **C.** Appearance of collagen in tissues. Part A reproduced from former site http://cwx.prenhall.com/horton/medialib/media_portfolio/text_images/FG04_34.jpg. Part B reproduced from http://cwx.prenhall.com/horton/medialib/media_portfolio/text_images/FG04_34.jpg.

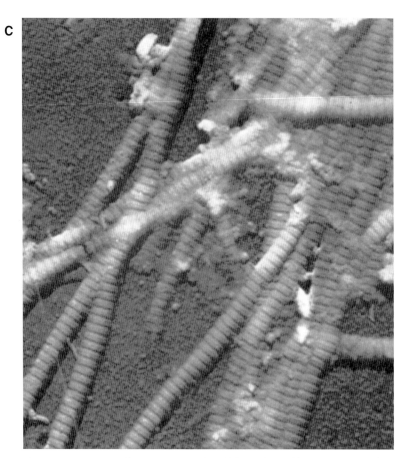

Figure 13-27, cont'd Part C reproduced from http://users.rcn.com/jkimball.ma.ultranet/BiologyPages/C/Collagen.gif.

four electrons participate in the reaction, so the enzyme must wait for the accumulation of four electrons because cytochrome c can transfer only one electron at a time (Figure 13-29). As the four electrons, needed to reduce oxygen to produce water, pass through the respiratory chain, 20 protons are pumped into the intermembrane space (against their concentration gradient by active transport). The protons can flow back down this gradient to reenter the matrix through another complex in the inner membrane, the **ATP synthase system.** As the concentration of protons increases in the intermembrane space (causing a decrease in the pH), a strong diffusion gradient develops. These protons can exit through the ATP synthase complex, and the energy released during the flow of these protons down their concentration gradient is used for the synthesis of ATP in a process known as **chemiosmosis** (Figure 13-30).

Another enzyme containing copper as a cofactor (two copper ions per subunit) is **dopamine β-hydroxylase,** which converts DOPA to norepinephrine, as shown in Figure 13-31. Ascorbic acid is a cosubstrate, and fumarate is an activator. This enzyme is important in the synthesis of catecholamines (neurotransmitters) from phenylalanine or tyrosine (Figure 13-32). Copper plays many other roles in the body, including

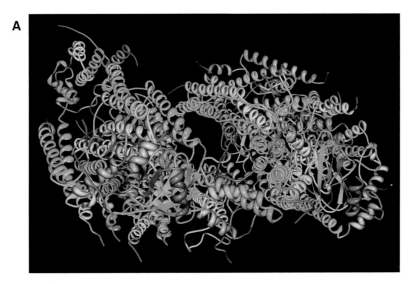

Figure 13-28. **A.** Structure of bovine heart cytochrome c oxidase (oxidoreductase). **B.** Reaction catalyzed by cytochrome c oxidase. Part A reproduced from RCSB PDB ID:2occ. S. Yoshikawa, K. Shinzawa-Itoh, R. Nakashima, R. Yaono, E. Yamashita, N. Inoue, M. Yao, M.J. Fei, C.P. Libeu, T. Mizushima, H. Yamaguchi, T. Tomizaki, T, Tsukihara. Redox-Coupled Crystal Structural Changes in Bovine Heart Cytochrome c Oxidase. *Science* **280** pp. 1723–1729 (1998).

blood clotting and maintaining low LDL levels and high HDL levels, and it is active in the immune system.

The best dietary sources of copper are calf liver, cremini mushrooms, greens, and blackstrap molasses, but many other foods are good sources of copper. There is no recommended daily allowance for copper, but a normal intake for an adult would be 2 mg. For other metals, the intakes are about 18 mg of iron, 2 mg of manganese, 120 μg of chromium, and 75 μg of molybdenum. Iron and (and ascorbic acid) inhibit the absorption of copper, and dietary copper forms complexes with sulfur and molybdenum. Zinc is absorbed by the same transporter (divalent metal transporter) as copper, so it competes with copper for absorption. High doses of zinc may, therefore, cause a deficiency of copper (and vice versa). Deficiency of copper reduces the activity of selenium-dependent enzymes because a number of selenium-dependent enzymes also contain copper.

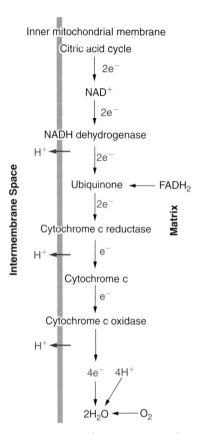

Figure 13-29. The respiratory chain, located on the inner mitochondrial membrane, consisting of NADH dehydrogenase, cytochrome c reductase, and cytochrome c oxidase, catalyze the stepwise transfer of electrons from NADH (and FADH$_2$) to oxygen molecules to form water (with protons). Redrawn from http://users.rcn.com/jkimball.ma.ultranet/BiologyPages/C/CellularRespiration.html.

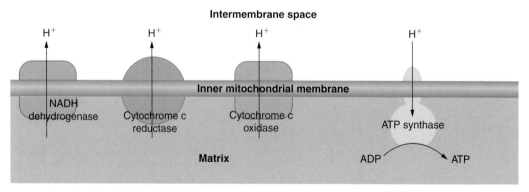

Figure 13-30. Components of the chemiosmosis process in mitochondria. Redrawn from http://users.rcn.com/jkimball.ma.ultranet/BiologyPages/C/CellularRespiration.html.

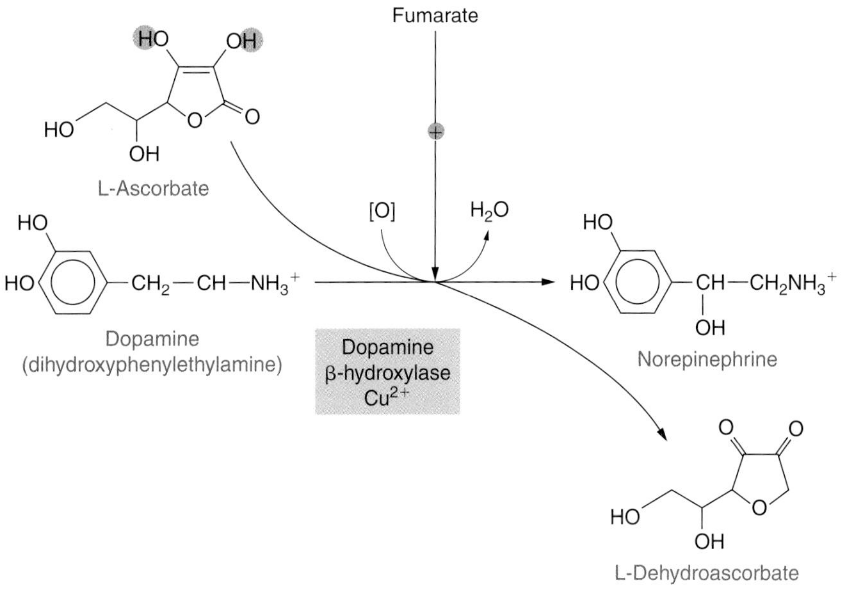

Figure 13-31. Reaction catalyzed by dopamine β-hydroxylase, a copper-containing enzyme.

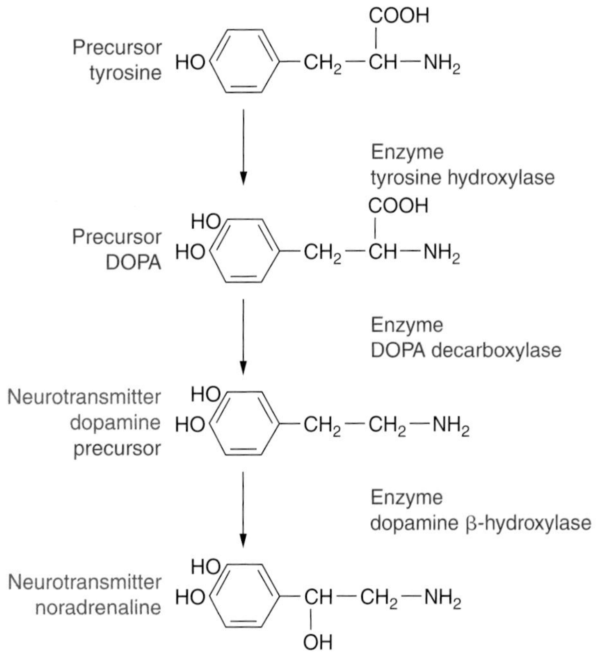

Figure 13-32. Synthesis of norepinephrine (noradrenaline) from tyrosine.

Selenium

Eleven or so selenoproteins have been characterized. Among these are antioxidant enzymes: glutathione peroxidases, iodothyronine deiodinases, selenophosphate synthetase, and thioredoxin reductase. Thioredoxin reductase contains a selenocysteine residue in a conserved C-terminal sequence (Gly–Cys–SeCys–Gly). If selenium–cysteine (SeCys) is deleted from the enzyme (deletion mutant), the enzyme loses activity. A cartoon structure of the enzyme is shown in Figure 13-33, showing the SeCys dipeptide as part of the C-terminal extension of the human enzyme. This enzyme has FAD and NADPH as cofactors in the catalyzed reaction.

Dietary Metals 773

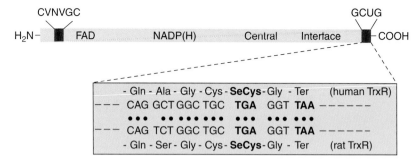

Figure 13-33. Cartoon structure of human thioredoxin reductase. *Ter,* Termination (stop) codon for mitochondria. Redrawn from L. Zhong and A. Holmgren, "Essential role of selenium in the catalytic activities of mammalian thioredoxin reductase revealed by characterization of recombinant enzymes with selenocysteine mutations," *J. Biol. Chem.*, 275: 18121–18128, 2000 and http://www.jbc.org/content/vol275/issue24/images/large/bc2406609001.jpeg.

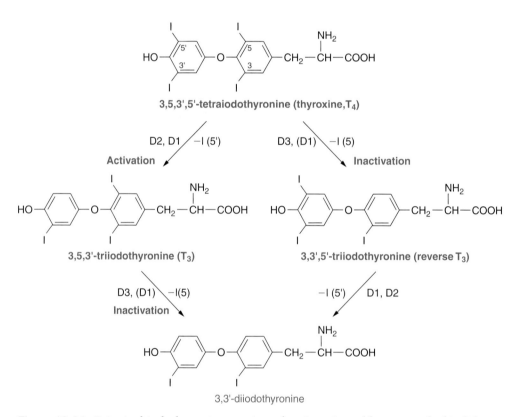

Figure 13-34. Principal iodothyronines activated or inactivated by removal of iodide by selenodeiodinases. *D1, D2,* and *D3* refer to isoforms of the same enzyme.

Iodothyronine deiodinases (there are three isoforms) catalyze the removal of iodide groups from the structure of T_4 and its relatives (Figure 13-34). Figure 13-35 shows the enzymatic reaction catalyzed by the deiodinase and the location of SeCys in the catalytic center of D1 deiodinase.

Another selenium enzyme is **selenophosphate synthetase (SePS)**, which catalyzes a reaction between HSe^- and ATP to produce selenophosphate

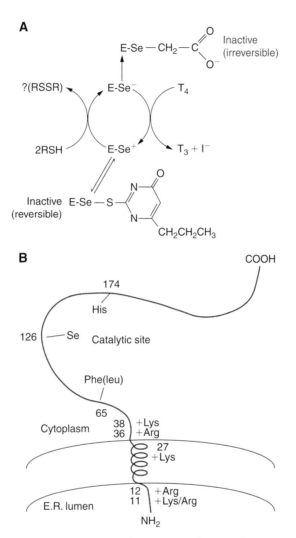

Figure 13-35. **A.** Deiodination mechanism for D1 catalyzed T4 to T3 conversion. **B.** The topology of the rat D1 as determined by protease sensitivities of the *in vitro*-translated rat Sec126Cys D1 mutant in the presence of pancreatic microsomes. Shown are locations of Phe65, important for rT_3 but not T_4 interaction with the active center, Sec 126, and His 174, which may be involved in maintaining Sec in a reduced state.

Dietary Metals   775

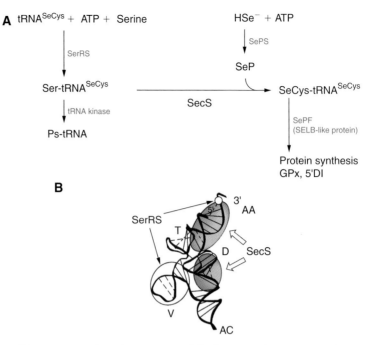

Figure 13-36. **A.** Schematic model of mammalian Sec synthesis. **B.** Tertiary structure model of tRNA^{Sec}. *White arrows* indicate the identity elements *(shaded area)* for SecS and *black arrows* indicate the identity elements *(white area)* for SerRS. Redrawn from Figures 1 and 2 of T. Mizutani, C. Goto and T. Totsuka, "Mammalian selenocysteine tRNA, its enzymes and selenophosphate," *J. Health Sci.*, 46: 399–404, 2000.

(SeP). SePS catalyzes the phosphorylation of selenide with magnesium ATP to form AMP, P_i, and $SePO_3$ (SeP). SeP serves as a selenium donor for tRNAs that incorporate selenium into specific proteins. Selenium–cysteine synthase (Sec synthase) converts Ser-tRNA^{Sec} to Sec-tRNA^{Sec} (selenocysteine is a *bona fide* amino acid with all of the components of protein synthesis specific to it) by addition of Se from SeP. Sec-tRNA is recognized by a specific elongation factor and is brought to Se-protein mRNA for the synthesis of selenoprotein. The set of reactions using SeP and the positioning of Sec synthase (SecS) with tRNA^{Sec} are shown in Figure 13-36. Rich sources of dietary selenium are oysters, Brazil nuts, beef, shrimp, crab meat, salmon, halibut, enriched noodles, pork, liver, yeast, brown rice, cereals, and eggs. Many other foods are good sources of selenium. Adults are expected to consume 5–20 μg of selenium per day.

Zinc

Some enzymes have zinc as a cofactor. Among these are carbonic anhydrase and cytosine deaminase. Some of these enzymes have zinc coordinated by three histidine residues and interacting with water or hydroxyl groups: Zn^{2+} $(N-His)_3(OH_2)$. The reaction catalyzed by carbonic anhydrase is $H_2CO_3 = CO_2 + H_2O$. The structure of human carbonic anhydrase complexed with an inhibitor is shown in Figure 13-37. *This reaction is*

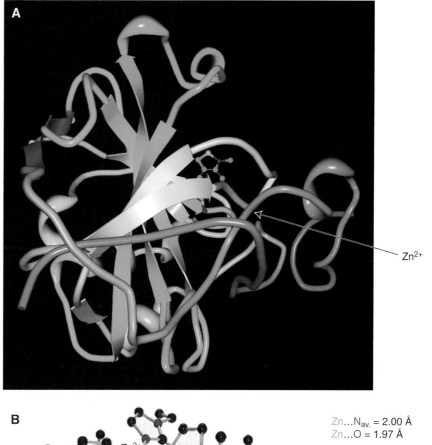
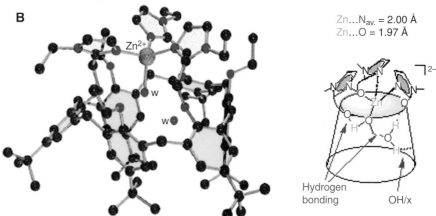

Figure 13-37. **A.** Structure of human carbonic anhydrase with an inhibitor. **B.** A model of the zinc coordination site in enzymes like carbonic anhydrase These histidine residues coordinate Zn^{2+} with interaction with an hydroxyl. Part A reproduced from http://www.ebi.ac.uk/thornton-srv/databases/cgi-bin/pdbsum/GetPage.pl?pdbcode=1kwr. PBD ID: 1kwr. S. Grüneberg, M.T. Stubbs, G. Klebe. Successful Virtual Screening for Novel Inhibitors of Human Carbonic Anhydrase: Strategy and Experimental Confirmation. *J. Med. Chem.* **45** pp. 3588–3602 (2002). Part B reproduced from http://www.biomedicale.univ-paris5.fr/umr8601/ORHTML/Zinc.html.

key for the elimination of carbon dioxide produced by metabolism. Cadmium and mercury, which can be toxic, easily can replace zinc in this enzyme and produce enzyme complexes that are more stable than those with zinc, resulting in the decreased activity of this enzyme. Another metalloenzyme, **cytosine deaminase,** catalyzes the conversion of cytosine to uracil, as shown in Figure 13-38. The structure of cytosine deaminase is shown in Figure 13-39. In addition to enzymes, zinc is found in certain DNA-binding proteins that contain the characteristic zinc finger structure. Two zinc fingers are in the DNA-binding domain of the glucocorticoid receptor (Figure 7-43) and have the general structure shown in Figure 13-40.

The recommended daily dose of zinc for adults is 11 mg for males and 8 mg for females, but a higher dosage is required during pregnancy (11 mg) and lactation (12 mg). Children require 3–8 mg, depending on their ages. Many people take oral zinc when they have a cold, but, so far, there is contradictory evidence on whether it is effective. Zinc is absorbed in the small intestine by **zinc family transporters (Zip),** by Zip4. Zip5 (Figure 13-41) is similar to Zip4, but Zip5 is found on the basolateral membrane of intestinal enterocytes and Zip4 is found on the apical membrane. These transporters should account for the movement of zinc into the intestinal cells from the intestinal lumen and across the cell into the extracellular space and the bloodstream. These are eight membrane-spanning subunit proteins similar to transporters discussed in Chapter 12. There is some evidence that zinc uptake in the intestine is inhibited by the vitamin folic acid. Zinc and folic acid form a complex at acid pH and the complex dissolves at pH 6.0, suggesting that there is a competition between zinc and folic acid at the site of intestinal transport.

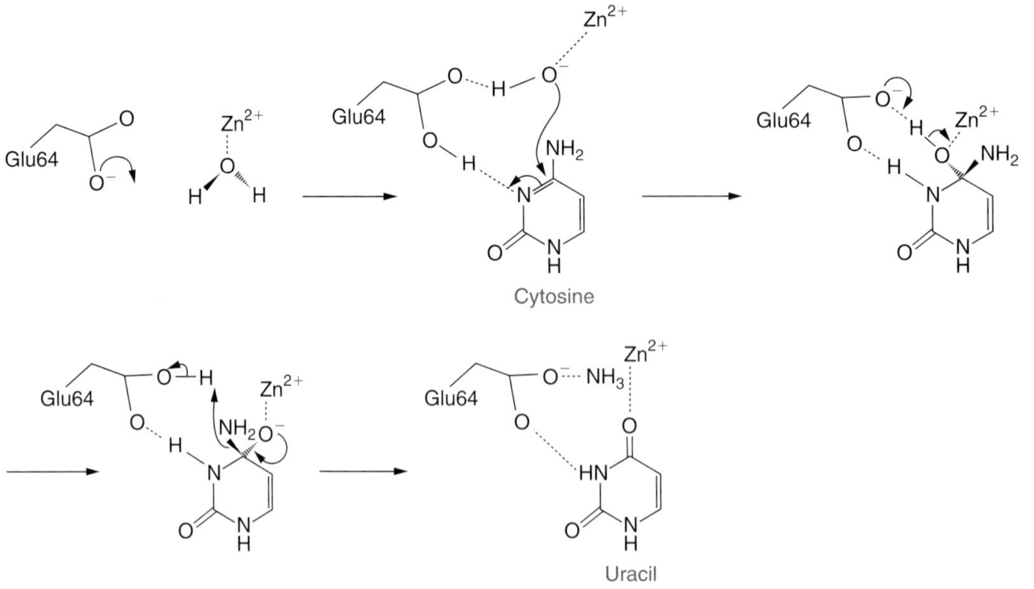

Figure 13-38. Proposed catalytic mechanism for yeast cytosine deaminase.

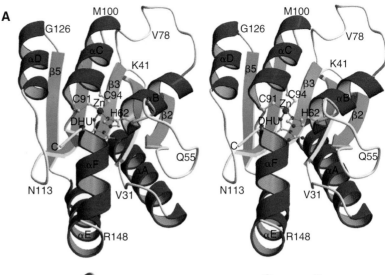

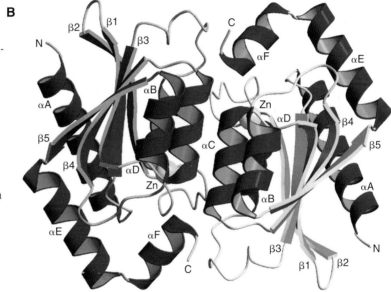

Figure 13-39. Structure of yeast cytosine deaminase. **A.** Stereo view of the monomer, which is a three-layered α/ß/α structure with a central ß-sheet sandwiched on either side by α-helices. The tightly bound zinc ion is shown as a *magenta sphere*, with its ligands and the inhibitor (3,4-dihydrouracil [DHU]) as ball-and-stick representations. **B.** The dimer has one monomer colored in *red* (helices) and *green* (strands), whereas the other is colored in *magenta* (helices) and *blue* (strands). The zinc ions are shown as *yellow spheres*. Reproduced from Figure 1 of T.-P. Ko et al., "Crystal structure of yeast cytosine deaminase: Insights into enzyme mechanism and evolution," *J. Biol.Chem.*, 278: 19111–19117, 2004 and http://www.jbc.org/cgi/content/full/278/21/19111.

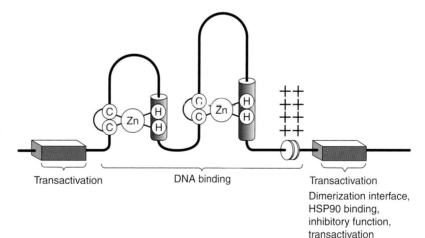

Figure 13-40. Structure of zinc finger motifs. Functional domains of zinc finger transcription factors. Cysteine *(C)* and histidine *(H)* coordinated by a zinc atom, causing the looping out of the "zinc fingers." The WT-1 transcription factor contains four zinc finger regions and is usually expressed in fetal kidney and gonads. Redrawn from http://web.archive.org/web/20030818055336/www.devbio.com/chap05/link0504.shtml.

Dietary Metals

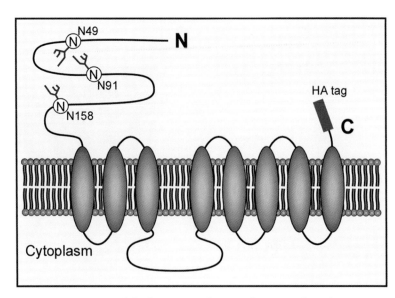

Figure 13-41. A model of Zip5 topology is shown with eight transmembrane domains. The positions of potential sites of *N*-glycosylation and the carboxyl-terminal HA tag are also shown. Both amino- and carboxyl-terminal domains are predicted to be extracytoplasmic. Reproduced from F. Wang et al., "The mammalian Zip5 protein is a zinc transporter that localizes to the basolateral surface of polarized cells," *J. Biol. Chem.*, 279: 51433–51441, 2004 and Figure 3 or http://www.jbc.org/cgi/content/full/279/49/51433/FIG3.

Magnesium

Magnesium is plentiful throughout the body and is involved in more than 300 metabolic reactions. Some examples are the ATP-synthesizing protein in mitochondria and ATP synthase, discussed here. Magnesium also is involved in the synthesis of nucleic acids, especially because magnesium forms a complex with many nucleotides (for example, MgATP). It is a cofactor for several enzymes involved in the synthesis of carbohydrates and lipids, and glutathione synthesis involves magnesium. Muscular contraction and relaxation requires magnesium. Many magnesium sites in enzymes can be replaced by manganese, another trace metal in nutrition. Magnesium also plays a structural role in bone, cell membranes, and chromosomes. The transport of magnesium and other divalent ions is an emerging field, and this topic has been addressed in Chapter 12.

An example of the functioning of magnesium is its role in ATP synthase, where the energy produced by protons being transported down their concentration gradient is used to couple ADP and inorganic phosphate to form ATP, the cell's primary energy source. The process is shown in Figure 13-42. Although the mechanism is not completely understood, the pivotal role of Mg^{2+} is evident.

The adult human body contains about 25 g of magnesium, more than 60% of which is found in the skeleton. In addition, more than 25% is found in muscle, about 7% is found in other tissue cells, and 1% is found outside cells. High doses of dietary zinc will interfere with the absorption

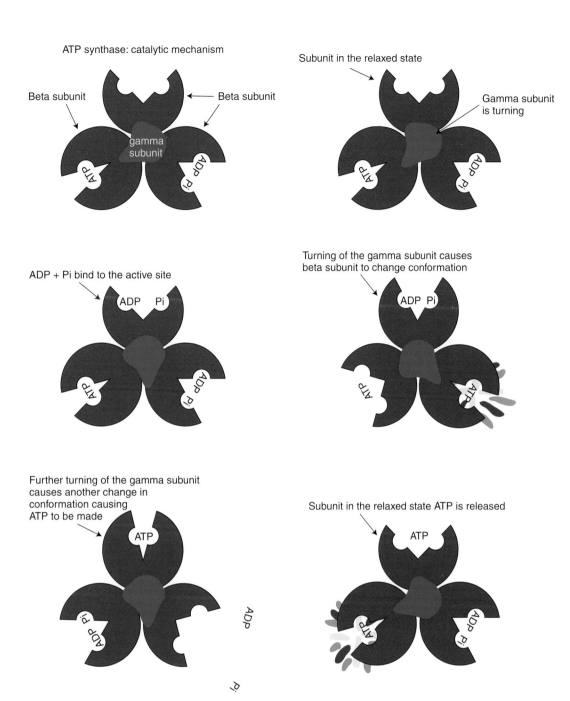

Figure 13-42. Steps (**A** through **F**) in the ATP synthase mechanism. Reproduced from http://www.stolaf.edu/people/giannini/flashanimat/metabolism/atpsyn2.swf.

of magnesium, indicating competition for the same transporter in the intestine. The recommended daily allowances for magnesium are shown in Table 13-6. The intake in males over age 31 should be about 420 mg/day. In the elderly, less magnesium seems to be absorbed in the diet and more excreted, so they risk lowered magnesium levels in the body. Early deficiency symptoms are irritability, anorexia, fatigue, insomnia, muscle twitching, poor memory, apathy, confusion, and reduced learning ability. Severe deficiency includes tingling, numbness, prolonged muscle contraction, hallucinations, and delirium. Food sources of magnesium are shown in Table 13-7.

Table 13-6
Recommended Daily Allowances for Dietary Magnesium

Life Stage	Age	Males (mg/day)	Females (mg/day)
Infants	0–6 months	30 (AI)	30 (AI)
Infants	7–12 months	75 (AI)	75 (AI)
Children	1–3 years	80	80
Children	4–8 years	130	130
Children	9–13 years	240	240
Adolescents	14–18 years	410	360
Adults	19–30 years	400	310
Adults	31 years and older	420	320
Pregnancy	18 years and younger	—	400
Pregnancy	19–30 years	—	350
Pregnancy	31 years and older	—	360
Breastfeeding	18 years and younger	—	360
Breastfeeding	19–30 years	—	310
Breastfeeding	31 years and older	—	320

AI, adequate intake, which is established when an RDA cannot be determined.
Reproduced from http://lpi.oregonstate.edu/infocenter/minerals/magnesium/.

Table 13-7
Food Sources of Magnesium

Food	Serving	Magnesium (mg)
100% bran cereal (e.g., All Bran)	1/2 cup	128.7
Oat bran	1/2 cup dry	96.4
Shredded wheat	2 biscuits	54.3
Brown rice	1 cup cooked	83.8
Almonds	1 ounce (22 almonds)	81.1
Hazelnuts	1 ounce	49.0
Peanuts	1 ounce	49.8
Lima beans	1/2 cup cooked	62.9
Black-eyed peas	1/2 cup cooked	42.8
Spinach, chopped	1/2 cup cooked	78.3
Swiss chard, chopped	1/2 cup cooked	75.2
Okra, sliced	1/2 cup cooked	45.6
Molasses, blackstrap	1 tablespoon	43.0
Banana	1 medium	34.2
Milk 1% fat	8 fluid ounces	33.7

Reproduced from http://lpi.oregonstate.edu/infocenter/minerals/magnesium/.

Calcium: A Micronutrient

Dietary calcium is the most abundant cation in the body, and most of the body's calcium (more than 99%) is in the structure of the bones and teeth. The remainder is in blood, muscle, and interstitial fluid. Calcium is used in muscle contraction; in contraction and expansion of blood vessels; as a second messenger in signaling systems, some of which are hormonal (Figure 12-26 and Figure 12-28); and in the functioning of the nervous system. Bone remodeling is a continuous process, in which formation and breakdown of bone vary with age. Levels of calcium in the body are well controlled, and the genes controlling the apical entry of calcium and the basolateral expulsion, eventually into the blood, are tightly linked.

Calcium is absorbed by way of the calcium transport channel (CaT1; also known as TRPV6) in the brush border membrane of the small intestine that facilitates the absorption of calcium ion from the intestinal lumen into the intestinal epithelial cells. This transporter (facilitated diffusion) is constitutive but is negatively regulated by calcium inside the cell; thus, the transporter is activated by calcium storage depletion. Calcium also can be transported outwardly toward the lumen when calcium exocytosis occurs. CaT1 is a member of the **TRPV family.** CaT1 and the epithelial calcium channel (ECaC; also known as TRPV5, the form found in the kidney) are calcium-selective channels that facilitate the entry of calcium in absorptive and secretory tissues. CaT1 is expressed largely in proximal intestine, placenta, and exocrine tissues. ECaC is mainly expressed in distal convoluted and connecting tubules of the kidney. In the intestine, CaT1 is responsive to activated vitamin D (1,25-dihydroxyvitamin D_3) and is regulated by fast and slow calcium-dependent feedback inhibition to prevent the overload of calcium in the cell. ECaC is only slowly inactivated and is regulated by calcium load in the kidney. CaT1 also is highly expressed in pancreas, prostate, and salivary gland. An overview of the uptake of calcium from the intestinal lumen is shown in Figure 13-43.

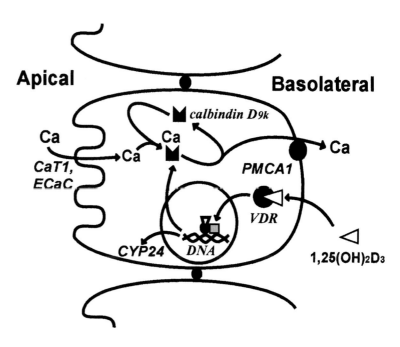

Figure 13-43. Calcium uptake in the intestinal epithelial cell through CaT1 (or distal convoluted or connecting kidney tubules, in the case of ECaC). The facilitated diffusion model proposed to explain vitamin D–regulated intestinal calcium absorption. *CaT1*, calcium transport channel; *ECaC*, epithelial calcium channel; *CYP24*, 25-hydroxyvitamin D_3 24-hydroxylase; *PMCA1*, plasma membrane calcium ATPase 1; *VDR*, vitamin D receptor. Reproduced from http://ajpgi.physiology.org/cgi/content/full/283/3/G618. J.C. Fleet et al., "Vitamin D-inducible calcium transport and gene expression in three Caco-2 cell lines," *Am. J. Physiol. Gastrointest Liver Physiol*, 283: G618–G625, 2002.

Calcium is absorbed through the CaT1 channel and complexed with calbindin inside the cell, carried across the cell, and pumped to the basolateral extracellular space by a calcium transporter driven by ATP (Ca-ATPase; Figure 13-44). The vitamin D receptor (VDR), inside the intestinal epithelial cell, binds vitamin D_3 and acts transcriptionally to increase the levels of **CYP24** (a unique cytochrome P450 involved in the degradation of vitamin D) and calbindin. The structures of calbindin, the VDR ligand–binding domain, and cytochrome P450 are shown in Figure 13-45. Cytochrome P450 family members are heme-containing monooxygenases acting on paired donors with the incorporation or reduction of molecular oxygen; reduced flavin or flavoprotein can be one donor. Often, a hydroxyl group is incorporated into the substrate structure. The architecture of the intestinal CaT1 channel is shown in Figure 13-46. The channel is composed of six transmembrane domains, and the N-terminus is anchored on the cytoplasmic side to an ankyrin repeat (Figure 12-49). An aspartate residue (D542) is the important residue for permeation of the calcium cation (Ca^{2+}). The process of calcium import through this channel is explained in the legend of Figure 13-46. Regulation of the activity of the CaT1 channel has been hypothesized to involve a constitutively active channel, a channel that is inactive but is activated by intracellular depletion of the

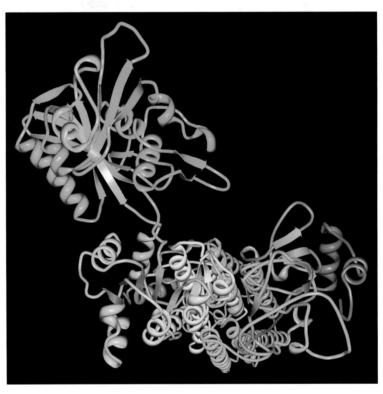

Figure 13-44. Crystal structure of calcium ATPase (rabbit muscle) with two bound calcium ions *(upper left)* and one Na^+ *(blue atom)*. This enzyme provides energy for the expulsion of Ca^{2+} on the basolateral membrane:

$$ATP + H_2O \rightleftharpoons ADP + P_i$$
$$Ca^{2+}_{Cis} \longrightarrow Ca^{2+}_{trans}$$

Reproduced from http://www.ebi.ac.uk/msd-srv/msdlite/atlas/visualization/1su4.html.

PDB ID: 1su4. C. Toyoshima, M. Nakasako, H. Nomura, H. Ogawa. Crystal structure of the calcium pump of sarcoplasmic reticulum at 2.6 A resolution. *Nature* **405** pp. 647 (2000).

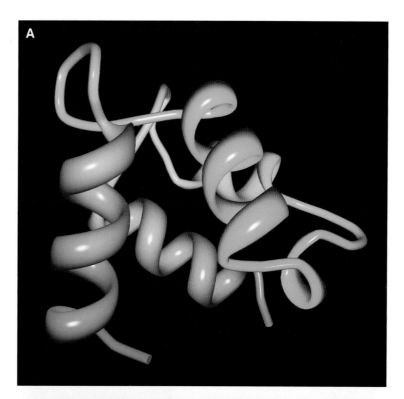

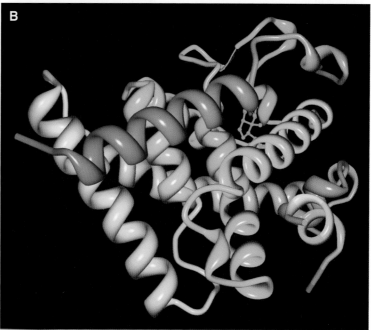

Figure 13-45. **A.** Bovine calbindin D, 9k binding Mg^{2+}. **B.** Crystal structure of the rat vitamin D receptor ligand binding domain complexed with 1,25-dihydroxyvitamin D_3 (ball-and-stick structure) and a synthetic peptide containing the MR2 box of DRIP 205. **C.** Crystal structure of human cytochrome P450 3A4. Part A reproduced from http://www.rcsb.org/pdb/cgi/explore.cgi?pdbId=lig5. PDB ID: lig5. M. Andersson, A. Malmendal, S. Linse, I. Ivarsson, S. Forsén, L.A. Svensson. Structural Basis for the Negative Allostery Between Ca(2+)- and Mg(2+)-Binding in the Intracellular Ca(2+)-Receptor Calbindin D9k. *Protein Science* **6** pp. 1139 (1997). Part B reproduced from http://www.rcsb.org/pdb/explore/images.do?structureId=lrk3. PDB ID: lrk3. J.L. Vanhooke, M.M. Benning, C.B. Bauer, J.W. Pike, H.F. DeLuca. Molecular Structure of the Rat Vitamin D Receptor Ligand Binding Domain Complexed with 2-Carbon-Substituted Vitamin D3 Hormone Analogues and a LXXLL-Containing Coactivator Peptide. *Biochemistry* **43** pp. 4101 (2004).

Continued

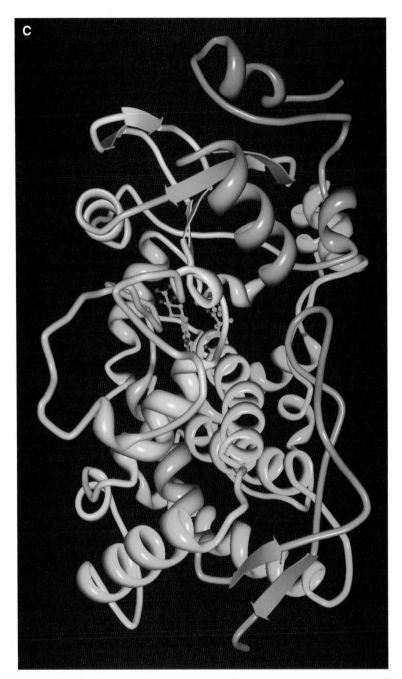

Figure 13-45, cont'd Part C reproduced from PDB ID: 1w0g. P.A. Williams, J. Cosme, D.M. Vinkovic, A. Ward, H.C. Angove, P.J. Day, C. Vonrhein, I.J. Tickle, H. Jhoti. Crystal Structures of Human Cytochrome P450 3A4 Bound to Metyrapone and Progesterone. *Science* **305** pp. 683 (2004).

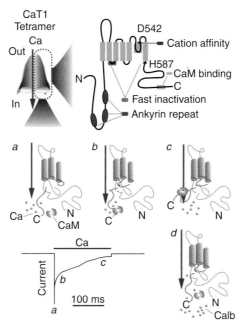

Figure 13-46. Molecular architecture and feedback inactivation of CaT1. *Upper panel,* key domains and residues in CaT1. A CaT1 tetramer, the functional unit of the channel, is shown on the left. Ankyrin repeats may mediate protein-protein interaction; D542 is a key residue cation affinity or permeation. The first intracellular loop and H587 are involved in the fast phase of inactivation. The calmodulin (CaM) binding site close to the carboxyl terminus is involved in the slow phase of Ca^{2+}-dependent inactivation. *Lower panel,* fast and slow phases of inactivation and involved residues and domains in CaT1. A typical Ca^{2+}-dependent inactivation process of Ca^{2+} current of CaT1 is shown in the box. The currents and the presumed corresponding CaT1 states are labeled *a, b,* and *c; d* shows the putative role of calbindin D_{9k} (Calb) on Ca^{2+} influx; buffering the Ca^{2+} underneath the channel thereby releases the channel from Ca^{2+} feedback inhibition. Redrawn from J.-B. Peng, E.M. Brown, and M.A. Hediger, "Epithelial Ca^{2+} entry channels: transcellular Ca^{2+} transport and beyond," *J. Physiol.,* 551.3: 729–740, 2003 and http://jp.physoc.org/content/vol551/issue3/fulltext/729/729-F2.html.

calcium store or a secretion-coupled mode, in which inactivated CaT1 may be localized within the cell and then exocytosed to the membrane after an appropriate signal. These processes may exist in different tissues to fulfill specific needs (Figure 13-47).

The relative of CaT1 in the kidney is ECaC1, which acts similarly to CaT1 (Figure 13-48). Calcium transport occurs in the distal nephron, exclusively, in a three-step process. Calbindin, again, plays a role in controlling intracellular concentrations of calcium, and calcium ion is pumped out of the cell through the calcium-ATPase (PMCA1b) on the basolateral surface and by a Na^+ exchanger (antiporter, NCX1). With its ligand, the VDR stimulates the expression of ECaC1 and calbindin. The predicted structure of ECaC1 is shown in diagrammatic form in Figure 13-49.

An adult male or female, 19–50 years, needs to ingest 1000 mg of calcium per day in the diet. The need increases in ages over 51 (Table 13-8). Food sources of calcium are shown in Table 13-9.

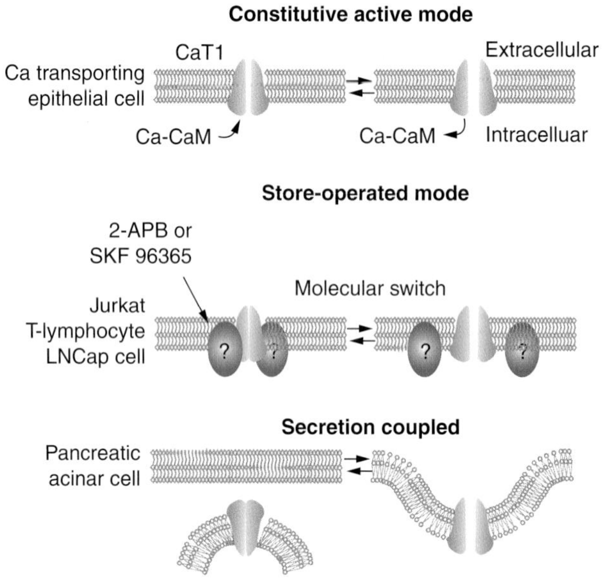

Figure 13-47. Modes of CaT1 activation. Constitutive active mode: CaT1 alone is constitutively active and works in this mode in Ca^{2+}-transporting epithelia such as the brush-border membrane of small intestine. The channels are subject to Ca^{2+}-dependent feedback regulation and this process is partially mediated by calmodulin (CaM). Store-operated mode: CaT1 is not active in unstimulated cells and is activated by calcium store depletion. Two putative mechanisms are shown. Molecular switch mode: another cellular component serves as a molecular switch to shut CaT1 off when it is associated with CaT1 or to activate CaT1 upon its dissociation. This component might be the target of 2-APB or SKF 96365, which may promote the association state, as suggested by studies in Jurkat T-lymphocytes or LNCaP (prostate cancer) cells. Secretion-coupled mode: inactivated CaT1 may localize in the membrane of intracellular vesicles. Upon exocytosis, CaT1 is incorporated into plasma membrane. The plasma membrane-associated CaT1 takes up Ca^{2+} released from the vesicle to refill the calcium store or to trigger further exocytosis. Redrawn from Figure 3 of reference to Figure 13-46.

Molybdenum

Molybdenum (Mo^{2+}) functions as a coenzyme in the form of a pterin (molybdoenzymes) for xanthine oxidase, sulfite oxidase, and aldehyde oxidase and for proteins like human **gephyrin,** which is located in the central nervous system, where it facilitates the clustering of inhibitory neuroreceptors in the postsynaptic membrane. The intestinal transporter for molybdenum is not well known; however, in bacteria, molybdenum transport is accomplished by ABC-like transporters (Figure 12-4A). The molybdenum cofactor (a pterin) is synthesized from GTP or guanosine-X nucleotide to form precursor Z. Molybdopterin synthase catalyzes the conversion of precursor Z to form molybdopterin, then molybdenum is inserted into this structure to form the molybdenum cofactor (moco), as shown in Figure 13-50. The structure of molybdopterin synthase from *Escherichia coli* is shown in Figure 13-51. The structures of molybdate-containing enzymes, as well as human gephyrin, are shown in Figure 13-52.

(Text continues on p. 795.)

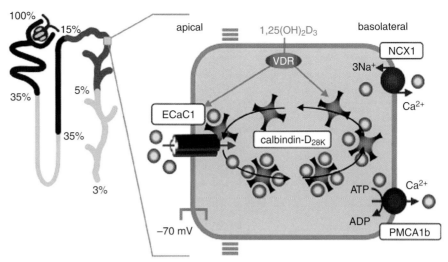

Figure 13-48. Ca^{2+} reabsorption along the tubule. The remaining Ca^{2+} at different sites of the nephron is indicated in percentages. Transcellular Ca^{2+} transport only takes place in the distal nephron *(pink)* and is carried out as a three-step process. Following entry of Ca^{2+} through the epithelial Ca^{2+} channel, ECaC1, cytosolic Ca^{2+} is buffered by calbindin-D_{28k}. At the basolateral membrane, Ca^{2+} is extruded by a Ca^{2+}-ATPase (PMCA1b) and a sodium-calcium exchanger (NCX1). $1,25(OH)_2D_3$ regulates this process by stimulating the expression of ECaC1 and calbindin-D_{28k}. Reproduced with permission from Figure 1 of D. Müller et al., "The epithelial calcium channel, ECaC1: molecular details of a novel player in renal calcium handling," *Nephrol. Dial. Transplant.*, 16: 1329–1335, 2001 and http://ndt.oxfordjournals.org/cgi/content/full/16/7/1329/F1.

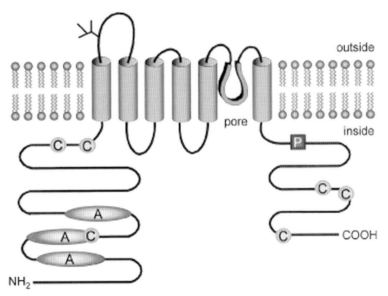

Figure 13-49. Predicted topology of ECaC1 consisting of six transmembrane-spanning domains with a short hydrophobic stretch between transmembrane domains 5 and 6. Cytosolic C- and N-terminal tails contain several potential regulatory sites for PKC phosphorylation, ankyrin repeats, and a PDZ domain. Reproduced with permission from Figure 2A of D. Müller et al., "The epithelial calcium channel, ECaC1: molecular details of a novel player in renal calcium handling," *Nephrol. Dial. Transplant.*, 16: 1329–1335, 2001 and http://ndt.oxfordjournals.org/cgi/content/full/16/7/1329/F2.

Table 13-8
Recommended Daily Intakes of Calcium

Male and Female Age	Calcium (mg/day)	Pregnancy & Lactation
0–6 months	210	N/A
7–12 months	270	N/A
1–3 years	500	N/A
4–years	800	N/A
9–13 years	1300	N/A
14–18 years	1300	1300
19–50 years	1000	1000
51+ years	1200	N/A

*mg, milligrams.
Reproduced from http://ods.od.nih.gov/factsheets/calcium.asp.

Table 13-9
Food Sources of Calcium

Food	Serving	Calcium (mg)	% DV
Yogurt, plain, low fat,	8 ounces	415	42
Yogurt, fruit, low fat,	8 ounces	245–384	25–38
Sardines, canned in oil, with bones,	3 ounces	324	32
Cheddar cheese,	1 ½ ounces shredded	306	31
Milk, nonfat,	8 fluid ounces	302	30
Milk, reduced fat (2% milk fat), no solids,	8 fluid ounces	297	30
Milk, whole (3.25% milk fat),	8 fluid ounces	291	29
Milk, buttermilk,	8 fluid ounces	285	29
Milk, lactose reduced,	8 fluid ounces*	285–302	29–30
Mozzarella, part skim	1½ ounces	275	28
Tofu, firm, made w/calcium sulfate,	½ cup†	204	20
Orange juice, calcium fortified,	6 fluid ounces	200–260	20–26
Salmon, pink, canned, solids with bone,	3 ounces	181	18
Pudding, chocolate, instant, made w/ 2% milk,	½ cup	153	15
Cottage cheese, 1% milk fat,	1 cup unpacked	138	14
Tofu, soft, made w/ calcium sulfate,	½ cup†	138	14
Spinach, cooked,	½ cup	120	12
Instant breakfast drink, various flavors and brands, powder prepared with water,	8 fluid ounces	105–250	10–25
Frozen yogurt, vanilla, soft serve,	½ cup	103	10
Ready to eat cereal, calcium fortified,	1 cup	100–1000	10–100
Turnip greens, boiled,	½ cup	99	10
Kale, cooked,	1 cup	94	9
Kale, raw,	1 cup	90	9
Ice cream, vanilla,	½ cup	85	8.5
Soy beverage, calcium fortified,	8 fluid ounces	80–500	8–50
Chinese cabbage, raw,	1 cup	74	7
Tortilla, corn, ready to bake/fry,	1 medium	42	4
Tortilla, flour, ready to bake/fry,	16″ diameter	37	4
Sour cream, reduced fat, cultured,	2 Tablespoons	32	3
Bread, white,	1 ounce	31	3
Broccoli, raw,	½ cup	21	2
Bread, whole wheat,	1 slice	20	2
Cheese, cream, regular,	1 Tablespoon	12	1

DV, daily value.
*Content varies slightly according to fat content; average = 300 mg calcium.
†Calcium values are only for tofu processed with calcium salt. Tofu processed with a noncalcium salt will not contain significant amounts of calcium.
Reproduced from http://ods.od.nih.gov/factsheets/calcium.asp.

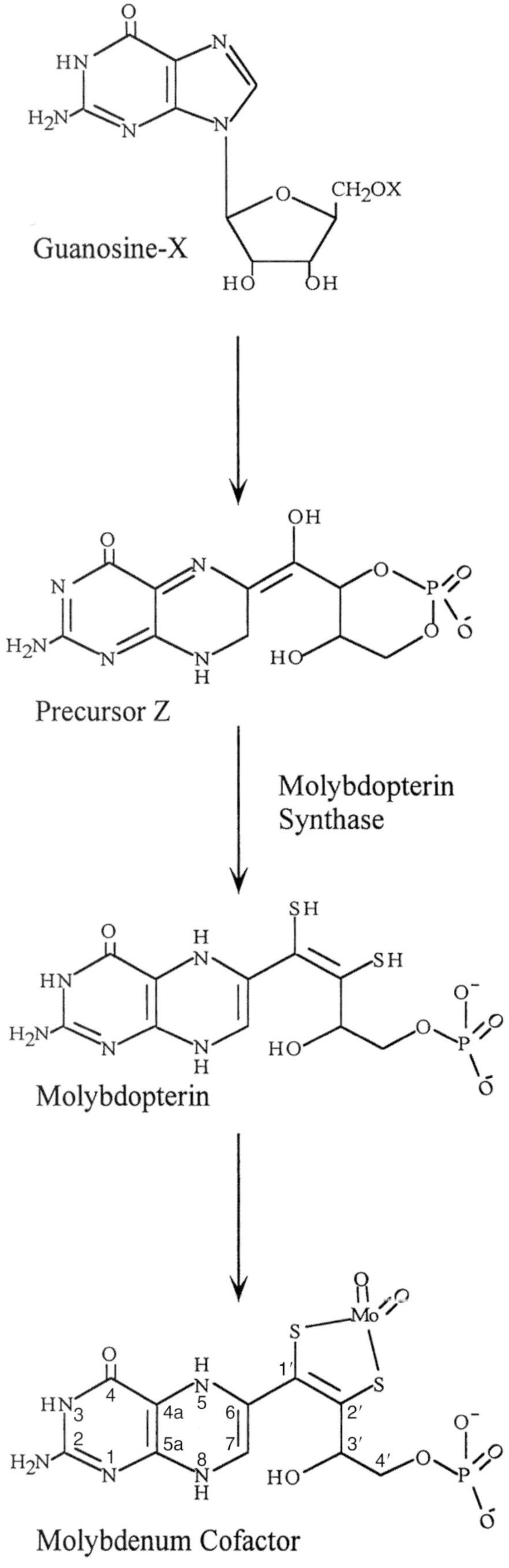

Figure 13-50. Synthesis of molybdenum cofactor from GTP or a guanosine derivative. Reproduced from S. Unkles et al., "Eukaryotic molybdopterin synthase: biochemical and molecular studies of *Aspergillus nidulans cnxG* and *cnxH* mutants," J. Biol. Chem., 274: 19286–19293, 1999 and http://www.jbc.org/content/vol274/issue27/images/large/bc2691088001.jpeg.

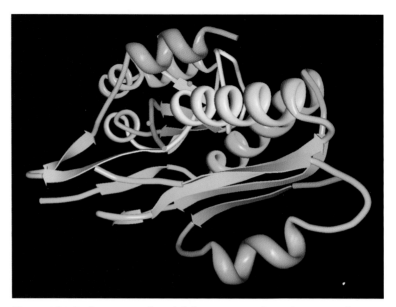

Figure 13-51. Structure of molybdopterin synthase from *E. coli*. PDB ID: 1nvi. M.J. Rudolph et al., "Crystal structure of molybdopterin synthase and its evolutionary relationship to ubiquitin activation," *Nat. Struct. Biol.*, 8: 42–46, 2001 and http://www.rcsb.org/pdb/cgi/explore.cgi?pdbId=1fmo.

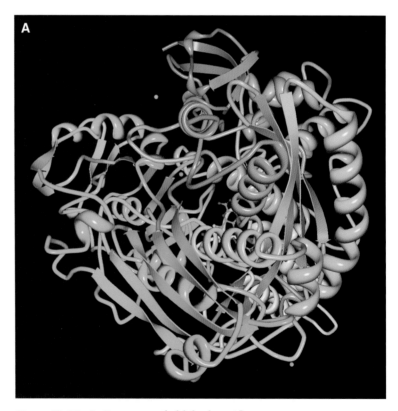

Figure 13-52. **A.** Structure of aldehyde oxidase.

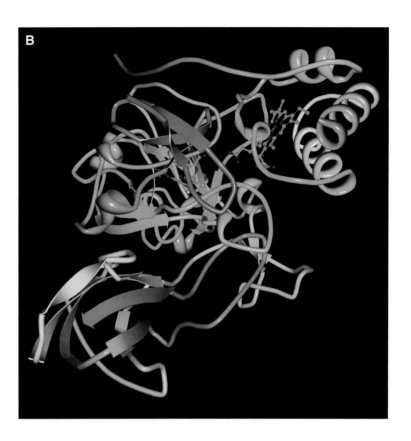

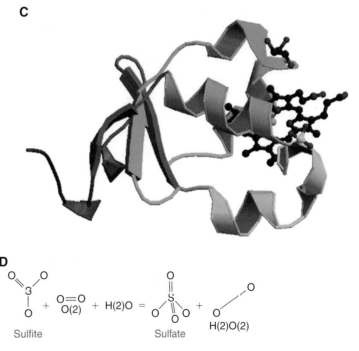

Figure 13-52, cont'd **B.** Structure of sulfite dehydrogenase. **C.** Crystal structure of human sulfite oxidase (cytochrome B5 domain). **D.** Reaction catalyzed by sulfite oxidase. **E.** X-ray structure of gephyrin N-terminal domain. **F.** Crystal structures of human gephyrin and plant Cnx1 G domains. Comparative analysis and functional implications. Part A reproduced from http://www.rcsb.org/pdb/cgi/explore.cgi?pdbId=1sij. PDB ID: 1sij. D.R. Boer, A. Thapper, C.D. Brondino, M.J. Romao, J.J.G. Moura. X-ray Crystal Structure and EPR Spectra of "Arsenite-Inhibited" Desulfovibriogigas Aldehyde Dehydrogenase: A Member of the Xanthine Oxidase Family. *J. Am. Chem. Soc.* **126** pp. 8614 (2004).

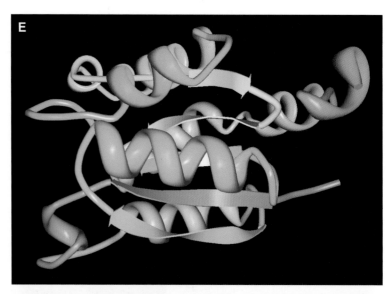

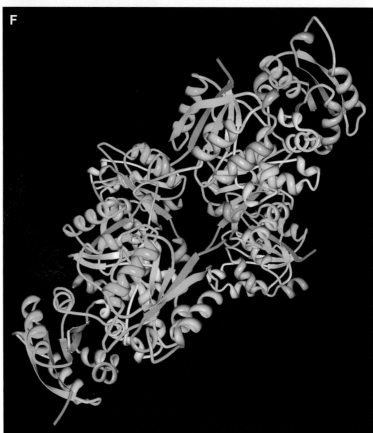

Fig. 13-52, cont'd Part B reproduced from U. Kappler and U.S. Bailey, "Molecular basis of intramolecular electron transfer in sulfite-oxidizing enzymes is revealed by high resolution structure of a heterodimeric complex of the catalytic molybdopterin subunit and c-type cytochrome subunit," *J. Biol. Chem.*, V280: pp. 24999, 2005 and http://www.rcsb.org/pdb/cgi/explore.cgi?pdbId=2blf. Part C reproduced from http://www.rcsb.org/pdb/explore/images.do?structureId=1mj4. PDB ID: 1mj4 M.J. Rudolph, J.L. Johnson, K.V. Rajagopalan, C. Kisker. The 1.2 Å Structure of the Human Sulfite Oxidase Cytochrome b(5) Domain. *Acta. Crystallogr.*, Sect. D, **59** pp. 1183 (2003). Part D reproduced from http://www.ebi.ac.uk/thornton-srv/databases/cgi-bin/pdbsum/GetPage.pl?pdbcode=1mj4. Part E reproduced from http://www.rcsb.org/pdb/cgi/explore.cgi?pdbId=1ihc. PDB ID: 1ihc. M. Sola, M. Kneussel, I.S. Heck, H. Betz, W, Weissenhorn. X-ray crystal structure of the trimeric N-terminal domain of gephyrin. *J. Biol. Chem.* **276** pp. 25294 (2001). Part F reproduced from PDB ID: 1eav. G. Schwarz, N. Schrader, R.R. Mendel, H.J. Hecht, H. Schindelin. Crystal structures of human gephyrin and plant Cnx1 G domains: comparative analysis and functional implications. *J. Mol. Biol.* **312** pp.405 (2001).

Table 13-10
Recommended Daily Allowances for Molybdenum

Life Stage	Age	Males (µg/day)	Females (µg/day)
Infants	0–6 months	2 (AI)	2 (AI)
Infants	7–12 months	3 (AI)	3 (AI)
Children	1–3 years	17	17
Children	4–8 years	22	22
Children	9–13 years	34	34
Adolescents	14–18 years	43	43
Adults	19 years and older	45	45
Pregnancy	All ages	—	50
Breastfeeding	All ages	—	50

AI, adequate intake, which is established when an RDA cannot be determined.
Reproduced from http://lpi.oregonstate.edu/infocenter/minerals/molybdenum.

Dietary deficiency of molybdenum is rare. A genetic disease known as sulfite oxidase deficiency or molybdenum cofactor deficiency has been characterized but is extremely rare. It is a recessive disorder in which a recessive, mutated gene must be contributed from each parent (the carrier of a recessive gene does not display the disorder). Severe brain damage is the result of this disorder and may relate to the loss of function of gephyrin.

Recommended daily allowances for molybdenum are shown in Table 13-10. An adult requires ingestion of 45 µg (10^{-6}g)/day and a 170-pound adult has 7.7 mg of bodily molybdenum. People with a reasonably normal diet do not have a problem in getting enough molybdenum. Good sources of this metal are beans, lentils, peas, grains, and nuts.

Most of the trace minerals (copper, molybdenum, zinc, and selenium) have been discussed; manganese, chromium, lithium, and vanadium have only been touched upon or have not been included in this discussion. Aluminum is known for its toxic effects when too much is absorbed in the body. It is unclear whether it causes Alzheimer's disease, but it can be toxic, especially by competing with other required metals, such as copper.

The major micronutrients, calcium, cobalt, and iron, have been discussed; phosphorous, potassium, sodium, and sulfur have not been presented here.

Iodine: A Micronutrient

Iodine, another micronutrient, is essential for the formation of thyroid hormone, which is needed in all cells to maintain metabolic rate. The synthesis of the thyroid hormone in the follicular cells of the thyroid gland is under the control of a hypothalamic hormone, an anterior pituitary hormone, and a sufficiency of iodine. The hypothalamus secretes **thyroid-releasing hormone (TRH),** which is connected to the anterior pituitary by a closed portal system. TRH activates the release of thyroid-stimulating hormone (TSH) from the anterior pituitary by binding to its receptor on the cell membrane of the anterior pituitary thyrotropic cell. TSH binds to its receptor on the thyroid follicular cell and generates biochemical changes within the cell (increases cyclic AMP) that culminate in the synthesis and release of T_4 (and a small amount of T_3). The control of T_4 secretion is shown in Figure 13-53 under normal conditions and under conditions of insufficient dietary iodine. The events occurring in the thyroid follicular cells under the influence of TSH are shown in Figure 13-54. Iodide enters the cell through the sodium/iodide symporter (NIS; Figure 13-55), that is, under the positive control of TSH. The symporter, a glycoprotein, moves two sodium ions and one iodide ion into the cell (Figure 13-56). NIS also pumps iodide into salivary glands, gastric mucosa, and lactating mammary gland (translocating I^- into the milk for the newborn's synthesis of thyroid hormone). The iodide ion inside the thyroid follicular cell, transported from the outside and from liberated iodide from free iodotyrosines by way of deiodinase, forms iodine. With the action of peroxidase, the thyronines on the colloid, thyroglobulin,

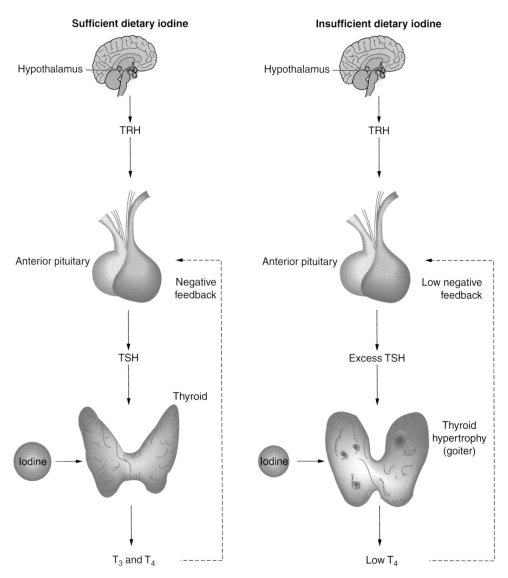

Figure 13-53. Iodine intake and thyroid function. Insufficient dietary iodine results in lowered production of thyroxine (T_4) and hypertrophy of the thyroid gland. Redrawn with permission from http://lpi.oregonstate.edu/infocenter/minerals/iodine/thyroid.html.

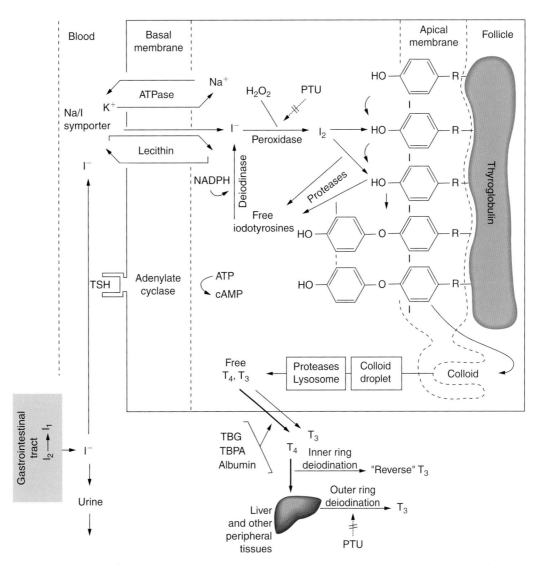

Figure 13-54. **Biosynthesis of thyroid hormones.** *ATP*, Adenosine triphosphate; *cAMP*, adenosine 3′:5′–cyclic phosphate; *I*, iodide; *NADPH*, the reduced form of nicotinamide-adenine dinucleotide phosphate; *PTU*, propylthiouracil; T_3, triiodothyronine; T_4, thyroxine; *TBG*, thyroxine-binding globulin; *TBPA*, thyroxine-binding prealbumin (transthyretin); *TSH*, thyroid-stimulating hormone. Redrawn with permission from http://www.merck.com/mrkshared/mmanual/figures/8fig1.jsp.

become iodinated (Figure 13-57). Follicles are filled with colloid, as shown in Figure 13-58. A small amount of T_4 is deiodinated in the outer thyronine ring by type I 5′-deiodinase, a selenoenzyme, to form T_3. The colloid droplet is acted upon by a protease in the lysosome that liberates free T_4 and T_3 into the bloodstream, where T_4 associates with the plasma protein, **thyroid-binding globulin (TBG)** that binds 75% of the hormone, **thyroid-binding prealbumin (transthyretin, TBPA)** and albumin. About 20% of circulating T_3 is produced in the thyroid gland; the other 80% of T_3 is the product of deiodination (5′-deiodinase) of the outer ring of T_4, primarily in the liver. The hormone is carried to the cells by these complexes, and the unbound (free) thyroid hormone (0.03% of total serum T_4 and T_3) is taken up into the cells, where T_4 is deiodinated to T_3 (active form of the hormone). T_3 is then moved into the nucleus, where it binds

Dietary Metals

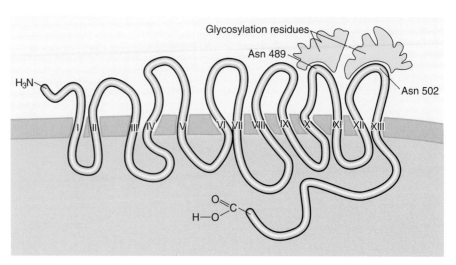

Figure 13-55. Current NIS secondary structure model. The model contains 13 putative transmembrane segments. The NH$_2$ terminus faces the extracellular milieu, and the COOH terminus faces the cytosol. Reproduced from Figure 2 of O. Dohán et al., "The sodium/iodide symporter (NIS): characterization, regulation, and medical significance," *Endocr. Revs.*, 24: 48–77, 2003 and http://edrv.endojournals.org/cgi/content/full/24/1/48.

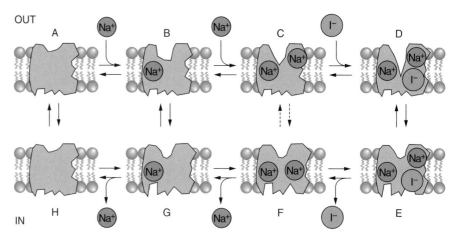

Figure 13-56. Schematic representation of a NIS mechanistic model. The kinetic data suggest that both Na$^+$ ions bind to NIS before I$^-$ (A → B → C). In the presence of I$^-$, the complex NIS-Na$_2$I is formed (symport mode) (C → D), which undergoes a conformational change to expose the bound two Na$^+$ and I$^-$ ions to the interior of the cell (D → E). Both Na$^+$ ions and I$^-$ are released into the cytoplasmic compartment (E → F → G → H), and the empty carrier (H) undergoes another conformational change to expose the binding sites to the external solution again (A). Charge movement data suggest that the Na$^+$ binding dissociation does not contribute greatly to the total observed charge. Thus, it is proposed that NIS charge movements arise primarily from conformational changes of the empty carrier (H → A). In the Na$^+$ uniport mode (B → G), one Na$^+$ ion binds to NIS (A → B) and may cross the membrane via NIS. Release of Na$^+$ into the cytoplasm (G → H) is followed by the return of the empty binding site to complete the pathway (H → A). Redrawn from Figure 13 of Orsolya Dohán et al., "The sodium/iodide symporter (NIS): characterization, regulation, and medical significance," *Endocr. Revs.* 24: 48–78, 2003 and http://edrv.endojournals.org/cgi/content/full/24/1/48.

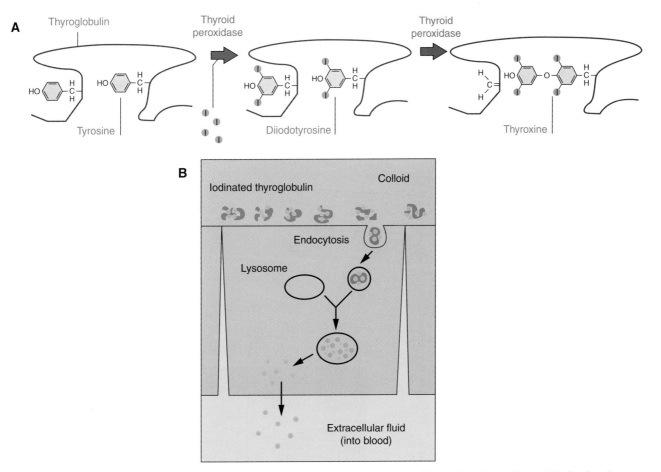

Figure 13-57. **A.** Formation of thyroxine residues on thyroglobulin. **B.** Release of T_4 from thyroglobulin by the action of lysosomal protease and release of liberated T_4 into the bloodstream. Redrawn from http://www.vivo.colostate.edu/hbooks/pathphys/endocrine/thyroid/synthesis.gif.

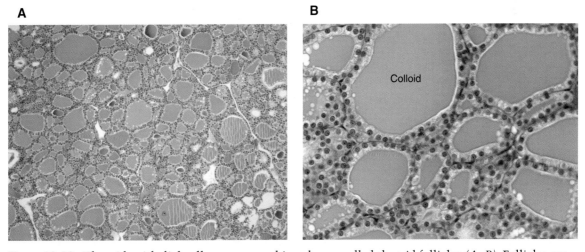

Figure 13-58. Thyroid epithelial cells are arranged in spheres, called thyroid follicles *(A, B)*. Follicles are filled with colloid *(pink)*, which is the repository of thyronines destined to become iodinated and synthesized into T_4. Some T_4 is deiodinated in the thyroid gland to form T_3. Photos courtesy of Richard Bowen, Colorado State University.

to the thyroid hormone receptor (TR). The TR activates transcription of mRNAs encoding enzymes that are part of the cell's metabolism for the formation of ATP energy. T_3 increases oxygen consumption in liver, kidney, heart, and skeletal muscle, primarily by increasing the amount of the sodium/potassium ATPase and increasing amounts of some enzymes (through transcription) in the citric acid cycle. The structures of T_4, T_3, and reverse T_3 (rT_3, inactive) are shown in Figure 13-59. The ligand-binding domain of the human thyroid hormone receptor (TR_β) complexed with a thyromimetic (a compound-like thyroid hormone) and the TR_β-retinoid X receptor complex DNA-binding domain are pictured in Figure 13-60.

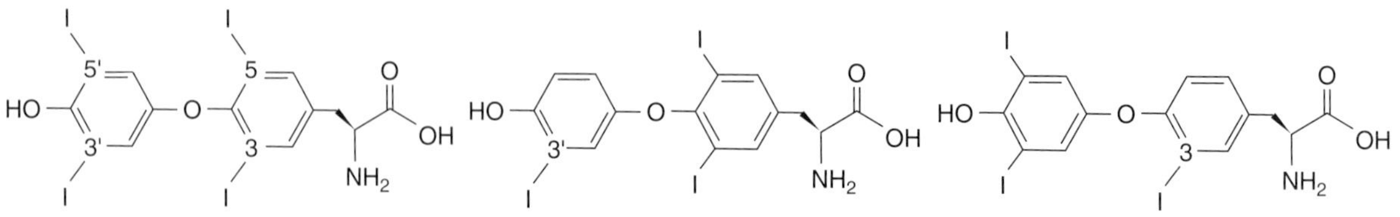

Figure 13-59. Structures of thyroid hormones (T_4 and T_3) and inactive T_3 called reverse T_3.

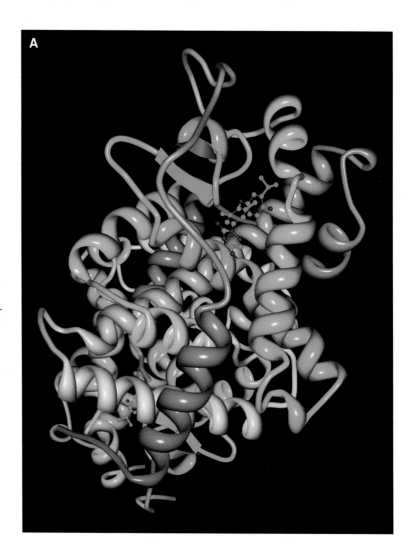

Figure 13-60. **A.** Crystal structure of human TR_β ligand–binding domain complexed with a potent subtype-selective thyromimetic. **B.** Complex retinoid X receptor–thyroid hormone receptor DNA-binding domain heterodimer bound to thyroid response element DNA. Part A reproduced from PDB ID: 1n46. R.L. Dow, S.R. Schneider, E.S. Paight, R.F. Hank, P. Chiang, P. Cornelius, E. Lee, W.P. Newsome, A.G. Swick, J. Spitzer, D.M. Hargrove, T.A. Patterson, J. Pandit, B.A. Chrunyk, P.K. LeMotte, D.E. Danley, M.H. Rosner, M.J. Ammirati, S.P. Simons, G.K. Schulte, B.F. Tate, P. DaSilva-Jardine. Discovery of a novel series of 6-azauracil-based thyroid hormone receptor ligands: potent, TR beta subtype-selective thyromimetics. *Bioorganic & Medicinal Chemistry Letters* **13** pp. 379 (2003).

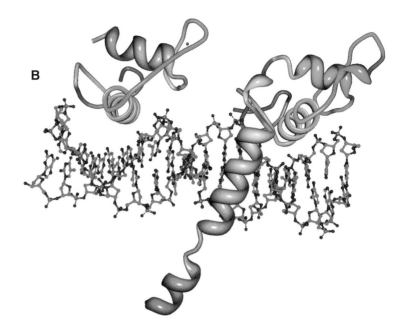

Figure 13-60, cont'd Part B reproduced from PDB ID: 2nll. F. Rastinejad et al. Structural determinants of nuclear receptor assembly on DNA direct repeats. *Nature* **375** pp. 203 (1995).

Vitamins

Water-Soluble Vitamins

Vitamins are not synthesized in the body and must be obtained in the diet. Tissues in the body are able to convert many of the water-soluble vitamins into coenzymes. Some major water-soluble vitamins are thiamine (vitamin B_1, aneurine); riboflavin (vitamin B_2); niacin (vitamin B_3); pantothenic acid (vitamin B_5); pyridoxine, pyridoxal, and pyridoxamine (vitamin B_6); biotin (vitamin H); cobalamin (vitamin B_{12}); folic acid (vitamin B_9); and ascorbic acid (vitamin C).

Thiamine (Vitamin B_1, Aneurine)

Thiamine is rapidly converted to **thiamine pyrophosphate (TPP)**, its active form, in liver and brain, by thiamine diphosphotransferase (thiamine diphosphokinase; Figure 13-61). TPP is a coenzyme of pyruvate dehydrogenase, α-ketoglutarate dehydrogenase, and transketolase (pentose phosphate pathway). The bacterial pyruvate dehydrogenase crystal structure and reaction are shown in Figure 13-62. A summary of the roles of TPP in the Krebs cycle is shown in Figure 13-63. *The common mechanism converting pyruvate to acetyl-CoA and α-ketoglutarate to succinyl-CoA involves the proton on C2 of TPP that dissociates to give a carbanion.* There follows a nucleophilic addition by the carbanion to the carbonyl of the α-keto acid (pyruvate or α-ketoglutarate) to the protonated forms of the activated α-hydroxy acid, which, then, undergoes decarboxylation. During the decarboxylation, the positively charged nitrogen of TPP is the critical electron sink and contributes to the resonance stabilization of the product. The hydroxy alkyl group is transferred by other proteins in the complex to CoA from pyruvate or to succinyl-CoA from α-ketoglutarate (Figure 13-64).

The structure of transketolase, the role of the enzyme in the pentose phosphate pathway, and the reaction mechanism are shown in Figure 13-65. As noted earlier, the carbanion at C2 of TPP is produced. It attacks the carbonyl of the α-ketose, forming an addition product. A hydroxyl of glyceraldehyde-3-phosphate is released after deprotonation, and an

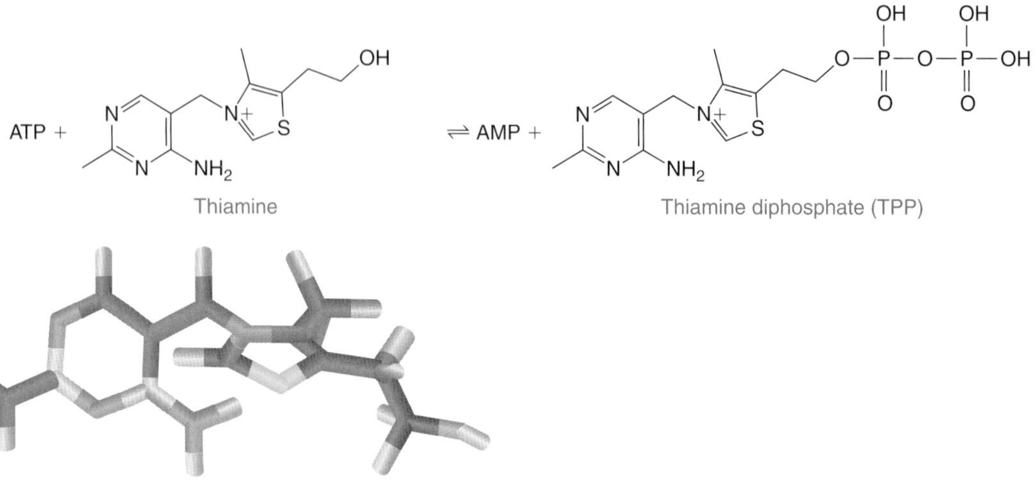

Figure 13-61. Conversion of thiamine (vitamin B_1) to its coenzyme form (thiamine diphosphate [TPP]) by the action of thiamine diphosphotransferase (thiamine diphosphokinase).

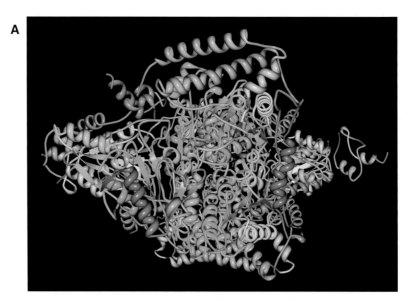

Figure 13-62. **A.** Crystal structure of pyruvate dehydrogenase e1 (d180n e183q), bound to the peripheral subunit binding domain of e2. **B.** Overall reaction catalyzed by pyruvate dehydrogenase. Part A reproduced from PDB ID: 1w88. R.A. Frank, C.M. Titman, J.V. Pratap, B.F. Luisi, R.N. Perham. A Molecular Switch and Proton Wire Synchronize the Active Sites in Thiamine Enzymes. *Science* **306** pp. 872 (2004).

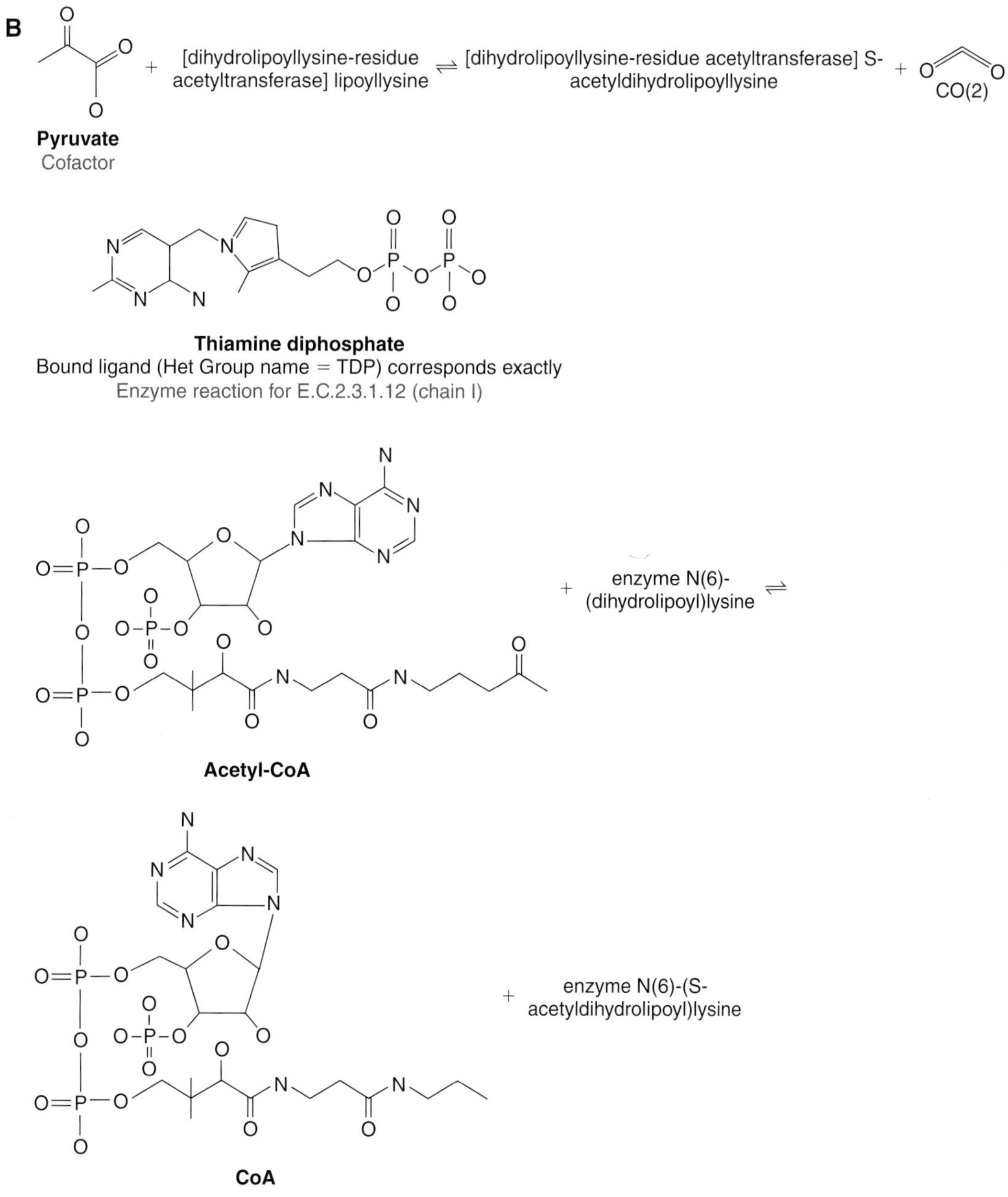

B

Pyruvate + [dihydrolipoyllysine-residue acetyltransferase] lipoyllysine ⇌ [dihydrolipoyllysine-residue acetyltransferase] S-acetyldihydrolipoyllysine + CO_2

Pyruvate
Cofactor

Thiamine diphosphate
Bound ligand (Het Group name = TDP) corresponds exactly
Enzyme reaction for E.C.2.3.1.12 (chain I)

Acetyl-CoA + enzyme N(6)-(dihydrolipoyl)lysine ⇌

CoA + enzyme N(6)-(S-acetyldihydrolipoyl)lysine

Figure 13-62, cont'd

activated glyceraldehyde bound to TPP is formed. Again, the thiazole nitrogen (of TPP) serves as the electron sink and contributes to the stabilization of the activated glycolaldehyde produced. It undergoes nucleophilic addition to the carbonyl of erythrose-4-phosphate, and the nascent α-ketose is released from TPP after another deprotonation.

Thiamine is thought to have a role in the brain and nerves independent of its role as coenzyme, and it may be involved in nerve impulses through the sodium/potassium gradient. These ideas stem from the fact that thiamine deficiency causes neurological disorders. Stimulation of

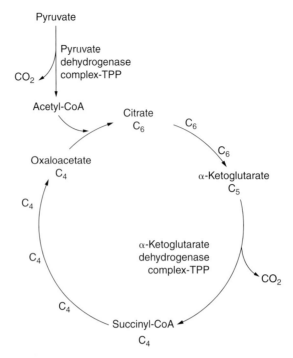

Figure 13-63. Summary of the roles of thiamine diphosphate as coenzyme of pyruvate dehydrogenase and α-ketoglutrate dehydrogenase in the Krebs cycle. Redrawn from http://chemistry.gsu.edu/glactone/vitamins/b1.

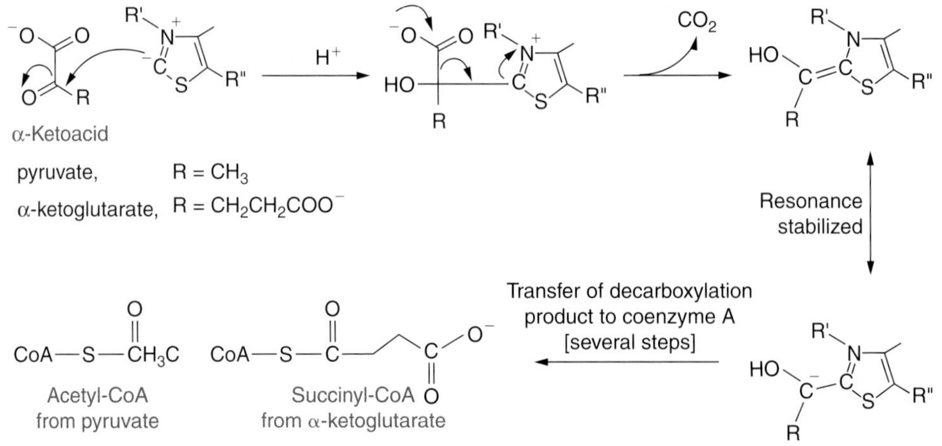

Figure 13-64. Mechanism of thiamine pyrophosphate (TPP) participation in pyruvate dehydrogenase reaction or α-ketoglutarate dehydrogenase reaction (see text for explanation).

nerves by electrical means or by acetylcholine causes the release of thiamine monophosphate and free thiamine into the medium. There is a concomitant decrease of TPP and thiamine triphosphate in the cell under this type of stimulation. Thiamine deficiency causes constipation, appetite suppression, nausea, mental depression, peripheral neuropathy, and fatigue. When the deficiency persists for a long period, there are severe neurological abnormalities, ataxia, mental confusion, and loss of eye coordination, as well as cardiovascular and defective musculature.

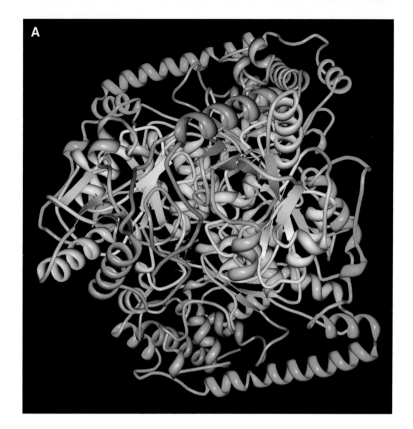

Figure 13-65. **A.** Yeast transketolase. TPP is depicted in atomic model, as is a calcium ion. **B.** Overall reaction catalyzed by transketolase in the pentose phosphate pathway. **C.** Reaction mechanism of ketose substrate and TPP in the transketolase catalyzed reaction. Part A reproduced from http://www.rcsb.org/pdb/cgi/explore.cgi?pdbId=1ay0. PDB ID: 1ay0. C. Wikner, U. Nilsson, L. Meshalkina, C. Udekwu, Y. Lindqvist, G. Schneider. Identification of Catalytically Important Residues in Yeast Transketolase. *Biochemistry* **36** pp. 15643 (1997). Part C redrawn from http://chemistry.gsu.edu/glactone/vitamins/b1/.

Chronic alcoholics often do not get sufficient thiamine in their abnormal diet, and this condition is known as the **Wernicke-Korsakoff syndrome. Beriberi** is the classic disease of thiamine deficiency wherein certain populations eating rice as the mainstay of their diet use polished rice instead of unpolished rice, whose covering contains the thiamine. This disease is still encountered in parts of Southeast Asia.

Riboflavin (Vitamin B_2)

Riboflavin is a vitamin precursor of flavin mononucleotide and flavin adenine dinucleotide (Figure 13-66). The synthesis of these coenzymes is shown in Figure 13-67. Riboflavin-5′-phosphate (FMN) is formed first by the action of riboflavin kinase. FMN is converted to flavin adenine dinucleotide (FAD) with ATP by FAD pyrophosphorylase. The crystal structure of human riboflavin kinase is shown in Figure 13-68. Flavoproteins usually contain metals, such as Mg^{2+}; examples of enzymes containing flavins are succinate dehydrogenase (Figure 13-69) and xanthine dehydrogenase (Figure 6-48). The enzymatic reactions are shown in

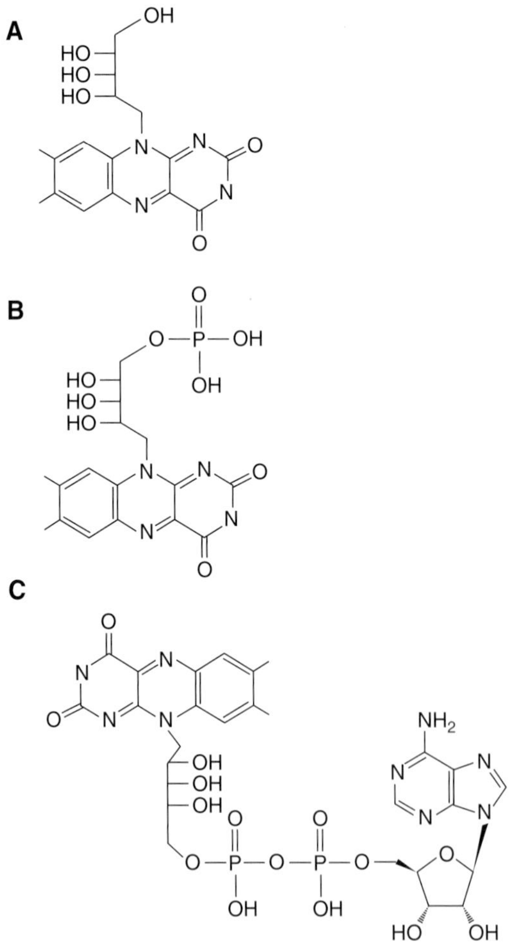

Figure 13-66. Structure of riboflavin (A), flavin mononucleotide in B, flavin adenine dinucleotide in C.

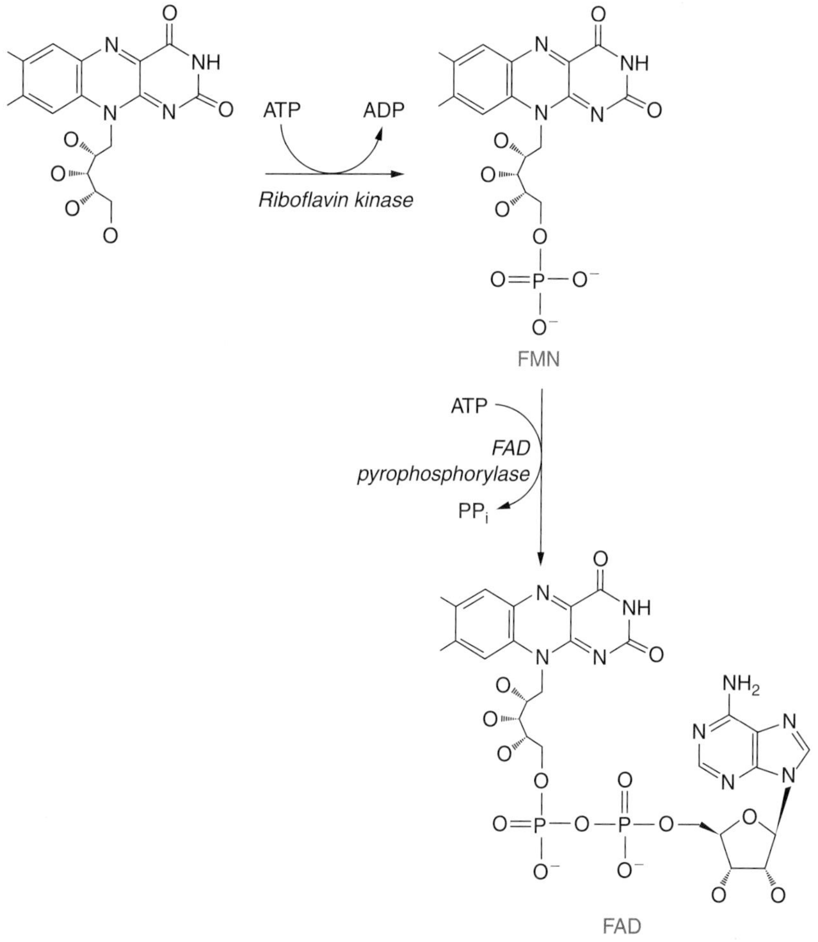

Figure 13-67. Synthesis of FMN and FAD by human tissues.

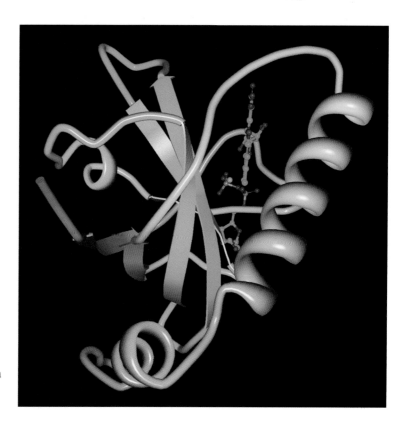

Figure 13-68. Crystal structure of human riboflavin kinase with bound riboflavin, Mg^{2+}, and ADP. Reproduced from http://www.rcsb.org/pdb/cgi/explore.cgi?pdbId=1nb9. PDB ID: 1nb9. S. Karthikeyan, Q. Zhou, F. Mseeh, N.V. Grishin, A.L. Osterman, H. Zhang. Crystal Structure of Human Riboflavin Kinase Reveals a Beta Barrel Fold and a Novel Active Site. *Arch Structure* **11** pp. 265 (2003).

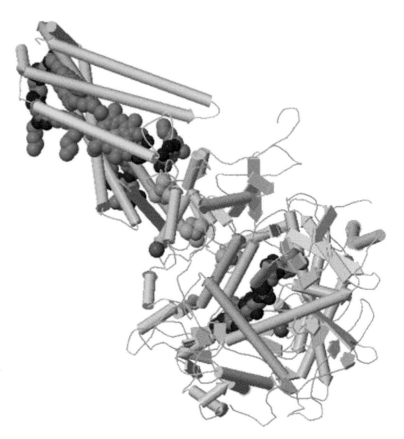

Figure 13-69. Structure of succinate dehydrogenase complex from *E. coli*. Atomic structures are Fe^{3+}–S4 cluster, phosphatidylethanolamine derivative, FAD, and oxaloacetate. Reproduced from http://www.imb-jena.de/cgibin/htmlit.pl?color=white&id=GIF&src=1nen.gif&name=Image%20Library%20Thumb%20Nail%201NEN. Institute for Age Research–Fritz Lippman Institute (formerly IMB-Jena) Beutenbergstr, Jena Germany. From V. Yankoskaya et al. "Architecture of succinate dehydrogenase and reactive oxygen species generation," *Science* 299: 700–704, 2003.

Figure 13-70. In these reactions, the reduced form of the coenzyme, $FADH_2$, appears as a product.

Riboflavin is plentiful in eggs, milk, meat, and cereals; therefore, riboflavin deficiency rarely is seen. However, it does occur in chronic alcoholics who have inadequate diets. Riboflavin breaks down when exposed to visible light (Figure 13-71), and in newborns treated for hyperbilirubinemia with phototherapy, riboflavin can become deficient. Riboflavin deficiency causes photophobia (aversion to light), glossitis (inflammation of mouth, face, and tongue), seborrhea (excessive oiliness of face and scalp), and angular stomatitis (also cheilosis, a condition characterized by fissures and inflammation of lower lip).

Niacin (Vitamin B_3)

The vitamin forms of niacin are nicotinic acid and nicotinamide (Figure 13-72). The coenzyme forms of niacin are the oxidized and reduced forms of nicotinamide adenine dinucleotide (NAD^+, NADH) and the oxidized and reduced forms of the phosphate derivative, nicotinamide adenine dinucleotide phosphate ($NADP^+$, NADPH); these forms are shown in Figure 13-73. The formation of the coenzymes from the vitamin forms is shown in Figure 13-74. Nicotinate phosphoribosyltransferase structure and reaction are shown in Figure 13-75. Figure 13-76 shows the structure of NAD^+ synthase and the reaction it catalyzes.

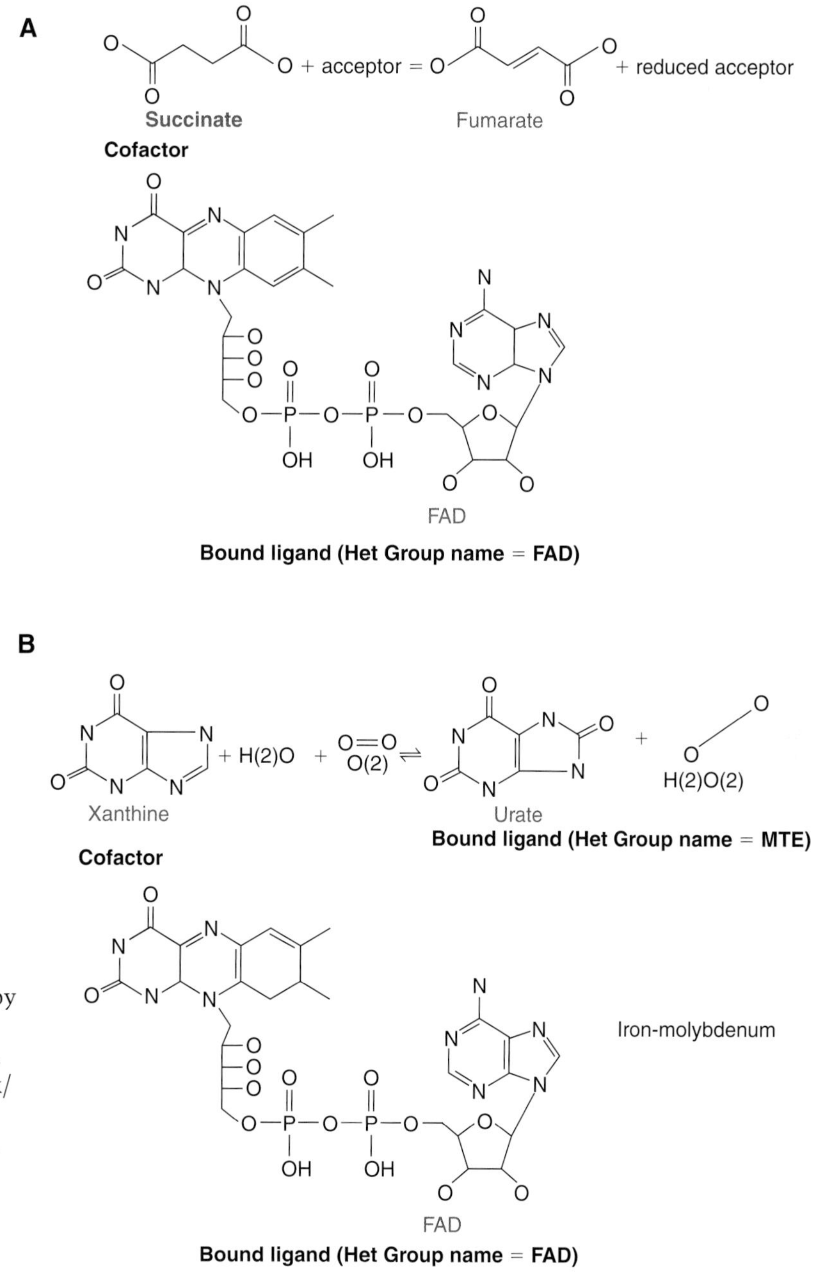

Figure 13-70. Reactions catalyzed by succinate dehydrogenase *(A)* and xanthine dehydrogenase *(B)*. Part A redrawn from http://www.ebi.ac.uk/thornton–srv/databases/cgi–bin/pdbsum/GetPage.pl?pdbcode=1I0v. Part B redrawn from http://www.ebi.ac.uk/thornton–srv/databases/cgi-bin/pdbsum/GetPage.pl?pdbcode=1fo4.

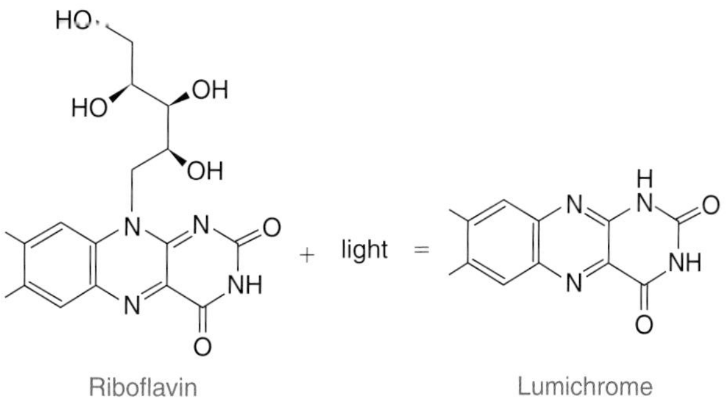

Figure 13-71. Riboflavin is degraded by light to form lumichrome.

Vitamins 809

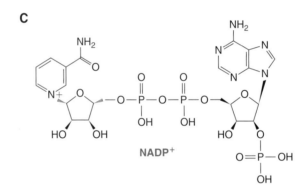

Figure 13-72. Structures of nicotinic acid *(A)* and nicotinamide *(B)*.

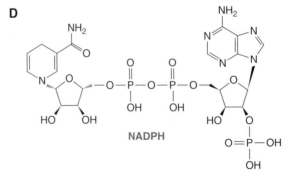

Figure 13-73. Structures of NAD⁺ *(A)*, NADH *(B)*, NADP⁺ *(C)*, and NADPH *(D)*.

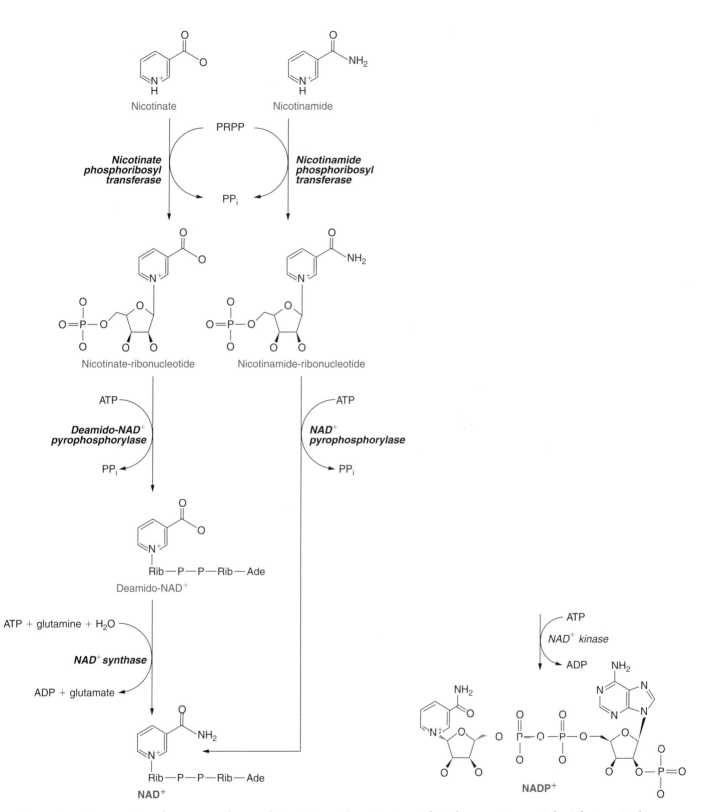

Figure 13-74. Formation of coenzyme forms of nicotinic acid or nicotinamide in human tissues. *Ade,* Adenine; *P,* phosphate; *Rib,* ribose.

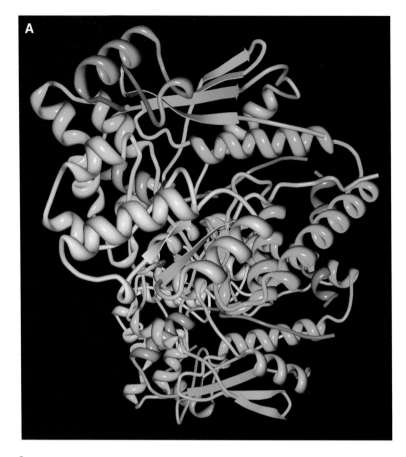

Figure 13-75. **A.** Crystal structure of a nicotinate phosphoribosyltransferase. Reproduced from http://bip.weizmann.ac.il/oca–bin/ccpeek?id=1VLP. **B.** Reaction catalyzed by nicotinate phosphoribosyltransferase *(right to left)*. Part A reproduced from http://www.ebi.ac.uk/thonton-srv/databases/cgibin/pdbsum/GetPage.pl?pdbcode=1ybe. PDB ID: 1ybe. J. Seetharman, S. Swaminathan. Crystal Structure of a Nicotinate Phosphoribosyltransferase. To be published. Part B redrawn from http://www.ebi.ac.uk/thornton–srv/databases/cgi–bin/pdbsum/GetPage.pl?pdbcode=1vlp.

Many dehydrogenases have NAD$^+$ or NADP$^+$ as coenzymes, including lactate dehydrogenase and malate dehydrogenase. Their structures and the reactions they catalyze are shown in Figure 13-77. Some niacin can be derived from tryptophan in the body, but this is an inefficient process (60 mg of tryptophan per 1 mg of niacin) and does not fulfill the dietary vitamin requirement. Also, other vitamins are involved in this conversion, such as vitamins B_1, B_2, and B_6 (pyridoxine). The overall conversion of tryptophan to niacin and its major derivative, NAD, occurs in the intestine (Figure 13-78). About 15 mg of dietary niacin is required per day for a normal adult.

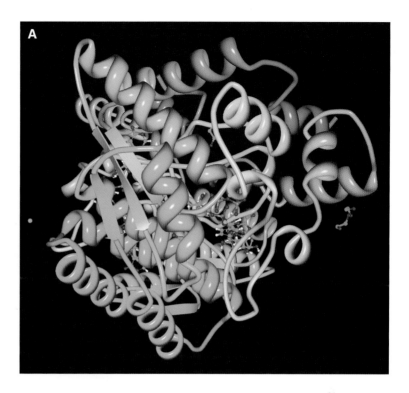

Figure 13-76. **A.** Structure of NAD⁺ synthetase (from *B. subtilis*). **B.** Reaction catalyzed by NAD⁺ synthetase. Part A reproduced from http://www.rcsb.org/pdb/cgi/explore.cgi?pdbId=1kqp. PDB ID: 1kqp. J. Symersky, Y. Devedjiev, K. Moore, C. Brouillette, L. DeLucas. NH3-Dependent NAD⁺ Synthetase from *Bacillus subtilis* at 1 A Resolution. *Acta Crystallogr.*, Sect. D, **58** pp. 1138 (2002). Part B redrawn from http://www.ebl.ac.uk/thornton-srv/databases/cgi-bin/pdbsum/GetPage.pl?pdbcode=1kqp.

Vitamins

Niacin was discovered by its deficiency in dogs, causing black tongue. In humans, a deficiency of niacin causes glossitis (inflammation) of the tongue, dermatitis, and diarrhea. The classical deficiency in humans is known as **pellagra.** If the absorption of tryptophan is compromised, as in **Hartnup disease** and in malignant carcinoid syndrome, niacin deficiency can follow. In malignant carcinoid syndrome, there is also an excess synthesis of serotonin. About 10% of patients with carcinoid tumor have the malignant carcinoid syndrome. In these cases,

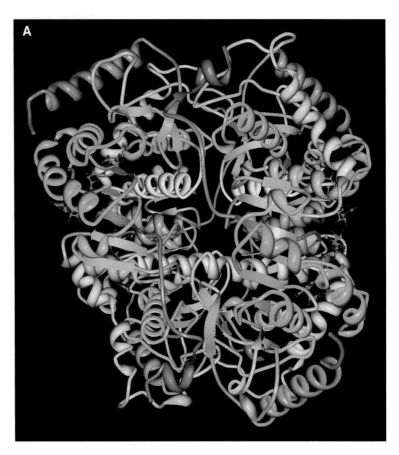

Figure 13-77. **A.** Structure of human heart lactate dehydrogenase. **B.** Lactate dehydrogenase catalyzed reaction. **C.** Structure of malate dehydrogenase (from *E. coli.*) containing bound NAD. **D.** Reaction catalyzed by malate dehydrogenase. Part A reproduced from http://www.rcsb.org/pdb/cgi/explore.cgi?pdbId=1t2f. PDB ID: 1t2f. A. Cameron, J. Read, R. Tranter, V.J. Winter, R.B. Sessions, R.L. Brady, L. Vivas, A. Easton, H. Kendrick, S.L. Croft, D. Barros, J.L. Lavandera, J.J. Martin, F. Risco, S. Garcia-Ochoa, F.J. Gamo, L. Sanz, L. Leon, J.R. Ruiz, R. Gabarro, A. Mallo, F.G. De Las Heras. Identification and Activity of a Series of Azole-Based Compounds with Lactate Dehydrogenase–Directed Anti-malarial Activity. *Journal of Biological Chemistry* **279** pp. 31429 (2004). Part B reproduced from http://www.ebi.ac.uk/thornton-srv/databases/cgi-bin/pdbsum/GetPage.pl?pdbcode=1i0z.

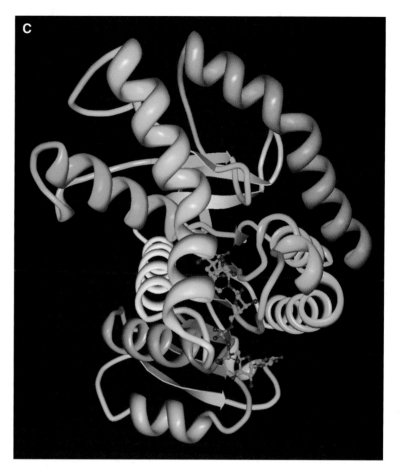

(S)-Malate + NAD(+) ⇌ Oxaloacetate +

NADH

Figure 13-77, cont'd Part C reproduced from http://www.rcsb.org/pdb/cgi/explore.cgi?pdbId=1emd. PDB ID: 1emd. M.D. Hall, L.J. Banaszak. Crystal Structure of a Ternary Complex of *Escherichia coli* Malate Dehydrogenase Citrate and NAD at 1.9 Å resolution. *Journal of Biological Chemistry* **232** pp. 213 (1993). Part D from http://www.ebi.ac.uk/thornton-srv/databases/cgi-bin/pdbsum/GetPage.pl?pdbcode=1emd.

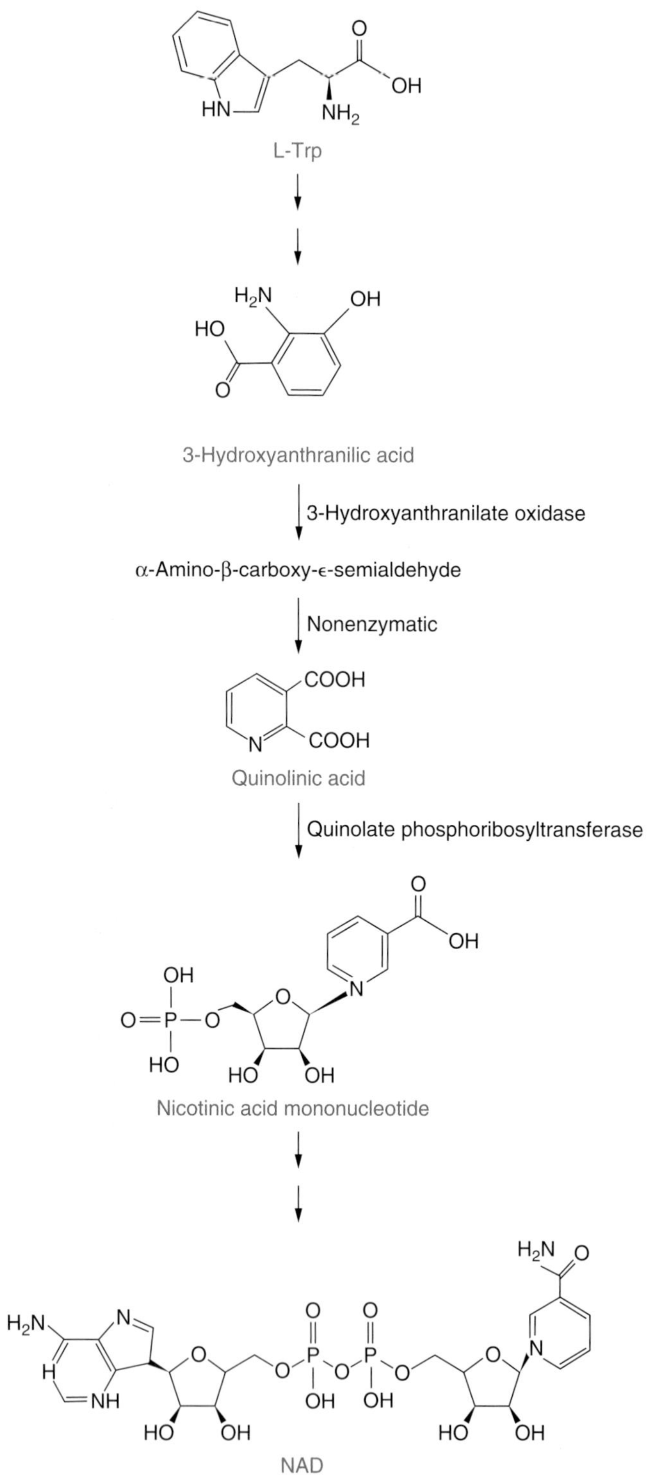

Figure 13-78. Conversion of L-tryptophan to niacin derivatives, especially NAD in the intestine.

the original tumor has metastasized and produces large amounts of serotonin, having been derived from neuroendocrine cells located in the small bowel or derived from the appendix. Symptoms include flushing of the face, severe diarrhea, and asthma attacks. Hartnup disease is an autosomal recessive disorder in which the Na^+-dependent transporter (Figure 13-79) of neutral amino acids, including tryptophan,

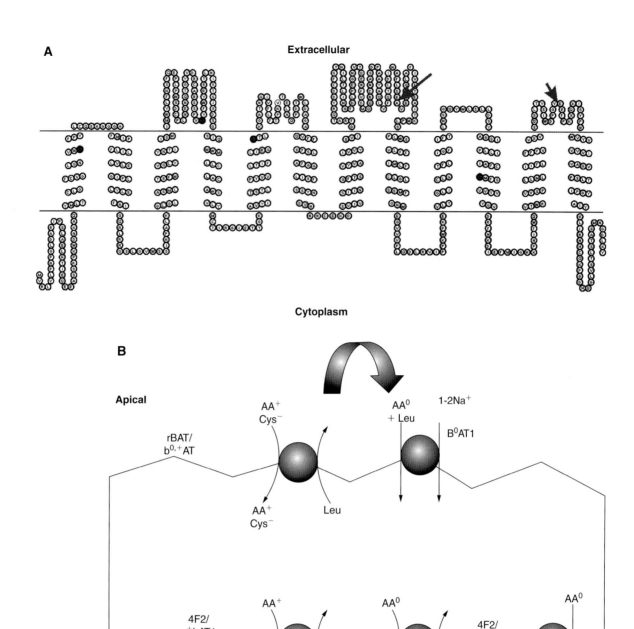

Figure 13-79. **A.** A topological model of Na$^+$-dependent neurotransmitter transporter family (B^{0}AT1) derived from the combined analysis of several topology predictions. The location of the amino and carboxyl terminus were both postulated to be intracellular in accordance with other members of this family. The transporter is predicted to have 12 transmembrane-spanning domains. Exon/intron boundaries, translated into the peptide sequence, are indicated by *arrows*. **B.** Resorption of neutral and cationic amino acids in kidney and intestine. Cationic amino acids are transported across the apical membrane by exchange against neutral amino acids via the heteromeric amino acid antiporter rBAT/b^{0}AT. Leucine, the preferred neutral amino acid released by rBAT/b^{0}AT, plus other neutral amino acids already present in the urine are subsequently removed by B^{0}AT1. B^{0}AT1 accumulates neutral amino acids by cotransport with 12 Na$^+$. Cationic amino acids are released on the basolateral side by the heteromeric amino acid transporter 4F2/y$^+$LAT1 in exchange for neutral amino acids; neutral amino acids are released by LAT2 and a yet unidentified uniporter. Part A reproduced from Figure 3 of S. Bröer et al., "Neutral amino acid transport in epithelial cells and its malfunction in Hartnup disorder" *Biochem. Soc. Trans.*, 33: 233–236, 2005. Part B redrawn from Figure 12 of A. Bröer et al., "Molecular cloning of mouse amino acid transport system B^0, a neutral amino acid transporter related to Hartnup disorder," *J. Biol. Chem.*, 279: 24467–24476, 2004 and from http://www.jbc.org/cgi/content-nw/full/279/23/24467/FIG12.

Vitamins 817

(monoamine monocarboxylic amino acids) is defective in the intestine and kidney. It is assumed that mutations in this transporter underlie Hartnup disease.

Pantothenic Acid (Vitamin B_5)

Pantothenic acid is formed from β-alanine and pantoic acid in microorganisms. It is a component of CoA and the **acyl carrier protein (ACP)** of fatty acid synthase (Figures 5-33 through 5-35). Seventy or so enzymes require CoA or ACP; therefore, pantothenate is critical for carbohydrate, fat, and protein metabolism and the functioning of the citric acid cycle. The synthesis of CoA from pantothenic acid is shown in Figure 13-80.

Pantothenate is plentiful in cereals, meat, and legumes, and deficiency rarely occurs. If it did occur, it would be difficult to diagnose because of symptoms that are similar to those in deficiencies of the other B vitamins. Pantothenate deficiency is reputed to be involved in premature gray or white hair. This may result from depressed synthesis of melanin that requires several vitamins in the process: pantothenate, folic acid, vitamin B_{12}, and ascorbic acid.

Pyridoxine, Pyridoxal, and Pyridoxamine (Vitamin B_6)

These are forms of the dietary vitamin, all of which are convertible to the coenzyme form, **pyridoxal phosphate (PLP)**. Structures of the three vitamin forms and their conversion to PLP are shown in Figure 13-81. The kinase (pyridoxal, pyridoxine, or pyridoxamine kinase) phosphorylates the three vitamin forms to their respective 5'-phosphate derivatives. Thus, pyridoxal is converted directly by phosphorylation (ATP) to the coenzyme form (PLP). Pyridoxine-5'-phosphate is converted to PLP by an oxidase (pyridoxine phosphate or pyridoxamine phosphate, or PMP, oxidase), and pyridoxamine-5'-phosphate is converted to PLP by two enzymes, the oxidase and amino acid transaminase (PLP or PMP dependent). Crystal structures have revealed the molecularity of the phosphorylation conducted by pyridoxal kinase (Figure 13-82). The overall pyridoxal kinase reaction mechanism is shown in Figure 13-83.

PLP is a coenzyme for aminotransferases, amino acid racemases, amino acid decarboxylases, and other enzymes of amino acid metabolism and is a coenzyme of glycogen phosphorylase (Figures 4-38, 4-42, 4-43, 4-53, and 10-97). The aminotransferases have been described, along with their reaction mechanisms (Figures 10-12 and 10-13). Adults should ingest 1.4 to 2.0 mg/day of vitamin B_6. There is a greater requirement for this vitamin during pregnancy, so a pregnant or lactating female should ingest about 2.5 mg/day. Deficiencies of this vitamin are uncommon and relate to overall B vitamin deficiency. There have been claims that ingestion of large amounts (up to 200 mg/day) of this vitamin for as long as 6 months can palliate certain conditions, such as carpal tunnel syndrome; however, it would be important to measure the blood level of the vitamin because it can become toxic and cause nerve damage.

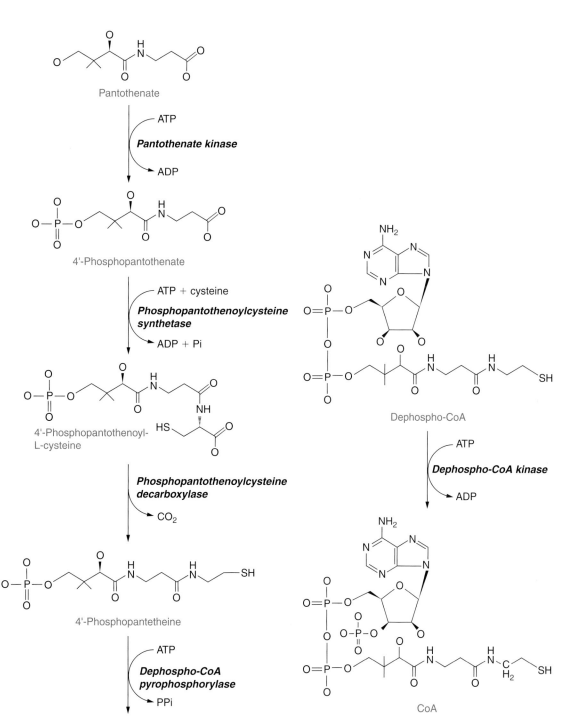

Figure 13-80. Synthesis of coenzyme A (CoA) from pantothenic acid.

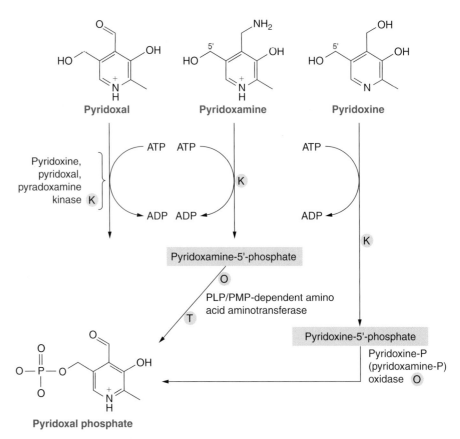

Figure 13-81. Conversions of the vitamin forms of vitamin B_6 pyridoxal, pyridoxamine and pyridoxine to the coenzyme form, pyridoxal-5'-phosphate (PLP) in human tissues, especially liver. *K*, kinase for the 3 vitamin forms; *O*, pyridoxine phosphate/pyridoxamine-phosphate oxidase; *T*, PLP/PMP-dependent amino acid aminotransferase; *P*, phosphate; *PMP*, pyridoxamine phosphate.

Biotin (Vitamin H)

Biotin (Figure 13-84) is a cofactor for enzymes of carboxylation, such as acetyl-CoA carboxylase (Figures 5-27 through 5-29) and pyruvate carboxylase (Figure 13-85). A illustration of the active site of pyruvate carboxylase is shown in Figure 13-86. The position of the biotin coenzyme is thought to react with enzyme for the first part of the reaction then swivel over to the second part of the reaction. Biotin acts as a carrier of carbon dioxide in this enzymatic reaction, and biotin is linked to a lysine ε-amino group in the enzyme (Figure 13-86C). In humans, biotin is a coenzyme for pyruvate carboxylase (as already shown), β-methylcrotonyl-CoA carboxylase, proprionyl-CoA carboxylase that converts proprionyl-CoA to succinyl-CoA, and acetyl-CoA carboxylase for the carboxylation of acetyl-CoA to malonyl-CoA. Pyruvate carboxylase is an important function in gluconeogenesis in the liver. β-Methylcrotonyl-CoA carboxylase is involved in leucine catabolism (Figure 10-66). Acetyl-CoA carboxylase functions in lipid synthesis from acetate.

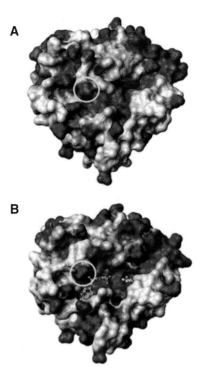

Figure 13-82. **A.** The electrostatic potential surface of human pyridoxal kinase compared to sheep brain pyridoxal kinase in *B*. On each enzyme surface there is a cavity with negative charge located along one edge of the central beta-sheet that is favorable for attracting and binding substrates with a positive charge, such as the pyridine ring of vitamin B6 and the adenine ring of ATP. A key 12-residue peptide over the active site in the human enzyme is a beta-strand/loop/beta-strand flap compared to the sheep enzyme that has a loop conformation. The human enzyme possesses a more hydrophobic ATP-binding pocket. Positive potential regions are *blue* and negative potential regions are *red*. Residues-121 at the surfaces are indicated by *yellow circles*. ATP and pyridoxal are indicated by ball and stick representation in B (ATP, *green*; pyridoxal, *yellow*). Reproduced from Figure 2 of P. Cao et al., *J. Struct. Biol.*, 154: 327–332, 2006.

The holocarboxylases (with bound biotin) can be acted upon by the enzyme **biotinidase,** which cleaves biotin from its binding to the ε-amino group of a lysine residue in the active site of the enzyme. The residue of biotin bound to the enzyme lysine is called **biocytin.** The released biotin can then be recycled to bind to other apocarboxylases. This biotin bicycle is shown in Figure 13-87. Biotin also binds to other proteins, notably avidin (in egg white), streptavidin, homocitrate synthase, and isopropylmalate synthase. Biotin is found in liver, kidneys, soy flour, egg yolk, cereal, and yeast. It is also synthesized by bacteria in the intestine.

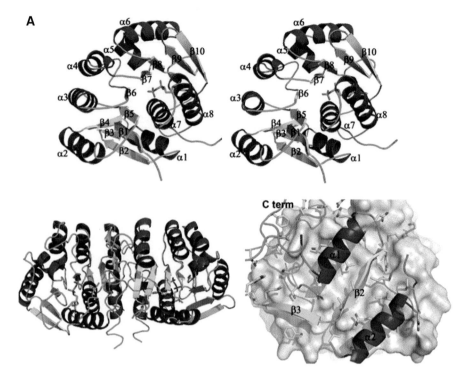

Figure 13-83. The pyridoxal kinase (PK) catalytic mechanism. **A.** (a) Stereo view of PK. Alpha helices are in *red*, beta strands in *yellow*, loops in *green*, and the ADP molecule is in *stick format*. (b) PK dimer viewed with the 2-fold axis vertical. (c) Residues involved in forming the dimer interface are shown in *stick representation* and contacting the surface of a 2-fold re‑ lated partner. The interface lies away from the active site and is highly co mplementary in shape and accounts for 16% of the total accessible surface area of each subunit. **B.** (a) Stereo view of PK nucleotide binding site; key interacting residues are labeled and hydrogen bonds between protein and ADP are shown as *dotted black lines.* (b) Stereo view of the superim‑ position of PK *(salmon with red loops)* and *S. typhimurium* HMPP (4-amino-5-hydroxymethyl-2-methylpyrimidine phos‑ phate) kinase *(blue with dark blue loops*, relating pyridoxal kinase to this family of enzymes. ADP and substrate HMP (4-amino-5-hydroxymethyl-2-methylpyrimidine) are shown for reference in *green*. Highlighted region A indicates residues that become ordered following nucleotide binding. Loops marked B and C show the differences in conformation in two loops that lie close to the substrate binding site. Loop B may act as an active site flap shielding the substrate from solvent and is trapped in an open conformation in PK by an intrachain disulfide bond. Loop C contains residues involved in sub‑ strate recognition and may be important in distinguishing the different substrate specificities of these two enyzmes. (C) Comparison of PK ADP complex *[green]* with the nucleotide positions from other ribokinase superfamily members (*E. coli* ribokinase ACP complex *[pink]*, *T. gondi* adenosine kinase ACP complex *[yellow]*, sheep brain PK ATP complex *[dark blue]*, sheep brain PK ACP complex *[light blue]*, and *T. thermophilus* 2-keto-3deoxygluconate kinase ATP complex *[red]*, superimposed on the basis of the putative anion hole residues (N terminus of alpha 7), which are shown in *blue gray*. **C.** Stereo diagram of a comparison between *B. subtilis* PK substrate-binding site *(red throughout)* with (a) sheep brain PK *(brown)* and *E. coli* PK *(dark blue)*, (b) *S. Typhimurium* HMPP kinase *(green)* and *T. thermophilus* HMPP kinase *(orange)*. The substrate pyridoxal is shown in its expected position throughout in the ball and stick representation *(light blue)*. It is expected that a metal ion is coordinated to the carboxyl oxygen atoms of Asp107 and Glu144 and the main-chain carbonyl oxygen atom of Thr139. The metal ion would lie about 3.5 Å from the beta phosphate of the ADP. Parts A, B, and C are reproduced from Figures 1, 3, and 4 respectively of J.A. Newman, S.K. Das, S.E. Sedelnikova, and D.W. Rice, "The crystal structure of an ADP complex of Bacillus subtilis pyridoxal kinase provides evidence for the parallel emergence of enzyme activity during evolution," *J. Mol. Biol.*, 363: 520–530, 2006.

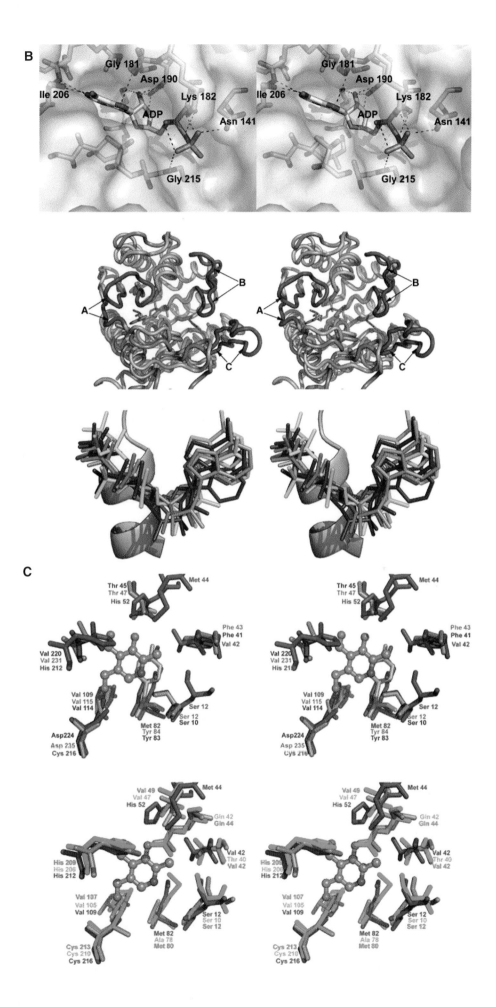

Figure 13-83, cont'd

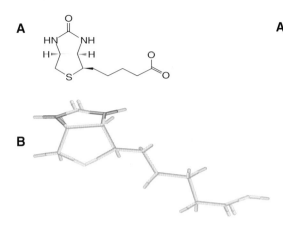

Figure 13-84. **A.** Chemical structure of biotin. **B.** Stick model of biotin.

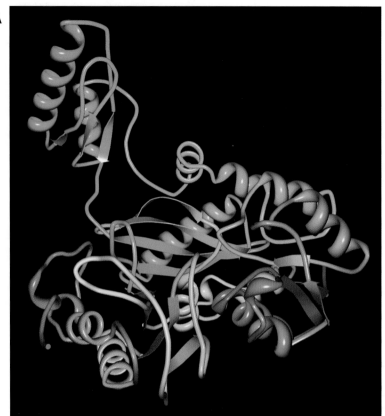

Figure 13-85. **A.** Crystal structure of biotin carboxylase subunit of pyruvate carboxylase (from a bacterium). **B.** Reaction catalyzed by pyruvate carboxylase. Part A reproduced from http://www.rcsb.org/pdb/explore/images.do?structureId=1ulz. PDB ID: 1ulz. S. Kondo, Y. Nakajima, S. Sugio, J. Young-Biao, S. Sueda, H. Kondo. Structure of the Biotin Carboxylase Subunit of Pyruvate Carboxylase from *Aquifex aeolicus* at 2.2 Å Resolution. *Acta Crystallogr.*, Sect. D, **60** pp. 486 (2004). Part B redrawn from http://www.ebi.ac.uk/thornton-srv/databases/cgi-bin/pdbsum/GetPage.pl?pdbcode=1ulz.

Cobalamin (Vitamin B$_{12}$)

The structure of cobalamin is shown in Figure 13-88. It has a complex tetrapyrrol ring structure with a cobalt atom in the center. The vitamin is synthesized by microorganisms and is found in animal meat, where it is bound by protein. The vitamin is released by hydrolysis in the stomach or by trypsin in the intestine. **Intrinsic factor** is a protein secreted by parietal cells of the stomach, and cobalamin is bound to it. The cobalamin–intrinsic factor complex is absorbed in the ileum. There is a receptor for the intrinsic factor–cobalamin complex called **cubilin,** which facilitates the absorption of the complex. Mutation of this receptor (Figure 13-89) causes impaired absorption of the intrinsic factor–cobalamin complex, which results in megaloblastic anemia, a rare disease. The vitamin is transported to the liver in the bloodstream, in which it is bound by transcobalamin II. The scenario of the ingestion and use of cobalamin is shown in Figure 13-90. Cobalamin, as the 5'-deoxyadenosine derivative, is a coenzyme of methylmalonyl CoA mutase that converts methylmalonyl-CoA to succinyl-CoA in the fatty acid catabolic system. It also is involved in methionine synthase

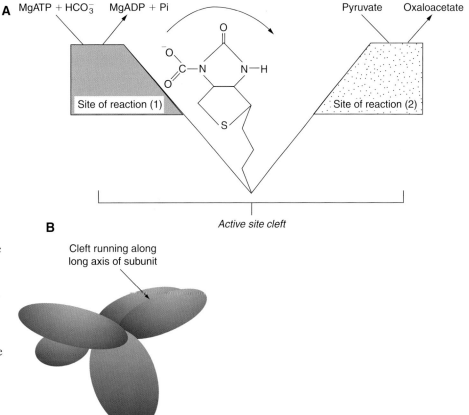

Figure 13-86. **A.** Schematic diagram of the active site cleft of one subunit of pyruvate carboxylase showing the biotin prosthetic group acting to carry the carboxyl group from the site of reaction [1] to that of reaction [2]. **B.** Diagram showing the tetrahedron-like arrangement of subunits of pyruvate carboxylase and illustrating the cleft that runs along the long axis of each subunit and which has been proposed to be the location of the active site.

Continued

Figure 13-86, cont'd **C.** The pyruvate carboxylase reaction (pyruvate + CO_2 + ATP → oxaloacetate + ADP + Pi) in which biotin is coenzyme and Mg^{2+} is cofactor. Carboxylases are synthesized as apo-enzymes without the cofactor biotin. The active form (holoenzyme) appears when biotin binds to an ε-amino group of the apoenzyme. The reaction in C, above, is catalyzed by the holoenzyme. When carboxylase holoenzymes are degraded, the biotin is cleaved from the ε-amino group of the enzyme by another enzyme, biotinidase, causing the release of free biotin that can be recycled. Part A reproduced from Figure 1 and part B reproduced from Figure 11 of P.V. Attwood, "The structure and the mechanism of action of pyruvate carboxylase," *Int. J. Biochem. Cell Biol.*, 27: 231–249, 1995. Elsevier Science Ltd. Printed in Great Britain. Part C redrawn from http://images.google.com/imgres?imgurl=http://www.chem.uwec.edu/...%3D10%26hl%3Den%26lr%3D%26client%3Dsafari%26rls%3Den–us%26sa%3DN.

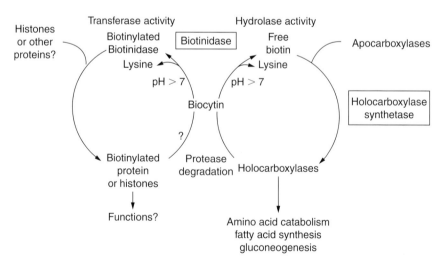

Figure 13-87. The biotin bi-cycle, which includes biotinyl-hydrolase activity and biotinyl-transferase activity. Redrawn from Figure 1 of B. Wolf, "Biotinidase: its role in biotinidase deficiency and biotin metabolism," *J. Nutr. Biochem.*, 16: 441–445, 2005.

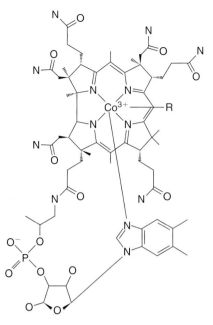

Figure 13-88. Structure of cobalamin, vitamin B_{12}.

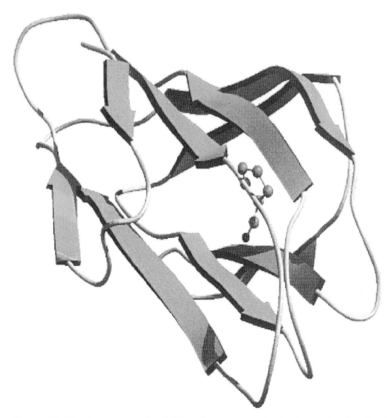

Figure 13-89. Structure of cubilin, the receptor in the intestine for the intrinsic factor–cobalamin complex. This structure is actually a mutant (FM1) of the protein. Reproduced from Figure 4 of M. Aminoff et al., "Mutations in CUBN, encoding the intrinsic factor-vitamin B 12 receptor, cubilin, cause hereditary megaloblastic anaemia," *Nat. Genet.* 21: 309, 1999 and http://www.nature.com/ng/journal/v2l/n3/full/ng0399_309.html.

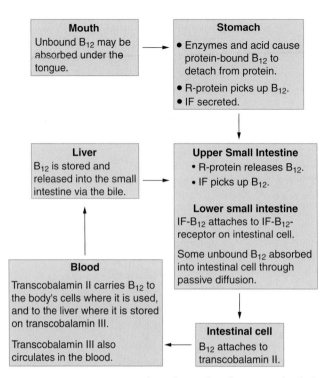

Figure 13-90. Summary of intake and utilization of cobalamin (vitamin B_{12}). R-protein is a protein in the stomach and small intestine that binds cobalamin. *IF*, Intrinsic factor; receptor on intestinal cell is cubilin. Redrawn from http://www.veganhealth.org/b12/images/b12absorption1.gif.

that catalyzes the conversion of homocysteine to methionine. In this reaction, a methyl group is transferred from N^5-methyltetrahydrofolate to hydroxycobalamin.

The liver is a long-term store of cobalamin. Intrinsic factor can be missing or inactive in some patients with pernicious anemia resulting from a deficiency of cobalamin, because without intrinsic factor there is no complex to bind to intestinal cubilin for absorption. With a deficiency of cobalamin, folic acid becomes trapped as the N^5-methyltetrahydrofolate because methionine synthase is no longer active. Also, other tetrahydrofolate derivatives are not formed, and these are required for purine and thymidine synthetic pathways. In the nervous system there is an increase in methylmalonyl-CoA, which inhibits the fatty acid synthesis required for the turnover of the myelin sheath, and progressive demyelination results.

Folic Acid (Vitamin B_9)

Folic acid is synthesized in plants and yeasts. The body cannot synthesize one component of folic acid, para-aminobenzoic acid (PABA). In the synthesis of folic acid by organisms outside the body, a pteridine ring structure is bonded to PABA and glutamate is linked to PABA to form folic acid (Figure 13-91). Dietary folic acid occurs as mixtures of polyglutamates, with as many as seven glutamic acid moieties (Figure 13-92). These may be hydrolyzed in the intestine by peptidases (and by lyso-

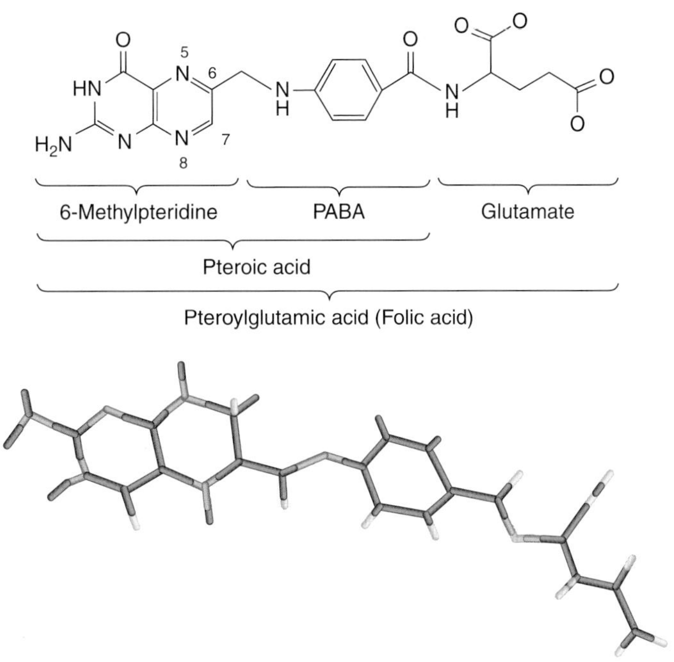

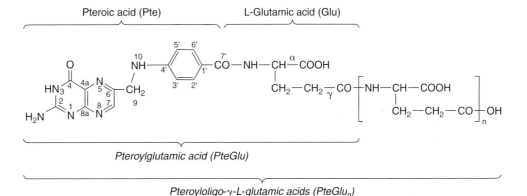

Figure 13-91. Structure of folic acid.

Figure 13-92. Structure of pteroyl-γ-L-glutamic acids. n can be as many as six glutamate residues.

somal conjugase) to the natural form of folic acid (or one with reduced number of glutamate residues) that can be recognized by the intestinal folic acid receptor, the RFC-1 gene product called the **RFC-1 transporter** (Figure 13-93). This transporter appears to be structurally related to the intestinal thiamine transporter. Removal of conjugated glutamate residues renders the folate molecule less negatively charged and more suitable for passage through the basolateral intestinal cellular membrane to gain access to the bloodstream.

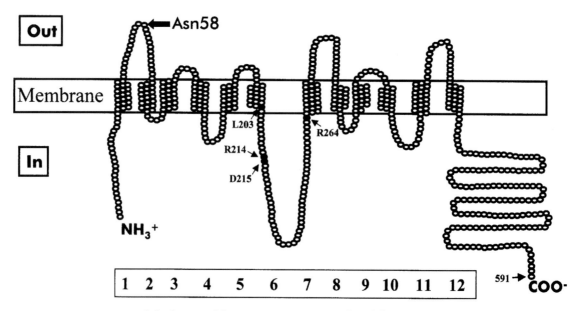

Figure 13-93. Diagram of the human folate transporter. Reproduced from Figure 1 of X.Y. Liu, T.L. Witt, and L.H. Matherly, "Restoration of high level transport activity by human reduced folate carrier/ThTrl thiamine transporter chimeras: role of the transmembrane domain 6/7 linker region in reduced folate carrier function," *Biochem. J.*, 369: 31–37, 2003.

When folic acid enters the liver cell, it is reduced to tetrahydrofolate (THF) by **dihydrofolate reductase (DHFR).** Also in the liver, folic acid in its storage form is in a polyglutamate form. THF derivatives transfer one carbon unit (methyl, methylene, formyl, or formimino groups) during biosynthetic reactions. Some of these intermediary structures of THF are shown in Figure 13-94. One-carbon transfer reactions are a feature of the biosyntheses of the serine, methionine, glycine, choline, and purine nucleotides and dTMP. *The role of N^5,N^{10}-methylene–THF in the regeneration of dTMP may be the most important role of folate* (see Figure 6-52). Cobalamin is also involved with N5,N10-methylene–THF in the formation of methionine from homocysteine (Figure 13-95). Folate deficiency (usually in alcoholics with inadequate diets) impairs dTMP synthesis, leading to cell cycle arrest in S-phase of rapidly dividing hemopoietic cells. In nonalcoholics, folate deficiency might be a feature of inadequate intestinal absorption of the vitamin. *Anticoagulants and contraceptives interfere with folate absorption.* Folate deficiency can lead to megaloblastic anemia, which is identical to that produced by cobalamin deficiency. The recommended daily allowance of folic acid for a normal adult is 400 μg/day. Common food sources of folic acid are given in Table 13-11.

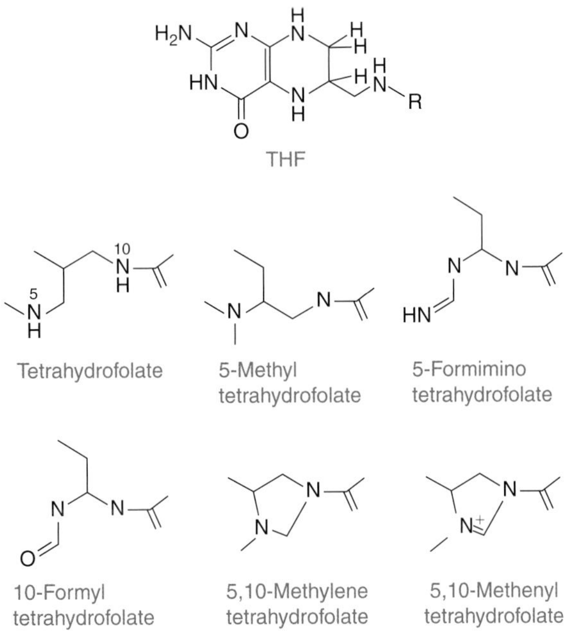

Figure 13-94. Structure of tetrahydrofolate *(top)* and some fractional one-carbon derivatives.

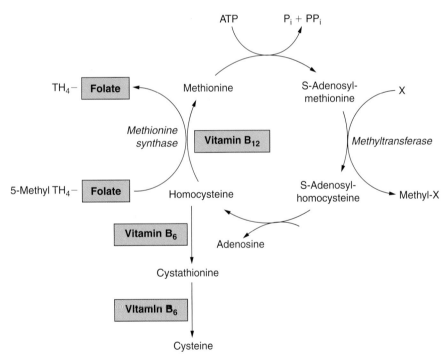

Figure 13-95. Homocysteine metabolism. S-adenosylhomocysteine is formed during S-adenosylmethionine–dependent methylation reactions, and the hydrolysis of S-adenosylhomocysteine results in homocysteine. Homocysteine may be remethylated to form methionine by a folate-dependent reaction that is catalyzed by methionine synthase, a vitamin B_{12}–dependent enzyme. Alternatively, homocysteine may be metabolized to cysteine in reactions catalyzed by two vitamin B_6–dependent enzymes.

Table 13-11
Common Food Sources of Folic Acid

Food	Serving	Folate (µg)
Fortified breakfast cereal	1 cup	200–400
Orange juice (from concentrate)	6 ounces	82
Spinach (cooked)	1/2 cup	131
Asparagus (cooked)	1/2 cup (~ 6 spears)	131
Lentils (cooked)	1/2 cup	179
Garbanzo beans (cooked)	1/2 cup	141
Lima beans (cooked)	1/2 cup	78
Bread	1 slice	20 (folic acid)
Pasta (cooked)	1 cup	60 (folic acid)
Rice (cooked)	1 cup	60 (folic acid)

Reproduced from http://lpi.oregonstate.edu/infocenter/vitamins/fa/.

Ascorbic Acid (Vitamin C)

The structure of ascorbic acid is shown in Figure 13-96. Ascorbic acid is an antioxidant, reducing other compounds while being converted to its oxidized form (dehydroascorbic acid). Although many species are capable of synthesizing ascorbate, humans are unable to do so. One of the main functions of this vitamin is in the hydroxylation of proline (in collagen), where the vitamin is a cofactor for the hydroxylation reaction (Figure 13-97). The peptide-bound proline-4-hydroxylase reaction is coupled to the oxidative decarboxylation of α-ketoglutarate, and the enzyme-bound Fe^{2+} is rapidly converted to enzyme-bound Fe^{3+}, causing inactivation of the enzyme. Ascorbate reduces enzyme-bound Fe^{3+} to enzyme-bound Fe^{2+}, reactivating the enzyme. This hydroxylation is required for the collagen molecule to aggregate into a triple helix. Thus, ascorbate is necessary for the maintenance of connective tissue. Also, it is important in wound healing, which involves the synthesis of connective tissue as an early step of the process. Because collagen is in the organic matrix of bone, ascorbate is necessary for bone remodeling. Ascorbate is involved in the synthesis of epinephrine from tyrosine and possibly in the synthesis of hormonal steroids, especially cortisol. Severe stress leads to increased need for ascorbate as the adrenal store of this vitamin is rapidly depleted. Vitamin C also may be involved in the signal transduction by NFκB. Dehydroascorbate (DHA), but not ascorbate, transported into the cell through a glucose transporter directly inhibits IκBα kinase (IKKβ). Ascorbate quenches reactive oxygen species intermediates involved in the activation of NFκB and is oxidized to DHA, which inhibits IKKβ and IKKα enzymatic activity; thus, ascorbate can modulate NFκB signaling downward (Figure 13-98). Ascorbate is degraded through dehydroascorbate to diketogulonic acid and then to other acids, such as oxalic acid. The average adult male needs 90 mg of ascorbate per day, and the adult female needs 75 mg ascorbate per day. Good sources of dietary vitamin C are listed in Table 13-12.

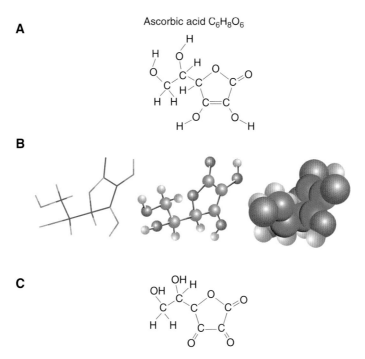

Figure 13-96. Structures of ascorbic acid (vitamin C). **A.** Chemical structure of reduced form. **B.** Models of the reduced form. **C.** Oxidized structure of ascorbate (dehydroascorbate).

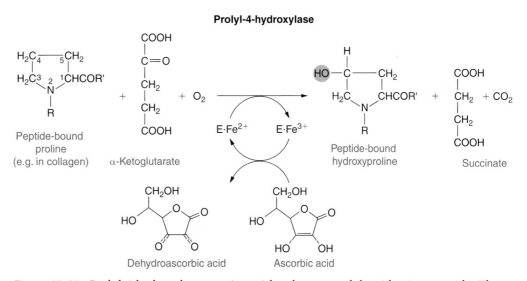

Figure 13-97. Prolyl-4-hydroxylase reaction with substrate prolyl residue in a peptide. The hydroxylation of peptide-bound proline is coupled to the oxidative decarboxylation of α-ketoglutarate and enzyme-bound Fe^{2+} is rapidly converted to Fe^{3+}. Enzyme-bound Fe^{3+} can be reduced again by ascorbate to reactivate the enzyme. E-enzyme; R and R' are amino acid residues.

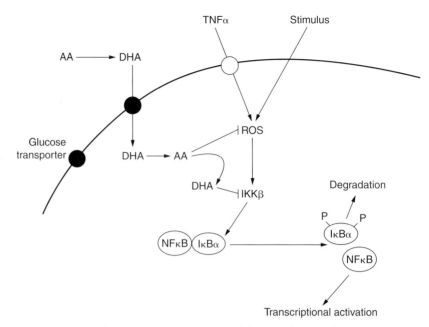

Figure 13-98. Schematic representation of the regulation of signaling responses by vitamin C. Vitamin C enters the cell through the glucose transporters as DHA and is rapidly reduced to AA. ROS induces NFκB signaling responses by activating IKKβ, and AA quenches ROS, inhibiting the activation of IKKβ. Throughout these processes, AA becomes oxidized to DHA, and DHA inhibits IKKβ. *AA*, Ascorbic acid; *DHA*, dehydroascorbic acid; *IκBα*, I kappa B; *IKKβ*, I Kappa B kinase; *NFκB*, nuclear factor kappa B; *ROS*, reactive oxygen species.

Table 13-12
Good Dietary Sources of Vitamin C

Food	Serving Size	Milligrams Vitamin C	% AI for Men	% AI for Women
Guava	1 medium	165	183	235
Red bell pepper	1/2 cup	95	94.7	135
Papaya	1 medium	95	94.7	135
Orange juice, from frozen concentrate	3/4 cup	75	83.3	107
Orange	1 medium	60	66.6	85.7
Broccoli, boiled	1/2 cup	60	66.6	85.7
Green bell pepper	1/2 cup	45	50	64.2
Kohlrabi, boiled	1/2 cup	45	50	64.2
Strawberries	1/2 cup	45	50	64.2
Grapefruit, white	Half	40	44.4	57.1
Cantaloupe	1/2 cup	35	38.8	50
Tomato juice	3/4 cup	35	38.8	50
Mango	1 medium	30	33.3	42.8
Tangerine	1 medium	25	27.7	35.7
Potato, baked with skin	1	25	27.7	35.7
Cabbage greens, frozen, boiled	1/2 cup	25	27.7	35.7
Spinach, raw	1 cup	15	16.6	21.4

AI, adequate intake.
Reproduced from http://ohioline.osu.edu/hyg-fact/5000/5552.html.

Fat-Soluble Vitamins

Vitamin A

Plant foods, especially carrots, contain β-carotene, which is the provitamin. β-Carotene contains two retinals linked head to tail, and it is cleaved in the intestine to yield two retinal molecules by the enzyme β-carotene **dioxygenase** (Figure 13-99). Retinal can be further metabolized to retinol or retinoic acid, which is the ligand for the retinoic acid receptor (RAR) that heterodimerizes with other receptors to become active at gene promoters (TR; VDR; retinoic X receptor, or RXR; and others). The metabolite, 9-*cis*-retinoic acid, is the ligand for the RXR. 9-*cis*-Retinoic acid is formed by retinol dehydrogenase that converts 9-*cis*-retinol to 9-*cis*-retinaldehyde. This is followed by an oxidation to 9-*cis*-retinoic acid. The metabolism of β-carotene in certain cells is shown in Figure 13-100, which gives an overview of the metabolism of β-carotene and indicates the roles of all-*trans* retinoic acid and 9-*cis*-retinoic acid as ligands for nuclear receptors. Retinol palmitate is incorporated into chylomicrons and enters the bloodstream. The liver absorbs chylomicron remnants, which store retinol as

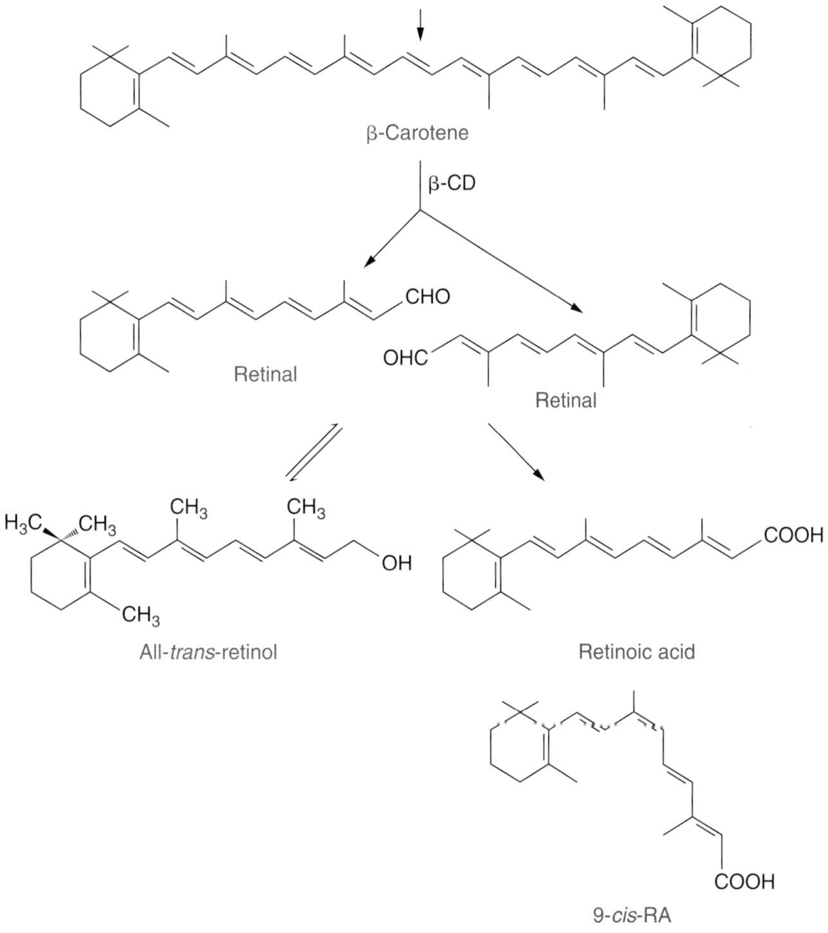

Figure 13-99. Conversion of β-carotene to retinal by β-carotene dioxygenase (β-CD). Retinal can be converted to all-*trans*-retinol or retinoic acid (RA). RA is the ligand for RAR. 9-cis-Retinoic acid (9-*cis*-RA) is the ligand for RXR.

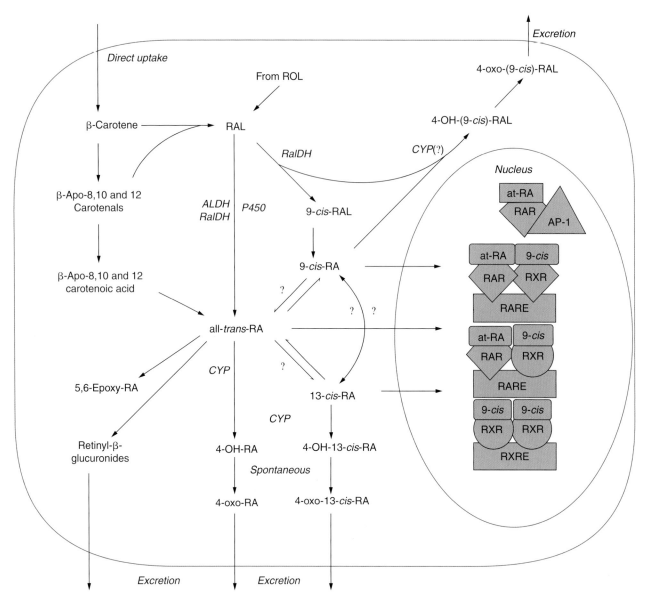

Figure 13-100. Overview of pathways of retinoid metabolism. Within the nucleus are shown homo- and heterodimers of RAR and RXR and their DNA binding sites for receptor-ligand complexes. *ALDH*, Aldehyde hydrogenase; *at-RA*, all trans retinoic acid; *9-cis*, 9-*cis*-retinoic acid; *9-cis-RA*, 9-*cis*-retinoic acid; *RA*, retinoic acid; *RAL*, retinal; *RalDH*, retinal dehydrogenase; *RAR*, retinoic acid receptor; *ROL*, retinol.

the ester. Retinol can be transported out of the liver to other tissues by first hydrolyzing the retinol ester so that free retinol can bind to the liver retinol-binding protein, which can be transported to the cell surface and secreted. This complex travels to other tissues, where free retinol enters and is bound to a **cellular retinol-binding protein (CRBP)**, whereas the transport of retinoic acid occurs through its binding to albumin. Retinoic acid and 9-*cis*-retinoic acid are like hormones in their ability to act as ligands for some heterodimeric receptors of the steroid hormone receptor superfamily, such as TR, VDR, RAR, and RXR. Retinoic acid is involved in early development.

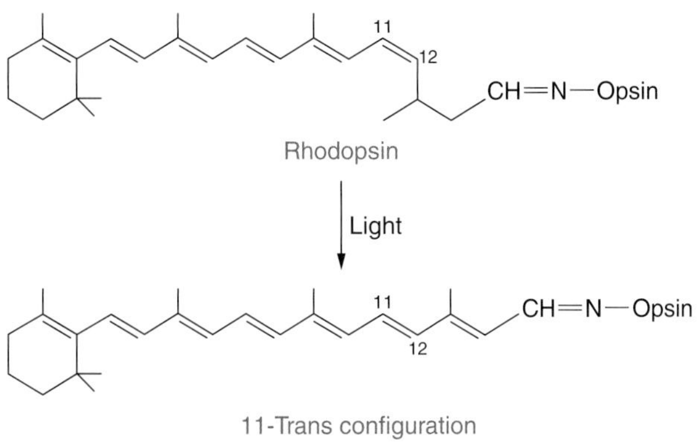

Figure 13-101. A primary chemical event in the visual process in rod cells of the eye is the absorption of light by rhodopsin that isomerizes 11-cis double bond to the 11-trans configuration.

Opsin, involved in the visual process, is a photosensitive pigment coupled covalently to 11-*cis*-retinal to form **rhodopsin.** Rhodopsin is a receptor incorporated into the rod cell membrane. Coupling of 11-*cis*-retinal occurs at three transmembrane domains of rhodopsin. The primary event of vision in rod cells of the eye is the absorption of light by rhodopsin (bleaching) by isomerization of 11-*cis* double bond to the 11-*trans* configuration (Figure 13-101). Opsin is released, resulting in a conformational change in the photoreceptor. This change activates a G protein, called **transducin,** that leads to increased GTP binding by its α-subunit, causing its release from the β- and γ-subunits. The activated α-subunit activates a local phosphodiesterase that hydrolyzes cyclic GMP to GMP. The fall in the concentration of cyclic GMP closes the Na^+ channels of the rod cells because cyclic GMP maintains the open position of these channels. Metarhodopsin (Figure 13-102) II, a conformational intermediate, initiates the closure of the Na^+ channels that hyperpolarizes the rod cell, causing propagation of nerve impulses to the brain.

Vitamin A deficiency leads to visual problems, including night blindness. Other problems are susceptibility to cancer, infection, and anemia. Susceptibility to cancer is related to the antioxidant activity of β-carotene to reduce free radicals that can cause cancer. Because fat-soluble vitamins like vitamin A are stored as esters in the liver, they are not catabolized and removed from the body. Consequently, it is possible to ingest too much vitamin A by supplementing the diet. Toxicity produces bone pain, enlargement of the liver and spleen, nausea, and diarrhea. Food sources for vitamin A are listed in Table 13-13.

Vitamin D

Vitamin D is a hormonal ligand for the VDR (Figure 10-36 and Figure 10-37). The active form, 1,25-dihydroxyvitamin D_3, regulates homeostasis of calcium and phosphorus. 7-Dehydrocholesterol is irradiated in the skin by sunlight to form the active ligand for the VDR (Figure 13-103). Dietary sources of cholecalciferol or ergocalciferol are absorbed from the intestine, transported to the liver, and bound to a

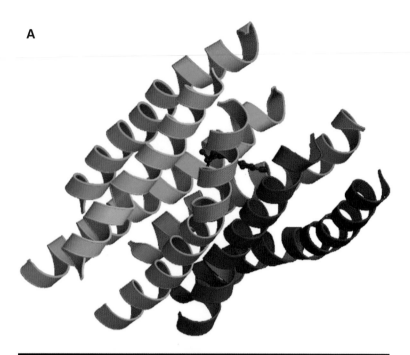

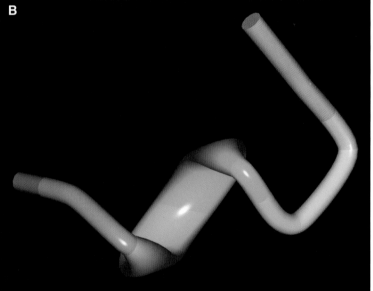

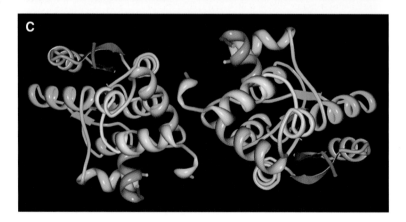

Figure 13-102. **A.** Bovine rhodopsin (7-helix bundle) with all-*trans* retinal, metarhodopsin II theoretical model. **B.** Rhodopsin-bound C-terminal peptide of the transducin α-subunit. **C.** Heterotrimeric complex of phosducin or transducin beta gamma. Part A reproduced from http://www.ebi.ac.uk/msd-srv/msdlite/atlas/visualization/1boj.html. PDB ID: 1boj. I.D. Pogozheva, A.L. Lomize, H.I. Mosberg. The Transmembrane 7-alpha-bundle of Rhodopsin: Distance Geometry Calculations with Hydrogen Bonding Constraints. *Biophysical Journal* **72** pp. 1963(1997). Part B reproduced from http://www.rcsb.org/pdb/cgi/explore.cgi?pdbId51aqg. PDB ID: 1aqg. O.G. Kisselev, J. Kao, J.W. Ponder, Y.C. Fann, N. Gautam, G.R. Marshall. Light-Activated Rhodopsin Induces Structural Binding Motif in G Protein Alpha Subunit. *Proceedings of the National Academy of Sciences USA* **95** pp. 4270 (1998). Part C reproduced from http://www.rcsb.org/pdb/cgi/explore.cgi?pdbId51a0r. PDB ID: 1a0r. A. Loew, Y.K. Ho, T. Blundell, B. Bax. Phosducin Induces a Structural Change in Transducin Beta Gamma.*Structure* **6** pp. 1007 (1998).

Table 13-13
Food Sources of Vitamin A (β-Carotene)

Food	Serving	IU/International Units	%DV*
Animal Sources			
Liver, beef, cooked,	3 ounces	30,325	610
Liver, chicken, cooked,	3 ounces	13,920	280
Egg substitute, fortified,	1/4 cup	1355	25
Fat free milk, fortified w/vitamin A,	1 cup	500	10
Cheese pizza,	1/8 of a 12″ diameter	380	8
Milk, whole, 3.25% fat,	1 cup	305	6
Cheddar cheese,	1 ounce	300	6
Whole egg,	1 medium	280	6
Swiss cheese,	1 ounce	240	4
Margarine, soft, corn oil,	1 teaspoon	165	4
Yogurt, fruit flavored, low fat,	1 cup	120	2
Plant Sources			
Carrot, 1 raw	(7 1/2″)	20,250	410
Carrots, boiled,	1/2 cup slices	19,150	380
Carrot juice, canned,	1/2 cup	12,915	260
Mango, raw, without refuse,	1 fruit	8,050	160
Sweet potatoes,	1/2 cup Junior mashed	7,430	150
Spinach, boiled,	1/2 cup	7,370	150
Cantaloupe, raw,	1 cup cubes	5,160	100
Kale, boiled,	1/2 cup	4,810	100
Vegetable soup, prepared with equal volume water,	1 cup	3,005	60
Pepper, sweet, red, raw,	1/2 cup sliced	2,620	50
Apricots, without skin, canned in water,	1/2 cup halves	2,055	40
Spinach, raw,	1 cup	2,015	40
Broccoli, frozen, chopped, boiled,	1/2 cup	1,740	35
Apricot nectar, canned,	1/2 cup	1,650	30
Oatmeal, instant, fortified, low sodium, dry,	1 packet	1,050	20
Tomato juice, canned,	6 ounces	1,010	20
Ready-to-eat cereal, fortified,	1 ounce (15% fortification)	750	15
Peaches, canned, water pack,	1/2 cup halves or slices	650	15
Peach, raw,	1 medium	525	10
Papaya, raw,	1 small	430	10
Orange, raw,	1 large	375	8

*DV, daily value. DVs are reference numbers based on the recommended daily dietary allowance (RDA). They were developed to help consumers determine if a food contains a lot or a little of a specific nutrient. The DV for vitamin A is 5000 IU. The percent DV (%DV) listed on the nutrition facts panel of food labels tells adults what percentage of the DV is provided by one serving. Percent DVs are based on a 2000-calorie diet. An individual's DV may be higher or lower depending on your calorie needs. Foods that provide lower percentages of the DV will contribute to a healthful diet.
Reproduced from http://ibscrohns.about.com/library/fda/blvita5.htm.

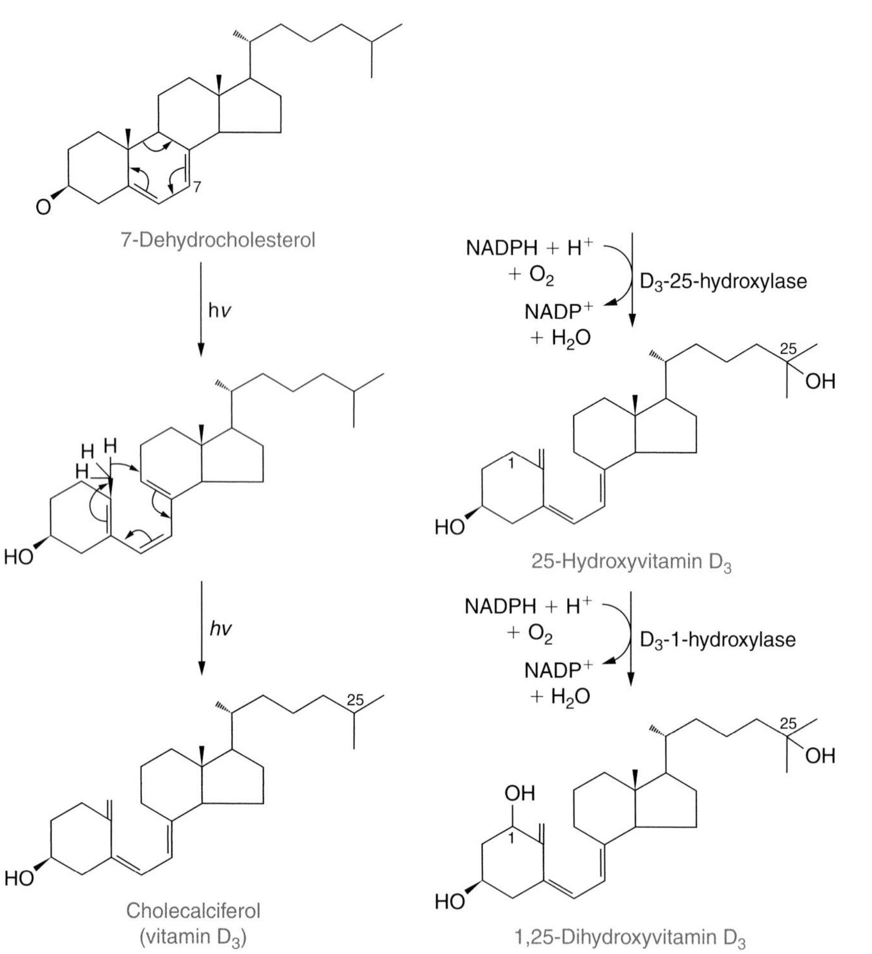

Figure 13-103. Conversion of 7-dehydrocholesterol to 1,25-dihydroxyvitamin D_3 in the skin.

specific vitamin D–binding protein. D_3-25-hydroxylase acts on liver cholecalciferol to produce 25-hydroxy-D_3, the major circulating form (also, D_3-24-hydroxylase in the kidneys, intestine, placenta, and cartilage can produce the 24-hydroxylated form). D_3-1-hydroxylase in the proximal convoluted kidney tubules (and in bone and placenta) produces 1,25-dihydroxyvitamin D_3 (calcitriol), the biologically active form of the vitamin.

The parathyroid gland releases **parathyroid hormone (PTH)** when the level of serum calcium falls, and PTH stimulates the synthesis of calcitriol. When the levels of PTH fall, the synthesis of inactive 24,25-dihydroxy D_3 is stimulated. Calcitriol acts through the vitamin D_3 receptor to induce **calbindin D** in the intestinal epithelium (and kidney), and calbindin D (Figure 13-48) is involved in the intestinal absorption of calcium. A fall in plasma calcium, calcitriol, and PTH stimulates bone resorption; they act on the kidneys to inhibit excretion of calcium through the stimulation of bone resorption. Because vitamin D is added to milk and only 10 minutes of sunshine is needed to activate 7-dehydrocholesterol, deficiency of the vitamin is rare. Osteomalacia in adults and rickets in children, however, result from a deficiency of this vitamin. Recommended intakes of the

Table 13-14
Food Sources of Vitamin D

Food	Serving	Vitamin D
Milk	1 cup	100 IU
Fortified rice or soy beverage	1 cup	100 IU
Fortified margarine	2 teaspoons	53 IU
Salmon, canned, pink	3 ounces	530 IU
Tuna, canned, light	3 ounces	200 IU

IU, international unit.
Reproduced from http://www.bchealthguide.org/healthfiles/hfile68e.stm.

vitamin are 200 international units (IU) per day (1–50 years), 400 IU (51–70 years), and 600 IU (over 70 years); in the first year of life, 400 IU/day is required. Food sources of vitamin D are given in Table 13-14.

Vitamin E

Vitamin E, an antioxidant, is known as α-tocopherol, the most active member of a number of compounds with vitamin E activity. These compounds are α-tocopherol (5,7,8-trimethyltocol), β-tocopherol (5,8-trimethyltocol), γ-tocopherol (7,8-trimethyltocol, δ-tocopherol (8-trimethyltocol), α-tocotrienol, β-tocotrienol, γ-tocotrienol, and δ-tocotrienol. The structures of α-tocopherol and α-tocotrienol are shown in Figure 13-104. It is becoming apparent that these different relatives of α-tocopherol may have different unique activities. γ-Tocopherol, for example, has been found to inhibit cyclooxygenase activity. α-Tocotrienol, but not α-tocopherol, modulates 12-lipoxygenase (12-LOX). The difference between these two compounds (Figure 13-104) is that α-tocopherol has a saturated phytyl side chain, whereas α-tocotrienol has an isoprenoid (farnesyl) side chain. 12-LOX is a mediator of glutamate-induced neurodegeneration, and α-tocotrienol inhibits 12-LOX and thus is a neuronal protector. Figure 13-105 shows the interaction of this form of vitamin E with 12-LOX.

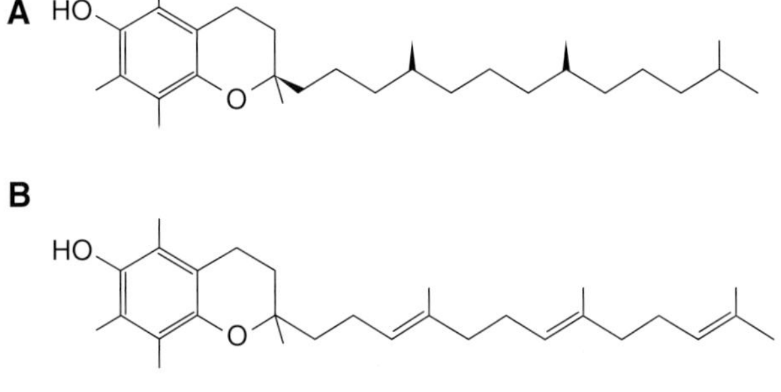

Figure 13-104. Structures of α-tocopherol *(A)* and α-tocotrienol *(B)*.

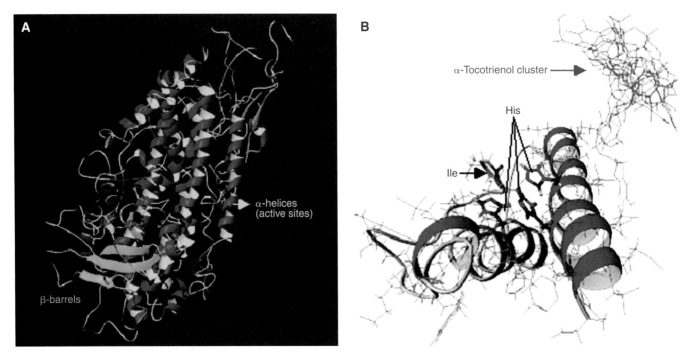
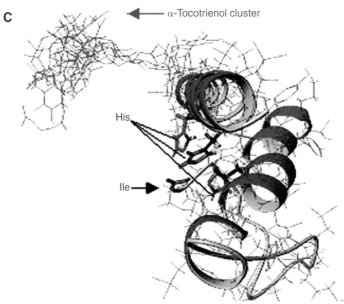

Figure 13-105. Three-dimensional modeling of 12-LOX and α-tocotrienol docking analysis. **A.** three-dimensional structure of 12-LOX. Homology model construction was carried out using a Silicon Graphics O2 workstation with 300-MHz MIPS R5000 (OS IRIX Version 6.5). The theoretical model of 12-LOX was built using the Sybyl GeneFold module (Version 6.8). **B** and **C,** theoretical model and α-tocotrienol docking (two positions in *B* and *C* are shown with 10 different docking positions). His-360, His-365, His-540, and Ile-663, flanking the iron atom, are visualized in red. Reproduced from Figure 7 of S. Khanna et al., "Molecular basis of vitamin E action: tocotrienol modulates 12-lipoxygenase, a key mediator of glutamate-induced neurodegeneration," *J. Biol. Chem.*, 278: 43508–43515, 2003 and http://www.jbc.org/cgi/content/full/278/44/43508.

Table 13-15
Some Dietary Sources of Vitamin E

Food	Serving	Milligrams (mg) Alpha-Tocopherol (ATE) Per Serving	Percent DV*
Wheat germ oil,	1 tablespoon	20.3	100
Almonds, dry roasted,	1 ounce	7.4	40
Sunflower seed kernels, dry roasted,	1 ounce	6.0	30
Sunflower oil, over 60% linoleic,	1 tablespoon	5.6	30
Safflower oil, over 70% oleic,	1 tablespoon	4.6	25
Hazelnuts, dry roasted,	1 ounce	4.3	20
Peanut butter, smooth style, vitamin and mineral fortified,	2 tablespoons	4.2	20
Peanuts, dry roasted,	1 ounce	2.2	10
Corn oil (salad or vegetable oil),	1 tablespoon	1.9	10
Spinach, frozen, chopped, boiled,	1/2 cup	1.6	6
Broccoli, frozen, chopped, boiled,	1/2 cup	1.2	6
Soybean oil,	1 tablespoon	1.3	6
Kiwi,	1 medium fruit without skin	1.1	6
Mango, raw, without refuse,	1/2 cup sliced	0.9	6
Spinach, raw,	1 cup	0.6	4

*DV, daily value. DVs are reference numbers developed by the Food and Drug administration (FDA) to help consumers determine if a food contains a lot or a little of a specific nutrient. The DV for vitamin E is 30 international units (or 20 mg ATE). Most food labels do not list a food's vitamin E content. The percent DV listed on the table indicates the percentage of the DV provided in one serving. A food providing 5% of the DV or less is a low source; a food that provides 10–19% of the DV is a good source. A food that provides 20% or more of the DV is high in that nutrient. It is important to remember that foods that provide lower percentages of the DV also contribute to a healthful diet. For foods not listed in this table, please refer to the U.S. Department of Agriculture's Nutrient Database Web site: http://www.nal.usda.gov/fnic/cgi-bin/nut_search.pl.
Reproduced from Table 1 of http://ods.od.nih.gov/factsheets/vitamine.asp.

Ingested vitamin E is absorbed in the intestine and incorporated into chylomicrons for delivery to the liver (chylomicron remnant uptake). Export of vitamin E from the liver is in the form of VLDLs. The vitamin is found in cell membranes, circulating lipoproteins, and adipose tissue, the major storage site. As an antioxidant, this vitamin prevents peroxidation of unsaturated fatty acids of the cell membrane. Following scavenge of a peroxy free radical, the vitamin can be regenerated by interacting with ascorbic acid. When it is not reacting with ascorbic acid, a molecule of α-tocopherol can scavenge two peroxy free radicals and then be excreted in the bile as a glucuronate.

Vitamin E deficiency is rare in a healthy diet, but the deficiency leads to increased red blood cell fragility. Adults need about 15 mg/day. Some food sources of vitamin E are listed in Table 13-15.

Vitamin K

There are two forms of naturally occurring vitamin K: vitamin K_1, which is phytylmenaquinone, and vitamin K_2, which is multiprenylmenaquinone. The K vitamins have in common a methylated naphthoquinone ring structure varying in the aliphatic side chain at the 3-position. These structures are shown in Figure 13-106. Vitamin K_1 occurs in plants, and vitamin K_2 is synthesized by intestinal bacteria. Vitamin K_3 is a synthetic molecule (menadione) with vitamin

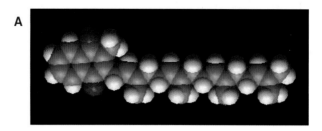

Figure 13-106. Structures of molecules with vitamin K activity. **A.** Atomic model of α-tocopherol. **B, C,** and **D** are chemical structures of vitamin K_1, K_2, and K_3. Part A reproduced from http://www.worldofmolecules.com/supplements/vitamin_K.htm. Parts B through D reproduced from http://www.med.unibs.it/~marchesi/vitamins2.html.

K activity. Vitamin K is a stimulator of blood clotting and achieves this activity by maintaining normal levels of the active blood-clotting protein factors II, VII, IX, and X and proteins C and S. Clotting factors are modified postranslationally as they pass through the endoplasmic reticulum to become active. *This modification is the carboxylation of multiple glutamate residues in the clotting factor protein by a vitamin K–dependent γ-carboxylase that converts clotting factors into γ-carboxyl glutamate (Gla)-containing proteins.* The structure of Gla is shown in Figure 13-107. The system is located in the membrane of the endoplasmic reticulum (ER) and includes both the γ-carboxylase and the enzyme **2,3-epoxide reductase (VKOR)** that converts vitamin K to the 2,3-epoxide form. VKOR provides the reduced vitamin K cofactor for the γ-carboxylase. A reductase regenerates the hydroquinone form of the vitamin. Another factor in the system is an ER chaperone protein called **calumenin** that associates with the γ-carboxylase and inhibits its activity. The system is shown in Figure 13-108. Carboxylation of multiple glutamate residues on clotting factors

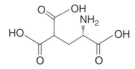

Figure 13-107. Structure of γ-carboxyglutamic acid.

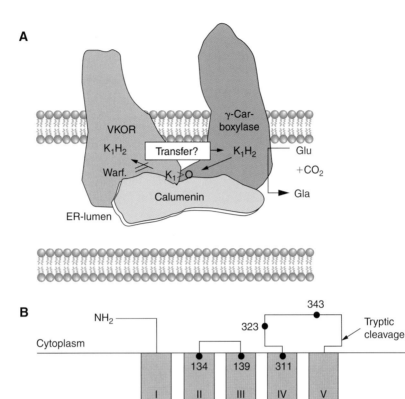

Figure 13-108. **A.** A putative molecular model of the γ-carboxylation system. VKOR and γ-carboxylase are parts of an enzyme complex in the ER lipid bilayer. Calumenin binds to γ-carboxylase as an inhibitory chaperone and also affects the activity and warfarin sensitivity of VKOR. **B.** The membrane topology (*B*) of human γ-glutamyl carboxylase. Cysteine residues are marked, and the *gray areas* correspond to the transmembrane domains. *Warf*, warfarin. Part A redrawn from N. Wajih et al., "The inhibitory effect of calumenin on the vitamin K–dependent γ-carboxylation system: characterization of the system in normal and warfarin-resistant rats," *J. Biol. Chem.*, 279: 25276–25283, 2004. Part B redrawn from Figure 8 of J.-K. Tie et al., "Chemical modification of cyteine residues is a misleading indicator of their status as active site residues in the vitamin K–dependent γ-glutamyl carboxylation reaction," *J. Biol. Chem.*, 279: 54079–54087, 2004. http://www.jbc.org/cgi/content/full/279/52/54079.

Vitamins 845

involves a processive mechanism and results in the carboxylation of the γ-carboxylase itself. The enzyme has a high-affinity binding site for vitamin K–dependent (VKD) proteins. The contact between the γ-carboxylase and the VKD protein increases the affinity of the carboxylase for vitamin K. The anticoagulant **warfarin** (Figure 13-109) indirectly blocks carboxylation of VKD proteins and increases accumulation of VKD protein precursors, leading to undercarboxylated inactive clotting proteins and preventing blood clotting. These ideas are conveyed in Figure 13-110. A case in point is prothrombin (factor II) derived from preprothrombin. Prothrombin, which contains Glas, chelates calcium and interacts with membrane phospholipids. Activated factor X (Xa), a protease, converts prothrombin to thrombin (factor IIa). Thrombin promotes the conversion of fibrinogen to the fibrinogen monomer that

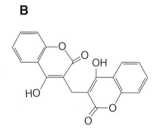

Figure 13-109. **A.** Structure of warfarin, also known as Coumadin. **B.** Structure of dicoumarol from which warfarin was derived.

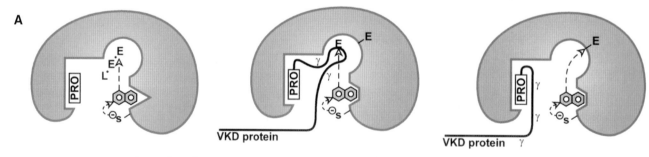

Figure 13-110. **A.** The carboxylase active site. Most vitamin K–dependent (VKD) proteins bind to the carboxylase through the propeptide (PRO), and there may be a second point of contact for the glutamyl (glu; E) residues, as illustrated by the pocket. Covalent attachment of the propeptide and gla domain increases the affinity of KH_2 binding, as contrasted in A and B. As shown in B and C, multiple carboxylations are accomplished by "tethered processivity," in which the propeptide remains bound throughout the reaction, and the Gla domain undergoes intramolecular movement to reposition the Glu's for conversion to gla's (γ). During the reaction, the carboxylase is also carboxylated.

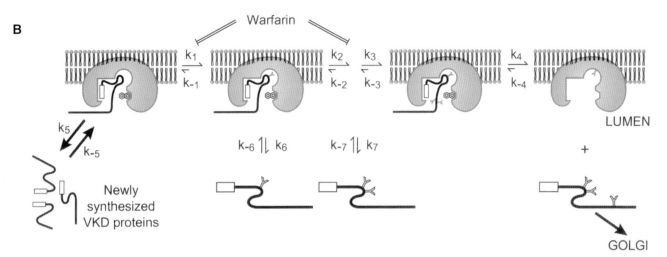

Figure 13-110, cont'd **B**. Carboxylation and secretion of VKD proteins. Normally, VKD proteins undergo comprehensive carboxylation [K_{1-3} in this example for a VKD protein with 3 γ-carboxylated glutamyl (gla) residues], release (k_4), and subsequent propeptide processing in the Golgi. Decarboxylation (i.e., k_{-1} to k_{-3}) is unlikely to be physiologically relevant. Warfarin effects premature release (k_{5-7}) of undercarboxylated forms. This effect could be due to a decrease in k_{1-3} as well as a buildup of VKD precursors that accelerate release. Most VKD proteins also undergo other posttranslational modifications, including N-glycosylation, O-glycosylation, β-aspartyl hydroxylation, and proteolytic cleavage at internal sites within the mature protein. Reproduced from Figures 1 and 2 of K.L. Berkner, "The vitamin K–dependent carboxylase," *J. Nutr.* 130: 1877–1880, 2000 and http://www.jn.nutrition.org/cgi/content/full/130/8/1877.

polymerizes to form a cross-linked fibrin polymer. The conversion of fibrin polymer to cross-linked fibrin polymer also is positively regulated by factor XIIIa. The conversion of factor XIII to factor XIIIa is accomplished by thrombin as well.

Vitamin K is absorbed in the intestine in the presence of bile salts and interacts with chylomicrons. Diseases in which fats are not absorbed well (such as bile duct obstruction) can lead to deficiency of vitamin K. However, deficiency (leading to bleeding or hemorrhagic problems) in adults is rare because the intestinal bacteria synthesize vitamin K_2. The recommended daily intake of vitamin K is 80 μg for men and 65 μg for women. Common food sources of vitamin K are listed in Table 13-16.

Table 13-16
Common Food Sources of Vitamin K

Food	Serving	Vitamin K (μg)
Olive oil	1 tablespoon	6.6
Soybean oil	1 tablespoon	26.1
Canola oil	1 tablespoon	19.7
Mayonnaise	1 tablespoon	11.9
Broccoli, cooked	1 cup (chopped)	420
Kale, raw	1 cup (chopped)	547
Spinach, raw	1 cup (chopped)	120
Leaf lettuce, raw	1 cup (shredded)	118
Swiss chard, raw	1 cup (chopped)	299
Watercress, raw	1 cup (chopped)	85
Parsley, raw	1 cup (chopped)	324

Reproduced from http://lpi.oregonstate.edu/infocenter/vitamins/vitaminK/.

Food Guide Pyramid
A Guide to Daily Food Choices

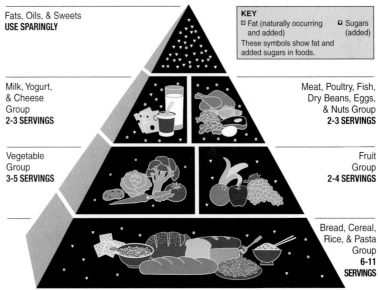

Source: U.S. Department of Agriculture/U.S. Department of Health and Human Services

Table 13-17
Recommended Diet

Group	Food Included	Daily Servings	Nutrition
Grains	Breakfast cereals, muesli, rice, pasta, bread, oats, noodles, and all other foods made from flour and grains. **Best** are less refined types.	6–11 servings	Good sources of starchy carbohydrates, fiber, plus calcium, iron, and B vitamins. Very filling and low in fat.
Vegetables	Includes fresh, frozen or canned, and blended vegetable juice drinks. **Best:** spinach, cabbage, bean sprouts, peas, carrots, broccoli, peppers, leeks, potatoes, onions, garlic, tomatoes.	4 servings	Good sources of antioxidant vitamins A and C, folates, fiber, and some carbohydrates. Very low in fat and calories.
Fruits	Includes fresh, frozen, or canned, plus blended 100% fruit drinks. **Best:** strawberries, grapes, kiwis, pineapple, blackcurrants, oranges, lemons, grapefruit, peaches, apricots.	3–4 servings	Good sources of antioxidant vitamins A and C, folates, fiber and some carbohydrates. Very low in fat and calories.
Meat & dairy	Fish, all meat, poultry, eggs, milk, cheese, yogurt, nuts. **Best:** any fish (esp. oily fish), turkey, free-range chicken, lean beef, free-range eggs, low fat dairy products, fat-free milk and lower fat cheese.	2–3 servings. You should try to eat at least two portions of fish per week	Lean red meat contains a wide variety of essential nutrients. Doctors advise us to eat less red meat and more oily fish.
Sugars & fats	All sugars, oils, butter, margarine, cream. **Best:** choose unrefined oils low in saturated fat, unrefined sugar, and low-fat spread (low-fat soft margarine).	Tiny amounts only	Good source of essential fatty acids and fat soluble vitamins A, D, E, and K.

Reproduced from http://www.sciencepages.co.uk/keystage3/year8/module13/images/foodpyramid.jpg.

The Diet

A balanced diet ensures that nutrients, vitamins, and calories will be provided in optimal proportion to ensure health. The food pyramid is represented in Table 13-17. Underlying the food pyramid is exercise, which is essential to health. Genetic variation plays a major part in the way people eat and exercise, and variations on this basis are not factored into daily recommendations. A recommended daily diet can be constructed from Table 13-17, and the optimal caloric intake can be calculated from Table 13-18. A summary of the daily requirements of micronutrients is given in Table 13-19.

Protein Nutrition

Proteins in foods obviously are needed to supply amino acids and build proteins and tissues in the body. Recommendations are only general because the requirements for protein and the type of protein ingested vary from one individual to the next. Plant proteins may be somewhat less well digested than animal proteins, such as those from eggs or milk.

Table 13-18
Calculations of Calorie Requirements for Women and Men

Calories for Women

- **Work out your weight in pounds**
  Example: Hilary is 150 pounds
- **Multiply your weight according to your lifestyle**
  Sedentary: multiply by 12
  A very inactive Hilary needs $150 \times 12 = 1800$ calories
  Light Exercise: multiply by 13
  If Hilary takes light exercise she needs about $150 \times 13 = 1950$ calories
  Moderate Exercise: multiply by 14
  If Hilary takes moderate exercise she needs about $150 \times 14 = 2100$ calories

Calories for Men

- **Work out your weight in pounds**
  Example: Bill is 165 pounds
- **Multiply your weight according to your lifestyle**
  Sedentary: multiply by 13
  A very inactive Bill needs $165 \times 13 = 2145$ calories
  Light Exercise: multiply by 14
  If Bill takes light exercise he needs about $165 \times 14 = 2310$ calories
  Moderate Exercise: multiply by 15.25
  If Bill takes moderate exercise he needs about $165 \times 15.25 = 2516$ calories
  Moderate-Heavy Exercise: multiply by 16.5
  If Bill takes moderate-heavy exercise he needs about $165 \times 16.5 = 2722$ calories
  Regular Heavy Exercise: multiply by 18
  If Bill takes regular heavy exercise he needs about $165 \times 18 = 2970$ calories

Please note: Calorie needs vary according to many things, including gender, age, height, level of activity, body fat percentage and type of food eaten. This calorie guide offers a ballpark estimation only.
Reproduced from http://www.diet-i.com/weight_loss/calories.htm.

Table 13-19
Daily Values of Nutrients Based on a 2000-Calorie/Day Diet

Nutrient	Unit of Measure	Daily Values
Total fat	grams (g)	65
Saturated fatty acids	grams (g)	20
Cholesterol	milligrams (mg)	300
Sodium	milligrams (mg)	2400
Potassium	milligrams (mg)	3500
Total carbohydrate	grams (g)	300
Fiber	grams (g)	25
Protein	grams (g)	50
Vitamin A	International Unit (IU)	5000
Vitamin C	milligrams (mg)	60
Calcium	milligrams (mg)	1000
Iron	milligrams (mg)	18
Vitamin D	International Unit (IU)	400
Vitamin E	International Unit (IU)	30
Vitamin K	micrograms (µg)	80
Thiamin	milligrams (mg)	1.5
Riboflavin	milligrams (mg)	1.7
Niacin	milligrams (mg)	20
Vitamin B_6	milligrams (mg)	2.0
Folate	micrograms (µg)	400
Vitamin B_{12}	micrograms (µg)	6.0
Biotin	micrograms (µg)	300
Pantothenic acid	milligrams (mg)	10
Phosphorus	milligrams (mg)	1000
Iodine	micrograms (µg)	150
Magnesium	milligrams (mg)	400
Zinc	milligrams (mg)	15
Selenium	micrograms (µg)	70
Copper	milligrams (mg)	2.0
Manganese	milligrams (mg)	2.0
Chromium	micrograms (µg)	120
Molybdenum	micrograms (µg)	75
Chloride	milligrams (mg)	3400

Based on a 2000 calorie intake; for adults and children 4 or more years of age.
Reproduced from http://www23.netrition.com/rdi_page.html.

Thus, vegetarians should multiply the Reference Nutrients Intakes (recommended in the United Kingdom) by a factor of 1.1. These recommendations are listed in Tables 13-20 and 13-21. The latter table provides the protein content of foods that a vegan (diet devoid of any animal protein) might ingest in the diet.

Some proteins are incomplete, that is, they lack sufficient amounts of one or more amino acids. Grain cereals, for example, may be low in lysine, a deficiency that is usually satisfied by milk that is combined with it. Too much protein (relative to carbohydrates, for example), such as in the case of the Atkins diet, can stress the kidneys as a larger amount than usual of urea and ammonium ion is required to be cleared.

Table 13-20
The United Kingdom Reference Nutrient Intakes (RNI) for Protein

Type of Person	Amounts Required (g/day)
Infants/Children	
0–12 months	12.5–14.9
1–3 yrs	14.50
4–10 yrs	19.7–28.3
Boys	
11–14 yrs	42.1
15–18 yrs	55.2
Girls	
11–14 yrs	41.2
15–18 yrs	45
Men	
19–50 yrs	55.5
50+ yrs	53.3
Women	
19–50 yrs	45
50+ yrs	46.5
During pregnancy	extra 6g/day
Breast feeding 0–6 mths	extra 11g/day
Breast feeding 6+ mths	extra 8g/day

- The RNI is a daily amount that is enough or more than enough for 97% of people. The RNI is similar to the Recommended Daily Amount used previously in the UK.
The US Recommended Dietary Allowances introduced in 1989 are similar to the UK values.
Reproduced from The Vegan Society http://www.vegansociety.com/html/food/nutrition/protein.php

Table 13-21
Examples of Amounts of Foods Providing 10 g of Protein

Type of Food	Quantity Providing 10 g Protein (g)
Soya flour	24
Peanuts	39
Pumpkin seeds	41
Almonds	47
Brazil Nuts	50
Sunflower seeds	51
Sesame seeds	55
Hazel Nuts	71
Wholemeal bread	95
Whole lentils dried & boiled	114
Chickpeas dried & boiled	119
Kidney beans dried & boiled	119
Wholemeal spaghetti boiled	213
Brown rice boiled	385

Reproduced from The Vegan Society http://www.vegansociety.com/html/food/nutrition/protein.php

Herbs and Nutraceuticals

Asian practices of using herb remedies to cure diseases have been popular in the West. Many of these remedies are available in health food stores and from the Internet. The major problem with herbal preparations is that these remedies have not been proven by clinical testing. Also, there is a concern about reproducibility from one batch to another and about bacterial or other type of contamination. Until there are uniform controls on these products and clinical testing on significant numbers of patients, conclusions about the use of these products are uncertain.

Further Reading

Books

Gropper, S., *The Biochemistry of Human Nutrition: A Desk Reference*, 2nd edition, Thomson Press, 2000.

Mann, J., and Truswell, S. (eds.), *Essentials of Human Nutrition*, Oxford University Press, 2002.

Reviews

Ghishan, F.K., Said, H.M., Wilson, P.C., Murrell, J.E., and Greene, H.L., "Intestinal transport of zinc and folic acid: a mutual inhibitory effect," *Am. J. Clin. Nutr.*, **43**: 258–262, 1986.

Liu, Y.L., Ang, S.O., Weigent, D.A., Prchal, J.T., and Bloomer, J.R., "Regulation of ferrochelatase gene expression by hypoxia," *Life Sci.*, **10**: 2035–2043, 2004.

CHAPTER 14

Blood and Lymphatic System

Deep Vein Thrombosis: A Major Health Problem

Blood clotting is a normal mechanism in wound healing. However, blood clots, unrelated to wound healing, can occur that can be dangerous and life threatening. Adventitious (accidental) clotting can occur in older people. Sometimes spending long periods without moving around, such as sitting on a plane or in a car or lying in bed, is conducive to clot formation. Other causative factors may be oral contraceptives, surgery, childbirth, massive trauma, burns, and fractures of the hips or femur. Excessive clotting usually occurs in vessels that carry blood to the heart **(deep vein thrombosis, DVT)**. DVT occurs with a frequency of 1 in a 1000 people/year. In these cases, a clot (thrombus) occurs in a deep vein. The clot can break up at some point, and part of the clot can travel in the bloodstream and come to rest in the lungs. Pulmonary embolism produces symptoms such as cough, bloody saliva, shortness of breath (especially on exertion), rapid breathing, and rapid heart rate. Deep veins are those pictured in Figure 14-1.

Clots can occur in the legs, hips, or pelvis. When a clot occurs in the arm (subclavian vein), usually detected by Doppler ultrasound scan, physicians are prone to scrupulously examine for the presence of a cancer (magnetic resonance imaging scan, magnetic resonance angiography, and computerized axial tomography, or CAT, scan). The result of a clot in the leg, also detected by Doppler ultrasound scan, is a clot that can travel to the lung (pulmonary embolus) and interfere with the flow of blood into the lungs and/or block vessels in the heart, sometimes causing a heart attack. Occasionally, the clot can travel further to the brain and cause a stroke. The tendency to form blood clots, especially in older people, can be caused by hypercoagulability of the blood. Hypercoagulability can be derived from various factors. A natural protein called protein C that functions to inhibit clotting can be underexpressed. Protein S is protein cofactor of protein C, and a shortage of it may be involved. Antithrombin III could be limiting. The levels of these proteins in blood can be determined and when found to be low may be the cause of hypercoagulability. The presence of cancer is also a cause of adventitious blood clotting. Possibly the tumor interferes physically with the flow of the blood. Heparin (which has a short half-life) is injected until the correct level of blood thinners, such as **Coumadin**

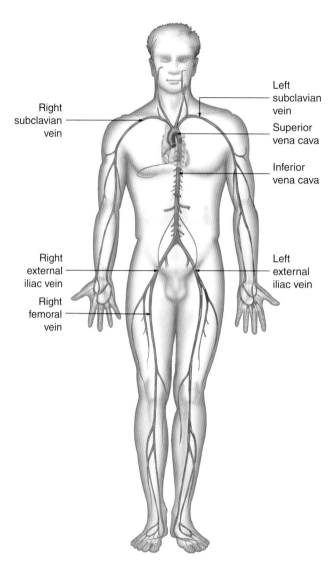

Figure 14-1. Deep veins. Redrawn from http://web.archive.org/web/20020815082456/http://www.nlm.nih.gov/medlineplus/ency/imagepage/8733htm (Oct. 21, 2002).

(**warfarin**) (Figure 13-109) is determined, and it is taken orally for a period or for life to prevent further deep vein thrombi. Taking an anticoagulant nearly doubles the clotting time, so excessive bleeding will always be a caution. With long-term anticoagulant therapy, there is about a 3% per year chance of having a major hemorrhage; 20% of these are fatal. A constantly surveilled dosage is followed that will generate an international normalized ratio of 2:3, based on prothrombin time determinations. This produces a relatively safe condition. If subsequent surgery or any procedure that will produce bleeding is required, the intake of Coumadin is stopped for 5 days before the procedure, enough time to clear the Coumadin from the blood. The prior intake of Coumadin is resumed after the procedure.

Some symptoms of DVT are pain, swelling, redness, and warmth of the affected area. In the groin, for example, the blood flow can be blocked and the leg may look like that in Figure 14-2. Figure 14-3 shows a deep

Figure 14-2. Red and swollen thigh resulting from a thrombus in deep veins of the groin (leofemoral levis), which prevents normal return of blood from the leg to the heart. Reproduced from http://www.nlm.nih.gov/medlineplus/ency/imagepages/2549.htm.

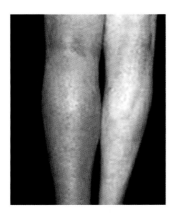

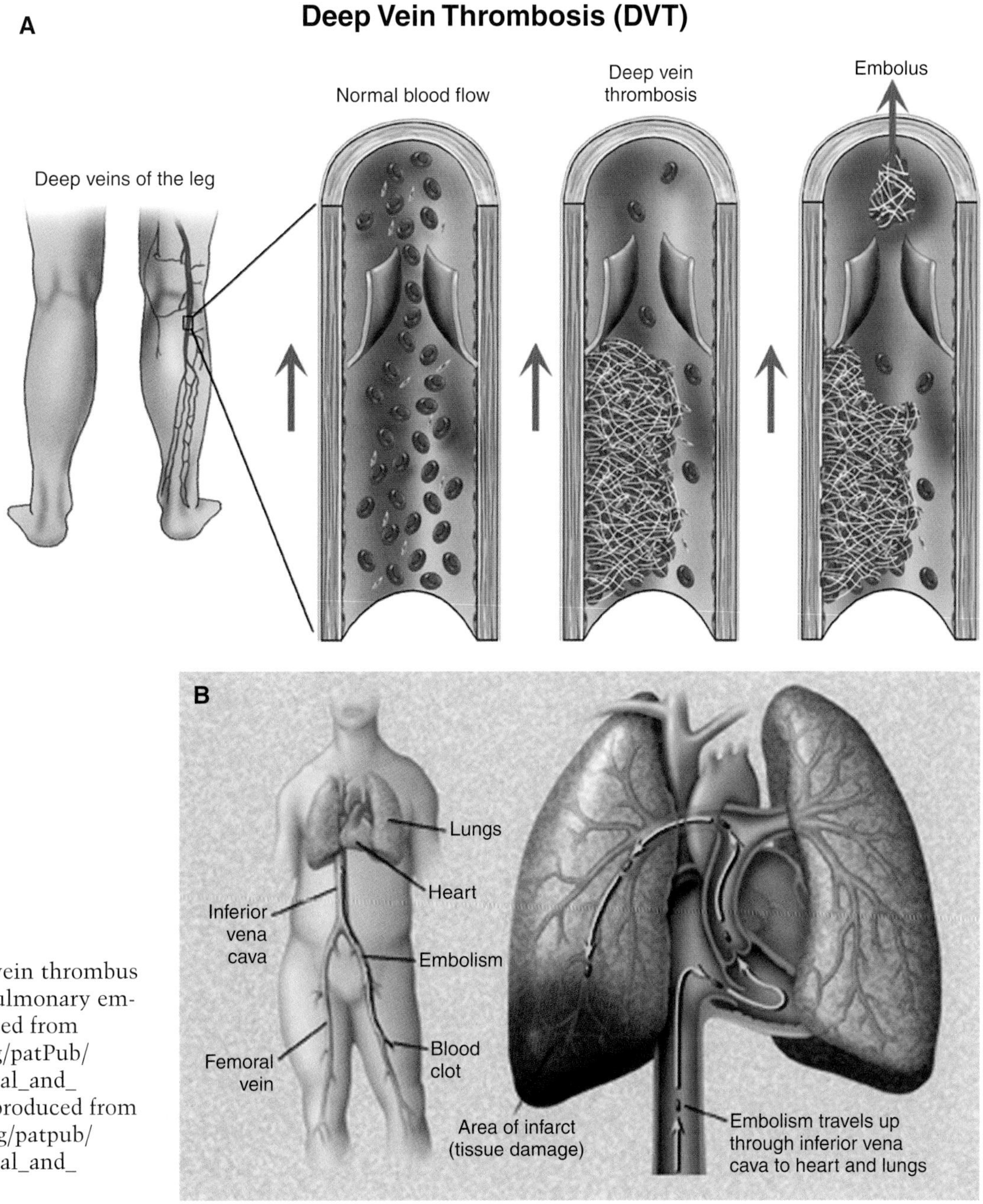

Figure 14-3. **A.** Deep vein thrombus in the leg. **B.** Site of pulmonary embolus. Part A reproduced from http://www.sirweb.org/patPub/dvtimages/DVT_normal_and_embolus.gif. Part B reproduced from http://www.sinweb.org/patpub/dvtimages/DVT_normal_and_embolus.gif.

vein thrombus in the leg and the position of a resulting pulmonary embolus. The frequency of hypercoagulation is about 2 million people per year globally.

As mentioned earlier, there are natural factors that prevent blood clotting. Protein C is a physiological anticoagulant, and protein C deficiency is an inherited trait (rare), or an acquired one, that predisposes to venous clots (and habitual abortion). The most common hereditary form is factor V Leiden. Acquired forms involve elevated factor VIII concentrations. Protein C functions to cleave and inactivate factor Va ("a" denotes the activated form) and factor VIIIa. The activation of protein C is accomplished by the activation of thrombomodulin by thrombin, which then activates protein C. Activated protein C (APC) then combines with its cofactor, protein S, and the complex binds to a platelet membrane that contains a receptor for APC. APC on the platelet membrane is then able to cleave and inactivate factor Va and factor VIIIa. Factor V Leiden is an inherited gene encoding a variant of factor V. This phenomenon is called **activated protein C resistance (APCR),** in which factor V Leiden is inactivated by APC at a subnormal rate. Expression of factor V Leiden makes blood more likely to clot (thrombophilia). *The activated altered factor V (Leiden) is resistant to inactivation by protein C.* At least 95% of patients with APCR express factor V Leiden. The clotting process in Figure 14-4 shows the function

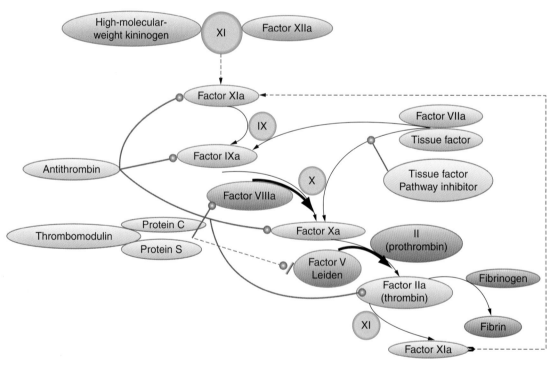

Figure 14-4. Clotting mechanism showing the consequence of expression of factor V Leiden. Factor V activates prothrombin to form activated factor II (thrombin) that forms a fibrin clot by activation of fibrinogen. Normally protein C and protein S complexed with thrombomodulin inhibit factor V but factor V Leiden is resistant to this inhibition. Consequently, factor V Leiden, when expressed, produces blood that is more prone to form clots.

Table 14-1
The Role of Factor V Leiden in Venous Thromboembolic Disease. Risk Is Shown Relative to a Normal Person Without Factor V Leiden

Thrombophilic Status	Relative Risk of Venous Thrombosis
Normal	1
OCP use	4
Factor V Leiden, heterozygous*	5–7
Factor V Leiden, heterozygous + OCP	30–35
Factor V Leiden, homozygous*	80
Factor V Leiden, homozygous + OCP	??? > 100
Prothrombin gene mutation, heterozygous	3
Prothrombin gene mutation, homozygous	??? possible risk of arterial thrombosis
Prothrombin gene mutation, heterozygous + OCP	16
Protein C deficiency, heterozygous	7
Protein C deficiency, homozygous	Severe thrombosis at birth
Protein S deficiency, heterozygous	6
Protein S deficiency, homozygous	Severe thrombosis at birth
Antithrombin deficiency, heterozygous	5
Antithrombin deficiency, homozygous	Thought to be lethal before birth
Hyperhomocysteinemia	2–4
Hyperhomocysteinemia combined with factor V Leiden, heterozygous	20

OCP, oral contraceptive pill.

*The terms heterozygous (*hetero* = different) and homozygous (*homo* = same) are terms used in genetics. The human genome contains two copies of the information. If the copies are the same, they are homozygous; if the copies are different, they are heterozygous. For example, take a protein called A. The normal genome would code for the protein as AA. This is homozygous for the normal protein. If there is a variation of the protein called a, there are two possible ways to get the a. The genome could be Aa, which is called heterozygous, or the genome could be aa, which is called homozygous.

Reproduced from http://web.archive.org/web20070425232308/http://www-admin.med.uiuc.edu/hematology/PtFacV2.htm.

of factor V and the consequences of expression of factor V Leiden (Table 14-1).

There are cases of hypercoagulation that are not related to deficiencies of protein C or protein S, to factor V Leiden, or even to other factors in the blood coagulation mechanism. Such situations are considered idiopathic (without a known cause); at least, the cause has not yet been discovered.

The gene encoding the message for protein C is located on chromosome 2 (2q13-q14). Protein C is a vitamin K–dependent serine protease that becomes activated by thrombin to APC. The primary structure of human protein C is shown in Figure 14-5, as well as its similarity with other serine proteases (Figure 14-5B). The crystal structures of the precursor (zymogen) of protein C and the activated form are shown in Figure 14-6. Figure 14-7 shows the **endothelial protein C receptor (EPCR)** bound with a portion of the protein C γ-carboxy glutamate (Gla) domain and a lipid molecule. The EPCR accelerates the thrombin- and thrombomodulin-dependent generation of APC. Protein C contains a phospholipid domain that is involved in the EPCR binding of protein C. EPCR binds to both protein C and APC with about the same affinity. It appears that the Gla domain of protein C is also key to its binding to the EPCR. A surface representation of the EPCR is shown in Figure 14-8. The functions of APC are shown in Figure 14-9. Besides inactivating factors Va and VIIIa, APC (with thrombomodulin)

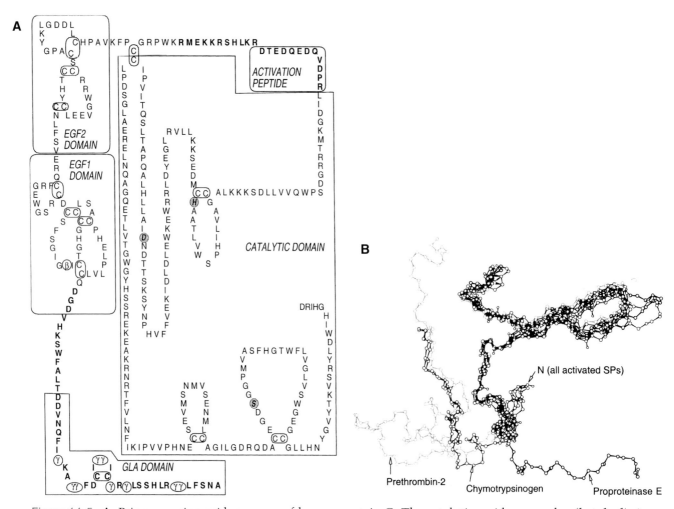

Figure 14-5. **A.** Primary amino acid sequence of human protein C. The catalytic residues are also *(hatched)* circled. Boldfaced letters indicate the residues whose positions were initially modeled. γ, γ-carboxyglutamic acid residues; β, β-hydroxyaspartic acid. **B.** Similarity of protein C to other serine proteases. The C_α alignment of the N-terminal residues of activated serine proteases: chymotrypsin, α-thrombin, bovine pancreatic elastase, and protein C. All N-termini are well superimposed for the preceding active serine proteases (SPs); N labels the N-terminal C_α of the active forms. The C_α values of zymogens (prethombin-2, chymotrypsinogen, and proproteinase E) are also shown here. Alignment uses residues 20–27(cn) for all proteins. In chymotrypsinogen and prethrombin-2, the residues corresponding to the activated protein of protein C are found to the left in the figure; the corresponding residues of proproteinase E exhibit a rotation of 120 degrees to the right. Parts A and B are reproduced from Figures 1 and 3, respectively, from L. Perera et al., "Modeling zymogen protein C," *Biopys. J.,* 79: 2925–2943, 2000 and http://www.biophysj.org/content/vol79/issue6/images/large/bj1204656001.jpeg.

inhibits inflammation. An idea of how these molecules are oriented on the vascular endothelial cell membrane is shown in Figure 14-10.

Protein S is a cofactor for protein C and binds to another binding protein, C4b-binding protein. The schematic structure and the three-dimensional structure of protein S are shown in Figure 14-11.

Factor Va is a cofactor in the prothrombinase complex that produces a huge increase in the rate of thrombin generation to form clots. A recently determined structure of a fragment of factor Va is shown in Figure 14-12.

(Text continues on p. 864.)

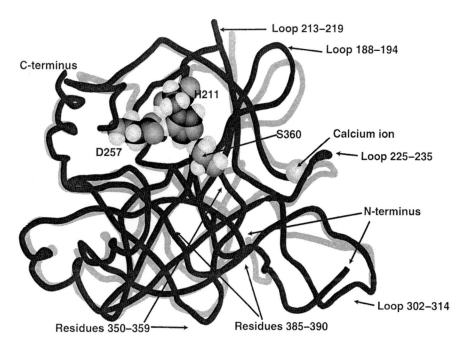

Figure 14-6. Comparison of the crystal structures of precursor of protein C (zymogen) and the activated protein C (APC). The catalytic domains of protein C in the zymogen model *(dark line)* and the active x-ray crystal structure *(light line)*. Catalytic residues in the zymogen model structure are rendered. Also, loops with observed movements (APC → protein C) in the backbone structure are labeled. Reproduced from Figure 11 of L. Perera et al., "Modeling zymogen protein C," *Biopys. J.*, 79: 2925–2943, 2000 and http://www.biophysj.org/content/vol79/issue6/images/large/bj1204656011.jpeg.

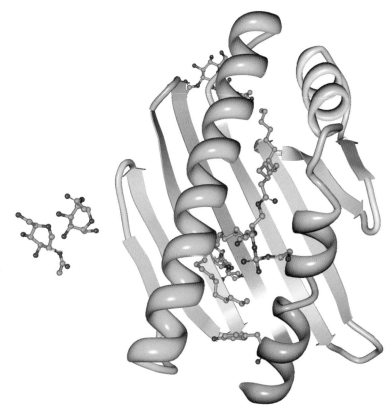

Figure 14-7. The recombinant soluble endothelial cell protein C receptor (EPCR) molecule with a portion of the protein C Gla domain and a lipid molecule. In EPCR *(yellow ribbon)*, two α-helices and an eight-stranded β-sheet create a groove that is filled with phospholipid (the space-filling *balls* in the *center*). Binding of Ca^{2+} ions *(magenta spheres)* to the protein C Gla domain *(green ribbon)* exposes the N-terminal ω loop, which in the absence of EPCR interacts with the phospholipid surfaces on the membrane. There do not appear to be direct interactions between the protein C Gla domain and the lipid molecule located in the groove of recombinant soluble EPCR. The model of the complex consists of residues 7–177 of recombinant soluble EPCR and the first 33 residues of the protein C Gla domain. Reproduced from Figure 1 of V. Oganesyan et al., "The crystal structure of the endothelial protein C receptor and a bound phospholipid," *J. Biol. Chem.*, 277: 24851–24854, 2002.

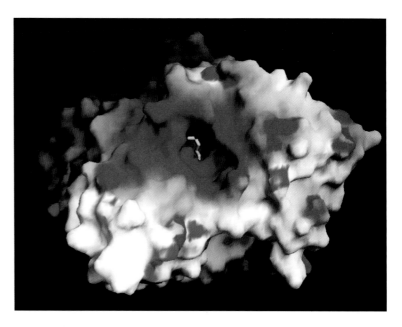

Figure 14-8. Surface representation of the recombinant soluble endothelial cell protein C receptor (EPCR) molecule. Electrostatic potentials are mapped on the surface. The head group of the lipid is solvent-exposed *(yellow stick model in the center)*, whereas the fatty acid chains are buried deep in the hydrophobic groove (and therefore unseen). Reproduced from Figure 3 of V. Oganesyan et al., "The crystal structure of the endothelial protein C receptor and a bound phospholipid," *J. Biol. Chem.*, 277: 24851–24854, 2002 and http://www.jbc.org/cgi/content-nw/full/277/28/24851/F3.

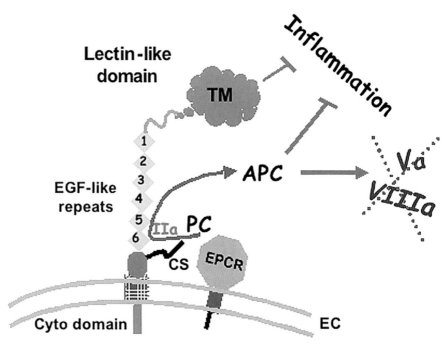

Figure 14-9. Actions of activated protein C (APC). Structure of thrombomodulin (TM) and activation of protein C (PC). TM is depicted with its five structural domains, including the cytoplasmic (cyto) and transmembrane domains, a serine/threonine-rich region with an attached chondroitin sulfate (CS) moiety, six EGF-like repeats, and the N-terminal lectin-like domain. EGF-like repeats 4 to 6 of TM provide cofactor function for thrombin (IIa)-mediated activation of PC, a step that is further amplified by the endothelial cell protein C receptor (EPCR). APC cleaves coagulation cofactors Va and VIIIa, thereby downregulating thrombin generation, and directly interferes with inflammation. The lectin-like domain of TM also suppresses inflammation. *EC*, vascular endothelial cell; *EGF*, epidermal growth factor. Reproduced from M. Van de Wouwer, D. Collen, and E.M. Conway, "Thrombomodulin-protein C-EPSCR system," *Arterioscler. Thromb. Vasc. Biol.*, 24: 1374, 2004 and http://atvb.ahajournals.org/content/vol24/issue8/images/large/10FF1.jpeg.

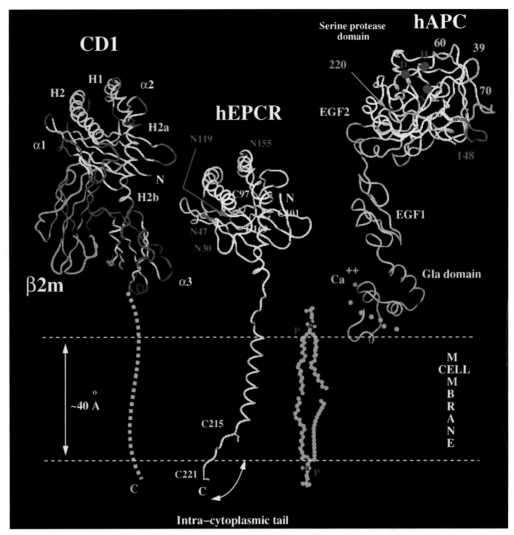

Figure 14-10. Concept of organization of CD1, the human endothelial protein C receptor and human activated protein C on the vascular endothelial cell. Ribbon diagram of mouse CD1, human EPCR, and human APC. The x-ray structures of mouse CD1 and β2m are shown *(left)*, with the secondary structure elements colored. The CD1 helices are in *yellow*, and the strands are *white* (*orange* in β2m). The structure of the CD1 transmembrane segment is not known and is shown here as a dashed line *(blue)*. The human endothelial cell protein C receptor (EPCR) model is presented in the middle with the Cys residues involved in a disulphide bond *yellow* (C101–C169). The free Cys (C97) is buried and is part of a β-strand. The N- and C-terminal residues in CD1 and EPCR are noted N and C, respectively. Potentially glycosylated Asn residues in EPCR are shown in *magenta*. An average conformation is presented for the last few C-terminal residues of EPCR, but the *white arrow* illustrates that several possibilities can be considered. The conformation of this region depends on the environment. The model of human full-length APC is presented *(right)* to show the relative size of the molecules. The Gla domain and the two EGF-like modules are in *yellow*, with the seven calcium ions shown as *blue spheres*. The serine protease (SP) domain is in *white*; some key surface loops are labeled and colored for orientation. The catalytic triad residues are in *red (filled circles)*, and from *left to right*, they correspond to Asp102, His57, and Ser195. The 60 and 39 areas of APC are rich in positively charged residues. Loop 70 is involved in calcium binding, whereas loop 220 is proposed to bind sodium ions. The numbering for SP domain residues and loops follows the chymotrypsinogen nomenclature. The overall orientation of these molecules with respect to the membrane plane has not been experimentally proven yet but is reasonable. A virtual membrane with a width of about 40 Å is displayed. Two phospholipid molecules were extracted from a lipid bilayer to help the reading of the figure. Phosphorus atoms are in *red*. Reproduced from B.O. Villoutreix, A.M. Blom, and B. Dahlbäck, "Structural prediction and analysis of endothelial cell protein C/activated protein C receptor," *Protein Eng.*, 12: 833–840, 1999 and http://peds.oxfordjournals.org/content/vol12/issue10/images/large/p1013.f2.jpeg.

A Vitamin K–dependent protein S

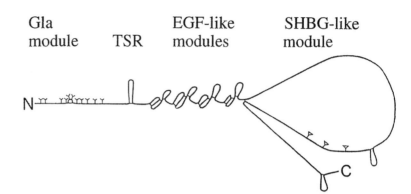

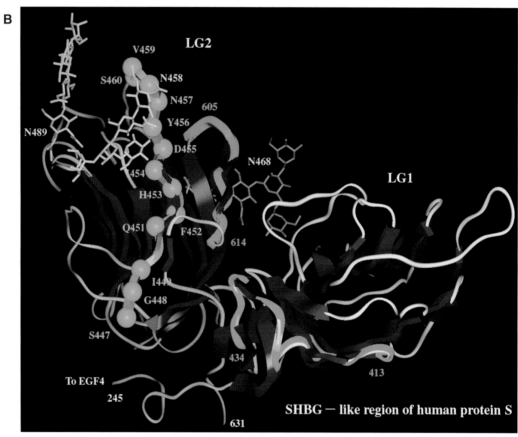

Figure 14-11. **A.** Schematic picture showing the modular arrangement of protein S. Protein S contains the N-terminal Gla-module, which is rich in γ-carboxyglutamic acid residues, the thrombin-sensitive disulfide loop, four EGF-like modules with high-affinity calcium-binding sites, and the SHBG-like domain, which is homologous to SHBG. **B.** Proposed three-dimensional structure for the SHBG-like region of protein S. The two LG modules are shown with their β-strands drawn as an *arrow*, whereas the remaining loops and short helices are presented as *tubes*. The glycosylated Asn residues are *yellow*, and a common sugar motif was grafted onto the structure to provide approximate information about the overall surface that could be covered by such side chains. Two peptide segments suggested to represent a binding region for C4BP are shown as *green ribbon*. The segment investigated in the present study in presented as a *blue ribbon*, with the C_α atom of all residues mutated to Ala depicted by a *blue sphere*. SHBG, sex hormone–binding globulin. Part A reproduced from Figure 1 of S. Linse et al., "A region of vitamin K–dependent protein S that binds to C4b binding protein (C4BP) identified using bacteriophage peptide display libraries," *J. Biol. Chem.*, 272: 14658–14665, 1997. © 1997 by The American Society for Biochemistry and Molecular Biology, Inc. http://www.jbc.org/content/vol272/issue23/images/large/bc8825001.jpeg. Part B reproduced from Figure 7 of T.K. Giri et al., "Structural requirements of anticoagulant protein S for its binding to the complement regulator C4b-binding protein," *J. Biol. Chem.*, 277: 15099–15106, 2002.

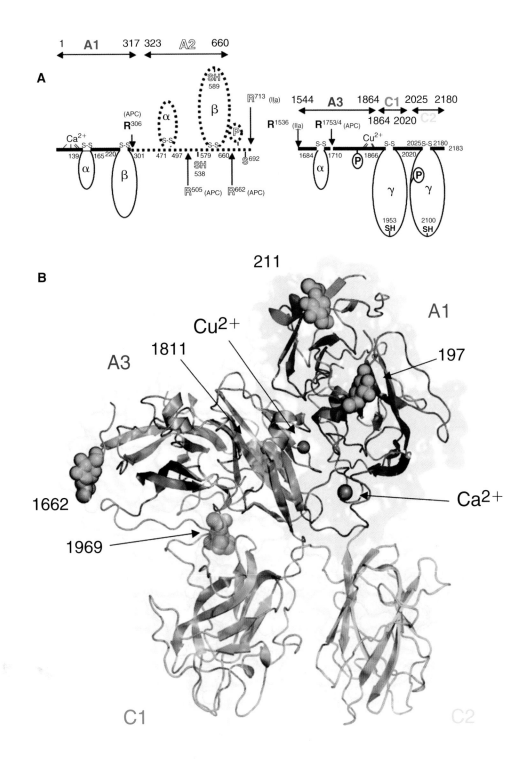

Figure 14-12. The structure of bovine factor Va$_i$. **A.** Schematic drawing of the structure of bovine factor Va. The extent and names of the five domains, metal-binding sites, and phosphorylation sites are indicated. *Dashed lines* and *outlined fonts* depict the A2 domain that is removed in factor Va$_i$. **B.** Ribbon diagram of bovine factor Va$_i$ indicating the positions of the carbohydrates *(orange)* and the metals (Ca^{2+}, *gray*; Cu^{2+}, *pink*). A van der Waals surface representation is shown in the background. Domains are color coded throughout all figures as follows: *A1*, red; *A3*, blue; *C1*, green; and *C2*, yellow. Reproduced from Figure 1 of T.E. Adams et al., "The crystal structure of activated protein C–inactivated bovine factor Va: implications for cofactor function," *PNAS USA*, 101: 8918–8923, 2004.

The Blood-Clotting Mechanism

The blood-clotting mechanism is a cascade of various proteins, most of which are proteases, that activate precursor proteins by proteolysis. The activated protein (serine protease) activates the next protein in the cascade, and so on, until fibrinogen becomes activated to fibrin (Figure 14-13) catalyzed by thrombin (Figure 14-14). Fibrin molecules

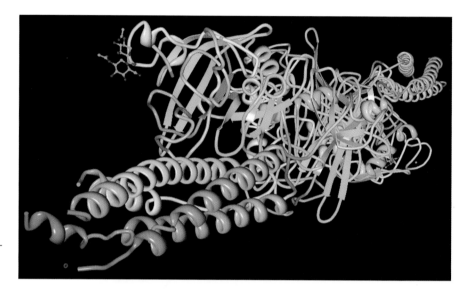

Figure 14-13. Crystal structure of fragment double-D from human fibrin with two different bound ligands. Reproduced from PDB ID: 1FZC. S.J. Everse et al. Crystal Structure of Fragment Double-D from Human Fibrin with Two Different Bound Ligands. *Biochemistry* **37** pp. 8637 (1998).

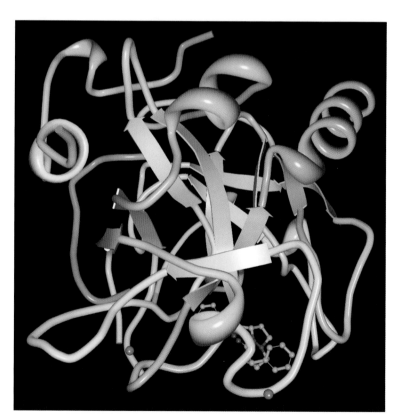

Figure 14-14. Human thrombin complexed with novel synthetic peptide mimetic inhibitor and hirugen complex (serine protease/inhibitor). Reproduced from PDB ID: 1a5g. R. St. Charles et al. Bound Structures of Novel P3-P1' Beta-Strand Mimetic Inhibitors of Thrombin. *Journal of Medicinal Chemistry* **42** pp. 1376 (1999).

are long and stringy and form the clot (Figure 14-15). Many proteins in the cascade are named for investigators or were named by original investigators to produce what appears to be a complicated scheme, but the cascade is straightforward. For example, factor XII is named the Hageman factor, factor IX is called the Christmas factor, and factor X is called the Stuart factor. The cascade is shown in Figure 14-16. Two events can signal the clotting pathway. In cases such as a deep vein thrombus from an undetermined cause inside the body, the intrinsic pathway is signaled. Events from the outside, such as trauma, trigger the extrinsic pathway; these pathways are shown in Figure 14-17.

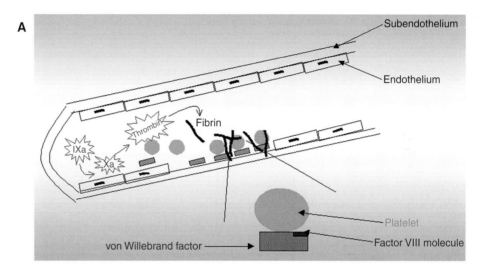

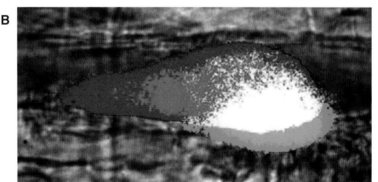

Figure 14-15. **A.** Formation of a blood clot in a blood vessel. Fibrin forms on the surface of platelets that are attached to the von Willebrand factor complexed with factor VIII attached to the blood vessel wall. **B.** Formation of a blood clot in the small blood vessel of an experimental animal. A laser-induced injury of a normal vessel wall initiates the clot. Blood flow is from right to left. *Red label* is platelets, and *green label* is tissue factor; these appear first. *Blue labeled* fibrin form and fill in the bulk of the thrombus. Tissue factor shows up in *turquoise, white,* and *yellow;* fibrin shows up in *turquoise, white,* and *magenta;* platelets show up in *yellow, white,* and *magenta.* Part A reproduced from http://medinfo.ufl.edu/year2/coag/bigpic.html. Part B reproduced from http://focus.hms.harvard.edu/2002/Nov8_2002/hematology.html.

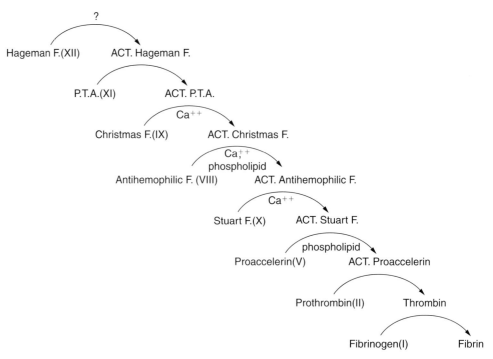

Figure 14-16. The blood-clotting cascade. The question mark at the top of the cascade indicates an unknown for the clotting mechanism to begin. *ACT*, activated; *F*, factor; *P.T.A.*, predicted transmitting ability. Reproduced from Figure 5 of E.W. Davie, "A brief historical review of the waterfall/cascade of blood coagulation," *J. Biol. Chem.*, 278: 50819–50832, 2003.

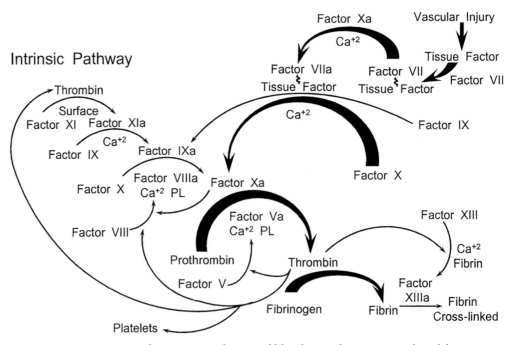

Figure 14-17. Intrinsic and extrinsic pathways of blood coagulation. Reproduced from http://www.jbc.org/content/vol278/issue51/images/large/bc5139454005.jpeg.

Fibrinogen, at the bottom of the pathway, is soluble and comprises about 3% of the blood proteins. In the center of the molecule, it has a sticky region covered by negatively charged short amino acid chains. To form a clot, thrombin (a serine protease activated from prothrombin by factor X) hydrolyzes off the short, charged amino acid chains, forming fibrin and exposing the sticky region. Molecules of fibrin start to stick together to form a clot. One reason clots do not normally form is the production of α1-antitrypsin, a serine protease inhibitor (Figure 14-18) produced by bodily tissues, and of another inhibitor called antithrombin that blocks thrombin. A plasma protein, plasminogen, when activated, has clot-dissolving action. The structures of

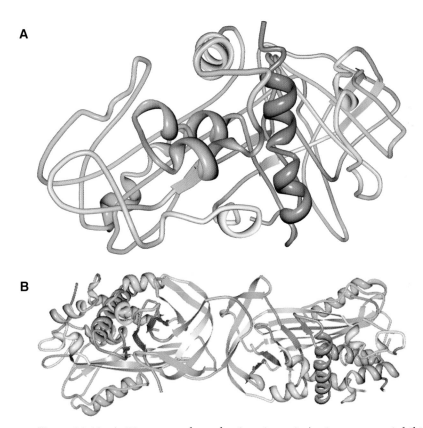

Figure 14-18. **A.** Human uncleaved α-1-antitrypsin (serine protease inhibitor). **B.** Structure of human antithrombin III. **C.** Crystal structure of human plasminogen catalytic domain. Part A reproduced from PDB ID: 1ATU.S.E. Ryu, H.J. Choi, K.S. Kwon et al. The Native Strains in the Hydrophobic Core and Flexible Reactive Loop of a Serine Protease Inhibitor: Crystal Structure of an Uncleaved α1-antitrypsin at 2. 7 A. *Structure* **4** pp. 1181 (1996). Part B reproduced from PDBsum ID: 1ATH. H.A. Schreuder, B. De Boer, R. Dijkema, et al. The Intact and Cleaved Human Antithrombin Iii Complex as a Model for Serpin-Proteinase Interactions. *Nature Structural and Molecular Biology* **1** pp. 48 (1994).

Figure 14-18, cont'd Part C reproduced from PDBsum ID: 1DDJ. X. Wang et al. Human Plasminogen Catalytic Domain Undergoes an Unusual Conformational Change upon Activation. *Journal of Molecular Biology* **295** pp. 903 (2000).

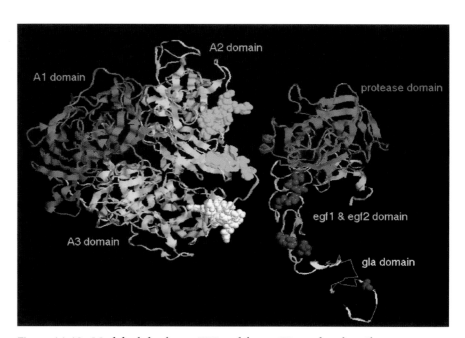

Figure 14-19. Model of the factor VIII and factor IXa molecules. Shown are representations of porcine factor IXa (Protein Data Bank accession code) and the triplicated A-domains of human factor VIII (hemophilia A web site, http://europium.mrc.rpms.ac.uk), which are derived from crystallography and homology modeling, respectively. Factor IXa binding region in the factor VIII A3 domain (residues 1811–1818) is shown in *white*, whereas the binding regions in the A2 domain (residues 558–565 and 698–710) are shown in *dark* and *light blue*, respectively (space-filling representations). These sites are in close vicinity and are exposed at the same side of the molecule. The factor VIII A2 domain is required to induce significant changes within the factor IXa protease domain, indicating that it binds to the factor IXa protease domain. The A3 domain of factor VIII has been proposed to interact with the factor IXa light chain. Within the factor IXa light chain, residues 12, 64, 69, 78, 92, and 94 are indicated (*red*, space-filling representation). These residues have been reported to be associated with an abnormal response to factor VIIIa in factor X activation. Reproduced from Figure 4 of P. S. Lenting, J.A. van Mourik, and K. Merteus, "The life cycle of coagulation factor VIII in view of its structure and function," *Blood*, 92: 3983–3996, 1998.

human α1-antitrypsin, antithrombin III, and the catalytic domain of human plasminogen (microplasmin) are shown in Figure 14-18.

Hemophilia is a familial bleeding tendency that affects 1 in 5000 males and is associated with a deficiency of factor VIII (antihemophilic factor). Factor VIII is the cofactor of activated factor IX in the factor X–activating complex (Figure 14-19) of the intrinsic pathway of coagulation. The formation, life cycle process, and degradation of factor VIII are shown in Figure 14-20.

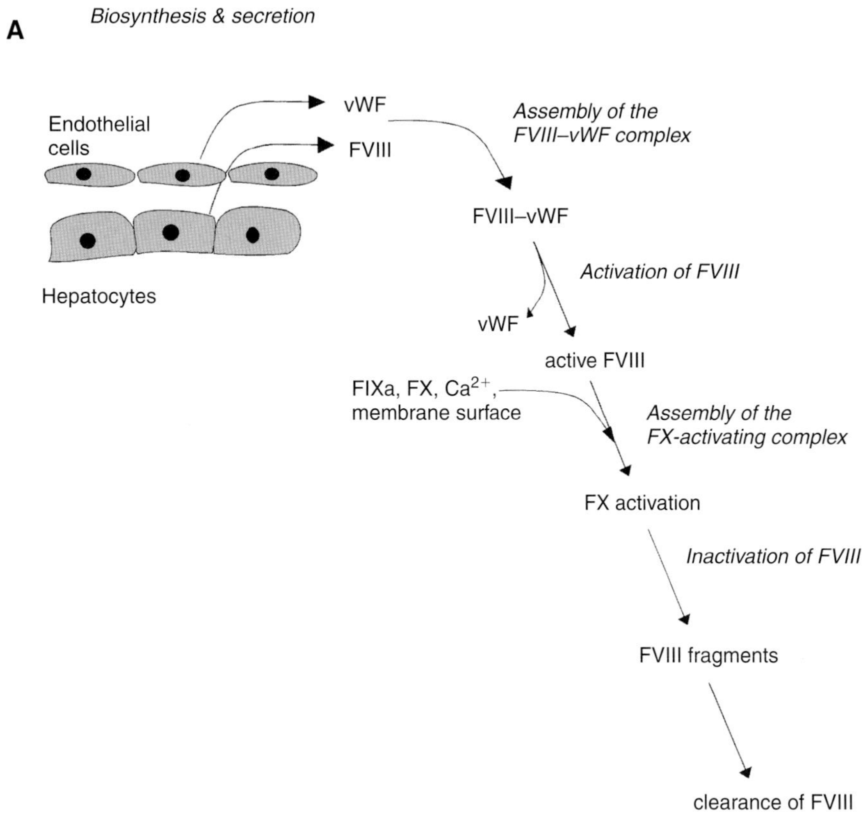

A *Biosynthesis & secretion*

```
Endothelial cells ──▶ vWF
              ──▶ FVIII   Assembly of the
Hepatocytes                FVIII–vWF complex
                              │
                              ▼
                          FVIII–vWF
                              │  Activation of FVIII
                              ▼
                    vWF ◀── active FVIII
          FIXa, FX, Ca²⁺,        │
          membrane surface       │  Assembly of the
                                 ▼  FX-activating complex
                            FX activation
                                 │  Inactivation of FVIII
                                 ▼
                           FVIII fragments
                                 │
                                 ▼
                           clearance of FVIII
```

Figure 14-20. **A.** The life span of factor VIII. Factor VIII is synthesized by various tissues, including liver, kidney, and spleen, as an inactive single-chain protein. After extensive posttranslational processing, factor VIII is released into the circulation as a set of heterodimeric proteins. This heterogenous population of factor VIII molecules readily interacts with vWF, which is produced and secreted by vascular endothelial cells. Upon triggering of the coagulation cascade and subsequent generation of serine proteases, factor VIII is subject to multiple proteolytic cleavages. These cleavages are associated with dramatic changes of the molecular properties of factor VIII, including dissociation of vWF and development of biological activity. After conversion into its active conformation and participation in the factor X–activating complex, activated factor VIII rapidly loses its activity. This process is governed by both enzymatic degradation and subunit dissociation.

Continued

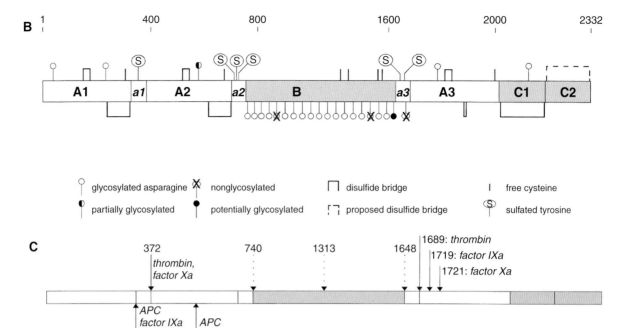

Figure 14-20, cont'd **B** and **C**. The factor VIII protein. Mature factor VIII consists of 2332 amino acids, which are arranged in a discrete domain structure: *A1* (residues 1–336), *A2* (373–710), *B* (741–1648), *A3* (1690–2019), *C1* (2020–2172), and *C2* (2173–2332). The A domains are bordered by acidic regions *a1* (337–372), *a2* (711–740), and *a3* (1649–1689). Disulfide bridges: Using B-domainless factor VIII, seven disulfide bonds have been identified: residues 153 and 179, 248 and 329 (A1 domain), 528 and 554, 630 and 711 (A2 domain), 1832 and 1858, 1899 and 1903 (A3 domain), and 2021 and 2169 (C1 domain). Within the C2 domain, residues 2174 and 2326 most likely also form a disulfide bridge. Free cysteine-residues have been identified at positions 310, 692, and 2000. Cys and Cys may be present as free cysteines, because these residues are reactive toward a sulfhydryl-specific fluorphor. With regard to the Cys residues in the B-domain, it is unknown whether they are free or linked. *N*-Linked glycosylation: Factor VIII contains 25 consensus sequences (Asn-Xxx-Thr/Ser) that allow N-linked glycosylation. Using either full-length or B-domainless factor VIII, most of these sites have been shown to be glycosylated: residues 42 and 239 (A1 domain); residues 757, 784, 828, 900, 963, 1001, 1005, 1055, 1066, 1185, 1255, 1259, 1282, 1300, 1412, and 1442 (B domain); residue 1810 (A3 domain); and residue 2118 (C2 domain). Nonglycosylated residues are present at positions 943 and 1384 (B domain) and at position 1685 (*a3* acidic region). Residue 582 (A2 domain) has been reported to be nonglycosylated in two studies, whereas one study reported this residue to be partially glycosylated. Finally, it remains to be investigated whether residue 1512 (B domain) is glycosylated. Tyrosine sulfation: The acidic regions contain consensus sequences that allow sulfation of Tyr residues at positions 346 (*a1* region); 718, 719, 723 (*a2* region); and 1664 and 1680 (*a3* region). Analysis using recombinant proteins established that all sites indeed can be sulfated. **C.** Limited proteolysis of factor VIII. The major part of factor VIII circulates as a set of heterogenous dimers, consisting of a light (*a3*-A3-C1-C2) and a heavy (A1-*a1*-A2-*a2*-B) chain. The heavy chain is variably sized because of limited proteolysis within the B domain. Some of these cleavages may occur intracellularly at positions 1313 and 1648 *(open downward arrows)*. Factor VIII can be converted into its active form by proteolysis in both the heavy and the light chains by various serine proteases *(closed downward arrows)*, including thrombin and factor Xa. Because proteolysis by factor Xa but not thrombin is inhibited by vWF, thrombin is probably the physiological activator of factor VIII. Proteolytic degradation of factor VIIIa proceeds through cleavages within the A1 and A2 domains by various serine proteases *(upward arrows)* and results in release of the *a1* acidic region and bisecting of the A2 domain. In contrast to what has previously been assumed, cleavages within the light chain by factor IXa or factor Xa do not result in inactivation of factor VIII but contribute to the development of factor VIII cofactor activity. *vWF*, von Willebrand factor; *F*, factor. Part A reproduced from P.J. Lenting, J.A. van Mourik, and K. Mertens, "The life cycle of coagulation factor VIII in view of its structure and function," *Blood*, 92: 3983–3996, 1998. Part B redrawn from http://www.bloodjournal.org/content/vol92/issue11/images/large/blod42351002y.jpeg. Part C redrawn from http://www.bloodjournal.org/cgi/content–nw/full/92/11/3983/F3.

Blood

The vascular circulatory system is responsible for transporting oxygen to the tissues and carbon dioxide from the tissues to be expired. It also carries hormones, enzymes, various nutrients, plasma proteins, and blood cells. In addition, it has a pH **buffering** system to maintain the critical pH range of blood in the range of 6.8 to 7.4. It carries water and the toxin urea for clearance. Circulating nutrients are mainly amino acids, glucose, vitamins, minerals, fatty acids, and glycerol. The volume of blood is about 8 pints, which circulates through the kidneys (to remove toxins to the urine) 36 times per day. Blood cells occupy 45% of the blood volume, of which 99% are erythrocytes and 1% consists of leukocytes and platelets. The other 55% is plasma that is 92% water and contains sodium, chlorine, potassium, manganese, and calcium ions; the blood plasma proteins; albumin; globulins; and fibrinogen—in addition to hormones. A diagram of the human systemic circulation is shown in Figure 14-21. The lungs oxygenate the blood, and the oxygenated blood enters the left auricle or atrium through the pulmonary vein. It is pumped to the left ventricle of the heart (arrow) and then out through the aorta (reverse arrow), the major artery from the heart, to the liver by way of the hepatic artery, small intestine (mesenteric artery), kidneys (renal artery), and the legs (iliac artery). Toward the upper body, oxygenated blood is carried to the arms (subclavian artery) and the head (carotid artery). After oxygenating the tissues, the deoxygenated blood is carried back to the right ventricle of the heart from the legs (iliac vein), kidneys (renal vein), and liver (hepatic portal vein from the small intestine and then the hepatic vein from the liver). These sources connect with the inferior vena cava into the right ventricle of the heart. From the upper body, the jugular vein carries deoxygenated blood from the head and the subclavian vein from the arms flowing into the superior vena cava that returns the blood to the right ventricle of the heart. Deoxygenated blood is then pumped out of the heart into the lungs through the pulmonary artery, and the newly oxygenated blood enters the aorta and recycles again.

Transport of Oxygen

The transport of oxygen in the blood is accomplished by hemoglobin. Oxygen is taken in through the lungs, and carbon dioxide is exhaled through the lungs. Hemoglobin is contained in the red blood cells (erythrocytes). Myoglobin is the counterpart of hemoglobin. Myoglobin is the small globular protein in muscle tissue that binds oxygen. Both myoglobin (Figure 13-6) and hemoglobin (Figure 13-21) bind oxygen, but the kinetics of oxygen binding differs with the two proteins (Figure 14-22). Myoglobin is a monomer and binds oxygen linearly, whereas hemoglobin is a tetramer and displays allosteric kinetics (Figure 3-17), in which an initial lag occurs as the oxygen concentration increases. When about 10 torr (10 mm of mercury, or mmHg) of oxygen pressure is reached, the rate of binding of oxygen increases to saturation. To bind oxygen, the heme iron of hemoglobin must be in the ferrous state (Fe^{2+}) and oxidation of iron to the ferric state (Fe^{3+}) will not allow oxygen binding to hemoglobin. Although the exact mechanism of the allosteric function of oxygen binding to hemoglobin is not known, one hypothesis states that when

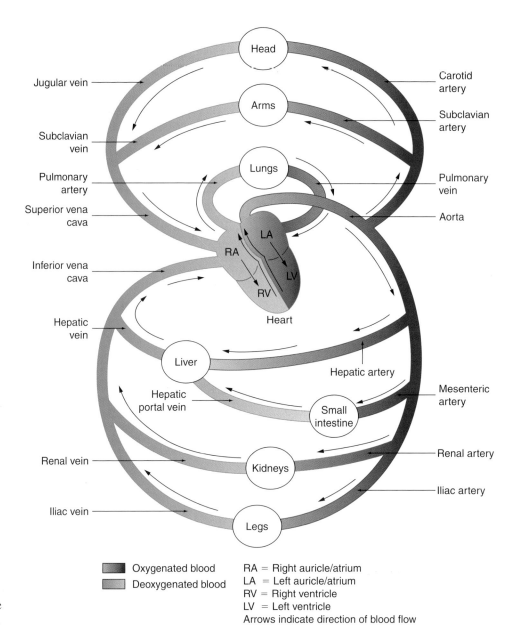

Figure 14-21. Diagram of the human systemic circulation.

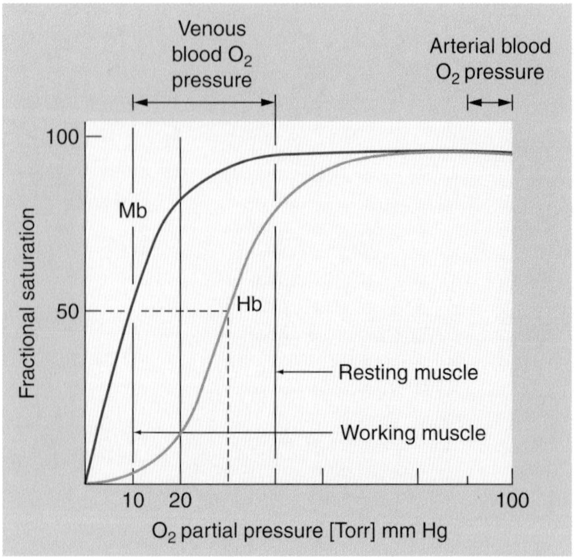

Figure 14-22. Differential binding characteristics between myoglobin and hemoglobin on binding oxygen. The different behavior of oxygen binding of myoglobin and hemoglobin can be summarized as follows. Myoglobin binds O_2 under conditions in which hemoglobin releases it. This is less than 20 Torr in muscle tissue. At this pressure, hemoglobin releases almost all of its oxygen, and myoglobin binds the freed oxygen at over 90%. The hyperbolic curve of Mb binding is typical for noncooperative processes, whereas the sigmoidal curve of Hb binding is typical for cooperativity. *Hb,* hemoglobin; *Mb,* myoglobin.

oxygen binds to unliganded (where oxygen is the ligand) hemoglobin (deoxyhemoglobin), the first molecule of oxygen binds with a relatively low affinity but the conformation of the binding subunit is changed. When the second molecule of oxygen binds, it binds with a higher affinity accompanied by a conformational change in the second subunit, and so on, until the four subunits are saturated with oxygen. This idea is diagrammed in Figure 14-23B. Figure 14-23A is a hypothesis positing that there are two forms of hemoglobin: one form is a low-affinity binder of oxygen, and the other form (altered conformation) is a high-affinity binder of oxygen; thus, oxygen will bind with high affinity to all four

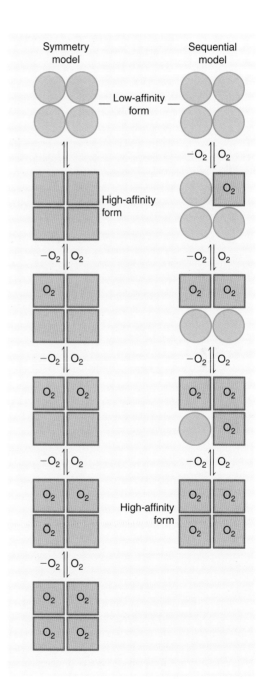

Figure 14-23. Models to explain the molecular mechanism of positive cooperativity when hemoglobin loads oxygen. In these models, hemoglobin exits in two forms: low affinity and high affinity. However, the difference is that in the Symmetry model, all subunits of hemoglobin exist in either the low-affinity or high-affinity form. In the Sequential model, the binding of an oxygen molecule leads to the conversion of that subunit to the high-affinity form, and so as each molecule of oxygen binds, it converts its subunit to the high-affinity form.

subunits. Presumably the two conformations would be in some sort of equilibrium. It also would be possible to combine the two hypotheses, whereby the binding of the first oxygen would convert the tetramer to the high-affinity form and all subsequent oxygens would bind with high affinity. The structural differences between fully oxygenated (R form, for "relaxed") and deoxygenated (T form, for "taut") hemoglobin have been solved (Figure 14-24), even though the precise intervening steps in the mechanism are unknown. Red blood cells, containing hemoglobin, transport oxygen to the tissue cells, where the oxygen is released. Oxyhemoglobin (R form) forms when oxygen concentrations are high, but at low oxygen concentrations, oxyhemoglobin dissociates into deoxyhemoglobin and oxygen.

When the level of glucose in the blood is high, for example, in uncontrolled diabetes, hemoglobin can be glycated (glucose bound to hemoglobin;

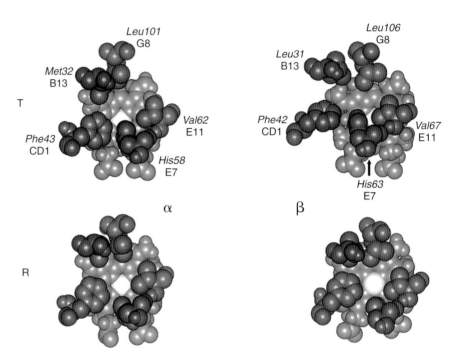

Figure 14-24. T is the deoxy state of hemoglobin; R is oxygenated hemoglobin. In the absence of the O_2 ligand, the T state has a tendency for dimer rotation, but steric hindrance of the joint region hinders dimer rotation. Binding of ligand to the α-subunit prevents this steric hindrance because of the coupling between this region and the α-proximal histidine, and the dimer rotation is completed. Near the end of this quaternary transition, the switch region adopts the R conformation, resulting in a shift of the β-proximal histidine. The β-heme slides, as a consequence, which opens the β-heme's distal side, increasing the accessibility of the Fe atom and thereby the affinity of the protein (for O_2). The distal side *(black)* of the α- and β- subunits in the T and R conformations. The heme is in *gray*, and the iron atom in *white*. The five residues (B13, CD1, E7, E11, and G8) that are in close contact with the heme on the distal side are shown as *hard spheres*. It can be seen that, in the T state, the Fe atom is more accessible to ligand in the α-chains than in the β-chains. Reproduced from Figure 14 of L. Mouawad et al., "New insights into the allosteric mechanism of human hemoglobin from molecular dynamics simulations," *Biophys. J.*, 82: 3224–3245, 2002 and http://www.biophysj.org/cgi/content–nw/full/82/6/3224/F14.

glycosylated); that is, glucose residues bind to amino acid residues on hemoglobin. Elevated levels of **glycated hemoglobin (HbA$_{1c}$)** are diagnostic for diabetes; levels above 10% of hemoglobin (HbA0) that are glycated indicate poor metabolic control of carbohydrates. The level of glycated hemoglobin can be used to monitor treatment. When blood glucose is high, not only hemoglobin but also other proteins (including insulin) can be glycated. High levels of HbA$_{1c}$ also can indicate kidney problems. The sugar moiety of the β1-subunit of glycated hemoglobin is directed toward the central cavity (Figure 14-25), and the sugar moiety of the β2-subunit enters the polypeptide chain. In the test tube, hemoglobin is extremely sensitive to damage by glucose, and the action of glucose leads to destruction of the heme

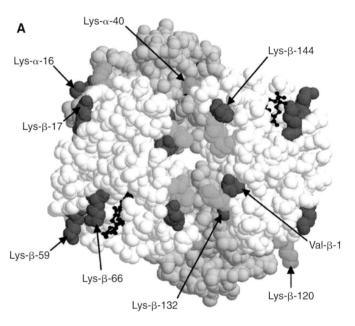

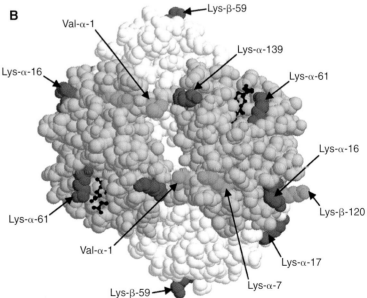

Figure 14-25. Three-dimensional structure of human oxy-hemoglobin showing the position of glycated residues. A space-filling rendition of the structure is shown. In *B*, the model shown in *A* has been rotated 180 degrees along the vertical axis and 115 degrees counterclockwise along the axis perpendicular to the plane of the figure. α- and β-chains are displayed in *gray* and *white*, respectively, and heme groups are in *black*. *Pink* residues correspond to the sites where fructosamines phosphorylated by FN3K have been identified, whereas *violet* residues were found to be glycated in previous studies. Residues interacting with 2,3-bisphosphoglycerate (besides Val β 1 and Lys-β-144 [i.e., His-β-2, Lys-β-82, and His-β-143]) are highlighted in *green*. FN3K, fructosamine-3-kinase. Reproduced from Figure 6 of G. Delpierre, D. Vertommen, and E. Van Schaftingen, "Identification of fructosamine residues deglycated by fructosamine-3-kinase in human hemoglobin," *J. Biol. Chem.*, 279: 27613–27620, 2004.

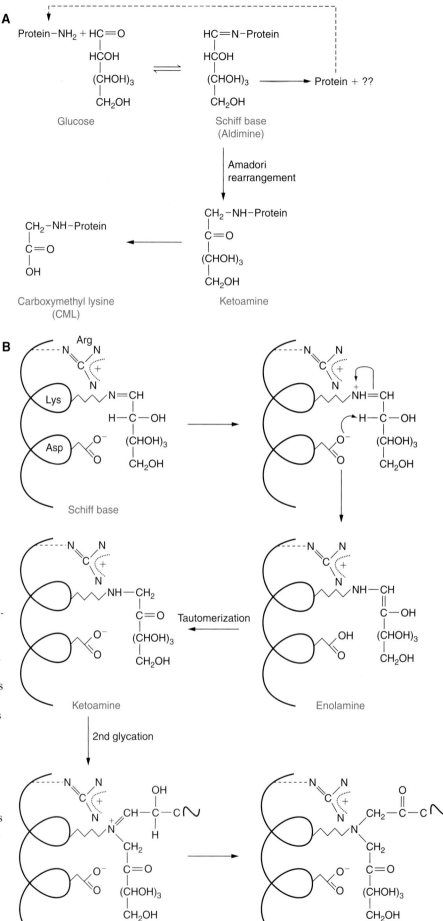

Figure 14-26. **A.** Schematic representation of the reactions involved in the glycation of proteins. The open chain form of the sugar (glucose) reacts with the ∈-amino group of lysine residues to form a Schiff base, which undergoes the Amadori rearrangement to form a ketoamine product. This ketoamine is subject to a series of reactions, which result in AGEs, one of which is CML. *AGE*, advanced glycosylation end products; carboxymethyl lysine.
B. Schematic representation summarizing the reactions occurring on the model peptide helices and the residues involved. The mechanism of Amadori rearrangement on Lys6 is shown as catalyzed by Asp10. The proximity of the arginine residue and its effect on the reactions involving Lys6 are also highlighted. Peptide RKD4 contains all three residues, whereas KD4 contains only Lys6 and Asp10.

group with the release of the iron atom; HbA_{1c} also is sensitive to catalase. In erythrocytes, there is a deglycosylation mechanism, although this mechanism must be overrun when circulating glucose is high. Hemoglobin is glycated primarily on valine and lysine residues *in vivo*. In an experimental system, the glycation positions were identified, and they are shown in Figure 14-26. An example of the chemistry of the glycation by glucose of a lysine residue in a peptide chain is shown in Figure 14-26B. HbA_{1c} has an altered oxygen saturation curve, indicating that glycation can modify the conversion of deoxyhemoglobin to oxyhemoglobin, and glycated hemoglobin has a greater affinity for oxygen than HbA0 and a decreased tendency to dissociate oxygen. This might decrease the availability of oxygen to tissues. Also, there are a number of possible genetic mutations in the hemoglobin molecule with serious consequences, as indicated in Table 14-2.

Table 14-2
Some Mis-Sense (Point) Mutations in Human Hemoglobin

Effect	Residue Changed	Change	Name	Consequence of Mutation	Explanation
Sickling	β 6 (A3)	Glu → Val	S	Sickling	Val fits into EF pocket in chain of another hemoglobin molecule
	β 6 (A3)	Glu → Ala	G Makassar	Not significant	Ala probably does not fit the pocket as well
	β 121 (GH4)	Glu → Lys	O Arab, Egypt	Enhances sickling in S/O heteroygotes	β 121 lies close to residue β 6; Lys increases interaction between molecules
Change in O_2 affinity	α 87 (F8)	His → Tyr	M Iwate	Forms methemoglobin, decreases O_2 affinity	The His normally ligated to Fe has been replaced by Tyr
	α 141 (HC3)	Arg → His	Suresnes	Increases O_2 affinity by favoring R state	Replacement eliminates bond between Arg141 and Asn126 in deoxy state
	β 74 (E18)	Gly → Asp	Shepherds Bush	Increases O_2 affinity by decreasing in BPG binding	The negative charge at this point decreases BPG binding
	β 146 (HC3)	His → Asp	Hiroshima	Increases O_2 affinity, reduced Bohr effect	Disrupts salt bridge in deoxy state and removes His that binds a Bohr effect proton
	β 92 (F8)	His → Gln	St. Etienne	Loss of heme	The normal bond from F8 to Fe is lost, and the polar glutamine tends to open the heme pocket
Heme loss	β 42 (CD1)	Phe → Ser	Hammersmith	Unstable, loses heme	Replacement of hydrophobic Phe with Ser attracts water into the heme pocket
Dissociation of tetramere	α 95 (G2)	Pro → Arg	St. Lukes	Dissociation	Chain geometry is altered in subunit contact region
	α 136 (H19)	Leu → Pro	Bibba	Dissociation	Pro disrupts helix H

BPG, 2,3-bisphosphoglycerate; *EF pocket*, acceptor site in hemoglobin; *S/O heterozygotes*, a rare compound heterozygous hemoglobinopathy.
Reproduced from http://www.chemsoc.org/exemplarchem/entries/2004/durham_mcdowall/prot-evo.html.

Carbon Dioxide

Hemoglobin can bind carbon dioxide, although to a lesser extent than oxygen, and some carboxyhemoglobin is formed. Oxygen is released from hemoglobin in the presence of carbon dioxide, a reaction known as the **Bohr effect.** The oxygen dissociation curves of hemoglobin in the absence or in the presence of carbon dioxide in blood are shown in Figure 14-27. In the presence of carbonic anhydrase, most of the carbon dioxide in the erythrocyte forms carbonic acid:

$$H_2O + CO_2 = H_2CO_3$$

Bicarbonate can form protons and hydrogen carbonate ions:

$$H_2CO_3 = H^+ + HCO_3^-$$

HCO_3^- diffuses from the erythrocyte cytoplasm into the plasma in exchange for chloride ions (Cl^-) to maintain the balance of negative charges within and without the erythrocyte **(chloride shift)**. Dissociation of carbonic acid (see the first equation) in the blood decreases the pH (release of protons), and protons react with oxyhemoglobin to release bound oxygen:

$$Hb.4O_2 + H^+ = HHb^+ + 4O_2$$

This reaction underlies the Bohr effect (release of oxygen from HbA0 in the presence of CO_2). When tissue cells respire, they form carbon dioxide so that its high concentration causes oxygen to be released from oxyhemoglobin to oxygenate the tissue cell. The changes occurring in the red blood cell are diagrammed in Figure 14-28. Carbon dioxide can react directly with hemoglobin to form a carbamate; in this case, carbon dioxide reacts with the α-amino group of an amino acid (AA) residue:

$$Hb\text{-AA residue-}NH_2 + CO_2 = Hb\text{-AA residue-NH-C} = O(O^-)$$

Erythrocyte hemoglobin carries carbon dioxide from the tissues, using the plasma, to the lungs, where it is expired. In the fetus, there is a different form of hemoglobin called **fetal hemoglobin** (**HbF,** $\alpha_2\gamma_2$). HbF binds oxygen with greater affinity than adult hemoglobin (HbA0). In this case, the oxygen saturation curve (Figure 14-27) for HbF would be shifted to the left of that for HbA0.

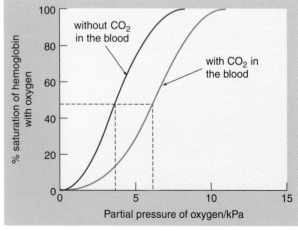

Figure 14-27. Oxygen dissociation curves of hemoglobin in the absence or presence of carbon dioxide in the blood. In the presence of CO_2 in the blood, the curve is pushed to the right. This means that more oxygen is required to saturate hemoglobin to (form oxyhemoglobin, HbAO). *KPa*, kilopascals; *1KPa*, 14.7 lbs/inch². Redrawn from http://www.chemsoc.org/networks/learnnet/cfb/transport.htm.

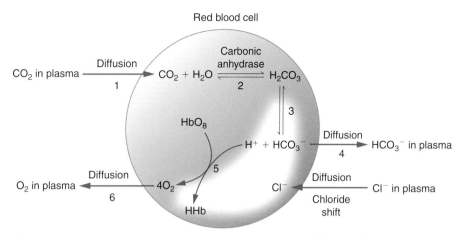

Figure 14-28. Reactions occurring in and around the red blood cell when carbon dioxide *(upper left)* increases in plasma. Redrawn from http://www.chemsoc.org/networks/learnnet/cfb/transport.htm.

Also, hemoglobin reacts strongly, more strongly than with oxygen, with carbon monoxide (CO), which has a 200-fold higher affinity for Hb than oxygen. CO forms carboxyhemoglobin, and the presence of the deadly toxin, CO, reduces the ability of hemoglobin to bind oxygen. Another factor that affects the function of hemoglobin is **2,3-bisphosphoglycerate (2,3-BPG)** which is derived from 1,3-bisphosphoglycerate (1,3-BPG) in the glycolytic pathway. 1,3-BPG is a negative allosteric effector on the oxygen-binding function of hemoglobin. Glucose is converted to two glyceraldehyde-3-phosphate molecules, and glyceraldehyde-3-phosphate is converted to 1,3-BPG by the action of glyceraldehyde-3-phosphate dehydrogenase. In turn, 1,3-BPG can be converted to 2,3-BPG by the action of BPG mutase. Both 1,3-BPG and 2,3-BPG can be converted to the normal pathway intermediate, 3-phosphoglycerate, that is metabolized down to pyruvate. The synthesis of 2,3-BPG is a major pathway for the use of glucose in red blood cells, and 2,3-BPG is important as a mechanism for controlling the affinity of hemoglobin for oxygen. 2,3-BPG binds to the T form (deoxy) of hemoglobin and tends to stabilize the T form, thus limiting the oxygen binding to hemoglobin. When 2,3-BPG is low or unavailable (lower glucose concentration in the red blood cell), hemoglobin can more readily convert to the oxygenated form (R state). HbF has a lower affinity for 2,3-BPG, thus giving the fetus greater access to oxygenated hemoglobin.

After 120 days at the end of the life of the red blood cell, it is degraded. Hemoglobin also is degraded, generating free iron, which is recycled, and bilirubin derived from the heme porphyrin is secreted in the bile.

Blood Cells

About 40% of the volume of blood is occupied by cells; the other 60% is plasma. Serum is the fluid remaining after the blood clots and the cells are removed. There are three main types of blood cells: red cells (erythrocytes), white cells (leukocytes), and platelets. Blood cells are derived from pluripotent stem cells, either myeloid or lymphoid stem cells from bone marrow, as shown in Figure 14-29. The red blood cell carries oxygen to

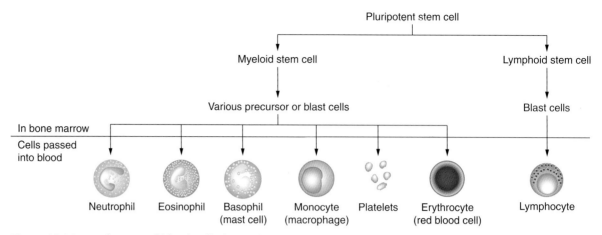

Figure 14-29. Production of blood cells from pluripotent stem cells.

the tissue cells and carries carbon dioxide from the tissue cells to the lungs for expiration. Hemoglobin in the red cell carries out these functions as described earlier. Red blood cells are shown in Figure 14-30; they have no nucleus and thus have a central faded spot that varies in size depending on the amount of (red) hemoglobin in the cell. The normal red cell is round and has a diameter of 6 to 8 μm.

Segmented neutrophils (polymorphonuclear leukocytes) are mature phagocytes (Figure 14-31) that can migrate through the tissues to engulf and destroy microorganisms and respond to inflammatory stimuli, such as IL-1 released from macrophages as a result of infection or tissue injury or histamines released by circulating basophils, tissues mast cells, and blood platelets. They are 40 to 75% of peripheral leukocytes and are 9–16 μm in diameter. They may have two to five nuclear lobes with connecting filaments. The pattern of chromatin in these cells is coarse or clumped, and they have a large cytoplasmic space.

The eosinophil is a mature granulocyte that responds to infections by parasites and allergic reactions. Eosinophils are 1 to 4% of peripheral leukocytes with diameters of 9 to 15 μm. There are one to three lobes in

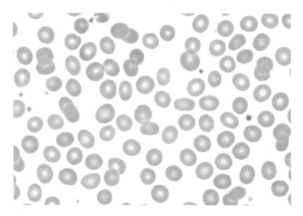

Figure 14-30. Appearance of erythrocytes (red blood cells). Reproduced from http://www.wadsworth.org/chemheme/heme/microscope/rbc.htm.

the nucleus, the pattern of chromatin is coarse and clumped, and the cytoplasmic space is large (Figure 14-32). Eosinophils can release major basic protein, cationic proteins, peroxidase, arylsulfatase B, phospholipase D, and histaminase from eosinophil granules. These substances can destroy the membrane of a parasite.

A basophil is a granulocyte containing granules of heparin and vasoactive compounds that are released when activated. It represents about 0.5% of the total leukocytes. Basophils (Figure 14-33) are involved in immediate hypersensitivity reactions (type I hypersensitivity reactions) and some delayed hypersensitivity reactions. An example of an immediate hypersensitivity reaction is a bee sting. They are the smallest of the circulating granulocytes, with diameters of 10 to 15 μm. The nucleus and the cytoplasm have about the same volume. The chromatin appears to be coarse. The cytoplasm is homogeneous with large dark granules.

The monocyte or large mononuclear phagocyte represents an immature stage of the macrophage. Monocytes are from 10 to 30 μm in diameter, and the nucleus of the monocyte can be twice as large as the

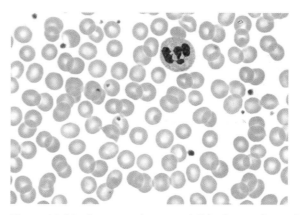

Figure 14-31. Segmented neutrophil (polymorphonuclear leukocyte). Reproduced from http://www.wadsworth.org/chemheme/heme/microscope/seg.htm.

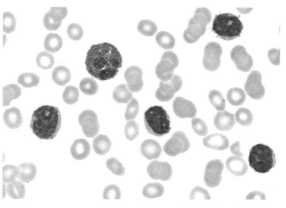

Figure 14-32. An eosinophil in a field of red blood cells. Reproduced from http://www.wadsworth.org/chemheme/heme/microscope/eos.htm.

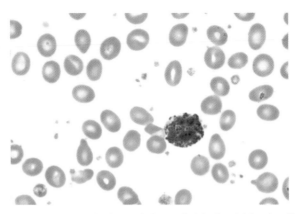

Figure 14-33. The basophil is a field of red blood cells. Reproduced from http://www.wadsworth.org/chemheme/heme/microscope/basophil.htm.

cytoplasm. The nucleus can be shaped like a horseshoe and may fold over on itself, showing convolutions. Nucleoli are not visible, and the chromatin pattern is arranged in strands. The cytoplasm may contain vacuoles (Figure 14-34). Monocytes circulate in the peripheral blood before emigrating into the tissues. In the liver, they are known as **Kupfer cells;** in the brain, they are known as microglia; in the kidney, they are known as mesangial cells; and in the bone, they are known as osteoclasts.

Platelets (Figure 14-35) are cytoplasmic fragments of megakaryocytes that circulate as small discs in the peripheral blood. They play an important role in forming blood clots (described earlier) and in maintaining the blood vessel lining. The diameter of a platelet is 1 to 4 μm. The platelet does not contain a nucleus. There are 130,000 to 450,000 platelets per microliter of normal blood.

Lymphocytes (Figure 14-36) vary in size and granularity. The most common lymphocytes in the peripheral blood are small (6 to 10 μm) with nuclei about 7 μm in diameter. A large part of the nuclear chromatin is condensed. The nucleus may be three to five times larger than

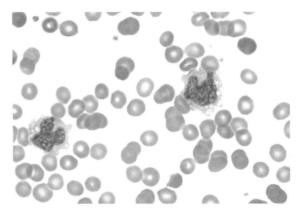

Figure 14-34. Monocytes *(center right and center left)*. Reproduced from http://www.wadsworth.org/chemheme/heme/microscope/monocyte.htm.

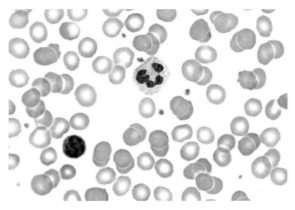

Figure 14-35. Platelets *(blue)* is a field of red blood cells. Reproduced from http://www.wadsworth.org/chemheme/heme/microscope/platelets.htm.

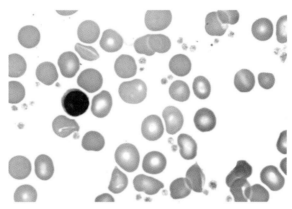

Figure 14-36. A lymphocyte *(center left)* is shown in a field of red blood cells with a few platelets *(lower right)*. Reproduced from http://www.wadsworth.org/chemheme/heme/microscope/lymphocytes.htm.

the cytoplasm, which may appear only as a ring around the nucleus. They are known as B cells if they achieve immune competence within the bone marrow or as T cells in the thymus. Elsewhere, organized lymphoid tissue is known as secondary lymphoid tissue and includes lymph nodes, adenoids, tonsils, and mucosa-associated lymphoid tissue (MALT). MALT includes bronchus-associated lymphoid tissue, gut-associated lymphoid tissue, nasopharyngeal-associated lymphoid tissue, and urogenital-associated tissue. Lymphocytes (B cells) respond to antigens by producing antibodies or lymphokines by producing T cells and B cells.

Band neutrophils (Figure 14-37) represent 1 to 3% of peripheral leukocytes. The nucleus appears as a U-shaped rod before segmentation, and the pattern of chromatin is coarse and clumped.

A picture of red and white cells in a small blood vessel is shown in Figure 14-38. A **complete blood (cell) count (CBC)** is usually determined on a patient to assess status. An overview of the CBC is shown in Table 14-3.

Blood 883

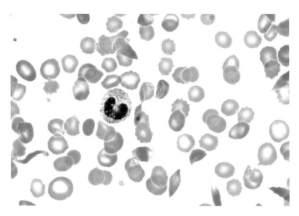

Figure 14-37. A band neutrophil is shown in the *left center* of the field. Reproduced from http://www.wadsworth.org/chemheme/heme/microscope/band.htm.

Figure 14-38. White cell among several red blood cells. Reproduced from http://images.encarta.msn.com/xrefmedia/sharemed/targets/images/pho/35a5c/35A5C297.jpg

Blood Proteins

Circulating human blood contains many proteins. Some major groups are serum albumins (Figure 14-39) that make up 55% of plasma proteins; globulins, especially γ-globulins (Figure 14-40); fibrinogens (Figure 14-41); and hemoglobin. γ-Globulins are composed mainly of antibodies. Some other proteins are haptoglobin in serum, which binds hemoglobin in the blood; protein C, a vitamin K–dependent plasma protein that cleaves activated factors V and VIII and thus inhibits blood coagulation and clot formation; and thromboplastin (factor III, platelet tissue factor, and thrombokinase; Figure 14-42), a plasma protein in tissues, platelets, and white blood cells required for the conversion of prothrombin to thrombin

(Text continues on p. 888.)

Table 14-3
Components of the Complete Blood Count (CBC)

Test	Name	Measuring	Use	Increased/Decreased
WBC	White blood cell	Total number of WBCs per volume of blood (sum of all types of WBCs)	The body uses WBCs to fight infection. Each type has a slightly different job. WBC is measured to make sure there are a sufficient number and to help detect and monitor conditions that lead to increases or decreases in total WBCs and/or to increases in one or more types of WBCs.	May be increased with infections, inflammation, cancer, leukemia; decreased with some medications (such as methotrexate), some autoimmune conditions, some severe infections, bone marrow failure, and congenital marrow aplasia (marrow does not develop normally).
% Neutrophil % Lymphs % Mono % Eos % Baso	Neutrophil/band/seg Lymphocyte Monocyte Eosinophil Basophil	Measures the percentage of each of five types of WBC, compared to total WBC count		This is a dynamic population that varies somewhat from day to day depending on what is going on in the body. Significant increases in particular types are associated with different temporary and acute and/or chronic conditions. An example of this is the increased number of lymphocytes seen with lymphocytic leukemia.
Neutrophil Lymphs Mono Eos Baso	Neutrophil/band/seg Lymphocyte Monocyte Eosinophil Basophil	Measures the actual number of each type of WBC per volume of blood		
RBC	Red blood cell	Total number of RBCs per volume of blood	Primarily measured to detect decreased production, increased loss, or increased destruction of RBCs, to detect anemia, and sometimes to help detect erythrocytosis (too many RBCs).	Decreased with anemia; increased when too many made and with fluid loss due to diarrhea, dehydration, burns.
Hgb	Hemoglobin	Total amount of oxygen-carrying protein inside RBCs		Mirrors RBC results.
Hct	Hematocrit	Percentage of blood volume made up of RBCs (solid versus liquid portion of blood)		Mirrors RBC results.
MCV	Mean corpuscular volume	Average size of RBCs	The size of RBCs and the average amount of hemoglobin inside them can help classify different types of anemia.	Increased with B_{12} and folate deficiency; decreased with iron deficiency and thalassemia.
MCH	Mean corpuscular hemoglobin	Average amount (weight) of hemoglobin inside each RBC		Mirrors MCV results.
MCHC	Mean Corpuscular Hemoglobin Concentration	Average concentration (%) of hemoglobin inside each RBC		May be decreased when MCV is decreased; increases limited to amount of Hgb that will fit inside a RBC.
RDW	RBC distribution width	Measures variation in size of RBCs; most normal RBCs are the same size	Help classify anemia.	Increased RDW indicates mixed population of RBCs. Immature RBCs tend to be larger.

Continued

Table 14-3
Components of the Complete Blood Count (CBC)—cont'd

Test	Name	Measuring	Use	Increased/Decreased
Platelet	Platelet	Total number of platelets per volume of blood; platelets are special cell fragments that are important in blood clotting	Determine whether number is adequate to control bleeding.	Decreased or increased with conditions that affect platelet production; decreased when greater numbers are used, as with bleeding; decreased with some inherited disorders (such as Wiskott-Aldrich and Bernard-Soulier), with systemic lupus erythematosus, pernicious anemia, hypersplenism (spleen takes too many out of circulation), leukemia, and chemotherapy.
MPV	Mean platelet volume	Average size of platelets	Help evaluate decreased platelets.	Vary with platelet production. Younger platelets are larger than older ones.

© 2005 American Association for Clinical Chemistry. Downloaded from Lab Tests Online (http:www.labtestsonline.org)

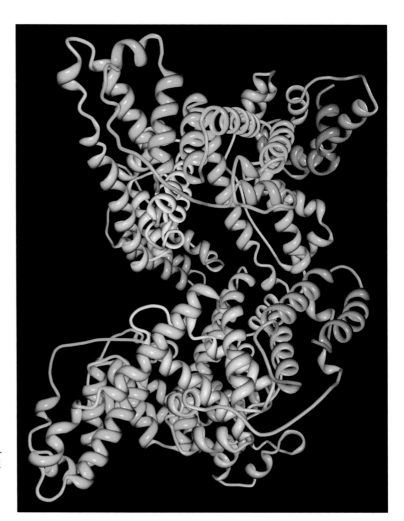

Figure 14-39. Crystal structure of human serum albumin. Reproduced from PDB RCSB ID: 1BMO. S. Sugio, A. Kashima, S. Mochizuki, et al. Crystal Structure of Human Serum Albumin at 2.5 A Resolution. *Protein Engineering Design and Selection* **12** pp. 439 (1999).

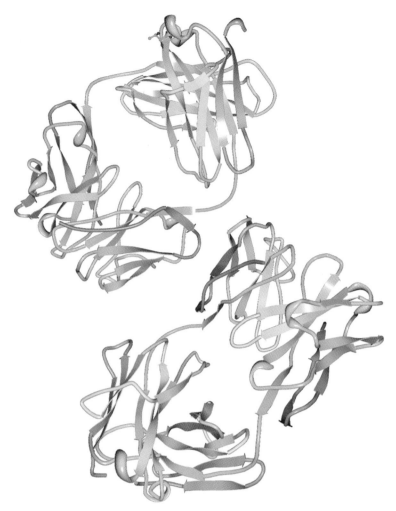

Figure 14-40. Structure of immunoglobulin 2E8 Fab fragment. Reproduced from PDBsum (Jena) ID: 12e8. S. Trakhanov et al. Crystal Structure of a Monoclonal 2E8 Fab Antibody Fragment Specific to the Low Density Lipoprotein Receptor Binding Region of Apolipoprotein E. *Acta. Crystallogr. D. Biol. Crystallogr.* **55** pp. 122 (1999).

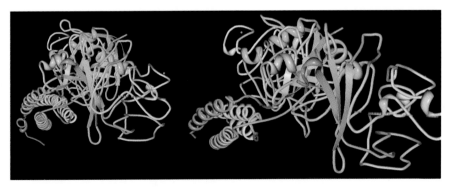

Figure 14-41. Crystal structure of recombinant human fibrinogen fragment D. Reproduced from PDB RCSB ID: 1LT9. M.S. Kostelanski et al. 2.8 Angstrom Crystal Structures of Recombinant Fibrinogen Fragment D with and without Two Peptide Ligands: GHRP Binding to the "b" Site Disrupts Its Nearly Calcium-Binding Site. *Biochemistry* **41** pp. 12124 (2002).

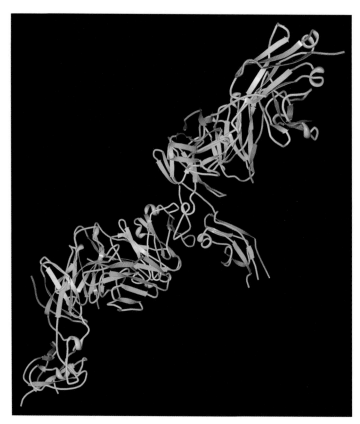

Figure 14-42. A complex of extracellular domain of tissue factor (thromboplastin) with an inhibitory Fab. Reproduced from PDB RCSB ID: 1AHW. M.H. Huang et al. The Mechanism of an Inhibitory Antibody on TF-initiated Blood Coagulation Revealed by the Crystal Structures of Human Tissue Factor, Fab 569 and TF.69 Complex. *Journal of Molecular Biology* **275** pp. 873 (1998).

in the presence of Ca^{2+}. An abnormal protein, paraprotein, can be a macroglobulin, cryoglobulin, or myeloma protein. Many enzymes in blood are important for clinical diagnosis, especially under disease conditions (Chapter 3). Regulatory peptides and protein fragments are also found in serum, and some of these are summarized in Table 14-4.

Blood Type and Rh

The human blood types are A, B, AB, and O. In the American population, 45% are type O, 42% are type A, 10% type B, and 3% type AB. These types derive from codominant alleles. There are three different alleles, I^A, I^B, and i, where the i allele is recessive. Types A and B can be either homozygous ($I^A I^A$ or $I^B I^B$) or heterozygous ($I^A i$ or $I^B i$). The AB phenotype has the codominant $I^A I^B$ alleles. Type O is homozygous recessive (ii). The blood type phenotype is related to the genotype, as shown in Table 14-5. For example, if the mother is type A and the father is type B, there are

Table 14-4
Proteins, Regulatory Peptides, and Protein Fragments Found in Plasma

Peptide hormones	Angiotensin I, guanylin (22–115), uroguanylin (89–112), atrial natriuretic factor (CDD/ANP99-126), GLP-1
Cytokines, growth factors	HCC-1, IGF-1, IGF-2, osteoinductive factor, PDGF, CTAP-III, pigment endothelium–derived factor
Defensins	β-Defensin 1, propeptides of neutrophil defensins 1 to 3
Plasma proteins	Albumin, fibrinogen A, fibrinogen B, β-2-microglobulin, zinc-α-2-glycoprotein, α-2-HS-glycoprotein (fetuin), serum amyloid protein A, haptoglobin, profilin, desmocollin, thymosin-β-4 and -β-10, apolipoprotein C-III, uteroglobin, ubiquitin, gelsolin
Transport proteins	Retinol-binding protein, α-1-microglobulin, transferrin, transthyretin, TGF β-binding protein, IGF-binding protein 2 and 3
Complement factors	Factor C3, factor D, factor C4A (anaphylatoxin)
enzymes, inhibitors	Lysozyme, cystatin C, α-1-antitrypsin, pancreatic trypsin inhibitor, plasminogen, α-2-antiplasmin, carboxypeptidase N, inter-α-trypsin inhibitor component II, somatomedin B, vitronectin
Matrix proteins	Collagens α-1-(I), α-2-(I), α-3-(IV), α-1-(XVIII), and osteopontin

Reproduced from Table 2 of http://www.abrf.org/jbt/1998/December98/dec98rcjurgens.html.

Table 14-5
Relationship Between Blood Type and Genotype

Phenotype (blood type)	Genotype
Type A	$I^A I^A$ or $I^A i$
Type B	$I^B I^B$ or $I^B i$
Type AB	$I^A I^B$
Type O	ii

Reproduced from http://www.biology.arizona.edu/mendelian_genetics/problem_sets/monohybrid_cross/11t.html.

possible four genetic crosses that will determine the blood type of offspring: $I^A I^A \times I^B i$, $I^A I^A \times I^B I^B$, $I^A i \times I^B I^B$, and $I^A i \times I^B i$. If the father is type AB (genotype $I^A I^B$) and the mother is type O (ii), half of the children would be type A and half type B ($I^A i$, $I^A i$, $I^B i$, or $I^B i$), but they would not be type AB or type O. Blood types represent macromolecules on the surface of red blood cells. If two different blood types are mixed, they could clump in the blood vessel. Blood type A has A antigens on the surface of the red blood cell and antibodies to type B in the blood plasma. Blood group B has B antigens on the red blood cell surface, and the plasma contains antibodies to blood group A. Blood group AB has both A and B antigens on the red blood cell and no antibodies to A or B in the plasma. Blood group O (null, universal donor) does not have A or B antigens on the red blood cell but has antibodies to both group A and group B antigens. These properties are important when blood transfusions are to be made. The allowable types that can be transfused are shown in Table 14-6.

Determination of a person's blood type can be made with antibodies to A, B, or Rh. Rh will be discussed later in this section. Agglutination

Table 14-6
The Blood Groups, Antigens on Red Blood Cells, Antibodies in Plasma, the Blood Groups That Can Be Donated to, and the Blood Groups That Can Be Received in Transfusion

Blood group	Antigens	Antibodies	Can give blood to	Can receive blood from
AB	A and B	None	AB	AB, A, B, O
A	A	B	A and AB	A and O
B	B	A	B and AB	B and O
O	None	A and B	AB, A, B, O	O

Reproduced from http://nobelprize.org/medicine/educational/landsteiner/readmore.html.

Sugars	Blood type	Antibodies to	Receive from	Donate to
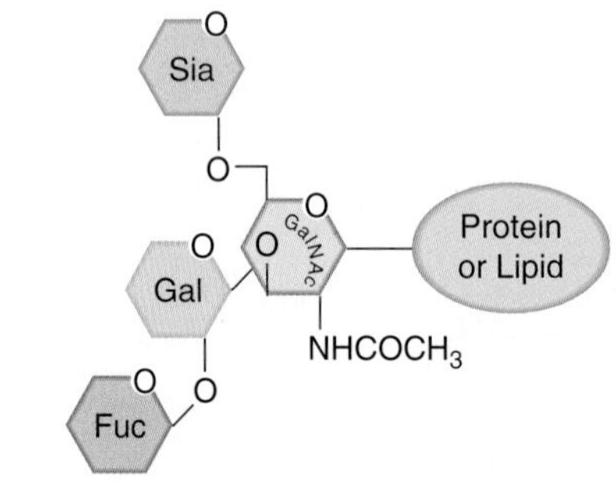 (Sia, Gal, GalNAc, Fuc — Protein or Lipid)	O	A B	O	O A B AB
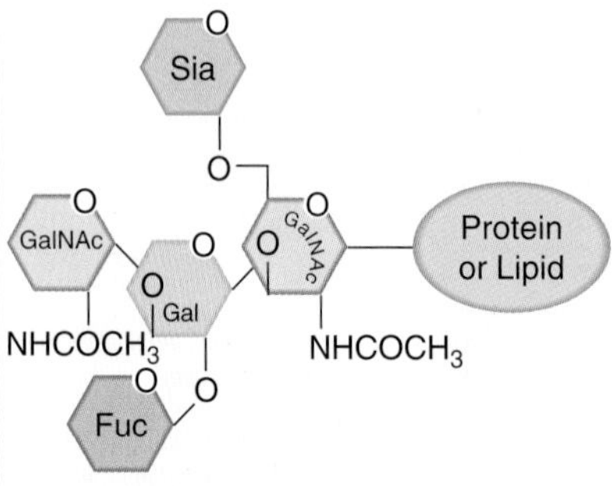 (Sia, GalNAc, Gal, GalNAc, Fuc — Protein or Lipid)	A	B	O A	A AB
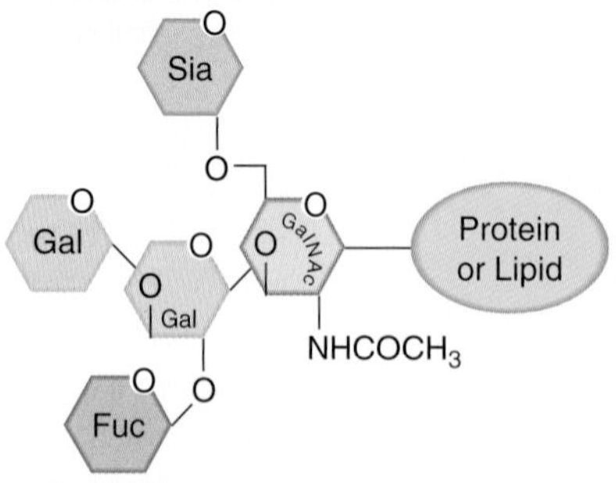 (Sia, Gal, GalNAc, Gal, Fuc — Protein or Lipid)	B	A	O B	B AB
Mix of the above	AB	None	O A B AB	AB

Figure 14-43. Blood types and their constituent carbohydrate substituents.

determines the presence of that antigen, and nonagglutination determines the absence of the antigen (AB will cross-react with either antibody to A or antibody to B; A will cross-react with antibody to A but not to B; B will cross-react with antibody to B but not to A; and O will not cross-react to either antibody). The blood type antigen on the surface of the red blood cell is a macromolecule attached to various carbohydrate groups that form the antibody reactive site (epitope). These are shown in Figure 14-43.

There is an antigen on the red blood cell in addition to the ABO system that is the **Rh antigen.** Individuals who have the Rh antigen on their red blood cells are typed Rh positive, and those who do not are Rh negative. Considering Rh, the ABO system is modified to indicate the presence or absence of the Rh antigen. Then blood types are indicated as O positive (36% of the U.S. population), AB negative (0.5%), AB positive (2.5%), B negative (2%), B positive (8%), A negative (8%), and A positive (34%). If a mother is Rh negative and the father is Rh positive, the baby has a 60% chance of being Rh positive. If the baby's blood mixes with the mother's blood, the mother could make antibodies (that could cross the placenta) against the baby's red blood cells and destroy them. There is an Rh immunoglobulin (RhoGAM) that can prevent the mother from forming antibodies unless the mother has already made antibodies. The Rh antigen is a 30,000 molecular weight protein, itself called the Rho D antigen, that is in the red cell membrane. It turns out that there are at least three membrane polypeptides encoded by *RHD-* and *RHCE-*related genes (Figure 14-44). The Rh antigen may have a 12-transmembrane structure similar to that shown in Figure 14-45.

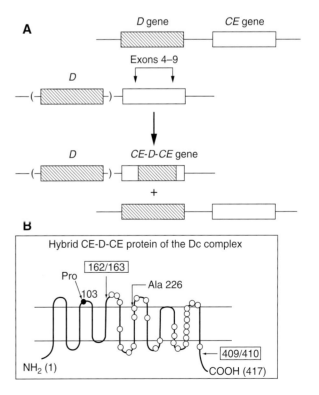

Figure 14-44. Model of gene conversion and predicted membrane topology for *De–* gene complex. **A.** A directional transfer of a homologous DNA segment encompassing axons 4 to 9 (indicated by the double arrows) from the *RHD* gene (donor) to the *RHCE* gene (recipient) generated the hybrid *RHCE-D-CE* gene structure of the *Dc–* complex. A similar directional transfer of axons 2 (or 3) to 9 generated the *DCW–* complex (not shown). While the recipient gene (either from an Rh-positive or an Rh-negative chromosome) is converted into a recombinant, the donor gene restores its native structure by the repair synthesis. **B.** Predicted membrane topology of the *Dc-* gene-encoded protein. The *bold line* represents the polypeptide sequence specific of the D protein. *Open circles* refer to amino acid substitutions that distinguish D from CE proteins. *Closed circle* indicates amino acids associated with the C/c (103 = Ser/Pro) polymorphisms. Alanine (Ala) at position 226 of the D protein is indicated. *Arrows* indicate the new junction sites of the hybrid protein. Redrawn from Figure 4 of B. Cherif-Zahar et al., "Molecular analysis of the structure and expression of the rh locus in individuals with *D-, Dc-,* and *DCW–*gene complexes," *Blood,* 84: 4354–4360, 1994.

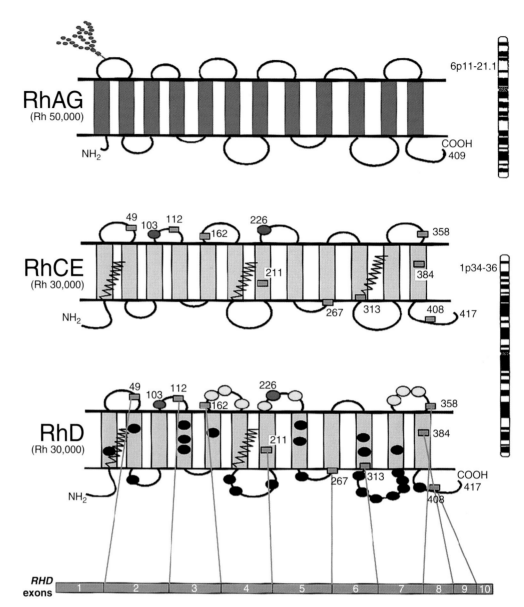

Figure 14-45. Predicted structure of three Rh antigens. http://www.bloodjournal.org/content/vol95/issue2/images/large/bloo0024401z.jpeg. Model of topology for RhAG, RhCE, and RhD. RhAG (M_r 50,000) consists of 409 amino acids and is encoded by *RHAG* on chromosome 6 p11-p21.1. RhCE and RhD (M_r 30,000) are predicted to have a similar topology and are encoded by *RHCE* and *RHD*, which are adjacent on chromosome 1 p34-p36. The domain of the RhD protein encoded by each exon is depicted by numbered boxes, which represent the start and finish of each exon. Of the D-specific amino acids, 8 are on the exofacial surface *(yellow ovals)* and 24 are predicted to reside in the transmembrane and cytoplasmic domains *(black ovals)*. *Red ovals* represent amino acids that are critical for C/c (Ser103Pro) and E/e (Pro226Ala) antigens; *purple ovals* represent Ser103 and Ala226 on RhD. The *zigzag lines* represent the Cys-Leu-Pro motifs that are probably involved in the palmitoylation sites. The N-glycan on the first loop of RhAG is indicated by the branched structure of *red circles*. Reproduced from Figure 1 of N.D. Avent and M.E. Reid, "The Rh blood group system: a review," *Blood*, 95: 375–387, 2000.

The Rh blood system is complex. As shown in Figure 14-44A, there are genes that carry two distinct Rh proteins, C (or c) or E (or e), together with the D antigen. An Rh glycoprotein has been found to be essential for the assembly of the Rh complex in the red cell membrane and for the expression of Rh antigens. The predicted appearance of three Rh antigens is shown in Figure 14-45. A person with Rh⁻ blood does not have Rh antibodies but can develop Rh antibodies in the blood plasma if transfused with Rh⁺ blood. A person with Rh⁺ blood can receive Rh⁻ blood with no problem. Of 100 blood donors, 84 will be Rh⁺ and 16 will be Rh⁻. Considering the Rh antigen and the blood type based on the ABO system, blood typing becomes more complex than shown in Table 14-6; this is indicated in Tables 14-7, 14-8, and 14-9.

Table 14-7
Transfusability of Various Blood Types Based on the ABO System and Rh Antigen

Type	You can give blood to	You can receive blood from
A+	A+ AB+	A+ A− O+ O−
O+	O+ A+ B+ AB+	O+ O−
B+	B+ AB+	B+ B− O+ O−
AB+	AB+	Everyone
A−	A+ A− AB+ AB−	A− O−
O−	Everyone	O−
B−	B+ B− AB+ AB−	B− O−
AB−	AB+ AB−	AB− A− B− O−

Reproduced from http://chapters.redcross.org/br/northernohio/INFO/bloodtype.html.

Table 14-8
Inheritance of Blood Types: Results for Offspring

Blood Type		Mother's Type			
		O	A	B	AB
Father's Type	O	O	O, A	O, B	A, B
	A	O, A	O, A	O, A, B, AB	A, B, AB
	B	O, B	O, A, B, AB	O, B	A, B, AB
	AB	A, B	A, B, AB	A, B, AB	A, B, AB

Reproduced from http://www.mistupid.com/health/bloodinherit.htm.

Table 14-9
Inheritance of Rh Factor

Rh Factor		Mother's Type	
		Rh⁺	Rh⁻
Father's Type	Rh⁺	Rh⁺, Rh⁻	Rh⁺, Rh⁻
	Rh⁻	Rh⁺, Rh⁻	Rh⁻

Reproduced from http://www.mistupid.com/health/bloodinherit.htm.

Lymphatic System

The lymphatic system is made up of thin-walled vessels that carry flow in one direction (contain valves that promote unidirectional flow), not unlike the venous system of the blood circulation. Skeletal muscle contraction causes movement of the lymph fluid through valves. The lymphatic vessels join with the thoracic duct and the right lymph duct, and these empty near the heart. An overview of the lymphatic system is shown in Figure 14-46.

The water that is lost in the blood capillaries and water from tissues is collected by the lymphatic system and emptied into the blood circulation. The lymphatic system also transports molecules, such as proteins and lipids that are too large for uptake by the capillaries. Some of the interstitial fluid (extracellular fluid) is absorbed by the lymph capillaries, which drain into the larger vessels of the lymphatic system. Figure 14-47 shows lymph capillaries as they interface with tissue cells and blood capillaries. Lymph is collected from the left side of the body, the digestive tract, and the right side of the lower body, and all flow into the thoracic duct, the single major vessel. About 100 ml of lymph are emptied from the thoracic duct per hour into the left subclavian vein, and the right subclavian vein collects lymph from the right side of the head, neck, and chest. Increased capillary blood pressure and a decreased concentration of plasma proteins contribute to increased production of lymph. If an increase in the production of lymph is great enough, the lymphatic system may not be able to accommodate, in which case the lymph can accumulate and distend the tissues producing edema. Because white blood cells congregate in the lymph nodes, this system serves as part of the immune system and antibodies can be produced in lymph nodes. Antibodies and lymphocytes can enter the blood through the subclavian veins.

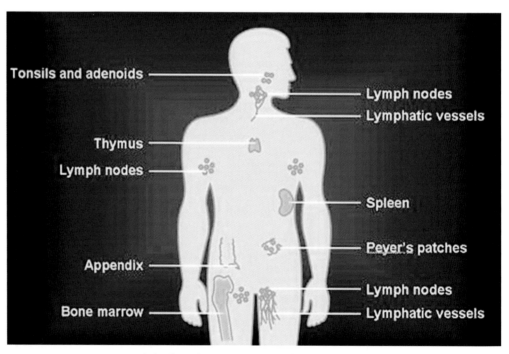

Figure 14-46. Overview of the lymphatic system. Redrawn from http://www.web-books.com/eLibrary/Medicine/Physiology/Immune/Organs.gif

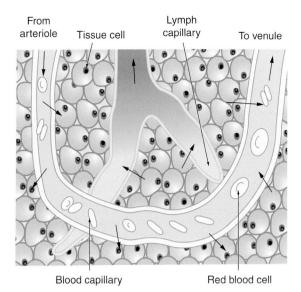

Figure 14-47. Lymph capillary in contact with tissue cells.

There are many lymph nodes scattered throughout the lymphatic system. These are mainly located in the armpits, neck, groin, and abdomen. The lymph nodes contain sinuses that collect the lymph, and the walls of these sinuses are lined with phagocytic cells (macrophages) that are able to engulf bacteria (Figure 14-48) or other foreign particles. Virtually all bacteria are destroyed before the lymph leaves a node to return to the blood.

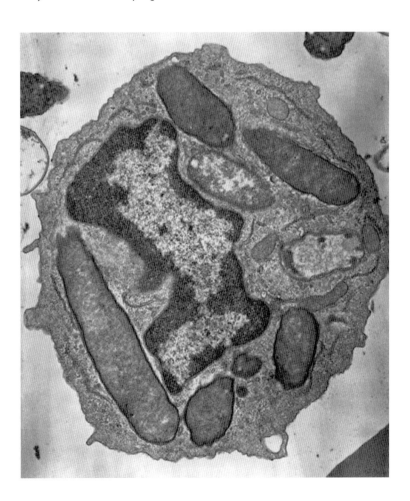

Figure 14-48. Electron micrograph of a macrophage cell that has endocytosed foreign particles. Reproduced with permission from http://missinglink.ucsf.edu.

Macrophages (white blood cells) identify bacteria or viruses as foreign particles (nonself), engulf them, and destroy them. Antibodies attack and bind to the foreign particle (for example, bacterium), and a receptor for the antibody on the macrophage membrane binds the antibodies and bacterium, allowing the complex to be engulfed by the macrophage (Figure 14-49).

Among the lymph organs are the spleen, bone marrow, and thymus, in addition to lymph nodes. Bone marrow produces lymphocytes (B cells mature in the bone marrow, and T cells mature in the thymus gland). The spleen, which filters and purifies the blood and lymph flowing through it, resembles a lymph node, but it is also a reservoir for blood.

Thymosin, a 5-kDa molecular weight hormone, is secreted by the thymus gland, and thymosin is a maturing hormone for precursor T cells in the thymus, stimulating their development into mature T cells (Figure 14-50). Thymosin β4 is the most abundant member of a family of thymosin peptides, originally isolated from the thymus gland but now known to be expressed in most cell types. Thymosin helps organize the cytoskeleton by sequestering actin (Figure 14-51). Thymosin (β-thymosins) have other

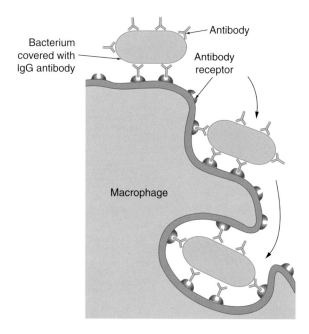

Figure 14-49. Macrophage engulfs bacteria coated with antibodies (bacteria are recognized as nonself). Receptors for antibodies on the cell surface bind the antibody-bacterium complex, facilitating engulfment and internal destruction of the bacterium. Redrawn from W.K. Purves et al., *Life: The Science of Biology*, 4th ed, Sinauer Associates and WH Freeman, Gardansville, VA, 2002 and from http://www.emc.maricopa.edu/faculty/farabee/BIOBK/BioBookIMMUN.html#Table%20of$20Contents.

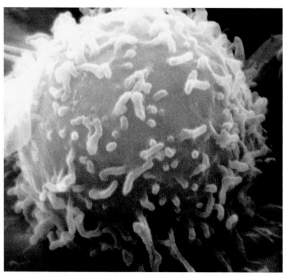

Figure 14-50. Human T lymphocyte. Reproduced from http://en.wikipedia.org/wiki/Image:SEM_Lymphocyte.jpg

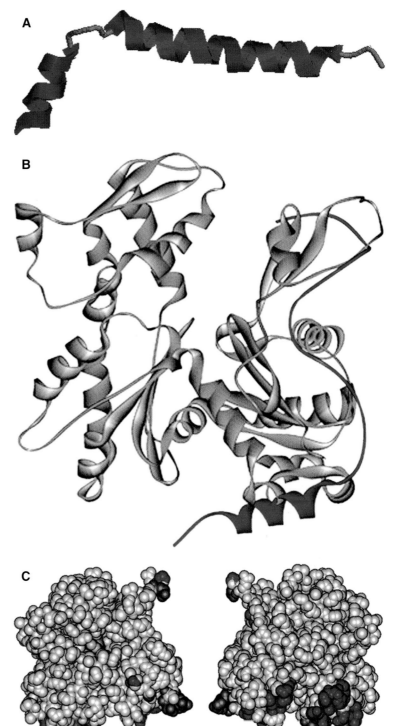

Figure 14-51. **A.** A structure of thymosin β9. **B.** Modeled complex of thymosin β4 and actin monomer. (Thymosin β4 is in *red*; actin is in *gray*). **C.** Molecular contact sites between actin and thymosin β4 *(dark green). Left image* is "front" view, and *right image* is back view. Part A reproduced from PDB ID: 1hjo. R. Stoll, W. Voelter, T.A. Holak. Conformation of Thymosin Beta9 in Water/Fluoroalcoh Solution Determined by NMR Spectroscopy. *Biopolymers* **41** pp. 623 (1997). Part B reproduced from Figure 14 of C.G. Dos Remedios et al., "Actin binding proteins: regulation of cytoskeletal microfilaments," *Physiol. Rev.*, 83: 433–473, 2003. Copyright ©2003 by the American Physiological Society. http://physrev.physiology.org/content/vol83/issue2/images/large/9j0230240114.jpeg. Part C reproduced from Figure 6 of C.G. Dos Remedios et al., "Actin binding proteins: regulation of cytoskeletal microfilaments," *Physiol. Rev.*, 83: 433–473, 2003. Copyright ©2003 by the American Physiological Society. http://physrev.physiology.org/content/vol83/issue2/images/large/9j0230240114.jpeg.

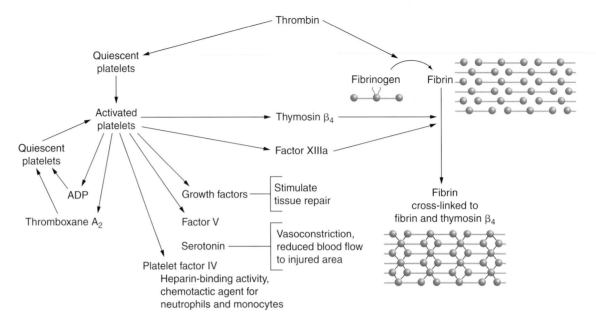

Figure 14-52. Thymosin β4 is co-released from activated platelets together with factor XIIIa (transglutaminase) to form cross links with fibrin during clot formation. Redrawn from http://www.fasebj.org/content/vol16/issue7/images/large/380421351s01.jpeg.

activities attributed to them. One example is the release of thymosin β4, from human blood platelets, which can be cross-linked to fibrin in a time- and calcium-dependent manner. Thymosin β4 cross-linking may be mediated by factor XIIIa, a transglutaminase, released with thymosin β4 from stimulated platelets. Thymosin β4 stabilizes the blood clot in this manner (Figure 14-52).

Further Reading

Daniels, Geoff, *Human Blood Groups*, 2nd edition, Blackwell Publishing, 2002.

Li, John K-J., *Arterial Circulation: Physical Principles and Clinical Applications*, Humana Press, 2000.

McDonnell, Julie, and Windelspecht, Michael, *The Lymphatic System*, Greenwood Press, 2004.

Mehler, Robert E., and Lauren Sompayrac (eds.), *How the Circulatory System Works*, Blackwell Science, 2000.

CHAPTER 15

Immunobiochemistry

The Surveillance System and Cancer

For decades, the immune surveillance system (a term coined by Lewis Thomas and Frank Macfarlane Burnet in the 1950s) has been credited with protecting the body against cancer. Presumably, cancer develops occasionally during a lifetime and the surveillance system is able to eradicate it from the body. It is only when a cancer escapes this system or the system becomes weakened that cancer can become established. In some cases, the cancer cells are able, biochemically, to escape the immune surveillance system. The key player in this system is the **natural killer (NK) cell.** NK cells are a subset of T cells that can kill tumor cells (or virus-infected cells) but do not harm normal cells. They respond to a variety of agents with cytolytic functions through the secretion of various cytokines and can affect even tumor growth and metastases. NK cells also are heterogeneous; they contain subsets of cells, at least one of which (A-NK cells) is capable of killing tumor cells and cells infected with viruses. The A-NK cell has a variety of receptors on its cellular surface (Figure 15-1). One or more of these will recognize an antigen on the surface of the tumor cell or virus-infected cell. *This antigen, not expressed on most normal cells, is an MHC class I chain-related antigen, which binds to NKG2D (Figure 15-2), a c-lectin–like receptor on the surface of human NK cells that is noncovalently associated to the DAP10 transmembrane signaling adaptor. The complex with the antigen can initiate the killing program through PI3K (normal cells express MHC class I peptides that are specific to normal cells, and these normal peptides prevent the killing program of NK cells).* Some other activating receptors on the cell surface operate through protein tyrosine kinase–dependent pathways through noncovalent associations with transmembrane signaling adaptors. There also are cell surface inhibitory receptors that antagonize the killing activating pathways through protein tyrosine phosphatases. When the killing program is activated, the contents of secretory granules are released from the activated A-NK cell and gain entrance to the tumor cell, resulting in the killing of the tumor cell by the secreted proteins and cytokines.

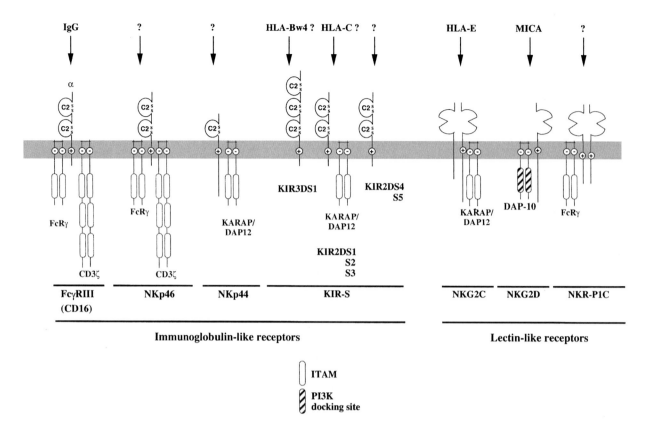

Figure 15-1. Activating oligomeric receptors expressed on human NK cells. Extracytoplasmic immunoglobulin domains and C-type lectin domains are depicted. Known ligands for activating NK cell receptors are indicated at the top of the *arrows*. Transmembrane charged amino acid residues and disulfide bridges (—) are also indicated. Reproduced from Figure 1 of E. Tomasello et al., "Signaling pathways engaged by NK cell receptors: double concerto for activating receptors, inhibitory receptors and NK cells," Semin. Immunol., 12: 139–147, 2000. doi: 10.1006/smim.2000.0216, available online at http://www.idealibrary.com on IDEAL®.

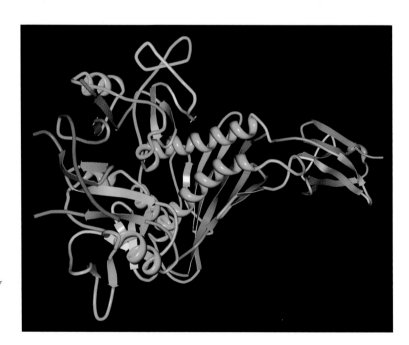

Figure 15-2. Crystal structure of human MHC class I chain–related gene A in complex with NK cell receptor NKG2D. Reproduced from PDB ID: 1hyr. P. Li, D.L. Morris, B.E. Willcox, et al. Complex Structure of the Activating Immunoreceptor NKG2D and Its MHC Class I-Like Ligand MICA. *Nature Immunology* 2 pp. 443 (2001) (PubMed entry 11323699) and http://www.ebi.ac.uk/msd-srv/msdlite/images/1hyr600.jpg.

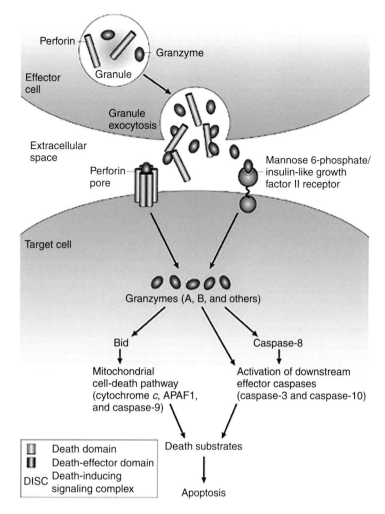

Figure 15-3. Perforin is expressed mainly by cytotoxic T cells and NK cells and is stored in cytotoxic granules, together with granzymes and other proteins. On recognition of a target cell, the granules are released, and perforin monomers insert themselves into the target cell membrane and polymerize into channel-forming aggregates. These perforin pores can cause osmotic lysis of the target cell and allow granzymes to enter the target cell and induce apoptosis through various downstream effector pathways, including effector caspases, Bid, and the mitochondrial cell-death pathway. Recent studies suggest that granzyme B can also enter the target cell without the need for a perforin channel, possibly through its binding to the mannose 6-phosphate/insulin-like growth factor II receptor, followed by endocytosis. Reproduced from M.R.M. van den Brink and S.J. Burakoff, "Cytolytic pathways in haematopoietic stem-cell transplantation," Nat. Rev. Immunol., 2: 273, 2002.

Killer cells have at least two mechanisms to destroy tumor cells: **perforin-granzyme–mediated necrosis** (Figure 15-3) and induction of apoptosis of the cancer cell induced by binding TNF family ligands (Figure 15-4) to receptors on the surface of the NK cell. *Perforin binds to the tumor cell membrane–producing pores* (Figure 15-5), *allowing the entry of granzymes and changing cell permeability to produce*

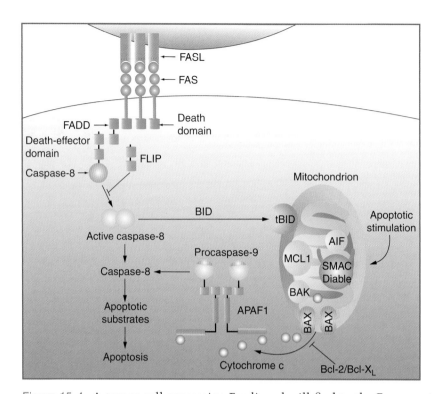

Figure 15-4. A tumor cell expressing Fas ligand will find to the Fas receptor on a natural killer cell membrane followed by trimerization of the liganded receptor *(top)*. This causes cell activation, allowing death domains to interact, including FADD. FADD recruits and activates procaspase-8 to its active form. If caspase-8 levels are high, it directly cleaves caspase-3 in the intrinsic pathway. If caspase-8 levels are low, it cleaves tBID into BID, which releases cytochrome c from the mitochondria. Cytochrome c interacts with Apaf-1 and dATP and procaspase-9 is converted to caspase-9, which cleaves procaspase-3 to the active form producing apoptosis through the cleavage of key proteins and DNA in the cell. *AIF*, apoptosis inducing factor; *APAF1*, apoptotic peptidase-activating factor; *BAK*, mediator p53–induced apoptosis; *BAX*, Bcl-2–associated x protein; *Bcl*, B-cell lymphoma (factor); *BID*, a proapoptotic member of Bcl-2 family; *FADD*, Fas-associated protein with death domain; *FAS*, apoptosis-mediating surface antigen; *FASL*, Fas ligand = apoptosis-mediating surface ligand = APO1 antigen, activates Fas; *FLIP*, FLICE-like inhibitory protein; *MCL1*, an antiapoptotic member of the Bcl-2 family; *SMAC*, second mitochondrial-derived activator of caspase; *tBID*, c-terminal BID fragment.

osmotic lysis—in addition to intracellular events caused by granzymes that lead to apoptosis (Figure 15-6). Perforin is synthesized as an inactive 70-kDa precursor, which is cleaved to the 60-kDa active form. The propeptide piece occurs at the boundary of a C2 domain (Figure 15-7) that can bind to phospholipid membranes in a calcium-dependent manner and begin pore formation. The crystal structure of human granzyme B is shown in Figure 15-8.

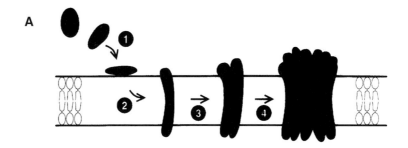

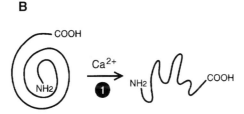

Figure 15-5. The postulated mechanism of action of perforin. **A.** In the presence of Ca^{2+}, perforin monomers may undergo conformational changes and bind to the membrane lipid bilayer *(step 1)*, insert into it *(step 2)*, and subsequently aggregate to form homopolymeric pore structures *(steps 3 and 4)*. This structure appears to perturb membrane permeability and result in osmotic lysis.
B. The hypothetical conformational change occurring in the presence of Ca^{2+} *(step 1)* probably unmasks the hidden hydrophobic and/or amphipathic regions of the originally hydrophilic perforin molecule. Reproduced from Figure 1 of C.-C. Liu, C.M. Walsh, and J.D.-E. Young, "Perforin: structure and function," *Immunol. Today*, 16: 194, 1995.

Two *of receptors on the surface of the NK cell are a killer cell inhibitory receptor, KIR, (KIR2DL2), an example of an immunoglobulin-like receptor, and Ly-49A, which binds an MHC class I ligand (Figure 15-9).* Pre-A-NK cells express a receptor (IL-2R-β/γ) through which IL-2 rapidly converts the pre–A-NK cell to the functionally competent A-NK cell. On the other hand, normal cells express an MHC class I chain–related antigen either at very low levels or not at all, and A-NK cells that are surveying the microenvironment do not attack them. Thus, NK cells have receptors that, when complexed with a tumor antigen, produce killing and, on the other hand, have receptors that recognize normal antigens on normal cells; this complex inhibits the killing program (Figure 15-10). Various photographs of NK cells are shown in Figure 15-11. The signal transduction pathway leading from the killer receptors on the surface of the NK cell to killing of tumor cells through the secretion of granular perforin and granzymes is shown in Figure 15-12. Figure 15-13A summarizes the activating cell surface receptors expressed by human NK cells and their potential ligands, and Figure 15-13B summarizes inhibitory receptors expressed on the surface of human NK cells and their potential ligands. Probably, individuals who do not develop cancer in their lifetimes have high levels of A-NK cells, whereas those who do succumb to cancer have low levels of A-NK cells.

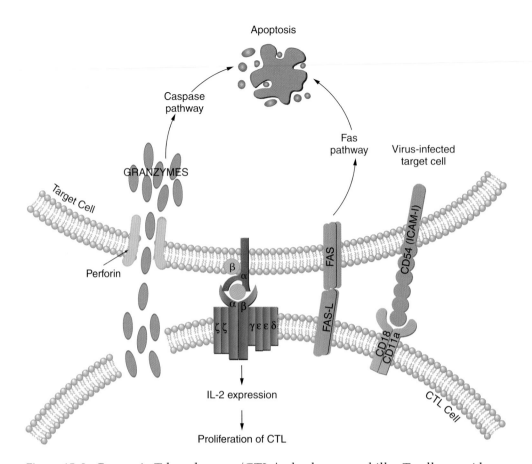

Figure 15-6. Cytotoxic T lymphocytes (CTLs), also known as killer T cells, provide a cell-mediated response to specific foreign antigens associated with cells. CTLs only respond to foreign antigen when it is presented bound to the MHC-1 expressed on the surface of all cells. CTLs do not respond to soluble antigen, but induce apoptosis in viral-infected cells and in cancer cells. When the complex of antigen bound to MHC-1 is bound to antigen-specific T cell receptor, the cytotoxic T cell induces apoptosis in the target cell primarily by two pathways, one involving perforin-mediated apoptosis and the other involving Fas/Fas-ligand interaction. When CTLs are activated by recognition of specific antigen on a cell, they release perforin proteins that integrate into the membrane of the target cell and organize to form a membrane pore. This allows the protease granzyme to enter the cell and activate the apoptotic caspase proteolytic cascade and also allows other molecules to cross the cell membrane and trigger osmotic lysis of the membrane. The interaction of T-cell Fas ligand with the Fas receptor in the target cell can also activate the caspase cascade and other pathways involved in apoptosis (see the Fas Signaling pathway). The interaction of a CTL with antigen-MHC I complex activates the CTL to proliferate and amplify the clone of T cells that respond to that antigen, amplifying the immune response against that specific antigen. *CTL*, cytotoxic T lymphocyte; *Fas*, APO-1 = CD95 = receptor for extrinsic apoptotic ligands, such as Fas ligand (Fasl); *ICAM*, intercellular adhesion molecule; *IL2*, interleukin-2. Redrawn from http://www.biocarta.com/pathfiles/ctlPathway.asp.

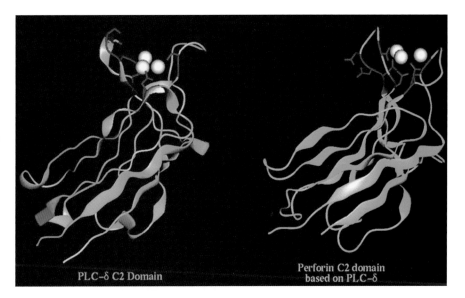

Figure 15-7. Ribbon diagram of C2 domain of PLC-δ and the perforin C2 domain, based on this structure. Aspartic acids in the calcium-binding site are shown in *red*, calcium in *white*. PLC, phospholipase C. Reproduced from Figure 7 of R. Uellner et al., "Perforin is activated by a proteolytic cleavage during biosynthesis which reveals a phospholipids-binding C2 domain," *EMBO J*, 16: 7287–7296, 1997.

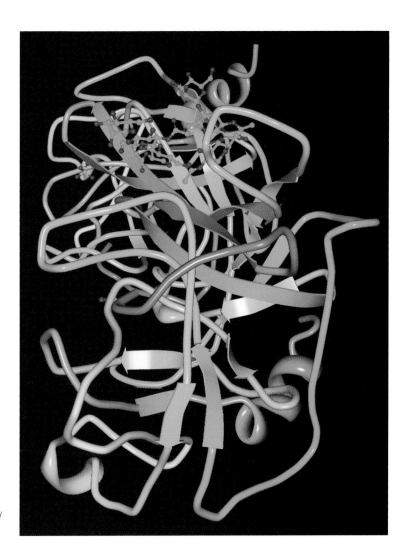

Figure 15-8. Crystal structure of human granzyme B. Reproduced from http://www.ebi.ac.uk/msd-srv/msdlite/images/1fq3600.jpg.

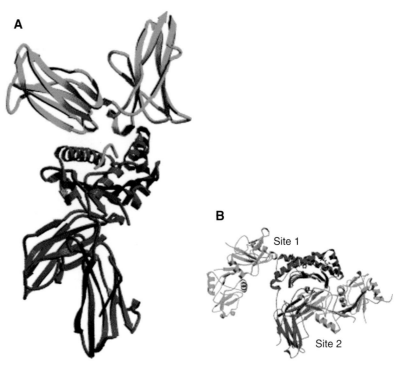

Figure 15-9. **A.** The first crystal structure of a KIR (KIR2DL2) with its ligand (HLA-Cw3) at 3-Å resolution. **B.** The first crystal structure of the Ly-49A homodimer bound to its class I ligand H-2D^d at 3-Å resolution. Reproduced from Figures 1 and 2, respectively, from M. Colonna et al., "A high-resolution view of NK-cell receptors: structure and function," *Immunology Today*, 21: 428–431, 2000.

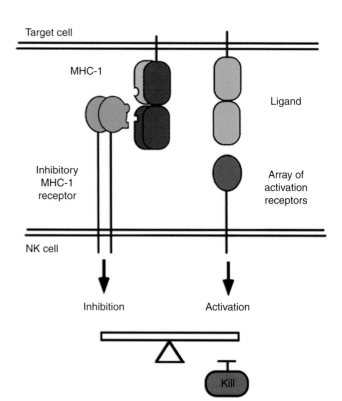

Figure 15-10. NK cells (lower part of figure) have receptors that recognize antigens on tumor cells that induce a killing program, and the NK cell also has inhibitory receptors that recognize normal MHC-1 peptides leading to inhibition of the killing program. Reproduced from http://www.lau.licr.org/images/NKTCDGFig1.gif

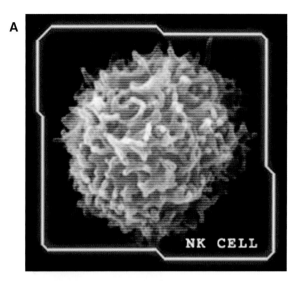

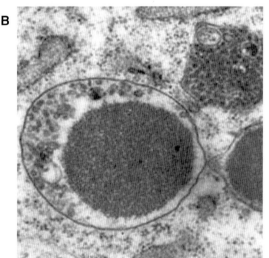

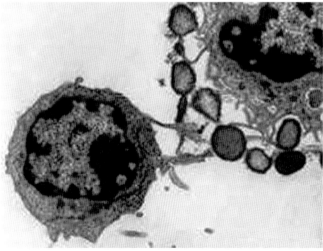

Figure 15-11. **A.** Scanning electron micrograph of an NK cell. **B.** High-power view of an NK cell granule containing perforin, granzymes, and other mediators. **C.** Photo of an NK cell *(left)*, preparing to destroy a cancer cell by attaching a tentacle with killer proteins to the cancer cell. **D.** An NK cell *(N)* attached to a target cell *(T)*. Part A reproduced from http://library.thinkquest.org/03oct/00520/gallery/photos/nkcell.jpg. Part B reproduced from http://www.staff.nci.ac.uk/colin.brooks/research.html. Part C reproduced from http://www.healthrecipes.com/images/nckcell1.jpg.

Continued

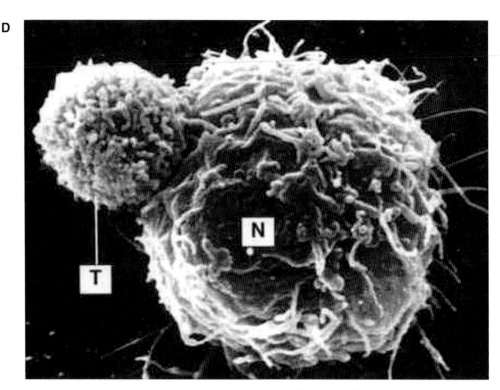

Figure 15-11, cont'd Part D reproduced from http://www.healingcancernaturally.com/Maars_image3.jpg.

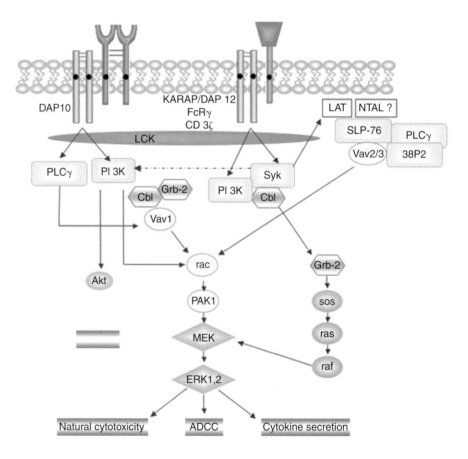

Figure 15-12. NK cell signaling pathways. A schematic representation of the transduction pathways initiated upon engagement of oligomeric activating receptors coupled to DAP10 or ITAM-bearing polypeptides is proposed. Charged transmembrane amino acids are represented as *filled circles*. Reproduced from Figure 2 of S. Chiesa et al., "Coordination of activating and inhibitory signals in natural killer cells," *Mol. Immunol.* 42: 477–484, 2005. del:10.1016/j.molimm.2004.07.030. Copyright © 2004 Elsevier Ltd All rights reserved. http://www.sciencedirect.com/science?_ob=ArticleURL&_udi=B6T9R-4F...=1&_urlVersion=0&_userid=10&md5=f33f8690a2b9667ac8b75fof321a47.

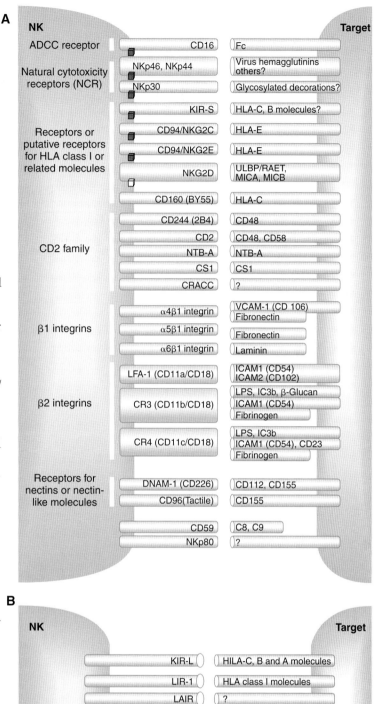

Figure 15-13. **A.** Activating cell surface receptors expressed by human NK cells. NK cell receptors are indicated with their transmembrane adapters when applicable. ITAM-bearing adapters are represented by a *black box* and include KARAP/DAP12, CD3ζ, and FCRγ. The YINM-containing adaptor, DAP10, is represented by an *open box*. Known and potential ligands are listed. **B.** inhibitory cell surface receptors expressed by human NK cells inhibitory receptors are indicated with their known ligands expressed on target cells. All these molecules harbor at least one ITIM in their intracytoplasmic domain. Parts A and B are reproduced from Figures 1 and 3, respectively, of S. Chiesa et al., "Coordination of activating and inhibitory signals in natural killer cells," *Mol. Immunol.*, 42: 477–484, 2005. del:10.1016/j.molimm.2004.07.030. Copyright © 2004 Elsevier Ltd All rights reserved. http://www.sciencedirect.com/science?_ob=ArticleURL&_udi=B6T9R-4F...=1&_urlVersion=0&_userid=10&md5=f33f8690a2b9667ac8b75fof321a47.

Types of Antibodies

Antibodies (Abs) are specific molecules that bind specific antigens (Ans) in a fashion resembling a semienzymic reaction

$$Ab + An \rightleftharpoons Ab\text{-}An$$

in which a high-affinity complex is formed. There are five classes of immunoglobulins (IgG, IgA, IgD, IgE, and IgM) that comprise the antibody repertoire. In human serum, IgG predominates and represents about 75% (approximately 1000 mg/dl) of the immunoglobulins; IgA is about 15% (200 mg/dl), IgM about 9% (120 mg/dl), and IgD and IgE both much less than 1%. IgG, the predominating antibody class, is 150,000 molecular weight (light chain = 25,000; heavy chain = 50,000); IgA has a higher molecular weight of 170,000 to 400,000; IgM is 900,000; IgD is 180,000; and IgE is 190,000. *Plasma antibodies (immunoglobulins) are derived from plasma cells that originate from B cells.* Antibodies are Y-shaped molecules with two long polypeptide chains (heavy chains) and two identical short polypeptides (light chains), as shown in Figure 15-14. The constant regions of the light chains are either κ or λ, and the light chains of an immunoglobulin will be either κ or λ. The heavy chains of the antibody are described in this section. The variable portions of the light chains will have distinctive sequences tailored for the binding of a specific antigen.

The complex of antibody and antigen involves specific recognition and inactivation of the antigen. The binding of the antigen takes place on the Fab region of the antibody. **Fab** is the fragment (F) of the antibody (Ab) that binds antigen (An), and it is made up of the N-terminal domains of the heavy and light chains that are variable (in terms of the sequence of amino acids) and are called VH for the variable heavy chain and VL for the variable light chain (Figure 15-15). The Fab region contains highly conserved amino acids, as well as *a region of greatly variable amino acids (Fv). The variable amino acids are located in six protein loops that join at the end of the Fab fragments to form a hypervariable surface that provides the specificity for binding the antigen.* The structure of the first antibody, an IgG type (IgG2A), crystallized is shown in Figure 15-16, and the crystal structure of a **catalytic antibody** (a few antibodies are able to function like enzymes) is shown in Figure 15-17. Crystal structures of the antigen–antibody complex between antibody and hen egg **lysozyme**

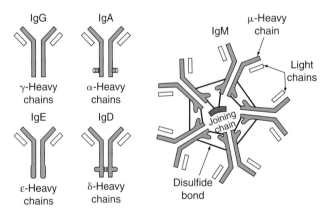

Figure 15-14. The five classes of antibodies.

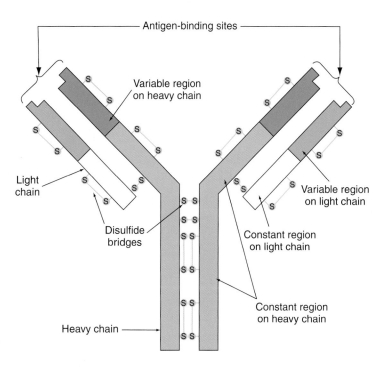

Figure 15-15. Structural regions of the antibody molecule.

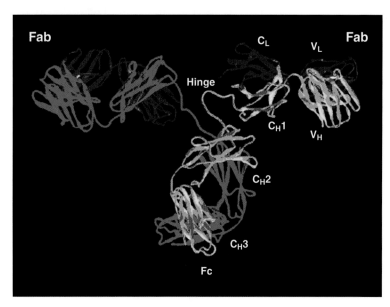

Figure 15-16. Ribbon drawing of the first antibody (IgG2A) crystallized. Reproduced from http://www.antibodyresource.com/intactab.html.

(lysozyme is an enzyme that splits linkages in the primarily gram-positive bacterial cell wall) are shown in Figure 15-18. An antigen–antibody complex formed in a test tube falls out of solution and is known as the precipitin. An extensive antibody–antigen reaction is shown in Figure 15-19.

Antibodies are produced by certain B lymphocytes (plasma cells) and by T lymphocytes (T cells). Antibodies must be extremely diverse to defend against a huge array of foreign substances (recognized as nonself). *B cells (Figure 15-20) can produce 1×10^8 to 1×10^{10} IgG antibodies that differ in the sequence of amino acids at the antigen-binding site, thus specifically tailoring binding.* An antibody is produced in response to an antigen

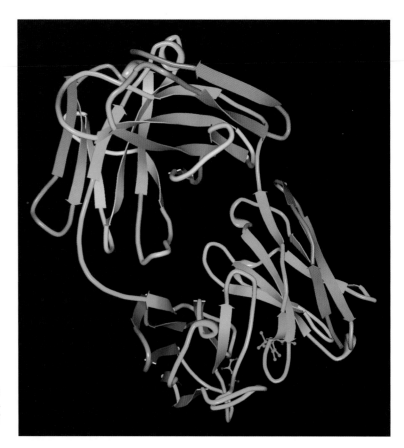

Figure 15-17. Crystal structure of a Diels-alderase catalytic antibody 1E9 in complex with its hapten *(stick models, bottom center)*. Reproduced from PBD ID: 1cle, http://www.ebi.ac.uk/msd-srv/msdlite/images/1cle600.jpg.

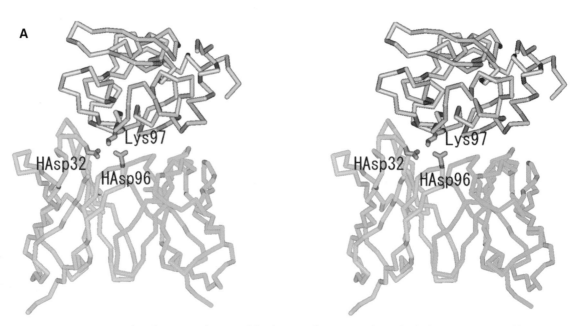

Figure 15-18. **A.** Interface between the variable domain fragment of antibody (HyHEL-10) and hen egg lysozyme. Wild-type lysozyme complex; variable light chain, VL, *green;* VH in *sky blue,* and lysozyme in *pink*. Residues participating in salt bridge formation are in orange. Part A reproduced from Figure 1 of M. Shiroishi et al., "Structural evidence for entropic contribution of salt bridge formation to a protein antigen-antibody interaction: the case of hen lysozyme-HyHEL-10Fv complex," *J. Biol., Chem.,* 276: 23042–23050, 2001.

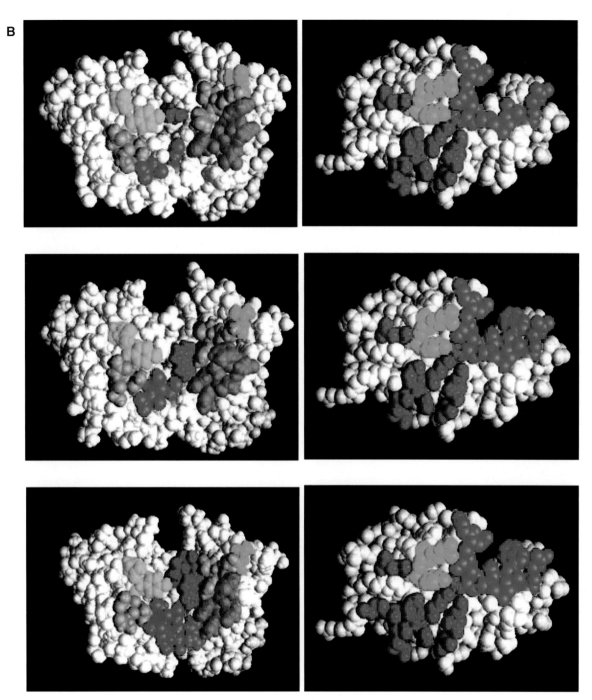

Figure 15-18, cont'd **B.** Composite figure of contact residues in antibodies and antigen, with figure showing the contact residues of the antibodies and HEL. All atoms are shown in space-filling representations, and those involved in electrostatic and van der Waals interactions alone are shown in color. In the *left panels*, antibodies are shown. Residues belonging to CDRs L1, L2, and L3 are shown in *light blue, cyan,* and *dark blue,* whereas those belonging to CDRs H1, H2, and H3 are shown in *orange, magenta,* and *red,* respectively. The framework residues at positions 30_{VH} in all antibodies are shown in *green*. In *right panels,* contact residues of HEL are shown: residues contacted by antibody light-chain residues are shown in *red;* those contacted by heavy-chain residues in *dark blue;* and those by residues of both chains in *green*. The *top, middle,* and *bottom panels* correspond to HH8-HEL, HH10-HEL, and HH26-HEL complexes, respectively. HH8 can be seen to have the fewest contact residues (which are also scattered), whereas HH26 has the most (and tightly clustered) contact residues. Correspondingly, the HEL in complex with HH8 has fewest contact residues, whereas the one in contact with HH26 has the most. HH8, HH10, and HH26 are three different antibodies to lysozyme. *HEL,* hen egg lysozyme; *CDRs,* complementarity-determining regions of the IgG antibody. L1, L2, and L3 and H1, H2, and H3 are CDRs. Part B reproduced from Figure 7 of S. Mohan, N. Sinha, and S.J. Smith-Gill, "Modeling the binding sites of anti-hen egg white lysozyme antibodies HyHEL-8 and HyHEL-26: an insight into the molecular basis of antibody cross-reactivity and specificity," *Biophys. J.*, 85: 3221–3236, 2003.

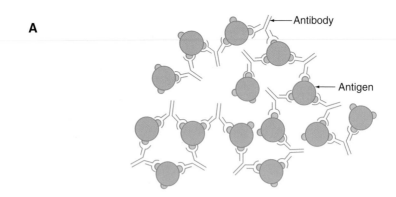

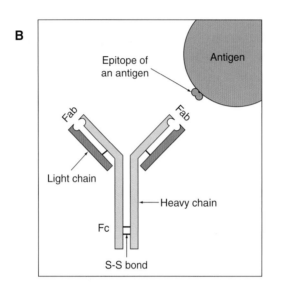
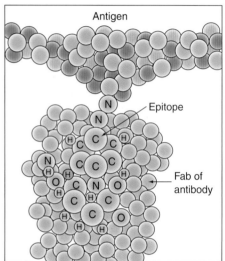

Figure 15-19. **A.** Formation of an antigen-antibody complex that can be extensive when the antigen contains several sites for interaction with antibody. **B.** The Fab portion of the antibody has specificity for binding an epitope of an antigen. The Fc portion directs the biological activity of the antibody. **C.** Epitope of an antigen binding to Fab of an antibody.

(foreign molecule) that has entered the body. The body contains hundreds of thousands of different B cells, and each cell can produce one type of antibody. Each B cell also has membrane sites that interact with a specific antigen. When an antigen reacts with a membrane site on a B cell, the B cell is signaled to reproduce to form a clone of that B cell (plasma cell) that can synthesize a huge amount of its antibody. The interactions between the Fab region of an antibody and the antigen are accomplished through different binding modes consisting of charge–charge interactions, dipole–dipole interactions, hydrogen bonding, and van der Waals interactions (for example, see Figure 15-18). The antibody pocket that binds the antigen accommodates the shape, charge, and hydrophobicity of the antigen by using the different chemical properties of the 20 amino acids that are varied and can be ordered in sequence specifically for any given

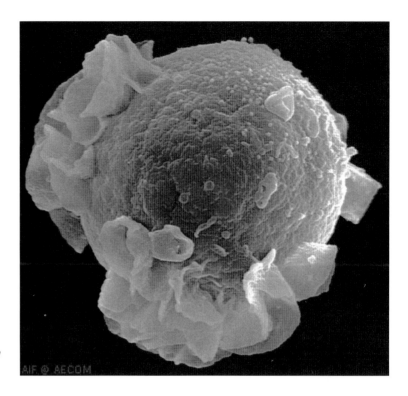

Figure 15-20. B-cell budding virus. Reproduced from http://www.aecom.yu.edu/aif/gallery/sem/sem.htm.

antigen. The high affinity of the antibody for the antigen is in the range of 10^{10} M^{-1}. This variable specificity is made possible through a complex process of gene splicing (see the section on antibody formation later in this chapter).

Polyclonal and Monoclonal Antibodies

A polyclonal antibody is one obtained from an immunized animal, usually an animal that is a different species than the source of the antigen. Thus, a human would be expected to make antibodies if injected with pig insulin, providing that the insulin from the pig is different in sequence at the **epitope** (antigenic determinant) from human insulin. A polyclonal antibody contains a mixture of antibodies that are active against a specific antigen, each antibody recognizing a different epitope or region of the antigen. Because of the multiple interactions of polyclonal antibodies, their binding to the antigen is strong and usually results in a precipitate (in the test tube).

A **monoclonal antibody** *is a single type of antibody against only one epitope of the antigen because it was produced by one type of immune cell and all clones derive from a single parental cell.* To obtain a monoclonal antibody, an inbred mouse (usually) is immunized repeatedly with the antigen. Because the spleen produces B cells that form the antibodies to the antigen, the spleen is removed and the spleen cells (that are not culturable) are fused with immortal myeloma cells. The resulting fused cells are called hybridoma cells. Hybridoma cells can be cultured or can be injected into animals to produce a tumor that can be used as a rich source of the monoclonal antibody. The simplified procedure for obtaining monoclonal antibodies is shown in Figure 15-21.

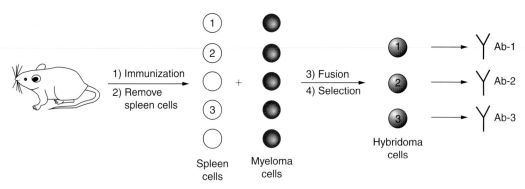

Figure 15-21. Outline of a procedure to produce monoclonal antibodies. In this scheme, three different antigen-specific antibodies are produced by the immunization process. Some hybridoma cells will produce nonspecific antibodies, and they can be screened out by an ELISA (enzyme-linked immunoabsorbent assay) procedure. Redrawn from http://www.antibodyresource.com/antibody.html.

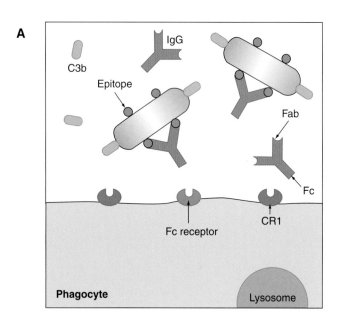

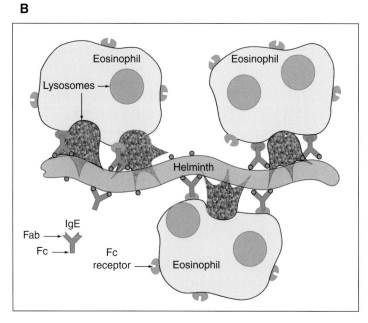

Figure 15-22. **A.** Enhanced attachment (opsonization). The Fab portion of IgG binds to epitopes of an antigen. The Fc portion can now attach the antigen to Fc receptors on phagocytes for enhanced attachment. This is especially important against encapsulated microbes. C3b and C4b from the complement pathways can also attach antigens to phagocytes. **B.** Opsonization of a helminth by IgE and eosinophils. A major function of the cytokines produced by T_h2 cells is to enable B-lymphocytes to activate eosinophils and produce increased amounts of a class of antibodies called IgE against helminths (parasitic worms) and arthropods. IgE act as an opsonizing antibody to stick phagocytic eosinophils to helminths for extracellular killing of the helminths. The Fab portion of IgE reacts with epitopes on the helminth while the Fc portion binds to Fc receptors of activated eosinophils. The lysosomal proteases of eosinophils are able to destroy the tough integument of helminths. IgE also promotes inflammation to recruit phagocytes. Part A reproduced from Figure 7 of http://student.ccbcmd.edu/courses/bio141/lecguide/unit5/humoral/abydefense/opsonization/images/u3fg18a.jpg. Part B reproduced from Figure 9 of http://student.ccbcmed.edu/courses/bio141/lecguide/unit5/humoral/abystructure/u3fg2t.html.

Opsonization

Opsonization is the process by which antibodies bind to an antigen, especially on the surface of an invading microbe, to enhance attachment of the microbe to a phagocytic cell (Figure 15-22).

Antibody Formation

There are two types of light chains, called χ (or κ) and λ, and five types of heavy chains, called α, δ, ϵ, γ, and μ (isotypic variations; Figure 15-14). The letters C and V are used to indicate constant and variable regions. IgG, for example, resembles the other immunoglobulins in that it is composed of two identical light chains and two identical heavy chains. The light chains are composed of two domains, C_L and V_L. The heavy chains have four domains, one variable and three constant: V_H, C_H1, C_H2, and C_H3. The variable regions of the light chains (V_L) and heavy chains (V_H) have different sequences, and any two variable chains obtained from different patients never have been found to have identical amino acid sequences. Hypervariable regions also are referred to as complementarity-determining regions (CDRs) because 5–10 amino acids in these regions contribute to the antigen-binding site. Combinatorial diversity refers to the various combinations of CDRs that give rise to many different antibodies with differing specificities. Heavy chains are interconnected by disulfide bonds, and light chains are connected to heavy chains by disulfide bonds. Disulfide bridges also stabilize structures within the domains. The Fab arms (variable regions, Figure 15-15) are connected to the F_C part (Figure 15-15) by a *hinge region* that generates flexibility.

Gene switching (for example, between χ or κ and λ of a light chain or between α and δ of a heavy chain) between one isotype and another occurs during biosynthesis of the immunoglobulins. In the words of Susumu Tonegawa: "In the genome of the germ-line cell, the genetic information for an immunoglobulin polypeptide chain is contained in multiple gene segments scattered along a chromosome. During the development of bone marrow–derived lymphocytes [B cells], these gene segments are assembled by recombination which leads to the formation of a complete gene. In addition, mutations are somatically introduced at a high rate into the amino-terminal region. Both somatic recombination and mutation contribute greatly to an increase in the diversity of antibody synthesized by a single organism." A joining region of DNA is able to connect the constant region of a chain with a suitable variable light chain region to produce a specific antibody. There are many variable regions in the genome, one of which would be needed to tailor an antibody to bind a specific antigen. The DNA containing the constant region would search for the appropriate DNA constituting the variable region, and the two would be combined by a joining region to produce the gene for the specific immunoglobulin. There are many genes encoding the variable domains of antibodies. When an antibody is being produced, only one of these variable regions is selected and expressed. An idea of the means by which the constant region might search out the variable region is shown in Figure 15-23. A later model was developed to show how the joining

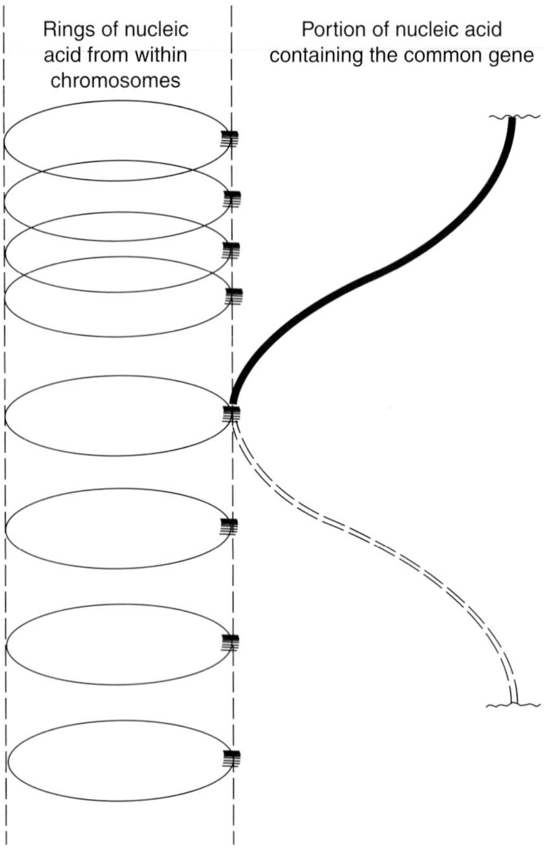

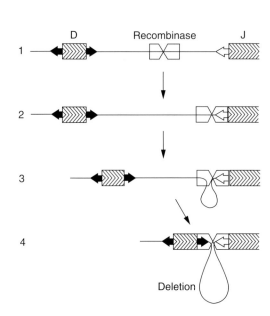

Figure 15-23. Diagram of the proposed genetic mechanism that accounts for the amino acid sequence variations found in L-chains. Genetic material that codes for the "variable" portion of L-chain molecules is inserted into that which codes for the "common" region of amino acid sequence by a mechanism similar to the insertion of the λ-virus into a bacterial chromosome. Redrawn from Figure 3 of W.J. Dreyer and J.C. Bennett, "The molecular basis of antibody formation: a paradox," *PNAS*, 54: 864–869, 1965.

Figure 15-24. Associative tracking model. D and J_H segments are in the same orientation. ⇦, Signal sequence with a 23-bp spacer; ⬅ signal sequence with a 12-bp spacer. Redrawn from Figure 5 of C. Wood and S. Tonegawa, "Diversity and joining segments of mouse immunoglobulin heavy chain genes are closely linked and in the same orientation: implications for the joining mechanism," *PNAS USA*, 80: 3030–3034, 1983.

region (J) combined with a diverse segment (D) of a heavy-chain gene (Figure 15-24).

The various regions of the gene encoding light chains are light (L), variable (V), joining (J), and constant (C); these regions are separated by introns of variable lengths and are located on chromosome 2 in the human. There are about 150 identical L gene segments. There are 150 V segments located next to the L segments. *L and V segments always occur together.* There are five variants of the J segment that encode a peptide of 13 amino acids that link the variable portion to the constant portion of the χ-chain. The 84 residues comprising the constant part of the light chain is encoded by a single C segment (Figure 15-25). Unique V–J combinations occur during the differentiation of the B cell. First, 1 of the 150 L–V tandem segments is selected, and this is linked to 1 of the 5 J segments, giving rise to a somatic gene that is smaller than the germ line gene. A heteronuclear RNA for the χ-chain derives from the transcription of the somatic gene, and then the extra J segments and introns are spliced out. The mature RNA containing the L, V, J, and C segments is transported to

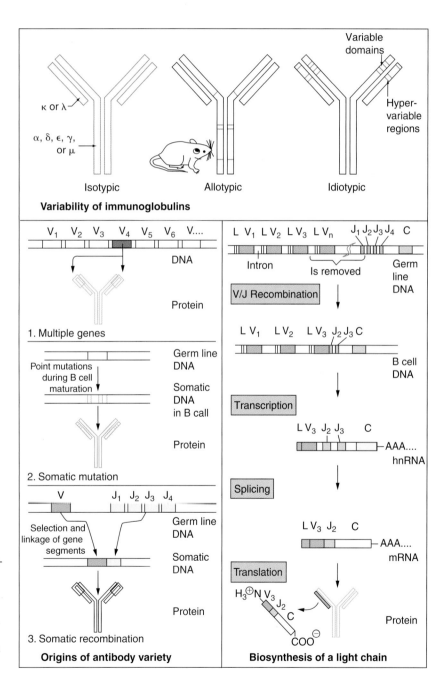

Figure 15-25. Variability of immunoglobulins. Origins of antibody variety and biosynthesis of a light chain. Redrawn from page 271 of *Color Atlas of Biochemistry*, Jan Koolman, Klaus-Heinrich Röhm, translated by Kathryn Schuller, 194 color plates by Jürgen Wirth Thieme Stuttgart, New York, 1996.

the cytoplasm and translated. The final steps in the biosynthetic process of the immunoglobulin are similar to those of any secretory protein.

Autoimmunity

Normally, the immune system distinguishes between self (molecules inside the body) and nonself (molecules outside the body that are different). Sometimes the immune system attacks proteins or other molecules inside the body, known as an autoimmune reaction, during a lifetime, but these are short lived and become resolved. This sometimes occurs following infection. Occasionally, in about 5% of people, the reaction exists over a long period and can even lead to an **autoimmune disease.** These diseases occur when antibodies are produced that attack normal bodily proteins. These diseases range from attack of an organ to a widespread attack that is manifested throughout the body.

Because the frequency of occurrence of certain autoimmune diseases is greater in one sex compared to the other, endocrine factors may be involved. For example, in females, Graves' disease and Hashimoto's disease occur 4–5 times more often and 10 times more often, respectively, than in males. **Systemic lupus erythematosus (SLE)** occurs 10 times more frequently in females than in males, and ankylosing spondylitis is about 4 times more common in males than in females. For rheumatoid arthritis, 3 times as many women are affected as men, and women are 9 times more affected than men by SLE and 25 times more affected than men for autoimmune thyroiditis. Women outlive men and have a more aggressive immune system than males; this is probably the reason for the greater occurrence of these autoimmune diseases in females. Although unclear as yet, estrogen may have a role in activating the immune system. Rheumatoid arthritis and other autoimmune diseases are often reduced in severity during pregnancy (during which there are profound hormonal changes). Genetic factors, as well as viral infections, also are thought to be involved sometimes in the generation of autoimmune disease.

Examples of autoimmune diseases are shown in Figure 15-26. Some of these diseases affect specific organs and are systemic, others affect a specific organ only, and still others are primarily systemic.

Graves' Disease

Graves' disease results from the development of an autoimmune antibody that recognizes thyroid-stimulating hormone receptor or thyrotropin receptor (TSHR) as its antigen. The antibody is sometimes known as the **long-acting thyroid stimulator (LATS).** As shown in Figure 8-53, which summarizes the hypothalamic–pituitary–thyroid axis, TSH stimulates the thyroid follicle to produce and secrete the thyroid hormone: tetraiodothyronine (T_4) and some triiodothyronine (T_3). One function of the end product hormone is to feed back negatively on the further production of TRH and TSH by the hypothalamus and the thyrotropic cell of the anterior pituitary. *An antibody that binds to TSHR on the thyroid follicle cell membrane activates the receptor constitutively (Figure 15-27) so that the*

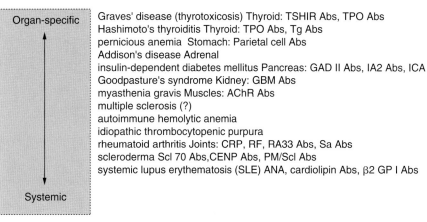

Figure 15-26. Examples of autoimmune diseases that affect specific organs or that are systemic affecting many bodily systems. *Abs,* antibodies; *AChR,* acetylcholine receptor; *ANA,* antinuclear antibodies; *TSHR,* thyrotropin receptor. Redrawn from http://www-immuno.path.cam.ac.uk/~immuno/part1/lec12/spectrum.gif.

TSHR = thyrotropin receptor; Abs = antibodies; TPO = thyroid peroxidase; Tg = thyroglobulin; GAD = glutamic acid decarboxylase; IA2 = protein tyrosine phosphatase-like protein; ICA = islet cell antibodies; GBM = glomerular basement membrane; CRP = C-reactive protein; RF = rheumatoid factor; RA33 = heteronuclear RNPA2; Sa = antibody in arthritis; Scl = scleroderma antibody; CENP = centromere antigen; PM/Scl = nuclear/nucleolar particle; GP = β2 glycoprotein; AChR = acetylcholine receptor; ANA = antinuclear antibodies

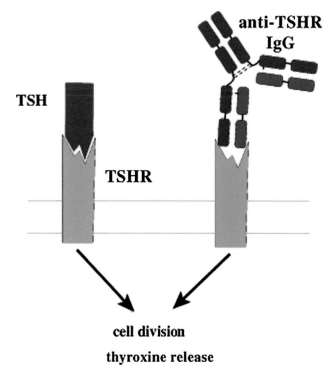

Figure 15-27. Activation of the TSH receptor (TSHR). On the left, the receptor is activated normally by TSH, which stimulates secretion of thyroid hormone (T4/T3). The thyroid hormone feeds back negatively on the thyrotrophic cell of the anterior pituitary to inhibit further release of TSH; therefore, the ligand for TSHR will disappear and the production of T4/T3 will stop until the next release of TSH, when needed. By contrast, on the right, the autoimmune antibody (LATS) binds directly to the TSHR turning the receptor on without regulation, so T4/T3 continue to be secreted without control, causing a hyperthyroid condition and thyrotoxicosis. Reproduced from http://www-immuno.path.cam.ac.uk/~immuno/part1/lec12/lec12_97.html.

follicle is chronically stimulated to produce thyroid hormones. This will cause the upper part of the pathway to shut down because of the negative feedback caused by the excess thyroid hormones, but the thyroid follicle will be turned on most of the time and hyperthyroid activity, and sometimes thyrotoxicosis, will result. An opposite condition is Hashimoto's disease, which produces hypothyroidism. In this condition, autoantibodies are produced in the body that bind thyroglobulin (precursor of thyroid hormones) and thyroid peroxidase (see Figure 8-63). This disease is complex, and there are several theories to explain effects on thyroid follicles (thyrocytes).

MHC Involvement

Genetics plays a role in many autoimmune diseases. The MHC is key in that linkages exist between MHC allotypes (genetically fixed variations from one individual to the next in the same species) and several specific autoimmune diseases. Those diseases that are not affected by MHC genes may be affected by other type genes in close linkage. The idea of a linkage disequilibrium is pictured in Figure 15-28. Many genetic loci are involved in susceptibility of autoimmune diseases, such as insulin-dependent diabetes and rheumatoid arthritis. Generally, there has to be a combination of susceptibility alleles, rather than a single allele, to predispose toward a specific disease. These autoimmune diseases are not the same as a truly inherited disease. In some autoimmune diseases, such as diabetes, there is an important role of environmental factors; it is known that dietary factors are important. Supporting this idea is the knowledge that identical twins have only a 20–40% chance to develop the same autoimmune disease. A major environmental factor, in many cases, is infection, especially by viruses.

Many human autoimmune diseases concern the MHC region on chromosome 6 p21. The particular alleles of HLA-DR or HLA-DQ genes (MHC genes) are associated with rheumatoid arthritis, multiple sclerosis, and type 1 diabetes (insulin-dependent diabetes). MHC proteins bind candidate peptide autoantigens, and it is thought that a complex of this type is responsible for autoimmunity. The interaction of MHC protein with an autoantigen peptide results from critical polymorphic residues in the MHC protein that determine the charge and shape of the binding site for the peptide. Figure 15-29 shows how the binding domain of an MHC protein has adapted its pocket to the binding of a specific peptide. MHC class II molecules that may be involved in promoting autoimmune disease could promote the selection of potentially pathogenic T cells in the

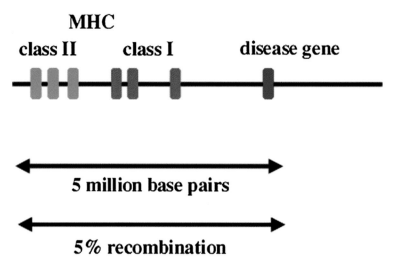

Figure 15-28. A genetic map showing linkage disequilibrium. Reproduced from http://www-immuno.path.cam.ac.uk/~immuno/part1/lec12/lec12_97.html.

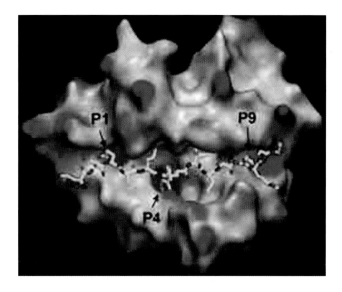

Figure 15-29. Peptide *(stick model)* binding by an MHC protein. Reproduced from http://www.eurekah.com/dbimages/Oksenberg03Wuncherpfening.jpg.

thymus that later could be involved in the presentation of peptides from a target organ. In relation to many autoimmune diseases, the target autoantigens associated with a specific autoimmune disease are unknown. There are general mechanisms for autoimmune pathology, including direct antibody mediated effects (an example is Graves' disease; see Figure 15-27), T cell–mediated effects, and immune complex effects. In addition to Graves' disease, where autoantibodies are formed against the TSH receptor, myasthenia gravis is an example in which autoantibodies are developed against a receptor, the acetylcholine receptor (shown in Figure 15-30), that block neuromuscular transmission from cholinergic neurons (the binding of acetylcholine to the receptor is blocked, causing downregulation of the receptor). This is the opposite effect of Graves' disease, where the autoantibody to the TSH receptor binds to the receptor, causing chronic activation and overactivity of the thyroid gland.

T cell–mediated damage implies that certain T cells recognize an autoantigen that leads to tissue damage without production of an autoantibody. Cytotoxicity can occur by CD8+ cytotoxic T cells (CTLs) and induce apoptosis (programmed cell death) of the target tissue by Fas ligand

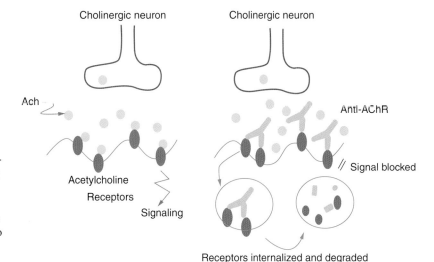

Figure 15-30. Autoantibodies to the acetylcholine receptor (AChR) block the binding of acetylcholine (Ach) to the receptor in cholinergic neurons, causing downregulation and development of myasthenia gravis. Redrawn from http://www-immuno.path.cam.ac.uk/~immuno/part1/lec12/myasthenia.gif.

(FasL) on the T cell membrane. Activated CTLs usually express high levels of FasL on their membranes that complex with the Fas receptor on the membrane of a target cell (Figure 15-31). A model of the FasL–receptor from its crystal structure is shown in Figure 15-32. Tissue cells also can self-destruct by TNFα or other cytokines (Figure 15-33). The structure of TNFα is shown in Figure 15-34, and the TNFα receptor (TNFαR1) is shown in Figure 15-35. Macrophages can be recruited and activated and

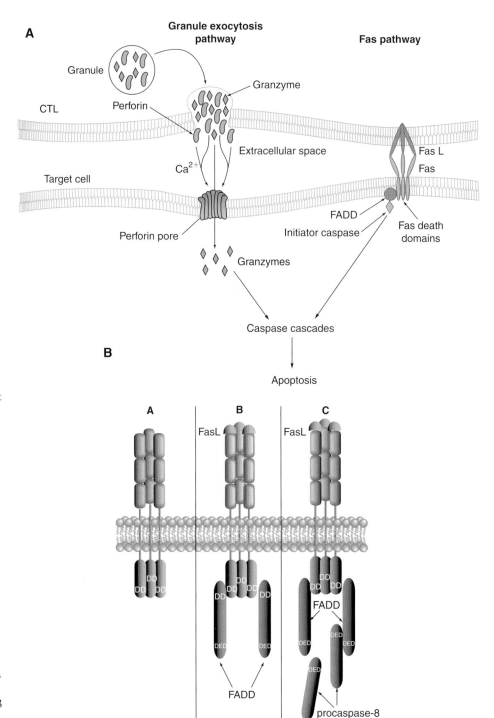

Figure 15-31. **A.** Fas pathway showing the Fas ligand on the surface of a cytotoxic T lymphocyte binding to the Fas receptor on the surface of a target cell. The complex of Fas/FasL leads to apoptosis of the target cell. **B.** Close up model of the Fas receptor *(A)*. In *B*, FasL *(orange trimer at top)* is shown bound to Fas and FADD (Fas-associated death domain) on the cytoplasmic domain of Fas receptor; *C* shows the receptor complex and the interaction of FADD with procaspase-8, which leads to apoptosis *(A, above)*. Part A redrawn from http://homepages.uel.ac.uk/cho5077v/Image2.jpg and Part B redrawn from http://images.google.com/imgres?imgurl=http://www.sghms.ac.uk/de...num%3D10%26hl%3Den%26lr%3D%26client%3Dsafari%26rls%3Den%26sa%3DN.

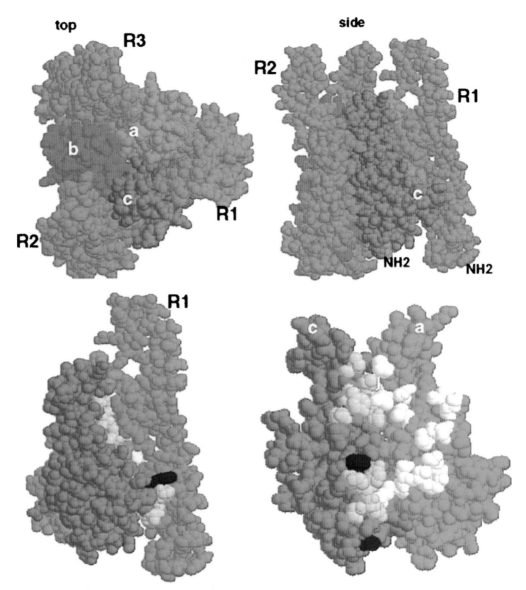

Figure 15-32. *Upper figures* show FasL–receptor complex. Three receptor chains are in *gold* and FasL, the trimer is in *green*. *Bottom* shows different views of the complex. Specific subunits of the ligand are in *dark* and *light gray* (phe-110 is in *red*; Ser-70 is in *blue* in polymorphisms). Reproduced from http://www.pnas.org/content/vol94/issue12/images/large/pq0870523002.jpeg.

can lead to local tissue destruction. Many autoimmune diseases are accompanied by inflammation, and antiinflammatories are sometimes helpful. In systemic lupus erythematosis (SLE), for example, glucocorticoid treatment can palliate the symptoms and sometime reverse the progress of the disease. The occurrence of SLE is 1 in 2000 Americans and as many as 1 in 250 African American women.

Genetic susceptibility to SLE involves immune function (MHC), as well as many other factors. Some of these other factors are thought to be nutritional susceptibility (dietary fat), hormonal susceptibility (menstrual cycle patterns, hormone replacement medications, oral contraceptives, menopausal age, bodily sex hormone, and prolactin levels), environmental

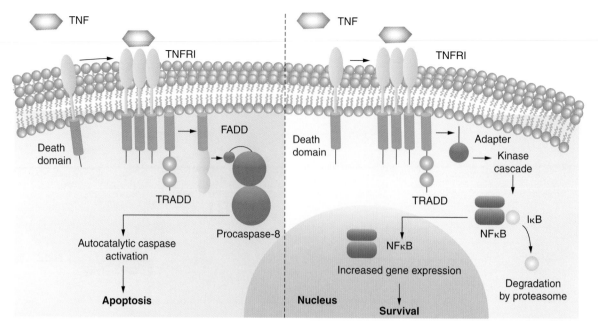

Figure 15-33. Cell death caused by TNF-α in combination with the TNF receptor *(left)* with subsequent signaling leading to cell death (apoptosis).

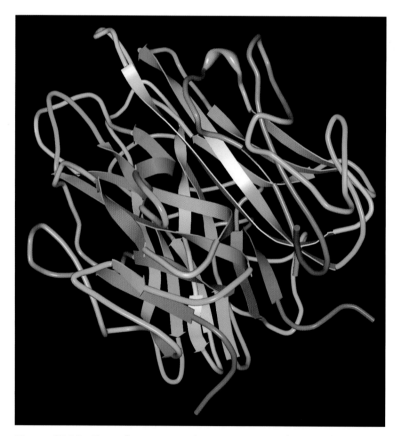

Figure 15-34. Crystal structure of tumor necrosis factor α (TNFα) R3ID mutant. Reproduced from PDB ID: 1a8m. C. Reed, Z.Q. Fu, J. Wu, et al. Crystal Structure of the TNF-Alpha Mutant R31D with Greater Affinity for Receptor R1 Compared with R2. *Protein Engineering* **10** pp. 1101 (1997).

exposures (silica, solvents, and smoking), and possibly infectious exposures (Epstein-Barr viral infection, cytomegalovirus, herpes zoster, and retroviral infections). The most important members of the immunoglobulin gene superfamily, including the MHC proteins, are shown in Figure 15-36. Antibody, T cell receptor, and MHC protein function together in the immune response. T cell receptors are located on the cell membranes of T cells, and they consist of two chains, an α-chain and a β-chain. The receptor has one specific binding site. The binding site can be diverse because there are multiple coding exons for this site and somatic recombination (as is the case with generating immunoglobulin specificity) makes possible a tailored binding site. The site to which the T cell receptor binds to an MHC protein (for example, type HLA-A2) involves the site

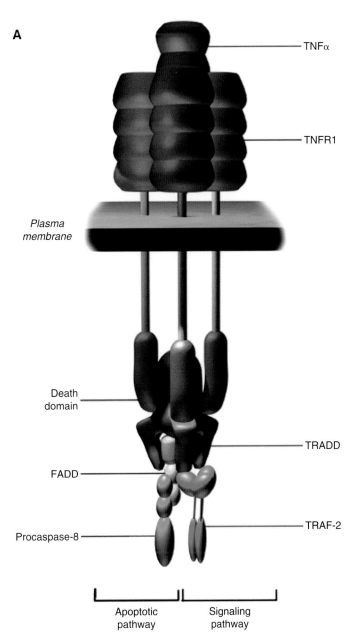

Figure 15-35. **A.** Cartoon of TNFα and TNFαR1, whose complex leads to apoptosis *(left legend)* or nonapoptotic pathway *(right legend)*.
B. Model of TNFR1 *(in orange)* with TNFα *(in blue)*. Part A redrawn from http://www.sgul.ac.uk/depts/immunology/~dash/apoptosis/tnfr.jpg.

Continued

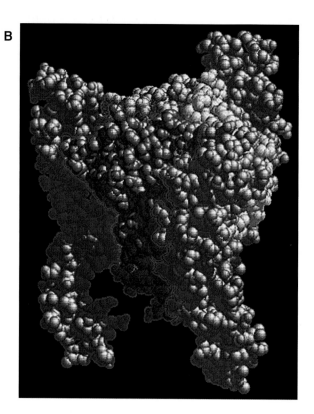

Figure 15-35, cont'd Part B reproduced from http://asterix.cs.gsu.edu/weber/model.html.

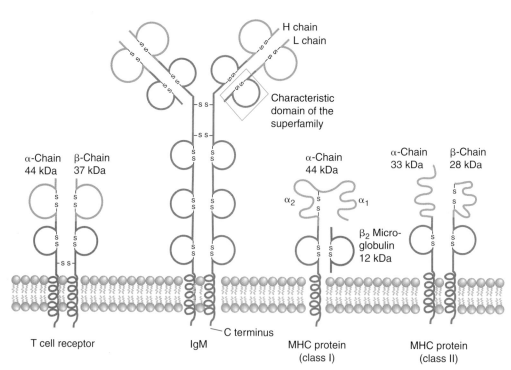

Figure 15-36. Major components of the immunoglobulin gene superfamily. Members of this family have characteristic domains *(blue box)*. Constant regions are shown in *black* and *green*. Variable regions are in *orange*. Homologous regions are in the same color. Most members have transmembrane regions near the C-terminus and are integral membrane proteins (exceptions are immunoglobulins IgA, IgD, IgE, IgG, and IgM). Disulfide bonds occur in the same molecule or between molecules. Reproduced from page 273 of *Color Atlas of Biochemistry*, Jan Koolman, Klaus-Heinrich Röhm, translated by Kathryn Schuller, 194 color plates by Jürgen Wirth Thieme Stuttgart, New York, 1996.

where the MHC protein has bound a foreign peptide. The foreign peptide derives from an interaction between a B cell and the foreign structure; the foreign structure is bound by B cell antibodies (or antibodies on the surface of another cell type), and it can be internalized and broken down to produce a peptide (foreign peptide) to which the T cell receptor (on a T cell) makes contact with an MHC protein on the surface of a B cell. These interactions are shown in Figure 15-37. The structure of a complex among a human T cell receptor, a viral peptide (foreign peptide), and a member of the MHC family is shown in Figure 15-38. To understand these components more clearly, Figure 15-39 shows a high-resolution crystal structure of the "foreign peptide" (JM22-MP; 58–66) in complex with MHC protein

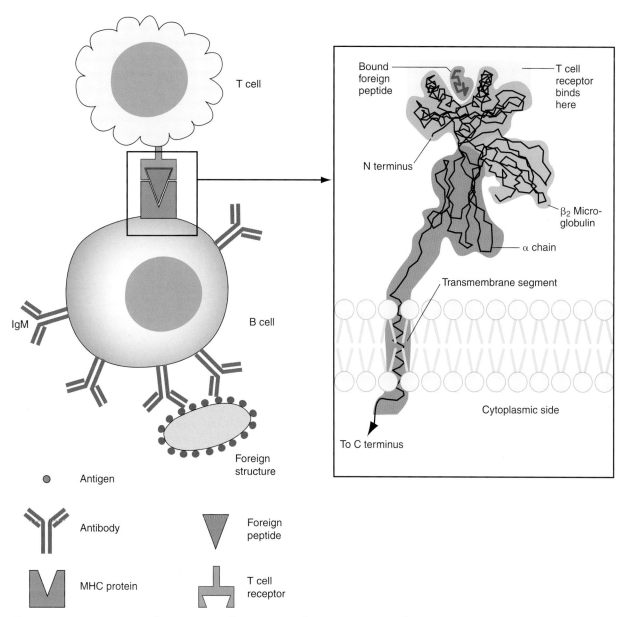

Figure 15-37. Interactions between T cell receptor and an MHC protein (bound to a foreign peptide) on a B cell surface (A). The foreign substance is recognized by IgM antibodies on the B cell membrane, and its structure is degraded to avail a foreign peptide bound to the B cell MHC protein, which, in turn, is bound by the T cell receptor. In B, the site of binding of the T cell receptor on the MHC protein is amplified. Reproduced from page 273 of the reference listed for Figure 15-36.

Autoimmunity

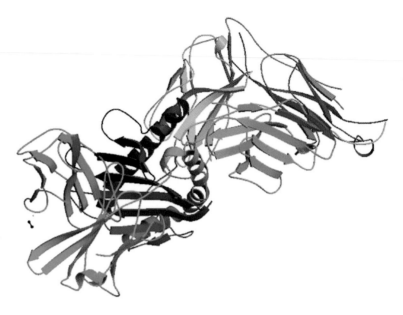

Figure 15-38. Complex among human T cell receptor, viral peptide (TAX), and HLA-A0201. Reproduced from PDB ID: 1a07. D.N. Garboczi, P. Ghosh, U. Utz, et al. Structure of the Complex Between Human T-cell Receptor, Viral Peptide and HLA-A2. *Nature* **384** pp. 134 (1996).

(HLA-A2). The foreign peptide is shown in yellow in Figure 15-39A, in green in part B, and yellow in part C.

MHC proteins also are referred to as **human leukocyte-associated (HLA) antigens,** and these proteins are so diverse that no two people, except for identical twins, are likely to have the same MHC proteins. Class I MHC proteins are located on the surfaces of nucleated cells, and they have one α-chain in association with a $β_2$-microglobulin (a small protein). Class II MHC proteins have two membrane-spanning regions, together with an α-chain and a β-chain. With these proteins, the body can differentiate between immune system cells and other cell types. Models of these structures are shown in Figure 15-36.

Theory on the Development of Type I Diabetes (Insulin-Dependent Diabetes Mellitus)

Autoimmune type I diabetes mellitus occurs in 1 in 800 people, usually in children or young adults, and is, therefore, an important health problem. The cause of type I diabetes has escaped researchers for years. A recent theory is *that somehow a "foreign peptide" resembles proinsulin or insulin enough to generate (or that insulin itself generates) an immune response to bodily insulin and that the autoantibodies produced by autoreactive T cells destroy the beta cells of the pancreas (source of insulin in the body). Alternatively, T cells might be generated in the thymus that produce anti-insulin antibodies, and a level of tolerance might prevail in certain genetic backgrounds that allows these T cells to survive (instead of being selected out) and destroy the beta cells of the pancreas.* Major mediators of this destruction are autoreactive CD4 and CD8 T cells, which recognize target antigens as peptide fragments presented by MHC molecules (Figure 15-37).

Some investigators now believe that insulin, itself, is the target of T cell attack. Experimental evidence comes from nonobese diabetic (NOD) mice, which are deficient in both preproinsulin 1 and preproin-

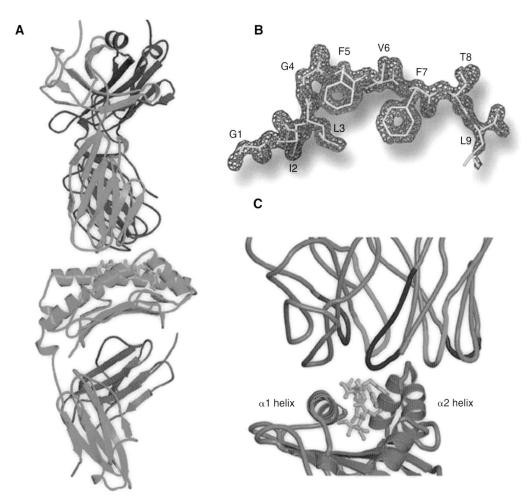

Figure 15-39. High-resolution crystal structure of the JM22-MP(58-66)-HLA-A2 complex. **A.** Ribbon representation of the HLA-A2 heavy chain *(green)*, β_2M *(coral)*, TCR α-chain *(cyan)*, and TCR β-chain *(magenta)*. The MP(58-66) peptide *(yellow)* is shown in *stick representation*. **B.** F_o–F_c electron density for the MP(58–66) peptide, shown as *green chicken wire* (contoured at 2.5σ). The view is the same as in *A*. **C.** Diagram highlighting the CDR loops of the JM22 in complex with MP (58–66)-HLA-A2. The CDR loops are colored as follows; V_α CDR1, *green;* V_α CDR2, *red;* V_α CDR3, *blue;* V_β CDR1, *magenta;* V_β CDR2, *orange;* V_β CDR3, *cyan*. The view is along the peptide-binding groove—that is, orthogonal to the view in *A*. Reproduced from Figure 1 of G.B.E. Stewart-Jones, A.J. McMichael, and J.I. Bell, "A structural basis for immunodominant human T cell receptor recognition," *Nat. Immunol.*, 4: 657–663, 2003.

sulin 2 (mice express two forms of insulin, whereas humans express only one form), as shown in Figure 15-40. The NOD mice, being deficient in insulin, die at an early age. However, a transgene could be introduced into these animals to promote the expression of a mutant insulin 2 molecule (amino acid 16 in the β chain was changed from a tyrosine to an alanine [Y-16 B-A]). An insulin molecule, which was not recognized by autoreactive T cells as a target, was therefore expressed that was metabolically active (allowing the mice to survive), and no autoimmune symptoms developed. The transgene (a gene from one

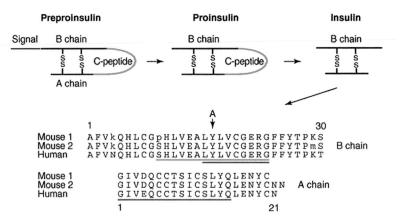

Figure 15-40. Sequence of insulin production from preproinsulin. Insulin is produced as a much larger molecule, preproinsulin. The signal peptide is cleaved to generate proinsulin. Finally, the c-peptide is cleaved to insulin, which consists of a B-chain linked to the A-chain by two disulfide bonds. The sequence of the two insulins, 1 and 2, expressed in mice, compared with insulin expressed in humans, is shown. The area underlined in *green* shows the peptide B9-23 of mouse insulin 2 recognized by pathogenic CD4 T cells. The area underlined in *red* shows the peptide B15-23, recognized by pathogenic CD8 T cells. The area in the A chain, underlined in *blue*, showing amino acids 115, is recognized in the human T cells. The *arrow* indicates the tyrosine residue in the B chain that is substituted with alanine. Reproduced from Figure 1 of F.S. Wong, "Insulin: a primary autoantigen in type 1 diabetes?" *Trends Mol. Med.*, 11: 445–448, 2005.

organism inserted into the genome of another) for mutant insulin 2 also was expressed in the thymus. Absence of autoimmunity indicated that the mutated region of the molecule (region of B16) is the focus of the immune attack in the pancreas. By mutating an amino acid, B16, in the main MHC-binding residue, the ability of CD8 T cells (as well as CD4 pathogenic T cells) to bind to an MHC class 1 molecule is extinguished. Thus, a theory of how islets may be destroyed, developed from this recent research, when an insulin-like peptide is recognized as a foreign peptide by autoreactive T cells in the thymus is shown in Figure 15-41. In the human, the situation is more difficult to analyze; however, it has been shown recently that, in the peripheral blood of a diabetic patient, T cells reactive to insulin have been propagated and insulin can be recognized by T cells in lymph nodes near the pancreas (but not in nondiabetic patients). Further, in diabetic patients expressing HLA DRB1 0401 (a susceptibility allele for diabetes), T cells were derived that recognized insulin chain A amino acids 1–15 (Figure 15-40). The genetic background for the development of type 1 diabetes has been narrowed to susceptibility loci on chromosomes 2 (2q31-q330, 6 (6q210, 10 (10p14-q11), and 16 (16q22-q24). The idea is that a limited number of autoantigens are targeted early but that later in the progression of diabetes "epitope spreading" may occur and different cells could target several other (than insulin) islet molecules. Although these ideas are embryonic, they may lead the way to successful treatment of insulin-dependent diabetes.

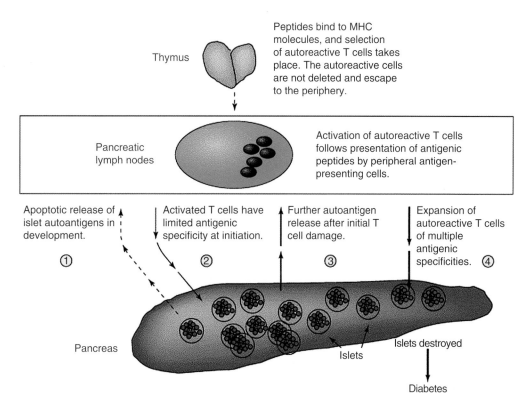

Figure 15-41. A possible scheme for the development of diabetes. Autoreactive CD4 and CD8 T cells that have not been deleted in the thymus, influenced by genetic susceptibility, are released to the periphery. Islet autoantigens, released by apoptosis during development, are presented to these T cells in local draining pancreatic lymph nodes (1). This might be influenced by environmental factors. Cells of limited specificity, such as to insulin, become activated (2). These cells travel back to the islets and enter and begin to destroy insulin-producing β cells. This releases more autoantigens, which are further taken up and activate many different types of islet-reactive T cells with different specificities (3). Finally, this diversified set of activated cells causes major damage to the β cells and diabetes occurs (4). Abbreviation: MHC, major histocompatibility complex. Reproduced from Figure 2 of F.S. Wong, "Insulin: a primary autoantigen in type 1 diabetes?" *Trends Mol. Med.*, doi:10.1016/1.molmed. 2005.08.005. Copyright © 2005 Elsevier Ltd.

Complement System

When foreign organisms attack the body, inflammation results. If the inflammatory response cannot stop the spread of disease-producing agents, such as viruses, bacteria, and fungi, there is a release of chemical agents by the body, ending with a final cleanup by monocytes (phagocytes). If this response is still insufficient to kill the invaders, the complement system and the immune system come into play. The complement system of about 30 proteins is produced in the liver and circulates in the blood plasma. These proteins are able to bind to a bacterial cell (gram-negative cells are high in lipid, as opposed to gram-positive cells that have rigid membranes high in carbohydrate derivatives) and open pores in its cell wall, allowing fluids and salts to move in and cause the bacterium to swell and burst (Figure 15-42). This system complements the actions of the immune system. Most complement proteins need to be activated by a protease. Once activated in this way, the complement protein also becomes an active protease. The activated proteins, in turn, activate other proteins of the system into a cascade resembling the blood-clotting mechanism (Figure 14-16). This system of complement molecules is activated when there is an infection and there is a sequence of events that take place on the surface of the infecting agent, resulting in its destruction and elimination. There are two major mechanisms: one, called the classical pathway, involves an antigen–antibody complex, and the other, called the

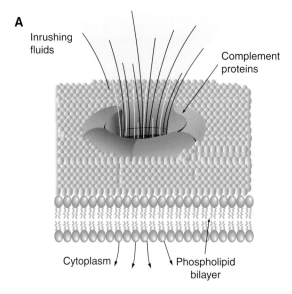

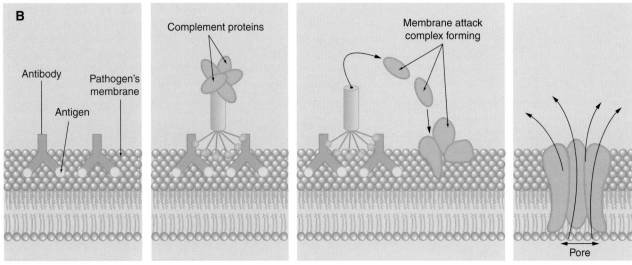

(1) Antibody molecules attach to antigens on pathogen's plasma membrane.

(2) Complement proteins link two antibody molecules.

(3) Activated complement proteins attach to pathogen's membrane in step-by-step sequence, forming a membrane attack complex (MAC).

(4) MAC pores in the membrane causes cell lysis.

Figure 15-42. **A.** Functioning of the complement system of proteins creating a pore in an invading pathogenic cell. **B.** Course of action of complement proteins to create a pore in invading cell membrane. Part A redrawn from the complement system of proteins and their functioning. Image from Purves et al., *Life: The Science of Biology*, 4th ed., by Sinauer Associates (www.sinauer.com) and WH Freeman (www.whfreeman.com).

alternative pathway, has events occur on the surface of the pathogen in the absence of antibody. These pathways are summarized in Figure 15-43. In the alternative pathway, there is a spontaneous conversion of complement protein C3 to an active protease (C3 has a thioester group that is hydrolyzed spontaneously at a slow rate). Participation of binding factor B and activation by factor D results in the cleavage of factor B to form an active C3 protease. This enzyme cleaves C3 to form C3b that can become C5 activating convertase. Here, the two pathways converge and additional complement factors (C6, C7, C8, and C9) are recruited to the growing **membrane attack complex (MAC)** that forms on the invading cell membrane and lyses the invader. A crystal structure of the heterotrimeric globular head of the C1q complement protein participating in the classical pathway by forming a complex with an antibody (Figure 15-43, upper right) is shown in Figure 15-44. It also shows a model of the interaction between this complement protein (head) and an IgG antibody.

In the classical pathway, complement activation mediates the specific antibody response. The C1 complement component binds to a specific part of the antibody (Figure 15-43) and initiates the pathway. C1q and C1r2s2 are two reversibly interacting complement subunits that combine by a calcium-dependent mechanism to form the initial enzyme, C1. Most of the C1 occurs as this complex. There is a C1 inhibitor in serum that controls its activity until C1 binds to an antibody (the subunit, C1q, has an affinity for IgM and IgG) when the inhibition is overcome and C1q activates the

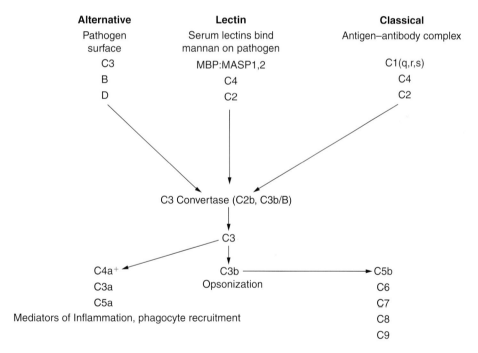

+C4a is a product of the cleavage of C4 by MASP or C1s, not C3, and it serves as a mediator of inflammation.

Figure 15-43. Overview of complement pathways in the face of infection. The classical pathway involves and antibody-antigen complex, whereas the alternative pathway does not involve antibody. Letters and numbers refer to members of the complement system. *MASP*, mannose-binding lectin-associated serine protease; *MBP*, mannose-binding protein. Redrawn from page 4 of http://www.talkorigins.org/faqs/behe/icsic.html.

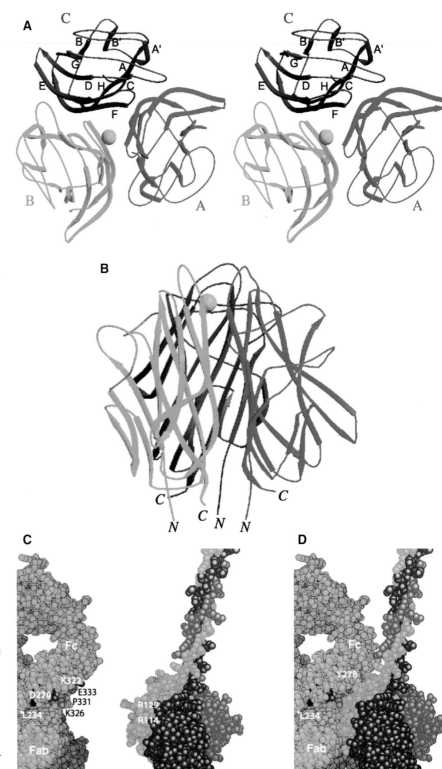

Figure 15-44. **A.** Structure of the heterotrimeric C1q globular domain. A stereographic ribbon representation of the assembly seen from the top. Modules *A*, *B*, and *C* are shown in *blue, green*, and *red*, respectively. β-strands are labeled according to the TNF nomenclature, and the calcium ion is represented as a golden sphere. **B.** Side view, *N* and *C* indicate the N- and C-terminal ends of each module. *C* and *D* show the C1q–IgGI interaction. **C.** Space-filling representation of the proposed interacting faces of human IgG1, G12, and C1q head, showing their shape complementarity. The IgG Fc (constant fragment) domain an the Fab (antibody-binding fragment) arms are indicated. Color coding is as follows: major charged residues crucial for the interaction (D270, K322) are in *red*; charged residues impairing the interaction (K326, E333) are in *blue*; other crucial residues in the contacting zone (P329, P331) are in *magenta*; and residues in the hinge region (L234, L235) are in *black*. The arginine residues of C1qB proposed as a possible interaction site are displayed in *light blue*. **D.** Same representation of the C1q/IgG1 assembly. Parts A and B reproduced from Figure 1 and parts C and D are reproduced from Figure 5 of C. Gaboriaud et al., "The crystal structure of the globular head of complement protein C1q provides a basis for its versatile recognition properties," *J. Biol. Chem.*, 278: 46974–46982, 2003.

other subunit, C1r2s2. When C1q binds an immune complex, the subunits of C1 are tightly bound and autoactivation begins. The two activated C1s subunits can catalyze the formation of C3 convertase (Figure 15-43) (C4b2a) that is formed from C2 and C4 (C4 contains three polypeptide chains and is a substrate of C1s*). Other reactions follow, as shown in Figure 15-43, that culminate in the MAC, which kills the invading cell. Details of the two complement pathways are diagrammed in Figures 15-45 and 15-46.

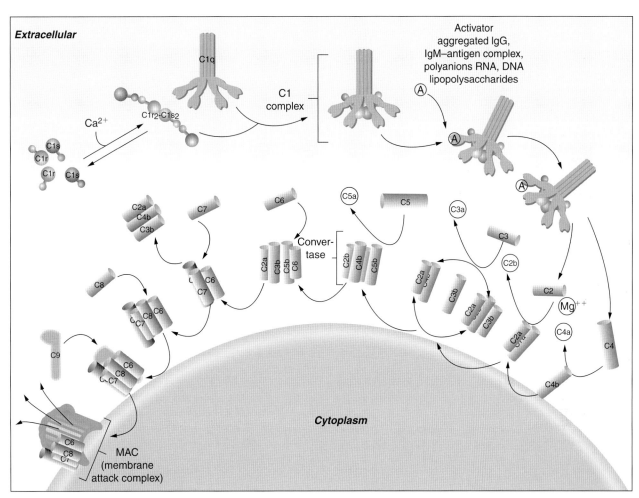

Figure 15-45. Classical complement pathway. The complement system is part of the defense against invading cells and is composed of about 20 different proteins found in the plasma. When activated, complement proteins form a pathway of proteolytic reactions that culminates in the lysis of foreign cells. The complement system also stimulates phagocytosis of foreign cells and an inflammatory response. There are two different complement systems, the classical complement pathway initiated by antibody complexes on the cell surface, and an alternative complement pathway that is initiated without antibodies. The complement system proteins are named with a capital C followed by a number. A small letter after the number indicates that the protein is a smaller protein resulting from the cleavage of a larger precursor by a protease. In the classical pathway, the first step is the initiation of the pathway triggered by recognition by complement factor C1 of antigen–antibody complexes on the cell surface. When C1 complex interacts with aggregates of IgG with antigen on a cell's surface, two C1-associated proteases, C1r and C1s, are activated. Other factors like lipopolysaccharide also activate C1s. Once C1s is activated, it cleaves C4 to form C4b that then binds to the cell membrane of the cell being attacked. The proteolytic complement cascade is then amplified on the cell membrane through sequential cleavage of complement factors and recruitment of new factors until a cell surface complex containing C5b, C6, C7, and C8 is formed. The addition of a multiple C9 proteins creates the membrane attack complex results in a large pore that spans the membrane of the cell being attacked, allowing ions to flow freely between the cellular interior and exterior, Ions flow out, but large molecules stay in, causing water to flood into the cell and ultimately burst the cell from osmotic pressure. Redrawn with permission from http://www.biocarta.com/pathfiles/h_classicPathway.asp.

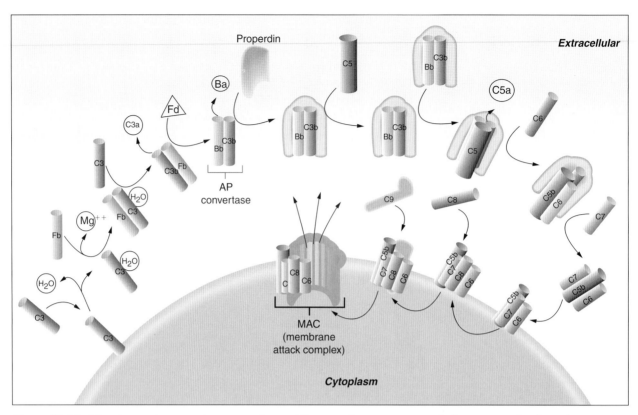

Figure 15-46. The alternative complement pathway. The complement system of plasma proteins is an important part of the immune system that forms a cascade of factors that lyses foreign cells. There are two branches of the complement system, the classical pathway that is initiated by antibody–antigen complexes on a cell and the alternative pathway that is antibody independent. The ultimate result in either pathway is the creation of the membrane attack complex, a large pore in the cell membrane that results in cell lysis. The alternative pathway starts with the spontaneous conversion of C3 to an active protease. C3 contains a thioester group that is spontaneously hydrolyzed at a slow rate to create C3(H2O). From there, binding of factor B (Fb) and activation by factor D (Fd) cleaves factor B to create the active protease C3 convertase (AP convertase). This enzyme cleaves C3 to form C3b, which can go on to form a C5 activating convertase. At this point the alternative pathway proceeds in the same manner as the classical pathway, recruiting additional complement factors (C6, C7, C8, and C9) to ultimately form the membrane attack complex and lyse the associated cell. One question about the alternative pathway is how the spontaneous activation of C3 in plasma leads to the lysis of specific cells in the absence of antibody on the cell surface. Active C3b binds to the cell surface, particularly to complement activators like cell wall components and lipopolysaccharide. A constant low level of spontaneous C3b formation ensures that C3b can bind to invading cells and trigger the rest of the alternative complement pathway to lyse the cells even in the absence of an antibody response. The constant low level of C3b activation and potential activation of the alternative pathway is kept in check by a natural damper, factor H, and factor I. Factors H and I in plasma inactivate C3b enzyme in solution. Factors H and I cannot inactivate C3b on the cell surface due to protection by properdin, ensuring that the alternative pathway is primarily inactive in plasma and specifically activated on the surface of invading. Redrawn with permission from http://www.biocarta.com/pathfiles/h_alternativePathway.asp.

The crystal structures of the C4d and the C3d fragments of C4 and C3 human complement factors are shown in Figures 15-47 and in Figure 15-48. The various faces of the C4d molecule show differences from the corresponding faces of C3d that probably reflect functional differences between C3 and C4. C3d and C4d have an approximately 30% sequence identity that produces a virtually identical fold; however, the electrostatic natures of the surfaces of the C4d molecule are different from those of C3d, again, probably reflecting the difference in function between C3 and C4.

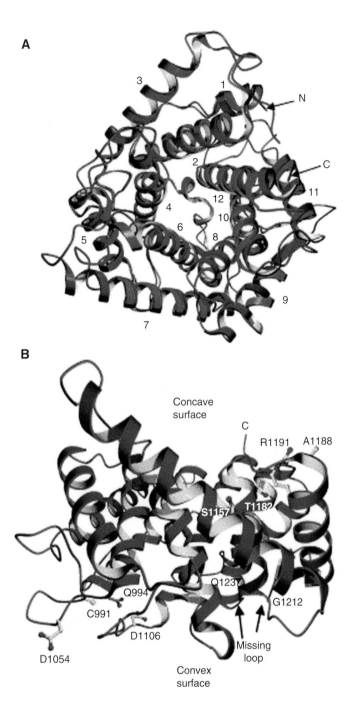

Figure 15-47. **A.** *Top view* superposition of the structures of C3d and C4Ad rendered in *magenta* and *gold*, respectively, showing the α–α 6 barrel topology of the molecules with 12 helices, consecutively alternating from the outside to the inside of the barrel. **B.** *Side view* ribbon representations of the C4Ad structure showing the positions of the thioester-forming residues, Cys991 and Gln994, and the C-terminal isotopic residue Asp1106 *(gold ball* and *stick)*, all located at the convex face of the molecule. Also shown as *ball and stick* are the side chains of the polymorphic amino acid residues Ser1157, Thr1182, Ala1188, and Arg1191. These are proximately located on the concave surface and with the exception of Thr1182 contribute the major Ch/Rg (epitope region). Reproduced from Figure 1 of J.M.H. van den Elsen et al., "X-ray crystal structure of the C4d fragment of human complement component C4," *J. Mol. Biol.*, 322: 1103–1115, 2002.

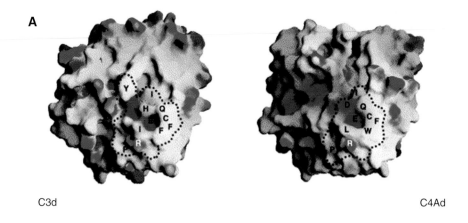

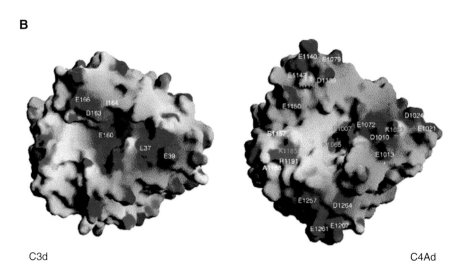

Figure 15-48. Molecular surface representation of four faces of the respective C3d and C4Ad molecules colored for electrostatic potential (highly negative, *red*; highly positive, *blue*). **A.** The convex face of C3d *(left)* and C4Ad *(right)*. Conserved residues surrounding the thioester residues are labeled in both molecules. The putative domain interfaces in intact C3 and C4A that would respectively sequester the thioester from the solvent are denoted by the dotted line boundaries. This domain interface is inferred from the sequence and charge conservation among diverse species C3 and C4 molecules. **B.** The concave faces of the structures of C3d *(left)* and C4Ad *(right)*. Labeled are the surface-exposed residues in C3d, which according to mutagenesis data are involved in interactions with CR2. Acidic residues are labeled in *yellow*, basic residues in *green*, and hydrophobic residues in *khaki*. Surface-exposed polymorphic residues of C4Ad, which in this case represent the antigenic determinates for Chl, Ch6, and Ch3 alloantibodies, are marked in *white*.

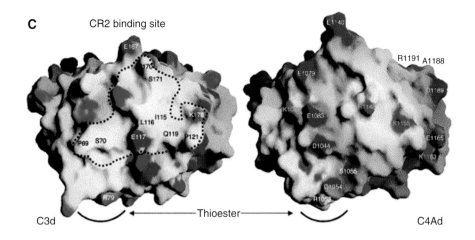

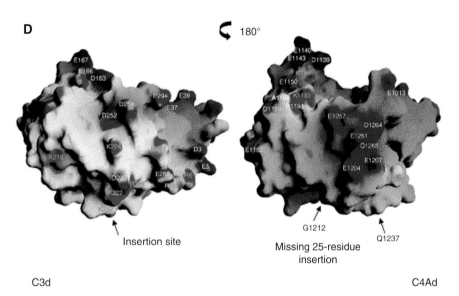

Figure 15-48, cont'd **C** and **D**. A comparison of C3d *(left)* and C4Ad *(right)* on a side face of the α–α barrel, which in the case of C3d has recently been implicated by a cocrystal structure as forming the binding interface for CR2. The buried surface of the C3d-CR2 interface is indicated by the *dotted line* and the labeled residues within the visualized interface are ones for which backbone carbonyl oxygen atoms are involved in hydrogen bonding (except for N170, where the side chain contributes to interactions within the interface). The thioester region for each molecule is indicated for orientation purposes. **D.** Molecular surface representations of C3d *(left)* and C4Ad *(right)* showing a side view of the α–α barrel that represents a 180 degree rotation from that shown in *(C)*. Labeled are charged surface-exposed residues. *Arrows* indicate the position of the 25-residue insertion in C4AD, relative to the C3d amino acid residue sequence, that is missing in the structure and that contains within it S1217, which is the site of covalent attachment of C3b in forming the classical pathway C5 convertase. Reproduced from Figure 4 J.M.H. van den Elsen et al., "X-ray crystal structure of the C4d fragment of human complement component C4," *J. Mol. Biol.*, 322: 1103–1115, 2002.

Regulators of Complement Pathways

There are several regulatory proteins that control the complement pathways and hold them in check until they are needed to function. When a pathway is needed to be turned on, suppression by the regulators is overcome by increased concentrations of members of the complement pathway. The regulators bring about the decay or degradation of certain complement factor pathway members, as shown in Figure 15-49. In this figure, the regulators (CR1, CD55, C4BP, CD46, and FH/FHL-1) are listed in the boxes, and their functions are indicated. From the figure, it can be seen that these regulators (in the fluid and membrane phases) of complement activation (RCA) control the C3 convertases of both the classical and the alternative pathways. In an attempt to escape attack by the complement system, many pathogens bind to the fluid phase RCA. Many of these interactions are known and are listed in Table 15-1.

The mechanism of escape could involve the microorganism binding to a plasma RCA protein and escaping attack by blocking antigenic

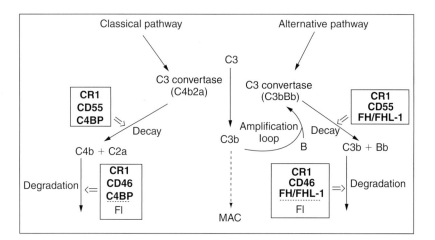

Figure 15-49. Overview of the human complement system and its regulation by RCA proteins. The conversion of C3 to C3b can be seen as the key reaction during complement activation. This reaction is catalyzed by the two C3 convertases, which are formed in the classical and alternative pathways, respectively. The classical pathway C3 convertase (C4b2a) is derived from C4 and C2 and the alternative pathway C3 convertase (C3bBb) is derived from C3 and factor B. An amplification loop enhances formation of the alternative pathway C3 convertase, as indicated. Surface-bound C3b causes opsonization and formation of surface-bound convertase, which recruits the membrane attack complex (MAC) that causes lysis of target cells. Not indicated in the figure is the formation of anaphylatoxins, for example C5a. The activity of the C3 convertases is regulated by RCA proteins (boxed and marked in bold), which cause decay and/or degradation of the convertases, as indicated. In causing degradation, the RCA proteins act as cofactors to the plasma protease factor I (FI). CR1 and the fluid-phase regulators C4BP and FH/FHL-1 cause both decay and degradation. CD55 only causes decay, whereas CD46 only acts as a cofactor. *RCA*, regulators of complement activation. Redrawn from Figure 1 of B. Lindahl, U. Sjöbring, and E. Johnsson, "Human complement regulators: a major target for pathogenic microorganisms," *Curr. Opin. Immunol.*, 12: 44–51, 2000.

Table 15-1
Interactions Between RCA Proteins and Microorganisms

RCA protein	Microorganism	Ligand in microorganism	SCRs required for complement regulation	SCRs required for binding to microorganism
Membrane RCA proteins				
CR2 (CD21)	EBV	gp350/220	1 and 2	1 and 2
CD46 (MCP)	MV	Hemagglutinin	2–4	1 and 2
	S. pyogenes ?	M6 protein		?
	N. gonorrhoeae	Pili		?
	N. meningitidis	Pili		?
	H. pylori	BabA protein		1?
CD55 (DAF)	E. coli	Dr-like antigens	2–4	2 and 3; or only 3
		X adhesin		3 or 4
	Picornaviruses:			
	echovirus 7	Capsid		2–4
	coxsackievirus A21	Capsid		1?
Fluid-phase RCA proteins				
C4BP	S. pyogenes	Some M proteins	1 and 2 (α-chain)	1 and 2 (α-chain)
	B. pertussis	FHA?		?
	N. gonorrhoeae	Porin		?
FH	S. pyogenes	Some M proteins	1–4	7
	N. gonorrhoeae	LOS		16–20
		Porin		?
	S. pneumoniae	?		?
	HIV-1	gp41, gp120		?
FHL-1	S. pyogenes	Some M proteins	1–4	7

DAF, decay-accelerating factor; LOS, lipooligosaccharide; MCP, membrane cofactor protein; MV, measles virus; SCR, short complement-like repeat. Reproduced from Table 1 of B. Lindahl, U. Sjöbring, and E. Johnsson, "Human complement regulators: a major target for pathogenic microorganisms," Curr. Opin. Immunol., 12: 44–51, 2000.
EBV, Epstein-Barr virus; FHA, filamentous hemagglutinin.

sites (where the antibody would bind) in the complex. The RCA proteins stem from genes linked closely together on chromosome 1. These proteins are constructed from short consensus repeats (SCRs), each of which contains about 60 amino acids with four cysteine residues. These are compact domains with a α-barrel structure (Figure 15-50). The SCRs shown in this figure are from the RCA, CD46 (Table 15-1). The manner in which two SCR domains interface with each other is shown in Figure 15-51. The docking site on the surface of CD46 for measles virus and the site for antibody binding to CD46 are shown in Figure 15-52. CD46 has a highly hydrophobic and protruding loop at the base of the first repeat (Figure 15-50B), providing a virus-binding site that defines the recognition epitope. The binding of antibody and virus involves overlapping sites (Figure 15-52). Presumably, molecules that could be synthesized to mimic the conformation of this loop might be effective antiviral agents that could prevent the binding of measles virus to CD46.

RCA proteins inhibit C3 convertases (Figure 15-49) either by causing the decay of these enzymes or or by acting as cofactors for the serine protease 1 that degrades C3b and C4b. Among the RCAs, CR1 is the most important as a receptor for complement–antibody complexes.

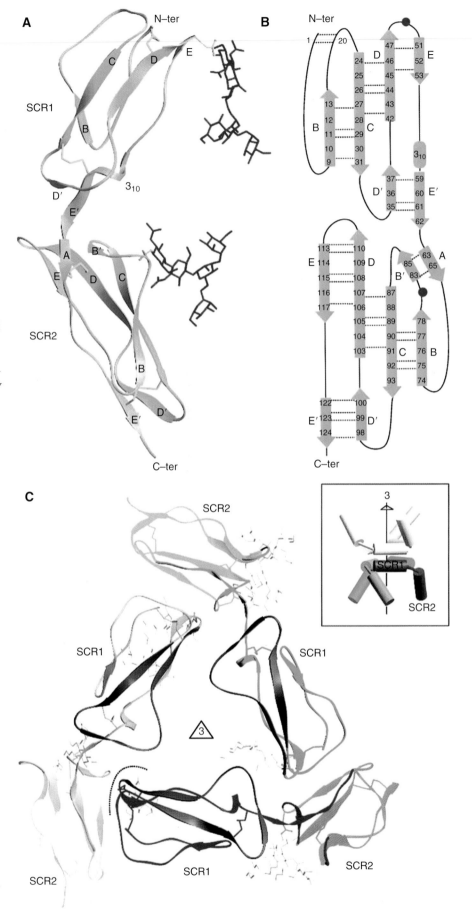

Figure 15-50. Structure of the N-terminal two short consensus repeats, SCR1 and SCR2, of CD46, and arrangement of the molecules in the crystal. **A.** Ribbon drawing of the molecule. Disulfide bonds and carbohydrate residues are shown in *yellow* and *red*, respectively. **B.** β-Sheet topology and secondary structure assignment. Main-chain hydrogen bonds are indicated with *dashed lines*. Residues in α-strands were assigned based on main-chain hydrogen bond formation with a donor-acceptor distance of 3.5 Å. The glycosylation sites at Asn49 and Asn80 are marked with *red spheres*. **C.** Arrangement of the six independent copies of the molecule in the crystal. Ribbon drawing of a trimer and schematic representation *(inset)* of two trimers present in the asymmetric unit of the crystals. The segments between strands C and D in the three SCR1 domains are located at the *top* of the trimer and are in *black*. Each contact between two molecules buries a relatively large surface area of 1190 Å and involves numerous hydrophobic residues (Ile22, Ile37, Pro38, Pro39, Leu40, Ile45, and Leu53). One such contact is marked with a broken line *(lower left)*. Part C reproduced from Figure 2 of J.M. Casasnovas, M. Larvie, and T. Stehle, "Crystal structure of two CD46 domains reveals an extended measles virus-binding surface," *EMBO J.*, 18: 2911–2922, 1999 and http://www.nature.com/embo/journal/v18/n11/full/7591712a.html.

CR2 provides a connection between innate and adaptive immune responses (innate immune responses are defined by bodily barriers, for example, skin, and by cells within the body, such as phagocytes and barriers set up by the inflammatory process; adaptive immunity refers to the response to pathogens, nonself recognition). CD46 and CD55, which are short molecules, are RSAs that have a wide occurrence and play a major role in protecting cells against attack by complement. C4b-binding protein (C4BP) and factor H (FH) are mainly in the fluid

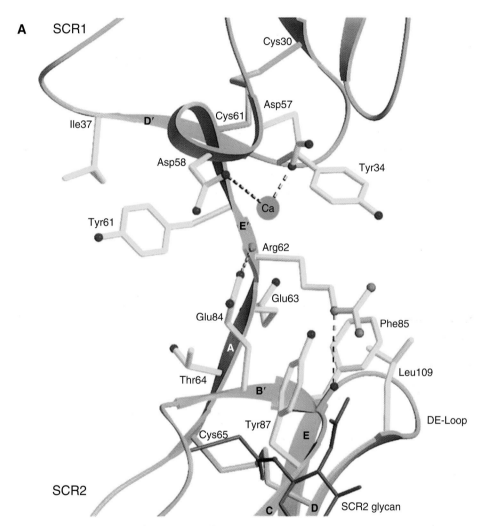

Figure 15-51. Interdomain interface and movement between two CD46 domains. **A.** Detailed view of the interdomain interface. The main chain of a monomer is shown as a ribbon drawing with residues at the interface colored *orange* and Asn80 with the attached glycan colored *red*. Disulfide bonds are shown in *blue*, and hydrogen bonds are represented with *broken lines*. The calcium ion located at the interface is shown in *green*. Oxygen and nitrogen atoms are *red* and *blue*, respectively. Arg62 is hydrogen bonded to Glu84 and Phe85 in SCR2, and it is also within 4 Å of the aromatic ring of Tyr34 in SCR1. The remaining interactions at the interface involve hydrophobic contacts between *(i)* Tyr61 and Ile37, *(ii)* Phe85 and Leu109, and *(iii)* the methyl group of the first *N*-acetyl glucosamine of SCR2 and Tyr87.

Continued

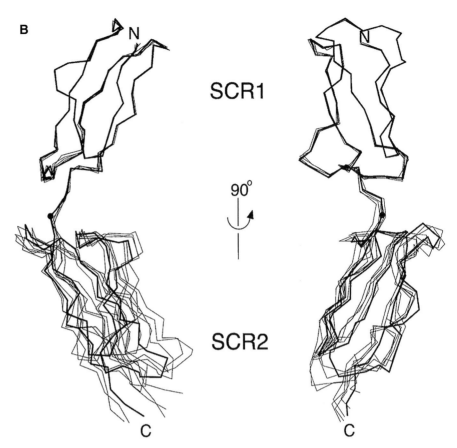

Figure 15-51, cont'd **B.** Superposition of the six crystallographically independent molecules, showing the interdomain movement and the flexibility of the molecule at the domain interface. Two orthogonal views are given. The view on the left side is approximately parallel to plane of maximum variation, and the view on the right side is approximately perpendicular to it. The superposition is based on residues within SCR1. The *black sphere* in the molecule drawn with *thicker lines* marks the kink in the polypeptide chain at Glu63, which is primarily responsible for the interdomain bend. Reproduced from Figure 3 of J.M. Casasnovas, M. Larvie, and T. Stehle, "Crystal structure of two CD46 domains reveals an extended measles virus-binding surface," *EMBO J.,* 18: 2911–2922, 1999 and http://www.nature.com/embo/journal/v18/n11/full/7591712a.html.

phase and act primarily on the C3 convertases of the classical and alternative pathways. When the pathogen *Streptococcus pyogenes* infects, it can bind to the RCA, C4BP (Table 15-1), and downregulate complement activation. To accomplish this, C4BP binds to the amino terminal region of the M protein on the surface of the *S. pyogenes* cell (Figure 15-53). The *Streptococcal* M protein can cause endothelial inflammation (Figure 15-54). The RCAs are summarized in Figure 15-55. There can be disease problems with the complement system, mostly of a hereditary nature. An inherited deficiency of C3 predisposes to bacterial infections. A deficiency of C2 or other early members of the pathway (C1q, C1r, C1s, or C4; Figure 15-43) causes immune complex disorders in the classical pathway. Early factor deficiency can be found in

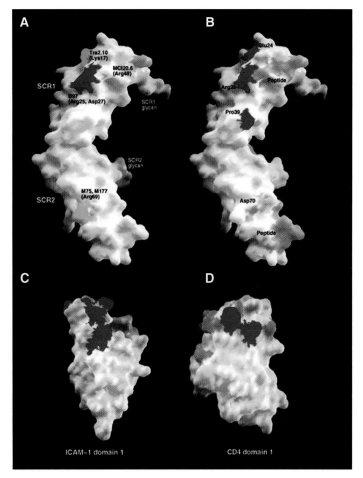

Figure 15-52. Virus-binding surfaces in human virus receptors CD46, ICAM-1, and CD4. *(A and B)* Surface representations of the N-terminal two repeats of CD46 with epitopes for antibody and measles virus binding. **A.** Epitopes for anti-CD46 monoclonal antibodies that significantly blocked virus binding to the receptor (inhibitory effect > 50%). Residues recognized by B97, Tra2.10, MCI20.6, M75, and M177 antibodies are colored. **B.** Regions involved in measles virus binding. Residues Glu24, Arg25, and Pro39 had a severe effect on hemagglutinin binding according to mutagenesis experiments and are shown in *red*. Residue Asp70 had a moderate effect on binding of CD46 to both measles virus and soluble hemagglutinin and is marked in *orange*. Stretches of amino acids 45–48 and 85–104 that were identified by peptide inhibition studies are shown in *orange*. The carbohydrate bound to SCR2 is shown in *yellow*. **C.** Surface representation of the N-terminal domain of ICAM-1. Residues Gln27, Pro28, Lys29, Leu30, Tyr66, Pro70, and Asp71 that are involved in interaction with several human rhinoviruses are shown in *red*. **D.** Surface representation of the N-terminal domain of CD4. Phe43 and Arg59, the two residues that interact primarily with human immunodeficiency virus (HIV), are shown in *red*; residues Gln25, Lys29, Lys35, Gln40, Ser42, and Ser60 make additional contacts and are shown in *orange*. The coordinates for ICAM-1 and the CD4–gp120 complex were obtained from the Brookhaven Protein Data Bank (accession codes 1IC1 and 1GC1). Only the side chains of residues are colored in (A–D). Reproduced from Figure 5 of J.M. Casasnovas, M. Larvie, and T. Stehle, "Crystal structure of two CD46 domains reveals an extended measles virus-binding surface," *EMBO J.*, 18: 2911–2922, 1999 and http://www.nature.com/embo/journal/v18/n11/full/7591712a.html.

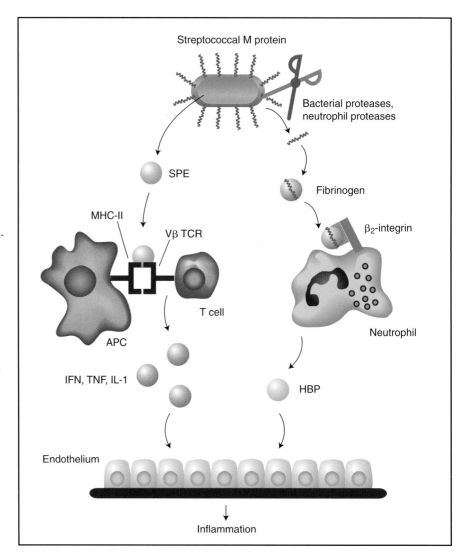

Figure 15-53. Model for the role of C4BP in *S. pyogenes* pathogenesis. C4BP binds to the amino-terminal region of a surface M protein and downregulates complement activation, thereby protecting the bacterium against opsonization and phagocytosis. This binding property makes the amino-terminal region of the M protein a key target for immune attack, which explains the hypervariability in this region. Redrawn from Figure 3 of B. Lindahl, U. Sjöbring, and E. Johnsson, "Human complement regulators: a major target for pathogenic microorganisms," *Curr. Opin. Immunol.*, 12: 44–51, 2000.

Figure 15-54. Bacterial M protein induces shock. M protein is released from the streptococcal surface by neutrophil proteases and bacterial-derived cysteine proteases. M protein binds to fibrinogen, and these complexes, as shown by Björck and colleagues, subsequently bind to β2-integrins, adhesion molecules on the neutrophil surface. This binding results in neutrophil activation and the release of heparin-binding protein *(HBP)*, a potent inflammatory mediator. *S. pyogenes* also releases potent pyrogenic exotoxins *(SPE)*, superantigenic toxins that can cross-link MHC class II molecules on antigen-presenting cells *(APC)* and the V_β domains of T-lymphocyte receptors on T cells *(T)*. This results in T-cell activation and the release of proinflammatory monokines. Redrawn from http://www.nature.com/nm/journal/v10/n4/thumbs/nm0404-342.f1.jpg.

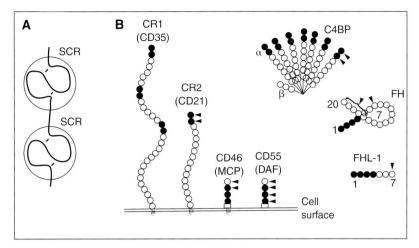

Figure 15-55. The structure of human complement regulators in the RCA family. **A.** Schematic representation of two SCR domains, with disulphide bridges indicated. **B.** Membrane-bound and fluid-phase RCA proteins. Membrane-bound proteins are shown to the left, fluid-phase proteins to the right. Each circle represents one SCR domain. Filled circles indicate domains required for complement regulation and/or for binding of complement components. *Arrowheads* indicate SCRs implicated in the binding of microorganisms (see Table 15-1). The open boxes at the membrane-proximal end of CD46 and CD55 represent serine/threonine-rich regions that are extensively O-glycosylated. *DAF*, decay-accelerating factor; *FH*, factor H; *MCP*, membrane cofactor protein; *SCR*, short consensus repeat. Reproduced from Figure 2 of B. Lindahl, U. Sjöbring, and E. Johnsson, "Human complement regulators: a major target for pathogenic microorganisms," *Curr. Opin. Immunol.*, 12: 44–51, 2000.

patients with SLE. C9 deficiency is usually not grave because the C5b, C6, C7, C8 complex can lyse bacteria, but not as efficiently as C9. Hereditary deficiency of **hereditary angioneurotic edema (C1INH)** is marked by occasional triggering of the complement system, generating massive release of anaphylatoxins (C3a, C5a) that can cause edema of the airways (swelling) and of the intestine and skin.

Properdin

The protein **properdin** is complement factor B, a positive regulator of the alternative complement pathway. Properdin is cleaved by complement factor D to form noncatalytic chain Ba and catalytic subunit Bb, a serine protease, which associates with C3b to form the alternative pathway C3 convertase (Figure 15-46). Properdin contains six thrombospondin repeat type I domains (TSR-1 to TSR-6). **Thrombospondin** itself is a glycoprotein that can interact with blood coagulation and anticoagulant factors, and it can be involved in a number of processes, including cell adhesion, platelet aggregation, and cell proliferation. Properdin can take the form of a dimer or trimer (with TSR domains), as shown in Figure 15-56.

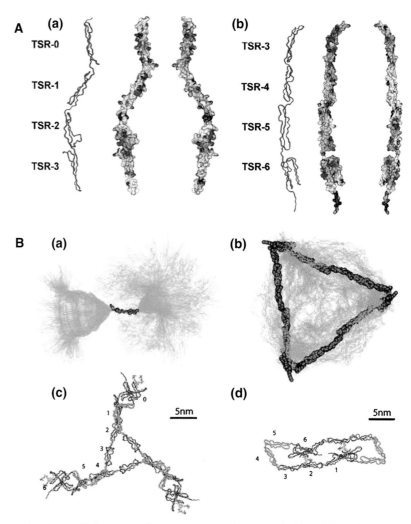

Figure 15-56. Structure of properdin. **A.** Electrostatic views of the properdin monomer. The extended model is shown for clarity. The *blue* and *red* colors denote positively and negatively charged surfaces, respectively. *(a)* Three views are shown of the four-domain N-terminal half, which is composed of TSR-0 to TSR-3. The ribbon view showing the location of the TSR domains is shown to the left. The electrostatic surface view to the left corresponds to the view of the ribbon representation, and that to the right corresponds to a 180-degree rotation about the vertical axis. *(b)* Three views are shown likewise for the four-domain C-terminal half, which is composed of TSR-3 to TSR-6. **B.** Summary of the conformational search to identify best-fit models for the properdin trimer and dimer. *(a)* A random selection of 200 of the 3125 models for properdin is superimposed on TSR-3, which is shown as a *blue ribbon* at the center, with six *red spheres* denoting the Cys residues. This shows the extent to which the TSR inter-domain arrangements have been conformationally sampled within a properdin monomer. *(b)* A random selection of 80 trimer models is superimposed at the center of mass of one of these 80 trimer models represented by *red spheres*. This shows the extent to which the TSR inter-domain arrangements have been sampled within a properdin monomer. *(c)* The final best-fit trimer model for properdin is shown in *green* and *dark blue,* with its extended carbohydrate chains in red. The TSR domains are numbered from 0 to 6. *(d)* The final best-fit dimer model for properdin is shown, with its TSR domains numbered from 0 to 6. *TSR,* thrombospondin repeat type domain. Parts A and B reproduced from Figures 11 and 9, respectively, from Z. Sun, K.B.M. Reid, and S.J. Perkins, "The dimeric and trimeric solution structures of the multidomain complement protein properdin by x-ray scattering, analytical ultracentrifugation, and constrained modeling," *J. Mol. Biol.*, 343: 1327–1343, 2004.

C-Reactive Protein

The **C-reactive protein (CRP)** is a highly conserved plasma protein that participates in the systemic response to inflammation. During inflammation, the concentration of CRP increases in the blood plasma. It binds to specific configurations on the surfaces of pathogens. More recently, a small increase in the level of the CRP has been used as an indicator of potential heart disease risk. In humans, the role of CRP can be either proinflammatory or anti-inflammatory, depending on the conditions under which it is acting. It may even participate in the pathogenesis of disease.

Its name derives from its ability to precipitate the "C" polysaccharide of the *pneumococcal* cell wall, but the specific ligand for CRP was shown to be phosphocholine, part of the **teichoic acid** of the cell wall structure (Figure 15-57). Teichoic acids are major components of gram-positive bacterial cell walls. They are complex polymers of polyols and phosphate, with a backbone of repeating units of glycerol or ribitol phosphate linked to residues of amino sugars (or unsubstituted sugars) and D-alanine. Teichoic acids are linked to glycolipids through the terminal glycerol phosphate residue of the teichoic acid chain to form lipoteichoic acids.

Human plasma levels of CRP can increase, by synthesis in the liver, 1000-fold or more after an acute inflammatory event. This is part of the acute phase response generated by gene expression in the liver during inflammation. The acute phase response reflects the synthesis in hepatocytes of some 40 plasma proteins, such as CRP and mannose-binding protein (lectin pathway, Figure 15-43) that promote inflammation, activate the complement system, and stimulate chemotaxis of phagocytes. Induction of CRP results from the transcriptional effects of interleukin-6 (IL-6) with an enhancing effect by IL-1β. Transcription factors, STAT3, C/EBP family members, and NFκB are part of the signal transduction process.

CRP is made up of five 23-kDa protomers arranged around a central pore, representing part of a family of proteins called pentraxins. Each protomer has a face with a phosphocholine-binding site containing two coordinated calcium ions next to a hydrophobic pocket. This is shown in Figure 15-58, where CRP has been cocrystallized with phosphocholine. Key residues important for the binding of phosphocholine are Phe-66 and

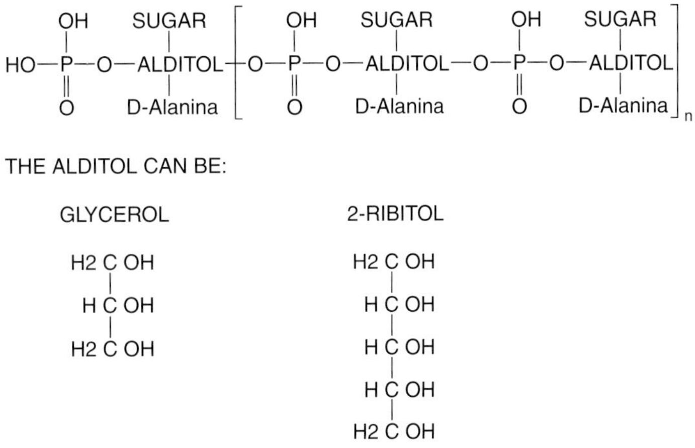

Figure 15-57. **A.** Structure of teichoic acid.

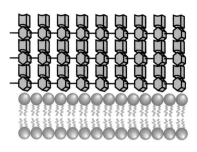

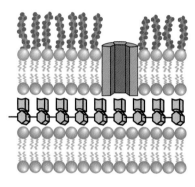

Figure 15-57. **B.** Constituents of gram-positive and gram-negative bacterial cell walls. Reproduced from http://images.google.com/imgres?imgurl=http://microvet.arizona.ed...vnum%3D10%26hl%3Den%26Ir%3D%26client%3Dsafari%26ris%3Den%26sa%3DN.

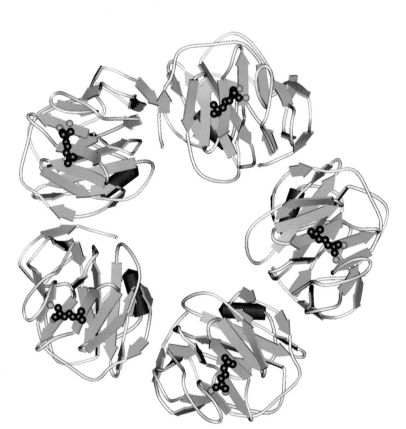

Figure 15-58. Crystal structure of C-reactive protein complexed with phosphocholine. The calcium ions are *yellow*, and phosphocholine is *green*. Reproduced from Figure 1 of S. Black, I. Kushner, and D. Samols, "C-reactive protein," *J. Biol. Chem.*, 279:48487–48490, 2004 and http://www.jbc.org/content/vol279/issue47/images/large/zbc0470450170001.jpeg.

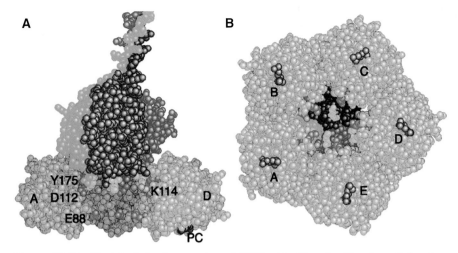

Figure 15-59. Model of the interaction of CRP with C1q. *A*, side view. Subunits B and C of CRP have been omitted for clarity. *B*, perpendicular bottom view. Modules A, B, and C of the C1q subunit are shown in *blue, green,* and *red*, respectively. The lysines at the top of the C1q head (Ala-173, Ala-200, Ala-201, Cys-170) and Tyr^{B175} are in *light blue*. A-E designate the CRP protomers. The phosphocholine *(PC)* ligand is in *red*, and the nearby Ca^{21} ion is in *green*. Color coding for CRP mutations is as follows: Mutations impairing complement activation (Glu-88, Asp-112, Tyr-175) are *magenta*, and mutations enhancing complement activation (Lys-114) are *blue*. Reproduced from Figure 2 of S. Black, I. Kushner, and D. Samols, "C-reactive protein," *J. Biol. Chem.*, 279: 48487–48490, 2004 and http://www.jbc.org/content/vol279/issue47/images/large/zbc0470450170002.jpeg.

Glu-81. Phe-66 interacts hydrophobically with the methyls of phosphocholine, and Glu-81 (opposite end of the binding pocket) interacts with the positive charge on the nitrogen of choline. The opposite face of the pentamer binds complement C1q (and Fcγ receptors). Residues, such as Asp-112 and Tyr-175, are involved in the binding of C1q. The positively charged head of the C1q interacts with negative charges of the central pore of the CRP pentamer, and optimal C1q binding is accommodated by certain conformational changes in the structure of CRP (Figure 15-59).

Further Reading

Books

Bertagnoli, M.M., *Cytokines and T Lymphocytes: Therapeutic Manipulation of the Immune System*, R.G Landes, 1993.

Dimitrios, M., and Lambris, J.D. (eds.), *Structural Biology of the Complement System*, CRC Press, 2005.

Schwartz, R.S., and Rose, N.R. (eds.), *Autoimmunity: Experimental and Clinical Aspects*, Annals of the New York Academy of Sciences, volume 475, 1986.

Szeleni, J., *The Complement System: Novel Roles in Health and Disease*, Kluwer Academic Publishers, 2004.

Reviews

Concannon, P. Erlich, H.A., Julier, C., Morahan, G., Nerup, J., Pociot, F., Todd, J.A., Rich, S.S., Type 1 Diabetes Genetics Consortium, "Type 1 diabetes: evidence for susceptibility loci from four genome-wide linkage scans in 1,435 multiplex families," *Diabetes*, **54**: 2995–3001, 2005.

Tonegawa, S., "Somatic generation of antibody diversity," *Nature*, **302**: 575–581, 1983.

CHAPTER 16

Neurobiochemistry

Pain: A Constant Health Problem

The people who suffer from either persistent or recurrent pain make up more than one third of the world's population. In the United States alone, the yearly cost associated with health care and financial considerations of compensation and litigation is about $100 billion dollars. Untreated pain suppresses the immune system, and it can predispose people to subsequent disease.

Nociceptor is the term for a nerve ending (of a neuron) that senses pain (the first syllable of the term derives from the Latin, *nocere*, which means injury). From the nociceptor, the pain signal travels through a complex of peripheral nerves distributed throughout the body and through the central nervous system, including the brain and the spinal cord. These nerves join in the dorsal horn of the spinal cord, and signals are transmitted from this region to the thalamus of the brain. There, the signals are sorted and sent on to the cerebral cortex, where the pain message is expressed so that the body perceives the pain (Figure 16-1).

There are two types of pain, acute and chronic. *Acute pain* is immediate and derives from tissue injury, bodily malfunction, or severe illness. The injury is sensed by nociceptors in the peripheral nerves, and injured tissues release chemical messengers, such as bradykinins and prostaglandins (Figure 16-2) that activate nociceptors. Structures of the receptors for bradykinin and prostaglandin E_2 that would be present on the nociceptor neuronal cell are shown in Figure 16-3. The pain message travels to the spinal cord and then to the brain, as outlined earlier. *Chronic pain* can derive from nerve disease or back pain. The messages take a more circuitous route to the cerebral cortex than acute pain messages.

There are parallel independent pain pathways that arise from different classes of nociceptors. These populations of nociceptors contain different neurotransmitters, and they may contain different receptors and ion channels. In this way, different populations may respond to different noxious stimuli and they may respond to different situations during injury, disease, and inflammation. More information needs to be gathered on the heterogeneity of nociceptors. There are endogenous opioids, such as enkephalins (methionine enkephalin, YGGFM, and

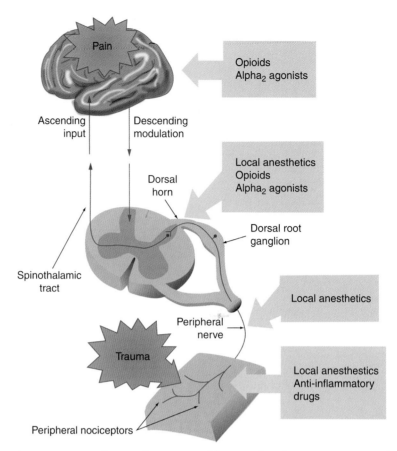

Figure 16-1. Pathway of pain sensed by peripheral nociceptors. Peripheral nociceptors in tissues transmit pain messages to peripheral nerve and collected in the dorsal horn of spinal column. Ascending nerves that transmit pain signals to the cerebral cortex of the brain foci, at which drugs are active, are indicated. Redrawn from A. Gottschalk, D.S. Smith, "New concepts in acute pain therapy: preemptive analgesia, *American Family Physician*, May 15, 2001.

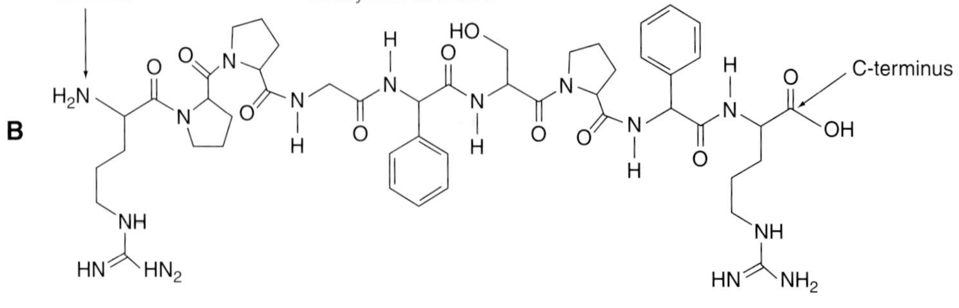

Figure 16-2. Structures of a prostaglandin (PGE$_2$) *(A)* and of bradykinin *(B)*. Amino acid sequence of bradykinin is: Arg-Pro-Pro-Gly-Phe-Ser-Pro-Phe-Arg.

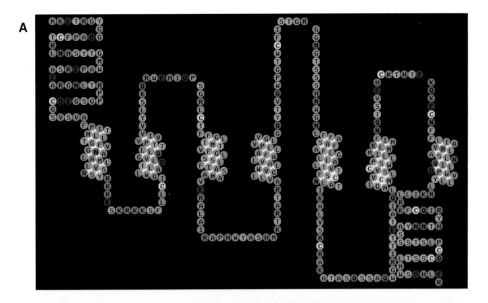
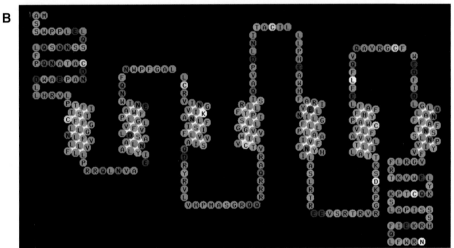

Figure 16-3. **A.** Suggested structure of receptor for prostaglandin E$_2$. **B.** Suggested structure of receptor A for bradykinin (B), a G protein−coupled receptor. Part A and Part B redrawn from http://www.wdv.com/CellWorld/Receptors/small/prostaglandinE2subtypeEP3Receptor_gif.gif.

leucine enkephalin, YGGFL) (Figure 16-4) and endorphins, that inhibit pain in the afferent (toward the brain) pathways.

The nociceptor is a neuron that can sense, through receptors, the release of chemical messengers, such as prostaglandins or bradykinins, from injured or inflamed tissues. A typical neuron and an interneuron are shown in Figure 16-5. An interneuron is a neuronal cell that can transmit a chemical or electrical signal to another neuron at its cell body or to a **dendrite** (branch of a neuron) extending from the cell body. In Figure 16-5B, the sending neuron is the interneuron that transmits signals to the cell body or to a dendrite of the receiving neuron,

Met-enkephalin, YGGFM, and Leu-enkephalin, YGGFL

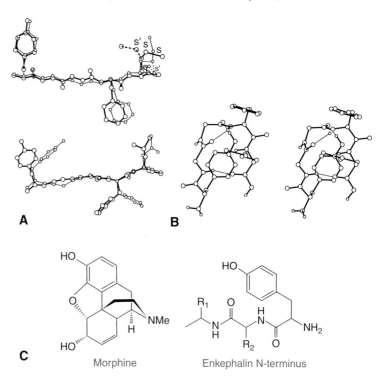

Figure 16-4. **A.** Superposition of the two crystallographically independent molecules of Met-enkephalin *(upper)* and Leu-enkephalin *(lower)*. **B.** Stereoview of the global minimum energy structure of Met-enkephalin. **C.** Similarity between enkephalin structure and that of morphine. Part A redrawn from Figure 2 of J. F. Griffin et al., "The crystal structures of [Met^5]enkephalin and a third form of [Leu^5]enkephalin: observations of a novel pleated β-sheet (enkephalin conformation/opioid peptides/anti-parallel planar β-sheet)," *PNAS USA*, 83:3272–3276, May 1986. Part B redrawn from Figure 1 of Z. Li and H.A. Scheraga, "Monte Carlo-minimization approach to the multiple-minima problem in protein folding," *PNAS USA*, 84: 6611–6615, 1987.

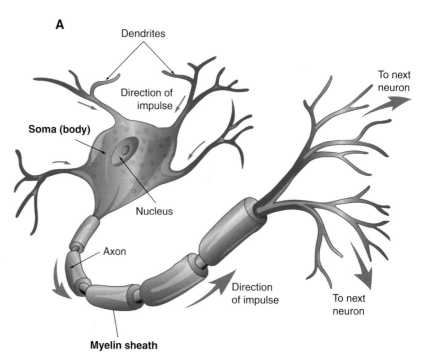

Figure 16-5. **A.** A typical neuron.

958   CHAPTER 16   Neurobiochemistry

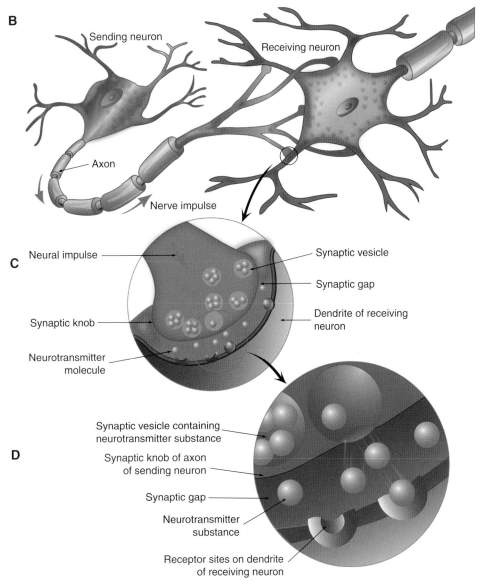

Figure 16-5, cont'd **B.** Interneuron *(left)* interacting with the cell body and dendrites of the receiving (target) cell. **C.** Magnification of the synapse. **D.** Magnification of the release of neurotransmitter from nerve ending of the sending cell (interneuron), the migration of the neurotransmitter across the synaptic gap and binding of the neurotransmitter to its cognate receptor on the surface of the dendrite of the target neuron.

creating a **synapse** (functional connection between a nerve cell and its target cell). The chemical signal, a neurotransmitter (a chemical messenger such as substance P, acetylcholine, dopamine, norepinephrine, or serotonin), is released from the nerve ending of the interneuron (sending neuron) into the synaptic gap (or synaptic cleft) and then combines with its cognate receptor (cognate means that the receptor is related to the ligand by recognizing its structure) on the dendrite of the receiving neuron. The balance of ions on the outer and inner surfaces of a neuron is altered when a chemical contacts the surface of the

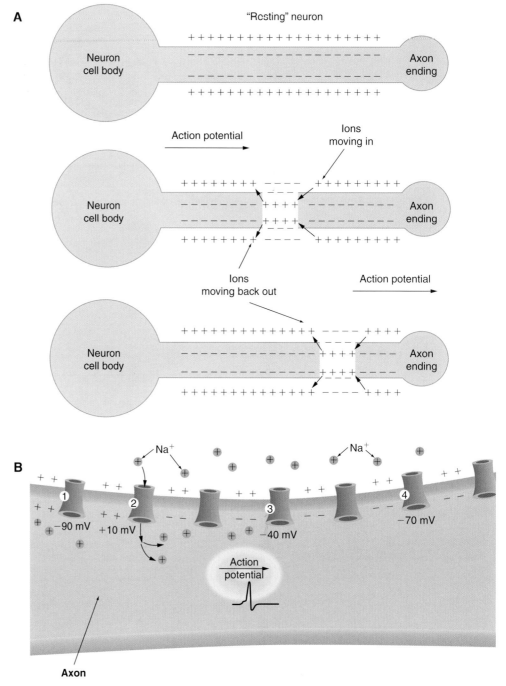

Figure 16-6. **A.** Transmission of an action potential along the axon of a neuron. Positive charges cover the surface of the axon (usually, Na$^+$). When Na$^+$ moves into the resting membrane, an action potential is developed (positive charges replace some of the negative charges inside the membrane). This process continues down the axon to the axon terminal (nerve ending) that culminates in the release of neurotransmitter from the nerve ending. Ca^{2+} ions also can generate an action potential. **B.** Ions moving into the membrane as the process continues toward the nerve terminal. Part A is redrawn from http://www.ship.edu/~cgboeree/theneuron.html.

membrane and reaches a threshold level, coursing across the membrane to the axon. Upon reaching the axon, this alteration produces an action potential. The action potential is a brief electrical pulse that travels along the axon. When the action potential reaches the nerve ending (axon terminal), it triggers the release of a neurotransmitter. Transmission of an action potential along a polarized membrane is shown in Figure 16-6.

Another type of receptor is the vanilloid receptor (VR1) that detects noxious heat. This receptor is also referred to as the **capsaicin** receptor because, in addition to heat, it senses capsaicin, the substance that causes the burning sensation in hot peppers. Capsaicin is a member of the capsaicinoid family (Figure 16-7). Capsaicin contains a double bond that can be reduced to dihydrocapsaicin. Capsaicinoids derive from the family of vanilloids (thus the name of vanilloid receptor) because they contain the vanilloid group (Figure 16-7). Interestingly, capsaicin aids

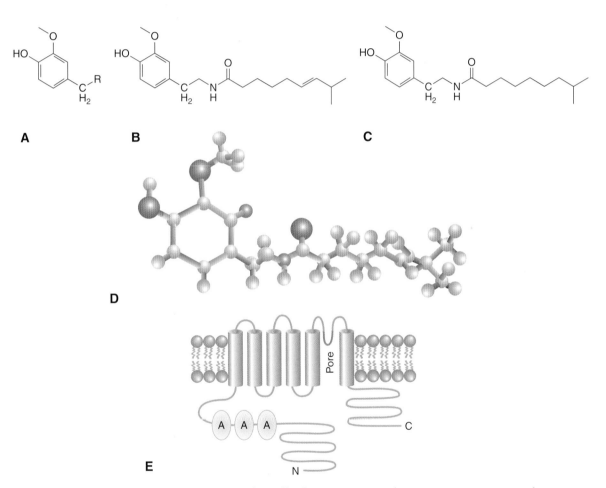

Figure 16-7. **A.** Vanillyl group component of vanilloids. **B.** Structure of capsaicin. **C.** Structure of dihydrocapsaicin. **D.** Ball- and stick-model of capsaicin [N-(4-hydroxy-3-methoxybenzyl)-8-methyl-non-*trans*-6-enanride]. **E.** Structure of the capsaicin receptor (VR1). The pore region admits Ca^{2+} into the cell when the receptor is activated by capsaicin. The receptor also responds to heat and dilute acid. Deficiency of VR1, in experimental mice, renders them relatively insensitive to heat.

in the metabolism of epoxide aromatic hydrocarbons (carcinogens), rendering them unable to bind to DNA and cause mutations. The structure of the vanilloid receptor is shown in Figure 16-7E.

Substance P

Substance P is an 11–amino acid residue polypeptide (11-mer) that functions as a neurotransmitter and a neuromodulator. It is a regulator of the transmission of pain (and is associated with regulation of neurotoxicity, stress, anxiety, nausea, neurogenesis, and mood disorders). It is a member of the **tachykinin** neuropeptide family. Substance P (Arg-Pro-Lys-Pro-Gln-Gln-Phe-Phe-Gly-Leu-Met-NH_2) is derived from a preprotachykinin and is matured to the active form by proteolysis. Other members of the family are neurokinin A (His-Lys-Thr-Asp-Ser-Phe-Val-Gly-Leu-Met-NH_2) and neurokinin B (Asp-Met-His-Asp-Phe-Phe-Val-Gly-Leu-Met-NH_2). Tachykinins are a family of amidated neuropeptides that share the sequence Phe-X-Gly-Leu-Met-NH_2 at the carboxy terminal. Receptors that recognize these tachykinins are NK_1, NK_2, and NK_3, as shown in Table 16-1. The structure of substance P and its receptor, NK_1, are shown in Figure 16-8. Detailed structures are shown in Figure 16-9. The signaling mechanism of substance P is shown in Figure 16-10. The NK_1 receptor is a G protein–coupled receptor, and substance P binding is followed by GDP–GTP exchange and subsequent activation of phospholipase-β. This leads to the formation

Table 16-1
Tachykinins; Neurotransmitters and Their Receptors

Transmitters	
Precursors:	Preprotachykinins
Synthesizing enzymes:	Peptidases
Metabolizing enzymes:	
Metabolite:	Peptide fragments

Receptors			
Receptor subtypes	**Agonists**	**Antagonists**	**Second messenger**
NK_1	SPOMe, GR 73632, $Sar^9Met(O_2)^{11}SP$; substance P*	GR82334, RP67580, CP-99994	PI; *G protein*
NK_2	GR 64349; neurokinin A*	L-659877, SR48968, MEN10207, GR94800 GR100679	PI; *G protein*
NK_3	Senktide; neurokinin B*	GR138676	PI

*Sequence in Figure 16-8A.
NK, neurokinin; *PI*, phosphatidylinositol.
Reproduced from http://www.neuro.wustl.edu/neuromuscular/lab/sp.htm.

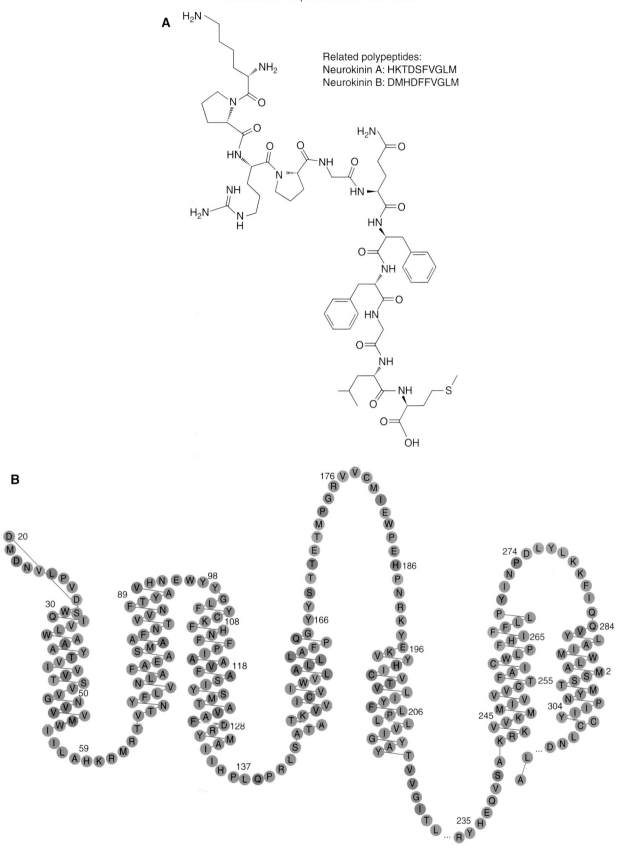

Figure 16-8. **A.** Sequence *(upper right)* formulas of related peptides, neurokinin A and neurokinin B are also shown and chemical formula of substance P. **B.** Substance P (NK1) receptor. Residues of the same color remain conserved or are mutated together. Some deletions of the sequence are introduced to avoid long loops. Part A redrawn from http://www.wdv.com/CellWorld/Biochemistry/SubstanceP/SubstancePmin.gif.

Substance P   963

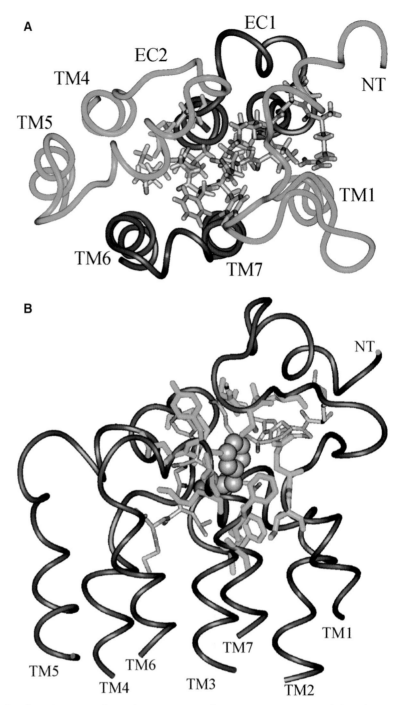

Figure 16-9. **A.** Model of the NK1R with substance P, resulting from MD simulations using structural data from NMR and homology modeling and contact points from photoaffinity labeling experiments. The ligand substance P is depicted as *sticks*. The N-terminus *(NT)*, TM1, and TM4-EC2-TM5 are *light green*, whereas TM2-EC1-TM3 and TM6-EC3-TM7 are *dark green*. The disulfide bridge between EC1 and EC2, $Cys^{105}-Cys^{180}$, is shown in *yellow*. *EC*, extracellular domain with number; *NK1R*, NK1 receptor; *TM*, transmembrane segment with number. **B.** Residues of the NK1R creating the binding pocket for Phe^7 of substance P in the ligand–receptor complex. The backbone of the receptor is shown as *dark-green ribbons*. The residues of the receptor contributing to the binding pocket for Phe^7 are illustrated. The ligand is displayed as *lines* in *gray* (carbon), *red* (oxygen), and *blue* (nitrogen); Phe^7 is depicted as Corey-Pauling-Koltun spheres. *NT*, N-terminus. **C.** Representation of the binding pocket for the C-terminus of substance P, Met^{11}-NH_2, in the NK1R. The portions of the receptor contributing to the pocket are shown as *dark-green ribbons*, with the residues labeled using the one-letter code. The residues of substance P are labeled using the three-letter code. Part A reproduced from http://www.jbc.org/cgi/content/full/276/25/22862/F1 and from Figure 1 of M. Pellegrini et al., "Molecular characterization of the substance P·Neurokinin-1 receptor complex: development of an experimentally based model," *J. Biol. Chem.*, 276: 22862–22867, 2001. Part B reproduced from http://www.jbc.org/cgi/content /full/276/25/22862/F4 and Figure 4 of M. Pellegrini et al., "Molecular characterization of the substance P·Neurokinin-1 receptor complex: development of an experimentally based model," *J. Biol. Chem.*, 276: 22862–22867, 2001.

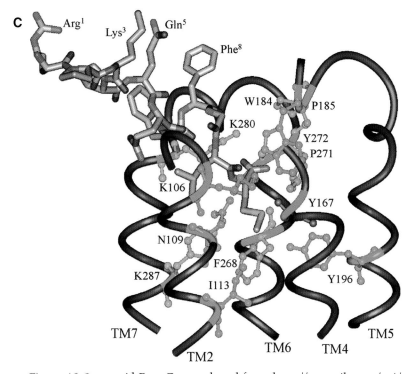

Figure 16-9, cont'd Part C reproduced from http://www.jbc.org/cgi/content/full/276/25/22862/F5 and Figure 5 of M. Pellegrini et al., "Molecular characterization of the substance P·Neurokinin-1 receptor complex: development of an experimentally based model," *J. Biol. Chem.*, 276: 22862–22867, 2001.

of inositol triphosphate (IP$_3$), which evokes the internal release of calcium ions from the calcium store in the endoplasmic reticulum. Because diacylglycerol (DAG) is produced in the initial reaction, protein kinase C (PKC) is activated, producing a cellular response. Activation of tyrosine kinase by G proteins leads to a nuclear pathway that results in gene expression and the cellular response. Details are necessarily vague because all products of the pathway are yet to be worked out. However, the inward rectifier K$^+$ channels are suppressed, chloride channels are opened, and arachidonic acid is mobilized from the cell membrane by phospholipase A$_2$. Substance P is generated from C-type sensory afferent neurons and coexists with the excitatory neurotransmitter, glutamate. These are located in primary afferent nerves that respond to painful stimulation. Glutamate receptor (Figure 16-11) action couples with substance P on dorsal horn neurons to generate the pain message and the release of substance P (and neurokinin A) from nociceptors in primary afferents and is essential to the production of moderate to intense pain.

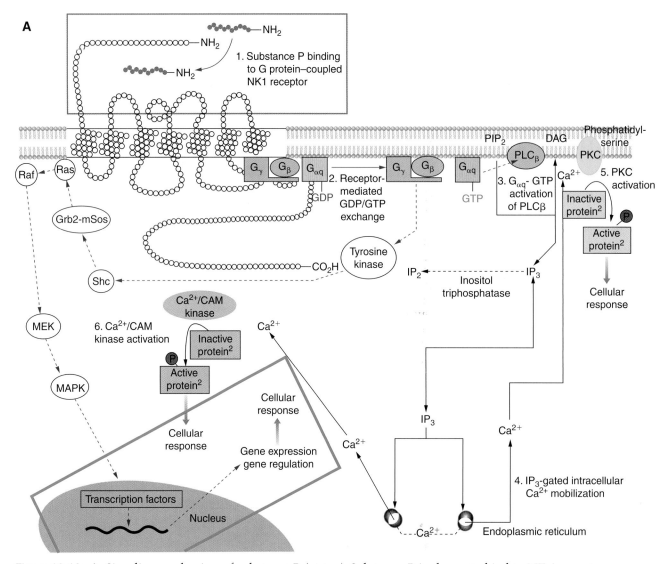

Figure 16-10. **A.** Signaling mechanism of substance P *(at top)*. Substance P is shown to bind to NK-1 receptor. Signaling/effector systems stimulated by NK-1 receptor activation:
- stimulation of P_i turnover via PLC, leading to Ca^{2+} mobilization from both intra- and extracellular sources (PTX-resistant)
- cAMP accumulation via stimulation of AC
- suppression of inward rectifier K^+ channels (PTX-resistant)
- opening of chloride channels (PTX-sensitive)
- arachidonic acid mobilization via PLA_2 (PTX-sensitive)
- mitogenic stimulation via regulation of gene expression

Figure 16-10, cont'd **B.** Steps involved in the activation of L-arginine–nitric oxide pathway after the release of substance P and excitatory amino acids (EAA) from the primary afferent terminals. Evidence suggests that nitric oxide formed in the stimulated neuron may not activate the formation of cyclic guanosine 3′, 5′-monophosphate (cGMP) in the same neuron because the elevated intracellular calcium in the neuron may inhibit the activation of guanylate cyclase. Nitric oxide, an easily diffusible gas, may diffuse into the neighboring neurons and activate guanylate cyclase in those target neurons to form cGMP from guanosine triphosphate. +, activation; −, inhibition; AC, adenylate cyclase; EAA, excitatory amino acid, such as glutamate or aspartate; GC, soluble guanylate cyclase; NMDA, N-methyl-D-aspartate receptor; NOS, nitric oxide synthase; PLA_2, phospholipase A_2; PLC, phospholipase C.

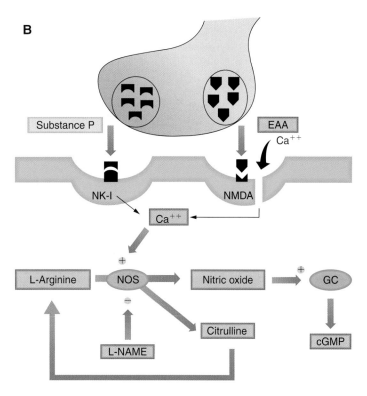

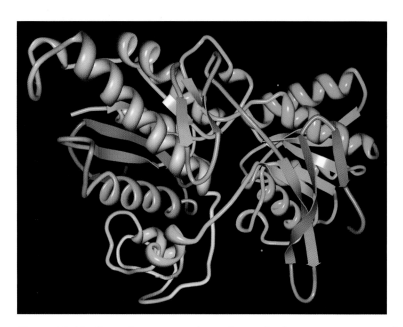

Figure 16-11. Crystal structure of metabotropic glutamate receptor subtype 1 complexed with glutamate and gadolinium ion. Reproduced from PDB ID: 1isr. D. Tsuchiya, N. Kunishima, N. Kamiya, H. Jingami, K. Morikawa. Structural Views of the Ligand-binding Cores of a Metabotropic Glutamate Receptor Complexed with an Antagonist and Both Glutamate and Gd3. *PNAS USA* **99** pp. 2660(2002).

Opioids and Morphine

Interference with the transmission of pain occurs endogenously with the opioid hormones (β-endorphin, leucine enkephalin, methionine enkephalin, dynorphin A, dynorphin B, endomorphin-1, and endomorphin-2). The term "opioid" refers to a substance that is an agonist of an opioid receptor, whereas the term "opiate" refers to substances derived from opium, some of which, but not all, are opioids. More recently, opioid has been used to designate an opioid receptor agonist regardless of its origin. New peptides have been discovered that are derived from the same precursor: nociceptin or orphanin FQ (OFQ) and nocistatin; as their names imply, these have mutually antagonistic actions for analgesia and hyperalgesia (analgesia is the absence of pain when a stimulus that normally causes pain is applied, and hyperalgesia is increased response to a stimulus that is normally painful). These substances do not operate through the opioid receptor system, but nociceptin or OFQ bind to the **orphanin FQ receptor (OFQR),** as discussed later. The amino acid sequences of some opioid peptides are presented in Table 16-2. *Opioids and endorphins* (endorphin means "endogenous morphine") *bind to opiate receptors to terminate pain signals by inhibiting release of substance P from presynaptic neurons,* as shown in Figure 16-12. Binding of an opioid to a receptor produces a cellular conformational change, resulting in the inhibition of the release of G proteins from the internal plasma membrane that affects the opened or

Table 16-2
Amino Acid Sequences of Some Opioid Peptides

Peptide	Amino acid sequence
β-Endorphin	**YGGFM**TSEKSQTPLVTLFKNAIIKNAHKKGQ
Methionine enkephalin	YGGFM
Leucine enkephalin	YGGFL
Dynorphin A	**YGGFL**RRIRPKLKWDNQ
Dynorphin B	**YGGFL**RRQFKVVT
Endomorphin-1	YPWF
Endomorphin-2	YPFF
(Nociceptin/Orphanin FQ; not an opioid; does not alleviate pain)	*FGGFTGARKSARKLANQ*
Nocistatin	SVFSSCQRDCLTCQEKLHPALDSFDLEVCILE CEEKVFPSPLWTPCTKVMARSSWQLSPAAPE HVAAALYQPRASEMQHL**RR**MPRVRSLFQEQ EEPEPGMEEAGEMEQKQLQ**KR***FGGFTGARKS ARKLANQ***KR**FSEFMRQYLVLSMQSSQ**RRR**TL HQNGNV (176 amino acid residues)

The first five amino acids *(highlighted)* at the N-terminal of β-endorphin are those of methionine enkephalin. The first five amino acids *(highlighted)* in the N-terminal of dynorphin A and dynorphin B are those of leucine enkephalin. The activities of these peptides are due to the first five amino acids. **Enkephalinase** enzymes inactivate enkephalins by cleaving between the tyrosine and the glycine residues. A unique 55-kDa endopeptidase converts dynorphin B to its bioactive fragment, YGGFLR[6], but does not cleave dynorphin A (J. Silberring and F. Nyberg. *J. Biol. Chem.*, 264: 11082–11086, 1989.) Nocistatin/orphanin FQ, which has the opposite effect of nociception, contains the amino acid sequence *(italics)* of nociception. The N-terminal amino acids of nociception orphanin FQ are FGGF, whereas some active opioids begin with the sequence YGGF. The difference between Y and F is the hydroxyl group of tyrosine, which is important for opioid activity (Figure 16-4c).

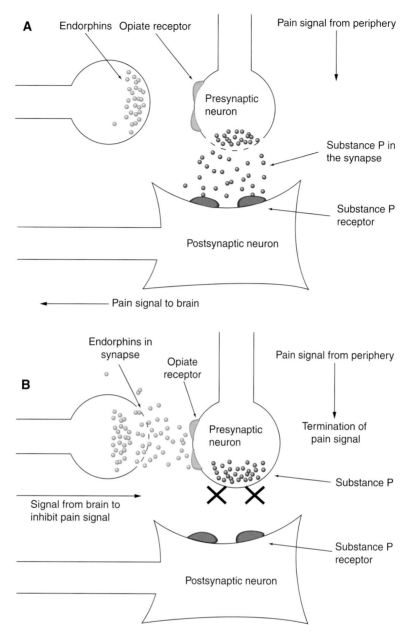

Figure 16-12. **A.** Substance P travels across the synapse between neurons to transmit a pain signal. Release of substance P results from electrical impulses that travel down the neuronal axon. **B.** Activation of opiate receptors of the presynaptic neuron produces a conformational change preventing the release of substance P. Opiate receptors are activated by release of endorphins (or other opioid) from a neighboring neuron. Release of endorphins is generated in the brain (anterior pituitary).

closed status of ion channels. Without the movement of ions to produce the appropriate propagation gradients, neurons that have activated opiate receptors are unable to release substance P.

There are three major opioid receptors: μ, δ, and κ. μ-Receptors are involved in most of the analgesic effects of opioids. δ-Receptors operate in the periphery but may contribute to analgesia. κ-Receptors are involved in analgesia at the spinal level, and there are certain analgesic drugs that are κ-receptor specific. All opioid receptors are G protein linked, and they signal the inhibition of adenylate cyclase (resulting in the reduction of cyclic AMP). Opioid receptors bring about the opening of potassium ion channels, causing hyperpolarization, and inhibit the opening of calcium ion channels, depressing transmitter release. The effects on the membrane are not, however, explained by the lowering of the cyclic AMP concentration. The affinities of opioid receptors for opioid peptides and morphine are summarized in Table 16-3. The μ, δ, and κ receptors are seven transmembrane proteins, as shown in Figure 16-13.

A system has been discovered with the agonist nociceptin, also called OFQ. The nociceptin or OFQ structure resembles dynorphin A but has some different amino acid residues that cause it to be an agonist for OFQR but not an agonist for the opioid κ-receptor (Figure 16-14) that binds dynorphin A avidly. Nociceptin or OFQ binds to OFQR, a G protein–coupled receptor, which stimulates the inositol triphosphate receptor (IP_3R) with the cytoplasmic increase of calcium ions and inhibits adenylate cyclase. *OFQ stimulates nerve endings of nociceptive primary afferent neurons indirectly by a local substance P release.* Theoretically, nocistatin, which contains the OFQ sequence within its structure (Table 16-2), should bind to OFQR and inhibit its function, provided that the OFQ sequence is at the surface of the nocistatin 276-amino acid polypeptide and available for receptor binding.

Morphine (Figure 16-4C) is an agonist of the μ-opioid receptor and serves as a model for the signaling pathways of this analgesic receptor. As a G protein–linked receptor, it stimulates adenylate cyclase (increases concentration of cyclic AMP), the export of potassium ions, and the import of sodium ions through the activation of protein kinase A (PKA) and phosphorylation of proteins by protein kinase A culminates in the stimulation of gene expression (Figure 16-15A). Morphine also can stimulate the increase of calcium ion, leading to the activation of constitutive nitric oxide synthase (cNOS). *This leads to an increase of nitric oxide that stabilizes the IκBα–NFκB/RelA complex, thus preventing NFκB from entering the nucleus so that the induction of inducible nitric oxide synthase (iNOS) and the production of inflammatory cytokines (and pain) would not occur,* as shown in Figure 16-15B.

Table 16-3
Selectivities of Opioid Peptides and Morphine for Specific Opioid Receptors

Opioid	Affinity of receptor		
	μ	δ	κ
β-Endorphin	+++	+++	+++
Leu-enkephalin	+	+++	−
Met-enkephalin	++	+++	−
Dynorphin	++	+	+++
Morphine	+++	+	++

Binding affinities of opioids for specific opioid receptors. A + sign indicates the affinity of the receptor for the specific opioid. A − sign indicates no productive binding to the specific receptor. Reproduced from H.P. Rang, M.M. Dale, J.M. Ritter, and P. Gardner. *Pharmacology,* Churchill Livingstone: New York, p. 621, 1995.

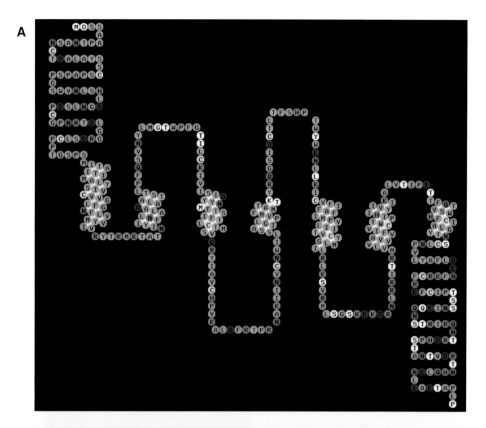
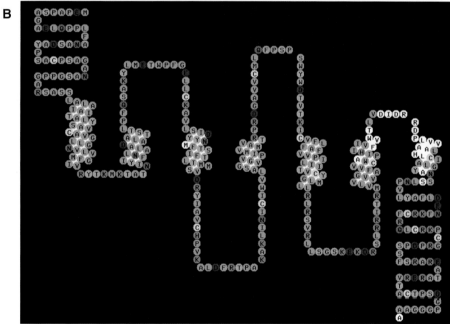

Figure 16-13. **A.** Diagram of the μ-opioid receptor indicating the amino acid residues. **B.** Diagram of the δ-opioid receptor. Part A redrawn from http://www.wdv.com/CellWorld/Receptors/pages/OpioidReceptorTypeM_gif.htm. L. Van Warren, Cell World. Part B redrawn from http://www.wdv.com/CellWorld/Receptors/pages/OpioidReceptorTypeM_gif.htm. L. Van Warren, Cell World.

Continued

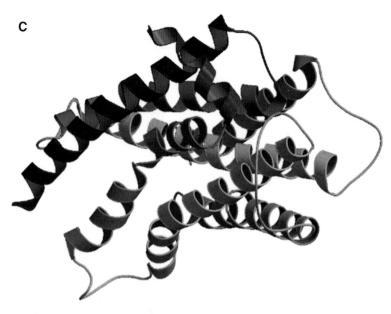

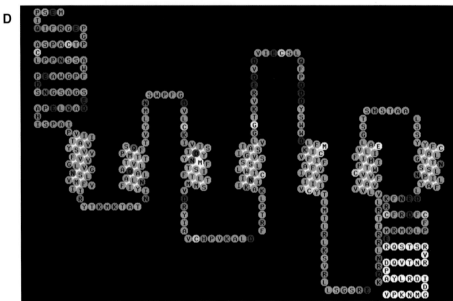

Figure 16-13, cont'd **C.** Structure of human δ-opioid receptor (ribbon structure). **D.** Diagram of the κ-opioid receptor. Part C reproduced from http://www.ebi.ac.uk/msd-srv/msdlite/atlas/visualization/1ozc.html. PDB ID: 1ozc. F.M. Décaillot, K. Befort, D. Filliol, S. Yue, P. Walker, B.L. Kieffer. Opioid Receptor Random Mutagenesis Reveals a Mechanism for G Protein-coupled Receptor Activation. *Nat. Struct. Biol.* **10** p. 629 (2003). Figure 16-13. Part D redrawn from http://www.wdv.com/CellWorld/Receptors/pages/OpioidReceptorTypeM_gif.htm. L. Van Warren, Cell World.

Another factor that can alter the extent of pain is so-called neuronal plasticity. This refers to changes that can occur in the nervous system, including structural changes in neurons (particularly nociceptors), alterations in the amounts of neurotransmitters, and changes in amounts and properties of receptors and ion channels, the summation of which can lead to increased activity of neurons in the pain pathway. Plasticity also can alter the body's pain inhibitory systems and increase the

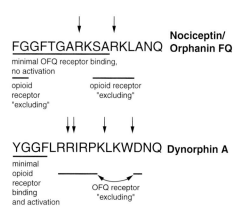

Figure 16-14. Amino acid sequences of orphanin FQ (OFQ) and dynorphin A, its closest structural relative. OFQ binds to the OFQ receptor based upon the sequence of the first 11 amino acid residues. Dynorphin A, which binds most strongly to the kappa opioid receptor (Table 16-3) based on the N-terminal sequence, does not bind to the OFQ receptor owing to the amino acid residues 11-14 (KLKW). At the same position, the amino acid residues 11-14 of OFQ are ARKL. It is concluded that there was a coordinated mechanism of evolution that separated the orphanin FQ system from the opioid system. *Arrows* indicate amino acid residues, in addition to N-terminal tetrapeptide, which are most critical for biological activity. Redrawn from http://www.jbc.org/content/vol273/issue3/images/large/bc0481819005.jpeg and from Figure 5 of R.K. Reinscheid et al., "Structures that delineate orphanin FQ and dynorphin A pharmacological selectivities," *J. Biol. Chem.*, 273: 1490–1495, 1998.

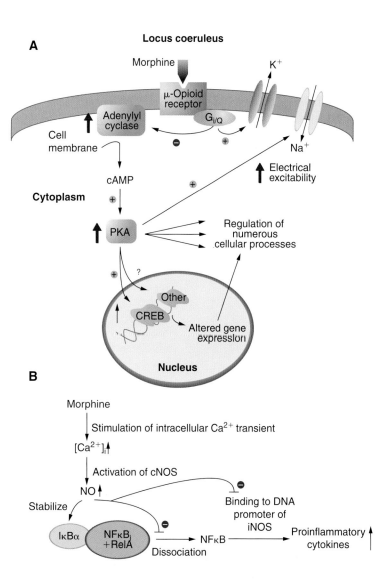

Figure 16-15. **A.** Action of morphine on neurons to the spinal cord *(locus coeruleus)* and the signaling pathway. **B.** Action of morphine to inhibit the action of NFκB in the cell nucleus to prevent the generation of proinflammatory cytokines.

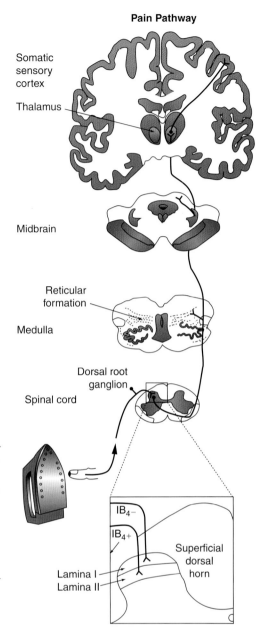

Figure 16-16. Diagram of the pain pathway. Painful stimuli such as intense heat activate the peripheral terminals of nociceptors. Action potentials are transmitted along the afferent axons to the spinal cord. The central terminals of IB$_4$-negative unmyelinated nociceptors synapse in lamina I and outer lamina II, whereas IB$_4$-positive unmyelinated nociceptors terminate in inner lamina II. By means of chemical transmission, nociceptors activate spinal neurons that send axons across the spinal cord and up fiber tracts and terminate in the medulla, midbrain, and thalamus. Thalamic neurons project to regions of the cortex, including the somatosensory cortex. Reproduced from Figure 1 of C.L. Stucky, M.S. Gold, and X. Zhang, "Mechanisms of pain," *PNAS USA*, 98: 11845–11846, 2001. From the Academy Chinese American Frontiers Of Science Symposium, http://www.pnas.org/content/vol98/issue21/images/large/pq2113733001.jpeg.

pain in this manner. These changes can be brief or lasting. Figure 16-16 shows the pain pathway in detail, indicating the locations of the IB$_4^-$ and IB$_4^+$ nociceptors.

Anandamide

Anandamide is a substance derived from arachidonic acid in the body. It resembles the drug class of cannabinoids in its actions and therefore is considered, along with analogous compounds, to be an endocannabinoid.

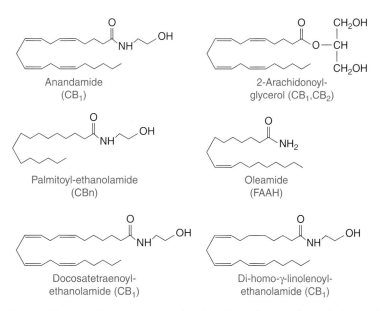

Figure 16-17. Substances considered to be endocannabinoids derived from arachidonic acid. The proposed target of each metabolite is shown in parentheses. *CB1*, central cannabinoid receptor; *CB2*, 'peripheral' cannabinoid receptor; *CBn*, non-CB1-non-CB2 cannabinoid receptor(s); *FAAH*, fatty acid amide hydrolase.

These are shown in Figure 16-17. Arachidonic acid is split out of phospholipids in the cell membrane by the action of phospholipase A_2. The action of anandamide synthase converts arachidonate and ethanolamine (split from a phosphatidic acid in the membrane by phospholipase D) to anandamide.

Anandamide can be released from the cell to be taken up by a specific transporter and carried inside a neighboring cell, where it becomes hydrolyzed (and inactivated) by **fatty acid amide hydrolase (FAAH)** to produce ethanolamine and fatty acid. *FAAH activity determines the extent to which anandamide is active.* FAAH might be a therapeutic target so that by its inhibition, the activity of anandamide could act with longer duration. It appears that anandamide is taken up to the cell interior by caveola or lipid raft endocytosis and later is metabolized and inactivated by FAAH. The metabolites (arachidonic acid and ethanolamine) are recycled to caveolin-rich membrane domains. These reactions are summarized in Figure 16-18. A more straightforward view of the biosynthetic process only is shown in Figure 16-19.

A close relative of anandamide is 2-arachidonylglycerol, which binds not only to the central cannabinoid receptor, CB-1, but also to the peripheral cannabinoid receptor, CB-2. *Both arachidonylglycerol and anandamide can be converted into prostaglandins, thromboxane, and prostacyclin glycerol esters through oxygenation of cannabinoids by inducible cyclooxygenase-2 (COX2), but not by a constitutive COX1.* These conversions are summarized in Figure 16-20. Structures of COX1 and COX2 are compared in Figure 16-21.

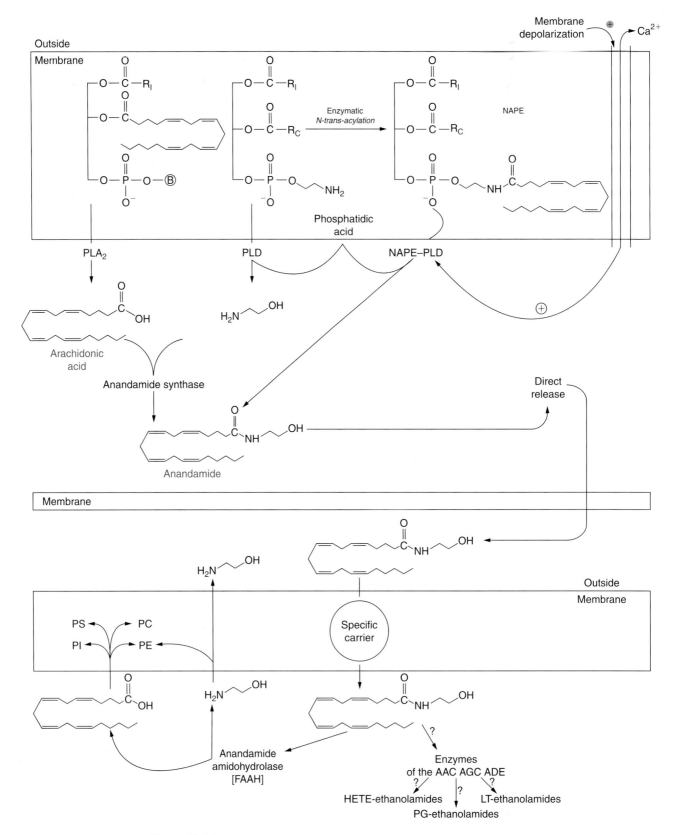

Figure 16-18.

Figure 16-18. Possible biosynthetic and catabolic pathways for anandamide and related acylethanolamides. Anandamide (and palmitoyl-ethanolamide) produced by either a synthase enzyme or N-acyl-phosphatidyl-ethanolamine-specific phospholipase D (NAPE-PLD) following membrane depolarization is released outside the cell and acts at neighboring cells. In order to be catalyzed by the synthase enzyme the formation of anandamide must be preceded by phospholipase A_2 (PLA_2) activation. Once released by cells, anandamide or palmitoyl-ethanolamide can be taken up by selective carrier mechanisms and degraded by anandamide amidohydrolase (*FAAH*, fatty acid amide hydrolase), thereby producing ethanolamine and fatty acids. The latter are readily incorporated into membrane phospholipids *(PC, PS, PE, PC)*. Alternative catabolic pathways for anandamide may utilize the enzymes of the arachidonic acid (AA) cascade; however, the formation of the ethanolamides of hydroxy-eicosatetraenoic acids (HETE), prostaglandins (PG) or leukotrienes (LT) has never been demonstrated in intact cells. *B*, phospholipid bases.

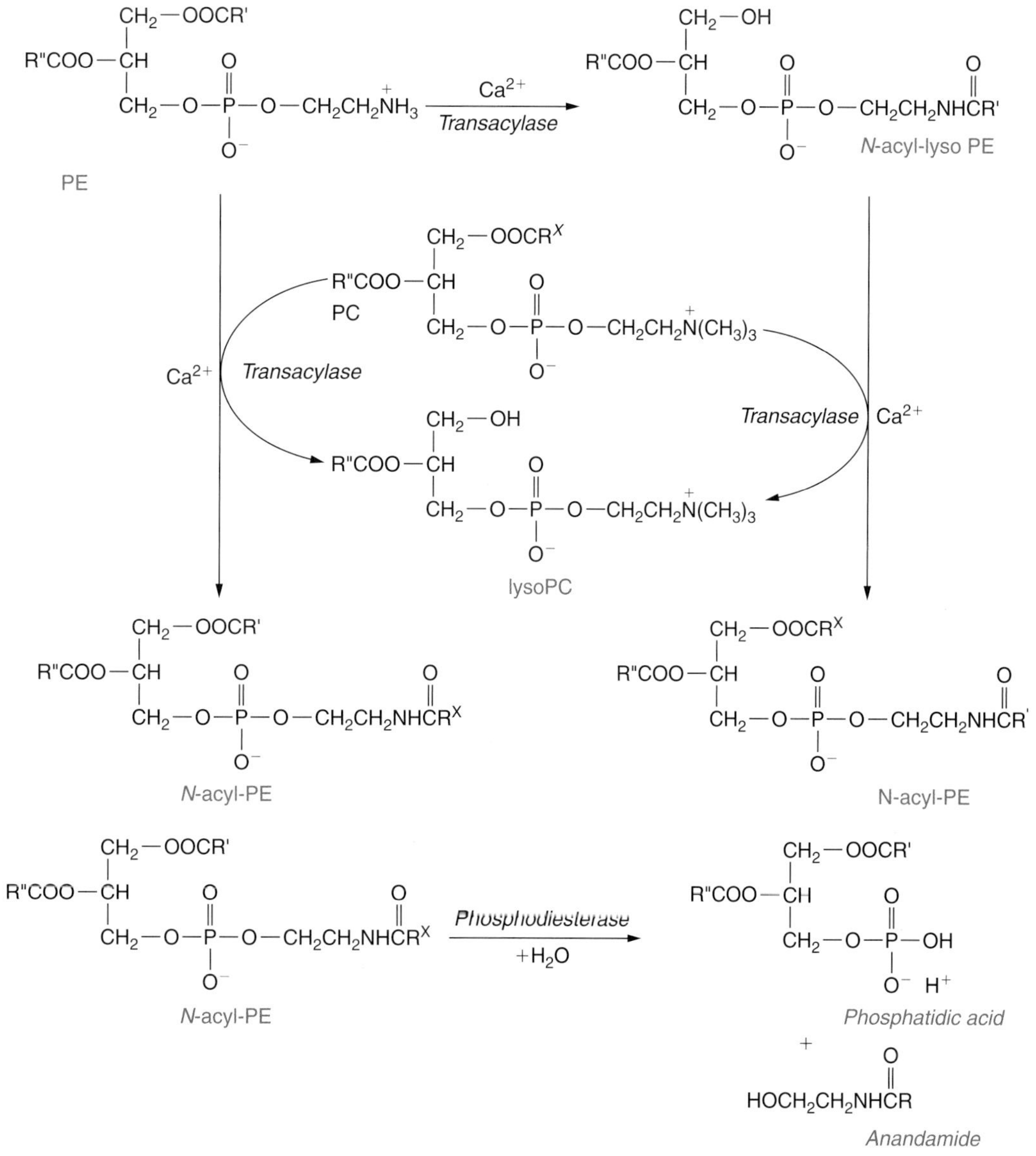

Figure 16-19. Biosynthesis of anandamide from phosphatidylethanolamine (PE). *LysoPC*, lysophosphatidyl choline.

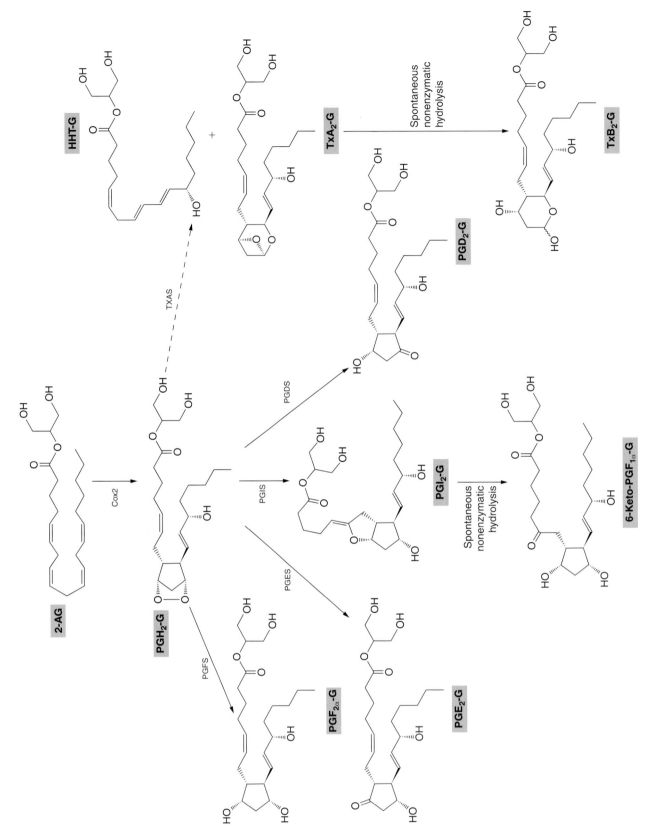

Figure 16-20. Conversion of the cannabinoid, 2-arachidonylglycerol (2-AG) into various prostaglandins, thromboxanes, and prostacyclin by action of cydocxygenase-2 (COX2), the inducible cyclooxygenase, followed by enzymatic conversions to the products. *PGD₂-G*, prostaglandin D₂-glycerol; *PGDS*, prostaglandin D synthase; *PGE₂G*, prostaglandin E₂-glycerol; *PGES*, prostaglandin E₂ synthase; *PGF₂α-G*, prostaglandin F₂α-glycerol; *PGFS*, prostaglandin F₂α synthase; *PGH₂-G*, prostacyclin H₂-glycerol; *PGI₂-G*, prostacyclin glycerol; *PGIS*, prostaglandin synthase; *TXA₂-G*, thromboxane A₂-glycerol; *TXAS*, thromboxine A synthase.

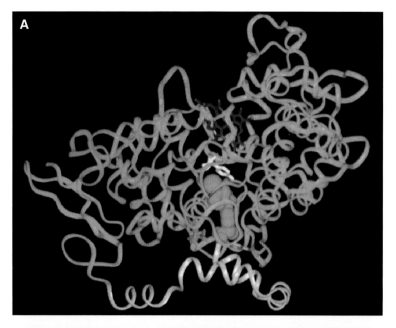

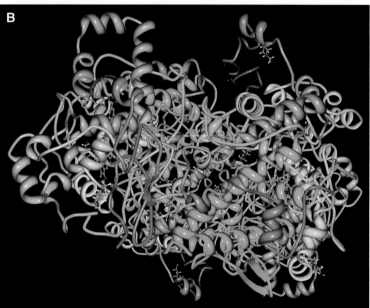

Figure 16-21. **A.** Subunit structure of COX1. Each cox subunit comprises three domains, an epidermal growth factor domain *(yellow)*, a membrane-binding domain *(lavender)*, and a catalytic domain *(blue)*. The catalytic domain contains the cyclooxygenase active site and the peroxidase active site separated by the heme prosthetic group *(red)*. In the present structure, the cyclooxygenase active site is occupied by a molecule of iodosuprofen *(lime)*. Arg-120, Tyr-355, and Glu-524 comprise an H-bonding network that introduces a constriction at the base of the cyclooxygenase active site. They are depicted in *gold*. The volume beneath this constriction is termed the lobby and is bordered on three sides by the membrane-binding domain. The catalytically important Tyr-385 residue is depicted in *white*. The peroxidase active site is at the top of the protein in this drawing and is visible as the wide opening to the heme prosthetic group.
B. Structure of COX2 complexed with a selective inhibitor (SC-558). Part A reproduced from http://www.jbc.org/cgi/content/full/274/33/22903/F2 and from Figure 2 of L.J. Marnett et al., "Arachidonic acid oxygenation by COX-1 and COX-2: Mechanisms of catalysis and inhibition," *J. Biol. Chem.*, 274: 22903–22906, 1999. Part B reproduced from http://www.rcsb.org/pdb/cgi/explore.cgi?pdbId=1cx2 and PDB ID: 1cx2. R.G. Kurumbail, A.M. Stevens, J.K. Gierse, J.J. McDonald, R.A. Stegeman, J.Y. Pak, D. Gildehaus, J.M. Miyashiro. Penning, T.D., Seibert, K., Isakson, P.C., Stallings, W.C. Structural Basis for Selective Inhibition of Cyclooxygenase-2 by Anti-inflammatory Agents. *Nature* **384** pp. 644 (1996).

Again, anandamide binds to the neuronal CB-1, a G protein–coupled receptor producing a cannabinoid-like effect. This is the receptor through which marijuana acts to produce its effects. One of the components of marijuana is tetrahydrocannabinol (THC). The structure of THC is shown in Figure 16-22. Interestingly, the structures of THC and anandamide are different, but there must be a structural similarity to merit a common binding site or dual recognition sites within the binding pocket. In addition, CB-2 is located mainly on the membranes of cells of the immune system. Diagrammatic structures of these two receptors are shown in Figure 16-23. Activated CB-1 leads to the inhibition of cyclic AMP

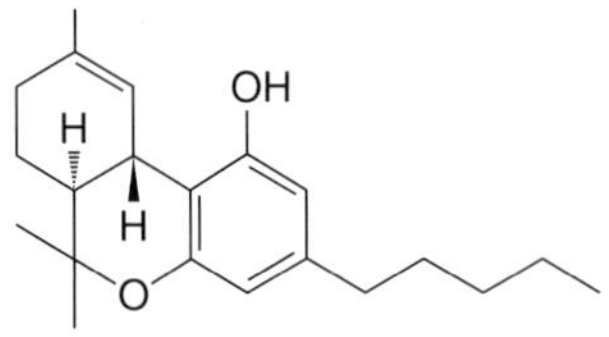

Figure 16-22. The structure of tetrahydrocannabinol (THC).

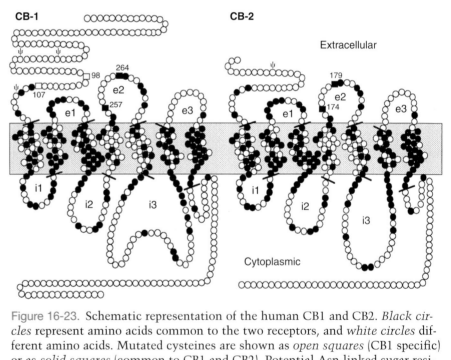

Figure 16-23. Schematic representation of the human CB1 and CB2. *Black circles* represent amino acids common to the two receptors, and *white circles* different amino acids. Mutated cysteines are shown as *open squares* (CB1 specific) or as *solid squares* (common to CB1 and CB2). Potential Asn-linked sugar residues are shown as ψ. The bars represent the sites of fusion to create the chimeras. *e*, extracellular domain; *i*, intracellular domain. Reproduced from Figure 1 of D. Shire et al., "Structural features of the central cannabinoid CB1 receptor involved in the binding of the specific CB1 antagonist SR 141716A," *J. Biol. Chem.*, 271: 6941–6946, 1996. http://www.jbc.org/cgi/content/full/271/12/6941/f1

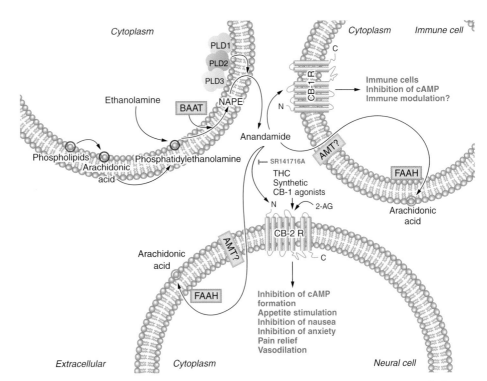

Figure 16-24. The lipid anandamide derived from arachidonic acid binds to the cannabinoid G protein–coupled receptors CB1 and CB2 to exert its biological effects. CB1 is also the site of action of the psychoactive components of marijuana, including tetrahydrocannabinol (THC), and responsible for the effects of cannabinoids to repress learning, memory, and anxiety; relieve pain and nausea; and stimulate appetite. CB2 is expressed mainly in cells of the immune system and appears to modulate immune functions but not to be involved in the effects of cannabinoids in the central nervous system. In addition to the illicit use of cannibinoids as drugs of abuse, the pharmaceutical use of natural and synthetic cannabinoids is being intensely studied, including a possible role in neurodegenerative diseases. Another therapeutic approach is the pharmacological modulation of levels of the endogenous cannabinoid anandamide. Although the subject of some controversy, the most likely route for the biosynthesis of anandamide starts with the formation of phosphatidylethanolamine from arachidonic acid and ethanolamine. Phosphatidylethanolamine is converted by N-acyltransferase to the phospholipid N-arachidonoyl-phosphatidylethanolamine (NAPE) and hydrolysis by phospholipase D releases phosphatidic acid and anandamide, which binds to the cannabinoid receptors. The signal it induces is terminated in part by degradation of anandamide by anandamide hydrolase (FAAH). 2-Arachidonoylglycerol (2-AG), and 2-arachidonoylglyceryl ether have also been shown to bind the cannabinoid receptors and have been suggested to be important endocannabinoids. Also complicating cannabinoid pharmacology is the suggestion that some effects of cannabinoids might be moderated by additional receptor subtypes that have not yet been identified or by the vanilloid receptors. Redrawn from http://www.biocarta.com/pathfiles/h_cb1rPathway.asp. Biocarta, Inc., San Diego, CA.

formation in the neuronal cell, and activated CB-2 leads to the inhibition of cyclic AMP formation in immune cells (Figure 16-24). CB-1 resides in the neuronal cell membrane in equilibrium between active (bound with ligand) and inactive (absence of ligand) states. The activated receptor becomes internalized (endocytosed) through a Rab5 pathway, and the receptor courses through a recycling pathway, ending in the reappearance of the inactive receptor on the membrane. This receptor trafficking is shown in Figure 16-25. CB-1 cycles between the plasma membrane and

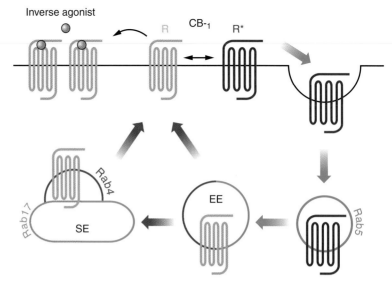

Figure 16-25. Proposed model for the constitutive trafficking of the CB1R. Plasma membrane–localized CB1Rs are in an equilibrium between active R* *(red)* and inactive R *(orange)* states. Once activated, either constitutively or by an agonist ligand, the receptor is internalized through a Rab5-mediated pathway *(green arrow)*. The receptor then travels through early endosomes *(EE)* and sorting endosomes *(SE)* and is recycled back to the plasma membrane by Rab4-mediated recycling pathways *(blue arrows)*. Since recycling is slower than endocytosis, the majority of receptors are localized to endosomes at steady state. The inverse agonist, which stabilizes the inactive conformation of the receptor, leads to externalization by sequestering recycled receptors on the plasma membrane. *CB1R*, CB1 receptor. Redrawn from Figure 7 of C. Leterrier et al., "Constitutive endocytic cycle of the CB1 cannabinoid receptor," *J. Biol. Chem.*, 279: 36013–36021, 2004. http://www.jbc.org/cgi/content-nw/full/279/34/36013/FIG7.

the internal endosomes whereby it is located intracellularly predominantly (intracellular activities are slower than membrane activities) at the steady state.

Binding of anandamide to the cannabinoid receptors in endothelial cells leads to activation of nitric oxide synthase (NOS), which increases intracellular nitric oxide concentration. Nitric oxide activates the cellular transporter which moves anandamide to the cell interior without reacting with its receptor and internal anandamide is cleaved by FAAH to arachidonic acid and ethanolamine (Figure 16-26). Anandamide is thought to increase food intake through activation of the G protein–coupled cannabinoid receptors. Interestingly, a relative of anandamide, oleoylethanolamide, appears to be an antiobesity agent by reducing food intake and bodily weight gain through activation of the nuclear receptor, peroxisome proliferator–activated receptor-α) and consequent stimulation of fat use.

In addition to binding to cannabinoid receptors, *anandamide can bind to vanilloid receptors (capsaicin receptors) and produce programmed*

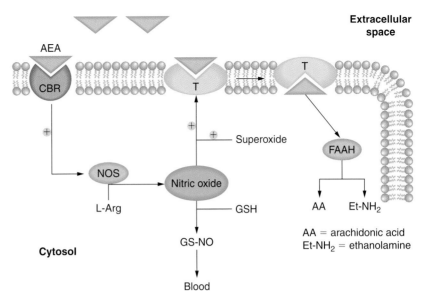

Figure 16-26. Regulatory loop between cannabinoid receptor and AEA transporter in endothelial cells. Binding of extracellular AEA to cannabinoid receptors *(CBR)* leads to activation of NOS and intracellular nitric oxide production from L-arginine *(L-Arg)*. Nitric oxide, or better, peroxynitrite, derived from its reaction with superoxide, activate, transporter *(T)*-mediated uptake of AEA. Once taken up, AEA can be cleaved rapidly to arachidonic acid and ethanolamine by membrane-bound FAAH. GSH can bind nitric oxide leading to *S*-nitrosoglutathione *(GS-NO)*, thus inhibiting its effect. AEA, anandamide; FAAH, fatty acid amide hydrolase; T, cannabinoid transporter.

cell death (apoptosis) by increasing the release of cytochrome c from mitochondria and activating caspase enzymes. In the same cell, anandamide can bind to the cannabinoid receptor, which allows calcium concentration to increase in the cell interior and activates the anandamide transporter on the cell membrane. This leads to hydrolysis and inactivation of anandamide inside the cell. Activation of the transporter admits anandamide into the cell interior and can limit the effects of anandamide in producing apoptosis. In inducing the membrane transporter for anandamide, the cannabinoid receptor can protect the cell against the apoptosis produced through the vanilloid receptor. These effects must be weighed against the individual effects of anandamide on the vanilloid receptor and the cannabinoid receptor (Figure 16-27). The binding site of the vanilloid receptor in complex with two agonists is shown in Figure 16-28.

The vanilloid receptor is controlled by phosphorylation by a kinase that is calcium ion–calmodulin dependent, as shown in Figure 16-29. It appears that *anandamide complexes with the vanilloid receptor by acting at an intracellular site and that this activity is limited by the extent to which FAAH inactivates anandamide.* The actions of anandamide at

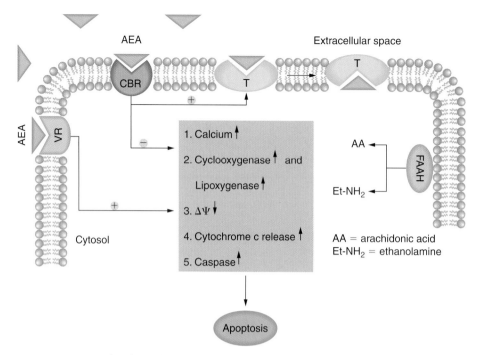

Figure 16-27. Role of vanilloid receptor and cannabinoid receptor in AEA-induced programmed cell death. Binding of extracellular AEA to VR triggers a sequence of events starting with a rise in intracellular calcium and followed by activation of cyclooxygenase and lipoxygenase, drop in mitochondrial membrane potential ($\Delta\psi$), release of cytochrome c, and activation of caspases, ultimately leading to programmed cell death (apoptosis). Binding of AEA to cannabinoid receptors *(CBR)* activates transporter *(T)*-mediated uptake of AEA and its subsequent cleavage to arachidonic acid and ethanolamine by membrane-bound FAAH. These latter events inhibit the proapoptotic activity of AEA. *AEA*, anandamide; *FAAH*, fatty acid amidohydrolase; *VR*, vanilloid receptor.

CB-1 are summarized in Figure 16-30, where it is shown that cyclic AMP concentration is reduced, potassium ion is expelled, and calcium ion is taken up from the cell exterior, leading to protein kinase activity (although not protein kinase A).

There are other known regulators of anandamide action. In the T lymphocyte, the hormone leptin (Figure 16-31), through a transcriptional effect, can increase the concentration of FAAH and decrease the amount of biologically active anandamide (Figure 16-32). Another activator of FAAH in human T lymphocytes is progesterone, which stimulates the enzyme nearly 300-fold *in vitro* through upregulation of gene expression at both the transcriptional and the translational levels. Thus, both leptin and progesterone can fine-tune the effects of anandamide on immune cells through the upregulation of the enzyme, FAAH, which hydrolyzes and inactivates anandamide.

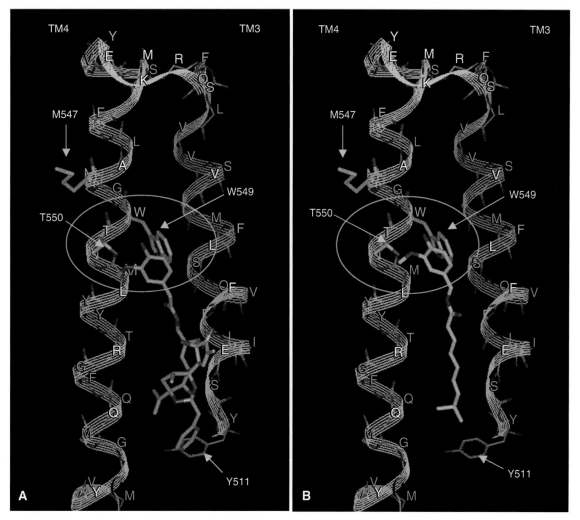

Figure 16-28. Structural model of RTX *(A)* and capsaicin *(B)* interacting with transmembrane helices TM3 and TM4 of TRPV1. The backbone of the complete structural model, along with the side chains of residues considered to be involved in interactions, is shown. The side chains of Met[547], Trp[549], and Thr[550] are shown as *sticks (thick lines)*. Indicated interactions of vanillyl moiety with Thr[550] and Trp[549] are highlighted in the *green ellipses*. Residues considered to be involved in interactions with the substituted phenyl portions of the two ligands are shown in *cyan*. Modeled hydrophobic contacts of Tyr[511] with the hydrophobic ends of RTX and capsaicin are shown. RTX and capsaicin are shown in *pink (A)* and *green (B)*, respectively. *RTX*, resiniferatoxin (from *Euphorbia resinifera*) a potent agonist of vanilloid receptors; *TM*, transmembrane. Reproduced from Figure 7 of N.R. Gavva et al., "Molecular determinants of vanilloid: sensitivity in TRPV1," *J. Biol. Chem.*, 279: 20283–20295, 2004.

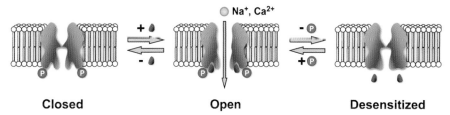

Figure 16-29. Regulation of the vanilloid receptor by phosphorylation by $Ca^{2+}/$calmodulin dependent kinase. At the *left* is the phosphorylated, inactive receptor, which can bind ligand (e.g., capsaicin), shown as *red oval*. The activated receptor *(middle)* admits Na^+ and Ca^{2+} into the cell. Dephosphorylation produces the desensitized receptor *(right)* that releases the ligand. Redrawn from Scheme 1 of J. Jung et al., "Phosphorylation of vanilloid receptor 1 by $Ca^{2+}/$calmodulin-dependent kinase II regulates its vanilloid binding," *J. Biol. Chem.*, 279: 7048–7054, 2004.

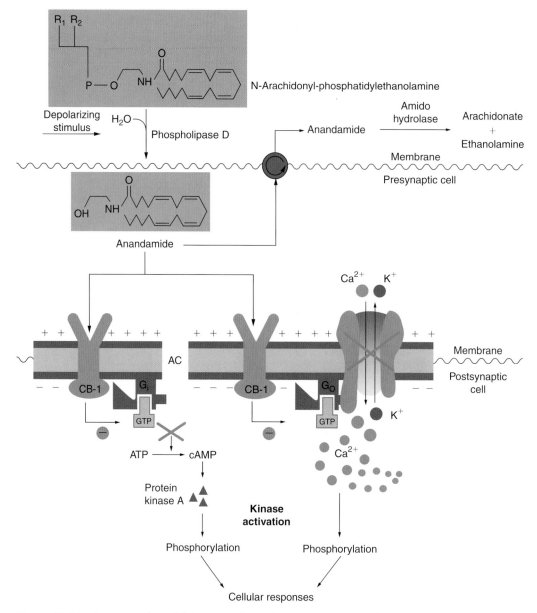

Figure 16-30. A proposed model for cannabinoid transmission involving a presynaptic and a postsynaptic element. Upon the input of a depolarizing stimulus, anandamide is released from lipid precursors in the presynaptic cell by enzymatic cleavage. Anandamide acts then through the stimulation of CB1 receptors coupled to either the inhibition of cAMP production or the opening of K^+ or Ca^{2+} channels. These signaling elements ultimately activate phosphorylation mechanisms modulating cell activity. Other transduction pathways coupled to cannabinoid receptors, including the mitogen-activated protein kinase cascade, are not depicted for increased clarity. Anandamide is degraded, after a specific uptake process, by intracellular enzymatic reactions. An additional model may colocalize all these elements in a single cell (autacoid/local mediator). Redrawn from Figure 2 of M. Navarro and F. Rodriguez de Fonsecaea. "Introduction: the neurobiology of cannabinoid transmission: from anandamide signaling to higher cerebral functions and disease." *Neurobiol. Dis.*, 5: 379–385, 1998. Article No. NB980216.

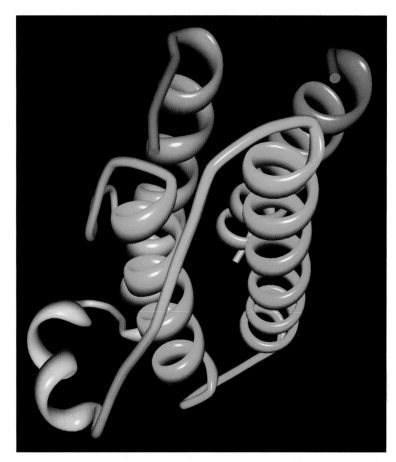

Figure 16-31. Structure and amino acid sequence of leptin. Note: The sequences shown here are the sequences observed in the structure. They may differ from the sequences found in the associated UniProt entry for entries. Some chains listed may not appear in the original PDB entry but are part of the assemblies for this entry. Sequence from http://www.ebi.ac.uk/msd-srv/msdlite/atlas/sequence/1ax8.html. Reproduced from http://www.rcsb.org/pdb/cgi/explore.cgi?pdbId=1ax8. PDB ID: 1ax8. F. Zhang, M.B. Basinski, J.M. Beals, S.L. Briggs, L.M. Churgay, D.K. Clawson, R.D. DiMarchi, T.C. Furman, J.E. Hale, H.M. Hsiung, B.E. Schoner, D.P. Smith, X.Y. Zhang, J.P. Wery, R,W, Schevitz. Crystal Structure of the Obese Protein Leptin-E100. *Nature* **387** pp. 206 (1997). Sequence Chain A,B,C (Uniprot entry p41159). >1a8_A VPIQKVQDDTKTLIK-TIVTRINDISHTQSVSSKQKVTGLDFIPGLHPILTLSKMDQTLAVYQQILTSMP-SRNVIQISNDLENLRDLLHVLAFSKSCHLPEASGLETLDSLGGVLEASGYSTE-VVALSRLQGSDMLWQLDLSPGC

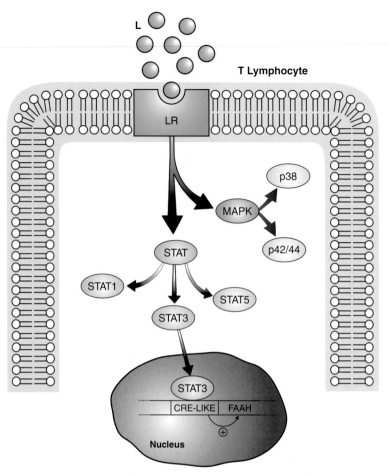

Figure 16-32. Model of the activation of FAAH promoter by leptin in human T cells. L binds to its receptor, LR, in human T lymphocytes, leading to the activation of STAT3 but not of other typical LR-dependent downstream signals like STAT1, STAT5, MAPK, p38, p42, or p44. Phosphorylation of STAT3 results in the activation of a CRE-like region in the FAAH promoter, thus upregulating gene expression. *L*, leptin; *LR*, leptin receptor.

Excitatory Amino Acids

Although excitatory amino acids, in particular, glutamic acid (also aspartate and a few other amino acids), play important roles in neural development, learning, and memory, *these amino acids also are associated with neural toxicity and neuronal damage.* They can operate in tandem with substance P in the transmission of pain (Figure 16-33). There are several excitatory amino acid receptors, and most bind glutamate or glutamate or aspartate analogs (Table 16-4). The structures of excitatory amino acids are shown in Figure 16-34.

In addition to its role as a classical amino acid in protein synthesis, glutamate acts as a neurotransmitter because it is localized in presynapses

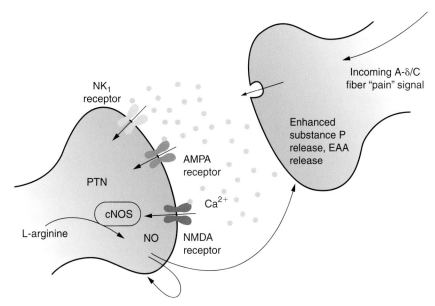

Figure 16-33. A synapse between two neuronal cell axon endings. The presynaptic nerve ending *(right)* is secreting substance P and excitatory amino acids in response to a pain signal. Excitatory amino acid receptors, in this case, include the AMPA receptor and the NMDA receptor. The NK-1 receptor is the substance P receptor. Activation of the NMDA receptor causes Ca^{2+} uptake and the activation of constitutive nitric oxide synthase (cNOS) in the postsynaptic nerve ending, which converts arginine to nitric oxide, which acts further on both the presynaptic and post-synaptic nerve endings.

Table 16-4
Excitatory Amino Acid (Glutamate) Receptors

NMDA	AMPA	Kain	Metabotropic
NMDAR1	GluR1	GluR5	mGluR1
NMDAR2A–2D	–GluR4	–GluR7	–mGluR8
		KA1, KA2	
NMDA	AMPA	Kain	1S,3R-ACPD
	Quis		L-AP4 (L-APB)
	Kain		CCG-1
			Quis
			3,5-DHPG
AP5 (APV)	CNQX		MCPG
CPP	DNQX		4CPG
			MAP4
Kynurenate	Kynurenate		MPPG
			MCCG

The four main receptors are NMDA, AMPA, Kainic Acid (kain), and metabotropic.
In *green* are the subunits of the receptors: NMDA has five subunits; AMPA, four subunits; kainate receptor, five subunit; and metabotropic, eight subunits. The ligands *(in cyan blue)* are amino acid analogs, but all receptors bind glutamate. Structures of excitatory amino acids are given in Figure 16-34.
Green, cloned Receptors; *cyan*, agonists; *red*, antagonists.
Reproduced from http://www.ucl.ac.uk/~smgxt01/frameh.htm?page=glutamat.htm.

in the cerebellum and hippocampus; it is released from the nerve ending by calcium ions; it has a sodium ion–dependent reuptake system; and it binds and agonizes the major excitatory amino acid receptors. The glutamate receptors are of two classes, the ligand-gated ion channels and the metabotropic G protein–coupled receptors. *The excitatory amino acid receptors (NMDA, kainate, and AMPA) elevate intracellular calcium ion and sodium ion concentrations.* The names of the receptors are derived

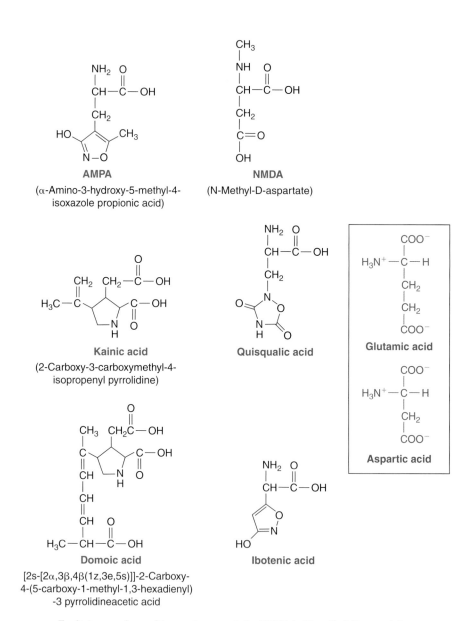

Excitatory amino acid receptor agonists. NMDA is N-methyl-D-aspartate.

Figure 16-34. Excitatory amino acid receptor agonists and antagonists. Quinolinic acid. An excitatory amino acid formed from tryptophan. When administered in high doses in the striatum, quinolinic acid is a neurotoxin used to produce a model of Huntington's disease. *NMDA*, N-methyl-D-aspartate.

Excitatory Amino Acid Receptor Antagonists

Metabotropic Receptor Agonists and Antagonists

Figure 16-34, cont'd

from their classical agonists. Thus, the NMDA receptor binds N-methyl-D-aspartate, as well as glutamate; the AMPA receptor binds **amino hydroxymethylisoxazole propionic acid (AMPA)** and glutamate; and the kainite receptor binds kainic acid (Figure 16-34) and glutamate. Ibotenic acid (Figure 16-34) also is a potent agonist of the NMDA receptor. Binding of agonists to the NMDA receptor is stimulated by Mg^{2+} and decreased by Zn^{2+}. Glycine enhances agonist binding through a secondary binding site, and polyamines (spermine, spermidine, and polycyclic phencyclidine) regulate ligand binding to the receptor (Figure 16-35).

The structure of the drug phencyclidine is shown in Figure 16-36. Models of the closed and open forms of the glutamate-binding site of the NMDA receptor are shown in Figure 16-37. The crystal structure of the glutamate receptor-2 (GluR2) binding core of the AMPA receptor (Table 16-4) is shown in Figure 16-38. Figure 16-39 shows the kainate receptor GluR6 agonist binding domain complexed with domoic acid (Figure 16-34). AMPA is the most commonly found receptor for glutamate of the non-NMDA type that mediates fast synaptic transmission in

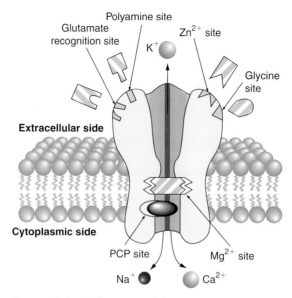

Figure 16-35. Schematic of the NMDA receptor complex. *PCP*, polycyclidine. Redrawn from http://www.chemistry.emory.edu/justice/seminar_borders/crest.gif.

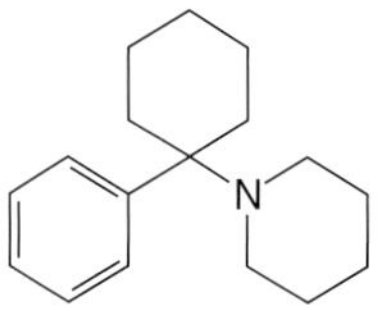

Figure 16-36. Structure of phencyclidine (PCP). It is an anesthetic (given intravenously) and analgesic.

the central nervous system. *The three glutamate receptors, AMPA, NMDA, and metabotropic glutamate receptor (mGluR) are highly expressed at the synapses of nociceptors.* When glutamate is released from a presynaptic nerve ending, it crosses the synaptic cleft and binds to glutamate receptors on the postsynaptic membrane (Figure 16-33), such as AMPA receptor and NMDA receptor. The NMDA receptor acts as a calcium channel, admitting calcium ions into the postsynaptic cell. NMDA receptors are attached in the postsynaptic membrane by interactions between their NR2 subunits (cytoplasmic C-terminal tails) and the "PDZ" domains of the postsynaptic membrane forming a lattice structure that underlies the membrane. The PDZ domains also bind potassium channels (K^+Ch), tyrosine kinases (ErbB4), and cell adhesion molecules **(neuroligin).** The postsynaptic membrane (PSD, for postsynaptic density) functions as a scaffold to assemble signaling proteins in the pathway of the NMDA receptor. The PSD interacts with GKAP and Shank scaffold proteins in the internal part of PSD. Shank complexes with Homer that interacts with the mGluR cytoplasmic tail. Homer also binds to IP_3R to cause the release of internal calcium ion stores in the smooth endoplasmic reticulum. These pathways from the NMDA receptor are summarized in Figure 16-40.

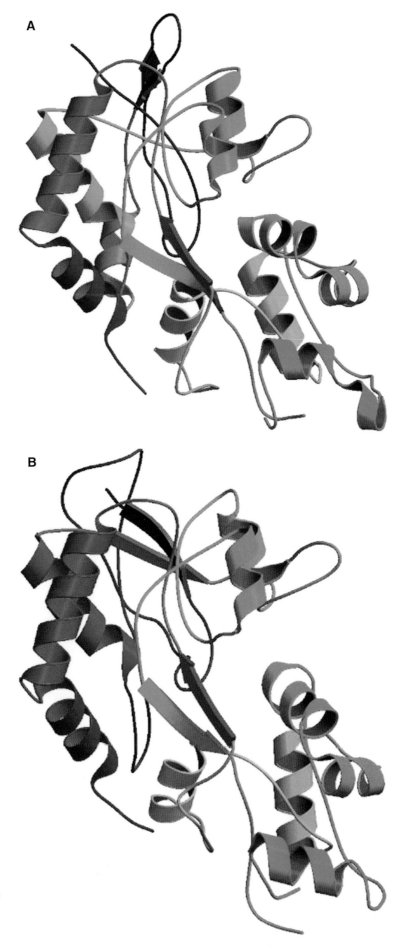

Figure 16-37. **A.** Closed form of the glutamate-binding site of the NMDA receptor. **B.** Model of the open form of the glutamate-binding site of the NMDA receptor. Part A reproduced from PDB ID: ls1l. I.G. Tikhonova, I.I. Baskin, V.A. Palyulin, N.S. Zefirov, S.O. Bachurin. Structural Basis for Understanding Structure-Activity Relationships for the Glutamate Binding Site of the NMDA Receptor. *Journal of Medicinal Chemistry* **45** pp. 3836 (2002). Part B reproduced from PDB ID: ls2s. I.G. Tikhonova, I.I. Baskin, V.A. Palyulin, N.S. Zefirov, S.O. Bachurin. Structural Basis for Understanding Structure-Activity Relationships for the Glutamate Binding Site of the NMDA Receptor. *Journal of Medicinal Chemistry* **45** pp. 3836 (2000).

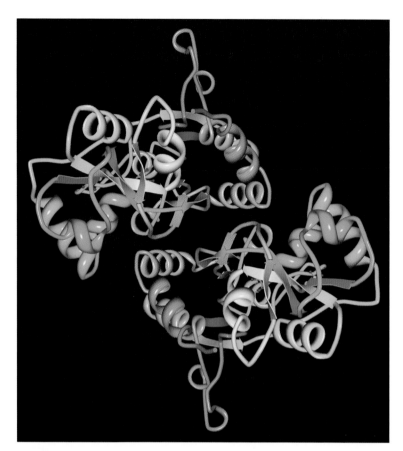

Figure 16-38. Crystal structure of the GluR2 ligand–binding core (S152J) with the L483Y and L650T mutations and in complex with AMPA (Figure 16-34). Reproduced from PDB ID: 1p1w. N. Armstrong, M. Mayer, E. Gouaux. Tuning Activation of the AMPA-Sensitive GluR2 Ion Channel by Genetic Adjustment of Agonist-induced Conformational Changes. *Proc. Natl. Acad. Sci. USA* **100** pp. 5736 (2003).

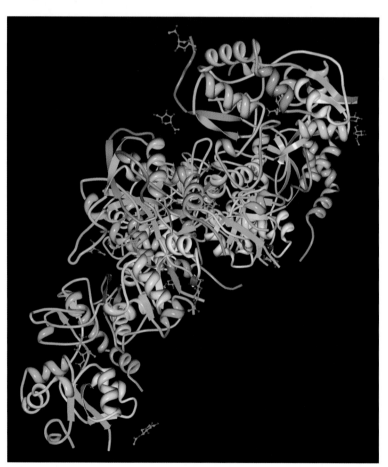

Figure 16-39. Structure of the kainate receptor subunit GluR6 agonist–binding domain complexed with domoic acid (Figure 16-34). Reproduced from PDB ID: 1yae. M.H. Nanao, T. Green, Y. Stern-Bach, S.F. Heinemann, S. Choe. Structure of the Kainate Receptor Subunit GluR6 Agonist-binding Domain Complexed with Domoic Acid. *Proc. Natl. Acad. Sci. USA* **102** pp. 1708 (2005).

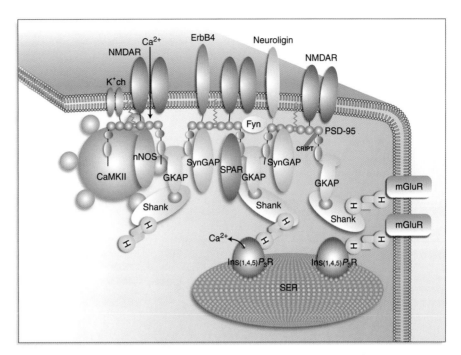

Figure 16-40. The postsynaptic NMDAR-PSD-95 signaling complex in excitatory synapses of the brain. PDZ domains are in *orange*. CaMKII, calcium-calmodulin protein kinase II; H, homer; mGlu R, metabotropic glutamate receptor; nNos, neuronal nitric oxide synthase; SER, smooth endoplasmic reticulum. Reproduced from M. Sheng, "Cell science at a glance the postsynaptic NMDA-receptor−PSD-95 signaling complex in excitatory synapses of the brain," *J. Cell. Sci.*, 114, 1251, 2001.

When polycyclic phencyclidine (Figure 16-36) is administered chronically to experimental animals, there appears to be an up-regulation of NMDA receptors that result in apoptosis (programmed cell death) during which as many as one third of the neurons die. Exposure of cultures of neurons to NMDA led to increased expression of Bax and decreased **Bcl-X**$_1$ expected during activation of apoptosis. Reactive oxygen species were intermediate in this process. Like the NMDA receptor, the AMPA receptor is located in the postsynaptic membrane and connects to a lattice (PDZ) structure that, in turn, connects other proteins, particularly protein kinases, in the pathway of signal transduction, as shown in Figure 16-41.

The topology of GluR6 subunit of the kainate receptor is shown in Figure 16-42. The kainite receptor subunit, GluR6, has five transmembrane domains. Like the other receptors of the excitatory amino acid class, the kainite receptor is associated with cytoplasmic proteins (synapse-associated proteins, SAPs) as shown in Figure 16-43. Kainate receptors are associated with members of the SAP90/PSD-95 group. SAP90 binds tightly to GluR6/PSD-95 kainite receptor subunits and modifies its electrophysiological properties. As indicated in the figure, SAP97 weakly associates in comparison to SAP90. It is suggested that the differential association of kainite receptors with the SAP family members may be a mechanism that determines the subcellular localization of these proteins.

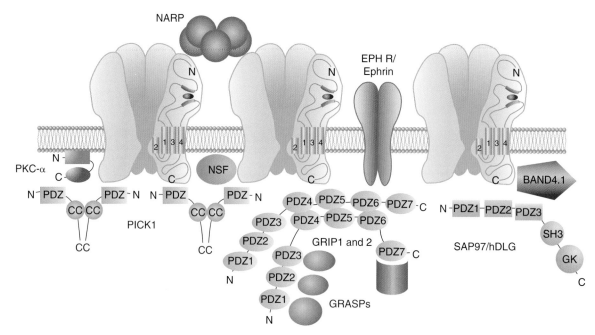

Figure. 16-41. The AMPA receptor complex contains proteins that interact with the subunits of the receptor and regulate function. Signaling proteins, such as protein kinases and other proteins associate with this complex and regulate receptor function as well as downstream signal transduction. Redrawn from http://neuroscience.jhu.edu/picdetail.asp?ID=22&ResPicNo=ResPic1.

Although volatile anesthetics, such as halothane, have been found to inhibit the function of GluR3 subtype of glutamate receptors (including the AMPA receptor; see Table 16-4), kainite receptor (GluR6) function is enhanced. Halothane interacts with a transmembrane region (TM4) of GluR6 and involves a glycine residue (glycine 819) that is important for the enhancement of receptor function. Although ethanol and phenobarbital both inhibit the function of this receptor, their actions do not involve this specific glycine residue. The interaction of halothane with TM4 of GluR6 is modeled in Figure 16-44, which shows the halothane-binding pocket formed by glycine 819.

Another activity of kainite receptors is to down regulate the release of the neurotransmitter γ-aminobutyric acid (GABA) in the hippocampus. Activation of the kainite receptor, by the binding of glutamate, generates a second messenger cascade, resulting in the stimulation of protein kinase C. This, in turn, results in the metabotropic inhibition (through G protein–coupled receptors) of GABA release. GABA is an important inhibitory neurotransmitter in the central nervous system. The activated postsynaptic GABA receptor opens a chloride channel of the receptor that leads to hyperpolarization of the postsynaptic cell. The drug benzodiazepine (Figure 16-45) binds to a modulatory site (different from the GABA-binding site) of the GABA receptor and generates hypnotic, anxiolytic, and anticonvulsant effects.

Figure 16-46 shows a glutaminergic neuronal ending, where circulated glutamine is converted to glutamic acid through the mitochondrial glutaminase and glutamate is released from the presynaptic nerve ending across the synaptic cleft to bind to an AMPA or kainite receptor in the

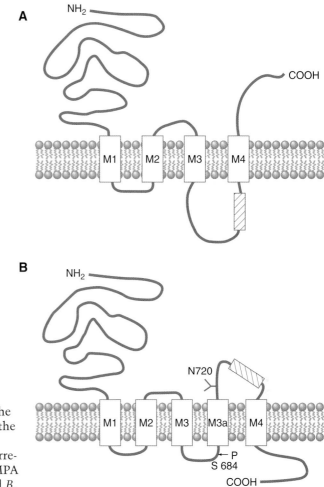

Figure 16-42. The GluR6 subunit of the kainate receptor. **A.** Schematic of the presently accepted GluR6 topology in the lipid bilayer. **B.** Schematic of the new proposed topology of the GluR6 subunit, including a new transmembrane domain, TM3a, located between Ser-684 and Asn-720. The region corresponding to the alternatively spliced module flip/flop in AMPA receptor subunits is pictured as a *hatched bar* in both *A* and *B*.

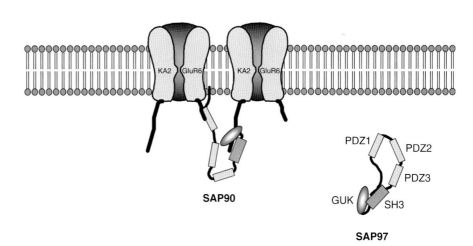

Figure 16-43. Mechanism regulating the differential interaction of the KA2 subunit of the kainate receptor with SAPs. SAP90 can cluster KA2 receptors by interacting with the SH3 and GUK domains. Although KA2 can potentially bind to the SH3 and GUK domains of SAP97 with, respectively, high and lower affinities, the SH3 site is occluded by binding of the SAP97 N-terminus and the GUK site is occluded by SH3 binding. Effectively, these intramolecular interactions block SAP97 association with KA2. *GluR6*, subunit of kainate receptor (Table 16-4); *GUK*, catalytically inactive guanylate kinase; *KA2*, indicated subunit of the kainate-receptor. Reproduced from S. Mehta et al., "Molecular mechanisms regulating the differential association of kainate receptor subunits with SAP90/PSD-95 and SAP97," *J. Biol. Chem.*, 276: 16092–16099, 2001 and from Figure 7 from http://www.jbc.org/cgi/content/full/276/19/16092/F7.

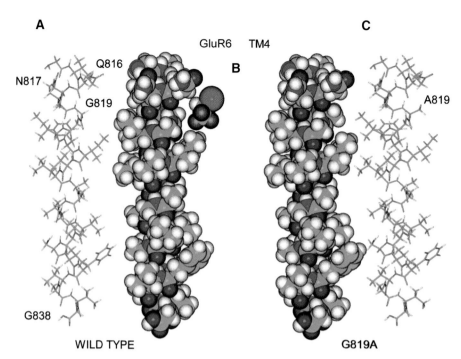

Figure 16-44. Interaction of halothane (atomic model at *B*; chemical structure of halothane above atomic model). The amino acid sequence of TM4 for GluR6 was modeled as an α-helix, with the pore of the ion channel toward the left *(A)*. A molecule of halothane was modeled, positioned near the pocket formed by Gly-819 in GluR6, and then docked onto TM4 *(B)*. The effect of the G819A mutation was modeled by replacing Gly-819 with alanine and then reoptimizing the side chain packing *(C)*. Reproduced from Figure 10 of K. Minami et al., "Sites of volatile anesthetic action on kainate (glutamate receptor 6) receptors," *J. Biol. Chem.*, 273: 8248–8255, 1998 and http://www.jbc.org/cgi/content/full/273/14/8248/F10.

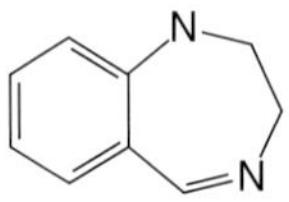

Figure 16-45. Structure of benzodiazepine.

postsynaptic membrane. After its action at the receptor is performed, the glutamic acid dissociates and is recycled both to the presynaptic neuron and to a neighboring glial cell (glial cells surround neurons and support them structurally and nutritionally), where it can again be converted to glutamine and transported to a neuron. There are five general types of glia: **astrocytes** (star shaped), ependymal cells (line the surfaces of brain ventricles and the spinal central canal), microglia (smallest central nervous cells that phagocytize debris), oligodendrocytes (myelinate axons in brain and spinal cord), and Schwann cells (myelinate axons in the peripheral nervous system). A picture of an astrocyte is shown in Figure 16-47. As shown in Figure 16-46, glutamate can be converted to GABA by the action of **glutamic acid decarboxylase (GAD)** in a GABAergic neuron and GABA can be released at the nerve ending to bind

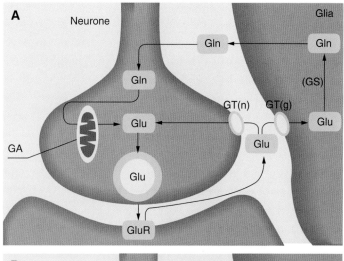

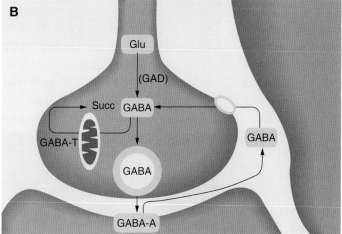

Figure. 16-46. AMPA/kainate receptors. Pathways of glutamate and GABA metabolism in AMPA and Kainate receptors. **A.** Glutamate (Glu) is released into the synaptic cleft and then retrieved by Na^+-coupled glutamate transporters (neuronal-type GT(n) and glial-type GT(g)). Glutamate in the glial cell is converted to glutamine (Gln) by glutamine synthetase (GS). When Gln is in high concentration (ca 0.5 mM) in the central nervous system, it can be taken up by the neuron to replenish glutamate after hydrolysis by mitochondrial glutaminase (GA). **B.** Glutamate forms GABA in the cytoplasm of the GABAergic neuron. The majority of the GABA that is released is retrieved into terminals in addition to some GABA being taken up by glial cells. Mitochondrial GABA-transaminase(GABA-T) converts GABA to succinic semialdehyde, which is oxidized to succinate.

to a GABA receptor in the membrane of a postsynaptic neuron. After dissociating from the GABA receptor, GABA can be recycled to the GABAergic neuron (or to a glial cell), where it can be reused as a neurotransmitter. GABA is also degraded to succinate by GABA transaminase in the GABAergic cell mitochondrion.

The fourth general type of excitatory glutamate receptor is the metabotropic receptor that has eight forms, mGluR1 through mGluR8 (Table 16-4). Structures of mGluR1 and mGluR3 are shown in Figure 16-48. mGluRs regulate NMDA synaptic transmission in the

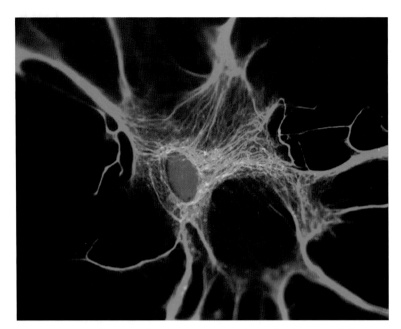

Figure 16-47. Photograph of an astrocyte *(green)*. Photograph by Drs. Edward Nyatia and Dirk Lang, S. African Agency for Science and Technology Advancement.

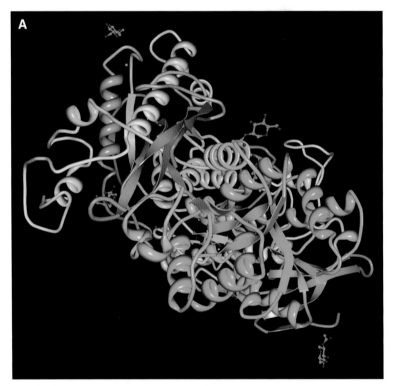

Figure 16-48. **A.** Crystal structure of metabotropic glutamate receptor subtype 1 (mGluR1) complexed with glutamate. **B.** Theoretical model of metabotropic glutamate receptor mGluR3 in open form. Part A reproduced from PDB ID: 1ewk. N. Kunishima, Y. Shimada, Y. Tsuji, T. Sato, M. Yamamoto, T. Kumasaka, S. Nakanishi, H. Jingami, K. Morikawa. Structural Basis of Glutamate Recognition by a Dimeric Metabotropic Glutamate Receptor. *Nature* **407** pp. 971 (2000). .

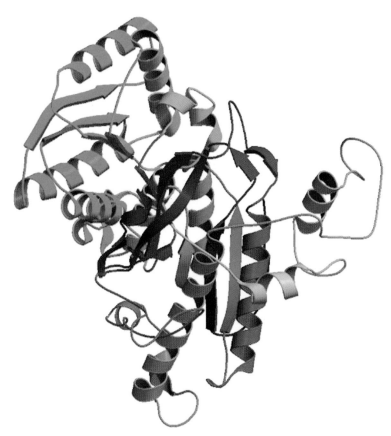

Figure 16-48, cont'd Part B reproduced from PDB ID: ls8m.
M.S. Belenikin, I.I. Baskin, G. Costantino, V.A. Palyulin, R. Pellicciari,
N.S. Zefirov. Comparative Analysis of the Ligand-Binding Sites of the Metabotropic Glutamate Receptors mGluR1-mGluR8. Doklady. *Biochemistry and Biophysics* **386** pp. 251.

nucleus accumbens (from the Latin meaning nucleus leans against the septum; *nucleus accumbens* is also the ventral striatum), which is a collection of neurons located just lateral to the *septum pellucidum* (Figure 16-49). This brain structure is responsible for reward, motivation, and addiction. Dopamine is released by dependent drugs, such as cocaine and nicotine (not caffeine). Aside from NMDAergic neurons, the *nucleus accumbens* contains mostly spiny GABAergic neurons (about 95%). A glutaminergic neuron releasing glutamate is diagrammed in Figure 16-50, where the release of glutamate is negatively regulated by GluR2 or GluR3 receptor (bound with glutamate) that sends a negative signal through a G protein, inhibiting the presynaptic release of glutamate. On the postsynaptic cell, glutamate normally binds to the NMDA receptor that admits calcium and sodium ions. The function of this receptor is negatively regulated by mGluR4 on the postsynaptic membrane that, through a G protein, signals protein kinase A to negatively regulate the NMDA receptor ion channel.

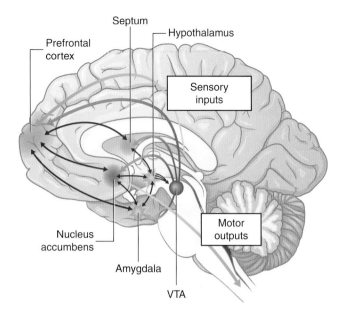

Figure 16-49. Location of the nucleus accumbens *(white space)* in the brain. Reproduced from http://library.thinkquest.org/04oct/01639/vn/health/popup/nucleus.html. ThinkQuest, Oracle Education Foundation, Redwood Shores, CA.

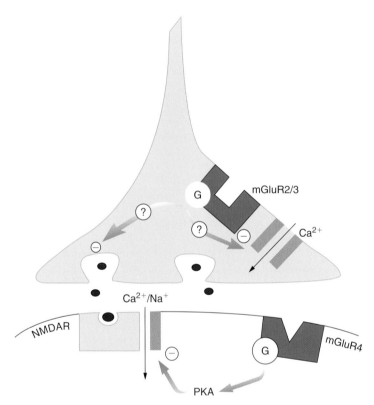

Figure 16-50. Schematic of hypothetical pre- and postsynaptic locations of mGluRs at a glutamatergic synapse on a NAcc neuron. Evidence suggests that group 2 mGluRs reduce glutamate release presynaptically, whereas the group 3 mGluR4 receptor may postsynaptically reduce currents passing through NMDA receptors. −, inhibitory effect; ?, unknown linkages; *G*, GTP-binding protein; *G inside a circle*, G protein, *NMDAR*, NMDA receptor; *PKA*, protein kinase A. Redrawn from Figure 10 of G. Martin, Z. Nie, and G.R. Siggins, "Metabotropic glutamate receptors regulate *N*-methyl-D-aspartate-mediated synaptic transmission in nucleus accumbens," *J. Neurophys.*, 78, 3028–3038, 1997 and http://jn.physiology.org/cgi/content-nw/full/78/6/3028/F10.

The Classical Neurotransmitters

The classical neurotransmitters are acetylcholine; the amino acids (in addition to the excitatory amino acids, glutamate and aspartate, which were discussed earlier); GABA and glycine; peptides, such as vasopressin, somatostatin, and neurotensin; monoamines, such as norepinephrine (and epinephrine, which is often considered to be a hormone); dopamine; and serotonin. The major neurotransmitters in the brain are glutamate and GABA. The monoamines and acetylcholine serve specialized modulating functions and are confined to specific structures. The peptides serve specialized functions in the hypothalamus (see Chapter 8) and act in other regions in the brain.

Members of the so-called nicotinicoid receptor superfamily (they usually have a binding site for nicotine) are the inhibitory glycine receptor, the inhibitory GABA type A receptor (GABA$_A$R), the **excitatory nicotinic acetylcholine receptor (nAChR)**, and the serotonin type 3 receptor (5-HT$_3$R). When a neurotransmitter binds to this class of receptors, the receptors transiently open selective pores for small ions that move down their electrochemical gradient, changing the potential across the membrane and affecting the activity of voltage-gated channels and cellular electrical activity. This family of receptors also is known as signature Cys-loop proteins because of a conserved 15-amino-acid–spaced disulfide loop that occurs in their extracellular ligand-binding domain. This family expresses heteropentameric receptors derived from different gene products, splice variants, or both. These receptor structures are diagrammed in Figure 16-51. A ribbon model of the extracellular domain of the glycine receptor, which functions in fast synaptic transmission in the brain stem and spinal cord, is shown in Figure 16-52. Binding of glycine to the ligand-binding domain is followed by a conformational change of the receptor that is mediated through the transmembrane domains and is followed by the opening of the channel pore. The location of the conserved Cys-loop is within the loops 2 and 7, as indicated in the figure. These loops of the extracellular domain play an important role in the activation of the glycine receptor. The extracellular domain contains the ligand-binding site that is separated spatially from the M2 domain (Figure 16-51A), lining the ion channel pore.

Besides glycine, other amino acids can activate the glycine receptor, including β-alanine, taurine, L-alanine, L-serine, and proline. The glycine receptor does not bind GABA. A highly selective inhibitor of the glycine receptor is strychnine, which binds the receptor with nanomolar affinity. It does so without opening the associated chloride ion channel. The structure of strychnine is shown in Figure 16-53. The binding site for strychnine is either the same as that for glycine or close to it

Glycine has dual roles as an inhibitory ion channel neurotransmitter and a neuromodulator and stimulator of the NMDA receptor. When glycine is released into a synapse, it binds to the glycine receptor on the postsynaptic membrane, making the postsynaptic membrane more permeable to chloride ion. *The uptake of Cl$^-$ hyperpolarizes the membrane, decreasing its ability to depolarize. This is the process by which glycine is inhibitory as mediated by its receptor.* A view of the GABA receptor associated with a chloride channel is shown in Figure 16-54. Strychnine binds to the glycine receptor without opening the chloride ion channel, thus inhibiting the inhibitory action of the liganded glycine receptor. The

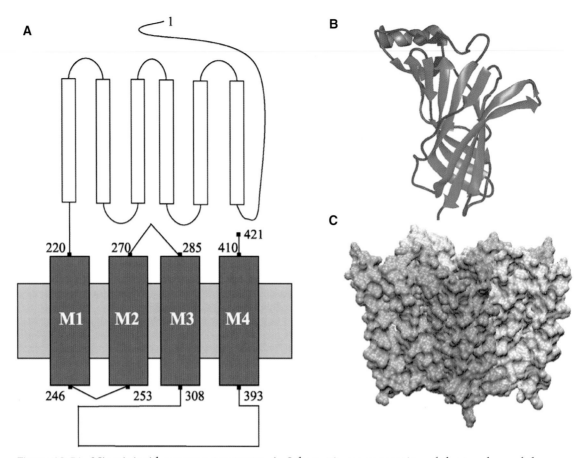

Figure 16-51. Nicotinicoid receptor structures. **A.** Schematic representation of the topology of the α_1 subunit of GlyR. **B.** Ribbon model of a single subunit of the pentameric AChBP. **C.** Surface representation of the top and side view of the transmembrane domains of the heteropentameric nAChR. *AChBP*, acetylcholine-binding protein; *GlyR*, glycine receptor; *nAChR*, neuronal acetylcholine receptor. Reproduced from Figure 1 of M. Cascio, "Structure and function of the glycine receptor and related nicotinicoid receptors," *J. Biol. Chem.*, 279: 19383–19386, 2004 and http://www.jbc.org/cgi/content/full/279/19/19383/F161.

hyperexcitability produced by strychnine in the ventral spinal cord is what makes it a poison. In the synapse, glycine can be deactivated by its active transport (reabsorption) back into the presynaptic membrane.

The GABA receptor is another member of the nicotinicoid superfamily of neurotransmitter receptors. GABA is second only to glutamate as a major brain neurotransmitter; it occurs in nearly 40% of all synapses. GABA concentrations in the brain are 200 to 1000 times those of acetylcholine or the monoamines. Unlike other neurotransmitters, GABA, after release into a synapse, can be inactivated by transport actively from the synapse into astrocyte glial cells near the synapse. GABA, as well as glutamate, is synthesized in the brain from α-ketoglutarate by way of the GABA shunt, and GABA can be metabolized back into the citric acid cycle (Figure 16-55). The enzyme forming GABA from glutamate is GAD, a pyridoxal phosphate–containing enzyme. In vitamin B_6 deficiency, seizures can occur because insufficient GABA is produced to generate the needed inhibitory activity of GABA (the inhibitory effects of GABA on neurons in the brain are protective during ischemia or hypoxia).

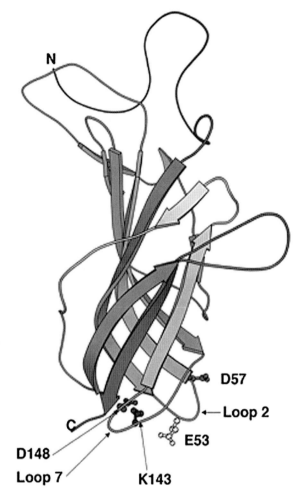

Figure 16-52. A ribbon diagram of the modeled GlyR extracellular domain. The diagram shows the model of the GlyR extracellular domain from outside of the pentameric ring, perpendicular to the fivefold axis. The side chains of residues Glu-53 and Asp-57 in loop 2 and of Lys-143 and Asp-148 in loop 7 are labeled. The N- and C-termini of the extracellular domain sequence are labeled *N* and *C*, respectively. *GlyR*, glycine receptor. Reproduced from Figure 1 of N.L. Absalom et al., "Role of charged residues in coupling ligand binding and channel activation in the extracellular domain of the glycine receptor," *J. Biol. Chem.*, 278: 50151–50157, 2003 and http://www.jbc.org/cgi/content-nw/full/278/50/50151/FIG1.

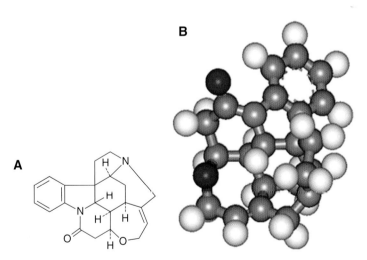

Figure 16-53. **A.** Chemical structure of strychnine. **B.** Atomic model of the plant alkaloid, strychnine. Oxygen atom in *red*. Nitrogen in *blue*. Part A reproduced from http://www.the-piedpiper.co.uk/graphics1/strychnine.jpg. Part B reproduced from http://jchemed.chem.wisc.edu/JCEWWW/Features/MonthlyMolecules/2004/Sep/JCE200 4p1366fig1.gif and J.W. Nicholson and A.J. Wilson, "Featured molecules," *Chem. Educ.*, 81: 1362, 2004.

The Classical Neurotransmitters   1005

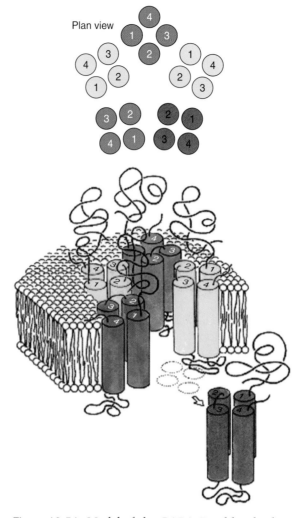

Figure 16-54. Model of the GABA$_A$R–chloride channel protein complex. Reproduced form http://www.molres.org/delorey/. Timothy M. DeLorey, Molecular Research Institute, Mountain View, CA.

In the presynaptic nerve terminal, GABA is formed from glutamate and stored in synaptic vesicles, awaiting the signal for release of GABA into the synaptic space. For the synthesis of GABA from L-glutamate, glutamate decarboxylase is anchored to the synaptic vesicle by attachment to heat shock cognate protein 70 (HSC70) and then to the **cysteine string protein (CSP)** to the vesicle, as shown in Figure 16-56. *GABA binds to GABAAR on the postsynaptic membrane, where it opens a chloride channel connected to the receptor (similarly to the glycine receptor). The influx of chloride ion into the postsynaptic cell causes the membrane to reduce its depolarization.* Certain drugs, the benzodiazepines (of which the tranquilizer, Valium, is an example; caffeine reduces the effects of benzodiazepines), increase the rate at which the channels open. Thus, they bind to GABA$_A$R and enhance the effects of GABA on the receptor. These structures are shown in Figure 16-57. GABA receptors and glycine receptors share some similarities in their actions of opening associated chloride ion channels (glycine receptor predominantly in the spinal cord

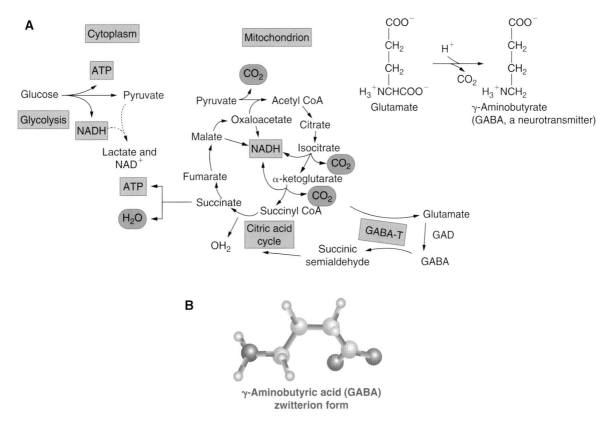

Figure 16-55. **A.** Brain biochemistry of γ-aminobutyric acid (GABA). *GABA-T*, GABA transaminase; *GAD*, glutamic acid decarboxylase. The GAD reaction is shown in the *upper right*. **B.** Zwitterion structure of GABA. Part A redrawn from page 6 of http://www.benbest.com/science/anatmind/anatmd10.html Part B redrawn from Timthy M. DeLorey (The DeLorey lab) webpage. http://images.google.com/imgres? imgurl=http://198.144.204.232/del...num%3D10%26hl%3Den%26lr%3D%26client%3Dsafari%26rls% 3Den%26sa%3DN

and GABA receptor predominantly in the brain) and in their attachment to the cell membrane and subcellular scaffolding. It has been known that three glycine molecules were required to activate the glycine receptor, and this may be because of the need for clustering of the receptors to generate the activated state. The mechanism of clustering of glycine and GABA$_A$Rs involves a protein (93 kDa) called **gephyrin** (Greek for bridge). Gephyrin binds directly to glycine receptors but not directly to GABA receptors. In the latter case, gephyrin binds first to a small protein (14 kDa) called the **GABA$_A$ receptor–associated protein (GABARAP)**, localized on intracellular membranes, that then can associate with the GABA receptor. A model illustrating the clustering of these receptors and their interactions with gephyrin is shown in Figure 16-58. The x-ray structure of the gephyrin N-terminal domain is shown in Figure 16-59, and the structure of GABARAP is shown in Figure 16-60. The N-terminal basic helical region of GABARAP associates with tubulin. The surface features of GABARAP are shown in Figure 16-61. This protein also could be important in receptor intraneural sorting and targeting of GABA receptors.

Acetylcholine is usually an excitatory neurotransmitter. The other peripheral neurotransmitter is norepinephrine, which is inhibitory. The structure of acetylcholine is shown in Figure 16-62. It is synthesized in

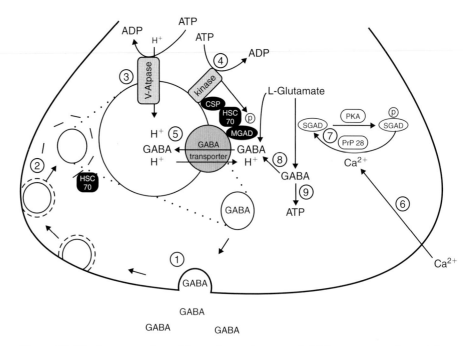

Figure 16-56. A proposed model on the anchorage of GAD_{65} to synaptic vesicles at the synaptic terminal. GAD_{65} is subcellularly targeted and anchored to synaptic vesicles first through the chaperone function of HSC70 to form an HSC70–GAD_{65} complex, followed by association of HSC70–GAD_{65} complex to CSP on synaptic vesicles. *CSP*, cysteine string protein; *GAD_{65}*, glutamic acid decarboxylase, 65 kDa; *HSC70*, 70-kDa heat shock cognate protein.

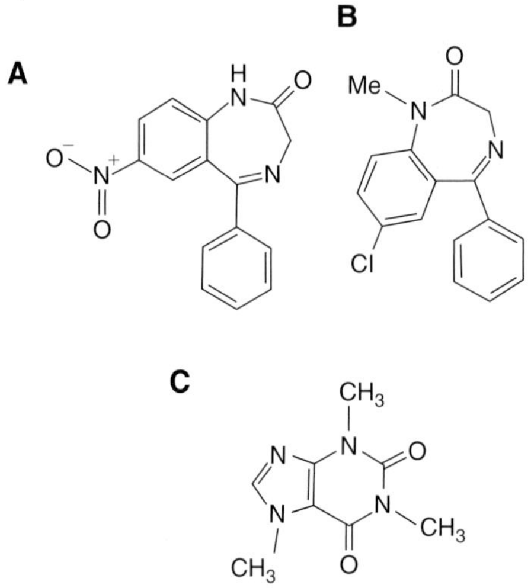

Figure 16-57. **A.** Chemical structure of a benzodiazepine. **B.** Chemical structure of valium. **C.** Chemical structure of caffeine.

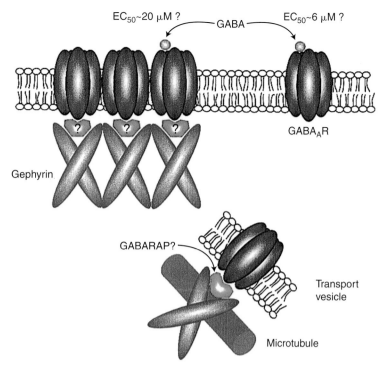

Figure 16-58. Gephyrin (93 kDa) binds glycine receptors and tubulin *in vitro* and is required for clustering of glycine and GABA$_A$Rs at synapses, yet it does not associate with GABA$_A$Rs *in vitro*. GABARAP (14 kDa) binds a subunit of the GABA$_A$R, gephyrin, and tubulin *in vitro*. Furthermore, GABARAP recruits GABA$_A$Rs into clusters at the plasma membrane in heterologous quail fibroblast cells. However, GABARAP does not colocalize with gephyrin or GABA$_A$Rs in primary neurons. Instead, it is associated with intracellular vesicles. The data suggest that GABARAP may mediate transport or targeting of gephyrin and GABA$_A$Rs, but another unknown protein may link gephyrin and the GABA$_A$R at synapses. Clustered receptors recruited to the membrane by GABARAP in quail cells have a fourfold reduced affinity for GABA. If the GABARAP promoted clustering resembles that at synapses, the implication is that clustering of GABA$_A$Rs and association with the subcellular cytoskeleton can dramatically alter their kinetic properties. Reproduced from Figure 1 of M.B. Kennedy, "Sticking together," *PNAS USA*, 97: 11135–11136, 2000 and http://www.pnas.org/cgi/content-nw/full/97/21/11135/F1.

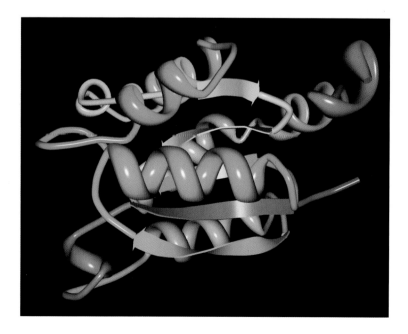

Figure 16-59. X-ray structure of gephyrin N-terminal domain. Reproduced from PDB ID: 1ihc. M. Sola, M. Kneussel, I.S. Heck, H. Betz. W. Weissenhorn, W. X-ray Crystal Structure of the Trimeric N-terminal Domain of Gephyrin. *J. Biol. Chem.* **276** pp. 25294 (2001).

The Classical Neurotransmitters

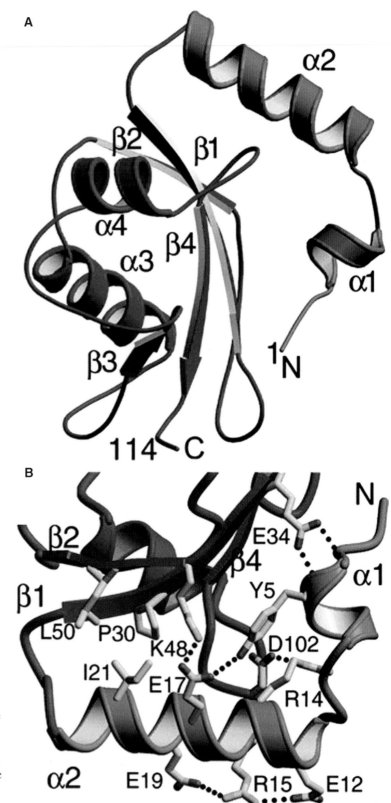

Figure 16-60. **A.** Ribbon diagram of GABARAP. The N-terminal region is shown in *gray,* and the core structure displaying a conserved ubiquitin-like fold is shown in *red.* Secondary structure elements are labeled. **B.** Close-up view of the interaction between the N-terminal region and the core structure. Hydrogen bonds are shown as *dashed lines.*

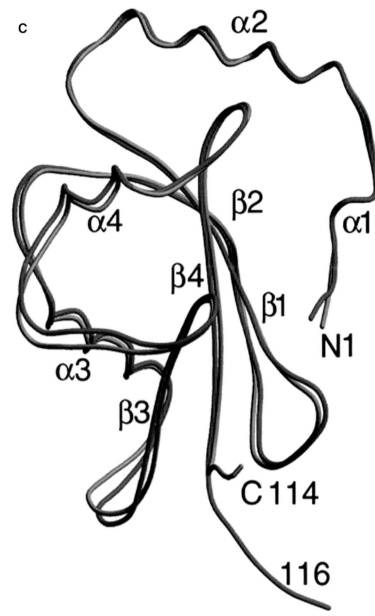

Figure 16-60, cont'd **C.** Superposition of the Cα atoms of GABARAP *(gray)* (residues 1–114) and corresponding residues of GATE-16. Reproduced with permission from Figure 1 of V.N. Bavro et al., "Crystal structure of the GABA$_A$-receptor-associated protein, GABARAP," *EMBO Reports*, 3: 183–189, 2002 and http://www.nature.com/embor/journal/v3/n2/full/embor233.html.

the nerve ending, stored in the presynaptic vesicles, and released into the synaptic space when an electric signal is transmitted down the axon to the terminal. Synthesis occurs in the mitochondrion from acetyl SCoA and choline, as shown in Figure 16-63. The action of acetylcholine liganded to the acetylcholine receptor, on the postsynaptic membrane, is terminated by the action of the enzyme cholinesterase, which cleaves acetylcholine into acetate and choline, rendering the ligand inactive. Two views of the acetylcholine receptor are shown in Figure 16-64. As indicated in Figure 16-64A, acetylcholine is bound to nAChR in its extracellular domain to activate the receptor (a ligand-gated cation channel). This domain is shown more clearly in Figure 16-65. The five subunits are arranged clockwise as a cylinder to surround the ion channel. Each subunit contains transmembrane domains, as well as extracellular and

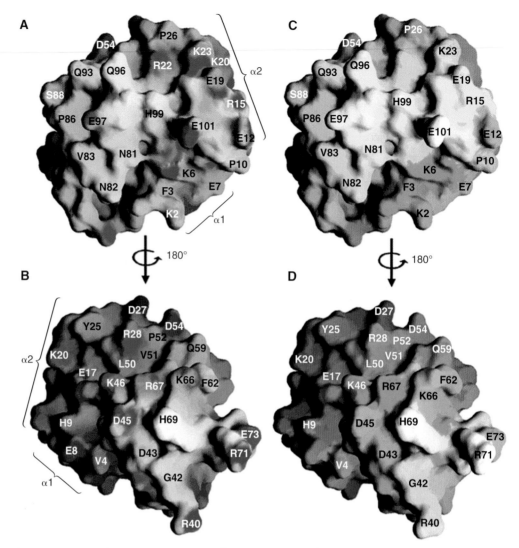

Figure 16-61. Surface representations of GABARAP. **A.** Electrostatic potential map of GABRAP surfaces (*B*, rotated 180 degree). Regions of electrostatic potential less than 20 k_b are shown in *red*, and those greater than +20 k_b are shown in *blue* (k_b, Boltzmann constant; *T*, absolute temperature). *(C* and *D)* Representation of GABARAP in which identical *(green)* and homologous *(yellow)* residues are mapped to the protein surface using the sequence alignment shown in Figure 16-61. Orientations are identical to those shown in *A* and *B* and show that surface residues are conserved on one side of the molecule *C*, but not on the other *D*. Note the charged surface around α-helix 2 (region 1) *A* and a cleft lined by basic residues on the opposite side of the molecule (region 2) *B*, which are highly conserved. Positions of α-helices 1 and 2 are marked. Figure 3 of V.N. Bavro et al., "Crystal structure of the GABA$_A$-receptor-associated protein, GABARAP," *EMBO Rep.*, 3: 183–189, 2002 with permission and http://www.nature.com/embo/journal/v3/n2/full/embor233.html.

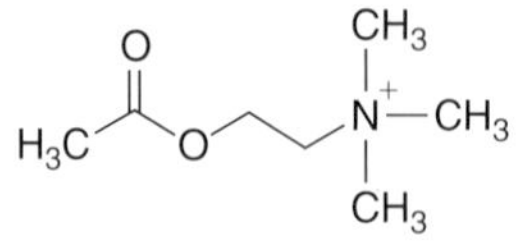

Figure 16-62. Chemical structure of acetylcholine.

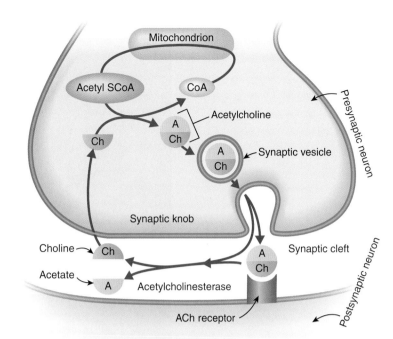

Figure 16-63. Synthesis of acetylcholine from acetyl-SCoA and choline. *A,* acetyl group; *ACh,* acetylcholine; *Ch,* choline. Abel Menacer Institute of Chemistry, Pharmacy and Biomedical Sciences University of Sunderland, UK.

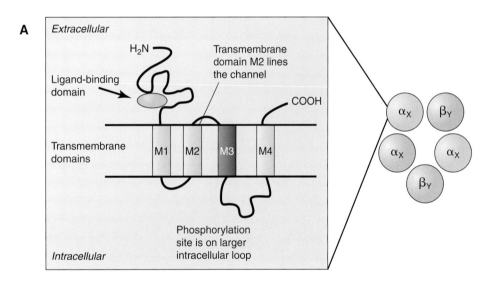

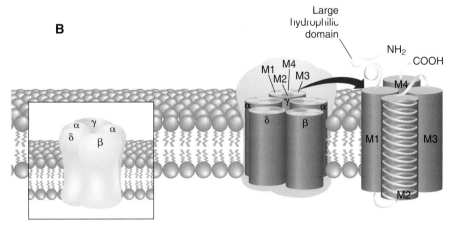

Figure 16-64. **A.** Diagram of the nicotinic acetylcholine receptor (nAChR). **B.** Schematic models of the acetylcholine receptor. *Left,* The AChR is a membranal heteropentameric glycoprotein. *Middle* and *right,* the topology of the subunits. M1, M2, M3, and M4 represent the four transmembrane domains, respectively.

The Classical Neurotransmitters 1013

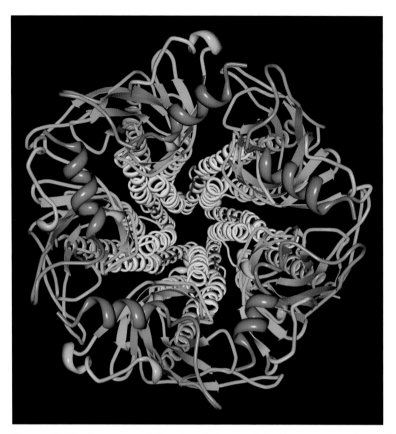

Figure 16-65. Structure of the nicotinic acetylcholine receptor. Reproduced from PDB ID:2bg9. N. Unwin. Refined Structure of the Nicotinic Acetylcholine Receptor at 4A Resolution. *J. Mol. Biol.* **346** pp.967 (2005).

intracellular domains. The acetylcholine receptor is located on the postsynaptic membrane of a neuromuscular junction, and the binding of acetylcholine is translated into an electrical signal that leads to muscle contraction.

There are four major classes of nAChRs: those occurring at postsynaptic neuromuscular junctions (just described); neuronal receptors that are homomeric and occur in the central and peripheral nervous systems; neuronal and autonomic receptors that occur in the brain and have a great affinity for nicotine; and epithelial and neuronal receptors that are homomeric and occur in hair cells and in cochlea. All classes are sensitive to α-bungarotoxin inhibition except neuronal and autonomic receptors, which are insensitive to the toxin. However, neuronal and autonomic receptors are inhibited by χ-bungarotoxin. Epithelial and neuronal receptors are also inhibited by strychnine and atropine. Structures of bungarotoxin and relatives are shown in Figure 16-66. A complex between α-bungarotoxin and the α7 subunit of the neuronal nAChR is shown in Figure 16-67. This binding site obscures the acetylcholine-binding site, accounting for the inhibition. *The second messengers of these receptors are transport of calcium and sodium ions.* The allosteric states of these receptors, with regard to opening and closing of the receptor-associated ion channel, are shown in Figure 16-68. The nAChRs are responsible for the peripheral effects of acetylcholine in the neuromuscular junctions and

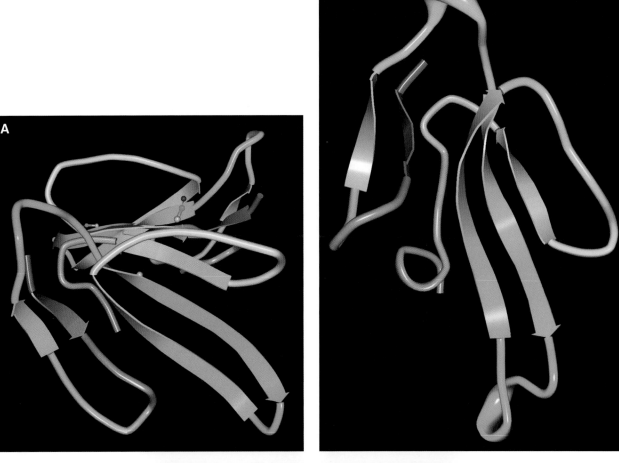
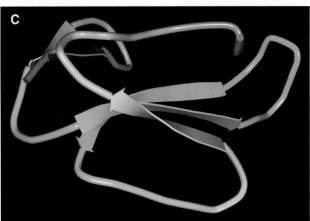

Figure 16-66. **A.** Structures of three α-neurotoxins: recombinant erabutoxin (s8T mutant). **B.** Superimposed structures of three α-neurotoxins: α-bungarotoxin. **C.** Superimposed structures of three α-neurotoxins: α-cobratoxin. Part A reproduced from PDBID: 3era. J.F. Gaucher, R. Menez, B. Arnoux, A. Menez, A. Ducruix. High Resolution X-ray Analysis of Two Mutants of a Curaremimetic Snake Toxin. *Eur. J. Biochem.* **267** pp.1323 (2000). Part B reproduced from PDB ID: 1kfh. L. Moise, A. Piserchio, V.J. Basus, E. Hawrot. NMR Structural Analysis of Alpha-bungarotoxin and Its Complex with the Principal Alpha-neurotoxin-binding Sequence on the Alpha 7 Subunit of a Neuronal Nicotinic Acetylcholine Receptor. *J. Biol. Chem.* **277** pp. 12406 (2002). Part C reproduced from PDB ID: 2ctx. C. Betzel, G. Lange, G.P. Pal, K.S. Wilson, A. Maelicke, W. Saenger. The Refined Crystal Structure of Alpha-cobratoxin from Naja Naja Siamensis at 2.4-Å resolution. *J. Biol. Chem* **266** pp. 21530 (1991).

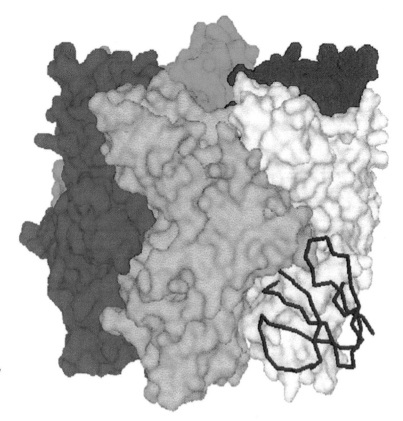

Figure 16-67. A surface model of AChR is shown with each subunit colored differently. The synaptic side is on *top,* and the membrane side is *below.* Bungarotoxin is shown as a *red stick model* with the unstructured C-terminal tail removed. The bungarotoxin-binding site is in *white.* Reproduced from Figure 9 of L. Moise et al., "NMR structural analysis of α-bungarotoxin and its complex with the principal a-neurotoxin-binding sequence on the α7 subunit of a neuronal nicotinic acetylcholine receptor," *J. Biol. Chem.,* 277: 12406–12417, 2002 and http://www.jbc.org/cgi/content/full/277/14/12406/F9.

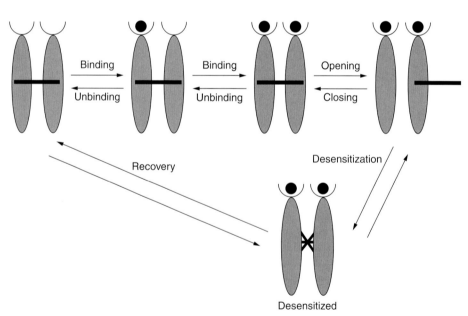

Figure 16-68. Allosteric states of the nicotinic acetylcholine receptor. When the receptor binds agonist, its global conformation changes. After binding two molecules of agonist, the protein may enter a conducting conformation. The open channel may adopt a desensitized state in which the channel is nonconducting and the agonist is bound with high affinity. Redrawn from http://www.medicine.megill.ca/mjn/v05n02/v05p090/v05p090fgr2.htm

autonomic ganglia. There is a cholinergic system in the *nucleus basalis* of Meynert that involves the cortex. This nucleus may play a role in learning and memory. There are cholinergic neurons in the basal forebrain involved in integration of information.

In addition to the nAChRs, there are muscarinic acetylcholine receptors. These acetylcholine receptors are named for artificial toxins that activate them selectively. Fast-acting acetylcholine receptor is labeled nicotinic because it is specifically activated by nicotine, and the slow-acting receptor is labeled muscarinic for the toxin, **muscarine,** found in poisonous mushrooms, which activates the receptor; nicotine will not do so. Muscarinic acetylcholine receptors consist of five receptor subtypes: M1 through M5. All bind the ligand, methacholine (Figure 16-69). Methacholine differs from acetylcholine (Figure 16-63) only by the addition of a methyl group to acetylcholine. Signaling from the muscarinic receptors is directly to a G protein or directly to phosphatidylinositol and then to a G protein in the cases of M1, M3, and M5. M2 modulates a potassium ion channel, as does M4. A theoretical model of the rat M3 muscarinic acetylcholine receptor is shown in Figure 16-70.

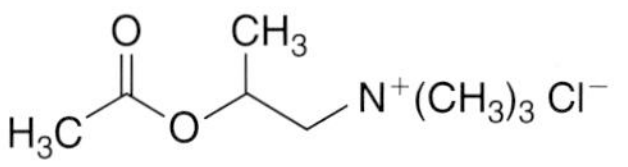

Figure 16-69. Structure of methacholine.

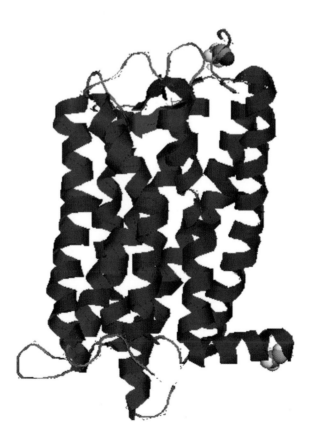

Figure 16-70. Theoretical model of rat M3 muscarinic acetylcholine receptor. Reproduced from PDB ID: 2amk. B. Li, N. M. Nowak, S. K. Kim, K. A. Jacobson, A. Bagheri, C. Schmidt, J. Wess. Random Mutagenesis of the M3 Muscarinic Acetylcholine Receptor Expressed in Yeast. *J. Biol. Chem.* **280** pp. 5664 (2005) and http://www.imb-jena.de/cgi-bin/ImgLib.pl?CODE=2amk.

Figure 16-71. Chemical structure of nicotine.

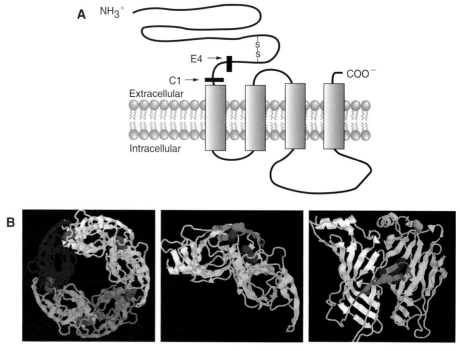

Figure 16-72. **A.** Schematic drawing of the chimeric 5-HT$_3$R subunit. *Arrows* indicate location of the "switch points" of the amino acid sequence. **B.** Structural model of the N-terminal domain of the 5-HT$_3$R based on the structure of AChBP. The N-terminal domain composed of five subunits (indicated by the different *colors, ribbons*) of 5-HT$_3$R is arranged in a *rosette* around a central cavity, here viewed from the membrane upward or bottom view (i.e., from the C to the N terminus) *(left panel)*. The ligand-binding site is present at the interface between adjacent subunits and is illustrated for clarity by the HEPES molecule *(purple, space fill)* found to be present in the AChBP structure. Detailed views of two subunits are shown *(middle panel*, bottom view; *left panel*, viewed from outside). Indicated in *white* and *yellow ribbons* are a principal and an adjacent subunit, respectively, which contribute to the binding site by the loops AC and DF, respectively. Colored *brown*, loop C reaches from the principal to the adjacent subunit closing the ligand-binding site. The backbone atoms of Glu residues investigated are indicated in *space fill*: Glu^{97} *(blue)*; Glu^{224} *(green)*; Glu^{235} *(red)*. Notice that Glu^{224} and Glu^{235} are juxtaposed on loop C, which forms an antiparallel β-strand, and are in close proximity of the ligand-binding site indicated by the HEPES molecule *(purple, space fill)*. Glu^{97}, here shown on the adjacent subunit, is more distant, located at the *top* of the subunit. *Left panel*, 80 × 80 Å; *middle* and *right panels*, 60 × 60 Å. Part A redrawn from Figure 9 of S. Lankiewicz et al., "Molecular cloning, functional expression, and pharmacological characterization of 5-hydroxytryptamine$_3$ receptor cDNA and its splice variants from guinea pig," *Mol. Pharmacol.*, 53: 202–212 1998. Part B reproduced from Figure 7 of C. Schreiter et al., "Characterization of the ligand-binding site of the serotonin 5-HT$_3$ receptor: the role of glutamate residues," *J. Biol. Chem.*, 278: 22709–22716, 2003.

5-HT$_3$R also belongs to the class of nicotinicoid receptors. This class of receptor usually has a binding site for nicotine (Figure 16-71). The subunit structure is diagrammed in Figure 16-72 with views of the receptor subunits forming a channel and the **serotonin (5-hydroxytryptamine; 5-HT,** Figure 16-73) binding site. Serotonin is synthesized from tryptophan by the action of tryptophan hydroxylase and aromatic amino acid decarboxylase, as shown in Figure 16-74. Serotonin is metabolized by monoamine oxidase that removes the amino group of serotonin and converts the remaining methyl group to an aldehyde; finally, this intermediate is acted on by aldehyde dehydrogenase to form the corresponding acid, **5-hydroxyindoleacetic acid (5-HIAA).** The serotonergic system extends from ascending axons from cell bodies in the raphe nuclei (Figure 16-75) that project into the frontal cortex, basal ganglia, thalamus, cerebellum, and hypothalamus. Serotonergic neurons are responsible for the stimulation of the hypothalamus that control the constitutive (baseline, not resulting from stress) levels of cortisol from the adrenal cortex. Serotonin is secreted from vesicles into the synaptic space from the axonal nerve ending, and it crosses the synapse to bind to serotonin receptors on the membrane of the postsynaptic cell (Figure 16-76). There is a reuptake mechanism for serotonin in the synaptic space so that it can be stored again in the presynaptic vesicles and reused. Nearly all tryptophan in the brain is converted to serotonin. Because the pineal body synthesizes melatonin

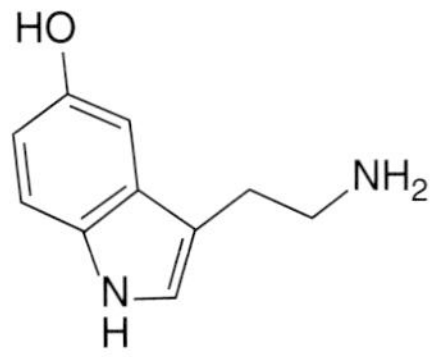

Figure 16-73. Chemical structure of serotonin, 5-hydroxytryptamine.

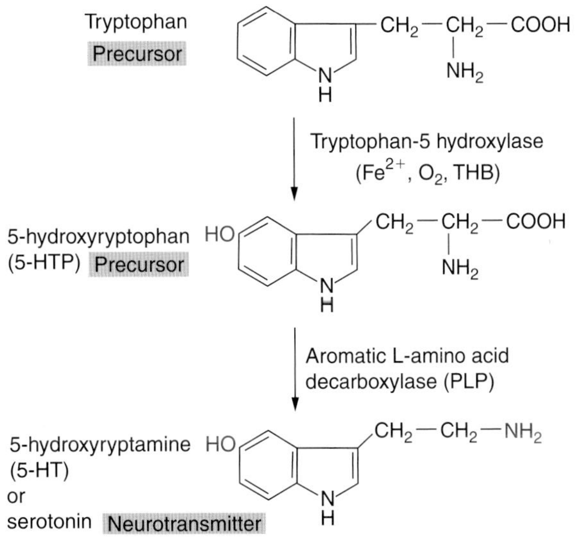

Figure 16-74. Synthesis of serotonin (5-hydroxytryptamine). *PLP*, pyridoxal phosphate (coenzyme); *THB*, tetrahydrobiopterin.

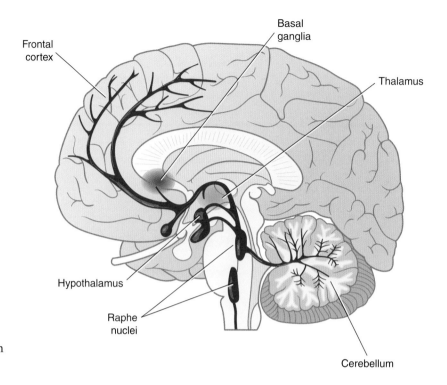

Figure 16-75. Serotonergic system in the brain.

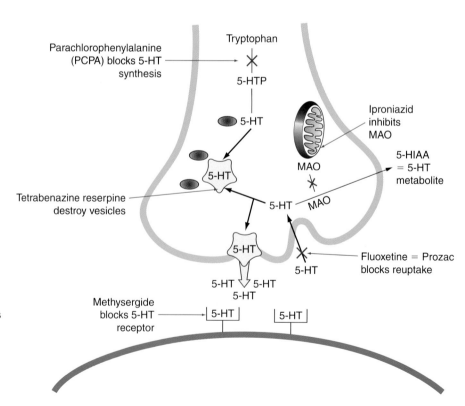

Figure 16-76. Synthesis of serotonin from tryptophan and incorporation of serotonin (5-HT) into vesicles. Serotonin is released in response to a signal into the synaptic space and binds to serotonin receptors on the membrane of the postsynaptic cell. Shown are various inhibitors of steps in the pathway including the reuptake mechanism. *HIAA*, hydroxyindole acetic acid; *HT*, hydroxytryptophan; *MAO*, monoamine oxidase.

from serotonin, it is the richest source of serotonin, although in the pineal gland it serves as a precursor rather than a neurotransmitter.

There are many types of serotonin receptors, and all, with the exception of 5-HT$_3$R (a ligand-gated ion channel), are seven transmembrane receptors coupled to G proteins that activate an intracellular second messenger signaling cascade. 5-HT$_1$Rs are coupled to G$_i$ or G$_0$ and mediate cellular changes by decreasing the level of cyclic AMP. 5-HT$_2$Rs (there are three subtypes: 5-HT$_{2A}$R, 5-HT$_{2B}$R, and 5-HT$_{2C}$R) are coupled to G$_q$ or G$_{11}$ and mediate effects through increasing IP$_3$ and DAG, activating protein kinase A and releasing intracellular calcium ions. 5-HT$_4$R is coupled to G$_S$ and increases levels of cyclic AMP. 5-HT$_7$R also is coupled to G$_S$ and increases the level of cyclic AMP. Some other information pertaining to these receptors is presented in Table 16-5. A diagram of 5-HT$_3$R, its function in transporting sodium ions, the receptor function in the presence of an antagonist, and the mechanism of serotonin reuptake are shown in Figure 16-77.

Table 16-5
Characteristics of 5-HT Receptors

Receptor	Actions	Agonists	Antagonists
5-HT$_{1A}$	CNS: neuronal inhibition, behavioral effects (sleep, feeding, thermoregulation, anxiety)	buspirone	spiperone, methiothepin, ergotamine, yohimbine
5-HT$_{1B}$	CNS: presynaptic inhibition, behavioral effects; vascular: pulmonary vasoconstriction	ergotamine, sumatriptan	methiothepin, yohimbine, metergoline
5-HT$_{1D}$	CNS: locomotion; vascular: cerebral vasoconstriction	sumatriptan	methiothepin, yohimbine, metergoline, ergotamine
5-HT$_{2A}$	CNS: neuronal excitation, behavioral effects; smooth muscle: contraction, vasoconstriction/dilatation; platelets: aggregation	α-methyl-5-HT, LSD (CNS)	ketanserin, cyproheptadine, pizotifen, LSD (PNS)
5-HT$_{2B}$	stomach: contraction	α-methyl-5-HT, LSD (CNS)	yohimbine, LSD (PNS)
5-HT$_{2C}$	CNS, choroid plexus: cerebrospinal fluid (CSF) secretion	α-methyl-5-HT, LSD (CNS)	mesulergine, LSD (PNS)
5-HT$_3$	CNS, PNS: neuronal excitation, anxiety, emesis	2-methyl-5-HT	metoclopramide (high doses), renzapride, ondansetron, alosetron
5-HT$_4$	GIT, CNS: neuronal excitation, gastrointestinal motility	5-methoxytryptamine, metoclopramide, renzapride, tegaserod	GR113808
5-ht$_5$	CNS: unknown	unknown	unknown
5-ht$_6$	CNS: unknown	unknown	unknown
5-HT$_7$	CNS, GIT, blood vessels: unknown	5-carboxytryptamine, LSD	methiothepin

CNS, central nervous system; *GIT*, gastrointestinal tract; *PNS*, peripheral nervous system.
Note that there is no 5-HT$_{1C}$R because, after the receptor was cloned and further characterized, it was found to have more in common with the 5-HT$_2$ family of receptors and was redesignated as 5-HT$_{2C}$R.
Reproduced from http://www.answers.com/topic/5-ht-receptor.

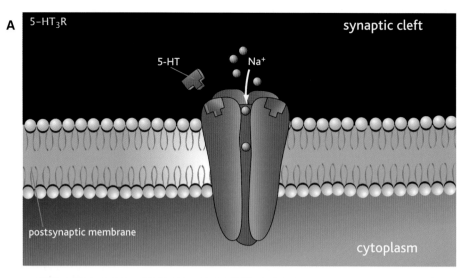
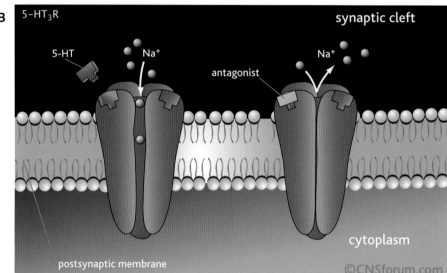

Figure 16-77. **A.** 5-HT$_3$R, a ligand-gated ion channel permeable to sodium and potassium ions. The receptor is structurally similar to the nicotinic acetylcholine receptor, composed of five subunits. **B.** Binding of an agonist at the 5-HT-binding site generates a conformational change and activation of 5-HT$_3$R, permitting movement of positively charged ions from the synaptic cleft (space) into the postsynaptic cell cytoplasm. Binding of an antagonist of 5-HT- binding prevents receptor activation and inhibits cell depolarization.

Based on experimental animal models, one theory holds that adult anxiety (panic attacks, obsessive compulsive disorder, social phobia, etc.) could be the result of a failure to express 5-HT$_{1A}$R during development. 5-HT$_{1A}$R is diagrammed in Figure 16-78. Many drugs are available for treating anxiety disorder, and these are categorized as **selective serotonin reuptake inhibitors (SSRIs).** An example is fluoxetine (Prozac;

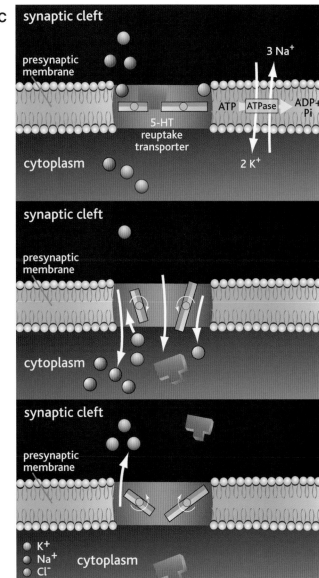

Figure 16-77, cont'd **C.** Mechanism of action of 5-HT reuptake transporters. Action of 5-HT at the synapse is terminated by its reuptake across the presynaptic membrane in an energy-dependent process. Hydrolysis of ATP by Na$^+$/K$^+$ ATPases provides energy that creates a concentration gradient of ions across the presynaptic membrane that drives the opening of the transporter and cotransport of sodium and chloride ions and 5-HT from the synaptic cleft. Potassium ions bind to the transporter and enable it to return to the outward position. Release of the potassium ions into the synaptic cleft equilibrates the ionic gradient across the presynaptic membrane. The 5-HT reuptake transporter is then available to bind another 5-HT molecule for reuptake. Parts A, B, and C reproduced from the CNS forum, Lundbeck Institute, Skodsborg, Denmark and http://www.cnsforum.com/imagebank/section/receptor_systems_Serotonergic/default.aspx.

Figure 16-79. SSRIs downregulate the **serotonin transporter** (**SERT**; Figure 16-77C) that moves serotonin back from the synaptic cleft to the presynaptic nerve ending in the reuptake process. The downregulation apparently is not the result of reduced SERT gene expression. SSRIs, by inhibition of the reuptake process, enhance serotonergic neurotransmission, but it has been observed that they also enhance the neurotransmission of norepinephrine; presumably, both effects contribute to the antidepressant action of SSRIs. In addition, there may be regional differences in 5-HT$_{1A}$R regulatory mechanisms.

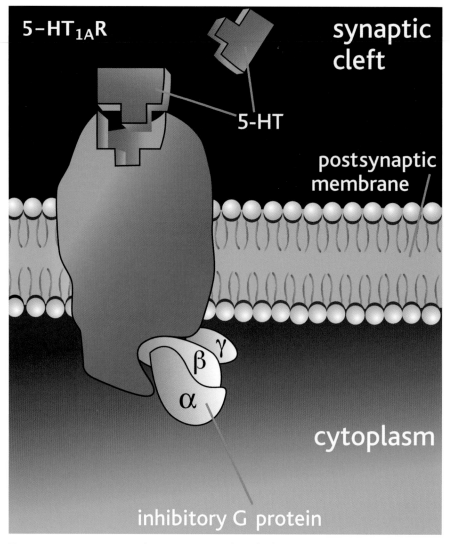

Figure 16-78. 5-HT$_{1A}$R. This receptor is classified into A, B, and D subtypes, which are found in the central nervous system and in blood vessels. This receptor is coupled to inhibitory G proteins, which have an inhibitory effect on neurotransmission when bound by an agonist. CNS Forum, Lundbeck Institute, Skodsborg, Denmark. Reproduced from http://www.cnsforum.com/imagebank/item/Rcpt_sys_SN1A/default.aspx.

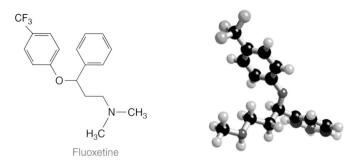

Figure 16-79. Structure of fluoxetine (Prozac) an example of a selective serotonin reuptake inhibitor.

Figure 16-80 summarizes the classes of 5-HTRs, and their locations, second messenger effects, agonists, and antagonists. A concept, based on data, of the SERT that mediates the reuptake process is shown in Figure 16-81. SERT is found in the plasma membrane of serotonergic neurons (presynaptic neuron). One or more molecules of 5-HT is moved from the synaptic cleft back into the presynaptic axon terminal per cycle (after a conformational change in the protein). SERT does not resemble an ion channel but is

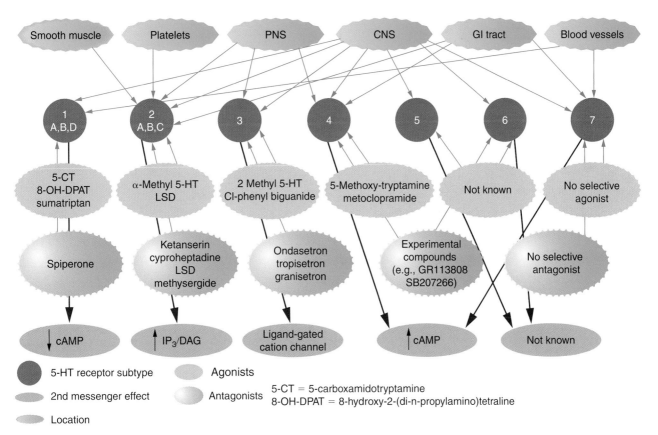

Figure 16-80. Classes of 5-HT receptors, their locations, agonists, antagonists, and second messenger effects. *CNS*, central nervous system; *PNS*, parasympathetic nervous system. Redrawn from http://www.cnsforum.com/content/pictures/imagebank/hirespng/5HT_rcpt_subtypes.png. CNS Forum, Lundbeck Institute Skodoborg, Denmark.

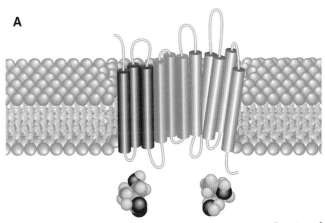

Figure 16-81. **A.** Model of the 5-HT transporter (SERT) responsible for reuptake of serotonin (5-HT) from the synaptic cleft into the axonal ending of a serotonergic (presynaptic) neuron.

Continued

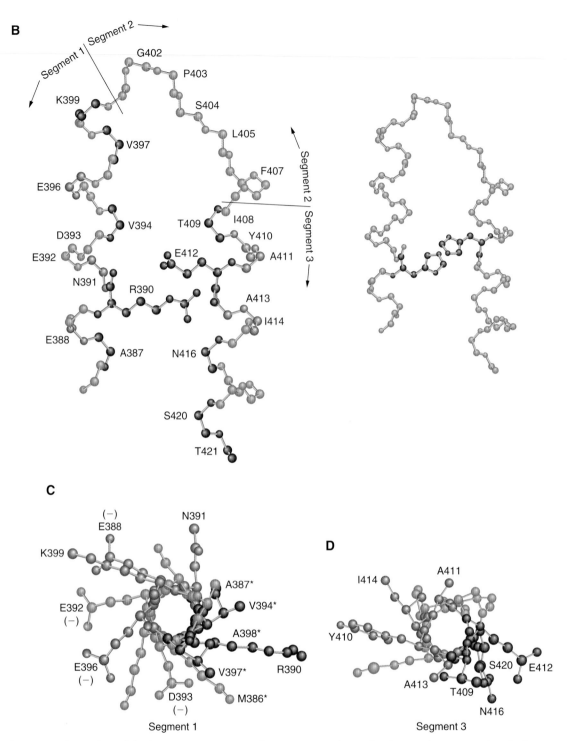

Figure 16-81, cont'd **B.** Hypothetical model of EL4 structure. Residues Met-386 through Ala-399 *(segment 1)* and Thr-409 through Ser-421 *(segment 3)* were modeled as standard α-helices. The residues lying in between *(segment 2)* are shown as random coil. The upper figures *(A* and *B)* show only the backbone atoms for clarity. *A*, entire EL4 loop with proposed α-helical segments indicated; *B*, entire loop with Arg-390 and Glu-412 substituted with histidines to show how the proposed model allows the proximity of these side chains as needed to form a Zn^{2+} binding site. *C* and *D*, segment 1 *(C)* and segment 3 *(D)* helices viewed end on. In *C, black* indicates functionally important, highly accessible positions, and *gray* indicates fully or partially occluded positions. In *D, black* indicates native hydrophilic positions, and *gray* indicates fully or partially occluded positions. These shading patterns are also used for segment 1 and segment 3 in *A*. EL4, extracellular loop 4.

a carrier; however, both sodium and chloride ions are associated with active 5-HT translocation. SERT resembles norepinephrine and dopamine transporters and is composed of 12 transmembrane domains with an extracellular loop between transmembrane helices (TMs) 3 and 4, as shown in Figure 16-81A. Extracellular loop 4 plays a role in the conformational changes involved with the translocation of serotonin from the synaptic cleft into the axonal nerve ending of a serotonergic neuron. The N- and C-terminals of SERT polypeptide are located in the cytoplasm, where there are six putative phosphorylation sites for protein kinase A and protein kinase C. 5-HT binding takes place within helices 1 through 3 and helices 8 through 12. A model of extracellular loop 4, which plays a role in conformational change, mediating transport, is shown in Figure 16-81B.

Catecholamines and Monoamines

Skeletal muscles are under voluntary control, whereas the involuntary smooth muscles are controlled by the autonomic nervous system. The autonomic nervous system has two parts, the sympathetic and parasympathetic (Figure 16-82). Both skeletal muscles and smooth muscles of the

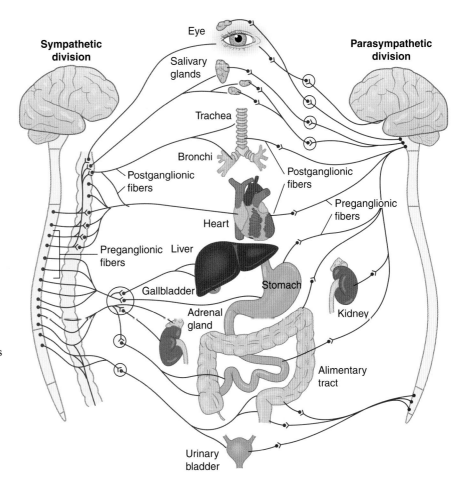

Figure 16-82. Sympathetic and parasympathetic nervous systems, parts of the autonomic nervous system, which is not under voluntary control. The skeletal muscles (voluntary) and the smooth muscles (involuntary) are innervated by acetylcholine. The sympathetic nervous system (except for the sweat glands) is innervated by norepinephrine (norepinephrine plus epinephrine in the adrenal medulla). Redrawn from http://ardb.bjmu.cn/image/ANS.jpg. Adrenergic Receptor Database, 973 Project in China and 985 Project of Peking University.

parasympathetic nervous system are innervated by acetylcholine (there are relatively few acetylcholine receptors in the brain). Norepinephrine is the main neurotransmitter in the direct innervation of the sympathetic nervous system (except for the sweat glands), and epinephrine and norepinephrine innervate the adrenal medulla. Norepinephrine is considered to be the main neurotransmitter, whereas epinephrine often is considered to be a hormone. Seventy-five percent of all parasympathetic fibers arise from the cranial vagus nerve (parasympathetic nerves are cranial and sacral) and travel to end organs containing ganglia. The postganglionic nerves to the smooth muscles in the end organs are muscarinic. The parasympathetic and the sympathetic preganglionic fibers are nicotinic, as is the neuromuscular junction of the skeletal muscles. Because administration of muscarinic blocking agents results in memory loss in normal people, it is theorized that cholinergic loss might be significant for Alzheimer's disease.

The most important monoamine neurotransmitters are dopamine, norepinephrine (catecholamines, Figure 10-32), and serotonin (Figure 16-73). The synthesis of the catecholamines has been presented in the chapter on metabolism (Chapter 10). *There are at least six G protein–coupled dopamine receptors of two major types, D_1 and a D_2-like subfamily. The D_1-like receptors are D_1 and D_5; these couple to Gs and activate adenylate kinase. D_{2a}, D_{2h}, D_3, and D_4 (members of the D_2-like subfamily) are coupled to G proteins that inhibit adenylate cyclase and activate potassium ion channels.* D_2 receptors are expressed in neurons of the midbrain, caudate and limbic systems (*nucleus accumbens*, amygdala, and hippocampus) and parts of the cerebral cortex. These receptors have a high affinity for antipsychotic drugs. A defect in D_2 receptors may predispose to alcoholism in people who drink alcohol for its euphoric affects rather than for its antianxiety properties. The general structure of the dopamine receptors is indicated in Figure 16-83. The D_1 type receptor is shown in the figure. The D_2-like receptors are similar except that they have a shorter C-terminal cytoplasmic tail and a larger third intracellular loop (I3). Dopamine bound to its binding site in the human D_2 dopamine receptor is shown in Figure 16-84. The predicted binding site for dopamine is located between transmembrane (TM) helices 3, 4, 5, and 6 (the same site is occupied by all dopamine agonists), whereas the most potent antagonists bind to a site composed of transmembrane helices 2, 3, 4, 6, and 7. The gene encoding the human D_2 receptor is the *DRD2* gene, shown in Figure 16-85A, and it is located on chromosome 11 (11 q22-23), shown in Figure 16-85B.

Dopamine receptors are the predominant catecholamine receptors expressed in the central nervous system, where they are involved in control of locomotion, cognition, emotion, and hormone secretion in the pituitary. Dopamine receptors also affect sodium homeostasis in kidney and vascular tone in the blood vessels. Although the signaling pathways of the activated dopamine receptors are not all completely understood, it is clear that the D_1-like receptors stimulate adenylate cyclase, inhibit the release of arachidonic acid, inhibit the sodium ion–proton exchanger, stimulate calcium ion uptake (calcium ion channels), may affect potassium ion export, and stimulate phospholipase C. The D_2-like receptors have opposite effects; they inhibit adenylate cyclase through an inhibitory G protein, stimulate sodium or potassium ATPase, stimulate the efflux of potassium ion (potassium ion channels) leading to cell

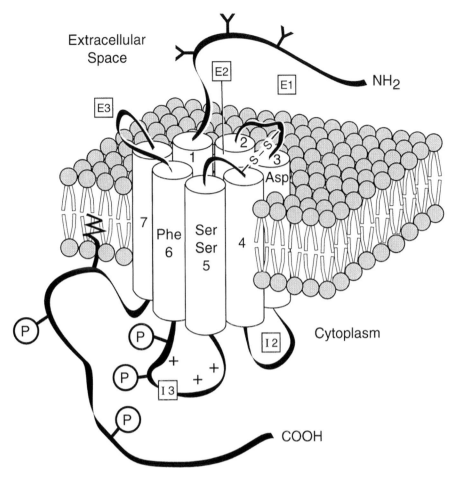

Figure 16-83. Dopamine receptor structure. Structural features of D_1-like receptors represented. D_2-like receptors are characterized by a shorter COOH-terminal tail and by a bigger third intracellular loop. Residues involved in dopamine binding are highlighted in transmembrane domains. Potential phosphorylation sites are represented on third intracellular loop and on COOH terminus. Potential glycosylation sites are represented on NH₂ terminal E1–E3, extracellular loops; 1–7, transmembrane domains; 12–13, intracellular loops. Reproduced with permission from Figure 1 of C. Missale et al., "Dopamine receptors: from structure to function," *Physiol. Rev.*, 78: 189–225, 1998.

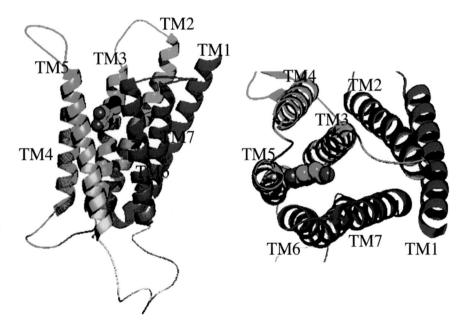

Figure 16-84. Predicted binding site of dopamine *(shown in spheres)* in the predicted structure of human dopamine D_2 receptor. The figure at the *left* shows a side view, and that on the *right* gives a top view. Dopamine is shown in atomic model. Reproduced with permission from Figure 2 of M. Yashar et al., "The predicted 3D structure of the human D_2 dopamine receptor and the binding site and binding affinities for agonists and antagonists," *PNAS*, 101: 3815–3820, 2004 and http://www.pnas.org/cgi/contant/full/101/11/3815/FIG2.

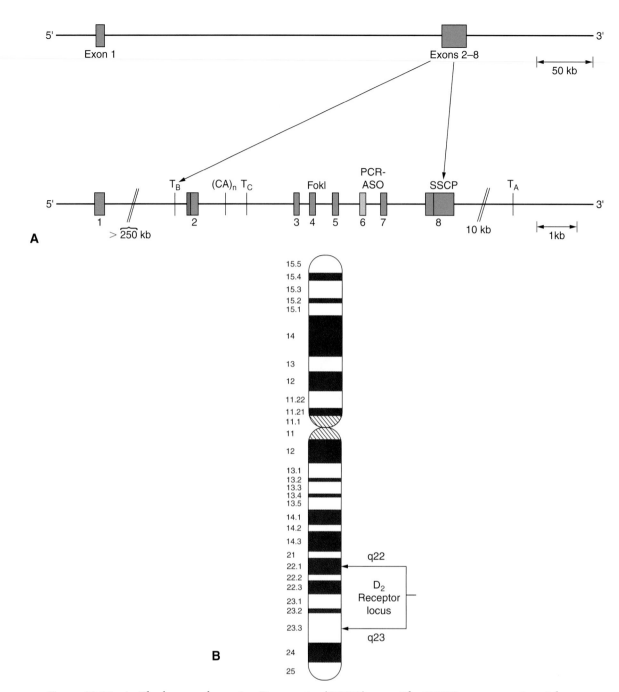

Figure 16-85. **A.** The human dopamine D_2 receptor (DRD2) gene. The DRD2 gene contains eight exons, spans 270 kb, and includes an intron of approximately 250 kb separating the first exon from the exons encoding the receptor protein. Exons 2–8 span 14 kb. The transcript undergoes alternative splicing of exon 6 to produce two mRNAs. The translation products differ by 29 amino acids. Also shown are the locations of other polymorphic markers. **B.** The chromosomal location of the human dopamine D2 receptor *(DRD2)* gene. The *DRD2* gene is located at chromosome 11 q22-23.

hyperpolarization, inhibit calcium ion uptake through the calcium ion channel, and stimulate the release of arachidonate. These signaling mechanisms are summarized in Figure 16-86.

Two types of neurons involve dopamine receptors. One population of neurons projects to the entopeduncular nucleus (a major outflow nucleus of the basal ganglia) and the *pars reticulata* of the *substantia nigra* (Figure 16-87) that expresses substance P and dynorphin. The other neuronal set projects to the external segment of the *globus pallidus* and contains enkephalin. The first set express D_1 receptors preferentially (Figure 16-88, left), wherein dopamine stimulates the expression of substance P and dynorphin. The second set of neurons (Figure 16-88, right) expresses mainly D_2 receptors and inhibits the expression of **proenkephalin A (PPA).** Also, there is an interrelation between D_2 dopamine receptors and the endocannabinoid system, producing anandamide (Figure 16-17), in the dorsal striatum (Figure 16-89). Dopamine and dopamine receptor imbalance are thought to play a role in neurological diseases such as Parkinson's disease and schizophrenia.

Norepinephrine (also called noradrenaline) is the other major neurotransmitter of the central nervous system in addition to dopamine. Norepinephrine is the major neurotransmitter of the peripheral sympathetic nervous system (Figure 16-82), and epinephrine is the primary hormone secreted by the adrenal medulla. There are nine adrenergic receptors, six α-receptors and three β-receptors (Figure 16-90). *The net effects of these receptors are to increase cellular calcium ion concentration in the case of the α-1 receptors, to inhibit adenylate cyclase and the cellular level of cyclic AMP in the case of the α-2 receptors, and to*

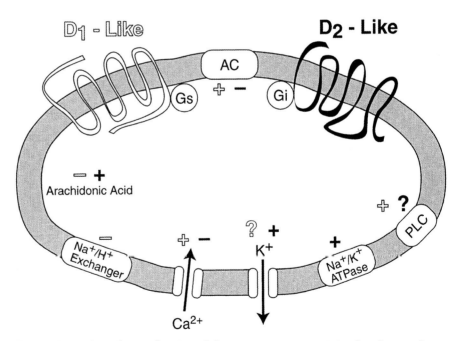

Figure 16-86. Signal transduction of dopamine receptors. *AC*, adenylate cyclase; *PLC*, phospholipase C. Reproduced from Figure 2 of http://physrev.physiology.org/cgi/content/full/78/1/189/F2. C. Missale et al., "Dopamine receptors: from structure to function," *Physiol. Rev.*, 78: 189–225, 1998.

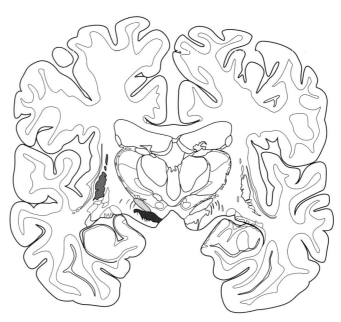

Figure 16-87. *Substantia nigra* and other brain structures. The subthalamic nucleus *(shown in yellow)* and the *substantia nigra (in red)* are also included in the basal ganglia. The putamen *(in blue)* can also be seen in this diagram. Although the subthalamic nucleus and *substantia nigra* appear separate from the other basal ganglia components, they have a clear anatomical relationship.

stimulate adenylate cyclase and the level of cellular cyclic AMP in the case of the β-receptors. The subclasses generally operate through the same signaling mechanisms.

Rather little has been confirmed on the crystal structures of the α-receptors, but a homology model of $α_{1A}$-adrenoreceptor, based on the crystal structure of rhodopsin, another G protein–coupled receptor (Figure 16-91), has been developed (Figure 16-92). Rhodopsin, the only G protein–coupled receptor whose crystal structure has been resolved, is a part of the visual cycle, where the sensing of light is initiated by this protein in rod cells. 11-*cis*-Retinal, from vitamin A, is bound by rhodopsin in the rod cell discs. 11-*cis*-Retinal is the light-sensitive molecule that absorbs light to initiate the process of visual transduction (Figure 16-93).

The predicted structure of the $β_1$-adrenergic receptor, from the structure of rhodopsin, is shown in Figure 16-94. A neuron synthesizing norepinephrine from tyrosine (or phenylalanine) packages norepinephrine in presynaptic vesicles in the nerve ending. An appropriate signal causes the release of the norepinephrine from the synaptic vesicle into the synaptic cleft, and norepinephrine binds to adrenoreceptors on the membrane of the postsynaptic cell (Figure 16-95). In some nerve terminals, norepinephrine is packaged alone or with neuropeptide Y. When norepinephrine is released into the synaptic cleft, it can, in addition to binding an appropriate receptor on the postsynaptic cell membrane, bind to an $α_2$-receptor on the membrane of the nerve terminal itself. Neuropeptide Y

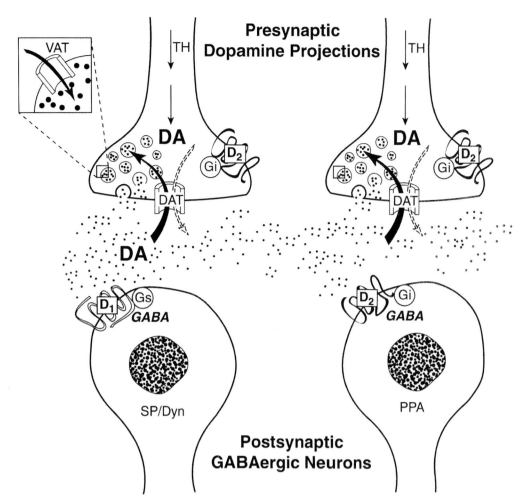

Figure 16-88. Organization of striatal dopaminergic synapses. D_1 receptors are preferentially expressed by γ-aminobutyric acid (GABA)ergic neurons coexpressing substance P (SP) and dynorphin (Dyn) and projecting to entopeduncular nucleus and *substantia nigra*, whereas D_2 receptors are segregated on GABAergic neurons containing enkephalin and projecting to *globus pallidus*. D_2-like autoreceptors are present on dopaminergic terminals. *DAT*, dopamine transporter; *PPA*, preproenkephalin A; *TH*, tyrosine hydroxylase; *VAT*, vesicular transporter. Reproduced with permission from http://physrev.physiology.org/cgi/content/full/78/1/189/F3 and Figure 3 from C. Missale et al., "Dopamine receptors: from structure to function," *Physiol. Rev.*, 78: 189–225, 1998.

released into the synaptic cleft also can bind to the Y_2 receptor on the membrane of the nerve terminal. These events cause reciprocal regulation of neurotransmitter release because norepinephrine binding to the α_2-receptor of the nerve ending inhibits the further release of norepinephrine and neuropeptide Y into the synaptic cleft and binding of neuropeptide Y from the synaptic cleft to the presynaptic Y_2 receptor causes inhibition of the release of norepinephrine (Figure 16-96).

When β-receptor agonist levels are high, there is a mechanism to shut down the receptor called desensitization that is accomplished by the phosphorylation of the receptor by protein kinase A of amino acid residues in the cytoplasmic domain of the receptor. A second phosphorylating

Figure 16-89. An hypothetical model of the functional interaction between endocannabinoid and dopaminergic systems in dorsal striatum. Activation of dopamine D_2-like receptors by dopamine may elicit the extracellular release of anandamide through a mechanism that remains undetermined. Anandamide, in turn, may counteract D_2-mediated motor behaviors by engaging CB1 cannabinoid receptors. Redrawn from http://web.archive.org/web/20060518031521http://www.uclm.es/inabis2000/symposia/files/119/fig1.gif.

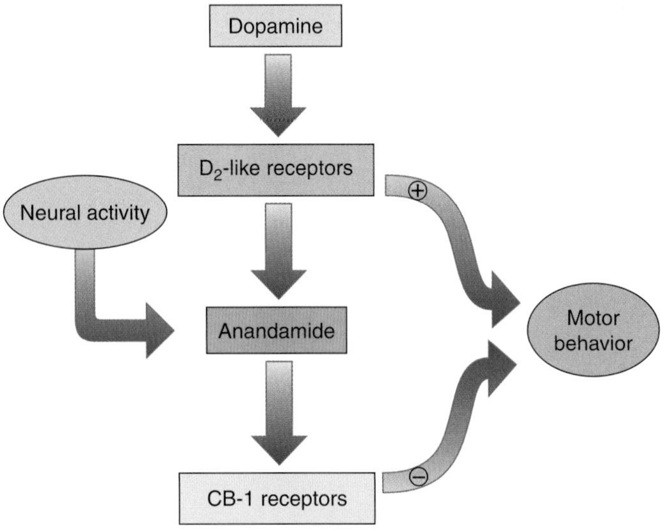

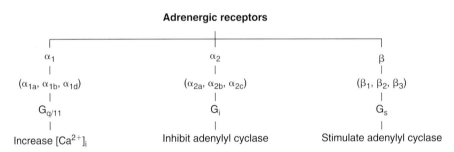

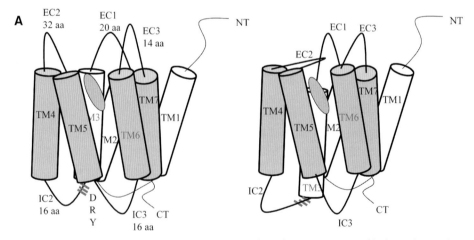

Figure 16-90. The adrenergic receptor family. Subtypes within a subfamily activate the same signaling mechanism in general. Alpha1 receptors operate through the G protein $G_{q/11}$ and increase cellular calcium ion. Alpha2 subtypes operate through the G_i inhibitory G protein to inhibit adenylate cyclase (reduce cyclic AMP levels). Beta receptors operate through the G protein, G_s, a stimulatory G protein to stimulate adenylate cyclase (increase the cellular level of cyclic AMP).

Figure 16-91. **A.** Schematic for a possible signaling mechanism in rhodopsin. Note that the movement of helix 3 (caused by interaction with the *trans*-isomer of retinal) exposes the *DRY* sequence to G protein activation and as a result closes the EC-II loop to maintain the ligand inside the bundle sequence. **B.** The 13 regions shown as boxes used in scanning the entire protein for the 11-*cis*-retinal putative binding site. The two boxes chosen as binding sites by HierDock are shown in red. *Front view* with N-terminus at the bottom. *Top view* obtained by rotating 90 degrees around the horizontal axis in A so that the N-terminus is out of view. These two orientations are used for all structures. *CT*, C-terminal; *EC*, extracellular domain; *IC*, intracellular domain; *NT*, N-terminal; *TM*, transmembrane (domain). Part A reproduced from Figure 4 of R.J. Trabanino et al., "First principles predictions of the structure and function of g-protein-coupled receptors: validation for bovine rhodopsin," *Biophys. J.*, 86:1904–1921, 2004 and http://www.biophysj.org/cgi/content/full/86/4/1904/FIG4.

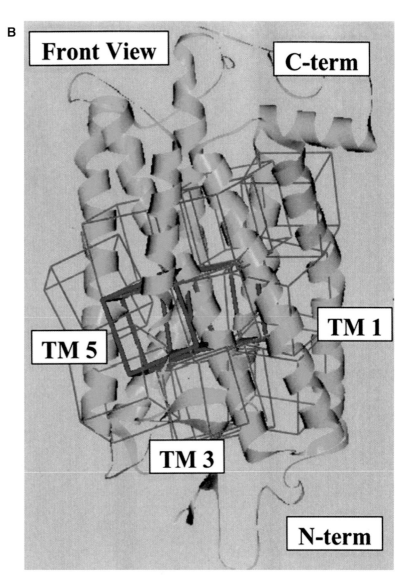

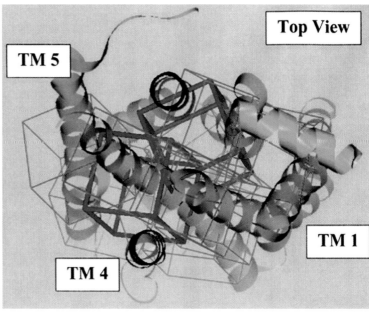

Figure 16-91, cont'd Part B reproduced from Figure 5 of R.J. Trabanino et al., "First principles predictions of the structure and function of g-protein-coupled receptors: validation for bovine rhodopsin," *Biophys. J.*, 86: 1904–1921, 2004 and from http://www.biophysj.org/cgi/content/full/86/4/1904/FIG5.

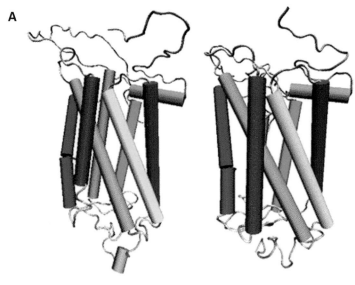

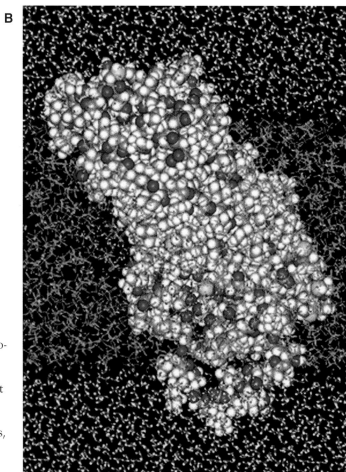

Figure 16-92. **A.** α1A adenoreceptor model I *(left)* and model II *(right)*, with TM-1 *(red)*, TM-2 *(yellow)*, TM-3 *(mauve)*, TM-4 *(purple)*, TM-5 *(blue)*, TM-6 *(orange)*, TM-7 *(green)*, TM-8 *(cyan)*, and loop regions *(pink)*. *TM,* transmembrane domain. **B.** Simulation of α1A adenoreceptor in a membrane (hydrophilic/hydrophobic interface). Part A reproduced from Figure 1 of G.K. Kinsella, I. Rozas, and G.W. Watson, "Computational development of an α1A-adrenoceptor model in a membrane mimic," *Biochem. Biophys. Res. Comm.,* 324: 916-992, 2004. Part B reproduced from Figure 2 of G.K. Kinsella, I. Rozas, and G.W. Watson, "Computational development of an α1A-adrenoceptor model in a membrane mimic," *Biochem. Biophys. Res. Comm.,* 324: 916–992, 2004.

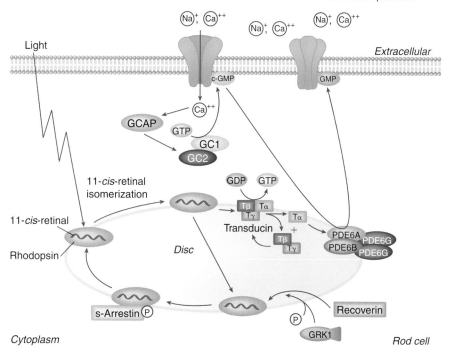

Figure 16-93. The signal transduction cascade responsible for sensing light in vertebrates is one of the best studied signal transduction processes and is initiated by rhodopsin in rod cells, a member of the G protein–coupled receptor gene family. Rhodopsin remains the only GPCR whose structure has been resolved at high resolution. Rhodopsin in the discs of rod cells contains a bound 11-*cis*-retinal chromophore, a small molecule derived from vitamin A that acts as the light sensitive portion of the receptor molecule, absorbing light to initiate the signal transduction cascade. When light strikes 11-*cis*-retinal and is absorbed, it isomerizes to all-*trans* retinal, changing the shape of the molecule and the receptor it is bound to. This change in rhodopsin's shape alters its interaction with transducin, the member of the G protein gene family that is specific in its role in visual signal transduction. Activation of transducin causes its α-subunit to dissociate from the trimer and exchange bound GDP for GTP, activating in turn a membrane-bound cyclic GMP–specific phosphodiesterase that hydrolyzes cGMP. In the resting rod cell, high levels of cGMP associate with a cyclic GMP–gated sodium channel in the plasma membrane, keeping the channels open and the membrane of the resting rod cells depolarized. This is distinct from synaptic generation of action potentials, in which stimulation induces opening of sodium channels and depolarization. When cyclic GMP-gated channels in rod cells open, both sodium and calcium ions enter the cell, hyperpolarizing the membrane and initiating the electrochemical impulse responsible for conveying the signal from the sensory neuron to the CNS. The rod cell in the resting state releases high levels of the inhibitory neurotransmitter glutamate, while the release of glutamate is repressed by the hyperpolarization in the presence of light to trigger a downstream action potential by ganglion cells that convey signals to the brain. The calcium that enters the cell also activates GCAP, which activates guanylate cyclase (GC-1 and GC-2) to rapidly produce more cGMP, ending the hyperpolarization and returning the cell to its resting depolarized state. A protein called recoverin helps mediate the inactivation of the signal transduction cascade, returning rhodopsin to its preactivated state, along with the rhodopsin kinase Grk1. Phosphorylation of rhodopsin by Grk1 causes arrestin to bind, helping terminate the receptor activation signal. Dissociation and reassociation of retinal, dephosphorylation of rhodopsin, and release of arrestin all return rhodopsin to its ready state, prepared once again to respond to light. Redrawn from http://www.biocarta.com/pathfiles/h_rhodopsinPathway.asp. BioCarta, Inc., San Diego, CA.

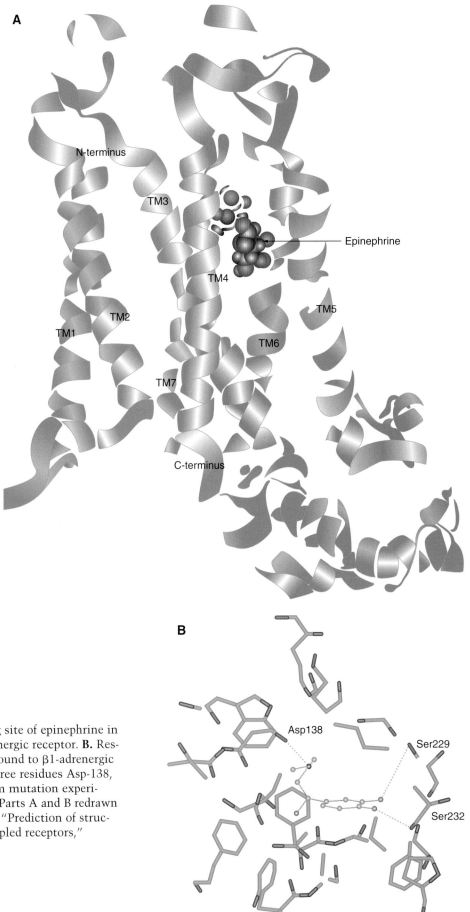

Figure 16-94. **A.** Predicted binding site of epinephrine in the predicted structure of β1-adrenergic receptor. **B.** Residues within 5 Å of epinephrine bound to β1-adrenergic receptor. Shown in bold are the three residues Asp-138, Ser-229, and Ser-232 (deduced from mutation experiments to be involved in binding). Parts A and B redrawn from Figure 2 of N. Vaidehi et al., "Prediction of structure and function of g-protein-coupled receptors," *PNAS*, 99: 12622–12627, 2002.

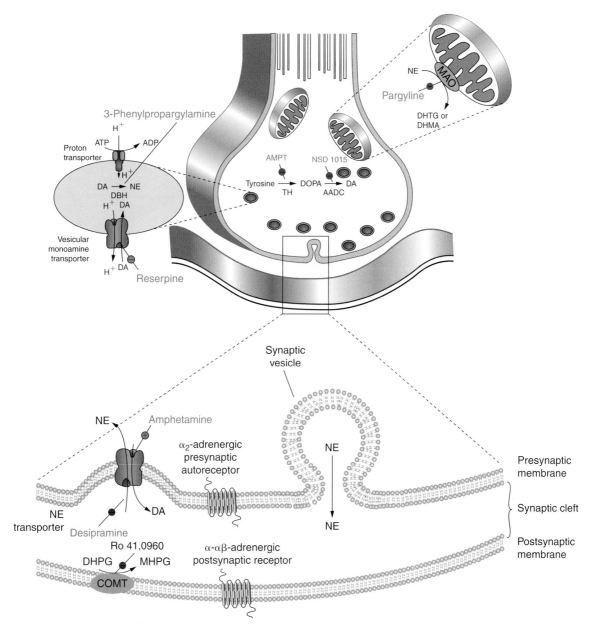

Figure 16-95. The nerve terminal of a neuron synthesizing norepinephrine (NE). *AADC*, amino acid decarboxylase; *COMT*, catecholamine *O*-methyl transferase *DA*, dopamine; *DOPA*, dihydroxyphenylalanine; *MAO*, monoamine oxidase; *NE*, norepinephrine; *TH*, tyrosine hydroxylase. Also shown are various drugs that bloc the indicated steps.

enzyme, β-**adrenergic receptor kinase** (β**ARK**), phosphorylates other amino acid residues in the cytoplasmic domain of the receptor, allowing a protein (β-arrestin) to associate with this domain and inactivate (as its name implies) or desensitize the receptor (Figure 16-97A), sequestering it from the active environment. The structure of β-arrestin and its association with the receptor (rhodopsin) are shown in Figure 16-98. The desensitized receptor can be resensitized by the removal of the phosphorylations by a

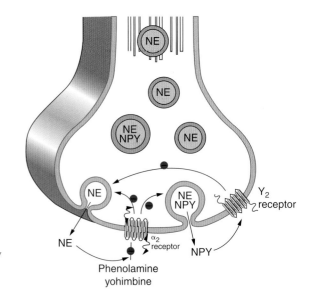

Figure 16-96. Reciprocal regulation of neurotransmitter release by norepinephrine and neuropeptide Y. *NE*, norepinephrine. Reproduced from http://www.chemistry.emory.

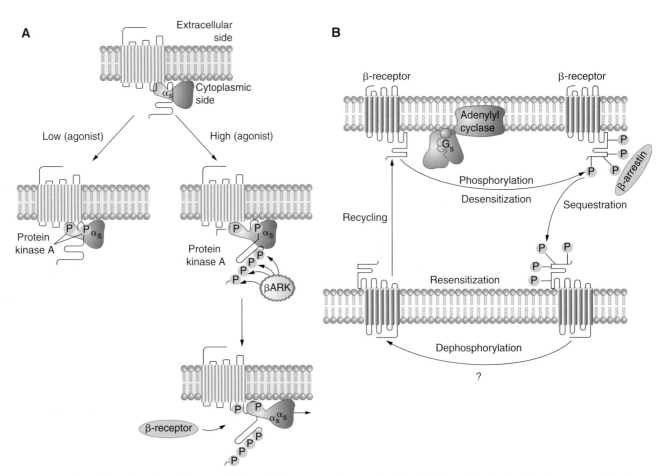

Figure 16-97. **A.** Mechanism of agonist-induced β-receptor desensitization. **B.** Hypothesized role of sequestration in β-receptor desensitization and resensitization. Reproduced from Figures 8.36 and 8.37 from http://www.chemistry.emory.edu/justice/seminar/adrenergic_receptors.htm.

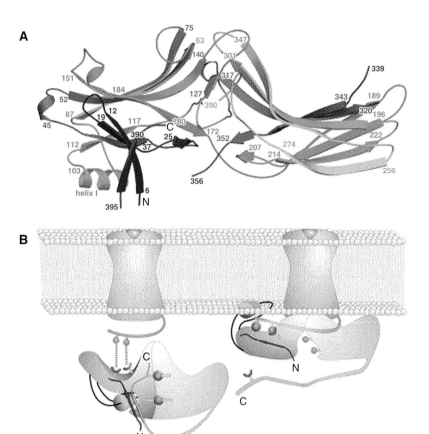

Figure 16-98. **A.** Ribbon diagram of β-arrestin. The N- and C-termini are indicated by N and C, respectively. The color varies from blue to red from the N-terminus to the C-terminus. **B.** Proposed receptor-binding mechanism of β-arrestin. The cartoon illustrates the proposed receptor-binding mechanism of arrestins. The N and C domain are shown in pink and cyan, respectively. The polar core is located at the interface between the N and C domains and is represented by charge–charge interactions (*blue–red* dots). The C tail is drawn in *yellow–green* hues; the N-terminal strand I is in *magenta*. Positive and negative charges are represented by *blue* and *red*; hydrophobic portions are in *yellow*. The activated receptor is depicted in *green*. Part A reproduced from Figure 1 of M. Han et al., "Crystal structure of β-arrestin at 1.9Å, possible mechanism of receptor binding and membrane translocation," *Structure*, 9: 869–880, 2001. Part B reproduced from Figure 6 of M. Han et al., "Crystal structure of β-arrestin at 1.9 Å, possible mechanism of receptor binding and membrane translocation," *Structure*, 9: 869–880, 2001.

phosphoprotein phosphatase, and the receptor assumes a functional state in the active environment of the membrane (Figure 16-98B).

The signal transduction pathways from the activated adrenergic receptors are complex. Figure 16-99 shows the pathways from the activated β_1-adrenergic receptors and the β_2-adrenergic receptors. In addition to these canonical pathways, there are crosstalk pathways to MAPK, shown in Figure 16-100. Similar treatments are presented for the α_1- and α_2-adrenergic receptors in Figures 16-101 and 16-102. These are

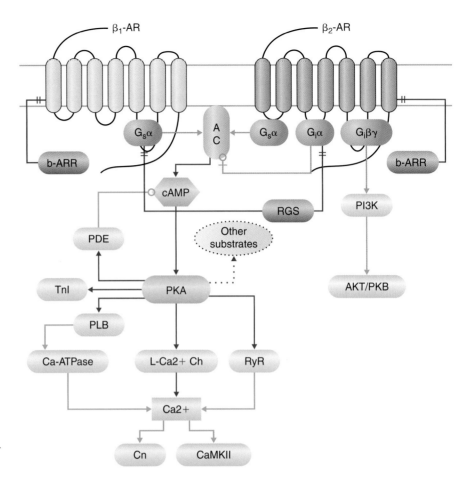

Figure 16-99. Canonical signal pathways of β_1- and β_2-adrenergic receptors(ARs).

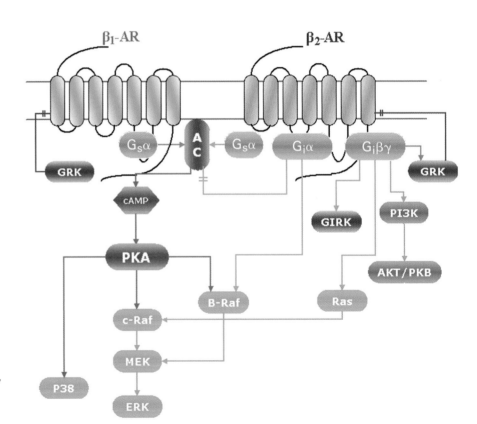

Figure 16-100. Crosstalk pathways of β_1- and β_2-adrenergic receptors (ARs) and the MAPK system. MEK, MAPK/ERK kinase.

1042   CHAPTER 16   Neurobiochemistry

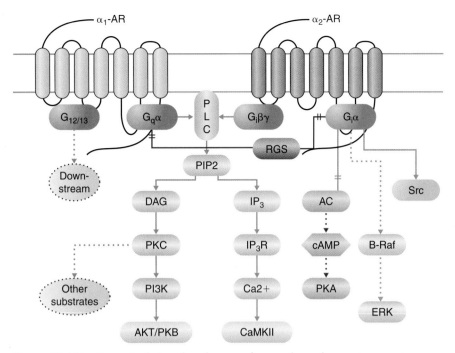

Figure 16-101. Canonical signal pathways of α_1- and α_2-adrenergic receptors (α_1-AR, α_2-AR).

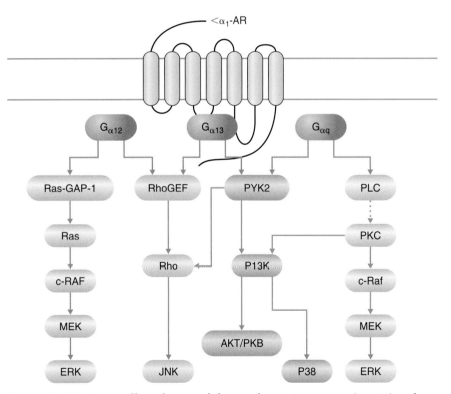

Figure 16-102. Crosstalk pathways of the α_1-adrenergic receptor (α_1-AR) with the MAPK system. MEK-MAP kinase/ ERK kinase.

Catecholamines and Monoamines 1043

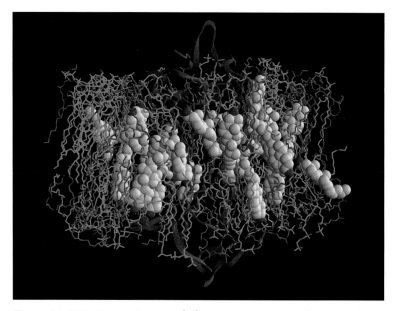

Figure 16-103. G protein–coupled receptors in a membrane environment. Reproduced from http://domino.research.ibm.com/comm/pr.nsf/pages/rscd.bluegene-picaa.html/$FILE/GPCR.jpg.IBM.

G protein–coupled receptors, and they are modeled as they might appear in the cell membrane (Figure 16-103).

Characteristics of the Brain

Table 16-6 gives the volumes of the central nervous system occupied by various structures. The composition of the brain compared to muscle is given in Table 16-7. The lipid component is important in the

Table 16-6
Structures in the Central Nervous System and Their Volumes

	Proportion by Volume (%)
	Human
Cerebral cortex	77
Diencephalon	4
Midbrain	4
Hindbrain	2
Cerebellum	10
Spinal cord	2

Reproduced from http://faculty.Washington.edu/chudler/facts.html.

Table 16-7
Composition of Brain and Muscle

	Skeletal Muscle (%)	Whole Brain (%)
Water	75	77 to 78
Lipids	5	10 to 12
Protein	18 to 20	8
Carbohydrate	1	1
Soluble organic substances	3 to 5	2
Inorganic salts	1	1

Components of brain compared to muscle. Of the solids, neglecting water content, the brain has twice the lipid content of muscle and is nearly half of the solid components, whereas in muscle, lipid is slightly above 15% of the solid matter.
(Reference: McIlwain, H. and Bachelard, H.S., *Biochemistry and the Central Nervous System*, Edinburgh: Churchill Livingstone, 1985.)
Reproduced from http://faculty.Washington.edu/chudler/facts.html.

Table 16-8
Comparison of the Composition of Serum and Cerebrospinal Fluid

Composition of Serum and Cerebrospinal Fluid (CSF)		
	CSF	Serum
Water (%)	99	93
Protein (mg/dl)	35	7000
Glucose (mg/dl)	60	90
Osmolarity (mOsm/l)	295	295
Na (mEq/l)	138	138
K (mEq/l)	2.8	4.5
Ca (mEq/l)	2.1	4.8
Mg (mEq/l)	0.3	1.7
Cl (mEq/l)	119	102
pH	7.33	7.41

(Reference: Fishman, R.A. Cerebrospinal Fluid in Disease of the Nervous System. Philadelphia: Saunders, 1980.)
Reproduced from http://faculty.Washington.edu/chudler/facts.html.

brain. Many of these lipids are sphingomyelin (Figure 5-44) and related structures. The newborn brain weighs 300–400 g, whereas the adult brain is 1300–1400 g. The brain partitions nutrients and needed substrates from the blood by the blood-brain barrier, described in Figure 12-55. A comparison of serum with cerebrospinal fluid is shown in Table 16-8.

Further Reading

Books

Glees, P. *The Human Brain*, Cambridge University Press, 1990.

Teelkeen, A., and Korf, J. (eds.), *Neurochemistry: Cellular, Molecular, and Clinical Aspects*, European Society of Neurochemistry, 1998.

Reviews

De Petrocellis, L., Bisogno, T., Maccarrone, M., Davis, J.B., Finazzi-Agro, A., and Di Marzo, V., "The activity of anandamide at vanilloid VR1 receptors requires facilitated transport across the cell membrane and is limited by intracellular metabolism," *J. Biol. Chem.*, **276**: 12,856–812, 863, 2001.

Guzman, M., Lo Verme, J., Fu, J., Oveisi, F., Blazquez, C., and Piomelli, D., "Oleoylethanolamide stimulates lipolysis by activating the nuclear receptor peroxisome proliferator-activated receptor alpha (PPAR-α)," *J. Biol. Chem.*, **279**: 27,849–827,854, 2004.

Inoue, M., Kobayashi, M., Kozaki, S., Zimmer, A, and Ueda, H., "Nociceptin/orphanin FQ-induced nociceptive responses through substance P release from peripheral nerve endings in mice," *Proc. Natl. Acad. Sci. USA*, **95**: 10,949–10,953, 1998.

Maccarrone, M., Bari, M., Di Rienzo, M., Finazzi-Agro, A., and Rossi, A., "Progesterone activates fatty acid amide hydrolase (FAAH) promoter in human T lymphocytes through the transcription factor Ikaros. Evidence for a synergistic effect of leptin," *J. Biol. Chem.* **278**: 32,726–732,732, 2003.

McFarland, M.J., Porter, A.C., Rakhshan, F.R., Rawat, D.S., Gibbs, R.A., and Barker, E.L., "A role for caveolae/lipid rafts in the uptake and recycling of the endogenous cannabinoid anandamide." *J. Biol. Chem.*, **279**: 41,991–41,997, 2004.

Stucky, C.L., Gold, M.S., and Zhang, X., "Mechanisms of pain," *Proc. Natl. Acad. Sci. USA*, **98**: 11,845–11,846, 2001.

Wang, C., Kaufmann, J.A., Sanchez-Ross, M.G., and Johnson, K.M. "Mechanisms of N-methyl-D-aspartate-induced apoptosis in phencyclidine-treated cultured forebrain neurons," *J. Pharmacol. Exp. Ther.*, **294**: 287–295, 2000.

CHAPTER 17

Microbial Biochemistry

AIDS: A Deadly Viral Disease

Currently, there are millions of people infected with human immunodeficiency virus (HIV). The virus was thought to originate from a west African subspecies of the chimpanzee. The virus apparently jumped species (presumably through a mutation) to infect the human (possibly through a bite by an animal). HIV can generate the syndrome known as **acquired immunodeficiency syndrome (AIDS),** or the virus can be carried but not expressed or minimally expressed; however, a carrier can infect another individual. The virus is spread through the openings in the human body: penis, vagina, anus, and mouth. It is possible to become infected by kissing an infected person or carrier on the mouth. This devastating disease was observed first in the early 1980s, and it is killing huge numbers of people, especially in underdeveloped countries. Because the viral genome has "hot spots," it is likely that chemotherapies can continue successfully for a while but that the virus could mutate to a form that likely will be even more difficult to treat.

There are two strains of the virus, HIV-1 and HIV-2. HIV-1 is the predominant strain, whereas HIV-2 is concentrated in west Africa and rarely seen elsewhere. There are three recognized subtypes of HIV-1: group M, group O, and group N. There are nine genetically distinct subtypes of HIV-1 group M: A, B, C, D, F, G, H, and K. HIV-1 group O is restricted to west-central Africa, and HIV-1 group N is extremely rare. *More than 90% of HIV-1 infections are of HIV-1 group M.*

HIV-1M subtype C predominates in southern and east Africa, India, and Nepal and has caused the worst HIV epidemics. Subtype B is most common in Europe, North and South America, Japan, and Australia; however, there are other subtypes causing new infections in Europe. Subtype A has caused the Russian epidemic, although it predominates in west and central Africa. Subtype D is found only in east and central Africa. Subtype F occurs in central Africa, South America, and eastern Europe. Subtype G has been seen in east and west Africa and central Europe. Subtype H occurs only in central Africa, so far. Subtype J occurs only in central America, and subtype K occurs in the Congo and Cameroon. It is possible for two viruses of different subtypes to infect the same cell, and the mixing of their genetic material sometimes generates a new hybrid virus that can survive if it infects more than one person. It is likely that new forms of the virus will be generated continually. Transmission of the virus from mother to child seems to be more common primarily with subtype C than with subtype D.

Testing for HIV infection is done by **enzyme-linked immunosorbent assay (ELISA)**, which uses specific antibodies to detect HIV RNA or HIV antigens in blood plasma or serum. Not every subtype is detected, especially the rarer HIV-2 and HIV-1 group O types.

HIV is a retrovirus, indicating that its genome consists of RNA rather than DNA—actually, two identical strands of RNA. After infection of the human cell, the RNA is converted to DNA by the enzyme reverse transcriptase, encoded in the viral RNA, and the transcribed DNA is incorporated into the host cell genome with the aid of the retroviral protein integrase. The DNA "gene" acts in an uncontrolled fashion, recruiting the host cell's machinery, to produce more RNA and the encoded proteins needed for assembly of new virus particles.

Initial infection with HIV passes through an acute phase that can last some 12 weeks and then enters a chronic phase lasting more than 8 years, when full-blown AIDS is developed, usually resulting in death in a few years. These stages are characterized by the levels of multiplication of the virus and the CD4 cell count, shown in Figure 17-1.

Two diagrams of the virus particle are shown in Figure 17-2. The "knobs" or "spikes" on the surface of the virus are constructed from two proteins encoded in the viral RNA gene *env*, standing for envelope. The proteins are glycoproteins, gp120 (120 indicates the molecular weight of

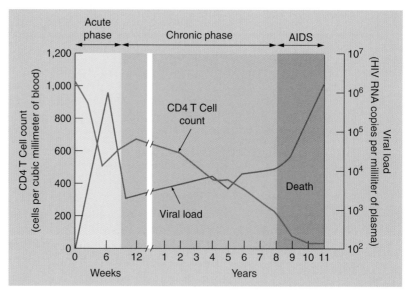

Figure 17-1. HIV infection begins with a sharp rise of virus in the blood *(orange line)* and a consequent drop in CD4 T cells *(blue line)*. The immune system soon recovers somewhat, however, and keeps HIV levels fairly steady for several years. Eventually, though, the virus gains the upper hand. AIDS is diagnosed when the CD4 T cell level drops below 200 cells per cubic millimeter of blood, or when opportunistic infections arise. Redrawn from Wikipedia, http://en.wikipedia.org/wiki/aids.

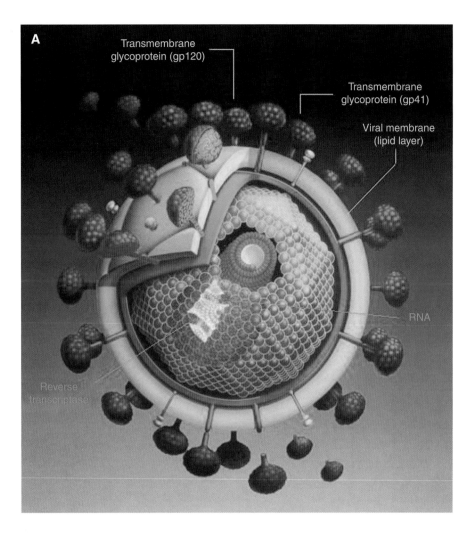
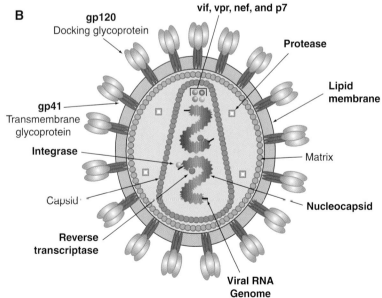

Figuro 17-2. **A.** Diagram of HIV. **B.** Diagram of an HIV particle indicating the constituent proteins encoded by the viral RNA genome. Part A reproduced from Virco Lab, Inc., at http://www.vircolab.com/content/backgrounders/www.vircolab.com/hiv_virus.gif. Part B reproduced from http://en.wikipedia.org/wiki/HIV.

120 kDa) and gp41. Gp120 forms the knob, and gp41 forms the stem. The knobs enable the virus to attach to the host cell and fuse with it. The viral membrane is a lipid layer. Beneath the viral membrane is a layer made up of a membrane-associated protein (matrix protein), MAp17, encoded by the **gag** (group-specific antigen) gene of the RNA genome. In addition to the matrix protein, p17, **ENV** codes for p24, constituting the viral capsid, and for p6 and p7, which are nucleocapsid proteins. Reverse transcriptase is encoded by the RNA gene **pol** (polymerase). Reverse transcriptase converts the two identical genomic RNAs into corresponding DNA strands.

There are at least six other RNA genes in the viral genome that encode single proteins: *tat, rev, nef, vif, vpr,* and *vpu*. Each gene encodes a single protein that has the same name as its gene. The protein **TAT** stands for transactivator of transcription (Figure 17-3), and it contains from 86 to 101 amino acid residues, depending on the subtype. HIV RNA contains a hairpin loop (Figure 17-4) that prevents full transcription but allows the TAT protein to be produced. TAT phosphorylates some cellular factors, which eliminate the hairpin structure, and allows the transcription of the full HIV DNA. Production of TAT is key to the enhanced transcription of HIV DNA. The protein **REV** (for regulator of virion) allows the export of small RNAs from the nucleus to the cytoplasm, which, in the absence of REV, would be spliced, preventing the formation of HIV structural proteins and allowing only the formation of smaller regulatory proteins. The REV-binding element is shown in Figure 17-5. The **NEF** protein (negative regulatory factor; Figure 17-6) is expressed early in the life cycle of the virus, and it promotes T cell activation and a persistent state of infection. In addition, it down regulates expression of some surface molecules (MHC I, MHC II, and CD4 and CD28 on CD$^+$ T cells) on the host cell that are important for host immune function. **VIF,** the viral infectivity factor, has a molecular weight of 23 kDa and is essential for

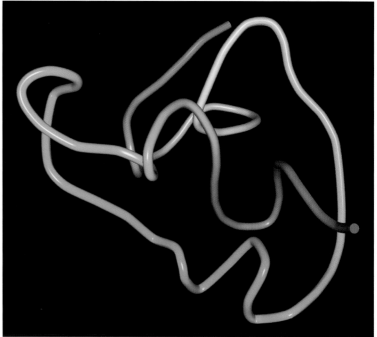

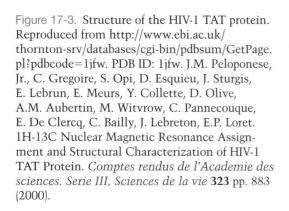

Figure 17-3. Structure of the HIV-1 TAT protein. Reproduced from http://www.ebi.ac.uk/thornton-srv/databases/cgi-bin/pdbsum/GetPage.pl?pdbcode=1jfw. PDB ID: 1jfw. J.M. Peloponese, Jr., C. Gregoire, S. Opi, D. Esquieu, J. Sturgis, E. Lebrun, E. Meurs, Y. Collette, D. Olive, A.M. Aubertin, M. Witvrow, C. Pannecouque, E. De Clercq, C. Bailly, J. Lebreton, E.P. Loret. 1H-13C Nuclear Magnetic Resonance Assignment and Structural Characterization of HIV-1 TAT Protein. *Comptes rendus de l'Academie des sciences. Serie III, Sciences de la vie* **323** pp. 883 (2000).

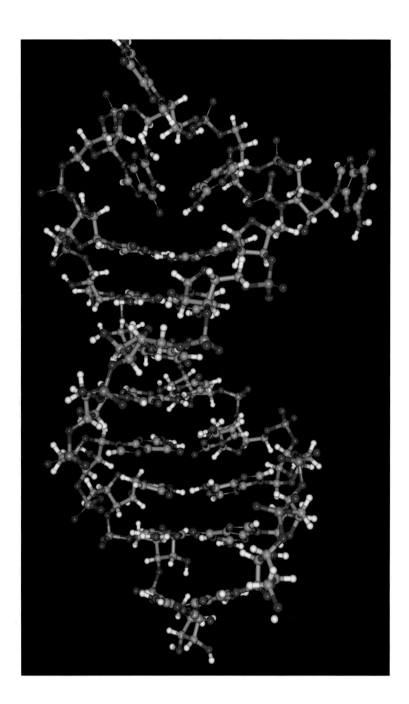

Figure 17-4. SL3 hairpin from the packaging signal of HIV-1 determined by NMR. Reproduced from PDB ID: 1bn0. L. Pappalardo, D.J. Kerwood, I. Pelczer, P.N. Borer. Three-dimensional Folding of an RNA Hairpin Required for Packaging HIV-1. *The Journal of Molecular Biology* **282** pp. 801 (1998).

viral replication. **VPR,** the viral protein R, is 10 kDa and regulates the nuclear import by the host cell of the HIV-1 preintegration complex. It also induces cell cycle arrest in proliferating cells. **VPU** (VPX in HIV-2 infection), the viral protein U factor, is involved in viral budding that enhances release of the virion from the cell. Integrase is expressed by the HIV genome, and this enzyme is responsible for integrating the DNA transcripts from the viral RNA into the host genome. A protease also is expressed, and it cleaves the precursors of *gag* and *pol* gene products (proteins) into their final active forms.

The viral genomes of HIV-1 and HIV-2 are diagrammed in Figure 17-7. The complete HIV genome is encoded on one long strand of RNA, and there are two separate identical strands of RNA in each viral particle. *Tat,*

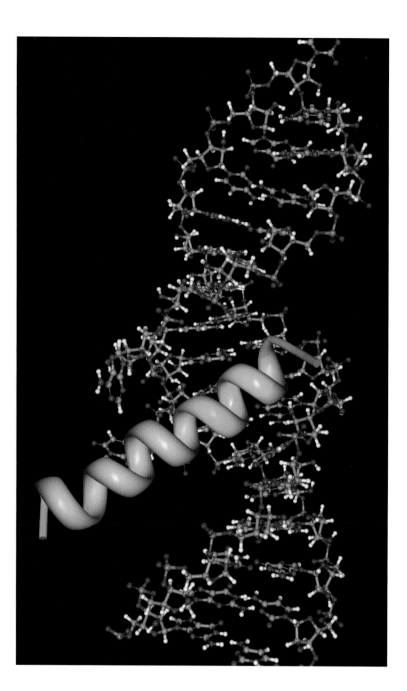

Figure 17-5. REV response element (RRE) RNA complexed with REV peptide determined by NMR. Reproduced from http://www.rcsb.org/pdb/cgi/explore.cgi?pdbId=1etf. PDB ID: 1etf. J.L. Battiste, H. Mao, N.S. Rao, R. Tan, D.R. Muhandiram, L.E., Kay, A.D. Frankel, J.R. Williamson. Alpha Helix-RNA Major Groove Recognition in an HIV-1 REV Peptide-RRE RNA Complex. *Science* **273** pp. 1547 (1996).

rev, and *nef* are expressed as early gene products (Figure 17-8). The **long terminal repeats (LTRs)** at the ends of the RNA strands are "sticky," and integrase uses the LTR to insert the HIV genome (as reverse-transcribed DNA) into the host cell DNA. LTRs also act as promoters when they are integrated into the host genome, influencing the transcription of DNA, and the LTRs have protein-binding sites (for constitutive SP-1, Lef, Ets, and inducible transcription factors NF-AT and AP-1) involved in RNA initiation.

Reverse transcription is unique to retroviruses. HIV reverse transcriptase has two subunits, p66 and p51 (p51 is a truncated form of p66). The structure of reverse transcriptase is shown in Figure 17-9. Regions of the enzyme are denoted as "fingers," "palm," and "thumb," in addition to

(Text continues on p. 1056.)

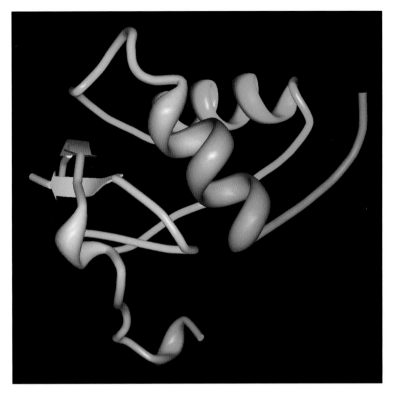

Figure 17-6. HIV-1 NEF protein, unliganded core domain. Reproduced from http://www.rcsb.org/pdb/explore/images.do?structureId=1avv. PDB ID: 1avv. S. Arold, P. Franken, M.P. Strub, F. Hoh, S. Benichou, R. Benarous, C. Dumas. The Crystal Structure of HIV-1 Nef Protein Bound to the Fyn Kinase SH3 Domain Suggests a Role for This Complex in Altered T Cell Receptor Signaling. *Structure* **5** pp. 1361 (1997).

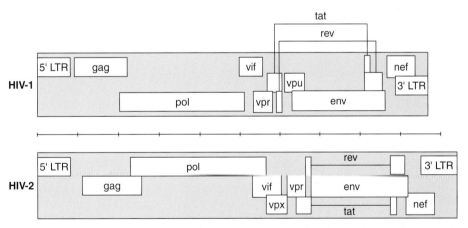

Figure 17-7. Genomes of HIV-1 and HIV-2. **gag** (coding for the viral capsid proteins); **pol** (notably, coding for **reverse transcriptase**); **gag** and **pol** together can be expressed in one long strand called **gag-pol**); **env** (coding for HIV's envelope-associated proteins). And the regulatory genes: **tat, rev, nef, vif, vpr, vpu** (N.B. not present in HIV-2), **vpx** (N.B. not present in HIV-1). *LTR,* long terminal repeat. Reproduced from http://hiv-web.lanl.gov/content/immunology/pdf/2000/intro/GenomeMaps.pdf at http://www.mcld.co.uk/hiv/?q=HIV%20genome.

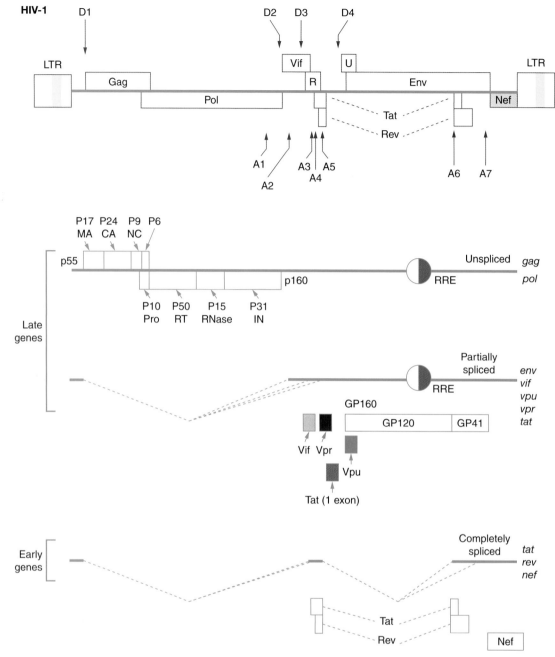

Figure 17-8. Expression of the HIV-1 genome. *At top*, a schematic view of the linear proviral genome, with coding sequences of the HIV genes depicted as open rectangles. Potential splice donor sites are designated as D1–4 and potential splice acceptor sites as A1–7. The three major classes of viral RNA and the viral proteins that they encode are shown below. The *Gag*, *Pol*, and *Env* genes are expressed as precursor polyproteins, which are then cleaved to yield mature viral proteins. *CA*, capsid; *gp*, glycoprotein; *IN*, integrase; *LTR*, long terminal repeat; *MA*, matrix; *NC*, nucleocapsid; *p*, protein; *Pro*, protease; *RNase*, ribonuclease H; *RRE*, Rev-Responsive Element; *RT*, reverse transcriptase.

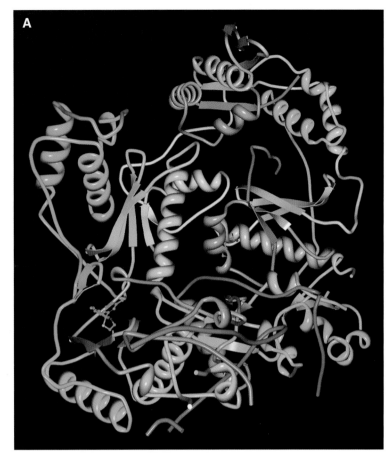

Figure 17-9. **A.** Crystal structure of HIV-1 reverse transcriptase. **B.** Model of HIV-1 reverse transcriptase with NNRTI, DNA primer/template, and incoming dNTP. The NNRTI from the structure described here (CP-94,707) is shown superimposed on the ternary complex of HIV-1 reverse transcriptase bound to DNA substrates. The incoming dNTP *(green)* and CP-94,707 *(yellow)* are in space-filling representation. The DNA primer *(light gray)* and template *(dark gray)*; fingers *(blue)*, palm *(purple)*, thumb *(green)*, connection *(yellow)*, and RNaseH *(red)* subdomains of the p66 subunit of HIV-1 reverse transcriptase; and p51 subunit *(white)* are in ribbons representation. The region circled includes the polymerase active site and NNRTI-binding pocket. **C.** The structure of a RNA/DNA hybrid: a substrate of the ribonuclease activity of HIV-1 reverse transcriptase. *NNRTI,* nonnucleotide reverse transcriptase inhibitor. Part A reproduced from http://www.rcsb.org/pdb/cgi/explore.cgi?pdbId=1c0t. PDB ID: 1c0t. J. Ren, R.M. Esnouf, A.L. Hopkins, D.I. Stuart, D.K. Stammers. Crystallographic Analysis of the Binding Modes of Thiazoloisoindolinone Nonnucleoside Inhibitors to HIV-1 Reverse Transcriptase and Comparison with Modeling Studies. *The Journal of Medicinal Chemistry* **42** p. 3845 (1999). Part B reproduced from http://www.rcsb.org/pdb/cgi/explore.cgi?pdbId=1c0t. J.D. Pata et al., "Structure of HIV-1 reverse transcriptase bound to an inhibitor active against mutant reverse transcriptases resistant to other nonnucleoside inhibitors," *Proc. Natl. Acad. Sci. USA,* 101: 10548–10553, 2004.

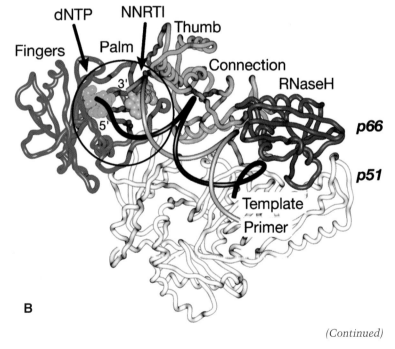

(Continued)

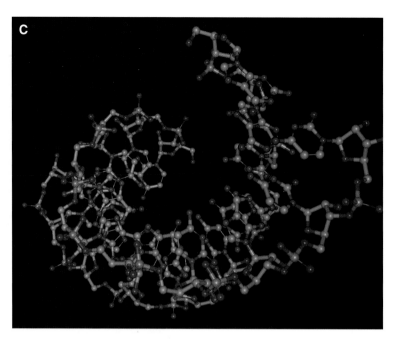

Figure 17-9, cont'd Part C reproduced from http://www.rcsb.org/pdb/cgi/explore.cgi?pdbId=1fix. PDB ID: 1fix. N.C. Horton, B.C. Finzel. The Structure of an RNA/DNA Hybrid: A Substrate of the Ribonuclease Activity of HIV-1 Reverse Transcriptase. *The Journal of Molecular Biology* **264** p. 521 (1996).

the RNaseH domain of the enzyme, as shown in Figure 17-9B. Figure 17-9C shows the RNA–DNA hybrid that is a substrate of the ribonuclease (RNase) activity of reverse transcriptase. This is clarified by a summary of the reverse transcriptase mechanism shown in Figure 17-10.

The viral integrase enzyme catalyzes the insertion of the transcribed viral RNA (DNA) into the host cell DNA. This process and the mechanism are described in Figure 17-11. Viral DNA stays associated with both the viral proteins and the host cellular proteins as a preintegration complex. Viral integrase is an important ingredient of the preintegration complex. Integrase performs the critical cutting of host DNA and the subsequent joining events that cause the viral DNA to become part of the host cell DNA. The information from the integrated viral DNA generates transcribed RNAs, which become translated into new components of new virions that eventually become released from the host cell when it lyses. Structures of the catalytic core domain and the N- and C-terminal domains of integrase are shown in Figure 17-12.

The HIV protease is an aspartyl protease (similarly to the caspases and renin) with an aspartyl residue in the active site. The enzyme attacks the carbonyl group of the peptide bond, cleaving the peptide substrate, and the enzyme is regenerated with one molecule of water (Figure 17-13). The protease cleaves the GAG–POL polypeptide to give rise to various products that are incorporated into the newly forming virion. The products of the GAG polyprotein are p17, p24, p9, and p6, and the products of the POL polyprotein are protease, reverse transcriptase, and integrase. Cleavage of the ENV polyprotein is carried out by a host cellular protease to

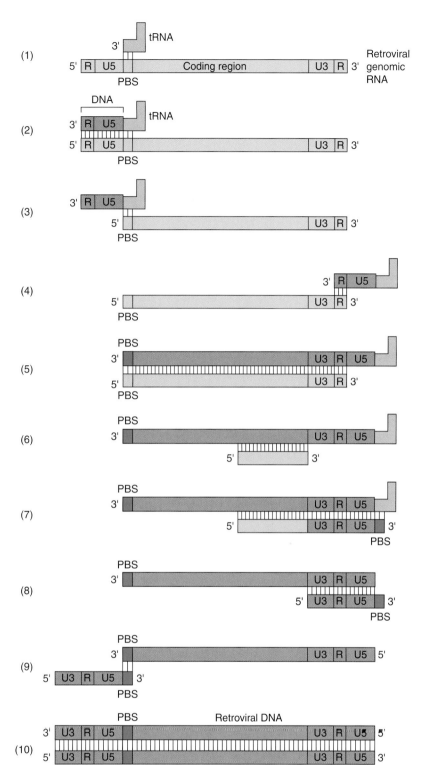

Figure 17-10. Mechanism of reverse transcription. The entire process is catalyzed by reverse transcriptase, which has both DNA polymerase and RNase H activities.
(1) A retrovirus-specific cellular tRNA hybridizes with a complementary region called the primer-binding site (PBS).
(2) A DNA segment is extended from tRNA based on the sequence of the retroviral genomic RNA. (3) The viral R and U5 sequences are removed by RNase H. (4) First jump: DNA hybridizes with the remaining R sequence at the 3' end. (5) A DNA strand is extended from the 3' end. (6) Most viral RNA is removed by RNase H. (7) A second DNA strand is extended from the viral RNA. (8) Both the tRNA and the remaining viral RNA are removed by RNase H. (9) Second jump: The PBS region of the second strand hybridizes with the PBS region of the first strand. (10) Extension on both DNA strands. LTR, long terminal repeat. Redrawn with permission from http://www.web-books.com/MoBio/Free/Ch4J1.htm.

AIDS: A Deadly Viral Disease

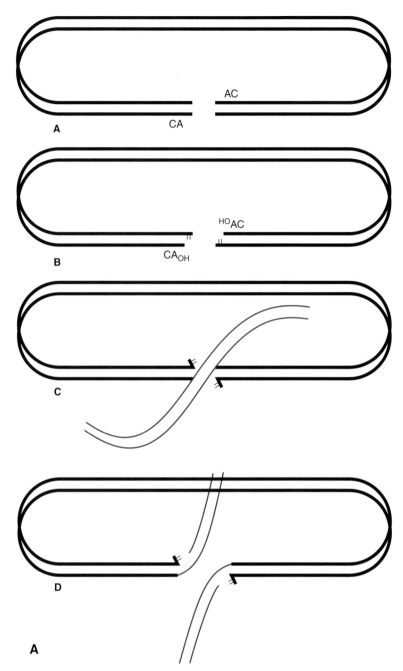

Figure 17-11. **A.** DNA cutting and joining steps in retroviral integration. *A,* The viral DNA *(orange)* made by reverse transcription is linear and blunt ended. *B,* In the first step of the integration process, 3′-end processing, two nucleotides are cleaved from each 3′-end of the viral DNA. *C,* In the next step, DNA strand transfer, the hydroxyl groups at the 3′-ends of the processed viral DNA attack a pair of phosphodiester bonds in the target DNA *(blue).* The spacing between the sites of attack on each target DNA strand is fixed and characteristic for each retrovirus. *D,* The resulting integration intermediate is redrawn to clarify the connections between viral and target DNA. Integrase is responsible for both the 3′-processing and the DNA strand transfer reactions that give rise to the integration intermediate. Completion of DNA integration requires removal of the two unpaired nucleotides at the 5′-ends of the viral DNA, filling in the single-strand gaps between host and viral DNA by a DNA polymerase, and finally ligation. These steps are likely to be carried out by cellular enzymes.

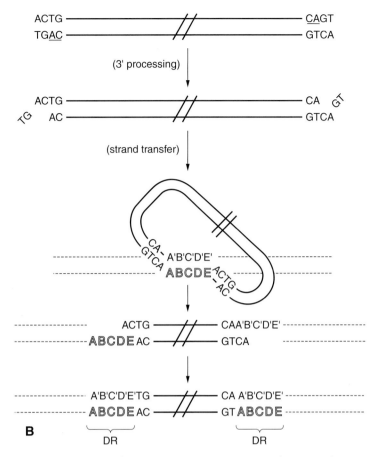

Figure 17-11, cont'd **B.** HIV-1 IN activities. A schematic diagram of HIV-1 IN activities depicts the double-stranded DNA viral genome at the top as parallel black lines with the terminal nucleotides CAGT. The conserved 3' CA dinucleotide is underlined at each viral end. IN first acts in the cytoplasm to remove the two 3' nucleotides (3' processing), leaving a 2-nt overhang at each 5' end. In the nucleus, IN mediates a concerted integration (strand transfer) by ligating each 3' end of the viral DNA (looped structure) to the host DNA *(striped lines)*. This generates a "gapped intermediate" with free viral 5' ends that are repaired to generate the fully integrated provirus. The characteristic HIV-1 5-bp staggered strand transfer is depicted by the letters A–E in the target DNA, and the resulting 5-bp direct repeats (DR) of host DNA flanking the provirus are indicated. *IN,* integrase. Part B reproduced from J.C. Chen, J. Krucinski, L.J. Miercke, J.S. Finer-Moore, A.H. Tang, A.D. Leavitt and R.M Stroud, "Crystal structure of the HIV-1 integrase catalytic core and C-terminal domains: a model for viral DNA binding," *PNAS USA*, 97: 8233–8238.

yield gp120 and gp41. The overall scheme is shown in Figure 17-14. Views of the structure of HIV protease are shown in Figure 17-15.

Inhibitors have been designed for the HIV protease as a possible therapeutic intervention. The design of inhibitors is similar to those in the normal substrate, with groups added to provide steric hindrance. One such inhibitor is **Saquinavir,** shown in Figure 17-16. Combining a protease inhibitor and an inhibitor of reverse transcriptase proves to be an effective method for controlling virus propagation in patients. Gp120

(Text continues on p. 1064.)

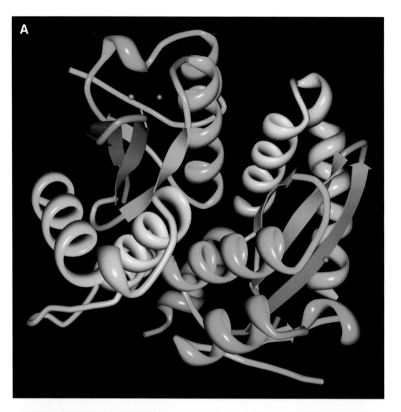

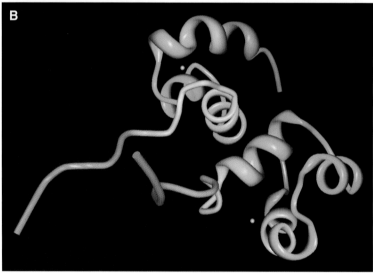

Figure 17-12. **A.** HIV-1 integrase core domain. **B.** Solution structure of H12C mutant of the N-terminal domain of HIV-1 integrase complexed to cadmium. Structure determined by NMR. **C.** C-terminal DNA-binding domain of HIV-1 integrase. Determined by NMR. **D.** Electrostatic potential map of the HIV-1 IN$^{52\text{-}288}$ dimer. *(A)* Potentials range from – 15/ kT *(red)* to + 15/kT *(blue)*. The strip of positive charge *(blue)* coursing up and to the left contains residues from both monomers of the dimer, K211, K215, and K219 from monomer A and K159, K186, R187, K188 from monomer B. The active site pocket of monomer B (*) includes catalytic residues D64, D116, and E152. *(B)* An 18-bp viral DNA end is modeled onto the IN dimer with the positively charged residues in contact with the DNA phosphodiester backbone. The adenine base of the conserved viral 3' CA dinucleotide contacts K159. *IN*, integrase. Part A reproduced from PDB ID: 1biz. Y. Goldgur, F. Dyda, A.B. Hickman, T.M. Jenkins, R. Craigie, D.R. Davies. Three New Structures of the Core Domain of HIV-1 Integrase: An Active Site That Binds Magnesium. *Proceedings of the National Academy of Sciences of the United States of America* **95** p. 9150 (1998). Part B reproduced from http://www.rcsb.org/pdb/cgi/ explore.cgi?pdbId=1wjf. PDB ID: 1wjf. M. Cai, Y. Huang, M. Caffrey, R. Zheng, R. Craigie, G.M. Clore, A.M Gronenborn. Solution Structure of the His 12–>. Cys Mutant of the N-terminal Zinc Binding Domain of HIV-1 Integrase Complexed to Cadmium. *Protein Science* **7** p. 2669 (1998).

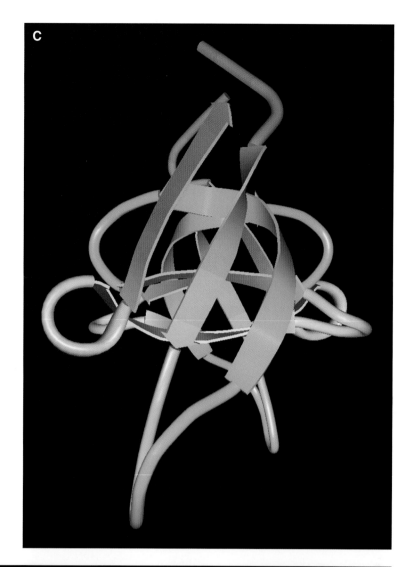

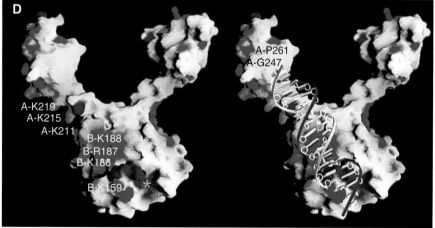

Figure 17-12, cont'd Part C reproduced from http://www.rcsb.org/pdb/cgi/explore.cgi?pdbId=1qmc. PDB ID: 1qmc. A.P. Eijkelenboom, R. Sprangers, K. Hard, R.A. Puras Lutzke, R.H. Plasterk, R. Boelens, R. Kaptein. Refined Solution Structure of the C-terminal DNA-binding Domain of Human Immunovirus-1 Integrase. *Proteins* **36** p. 556 (1999). Part D reproduced from J.C.-H. Chen, J. Krucinski, L.J.W. Miercke, J.S. Finer-Moore, A.H. Tang, A.D. Leavitt, R.M. Stroud. Crystal Structure of the HIV-1 Integrase Catalytic Core and C-terminal Domains: A Model for Viral DNA Binding. *Proceedings of the National Academy of Sciences of the United States of America* **97** p. 8233 (2000), Figure 3, and from http://www.pnas.org/cgi/content/full/97/15/8233/F4.

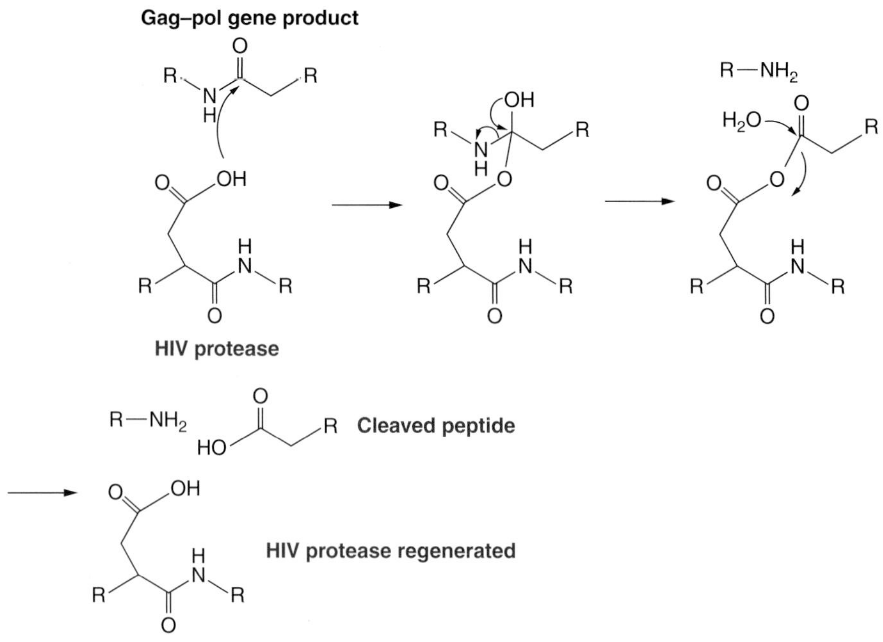

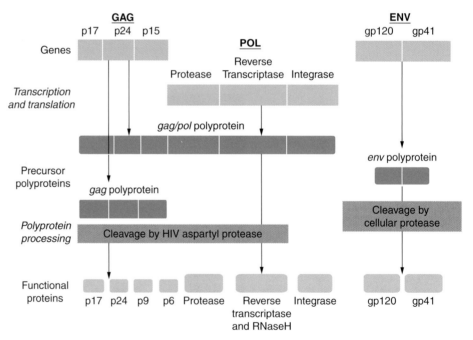

Figure 17-13. HIV protease mechanism. Shown here is the cleavage of the gag-pol gene product. The GAG polypeptide, upon cleavage by protease, gives rise to P17, P24, P9, and P6 polypeptides while the cleavage of the POL polypeptide gives rise to protease, reverse transcriptase, and integrase.

Figure 17-14. Overview of the maturation of polyproteins of the *GAG*, *POL*, and *ENV* genes. GAG and POL polyproteins are cleaved to final products by HIV protease, whereas maturation of the ENV polyprotein is done by a host cellular protease.

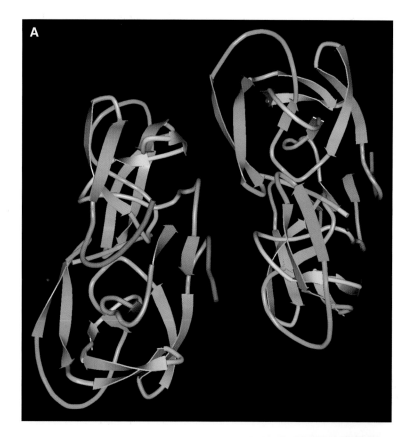

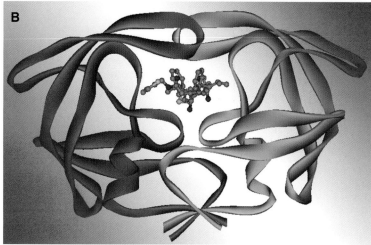

Figure 17-15. **A.** Crystallographic analysis of human immunodeficiency virus 1 protease with an analog of the conserved CA-P2 substrate: interactions with frequently occurring glutamic acid residue at P2' position of substrates. **B.** HIV-1 protease showing its dimeric structure. Part A reproduced from http://www.ebi.ac.uk/msd-srv/msdlite/images/1a8k600.jpg. PDB ID:1a8k. I.T. Weber, J. Wu, J. Adomat, R.W. Harrison, A.R. Kimmel, E.M. Wondrak, J.M. Louis. Crystallographic Analysis of Human Immunodeficiency Virus 1 Protease with an Analog of the Conserved CA-p2 Substrate: Interactions with Frequently Occurring Glutamic Acid Residue at P2' Position of Substrates. *European Journal of Biochemistry* **249** p. 523 (1997).

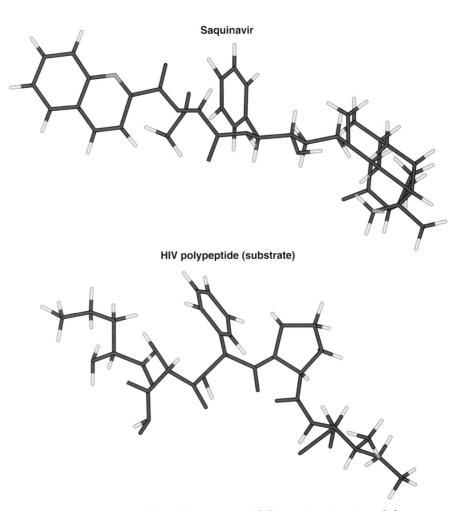

Figure 17-16. Structure of the HIV protease inhibitor, Saquinavir, and the normal HIV polypeptide (substrate).

and gp41 are the docking glycoprotein and the transmembrane glycoprotein that constitute the knobs or spikes on the surface of the virus (Figure 17-2B). The C5 domain of gp120 is thought to interact directly with gp41, and the structure of this domain is shown in Figure 17-17. The structure of a trimer of gp41 is shown in Figure 17-18. One spike is made of three gp120 molecules and three gp41 molecules (Figure 17-19). The overall HIV cycle is shown in Figure 17-20. The major events are the binding of the virus particle to the host cell followed by penetration, endocytosis, and uncoating to liberate the RNA viral genome. HIV reverse transcriptase produces a double strand of DNA from the RNA template, and the DNA is transported to the host cell nucleus, where it is integrated into the host cell DNA. mRNAs are produced through the usual mechanism and translated by the cellular machinery to form the constituents of new viral particles. These are assembled and released from the host cell by budding. Eventually, the host cell fills with virus particles and lyses to release more virions (virus particles), which can infect other host cells.

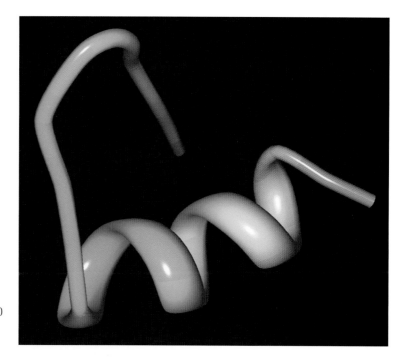

Figure 17-17. Solution structure of the HIV gp120 C5 domain. Reproduced from http://www.rcsb.org/pdb/cgi/explore.cgi?pdbId=1meq. PDB ID:1meq. L. Guilhaudis, A. Jacobs, M. Caffrey. Solution Structure of the HIV gp 120 C5 Domain. *European Journal of Biochemistry* **269** pp. 4860.

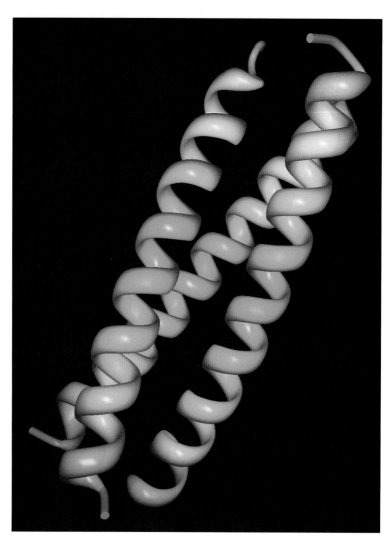

Figure 17-18. Trimerization specificity in HIV gp41: analysis with a GCN4 leucine zipper model. Reproduced from http://www.rcsb.org/pdb/cgi/explore.cgi?pdbId=1ce0. PDB ID: 1ce0. W. Shu, H. Ji, M. Lu. Trimerization Specificity in HIV-1 gp41: Analysis with a GCN4 leucine zipper model. *Biochemistry* **38** pp. 5378 (1999).

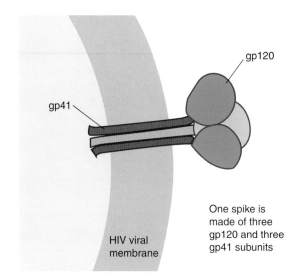

Figure 17-19. Structure of an HIV spike. There is some confusion over the number of spikes on an HIV particle. Estimates vary from 10 to 72.

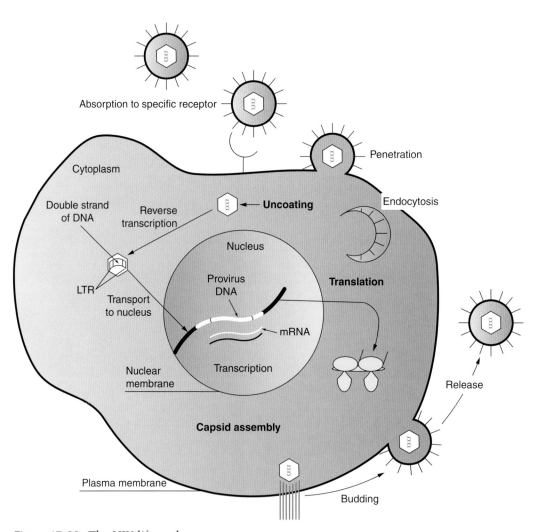

Figure 17-20. The HIV life cycle.

1066   CHAPTER 17   Microbial Biochemistry

HIV infection eventually destroys the immune system, opening the infected person to undefended infections. Within 2 to 4 weeks of exposure to the virus, the acute infection will resemble flulike symptoms, and **seroconversion** from HIV negative to HIV positive will occur within 3 months. Although infected carriers can transmit the infection to other people, carriers may not show any symptoms for as long as 10 years. During this period, there is a weakening of the immune system, and eventually there is a diagnosis of AIDS. When the infection is acute, there is a progression to asymptomatic HIV infection followed by early symptomatic infection and full-blown AIDS. There is a certain subset of patients (nonprogressors) who develop AIDS very slowly or not at all. HIV infection may be diagnosed without occurrence of any symptoms; however, the infection may be followed by such seemingly nonspecific symptoms as sore throat, muscular stiffness, headache, diarrhea, swollen lymph glands, fever, fatigue, rash, and vaginal yeast infection. There are a variety of opportunistic infections that can occur, as well as other conditions, such as HIV dementia, HIV lipodystrophy (refers to disturbed fat metabolism, including wasting of the face and extremities and accumulation of abdominal fat and breast enlargement), chronic wasting, and malignancies, one of which is Kaposi's sarcoma (Figure 17-21).

Treatment of AIDS is generally based on the combination of two inhibitors of HIV reverse transcriptase and HIV protease. These drugs reduce the viral load, increase $CD4^+$ T cell count (Figure 17-1), and prolong life. Drugs used to treat HIV-1 are not necessarily effective against HIV-2. There are many drugs now available, including nucleoside reverse transcriptase inhibitors, nonnucleoside reverse transcriptase inhibitors, and protease inhibitors, as shown in Table 17-1. Examples of these drugs are **AZT** (a nucleoside reverse transcriptase inhibitor) and **nevirapine (nonnucleoside reverse transcriptase inhibitor, NNRTI),** shown in Figure 17-22, and the protease inhibitor Saquinavir, shown in Figure 17-16. The binding site for NNRTIs is shown in Figure 17-9B. Currently, attempts are being made to develop vaccines against various strains of HIV.

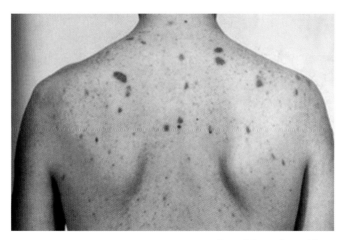

Figure 17-21. Kaposi's sarcoma. Reproduced from http://history.nih.gov/NIHInOwnWords/docs/page_01d.html. The National Institutes of Health.

Table 17-1
Drugs Used to Treat HIV Infections (AIDS)

Drug	Brand Name	Manufacturer
Nucleoside reverse transcriptase inhibitors (NRTIs)		
Zidovudine (AZT)	Retrovir	Glaxo SmithKline
Didanosine (ddI)	Videx	Bristol-Myers Squibb
Zalcitabine (ddC)	Hivid	Roche
Stavudine (d4T)	Zerit	Bristol-Myers Squibb
Lamivudine (3TC)	Epivir	Glaxo SmithKline
Abacavir (ABC)	Ziagen	Glaxo SmithKline
Nucleotide reverse transcriptase inhibitors (NtRTIs)		
Tenofovir disoproxil fumarate (TDF)	Viread	Gilead Sciences
Non-nucleoside reverse transcriptase inhibitors (NNRTIs)		
Nevirapine	Viramune	Roxane/Boehringer Ingelheim
Delavirdine	Rescriptor	Pharmacia
Efavirenz	Sustiva, Stocrin	DuPont Merck
Protease inhibitors (PIs)		
Saquinavir	Fortovase, Invirase	Roche
Ritonavir	Norvir	Abbott
Indinavir	Crixivan	Merck
Nelfinavir	Viracept	Agouron/Pfizer
Amprenavir	Agenerase	Glaxo SmithKline
Lopinavir (with ritonavir)	Kaletra	Abbott

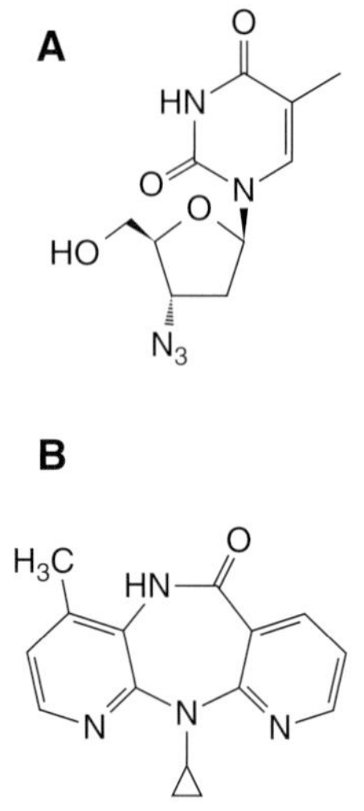

Figure 17-22. **A.** Structure of the nucleoside reverse transcriptase inhibitor, AZT(3′-azido-3′-deoxythymidine). **B.** Structure of the nonnucleoside reverse transcriptase inhibitor (NNRTI) nevirapine.

Other Viruses of Current Interest

Viruses that infect humans take a variety of forms, but the basic biology of the infection of a cell is similar among them. They attach to a host cell, become endocytosed into the cell, and avail their genome—either RNA (reverse transcribed into DNA in retroviruses) or DNA—into the host cell nucleus. DNA from the virus is incorporated into the DNA of the host. The transcribed mRNAs are processed by the host cell to form the components of the virus particle. The new virion is extruded from the host cell, usually by a budding process (in some cases, a lytic process without budding), and the virus circulates to infect another host cell. Meanwhile, the copied virus particles accumulate in the original host cell and eventually cause the lysis of that cell, releasing many virus particles to proliferate the infection.

Viruses are not living organisms in that they, by themselves, do not carry out metabolism or reproduce, functions that depend on their infection of a host cell. In this sense, they are parasitic symbionts. It is not known how viruses evolved, although some believe they derived from cells and gradually lost much genomic information. Viruses, at least those with RNA genomes, may be relics of the switch from genetics based on RNA (the probable initial genetic form) to genetics based on DNA that required the activity of reverse transcriptase. In any case, there is a huge variety of viruses. Those that infect man are generally spherical with a structure based on **icosahedral** (a polyhedron with 20 faces; a many-sided three-dimensional hexagonal shape composed of many small triangles) symmetry, like the geodesic dome of Fuller. Herpes virus is an example (Figure 17-23).

Viruses, in addition to their genetic information, contain a membrane envelope and capsid consisting of a layer of fatty acids that derives from the host cell. The viral ligands are spikes that protrude from the viral surface. These are keys to the recognition of the cell they will invade. The adenovirus, causing sore throat and serious infections of the lung in humans, is shown in various formats in Figure 17-24. The poliovirus particle (causing poliomyelitis) is shown in Figure 17-25. The virus causing the common cold (minor lung infection) is the rhinovirus, and its x-ray structure is shown in Figure 17-26A. Figure 17-26B shows regions of the virus attaching to the binding site of the host cell receptor. The icosahedral rhinovirus particle is 300 angstroms in diameter with a

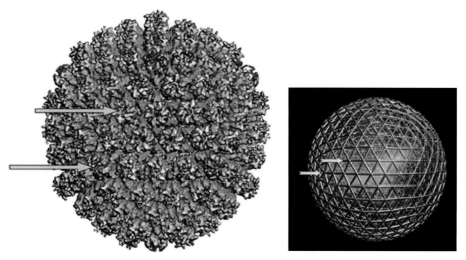

Figure 17-23. The herpes virus *(left)* and icosahedral symmetry, the key to encoding the proteins that repeat. Reproduced from http://pr.caltech.edu/periodicals/EandS/articles/LXVIII1/viruses.html. Photo by Z. Hong Zhou, U. of Texas Medical School, Houston. Appears in David Baltimore, Viruses, Viruses, Viruses. Engineering & Science, 626-395-3630, published by the California Institute of Technology.

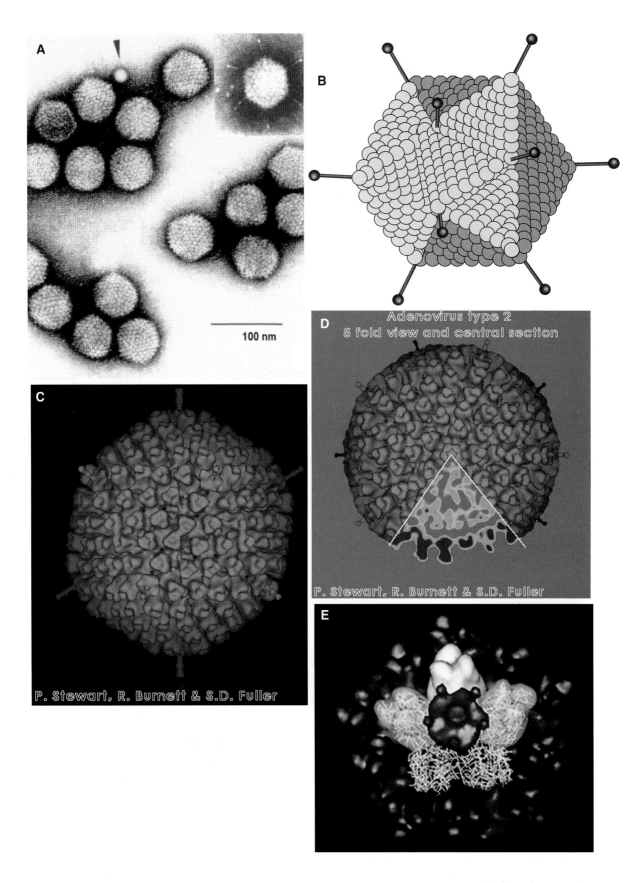

For legend see opposite page.

Figure 17-24. Various images of the adenovirus. **A.** An electron micrograph of human adenovirus. **B.** A diagram of an adenovirus virion made up of 252 identical protein building blocks wrapped around a molecule of DNA. Spikes project from the viral particle, which help the virus attach to a host cell (throat or lung). **C.** A twofold view of adenovirus type 2. **D.** A fivefold view of adenovirus type 2 with a central section. **E.** The penton region of adenovirus showing the fitting of hexon polypeptide into cryo-em map. Part A courtesy of C. Buchen-Osmond. Part B reproduced from http://www.che.utah.edu/images/slideShow/popUpSlideShow/popUpSlideShow.php?directory=.&currentPic=6. University of Utah, Department of Chemical Engineering, Salt Lake City, UT. Part C reproduced from http://www-db.embl-heidelberg.de/jss/servlet/de.embl.bk.wwwTools.GroupLeftEMBL/ExternalInfo/fuller/ad2fb.gif. Fuller Group, c/o Division of Structural Biology, Wellcome Trust Centre for Human Genetics, Henry Wellcome Building for Genomic Medicine, University of Oxford, UK. Part D reproduced from http://www-db.embl-heidelberg.de/jss/servlet/de.embl.bk.wwwTools.GroupLeftEMBL/ExternalInfo/fuller/ad5f_seca.gif. Fuller group left EMBL Heidelberg. Contact Division of Structural Biology, Wellcome Trust Centre for Human Genetics, Henry Wellcome Building for Genomic Medicine, University of Oxford, Roosevelt Drive, Oxford OX3 7BN, UK. Part E reproduced from http://www.ncbi.nim.nih.gov/ICTVdb/Wintkey/Images/3d_adenovirus.jpg.

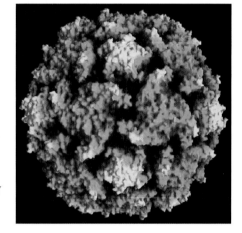

Figure 17-25. Three-dimensional reconstruction of a poliovirus particle. Polio is a small RNA-containing virus, a retrovirus. This structure was determined by bombarding crystals of virus particles with x-rays (x-ray diffraction).

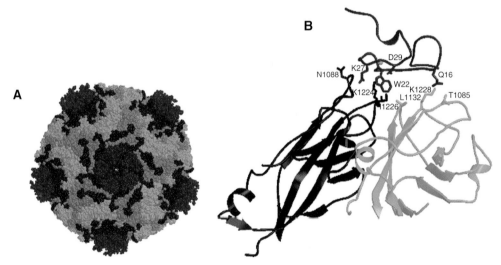

Figure 17-26. **A.** Structure of HRV2-V3 (human rhinovirus 2-V3) complex shown in a space-filling representation. Viral subunits VP1, VP2, and VP3 are in *blue, cyan* and *black*, respectively, and the bound V_r; modules are displayed in *red*. V3 binds outside the "canyon" and close to the virus fivefold vertex. The proximity between the N-terminal and C-terminal ends of neighbor V3 modules is apparent. **B.** Intermolecular interactions between the V3 module of the receptor *(red)* and the viral VP1 proteins from two adjacent protomers *(dark blue* and *turquoise*, respectively). The Ca^{2+} ion is shown as a *yellow sphere* between D29 and 127 of the receptor in *red* while chains of the residues directly involved in the virus-receptor interactions are shown as ticks. Part A reproduced from http://www.esrf.fr/UsersAndScience/Publications/Highlights/2004/images/FIG071.jpg. European Synchroton Radiation Facility. X-ray Structure of a Minor Group Human Rhinovirus Bound to a Fragment of its Cellular Receptor Protein, Grenoble, France. Part B reproduced from http://www.esrf.fr/UsersAndScience/Publications/Highlights/2004/images/FIG072.jpg. European Synchroton Radiation Facility, "X-ray structure of a minor group human rhinovirus bound to a fragment of its cellular receptor protein," Grenoble, France.

protein shell that envelops a single positive RNA strand (making it a retrovirus). The envelope, or capsid, is constructed from 60 copies, each containing four viral proteins, VP1, VP2, VP3, and VP4. The external surface is formed by VP1, VP2, and VP3. VP4 is internal and is in contact with the RNA genome.

Human Rhinoviruses

Human rhinoviruses (HRVs) form the largest genus among the Picornaviridae family, and HRVs have more than 100 immunologically distinct serotypes. Members of the major group of HRV serotypes (about 90%) bind to ICAM-1 (a member of the Ig superfamily) for entry into the host cell. Members of the minor group (about 10%) bind to various members of the extracellular membrane low-density lipoprotein (LDL) receptors, unrelated to ICAM-1. Features of ICAM-1 are shown in Figure 17-27. Features of the LDL receptor are shown in Figure 17-28. The figures describe the external ligand-binding domains that come in contact with respective virus particles. Both HRV14 and poliovirus have deep clefts, or "canyons," which are the attachment sites used for binding to host cell receptor Ig-like domains. The canyon surrounds each pentagonal vertex of the icosahedral shells.

Influenza Viruses

The influenza viruses, generally derived from birds, have jumped to the human species and are transmissible from one person to another. A bird virus can jump to a human and become infectious to another human when the individual is infected with a bird flu and at the same time is infected with a strain of flu that already has been adapted to the human. The two viruses infecting the same cell can exchange genetic information. The pandemic of flu virus in 1918 seems to be similar to the bird flu looming as a potential current pandemic. Many strains of influenza are variations of the same few avian viruses that have been with us for many decades. A feature of these viruses is that they have successfully mutated and continue to do so, requiring that vaccines be updated frequently.

A current potential problem is a new bird flu that is infectious to birds but has caused flu and death in some individuals who work in raising chickens, especially in Asia. This virus is identified as H5N1 avian influenza. There are 15 subtypes of influenza A virus, designated H1 to H15. All of these are derived from avian species. H1 is the strain of the 1918 pandemic that also occurred in 1977. H3 was the virus in the outbreak of 1968, and H5 produced respiratory disease in 1999 in Hong Kong. On the surface of the flu virus there are molecules of neuraminidase (sialidase) and **hemagglutinin** (**HA,** Figure 17-29).

N-acetylneuraminic acid (sialic acid) is linked through a hydroxyl group to the host cell glycoprotein. The viral neuraminidase cleaves the bond between sialic acid and glycoprotein, permitting the host cell wall to rupture and release the plethora of replicated viral particles. A recent effective inhibitor of viral neuraminidase prevents the enzyme from cleaving the sialic acid–glycoprotein bond, thus essentially preventing the lysis of the viral loaded cell and decreasing the spread of the virus.

(Text continues on p. 1076.)

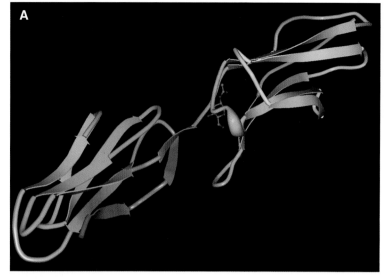

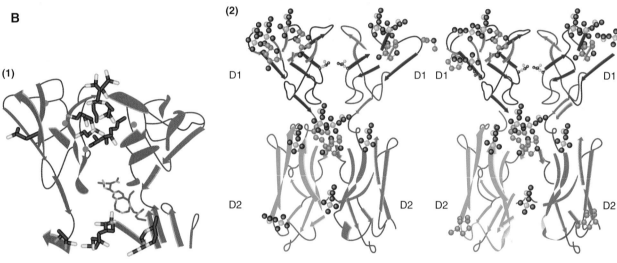

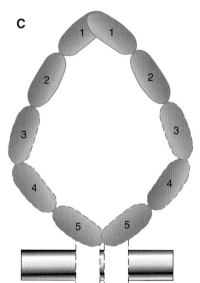

Figure 17-27. **A.** Crystal Structure of the N-terminal Two-domain of ICAM-1. *(A)* Ribbon diagram with beta-strands in red, alpha-helix in *blue*, and coil in *orange*. The last residue of domain 1 (Tyr-83) is *rose*. N-linked sugars *(yellow)*, Glu-37 in strand C of domain 1 *(black)*, and disulfide bonds *(green)* are included. The last two residues of the structure are omitted. *(B)* Movement between domains 1 and 2 and comparison of domain 1 of molecules A and B. The C alpha backbones of ICAM-1 molecules A *(magenta)* and B *(yellow)* are shown after superposition on domain 1. *(C)* Hydrogen bonds at the boundary between domains 1 and 2. **B.** The dimer interface and ligand-binding residues. Interacting residues in domain 1. Side chains are shown for residues that interact across the dimer interface in domain 1. The conserved central Val-51 residue is *blue*, and Glu-34 is *red*. Salt bridges between residues at the periphery of the interface are *dashed lines*. Stereoview of the dimer. Side chains and alpha carbons are shown for residues important in binding to LFA-1 *(red and orange)*, human rhinoviruses 3, 14, 15, 36, and 41 *(yellow and orange)*, and *Plasmodium falciparum (blue)*. Only single amino acid substitutions that reduced binding 50% or 2 SD below control are shown. **C.** A model for the ICAM-1 dimer on the cell surface. Domains 1 and 2 and their orientation in the dimer are from the crystal structure. The rod-like shape of domains 1-5 in the monomer and the bend between domains 3 and 4 are from electron microscopy. Dimerization or proximity between domain 5 is based on hindrance of antibody binding to this domain in the dimer, and association at the transmembrane domain is based on its role in dimerization. Part A reproduced from PDB ID: 1iam. J. Bella, et al. The Structure of the Two Amino-Terminal Domains of Human ICAM-1 Suggests How It Functions as a Rhinovirus Receptor and as an LFA-1 Integrin Ligand. *Proceedings of the National Academy of Sciences* **95** pp. 4140 (1998).

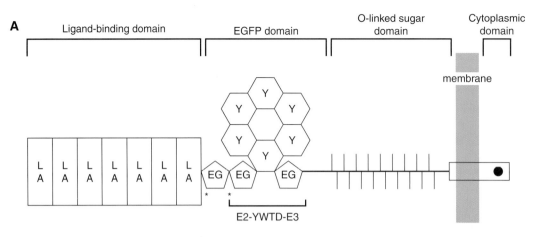

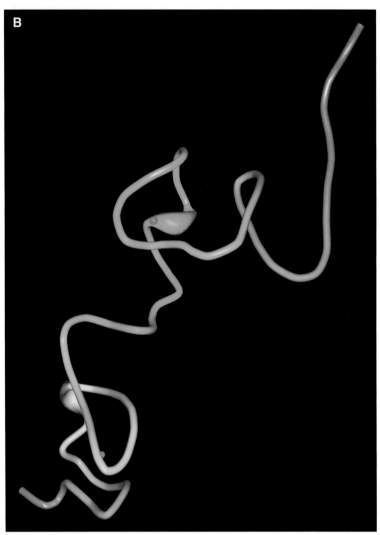

Figure 17-28. **A.** Modular domain organization of the LDL receptor. The black dot on the cytoplasmic tail represents the Asn-Pro-Xaa-Tyr motif required for receptor internalization. **B.** NMR structure of a concatemer of the first and second ligand-binding modules of the human low-density lipoprotein receptor. **C.** NMR study of a pair of LDL receptor calcium-binding epidermal growth factor-like domains. *EG*, epidermal growth factor-like modules; *EGFP*, epidermal growth factor precursor; *LA*, LDL-A ligand-binding modules; *Y*, repeats containing a YWTD consensus motif. Part A redrawn from Figure 1A of H. Jeon et al., "Implications for familial hypercholesterolemia from the structure of the LDL receptor YWTD-EGF domain pair," *Nat. Struct. Biol.*, 8: 499–504, 2001 and directly from http://www.nature.com/nsmb/journal/v8/n6/fig_tab/nsb0601_499_F1.html. Part B reproduced from http://www.ebi.ac.uk/msd-srv/msdlite/images/1f5y600.jpg. PDB ID: 1 f5y. N.D. Kurniawan, A.R. Atkins, S. Bieri, C.J. Brown, I.M. Brereton, P.A. Kroon, R. Smith. NMR Structure of a Concatemer of the First and Second Ligand-binding Modules of the Human Low-density Lipoprotein Receptor. *Protein Sci.* **9** pp. 1282 (2000).

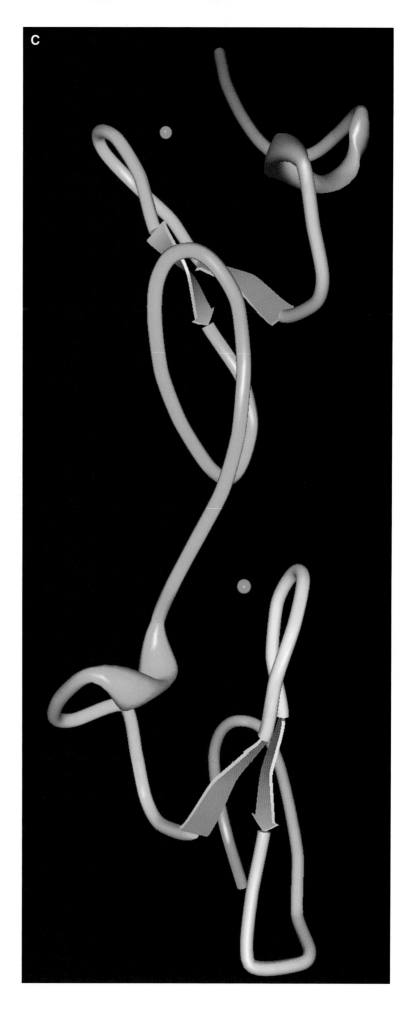

Figure 17-28, cont'd Part C reproduced from http://www.ebi.ac.uk/thornton-srv/databases/cgi-bin/pdbsum/GetPage.pl?pdbcode=1hj7. PDB ID: 1hj7. S. Saha et al. Solution Structure of the LDL Receptor EGF-AB Pair: A Paradigm for the Assembly of Tandem Calcium Binding EGF Domains. *Structure* **9** pp. 451 (2004).

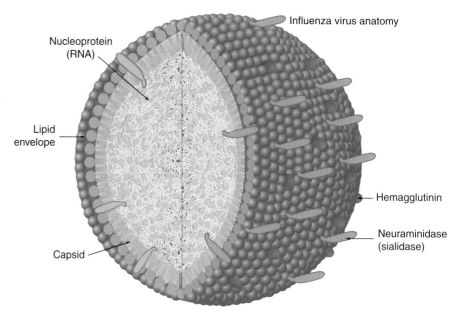

Figure 17-29. Diagram of a flu virus particle.

Neuraminidase inhibitors are Tamiflu (Hoffmann-La Roche) and its competitor Relenza (GlaxoSmithKline), whose structures are shown in Figure 17-30, along with the neuraminidase reaction. The location of the molecules of neuraminic acid (sialic acid) on the surface of the viral particle was shown in Figure 17-29. By blocking the bond between neuraminic acid and cellular glycoprotein, the replicated viral particles cannot be released into the extracellular fluid to infect other cells by way of the bloodstream; instead, the viral particles are aggregated into bundles and individual particles are not released (Figure 17-30C).

The HA molecule on the surface of the virus is a binding site for the receptor on the host cell membrane. The HA of the 1918 pandemic bird virus and its interaction with the receptor-binding site is shown in Figure 17-31. Influenza HA structures and interactions with the binding site of host cell receptor for various species are shown in Figure 17-32.

West Nile Virus

Another virus of recent interest is the West Nile virus (one of the flaviviruses), carried by mosquitoes (contracted by biting birds) and extremely dangerous to elderly and very young humans. Pictures of the virus (50 nm in diameter) showing its surface and envelope protein molecules are shown in Figure 17-33. The envelope protein is made up of two proteins, envelope E and membrane M proteins embedded in a lipid layer (Figure 17-33B). Protein E is involved in cell receptor recognition, fusion with endosomal membranes of the host cell, and virion assembly. The genome consists of a single-stranded RNA packaged inside of the core protein C (Figure 17-34). It is theorized that the virus binds through its E protein to highly sulfated heparan sulfate residues on the surface of the host cell. Once inside the host cell, the lowered pH produces a conformational alteration of the E protein,

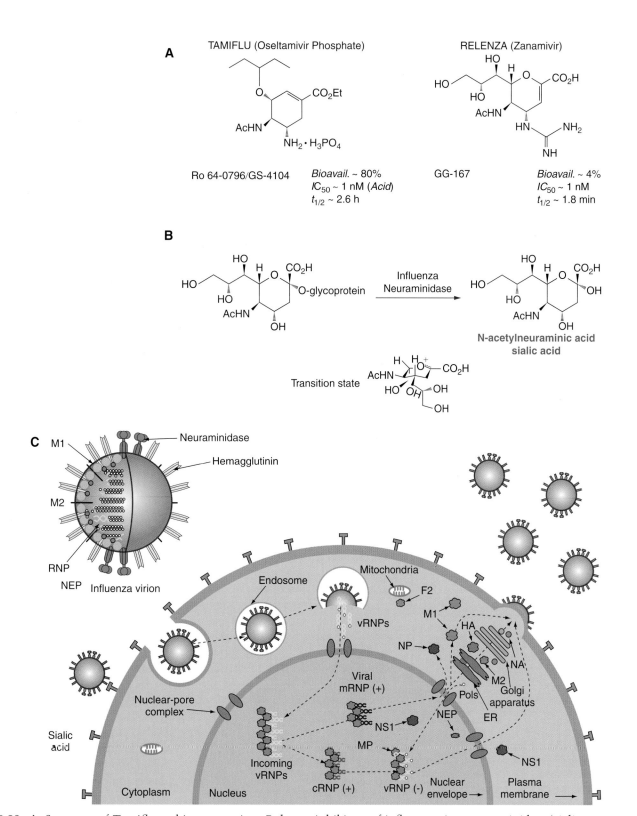

Figure 17-30. **A.** Structure of Tamiflu and its competitor, Relenza, inhibitors of influenza virus neuraminidase (sialidase). **B.** The viral neuraminidase reaction resulting in the cleavage of the bond between sialic acid and host cell glycoprotein. **C.** Influenza virus replication cycle. Virus attachment through hemagglutinin (HA) to receptors containing terminal neuraminic acid residues and penetration into host cell; transcription of viral RNA and translation of viral proteins; replication of viral RNA and assembly of virion, budding, and subsequent release from host cell. **D.** Electron micrographs of MDCK cells infected with influenza A virus. *(A)* Normal assembly and budding of virus in absence of NA inhibitor. *(B)* Lateral aggregation and formation of large bundles by virus in presence of NA inhibitor. Respiratory tract epithelial cells before and after infection with influenza A virus. *Bar*, 1 μm. Parts A and B are reproduced from http://www.siegtried.ch/pdf/syneposium/karpf_Tamiflu_booklet.pdf. Part C reproduced from http://www.reactome.org/figures/influenza_life_cycle_overview.jpg. Reactome project of the Cold Spring Harbor Laboratory, The European Bioinformatics Institute and the Gene Ontology Consortium. help@reactome.org.

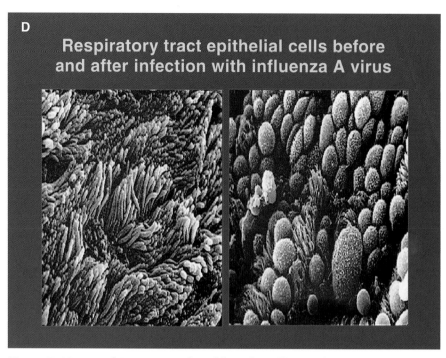

Figure 17-30, cont'd Part D reproduced from http://www.sbimc.org/2000/spring/fiddian-slides/img003.gif. A.P. Fiddian, SBIMC-BVIKM.

exposing a hydrophobic domain. This domain produces a fusion between viral and host vesicle membranes.

Genomic positive-strand RNA (which can function directly as mRNA) is released into the host cell cytoplasm, where it is translated ribosomally into viral polyprotein (this is a different mechanism from that of HIV, which initially requires reverse transcription and incorporation into host cell DNA). Viral polyprotein is cleaved proteolytically by host cell protease and by viral serine protease into 10 mature viral proteins (3 structural and 7 nonstructural proteins), as shown in Figure 17-35. The translated proteins assemble and form copies of the complementary negative-strand RNA that is copied to form positive-strand RNA. The positive strands serve as templates for more replication or are moved to the ribosome to generate more polyproteins. Proteins E, pe M (pr M in Figure 17-35), and C and new RNA positive strands assemble into new viruses that travel to the surface of the cell and are released to the blood and lymphatics. An electron photomicrograph of West Nile virus particles is shown in Figure 17-36. Appearing first in New York City, West Nile virus has spread to 44 states, with deaths approaching 100 people in the United States.

Severe Acute Respiratory Syndrome

Finally, a discussion of the **severe acute respiratory syndrome (SARS)** coronavirus (CoV) follows. The SARS virus is spherical, like many other viruses that infect humans. It has a diameter of about 100 nm. An

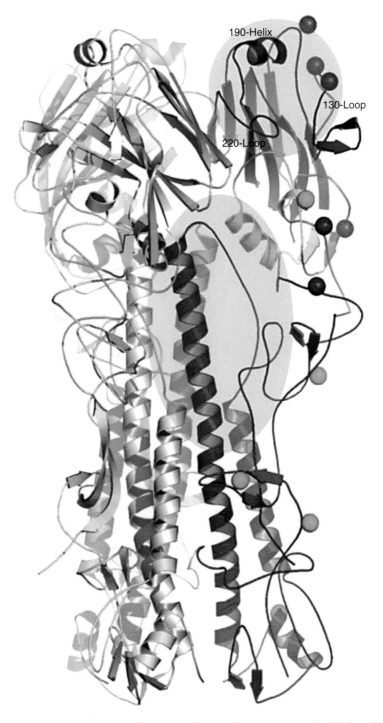

Figure 17-31. Structure of the 1918 hemagglutinin protein highlights the receptor-binding site *(blue)* and the membrane fusion domain *(magenta)*. Redrawn with permission from S.J. Gamblin et al., "The structure and receptor binding properties of the 1918 influenza hemagglutinin," *Science*, 303: 1838–1842, 2004.

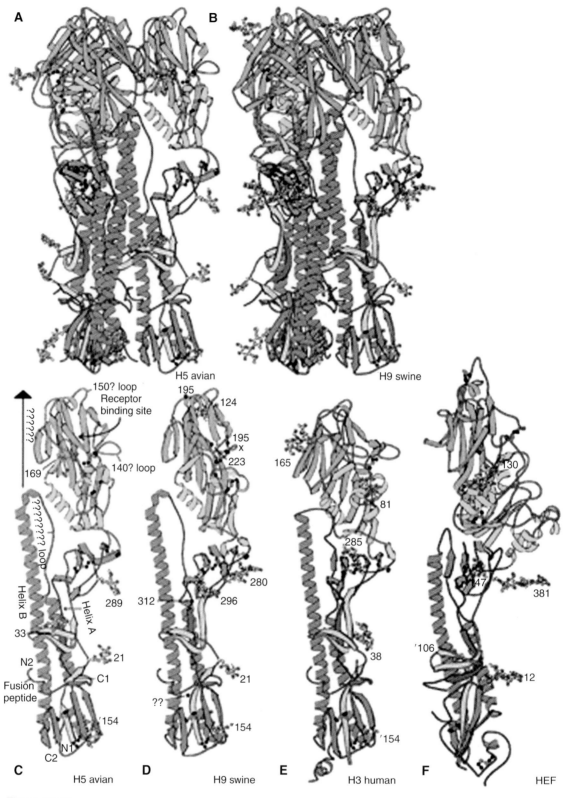

Figure 17-32.

1080 CHAPTER 17 Microbial Biochemistry

Figure 17-32. H5 avian and H9 swine influenza virus hemagglutinin structures and their interactions with receptor-binding sites of various species. H5 avian and H9 swine HA structures compared with H3 HA and HEF. *(A)* Ribbon diagram of the trimer of H5 avian HA colored by subdomains: receptor subdomain R *(blue)*, vestigial enzyme subdomain E' *(yellow)*, HA2 stem F subdomain *(red)*, F' subdomain HA1 1–52 *(pink)*, F' subdomain HA1 275–307 *(purple)*. Oligosaccharides are colored by atom type. *(B)* Trimer of H9 swine (A/Sw/9/98) HA. *(C)* Monomer of H5 (A/Dk/Sing/97) avian HA. Helix A and B, the interhelical loop between them, and the fusion peptide in HA2 are labeled. N1 and C1 indicate the termini of HA1; N2 and C2 are those of HA2. The trimer axis, receptor-binding site, and two prominent antigenic loops, 140s and 150s, are labeled. Residue numbers mark oligosaccharides, which are colored by atom type. Carbohydrate attachment sites are at HA1 21, 33, 169, and 289, and at HA2 154. *(D)* Monomer of the H9 (A/Sw/9/98) swine HA. The x marks the deleted 140s loop. Carbohydrate attachment sites are at HA1 21, 128, 210, 289 and 296 and at HA2 154. *(E)* Monomer of trimeric H3 (A/Aichi/2/68) human HA. Carbohydrate attachment sites are at HA1 8, 22, 38, 81, 165, and 285 and at HA2 154. *(F)* Monomer of trimeric influenza C virus hemagglutinin−esterase fusion (HEF) protein. *HA*, hemagglutinin; *HEF*, HA−esterase fusion; blue region (*upper part of Figures*, is receptor-binding domain). Reproduced from Figure 2 of Y. Ha et al., "H5 avian and H9 swine influenza virus hemagglutinin structures; possible origin of influenza subtypes," *EMBO J.*, 21: 865–875, 2002 and directly from http://www.nature.com/emboj/journal/v21/n5/full/7594305a.html.

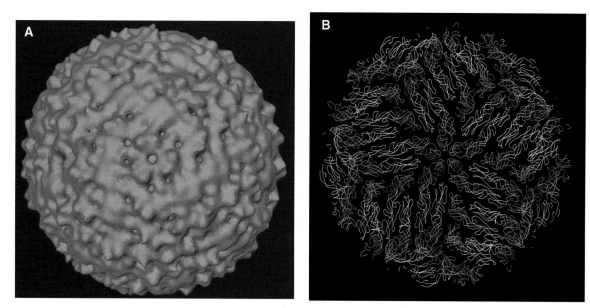

Figure 17-33. **A.** Surface-shaded image of West Nile virus. **B.** The West Nile virus showing envelope protein molecules. Part A reproduced with permission from http://www.newsroom.ucr.edu/images/releases/1033_1.jpg. University of California Riverside. Image from Purdue Department of Biological Sciences. Part B reproduced with permission from http://news.uns.purdue.edu/UNS/images/kuhn.westnile2.jpeg.

electron micrograph of the virus particle and a micrograph of lung tissue showing infection with SARS are shown in Figure 17-37. Its genome is composed of a single positive-strand RNA containing approximately 29,740 bases. It is shown in Figure 17-38. The genomic RNA is contained in a complex spherical structure composed of several proteins, as shown in Figure 17-39. This spherical structure attaches to the membrane of the host cell and becomes absorbed inside, where the single-stranded RNA is

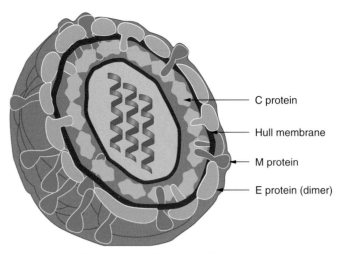

Figure 17-34. Diagram of the west Nile virus.

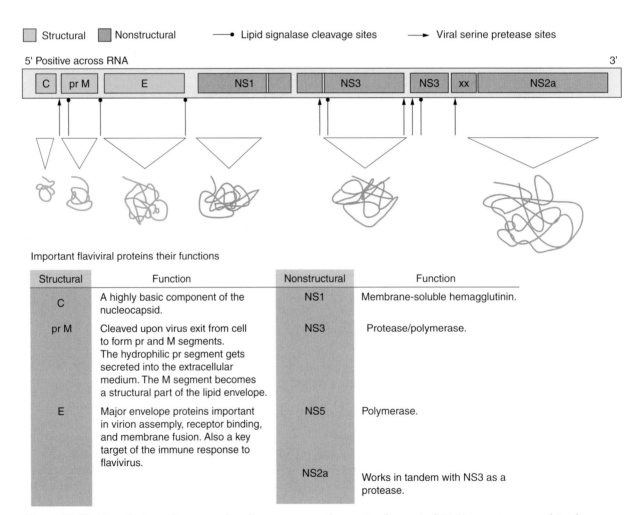

Figure 17-35. Translation of structural and non-structural proteins from viral RNA positive strand in the cytoplasm of the infected cell. Redrawn from Stanford University data.

1082   CHAPTER 17   Microbial Biochemistry

Figure 17-36. Electron micrograph of West Nile virus. Reproduced from http://www.lib.uiowa.edu/hardin/md/cdc/2290.html. Public Health Image Library Database, U.S. Centers for Disease Control.

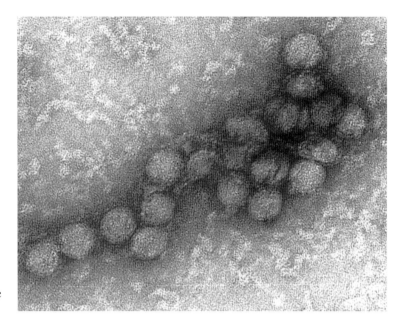

Figure 17-37. **A.** Scanning electron microscope image of SARS virus particles. **B.** Atomic force microscopy of Vero cells infected with SARS virus at 15 hours after infection *(A)* at much higher resolution imaging of the edge of a cell, a virus particle *(thick arrow)* in the process of extruding from the cell plasma membrane (PM) after fusion of the transport vesicle with the cell membrane. PM shows some loss of integrity *(thin arrows)* during this exit process. *(B)* A three-dimensional reconstruction of the extruding virus particle from panel B. *Arrow* indicates the thickened cell edge. *(C)* Arrow indicates the knoblike structures on the virus particles. Part A reproduced from http://www.cell-research.com/20033/20033COVER.htm. Prof. Yi Xue LI and Ye CHEN of Bioinformation Center, Shanghai Institutes for Biological Sciences, CAS, cover photo from Cell-Research, 13, 3, 2003. Part B reproduced from http://www.cdc.gov/ncidod/EID/vol10no11/04-0195-G4.htm. From M.L. Ng et al., National University of Singapore, Singapore General Hospital, and Singapore and National Environment Agency, Singapore, "Topographic changes in SARS coronavirus–infected cells at late stages of infection," *Emerg. Infect. Dis.*, 10, 2004.

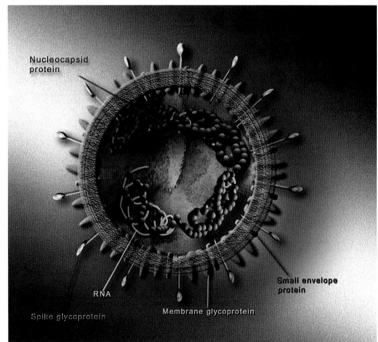

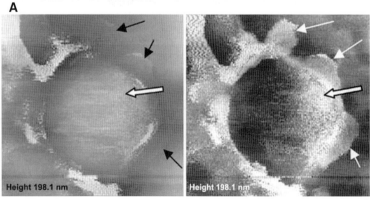

(Continued)

Other Viruses of Current Interest 1083

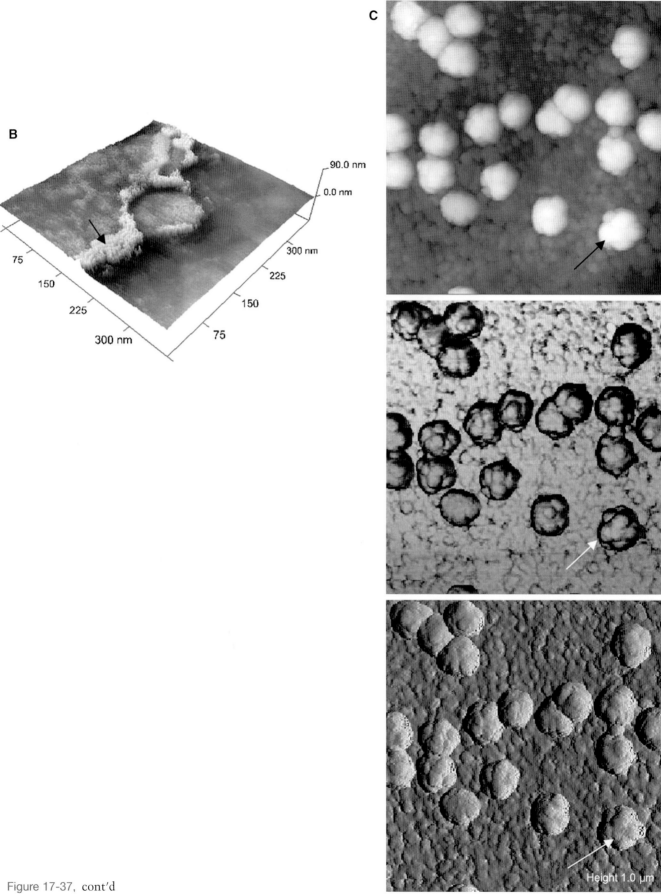

Figure 17-37, cont'd

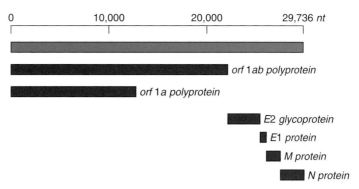

Figure 17-38. Representation of the SARS genome consisting of a single strand of RNA. The message is translated to a polyprotein that becomes cleaved to form several proteins including E2 glycoprotein, E1 protein, M protein and N protein.

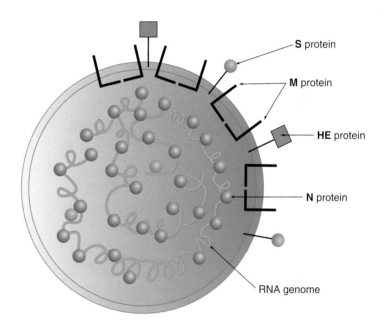

Figure 17-39. Internal SARS structure containing positive single-stranded RNA. The capsule is comprised by several proteins including, S, M, HE, and N. *HE*, hemagglutinin.

released. The N (neocapsid) protein interacts with the viral genome to form the viral core; it is a phosphoprotein of about 55 kDa.

Inside the host cell, the expression of the viral RNA is more complicated than the cases already reviewed (Figure 17-40). The hypothesized SARS CoV genome is shown in Figure 17-41. CoV RNA synthesis takes place in the host cell cytoplasm by way of a negative-strand RNA intermediate (Figure 17-40). The SARS CoV has a polycistronic genome and a nested set of multiple subgenomic RNAs (overlapping at the 3' end), each with the same 5' leader sequence (from the 5' end of the genome). Of the nine total RNAs, each is translated to generate a protein product, as shown in Figure 17-41B. Genomic RNA 1a and 1b encode replicase 1a and replicase 1b.

Structural and nonstructural proteins are encoded in the various derivative RNAs: S protein (spike protein), E protein, M protein, N protein, and others. Replication is carried out by a process of discontinuous

Other Viruses of Current Interest   1085

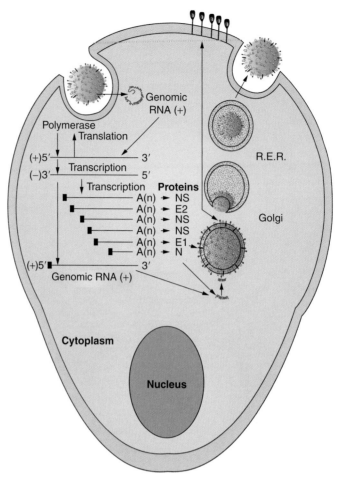

Figure 17-40. SARS replication concepts. Once inside the cell the ssRNA is able to make use of existing molecular structures to fabricate copies of the ssRNA and also build the proteins needed for construction of new shells. As the new viruses are constructed they are encapsulated, accumulate near the outer membrane, then released to attack other cells. Notably each time a new virus package is created it might be altered somewhat from the original. Redrawn from http://czws.gov.cn/jkzn/index1.files/SARS/SARS3.gif.

transcription (in the cytoplasm) that is incompletely understood. **Transcriptional regulatory sequences (TRSs)** at the 5′ end of each transcriptional unit regulate the discontinuous transcription of subgenomic RNAs. CoV genes are arranged in the order 5′-replicase-(HE)-S-E-M-N-3′ with other nonessential (so far) genes. The proteins S (spike), M (matrix), E (envelope), and, in the case of group II CoVs, HE (hemagglutinin esterase; not shown in Figure 17-41B), make up the virion envelope surrounding the neocapsid. The S protein (180 kDa) is a peplomer (knoblike projection on viruses) glycoprotein on the viral envelope and on the plasma membrane of infected cells. It is responsible for attachment to the host cellular receptor, virus–cell fusion during viral entry. The S protein interacts with the M and N proteins in neocapsid formation and viral assembly (Figure 17-42). The M protein is a transmembrane glycoprotein integrated with the virion core through its carboxy terminus. It maintains the core structure. The E protein (9.6 kDa) is membrane-associated protein needed for assembly of the virion. The

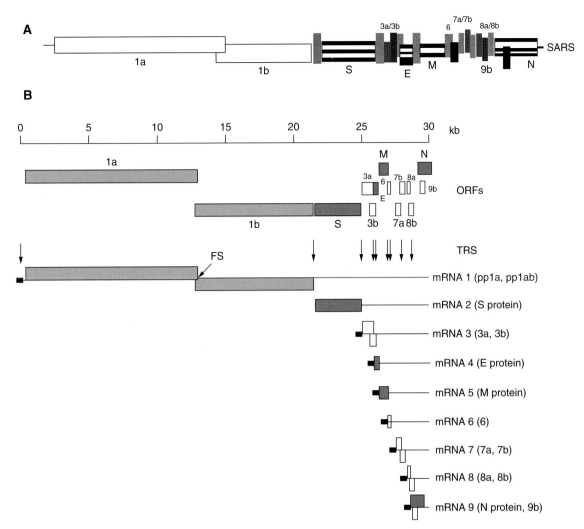

Figure 17-41. **A.** A hypothesized SARS-CoV genome. The replicase gene (ORFs 1a, 1b) is shown by *open bars*; structural genes (S, E, M, N, and HE) are depicted with *striped bars*; nonstructural genes *(black bars)* are variable in number and location in the coronavirus genome among the different viral groups. Small open reading frames (ORFs) are depicted in *solid bars*. **B.** SARS-CoV genome organization and expression. The SARS-CoV ORFs, frameshift (FS) and TRS elements, and genomic and subgenomic mRNAs, are shown. *Black boxes* represent the 72 nt leader RNA sequence located at the 5' end of each viral mRNA. Also indicated are the viral proteins predicted to be expressed from a given mRNA's 'unique' region (i.e., the region not present on smaller mRNAs). *HE*, helicase; *ORF*, open reading frame; *TRS*, transcription regulatory sequence. Part A redrawn in part from Figure 1 of S. Navas-Marting and S.R. Weiss, "Coronavirus replication and pathogenesis: Implications for the recent outbreak of severe acute respiratory syndrome (SARS), and the challenge for vaccine development," *J. Neurovirol.*, 10: 75–85, 2004. Part B reproduced in part from Figure 1 of T. Volker et al., "Mechanisms and enzymes involved in SARS coronavirus genome expression," *J. Gen. Virol.*, 84: 2305–2315, 2003 and http://vir.sgmjournals.org/cgi/content/full/84/9/2305.

N (neocapsid) protein (60 kDa) complexes with the RNA genome to form the neocapsid (see Figure 17-38). Newly formed virions bud into intracellular membranes and are released from the infected cell through vesicles of the secretory pathway. The virus infects the host cell by binding to a CAM receptor on its surface, as shown in Figure 17-43. The main proteinase produced from viral RNA is shown in Figure 17-44.

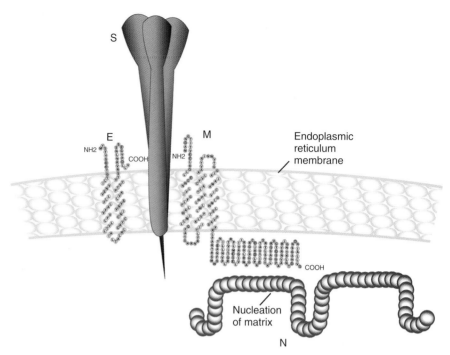

Figure 17-42. Proposed schema of interactions between S, M, and N-proteins and their roles in nucleocapsid formation and viral assembly. The newly synthesized M protein is anchored in intracellular membrane through its membrane-spanning region. The COOH-terminal domain of M interacts with N to facilitate formation of nucleocapsids. In the membrane, the transmembrane domain of S may interact with M to promote viral particle budding. E interacts with S independent of the S interaction with M. Reproduced from Figure 6 of http://jvi.asm.org/cgi/content/full/78/22/12557/F6 and Y. Huang et al., "Generation of synthetic severe acute respiratory syndrome coronavirus pseudoparticles: Implications for assembly and vaccine production," *J. Virol.*, 78: 12557–12565, 2004.

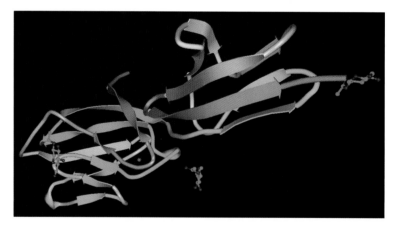

Figure 17-43. CAM molecule similar to the host molecule that binds SARS. Reproduced from http://www.rcsb.org/pdb/cgi/explore.cgi?pdbId=1hnf. PDB ID: 1hnf. D.L. Bodian, E.Y. Jones, K. Harlos, D.I. Stuart, S.J. Davis. Crystal Structure of the Extracellular Region of the Human Cell Adhesion Molecule CD2 at 2.5 Å Resolution. *Structure* **2** pp. 755 (1994).

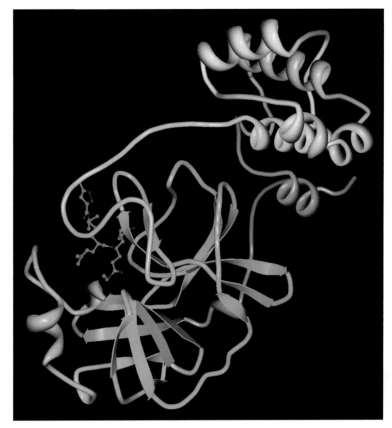

Figure 17-44. Coronavirus main protease. Reproduced from http://www.rcsb.org/pdb/cgi/explore.cgi?pdbId=2alv. PDB ID: 2alv. B.H. Harcourt, P.A. Rota, S.C. Baker, M.E. Johnson, A.D Mesecar. Design and Synthesis of Peptidomimetic Severe Acute Respiratory Syndrome Chymotrypsin-like Protease Inhibitors. *J. Med. Chem.* **48** pp. 6767 (2005).

Bacteriophage

A **bacteriophage** is a virus that infects bacterial cells. There are many viruses that infect plant cells, but they will not be discussed here except to say that because bacteriophages cannot penetrate the plant cell wall, the **phage** is carried by insects that feed on the plants. The general structure of a bacteriophage is shown in Figure 17-45. Bacteriophages have either an RNA genome or a DNA genome, similarly to viruses that infect humans. The general structure shown in Figure 17-45 is typified by the electron micrograph of T4 phage (phage derives from the Greek word *phagein*, meaning to eat) shown in Figure 17-46. The tail of the virus attaches to the bacterial surface by means of the pins (Figure 17-45). After the contraction of the tail, the tail plug penetrates the cell wall and the inner bacterial membrane. The viral nucleic acid is then injected into the bacterial cell (Figure 17-47). The life cycle of a bacteriophage, typified by bacteriophage-λ, is shown in Figure 17-48.

A great deal of research has been carried out using the T phages. The T phages can be characterized by the dimensions of the plaque formed, the head size, the length of the tail, the latent period, and the burst size. This summary is shown in Table 17-2. All T phages listed in Table 17-2

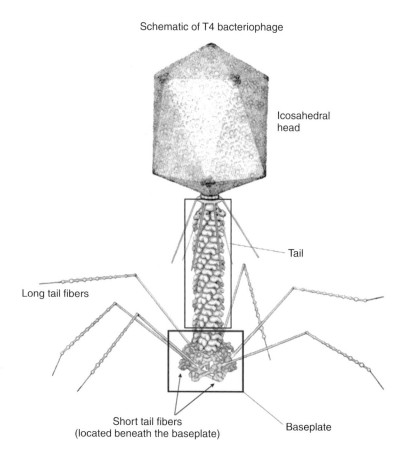

Figure 17-45. Structure of a bacteriophage. Reproduced from http://www.nsf.gov/od/lpa/news/02/pr0207images.htm.

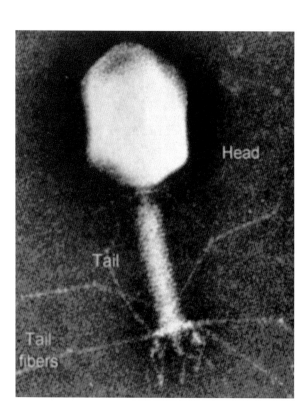

Figure 17-46. Electron micrograph of the T4 phage (aT-even phage consisting of T2, T4, and T6 phages).

1090   CHAPTER 17   Microbial Biochemistry

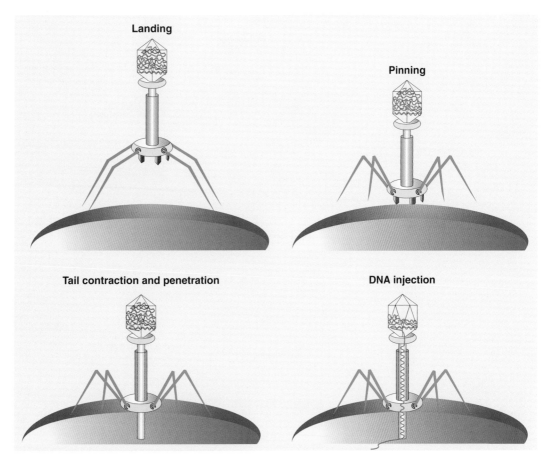

Figure 17-47. T-even bacteriophages (in this case T4) land on a bacterial cell, contract the tail so the pin interacts with the surface and penetration is made through the cell wall and the inner membrane and the nucleic acid (DNA) is injected into the cell.

have a single linear molecule of double-stranded DNA as the genome. Circular forms of the genome can exist, and all of these phages undergo lytic growth (new phages accumulate in the infected cell and cause the cell to burst). The T-even phages have large genomes and are related serologically; for example, the T4 phage has a genome length of 168,895 base pairs. Of the T-odd phages, T3 and T7 are related; however, T1 and T5 are unrelated. T7 has a genome of 39,937 base pairs. The commercially available **Sequenase** is a modified form of T7 DNA polymerase. The structure of T7 DNA polymerase is shown in Figure 17-49, and the residues of the polymerase that bind single-stranded DNA are shown in Figure 17-50. A model for primer synthesis and initiation of DNA synthesis by phage T7 is shown in Figure 17-51.

Some phages can become dormant (temperate bacteriophages) after infecting a cell and undergoing a lytic growth cycle. When dormant, a phage remains as a **prophage** in the host cell and can become active again, resuming a typical lytic growth cycle, when it receives a signal, such as ultraviolet radiation. Bacteriophage-λ is a typical example of a temperate bacteriophage. This phage, which infects *Escherichia coli (E. coli)*, has a linear double-stranded DNA genome containing 48,502 base pairs that becomes circularized after infection.

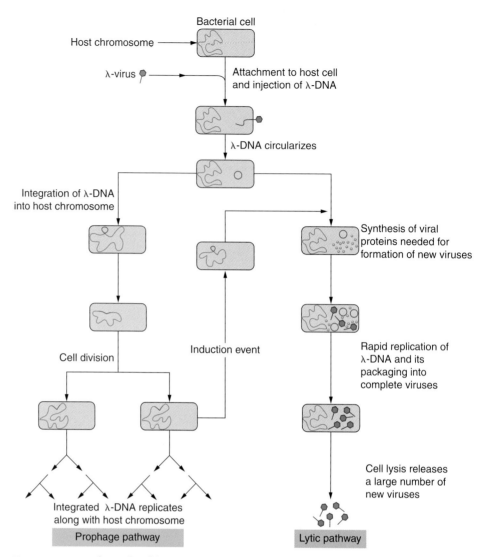

Figure 17-48. Life cycle of bacteriophage lambda. The linear double-stranded DNA lambda genome contains about 50,000 nucleotide pairs and encodes 50 to 60 different proteins. When the lambda DNA enters the cell the ends join to form a circular DNA molecule. The bacteriophage can multiply in *E. coli* by a lytic pathway, which destroys the cell, or it can enter a latent prophage state. Damage to a cell carrying a lambda prophage induces the prophage to exit from the host chromosome and shift to lytic growth. The entrance and exit of the lambda DNA from the bacterial chromosome are site-specific recombination events. Redrawn from *The Life Cycle Of Bacteriophage Lambda* ©1998 by Alberts, Bray, Johnson, Lewis, Raff, Roberts, Walter. http://www.essentialcellbiology.com published by Garland publishing, a member of the Taylor & Francis Group. http://www.accessexcellence.org/RC/VL/GG/bacts_Lambda.html.

Not all phages resemble the T-even phages and some are spherical in shape, such as φX174 and Ms2, and these phages cause lysis of the host cell. The appearances of these phages are shown in Figures 17-52 and 17-53. The genome of φX174 is a circular single-stranded DNA of 5386 base pairs coding for 11 genes. Its life cycle after infecting an *E. coli* cell is shown in Figure 17-54. To package viral genomes into their capsid, the

Table 17-2
Morphological Characteristics of Several T Bacteriophages

Name	Plaque size	Morphology		Latent period (min)	Burst size[*]
		Head (nm)	Tail (nm)		
T1	Medium	50	150 × 15	13	180
T2	Small	65 × 80	120 × 20	21	120
T3	Large	45	Invisible	13	300
T4	Small	65 × 80	120 × 20	23.5	300
T5	Small	100	Tiny	40	300
T6	Small	65 × 80	120 × 20	25.5	200–300
T7	Large	45	Invisible	13	300

[*]*Burst size*, Mean number of phage particles released per infected bacterial cell.

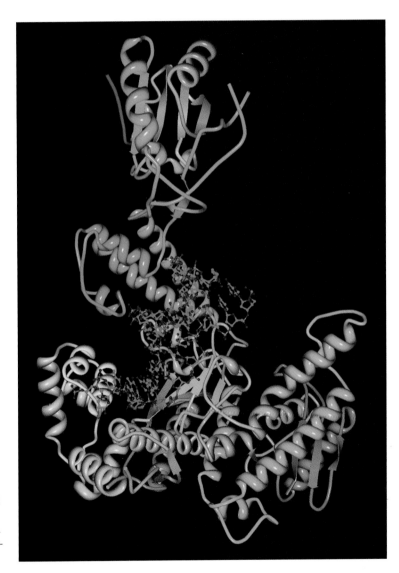

Figure 17-49. T7 DNA polymerase complexed to DNA primer/template and ddATP. From PDB ID: 1SKR. Y. Li et al. Nucleotide Insertion Opposite a *Cis-Syn* Thymine Dimer by a Replicative DNA Polymerase from Bacteriophage T7. *Nature Structural & Molecular Biology* **11** pp. 784 (2004).

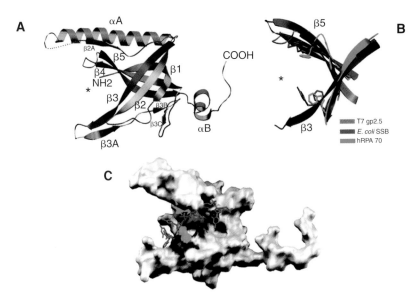

Figure 17-50. gp2.5 is an OB-fold protein. **A.** The structure of gp2.5 is a single domain consisting of a 5-stranded antiparallel β-barrel (the OB-fold in *gold*) capped by a α-helix (α, *blue*). A 28-amino acid tail protrudes away from the β-barrel at the carboxyl terminus of the gp2.5delta26C protein. An additional 26 residues, which are missing from the C terminus of gp2.5delta26C, participate in dimer formation and interactions with other replication proteins. **B.** The proposed DNA binding surface (asterisk) contains a pair of structurally conserved aromatic residues surrounded by basic residues. This region of the gp2.5 structure *(gold)* is superimposed on structures of the *E. coli* SSB protein *(magenta)* and human RPA70 *(green)*. In the *E. coli* SSB protein and human RPA70 structures these residues contact 3 nucleotides of the bound ssDNA. **C.** A DNA docking model. The DNA from *E. coli* SSB structure overlies the conserved aromatic residues *(green)* described in *B* and is near a series of conserved basic residues *(blue)*. The DNA is closely apposed to these residues without significant steric clashes. Reproduced from Figure 1 of T. Hollis et al., "Structure of the gene 2.5 protein, a single-stranded DNA binding protein encoded by bacteriophage T7", PNAS USA, 98: 9557–9562, 2001.

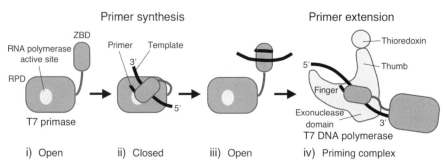

Figure 17-51. A mechanistic model for primer synthesis and the initiation of DNA synthesis by bacteriophage T7. *i*, T7 primase consists of a ZBD and RPD that are connected by a flexible linker. The ZBD is essential for the recognition of priming sites on a DNA template, and the RPD contains the active site for primer synthesis. However, these domains do not interact in the absence of substrates. *ii*, During primer synthesis, both the ZBD and the RPD bind to the template, inducing a closed conformation of the primase. *iii*, After primer synthesis, the primase opens again and only the ZBD remains bound to the primed template. *iv*, The ZBD delivers the primed DNA template to T7 DNA polymerase to stimulate primer extension. *RPD*, residues 60-225 comprising the active primase; *ZBD*, zinc binding domain. Reproduced from Figure 1 of M. Kato, T. Ito, G. Wagner, "A molecular handoff between bacteriophage T7 DNA primase and T7 DNA polymerase initiates DNA synthesis," *J. Biol. Chem.*, 279: 30554–30562, 2004 and http://www.jbc.org/cgi/content/full/279/29/30554/FIG1.

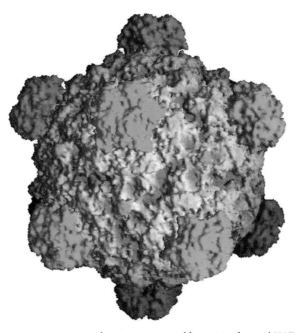

Figure 17-52. Surface structure of bacteriophage ϕX174. Reproduced from http://www.webpages.uidaho.edu/~snuismer/Nuismer_Lab/openeings_files/capsid.jpg. Dr. Scott L. Nuismer, Department of Biological Sciences, University of Idaho, Moscow, ID 83844.

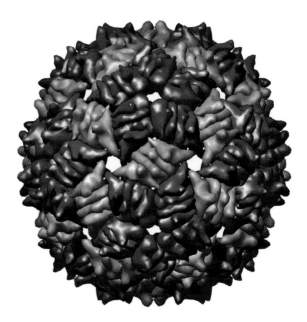

Figure 17-53. Surface appearance of bacteriophage Ms2.

nucleic acid (strongly negatively charged) genome needs to be partially neutralized. This is accomplished by the expression of a positively charged protein, protein J (encoded in the viral genome), which binds to the DNA. The J protein thus facilitates DNA packaging and mediates cell attachment and infection of a host cell. The structure of protein J is shown in Figure 17-55, and the overall process of packaging the single-stranded DNA into the procapsid to form the mature virion is shown in Figure 17-56. Protein J displaces the internal scaffolding protein B by

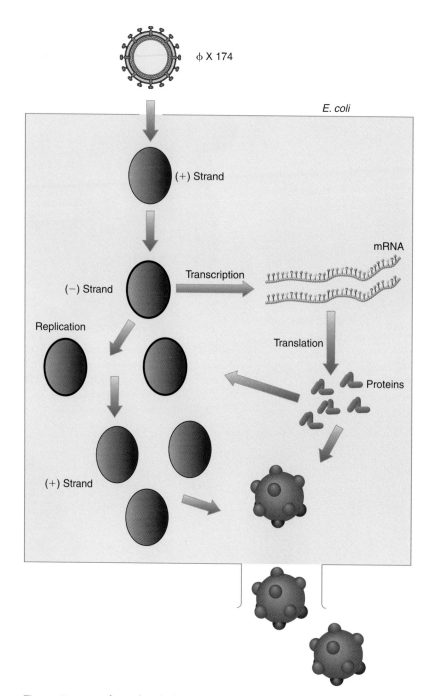

Figure 17-54. Life cycle of Phi X 174 (φX174).

competition for the binding pocket on the internal surface of the capsid protein F. The resulting provirion sheds the external scaffolding lattice to form the mature structure. The single-stranded circular DNA (5386 base pairs) of φX174 is diagrammed in Figure 17-57, where the 11 encoded genes are labeled A, A* (splice variant of A), C, D, J, F, G, H, B, K, and E. The genes are transcribed in the clockwise direction of the positive strand (in the 5' to 3' direction).

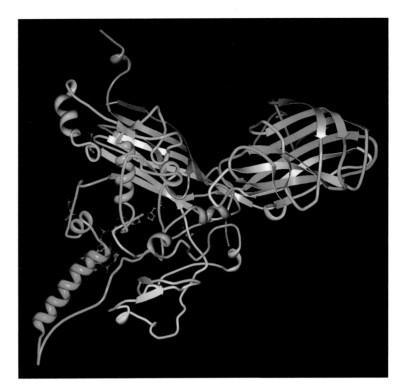

Figure 17-55. The φX174 DNA-binding protein J in two different capsid environments. Reproduced from PBD ID: 1rb8. R.A. Bernal, S. Hafenstein, R. Esmeralda, B.A. Fane, M.G. Rossmann. The φX174DNA Binding Protein J in Two Different Capsid Environments. *J. Mol. Biol.* **337** pp. 1109 (2004) and http://www.ebi.ac.uk/msd–srv/msdlite/images/1rb8600.jpg.

Various activities are ascribed to the φX174 gene-encoded proteins. The gene A protein, 59 kDa, cleaves the φX174 duplex replicative form of DNA and is bound in the replication complex. Of the 10 proteins required for the conversion of φX174 single-stranded DNA to the duplex form, the (DNA) B protein, (DNA) C protein (5.8 kDa), and (DNA) G protein (a primase), plus two other proteins, i and n (for which the genetic locus is not clear), derive from the φX174 genome with holo-DNA polymerase III and DNA unwinding protein. Gene E of φX174 is a lysis gene. Proteins B and D are scaffolding proteins. Proteins F, G, and H are involved in procapsid formation. The external scaffold is formed by 240 molecules of protein D. The procapsid contains 60 copies each of the internal scaffolding protein B, the capsid protein F (undergoes conformational changes during capsid maturation), and the spike protein G. During maturation, DNA and J proteins are packaged and protein B is excluded. *The mature virion requires removal of the external scaffold* (see Figure 17-56). The structure of the φX174 DNA-binding protein (protein J) is shown in Figure 17-58. The structure of a viral procapsid with scaffold is shown in Figure 17-59.

The life cycle of bacteriophage-λ is shown in Figure 17-48. A simplified genetic map of phage-λ is shown in Figure 17-60. P_L is the promoter for transcription of the left side of the genome (including N and cIII); O_L is the 50–base pair short noncoding region, which lies between the cI and N genes next to P_L; P_R is the promoter for the right side of the genome, including cro, cII, and the genes for structural proteins; and O_R is a 50–base pair noncoding region between cI and cro genes next to P_R. cI is transcribed from its own promoter and encodes a 236–amino acid

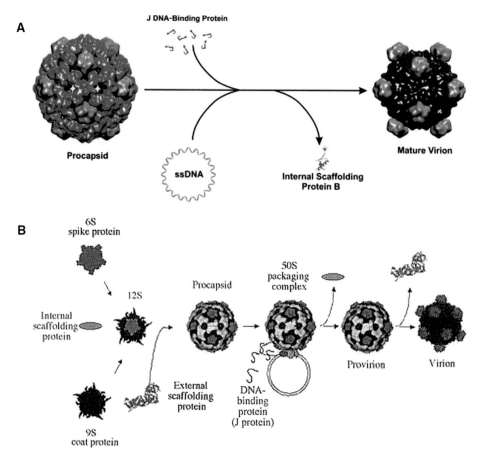

Figure 17-56. **A.** Assembly pathway of (φX174) *Microviridae*. The internal scaffolding protein B (*shaded red,* and barely visible) and external scaffolding protein D *(shaded green)* are required for the assembly of pentameric intermediates into an empty protein shell called the procapsid. The F capsid protein is *shaded blue,* and the G spike protein is *shaded yellow.* The single-stranded DNA (ssDNA) is concurrently synthesized and packaged with the basic DNA-binding protein J, which displaces the internal scaffolding protein B by competition for the same hydrophobic binding pocket on the internal surface of the F capsid protein. The resulting particle, or provirion, sheds the external scaffolding protein lattice to form the mature virion. **B.** More detailed morphogenic pathway of the ΦX174-like phages. Part A reproduced from Figure 1 of R. Bernal et al., "The φX174 protein J mediates DNA packaging and viral attachment to host cells," *J. Mol. Biol.*, 337: 1109–1122, 2004. Part B reproduced from Figure 1 of A. Burch and B. Fane, "Genetic analyses of putative conformation switching and cross-species inhibitory domains in microviridae external scaffolding proteins," *Virology*, 310: 64–71, 2003.

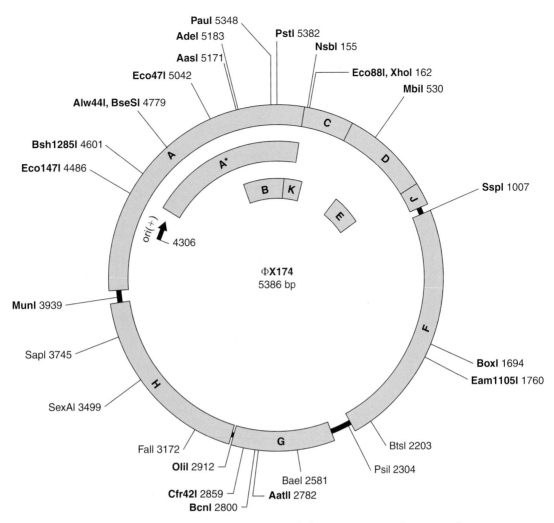

Figure 17-57. Representation of circular single-stranded DNA genome of φX174. The restriction enzymes used to cut the DNA are indicated along with the base position. *A*, 3981-136; *A**, 4497-136; *B*, 5075-51; *C*, 133-393; *D*, 390-848; *E*, 568-843; *F*, 1001-2284; *G*, 2395-2922; *H*, 2931-3917; *J*, 848-964; and *K*, 51-221.

repressor protein, which binds to O_R, preventing the transcription of cro but allowing transcription of cI (O_R to O_L) and preventing the transcription of N and the other genes in the left end of the genome; cII and cIII encode activator proteins, which bind to the genome and enhance the transcription of the cI gene. Cro (encodes a 66-amino acid protein) binds to O_R and blocks the binding of repressor to this site; N encodes an antiterminator protein (alternative rho factor) for host cell RNA polymerase, modifying its activity and permitting extensive transcription from P_L and P_R; and Q is an antiterminator similar to N, but it permits extended transcription only from P_R (definitions of these factors are taken from the reference in Figure 17-60).

N is transcribed from P_L and cro is transcribed from P_R in a newly infected cell, as shown in Figure 17-61. The N protein functions as a positive transcriptional regulator, allowing RNA polymerase to transcribe

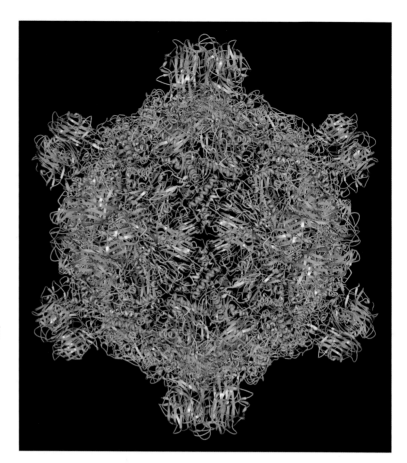

Figure 17-58. The ϕX174 DNA-binding protein J in two different capsid environments. Reproduced from PDB ID: 1rb8. R.A. Bernal, S. Hafenstein, N.H. Olson, V.D. Bowman, P.R. Chipman, T.S. Baker, B.A. Fane, M.G. Rossmann. Structural Studies of Bacteriophage-α3 Assembly. *J. Mol. Biol.* **325** pp. 11 (2003) and http://www.icsb.org/pdb/explore/explore.do?structureid=1RB8.

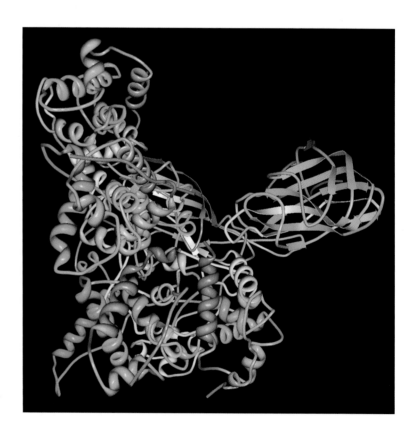

Figure 17-59. Structure of a viral procapsid with molecular scaffolding. Reproduced from L.L. Ilag et al., "DNA packaging intermediates of bacteriophage ϕX174," *Structure (London, England)*, 3: 353-363, 1995 and http://www.ebi.ac.uk/msd–srv/msdlite/images/1al0600.jpg.

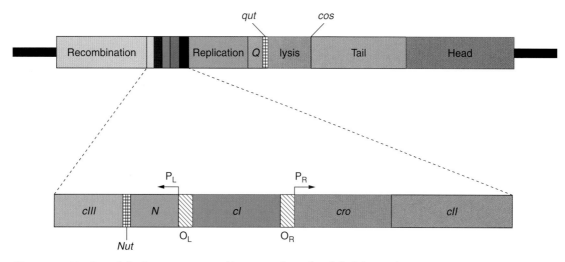

Figure 17-60. Simplified genetic map of bacteriophage lambda (phage λ).

many phage genes, including those responsible for DNA recombination and integration and cII and cIII. If the N protein is not expressed, RNA polymerase stops at certain sequences at the end of the N (*nut* site) and Q (*qut* site) genes. This restriction can be overcome by RNA polymerase–N protein complexes to permit full transcription from P_L and P_R. The complex of RNA polymerase–Q protein allows extended transcription only from P_R. The cI repressor gene is turned on as the *E. coli* cellular levels of cII and cIII proteins increase. This is a critical event determining the outcome of infection. *If cII remains below a critical level (cII is constantly degraded by cellular proteases), transcription from P_R and P_L continues (the usual case) and the phage is replicated and the cell lyses, releasing the particles.*

Occasionally, the concentration of cII increases, enhancing the transcription of the cI gene and increasing the concentration of the cI repressor (lower left of Figure 17-61). Binding of cI to O_R and O_L prevents transcription of all phage genes except cI, and this effective level (but not higher) is maintained by a negative feedback mechanism, keeping the cell in a stable state of lysogeny. At higher concentrations of cI, its own transcription is inhibited (Figure 17-62). When cI is not bound to O_R, cro is transcribed from P_R and cro binds preferentially to the left end of O_R (cI binds preferentially to the right end of O_R). The transcription of cI stops, and that of cro is enhanced (positive feedback loop). The phage is now committed to a lytic cycle and cannot return to the lysogenic state (prophage pathway, Figure 17-48). Some structures of proteins derived from phage-λ are shown in Figure 17-63.

The sequence of events of phage infection of a bacterial cell (in this case, *E. coli*; therefore, these phages are coliphages) can be summarized. First, there is the binding of the phage to the bacterial cell. In the case of phage-λ, the cell wall receptor is the high-affinity maltose receptor (Figure 17-64). This is followed by penetration, wherein the phage genome is injected and enters the host cell cytoplasm within seconds. The capsid and other structural elements do not penetrate the host cell. The

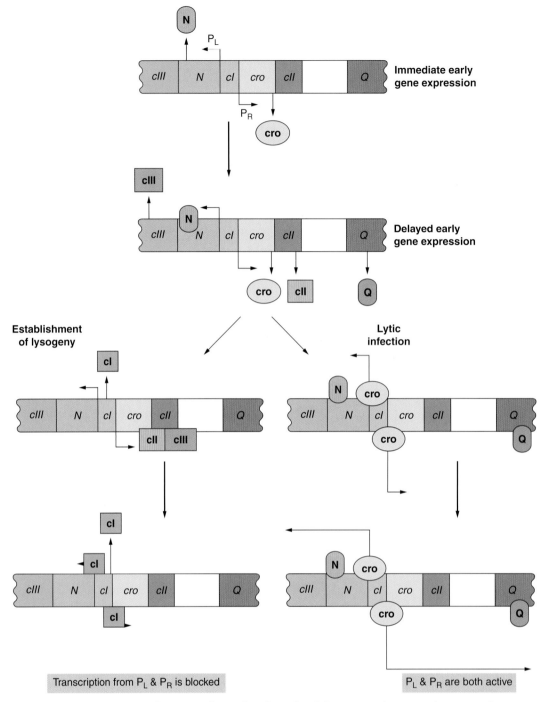

Figure 17-61. Expression of proteins from the phage lambda genome (via RNA) in immediate early gene expression, delayed early gene expression, establishment of lysogeny and lytic infection.

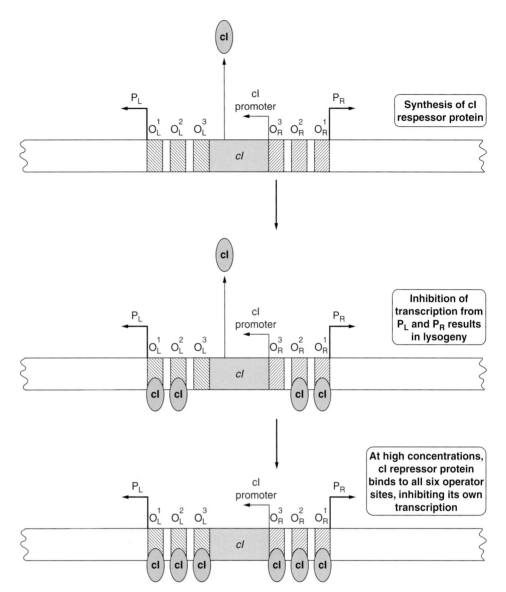

Figure 17-62. Expression of cI and consequences.

host cell is then converted to a phage-producing cell that requires the cessation of bacterial RNA and DNA syntheses. The nucleotide pool is used for replication of the virus. A number of biosynthetic steps follow: Phage DNA is replicated and transcribed into phage mRNAs that are translated into phage proteins. Phage mRNAs are synthesized as needed rather than all at once. The early messages encode enzymes necessary to take over the bacterial cell, and the proteins for phage DNA replication are made, as well as the proteins required for the production of late mRNAs. For late mRNAs, messages encode phage structural proteins, proteins for packaging phage nucleic acid into newly formed capsids, and proteins required for the lysis of the host cell. Finally, the host cell lyses

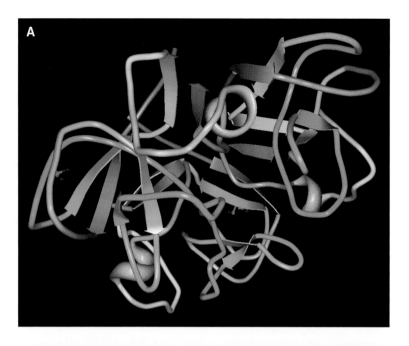

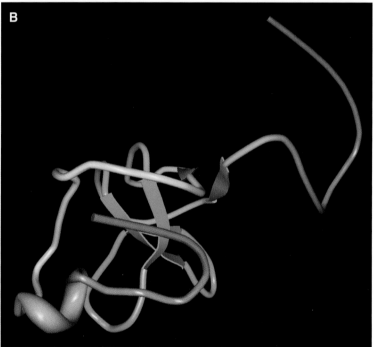

Figure 17-63. **A.** Crystal structure of gpD, the capsid-stabilizing protein of bacteriophage-λ. **B.** Head decoration protein of phage-λ. **C.** Fragment core-binding domain of phage-λ integrase. **D.** Bacteriophage-λ excisionase−DNA complex. **E.** Regulatory protein CII of bacteriophage-λ. Part A reproduced from PDB ID: 1e5e. C. Chang, A. Pluckthun, A. Wlodawer. Crystal Structure of a Truncated Version of the Phage Lambda Protein gpD. *Proteins* **57** pp. 866 (2004) and http://www.rcsb.org/pdb/explore/images.do?structure. Part B reproduced from http://www.rcsb.org/pdb/static.do?p=explorer/viewers/king.jsp. PDB ID: 1 vd0. H. Iwai, et al. NMR Solution Structure of the Monomeric Form of the Bacteriophage Lambda Capsid Stabilizing Protein gpD. *J. Biomol. NMR.* **31** pp. 351 (2005).

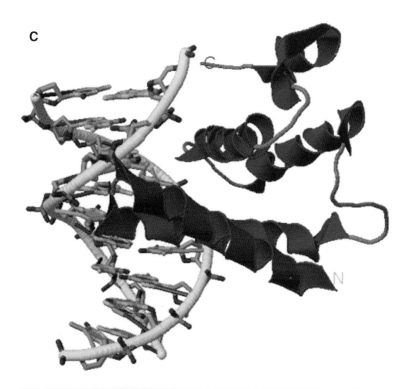

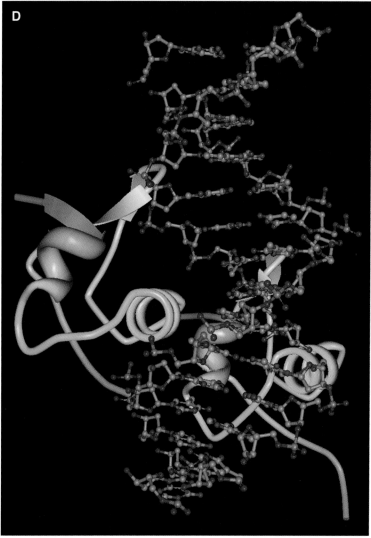

Figure 17-63, cont'd Part C reproduced from http://www.imb-jena.de/cgi-bin/ImgLib.pl?CODE=1M97. PDB ID: 1m97. B.M. Swalla, R.I. Gumport, J.F. Gardner. Conservation of Structure and Function Among Tyrosine Recombinases: Homology-based Modeling of the Lambda Integrase Core-binding Domain. *Nucleic Acids Research* **31** pp. 805 (2003). Part D reproduced from http://www.ebi.ac.uk/thornton-srv/databases/cgi-bin/pdbsum/GetPage.pl?pdbcode=1rh6. PDB ID: 1rh6. M.D. Sam, D. Cascio, R.C. Johnson, R.T. Clubb. Crystal Structure of the Excisionase-DNA Complex from Bacteriophage Lambda. *Journal of Molecular Biology* **338** pp. 229 (2004).

(Continued)

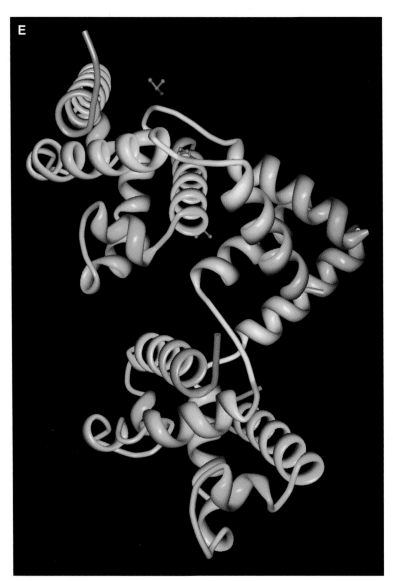

Figure 17-63, cont'd Part E reproduced from http://www.rcsb.org/pdb/cgi/explore.cgi?pdbId=1xwr. PDB ID: 1xwr. A.B. Datta, et al. Structure of Lambda CII: Implications for Recognition of Direct-repeat DNA by an Unusual Tetrameric Organization. *PNAS* **102** pp. 11242 (2005).

and progeny phage are released. Lysozyme or endolysin is synthesized late in the cycle to stimulate lysis of the cell membrane. The sequence of these steps occurs over a burst time of 10 to 60 seconds. Those phages that have a lysogenic cycle will enter this cycle (synthesis of prophages) if the environmental conditions are unfavorable. This cycle allows the phage genome to survive by being integrated into the chromosome of the host bacterium. Thus, the lysogenic cycle is selected when cII is high in concentration. When it is low, the lytic cycle takes place.

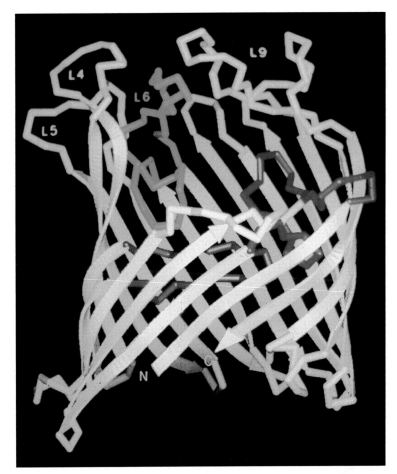

Figure 17-64. Crystal structure of phage-λ receptor (maltoporin). Crystal structure of LamB. A schematic drawing of the λ-receptor (maltoporin) monomer is shown. The cell exterior is at the top, and the periplasmic space at the bottom. The area of the subunit in trimer contacts is facing the viewer. The 18 antiparallel β-strands of the barrel are represented by *arrows*. Strands are connected to their nearest neighbors by loops or regular turns. Loops L1 *(blue)*, L3 *(red)*, and L6 *(green)* fold inward toward the barrel. L3 is the major determinant of the constriction site. The *yellow* bond symbolizes the disulfide bridge Cys22–Cys38 within loop 1. Loop 2, facing the viewer, latches onto an adjacent subunit in the trimer. Loops L4 to L6 and L9 form a large protrusion. The horizontal lines delineate the boundaries of the hydrophobic core of the membrane as inferred from the hydrophobic area found on the molecular surface. *LamB*, high affinity maltose receptor; *maltoporin*, Phage-λ receptor. Reproduced from Figure 5 of W. Boos and H. Shuman, "Maltose/maltodextrin system of *Escherichia coli*: Transport, metabolism, and regulation," Microbiol. Mol. Biol. Rev., 62, 204–229, 1998 and http://mmbr.asm.org/cgi/content/full/62/1/204/F5.

A Bacterial Cell, *E. coli*

E. coli, a gram-negative bacterium with a rod shape (0.5 μm wide; 1.5 μm long) that contains no nucleus but does contain chromatin (Figure 17-65), inhabits the intestinal tracts of most vertebrae. It has flagella (proteinaceous) that either propel or stop the forward motion of the cell. The cell has pili that are involved in sexual conjugation and may adhere the *E. coli* cell to surfaces. The major components of a bacterial cell are illustrated by a cartoon in Figure 17-66. Flagella and pili are pictured in Figures 17-67 and 17-68.

The Gram stain is a universal stain that separates bacteria into two classes: those that take up and retain the stain (gram-positive) and those that do not (gram-negative). The stain (invented by Hans Christian Gram in 1884) uses crystal violet as the primary stain, iodine as a mordant, alcohol (or acetone, or a mixture) as decolorizer, and safranin as a counterstain. The characteristics of the reactions of gram-positive cells and gram-negative cells are summarized in Table 17-3.

Gram-positive cells contain thick layers of peptidoglycan (also known as murein), constituting about 90% of the cell wall, and gram-negative bacteria contain peptidoglycan, constituting only about 10% of the cell wall (Figure 17-69). A peptidoglycan is synthesized in the bacterial cell membrane. Acetylmuramic acid–uridine diphosphate (UDP; attached to a peptide side chain) is transferred across the lipid cell membrane

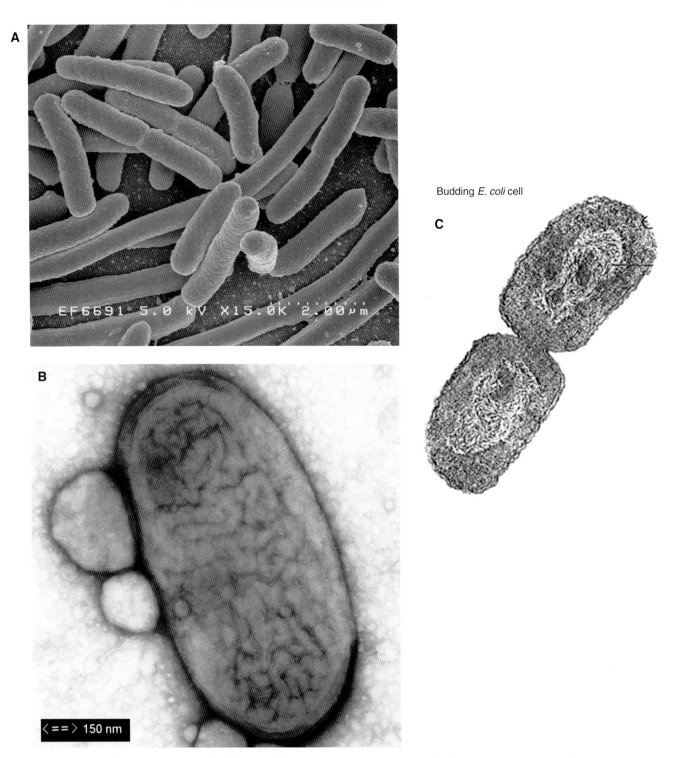

Figure 17-65. **A.** Electron micrograph of *E. coli*. This is an electron micrograph of common gram-negative bacteria that reside in the intestines of most vertebrates on the planet. In size it is approximately 1 to 2 μm in length by 0.5 to 1.0 μm in width. These cells are actively growing as can be seen by the number of cells in the process of cell division or binary fission. Some of them look ready to separate, and others appear to be just beginning to form their "cross walls." Their surface appears to be covered with a fuzzy material. This "fuzz" is composed of lipopolysaccharide and capsular material that covers the outer portion of the cell. These substances serve as armor to protect the cells. **B.** Transmission electron microscopy of an *E. coli* cell stained with uranylacetate. **C.** Transmission electron microscopy (TEM = 92,750) of rod-shaped *E. coli* dividing by binary fission. Part A reproduced from http://pl.wikipedia.org/wiki/Grafika:E_coli.jpg. Rocky Mountain Laboratories, National Institute of Allergy and Infectious Diseases, National Institutes of Health. Part B reproduced from http://www.mardre.com/homepage/mic/tem/samples/bio/ecoli/ecoli_3.html. Dr. rer. Nat. Markus Drechsler, University of Bayreuth. Part C reproduced with permission from Lehninger's Biochemistry Worth Publishers.

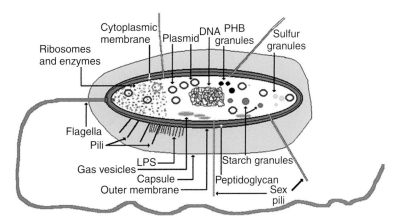

Figure 17-66. Composite cartoon showing the major structures found in bacteria. No bacterium contains all of these different components, but most bacteria contain the majority of them. *LPS*, lipopolysaccharide. Reproduced from Figure 3 of http://www.slic2.wsu.edu:82/hurlbert/micro101/pages/Chap3.html#cell_cartoon. Dr. R. E. Hurlbert, Washington State University.

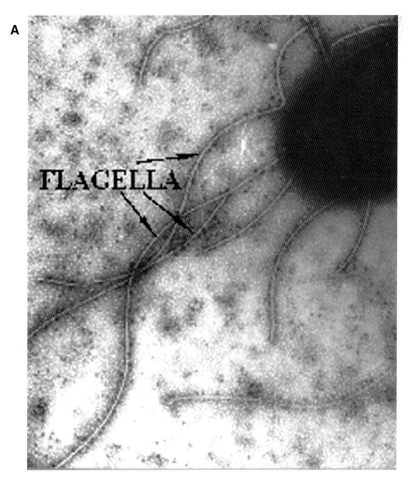

Figure 17-67. **A.** Bacterial flagella, the means of bacterial locomotion. Flagella are arranged in a manner (Figure 17-67B) descriptive of each bacteria.

(Continued)

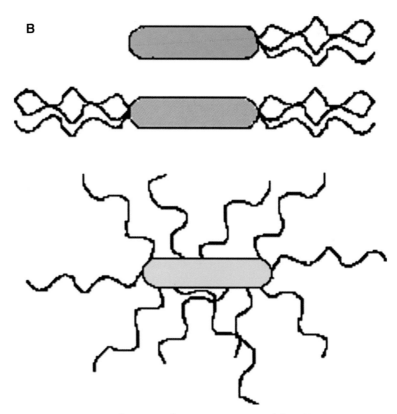

Figure 17-67, cont'd **B.** Specific arrangements of flagella that distinguish a specific bacterium. Parts A and B reproduced from pages 7 and 8 of http://www.slic2.wsu.edu:82/hurlbert/micro101/pages/Chap3.html#cell_cartoon. Dr. R. E. Hurlbert, Washington State University.

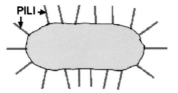

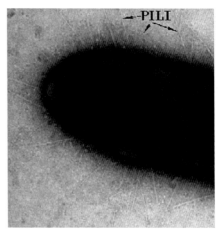

Figure 17-68. Pili. These fine, hairlike protein structures on the cell wall are pili. There are usually several hundred per cell. In most cases they have special binding proteins at the end of the stiff rods. These types of pili are often important in adhesion of the cell to surfaces, such as teeth. Reproduced from http://www.slic2.wsu.edu:82/hurlbert/micro101/pages/Chap3.html#cell_cartoon. Dr. R. E. Hurlbert, Washington State University.

Table 17-3
The Gram Stain Procedure

Staining solution	Gram +	Gram −
Crystal violet (primary stain)	Cells are stained violet	Cells are stained violet
Iodine (mordant)	Crystal violet–iodine complex forms	Crystal violet–iodine complex forms
Alcohol (decolorizer)	Cells remain violet	Cells are colorless
Safranin (counterstain)	Cells remain violet	Cells are red

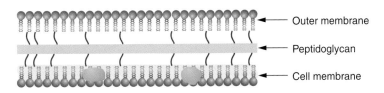

Figure 17-69. Cartoon illustrating the relative structure of gram-negative *(top)* and gram-positive *(bottom)* cell walls. The major differences lie in the thickness of the rigid peptidoglycan layer and in the presence of an outer membrane in gram-negative cells. In gram-negative cells the peptidoglycan layer is very thin, being only a few molecules thick, whereas in gram-positive cells this layer is very thick. CM, cell membrane.

containing a large C_{55} lipid by a pyrophosphate bridge (peptidomuramic acid–P–P–C_{55} lipid, in the membrane). N-Acetylglucosamine (NAG)–UDP, on the outside, is added to the acetylmuramic acid–P–P–C_{55} lipid on the inside with the release of UDP. In the final step, there is a transpeptidation in which the loose end of the (Gly)$_5$ is attached to the peptide side chain of the acetylmuramic acid peptide to form NAG–N-acetylmuramic acid (NAM) with a peptide side chain (and the glycine pentamer added to the peptide side chain laterally) to form the NAG–NAM peptide (containing Gly$_5$ added laterally)–P–P–C_{55} lipid, which is a peptidoglycan.

An enzyme involved in the cross-linking process by way of L-alanyl-pentaglycyl–peptidoglycan is a ligase, UDP–N-acetylmuramoyl: L-alanine synthetase, whose structure is shown in Figure 17-70. The

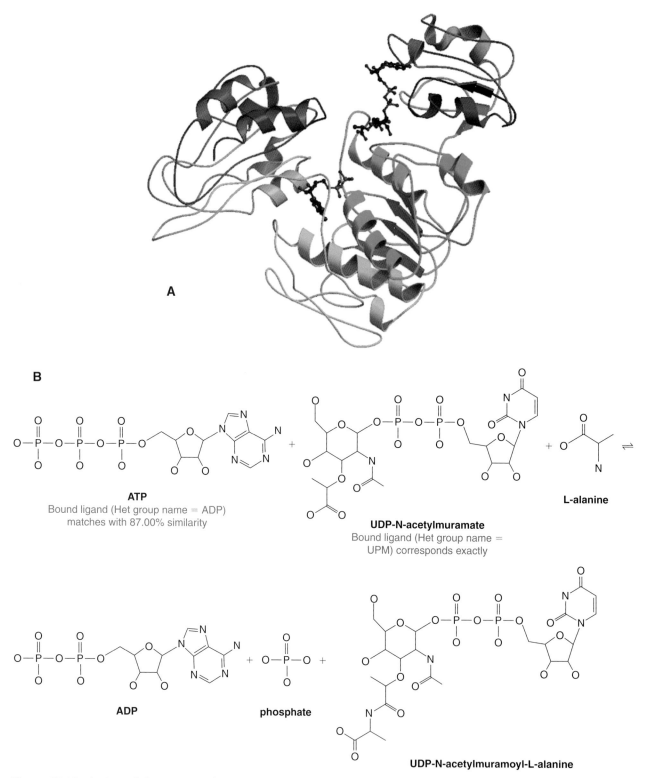

Figure 17-70. **A.** A model structure of UDP-N-acetylmuramoyl: l-alanine synthetase (murc) from *E. coli* based on homology and hydrophobic cluster analysis and its validation by amino acid modifications. **B.** Reaction catalyzed by this enzyme. Part A reproduced from http://www.ebi.ac.uk/msd−srv/msdlite/images/1cc9600.jpg. PDB ID: 1cc9. F. Nosal, et al. A Model Structure of UDP-N-acetylmuramoyl: 1-alanine Synthetase (murc) from *Escherichia coli* Based on Homology and Hydrophobic Cluster Analysis and Its Validation by Amino Acid Modifications. To be published.

pentapeptide from NAM is composed of L-alanine, D-glutamic acid, *meso*-diaminopimelic acid, and two D-alanines. Diaminopimelic acid is formed from L-aspartate, which is converted to L-aspartate semialdehyde in two steps; then, in four subsequent enzymatic reactions, L-aspartate semialdehyde is converted to dihydrodipicolinate, L-diaminopimelate, and *meso*-diaminopimelate. The structures of NAG and NAM are shown in Figure 17-71.

The overall structure of the *E. coli* peptidoglycan is shown in Figure 17-72. The peptidoglycan of the gram-negative *E. coli* is a polymer of the monomer, NAG–NAM–pentapeptide. Transpeptidase (Figure 17-73) enzymes then cross-link the chains, providing a strong structure that can resist osmotic lysis. A transglycosylase from family 2 of *E. coli* (there are several transglycosylase families), sMltA, is shown in Figure 17-74, which also shows the reaction catalyzed by the enzyme. This enzyme cleaves the β-(1-4)-glycosidic bonds in peptidoglycan, forming the non-reducing 1,6-anhydromuropeptides. Opening the chain is part of the maintenance and growth of the bacterial cell wall peptidoglycan.

In the cytosol are formed peptidoglycan monomers (Figure 17-75) that are attached to **bactoprenol** (a membrane carrier molecule). Bactoprenol, an abundant membrane lipid, is formed by the condensation of 10 unsaturated isoprene units and 1 saturated isoprene unit; it contains 10 double bonds per molecule with a molecular weight of 768. It contains a hydroxyl group, making it a C_{55} isoprenoid alcohol. Bactoprenol transports the peptidoglycan monomers across the cytoplasmic membrane and facilitates their insertion into the growing peptidoglycan chains (Figure 17-76). NAG is linked to UDP, forming UDP–NAG; some of the NAG is converted to NAM, and UDP–NAM can be formed.

The pentapeptide is formed from five amino acids (see Figure 17-75), of which the terminal two amino acids are D-alanine (D forms are not incorporated into proteins) formed from L-alanine by a racemase, a pyridoxal

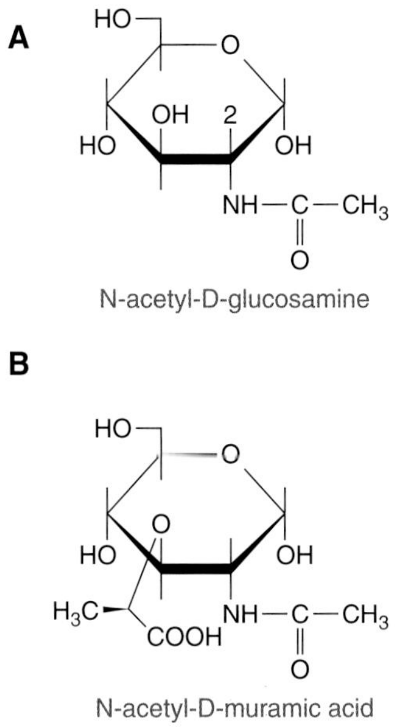

Figure 17-71. Structures of N-acetyl-D-glucosamine *(A)* and N-acetyl-D-muramic acid *(B)*.

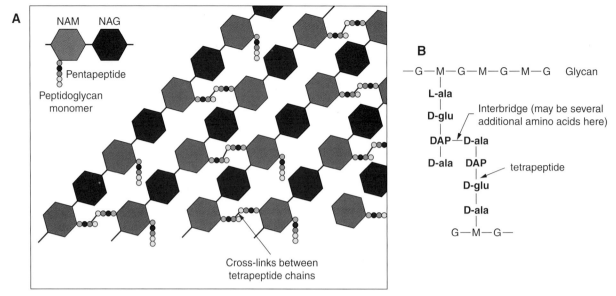

Figure 17-72. **A.** *E. coli* peptidoglycan composed of cross-linked chains of peptidoglycan monomers (NAG-NAM-pentapeptide). Transglucosidase enzymes join these monomers together to form chains. Transpeptidase enzymes then cross-link the chains to provide strength to the cell wall and enable the bacterium to resist osmotic lysis. In *E. coli*, the pentapeptide coming off the NAM is composed of the amino acids L-alanine, D-glutamic acid, *meso*-diaminopimelic acid, and two D-alanines. A gram-positive cell would contain peptidoglycan with more cross-links (in a stronger structure) than gram-negative cells (e.g., *E. coli*). *NAG*, N-acetylglucosamine; *NAM*, N-acetylmuramic acid. **B.** Structure of peptidoglycan showing side chain interbridge. *DAP*, diaminopimelic acid. Part A redrawn from Doc Kaiser's Microbiology Home Page. Copyright © Gary E. Kaiser. All Rights Reserved. Updated: February 8, 2005. http://www.cat.cc.md.us/courses/bio141/lecguide/unit1/prostruct/u1fig8_ec.html.

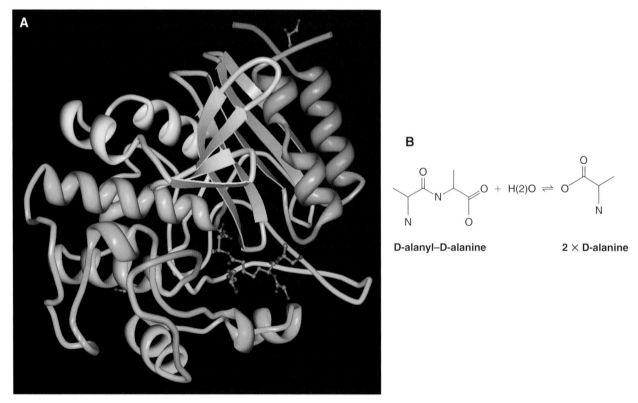

Figure 17-73. **A.** The crystal structure of phosphonate-inhibited D-Ala−D-Ala peptidase reveals an analog of a tetrahedral transition state. **B.** Reaction catalyzed by transpeptidase B. Part A reproduced from PDB ID: 1mpl. N.R. Silvaggi, J.W. Anderson, S.R. Brinsmade, R.F. Pratt, J.A. Kelly. *Biochemistry* **42** pp. 1199 (2003).

phosphate enzyme (Figure 17-77). Attachment of NAM–pentapeptide to bactoprenol is powered by a high-energy phosphate group of UDP. The NAG is attached to the NAM–pentapeptide on bactoprenol to form the completed peptidoglycan monomer. A bacterial enzyme, **autolysin** (Figure 17-78), cleaves the glycosidic bonds between the peptidoglycan monomers at the growth point of the existing peptidoglycan (Figure 17-79). The peptidoglycan monomers are inserted into the breaks in the peptidoglycan at the growing point of the cell wall by bactoprenol and

(Text continues on p. 1118).

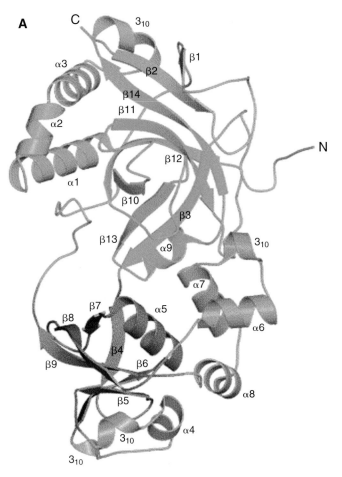

Figure 17-74. **A.** Crystal structure and topology of sMltA. The overall structure of sMltA shown as a ribbon diagram. Domain A is shown in *blue* and *green*, domain B is shown in *yellow* and *orange*, with the double-Ψ and TL5-related β-barrel folds highlighted in *green* and *orange*, respectively. Secondary structure elements are indicated. SmltA, soluble family 2 transglycosylase of *E. coli*. TL5 is an RNA-binding domain. **B.** Proposed mechanism of MltA *E. coli* transglycosylase. **C.** Surface representation of sMltA with the modeled glycan substrate. The modeled (MurNac-GlcNac)$_3$ glycan substrate is shown in stick representation, with the lactyl groups on the MurNAc residues omitted for ease of illustration. Residues that are strictly conserved among the different MltA sequences are colored *violet*. Part A reproduced from http://www.sciencedirect.com and Figure 1 of K.E. van Straaten et al., "Crystal structure of MltA from *Escherichia coli* reveals a unique lytic transglycosylase fold," *J. Mol. Biol.*, 352: 1068–1080, 2005.

(Continued)

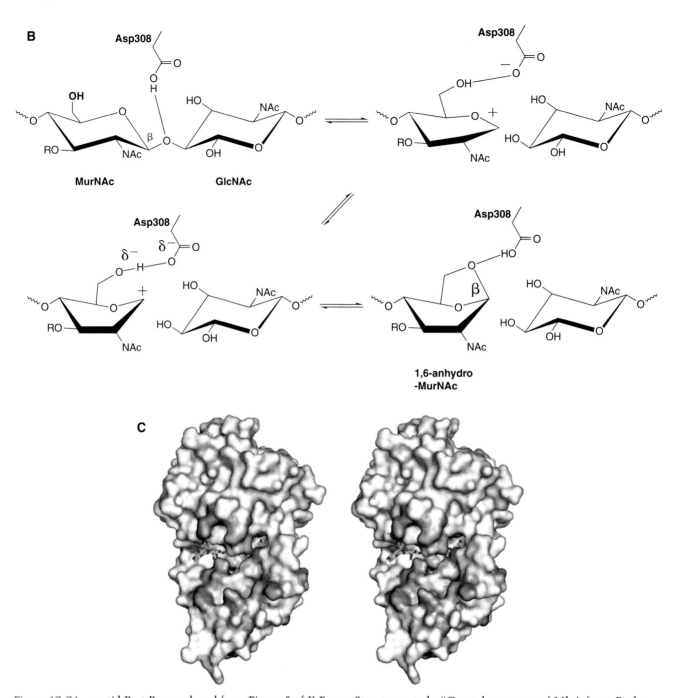

Figure 17-74, cont'd Part B reproduced from Figure 5 of K.E. van Straatena et al., "Crystal structure of MltA from *Escherichia coli* reveals a unique lytic transglycosylase fold," *J. Mol. Biol.*, 352: 1068–1080, 2005 and http://www.sciencedirect.com. Part C reproduced from Figure 6 of K.E. van Straaten et al., "Crystal structure of MltA from *Escherichia coli* reveals a unique lytic transglycosylase fold," *J. Mol. Biol.*, 352: 1068–1080, 2005.

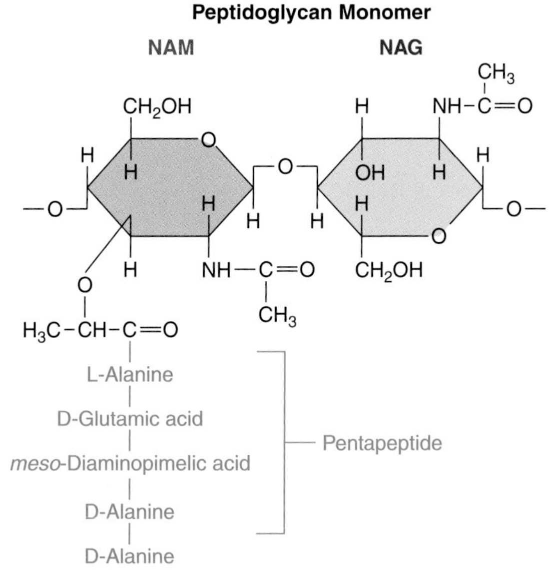

Peptidoglycan Monomer

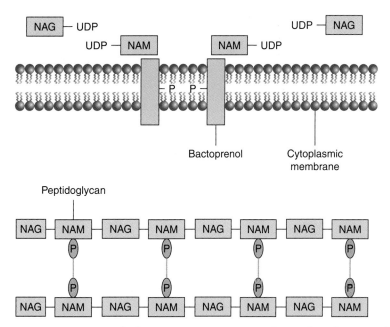

Figure 17-75. A peptidoglycan monomer consists of two joined amino sugars, N-acetylglucosamine (NAG) and N-acetylmuramic acid (NAM), with a pentapeptide coming off of the NAM. In *E. coli*, the pentapeptide consists of the amino acids L-alanine, D-glutamic acid, *meso*-diaminopimelic acid, and two D-alanines. Redrawn from Figure 2A of Doc Kaiser's microbiology Home Page. Copyright © Gary E. Kaiser. All Rights Reserved. http://www.cat.cc.md.us/courses/bio141/lecguide/unit1/images/prostruct/u1fig8a.html.

Figure 17-76. Peptidoglycan monomers are synthesized in the cytosol of the bacterium, where they attach to a membrane carrier molecule called **bactoprenol**. The bactoprenols transport the peptidoglycan monomers across the cytoplasmic membrane and help insert them into the growing peptidoglycan chains. Redrawn from Figure 3 of Doc Kaiser's Microbiology Home Page. Copyright © Gary E. Kaiser. All Rights Reserved. http://faculty.ccbcmd.edu/courses/bio141/lecguide/unit1/prostruct/images/ppgmonomer03.jpg.

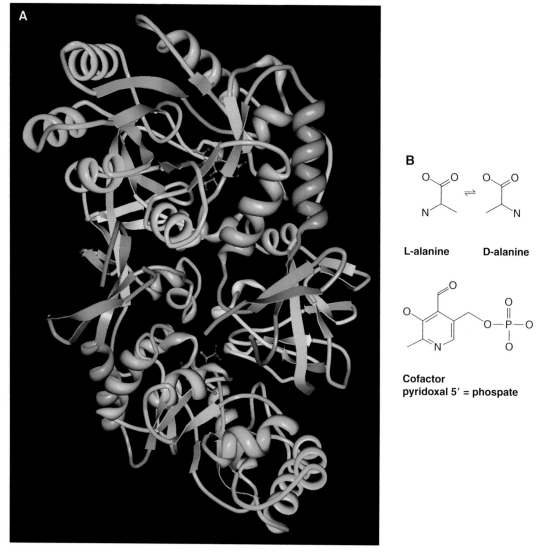

Figure 17-77. **A.** Alanine racemase complexed with alanine phosphonate. **B.** Racemase reaction converting L-alanine to D-alanine. Reproduced from PDB ID: 1bd0. G.F. Stamper, A.A. Morollo, D. Ringe, C.G. Stamper. Reaction of Alanine Racemase with 1-Aminoethylphosphonic Acid Forms a Stable External Aldimine. *Biochemistry* **37** pp. 10438 (1998) and http://www.rcsb.org/pdb/explore/images.do?structureId=1bd0.

transglycosidase (Figures 17-80 and 17-81). In this way, the inner part of the cell wall peptidoglycan of gram-negative bacteria is formed. A summary of the formation of peptidoglycan is shown in Figure 17-82.

Although the gram-positive bacterial cell has a thick proteoglycan layer as the outer part of its cell wall (and consequently binds the Gram stain), the *gram-negative bacterial cell, like E. coli, has an outer layer (selective permeability barrier, a lipopolysaccharide–phospholipid asymmetric bilayer) covering the proteoglycan* (see Figure 17-69B). This layer in *E. coli* has characteristic lipid modifications located at an amino-terminal cysteine, and the layer can be covalently bound to the peptidoglycan layer by way of a carboxy terminal lysine (Figure 17-83). Although the outer layer (Figure 17-84) is relatively impermeable to many types of small molecules, there are channels and receptors in the outer membrane

(Text continues on p. 1125.)

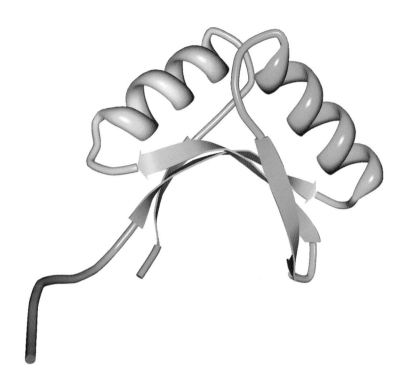

Figure 17-78. Solution structure of the peptidoglycan-binding domain of Bacillus subtilis cell wall lytic enzyme cwlc (autolysin). PBD ID: 1x60. M. Mishima, T. Shida, K. Yabuki, K. Kato, J. Sekiguchi, C. Kojima. Structure of the Peptidoglycan Binding Domain of *Bacillus subtilis* Cell Wall Lytic Enzyme CwIC: Characterization of the Sporulation-Related Repeats by NMR. *Biochemistry* **44** pp. 10153 (2005).

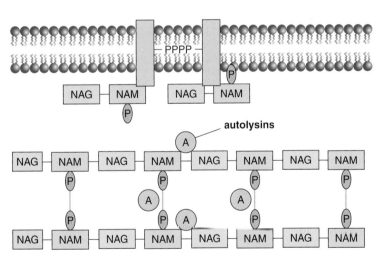

Figure 17-79. A group of bacterial enzymes called autolysins break the glycosidic bonds between the peptidoglycan monomers at the point of growth along the existing peptidoglycan. They also break the peptide cross-bridges that link the rows of sugars together. In this way, new peptidoglycan monomers can be inserted and enable bacterial growth. Redrawn from Figure 5 of Doc Kaiser's Microbiology Home Page. Copyright © Gary E. Kaiser. All Rights Reserved. http://faculty.ccbcmd.edu/courses/bio141/lecguide/unit1/prostruct/images/ppgmonomer03.jpg.

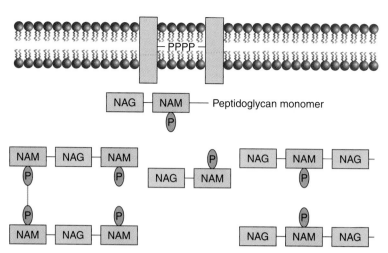

Figure 17-80. Bactoprenol and transglycosidase enzymes (not shown here) insert the peptidoglycan monomers into the breaks in the peptidoglycan at the growing point of the cell wall. Redrawn from Figure 4 of Doc Kaiser's Microbiology Home Page. Copyright © Gary E. Kaiser. All Rights Reserved. http://faculty.ccbcmd.edu/courses/bio141/lecguide/unit1/prostruct/images/ppgmonomer03.jpg.

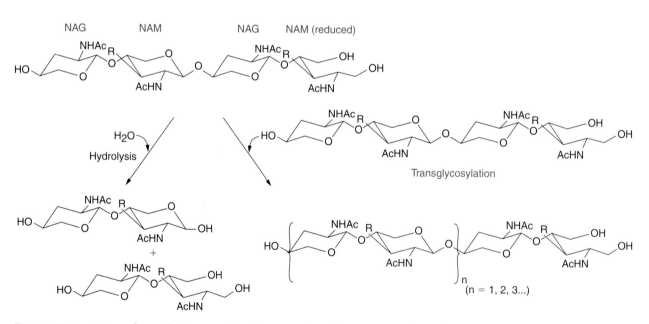

Figure 17-81. A transglycosylation reaction of a transglycosidase or transglycosylase *(pathway on right)*. Reproduced from R. Kuroki, L.H. Weaver, and B.W. Matthews, "Biochemistry structural basis of the conversion of T4 lysozyme into a transglycosidase by reengineering the active site," *PNAS*, 96: 8949–8954, 1999 and http://www.pnas.org/content/vol96/issue16/images/large/pq1692349001.jpeg.

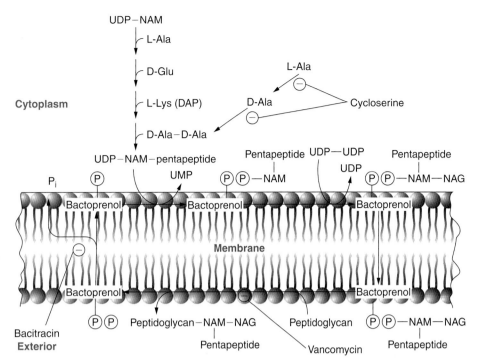

Figure 17-82. Overall summary of peptidoglycan synthesis. Locations of inhibition of cell wall formation by cycloserine, bacitracin, and vancomycin are shown. *DAP*, diaminopimelic acid; *NAG*, N-acetylglucosamine; *NAM*, N-acetylmuramic acid. Redrawn from L.M. Prescott, J.P. Harby, D.A. Klein, *Microbiology*, 4e. Copyright © 1999 The McGraw-Hill Companies, Inc. All rights reserved.

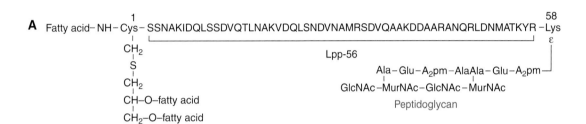

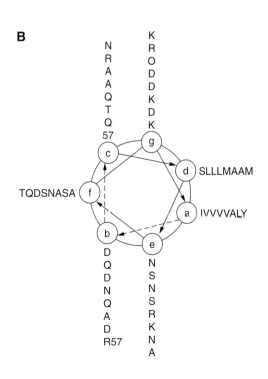

Figure 17-83.

For legend see next page.

A Bacterial Cell, *E. coli*

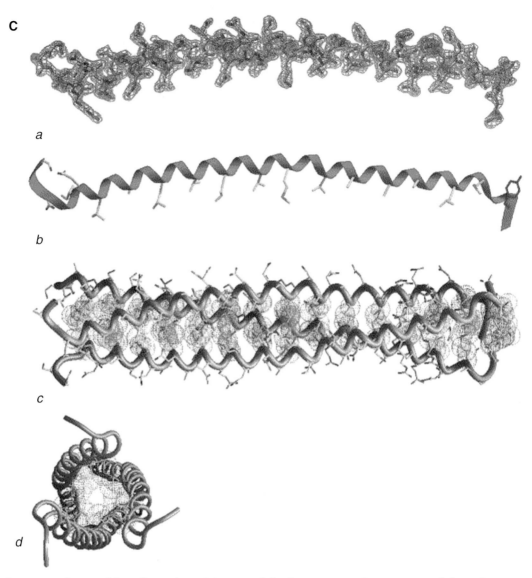

Figure 17-83, cont'd **A.** Outer membrane of *E. coli* envelope. Diagram of the lipoprotein. The sequence of the polypeptide chain, designated Lpp-56, is indicated. The N-terminal glycerylcysteine residue is attached to two fatty acids by two ester linkages and one fatty acid by an amide linkage. The ε-amino group of the C-terminal lysine is linked to the carboxyl group of every tenth to twelfth *meso*-diaminopimelic acid of the peptidoglycan. *(b)* Alignment of outer membrane lipoprotein sequences from four gram-negative bacteria. **B.** Folding of Lpp-56 as a stable, α-helical trimer. *(a)* helical wheel projection. The view is from the amino terminus looking toward the carboxyl terminus. Heptad positions are labeled *a* through *g*. **C.** The fold of the polypeptide chain. *(a)* $2F_o - F_c$ electron density map at 1.9-Å resolution, contoured at 1.0 σ. The Lpp-56 molecule is shown (residues 2 to 57). The overall fold is of a mostly α-helical segment from which extends amino- and carboxyl-terminal capping motifs. The side chains of residues at positions *a* and *d*, as well as key helix-capping residues are shown. *(c)* Side view of the Lpp-56 trimer. The van der Waals surfaces of residues at the *a (red)* and *d (green)* positions are superimposed on the helix backbone. The carboxyl-terminal tyrosine cap is shown in *magenta*. *(d)* Axial view of the Lpp-56 trimer. The view is from the amino terminus looking down the carboxyl terminus. Part A reproduced from http://www.sciencedirect.com/science?_ob=ArticleURL&_udi=B6WK7-45F51HTGD&_user=10&_handle=V-WA-A-W-WZ-MsSAYVW-UUA-U-AACDEAAEYD-AAC-CVEWDYD-CEVCVVDC-WZ-U&_fmt=full&_coverDate=06%2F16%2F2000&_rdoc=22&_orig=browse&_srch=%23toc%236899%232000%23997009995%23294249!&_cdi=6899&view=c&_acct=C000050221&_version=1&_version=1&_urlVersion=0&_userid=10&md5=fef74b8b0c1dbb6bd06cb333f5b96fdd and Figure 3 of W. Shua et al., "Core structure of the outer membrane lipoprotein from *Escherichia coli* at 1.9 å resolution," *J. Mol. Biol.*, 299: 1101–1112, 2000. Part B reproduced from Figure 2 of W. Shua, J. Liua, H. Jia and M. Lu, "Core structure of the outer membrane lipoprotein from *Escherichia coli* at 1.9 å resolution," *J. Mol. Biol.*, 299: 1101–1112, 2000. Part C reproduced from Figure 3 of W. Shua et al., "Core structure of the outer membrane lipoprotein from *Escherichia coli* at 1.9 å resolution," *J. Mol. Biol.*, 299: 1101–1112, 2000.

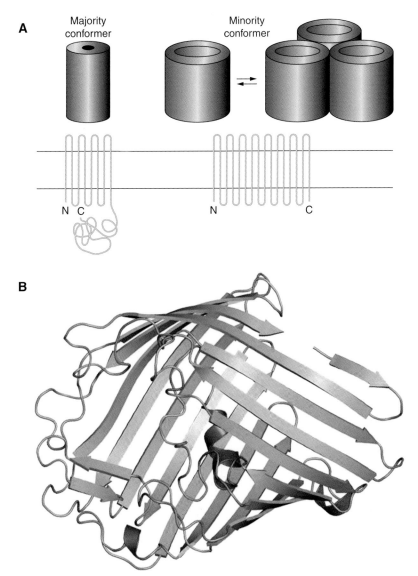

Figure 17-84. Cont'd **A.** Folding model of OmpA-OprF family slow porins. The major fraction of the population folds as a two-domain protein *(left)* and is important in binding the OM to the underlying peptidoglycan, since the C-terminal globular domain contains a peptidoglycan-binding motif. A minor fraction of the population, however, folds differently to produce an open β-barrel *(right)*. In *E. coli*, which produces trimeric, high-permeability porins, the presence of this fraction has no functional consequence. However, in fluorescent pseudomonads, which lack the high-permeability porin, this fraction functions as the major nonspecific porin. This fraction also tends to form a loosely associated oligomeric structure, as shown. The oligomer is shown as a trimer only for illustrative purposes. **B.** OmpF porin mutant D744. **C.** Electrostatic potential of porin Omp32 calculated at pH 7 and an ionic strength of 0.12 M. The size and the color of the grid dots characterize the potential at the respective points *(blue,* positive; *red,* negative). The upper threshold value was set to 2.5 kcal/mole (represented by the largest dots). Views are from the exoplasmic face *(a)*, periplasmic face *(b)*, longitudinal section showing the inner half of the porin channel with the arginine cluster in the construction zone *(c)*, and longitudinal section oriented toward the lipid membrane *(d)*. The strong positive potential in the top regions of *a* and *b* indicates the lysine girdle, the negative potential *(bottom)* the trimer contact regions. **D.** Isosurface of the electrostatic potential of Omp32, calculated for conditions of pH 7 and an ionic strength of .12M. View from the exoplasmic face. The isosurface was drawn just beyond the level of zero, leaving the regions with positive potential free from the isosurface network. Two positive funnel-shaped regions direct anions from the environment toward the channel. Part A redrawn from http://mmbr.asm.org/content/vol67/issue4/images/large/mr0430026003.jpeg. Part B redrawn from P.S. Phale et al., "Stability of trimeric OmpF porin: the contributions of the latching loop L2," *Biochem.*, 37: 15663–15670, 1998.

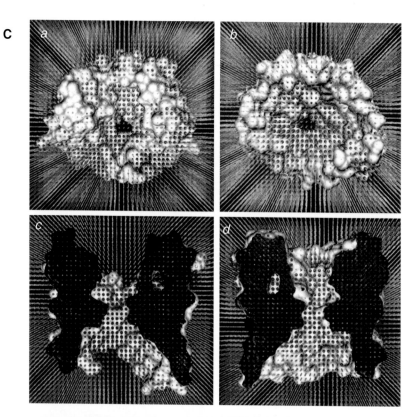
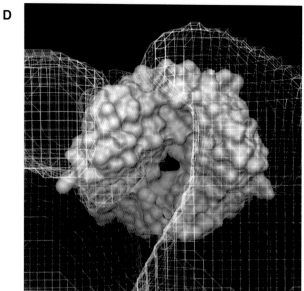

Figure 17-84, cont'd Part C reproduced from Figure 3 of U. Zachariae et al., "Electrostatic properties of the anion porin Omp32 from *Delftia acidovorans* and of the arginine cluster of bacterial porins," *Protein Sci.*, 11: 1309–1319, 2002 and http://proteinscience.org/cgi/content/full/11/6/1309/F3. Part D reproduced from Figure 4 of U. Zachariae et al., "Electrostatic properties of the anion porin Omp32 from *Delftia acidovorans* and of the arginine cluster of bacterial porins," *Protein Sci.*, 11: 1309–1319, 2002.

that allow the influx of nutrients and proteins. Of these, the **porins** (as many as 200,000 per cell) are made up of proteins that function as nonspecific diffusion channels (Figure 17-85). The channels are constricted by amino acid residues that develop a strong electrostatic field, as shown for the OmpF porin in Figure 17-86. A high-resolution topograph of crystals of the porin OmpF accentuates the characteristics of these structures (Figure 17-87) and other views of this channel are shown in Figure 17-88. A view of the outer membrane in context with the peptidoglycan layer and the inner membrane of gram-negative bacteria is shown in Figure 17-89. Other examples of porins are **LamB,** the maltodextrin channel, the **ferric citrate receptor (FecA)**, and **ScrY,** a sucrose-specific porin (Figure 17-90).

Proteins, such as lactoferrin (Lf) from leukocytes, modulate the functions of the porin channels. When lactoferrin is in close contact with the

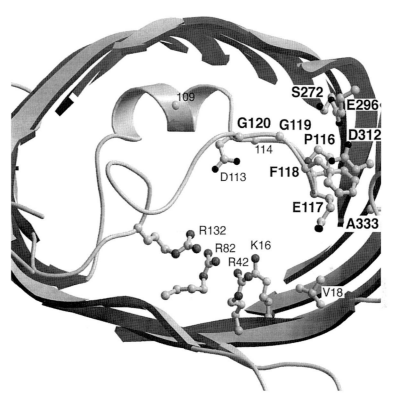

Figure 17-85. Structure of the channel constriction in wild-type OmpF porin. The cationic cluster (Arg-42, Arg-82, Arg-132, Lys-16) and the two carboxyl side chains (Asp-113, Glu-117) on loop L3, which are the main source of the strong electrostatic field across the constriction (1), are shown. The amino acid sequence of loop L3 from residue 112 to residue 121 is TDMLPEFGGD. The loop is linked to the β-barrel wall by a chain of hydrogen bonds (Ser-272···Glu-296···Asp-312···backbone Glu-117/Phe-118). Other contact areas between wall and L3 consist of hydrophobic residues. Residues indicated in boldface are those mutated in the present study. The strands of the β-barrel *(dark ribbons)* are clipped to allow a full view of the channel constriction. Reproduced from Figure 1 of P.S. Phale et al., "Voltage gating of *Escherichia coli* porin channels: Role of the constriction loop," *PNAS,* 94: 6741–6745, 1997 and http://www.pnas.org/content/full/94/13/6741/F1.

Figure 17-86. Comparison of high-resolution topographs of two-dimensional porin OmpF crystals *(brown-yellow)* and the atomic model rendered at 3 5 *(blue)*. Topographs were recorded with the AFM in 20 mM Tris-HCl (pH 7.8), 0.3 M KCl. The brightness range corresponds to 15 Å, and topographs are displayed as perspective views. *(A)* Periplasmic surface; *arrows* indicate short β-strand-connecting turns that can sometimes be seen in the topographs. *(B)* Extracellular surface; domains formed by long loops protrude by 13 Å away from the membrane and are flexible, leading to a disordered appearance. The twofold centers of rectangular unit cells determined by cross-correlation of the topograph with the atomic model are indicated by asterisks. Two porin trimers with their triangular vestibules are marked with circles. The two-dimensional OmpF crystals were reconstituted in the presence of dimyristoyl phosphatidylcholine at a lipid-to-protein ratio of 0.2. These crystals exhibited a rectangular ($A = 135$Å, $B = 82$Å) or a trigonal ($A = B = 82$Å) packing arrangement and were frequently double-layered. Crystals were adsorbed to mica in 20 mM Tris-HCl (pH 7.8), 0.3 M KCl and incubated for 30 minutes. Imaging was performed operating a Nanoscope III (Digital Instruments, Santa Barbara, CA) in the contact mode, using cantilevers from Olympus (Tokyo, Japan) with a length of 100 μm and a force constant of $k = 0.1$ N/m. In all cases, the force applied to the stylus was ≤100 pN to prevent sample distortion and the scan frequency was 16 Hz. To expose the extracellular surface, double-layered crystals were separated with the stylus by applying a larger force. Images recorded in trace and retrace direction simultaneously indicated that protein surfaces were not laterally distorted by the interaction with the AFM tip. Atomic models of trigonal or rectangular two-dimensional porin crystals were built by adjusting the angular orientation and lateral displacement of the three-dimensional atomic OmpF structure to topographs of the periplasmic surface by correlation methods using the SEMPER image processing program. The lipid surface was modeled as a planar layer fitted into the horizontal sections of appropriate height. Topographs were calculated by contouring the models at a threshold density determined by visual inspection. Reproduced from D. Muller and A. Engel, "Voltage and pH-induced channel closure of porin OmpF visualized by atomic force microscopy," *J. Mol. Biol.*, 285: 1347–1351, 1999.

OmpC channel, the ion flux is impeded, whereas the contact between human Lf and PhoE is weaker, allowing ion flux and contact with the lipid bilayer (OmpF and OmpC transport cations slightly better than anions, and OmpF allows passage of larger molecules than OmpC; PhoE prefers anions) (Figure 17-91). Between the peptidoglycan and the outer membrane is the periplasmic space which accounts for 20% to 40% of the total volume of the cell. It contains binding proteins for arabinose, maltose, galactose, histidine, leucine (as well as other amino acids), phosphate, sulfate, and the vitamins B_{12} and B_1. Also present in this space are asparaginase, acid phosphatase, alkaline phosphatase, carboxypeptidase-II, endonuclease-I, and 5'-nucleotidase. There are enzymes that inactivate antibiotics, such as β-lactamase and neomycin phosphotransferase, as well as membrane-derived oligosaccharides that increase in concentration in case of a fall in osmolarity.

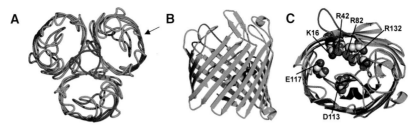

Figure 17-87. Structure of the OmpF porin of E. coli. **A.** View of the trimer from the top, that is, in a direction perpendicular to the plane of the membrane. Loop 2, colored *blue*, plays a role in interaction of the monomer with its neighboring unit. Loop 3, colored *orange*, narrows the channel. **B.** View of the monomeric unit from the side, roughly in the direction of the arrow in panel A. Loops 2 and 3 are colored as in panel A. **C.** View of the monomeric unit from the top, showing the "eyelet" or the constricted region of the channel. The eyelet is formed by Glu 117 and Asp 113 from the L3 loop, as well as four basic residues from the opposing barrel wall, Lys 16, Arg42, Arg82, and Arg132, all shown as spheres. Reproduced from Figure 2 of H. Hikaido, "Molecular basis of bacterial outer membrane permeability revisited," *Microbiol. Mol. Biol. Rev.*, 67: 593–656, 2003 and http://mmbr.asm.org/cgi/content/full/67/4/593/F2.

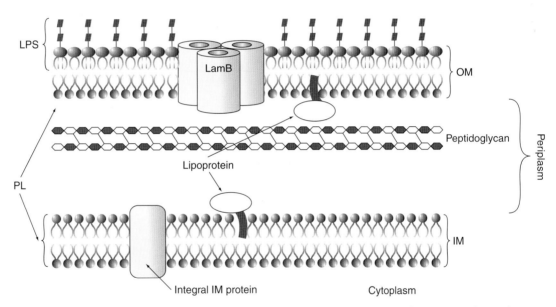

Figure 17-88. Outer membrane of gram-negative bacteria. *IM*, inner membrane; *LPS*, lipopolysaccharide; *OM*, outer membrane; *PL*, phospholipids. Redrawn from Figure 7 of T. Wu et al., "Identification of a multicomponent complex required for outer membrane biogenesis in *Escherichia coli*," *Cell*, 121: 235–245, 2005 and http://www.sciencedirect.com.

Substances are exported from the *E. coli* cell interior through export channels connecting the inner and outer membranes (Figure 17-91). A lipopolysaccharide structure is located in the outer membrane and is joined to the peptidoglycan layer (Figures 17-66 and 17-89). The active entity of the lipopolysaccharide is lipid A (Figure 17-92). Also indicated in Figure 17-89 is a lipoprotein residing both on the inner side of the

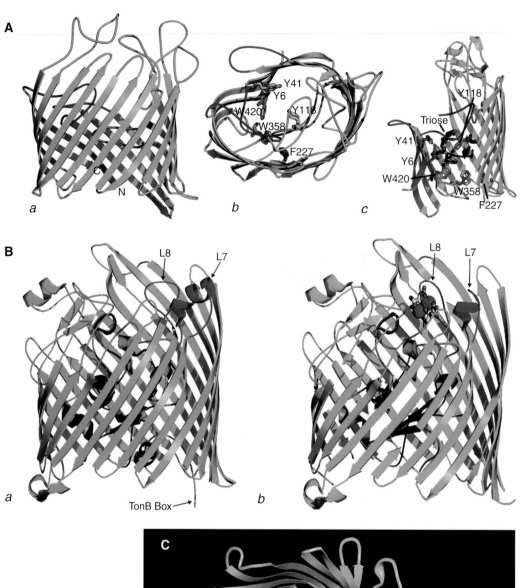

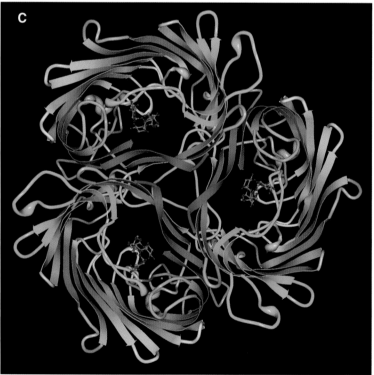

Figure 17-89.

For legend see opposite page.

Figure 17-89. **A.** X-ray crystallographic structure of LamB. *(A)* Side view of the monomeric unit. The β-barrel contains 18 strands in this protein, in contrast to the 16 strands seen in the trimeric porins. In addition to loop 3 *(orange)*, loop 1 *(red)* folds deeply into the channel. Loop 2 *(blue)* folds outward and interacts with the neighboring subunit in the trimer, as in OmpF. Other loops also are often large and tend to cover the entrance of the channel from the outside. *(B)* View of the monomeric unit from the top. The greasy slides (Tyr41, Tyr6, Trp420, Trp358, and Phe227) are shown as *blue stick diagrams,* and Tyr118, which constricts the diffusion channel from the opposite side, is shown as a *yellow stick diagram.* *(C)* View of the greasy slide and its interaction with maltotriose. This is a side view with the front of the β-barrel cut out for a better view of the slide. The aromatic residues that comprise the greasy slide and Tyr118 are shown as stick diagrams colored as in panel *B.* The maltotriose molecule (Triose) is shown as an *orange stick diagram.* The coloring of the loops is the same as in panel *A.* **B.** X-ray crystallographic structures of the ferric citrate receptor, FecA, of *E. coli.* *(A)* side view of the unliganded FecA. The "plug" domain inside the β-barrel is shown in *orange.* At its N-terminal end, the short sequence comprising the Ton box is shown in *light blue.* Loops 7 and 8 are shown in *deep blue* and *mauve,* respectively. *(B)* Liganded FecA. On binding of the ferric citrate (with two large *blue balls near the top* indicating the two iron atoms, and citrate molecules in stick diagrams), large displacements are seen at the N-terminal end of the plug domain, where residues 80 through 95 (including the Ton box of residues 80 to 84) become disordered and invisible. Loops 7 and 8 also undergo large conformational changes, with the loss of part of the helical structure in loop 7. **C.** Crystallization and preliminary x-ray diffraction analysis of ScrY, a specific bacterial outer membrane porin. Part A reproduced from Figure 4 H. Nikaido, "Molecular basis of bacterial outer membrane permeability revisited," *Microbiol. Mol. Biol. Rev.*, 67: 593–656, 2003 and http://mmbr.asm.org/cgi/content-nw/full/67/4/593/F7. Part B reproduced from Figure 4 of D. Muller and A. Engel, "Voltage and pH-induced channel closure of porin OmpF visualized by atomic force microscopy," *J. Mol. Biol.*, 285: 1347–1351, 1999. Part C reproduced from D. Forst et al., "Crystallization and preliminary x-ray diffraction analysis of ScrY, a specific bacterial outer membrane porin," *J. Mol. Biol.*, 229: 258–262, 1993 and http://www.rcsb.org/pdb/explore/images.do?structureId=1oh2.

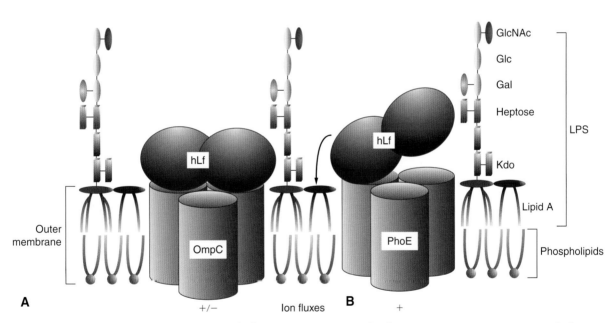

Figure 17-90. Schematic representation of Lf-porin interactions. The figure represents two OmpC and PhoE porin trimers *(barrels)* integrated in the outer membrane of gram-negative bacteria. **A.** Close interaction between hLf and OmpC impedes ion fluxes through the porin channels but probably does not allow further binding of hLf to the lipid bilayer. **B.** Weaker interactions between hLf and PhoE than for OmpC do not impede ion fluxes but probably favor further interactions with the lipid bilayer *(black arrow).* KDO, ketodeoxyoctulonic acid; *Lf,* lactoferrin from leukocytes. Redrawn from Figure 6 of F.R. Sallmann et al., "Porins OmpC and PhoE of *Escherichia coli* as specific cell-surface targets of human lactoferrin: binding characteristics and biological effects," *J. Biol. Chem.*, 274: 16107–16114, 1999 and http://www.jbc.org/content/vol274/issue23/images/large/bc2390844006.jpeg.

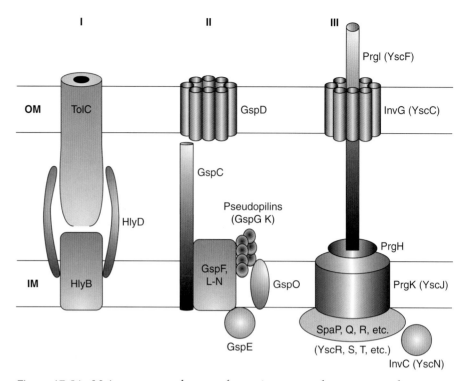

Figure 17-91. Major export pathways of proteins across the outer membrane (OM). Components of type I, type II, and type III secretion pathways are shown schematically. The type I pathway is composed of an ABC transporter (here HlyB for *E. coli* hemolysin), a membrane fusion protein (MFP) family protein (HlyD), and an OM channel (TolC). In the type II pathway, the proteins reach the periplasmic space via the Sec pathway and are then secreted across the OM by a machinery with many components, including pilin-like proteins (pseudopilins). The proteins are labeled according to the universal Gsp (general secretory pathway) nomenclature. The energy is apparently supplied by ATP hydrolysis by the GspE protein. The type III system is involved in the secretion (or perhaps injection) of virulence-related proteins into animal and plant host cells. Many of the components have homologies to the proteins of the flagellar hook and basal plate system. The figure is based mainly on a recent proposal that relies heavily on this similarity, as well as experimental studies of the "needle complex." The names of the proteins are those from the type III pathway in *Salmonella enterica* serovar typhimurium, with those from that of *Yersinia enterocolitica* shown in parentheses. The energy for export is thought to be supplied by the InvC (YscN) ATPase. *IM*, inner membrane; *OM*, outer membrane. Redrawn from Figure 7 of H. Nikaido, "Molecular basis of bacterial outer membrane permeability revisited," *Microbiol. Mol. Biol. Rev.*, 67: 593–656, 2003 and http://mmbr.asm.org/cgi/content-nw/full/67/4/593/F7.

outer membrane and on the outer side of the inner membrane. The structure of this lipoprotein is shown in Figure 17-94. *E. coli* also contains complex polysaccharides, as shown in Figure 17-95.

E. coli does not have a nucleus but contains genetic material in the form of highly compacted DNA as chromatin. When *E. coli* is lysed carefully and the proteins are removed, it can be spread on a grid and examined by electron microscopy (Figure 17-96). The DNA in this organism is compacted in the cell more than 1000-fold. The DNA must be flexible, changing conformation upon binding a protein to allow for expression (transcription). Some

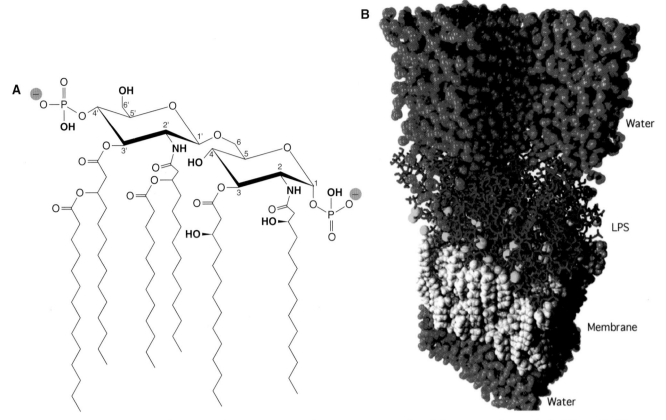

Figure 17-92. **A.** Lipid A from *E. coli* K-12. Groups that may act as H-bond donors are shown in boldface. **B.** The structure of lipid A in the lipopolysaccharide (LPS) component of the cell wall outer membrane. Part A Reproduced from Figure 9 of H. Nikaido, "Molecular basis of bacterial outer membrane permeability revisited," *Microbiol. Mol. Biol. Rev.*, 67: 593–656, 2003 and http://mmbr.asm.org/cgi/content-nw/full/67/4/593/F9.

DNA sequences are more flexible than others. The TATA motif region is very flexible, for example. **Integration host factor (IHF)** is used as an example of the bending of DNA in contact with a protein. Certain base pairs are greatly distorted because of this binding (Figure 17-97). The DNA in a bacterial cell is supercoiled in two ways: unrestrained and restrained (by proteins). Unrestrained supercoiling increases the concentration of DNA sites by a factor of about 100-fold, and branching of the unrestrained supercoil is driven by entropy.

In certain circumstances, there may be negatively supercoiled DNA; in others, there can be positively supercoiled DNA. Most DNA from *E. coli* has a level of supercoiling such that there would be about 20 supercoils per 3000 base pairs (10 unrestrained and 10 restrained). The restrained supercoils would contain five RNA polymerase (RNAP) molecules bound, three molecules of HU protein (histone-like) bound, and two molecules of H-NS protein (histone-like) bound, as shown in Figure 17-97. A bacterial cell may contain 100,000 copies of each histone-like protein (FIS, H-NS, HU, and IHF), amounting to 30% to 40% of the total proteins bound to DNA, because there is a total of about 1 million protein molecules in an *E. coli* cell. In isolated *E. coli* chromatin, there are equal masses of DNA and protein (in the 1000-fold compacted

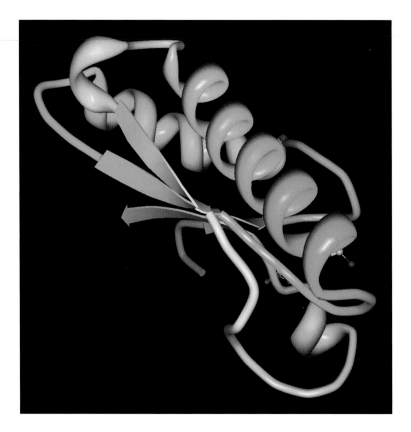

Figure 17-93. Peptidoglycan-associated lipoprotein. Chain: a. Fragment: periplasmic domain, residues 65–137. Reproduced from PDB ID: 1oap. C. Abergel, A. Walburger, E. Bouveret, J.M. Claverie. MAD Structure of the Periplasmic Domain of the E. coli Pal Protein. To be published.

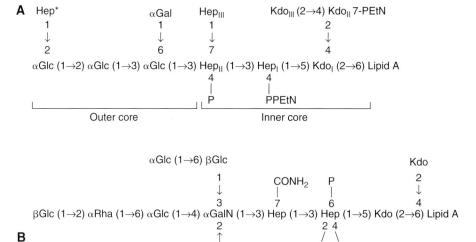

Figure 17-94. **A.** Structure of the R-core oligosaccharide in *E. coli* K-12 (the Hep residue indicated by the asterisk is replaced by GlcNAc in *S. enterica* serovar typhimurium) and in *Pseudomonas aeruginosa*. **B.** Individual LPS molecules may not necessarily contain all of the residues shown. Furthermore, the core structure in the O-antigen-containing LPS of *P. aeruginosa* is modified from the structure in the R-mutant LPS shown. The anomeric configurations of Hep and Kdo are always α. In the *P. aeruginosa* core, there are more formal negative charges (2 from Kdo residues plus about 4.5 from monophosphates, assuming roughly 1.5 negative charges per phosphatomonoester, i.e., a total of 6.5), especially concentrated in the trisubstituted heptose residue (although the negative charge is partially compensated by the presence of one positive charge in *N*-alanylgalactosamine). The *E. coli* core carries about 5.5 formal negative charges, but they are present in a more dispersed manner. *EtN*, ethanolamine; *Gal*, D-Galactose; *GalN*, D-galactosamine; *Glc*, D-glucose; *Hep*, L-*glycero*-D-*manno*-heptose; *Kdo*, 3-deoxy-D-*manno*-Oct-2-ulosonic acid; *Rha*, L-rhamnose. Redrawn from Figure 10 of H. Nikaido, "Molecular basis of bacterial outer membrane permeability revisited," *Microbiol. Mol. Biol. Rev.*, 67: 593–656, 2003 and http://mmbr.asm.org/content/vol67/issue4/images/large/mr0430026010.jpeg.

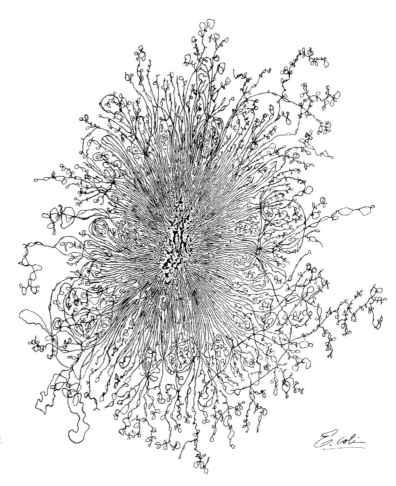

Figure 17-95. This is a famous electron micrograph of an *E. coli* cell that has been carefully lysed, then all the proteins were removed, and it was spread on an EM grid to reveal all of its DNA.

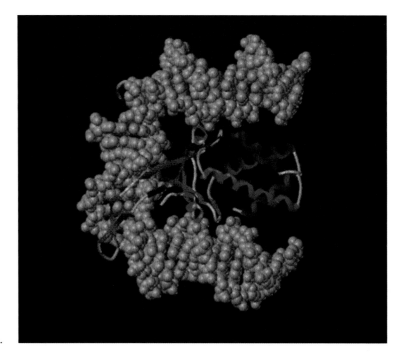

Figure 17-96. Binding of integration host factor (IHF), shown on the right center, to DNA, causing binding of DNA. Reproduced from http://gibk26.bse.kyutech.ac.jp/jouhou/image/dna-protein/all/small_N1ihf.gif. BioInfo Bank, Kyushu Institute of Technology, Fukuoka, Japan.

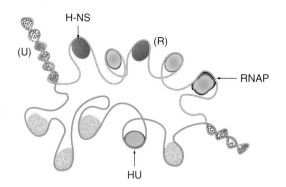

Figure 17-97. Approximate level of supercoiling of DNA in *E. coli*. Unrestrained supercoiling is indicated by *(U)* and restrained supercoiling is indicated by *(R)*. Darkened ovals represent proteins bound to DNA to produce restrained supercoiling. Proteins are RNAP (RNA polymerase), HU and H-NS.

DNA). Conversely, in eukaryotes, chromatin forms a stable DNA–protein complex in which only 2% of the genome codes for protein, in contrast to more than 90% of the chromatin in bacteria that codes for protein. Thus, the DNA in eukaryotes is not highly supercoiled and is generally wrapped around proteins, although here small regions in eukaryotic DNA are supercoiled around actively transcribed genes.

A typical bacterial operon (a set of genes under the control of an operator gene) may contain three genes, and if the operon B-DNA double-helical conformation were stretched out, it would be as long as the bacterial cell. The *lac* operon in *E. coli* is shown in Figure 17-98. This operon codes for three enzymes: β-galactosidase, permease, and transacetylase.

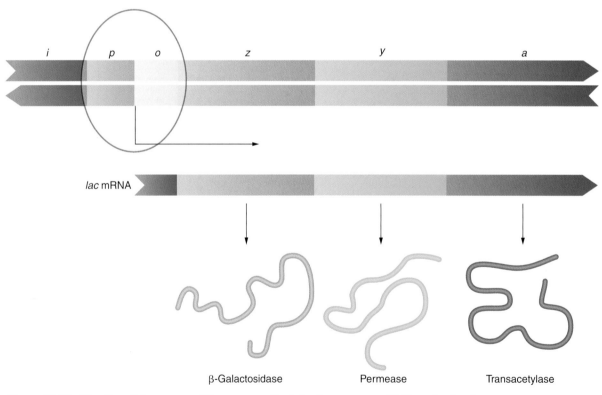

Figure 17-98. The *E. coli lac* operon. When transcribed the *lac* operon mRNA codes for three enzymes: β-galactosidase, permease, and transacetylase. The length of a bacterial cell might encompass position *p* through *a* of the stretched out B-DNA helix. Redrawn from http://www.cbs.dtu.dk/staff/dave/roanoke/fig11_03e.jpg.

The structure of β-galactosidase is shown in Figure 17-99 with the chemical reaction. *Hydrolysis of lactose generates galactose and glucose, the latter being the preferred energy substrate for E. coli.* β-Galactosidase acts exactly like intestinal lactase, but β-galactosidase is not strictly a lactase, which is specific for lactose, because β-galactosidase can hydrolyse substrates in addition to lactose.

The second gene (LacY) in the *lac* operon specifies the enzyme lactose permease, whose structure is shown in Figure 17-100a and b. The mechanism of the permease is shown in Figure 17-100c. Lactose permease is a symport of lactose and a proton that are transported inwardly or outwardly

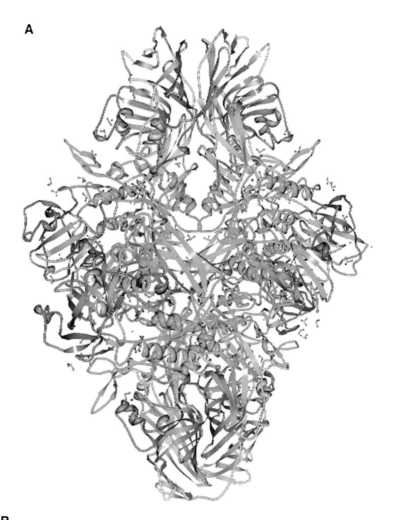

Figure 17-99. **A.** High-resolution refinement of the tetrameric β-galactosidase in a new crystal form reveals multiple metal-binding sites and provides a structural basis for α-complementation. **B.** Reaction catalyzed by β-galactosidase. Part A reproduced from PDB ID: 1jyn. D.H. Juers, B.W. Matthews. A Structural View of the Action of *Escherichia coli* (Lacz) Beta-Galactosidase. *Biochemistry* **40** pp. 14781 (2001).

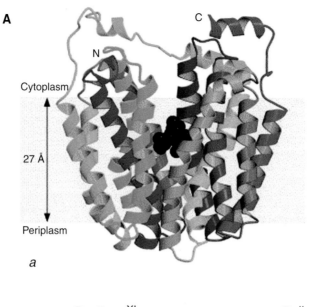
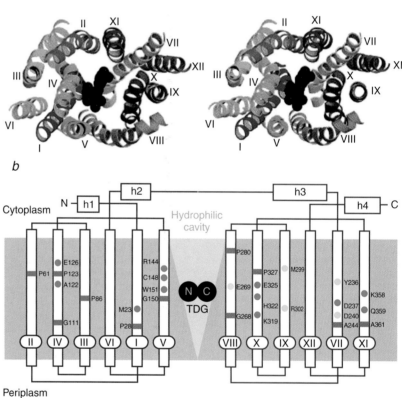

Figure 17-100. **A.** Overall structure of LacY (lactose permease). The figures are based on the C154G mutant structure with bound β-D-galactopyranosyl-1-thio-β-D-galactospyranoside (TDG). *(a)* Ribbon representation of LacY viewed parallel to the membrane. The twelve transmembrane helices are colored from the N-terminus (N) in *purple* to the C-terminus (c) in *pink* with TDG represented by *black spheres*. *(b)* Stereo view of the ribbon representation of LacY viewed along the membrane normal from the cytoplasmic side. For clarity, the loops have been omitted. The color scheme is the same as in *(a)*, and the twelve transmembrane helices are labeled with Roman numerals. *(c)* Secondary structure schematic of LacY. The N- and C-terminal portions of the enzyme are colored in *blue* and *red*.

1136   CHAPTER 17   Microbial Biochemistry

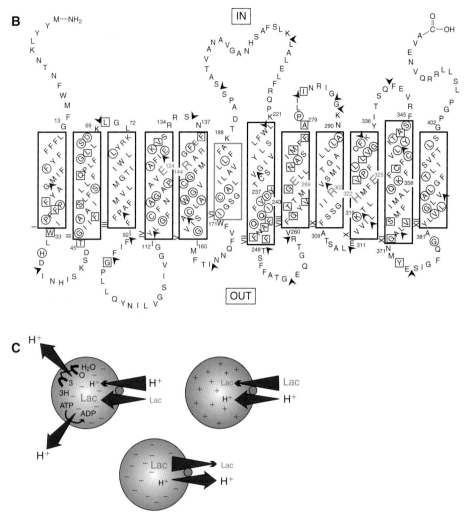

Figure 17-100, cont'd **B.** Secondary structure model of LacY derived from hydropathy and deletion analyses. The single-letter amino acid codes are used; residues irreplaceable for active transport are highlighted in *large green type*. Charge pairs Asp237/Lys358 and Asp240/Lys319 are shown in *small brown type*. *Solid rectangles* represent transmembrane regions defined by deletion analysis. The *dashed helix VI* represents the region defined by a decrease in downhill transport assessed by phenotype on indicator media. *Orange letters* represent ionizable residues predicted to be within the cytoplasmic ends of transmembrane helices II, III, IV, and V by deletion analysis. *Squared residues* represent positions where transport activity of single-cysteine replacement mutants is inhibited by N-ethylmaleimide (NEM) treatment. *Circled residues* represent positions where missense mutations have been shown to inhibit lactose accumulation. Residues in *gray circles* represent positions where both results have been observed. *Two-tone arrowheads* indicate locations where discontinuities in the primary sequence have been introduced, and *solid arrowheads* indicate regions where amino acids have been inserted into LacY. *Purple arrowheads* indicate good transport activity, and *red arrowheads* indicate little or no transport activity. **C.** Lactose/H$^+$ symport. In the absence of substrate, LacY does not translocate H$^+$; substrate gradients generate electrochemical H$^+$ gradients. Parts A, B, and C are reproduced from http://www.sciencedirect.com/science?_ob=ArticleURL&_udi=B6X1F-4G82Y3H-1&_user=4207628&_handle=V-WA-A-W-E-MsSAYZW-UUW-U-AACYUBUZZW-AACCZAAVZW-ZUEBABCZ-E-U&_fmt=full&_coverDate=06%2F30%2F2005&_rdoc=6&_orig=browse&_srch=%23toc%237241%232005%23996719993%23598285!&_cdi=7241&view=c&_acct=C000047720&_version=1&-urlVersion=0&_userid=4207628&md5=edbd45cf208687c3d44edfef0d69002b and Figures A, B, and C, respectively, of H.R. Kaback, "Structure and mechanism of the lactose permease," *Comptes Rendus Biologies*, 328: 557–567, 2005.

together. In the absence of lactose, the symport does not function but protons can be lost from the cell depending on the gradient of lactose inside.

A fragment of the transacetylase of *E. coli* is the dihydrolipoamide acetyltransferase domain of the pyruvate dehydrogenase complex enzyme of *E. coli*, which is shown in Figure 17-101.

In the absence of lactose outside the cell or when its concentration is low, the *lac* operon does not function because a repressor protein, the *lac* repressor, is in contact with the *lac* operator. The *lac* repressor is a product of a regulator gene upstream from the promoter site (Figure 17-102A).

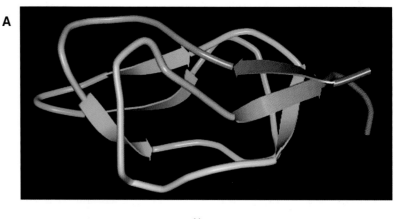

Figure 17-101. **A.** Dihydrolipoamide acetyltransferase. Innermost lipoyl domain of the pyruvate dehydrogenase from *E. coli*. Dihydrolipoamide acetyltransferase. Chain: a. Fragment: Lipoamide-binding domain of e2p. **B.** Reaction catalyzed by *E. coli* transacetylase. Part A reproduced from PDB ID: 1qjo. D.D. Jones, K.M. Stott, M.J Howard, R.N. Perham. *Biochemistry* **39** pp. 8448 (2000) and http://www.rcsb.org/pdb/explore/images.do?structureId=1qjo. Part B reproduced from http://www.ebi.ac.uk/thornton-srv/databases/cgi-bin/pdbsum/GetPage.pl?pdbcode=1qjo.

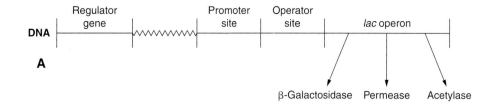
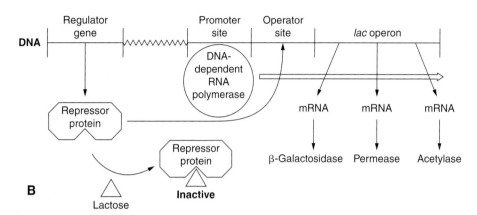

Figure 17-102. **A.** Diagram of the *lac* operon of *E. coli*. **B.** The *lac* operon showing that a repressor protein is encoded from information in an upstream regulator gene. Normally, when lactose is absent or in very low concentration, the regulator protein is bound to the operator site, repressing the lac operon. When lactose is present it binds to the repressor protein, preventing it from binding to the operator site. In consequence, the repressor does not bind to the operator site and the *lac* operon is derepressed, resulting in the synthesis of β-galactosudase, permease, and transacetylase.

A picture of the *lac* repressor in contact with the operator region of DNA is shown in Figure 17-103. Metabolism of the *E. coli* cell in the resting state depends mainly on glucose and yields products, such as succinate, lactate, acetate, ethanol, and carbon dioxide (Figure 17-104).

One of the curiosities of bacterial cells is the mechanism used to import proteins to the cell interior. Apparently a receptor—for example, the **BtuB receptor,** dedicated to the transport of vitamin B_{12} from the outside to the cell interior—has a mechanism to assist in moving an entire protein into the cell from the outside. A protein on the outside, colicin, in this case, crosses the *E. coli* outer membrane by tethering one of its ends into the BtuB receptor, and colicin searches for another receptor porous enough to admit it into the cell—in this case, the OmpF receptor, as shown in Figure 17-105.

Another interesting feature of many bacterial cells is the mechanism for regulating biosynthetic pathways. Pyrimidine synthesis in *E. coli* is a classical example wherein the first committed step in this pathway is catalyzed by the enzyme aspartate transcarbamoylase (ATCase). The pyrimidine synthetic pathway is shown in Figure 17-106. ATCase is a complex allosteric (from the Greek, *allos* = other; *stereos* = rigid) enzyme regulated positively by adenosine triphosphate (ATP) and negatively, through product feedback, by cytosine triphosphate (CTP). CTP can be further degraded to cytidine diphosphate (CDP; by nucleoside diphosphate kinase: CTP + ADP = CDP + ATP). The net negative feed-

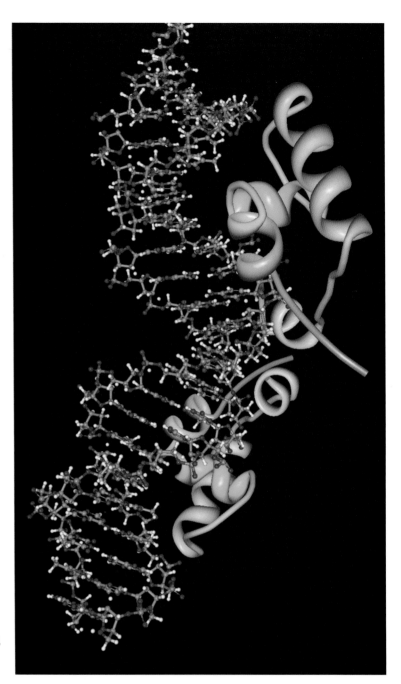

Figure 17-103. NMR Structure of lac repressor Hp62–DNA complex. Reproduced from http://www.rcsb.org/pdb/cgi/explore.cgi?pdbId=1cjg. PDB ID: 1cjg. C.A. Spronk, A.M. Bonvin, P.K. Radha, G. Melacini, R. Boelens, R. Kaptein. The Solution Structure of lac Repressor Headpiece 62 Complexed to a Symmetrical lac Operator. *Structure Fold. Des.* **7** pp. 1483 (1999).

back, then, is a product of the formation of CTP and the extent of its breakdown to CDP. The kinetic effects of these two regulators are shown in Figure 17-108.

The structure of ATCase is that of a 12-subunit enzyme (310 kDa) containing 6 catalytic subunits (as two trimers of catalytic chains) and 6 regulatory subunits (three dimers of regulatory chains), as shown in Figure 17-109. The catalytic chains are 33 kDa each, and the regulatory chains are 17 kDa each. The positively regulated enzyme under the influence of ATP reflects a conformation called the relaxed (R) state (high affinity for aspartate, high activity), and that reflected by the negatively regulated condition under the influence of CTP is called the taut (T) state

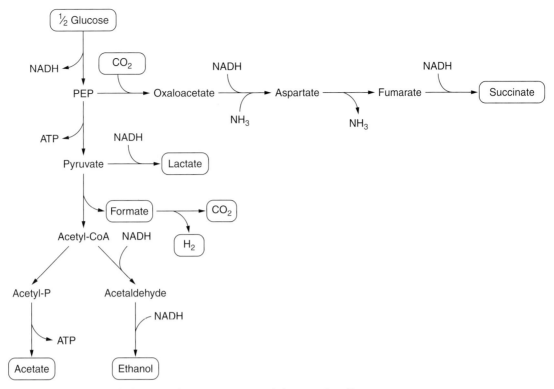

Figure 17-104. Metabolism in the resting state of the *E. coli* cell.

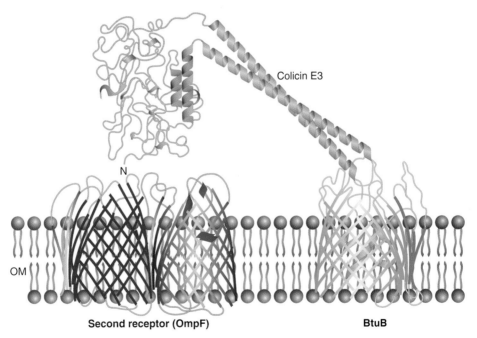

Figure 17-105. Shown in this graphic is the protein colicin and the receptors it uses to cross the cellular membrane of an *E. coli* bacterium. After lodging one of its ends in the "BtuB" receptor, usually used to admit vitamin B_{12} into the cell, the protein uses its extended "fishing pole" to search for a porous second receptor, "OmpF," that will allow its import across the membrane. (Kurisu et al., Nature Structural Biology, 2003). Redrawn from page 1 of http://www.innovations-report.com/bilder_neu/23696_cramer_ecoli.jpg.

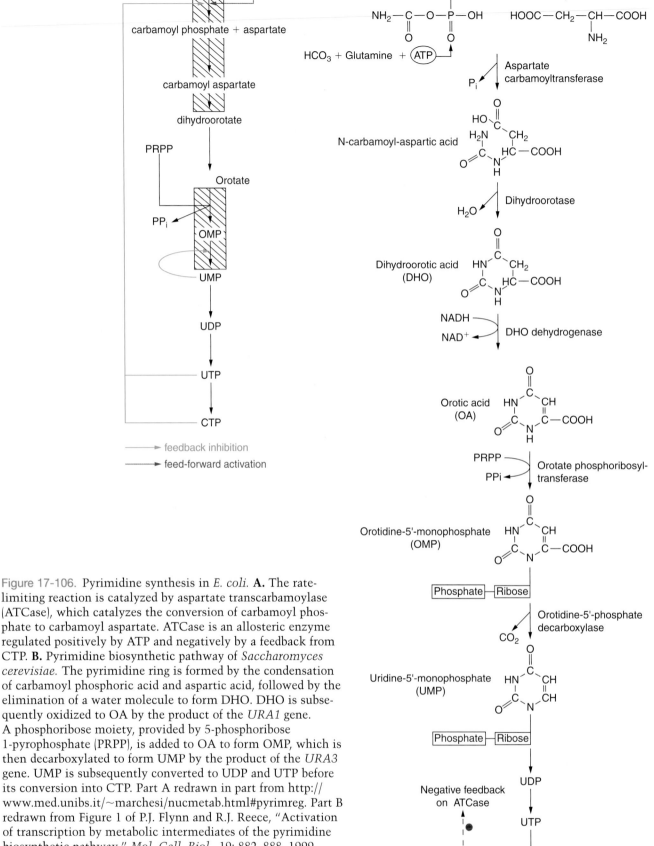

Figure 17-106. Pyrimidine synthesis in *E. coli*. **A.** The rate-limiting reaction is catalyzed by aspartate transcarbamoylase (ATCase), which catalyzes the conversion of carbamoyl phosphate to carbamoyl aspartate. ATCase is an allosteric enzyme regulated positively by ATP and negatively by a feedback from CTP. **B.** Pyrimidine biosynthetic pathway of *Saccharomyces cerevisiae*. The pyrimidine ring is formed by the condensation of carbamoyl phosphoric acid and aspartic acid, followed by the elimination of a water molecule to form DHO. DHO is subsequently oxidized to OA by the product of the *URA1* gene. A phosphoribose moiety, provided by 5-phosphoribose 1-pyrophosphate (PRPP), is added to OA to form OMP, which is then decarboxylated to form UMP by the product of the *URA3* gene. UMP is subsequently converted to UDP and UTP before its conversion into CTP. Part A redrawn in part from http://www.med.unibs.it/~marchesi/nucmetab.html#pyrimreg. Part B redrawn from Figure 1 of P.J. Flynn and R.J. Reece, "Activation of transcription by metabolic intermediates of the pyrimidine biosynthetic pathway," *Mol. Cell. Biol.*, 19: 882–888, 1999.

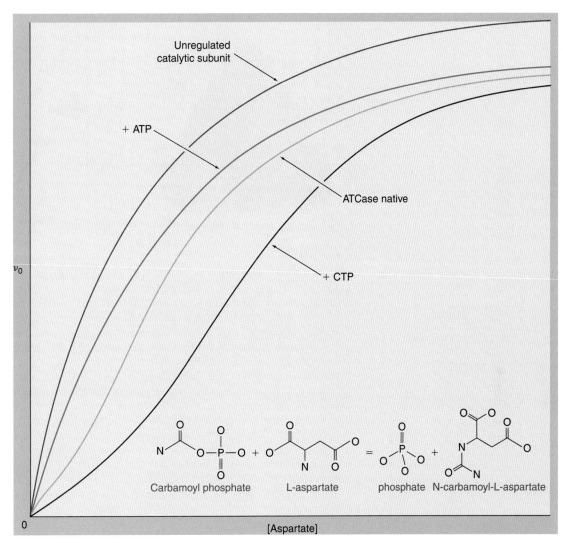

Figure 17-107. Rate of reaction (v_o) as a function of aspartate concentration. The native ATCase in the absence of regulators is shown in *yellow*. The *red* and *green lines* reflect the negative regulation by CTP and the positive regulation by ATP, respectively. Note that the *red* and *yellow curves* are S-shaped, indicative of allosteric behavior. Reproduced from http://www.unine.ch/bota/bioch/cours/enzymes/ATCase.jpg.

(low affinity, low activity). The conversion of the T-state to the R-state requires major changes of the three-dimensional structure. There is a 12 Å expansion along the threefold axis and rotation of the regulatory subunits about their respective twofold axes, as well as other changes. The resulting changes in the catalytic subunits, C1 and C4, between the T-state and the R-state are shown in Figure 17-110. The allosteric domain of the regulatory subunits binds ATP and CTP (or uridine triphosphate, UTP). Each regulatory dimer has one site of high affinity for ATP and CTP and another site with much lower affinity for these nucleotides. ATP binds preferentially to the high-affinity site and activates the enzyme, whereas CTP binding produces inhibition. If UTP is present, CTP binds to the high-affinity site and UTP binds to the low-affinity site. UTP binding increases the affinity for CTP at the high-affinity sites and

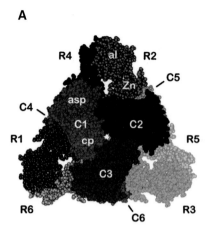
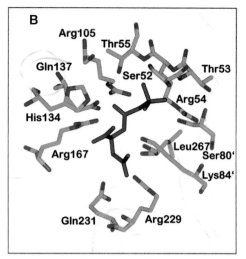

ATCase

Figure 17-108. Quaternary structure of *E. coli* ATCase. **A.** Holoenzyme viewed along the threefold axis. Catalytic chains are numbered C1 to C6, and regulatory chains are numbered R1 to R6. The different catalytic and regulatory subunits are indicated by different colors. The aspartate domain of the catalytic chain is designated *asp*, and the carbamoylphosphate domain is designated *cp*. The domains of the regulatory chain are named *Zn* for zinc domain and *al* for allosteric domain. **B.** Binding mode of the bisubstrate analog PALA (purple) to the active site of ATCase. Side chains are shown as sticks with atoms labeled by color (*green*, carbon; *blue*, nitrogen; *red*, oxygen). Primes after residue numbers indicate the position of the residue in an adjacent polypeptide chain. The figures are based on data for the CTP-liganded structure and the bisubstrate analog PALA-liganded structure, respectively. *PALA*, the bisubstrate analog, N-(phosphonoacetyl)-L-aspartate. Reproduced from Figure 1 of K. Helmstaedt, S. Krappmann, and G.H. Braus, "Allosteric regulation of catalytic activity: *Escherichia coli* aspartate transcarbamoylase versus yeast chorismate mutase," *Microbiol. Mol. Biol. Rev.*, 65: 404–421, 2001 and http://mmbr.asm.org/content/vol65/issue3/images/large/mr0310020001.jpeg.

causes inhibition up to 95%, whereas CTP binding alone, in the absence of UTP, produces 50% to 70% inhibition. ATP reduces cooperativity, and CTP enhances it as shown by the extent of sigmoidicity in Figure 17-107. A more complete view of ATCase in the T-state and the R-state is shown in Figure 17-110, and the binding sites are made clear, diagrammatically, in Figure 17-111.

In the *E. coli* cell cycle in which reproduction takes place by cell division, partitioning occurs while replication is occurring. First there is an increase in the mass of the cell with an orientation of the genomic material. Then replication begins, where the domains at the replication forks are decondensed. This is followed by spooling of the DNA through the replication machinery in the elongation process. The structures are terminated and resolved, and there is a final folding and separation of the chromosomes, followed by the ultimate step of cell division. A summary is shown in Figure 17-112.

In *E. coli*, the replication of chromosomal DNA is accomplished by the DNA polymerase III holoenzyme. This enzyme has three subassemblies. Subunits α, ε, and θ make up the catalytic core carrying out DNA syn-

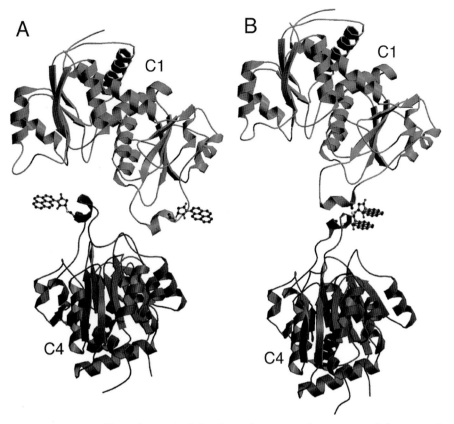

Figure 17-109. Ribbon diagram of the three-dimensional structure of the C1 and C4 catalytic chains of aspartate transcarbamoylase, with a model of Cys-241 labeled with pyrene, in the T-state *(A)* and R-state *(B)*. To highlight Cys-241 and pyrene, they are represented by *ball-and-stick* units. Pyrene is used as a fluorescent probe. Reproduced from Figure 1 of J.M. West, H. Tsurutal, and E.R. Kantrowitz, "A fluorescent probe-labeled *Escherichia coli* aspartate transcarbamoylase that monitors the allosteric conformational state," *J. Biol. Chem.*, 279: 945–951, 2004.

thesis and proofreading. The β-subunit is the sliding clamp that provides processivity. There is a clamp loader, a γ-complex, consisting of γ-, δ- (the wrench), and δ'-subunits that is needed to load the sliding clamp (β-subunit) onto DNA (in eukaryotes, the clamp loader consists of the proliferating cell nuclear antigen and five subunits, which together form replication factor C). Assembly of the sliding clamp on DNA requires the γ-complex and ATP. The ATPase motor is the γ-subunit, and the δ'-subunit is an inert stator, both being ATPases. The δ-subunit of the γ-complex is the only subunit alone that can bind to and open the β-clamp. The structure of DNA polymerase III at a replication fork is shown in Figure 17-113. In Figure 17-114, the replication process is shown from the point of view of replicating the DNA, in which two primary replicases are involved until halfway through the replication process, when four new replicases are expressed.

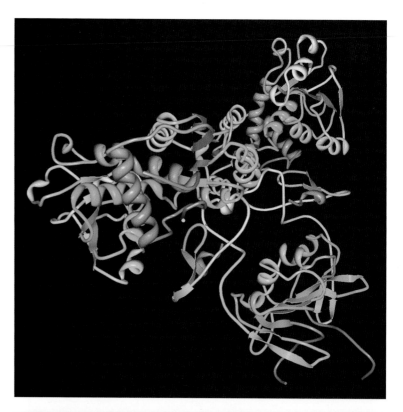

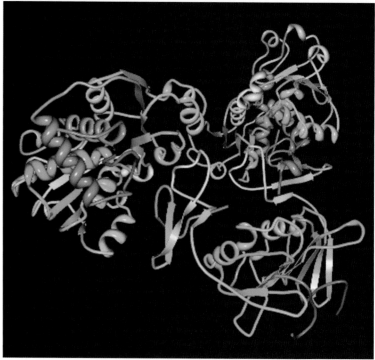

Figure 17-110. Structural changes in aspartate transcarbamoylase. *Upper view* is T-state, and *lower view* is R-state. Upper view reproduced from http://www.imb-jena.de/cgi-bin/ImgLib.pl?CODE=1ezz. PDB ID: 1ezz. L. Jin, B. Stec, E. R. Kantrowitz. A Cis-Proline to Alanine Mutant of *E. coli* Aspartate Transcarbamoylase: Kinetic Studies and Three-Dimensional Crystal Structures. *Biochem.* **39** pp. 8058 (2000). Lower view is reproduced from PDB ID: 1f1b. L. Jin, B. Stec, E. R. Kantrowitz A Cis-Proline to Alanine Mutant of *E. coli* Aspartate Transcarbamoylase: Kinetic Studies and Three-Dimensional Crystal Structures. *Biochemistry* **39** pp. 8058 (2000).

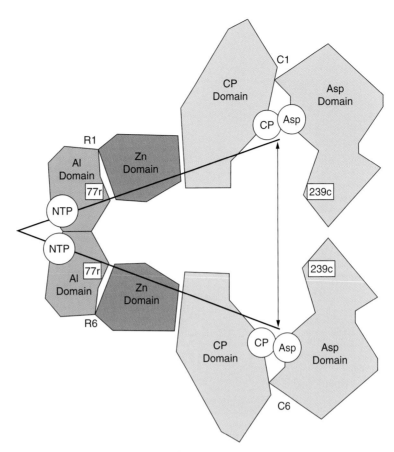

Figure 17-111. A schematic of one third of the aspartate transcarbamoylase enzyme showing one catalytic chain (C1) from the upper trimer, one catalytic chain (C6) from the lower trimer, and one regulatory dimer (R1-R6). Each catalytic chain is composed of the carbamoyl phosphate (CP) and aspartate domains, and each regulatory chain is composed of the allosteric (Al) and zinc domains. The active site is indicated by the locations of carbamoyl phosphate and aspartate and the regulatory biding site is indicated (NTP). The approximate position of residue 77 in the regulatory chain (77r) and of residue 239 in the catalytic chain (239c) are also shown.

Diseases Caused by *E. coli*

One strain of *E. coli*, O157:H7 (numbers refer to markers on the bacterial cell surface) is a major cause of illness from food. The organism lives in the intestines of healthy cattle (most *E. coli* strains are harmless and live in the intestinal tracts of humans, for example). Usually, the disease coming from *E. coli* O157:H7 (bearing one or more toxins) comes from eating undercooked, contaminated ground beef. Another mode of transmission is person-to-person contact in families and in child-care centers, according to the Centers for Disease Control and Prevention. In the United States, there are about 73,000 cases per year, with an annual death rate of 61 people. Infection with this strain of *E. coli* can lead to bloody diarrhea, cramps, and sometimes to kidney failure. Ingestion of raw milk (unpasteurized) and swimming in sewage-contaminated water are other causes of illness. A bloody diarrhea should be followed with testing for this organism (using

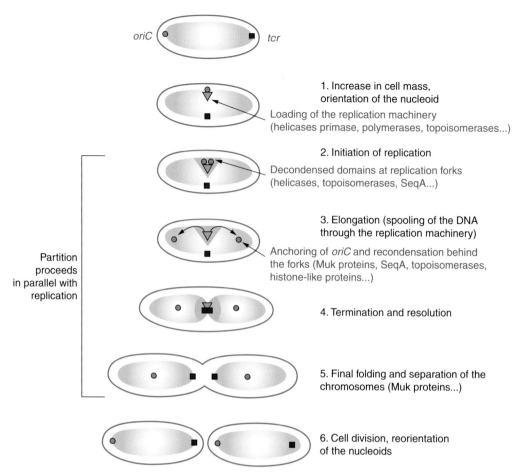

Figure 17-112. Model for nucleoid segregation through decondensation/condensation and supercoiling during the *E. coli* cell cycle. *Green circles* represent the origin of replication (oriC), *blue squares* the terminus, and the *orange triangle* the replication machinery. *Blue* represents the condensed, and *yellow* the decondensed, domains of the nucleoid or the bacterial chromosome. Refolding and partitioning of the chromosome occur continuously and in parallel with replication. According to this model, DNA decondenses in the vicinity of the replicating segment. Duplicate copies of oriC move rapidly apart from the center of the cell to quarter positions on opposite sides. The replicated DNA behind the fork recondenses and positions itself within the cell. Redrawn from Figure 1 of S. Dasgupta, S. Maisnier-Patin, and K. Nordström, "New genes with old modus operandi," *EMBO Rep.*, 1: 323–327, 2000.

sorbitol-MacConkey agar). It is always recommended to cook hamburger well and wash hands frequently. Ground meat can become contaminated during slaughter of an animal carrying the disease-producing strain in its intestines. It is also possible that the udder can become contaminated, thus infecting the raw milk and possibly the milking equipment. Some vegetables can be contaminated that are normally consumed in raw form, like lettuce, and these should be washed thoroughly before eating. Most infected individuals recover in 5 to 10 days without antibiotic treatment (some antibiotics may generate kidney complications).

There are more than 700 serotypes of *E. coli* based on O, H, and K antigens that can cause urinary tract infections, neonatal meningitis, and gastroenteritis, depending on a specific array of virulence determinants.

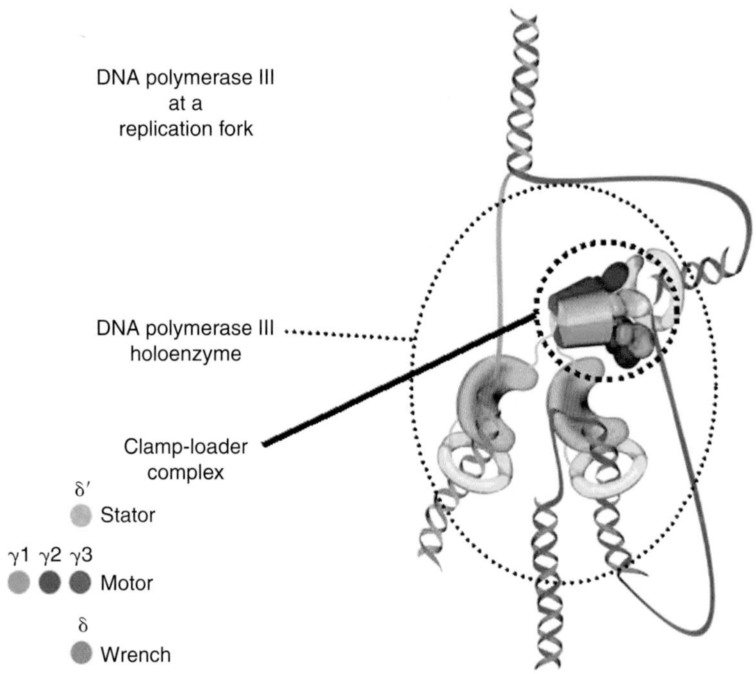

Figure 17-113. DNA polymerase III at a replication fork in the process of cell division of *E. coli*. Reproduced from http://jkweb.berkeley.edu/external/pdb/2001/clamp_wrench/replication_schematic.jpg and D. Jeruzalmi et al., "Mechanism of processivity clamp opening by the delta-subunit wrench of the clamp loader complex of *E. coli* DNA polymerase III," *Cell,* 106: 417–428, 2001.

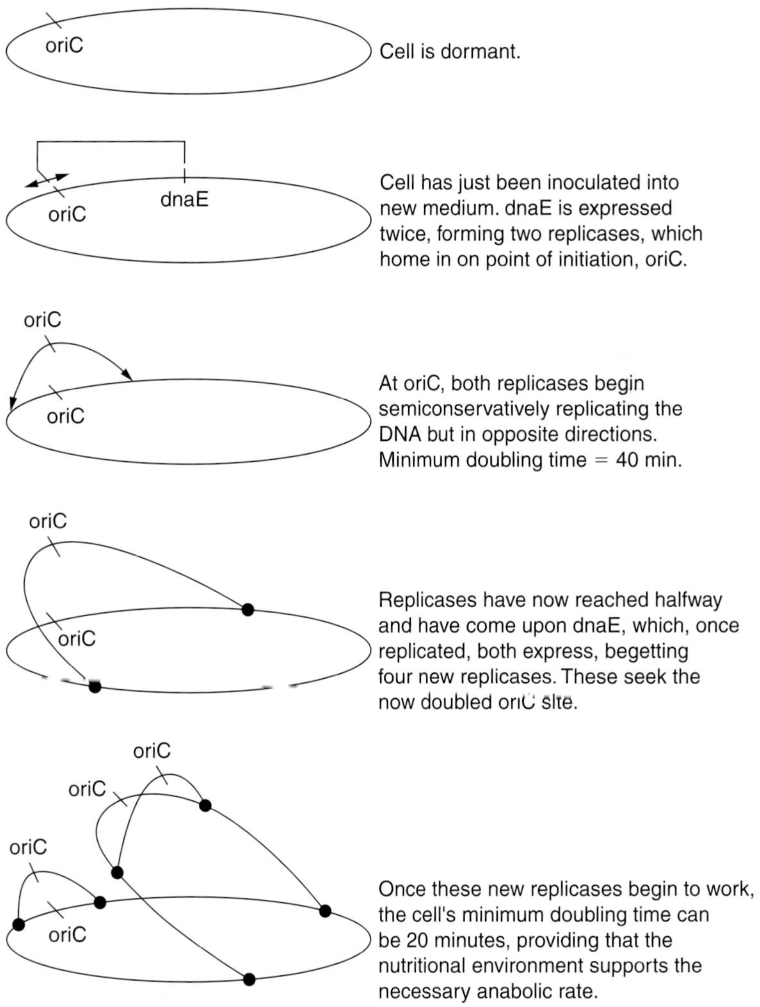

DNA polymerase III at a replication fork

DNA polymerase III holoenzyme

Clamp-loader complex

δ′ Stator

γ1 γ2 γ3 Motor

δ Wrench

oriC — Cell is dormant.

oriC dnaE — Cell has just been inoculated into new medium. dnaE is expressed twice, forming two replicases, which home in on point of initiation, oriC.

oriC / oriC — At oriC, both replicases begin semiconservatively replicating the DNA but in opposite directions. Minimum doubling time = 40 min.

oriC / oriC — Replicases have now reached halfway and have come upon dnaE, which, once replicated, both express, begetting four new replicases. These seek the now doubled oriC site.

oriC / oriC / oriC / oriC — Once these new replicases begin to work, the cell's minimum doubling time can be 20 minutes, providing that the nutritional environment supports the necessary anabolic rate.

Figure 17-114. Involvement of replicases in the replication of DNA.

The virulence determinants are summarized in Table 17-4. There are myriad other types of bacteria that are pathogenic, but discussion of these does not fall within the scope of this chapter.

Only a few topics in this chapter have been selected, mainly viruses of current interest; a short summary of viruses (coliphages) that infect bacteria, particularly *E. coli*; and one bacterium, *E. coli*. It is hoped that this will be sufficient for an introduction to microbial biochemistry. For interested readers, it is recommended to consult books listed in the "Further Reading" section or research articles listed in the permissions of figures and tables in this chapter.

Table 17-4
Virulence Determinants of Pathogenic *E. coli*

Summary of the Virulence Determinants of Pathogenic *E. coli*

Adhesins
  CFAI/CFAII
  Type 1 fimbriae
  P fimbriae
  S fimbriae
  Intimin (nonfimbrial adhesin)
Invasins
  hemolysins
  siderophores and siderophore uptake systems
  Shigella-like "invasins" for intracellular invasion and spread
Motility/chemotaxis
  flagella
Toxins
  LT toxin
  ST toxin
  Shiga-like toxin
  cytotoxins
  endotoxin lipopolysaccharide
Antiphagocytic surface properties
  capsules
  K antigens
  lipopolysaccharide
Defense against serum bactericidal reactions
  lipopolysaccharide
  K antigens
Defense against immune responses
  capsules
  K antigens
  lipopolysaccharide
  antigenic variation
Genetic attributes
  genetic exchange by transduction and conjugation
  transmissible plasmids
  R factors and drug-resistance plasmids
  toxin and other virulence plasmids

Reproduced from Table 1 of http://textbookofbacteriology.net/e.coli.html.

Further Reading

Chiu, Wah, and Johnson, John, *Virus Structure,* Academic Press/Elsevier, 2003.

Cohen, G.N., *Microbial Biochemistry,* Springer, 2004.

Schlegel, H.G., *General Microbiology,* Cambridge University Press, 7th edition, 1993.

APPENDIX 1

Abbreviations of the Common Amino Acids

Amino acid	Molecular weight	3-Letter	1-Letter	Character
Alanine	89.09	Ala	A	Neutral, aliphatic
Arginine	174.20	Arg	R	Basic
Asparagine	132.12	Asn	N	Neutral, amide
Aspartic acid	133.10	Asp	D	Acidic
Cysteine	121.16	Cys	C	Neutral, sulfhydryl group
Glutamic acid	147.13	Glu	E	Acidic
Glutamine	146.15	Gln	Q	Neutral, amide
Glycine	75.07	Gly	G	Neutral, aliphatic
Histidine	155.16	His	H	Aromatic ring, basic
Isoleucine	131.17	Ile	I	Neutral, aliphatic, hydrophobic
Leucine	131.17	Leu	L	Neutral, aliphatic, hydrophobic
Lysine	146.19	Lys	K	Basic
Methionine	149.21	Met		Neutral, sulfhydryl group
Phenylalanine	165.19	Phe	F	Neutral, contains aromatic benzene ring
Proline	115.13	Pro	P	Contains ring, imino acid
Serine	105.09	Ser	S	Contains hydroxyl, neutral
Threonine	119.12	Thr	T	Neutral, contains hydroxyl
Tryptophan	204.23	Trp	W	Aromatic indole ring, neutral
Tyrosine	181.19	Tyr	Y	Aromatic benzene ring, hydroxyl
Valine	117.15	Val	V	Neutral, aliphatic

APPENDIX 2

The Genetic Code

Base Pairing

In DNA, adenine (A) forms two hydrogen bonds with thymine (T), and guanine (G) forms three hydrogen bonds with cytosine (C). In RNA, the same base-pairing occurs, except that T is replaced by uracil (U). The DNA strand begins with the 5'-hydroxyl (or 5'-phospho) group of the beginning nucleotide and stretches to the final nucleotide's 3'-hydroxyl group. Thus, the upstream to downstream direction is from 5' to 3'. In double-stranded DNA, a complementary strand binds to the first strand in an antiparallel mode: the first strand runs in the 5' to 3' direction and the antiparallel strand runs in the opposite direction, 3' to 5', creating a double helical structure. Each nucleotide residue of double-stranded DNA is opposed by a complementary nucleotide (A:T and G:C). The sense strand of DNA contains the codons, and the antisense strand (template strand) contains anticodons. mRNA is synthesized from the antisense strand and therefore contains the same information as the sense template strand. A sample of this process is as follows:

Coding sense strand (5' to 3'):	ATGCTACGCAAGGCCCGA...
Template antisense strand (3' to 5'):	TACGATGCGTTCCGGGCT...
mRNA (5' to 3') from template strand:	AUGCUACGCAAGGCCCGA...
protein:	Met–Leu–Arg–Lys–Ala–Arg...

The amino acids are specified in triplets, as shown for DNA in the next table and for RNA is the subsequent table.

In the preceding table, the *ter* or end codons are equivalent to a stop codon. The preceding table is reproduced from page 1 of http://psyche.uthct.edu/shaun/SBlack/geneticd.html.

The same table for mRNA (where U replaces T) is shown here and is reproduced from http://users.rcn.com/jkimball.ma.ultranet/BiologyPages/C/Codons.html.

The codons in mRNA are bonded to anticodons in tRNAs, each of which carries the specific amino acid to the ribosomal site of protein synthesis, as illustrated in Figure 2-33.

		Second Position of Codon					
		T	**C**	**A**	**G**		
F i r s t P o s i t i o n	**T**	TTT Phe [F] TTC Phe [F] TTA Leu [L] TTG Leu [L]	TCT Ser [S] TCC Ser [S] TCA Ser [S] TCG Ser [S]	TAT Tyr [Y] TAC Tyr [Y] TAA *Ter* [end] TAG *Ter* [end]	TGT Cys [C] TGC Cys [C] TGA *Ter* [end] TGG Trp [W]	T C A G	**T h i r d P o s i t i o n**
	C	CTT Leu [L] CTC Leu [L] CTA Leu [L] CTG Leu [L]	CCT Pro [P] CCC Pro [P] CCA Pro [P] CCG Pro [P]	CAT His [H] CAC His [H] CAA Gln [Q] CAG Gln [Q]	CGT Arg [R] CGC Arg [R] CGA Arg [R] CGG Arg [R]	T C A G	
	A	ATT Ile [I] ATC Ile [I] ATA Ile [I] ATG Met [M]	ACT Thr [T] ACC Thr [T] ACA Thr [T] ACG Thr [T]	AAT Asn [N] AAC Asn [N] AAA Lys [K] AAG Lys [K]	AGT Ser [S] AGC Ser [S] AGA Arg [R] AGG Arg [R]	T C A G	
	G	GTT Val [V] GTC Val [V] GTA Val [V] GTG Val [V]	GCT Ala [A] GCC Ala [A] GCA Ala [A] GCG Ala [A]	GAT Asp [D] GAC Asp [D] GAA Glu [E] GAG Glu [E]	GGT Gly [G] GGC Gly [G] GGA Gly [G] GGG Gly [G]	T C A G	

		Second nucleotide				
		U	**C**	**A**	**G**	
U		UUU **Phenylalanine** (Phe) UUC Phe UUA **Leucine** (Leu) UUG Leu	UCU **Serine** (Ser) UCC Ser UCA Ser UCG Ser	UAU **Tyrosine** (Tyr) UAC Tyr UAA **STOP** UAG **STOP**	UGU **Cysteine** (Cys) UGC Cys UGA **STOP** UGG **Tryptophan** (Trp)	U C A G
C		CUU **Leucine** (Leu) CUC Leu CUA Leu CUG Leu	CCU **Proline** (Pro) CCC Pro CCA Pro CCG Pro	CAU **Histidine** (His) CAC His CAA **Glutamine** (Gln) CAG Gln	CGU **Arginine** (Arg) CGC Arg CGA Arg CGG Arg	U C A G
A		AUU **Isoleucine** (Ile) AUC Ile AUA Ile AUG **Methionine** (Met) or **START**	ACU **Threonine** (Thr) ACC Thr ACA Thr ACG Thr	AAU **Asparagine** (Asn) AAC Asn AAA **Lysine** (Lys) AAG Lys	AGU **Serine** (Ser) AGC Ser AGA **Arginine** (Arg) AGG Arg	U C A G
G		GUU **Valine** Val GUC (Val) GUA Val GUG Val	GCU **Alanine** (Ala) GCC Ala GCA Ala GCG Ala	GAU **Aspartic acid** (Asp) GAC Asp GAA **Glutamine acid** (Glu) GAG Glu	GGU **Glycine** (Gly) GGC Gly GGA Gly GGG Gly	U C A G

The mRNA code for amino acids in mitochondria differs in some respects from the nuclear-directed processes already discussed. UGA encodes tryptophan (Trp) rather than being a stop codon. AUA specifies methionine (Met) as opposed to isoleucine in the nuclear process, and AGA and AGG are the stop codons.

A rare amino acid, selenocysteine, is encoded by UGA. The translation machinery is able to distinguish the use of UGA for the seleno-amino acid from the stop codon, although UGA is also used for the stop codon.

APPENDIX 3

Weights and Measures

Abbreviation	Unit
M	Molar; molecular weight in grams per liter
mM	millimolar; Molar/1000; 10^{-3} M; 0.001 M
μM	micromolar; Molar/1,000,000; 10^{-6} M; 0.000001 M
nM	nanomolar; Molar/1,000,000,000; 10^{-9} M; 0.000000001 M
fM	femptomolar; Molar/1,000,000,000,000,000; 10^{-15} M; 0.000000000000001 M
m	meter
cm	centimeter; meter/100; 10^{-2} m; 0.01 m
mm	millimeter; meter/1000; 10^{-3} m; 0.001 m
μm	micrometer; meter/1,000,000; 10^{-6} m; 0.000001 m
angstrom	meter/10,000,000,000; 10^{-10} m; 0.0000000001 m
g	gram
mg	milligram; gram/1000; 10^{-3} g; 0.001 g
μg	microgram; gram/1,000,000; 10^{-6} g; 0.000001 g
ng	nanogram; gram/1,000,000,000; 10^{-9} g; 0.000000001 g
kg	kilogram; gram × 1000; 10^{3} g
liter	liter; 1000 cm^3; 1000 ml
dl	deciliter; liter/10; 100 cm^3; 100 ml
ml	milliliter; liter/1000; 1 cm^3; 10^{-3} liter
μl	microliter; liter/1,000,000; 10^{-6} liter

Glossary

AAAD: aromatic amino acid decarboxylase that converts L-dihydroxyphenylalanine (DOPA) to form dopamine

ABC superfamily: includes seven known transporters, including the multidrug resistance channel (MDR; P-glycoprotein) and the cystic fibrosis conductance regulator CFTR

ACDP2: ancient conserved domain protein; transports magnesium ion

ACP: acyl carrier protein

ACTH: adrenocorticotropic hormone

Action potential: a brief electrical pulse traveling along the axon of a neuron

Adenohypophysis: the anterior pituitary gland

Addison's disease: underproduction of cortisol by the adrenal gland

Adipocyte: a fat cell

ADH: antidiuretic hormone or vasopressin

AHSP: α-hemoglobin stabilizing protein

AIDS: acquired immunodeficiency syndrome

Albinism: a defect of pigmentation that results when the conversion of tyrosine to melanin is blocked

Aldosterone: the major salt-retaining hormone produced in the adrenal gland

Allopurinol: an inhibitor of xanthine oxidase; it is a drug used to treat gout

Allosteric enzyme: an enzyme that has a binding site or sites for a regulator that does not compete with the substrate but that affects the reaction rate, often through a change in conformation at the active site; such an enzyme displays an S-shaped velocity curve (as a function of substrate concentration), which can be moved to the left (K_m reduced) in the presence of a positive allosteric regulator or to the right (K_m increased) in the presence of a negative allosteric regulator

Allysine: aminoadipic-δ-semialdehyde

Amiloride: a drug inhibitor of one of the sodium–proton antiports, the amiloride-sensitive antiport

AMPA: amino hydroxymethylisoxazole propionic acid

Amylin: a 37–amino acid protein in hyaline

Amyloid: a protein deposit associated with tissue degradation

Amylopectin: a branched polysaccharide component of starch

Amylose: a linear polysaccharide in starch

Anandamide: a cannabinoid-like substance (an endocannabinoid) derived from arachidonic acid, produced in the body, that binds to receptors and stimulates feeding

Anaphase: when paired chromosomes dissociate and each member of the pair moves to opposite poles of the cell

Ankyrin: a cell skeletal protein

Annexin I (lipocortin 1): a protein induced by glucocorticoids; annexin I interacts with phospholipase A_2 (PLA_2) to inhibit the release of arachidonic acid from the cell membrane; annexin I is an endogenous inhibitor of inflammation because arachidonic acid gives rise to inflammatory prostaglandins, leukotrienes, and lipoxins

ANP (atriopeptin): atrionatriuretic peptide

Anticodon: as in tRNA, specific for a given amino acid, the three-letter anticodon combines with the sense codon on mRNA; the same relationship can exist in DNA in the sense and antisense strands

Antiport: a transporter moving two different substances in opposite directions

Apaf-1: an apoptosis factor; in combination with cytochrome c, the activation of caspase-9 occurs

APCR: activated protein C resistance

Apoenzyme: an enzyme requiring a coenzyme for activity that does not yet have the coenzyme bound to it

Apoptosis (programmed cell death): following appropriate signals, the process of a cell killing itself

APRT: adenine phosphoribosyltransferase

Arachidonic acid: a cell membrane constituent and precursor of prostaglandins, leukotrienes, and lipoxins; a 20-carbon fatty acid containing four double bonds

βARK: β-adrenergic receptor kinase

Aromatase: an enzyme, especially in female sexual tissues (e.g., mammary gland) that converts androgens (e.g., δ-4-androstenedione and testosterone) to estradiol or estrone

Aspartame: a chemical noncaloric sweetener (NutraSweet) that is a dipeptide of aspartic acid and the methyl ester of phenylalanine

Astrocyte: a star-shaped glial cell

ATP: adenosine 5′-triphosphate

Autocrine: a substance that affects the same cell that releases it

Autoimmune disease: a disease (e.g., lupus erythematosus, Grave's disease, Hashimoto's disease, rheumatoid arthritis, or myasthenia gravis) produced by the antibodies synthesized in the body that attack normal tissues; they recognize "self" as "nonself"

Autolysin: a bacterial enzyme that cleaves the glycosidic bonds between the peptidoglycan monomers at the growth point of the existing peptidoglycan polymer

AZT: a nucleoside reverse transcriptase inhibitor

Bacteriophage: a virus that infects a bacterial cell

Bactoprenol: a bacterial membrane carrier protein usually attached to peptidoglycan monomers; a membrane lipid formed by the condensation of 10 unsaturated isoprene units and 1 saturated isoprene unit, containing 10 double bonds per molecule and having a molecular weight of 768

Base excision repair mechanism: removal of an altered base in DNA after removal of the resulting sugar phosphate by DNA glyoxylase

Bax: a proapoptotic (producing apoptosis) factor

BBB: blood-brain barrier

Bcl-X: a proapoptotic molecule

B-DNA: the major physiological form of double-stranded DNA; right-handed helix

Beriberi: the classic thiamine deficiency disease among populations eating polished rice as the mainstay of the diet

β-Oxidation pathway: the mitochondrial system to degrade fatty acids and produce acetyl CoA

BH4: tetrahydrobiopterin

Bile acid: a derivative of cholesterol excreted through the intestine

Biotinidase: an enzyme (holodecarboxylases with bound biotin are acted upon by this enzyme) that cleaves biotin from its binding to the ε-amino group of a lysine residue in the active site of the decarboxylase

Bohr effect: when oxygen is released from hemoglobin in the presence of carbon dioxide

2,3-BPG: 2,3-bisphosphoglycerate

BRCA: an oncogene associated with breast cancer (*BRCA1* and *BRCA2*)

Bromodomain: the space to bind consecutive acetyllysines in the tail of histone H4 favoring propagation of acetylation to immediate neighbors

BtuB receptor: the vitamin B_{12} transporter of the *E. coli* outer membrane

Buffer: a solution of an ionic compound that resists changes of pH

CaaX motif: in *ras* that interacts directly with farnesyl in the active site of farnesyl transferase

Ca–CAM kinase: calcium–calmodulin-dependent protein kinase

Calnexin–calreticulin cycle: ensures the correct disulfide bond formation in proteins

CAM: cell adhesion molecule

Capsaicin: a substance that causes the burning sensation in hot peppers

CARM1: an arginine methyltransferase

Carnitine cycle: a system that carries cytoplasmic long fatty acid–CoA molecules into the mitochondrion complexed to carnitine (a derivative of lysine)

Caspase: an aspartate-specific cysteine protease; several of these enzymes are involved in the process of apoptosis

Catalase: an enzyme catalyzing the conversion of hydrogen peroxide to oxygen and water

Catalytic antibody: an unusual antibody that can function as an enzyme

CATH: a database for protein structure classification; CATH stands for Class, Architecture, Topology, and Homologous superfamily

CBC: complete blood (cell) count

CBP: CREB-binding protein

CCK: cholecystokinin

Cell cycle: stages of cell growth leading to division

Cell membrane potential: a differential charge on the membrane resulting from the flow of ions

Centriole: an organelle involved in spindle formation during cell division

Centromere: the central portion of the chromosome to which the spindle fibers attach during cell division

Ceramide: synthesized from serine and palmitoyl CoA; present in sphingomyelins and glycosphingolipids

Ceruloplasmin: a copper transporter

CF antigen: a gene on chromosome 1 encodes the cystic fibrosis antigen protein that is elevated in the serum of cystic fibrosis patients

CFTR: cystic fibrosis transmembrane regulator; involved in the transport of chloride

Chaperone: a protein that assists protein folding

Chaperonin: usually an HSP that functions as a chaperone

Chemiosmotic theory: states that protons are transferred through the mitochondrial membrane in a directed fashion, resulting in a pH gradient

Chemotaxis: movement of a cell toward a chemical gradient of a specific substance

Chief cell: a cell in the stomach that releases pepsinogen

CHIF: corticosteroid hormone-induced factor

Chirality: a group substituting a position linking to an asymmetrical atom and giving rise to right-handedness or left-handedness, depending on the position of the substituting group in space; can refer to the bending of light when light is passed through a solution of the molecule (e.g., *D* for *dextro* or *L* for *levo*)

Chloramphenicol: an antibiotic that inhibits the peptidyl transferase activity of the 50S bacterial ribosomal subunit

Chromaffin cell: a specialized postganglionic neuron, without axons, like those in the adrenal medulla

Chromatin: genetic material in the nucleus made up of chromosomes

Chylomicron: a small triglyceride-rich protein synthesized in the intestinal mucosa that is involved in the digestion of fat; also found in blood and lymph; name derived from *chylo*, meaning milky, and from *micron*, meaning small

Chyme: partially digested food

Cimetidine (Tagamet): a drug that blocks histamine action at the H_2 receptor

Circadian rhythm: the cyclic sleep-and-wake patterns governed by light and mediated by the pineal gland

Coenzyme: a substance that binds to an inactive apoenzyme to produce the active holoenzyme

Coenzyme A: a coenzyme in acetyl transfer reaction; made up of pantothenate, ATP, and cysteine

Competitive inhibitor: one that competes with binding of substrate to the active site

Conn's syndrome: the overproduction of aldosterone

Corticotroph (corticotrope): a specific cell in the anterior pituitary that produces and secretes ACTH

Cotransporter: a membrane exchanger that admits one ion into the cell in exchange for emitting another different ion

Coumadin (Warfarin): an anticoagulant used to thin the blood to prevent clotting

COX1 and COX2: two cyclooxygenase enzymes; COX1 is the constitutive enzyme, and COX2 is the inducible enzyme; COX2 is induced as part of the inflammatory response and is a popular drug target

C-reactive protein: a highly conserved plasma protein participating in the systemic response to inflammation

CRBP: cellular retinol-binding protein

CSF: colony-stimulating factor (M-CSF: macrophage colony-stimulating factor; G-CSF: granulocyte colony-stimulating factor; GM-CSF: granulocyte–macrophage colony-stimulating factor)

CSP: cysteine string protein

Cushing's disease: overproduction of cortisol by the adrenal gland

Cycloheximide: inhibits eukaryotic peptidyl transferase

Cyclooxygenase (COX): an enzyme involved in the synthesis of prostaglandins and development of inflammation

CYP24: a unique cytochrome P450 in the degradation of vitamin D

Cytokine: a member of a large family of compounds that undergo many activities, including inflammation

DAG: diacylglycerol

Dalton: a mass unit; macromolecular weights are often expressed as kilodaltons (kDa); a molecular weight of 100,000 can be expressed as 100 kDa

DD: death domain

Deamination: the process of removing an amino group from a compound

Deciliter: one-tenth of a liter, or 100 ml

Dendrite: a branch of a neuron extending from the cell body

Dexamethasone: a potent artificial glucocorticoid

DHEA: dehydroepiandrosterone

DiGeorge syndrome: a chromosome 22q11 microdeletion (characterized by lesions of the heart)

1,25-Dihydroxyvitamin D_3 (1,25-dihydroxycholecalciferol): the active hormonal form of vitamin D; ligand for the vitamin D receptor

Diphenhydramine (Benadryl): a drug that inhibits the H_1 receptor

Disulfide bond: the bond between two sulfhydryl groups (R_1-SH + R_2-SH) that become oxidized (R_1-S-S-R_2)

DMT1: a divalent ion transporter in the duodenum and upper jejunum

DVT: deep vein thrombosis

EAAT: excitatory amino acid transporter

EBP50: ezrin-radixin-moesin binding phosphoprotein

Ectoderm: the outer germ layer of the early embryo

Editing: the removal of introns from DNA allowing the exons to be spliced together so that mRNA can be generated

Effectors: usually small molecules that bind to sites other than the enzymatic catalytic center and produce conformational changes that affect the active center

EGF: epidermal growth factor

Electrostatic bonding: attraction between opposite charges

ELISA: enzyme-linked immunosorbent assay

ENaC: epithelial sodium ion channel; the sodium conductance channel in the renal tubular cell (stimulated by aldosterone)

Enantiomer: one of two mirror images of a chiral molecule (D-alanine and L-alanine are enantiomers)

Endocrine pancreas: the part of the pancreas that produces hormones

Endocytic vesicle: the vesicle formed at the cell membrane for moving materials from the outside of the cell to its interior

Endoderm: the innermost of three germ layers of the early embryo

Energy charge: usually refers to the ADP/ATP ratio in the cell

Enhancer: binds to DNA and facilitates recognition over long distances; actions involve protein–protein contacts; alterations of proteins, such as phosphorylation; changes in chromatin structure; nuclear localization; and others

Enkephalin: an endogenous opioid inhibiting pain in afferent (toward the brain) pathways; leucine enkephalin = YGGFL, methionine enkephalin = YGGFM

Enterochromaffin-like (ECL) cell: an endocrine cell in the acid-producing part of the stomach

Eosinophil: a cytotoxic white blood cell, granulocyte, that increases in allergy, inflammation, and infection as part of the inflammatory response

Epitope: sequence of amino acids in an antigen to which an antibody binds

Eplerenone: a drug related to spironolactone, an inhibitor of the mineralocorticoid receptor (aldosterone receptor)

ERα: estrogen receptor-α

ERK: extracellular signal-regulated kinase

Erythromycin: an antibiotic that binds to the 50S bacterial ribosomal subunit and inhibits translocation

Essential amino acids: those that must be obtained in the diet; not synthesized in the body (or minimally synthesized in the body): tryptophan, lysine, methionine, phenylalanine, threonine, valine, leucine, and isoleucine (arginine and histidine are essential in children)

Euchromatin: the open stringy form of chromatin

Eukaryote: an organism with a defined nucleus, such as animals, plants, and fungi

Exocrine pancreas: the part of the pancreas that produces digestive enzymes that are secreted into the intestinal tract

Exocytosis: release of hormones or other substances from a secretory granule to the cell exterior

Ex vivo **therapy:** cells are removed from a patient's body, treated in some way in vitro, and readministered to the patient

FAAH: fatty acid amide hydrolase; it hydrolyzes anandamide to yield fatty acid and ethanolamine

Fab: the fragment of an antibody that binds antigen and is made up of the N-terminal domains of the heavy and light chains that are variable in terms of amino acid sequence

Fabry's disease: a condition in which globotriaosylceramide accumulates

FAD: flavin adenine dinucleotide

FADD: fas-associated death domain protein

Farnesoid X receptor (FXR): a lipid-activated nuclear receptor

Farnesyl transferase: the enzyme-catalyzing interaction of *ras* protein and farnesyl

FATP: fatty acid transport protein

FecA: a bacterial porin; the ferric citrate receptor

Ferritin: an iron storage protein

FGF: fibroblast growth factor

Fibrous protein: poorly soluble protein, such as collagen that contains Gly-Pro-Pro repeated sequences

FKBP: one of several immunophilins

Frataxin: the iron donor for the iron–sulfur cluster and for the heme pathway

FSH: follicle-stimulating hormone

GABA: γ-aminobutyric acid

GABARAP: $GABA_A$ receptor-associated protein

GAD: glutamic acid decarboxylase

Gastric inhibitory peptide (GIP or enterogastrone): a 42–amino acid peptide (YAEGTFISDYSIAMDKIHQQDFVNWLLAQKGKKNDWKHNITQ); a member of the secretin family from the epithelial cells of intestinal mucosa that inhibits gastric motility and secretion of gastric acid

Gaucher's disease: a disease in which there is an excess accumulation of glucocerebrosides

G cell: a cell in the antrum of the stomach that releases gastrin

Gene therapy: replacement of a malfunctioning or nonfunctioning gene

Genome: all genes expressed in an organism

Gephyrin: facilitates the clustering of inhibitory neuroreceptors in the postsynaptic membrane

Ghrelin: a hormone from the stomach that stimulates the secretion of growth hormone from the anterior pituitary

Globoside: a cerebroside that contains more than one carbohydrate

GLP: glucagon-like peptide

Glucocorticoid response element: 5′ AGAACAnnnTGTTCT 3′

Glucogenic: a compound that can give rise, through metabolism, to glucose

GLU-R: glucagon receptor

Glut 4: a glucose transporter in the cell membrane that can be internalized and recycled to the membrane

Glutathione (GSH): glutamyl-cysteinyl-glycine; a tripeptide cellular store of reducing equivalents

Glycogenin: an enzyme that plays an initial role in the formation of glycogen from uridine diphosphate–glucose in muscle and liver

Glycogenolysis: the breakdown of glycogen for energy use

Glycolysis: anaerobic conversion of glucose to pyruvic acid in the cytoplasm

Glycophosphatidylinositol: part of a cell membrane to protein anchor

Glycoprotein: a protein containing a carbohydrate moiety or moieties

Gonadotroph (gonadotrope): a specific cell in the anterior pituitary gland that produces and secretes LH and FSH

G6PD: glucose-6-phosphate dehydrogenase

Graafian follicle: the residual follicle after ovulation

GRIP: glucocorticoid receptor–interacting protein

Growth factor: usually a hormone-like protein whose action causes a cell to enter mitosis

GSH reductase: the enzyme that reduces oxidized glutathione (GSSG $\rightleftharpoons$ 2GSH)

GTP: guanosine triphosphate

Gyrase: a topoisomerase controlling topological transitions of DNA, including unwinding of double-stranded DNA

HA: hemagglutinin

Hairpin: a secondary structure in RNA resembling a hairpin

HAP1: a specific brain protein

Hartnup disease: a disease in which the absorption of tryptophan is compromised

HAT: histone acetyltransferase; acetylates a lysine residue

HbA$_1$c: glycated hemoglobin; elevated levels are diagnostic for diabetes

HbF: fetal hemoglobin

hCG: human chorionic gonadotropin

Helix: a coil of peptide or DNA stabilized by internal bonds; the helix can be either right-handed or left-handed

Hematocrit: the volume of red cells as a percentage of total blood volume (43–49% in males; 37–43% in females)

Hemosiderin: an iron storage protein

Heterochromatin: condensed chromatin of the nucleus

HGPRT: hypoxanthine-guanine phosphoribosyltransferase

5-HIAA: 5-hydroxyindole acetic acid

HIF: hypoxia-inducible factor

High-density lipoprotein (HDL): contains lipids and protein in about equal amounts; promotes breakdown and removal of cholesterol from the body and carries cholesterol from peripheral tissues to the liver; high level of blood HDLs is related to prevention of a heart attack

HIOMT: hydroxyindole-O-methyltransferase

Hirsutism: masculinization in females; one development is unwanted hair

HIV: human immunodeficiency virus

HLA antigens: human leukocyte-associated antigens

HMG-CoA reductase: catalyzes the conversion of 3-hydroxy-3-methylglutaryl-CoA to mevalonate; rate-limiting enzyme in the biosynthesis of cholesterol

Holoenzyme: active form of an enzyme after an inactive apoenzyme is bound to factors or a coenzyme required for its activity

Hormone-sensitive lipase: certain hormones (glucagon, epinephrine, or β-corticotropin) operating through their cognate receptors, stimulate the level of cyclic AMP and the activation of protein kinase A, which phosphorylates HSL to its active form; with DAG lipase, the two enzymes hydrolyze triglycerides to monoacylglycerols (which are subsequently hydrolyzed by monoacylglycerol lipase to free fatty acids and glycerol)

HRE: hormone-responsive element

HRV: human rhinovirus

HSD: hydroxysteroid dehydrogenase

HSL: hormone-sensitive lipase

HSP90: a 90-kDa heat shock protein

5-HT: 5-hydroxytryptamine; serotonin

Huntingtin: the Huntington's disease gene protein product expressed in all types of neurons

Huntington's chorea: quick, jerky, and purposeless movements in Huntington's disease; *chores* is Greek for dance

Huntington's disease: caused by a mutant DNA sequence containing an expanded trinucleotide repeat of cytosine, adenine, and guanine within the coding region of the gene

Hyaline: a clear membranous material

Hydrolase: an enzyme that catalyzes the hydrolysis a bond by the addition of water

Hydrophobic bonding: attraction between nonpolar (uncharged) groups

Hyperammonemia: high levels of ammonia in the blood that can be toxic; usually a result of a defect in the urea cycle

Hypercholesterolemia: high levels of cholesterol in blood; can be caused by the dysfunction or limited expression of LDL receptors

Hyperhomocysteinemia (HHCE): elevated levels in blood of homocysteine; if high enough, the level can indicate premature coronary artery disease (15–100 μmol/L); more than 100 μmol/L is severe; occurrence is 5–7% of the population

IAP: inhibitor of apoptosis

ICAM: intercellular adhesion molecule

Icosahedral: a many-sided, three-dimensional hexagonal shape composed of many small triangles; a polyhedron with 20 faces

IGF-I: insulin-like growth factor

IGF-II receptor: the mannose-6-phosphate receptor that functions as a transport protein in embryonic and neonatal tissues

IGFBP: IGF-binding protein

IHF: integration host factor

IL: interleukin

Initiation factor (eIF, eukaryotic initiation factor): a protein (or proteins) that binds to the 40S ribosomal subunit that allows the binding of mRNA so that protein synthesis can occur

Interferon (IFN): a small protein cytokine released by tissue cells, macrophages, and lymphocytes in response to viral infection

Interneuron: a neuronal cell that, through a cellular process, affects another neuron

Intrinsic factor: a protein from the stomach parietal cells that has cobalamin (vitamin B_{12}) attached to it

Ion channel: a proteinaceous structure, usually in the cell membrane, that participates in the movement of ions into and out of the cell; it is often regulated

IP_3: inositol 1,4,5-trisphosphate

IRAK: interleukin-1 receptor–associated kinase

IRR: insulin receptor-like receptor

IRSI and IRSII: substrates of the insulin receptor

Islets of Langerhans: groups of pancreatic cells that produce glucagon (alpha cells), insulin (beta cells), and somatostatin (delta cells)

Isomerase: an enzyme that catalyzes a mutase reaction, such as conversion of an L-form to a D-form, a shift of a chemical group on a molecule, or movement of a double bond

Ketogenic: a compound that, through metabolism, can give rise to keto acids

Kinetochore: the point of attachment for microtubules of the spindle

Kupfer cell: a monocyte in the liver

Lactaid: used by people with lactose intolerance; contains lactase and is taken orally before consuming dairy products

Lactose intolerance: a failure to produce the enzyme lactase (β-galactosidase), which is essential for the digestion of lactose; people with this condition have an allergic reaction to foods containing lactose

Lactotroph (lactotrope): a specific cell of the anterior pituitary that produces and secretes PRL

LamB: a bacterial porin; the maltodextrin channel

LANR: low-affinity neurotrophin receptor

Laparoscope: a tube with a camera lens for viewing

Laparotomy: an incision through the abdominal wall

LATS: long-acting thyroid stimulator, an autoimmune antibody that recognizes the thyroid-stimulating hormone receptor as antigen

LBD: ligand-binding domain

LCF: lymphocyte chemoattractant factor

LDL receptor: a cell surface receptor that binds LDL and moves it into the cell, where cholesterol is released and often stored

Lecithin: a phosphatide found in plants and animals; a mixture of diglycerides of stearic, palmitic, and oleic acids linked to the choline ester of phosphoric acid

Lecithin cholesterol acyltransferase (LCAT): catalyzes mobilization of cholesterol from the plasma membrane to HDL; reaction in blood: cholesterol + lecithin $\rightleftharpoons$ cholesteryl ester + lysolecithin

Leptin: a 130–amino acid hormone produced in adipose cells; a ligand for the leptin receptor in the hypothalamus whose action promotes the synthesis of an α-melanocyte-stimulating hormone that suppresses appetite

Lesch-Nyhan syndrome: a recessive X-chromosome-linked disease generating severe gout and central nervous system disorder

LH: luteinizing hormone

LH spike: the sharp rise in the level of LH about midway through the ovarian cycle

Ligase: an enzyme whose action joins two moieties together

Limbic system: an important system in the brain in which key structures are the hippocampus and the amygdala

Lineweaver-Burk plot: a plot of 1/[S] (abscissa) against 1/v (ordinate) for an enzymatic reaction; the slope of the resulting line is K_m/V_{max}

Lipolysis: triglyceride breakdown

LTR: long terminal repeat

Liver X receptor (LXR): a lipid-activated nuclear receptor

Loratadine (Claritin): a drug that inhibits the H_1 receptor

Low-density lipoprotein (LDL): contains more lipid than protein; transports cholesterol in blood from the liver to the peripheral tissues; binds to the LDL receptor in some tissues to internalize cholesterol

LXXLL: the leucine zipper motif

Lyase: an enzyme that catalyzes the addition of water, ammonia, or carbon dioxide to a double bond, or these substances can be removed to create a double bond

Lysosome: a subcellular structure in the cytoplasm in which proteins and other molecules are degraded

Lysozyme: an enzyme that can break down gram-positive cell walls

MAC: membrane attack complex

MAPK: mitogen-activated protein kinase

Maple syrup disease: accumulation of branched chain amino acids in the blood and urine, imparting the odor of maple syrup in the urine

MEK: MAPK/ERK kinase

Melanin: a dark pigment (usually black or brown) formed from aggregates derived from tyrosine derivatives (dihydroxyphenylalanine-forming indolequinone)

Mesoderm: the middle of three germ layers of the early embryo

Mesothelioma: a rare cancer of the lung or stomach lining, often caused by asbestosis

Messenger RNA (mRNA): coding RNA formed from the template strand of DNA; triplets in RNA encode specific amino acids in protein formation (see Appendix 2)

Metabolic water: water produced as the result of metabolism, especially of fuels

Metaphase: when chromosomes are arrayed around the center of the cell by spindle fibers

MHC: major histocompatibility complex

Michaelis constant (K_m): the concentration of substrate in an enzymatic reaction that produces half-maximal velocity; it is not a strict measure of affinity but can be used often as an approximation; the K_m value may reflect the substrate concentration in the cell

Microsome: the fraction of the cytoplasm containing the ribosomes and the endoplasmic reticulum

Mineralocorticoid: an adrenocortical steroid hormone, especially aldosterone, which is active in salt retention

Mitogen: the stimulator of cell division

Mitochondrion: a subcellular structure containing enzymes of the citric acid cycle and some active genes

Monoclonal antibody: a single type of antibody against one epitope of an antigen

Motilin: a 22–amino acid gastrointestinal hormone (FVPIFTYGELQRMQEK ERNKGQ)

Multiprenylmenaquinone: vitamin K_2

Muscarine: a toxin found in poisonous mushrooms; used to name muscarinic acetylcholine receptors that can bind muscarine-like toxins

Mutant huntingtin protein (Htt): inhibits normal histone acetylation by blocking histone acetylases or recruiting histones into aggregates

nAChR: excitatory nicotinic acetylcholine receptor

NAD: nicotinamide adenine dinucleotide (reduced form is NADH; phosphorylated forms are $NADP^+$ and NADPH)

hNAT3: a human amino acid transporter that has 547 amino acid residues with the highest specificity for transport of L-alanine

Neuraminic acid: a parent acid of a family of amino sugars containing nine or more carbon atoms

Neurofascin: a cell skeletal protein

Neuroligin: a cell adhesion molecule

Neuropeptide Y (NPY): a feeding stimulant; the most abundant neuropeptide in brain; besides feeding behavior, it affects circadian rhythms, sexual function, anxiety responses, and vascular resistance

Neurophysin: a protein of a hypothalamic neuronal cell body that accompanies the hormone (e.g., oxytocin) of the cell from the cell body through the axon to the nerve ending

NFAT: nuclear factor of activated T cells

NFκB: nuclear factor kappa B

NGF: nerve growth factor

Niemann-Pick disease: a lysosomal storage disease

NK: natural killer (cell)

NMDA: N-methyl-D-aspartate

NNRTI (Nevirapine): nonnucleoside reverse transcriptase inhibitor

NO: nitric oxide

Nociceptive: pain transmitting

Noncompetitive inhibitor: one that produces enzyme inhibition by binding irreversibly; does not compete with substrate binding

Nonessential amino acids: those that are synthesized in the body or by the intestinal bacteria: glutamic acid, glutamine, aspartic acid, asparagine, alanine, cysteine, tyrosine, proline, serine, and glycine (and ornithine)

NOS: nitric oxide synthase (iNOS: cytokine-inducible nitric oxide synthase; cNOS: constitutive nitric oxide synthase; nNOS: neuronal nitric oxide synthase; eNOS: endothelial nitric oxide synthase)

NPEY motif: Asn-Pro-Glu-Tyr sequence that plays a role in contacting receptor tyrosine kinase to various signaling molecules

NSAID: nonsteroidal anti-inflammatory drug (examples are aspirin and naproxen)

Nuclear lamina: a layer of thin filaments stabilizing the inner nuclear membrane

Nuclear localization motif: NYKKPLK; directs proteins to enter the nucleus through the nuclear pore

Nucleolus: a small structure in the nucleus; site of action of RNA polymerase III

Nucleoplasm: the soluble cytoplasm within the cell nucleus

Nucleopore (nuclear pore): the port for the entry and egress of small and large molecules through the nuclear membrane

Nucleoporin: one of several cytoplasmic proteins that attach to a protein to facilitate transport into the nucleus

Nucleotide excision repair: a small region of a DNA strand surrounding a damaged nucleotide is removed as an oligonucleotide

Octreotide: a synthetic peptidic drug that inhibits growth hormone secretion; mimics somatostatin

Okazaki fragments: stretches of the lagging strand (starting with the 3′ end) of DNA synthesized by DNA polymerase to continue the action of the enzyme in the 5′ to 3′ direction

Oncogene: a gene that causes cancer; one or more oncogenes may be involved in the production of a given cancer

Oophorectomy: removal of one or both ovaries; when other tissues are removed in addition, it becomes a hysterectomy

OPG (osteoprotegerin): a soluble receptor for TRAIL that lacks a membrane anchor, a transmembrane domain, and a cytoplasmic domain

Opioid: a substance (e.g., β-endorphin, leucine enkephalin, methionine enkephalin, dynorphin A, or dynorphin B) that is an agonist of the opioid receptor

Opsin: a photosensitive pigment coupled covalently to 11-*cis*-retinal to form rhodopsin

Opsonization: the process in which an antibody binds to an invading microbe to enhance the microbe's attachment to a phagocytic cell

Orotic acid: a pyrimidine metabolite that does not occur in nucleic acids

Orphan receptor: a receptor (e.g., in the steroid receptor superfamily) for which the ligand is unknown or its function is obscure

Osmolarity: the sodium ion concentration, usually in plasma

Osmosis: passive membrane transport with no energy requirement above free diffusion

Oxidoreductase: an enzyme that catalyzes the addition or removal hydrogen atoms to or from a substrate, depending on the direction of the reaction

Panhypopituitarism: damage to the hypothalamus or to the stalk carrying hormones from the hypothalamus to the pituitary that results in a shortage of the hypothalamic-releasing hormones that eventually, through the actions of the pituitary gland, affect many target organs, causing a deficiency in the hormonal secretions from those target glands

Paracrine: a substance released by a cell that affects nearby cells

Parietal cell: a cell in the stomach that produces and secretes acid

PDGF: platelet-derived growth factor

Pellagra: a classic niacin deficiency disease

Pentose phosphate pathway: the anabolic pathway converting glucose to five carbon sugars (e.g., ribose-5-phosphate) for nucleic acid synthesis and reducing equivalents as NADPH

PEPCK: phosphoenolpyruvate carboxykinase

PEPT1: proton-dependent dipeptide transporter

Peptide bond: the linkage between a carboxyl group and an α-amino group, such as the linkage between two amino acids to form a dipeptide

Perforin-granzyme–mediated necrosis: an enzyme system from NK cells that causes cell necrosis; a mechanism against tumor cells

Perilipin: a protein involved in the translocation of HSL; one form coats lipid storage droplets in adipocytes; it becomes activated by protein kinase A

Periportal cell: a cell adjacent to the portal vein; periportal cells surround a portal vein

Perivenous hepatocyte: a liver cell that is adjacent to a vein; perivenous hepatocytes surround a vein

Peroxisome: a microbody in the cytoplasm in which organic molecules are degraded oxidatively to produce hydrogen peroxide that is subsequently converted to oxygen and water

Peroxisome proliferators: industrial and pharmaceutical chemicals such as plasticizers, herbicides, and hypolipidemic drugs

PEST sequence: Pro-Glu-Ser-Thr or another sequence, such as Pro-Gly-Ser-Cys, that renders some proteins more degradable by the ubiquitin pathway

pH: a logarithm of the reciprocal of the hydrogen ion concentration in gram atoms per liter; the value of pH is from 1 to 14, with 7 being neutral; values below 7 are acidic, and values above 7 are alkaline

Phage: from the Greek word *phagein*, meaning to eat; a virus, usually a bacteriophage

Phenylketonuria: an autosomal recessive inherited disease in which phenylalanine cannot be converted to tyrosine because of the lack of expression of phenylalanine hydroxylase activity

Phytic acid: an acid in plants (e.g., cereals and soybeans) that binds iron; iron in this form is not absorbed in the intestine

Phytylmenaquinone: vitamin K_1

Pica: a condition of advanced iron deficiency in which nonfood items (e.g., clay or starch) are eaten

PI3K: phosphatidylinositol-3-kinase

PIF: prolactin release–inhibiting factor

Pinocytosis: the process of ingesting fluid and macromolecules in small vesicles less than 150 nm in diameter

Point mutation: alteration of a single amino acid in a protein; it can arise from a single base change in DNA

Porin: a bacterial diffusion channel in the cell surface for the influx of nutrients and proteins

Posttranscriptional gene silencing (PTGS): double-stranded RNA is used to knock out the expression of specific genes; also referred to as RNA interference

POT: proton-dependent oligopeptide transporters; *pot* is a gene family

PPA: proenkephalin A

PPAR: peroxisome proliferator-activated receptor, usually an orphan receptor

Prediabetes: a condition of elevated blood glucose levels that are not high enough to be considered type 2 diabetes

Preproinsulin: a dimer of two polypeptide chains, chain A (21 amino acids) and chain B (30 amino acids), containing three disulfide bonds with a molecular weight of 5808 Daltons

Preproopiomelanocortin: a precursor of anterior pituitary hormones yielding proopiomelanocortin before final cleavages to produce the constituent hormones

Preprotein: a completed polypeptide chain with sequences added to the N-terminus that will be cleaved to form the proprotein (the proprotein is usually cleaved also to form the mature protein product)

Prion protein: a protein that has a normal function in the cell (PrP^c) but through a conformational change can give rise to an infectious protein particle, prion scrapie (PrP^{Sc}); in cattle, the prion disease is bovine spongiform encephalopathy ("mad cow disease"); in humans, it is Creutzfeldt-Jacob disease

PRL: prolactin

Prokaryote: an organism with cells that do not have a defined nucleus; refers mainly to bacteria

Prometaphase: when kinetochores are formed and begin to move to the central latitude of the cell

Properdin: a complement B-factor protein, a positive regulator of the alternative complement pathway

Prophage pathway: the course of a phage when it is committed to a lytic cycle and cannot return to the lysogenic state

Prophase: condensation of chromatin with disappearance of the nucleolus in preparation for cell division

Prosthetic group: a substance required for enzymatic activity that remains tightly bound to the enzyme throughout the catalytic reaction; it is different from a coenzyme, which may become dissociated from the enzyme during a reaction

Proteasome: a complex that degrades proteins, especially those tagged with ubiquitin

Protein kinase A (PKA): one of many protein kinases that phosphorylate proteins, mainly on serine or threonine residues; in the inactive form, two catalytic molecules of PKA are bound to two molecules of cyclic AMP–binding protein; upon activation by cyclic AMP, the subunits dissociate, yielding the active catalytic proteins

Prozac: fluoxetine; an SSRI antianxiety drug

PTH: parathyroid hormone

Puromycin: an antibiotic that blocks the peptide chain coming from the P site and terminates protein synthesis

Pyridoxal phosphate: the coenzyme form of the vitamin pyridoxal or pyridoxine

OFQR: orphanin FQ receptor; it binds nociceptin or orphanin FQ

Quaternary structure: a protein, for example, containing all of its subunits

Racemase: an enzyme that catalyzes the interchange between enantiomers (e.g., the conversion of D-alanine to L-alanine; this can be a pyridoxal phosphate–containing enzyme)

Ranitidine (Zantac): a drug that blocks histamine action at the H_2 receptor

RAR: retinoic acid receptor

Receptor: a protein on the cell surface or in the cell interior that specifically binds a ligand; the complex usually participates in cell signaling

Release factor: recognizes the stop codon (UGA)

Releasing hormone: one of several specific hormonal secretions from the hypothalamus that causes the release of a specific pituitary hormone

Restriction enzyme (restriction endonuclease): one of many bacterial enzymes that cut DNA at a specific sequence

RFC-1 transporter: an intestinal folic acid receptor

Rh antigen: an antigen on the red blood cell, in addition to those of the ABO system

Ribonucleotide reductase: an enzyme catalyzing the conversion of a ribose nucleotide to a deoxyribose nucleotide; the enzyme is coupled with thioredoxin reductase, which regenerates the oxidized coenzyme

Ribozyme: ribosomal RNA that can catalyze the formation of a peptide bond during protein synthesis

Rough endoplasmic reticulum (RER): membranes within the cytoplasm with ribosomes bound to the surface

RXR: retinoid X receptor

Salvage pathway: purine bases from the metabolism can form nucleotides again by phosphoribosylation by APRT and HGPRT

Saquinavir: an inhibitor of HIV protease

Sarcolemma: the plasma membrane covering the outer surface of a muscle fiber

SARS: severe acute respiratory syndrome

Scavenger receptor: a receptor on the surface of a macrophage; it takes up LDL particles, facilitating the transformation of a macrophage to a foam cell (rich in cholesterol)

Schiff base: an amino acid, such as lysine bonds to a carboxyl of an acid; usually an intermediate in pyridoxal phosphate (PLP) enzyme catalyzed reactions (e.g.,

Enz-lysN-PLP); the bond can shift to an incoming amino acid substrate to form another Schiff base (e.g., Enz-PLP-Nasp)

ScrY: a bacterial porin, the sucrose-specific channel

Sequenase: a commercial enzyme preparation; a modified form of T7 DNA polymerase

Sequence for N-glycosylation: Asn-X-Ser/Thr, where X is any amino acid except proline or aspartic acid

Seroconversion: when an HIV-negative test becomes HIV-positive

SERT: serotonin transporter

Sgk: serum and glucocorticoid-inducible kinase

Sheehan's syndrome: hypopituitarism occurring after profound blood loss during and after childbirth

Sialic acid: N- and O-substituted neuraminic acid

Sildenafil (Viagra): a drug that blocks the activity of phosphodiesterase, prolonging the vasodilation caused by NO through cyclic guanosine monophosphate

SLE: systemic lupus erythematosus

Smac/DIABLO: a proapoptotic protein released from mitochondria

Smooth endoplasmic reticulum (SER): a sac structure that is the same as the rough endoplasmic reticulum but without attached ribosomes; it does not participate in protein synthesis

Somatoliberin: the growth hormone–releasing hormone; also called somatocrinin or sermorelin

Somatomedin C: another name for IGF-I

Somatotroph (somatotrope): specific cells of the anterior pituitary that produce and secrete growth hormone

Spark chamber: an experimental apparatus used to simulate prebiotic conditions, leading to the formation of organic substances (e.g., amino acids, nucleic acids, and their precursors)

Sphingolipid: made up of cerebrosides, sulfatides, globosides, and gangliosides

Sphingomyelin: a sphingolipid (found in brain) that has a polar head group and two lipophilic tails; sphingosine, a long-chain amino alcohol, is the central molecule of the sphingolipids

Sphingomyelinase: the enzyme that hydrolyzes sphingomyelin to ceramide and choline phosphate

Spironolactone: an inhibitor of the mineralocorticoid receptor (aldosterone receptor)

Squalene: an intermediate in cholesterol biosynthesis; a polymer of six isoprene units

SREBP: sterol response element-binding protein

SSRI: selective serotonin reuptake inhibitor

StAR: steroid acute response protein in mitochondria

Start codon: AUG, begins the reading of mRNA

Statins: drugs that inhibit cholesterol biosynthesis by interacting with HMG-CoA reductase

Stereoisomer: the same molecule but with a different special arrangement of one or more atoms

Stop codon: UAG, terminates the reading of mRNA

Straight line equation: the general expression is $y = mx + b$, where y is the ordinate value, m is the slope, x is the abscissa value, and b is the x-axis intercept; in Michaelis-Menten terms: $y = 1/v_0$; $m = K_m/V_{max}$; $x = 1/[S]$; $b = 1/V_{max}$

Streptomycin: an antibiotic that inhibits bacterial growth by causing mRNA misreading; previously used to treat tuberculosis

Substance P: an 11–amino acid (RPKPQQFFGL-Met-NH$_2$) polypeptide that functions as a neurotransmitter and a neuromodulator

Symport: a transporter of two substances in the same direction

Synapse: the functional connection between a nerve cell and its target cell; the synaptic space is sometimes called a cleft

T$_3$: triiodothyronine; the most active form of the thyroid hormone

T$_4$: tetraiodothyronine (thyroxine)

Tachykinins: a family of amidated neuropeptides that share the sequence: Phe-X-Gly-Leu-Met-NH$_2$ at the carboxy terminal end

TAF: transcriptional activator factor

TATA-binding protein: a 27-kDa protein required by all three RNA polymerases to initiate transcription; it binds to the TATA box (TATAAA) on DNA

TBG: thyroid-binding globulin

TBPA (transthyretin): thyroid-binding prealbumin

Telophase: when a membrane forms around material that will become the daughter nucleus

Tetracycline: an antibiotic that binds to the bacterial 30S ribosomal subunit and inhibits subsequent binding of aminoacyl tRNAs

TFIIB: a core transcription element

Thrombospondin: a glycoprotein that can interact with blood coagulation and anticoagulant factors and can be involved in such processes as cell adhesion, platelet aggregation, and cell proliferation

Thymosin: a hormone with a molecular weight of 5 kDa secreted by the thymus gland; a maturing hormone for precursor T cells in the thymus

Thyroliberinase (pyroglutamyl peptidase II): a zinc-containing enzyme that cleaves the pyroglutamyl–histidine bond in TRH, inactivating the hormone

TIDA: tuberoinfundibular dopaminergic system

Tim complex: the translocator of the inner mitochondrial membrane

TNFα: tumor necrosis factor-α

TNFRSF: tumor necrosis factor receptor superfamily

α-Tocopherol: vitamin E

Tom complex: the protein translocator of the outer mitochondrial membrane (Tom is the abbreviation for translocator outer membrane)

TPP: thiamine pyrophosphate

TRAIL: tumor necrosis factor–related apoptosis-inducing ligand

Transamidation: a special case of transamination; for example, the ε- (C5) amide of glutamine contributes its *amide group* to fructose-6-phosphate to form glutamate and glucosamine-6-phosphate

Transamination: transposition of an amino group from one compound to another by a transaminase (also aminotransferase), usually requiring pyridoxal phosphate as coenzyme

Transcortin: a blood glucocorticoid–binding protein

Transferase: an enzyme that catalyzes the transfer of a group from one molecule to another

Transferrin: a carrier protein that regulates the transport of iron from the site of absorption (intestinal cells) to all tissues of the body

Transgene: a gene placed experimentally in the germline that functions as a normal gene

Transsulfuration pathway: syntheses from methionine to cysteine, carnitine, and taurine

TRBP: thyroid receptor–binding protein

TRH: thyroid-releasing hormone (L-pyroglutamyl-L-histidinyl-L-prolinamide) produced by the hypothalamus

Triamcinolone: a potent artificial glucocorticoid

Tricarboxylic acid cycle (TCA; citric acid cycle; Krebs cycle): metabolism of fuels in the mitochondrion, especially pyruvate, to form ATP, water, and carbon dioxide; the major supplier of energy to the cell

TRPV family: a family of channels including a calcium channel (CaT1)

TRS: transcriptional regulatory sequence

TSH: thyroid-stimulating hormone produced by the anterior pituitary thyrotrope

Type 1 diabetes: a disease produced by the destruction of the beta cells of the pancreas, creating a shortage of insulin

Type 2 diabetes: a disease in which insulin is in short supply or its use does not function properly (the beta cells of the pancreas are not destroyed)

Tyrosine kinase: a protein kinase that phosphorylates tyrosine residues

Ubiquitin: a protein that binds to certain amino acid residues on a protein destined for proteolysis

Uncompetitive inhibitor: an enzyme inhibitor that binds to a site other than the substrate-binding site and produces parallel lines in a double reciprocal plot; the value of $1/V_{max}$ is modified by the factor $(1 + [I]/K_i)$, but the slope is unchanged from the control

Uniport: a transporter of a single substance

Urocanate: a histidine deamination product (through the action of histidine ammonia lyase)

USF: upstream stimulatory (transcriptional) factor

Van der Waals forces: ionic charges usually occurring between two oppositely charged molecules

VDR: vitamin D receptor

Very low-density lipoprotein (VLDL): together with chylomicrons, VLDLs transport fatty acids from the liver to peripheral tissues

VKOR: 2,3-epoxide reductase (it converts vitamin K to the 2,3-epoxide form)

VMAT: vesicular monoamine transporter

Von Gierke's disease: a disease provoked by an excessive buildup of glycogen because of a failure to express glucose-6-phosphatase

Wernicke-Korsakoff syndrome: a condition of insufficient thiamine intake in chronic alcoholics

Z-DNA: a novel form of DNA in a left-handed double helix

Zinc finger: a DNA-binding motif that contains zinc

Zip: zinc family transporters

Zwitterion: a molecule with two types of charges (e.g., an amino acid is a zwitterion that can be positively, NH_4^+, or negatively, COO^- charged)

Index

A2486 amino acid, 70, 75f
AA. *See* Amino acid(s)
AAAD. *See* Amino acid-α decarboxylase
AADC. *See* Amino acid decarboxylase
ABC transporter superfamily, 732, 732f
ABC transporters/ABC-like transporters, 688, 788, 1130f
  models of, 689f
Abdominal swelling, 588
Aberrant ion transport, 685–694, 686f–694f
ABO system, 888–889, 891
Abortion, habitual, 856
Absorption. *See also* Uptake
  amino acid, 442
  calcium
    intestinal, 789f, 840
    ion, 783, 1028
  fat, 561
  folate, 830
  folic acid, 830
  ileum, cobalamin-intrinsic factor complex, 825, 827f
  iron, 743, 749f
  magnesium, 780, 782
  vitamin E, 843
ACAT. Coenzyme A cholesterol acyltransferase
AC+βHB, 717f
ACC. *See* Acetyl CoA carboxylase
Accessory factor I, 647f
ACDP2 (Ancient conserved domain protein), 714
Acetal grouping, aldosterone, 477f

Acetoacetate, 546, 553, 553f, 561
Acetoacetate plus β-hydroxybutyrate (AC+βHB), 717f
Acetone, 561
Acetyl CoA, 176, 176f, 178f, 546, 549f, 554
  ADP/ATP levels and, 566
  cholesterol biosynthesis from, 193, 194f
  conversion of fatty, 205–206, 205f–207f
  excess of, 561
  glucosamine-6-phosphate with, 512, 512f
  leucine conversion to, 553, 553f
  β-oxidation pathway for, 560, 560f
  structures of, 555f
Acetyl CoA carboxylase (ACC), 209, 213f, 554, 555f, 820
2-acetylaminofluorene, 301f
Acetylation
  histone, 358, 360, 360f
  NF$_k$B DNA, 327, 327f
Acetylcholine, 447, 454, 456, 804, 1003, 1007
  -binding protein, 1004f
  nervous system and, 1027–1028, 1027f
  structure of, 1012f
  synthesis of, 1013f
Acetylcholine binding protein (AChBP), 1004f
Acetylcholine receptor (AChR), 923f, 1011
  brain, 1028
  location of, 1014

Acetylcholine receptor (AChR), (*Continued*)
  models of, 1013f
  muscarinic, 1017, 1017f
  nicotinic, 1003, 1013f, 1014, 1014f, 1016f, 1017, 1019
  stress response of, 448
    calcium channels opened by, 448, 450f
  subunits of, 450f
  surface model of, 1016f
Acetyltransferases, 243f
AChBP. *See* Acetylcholine binding protein, 1004f
AChR. *See* Acetylcholine receptor
Acid phosphatase, 93
Acid secretion, histamine secretion and, 439
Acid sphingomyelinase, 221, 226
Acidification, cellular model of, 715f
ACP. *See* Acyl carrier protein
Acquired immunodeficiency syndrome (AIDS), 1047, 1067
  treatment of, 1067
ACTH. *See* Adrenocorticotropic hormone
Actin, 896, 897f
  filament system, 600–601
Actinolite, 323
Action potentials, 729
  transmission of, 959, 960f, 961
Activated factor X (Xa), 846
Activated protein C (APC), 857, 859f
  functions of, 860f
  in vascular endothelial cell membrane, 858, 861f

1177

Activated protein C resistance (APCR), 856, 860f
Activation energy enzymes, 96–97, 97f
Activation mechanism
 DNA cleavage, 494f
 hormone receptor, 478, 481f
Activator factor regions, 352–353
Activator protein-1 (AP-1), 326, 493, 494f, 609, 626
 diagram of, 328f
 transcription factors, 654
Active isoprene, 193, 194f
Activins, 612, 613
Acute pain, chronic pain v., 955
Acute phase response, 654, 951
Acute stress, 445, 446
Acyl carrier protein (ACP), 818
Acyl-CoA synthase, 723
Acyl-CoA synthetase, long-chain, 726f
Acylethanolamides, 977f
Adaptive immunity, 945
Addiction, 1001
Addison's disease, 474
Adenine, 118, 119t, 244, 247f, 248f, 262f. *See also* Bases; Flavin adenine dinucleotide
 base-pairing with, 250f
 deaminated to form hypoxanthine from, 271, 274f
Adenine mononucleotide. *See* AMP
Adenine phosphoribosyltransferase (APRT), 271
 structure of, 275f
Adeno-associated viruses, 1074t
 gene therapy with, 315t
Adenohypophysis, 368
Adenoids, 883
Adenoreceptor model, 1036f
Adenosine 5'-triphosphate (ATP), 13, 18–19, 20f, 176–177. *See also* DideoxyATP; UN coating ATPase
 adenylate kinase reaction with, 270
 amino acid metabolism with, 505
 -dependent helicase, 359
 E. coli with, 1112f, 1139–1140, 1140–1144, 1142f, 1143f
 electron transport chain generating, 740

Adenosine 5'-triphosphate (ATP) *(Continued)*
 fatty acid breakdown generating, 208
 glucose metabolism with, 142
 guanylate kinase reaction with, 270
 high-energy bond of terminal phosphate of, 182
 purine synthetic pathway regulation with, 263f
 supply, 740
 synthesis of, 770
Adenosine, conformation of syn- or anti-, 249f
Adenosine deaminase deficiency, 314, 315
Adenosine diphosphate (ADP), 18–19, 176–177. *See also* ADPr
 glucose metabolism with, 142
 purine synthetic pathway regulation with, 263f, 831f
Adenovirus, 1070f
 gene therapy with, 315t
Adenylate cyclase, 249, 387, 408, 410, 1031–1032
 activation of
  calcium/potassium channels from, 446
  cyclic AMP with, 970
  D2 receptor, 1028
 inhibition of, cyclic AMP reduction from, 970
Adenylate cyclase-activating peptide, 436
Adenylate kinase reaction, 270
Adenylosuccinate
 adenylosuccinate synthetase in IMP conversion to, 268, 270f
 conversion of IMP to AMP/GMP with, 268, 270f, 275f
Adenylosuccinate lyase, 255f
 adenylosuccinate converted to AMP by, 268
 structure of, 261f, 272f
Adenylosuccinate synthetase
 adenylosuccinate catalyzed by, 268, 270f
 structure of, 272f
ADH. *See* Antidiuretic hormone
Adhesion molecules, 658
Adipocytes, 204, 561, 723
 differentiation of, 616

Adipose tissue, 565
A-DNA, 289
ADP. *See* Adenosine diphosphate
ADP + Pi 1-(5'phosphoribosyl)-4-(N-succinocarboxamide)-5-aminoimidazole (SAICAR), 255f
ADP/ATP ratio, 566
ADPr, 710, 710f
Adrenal cortex, 453f
 blood supply of, 449f
 cell layers of, 449f, 452
  zona fasciculata, 387, 449f, 452, 453f, 454, 456
  zona glomerulosa, 449f, 452, 453f, 454
  zona reticularis, 449f, 452, 453f, 474
 serotonin from, 1019
Adrenal gland
 cross section of, 449f
 fetal, 452, 476
Adrenal medulla, 449f, 526, 526f
 epinephrine innervating, 1028, 1031
 in stressful event, 446, 448–451, 451f
Adrenaline, 448, 524
Adrenergic receptors, 163, 165f, 167, 168f, 448, 451t, 1031–1032, 1034f, 1038f, 1041, 1042f, 1043f
 signal pathways of, 1042f, 1043f
 α_1-adrenergic receptor, 163, 167, 168f, 1043f
 α_2-adrenergic receptor, 1043f
 α-adrenergic stimulation, 417
 β_1-adrenergic receptor, 1032, 1038f, 1041, 1042f
 β_2-adrenergic receptor, 1041, 1042f
 β-adrenergic receptor, 163, 168f
  structure of, 165f
 β-adrenergic receptor kinase (βARK), 1033, 1039
Adrenocorticotropic hormone (ACTH), 368, 373t, 378f, 452, 561
 action of, 391f
 cortisol pathway of CRH and, 387–392
 release of, 387, 447, 454, 456
 sequence of, 390f

1178 INDEX

α_{1A}adrenoreceptor, homology model of, 1032
ADRP, 561, 564f
AEA
  binding of, 984f
  CB-1 with, 984f
AEA transporter, CB-1 with, 983f
AEA-induced cell death, programmed, 984f
Aerobic oxidation, depression of, 500
Affinity(ies), 422–423
  antibody for antigen, 915
  ATP/UTP, 1143–1144
  nicotine, 1014
  opioid receptor, 970t
  taut state of, CTP as, 1140–1144
Aflatoxin B_1, 301f
Aging, DHEA-S and, 476
Agmatine, 536, 538f, 541
Agonist
  androgen receptor, 488f
  binding, 348f, 1033
  EAA receptor, 990f
  morphine as, 970
  NMDA receptor, 991
  OFQR, 970
  progesterone, 465–466
  β-receptor, 1033
Agouti receptor protein, 418
α-granules platelet, 600
α-hemoglobin stabilizing protein (AHSP), 760
  cellular, 762
  mechanism of, 763f
AHSP. See α-hemoglobin stabilizing protein
AHSP-α-globin complex, 760, 763f
AHSP-α-globin dimer, 760
AICAR. See 1-(5′phosphoribosyl)-5-amino-4-imidazole carboxamide
AICAR transformylase. See Phosphoribosyl aminoimidazole carboxamide formyltransferase
AIDS. See Acquired immunodeficiency syndrome
AIF. See Apoptosis inducing factor
AIR carboxylase. See Phosphoribosyl-aminoimidazole carboxylase

AIR synthase. See Phosphoribosyl-aminoimidazole synthase
AKT activation, 661f
ALA. See δ-aminolevulinic acid
ALA dehydrase, 753
ALA synthase, 766
  mutation, 756
Alanine, 43f, 44t
  E. coli, 1113, 1114f, 1118f
  α-helix formation with, 58, 81
β-alanine, 271, 277f, 1003
  pantothenic acid formed from, 818, 819f
  pyrimidine nucleotide catabolism leading to, 271, 277f
Alanine aminotransferase, 93, 546
Alanine racemase, 1113, 1114f, 1118f
Albinism, 531
Albumin, 418, 448, 797, 871
  free fatty acids complexed with, 561
  retinoic acid binding to, 836
Alcohol consumption, 717, 806
Alcoholics, chronic, 806, 808
Aldehyde crosslinks, 765
Aldehyde oxidase, 765, 788, 792f
Aldehyde reductase (AR), 537f
Aldimine linkage, 511f
Aldol condensation products, 768
Aldosterone, 387, 447, 452–466
  acetal grouping of, 477f
  ACTH and, 452
  action of
    ion transport, 458
    mechanism of, renal tubular, 458f, 459f
    sodium ion in, 456–457
    time course of, 459f
  cortisol v., 452, 463, 478
  ENaC synthesis with, 729
  fluctuations of, 452
  formation of, 453f
  11β HSD2 not being a substrate of, 463, 465f
  in hyperaldosteronism, 465
  inactivation of, 568
  Na^+/K^+-ATPase and, 458, 458f
  rapid effects of, 460f
  second receptor of, 454
  secretion, 454f
Alimentary system, 3, 4f
Alkaline phosphatase, 93

Alkaptonuria, 532
Allergy
  drugs for, 542, 543f
  histidine metabolism and, 542
Allo tetrahydrocortisone, 568, 569f
Allopurinol, 128f, 127
  structure of, 287f
  xanthine oxidase as inhibitor of, 283
Allosteric enzymes, 105–107, 106f, 107f
  effectors bound by, 105
  half of sites case with, 105
  protein kinase A as, 105, 106f
Allosteric kinetics, 107f, 871
Allosterism, 105–107, 106f, 107f
All-trans-retinal, 838f, 1037f
Allysine, aminoadipic-δ-semialdehyde, 765
Alpers syndrome, 33, 35
Alpha linolenic acid, 557f
Alpha-amylase, 93
Alternative pathway, 934–935, 935f, 937f, 938f
  properdin as regulator of, 949, 950f
Aluminum, 795
Alzheimer's disease
  aluminum toxicity causing, 795
  amyloid deposition in, 170, 172f
  cholinergic loss and, 1028
Amadori rearrangement, 876f
Amide, amino acid metabolism group transfers of, 503–508
Amiloride, 714, 715f
Amino acid(s) (AA), 39–54, 40f–43f, 44t, 45f–53f, 47t. See also Amino acid decarboxylase; Aminotransferase(s); Essential amino acids; Excitatory amino acid transporters; Transamination
  AAAD, 448
  absorption of, 442
  alanine, 43f, 44t, 58, 81, 93, 546
    E. coli, 996, 1113, 1114f, 1118f
  ammonia derived from, 497, 500f, 503
  anatomy of, 43f

Amino Acid(s) (AA) *(Continued)*
  antibody formation with, 917–920, 918f
  aromatic
    catabolism of, 531f, 537f
    decarboxylase, 516, 537f
  asparagine, 41f, 44t, 81, 182, 549, 547f, 550f
  aspartate, 510, 549, 547f
    aminotransferase with, 51
  aspartic acid, 42f, 44t, 47f, 990f
  asymmetric carbon in, 44–46
  carbon dioxide group of, 878
  catabolism of, 531f, 544f, 546–554, 547f, 551f, 554f, 820
  chains
    branched, 442, 442f, 551, 552f
    insulin, 932
    L-, sequence variations in antibody formation with, 918f
    side, 47f, 47t
  chirality of, 44–46, 45f
  cysteine, 41f, 44t, 47f, 519f, 520, 831f, 1118
  decarboxylation of, 541–542
  deprotonation with, 46
  enantiomer, 45
  in erythropoietin, 638
  glucogenic, 52, 546
  glutamate, 524, 536, 538f, 541, 549
    brain neurotransmitter, 544, 545f, 1004
    metabolism of, 500f, 505, 512, 512f, 544, 545f
  glutamic acid, 41f, 44t, 47f, 81, 829f, 858f
  glutamine, 41f, 44t, 58, 500f, 506, 507f, 549
    carbamoyl phosphate formation with, 263f
    glutamic acid conversion of, 503
    α-helix formation with, 58
    metabolism of, 500f, 505
  glycine, 43f, 44t, 58, 81, 545, 546f, 547f, 549, 753, 753f, 754f
    titration curves for, 48f, 49f
    transporters of, 721

Amino Acid(s) (AA) *(Continued)*
  histidine, 42f, 44t, 47f, 430, 513, 513f, 536, 541–544, 544f, 549, 550, 768, 1126
    decarboxylase reaction, 436f, 516t, 531–542
  isoleucine, 43f, 44t, 58, 550
  ketogenic, 52, 546, 552
  L/D nomenclature with, 45, 45f
  leucine, 43f, 44t, 81, 820, 955, 957, 958f, 1126
    acetyl CoA conversion to, 553, 553f
    transamination of, 552
    zipper, 327, 328
  lysine, 42f, 44t, 47f, 359, 510, 552, 877
    metabolism of, 500f, 503–508, 506f, 512f, 517–546, 519f, 527f–536f, 540f, 542f, 544f, 545f
  methionine, 41f
    monoamine, 1003, 1027–1044
    monocarboxylic, 816, 818, 817f
  neutral, 816, 818
    resorption of cationic and, 817f
  nonessential, 39, 50, 517
  oxidation of, 514, 514f, 515f
  oxidative deamination with, 52, 53f
  peptide bond formed in ribosomes from, 67–73, 71f–75f
    elongation factor for, 70, 74–77, 78f, 79f
    initiation factor for, 68, 73–74, 75f–77f
    termination factor for, 68, 77, 80f
  phenylalanine, 40f, 44t
  pK values for ionizable groups with, 47f, 47t
  plasma levels of, 500
  potential charges with, 46–50, 45f–50f, 47t
  proline, 40f, 44t, 58, 81, 541, 545, 550f
  protonation of, 46
  racemase/racemization, 46, 509–510
  residues in oxytocin/vasopressin of, 380

Amino Acid(s) (AA) *(Continued)*
  sequences of, 56–57, 57f
    dynorphin A, 973f
    human TGFβ1/TGFβ2, 613, 613f
    insulin with, 139f
    leptin, 987f
    OFQ, 973f
    opioid peptides, 968t
  serine, 40f, 44t, 545, 546f, 547f, 549
  stereoisomer, 46
  synthesis/degradation of, 50–54, 51f–53f
  taurine, 1003
  threonine, 40f, 44t, 549, 547f
  transport of, 442, 522f, 720–723, 720f, 721f, 722t
  tryptophan, 40f, 44t, 532–536, 532f–536f
  tyrosine, 42f, 44t, 47f
  valine, 44f, 44t, 550
  zwitterion, 46–50, 45f–50f, 47t
Amino acid decarboxylase (AADC), 516, 516t, 517f, 537f, 1039f
Amino acid proteins, 39–54, 40f–43f, 45f–53f, 47t
  sequences of, 56–57, 57f
Amino acid transporter family, 720
  hNAT3 in, 721
  Amino acid-α decarboxylase (AAAD), 448, 451f, 514, 514f, 515f, 720, 720f, 721
Amino hydroxymethylisoxazole propionic acid (AMPA), 989f, 990, 990f, 991
Amino sugar synthesis, 512, 512f
α-amino group, carbon dioxide and, 878
α-amino-β-ketobutyrate lyase, 549, 549f
Aminoacrylate hydration, 513f
Aminoacyl receptor site (A-site), 68, 70, 74f
Aminoadipic-δ-semialdehyde, al-lysine as, 765
γ-aminobutyric acid (GABA), 544–545, 545f, 996, 1003, 1033f. *See also* GABA transaminase
  brain biochemistry of, 1007f

1180 INDEX

γ-aminobutyric acid (GABA) (Continued)
  pathway for metabolism of glutamate and, 999f
  shunt, 1004
  succinate degrading of, 999, 999f
  synthesis of, 1006
2-aminoethylphosphoric acid, 713
Aminoimidazole ribonucleotide synthetase, crystal structure of, 259f
β-aminoisobutyrate, 271, 277f
β-aminoisobutyric aciduria, pyrimidine metabolism disorder with, 288t, 720f
δ-aminolevulinic acid (ALA), 753, 754f. *See also* ALA dehydrase; ALA synthase
Aminotransferase(s), 50, 508, 509f
  alanine, 93, 546
  aspartate, 51, 93, 107, 508
    structure of, 108f, 509f, 510f, 511, 549, 550f
  branched-chain, 551, 552f
  glutamine, 500f
  glycine, 538f
  PLP as, 818
Ammonia, 550f
  in amino acid metabolism, 497, 500f, 503
  blood level of, 497
  in histidine deamination, 513, 513f
  urea from, 499
AMP (adenine mononucleotide)
  adenylate kinase reaction with, 270
  conversion of IMP to, 252, 262f, 268, 270f, 275f
  cyclic, 249, 387, 395, 408, 410–410, 432
    aquaporin mechanism with, 698
    cellular levels of, 1031
    formation of, 561, 565f
    GLU-R receptor in, 559
    inhibition of, 970, 980–981, 981f
    insulin as decreasing, 554
    protein kinase A depended on by, 688
  TSH-TSH receptors and, 419

AMP (adenine mononucleotide) (Continued)
  nucleotidase in degradation of, 271, 274f
  purine nucleotide cycle with, 275f
  purine synthetic pathway regulation with, 263f
AMP kinase, 554
AMPA. *See* Amino hydroxymethylisoxazole propionic acid
AMPA receptor, 989f, 992
  complex, 996f
  kainate with, 999f
  location of, 995
AMP-activated Cl⁻ channel, 732
AMP-PCP, 821f
Amygdala, 446
Amylase, 441, 442–443
Amylin, 170–173, 172f, 173f
Amylo-(1,4-1,6)-transglycylase, 163
Amyloids, 170–173, 171f–173f, 172t
  disease associated with, 170, 172t
  electron micrograph of, 172f
  pancreatic islet with, 172t
Amylopectin, 153, 153f, 154f
  α-D-1,6 bonds with, 153, 154f
Amylose, 153, 153f
  α-D-1,4 bonds with, 153
Anaerobic glycolysis, 18
Analgesia, 968, 969–970
Anandamide, 219, 974–975, 975f–988f, 980–984
  binding of, 979, 980, 980f, 981f, 982
  CB-1 with, 986f
  extracellular release of, 1034f
  food intake increased by, 982
  immune cell effects of, 984
  pathways for, 977f
  phospholipase D conversion to, 975, 977f
Anandamide aminohydrolase, 977f
Anandamide hydrolase, 981f
Anandamide synthase, 975, 976f
Anaphase, 14
Anaphylatoxins, 942f
  edema with, 949
Ancient conserved domain protein (ACDP2), 714

Anderson's disease, 171t
Androgen pathway, 452, 453f
Androgen receptor
  DNA-binding domain of, 487f
  ligand-binding domain of, 486, 487f
  ligands of, agonist/antagonist, 488f
Δ^4-Androstenedione, 566
Androstenedione, 452, 566
  estrone conversion of, 567f
Androsterone, 568, 568f
Anemia, 119t, 636, 837
  causes of, 740
  diagnosis for, 740–741
  hematocrit/hemoglobin levels in, 742t
  iron-deficiency, 739–743, 746f
    chronic v., 741–742
  megaloblastic, 825, 830
  microcytic, 762
  pernicious, 828
Anesthetics, 992f, 996
ANF (atrionatriuretic factor), 465
Angiotensin (AT), renin-, 452, 454f, 457–458
Angiotensin I receptor (AT_{1a}), 456f
Angiotensin II, 447, 454, 456
Angiotensin II receptor, 452, 454
Angiotensin receptor (AT), 452, 454, 456f
Angiotensinogen, 452
Anhydroglucose, 153, 153f
A-NK cells, 899
  pre-, 903
Ankylosing spondylitis, 920
Ankyrin, 729, 730f
  repeat, 730f, 784, 787f, 789f
  domain of, 730f
Annexin I, 467, 469f
ANP. *See* Atrionatriuretic peptide
Anterior pituitary, 365
  hormones, 368, 373t–377t, 378, 379f
    actions of, 387–406
    gonadotropins, 406–415
    PRL, 430
    target end organs for, 379f
    thyrotropin, 417–428
    hypothalamus connection to, 369f
Anthophyllite, 323

INDEX 1181

Antiapoptotic factors, 644
Antibiotics
   E. coli disease recovery without, 1148
   protein synthesis inhibitors as, 83, 85f
Antibodies, 884, 890t, 910–916, 947f. See also Epitopes
   antigen-antibody complex with, 912f, 914f, 915, 934–935
   anti-insulin, 930
   auto-, 923f
   B cell, 910, 915, 929, 929f
   catalytic, 910, 912f
   classes of, 910f
   contact residues in antigens and, 913f
   definition of, 910
   diversity of, 917
   epitopes for binding viruses and, 947f
   Fab portion of, 887f, 910, 914f, 916f, 936f
   FC portion of, 914f
   formation of, 917–920, 919f
      amino acids in, 917–920, 918f
      combinatorial diversity in, 917
   IgG, 671, 910, 910f, 911f, 916f, 917, 935, 936f
   IgM, 910, 910f, 929f
   monoclonal, 915, 916f
   polyclonal, 915
   production of, 664, 911, 914
   receptors for, 896f, 916f, 927, 943
   semienzymic reaction of, 910
   T cell receptor, 927
   variable domains of, 910, 912f, 917, 918f, 919
   variable fragment of, 912f
Antibody pocket, 914
Antibody reactive site. See Epitopes
Anticancer agents, 491
Anticoagulants, folic acid absorption interference from, 830
Anticodon, 63
Antidepressant drugs, 533
Antidiuretic hormone (ADH), 367. See also Vasopressin

Antigen(s)
   antibody fragment binding to, 910
      hypervariable surface/specificity for, 910
   APO1, 902f
   binding of, 910, 914–915
      opsonization in, 916f, 917
      phagocytes in, 916f
   blood type, 889, 891
   CF, 685
   contact residues in antibodies and, 913f
   E. coli, 1148
   exogenous, 675
   gag, 1050, 1053f
   islet, 933f
   MHC I
      chain-related, 899, 900f, 903, 904f
      complex of, 904f
   MHC proteins as HLA, 928f, 930
   T cell, 725f
Antigen-antibody complex, 934–935
   formation of, 914f
   structures of, 912f
Antigen-presenting cells (APC), 639, 725f, 948f
Antihemophilic factor. See Factor VIII
Anti-inflammatory drugs, 283, 472f
Antioxidant(s), 832, 837, 841, 843
Antiport, 704
   Ca^{2+}/Na^+, 710, 712f
   cystine-glutamate, 721
   glutamate-proton, 722
   ion, 737f
   Na^+/H^+, 714, 715f
      cytoskeletal network with, 716f
   NCX1, 787
   sodium-proton, 729
   system, 702f
Antiporter(s)
   NCX1, 787
   vesicular, 722
Antisense DNA, 311
Antisense RNA, 311
Antithrombin, 867, 867f
Antithrombin III, 853, 867f, 869
α1-antitrypsin, 867, 867f

Anti-tumor activity, 594
   IL-18, 679f
Antiviral protein, 642–643
   guanylate-binding, 642–643
Antiviral response, 311
Anxiety, 962, 1022
Anxiolytic substance, 446
Aorta, 871, 872f
AP convertase, 938f
AP-1. See Activator protein-1
Apaf-1, 493, 494f
   cytochrome c interaction with, 902f
APC. See Activated protein C; Antigen-presenting cells
APCR. See Activated protein C resistance
Apical membrane
   calcium in, 783, 783f
   water permeation on, 699f
APO1 antigen, 902f
Apoenzymes, 111
Apolipoprotein E, 562, 564f
Apoptosis, 21, 224, 225f, 405, 679f, 923–924, 926f, 933f
   AEA-induced, 984f
   cytochrome c in, 982–983
   extrinsic pathway of, 492, 493, 494f, 495f, 593f, 594
   induced by glucocorticoids, 491–496, 494f–495f
   inhibition of, 658
   IAP, 593f
   intrinsic pathway of, 492, 493, 494f, 494f, 592, 593f, 594
   mediator p53 induced, 902f
   NMDA receptors causing, 995
   perforin pores in, 901f
   perforin-mediated, 904f
   regulator of, 654
   tumor necrosis factor in, 901, 902f, 927f
   of viral-infected cells, 904f
Apoptosis inducing factor (AIF), 902f
Apparent mineralocorticoid excess, 463–465
Appendix, serotonin in, 816
Appetite loss, 588
APRT. See Adenine phosphoribosyltransferase
AQP1. See Aquaporin water channel

AQs. *See* Aquaporin(s)
Aquaporin(s) (AQs), 699f
Aquaporin mechanism, 698, 699f
Aquaporin water channel (AQP1), 28, 29f
AR. *See* Aldehyde reductase
Arabinose, 1126
Arachidonate release, 1031
Arachidonic acid, 199–201, 200f, 521f, 1028
　anandamide derived from, 974, 975f, 981f
　phosphatidylethanolamine formation from, 981f
　phospholipase A_2 with, 975
　PLA_2 with, 965
2-AG. *See* 2-arachidonylglycerol
2-arachidonylglycero (2-AG), 978f
Arginase, 500–501, 503t, 538f
　human, 502f
Arginine, 42f, 44t, 47f, 500, 509, 507f, 549, 550f, 967f
　E. coli with, 1124f
　metabolism of, 536–541, 538f, 540f, 542f
　nitric oxide conversion of, 989f
Arginine decarboxylase, 516t, 538f, 541, 967f
Argininosuccinate, 500
Argininosuccinate lyase, 500, 503t
　arginine synthesis via, 541
　deficiency in, 497, 501, 503t
　structure of human, 502f
Argininosuccinate synthetase, 503t, 541
Argininosuccinic acid synthase, 500
　deficiency in, 497
Argininosuccinic aciduria, 497, 499, 501
Arginyl-tRNA synthetase, 538f
Arg-vasopressin (AVP), neuron synthesizing signals of, 382f
A-rings, estradiol, 476, 478
Aromatase, 566
　inhibitors of, 567f
　reaction, 567f
Aromatic amino acids
　catabolism of, 531f, 537f
　decarboxylase, 516, 537f
Aromatic L-amino acid decarboxylase, 516t

β-arrestin, 1039
　receptor-binding mechanism of, 1041f
　structure of, 1041f
Arteries
　endothelial cells of, 518
　vascular circulatory system and, 871, 872f
Arthritis, 678
Arylsulfatase B, 881
Asbestos, 323
　exposure to, 324f, 588
Asbestosis, 323–327
　lungs with, 325f
Ascending nerves, 956f
Ascites, 591
Ascorbate, 832, 834f
Ascorbic acid (vitamin C), 740, 770, 771, 818, 832–838, 833f
　dietary sources of, 834t
　signaling responses by, 834f
　structure of, 833f
　vitamin E with, 843
Ascorbic acid-iron complex, 740
A-site. *See* Aminoacyl receptor site
Asn-X-Ser/Thr, 184
Asparaginase, 549
Asparagine, 41f, 44t, 183, 549, 547f, 550f
α-helix structure disrupted with, 81, 550f
Aspartame (NutraSweet), 54, 149, 152f
Aspartate, 510, 549, 547f, 988, 1003
　analogs, 989t
　bacterial cell concentration of, 1143f
Aspartate α-decarboxylase, 516t
Aspartate aminotransferase, 51, 93, 95f, 508, 510f, 549, 550f
　coupling of, 511
　structure of, 108f, 509f, 510f, 511, 549, 550f
　transferases classification of, 107, 108f
Aspartate transcarbamoylase (ATCase), 456f
　bacterial cell, 1139, 1140, 1142f, 1140
　structure of, 1144f
　R-state, 1146f
　structure of, 253, 264, 265f
　T-state, 1146f

Aspartate-specific cysteine protease, 492
Aspartic acid, 42f, 44t, 47f, 990f
Aspartyl protease, 1056
Aspirin, 121, 739
Asp-X, 493f
Associative tracking model, antibody formation, 918f
Astrocytes, 654, 1000f, 1004
　types of, 998
Astroglia, 722, 723
Asymmetric carbon, 45
AT. *See* Angiotensin
AT_{1a} receptor, signaling domains in rat, 456f
AT_2 receptor, 452, 454
Ataxia, 503t
ATCase. *See* Aspartate transcarbamoylase
Atherogenic lipoprotein profile, 235t
Atherosclerosis, 198, 233
Atherosclerotic plaque, 189
Atkins diet, 850
Atoms, numbering of, 246f
Atorvastatin, HMG-CoA reductase inhibition with, 195f
ATP. *See* Adenosine 5'-triphosphate
ATP synthase, 768, 770, 772f, 780
ATP synthase-catalyzed reaction, 781f
ATPase
　calcium, 710, 712f, 784f
　E. coli, 1145
　sarcoplasmic reticulum Ca^{2+}, 712f
　sodium/potassium, 702, 702f, 703, 703f, 800
Atrial myocytes, 454
Atrionatriuretic factor (ANF), 465
Atrionatriuretic peptide (ANP), 454
　ANF and, 465
Atrionatriuretic receptor, dimerized hormone-binding domain of, 455f
Atriopeptin. *See* Atrionatriuretic peptide
AUG codon, 517
Auricle, left, 871, 872f

INDEX 1183

Autacoid mediator, 986f
Autoantibodies, 923f
Autoantigen peptide, 922
Autocrine, 402
Autoimmune disease(s), 678f, 920
　environmental factors in, 922
　examples of, 921f
　general mechanism for, 923
　inflammation with, 925
　MHC involvement in, 922
　susceptibility to, 922
Autoimmune reaction, 920
Autoimmune thyroiditis, 920
Autoimmunity, 920–933
　absence of, 932
　Graves' disease and, 920–922, 921f, 923
　MHC involvement in, 922–930, 929f
Autolysin(s), glycosidic bonds cleaved by, 1115, 1119f
Autonomic nervous system, 1027
Autoreactive CD4/CD8 cells, 930, 932f, 933f
Autosomal dominant genetics, 240
　pedigree for disease of, 241f
Autosomal recessive disorder pedigrees, 687f
Avidin, 820
AVP. See Arg-vasopressin
Axon, 959, 961, 969f
Axon terminal, 960f
AZT. See Zidovudine

B. See Bradykinins
B cells, 639, 654, 658, 665, 671, 675
　antibodies and, 910, 915, 929, 929f
　development of, 664
　immune competence and, 883
　lymphoma factor, 902f
　V-J combinations in differentiation of, 919
B Lymphocytes, 911
B^{0+} system, 721
B97, 947f
Bacterial cells. See also Bacterium; E. coli
　ATCase in, 1139, 1140, 1142f, 1143f

Bacterial cells (Continued)
　biosynthetic pathway regulation in, 1139–1140, 1142f
　CTP formation in, 1140, 1142f, 1143f
　cysteine protease derived from, 948f
　DNA in, 1130, 1131
　Gram stain for separating, 1107, 1111t
　gram-negative, 1107, 1108f, 1111t, 1118
　　gram-positive compared with, 1111f
　outer membrane of, 1127f
　major components of, 1109f
　operons in, 1134f, 1134, 1135
　periplasmic space of, 1126
　RNA polymerase in, 1131
　TATA motif region of, 1131
Bacteriophage(s), 1089–1107
　definition of, 1089
　genomes of, 1089
　　DNA, 1089, 1091–1092, 1092f–1094f, 1095–1097, 1096f
　LamB, 1107f, 1125, 1128f–1129f
　proteins of, 1095–1096, 1102f, 1104f
　RNA, 1102f, 1103
　structure of, 1090f
　temperate, 1091
　T-even, 1091f, 1092
Bacteriophage-λ, 1092f, 1101f
　capsid-stabilizing protein of, 1104f
　life cycle of, 1092f
Bacterium, internal destruction of, 896f
Bactoprenol, 1113, 1115, 1117f, 1118
　peptidoglycan monomers with, 1120f
BAD (Bcl-2-associated death), 654
Band neutrophils, 883, 884f
Baroreceptor, 380
Barrier function, 735f
Basal forebrain, 1017
Basal Ganglia, 1032f
　serotonergic system in, 1020f
Basal ganglia, Huntington's disease and functioning of, 239

Base excision repair, 245, 294
Base excision repair mechanism, 245
Base-pairing, 249–251, 250f, 251f
　bacterial cell, 1131
　hydrogen bonding with, 249, 250f
　Watson-Crick type of, 251f
Bases, 245. See also Schiff bases
　adenine, 244, 247f, 248f, 250f, 271, 274f
　anti orientation of, 245
　cytosine, 244, 247f, 248f, 250f, 268, 297f
　guanine, 244, 247f, 248f, 250f, 271, 274f
　hypoxanthine, 245, 248f, 249f, 271, 274f, 287f
　modified, 249
　orotic acid as, 245, 248f, 249f
　structures of, 248f
　syn orientation of, 245
　thymine, 244, 247f, 248f
　uracil, 244, 247f, 248f, 271, 297f
　xanthine, 245, 248f, 249f, 271, 274f
Basophil, 881, 882f
　progenitor, 630f
B°AT1. See Na^+-dependent neurotransmitter transporter family
BAX. See BCL-2-associated x protein
BB isozyme, 95
BBB. See Blood-brain barrier
Bc12, 670f
B-cell lymphoma factor, 902f
Bcl-2 family, antiapoptotic member of, 902f
Bcl-2-associated death (BAD), 654
Bcl-2-associated x protein (BAX), 224, 493, 494f, 902f, 995
　increased expression of, 995f
Bcl-X, 995
B-DNA, 289, 291f
Bee sting, 881
Beef, 776, 1147, 1148
Benadryl. See Diphenhydramine
Benign fibrosis, 588
Benign hyperphenylalaninemia, 523
Benzodiazepine, 996, 998f, 1006
　structure of, 1008f

1184　INDEX

Benzo(a)pyrene, 301f
Beriberi, 119t, 806
Betaine, 518
Betaine methyl group, 519f
Beta-oxidation. See β-oxidation
BH4. See Tetrahydrobiopterin
Bicarbonate
   ammonia combining with, 500f
   secretion of, 441
Bicarbonate ion (HCO$_3^-$)
   pancreatic duct homeostasis of, 693f
   regulation of, CFTR in, 688
BID, 902f
Big gastrin, 438
Bile
   heme porphyrin in, 879
   salts, 847
   secretion of, 506f
Bile acids
   circulation of, 197, 197f
   pigment, 743
   synthesis of, 193, 196f, 197–198, 197f, 554
Bile canaliculi, 506f
Bile ducts, 506f
   obstruction of, 847
Bilirubin, 743, 879
   heme degradation to, 745f
Bilirubin diglucuronide, 745f
Biliverdin, 743
   heme conversion to, 744f
Biliverdin reductase, 743, 745f
Binding. See also Binding sites; Ligand-binding domain; Protein(s)
   acetylcholine, 1004f, 1014
   AEA, 984f
   agonist, to cytoplasmic receptor, 348f
   AHSP, 760, 761f
   anandamide, 979, 980, 980f, 981f, 982
   antigen, 910, 914–915
      hypervariable surface for, 910
      opsonization in, 916f, 917
      specificity for, 910
   antiviral protein guanylate-, 642–643
   biotin, 820, 821
   C1q, 953
   CaM, 541
   class 1 receptor, 349

Binding *(Continued)*
   coactivator, 344, 345f
      TRBP to ERα/TRβ, 344, 345f
   complement components, 949f
   cortisol, 463–465, 470, 474
      transcortin, 466
   DNA, 1095
      C-terminal HIV-1 integrase, 1061f
      single-stranded, 1094f
   dopamine, 1029f
   EAA, 990f–991f, 991
   EGF, 604, 609
   estradiol, 478
   estrogen, 485f
   excess cortisol, 463–465, 465f
   factor B, 935
   FGF to FGFR, 623f
   foreign peptide, B cell MHC protein and, 929f
   gephyrin, 1007
   glucose residue, 874–875
   glutamate, NMDA site of, 993f
   glycine, 1003
   Gram stain, 1118
   heparan sulfate, 671f
   heterodimer of NF$_k$B, 326, 326f, 327f
   homer/IP$_3$R, 992
   homodimer p50, 326, 326f
   HRV serotype, 1072, 1073f
   human FGF receptor, 618t
   IGF-II receptor, perforin, 901f
   IHF, 1133f
   insulin, 648
   interleukin, 658, 666
   iron, 741, 760f
   LDL receptor/EG, 1075f
   lipocortin, 467
   measles virus, epitopes for antibody and, 947f
   microorganism, 943t, 949f
   norepinephrine, 1033
   oxygen, 871, 872f, 873, 873f
   P-box for, 461
   PDGF, 600
   peptide, MHC protein, 923f
   peptidoglycan, 1119f, 1123f
   perforin, 901, 903f
   phosphocholine, 953
   potassium channel, 992
   putative
      11-cis-Retinal, 1035f
      HIF-1α, 640f

Binding *(Continued)*
   receptor, β-arrestin, 1041f
   retinoic acid, 836
   SMRT, 344f
   STAT1, 647f
   substance P to NK-1, 966f
   TATA-binding protein, 333, 331f
   TBG, 797f
   testosterone, 478
   thyroglobulin/thyroid peroxidase, 922
   TRAIL, 591
   VDR ligand-, 784
   virus, 947
   virus particle, 1064, 1066f
Binding domains
   dimerized hormone-, atrion-atriuretic receptor, 455f
   DNA, 349, 350f, 461
      androgen receptor, 487f
   MHC protein, 922
   NBD1 CFTR, 688, 690f
   PDZ, 992
   thioredoxin-, 1093f
Binding proteins, 75f, 326, 335, 342
   acetylcholine, 1004f
   C4b-, 858, 937f, 945–946, 946, 948f
   CREB-, 342, 344, 346f, 356, 356f
   DNA-, 778, 1097f, 1100f
   Fatty acid-, 723
   GTP-, 713f
   heparin-, 948f
   IGF, 644, 648, 651f, 652f
   IL-22, 682f
   initiation factor for 4E, 74, 75f
   mannose-, 951
   matrix-, 726f
   serum-, 648
   single-stranded, 294, 295f
   SREBP-1, 218
   TATA-, 330, 330f, 333, 331f
   thyroid receptor-, 342, 345f
   vitamin D, 530, 837, 840
Binding sites
   bungarotoxin, 1016f
   dopamine, 1028
   epinephrine, 1038f
   glucose, 700f
   hemagglutinin, 1076

INDEX   1185

Binding sites *(Continued)*
  5-HT, 1022f, 1027
  nicotine, 1003, 1018f
  putative, 11-cis-Retinal, 1035f
  strychnine, 1003
  sugar, 700f
  thioredoxin-binding, 1093f
  virus, 943
  Zn^{2+}, 1026f
Biocytin, 820, 826f
Biological clock, 535
Biosynthesis
  anandamide, PE, 977
  bacterial cell, 1139–1140, 1142f
  ceramide of sphingomyelin, 220–221, 224f
  cholesterol, 193, 194f, 195f
    acetyl CoA in, 193, 194f
    active isoprene in, 193, 194f
    mevalonate in, 193, 194f
  DNA, 288–312, 290f–293f, 291t, 295f–305f, 301t, 306t, 307f–311f
    dinucleotide formation in, 290f
    next nucleotide added in, terminal phosphate of, 290f
  virus replication, 1103
  fatty acid, 554
  gangliosides, 229
  glycosphingolipids, 230f
  immunoglobulin, 917, 919–920
  IMP, 263f
  light chain, 919f
  pineal methoxyindoles, 537f
    serotonin in, 537f
  polyamine, 543f
  purine, 252–268, 252f–269f
  thyroid hormone, 797f
Biotin (vitamin H), 114f, 118, 119t, 820–821, 825f
  bi-cycle, 826f
  biotinyl-hydrolase/transferase of, 826f
  carboxylase subunit of pyruvate carboxylase, 824f
  chemical structure of, 824f
Biotinidase, 821, 826f
Bipolar disease, 718, 721
Bird flu, 1072, 1076, 1076f, 1081f
Bisphosphoglycerate (BPG), 879

1,3-bisphosphoglycerate (1,3-BPG), 879
2,3-bisphosphoglycerate (2,3-BPG), 879
Black tongue, 814
Bleeding, excessive, 854
Blood, 871–893
  carbon dioxide in, 878–879, 878f
  cortisol circulation in, 466
  glucose level in, high, 874–875
  hemoglobin/hematocrit values in, anemia, 742t
  hypercoagulability/hypercoagulation of, 853, 856
  IGFs in, 644
  iron-binding capacity of, anemia diagnosis as, 741
  oxygen transport in, 871–877, 872f–876f, 877t
  oxygenated/deoxygenated, 871
  peripheral, 932
  pH range of, 871
  thinners, 853–854
  volume, 871, 879
Blood cells, 879–884. *See also* Red blood cell(s)
  blood volume % of, 871, 879
  erythrocyte, 636f, 760, 871, 879
    appearance, 880f
    deglycosylation mechanism in, 877
  leukocytes, 470, 471f, 638–639, 871, 879
    integrins on surface of, 471f
    lactoferrins from, 1125, 1126, 1129f
    lymph node congregation of, 894
    peripheral, 880
  platelet, 600, 630f, 865f, 871, 879, 883f, 884, 898
    aggregation of, 949
    diameter of, 882
    PDGF with, 600–601, 602f 603f, 653f
  production of, 880f
  types of, 879
  white, 884
Blood clotting, 518, 844, 846, 853
  cascade in, 866f
  clot formation in, 865f
  detection of, 853

Blood clotting *(Continued)*
  mechanism, 856–857, 856f, 864–874, 864f–870f
  pathways of, 866f
  thymosin β4 in, 898, 898f
Blood concentrations
  decreased, 454f
  ghrelin, 396f
Blood concentrations/levels, testosterone, 486
Blood glucose, 565
Blood group proteins, 185–186, 187f
  blood group, 185–186
  posttranslational modifications of, 185
Blood groups, 890t
  blood transfusion receivable, 890t
Blood level
  ammonia, 497
  hemoglobin/hematocrit, anemia, 742t
  homocysteine, 518
Blood pressure
  increased
    epinephrine/norepinephrine secretion, 447
    excess kidney cortisol causing, 465
    sodium ion and, 457
  NO role in, 540
Blood protein(s)
  blood group, 185, 187f
  fibrinogen, 867, 884, 887f
  globulin, 884, 887f
  serum albumin, 884, 886f
Blood system, 4
  adrenal cortex, 449f
  Rh, 893
Blood transfusion, 889
  prion disease with, 33
  transfusability and, 889, 890t, 893, 893t
Blood type(s), 888–893, 890f
  ABO system of, 888–889, 891
  antigens in, 889, 891
  Rh, 891, 892f
  carbohydrate substituents of, 890f
  determination of, 888–889
  genotype and, 889t
  human, 888

1186   INDEX

Blood urea nitrogen, 497
Blood vessels, 783
Blood-brain barrier (BBB), 720, 734–737, 735f, 736f, 1045
 transport mechanisms of, 736, 737f
Bloodletting, 753
Blunt-ended products, 302, 302f
Boat configurations, 147, 148f, 150f
Bodily-growth path, GHRH/GH, 392–406
Body temperature, 740
Body weight, 588
Bohr effect, 878
Bone
 dietary metals and, 93, 780, 782
  breakdown of, 632
  growth, 402, 644
  mineral density of, 402
  mineralization of, 531
  osteoclastic resorption of, 632
  remodeling of, 782, 832
  resorption of, 840
Bone marrow, 637, 752f
 lymphocytes produced from, 894
Bone matrix, organic, 832
Bone morphogenic protein, 612
Bound ribosomes, 65
Bovine spongiform encephalopathy (BSE), 33
BPG. See Bisphosphoglycerate
1,3-BPG. See 1,3-bisphosphoglycerate
2,3-BPG. See 2,3-bisphosphoglycerate
BPG mutase, 879
Bradykinin receptor structure, 957f
Bradykinins (B), 955, 956f
Brain, 243f. See also Thalamus
 AChR in, 1028
 basal forebrain of, 1017
 characteristics of, 1044–1045, 1044t, 1045t
 components of muscle compared with, 1045t
 damage, 795
 GABA biochemistry in, 1007f
 -gut, 435
 infant, 1045
 information integration in, 1017
 injury trauma, 365–366

Brain (Continued)
 Kupffer cells of, 882
 mid, 974f
 neurotransmitters in, 544, 545f, 1003, 1004
 serotonergic system of, 1019, 1020f, 1027
 -type creatine kinase, 539f
Branched-chain α-keto acid dehydrogenase, 551, 552, 552f
 inactive, 551
Branched-chain amino acid(s), 442, 442f, 551, 552f
Branched-chain amino acid aminotransferase, 551, 552f
Branching enzyme, 154, 155f, 162–163
 amylo-(1,4-1,6)-transglycylase as, 163
Brazil nuts, 776
BRCA. See Breast cancer oncogenes
BRCA-1, 587–588
BRCA-2, 587
BRE. See TFIB recognition element
Breast cancer oncogenes (BRCA), 587–588
Breast development, 430
Broccoli, 740
Bromdomain, HAT, 359
Bronchoconstriction, 542
Bronchus-associated lymphoid tissue, 883
Brown rice, 776
Brush border membrane, 783
BSE. See Bovine spongiform encephalopathy
BtuB receptor, 1139, 1141f
Budding
 E. coli, 1108f
 virus particle, 1064, 1069
Buffering capacity, 48
Bulge loops, 317, 318f
Bulges, 317, 318f
α-bungarotoxin, 1014, 1015f
χ-bungarotoxin, 1014, 1015f
Bungarotoxin, 1016f
Buspirone, 1021t

C1, E. coli ATCase with, 1147f
C1 inhibitor, 935
Cl$^-$ permeating pore, 688

C1INH. See Hereditary angioneurotic edema
C1q
 binding of, 953
 heterotrimeric globular head of, 935, 936f
 interaction with CRP of, 953f
C1r2s2, 935
C2 deficiency, 946
C2 domain, 937
 perforin, 905f
 of PLC-δ, 905f
C3
 C3b conversion of, 942f
 deficiency of, 946
C3 convertase, 935f, 937, 942, 943, 946, 949
C3b, 935, 935f, 949
 C3 conversion to, 942f
C3d, 939f, 940f, 941f
C4, 677, 937
C4Ad, 939f, 940f, 941f
C4b-binding protein (C4BP), 858, 937f, 945–946, 946, 948f
C4BP. See C4b-binding protein
C5 activating convertase, 935, 938f, 941f
C5a, 942f
C9 deficiency of, 949
Ca^{2+}, 436
 antiports of Na$^+$ and, 710, 712f
 cellular concentration of, 418
 channel tetramer of, 711f
 cotransporter for, 708f
 cytosolic concentration of, 430, 438
 -dependent feedback, 788f
 endoplasmic reticulum, 708f
 ions of, 711f
 pumps, 712f
 release of
  intracellular, 710, 711f
  RYR, 707f
 -transporting epithelia, 788f
 uptake of, 789f
 vanilloid receptor regulation by, 985f
Ca^{2+}-calmodulin (CaM), 540, 541, 1088f
Ca-ATPase, 784, 784f
CaaX motif, 238
CaCAM kinase. See Calcium-calmodulin-dependent protein kinase

CACT. *See* Carnitine/acylcarnitine translocase
Cadmium, 778, 1060f
cADPr. *See* Cyclic adenosine diphosphate ribose
Caffeine, structure of, 1008f
CAG repeats, 239, 242f
Calbindin, 784, 785f, 787
Calcineurin homologous protein (CHP), 716f
Calcitrol837840f. *See* 1, 25-dihydroxyvitamin D_3
Calcium
   1,25-dihydroxyvitamin D3 regulating, 837, 840f
   absorption of, intestinal, 789f, 840
   apical entry of, 783, 783f
   in cell death, 707, 709
   excretion of, 840
   food sources of, 790t
   homeostasis, 712f
   import, 787f
   intracellular concentrations of, 787
   levels, 783
   as micronutrient, 783–788, 783f–788f
   release of, phosphatidylinositol stimulating, 609
   store of, 965
      depleted, 784, 787, 788f
   -transporting proteins, 530–531
Calcium channels
   acetylcholine receptor opening of, 448
   adenylate cyclase activation reducing, 446
   NMDA receptor as, 992
Calcium influx factor (CIF), 712f
Calcium ion(s)
   absorption, 783, 1028
   cyclic AMP with, 984, 986f
   cytoplasmic release of, 970
   glutamate with, 988, 990
   insulin secretion triggered by accumulation of, 142
   intracellular *vs.* extracellular concentrations of, 706t, 707
   IP_3 with, 709, 964–965
   in membrane transport, 707, 709, 710, 712f
   protein kinase A and, 411–413, 1001

Calcium ion channels, 731, 970, 1031
Calcium ion influx, 395, 430
Calcium transport channel (CaT1), 783, 783f, 784, 787, 787f
   activation of, 788f
   distal nephron location of, 787
   molecular architecture of, 787f
Calcium-ATPase (PMCA1b), 710, 712f, 787
Calcium-calmodulin protein kinase II (CaMKII), 995f
Calcium-calmodulin-dependent protein kinase (CaCAM kinase), 418
Calcium-dependent feedback inhibition, 783
Calmodulin (CaM), 716f
Calmodulin dependent kinase, 985f
Calnexin-calreticulin cycle, 184
Calorie requirements, calculations of, 849t
Calpactin II, 467
Calumenin, 844, 845f
CaM. *See* Ca^{2+}-calmodulin; Calmodulin
CaMKII. *See* Calcium-calmodulin protein kinase II
cAMP. *See* Cyclic AMP
cAMP phosphodiesterase, Insulin stimulating, 565f
cAMP-dependent phosphorylation, 689f
CAMs. *See* Cell adhesion molecules
Cancer, 236. *See also* Anticancer agents
   breast, 587, 588
   colon
      blood loss from, 743
      iron loss from, 739
   hematopoietic cell derivation of, 491
   immune surveillance system and, 899–909
   mammary, 566
   ovarian, 587–594, 589f, 590f
   prostate, 788f
   skin, Kaposi's, 1067, 1067f
   susceptibility to, vitamin A deficiency leading to, 837
   TGFα secretion in, 616

Cancer cells
   CTLs inducing apoptosis in, 904f
   NK destruction of, 901, 901f, 907f, 908f
Cancer drug target, farnesyl transferase as, 236
Cannabinoid(s), 974
   G protein-coupled receptors, 981f
   transmission, 986f
   transporter, 983f
Cannabinoid receptor (CB-1)
   AEA transporter with, 983f
   anandamide actions at, 986f, 1034f
   anandamide binding to, 980, 980f, 981f, 982
   cAMP with, 986f
   central, 975f
   G protein-coupled receptor, 982
   VR1 with, 983, 984f
Capillaries, lymph, 895f
Capsaicin, 961, 961f
   ball and stick model of, 961f
   model of RTX and, 985f
   structure of, 961f
Capsaicin receptor, 961, 961f
   activation of, 961f
   vanilloid receptors as, 982–983, 984f, 985f
Capsaicinoid family, 961f
Capsid-stabilizing protein, bacteriophage-λ, 1104f
Carbamoyl glutamate, 503t
Carbamoyl phosphate, 499
   formation from glutamine and bicarbonate, 263f
   mitochondrial, 503t
   pyrimidine synthesis from, 265f, 266f
Carbamoyl phosphate synthetase, 497, 499–500, 501f
Carbamoyl phosphate synthetase I (CPSI), 503t
Carbamoyl phosphate synthetase II (CPSII), 263f
   distinction from CPSI, 253
   structure of, 264, 264f
Carbohydrate(s), 131–188
   blood group antigens with, 185, 187f
   blood type substituent, 890f

Carbohydrate(s) *(Continued)*
  diabetes as disorder of, 131–146, 132f, 132t, 133t, 134f–146f, 170–173, 171f–173f, 172t
  glucose as, 138, 142, 145f, 146, 147f, 148f, 154, 155f, 157f, 162–163, 163f, 173–182, 174f–176f, 179f–182f, 211f
  glycerol as, 182, 182f, 201, 201f, 207
  glycobiology on, 188, 188f
  glycogen as, 138, 154–170, 155f–170f, 171t
  glycoproteins part of, 182–185, 183f–187f
  metabolism of, 565
  starch as, 153, 153f, 154f
  sugars, simple, as, 146–152, 147f–152f
Carbohydrate substitution, TSH molecule, 423f
Carbon(s)
  asymmetric, 45
  One-carbon transfer reactions of, 830
  pyrimidine in structure of, 253, 263f
Carbon dioxide, 878–879, 878f, 879f
  biotin as carrier of, 820
  elimination of, 776, 778
  hemoglobin with, 878
  in red blood cell reactions of, 879f, 880
Carbon monoxide (CO), 879
  removal of, 673f
ε-carbon, 510
Carbonic acid, 878
Carbonic anhydrase, 441, 441f, 776, 878
  reaction catalyzed by, 776
  structure of human, 777f
Carboplatin, 588
Carboxyhemoglobin, 878, 879
Carboxyl, pK values for, 47t
Carboxylase, active site of, 846f
γ-carboxyl glutamate (Gla), 844, 846, 846f
  -containing proteins, 844
  domain, 857
  structure of, 845f
γ-carboxyl glutamic acid, 858f
γ-carboxylase, 844, 845f

Carboxypeptidase, 442
5-carboxytryptamine (LSD), 1021t
Carcinogens
  direct-acting, 301f
    dimethyl sulfate as, 301f
    ethylmethane sulfonate as, 301f
    methyl nitrosourea as, 301f
    nitrogen mustard as, 301f
    β-propiolactone as, 301f
  DNA mutations/damage from, 298, 301f
  indirect-acting, 301f
    2-acetylaminofluorene as, 301f
    2-naphthylamine as, 301f
    aflatoxin B_1 as, 301f
    benzo(a)pyrene as, 301f
    dibenz (a,h) anthracene as, 301f
    dimethylnitrosamine as, 301f
    salrole as, 301f
    vinyl chloride as, 301f
  liver, 489
  metabolism of, 961–962
Carcinoid tumor, 814
Cardiac myocytes, 724
Cardiac remodeling, 632
Cardiomyopathy, end-stage dilated, 638f
Cardiovascular system, 3, 4f
Carnitine cycle, 206, 206f, 207f, 723
Carnitine palmitoyltransferase (CPT-1), 723
Carnitine palmitoyltransferase II, 723
Carnitine/acylcarnitine translocase (CACT), 726f
β-carotene, 835
  metabolism of, 836f
  retinal conversion of, 835f
β-carotene dioxygenase, 835f
Carpal tunnel syndrome, 818, 818
Cartoid artery, 871, 872f
Casein, 430
Caspase(s), 491–492, 492f, 591
  activation of, 492f, 492–493, 493f, 984f
  apoptosis with, 902f
Caspase 9, 493, 494f, 495f
  structure of dimeric, 495f

Caspase-1, 677, 678f
Caspase-3, 592, 902f
Caspase-8, 591
CaT1. *See* Calcium transport channel
Catabolism
  amino acid, 546–554
    aromatic, 531f, 537f
    aspartate, 549, 547f
    through citric acid cycle, 547f
    glutamine, 549
    histidine, 544f
    leucine, 820
    methionine, 550, 551f
    threonine, 549, 547f
    tryptophan, 554f
    valine, 550
  fatty acid, 546
Catalase, 25, 740, 875, 877
Catalytic antibody, 910, 912f
Catecholamine(s), 1027–1044
  copper with, 770, 773f
  dihydroxyindole, 527f
  in fat storage energy, 565f
  formation of, 524–526, 527f, 526f
  precursor of, 521, 524
  in stress reaction, 450, 450f, 451f, 526f
  synthesis of, 526
Catecholamine-O-methyl transferase DA dopamine (COMT), 1039f
CATH protein structure classification, 86, 88f–89f
Cations, 706
  resorption of neutral AAs and, 817f
CB-1. *See* Cannabinoid receptor
CB1R, constitutive trafficking of, 982f
CB-2. *See* Peripheral cannabinoid receptor
CBC. *See* Complete blood count
CBn, 975f
CBP. *See* CREB-binding protein
CBP-associated protein (p/CAF), 356
  crystal structure of, 362f
  human, 360
CCAAT box, 330
CCK2E3, 438f
CCR3, 677

INDEX 1189

CCR3-chemokine-mediated
  signaling, 677
CCR5, 671
CCs. *See* Chemokine(s)
CD4 cells, 947f
  autoreactive, 930, 932f, 933f
  migratory response in, 677
  pathogenic, 932
CD4 T cell, HIV levels of, 1048f
CD8 cells, 932
  autoreactive, 930, 932f, 933f
CD8+ CTLs, 923–924
CD8 T cells, 671, 675, 932
CD23, 671
CD31, 909f
CD45 protein, 723
CD46, 943, 943t, 944f, 947f
  domain movement between
    two, 945f
CD124, 664f
Cdc25A, 613–614
cDNA. *See* Complementary
  DNA
CDRs. *See* Complementarity-
  determining regions
CE. *See* Cholesteryl ester
Celecoxib, 127f, 472f
Cell(s), 5–6, 6f, 7f, 314, 762, 787,
  984. *See also* Apoptosis; Bac-
  terial cells; Blood cells; Cell
  death; Cell wall; *E. coli*; Red
  blood cell(s)
  acidification and, 715f
  adrenal cortex, 449f, 452
    zona fasciculata, 387, 449f,
      452, 453f, 454, 456
    zona glomerulosa, 449f, 452,
      453f, 454
    zona reticularis, 449f, 452,
      453f, 474
  A-NK, 899, 903
  antigen-presenting, 639, 725f,
    948f
  astrocytes, 654, 882, 998,
    1000f, 1004
  autocrine/paracrine, 402
  autoreactive, CD4/CD8, 930,
    932f, 933f
  B, 639, 654, 658, 665, 671, 675
    antibodies and, 910, 915, 929,
      929f
    development of, 664
    immune competence of, 883
    lymphoma factor, 902f

Cell(s) *(Continued)*
  V-J combinations in differen-
    tiation, 919
  beta
    diabetes, type 1 and, 930
    pancreas, 133, 135f, 138,
      142–145, 144f–145f, 173f,
      559
  brain, 882
  Ca^{2+} concentration of, 418,
    710, 711f
  cancer, 901, 901f, 904f, 907f,
    908f
  CD4, 677, 930, 932, 932f, 933f,
    947f
  CD8, 930, 932, 932f, 933f
  cerebellum, Purkinje, 723
  chief, 439
  chromaffin, 448
  corticotropic, 447
  cortisol and, 470
  crypt, 743, 749f
  cycle, 587–588
    S-phase progression in, 614
  cyclic AMP level in, 1031
  deoxytrinucleotide, 278
  ductal, 441
  duodenal mucosal S, 439
  endothelial, 470, 518, 616, 654,
    666, 858, 861f, 982, 983f
  enterochromaffin-like, 435–
    436, 439
  ependymal, 998
  epithelial, thyroid, 418f, 799f
  erythroid, 760
    progenitor, 633, 633
  fat (adipocytes), 204, 561
  foam, 233, 235
  follicular, 417, 418, 419, 795,
    797f
    thyroid, 417, 418, 418f, 419,
      797f
  gastric gland, 440f
  gastric mucosa, 439, 440f
  germ-line, 917, 919
  glial, 654, 998, 999f, 1004
  glucagon-secreting, 559f
  gram-negative, 1107, 1108f,
    1111t, 1118
    gram-positive compared
      with, 1111f
    outer membrane of, 1127f
  gram-positive, bacterial wall of,
    912f, 1107, 1111f

Cell(s) *(Continued)*
  hair/nAChR, 1014
  hematopoietic, 491, 494f, 627,
    629, 630f, 639f, 654, 830
  hepatobiliary, 466
  host, 1068–1069, 1078
    infection of, 1095, 1101,
      1103
    lysis of, 1092
    prophage in, 1091
    SARS, 1085
  hybridoma, 915
  IGF-1-producing, 402
  IL secretion, 654
  immune, 735f, 984
  inner membrane/PLA_2 interac-
    tion of, 470, 470f
  intestinal, 705f, 743
    enterochromaffin cells, 619
    zinc movement into, 778
  intracellular concentrations,
    calcium, 787
  kidney
    juxtaglomerular, 452
    mesangial, 882, 882f
  Kupffer, 882
  lactotrophic, 430
  Leydig, 408, 430
  liver, 6f
  magnesium entry into, 713–
    714
  mammary gland muscle, 380
  MDCK, 1078f
  mesangial, 882
  muscle, 380, 616
  myeloid, signaling paths/CSF-
    1R regulation in, 637f
  myeloma, 888, 915
  neuronal cell bodies, 380
  nucleus of, 6
  ovarian, 591
  pancreas
    acinar, 441
    alpha, 133, 135f, 173f, 559
    delta, 133, 134, 135f, 173f,
      559
  parietal, 436, 440f, 825
  pathogenic, 922, 932, 934f
  periportal, 506, 508f
  pituitary
    pars intermedia, 352, 366f,
      466
    PILC (*pars intermedia*-like),
      368, 378f

Cell(s) *(Continued)*
  plasma, antibody derivation from, 910, 911, 914
  pluripotent stem, 879–880, 880f
  pneumococcal wall of, 951
  receptors in, 26–27, 27f
  renal tubular, 458f, 459f
  rod, 837, 837f, 1037f
  Schwann, 998
  Sertoli, 409, 410
  sponge, 189
  steroidogenic, 561
  survival of, 654, 658, 666
  T, 654, 658, 661f, 665, 677, 725f, 883, 930f
    absence of, 670f
    CD4, 932, 1048f
    CD8, 671, 675
    cytotoxic, 901f, 923–924
    Graves' disease with, 923
    helper, 675
    killer, 904f
    maturation of, 894
    MHC protein interactions with, 929, 929f
    natural killer, 639, 899, 900f, 901–903, 901f, 904f, 906f–909f
    pathogenic, 922, 932
    target, 959f
    lysis of, 942f
  Th2, 658, 664f
  thyrotrophic, 417–418, 419, 921f
  tissue, diabetic, 134f
  unstimulated, calcium transport with, 788f
  uterine, estrogen receptor in, 484, 486
  viral-infected, apoptosis in, 904f
Cell adhesion
  mast, 677
  properdin with, 949
Cell adhesion molecules (CAMs), 224, 470
Cell cycle, 13, 13f
  arrest, 830
Cell death, 592. *See also* Apoptosis
  calcium in, 707, 709
  TNF-α_1TNΦ receptor causing, 926f

Cell differentiation, 6, 7f
  adipocytes, 616
  B cell, V-J combinations in, 919
  hematopoietic, 639f
  osteoclast differentiation factor, 632, 636f
  Th1, 669, 679f
Cell division
  DNA with, 289
  E. coli, 1144–1145, 1148f
  by FGFs, 617
  ligand-receptor complexes in, 607–608
  MAPK pathway of, 604
  mitogen stimulator of, 600
  nucleus in, 13–16, 12f–16f
Cell membrane, 6–7, 7f, 8f, 780. *See also* C2 domain; Membrane transport; Nuclear membrane
  apical, 699f
  cholesterol in, 7
  clathrin-coated pit on, 609, 610f
  cytoplasmic side of, 706
  de-gene complex topology of, 891f
  E. coli, 1107–1108, 1113, 1117f, 1118, 1121f
  inner, 470, 470f
  lactotrophic, 432
  lipid, 1107–1108, 1113, 1117f, 1118, 1126, 1131f
  lipid rafts in, 723, 724f, 725f
  muscle, 727f
  permeability of, 903f
  polar head of, 6–7, 7f
  polarized, 961
  pore formation in, 902, 904f
  pumps in, 702f, 703f, 707, 712f
  receptors of, 7
  rod, 837
  semipermeable, 697, 698
  T, 923–924
  transmembrane proteins in, 8f
  vascular endothelial, APC molecules in, 858, 861f
Cell membrane potential, 28
Cell proliferation, 707
  apoptosis balance to, 491
  FGF induction of, 616
  growth factors as causing, 627–628
  ligand binding leading to, 666

Cell proliferation *(Continued)*
  MAPK pathway in, 604
  somatostatins, pathways of, 406f
  STAT/MAPK/PI3K in, 405
  thrombospondin in, 949
Cell suicide. *See* Apoptosis
Cell wall
  peptidoglycan in, 1107, 1111f, 1114f, 1115, 1118
    binding of, 1119f
  pili on, 1110f
Cellular messenger, NO as, 540
Cellular prion protein (PrP^c)
  amino acids with, 39–54, 40f–43f, 44t, 45f–53f, 47t
  conversion diagram for, 38f
  copper-binding in model of, 35–36, 37f
  glycosylphosphatidylinositol protein anchor with, 36f
  knockout experiments in mice with, 35
  PrP^{Sc} propagation from, 35–54, 36f–43f, 44t, 45f–53f, 47t
  three-dimensional structure of, 37f
Cellular resistance, TRAIL, 591
Cellular retinol-binding protein (CRBP), 836
Central hepatic vein, 506
Central nervous system
  gephyrin in, 788, 794f
  structures in, 1044t
  volumes in, 1044t
Central nervous system disease, 240, 242f
  fragile X syndrome as, 240, 242f
  Huntington's disease as, 239–244, 240f–244f
  myotonic dystrophy as, 240, 242f
  spinobulbar muscular atrophy as, 240, 242f
Centrioles, 14
Centromere, 14
Cephalic phase, 439
Cephalosporins, distribution of, 719
Ceramide, 220–224, 223f–225f, 226t
  proapoptotic factor BAX with, 224
  sphingomyelin biosynthesis with, 220–221, 224f

INDEX 1191

Cereals, 776
Cerebellum
  presynapse of, 988, 990
  serotonergic system in, 1020f
Cerebral cortex
  dopamine receptors and, 1028
  pain signals in, 955
Cerebral edema, 499
Cerebrosides, 220, 226–227, 227f
Cerebrospinal fluid, 721, 1045t
Ceruloplasmin, 764
  structure of human, 765f
CF. See Cystic fibrosis
CF antigen, 685
C-Fos, 609
C-Fos, structure of, 328f
CFTR. See Cystic fibrosis transmembrane regulator
cGMP. See Guanosine 3', 5'-monophosphate
cGMP-gated channels, rod cell, 1037f
Chain(s)
  AA, 47f, 47t, 442, 442f, 551, 552f, 918f, 932
  E. coli, 1122f, 1143, 1147f
  electron transport, 740
  glucose—1, 6-biphosphate, 160f
  heavy, 910, 917
    HLA-A2, 931f
  interleukin receptor, 661f
  α-keto acid dehydrogenase, 551, 552, 552f
  light, 910, 917–918, 918f, 919, 919f
  long
    acetyl-CoA synthetase, 726f
    fatty acid, 554, 723
  phytyl side, 841
  -related antigen, 899, 900f, 903, 904f
  respiratory, 177, 178f, 180f, 770, 772f
  side chain cleavage enzyme system, 453f
  TCR, 931f
Chair/boat configurations, sugar carbon, 147, 148f, 150f
Chaperone proteins, 80, 764
Chaperone, yCCS, 766f
Chaperone-like proteins, 760
Chaperonin, 81
Charge repulsion, 59

Charge-charge interaction, 59
Charge-dipole interaction, 59, 60f
Cheilosis, 808
Chemical(s)
  peroxisome proliferator, 488–491, 489f, 490f
  reactions, heme synthesis, 754f–756f
Chemical compounds, estrogen-binding, 485f
Chemical gradient, PDGF, 600–601
Chemical messengers, pain signaling through, 957
Chemokine(s) (CCs), 658, 666, 677
Chemokine receptor family, 632f
Chemokine-mediated signaling, CCR3, 677
Chemosmosis, 770, 772f
Chemosmotic theory, 177, 182
Chemotaxis, 600–601, 951
Chief cells, 439
CHIF. See Corticosteroid hormone-induced factor
Childhood
  Cystic Fibrosis statistics, 686
  essential amino acids during, 536
Childhood craniopharyngioma, 366
Childhood mumps, 588
Childhood phenylketonuria, 523
Chimeric 5-HT 3R subunit, 1018f
Chirality, 44–46, 45f
Chloramphenicol, 85, 85f
Chloride
  entering of, 691
  normal concentration of, 685
Chloride channel protein complex (GABA$_A$R), 1003, 1006, 1006f, 1009f
Chloride channels, 965, 996
Chloride ion(s), 688, 732, 1003
Chloride ion transport, 685
Chloride shift, 878
Cholecalciferol, 530, 837, 840
Cholesterol, 189–198, 190f–192f, 194f–197f, 203, 204, 204f. See also 7-Dehydrocholesterol
  bile acids synthesized from, 193, 196f, 197–198, 197f

Cholesterol (Continued)
  biosynthesis of, 193, 194f, 195f
    acetyl CoA in, 193, 194f
    active isoprene in, 193, 194f
    mevalonate in, 193, 194f
    squalene in, 193, 194f
  cell membrane with, 7, 723, 724f
  drugs for lowering, 195f, 197
  lipid droplet freeing of, 390, 406–409
  UDP glucuronyl transferase with, 743
Cholesteryl ester (CE), 189, 191f, 561–562
  conversion of, 390
  endosome of, 189
  lipoproteins with, 233, 233f
Cholesteryl ester hydrolase, 406–409
Cholesteryl esterase, 390, 452, 454
Choline, 220, 221f, 1028
Choline plasmalogen, 222f
Cholinergic neurons, 1017
Cholinergic reflexes, 439
Cholinesterase, 1011
Chondrocytes, 616
Chorea, 240
CHP. See Calcineurin homologous protein
Christmas factor, 865
Chromaffin cells, 448
Chromatin, 906–911, 10f–12f
  conformations of, 360f
  E. coli as containing, 1107, 1108f, 1130
  gene expression and, 357–359
  pattern of, 880–881
  relaxation of, 358
  representation of, 359
  structure/fiber of, 358f
Chromium, intake amount for, 771
Chromophore, 11-cis-Retinal, 1037f
Chromosome(s), 12f, 15f, 131, 132f, 191, 239, 240f, 283, 780, 906
  BRCA, 587
  dopamine location of, 1030f
Chromosome 1, 947f

1192   INDEX

Chromosome 2
    diabetes with, 131, 132f, 932
    gene encoding light chains on, 918
    protein C gene encoding on, 857
Chromosome 4, 239, 240f
Chromosome 5, 629
Chromosome 6 p21, 922
Chromosome 7, 685, 688
Chromosome 9, 639
Chromosome 11, 1028, 1030f
Chromosome 12, 639
Chromosome 13, 587
Chromosome 17, 587
Chromosome 22, 717
Chronic pain, acute pain v., 955
Chronic stress, 445–446
Chrysotile, 323
Chylomicron(s), 145, 202–203, 203f, 204f
    diagram of, 203f
    remnants, 835–836
        uptake of, 843
    retinol palmitate incorporated into, 835
    triglyceride fat absorption in, 561
    vitamin E with, 843
    vitamin K with, 847
Chyme, 439
Chymotrypsin, 441
Cl^- channel, AMP-activated, 732
CI repressor gene, E. coli, 1101, 1103f
CIF. See Calcium influx factor
Cl^-/HCO_3^- exchanger, 693f, 732
CII repressor gene, 1101, 1106, 1106f
Cimetidine (Tagamet), 542
Circadian rhythm, 448, 535
Circulation, human systemic, 872f
11-cis double bond, isomerization of, 837, 837f
9-cis-retinaldehyde, 835, 836f
9-cis-retinoic acid, 490, 835, 836, 836f
11-cis-Retinal, 1032
    chromophore, 1037f
    putative binding site, 1035f
11-cis retinal, 837
Cisplatin, 588, 837

Cis-urocanic acid, 513f
Citrate, 554
Citrate lyases, 109–111, 110f
Citrate-proton transporter, 717, 717f
Citric acid, 1004
Citric acid cycle, 138, 546. See also Tricarboxylic acid cycle
    acetyl CoA metabolized through, 176, 178f
    amino acid catabolism through, 547f
    ATPase with, 800
    in fatty acid degradation, 560–561, 560f
    glycolysis with, 174, 175f
Citrulline, 499–500
    arginine giving rise to, 536, 538f, 540f
    increased plasma levels of, 500, 501, 503t
Citrullinemia, 503t
Citrullinuria, 497
c-JUN, 328f, 669
CL channel, 689f, 693f
C_L light chain domain, 917
Clamp mechanism, 335
    structure of, 335f
Clams, 740
Claritin. See Loratadine
Classical pathway, 934, 935, 935f, 937f
    disorders in, 946
Clathrin, 608–609
    network, 609
Clathrin triskelion, 610f
Clathrin triskelia, 610f
    legs of, 610f
Clathrin-coated pit, 609, 610f
Clathrin-coated vesicles, 608, 611f
Clathrin-scaffold layer, 610f
Cleavage
    autolysin, glycosidic bonds by, 1115, 1119f
    DNA, 492, 494f
    factor B, 935, 1097f
    GAG/POL polyproteins, 1062f
    procaspase-9, 592
    protein C, 884
    side chain cleavage enzyme system, 453f
    viral polyproteins, 1078
    ΦX174, 1097f

Cleft palate, 717
CLIP. See Corticotropic-like inhibitory peptide
Cloning, gene, 309f, 310f
cNOS. See Constitutive nitric oxide synthase
CO. See Carbon monoxide
CoA
    derivatives, 723
    malonyl, 554
    pantothenic acid component of, 818, 819f
    succinyl, 546, 550, 551f, 753, 753f, 802, 804f
Coactivators, 337–344, 341f
    activities associated with, 347f, 358, 360
    HAT, 360
    classes of, 342t
    complexes of, 360f
    exchange of, for corepressors, 341f
    ligand binding of, 344
        recruitment in, 357f
    p160, 342, 346f
    SRC-1, crystal structure of, 342t
    TRBP, 342, 345f
Coagulation, intrinsic pathway of, 866f, 868f, 869. See also Blood clotting
CoA-SH, 551
Cobalamin (vitamin B_{12}), 825, 828
    deficiency of, 828, 830
    E. coli with, 1126
    intake/utilization of, 828f
    store of, 828
    structure of, 827f
Cobalamin-intrinsic factor complex, ileum absorption of, 825, 827f
Cobalt, 825
α-cobrotoxin, 1015f
Cocaine, 1001
Cochlea, 1014
Codon
    methionine, 517
    tryptophan, 518
Coenzyme(s), 111–119, 113f, 119t, 114f–118f, 191f, 214, 214f, 801, 818, 820f, 822f. See also Acetyl CoA
    apoenzymes needing, 111

Coenzyme(s) *(Continued)*
  biotin, 118, 119t, 114f
  coenzyme A, 118, 119t, 114f, 214, 214f
  coenzyme B$_{12}$, 119t, 117f
  flavins, 119, 119t, 116f
  lipoic acid, 118, 119t, 115f
  NADH, 119, 119t
  NADP, 119, 119t, 116f
  NADPH, 119, 119t
  nicotinamide, 119, 119t, 811f
  nicotinic acid, 811f
  pyridoxal-P, 112–118, 113f, 119t
  Schiff bases, 118, 119t
  tetrahydrofolic acid, 118, 119t, 115f
  thiamine pyrophosphate, 118, 119t
  vitamins with, 119t
Coenzyme A, 118, 119t, 114f, 214, 214f. *See also* Acetyl CoA
  structure of, 214f
Coenzyme A cholesterol acyl-transferase (ACAT), 191f
Coenzyme B$_{12}$, 119t, 117f
Cofactors. *See* Coenzyme(s)
Cognate, 959
Cognition, 1028
Colds, virus associated with, 1069
Colicin, 1141f
Coliphages, 1101
Collagen
  appearance of, 770f
  cross-linked, 768, 768f
  molecules, 832
  procollagen maturation process to, 768, 768f, 769f
  synthesis of, lysyl oxidase in, 765
Colloid, 796–797
  follicles as containing, 797, 799f
Colloid droplets, 422, 797
Colloidal space, 422
Colon cancer
  blood loss from, 743
  iron loss from, 739
Colony-stimulating factor (CSF), 601, 627–633, 630–633f. *See also* Granulocyte CSF; Macrophage-CSF
  multi-, 627

Colony-stimulating factor receptor (CSFR), 629, 632f, 634f
Color
  blood cells, 879, 880f, 883f
  eyes, 527
  hair/skin, 527
β-common receptor, 629, 630f, 631f, 633, 635f
  extracellular domain of, 630f, 635f
Competitive inhibition, 101–102, 101f
  noncompetitive v., 102
Complement activation (RCA), 942, 942f
  family, structure of, 949f
Complement antibody complex, 943
Complement system, 933–949
  adaptive immunity in, 945
  attack by complement in, 945
  factors in, 935, 938f, 942, 949
  occasional triggering of, 949
  overview of human, 942f
  pathways in
    alternative, 934–935, 935f, 937f
    classical, 934, 935, 935f, 937f, 946
  overview for, 935f
  regulators of, 942–949
  RCA family in, structure of, 949f
  RCA proteins/microorganism interactions in, 943t
Complementarity-determining regions (CDRs), 917
Complementary DNA (cDNA), 307, 310f
  production of, 310f
Complete blood count (CBC), 883, 889t–890t
COMT. *See* Catecholamine-O-methyl transferase DA
Conjugase, folic acid, 828–829
Connective tissue, 832
  matrix, 677
consensus sequence
  DPE, 337
  TATAAA, 330
Conservative model, 294–295
Constitutive nitric oxide synthase (cNOS), 970, 989f
Constitutive secretion, 695

Contraceptives
  absorption interference from, folate, 830
  oral, 853
Converting enzymes, 173, 677, 678f
  aldosterone and, 452
Coporphyrinogen III, 753
Copper, 527, 764–773, 765f–773f
  aluminum and, 795
  complexes formed from, 771
  dietary sources of, 771
  iron and, 740
  transfer mechanism for, 766f
  transport system, 714
Copper-zinc enzyme(s), 764
Coproporphyrinogen III oxidase, 753, 753f, 756f
Corepressors, 337–344, 341f
  activities associated with, 344, 347f
  complexes of, 360f
  displacement of, 342
  exchange corepressors for, 341f
  NCoR, 342, 343f
  nuclear receptors bound with, 348f
Cori's disease, 171t
Corneal transplants, prion disease with, 35
Coronary artery disease, 235t, 518
  premature, 518
Coronavirus main protease, 1089f
Corpus luteum, pregnancy with, 413
Corticosteroid hormone-induced factor (CHIF), 457, 458f
Corticosteroid-binding globulin, 352, 466
Corticotrophs, 368, 447
Corticotropic-like inhibitory peptide (CLIP), 368
β-corticotropin, fat storage breakdown stimulated with, 209
Corticotropin-releasing hormone (CRH), 368, 370t
  cortisol pathway of ACTH and, 387–392
  gene promoter regulation of, 387
  human (hCRH(F), 387, 388t
    diagram of, 389f
    subtypes of, 387, 389f

Corticotropin-releasing hormone (CRH)
  IL-13Rα, 676f
  regulation of, 390f
  sequences of, 387f
  stress causing release of, 447
Cortisol, 349–350, 352, 353f, 466–474, 658
  ACTH/CRH pathway of, 387–392, 393f
  aldosterone v., 452, 463, 478
  apoptosis and, 491
  ascorbate with, 832
  binding of, 463
    excess, 463–465, 465f
    transcortin, 466
  conversion to cortisone of, 465f
  cox1/cox2 and, 470, 472f
  depression and, 445–446
  epinephrine elevation causing release of, 446–447
  inactivation of, 568
  increased, 445–446, 449f
    adrenal gland in, 449f
  inflammation prevented by, 467, 469f, 470, 471f, 472f
  kidney, 465, 465f
  limbic system negatively impacted by, 445–446
  overproduction/underproduction of, 474
  PNMT induced by, 526
  pregnancy and, 430
  production, 393f
  progesterone and, 413, 430
  serotonin with, 1019
  stress and, 445–450, 446f, 449f, 451f, 466, 467
  stress-induced diabetes with, 133
  structure of, 351f
  surfactant protein induced by, 466, 467f
Cortisol-related compounds, 473t–474t
Cortisone, cortisol conversion to, 465f
Cosmic radiation, DNA mutations/damage from, 299
Cosmid, 304, 305f
Cotransporter, 421
Coumadin, 846, 853–854
  structure of, 846f

COUP-TF. See Ovalbumin upstream promoter transcription factor
Cox. See Cyclooxygenase
cox1, 120, 121f, 122f
  constitutive, 120
  endogenous, 470, 472f
  prostaglandin production with, 121
  subunit structure of, 979f
cox2, 120–123, 121f–122f, 472f, 675. See also Cycloxygenase-2; Inducible cyclooxygenase-2
  acid bound to, crystal structure of, 122f
  expression of, 645
  inducible enzyme of, 123, 470
  inhibitor of, 125–129, 127f, 128f
  macrophage induced, 123
  structure of, 120–123, 472f, 675
CPSI. See Carbamoyl phosphate synthetase I
CPSII. See Carbamoyl phosphate synthetase II
CPT-1. See Carnitine palmitoyltransferase (CPT-1)
CR1, 943, 945
CR2, 943, 945
Crab meat, 776
Cranial vagus nerve, 1028
Craniopharyngioma, childhood, 366
Craniosynostosis syndromes, 623f
CRBP. See Cellular retinol-binding protein
C-reactive protein (CRP), 951–953, 953f
  interaction with C1q of, 953f
Creatine, 540
  pathway, 538f
  precursor of, 536
Creatine kinase, 93, 95, 95f, 536, 539f
  BB isozyme of, 95
  brain-type, 539f
  MB isozyme of, 95
  MM isozyme of, 95
  muscle, 539f
Creatine-P, 95
Creatinine
  pathway, 538f
  precursor of, 536

CREB-binding protein (CBP), 326, 342
  activities of, 342, 344, 346f
  cointegrators p300 and, 356, 356f
  structural sites of, 346f
CRE-like region, 986f
Cretinism, 531–532
Creutzfeldt-Jakob disease, 33, 35
  amyloids with, 172t
CRH. See Corticotropin-releasing hormone
Crick, Francis, 289
Crocidolite, 323
Crohn's disease, amyloids with, 172t
Cross-links
  aldehyde, 765
  bifunctional, 765, 768
  fibrin, 847, 898
  formation of, 769f
  tetrafunctional, 768
CRP. See C-reactive protein
Cryoglobulin, 888
Crypt cells, 743, 749f
CSF. See Colony-stimulating factor
CSF receptor, 601
CSF-1, 627, 637f
CSF-1R, signaling paths regulated by, myeloid cell, 637f
CSF-2, 627
CSFR. See Colony-stimulating factor receptor
CSP. See Cysteine string protein
C-terminal BID fragment, 902f
C-terminal domain
  HIV
    glycoproteins in, 1049f, 1064, 1065f
    reductase, 541
  rhinovirus, 1071f
C-terminal globular domain
  E. coli, 1123f
  lamin A, 734f
C-terminus, substance P, 965, 965f
CTLs. See Cytotoxic T lymphocytes
CTP (Cytosine triphosphate)
  bacterial cell formation of, 1139, 1142f, 1143, 1143f
  inhibition by, 268, 269f, 1143
  synthesis route of, 269f
  UTP conversion to, 269f

CTP synthetase, 268, 269f
  GTP in activation of, 268
C-type lectin domains, 900f
Cubilin, 825, 827f, 828, 828f
Cushing's disease, 470
Cyclic adenosine diphosphate ribose (cADPr), 712f
Cyclic AMP (cAMP), 249, 387, 395, 408, 410–410, 432, 689f, 1021
  adenylate cyclase inhibition reduction of, 970
  adenylate cyclase with, 970
  aquaporin mechanism with, 698
  calcium ion with, 984, 986f
  CB-1 with, 986f
  CB-2 inhibition of, 981, 981f
  cellular levels of, 1031
  -dependent protein kinase A, 688
  elevated, CFTR channels activated by, 691f
  formation of
    adrenergic receptors in, 448, 451t
    fat storage energy and, 561, 565f
    GLU-R receptor in, 559
    TSH-TSH receptor complex and, 419
  inhibition of, 970, 980–981
    adenylate cyclase, 970
    insulin decreasing, 554
    iodine with, 795–796
Cyclic AMP-independent kinases, 557
Cyclic GMP, 249, 454, 500, 540, 837
  fall in concentration of, 837
  urea cycle disruptions and, 500
Cyclin D3, 613
Cycloheximide, 85, 85f
Cyclooxygenase (Cox), 120–123, 121f–122f, 841, 984f
  active site, 979f
Cyclooxygenase 2, 323
Cyclophilin 40, 478, 480f
Cydocxygenase-2 (cox-2), 978f
Cys-loop, conserved, 1003
Cys-loop protein signature, 1003
Cystathionase, 518, 520
Cystathionine, 518
  urine containing, 520

Cystathionine β-synthase, deficiency of, 518
Cystathionine synthase, 518, 520
  impaired functioning of, 520
Cysteine, 41f, 44t, 47f, 492, 831f
  E. coli with, 1118
  GSH and, 520
  synthesis of, 519f
Cysteine protease, bacteria-derived, 948f
Cysteine string protein (CSP), 1006, 1008f
Cystic fibrosis (CF)
  aberrant ion transport in, 685–694, 686f–694f
  as autosomal recessive disorder, 687f
  diagnostic chloride for, 685
  gene therapy for, 314
  overt, 688
Cystic fibrosis transmembrane conductance regulator, 691f
Cystic fibrosis transmembrane regulator (CFTR), 314, 632, 685, 688, 691, 692f–694f, 693t
  activation of, 693f
  CF antigen interaction of, 685
  chloride ion channel of, 732
  cyclic AMP elevation activating channels of, 691f
  ENaC link of, 691, 694f
  MDR compared with, 732f
  NBD1 of, 688, 690f
  nonfunctional, 685
Cystic fibrosis transmembrane sodium-proton exchanger increased by, 688, 691, 693f
Cystine-glutamate antiporter, 721
Cystines, chelate of iron with, 758
Cytochrome b5 reductase, 557
Cytochrome c, 492–493, 494f, 592, 593f, 770, 984f
  Apaf-1 interaction with, 902f
  apoptosis from, 982–983
  NADPH-dependent reduction of, 541
Cytochrome c oxidase, 768, 771f, 772f
Cytochrome c reductase, 768, 772f
Cytochrome P450, 567f, 784, 785f, 786f

Cytokine(s). See also specific ILs
  anti-inflammatory, 666, 671, 675
  interferon, 638–643, 645f, 646f
  receptors, 639, 647f, 658
  interleukin, 639f, 654–682, 655t, 656f, 657t, 659f, 665f, 670f, 672f, 673f, 674f
  proinflammatory, 654, 675, 677, 973f
  secretion of, 475, 899
Cytokine receptor superfamily, 629, 633f
Cytokine-inducible. See iNOS
Cytomegalovirus, 925, 927
Cytoplasm, 18, 706
  calcium ion release in, 970
  erythrocyte, 878
  soluble, 499
Cytoplasmic domain, 639
Cytoplasmic receptor, agonist binding to, 348f
Cytoplasmic steroid receptor, 340
  mechanism of, 339f
Cytosine, 244, 247f, 248f
  base-pairing with, 250f
  deamination of, 247f
  synthesis of, 268
  uracil from deamination of, 297f
Cytosine deaminase, 776, 778
  mechanism for, 778f
  structure of, 779f
Cytosine triphosphate. See CTP
Cytoskeleton, 896
  components of, 715, 716f
Cytosolic concentration, Ca^{2+}, 430, 438
Cytotoxic T lymphocytes (CTLs), 904f, 923–924
Cytotoxicity, T cells involved in, 923–924
Cytotoxins, 475, 923
D1. See Selenium-containing deiodinases
D1 receptors, 1031f
α-D-1,4 bonds, 153
α-D-1,6 bonds, 153, 154f
D2. See Selenium-containing deiodinases
D2 receptors, 1028
D3. See Selenium-containing deiodinases

D₃-24-hydroxylase, 840
D₃-25-hydroxylase, 840, 840f
DAG. *See* Diacylglycerol
D-Ala peptidase, phosphonate-inhibited D-Ala-, 1114f
D-alanine, 1113, 1114f
D-amino acid oxidase, 514
   mechanism of, 515f
DAP10 transmembrane, 908f
   signaling adaptor, 899
DAP12, 909f
Dark, sensitivity to, 532–533
DAT. *See* Dopamine transporter
DcR1, 593f
DcR2, 593f
DD. *See* Death domain
Deacetylases, histone, 640f
Deamination, 513, 513f
   adenine, 271, 274f
   amino acids oxidative, 52, 53f
   cytosine, 247f, 297f
   guanine, 271, 274f
   histidine, 513
   hypoxanthine, 271, 274f
   oxidative, 52, 53f
   serine, 513, 513f
   uracil, 297f
   xanthine, 271
Death
   E. coli rates of, 1147
   traumatic brain injury, 366
Death domain (DD), 591, 592f, 902f
Death receptor 4, 591
Debranching enzyme, 156, 161
   α-1,6 glucosidase activity of, 156
Decarboxylase, 516, 516t, 517f
   for some amino acids, 516t
Decarboxylation, 509
   amino acid, 516, 516t, 517f
Deciliter, 193
Decoy function, 591
Decoy receptor, 591, 593f, 594, 654
Deep vein thrombosis (DVT), 853–867, 854f, 855f
   frequency of, 853
   Leiden factor in, 856, 856f
Deficiency
   C2, 946
   C9, 949
   cobalamin, 828, 830
   cystathionine β-synthase, 518

Deficiency *(Continued)*
   dietary iron, 756
   factor VIII (antihemophilic factor), 869
   folate, 830
   inherited C3, 946
   magnesium, 714, 782
   molybdenum/sulfite oxidase deficiency, 795
   niacin, 814
   oxygen, 758
   PAH, 527f
   pantothenate, 818
   pigmentation, 531
   protein C, 856
   pyridoxamine, 1004
   riboflavin, 808
   thiamine, 803, 804
   vitamin A, 837
   vitamin B, 518, 818, 1004
   vitamin D, 840
   vitamin E, 843
   vitamin K, 847
   VR1, 961f
De-gene complex, membrane topology for, 891f
Deglycosylation, 877
7-dehydrocholesterol, 527, 529f, 837, 840f
Dehydroascorbic acid, 832
Dehydroepiandrosterone (DHEA), 387, 452, 453f, 474–476, 475f, 476t
   aromatase action on, 566
   inhibitory activity of, 476
Deiodinase, 796
   5'-, 797
5'-deiodinase, 797
Deiodination mechanism, 429f, 775f, 799f, 797, 799f
Delirium, 782
Dementia, HIV, 1067
Demyelination, 828
Denatured proteins, 61
Dendrite(s), 957, 958f
   receiving target cell, 959f
Deoxy nucleotide diphosphate (dNDP), NDP conversion to, 286f
Deoxycytidine, 276
Deoxycytidine kinase reaction, 276
Deoxyglucose, 244
Deoxyhemoglobin, 873, 874, 874f, 877

Deoxyribonuclease, 441
Deoxyribonucleic acid (DNA). *See also* DNA ligase; DNA polymerase; DNA synthesis; DNA-binding domain; Transcription
   acetylation of NF$_k$B, 326
   antisense, 311
   B form of, 289
   bacterial cell, 1130, 1131
   base excision repair mechanism with, 245, 294
   base-pairing in, 249–251, 250f, 251f
   binding, 1094f, 1095
      C-terminal HIV-1 integrase, 1061f
   biosynthesis of, 288–312, 290f–293f, 291t, 295f–305f, 301t, 306t, 307f–311f
      dinucleotide formation in, 290f
      next nucleotide added in, terminal phosphate of, 290f
   cell division with, 289
   cleavage of, 492, 494f
   complementary, 307, 310f
   defined, 288
   deoxyglucose with, 244
   deoxyribose-containing trinucleotides in, 280
   digestion of, 307f
   discovery of, 289
   double helix of, 289, 291f, 292f
   E. coli, 1130, 1131, 1133f, 1134f
   supercoiling of, 1131–1134, 1134f
   fragmentation of, 592, 593f
   genomic, 305–307, 310f, 311f
   glyoxalase, 294
   information transfer through RNA from, 316f
   major groove in, 289
   melting of, 251
   5-methylcytosine with, 249
   microarrays, 327
   minor groove in, 289
   mitochondrial, 295–298, 299f
   mutations/damage to, 298–300, 299f–301f, 301t
      carcinogens causing, 298, 301f
      chemicals causing, 298, 301f
      mistakes, frequency of, 298

Deoxyribonucleic acid (DNA) (Continued)
- radioactivity causing, 298–299
- types of, 301t
- ultraviolet light causing, 298–299, 299f
- nucleosome location on, 358
- nucleotide excision with, 294
- plasmid with introduction of foreign, 305, 309f
- proofreading/repair mechanism of, 294, 297f, 298
- recombinant, 305, 309f
- replication, 291
  - conservative model of, 294–295
  - dispersive model of, 295
  - helicase as enzyme for, 292
  - process of, 295f, 296f
  - semiconservative model of, 295
  - topoisomerase enzymes for, 291, 292
- restriction enzymes with, 300–305, 302f–305f, 306t
  - recognition sites of, 302f, 306f, 1058f
- retroviral integration of, cutting/joining steps, 1058f
- RNAP as bound to, 337f
- sequencing, 308–311, 310f
- single-stranded, 1094f
  - circular, 1099f
- spooling of, 1144, 1148f
- synthesis, error rate in, 294
- viral, 1056, 1056f
  - gapped intermediate in, 1059f
  - integrase of, 1056, 1060f
  - retroviral integration of, 1058f

Deoxyribose
- atomic model of, 286f
- attachment of, 246f

Deoxyribose-containing trinucleotides, synthesis of, 280

Deoxytrinucleotides, cellular concentrations of, 278

Depolarization
- exocytosis produced by, 707
- pancreas, 142

Depression
- cortisol and, 445–446
- immune system, stress as cause of, 445–446

Desaturases
- fatty acid, 557, 557f
- mixed-function oxidases as, 216
- Δ-5(4)-desaturase, 557
- Δ-6-desaturase, 557
- Δ-9-desaturase, 557

Desensitization
- β-receptor, 1040f
- receptor, 1033

Dexamethasone, 351f
D-fructose 6-phosphate, 512, 512f
D-glutamic acid, 115, 1113, 1113f
DHEA. See Dehydroepiandrosterone
DHEA-S, 452, 474, 475f
DHFR. See Dihydrofolate reductase
DHICA. See Dihydroxyindole catecholamine
DI deiodinase, 775

Diabetes, 131–146
- defined, 131
- effects of, 145–146, 146f
  - hyperglycemia as, 145, 146f
  - hyperinsulinemia as, 145, 146f
  - hyperlipidemia as, 145
  - hypertriglyceridemia as, 145
  - ketoacidosis as, 145
  - ketosis as, 145
- genetic factors for, 131, 132f, 132t
  - chromosome 2 with, 131, 132f
  - glucokinase mutation as, 131
  - glucose tolerance test for, 133, 133t
- HbA1c diagnostic for, 875
- hyperlipidemia with, 145
- peripheral blood in, 932
- prevalence of, 131
- stress-induced, 133, 445, 447–448
- type 1
  - defined, 131
  - development theory for, 930–933, 932f, 933f

Diabetes (Continued)
- genetic background for, 932
- insulin-dependent, 145, 922
- tissue cell characteristics with, 134f
- type 2
  - defined, 131
  - genetic factors for, 131, 132f, 132t
  - insulin independence of, 145
  - protein aggregation with, 170–173, 171f–173f, 172t
  - tissue cell characteristics with, 134f
- uncontrolled, 874–875

Diabetes insipidus, 367
Diabetic ketoacidosis, 209
DIABLO, 592
Diacylglycerol (DAG), 418, 430, 709, 710, 713f, 965, 1021
Diacylglycerol lipase, 209, 211f
Dibenz (a,h) anthracene, 301f
Dicoumarol, 846f
DideoxyATP (ddATP), 308
Dielsalderase catalytic antibody 1E9, 912f

Diet
- autoimmune disease and, 922
- calorie requirements and, 849t
- recommended, 848t, 849–850
Dietary copper, 771
Dietary fat, 588
- in SLE genetic susceptibility, 925
Dietary folic acid, 828, 829f
Dietary Iodine, insufficient, 796
Dietary iron, 739–743, 746f
- deficiency, 756
- uptake of, 743–753
Dietary metal(s), 764–801
- intake amounts for, 771
- iron, 739–743, 746f
  - RDA for, 742t
  - uptake of, 743–753
- iron-deficiency anemia and, 739–743, 746f
- magnesium, 780–782, 781f
  - RDAs for, 782t
- selenium, 773–776, 774f–776f
- zinc, 713, 776–780, 777f–780f
Dietary vitamin C, sources of, 834t

1198 INDEX

Diffusion
 facilitated, 697, 698–701, 699f, 737f
 free, 698
 paracellular, 737f
 passive, 697–698, 701f, 705
 simple, 701
 transcellular, 737f
DiGeorge syndrome, 717
Digestion, starch, 442–443
Digestive enzymes, 439, 441
Diglyceride, 201
3,4-dihydroxyphenylethylamine, 526. *See also* Dopamine
Dihomogammalinolenic acid, 557f
Dihydrobiopterin reductase, 522, 523f
Dihydrocapsaicin, structure of, 961f
Dihydrofolate reductase (DHFR), 125, 126f, 126f, 830
Dihydrolipoamide acetyltransferase, 1138f
Dihydroorotase, structure of, 264, 267f
Dihydroorotate dehydrogenase, structure of, 264, 267f
Dihydropteridine reductase, 526
5α-dihydrotestosterone, structure of, 410, 412f
Dihydrotestosterone, 478
1,25-dihydroxyvitamin D_3 (calcitrol), 530, 837, 840f
Dihydroxyindole catecholamine (DHICA), 527f
Dihydroxyphenylethylamine. *See* Dopamine
Dihydroxyvitamin D_3, 530, 530f
1,25-dihydroxyvitamin D_3. *See* Calcitrol
Di-iodotyrosines (DIT), 422
Diketogulonic acid, 832
Dimerization
 AHSP-α-globin, 760
 atrionatriuretic receptor, 455f
 caspase-9, 495f
 ferrochelatase, 758
 GH receptor for, 404–405
 IL-8, 666, 671f
 IL-12, 673f
 liganded receptor, 478
 NGF, 628f, 629f, 627f

Dimerization *(Continued)*
 NOS, 474
 P-box for, 461
 PDGF, 602f
 PDGF-CC, 604, 605f
  GFD-, 605f
 sites for, 349
 TGFβ1, 615f
 TNF-R1, 600f
 trkA receptor, 624, 625, 627f
Dimethyl sulfate (DMS), 301f
Dimethylnitrosamine, 301f
Dinucleotide, 245, 248f. *See also specific nucleotide*
 formation of first, 290f
 structure of, 248f
Dioxygen, 522
Diphenhydramine (Benadryl), 542
Diphenol oxidase, 527
Diphosphatidylglycerol, 220, 222f
Disaccharides, 147–149, 149f, 150f
Dispersive model, 295
Distal nephron, CaT1 location in, 787
Distal zinc finger, 463, 464f
Disulfide bond, 60, 60f, 917
Disulfide isomerase activity, 184
DIT. *See* Di-iodotyrosines
Divalent ion transporter (DMT) 1, 714, 746f
Divalent metal transporter, copper/zinc, 771
D-loops, 297
D-methylmalonyl CoA, 550
DMS. *See* Dimethyl sulfate
DMT. *See* Divalent ion transporter
DMT1
 expression, 749f
 -TM4, NMR structures of, 747f
DNA. *See* Deoxyribonucleic acid
DNA glyoxalase, 294
DNA ligase (Polydeoxyribonucleotide synthase), 289, 309f, 312
 structure of, 293
DNA polymerase, 289, 292, 293f
 DNA synthesis catalyzed by, 308
 E. coli, 1145, 1149f
 family of, 291

DNA polymerase *(Continued)*
 next nucleotide added with, terminal phosphate of, 290f
 parental double helix opening with, 292f
 proofreading capacity of, 294, 297f, 298
 sequenase form of, 1091, 1093f, 1094f
 3′,5′ exonuclease activity with, 294
DNA polymerase β (Pol β), 293f
DNA polymerase I (Pol I), 294, 295f, 312
DNA polymerase II (Pol II), 294, 295f
DNA polymerase III (Pol III), 292, 294
 E. coli, 1144, 1149f
DNA synthesis
 complementary template strand in, 311
 conservative model of, 294–295
 dispersive model of, 295
 DNA polymerase as catalyst of, 308
 error rate in, 294
 inhibition of, 311–312
 mistakes in, frequency, 298
 original template strand in, 311
 by phage T7, 1091, 1094f
 semiconservative model of, 295, 298f
DNA-binding domain, 461
 androgen receptor, 487f
 conserved, 349, 350f
 estrogen receptor, 485f
 thyroid hormone, 420
DNA-binding protein, 778
 ΦX174, 1097f, 1100f
dNDP. *See* Deoxy nucleotide diphosphate
Domoic acid, 990f
DOPA decarboxylase, 516, 526
 crystal structure of, 516, 517f, 720
Dopamine (DOPA), 432, 448, 479, 516, 526f, 1003
 binding of, 1029f
 binding site for, 1028
 drug release by, 1001
 endocannabinoids interaction with, 1034f
 in melanin synthesis, 527

INDEX 1199

Dopamine (DOPA) *(Continued)*
   as monoamine neurotransmitter, 1028
   norepinephrine conversion of, 770, 773f
   precursor to, 524
Dopamine D2 receptor, 432, 435
Dopamine D2 receptor gene, 1030f
Dopamine receptors, 432, 1028, 1029f–1031f
   structure, 1029f
Dopamine transporter (DAT), 1033f
Dopamine-β hydroxylase (DBH), 448, 451f, 770, 773f
Dopaminergic synapse, 1033f
Dopaquinone, 527
Doppler ultrasound scan, 853
Dorsal horn, 955, 956f
Dorsal horn neurons, 965
Dorsal striatum, dopaminergic/endocannabioid interaction in, 1034
Double helix, 289
   opening of parental, 292f
   parameters of, 291t
Double-stranded RNA (dsRNA), 311
Down's syndrome, amyloids with, 172t
Downstream caspases, 493
Downstream promoter element (DPE), 329, 329f
   consensus sequence of, 337
DPE. *See* Downstream promoter element
Drugs, 540. *See also* Multi drug-resistance protein
   allergy/H^1 receptor-inhibiting, 542, 543f
   analgesic, κ-receptor specific, 969–970
   anesthetic, 992f, 996
   antipsychotic, 1028
   benzodiazepine, 996, 998f, 1006
   cancer, 236
   cannabinoid, 974
   dopamine-releasing, 1001
   enzymes and, 125–129, 126f–129f
   HIV infection/AIDS, 1067, 1068t

Drugs *(Continued)*
   hypolipidemic, 488
   kainite receptor and, 996
   NSAIDs, 739
   ovarian tumor/cancer, 588, 591
   phencyclidine, 992f, 995
   resistance to, 588
dsRNA. *See* Double-stranded RNA
dTMP
   dUMP conversion to, 278, 278f
   regeneration of, 830
dTMP synthesis, impaired, 830
Duchenne muscular dystrophy, gene therapy for, 314
dUMP, dTMP conversion from, 278, 278f
Duodenal mucosal S cells, 439
Duodenum, iron uptake in, 743, 749f
Duplexes, 316, 318f
DVT. *See* Deep vein thrombosis
ΔΨ, 984f
Dyn. *See* Dynorphin
Dynorphin (Dyn), 968, 970, 1033f
Dynorphin A, 968, 970
Dynorphin B, 968
Dysbetalipoproteinemia, 235t
Dyslipidemic hypertension, 235t
Dystrophin, 314

4E binding protein (4E-BP), 74, 75f
E. coli (Escherichia coli), 1107–1150, 1108f, 1112f, 1124f
   ATP with, 1112f, 1139–1140, 1142f, 1143f
   contents of
      flagella, 1109f, 1110f
      peptidoglycan, 1107, 1111f, 1114f, 1115, 1117f, 1118, 1119f, 1121f
      pili, 1110f
   crystal structure of, 282f
   cycle of, 1144, 1148f
   diseases caused by, 1147–1150
   DNA, 1130, 1133f, 1131–1134, 1134f
      polymerase III of, 1144, 1149f
      replication, 1145, 1149f
   envelope, 1122f
   lysis, 1133f
   metabolism of, 1139, 1141f

E. coli (Escherichia coli) *(Continued)*
   NAG with, 1111, 1113, 1113f, 1117f, 1119f
   NAM with, 1113, 1114f, 1115, 1117f
   oligosaccharide, 1132f
   pathogenic
      transmission of, 1147–1148, 1150
      virulent determinants of, 1150t
   phage infection in, 1091, 1092, 1095, 1101, 1103
   polysaccharides in, 1130, 1131f
   porins, 1123f, 1124, 1125, 1125f–1127f, 1129f
      membrane transport of, 1126, 1126–1131
   proteins, 1125, 1126, 1130f, 1131–1134, 1134f
   pyrimidine synthesis in, 1139–1140, 1142f
   pyruvate dehydrogenase of, 1138, 1138f
   repressor genes and, 1101, 1103f
   restriction endonucleases action of, 302f
   transacetylase of, 1138, 1138f
   transglycosidase, 1120f
   transglycosylase, 1113, 1115f, 1116f, 1118
   trimerization, 1122f, 1127f
EAAC1, 722t
EAAs. *See* Excitatory amino acid(s)
EAAT1, 722, 722t
EAAT2, 722, 722t
EAAT3, 722, 722t, 723
EAAT4, 722–723, 722t
EAAT5, 722, 722t, 723
Early development, retinoic acid involved in, 836
Eating, four-phase response of, 439
4E-BP. *See* 4E binding protein
EBP50. *See* Ezrin-radixin-moesin binding phosphoprotein
ECaC. *See* Epithelial calcium channel
ECaC1, 787, 789f
   topology of, 789f

ECL. *See* Enterochromaffin-like
Eco RI, 302–304, 302f, 303f, 305f, 306t, 309f
  atomic models of, 303f
  map of sites of, 305f
Eco RV, 303, 304f, 306t
  structure of, bound to DNA, 304f
Ectoderm, 4, 5f, 619
Edema, 365, 714, 894
  anaphylatoxins causing, 949
EEATs. *See* Excitatory amino acid transporters
EF-1α-GTP, 74–77, 78f, 79f
EF-2-GTP, 74–77, 78f, 79f
Effector(s), 105
  caspases, 901f
EGF. *See* Epidermal growth factor
EGF-EGFR complexes, structure of, 608f
EGF-like repeats, 860f
EGFR, 616
Egg yolks, 740
Eggs, 776
eIF-2. *See* Eukaryotic initiation factor-2
eIF-4F, 74, 75f–77f
EL4 model, 1026f
Elastase, 441
Electrochemical gradient, 698, 706f
  inwardly directed, 704
Electron micrograph, melanosome fraction, 528f
Electron transport chain, 740
Electrophoretic gel, 311
Electrostatic bonds, 59, 60f
  charge-charge interaction in, 59
  charge-dipole interaction in, 59, 60f
ELISA. *See* Enzyme-linked immunoabsorbent assay
Elk-1, 609, 626
Elongation, 70, 74–77, 78f, 79f, 329, 337
  DNA, 1144, 1148f
  EF-1α-GTP as factor for, 74–77, 78f, 79f
  EF-2-GTP as factor for, 74–77, 78f, 79f
  RNAP complex of, 338f
Emotion, 1028
EMS. *See* Ethylmethane sulfonate

ENaC. *See* Epithelial sodium ion channels
Enantiomers, 45
Encapsulated microbes, 916f
Endocannabinoids
  anandamide, 219, 974–975, 975f–988f, 980–984
  dopaminergic interaction with, 1034f
  substances considered as, 975f
Endocrine pancreas, 133
  islets of Langerhans in, 135f
Endocrine receptors, 339
Endocrine system, 3
Endocytic vesicles, 22
  PrP^{Sc} propagation in, 35, 36f
Endocytosis, 607
  receptor-mediated, 695
Endoderm, 4, 5f
Endogenous morphine, 968
Endogenous opioids, 955, 957
Endomorphin-1, 968
Endonuclease, 493
Endoplasmic reticulum, 430
  Ca^{2+} in, 708f
  calcium ion release in, 965
  CFTR in, 688
  Golgi apparatus in relation to, 23f
  protein from ribosome released into, 67, 69f
  rough, 16, 21, 22f
  smooth, 21, 22f
Endorphin(s)
  defined, 968
  pain inhibiting of, 955, 957
  release of, 969f
  β-endorphin, 368, 968, 968t, 970t
Endosome, 189
Endothelial nitric oxide synthetase (eNOS), 474, 540–541, 540t
  infection, 946
  surrounding proteins of, 734, 734f
Endothelial cell(s), 616, 654, 666
  AEA transporter/CB-1 in, 983f
  anandamide binding to CB-1 in, NOS activated by, 982
  arterial, 518
  vascular, 470, 861f
    APC molecules with, 858, 861f
Endothelial cell membrane, 540–541, 858, 861f

Endothelial cell oxidant stress, nitric oxide bioavailability and, 476
Endothelial protein C receptor (EPCR), 857, 859f, 860f, 861f
Endotoxins, 475
Energy charge, 566
Energy, fat storage, 561–565, 562f–565f
Enhancers, 337
Enkephalinase, 968t
Enkephalins, 955, 957, 958f, 1031f, 1033f
  morphine structure similarity to, 958f
eNOS. *See* Endothelial
Enterochromaffin-like (ECL) cells, 435–436
  acid/histamine secretion by, 439
Enterocyte, 442
Enterogastrone. *See* Gastric inhibitory peptide
Enterokinase, 441
Entopeduncular nucleus, 1031
ENV gene, 1048, 1050, 1053f, 1054f
  rhinovirus, 1072
ENV polyprotein, 1062f
Envelope protein, 1076
  E. coli, 1122f
Environmental factors, in autoimmune diseases, 922
Enzyme(s), 531f, 916f, 1048. *See also* Coenzyme(s); Metalloenzymes; Restriction enzymes
  acid phosphatase, 93
  alanine aminotransferase, 93, 546
  alkaline phosphatase, 93
  allosteric, 105–107, 106f, 107f
    effectors bound by, 105
    half of sites case with, 105
    protein kinase A as, 105, 106f
  alpha-amylase, 93
  antibody semienzymic reaction of, 910
  aspartate aminotransferase, 93, 95f
  bone with, 93
  branching, glycogen synthase with, 154, 155f, 162–163
  caspase, 491–492, 492f, 493f

INDEX 1201

Enzyme(s) *(Continued)*
　catalase, 25, 740
　catalytic proteins as, 96–97, 97f
　classification of, 107–112, 108f–110f, 112f
　　apoenzymes, 111
　　coenzymes, 111
　　holoenzyme, 111
　　hydrolase, 108–109, 109f
　　isomerases, 110f, 111
　　ligase, 111, 112f
　　lyases, 109–111, 110f
　　oxidoreductases, 107
　　transferase, 107, 108f
　converting, 173, 452, 677, 678f
　cox2, 470, 472f
　creatine kinase with, 93, 95, 95f
　debranching, 156, 161
　deficiencies in, urea cycle, 497, 501
　diagnosis of disease with, 93–96, 94f, 95f
　digestive, 439, 441
　drugs and, 125–129, 126f–129f
　energy of activation with, 96–97, 97f
　G6PD, 474
　heart with, 93, 94f, 95f
　helicase, 292
　holo-, 111
　HSD2, 463–465, 465f
　inducible, 123, 470
　inhibition, 101–105, 101f–104f
　interleukin, 677, 678f
　kinetics of, 97–112, 98f, 100f, 106f–110f, 112f
　　allosterism in, 105–107, 106f, 107f
　　inhibition in, 101–105, 101f–104f
　　Lineweaver-Burk equation in, 99–100, 103f
　　Lineweaver-Burk plot in, 97–99, 103f
　　Michaelis-Menten constant in, 97, 98f, 99, 100f, 102
　　Michaelis-Menten equation in, 99–101, 100f
　　straight line equation in, 99
　　turnover number in, 100
　　velocity of reaction in, 98f, 102, 102f, 103f
　lactate dehydrogenase, 93, 94f

Enzyme(s) *(Continued)*
　liver with, 93
　lysozyme, 910–911, 912f
　metallo-, 120–125, 120f–125f
　Michaelis-Menten constant/substrate affinity for, 97
　molybdate-containing, 688, 792f
　pentose phosphate pathway, 284f–285f, 474
　peroxidase of, 740
　prostate gland with, 93
　proteolytic/renin, 452, 455f
　restriction, 300–305, 302f–305f, 306t, 309f
　selenium-dependent, 771
　side chain cleavage enzyme system, 453f
　topoisomerase, 291, 292
　transformylase/cyclohydrolase, 261f
　ubiquitin and, 82, 83, 83f, 84f
　urea cycle, 499, 501f, 502f
Enzyme-linked immunoabsorbent assay (ELISA), 916f, 1048
Eosinophil(s), 630f, 664, 665f, 677, 880–881, 881f
　progenitor, 630f
EPCR. *See* Endothelial protein C receptor
Ependymal astrocytes, 998
Epidermal growth factor (EGF), 479, 604–611, 608f, 614, 864. *See also* EGF-like repeats
　LDL receptor binding to, 1075f
　structure of human, 607f
Epidermal growth factor receptor, extracellular domain, 617f
Epinephrine, 1003
　adrenaline as, 448
　α_1-adrenergic receptor with, 163, 167, 168f
　β-adrenergic receptor with, 163, 168f
　AMP kinase with, 554
　binding site of, 1038f
　catecholamine formation with, 524, 526, 526f
　fat storage breakdown stimulated with, 209, 212f
　glycogen breakdown with, 163, 164f, 166f, 167, 168f

Epinephrine *(Continued)*
　glycogen synthesis with, 166f
　innervation by, 1028
　stress producing, 445, 526
　　cortisol release sequence for, 446–447
　structure of, 164f
　synthesis of, ascorbate in, 832
　triacylglycerol lipase activated by, 207
Episodic hyperammonemia, 503t
Epithelial calcium channel (ECaC), 783, 783f. *See also* ECaC1
Epithelial cells, 677
　Ca^{2+}-transporting, 788f
　intestinal, 705f, 743
　thyroid, 418f, 799f
Epithelial sodium ion channels (ENaC), 457, 458f, 460f, 691, 694f
　aldosterone with, 729
　inhibition of, 691
Epitopes, 891, 914, 915
　antibody/measles virus binding, 947f
　recognition of, 943
　spreading of, 932
Eplerenone, 465, 466f
EPO. *See* Erythropoietin
EPObp 2 complex, crystal structure of, 643f
EPOR. *See* Erythropoietin receptor
Epoxide aromatic hydrocarbons. *See* Carcinogens
2,3-epoxide reductase (VKOR), 844, 845f
Epstein-Barr, 925, 927
ER. *See* Estrogen receptor
ERα; binding to TRBP to TRβ and, 345f
Erabutoxin, 1015f
ErbB4, 992
ERcx. *See* Estrogen receptor
Ergocalciferol, 837, 840
Ergotamine, 1021t
ERK. *See* Extracellular signal-regulated kinase
ERM. *See* Ezrin, radixin, and moesin
ERp57, disulfide isomerase activity of, 184
Erythro-BH4, 522

Erythrocyte hemoglobin, 878
Erythrocytes. *See* Red blood cell(s)
Erythroid cells, hemoglobin formation and, 760
Erythroid progenitor, 630f
Erythroid progenitor cells, 633, 636
Erythromycin, 85, 85f, 669
Erythropoiesis, 633, 743
Erythropoietin (EPO), 633–638, 644f
　amino acids in, 638
　primary structure of, 642f
Erythropoietin receptor (EPOR), 638, 643f
Erythrose-4-phosphate, 803f
Escherichia coli. See E. coli
E-site. *See* Exit site
Essential amino acids (EAAs)
　arginine, 42f, 44t, 47t, 500, 509, 507f, 549, 550f, 967f, 1124f
　　metabolism of, 536–541, 538f, 540f, 542f
　　nitric oxide conversion of, 989f
　childhood, 536
　homocysteine conversion to, 518, 519f, 825, 828, 830, 831f
　methionine, 41f, 44t, 58, 517–521, 519f, 550, 551f, 958f
Essential fatty acids, 218, 218f
　linoleic acid, 199, 200f, 218, 218f
Estradiol, 430
　A-rings of, 476, 478
　binding to estrogen receptor of, 478
　formation of, 566
Estradiol receptor, 566
17β-estradiol, 566, 568f
Estriol, formation of, 566
Estrogen
　immune system and, 920
　ovarian cycle, 413
　oxytocin release by, 380
Estrogen receptor (ER), 349
　binding to
　　chemical compound, 485f
　　estradiol, 478
　DNA-binding domain of, 485f
　HRE for, 485f
　phosphorylation sites on, 486, 486f

Estrogen receptor-α (ERα), 342
Estrogen responsive element, 485f, 486
Estrone, 566, 567f
　androstenedione conversion to, 567f
Estrus, FSH receptor during, 414f
Ethanol, 606f, 996
Ethanolamine, 975, 975f, 976f, 977f, 981f
　formation of HETE/PG/LT with, 977f
Ethidium bromide, 304
Ethylmethane sulfonate (EMS), 301f
Etiocholanolone, 568, 568f
ets variant gene 5, 669
ETV5, 669, 669
Euchromatin, 10, 10f–11f
Eukaryotic elongation factors, 74–77, 78f, 79f
Eukaryotic initiation factor-2 (eIF-2), 73–74, 75f–77f
Eukaryotic promoter, basal, elements of, 329f
Eukaryotic ribosomes, 65
Eukaryotic transcription factors, 328, 329f
Excitatory amino acid(s) (EAAs), 988–1002
　glutamic acid, 41f, 47f, 858f
　　conversion to glutamine of, 503
　α-helix formation with, 81
　neurotransmitter, 1007
　quinolinic acid, 990f
　structures of, 990f
　substance P and, 967f
Excitatory amino acid receptors, 988, 989t, 990f–991f, 992
　agonists/antagonists, 990f
　listed, 990
　names of, 990–991
Excitatory amino acid transporters (EAATs), 722–723, 722t
　alternate names for, 722t
Excitatory nAChR, 1003
Exit site (E-site), 70, 74f
Exocrine pancreas, 133
Exocytosis, 142, 695, 697f, 698, 707
　calcium with, 788f
　depolarization to produce, 707
Exogenous antigens, 675

Exons, 305
3′,5′ exonuclease activity, 294
Exposure, asbestos, 324f
Extender-4, 558f, 559
Extracellular fluid, 894
Extracellular signal-regulated kinase (ERK), 604, 658
Extrinsic pathway, 492, 866f
　apoptosis, 492, 493, 494f, 495f, 593f, 594
　triggering of, 865
Eyes
　color of, 527
　pineal gland neural connection to, 533
Ezrin, radixin, and moesin (ERM), 716f
Ezrin-radixin-moesin binding phosphoprotein (EBP50), 691

FAAH. *See* Fatty acid amide hydrolase
Fab
　arms, 911f, 917
　fragments, 887f, 910, 914f, 916f, 936f
　region, 910
FABP. *See* Fatty acid-binding protein
FABPpm. *See* Peripheral membrane protein
Fabry's disease, 226t, 227
Facilitated diffusion, 697, 698–701, 699f, 737f
Factor B
　binding, 935
　properdin as, 949
Factor D
　activation by, 935
　properdin cleaved by, 949
Factor H, 938f, 949f
Factor I, 938f
Factor IIa. *See* Thrombin
Factor IX, 868f, 869
Factor IXa, 868f
Factor V, 856–857, 857t
Factor Va, fragmental structure of, 863f
Factor VIII, 865f, 866f, 870f
　deficiency of, 869
　life span of, 869f
Factor VIIIa, 856
Factor X, 866f, 867

Factor X-activating complex, 868f, 869
Factor XIII, 847, 865, 866f
Factor XIIIa, 847, 898, 898f
Factor-inhibiting HIF-1 (FIH-1), 636
Factors
  accessory, 647f
  complement system pathway, 935, 938f, 942
FAD. *See* Flavin adenine dinucleotide
FAD pyrophosphorylase, 806
FADD. *See* Fas-associated death domain protein
$FADH_2$, 808
FADL. *See* Fatty acid transporter
Fanconi-Bickel disease, 171t
Farber's lipogranulomatosis, 226t
Farnesoid X receptor (FXR), 198
Farnesyl, 236
Farnesyl pyrophosphate, 238
Farnesyl transferase, 236, 237f, 238f
  cancer drug target of, 236
Fas ligand, 679f, 902f, 904f, 923–924
Fas receptor, 902f, 904f, 924, 925f
Fas-associated death domain protein (FADD), 591–592
  immune surveillance system with, 902f
Fas/Fas-ligand interaction, 904f
FasL receptor, 924
  complex, 925f
  model of, 925f
Fasted motor activity, 443
Fasting, somatostatin and, 419f
Fat. *See also* Lipid(s)
  absorption of, 561
  dietary, 588, 925
  storage, 209, 561
    energy from, 561–565, 562f–565f
FAT/CD36. *See* Fatty acid translocase
FATP. *See* Fatty acid transport proteins
Fat-soluble vitamins, 835–847, 835f–838f, 839t, 840f–847f, 841t, 843t, 847t
  vitamin E, 841–843, 841f–847f

Fatty acid(s), 198–206, 199f–207f, 489–490, 723
  ACP synthase of, 818
  amino acid catabolism and, 546
  anandamide with, 977f
  arachidonic acid, 199–201, 200f, 521f, 1028
  ATP from breakdown of, 208
  biosynthesis of, 554
  carnitine cycle with, 206, 206f, 207f
  definition of, 198
  degradation of, 560–561, 560f
  desaturase, 557, 557f
  essential, 199, 200f, 218, 218f
  free, 561, 565f
  glycerol with, 201, 201f
  ketone bodies with, 208, 208f
  long-chain, 554, 723
  metabolism of, 565
  oxidation of, 202–205, 202f–205f, 554
  β-oxidation of, 204–205, 205f, 207, 208f, 489, 566
  peroxidation of, 843
  polyunsaturated, 200f, 218
  saturated, 198, 199f, 200f, 491
    lauric acid as, 200f
    myristic acid as, 200f
    palmitic acid as, 200f
    stearic acid as, 199f, 200f
  synthesis of, 209, 213f, 214, 215f–217f
    lipid metabolism and, 554
    reaction summary for, 554
    regulation of, 218
    steps in, 216f
  trans configuration of, 199, 201f
  translocase roles of, 727f
  transport of, 723, 727f
    mitochondrial activation and, 206, 206f, 207f, 566
  unsaturated, 198–199, 199f, 200f
    arachidonic acid as, 199–201, 200f
    linoleic acid as, 199, 200f, 208, 208f
    oleic acid, 199, 199f, 200f, 208, 208f
  uptake of, 723–728, 724f–728f
    mitochondrial, 726f

Fatty acid amide hydrolase (FAAH), 975f
  anandamide hydrolase as, 981f
  promoter, activation of, 984, 988f
Fatty acid spiral, 560, 560f
Fatty acid synthase, 214, 215f–217f
  insulin stimulation of, 218
  leptin in inhibition of, 218
  transcription of, 218
Fatty acid translocase (FAT/CD36), 723–724, 727f
Fatty acid transport proteins (FATP), 723
Fatty acid transporter (FADL), 727f
  structure of, 727f
Fatty acid-binding protein (FABP), 723
FC, 914f
FC receptors, 916f
FCRγ, 909f
Fe^{2+}. *See* Ferrous iron
Fe^{3+}, 740, 762
Fe^{3+}-S4 cluster, 808f
FecA. *See* Ferric citrate receptor
Feedback, negative, 818
Feeding, 559
Fenestration, 395, 736
Ferric citrate receptor (FecA), 1125, 1129f
Ferritin, 740, 750f
  mitochondrial/human, 751f
  saturation, 741
  serum, 741
Ferrochelatase, 753, 756, 756f, 757–758, 757f
  dimer for, 758
  gene promoter, 758
  reaction, 757f
Ferrous iron (Fe^{2+}), 762
  iron-deficiency and, 742
  oral administration of, 742
Fetal adrenal gland, 452, 476
Fetal hemoglobin (HbF,$\alpha_2\gamma_2$), 878
Fetus, 878
FGAM synthase. *See* Phosphoribosyl formylglycinamidine synthase
FGF. *See* Fibroblast growth factor
FGF receptors (FGFR), 617–618, 618t, 622f

FGF receptors (FGFR)
  (Continued)
  alternative splice variants of, 621f
  domain structure of, 620f
  extracellular region of, 623f
  FGF binding to, 623f
  human, binding of, 622f
  mutated, 620, 620f
FGF1, 622f
FGF-1, 619
FGF-2
  high molecular weight forms of, 618
  structure of, 619f
FGF-7, structure of, 618, 619f
FGFR. *See* FGF receptors
FGFR1, 618t
FGFR1c, 622f
FGFR2, 618t
FGFR2c, 622f
FGFR3, 618t
FGFR3c, complex, 622f
FGFR3c-FGF1, complex/structure of, 622f
FGFR4, 618t
Fibrin, 847, 856f, 864, 864f, 865, 898
  cross-linked, 847, 898
Fibrin clot, fibrinogen activation of, 856f
Fibrinogen, 846, 864, 867, 871, 884, 887f
  fibrin clot activated by, 856f
  percentage of blood protein, 867
  structure of human fragment D of, 887f
Fibroblast growth factor (FGF), 613, 616–623. *See also* FGF receptors
  binding to FGFR of, 623f
Fibroblasts, 638–639, 654, 678f
Fibronectin, 734, 735f
Fight or flight response, 445, 450
FiGlu. *See* Formiminoglutamate
Figlu formiminotransferase, 544f
FIH-1. *See* Factor-inhibiting HIF-1
Fischer projection, D-glucose, 147f, 148f
5-formimino-THF deaminase, 544f

FKBP51 immunophilin, 478, 479f
Flagella, 1109f
  specific arrangements of, 1110f
Flavin adenine dinucleotide (FAD), 116f, 119, 119t, 129, 540, 773, 806f
  hydrogenase, 560, 806, 806f
  synthesis of, 807f
Flavin mononucleotide (FMN), 475, 540, 806
  synthesis of, 807f
Flavin, succinate dehydrogenase as, 806
Flaviviruses, 1076
Flavoproteins, 514, 514f, 515f, 806
FLICE-like inhibitory protein (FLIP), 902f
FLIP. *See* FLICE-like inhibitory protein
Flt3, 629, 634f
Flu. *See* Influenza viruses
Fluid retention. *See* Edema
Fluoxetine (Prozac), 1022, 1024f, 1027
  structure of, 1024f
Fluvastatin, HMG-CoA reductase inhibition with, 195f
FMN. *See* Flavin mononucleotide; Riboflavin-5′-phosphate
FMS-interacting protein, 637f
FNIII, 730f
Foam cells, 233, 235
Folate, 518
  absorption, 830
  deficiency of, 830
  most important role of, 830
Folic acid (vitamin B_9), 119t, 818, 828–832, 829f
  in cobalamin deficiency, 828
  dietary, 828, 829f
  food sources of, 832t
  RDA, 830
  structure of, 829f
  zinc uptake inhibited by, 778
Folic acid receptor, intestinal, 829
Follicle(s)
  colloid in, 797, 799f
  thyroid, 799f
Follicle-stimulating hormone (FSH), 366, 373t, 408, 413
  in males, 411f

Follicle-stimulating hormone receptor, 410, 413f
  estrus expression of, 414f
  signal transductional pathways of, 414f
Follicular cells
  thyroid, 417, 418, 418f, 419, 797f
  TRH in, 795
  TSH-influenced, 797f
Food(s)
  calcium, 790t
  E. coli diseases from, 1147, 1148, 1150
  folic acid, 832t
  intake of, anandamide increasing, 982
  iron content in various, 741t
  magnesium content in various, 782t
  protein content in, 851t
  vitamin A, 839t
  vitamin D, 841t
  vitamin E, 847f
  vitamin K, 847f
Food guide pyramid, 852
Forbes' disease, 171t
Foreign peptide, 929–930, 929f, 931f
  structure of, 931f
Formate, 544
Formiminoglutamate (FiGlu), 544f
5-formiminotetra-hydrofolate (5-formimino-THF), 544f
10-formyl-THF, 544f
10-formyl-THF dehydrogenase, 544f
10-formyl-THF hydrolase, 544f, 550
FQ. *See* Orphanin
Fragile X syndrome, 240, 242f
Frataxin, 758
  regulation/structure of, 759f
Free fatty acids, 561, 565f
Free iron, 762
Free radicals
  peroxy, 843
  reactive oxygen, 762, 764
Free ribosomes, 65
Frontal cortex, serotonergic system in, 1020f
Fructose, 149f, 150f, 152f

FSH. *See* Follicle-stimulating hormone
Fucosidosis, 226t
Fumarate, 546, 770
ΦX174, 1092, 1095f, 1096, 1097, 1100f
   life cycle of, 1096f
   protein J with, 1097f
FXR. *See* Farnesoid X receptor

G protein, 430, 432, 713f, 1021
   activation, 1034f
   inhibited release of, 968–969
   inhibitory, 436
   muscarinic signaling to, 1017
   opioid receptors linked with, 970
   TK by, 965
G protein receptor, 677
G protein-coupled receptors, 408, 438, 957f, 962, 970
   cannabinoid, 982
   in membrane environment, 1044f
   metabotropic, 990, 991f
   rhodopsin as, 837, 837f, 1032, 1034f, 1037, 1037f
G6PD, 474
G6PDH, 476
GABA. *See* γ-aminobutyric acid
GABA receptor, 998–999, 1004, 1006f
GABA shunt, 1004
GABA transaminase (GABA-T), 999, 999f, 1007f
GABA type A receptor (GAB-A_AR), 1003
$GABA_A$ receptor-associated protein (GABARAP), 1007, 1009f, 1010f, 1011f
   electrostatic potential map of, 1013f
   surface representations of, 1013f
$GABA_A$R. *See* Chloride channel protein complex
GABAergic neuron, 999f
   spiny, 1001
GABARAP. *See* $GABA_A$ receptor-associated protein
GABA-T. *See* GABA transaminase
GAD. *See* Glutamic acid decarboxylase

Gag antigen, 1050
Gag gene, 1050, 1053f, 1054f
GAG polypeptide, HIV, 1062f
Gag polyprotein, 1056
GAG-POL polypeptide, HIV, 1056
Gag-pol strand, 1053f
Galactose, 149f, 150f
   E. coli with, 1126
β-galactosidase, 1134, 1135f, 1135
γ-aminobutyric acid (GABA), 996
Gamma-carboxyglutamic acid, 845f
Gamma-glutamyl cycle, 522f
Ganglia, superior cervical, 533f
Gangliosides, 220, 226, 227, 228f
   biosynthesis of, 229, 232f
GAP, 713f
Gapped intermediate, viral DNA, 1059f
GAR synthetase. *See* Glycinamide ribonucleotide synthetase
GAR transformylase. *See* Glycinamide ribonucleotide transformlyase; Phosphoribosyl glycinamide formyltransferase
Gastric acid secretion, 435–436
   histamine H_2 receptors mediating, 542
Gastric gland, cellular composition of, 440f
Gastric inhibitory peptide (GIP), 443, 559
   PreproGIP, 443–444
Gastric lumen, pH of, 435
Gastric motility, 444
Gastric mucosa, 796
   cellular composition of, 439, 440f
Gastric phase, 439
Gastrin, 435, 436
   big/mini, 438
Gastrin/CCK_B receptor, 438
Gastrointestinal hormones, 435–444, 436t
Gaucher's disease, 226t, 229
GC-1/GC-2. *See* Guanylate cyclase
G-CSF. *See* Granulocyte CSF
G-CSF-R, 633f

Gelatinase, 441
Gene(s)
   ALA synthase, mutation in, 756
   BAX, 494f
   conversion, 891f
   D2 receptor, 1030f
   ets variant 5, 669
   GH, 402, 403f
   for GHRH, 395
   HIV/viral RNA, 1048, 1050, 1052, 1053f, 1054f
   IFN-γ- responsive, 642
   IFN-stimulated, 639
   IL-7 receptor, 670f
   lysis, 1097
   MHC class I chain-related, 900f
   NOS isozymes derived from, 540, 540t
   repressor, 1101
   RHD-/RHCE-related, 891, 891f
   somatic, 917, 919
   for somatostatin, 401f, 398
   TGFβ family of, 613, 613f
   TRAIL, 594
   tumor suppressor, 587–588
   VP, 367
Gene chip array, 313
Gene cloning, 309f, 310f
Gene encoding
   in antibody formation, 917–919
   for GH receptor/mature protein, 403f
   ghrelin receptor (GHSRIa)/GHSR, 398f
   GHRH receptor, 398f
   glucocorticoid receptor, 351
   LDL receptor, chromosome 19, 191
   for protein C, chromosome, 857
   T_3 receptors, 403f
Gene env, 1048, 1050, 1053f, 1054f
Gene expression
   acute phase response of, 951
   AP-1 complex regulation of, 626
   chromatin in, 357–359
   protein kinase A culmination in, 973f
Gene pol (polymerase), 1050
   HIV reverse transcription, 1053f, 1054f

Gene processing, human ghrelin structure and, 398f
Gene splicing, 915
Gene switching, 917
Gene therapy, 314–315, 313f, 315f
  adeno-associated viruses for, 315t
  adenosine deaminase deficiency as target for, 314, 315
  adenoviruses for, 315t
  cystic fibrosis as target for, 314
  Duchenne muscular dystrophy as target for, 314
  familial hypercholesterolemia as target for, 314
  hemophilia as target for, 314
  herpes simplex virus for, 315t
  HGPRT gene with, 283
  lentiviruses for, 315t
  liposomes for, 315t
  plasmid therapy for, 315t
  retroviruses for, 315t
  SCID as target for, 314, 315
  sickle cell disease as target for, 314
  viral/nonviral vectors for, 315t
  in vivo therapy with, 314
Genetic crosses, blood type determined by, 888–889
Genetic diseases, 239. *See also* Autoimmunity; Huntington's disease
  diabetes and, 932
  hemochromatosis, 747, 753
  point mutation with, 239
Genetic susceptibility, SLE, 925
Genetics, virus, 1069
Genome, 9, 917
  bacteriophage, DNA, 1089, 1091–1092, 1092f–1094f, 1095–1097, 1096f
  germ-line cell, 919
    immunoglobulin polypeptide chain genetic information in, 917
  HIV, 1053f, 1054f
    encoding of, 1051–1052
  lambda, 1102f
  phage λ, 1092f
  retrovirus, 1048
  RNA, 1050
  SARS, 1085f, 1087f
    hypothesized CoV, 1087f

Genome *(Continued)*
  single-stranded DNA, circular, 1099f
  viral, HIV, 1051–1052, 1053f, 1054f, 1059f
  viral RNA, 1049f
Genomics, functional, 312–315, 313f, 315t. *See also* Gene therapy
Genotype, blood type and, 889t
Gephyrin, 788, 794f, 1007, 1009f
  function loss of, 795
  N-terminal domain of, 794f, 1009f, 1010f
  structure of, 1009f
Geranylgeranyl residues, 236
Germ-line cell, 917, 919
Gerstmann-Sträussler syndrome, amyloids with, 172t
Gerstmann-Sträussler-Scheinker syndrome, 33
GFAT. *See* L-glutamate: D-fructose 6-phosphate transaminase
GFD-PDGF-C, structure of, 605f
α-1,6 glucosidase activity, 156
γ-glutamyl cycle, 500f, 521, 522f
γ-glutamyl transpeptidase, 500f
GH. *See* Growth hormone
Ghrelin, 392, 395–395, 396f, 443
  blood concentrations of, 396f
  human, 398f
Ghrelin receptor(GHSRIa), 398f
GHRH. *See* Growth hormone-releasing hormone
GHSRIa. *See* Ghrelin receptor
GIP. *See* Gastric inhibitory peptide
Gla. *See* γ-carboxyl glutamate
Gla domain, 846f, 859f
GLAST, 722t
Glial cells, 654, 998, 999f
  astrocyte, 1004
α-globin, 760, 761f–763f, 764
  defect in synthesis of, 762
β-globin, 314, 760, 761f, 764
Globosides, 220, 226, 227
Globular class, 58
Globulin, 871, 884, 887f
  corticosteroid-binding, 352
Globus pallidus, 1031, 1033f
Glossitis, 808, 814
GLP. *See* Glucagon-like peptide
GLP-1, 558f

GLP-1 receptor, 559
GLT-1, 722t
Glucagon, 163, 554, 558f, 565f
  definition/function of, 559–560, 559f
  fat storage breakdown stimulated with, 209, 561
  glycogen breakdown with, 163, 166f, 167, 168f
  glycogen synthesis with, 166f
  inhibited release of, 559–560
  insulin secretion role of, 559, 559f
  islets of Langerhans' alpha cells producing, 133, 173f
  structure of, 164f, 558f
Glucagon receptor (GLU-R), 559f
  in cyclic AMP production, 559
Glucagon-like peptide (GLP), 558f, 559f
Glucocerebrosides, 227, 228f
Glucocorticoid(s), 492, 658. *See also* Cortisol
  apoptosis induced by, 491–496, 494f–495f
  as SLE treatment, 925
Glucocorticoid homodimeric receptor, 464f
Glucocorticoid pathway, 452, 453f
Glucocorticoid receptor-interacting protein 1 (GRIP1), 356, 356f
  functional domains of, 356f
Glucocorticoid receptors, 355f, 470, 474
  amounts of, 466
  gene encoding, 352
  human, 466
    ligand for, 349–350, 355f
    structure of, 355f
  ligand-binding domain of, 353f, 462f
  mouse, 353f
  rat liver, 354f
  lipocortin I produced through, 467
  as model transcription factor, 349–356, 353f, 354f, 355f
  in ovarian cycle, 413
  pathway of, 430, 452
  zinc fingers of, 349, 354f
Glucocorticoid resistance, 351
Glucocorticoid response element (GRE), 349, 349, 352

Glucogenic amino acids, 52
  metabolites of, 546
Glucokinase, 131
Gluconeogenesis, 174, 174f, 565
Glucosamine-6-phosphate, 512, 512f
D-glucosamine 6-phosphate, 512f
Glucose, 146, 147f, 153, 153f, 244, 700f
  blood, 565
    high level of, 874–875
  conversion to fats from, 211f
  damage by, 875, 877
  energy uses of, 173–182, 174f–176f, 178f–182f
  Fischer projection of, 147f, 148f
  glyceraldehyde-3-phosphate from, 207, 211f
  glycerol converted to, 182, 182f
  glycolysis with, 173–174, 174f, 175f
  insulin stimulation from, 145f, 444, 559
  metabolism, 142
  ring structure of, 147f, 148f
  serine synthesized from, 545, 545f
  stick structure of, 147f
  synthesis of, 506
  transport/membrane transport of, 138, 698–699, 699f, 704–705, 705f
    ascorbic acid in, 834f
  UDP-, 154, 155f, 157f, 162–163, 163f
Glucose 1-phosphate (Glc-1-P), 717f
Glucose 6-phosphate (Glc-6-P), 717f
Glucose tolerance test, 133, 133t
Glucose-6-phosphatase, 161, 161f
  purine metabolism disorder with, 288t
  von Gierke's disease from lack of, 169, 171t
Glucose-6-phosphate, 138, 156, 159f, 160f, 161
  ribose-5-phosphate from, 278, 281, 281f–283f
  travel in endoplasmic reticulum of, 169–170
Glucose-6-phosphate dehydrogenase (G6PD), 474
Glucose-dependent insulinotropic peptide, 444
Glucosylation, models for, 158f
Glucuronidation, 568, 568f
3-Glucuronide, of 17β-estradiol, 568f
Glucuronyl transferase, UDP, 743
GLU-R. See Glucagon receptor
GluR2 (glutamate receptor-2), 994f, 1001
  ligand-binding core, 994f
GluR3, 996, 1001, 1001f
GluR6, 995
  kainate receptor subunit, 994f, 997f
  TM4 of, 998f
Glut 2, 699
Glut 3, 699
Glut 4, 138, 143f, 699
Glut-1, 699, 700f, 737f
Glutamate, 536, 538f, 549, 1003. See also Formiminoglutamate
  analogs, 988, 989t
  brain neurotransmitter
    inhibitory, 544, 545f, 1037f
    major, 1004
  glutamine conversion of, 999f
  metabolism of, 500f, 505, 512, 512f, 544, 545f
  NMDA binding site for, 993f
  ornithine conversion to, 541
  PABA link to, 828, 829f
  pathways, GABA/glutamate metabolism, 999f
  removal of conjugated residues of, 829
  transaminated, 524
  xCT transporter, 721f
Glutamate 5-semialdehyde, 549
Glutamate, calcium ions and, 988, 990
Glutamate decarboxylase, 1006
Glutamate dehydrogenase, 52, 53f, 500f, 503, 504f, 505f, 545
  structure of, 505f
Glutamate γ−semialdehyde, 541, 549
Glutamate pyruvate aminotransferase, 511
Glutamate receptor(s), 990, 994f
  common, 991–992
  subtypes of, 995, 996
    metabotropic, 965, 967f
    subtype 1 in, 967f
Glutamate semialdehyde, 541
Glutamate synapse, 722
Glutamate synthetase, 503, 504f
Glutamate transporters, 442, 443f, 999f
Glutamate-aspartate aminotransferase. See Aspartate aminotransferase
Glutamate-induced neurodegeneration, 841, 1006, 512f
Glutamate-oxaloacetate transaminase (GOT), 508, 509f
Glutamate-proton antiport, 722
Glutamic acid, 41f, 44t, 47f, 990f
  conversion to glutamine of, 503, 996
  α-helix formation with, 81
Glutamic acid decarboxylase (GAD), 516t, 998–999, 1004, 1007f
  anchorage of, 1008f, 1113, 1113f
Glutaminase, 506, 507f, 549, 996
  phosphate-dependent, 500f
Glutamine, 41f, 44t, 998
  carbamoyl phosphate formation with, 263f
  catabolism of, 549
  conversion of
    glutamate, 999f
    glutamic acid, 503, 996
  α-helix formation with, 58
  hepatic glutamine cycle, 508f
  metabolism of, 500f, 505
  nitrogen flow importance of, 506
  pathways for utilization of, 507f
Glutamine aminotransferase pathway, 500f
Glutamine phosphoribosyl-pyrophosphate aminotransferase, 253f
  crystal structure of, 257f
Glutamine synthesis, 509, 507f
Glutamine synthetase (GS), 505, 506, 507f, 999f
  structure of, 505f
Glutaminergic neurotransmission, 722, 996, 1001
γ-glutamyl carboxylase, 845f
Glutathione (GSH), 55–56, 55f, 56f, 521f, 619
  efflux of, CTFR modulation of, 691

Glutathione (GSH) *(Continued)*
  peroxidases, 773
  S-transferase, 521
  synthesis of, 520, 520f, 721
    magnesium in, 780
Glutathione disulfide (GSSG), 56, 520f
Glycated hemoglobin (HbA1c), 875
  β-1-subunit of, sugar moiety in, 875
Glycation
  hemoglobin, 877
  protein, 876f
Glyceraldehyde-3-phosphate, 802–803, 879
Glyceraldehyde-3-phosphate dehydrogenase, 879
Glycerine receptors, gephyrin binding to, 1007
Glycerol
  fatty acids with, 201, 201f
  glucose converted from, 182, 182f
  glyceraldehyde-3-phosphate from, 207
  structure of, 201, 201f
  triacylglycerides breakdown to, 182
Glycerophosphatidylinositol bridge, 184
Glycinamide ribonucleotide synthetase (GAR synthetase), 253f
  crystal structure of, 258f
Glycinamide ribonucleotide transformlyase (GAR transformlyase), human, apo structure of, 258f
Glycine, 43f, 44t, 1003
  α-helix structure disrupted with, 58, 81
  in heme synthesis, 753, 753f, 754f
  in serine conversion, 545, 546f, 549, 547f
  threonine conversion to, 549, 547f
  titration curves for, 48f, 49f
  transporters, 721
Glycine aminotransferase, 538f
Glycine receptor, 1003, 1004f, 1005f
  AAs activating, 1003
  extracellular domain of, 1003
  inhibitor of, 1003

Glycobiology, 188, 188f
Glycogen, 138, 154–170, 155f–170f, 171t
  attack at non-reducing ends of, 162f
  branching enzyme in synthesis of, 154, 155f, 162–163
  breakdown of
    energy use, 156–161, 159f–162f
    glucose phosphatase, 161f
  defined, 154
  epinephrine in breakdown of, 163, 164f, 166f, 167, 168f
  epinephrine in synthesis of, 166f
  glucagon in breakdown of, 163, 166f, 167, 168f
  glucagon in synthesis of, 166f
  glycogen synthase in synthesis of, 154, 155f, 162–163, 168f
  glycogenin's role in formation of, 154, 155f, 162–163, 163f
  hormones' effect on, 163–169, 164f–170f
  particle, structure of, 158f
  phosphorylase, 156, 159f
  storage diseases, 169–170, 171t
    Cori's, 171t
    Fanconi-Bickel, 171t
    Forbes', 171t
    Hers', 171t
    McArdle's, 171t
    Pompe's, 171t
    Tarui's, 171t
    von Gierke's, 169, 171t
  structure of, 158f
  synthesis of, 162–163, 163f, 168f
  UDP-glucose's role in formation of, 154, 155f, 162–163, 163f
Glycogen phosphorylase, 818
Glycogen phosphorylase a, 163, 166f, 167, 168f
Glycogen phosphorylase b, 163, 166f, 167, 168f
Glycogen synthase, 154, 155f, 162–163, 168f
Glycogen synthase a, 163, 168f
Glycogen synthase b, 163, 167, 168f

Glycogenin (GN)
  branching enzyme with, 154, 155f
  dimer, crystal structure of, 156f
  glycogen formation with, 154, 155f, 162–163, 163f
  glycogen synthase with, 154, 155f, 162–163
  monomer, crystal structure of, 157f
Glycogenolysis, 156–161, 159f–162f
  debranching enzyme with, 156
  phosphorolysis in, 156
Glycolysis, 18, 20f, 50, 173–174, 174f, 175f, 565
  pentose phosphate pathway's connection with, 281, 282f
Glycolytic pathway, 50
Glycophoshatidylinositol membrane to protein anchor, 184, 186f
Glycophosphatidylinositol anchoring, 238
Glycophospholipid anchor, 591
Glycoprotein(s), 28, 182–185, 183f–187f
  carbohydrate substituents in, 185f
  HIV, 1048, 1049f, 1050, 1059, 1064
  Influenza virus, 1076
  N-acetylgalactosamine with, 183, 184f
  N-acetylglucosamine with, 183, 184f
  N-linked sugars of, 183, 183f, 184f
  O-linked sugars of, 183, 183f, 184f
  Rh, 893
Glycoprotein 130 (gp130), 664, 667f
Glycoprotein 130-JAK-STAT signaling, 638f
P-glycoprotein, 732, 732f, 737f
Glycosaminoglycans, matrix, 618
Glycosphingolipids, 220, 222f, 223f, 226–232, 227f–233f
  biosynthesis of, 230f
  catabolism of, 231f
  choline plasmalogen as, 222f
  diphosphatidylglycerol as, 222f

Glycosphingolipids
  four classes of, 226
    cerebrosides in, 226–227, 227f
    gangliosides in, 226, 227, 228f, 229, 232f
    globosides in, 226, 227
    sulfatides in, 226, 227, 227f
  lecithin as, 222f
  occurrence in membranes of, 223f
  phosphatidic acid as, 222f
  phosphatidylethanolamine as, 222f
  phosphatidylglycerol as, 222f
  phosphatidylinositol as, 222f
  phosphatidylserine as, 222f
  plasmalogen as, 222f
  structures of, 230f
  2-lysolecithin as, 222f
Glycosylation sites, 780f, 944f
Glycosylphosphatidylinositol protein (GPI), 36f, 724f
Glycosyltransferase, 154
Glycyrrhetinic acid, 152f
Glycyrrhizin, 152f
Gly-pro-hydroxypro, 768f
GlyR, 1004f
  extracellular domain, 1005f
GM-CSF. See Granulocyte-macrophage CSF
GM-CSF-Rb, 633f
GMP. See Guanine mononucleotide
GnRH. See Gonadotropin-releasing hormone
GN. See Glycogenin
GnRH. See Gonadotropin-releasing hormone
Golgi apparatus, 21–22, 23f
  endoplasmic reticulum in relation to, 23f
Golgi network, trans-, 697f
Gonadotrophs, 368
Gonadotropin-releasing hormone (GnRH), 368, 371t, 410, 416f
Gonadotropins, 406–415
  feedback system for, 416f
Gonads, melatonin receptors in, 533
GOT. See Glutamate-oxaloacetate transaminase, 508, 509f
Gout, purine metabolism disorder with, 288t

gp130, 664, 667f
gp130-JAK-STAT signaling, 638f
gpD protein, 1104f
GPI. See Glycosylphosphatidylinositol protein
Gq protein, 419–420
Graafian follicle, 413
Gradients, membrane transport, 706–713, 706f–713f, 706t
Gram stain, 1107, 1111t
  binding of, 1118
Granulocyte CSF (G-CSF), 627, 629, 631f, 632, 635f. See also G-CSF-R
  mature, 880
Granulocyte-macrophage CSF (GM-CSF), 630f, 631f, 633f, 635f, 658
Granulocytes, 633
  mononuclear, 632
Granzyme(s), 901–902
  perforin, 901f, 903
Granzyme B structure, 905f
Graves' disease, 920–922, 921f, 923
  opposite effect of, 923
Gray hair, premature, 818
GRβ isoform, 352, 355f
GRE. See Glucocorticoid response element
GRIP1, 356
  functional domains of, 356f
Groin (ileofemoral levis), thrombus/deep veins in, 855f
Growth factor(s), 74, 600–682, 601t, 602f, 603f
  cell proliferation caused by, 627–628
  CSF, 601, 627–633, 630–633f
  ECL cells, 436
  epidermal, 479, 604–611, 607f, 614, 1075f
  erythropoietin, 633–638
  hematopoietic, 637f
  insulin-like, 392, 394f, 395, 479, 643–654
  nerve growth, 623–627, 625f, 628f, 632t
  platelet-derived, 600–601, 602f, 603f, 653f
  signal transduction, 406f
  transforming, 612–616, 613f–616f
  tumor necrosis family of, 591

Growth hormone (GH), 366–367, 373t, 395f
  bodily-growth path of, 392–406, 405f
  gene for, 403, 403f
  receptors for, 403, 404f, 405
  prolactin, 432f
  structure of, 404f
  with tyrosine kinase activity, 404, 404f
Growth hormone injections, prion disease with, 35
Growth hormone-releasing hormone (GHRH), 368, 370t
  bodily-growth path of, 392–406
  gene for, 395
  secretion of, 392, 394f
  structure of, 397f
GS. See Glutamine synthetase
GSH. See Glutathione
GSH reductase, 56
GSH-containing leukotriene (LTC$_4$), 521, 521f
GS-NO. See S-nitrosoglutathione
GSSG. See Glutathione disulfide
GTP (Guanosine triphosphate), 68
  amino acid metabolism and, 505
  CTP synthetase activated by, 268
  molybdenum pterin synthesis from, 788
GTPase activity, 642–643
GTPase-activating protein (GAP), 713f
GTP-binding protein (G), 713f
Guanido, pK values for, 47t
Guanine, 244, 247f, 248f
  base-pairing with, 250f
  bases, 244, 247f, 248f, 250f, 271, 274f
  deaminated to form xanthine from, 271, 274f
Guanine mononucleotide (GMP)
  conversion IMP to, 252, 262f, 268, 270f
  cyclic, 249, 454, 500, 540
  guanylate kinase reaction with, 270
  nucleotide catabolism with, 271, 274f
  purine synthetic pathway regulation with, 263f

Guanosine 3′, 5′-monophosphate (cGMP), 967f
  hydrolyzed, 1037f
Guanosine triphosphate. See GTP
Guanosine-X, 788, 791f
Guanylate cyclase (GC-1/GC-2), 249, 540, 967f, 1037f
  activity, 454
Guanylate kinase reaction, 270
GUK domains, 997f
Gut-associated lymphoid tissue, 883
Gyrase, 294, 295f

H. See Homer
H^+. See Proton
H_2M_2 (LDH3), 93, 94f
H_2O_2, iodide and, 421–422
H_3M_1 (LDH2), 93, 94f
H_4 (LDH1), 93, 94f
HA. See Hemagglutinin
Hae III, 302, 302f, 306t
Hageman factor, 865
Hair
  color of, 527
  nAChR cells, 1014
  premature gray, 818
Hairpins, 316, 318f, 320
  HIV-1 signaling, 1051f
Half of sites, 105
Half-life, hnRNA, 320
Halibut, 776
Hallucinations, 782
Halogen substituents, 474
Halothane, 996, 998f
Halothane-binding pocket, 996
HAP1, 244
Haptoglobin, 884
Hartnup disease, 814, 816, 818
Hashimoto's disease, 920, 922
HAT. See Histone acetyltransferase
Haworth projection, 148f
HbA1c. See Glycated hemoglobin
HbF,$\alpha_2\gamma_2$. See Fetal hemoglobin
HBP. See Heparin-binding protein
hCG. See Human chronic gonadotropin
HCl. See Hydrochloric acid
HCO_3, 878, 879f
hCRH(F), 387
  properties/locations of, 388t

hCRH(F)$_1$, 387
hCRH(F)$_{2\alpha}$, 387
hCRH(F)$_{2\beta}$, 387
HDAC. See Histone deacetylases
HDL. See High-density lipoproteins
HDL receptor, 191f
HE. See Hemagglutinin esterase
Healing, wound, 616
Heart, 93, 94f, 95f
  aspartate aminotransferase in, 93
  atrial myocytes of, 454
    ANP secretion from, 454
  creatine kinase with, 95
Heart disease, 445
Heat shock cognate protein 70 (HSC70), 1006
Heat shock protein(s), 80–81, 340, 349, 478, 1006
  HSP60, 81
  HSP70, 80, 81, 478
  HSP90, 349
Heavy chains, 910, 917
  HLA-A2, 931f
  variable, 910, 917
hEGF molecules, 607f
Helicase, 292, 295f, 335
  ATP-dependent, 359
α-helix, 58, 58f, 81
Helminths, opsonization of, 916f
Hemagglutinin (HA), 1072, 1076, 1076f, 1079f, 1081f
Hemagglutinin esterase (HE), 1086
Hematocrit, anemia levels of, 742t
Hematopoiesis, IL-1 with, 654
Hematopoietic cells, 491, 494f, 627, 629, 654, 830
  formation of, 630f
  proliferation/differentiation of, 639f
Hematopoietic growth factors (HGFs), 637f
Heme, 120, 120f, 475, 744f
  in arginine metabolism, 540
  bilirubin degradation of, 745f
  chemical structure of, 120f
  crystal structure of, 120f
  -depleted nNOS, 541
  intracellular concentrations of iron and, 756

Heme (Continued)
  iron bound in, 120
  iron component of, 740, 743, 756
  with monooxygenases, 784
  pocket, 760
  prosthetic group, 979f
  synthesis of, 753–760, 753f, 754f
    chemical reactions in, 754f–756f
    final step in, 758
Heme arginate, 756
Heme iron, 120, 740, 743, 756
  in oxygen binding, 871
Heme oxygenase, biliverdin conversion of heme by, 744f
Heme oxygenase-1 (HO-1), 666, 673f
Heme porphyrin, 879
heme-globin complex, 760
Hemochromatosis, 753, 764
  genetic, 747
  treatment for, 753
Hemochromatosis gene (HFE), 747, 749f, 753
Hemodialysis, 499
Hemoglobin, 120, 636, 739, 752f, 880
  affinity for oxygen of, 873, 873f, 879
  in anemia diagnosis, 740–741
  blood proteins and, 884
  carbon dioxide with, 878
  CO with, 879
  counterpart of, 871
  deoxystate of, 874f
  formation of, 760–764, 761f–763f
    rate-limiting event in, 764
    red blood cell, 763f
  forms of, 873
    T form/R form, 874
  glycation of, 877
  iron incorporation into, 740, 743
  in normal red blood cells, 762
  oxygen dissociation curves of, 878, 878f
  oxygen with
    binding of, 871, 872f, 873
    transport of, 871
  production of, immature red blood cell, 743, 764

Hemoglobin *(Continued)*
  synthesis of, 740, 760
  tetrameric, 762, 763f
α-hemoglobin stabilizing protein (AHSP), mechanism of, 763f
Hemophilia, gene therapy for, 314
Hemopoietic cells, 654, 830
Hemorrhage, 452, 854
Hemosiderin, 740
Henderson-Hasselbalch equation, 48
Heparan, 671f
Heparan sulfate, 618–619, 666
  binding of, 671f
Heparan sulfate proteoglycan (HSPG), 623f
Heparin, 853, 881
Heparin-binding protein (HBP), 948f
Hepatic artery, 871, 872f
Hepatic glutamine cycle, 508f
Hepatic lobule anatomy, 506f
Hepatic portal vein, 871, 872f
Hepatobiliary cells, 466
Hepatocyte(s)
  amino acid metabolism and, 506, 506f
  bilirubin with, 743
  hypertrophy of, 489
  IL-1 secretion from, 654
  perivenous, 509
Hepatocyte nuclear factor (HNF), 131, 132t
HER-2/neu, 587
Herbal remedies, 852
Herbicides, 488
Herbs, 852
Hereditary angioneurotic edema (C1INH), 949
Herpes, 925, 927
Herpes simplex virus, gene therapy with, 315t
Herpes virus, 1069, 1069f
Hers' disease, 171t
HETE. *See* Hydroxy-icosatetraenoic acids
Heterochromatin, 9f, 10f, 906
Heterodimer
  of NF$_k$B, 326, 326f, 327f
  of Smad2/Smad4, 615f
  of Smad3/Smad4, 615f
  with Smad4, 613

Heterodimeric nuclear receptors, mechanism for, 340f
Heterogeneous nuclear ribonucleic acid (hnRNA), 320
  half-life of, 320
Heterozygosity (LOH), loss of, 587
Hexokinase, 138, 174, 174f
HFE. *See* Hemochromatosis gene
HFE protein, 749f
HGFs. *See* Hematopoietic growth factors
hGH. *See* Human growth hormone
HGPRT. *See* Hypoxanthine-guanine phosphoribosyltransferase
hGR. *See* Human glucocorticoid receptor
hGRα, nucleocytoplasmic shuttling of, 355f
HHCE. *See* Hyperhomocysteinemia
HIF-1. *See* Hypoxia-inducible factor
HIF-1α, 636
  putative binding sites for, 640f
HIF-1β, 636
High blood pressure, 445, 447
High-density lipoproteins (HDL), 189, 191f, 203–204, 204f. *See also* HDL receptor
  component distribution with, 234f
  copper maintaining, 771
  functions of, 235f
  sphingosine conversion blocked by, 224
High-energy bond, terminal phosphate of ATP with, 182
HIMOT. *See* Hydroxyindole-O-methyltransferase
Hind III, 302, 302f, 306t
Hinge region, 349, 911f, 917
Hippocampus, 446
  GABA in, 996
  presynapse of, 988, 990
Histaminase, 881
Histamine, 436
  secretion of acid and, 439
Histamine H$_2$ receptors, gastric acid mediated by, 542
Histamine receptors, 436, 437t
  inhibitors of, 542, 543f

Histidase, 544f
Histidase reaction, 513, 513f
Histidine, 42f, 44t, 47f, 430, 536, 768, 1126
  catabolism, 544f
  deamination of, 513, 513f
  degradation of, 542, 544, 544f
  metabolism of, 541–544, 544f
Histidine ammonia lyase, 549, 550
Histidine decarboxylase, 516t, 541–542
Histidinohydroxymerodesmosine, 768
Histone
  acetylation of, coactivators/corepressor complexes and, 358, 360f
  deacetylases, 640f
  -like proteins, 1131–1134, 1134f
  lysine residues in, 360
Histone acetyltransferase (HAT), 243f, 326, 357, 358, 360
  bromodomain, 360
  composition of, 362f
  human, 361t
  pCAF, 359
Histone chaperone CAFI, 360
Histone deacetylases (HDAC), 243f
HIV. *See* Human immunodeficiency virus
HIV dementia, 1067
HIV lipodystrophy, 1067
HIV protease, 1056, 1063f
  AIDS treatment based on, 1067
  inhibitor of, 1059, 1064f
  mechanism, 1062f
HIV-1
  protease, 1063f
  rare type of, 1048
  signaling, 1051f
  subtypes, 1047
    countries with, 1047
  viral genomes of, 1051, 1053f, 1054f
HIV-1 integrase, 1060f
  C-terminal DNA binding in, 1061f
HIV-2, 1047, 1048
HLA antigens. *See* Human leukocyte-associated (HLA) antigens

HLA class II, histocompatibility antigen, 421
HLA-A2 heavy chains, 931f
HLA-A0201, TAX interaction with human T cell and, 930f
HLA-DQ genes, 922
HLA-DR genes, 922
HM_3 (LDH4), 93, 94f
HMG. See Hydroxyl-methyl-glutarate
HMG-CoA lyase, 553
HMG-CoA reductase
  inhibitors of, 195f, 197–198
  structure of, 193, 195f
hNAT3, 721
HNF. See Hepatocyte nuclear factor
hnRNA. See Heterogeneous nuclear ribonucleic acid
H-NS protein, 1131–1134, 1134f
HO-1. See Heme oxygenase-1
Hodgkin's disease, amyloids with, 172t
Holoenzyme, 111
Holoferrochelatase, 758
Holofrataxin (Hftx), 758, 759f
Homeostasis, sodium, 1028
Homer (H), 992
Homocitrate synthase, 818, 825
Homocysteine, 518, 519f
  conversion to methionine of, 518, 519f, 825, 828, 830, 831f
  metabolism, 831f
Homodimer
  binding of p50, 326, 326f
  IL-10, 680f
Homogentisic acid, Hydroxyphenylpyruvate conversion to, 532
Homology, 604, 640f
  α_{1A}adrenoreceptor, 1032
Homotrimers, TRAIL, 592f
Hormonal response element (HRE), 341
  binding to, 461
  estrogen, 485f
  hypothalamus, releasing hormones of, 370t
Hormone(s). See also specific hormones
  ACTH, 368, 373t, 378f, 561
    action of, 391f

Hormone(s) (Continued)
  antidiuretic/vasopressin, 367, 380, 381f, 382f, 385f, 386f, 698, 699f, 1003
  gastrointestinal, 435–444, 436t
  gonadotropic, 367
    FSH (follicle-stimulating), 366, 373t
    LH (luteinizing), 366, 374t
  growth, 366–367, 373t
    human, 374t
  hypothalamic-releasing, 430
  lipolytic, 368, 378f, 561, 565f
  ovarian cycle changes in, 415f
  pituitary, 378
    anterior, 368, 373t–377t, 378, 379f, 387–406, 417–421
    posterior, 380
  polypeptide, 365–444
    neurofascin, 730f
    T_3, 366
    T_4, thyroxine, 366
    TSH, 366
  prolactin-releasing, 430
  releasing, 367f, 368
    ACTH, 368, 373t, 391f
    actions of, 387–406
    β-LPH, 368
    CRH, 368, 370t
    functions of, 368
    GHRH, 368, 370t, 392–406
    GnRH, 371t
    hypothalamus, 367f, 368, 370t
    γ–melanocyte-stimulating hormone (γ-MSH), 368
    TRH, 368, 372t
  secretion of, 1028
  sex, 483–488, 483f–488f
  signal transmission of, 368, 406f
  steroid, 445–496
    adrenal, 446, 448–451, 449f, 452, 453f, 566
    aldosterone, 387, 447, 452–466, 454f, 458f, 459f, 460f, 568, 729
    cortisol, 349–350, 352, 353f, 466–474, 568
    metabolism of, 566–570, 567f–569f
    synthetic mechanism for, 409

Hormone(s) (Continued)
  x-ray structures of, 476–478, 477f
  stress responses of, 447f
  thyroid, 417f, 800f
    biosynthesis of, 797f
    iodine in formation of, 795
    pathway of, 417f
    precursor, 418f
    secretion of, 428f
    tissue targets for, 423
Hormone receptors, activation mechanism for nuclear, 478, 481f
Hormone-binding domain, dimerized, arionatriuretic receptor, 455f
Hormone-sensitive lipase (HSL), 209, 561, 563f
  catalytic domain of, 563f
  regulation model for, 565f
  triacylglycerol as substrate of, 562f
Hormone-stimulated lipase. See Hormone-sensitive lipase
Hot peppers, burning sensation in, 961
HPETE. See Hydroperoxyicosatetraenoic acid
hPHT1, predicted structures of, 719f
hPHT2, predicted structures of, 719f
HRE. See Hormonal response element
HRVs. See Human rhinoviruses
11β-HSD2, 463–465, 465f
  mutations in, 465f
HSC70. See Heat shock cognate protein 70
HSD2. See Hydroxysteroid dehydrogenase 2
HSL. See Hormone-sensitive lipase
HSP. See Sodium conductance channel
HSP60. See Heat shock protein(s)
HSP70. See Heat shock protein(s)
HSP90
  dimer, 478
  human, partial structure of, 478, 480f
HSPG. See Heparan sulfate proteoglycan

hTSH, 376t, 422
Htt. See Mutant Huntington protein
H_2O_2. See Hydrogen peroxide
HU protein, 1131–1134, 1134f
Human chronic gonadotropin (hCG), 408
  luteal phase, 413
Human FGF receptor, 618t
Human fibrillin-1, secretion signal for, 695
Human glucocorticoid receptor (hGR)
  ligand/ligand-binding for, 349–350, 355f
  structure of, 355f
Human growth hormone (hGH), 374t
  structure of, 402f
Human immunodeficiency virus (HIV), 108–109, 109f, 947f, 1047–1068, 1065f
  acute/chronic phases of, 1048
  CD4 T cell levels in, 1048f
  AIDS treatment and, 1067
  diagnosis of, 1067
  drugs for infections of, 1067, 1068t
  HIV-1/HIV-2 strains of, 1047
  life cycle of, 1066f
  origination theory for, 1047
  reverse transcription in, 1052, 1053f, 1055f, 1056f, 1057f, 1066f, 1067
  RNA in, 1048, 1049f, 1050–1052, 1052f, 1054f
  viral, 1048, 1049f, 1050, 1064
  seroconversion in, 1067
  spread of, 1064
  symptoms of, 1067
  TAT proteins in, 1050, 1050f, 1051–1052
  testing for, 1048
  therapeutic intervention for, 1059
  transmission of, 1048, 1067
  vaccines against, 1067
  viral DNA in, 1056, 1056f, 1058f, 1059f, 1060f
  viral genomes of, 1051–1052, 1053f, 1054f, 1059f
  viral RNA, 1048, 1049f, 1050, 1052, 1053f, 1054f

Human leukocyte-associated (HLA) antigens, 928f, 930
Human Rhinoviruses (HRVs), 947f, 1069, 1071f, 1072
Humoral mechanism, 367–378
  hormonal signal transmission as, 368
  hypothalamus connection to anterior pituitary via, 369f
Huntington's chorea, 240
Huntington's disease, 239–244, 240f–244f
  basal ganglia functioning with, 239
  brain regions characteristic of, 243f
  chromosome 4 with, 239
  gene expression and, 243f
  pedigree for, 241f
Hyaline, 170
Hybrid. See RNA/DNA hybrid
Hybridoma cells, 915
Hydrochloric acid (HCl), 439
  secretion, 439f
Hydrogen bonding, protein structures with, 58, 59f, 60f
Hydrogen peroxide (H_2O_2), 25, 740
Hydrolase, 108–109, 109f
Hydrolysis
  bacterial cell, β-galactosidase, 1135
  cGMP, 1037f
  PIP_2, 438
  S-adenosylhomocysteine, 831f
Hydropathy, Lac Y derived from, 1137f
Hydroperoxyicosatetraenoic acid (HPETE), 521f
Hydrophobic bonding, 60, 60f, 250
  nonpolar groups in, 60
Hydrophobic groups, 60
5-Hydroxyindole acetic acid, 537f
5-Hydroxy-N-acetyltryptamine, 537f
25-Hydroxycholecalciferol, 530
3beta-hydroxysteroid dehydrogenase, 453f
5-HIAA, 1019
5-HT
  binding site, 1022f
  translocation, 1027

5-HT reuptake transporters, 1023f
5-HT_1AR, 1024f
  failure to express, 1022
5-HT_3R, 1019, 1022f
  chimeric subunit of, 1018f
  N-terminal domain of, 1018f
5-HTRs
  characteristics, 1021t
  classes of, 1025f
5-hydroxyindoleacetic acid (5-HIAA), 1019
5-hydroxytryptamine, 1019f
25-hydroxy-D_3, 840
Hydroxyallysine, 765
11β-Hydoroxyandrosterone, 568, 569f
Hydroxybiopterin, 522
β-hydroxybutyrate, 561
Hydroxycobalamin, 828
D_3-25-hydroxylase, 840, 840f
Hydroxy-icosatetraenoic acids (HETE), 977f
Hydroxyindole-O-methyltransferase (HIMOT), 533, 533f, 537f
Hydroxyl, pK values for, 47t
Hydroxyl ions (.OH + OH⁻), 762
Hydroxylation, proline, 832, 833f
Hydroxyl-methyl-glutarate (HMG), 195f
Hydroxylysinonorleucine, 765, 768
Hydroxyphenylpyruvate, homogentisic acid conversion of, 532
Hydroxysteroid dehydrogenase 2 (HSD2), 463–465, 465f
5-Hydroxytryptophol, 537f
HyHEL-10. See Variable domain fragment of antibody
Hyperaldosteronism, 465
Hyperalgesia, 623, 968
  inflammation associated with, 623, 625
Hyperalgesic response, NGF in, 623, 625
Hyperammonemia, 497–503
  episodic, 503t
  symptoms of, 503t
  treatment for, 499, 503t
  type 1/type 2, 503t
Hyperbilirubinemia, 808

Hypercholesterolemia, 189–198, 190f–192f, 194f–197f, 240
  atherosclerosis with, 198
  familial, 189, 191
    gene therapy for, 314
    LCAT deficiency with, 191
    mutated LDL receptor in, 191, 192f
    pedigree showing, 192f
    prognosis for, 198
  HMG-CoA reductase inhibitors treatment of, 195f, 197–198
Hyperchylomicronemia, 235t
Hypercoagulability, 853
Hypercoagulation, 856
Hyperexcitability, strychnine, 1003–1004
Hyperglycemia, diabetes with, 145, 146f
Hyperhomocysteinemia (HHCE), 518
Hyperinsulinemia, diabetes with, 145, 146f
Hyperlipoproteinemia, 235t
Hypermethylated BRCA1, 587
Hyperornithinemia, 499, 503t
Hyperphenylalaninemia, benign, 523
Hyperpolarization, 970
Hypersensitivity reactions
  delayed, 881
  immediate, 881
Hypertension, 465, 465f
  dyslipidemic, 235t
  new cause of, 466
  preclampsia, 466
  treatment of, 714
Hyperthyroid condition/activity, 921, 921f
Hypertriglyceridemia, diabetes with, 145
Hypertrophy, thyroid gland, 796f
Hypervariable regions. See Complementarity-determining regions
Hypervolemia, 454
Hypoalphalipoproteinemia, 235t
Hypoglycemia, 503t
Hypolipidemic drugs, 488
Hyponatremia, 366
Hypopituitarism, pregnancy and, 367
Hypothalamic-pituitary system, pituitary stalk in, 367f

Hypothalamic-pituitary-thyroid axis, 920
Hypothalamic-releasing hormones, 430
Hypothalamus, 1003
  connection to anterior pituitary, 369f
  CRH release by, stress and, 447
  in Graves' disease, 920
  human, 366f
  as IL-1 target, 654
  PIF/PRF release from neurons of, 430, 432
  releasing hormones of, 367f, 368, 370t
  serotonergic system in, 1020f
  stalk, 365
    trauma severing, 365
Hypothyroidism, 921
Hypotonic solution, 698
Hypovolemia, 457–458
Hypoxanthine, 245, 248f, 249f
  adenine deaminated to form, 271, 274f
  structure of, 287f
Hypoxanthine-guanine phosphoribosyltransferase (HGPRT), 271
  purine metabolism disorder with, 283, 288t
Hypoxia, 636, 640f, 758, 1004
Hypoxia-inducible factor (HIF-1), 636, 758
Hysterectomy, 588

I⁻. See Iodide ion
IAP. See Inhibitor of apoptosis
IB$_4$-negative unmyelinated nociceptors, 974f
Ibotenic acid, 990f
ICAM-1, 323, 947f
  human rhinovirus, 1072, 1073f
ICAMs. See Intercellular adhesion molecules
ICE. See Interleukin-1 converting enzyme
ICII64, 479
Icosahedral symmetry, viruses, 1069, 1069f, 1090f
Icosapentaenoic acid, 557f
I 5'-deiodinase, 797
IFN receptors, type I, 639
IFN-α, 638–639

IFNa-R1, 639
IFNa-R2, 639
IFN-β, 638–639
IFN-γ, 654, 669
  -responsive genes, 642
  structure of, 646f
IFNγ-R. See Interferon-γ receptor
IFNGR1, 647f
IFNGR2, 647f
IFNs. See Interferon(s)
IFN-stimulated genes, 639
Ig domains, 604, 900f
IgA, 910, 910f
IgD, 910, 910f
IgE, 910, 910f, 916f
  synthesis of, 671
IGF. See Insulin-like growth factors
IGF system, 651f
IGF-binding proteins (IGFBPs), 644, 645, 648, 651f, 652f
IGFBP-2, 648
IGFBP-3, 648
IGFBP-4, 652f
IGFBP-5, 652f
IGFBP-6, 652f
IGFBPs. See IGF-binding proteins
IGF-I, 402, 643, 644, 648
  half-lives of free, 648
  signaling pathway of, 650, 653f
  structure of, 648f
IGF-I receptor, 644
IGF-I receptor kinase, 650f
IGF-II, 643–644, 649f
IGF-II receptor
  binding to
    insulin, 648
    perforin, 901f
  human, 651f
  structure of, 645, 651f
IGF-IR receptor, signaling pathways mediated by, 653f
IgG antibodies, 671, 910, 910f, 911f, 916f, 917, 935, 936f
IgG FC, 936f
IgG2A antibodies, crystallized, 911f
IgG4 antibodies, 671
IgM, 910, 910f, 929f
IHF. See Integration host factor
I$_κ$B phosphorylation, 326, 658
I$_κ$Bα, 658
I$_κ$Bα kinase (IKKβ), 832, 834f
IKKβ. See I$_κ$Bα kinase

IL. *See* Interleukin(s)
IL-1, 470, 654, 655t, 657t, 880
    signal transduction pathway for, 659f
IL-1 receptor antagonist (IL-1ra), 654
IL-1 receptor I (IL-1RI), 654
IL-1 receptor II (IL-1RII), 654
IL-1 receptor superfamily, 632f
IL-1α, 656f
IL-1β, 656f, 675, 951
IL-1β precursor, 678f
IL-1R domain, 654
IL-1ra. *See* IL-1 receptor antagonist
IL-1RI. *See* IL-1 receptor I (IL-1RI)
IL-1RII. *See* IL-1 receptor II
IL-2, 654, 655t, 662f, 903
    signaling pathway for, 658, 661f
IL-2 receptor (IL-2R), 661f, 662f, 665, 677
    -IL-2R complex, 662f
IL-2 receptor beta chain, 661f
IL-2 receptor gamma chain, 661f
IL-2R. *See* IL-2 receptor
IL-2R-β/γ, 903
IL-3, 627, 635f, 639f
IL-4, 663f, 658, 664f, 677
    structures of, 658
IL-4 receptor, 663f, 665
IL-5, 635f, 658
    Inflammatory response with, 665f
IL-6, 664, 666f, 667f, 673f, 675
    CRP induction with, 951
    IL-6R interaction with, 667f
    transcription of IL-8 and, 327–327
IL-6 expression, FGF-2 increase of, 618
IL-6 production, 618
IL-6R, 664
    IL-6 interactions with, 667f
IL-7, 665, 670f
    model of, 669f
IL-7 receptor, 665–666
    deletion of, 670f
IL-7 receptor gene, 670f
IL-8, 591, 666, 675
    structure of, 671f
    transcription of IL-6 and, 327

IL-10, 666, 672f
    anti-inflammatory actions of, 673f
    homodimer, 680f
    signaling pathways of, 673f
IL-12, 669, 675
    signaling pathways for, 674f
    structure of dimeric, 673f
IL-13, 671, 675
    structure of, 677f
IL-13 precursor, 675
IL-13Rα1, 676f
IL-16, 581
    structure of, 677f
IL-17, 677
    signaling pathway of, 678f
IL-18
    anti-tumor activity, 679f
    macrophage, 677
    signaling pathway of, 679f
IL-18 receptor, 669
IL-19, 677, 680f
IL-20, 677, 680f
IL-20R1, IL-21R2 and, 680, 680f
IL-21R2, IL-20R1 and, 680, 680f
IL-22
    signaling pathways of, 682f
    structure of, 681f
IL-22 binding protein (IL-22BP), 682f
IL-22BP. *See* IL-22 binding protein
Ileofemoral levis. *See* Groin
Ileum, cobalamin-intrinsic factor complex absorbed by, 825, 827f
Iliac artery, 871, 872f
Iliac vein, 871, 872f
Imidazole pK values, 47t
4-Imidazole-5-propionate, 550
Imidazolepropionate amino hydrolase, 544f
Immune cells, 735f
    anandamide effects on, 984
Immune competence, 883
Immune complex effects, 923
Immune inflammatory stimuli, 639
Immune response, 628, 638, 904f
    BBB with, 734
    IL-1 modifier of, 654, 659f
    innate, 945
    MHC in, 927

Immune surveillance system
    cancer and, 899–909
    FADD in, 902f
    immunization process in, 916f
    perforin in, 901f
Immune system. *See also* Complement system
    cells, 735f
    cytokines produced by, 628
    depression of, 446
    estrogen influence on, 920
    functions, 981f
    HIV and, 1067
    regulation of, TNFRSF, 594
    skin immune reaction and, 678f
    suppression of, 445, 447
        pain causing, 955
Immunity, adaptive, 945
Immunization process, 916f
Immunoglobulin(s), 917
    biosynthesis of, 917, 919–920
    classes of, 910f
Immunoglobulin 2E8 Fab fragment, 887f
Immunoglobulin gene superfamily, 928f
Immunoglobulin-like receptor, 903
Immunophilin, 478, 480f
    FKBP51, 478, 479f
Immunoreactive trypsinogen (IRT), 685
IMP. *See* Inosine 5'-monophosphate
IMP cyclohydrolase (IMP synthetase), 256f
IMP dehydrogenase
    IMP conversion to xanthosine monophosphate with, 270, 270f
    structures of, 273f
IMP synthetase. *See* IMP cyclohydrolase
In vivo therapy, 314
Indolequinone, 527
Induced cyclooxygenase-2 (cox2), 470, 472f. *See also* cox2
Inducible cyclooxygenase-2 (cox2), 975, 978f. *See also* cox2
Infants, 836
    brain of, 1045

Infants *(Continued)*
  CF screening for, 685
  mental retardation in, 524, 527f
  urea cycle disorders in, 497
Infection(s), 837, 934
  bacterial
    C3 deficiency predisposing, 946
    phage sequence in, 1101, 1103
  in brain, 734
  complement pathways in, 935f
  E. coli, 1147–1149, 1150
  exposure to, 925, 927
  host cell, 1095, 1101, 1103
  SARS, 1085
  outcome of, cI/cII in, 1101
  phagocyte role in, 880
  stress from, GSH efflux in, 691
  viral, 638
    HIV, 1067
Inferior *vena cava*, 871, 872f
Inflammation, 639, 721, 933
  APC inhibition of, 857–858
  autoimmune disease accompanied by, 925
  cortisol protecting against, 467, 469f, 470, 471f, 474
  endothelial, *Streptococcal* M protein causing, 946
  hyperalgesia associated with, 623, 625
  IgE promotion of, 916f
  TNFRSF regulation of, 594
  tongue, 808, 814
Inflammatory response
  complement system pathways of, 937f
  IL-5 in, 665f
Influenza A virus
  HA with, 1076
  subtypes of, 1072
Influenza viruses, 1072, 1076, 1076f
  HA with, 1072, 1076, 1076f, 1079f, 1081f
Inheritance
  autosomal recessive disorder, 687f
  C3 infection, 946
Inhibins, 612, 613

Inhibition
  competitive, 101–102, 101f
    noncompetitive v., 102
  by CTP, 268, 269f
  CTP producing, 1143
  DNA synthesis, 312
  enzymes in, 101–105, 101f–104f
  HIV protease, 1059, 1064f
  HMG-CoA reductase, 195f
  leptin in fatty acid synthase, 218
  noncompetitive, 101–102, 103f, 104f
  UMP in OMP decarboxylase, 268
Inhibitor(s)
  NRTI, 1068t
  NtRTI, 1068t
  viral neuraminidase, 1076, 1077f
Inhibitor of apoptosis (IAP), 592, 593f
Inhibitors, histamine receptor, 542, 543f
Inhibitory cell surface receptors, 909f
Initiation, 68, 73–74, 75f–77f, 327, 331f
  4E binding protein with, 74, 75f
  eIF-4F as, 74, 75f–77f
  eukaryotic initiation factor-2 as, 73–74, 75f–77f
  phosphorylation with, 74, 76f
Initiation factor, 68
Initiator element (INR), 327, 329f
  binding protein, 335
  sequence of, 335
Innate immune response, 945
Inner cell membrane, PLA$_2$ interaction with, 470, 470f
iNOS, 474, 475f, 540–541, 540t
Inosine 5'-monophosphate (IMP), 256f
  biosynthesis of, 263f
  conversion to AMP/GMP of, 252, 262f, 268, 270f, 275f
  conversion to xanthosine monophosphate, 268–270, 270f
  nucleotidase in degradation of, 271, 274f
  purine nucleotide cycle with, 275f

Inositol 1, 4, 5-triphosphate, 418, 430, 712f, 1021
Inositol triphosphate (IP$_3$), 418, 430, 707, 709
  calcium ion evoking, 964–965
  formation of, 713f
Inositol triphosphate receptors (InsP$_3$Rs), 707, 707f, 708f, 709, 711f, 713f, 970, 992, 1021
INR. *See* Initiator element
INR box, RNA polymerase binding of, 335
InsP$_3$Rs. *See* Inositol triphosphate receptors
Insulin, 138–142. *See also* Proinsulin
  A-chain of, 138
  active, 138, 139f
  antibody anti-, 930
  B-chain of, 138
  binding of, 648
  calcium ion accumulation triggering secretion of, 142
  cAMP phosphodiesterase stimulated by, 565f
  cloning of gene for, 310f
  C-peptide of, 138
  cyclic AMP decreased by, 554
  dimer, 139, 141f
  epitope from, 915
  exocytosis of, 142
  fatty acid synthase stimulated by, 218
  GIP enhancing release of, 444
  glucose stimulation of, 145f, 444, 559
  glycated, 875
  glycogen synthesis/degradation with, 164, 167, 168f
  hexamer, 140f
  human, 138, 139f
    amino acid sequences of, 139f
  islets of Langerhans' beta cells producing, 133
  in lipid/carbohydrate metabolism, 565
  liver sensitivity to, 219
  monomer, 138–139, 141f, 143f
  protein phosphatase-1 in stimulation of, 164

INDEX 1217

Insulin *(Continued)*
  secretion, 142, 144f, 447, 559, 559f
  signal transduction pathways of, 137f
  signaling mechanism of, 137f
  in triacylglyceride synthesis, 556f
  zinc complex with, 140f
Insulin chain amino acids, 932
Insulin receptor, 133–138, 134f, 136f–138f, 168f, 648
  structure of, 134, 136f
  tyrosine kinase activity with, 134, 136f, 644, 650f
Insulin receptor substrate I (IRS-I), 650, 653f, 650
Insulin receptor substrate II (IRS-II), 650, 650
Insulin receptor-like receptor (IRR), 648
Insulin-dependent Diabetes (type I), 131, 134f, 145, 922
Insulin-dependent diabetes *mellitus* development theory, 930–933, 932f, 933f
Insulin-like growth factors (IGFs), 392, 394f, 395, 479, 643–650. See also IGF-I; IGF-II
  signal transduction pathways for, 650, 653f
Intake
  cobalamin, 828f
  dietary metals, 771
  food, anandamide increase of, 982
  iodine, 796f
  optimum calorie, 849t
  protein, 851t
  recommended daily
    calcium, 790t
    selenium, 776
    vitamin D, 840–841
    vitamin K, 847
Integrase. See Viral integrase
Integration host factor (IHF), 1131, 1131–1134
  binding of, 1133f
Integrins, 471f
β_2-integrins, 948f
Integumentary system, 3

Intercellular adhesion molecules (ICAMs), 323, 470, 947f, 1072, 1073f
Interferon(s) (IFNs), 638–643
  actions of, 645f
  type I, 645f
Interferon-γ, 646f
Interferon-γ receptor (IFNγ-R), 647f
  signaling pathway of, 647f
Interleukin(s) (ILS), 654–682, 656f, 659f, 665f, 670f, 672f, 673f, 674f
  activities/sources of, 655t
  hematopoietic cells with, 639f
  ligand family of, 657t, 658t
  receptors for, 632f, 639f, 654, 662f, 661f–663f, 665–666, 670f, 671, 677, 680, 680f, 903
  signal transduction pathways for, 658, 659f, 661f, 673f, 674f, 678f, 679f, 682f
Interleukin 1 receptor associated kinase (IRAK), 654
Interleukin-1 converting enzyme (ICE), 677, 678f
Intermembrane space, 770
Internal bleeding, 739
Internal loops, 317, 318f
Interneurons, 380, 959, 959f
Intestinal enterochromaffin cells, 619
Intestinal gas, 588
Intestinal phase, 439
Intestine(s)
  alkaline phosphatase in, 93
  brush border membrane of small, 783
  calcium absorption in, 789f, 840
  cells of, zinc movement into, 778
  E. coli presence in, 1107
  epithelial cells of, 705f, 743
  folic acid receptor of, 829
  GLP-1 in, 559
  Hartnup disease of, 816, 818
  membrane transport in, 778, 788
  obstruction in, CF as, 686f
Intrinsic factor, 825, 828

Intrinsic factor-cobalamin complex, impaired absorption of, 825, 827f
Intrinsic pathway
  apoptosis, 492, 493, 494f, 592, 593f, 594
  caspases in, 902f
  coagulation, 866f, 868f, 869
Introns, 305
Iodide, 774f
Iodide ion (I$^-$), 421–422, 796
Iodine
  active, 422
  deiodination mechanism, 429f, 775f, 797, 799f
  insufficient, 796
  intake of, 796f
  micronutrient, 795–800, 797f
  uptake of, 427f
Iodothyronine deiodinases, 773, 775, 775f
Iodothyronines, 774f
Iodotyrosines, 796
Ion-antiport channel, 737f
Ions/ion channels, 27–30, 30f, 31f, 458. See also Aberrant ion transport; Calcium ion(s)
  Ca^{2+}, 711f
  chloride, 685
  intracellular v. extracellular concentration of, 706t
  membrane transport, 706–713, 706f–713f, 706t
  Na$^+$, 704
  pain signals with, 955, 969f, 972
  receptor-associated, 1014
Ion-symport channel (Na$^+$/K$^+$/Cl$^-$), 737f
IP$_3$. See Inositol triphosphate
IP$_3$ receptor (IP$_3$R)
  binding core, 708f
  homer with, 992
IP$_3$R. See IP$_3$ receptor (IP$_3$R)
IRAK. See Interleukin 1 receptor associated kinase
Iron
  absorption of, 743, 749f
  atom, 758
  binding of, 741, 760f
  chelate of, 758
    cystines with, 758

Iron *(Continued)*
 ferrous, 742, 762
 free, 762
 heme, 120, 740, 743, 756, 871
Iron, dietary, 120, 713
 absorption of, 743, 749f
 content of in various, 741t
 copper and, 740
 excretion of excess, 743
 as heme component, 120, 740, 743, 756, 871
 heme insertion by, 756
 intake of, 771
 internal need for, 739
 intracellular concentrations of heme and, 756
 lack of, serum ferritin/stainable iron cause of, 741–742
 loss of
  menstruation, 739
  pregnancy, 739
 nonheme, 522, 740, 743
 normal bodily concentration of, 739
 overloading of, 747
 peroxypterin, 522
 RDA for, 742t
 stainable, 741
 stores of, 739
 uptake/ingestion of, 743–753, 749f
Iron transport, 743
Iron-deficiency anemia, 739–743, 746f
 chronic v., 741–742
 statistics, 739
Iron-sulfur biosyntheses, 759f
IRR. *See* Insulin receptor-like receptor
IRS-2, 653f
IRS-1, 653f
IRT. *See* Immunoreactive trypsinogen
Ischemia, 1004
Islet amyloid polypeptide, 170
Islet autoantigens, 933f
Islets of Langerhans
 cells types in, 135f
 delta cells of, 133, 134, 135f, 173f, 559
  somatostatin from, 133, 173f, 559
 endocrine pancreas with, 135f

Islets of Langerhans *(Continued)*
 glucagon from alpha cells of, 133, 173f
 insulin from beta cells of, 133, 144f
Isoelectric point, 48, 50f
Isoform GRβ, 352, 355f
Isoform I Nramp2, 746f
Isoform II Nramp2, 746f
Isoleucine, 43f, 44t, 550
 α-helix formation with, 58
Isomerases, 110f, 111
Isomerization, 11-cis double bond, 837, 837f
Isopentenyl diphosphate, cholesterol synthesis with, 193, 194f
Isopropylmalate synthase, 820, 821
Isovaleric acid, 499
Isovaleryl CoA, 552, 553
 dehydrogenase, 552
Isozyme, MM, Muscular dystrophy and, 95
ITAM, 908f
ITIM, 909f

J gene segment, 919
JAK. *See* Janus tyrosine kinase
JAK1, 647f, 661f
JAK2, 405, 431f, 588, 638f
JAK3, 661f, 670f
Janus tyrosine kinase (JAK)
 kinase activity of, 661f
 pathways of, 632, 638f, 658, 661f, 658, 669
Jejunum, upper, iron uptake in, 743
Jugular vein, 871, 872f
Junctions, 317, 318f
Jun-Fos dimer, 327, 328f
Jurkat T-lymphocytes, 788f
Juxtaglomerular cells, 452

K^+, 703t, 711t, 1023f
 inward rectifier, 965
K^+ ATPases, 1023f
KA2, kainate receptor subunit, 997f
Kainate, 990–991, 990f
Kainate receptor
 GABA with, 999f
 GluR6 subunit of, 994f, 997f
 KA2 subunit of, 997f

Kainic acid, 990f, 991
Kainite receptor, 996
 subunits of, 995, 997f
Kaposi's skin cancer, 1067, 1067f
K_{cat}. *See* Turnover number
K^+Ch, 992
Keratinocyte growth factor, 616
Keratinocytes, 616, 645, 654
Ketoacidosis, diabetes with, 145
α-ketoglutarate, 500, 503, 505, 518, 545, 546, 549
 dehydrogenase, 801, 804f
 GABA synthesis from, 1004
 lowered level of, 500
 transketolase, 801
α-ketoisocaproate, 552
Ketogenesis, 209, 212f
Ketogenic amino acids, 52, 546, 552
Ketone bodies, 208–209, 208f, 213f, 554
 excess acetyl CoA conversion to, 561
Ketosis, diabetes with, 145
Kidney
 calcium relative in, 787
 cells of
  juxtaglomerular, 452
  mesangial, 882, 882f
 cortisol, 465, 465f
 failure, 1147
 Hartnup disease in, 816, 818
 phosphate-dependent glutaminase, 499
Killer cell inhibitory receptor (KIR), 903, 906f
Killer T cells, 904f
Killing action, TRAIL, 591
Killing program, 899
Kinetics
 enzyme, 97–112, 98f, 106f–110f
 inhibition in, 101f–104f
 solute uptake, 701
Kinetochore, 14, 15f
KIR. *See* Killer cell inhibitory receptor
KIR2DL2, 903, 906f
Ki-RasA, 458f
Kirsten-RasA, 458f
K_m. *See* Michaelis-Menten constant
Krabbe's disease, 226t
K-*ras*, 587

Krebs cycle, 207f. *See also* Tricarboxylic acid cycle
Kupffer cells, 882
Kuru, 33

L gene segments, 918
L-5-hydroxytryptophan, 516
Lac operon, 1134f, 1135
  repressor of, 1135–1139, 1139f
Lac Y, 1136f, 1137f
Lactaid. *See* Lactase
β-lactalbumin, 430
Lactase, 188
Lactate, 507f, 509, 565
Lactate dehydrogenase (LDH), 93, 94f, 812
  catalyzed reaction of, 814f
  human heart, 814f
  oxidoreductases classification of, 107
Lactating mammary glands, 796
Lactation, 413
Lactic acid, iron deficiency increasing, 740
Lactoferrins (LF), 1125, 1126, 1129f
Lactose, 149f, 150f
Lactose intolerance, 188
Lactose permease (Lac Y), 1136f, 1135
Lactotrophic cell, 430
Lactotrophic cell membrane, 432
L-alanine, 1113, 1114f. *See also* UDP-N-acetylmuramoyl: L-anine synthetase
E. coli with, 1112f, 1113f
LamB
  as porin example, 1125, 1129f
  structure of, 1107f, 1128f–1129f
Lambda. *See* Phage λ
Lambda genome, 1102f
Lamin, 734, 734f
L-amino acid oxidases, 514, 514f, 515f
  deamination with, 52, 53f
L-amino acid transporter (LAT1), 720, 720f
  4F2hc, 720f
  system, 720–721
Langerhans, Islets of. *See* Islets of Langerhans, 1112f, 1113f

LANR. *See* Low-affinity neurotrophin receptor
Laparoscope, 588
Laparotomy, 588
L-arginine, 541
L-arginine-nitric oxide pathway, 967f
L-argininosuccinate, 541
L-asparagine conversion, 550f
L-aspartate, 550f
LAT1 L-amino acid transporter, 720f
LATS. *See* Long-acting thyroid stimulator
Lauric acid, 200f
LCAT. *See* Lecithin-cholesterol acyltransferase
LCF. *See* Lymphocyte chemoattractant factor
L-chains, antibody formation and, 918f
LDH. *See* Lactate dehydrogenase
LDH1. *See* H_4
LDH2. *See* H_3M_1
LDH3. *See* H_2M_2
LDH4. *See* HM_3
LDH5. *See* M_4
L-dihydroxyphenylalanine (L-DOPA), conversion of tyrosine to, 448
LDL. *See* Low-density lipoproteins
LDL receptor, 189–191, 190f–192f, 1074f
  coated pits with, 189, 191f
  EG binding of, 1075f
  encoding on chromosome 19 of, 191
  structure of, 190f
LDL transporter, 314
L-DOPA, 516, 720
Learning, 540, 1017
Lecithin, 222f, 518, 519f
Lecithin-cholesterol acyltransferase (LCAT), 191, 191f, 204
  familial hypercholesterolemia with deficiency in, 191
Lectin domains, C-type, 900f
Lectin-like domain, 864
Leg swelling, 588
Lentiviruses, gene therapy with, 315t

Leptin, 218–219, 219f, 419
  AA sequence of, 987f
  Agouti receptor protein inhibition by, 418
  FAAH activation of, 984, 988f
  fatty acid synthase inhibition with, 218
  SS pathways/TRH/fasting and, 419f
  structure of, 219f
Lesch-Nyhan syndrome, purine metabolism disorder with, 283, 288t
Lettuce, 1148
Leucine, 43f, 44t, 820, 955, 957, 958f, 1126. *See also* Leu-enkephalin
  α-helix formation with, 81
  transamination of, 552
Leucine zippers (LZ), 327, 328
Leu-enkephalin, 958f, 968, 968t
Leukocytes, 470, 638–639, 928f, 930
  integrins on surface of, 471f
  lactoferrins from, 1125, 1126, 1129f
  leakage of, 471f
  peripheral, 880
Leukotrienes (LTs), 468f, 489–490, 677
  arachidonic acid as substrate for, 201
  cortisol and, 470, 472f
  ethanolamides of, 977f
  synthesis of, 521f
Leydig cells, 408, 430
LF. *See* Lactoferrins
Lf-porin interactions, 1125, 1126, 1129f
L-glutamate, GABA synthesis from, 1006
L-glutamate: D-fructose 6-phosphate transaminase (GFAT), 512f
L-glycine, 549f
LH. *See* Luteinizing hormone
LH receptor, 430
LH spike, 413
LHβ promoter, repression, 412f
LH/HCG receptor, 407f
L-Histidine decarboxylase reaction, 436f

Ligand(s), 378
  agonist/antagonist, androgen receptor, 488f
  binding of, 665
    polyamines in, 991
    transmembrane domains in, 456f
  dependence on in recruitment of multiple coactivator complexes, 357f
  Fas, 679f, 902f
    T-cell, 904f, 923–924
  for human glucocorticoid receptor, 349–350, 355f
  interleukin, 654, 657t
    receptor overview for, 658t
  receptor conformation and, 483–488
  TNF family of, 594, 595t
    apoptosis through binding of, 901, 902f
Ligand-binding domain, 349, 461
  androgen receptor, 486, 487f
  glucocorticoid receptor, 353f, 354f, 462f
  GluR2 core of, 994f
  RXRα/PPARγ, 489, 490f
  TNF family, apoptosis through, 901, 902f
  TR_β, 800f
  VDR, 784
  vitamin D, 530
Ligand-receptor complexes, in mitosis, 607–608
Ligase, 82, 83f, 111, 112f, 294, 295f, 296f
Light
  sensitivity to, 532–533, 1032
  vertebrates sensing of, 1037f
Light chains, 910, 917–918
  biosynthesis of, 919f
  constant regions of, 910
  domains of, 917
  gene encoding for, 917–918
  variable, 910, 917, 918f, 919
  variable portions of, 910
Limbic system, cortisol negative feedback on, 445–446
Lineweaver-Burk equation, 99–100, 103f
Lineweaver-Burk plot, 97–99, 103f

Linkage disequilibrium, 922, 922f
Linoleic acid, 199, 200f, 218, 218f
α-Linoleic acid, 218, 218f
Linolenic acid, 557f
Lipase, 441, 443, 561
  diacylglycerol, 209, 211f
  hormone-sensitive, 209, 561, 563f
  hormone-stimulated, 562f
  pancreatic, 202
Lipid(s), 189–238. See also Anandamide
  cell membrane, E. coli, 1107–1108, 1113, 1117f, 1118, 1126, 1131f
  cholesterol as, 189–198, 190f–192f, 194f–197f
  chylomicrons with, 202–203, 203f, 204f
  fat/fatty acids, 198–206, 199f–207f, 207–219, 208f, 210f–220f
    activation/transport into mitochondria of, 206, 206f, 207f, 566
    oxidation of, 202–205, 202f–205f
  glycosphingolipids as, 220, 222f, 223f, 226–232, 227f–233f
  hypercholesterolemia as disease of, 189–198, 190f–192f, 194f–197f
  lipoproteins and, 233–235, 233f–235f, 235t
  metabolism of, 554–557, 555f–558f
    phosphatidylinositol, 725f
    regulation of, 565–566
  nuclear receptors activated by, 198
    farnesoid X receptor as, 198
    liver X receptor as, 198
  phospholipids, 220–226, 221f–225f, 226t
  proteins anchored to membranes with, 236–238, 236f–238f
  sphingomyelin, 220–221, 222f, 1045
  synthesis, 820

Lipid bilayer, 699, 701
Lipid droplet, 390, 406–409
Lipid rafts, 723, 724f, 725f
Lipid storage droplet, 561
Lipoate. See Lipoic acid
β-lipotropin (β-LPH), 368, 378f
  breakdown of, 368
Lipocortin, 467
Lipocortin I, 467, 469f
Lipodystrophy, HIV, 1067
Lipoic acid (Lipoate), 118, 115f, 119t
Lipolysis, 207, 447
Lipolytic hormones, 561, 565f
Lipopolysaccharide, 937f, 1127f
Lipopolysaccharide (LPS), E. coli, 1127f
Lipoproteins, 233–235, 233f–235f, 235t. See also High-density lipoproteins; Low-density lipoproteins; Very low-density lipoproteins
  cholesteryl esters with, 233, 233f
  classification of, 234f
  component distribution with, 234f
  coronary artery disease susceptibility with, 235t
  E. coli, 1127, 1132f
  functions of, 235f
  particle, 233, 233f, 562
  peptidoglycan-associated, 1132f
  triacylglycerides with, 233, 233f
Liposomes, gene therapy with, 315t
Lipoteichoic acids, 951
Lipoxins, 467, 468f
  arachidonic acid as substrate for, 201
12-lipoxygenase (12-LOX), 841
  modeling of, 842f
Lipoxygenase (LOX), 984f
Lithium, 718
Liver, 776
  aspartate aminotransferase in, 93
  bile acids in, 193, 197, 197f
  carcinogens, 489
  creatine kinase with, 95
  failure, 747
  fatty acid synthesis in, 213f

Liver, 776 *(Continued)*
  gene expression in, acute phase response, 951
  glutamine synthetase in, 506, 507f
  insulin sensitivity reduced for, 219
Liver cell, 6f. *See also* Hepatocyte(s)
Liver X receptor (LXR), 198
L-methylmalonyl CoA, 550
LNCaP. *See* Prostate cancer cells
Locomotion, 1028
LOH. *See* Heterozygosity
Long chains. *See* Chain(s)
Long terminal repeats (LTRs), 1052, 1054f
Long-acting thyroid stimulator (LATS), 920, 921f
Loratadine (Claritin), 542
Lovastatin, HMG-CoA reductase inhibition with, 195f
Low-affinity neurotrophin receptor (LANR), 625
Low-density lipoproteins (LDL), 189–191, 190f–192f. *See also* LDL receptor; LDL transporter
  cholesterol transport with, 203, 204f
  coated pits that bind, 189, 191f
  component distribution with, 234f
  copper with, 770–771
  functions of, 235f
  macrophages becoming sponge cells with, 189
  metabolism, 189, 191f
  uptake from bloodstream of, 191f
LOX. *See* Lipoxygenase
12-LOX. *See* 12-lipoxygenase
β-LPH. *See* β−lipotropin
γ-LPH, 368
Lpp 56, 1122f
LPS. *See* Lipopolysaccharide
L-pyroglutamyl-L-histidinyl-L-prolinamide, 372t, 418
L-Δ^1-Pyroline-5-carboxylate, 542f
LSD, 1021t
L-serine, 1003
  reaction of, 548f

LTC$_4$. *See* GSH-containing leukotriene
L-threonine, 549f
LTRs. *See* Long terminal repeats
L-Tryptophan, niacin conversion of, 812, 816f
LTs. *See* Leukotrienes
L-type channel, 436
L-tyrosine, 432, 522, 523f
  conversion to norepinephrine of, 451f
Lumichrome, 809f
Lung transplant, CF with, 686–687
Lungs, asbestosis, 325*f*
Lupus erythematosus. *See* Systemic lupus erythematosus
Luteal phase, 413, 414f
Luteinizing hormone (LH), 366, 408
  in males, 411f
  ovarian cycle levels of progesterone and, 413, 415f
  receptor sequence of, 374t
Luteinizing hormone/human chorionic gonadotrophin (LH/hCG), 407f
LXR. *See* Liver X receptor
LXXLL motif, 341, 341f, 357
  domains of, 347f
Ly-49A, 903
Lyases, 109–111, 110f
Lymph, production of, 894
  increased, 894–898
Lymph fluid, 894
Lymphatic system, 3, 4f, 894–896, 895f–898f, 898
  capillary in, 895f
  lymph nodes of, 883, 894
  organs of, 894
  overview of, 894f
Lymphocyte(s), 638, 654, 735f
  bone marrow produced from, 894
  common, 882, 883f
  cytotoxic T, 904f
  Jurkat T-, 788f
Lymphocyte chemoattractant factor (LCF), 675, 911
Lymphocyte-induced maturation protein, 637f
Lymphoid tissue, 883
Lymphokines, 654
Lymphoma factor, B-cell, 902f

Lysine (lys), 42f, 44t, 47f, 552, 877
  histone residues of, 360
  transamination with, 510
Lysine decarboxylase, 516t
Lysinonorleucine, 765, 768
Lysis
  E. coli cell, 1133f
  gene, 1097
  host cell, 1092
  osmotic, 901–902, 903f
  target cell, 942f
2-Lysolecithin, 222f
Lysogeny, 1101, 1102f, 1106
LysoPC, 977f
Lysophosphatidyl choline (LysoPC), 977f
Lysosomal proteases, 916f
Lysosomal storage disease, 224. *See also* Niemann-Pick disease
Lysosome(s), 21–25, 24f, 25f, 422, 645
  colloid droplet with, 797
  proteins broken down in, 81
Lysozyme, 910–911, 912f, 1120f
Lysyl oxidase, 765, 767f
LZ. *See* Leucine zippers
M protein
  bacterial, shock from, 948f
  streptococcal, 946, 948f
M3 muscarinic acetylcholine receptor, 1017f
M$_4$ (LDH5), 93, 94f
M22-MP;59-67 (foreign peptide), 929–930, 929f, 931f
M75, 947f
M177, 947f
MAC. *See* Membrane attack complex
β2 Macroglobulin, 675f
Macroglobulin, 888
Macromolecules
  blood types as, 889
  important components of, 713
Macrophage(s), 233, 235, 616, 633, 638, 896f
  activated, 666
  electron micrograph of, 895f
  IL-1s secreted by, 654, 880
  IL-18, 677
  immature stage of, 881
  induction of, 642
  M-CSF regulation of, 637f
  tissue destruction by, 924–925

Macrophage-CSF (M-CSF), 627, 629, 630f, 631f
  in osteoclasts, 632, 636f, 637f
Mad cow disease. *See* Bovine spongiform encephalopathy
Magnesium, 780–782, 781f
  cell entry of, 713–714
  deficiency, 714
  dietary sources for, 782t
  RDAs for dietary, 782, 782t
  transport of, 713–714, 780
Magnesium ion transporters, 714
MagT1, 714
Major groove, 289
Major histocompatibility complex (MHC), 904f, 906f
  allotypes, 922
  autoimmunity involvement of, 922–930, 929f
  class I, 675f, 899, 932
    chain-related antigen, 899, 900f, 903, 904f
  class II, 675
Malate dehydrogenase, 812, 815f
  catalyzed reaction of, 815f
Maleylacetoacetate, 532
Malignant carcinoid syndrome, 814, 816
Malonyl CoA, 554
  sensitive/insensitive carnitine palmitoyltransferase, 726f
MALT. *See* Mucosa-associated lymphoid tissue
Maltodextrin channel, as *E. coli* porin, 1125
Maltoporin, 1107f
Maltose, 149f, 150f, 1126
Mammary cancer, growth factor for, 566
Mammary gland
  estriol formation in, 566
  lactating, 796
  muscle cells, 380, 383f
  point mutation and, 465–466
  PRL action on, 430
Mammotrophic cell. *See* Lactotrophic cell
Manganese, 780
  intake amount for, 771
Mannose-6-phosphate, 645, 648, 901f
Mannose-binding protein, 951
MAO. *See* Monoamine oxidase
MAPK. *See* Mitogen-activated protein kinase

MAPK kinase (MEK), 606f, 626, 637f
MAPK/extracellular signal-regulated kinase (ERK) kinase (MEK), 604
Maple Syrup disease, 551–552
Marijuana, receptor producing effects of, 980, 981f
Mast cell adhesion, 677
Mating, 430
Matrix glycosaminoglycans, 618
Matrix metalloproteinases, 658
Matrix-binding protein, palmitoylcarnitine, 726f
Maturity onset diabetes of youth (MODY), 132t
MB isozyme, 95
MC120.6, 947f
McArdle's disease, 171t
MCL1, 902f
MCP. *See* Membrane cofactor protein
M-CSF. *See* Macrophage-CSF
MCT. *See* Monocarboxylate transporter
MDCK cells, 1078f
MDR. *See* Multi drug-resistance protein
MDR1 model, 732f
MDR-related protein, 732
Meats, 740
  E. coli disease from, 1147, 1148
Meconium, 685
  ileus, 686f
Median eminence, 432
Mediator(s), 470
  apoptosis, p53, 902f
Medulla, pain pathway in, 974f
Megakaryocyte, progenitor, 630f
Megaloblastic anemia, 825, 830
MEK. *See* MAPK kinase
Melanin
  formation of, 527–532, 527f–531f
  precursor of, 521–522
  synthesis of, 527, 527f
    depressed, 818
Melanocortin 4 receptor, 419
Melanocyte, in skin, 527, 528f
γ–melanocyte-stimulating hormone (γ-MSH), 368, 378f
γ-MSH. *See* γ-melanocyte-stimulating hormone
Melanosome fraction, electron micrograph of, 528f

Melatonin, 533, 537f, 1019, 1021
  PRL increased by, 535
  synthesis of, 532, 532f, 533f, 535
Melatonin receptors, 533, 535, 536f
  MT_2, 535, 536f
Melting of DNA, 251
Membrane attack complex (MAC), 935, 937, 942f
Membrane cofactor protein (MCP), 949f
Membrane potential, 706
Membrane transport, 685–737
  active, 702–704, 702f, 703f
  amino acid, 442, 522f, 720–723, 720f, 721f, 722t
  antiport system in, 702f
  blood-brain barrier in, 720, 734–737, 735f, 736f
  calcium ion, 707, 709, 710, 712f
  diffusion
    facilitated, 697, 698–701, 699f, 737f
    passive, 697–698, 701f, 705
  E. coli porin, 1126–1131
  exocytosis, 142, 695, 697f, 698, 707
  fatty acid, 723, 727f
    mitochondrial, 206, 206f, 207f, 566
  fatty acid uptake in, 723–728
  glucose, 698–699, 699f, 704–705, 705f
  Intestinal, 778, 788
  ions/gradients in, 706–713, 706f–713f, 706t
  iron, 743
  magnesium in, 713–714, 780
  MDR in, 688, 732, 732f, 733f
  osmosis, 697–698
  phagocytosis, 695, 695t, 696f
  pinocytosis, 695, 695t, 696f
  proton, 714–719, 715f 719f
  simple/coupled, 704–705, 704f–705f
  sodium conductance in, 729–731
  types of, 695–701, 695t, 696f, 697f, 699f–701f, 704f
Memory, 540, 1017
Memory loss
  short-term, 446
  stress and, 446

Menadione, 843–844
Menopause, early, 587
Menstruation, 413, 535
  loss of iron during, 739
Mental retardation, 503t, 520, 551–552
  in infants, 524, 527f
Mercury, 778
Mesangial cells, 882, 882f
Mesenteric artery, 871, 872f
Mesoderm, 4, 5f, 619
Mesothelial cells, plural, 327
Messenger RNA (mRNA), 9, 316–321, 316f, 317f, 320f, 1064, 1068–1069
  host-cell release of, 1078
  lac operon with, 1134f
  pathway of synthesis with, 317f
  phages in transcription of, 1103
  poly-A tail of, 75f, 320f
  protein coding with, 54, 54f
  sense strand in synthesis of, 62–63
  start codon in reading of, 67, 71f
  stop codon in reading of, 67, 70, 71f
Metabolic rate, 795
Metabolic water, 177
Metabolism, 497–589
  amino acid, 500f, 503–508, 517–546, 519f, 1004
    arginine, 536–541, 538f, 540f, 542f
    glutamate, 500f, 505, 512, 512f, 544, 545f
    glutamine, 500f
    histidine, 541–544, 544f
    melanin, 527–532, 527f–531f
    serine, 545, 546f
    tryptophan, 532–536, 532f–536f
  β-carotene, 836f
  carbohydrate, 565
  carcinogens, 961–962
  deamination in, 52, 53, 271, 274f, 297f, 513, 513f
  E. coli, 1139, 1141f
  fatty acid, 565
  homocysteine, 831f
  lipid, 554–557, 555f–558f
    phosphatidylinositol, 725f
    regulation of, 565–566

Metabolism (Continued)
  retinoid pathways of, 836f
  steroid hormone, 566–570, 567f–569f
  transamination in, 50, 503, 508–512, 510f, 512
Metabolites
  glucogenic amino acid, 546
  phenylalanine conversion to, 527f
  precursor of, 521–522
  steroid hormone, 566
Metabotropic glutamate receptor (mGluR)
  forms of, 999, 1000f, 1001, 1001f
  nociceptor synapses with, 992, 995f
  schematic of locations for, 1002f
  subtype 1, 967f, 1000f
Metalloenzymes, 120–125, 120f–125f
  cyclooxygenase as, 120–123, 121f–122f
  metals coordinated in ribozymes with, 125, 125f
  SOD as, 123–125, 122f, 124f
Metals, dietary, 739–852
Metaphase, 14, 15f
Metarhodopsin, 837, 838f
Metarhodopsin II, 838f
Met-enkephalin, 368, 958f
Methacholine, 1017, 1017f
Methionine, 41f, 44t, 831f, 958f
  catabolism of, 550, 551f
  codon of, 517
  conversion of homocysteine to, 518, 519f, 825, 828, 830, 831f
  α-helix formation with, 58
  metabolism of, 517–521, 519f
Methionine decarboxylase, 516t
Methionine enkephalin, 955, 957, 958f
Methionine synthase, 825, 828
5-methoxytryptamine, 1021t
Methyl group, 828, 1017
Methyl group donor, 518
Methyl nitrosourea (MNU), 301f
2-methyl-5-HT, 1021t
5,10-methylene THF. See Methylene tetrahydrofolate
5,10-methyl-THF, 544f

α-methyl-5-HT, LSD CNS, 1021t
Methylases, erythromycin-resistant, 669
Methylated naphthoquinone ring structure, 844f
β-methylcrotonyl CoA, 552
β-methylcrotonyl CoA carboxylase, 820
Methylcrotonyl CoA carboxylase, 553, 553f, 820
5-methylcytosine, 245, 249
  in DNA, 249
Methylene tetrahydrofolate (5,10-methylene THF), 278, 278f, 544f
β-methylglutaryl CoA, 553, 553f
Methylmalonic acid, 499
Methylmalonyl CoA, 825, 828
Methylmalonyl CoA mutase, 550, 825
Methylmalonyl CoA racemase, 550
Methyltetrahydrofolate reductase polymorphisms, 518
Metoclopramide, 1021t
Mevalonate, cholesterol synthesis with, 193, 194f
Mevastatin, HMG-CoA reductase inhibition with, 195f
Mg^{2+}, 786f, 991
  homeostasis, 714
MgATP, 780
mGluR. See Metabotropic glutamate receptor
mGluR1, 1000f
mGluR3, 1000f, 1001f
mGluR4, 1001, 1002
m-GOT, 508, 509f
MHC. See Major histocompatibility complex
MHC protein
  class 1, 932
    $β_2$-microglobulin with, 928f, 930
  class II, 928f, 930, 948f
  as HLA antigens, 928f, 930
  peptide fragments interacting with, type 1 diabetes with, 929f, 930
  T cell interactions with, 929, 929f
MHC-1 complex, 904f
MHC-1 peptides, 906f

Mice
  IL-7 receptor deletion in, 670f
  NOD (nonobese diabetic), 930–931
Michaelis-Menten constant (K_m), 97, 98f, 99, 100f, 102
  enzyme/substrate affinity not given by, 97
Michaelis-Menten equation, 99–101, 100f
Microarray analysis, 313, 313f
Microbes, encapsulated, 916f
Microcytic anemia, 762
Microglia astrocytes, 882, 998
β_2-microglobulin, MHC proteins class I with, 928f, 930
Micronutrient(s), 739–852
  calcium, 783–788, 783f–788f
  daily requirements for, 850t
  iodine, 795–800
  major, 795
Microorganism
  binding of, 943t, 949f
  RCA protein interactions with, 943t
Microplasmin. See Plasminogen
Microsomes, 21
Microviridae pathways, 1098f
Midbrain, pain pathway in, 974f
Milk
  cow, 739
  iodine ion NIS into, 796
  iron-deficient, 739
  raw, 1148
Milk proteins, 466
  synthesis of, 430
Mineralocorticoid, 447, 452
  excess of, 463–465
Mineralocorticoid pathway, 453f
Mineralocorticoid receptor (MR), 458, 458f, 461f, 462f
  cortisol binding to, 463, 470, 474, 478
  inhibitor of, 465
Mini gastrin, 438
Minor groove, 289
Misoprostol, 436
MIT. See Mono-iodotyrosines
Mitochondrial division, 295
Mitochondrial DNA, 295–298, 299f
  replication, 295–297, 299f
    mechanism of, 299f

Mitochondrial genes, inheritance, 295
Mitochondrial matrix space, 758
Mitochondrial membrane potential ($\Delta\Psi$), 984f
Mitochondrion, 9, 18–21, 19f, 20f, 554, 901f
  activation/transport of fatty acids into, 206, 206f, 207f, 566
  carbamoyl phosphate, 503t
  chemosmosis in, 772f
  cytochrome c from, 494f, 982–983
  FADD, 591–592
  fatty acid in, 206, 206f, 207f, 566
    degradation of, 560, 560f
    uptake with, 726f
  ferritin in human, 751f
  ferrochelatase associated with, 758
  in heme synthesis, 753, 753f
  LH testosterone synthesis and, 409–410
  outer membrane of, 566
  pathways of, cell-death, 901f
  positively charged regions of, 78
  proteins completed transferred to, 78–81
  receptor binding to preprotein in, 78
  ribosomes in, 65
  TIM complex of, 80
  TOM complex of, 80
  urea-producing enzymes in, 499, 501f, 502f
Mitogen, 600, 614
Mitogen-activated protein kinase (MAPK), 405, 405f, 430
  activation of, 620, 658, 661f
  cascade, 986f
  growth factors with, 604, 620
  pathways of, 609, 637f, 648, 710, 713f
    crosstalk, 1041, 1043f
    iron/p38 kinases with, 666
    signal transduction, 606f
Mitosis. See Cell division
Mixed-function oxidases, 216
MKP-1, 658
MltA E. coli transglycosylase, 1116f
MM isozyme, 95

MNDA receptor, 989f
MNU. See Methyl nitrosourea
Mo^{2+}. See Molybdenum
MODY. See Maturity onset diabetes of youth
Molecular chaperones, 184
Molecules
  adhesion, 658
  CaT1, 787f
  cellular/intercellular adhesion, 470
  cholesterol, 723, 724f
  collagen, 832
  FGF-2, high weight forms of, 618
  globin, 760, 761f
  large, absorption of, 695
  light-sensitive, 1032
  macro
    blood types as, 889
    important components of, 713
Molybdate enzymes, 788, 792f
Molybdenum (Mo^{2+}), 788–795, 791f–799f
  intake amount for, 771
  intestinal transporter for, 788
  RDA for, 795t
Molybdenum cofactor (moco), 788, 791f
  deficiency, 795
Molybdopterin
  precursor, 788, 791f
  synthase, 792f
Monoacylglycerol lipase, 209, 211f
Monoamine monocarboxylic amino acids, 816, 817f, 818
Monoamine oxidase (MAO), 537f, 1039f
Monoamines, 722, 1003, 1027–1044
Monocarboxylate transporter (MCT), 715, 716f, 718f
Monocarboxylate transporter (MCT) family, membrane topology of, 718f
Monoclonal antibodies, 915, 916f
Monocyte(s), 735f, 881, 882f. See also Phagocytes
  circulation of, 882
  diameter of, 881–882
  in inflammation, 933
  progenitor, 630f

INDEX 1225

Monoglyceride, 201
Mono-iodotyrosines (MIT), 422
Monokines, 948f
Monomer
  NGF, 625f
  peptidoglycan, 1115, 1117f, 1118, 1120f
  properdin, electrostatic views of, 950f
Monomer A, HIV integrase with, 1061f
Monomeric nuclear receptor mechanism, 340f
Mononuclear granulocytes, 632
Mononuclear phagocytes, 632, 637f. *See also* Monocyte(s)
  large, 881
Mononucleotide, 245, 248f. *See also specific nucleotide*
  adenine, 262f
  guanine, 262f
  structure of, 248f
Monooxygenase, heme-containing, 784
Monosaccharides, 146–147, 147f, 150f. *See also* Glucose
Mood disorders, 962
Morphine, 958f
  as agonist, 970
  endogenous, 968
  neuronal action of, 973f
  opioid receptor affinities for, 970t
  opioids and, 968–974, 968t, 969f
Mosquitoes, 1076
Motilin, 443, 444f
Motivation, 1001
Motor activity, fasted, 443
M-protein, 1088f
mRNA. *See* Messenger RNA
MRP1, human, 733f
Ms2, 1092, 1095f
α-MSH, 418
β-MSH, 368
MT_2 receptor, 535, 536f
Mucosa-associated lymphoid tissue (MALT), 883
Müllerian-inhibiting substance, 612, 614
Multi drug-resistance protein (MDR), 688, 732, 732f, 733f
Multi-CSF, 627

Multiple sclerosis, 922
Multiprenylmenaquinone, 843
Muscarine, 1017
  blocking agents of, 1028
Muscarinic acetylcholine receptors, 1017, 1017f
Muscle(s)
  components of brain compared with, 1045t
  creatine kinase with, 95
  growth, 518
  lactate dehydrogenase with, 93
  muscular performance, 540
  plasma membrane, 727f
  skeletal, 540
Muscle creatine kinase, 539f
Muscular contraction, 707, 780, 783
  acetylcholine binding in, 1014
  skeletal, 894
Muscular dystrophy, MM isozyme with, 95
Muscular relaxation, 780
Mushrooms, 1017
Mutant Huntington protein (Htt), 242, 243f
Mutation(s)
  ALA synthase gene, 756
  in antibody formation, 917
  BRCA1 gene, 588
  CFTR, 688, 693t
  E. coli, 1123f
  FGF receptors, 620t
  FGFR, 620
  HIV, 1047
  11β-HSD2, 465f
  point, 239, 465–466
  TNF-α, ribbon drawing of, 599f
Myasthenia gravis, 923, 923f
Myelin sheath, 220, 731, 828, 958f
Myeloid cells, signaling paths in, CSF-1R regulation of, 637f
Myeloma cells, 888, 915
Myeloma protein, 888
Myocardial infarction, 632
  LDH with, 95
Myocytes, cardiac, 724
Myoglobin, 740, 871
  dietary, 740
  in oxygen binding, 872f
  sperm whale, 748f
Myo-inositol hexaphosphoric acid, 741f

Myotonic dystrophy, 240, 242f
Myristic acid, 200f
Myristoyl, protein anchors with, 236f
Myristoylation, 236, 475

N protein (neocapsid), 1085, 1087, 1087f, 1088f
  RNA polymerase transcription with, 1099, 1101
  S/M/N-proteins in formation of, 1088f
N-(4-hydroxy-3-methoxybenzyl)-8-methylnon-trans-6-enan-ridel, 961f
N^5-formimino-tetrahydrofolate, 550
N^5-methyltetrahydrofolate, 828
N^5,N^{10}-methylene-tetrahydrofolate, 830, 831f
N^5,N^{10}-methylene-tetrahydrofolate cofactor, 546
Na^+, 1023f
  action potentials with, 960f
  antiport Ca^+ and, 710, 712f
  cotransporter, 708f
  ions, 704
Na^+ channels, closing of, 837
Na^+ concentration. *See* Plasma osmolarity
Na^+ exchanger (antiporter, NCX1), 787
Na^+/H^+ antiport, 714, 715f
NAADP. *See* Nicotinic acid adenine dinucleotide phosphate
N-acetylgalactosamine, 183, 184f
N-acetylglucosamine (NAG), 183, 184f, 1113f
  E. coli with, 1111, 1113, 1113f, 1117f, 1119f, 1121f
N-acetylglutamate, 500f
N-acetylglutamate synthetase, 503t
  deficiency in, 503t
N-acetylmuramic acid (NAM), *E. coli* with, 1107–1108, 1111, 1113, 1113f, 1114f, 1115, 1117f
  peptidoglycan synthesis in, 1121f
N-acetylneuraminic acid (NANA), 227, 228f, 1072
  structure of, 229, 233f

N-acetylserotonin, 533, 537f
N-acetyltransferase
   structure of, 534f
   synthesis of, 533, 533f
nAChR. See Nicotinic acetylcholine receptor
Na^+/Cl^--dependent neurotransmitter transporter, 721
NaCT, 718
N-acyl-phosphatidyl-ethanolamine-specific phospholipase D (NAPE-PLD), 977f
NAD. See Nicotinamide adenine dinucleotide
NAD+. See Threonine dehydrogenase
NAD+ synthetase, 813f
Na^+-dependent neurotransmitter transporter family (B°AT1), 817f
NADH, 119, 119t, 174, 176–177, 176f, 178f, 810f
   dehydrogenase
      cytochrome c as, 768, 772f
      nonheme iron with, 740
   reduced, 503
NADP, 119, 119t, 116f, 710
$NADP^+$. See Nicotinamide adenine dinucleotide phosphate
NADPH. See Nicotinamide adenine dinucleotide phosphate
$NADPH + H^+$, 503
NADPH:P450 reductase, 541
NAG. See N-acetylglucosamine
NAG-NAM pentapeptide, 1113
Na^+/H^+, antiport, cytoskeletal network with, 716f
Na^+/H^+ exchanger isoform 1 (NHE1), 716f
Na^+/K^+-ATPase, 458, 458f, 459f
Na^+/K^+ ATPase, 702, 702f, 703, 703f
   pumping cycle of, 702f
Na^+/K^+ ATPase pump, 703f
   mechanism of, 703f
$Na^+/K^+/Cl^-$ ion-symport channel, 737f
NAM. See N-acetylmuramic acid
NANA. See N-acetylneuraminic acid
2-naphthylamine, 301f
NAPE. See N-arachidonoyl-phosphatidylethanolamine

NAPE-PLD. See N-acyl-phosphatidyl-ethanolamine-specific phospholipase D
N-arachidonoyl-phosphatidylethanolamine (NAPE), 981f
Nasopharyngeal-associated lymphoid tissue, 883
Natural killer cells (NK), 639, 899, 901–903, 901f, 904f. See also NK cell receptors
   human, 903
   oligomeric receptors expressed on, 900f
   perforin contained in, 907f
   receptors on surface of, 903, 906f
   signaling pathways of, 908f
   tumor cell destruction by, 901, 901f, 907f, 908f
Nausea, 588
   substance P with, 962
NBD1. See Nucleotide-binding domain 1
NBD1-NBD2 interaction, 691f
NBD2, 691f
NCoR, 342, 343f
NCX1, 787
NDP. See Nucleotide diphosphate
NE. See Norepinephrine
NEF protein, HIV with, 1052, 1053f, 1054f
Negative feedback, 818, 974f
Neocapsid protein, 1085, 1087, 1087f
   S/M/N-proteins in formation of, 1088f
Neonatal meningitis, 1148
Nerve(s)
   afferent, primary, 965
   ascending, pain signaling in, 956f
Nerve ending. See Axon terminal
Nerve growth factor (NGF), 623–627
   dimer for, 628f
   monomer, 625f
   p75 receptor for, 626, 628f
   source of, 625
Nervous system, 4. See also Central nervous system; Central nervous system disease
   autonomic, 1027
   parasympathetic, 1027f
   sympathetic, 447f, 448, 1027f

Neural connections, pineal-eyes, 533
Neural toxicity, 988
Neuraminic acid, virus particle surface, 1076, 1076f
Neurobiochemistry, 955–1045
Neurodegenerative diseases, 723
Neurofascin, 729, 730f
Neurogenesis, 962
Neurokinin A, 962, 963f
Neurokinin B, 962, 963f
Neuroligin, 992
Neurological disorders, thiamine deficiency, 803
Neuromodulator, 962
   substance P as, 962
Neuromuscular junction, acetylcholine receptor on, 1014
Neuron(s). See also Interneurons
   activity of, 707
   catecholamine-secreting, 526
   cholinergic, 1017
   C-type sensory afferent, 965, 965f
   damage to, 988
   dorsal horn, 965
   GABAergic, 999f
   morphine action on, 973f
   nerve ending of, 955
   nociceptive, 623, 955, 956f, 957
      primary afferent, 970
   oxytocinergic, 380
   pain sensing/signaling with, 957
   plasticity altering, 972
   presynaptic, 968
   sending, 959
   serotonergic, 1019, 1020f, 1027
   signals on AVP-synthesizing, 382f
   somatostatinergic, 392
   spinal, 974f
   typical, 958f
   vasopressinergic, 380
Neuronal acetylcholine receptor, 1004f
Neuronal axons, 367f
Neuronal cell bodies, 380
   survival of, 616
Neuronal nAChR, 1016f
Neuronal nitric oxide synthase (nNos), 995f
Neuropeptide(s), 219, 446
   family of aminated, 962, 962t

INDEX   1227

Neuropeptide Y (NPY), 219, 443, 1032, 1033
  neurotransmitter release by, 1040f
  orphanin, 446
Neurophysin, 380
Neuroreceptors, inhibitory, gephyrin facilitating cluster of, 788, 794f
Neurotensin, 432, 1003
Neurotoxicity, 962
α-neurotoxins, 1015f
Neurotransmission
  neurotransmitter release in, 960f
  NO role in, 540
  pain signal, 959f
Neurotransmitters, 996. See also specific neurotransmitters
  amounts of, 972
  brain, 544, 545f, 1003
    major, 1004
  classical, 1003–1027, 1028
  excitatory, 1007
  monoamine, 1003, 1027–1044
    most important, 1028
  release, 1040f
  substance P as, 962, 962t
Neurotrophins, 625
Neutrophils, 654, 735f, 948f
  activation of, 948f
  band, 883, 884f
Nevirapine, 1067, 1068t
NF-1. See Nuclear factor-1
NFAT. See Nuclear factor of activated T cells
NFκB, 973f
  signaling responses, 834f, 951
NF-κB. See Nuclear factor kappa B
NF-κB ligand (RANKL), 632, 636f, 637f
N-formiminoglutamate, 550
NGF. See Nerve growth factor
N-Glycosylation
  sites of, 780f
  target sequence for, 184
NH_4^+, 505, 513f
NHE1. See Na^+/H^+ exchanger isoform 1
Niacin (Vitamin B_3), 119t, 808–818, 809f–817f
  deficiency, 814
  RDA for, 812
  tryptophan conversion to, 812, 816f

Nick-interacting kinase (NIK), 716f
Nicotinamide, 808, 810f
  coenzyme form of, 811f
Nicotinamide adenine dinucleotide (NAD), 119, 119t, 812, 816f
Nicotinamide adenine dinucleotide phosphate ($NADP^+$, NADPH), 119, 119t, 476, 743, 773, 808, 810f, 812
Nicotinate phosphoribosyltransferase, 812f
Nicotine, 1001, 1028
  acetylcholine receptor activation by, 1017
  affinity, 1014
  binding site for, 1003
  structure of, 1018f
  vasopressin released by, 380
Nicotinic acetylcholine receptor (nAChR), 1013f, 1014f, 1017, 1019
  allosteric states of, 1016f
  classes of, 1014
  excitatory, 1003
  neuronal, 1016f
Nicotinic acid, 808, 810f
  coenzyme form of, 811f
Nicotinic acid adenine dinucleotide phosphate (NAADP), 710, 710f
Nicotinicoid receptor(s)
  5-HT_3R in, 1019
  structures, 1004f
  superfamily
    as classical neurotransmitters, 1003
    GABA receptor member of, 1004
Niemann-Pick disease, 226, 226t
NIK. See Nick-interacting kinase
NIS. See Sodium/iodide symporter (NIS)
Nitric oxide (NO), 536, 540. See also Neuronal nitric oxide synthase
  bioavailability of, 476
  as cellular messenger, 540
  L-arginine with, 967f
Nitric oxide synthase (NOS), 323, 474, 538f, 675. See also Constitutive nitric oxide synthase

Nitric oxide synthase (NOS) (Continued)
  activation of, 983f
  anandamide binding to CB-1 activating, 982
  arginine formation thru, 540
  crystal structure of induced, 475f
  crystal structure of inducible, 541f
Nitric oxide synthase isozymes, 540, 540t
Nitrogen, blood urea, 497
Nitrogen flow
  amino acid metabolism, 503–508
  glutamine important to, 506
Nitrogen mustard, 301f
NK cell receptors, 909f
NK cells. See Natural killer cells
NK1. See Substance P
NK_1, 962
NK1R. See Substance P receptor
NK_2, 962
NK_3, 962
NKG2D, 899, 900f
N-linked sugars, 183, 183f, 184f
NMDA. See N-methyl-D-aspartate
NMDA receptor, 500, 990–991, 995
  complex, 992f
  glutamate-binding site of, 993f
  glycine with, 1003
  pathways of, 995f
  protein kinase A with, 1001
  up-regulation of, 995
NMDAR-PSD-95 signaling complex, 995f
N-methyl-D-aspartate (NMDA), 500, 990–991, 990f
  activation of, 989f
NMR structures, DMT1-TM4, 747f
nNOS, 474, 540–541, 540t
  heme-depleted, 541
nNos. See Neuronal nitric oxide synthase
NNRTI. See Nonnucleoside reverse transcriptase inhibitor
NO. See Nitric oxide
Nociceptin receptor
  OFQ, 970
  orphanin binding to, 446, 446f

Nociceptive stimuli, 623
Nociceptor(s), 623, 955, 968, 972
 classes of, 955
  IB_4-negative unmyelinated, 974f
 peripheral, 956f, 974f
 primary afferent, 970
 synapses of, mGluR with, 992, 995f
Nocistatin, 970
NOD mice (nonobese diabetic), 930–931
Nodes of Ranvier, 731, 731f
Non-CB1-non-CB2 cannabinoid receptors, 975f
Noncompetitive inhibition, 101–102, 103f, 104f
 competitive v., 102
Nonessential amino acids, 39, 50, 517
Nonheme iron, 522, 740
Nonnucleoside reverse transcriptase inhibitor (NNRTI), 1067, 1068f, 1068t
Nonprogressors, 1067
Nonsteroidal anti-inflammatory drugs (NSAIDs), 472f, 739
Noodles, enriched, 776
Noradrenaline. See Norepinephrine
Norepinephrine (NE), 380, 524, 533, 533f, 1003, 1027f, 1031
 DOPA conversion to, 770, 773f
 in fat storage conversion, 561
 as main neurotransmitter, 1028
 neurotransmitter release by, 1040f
 stress response of, 445, 446–447
  adrenal medulla in, 448, 451f
 synthesis of, 1039f
Northern blotting, 313
NOS. See Nitric oxide synthase
NPY. See Neuropeptide Y
NR2 subunits, 992
Nramp2 (DMT1) isoform I/isoform II, 746f
NSAIDs. See Nonsteroidal anti-inflammatory drugs
 N-terminal domain1018f
 gephyrin, 794f, 1009f, 1010f
 HIV-1 integrase, 1060f
 5-HT_3R, 1018f
 rhinovirus, 1071f

N-terminal myristoylation, 540–541, 541f
N-terminal oxygenase domain, 541, 541f
NtRTIs. See Nucleotide reverse transcriptase inhibitors
Nuclear factor kappa B (NF-κB), 224, 323, 654, 654, 660f
 acetylation of DNA in, 326
 activation of, 325f, 326, 654, 659f
 heterodimer of, binding of, 326
 modulation of, ascorbate, 832, 834f
Nuclear factor of activated T cells (NFAT), 327
Nuclear factor-1 (NF-1), 330
Nuclear hormone receptors, 198
 activation of, 481f
Nuclear lamina, 16
Nuclear localization motif, 619
Nuclear membrane, 16–18, 17f
 nucleopores of, 16, 17f
Nuclear receptor(s)
 activation of hormone, 481f
 bound with corepressor, 348f
 glucocorticoid, 349–356, 353f, 354f, 355f
Nuclear receptor gene family, vitamin D receptor in, 527
Nuclear receptor superfamily, 339f
 classes of receptors in, 351f, 488
Nuclear translocation, 349
Nuclear translocation signal, 352
Nucleic acids, 239–311. See also Deoxyribonucleic acid
 base-pairing with, 249–251, 250f, 251f
 deoxyribonucleic acid as, 244
  base-pairing in, 250f, 251f
  mutations/damage to, 298–300, 299f–301f, 301t
  replication of, 292, 294–295, 295f, 296f
  restriction enzymes with, 300–305, 302f–305f, 306t
  sequencing of, 308–311, 310f
  Huntington's disease as condition of, 239–244, 240f–244f
 purines/pyrimidines as, 244–288, 245f–270f, 272f–287f, 288t

Nucleic acids (Continued)
 ribonucleic acid as, 11, 244, 271, 311, 316–321, 316f–321f
 synthesis of, magnesium in, 780
 viral, 1089, 1091f
 viral genome packaging by, 1092, 1095
Nucleocytoplasmic shuttling, hGRα, 355f
Nucleoplasm, 9
Nucleopores, 9
 nuclear membrane, 16, 17f
Nucleoporins, 18
Nucleoside diphosphate kinase, 264, 270
Nucleoside monophosphate kinase, 264
Nucleosome, 358
 DNA location of, 358
 remodeling of, 359
 structure of, 359f
Nucleotidase degradation of AMP/IMP by, 271, 274f
Nucleotide(s), 245
 guanosine-X, 788
 magnesium with, 780
Nucleotide diphosphate (NDP), dNDP conversion from, 286f
Nucleotide excision, 294
Nucleotide pool, virus replication with, 1103
Nucleotide reverse transcriptase inhibitors (NtRTIs), 1068t
Nucleotide-binding domain 1 (NBD1), CFTR, 688, 690f. See also NBD2
Nucleus, 9, 9f, 10f, 12f
 cell division with, 13–16, 12f–16f
 chromatin of, 9
 genome of, 9
 mitochondrion of, 9
 mRNA of, 9
 nucleoplasm of, 9
 nucleopores of, 9
 paraventricular, 380, 533f
 subthalamic, 1032f
 suprachiasmatic, 533f
 supraoptic, 366f, 380
Nucleus accumbens, 999, 1001, 1002f

Nucleus basalis of Meynert, cholinergic system in, 1017
Nutraceuticals, 852
NutraSweet. *See* Aspartame
Nutrition. *See also* Diet; Micronutrient(s)
  protein, 849–850
N^ω-hydroxy-L-arginine, 540

Obesity, perilipin for, 562
Octreotide, 371t, 398
  structure of, 400f
OFQ. *See* Orphanin FQ
OFQR. *See* Orphanin FQ receptor
OH_2D_3 1, 25, 789f
OH + OH⁻. *See* Hydroxyl ions
Okazaki fragments, 294, 296f
Oleic acid, 199, 199f, 200f, 208, 208f
Oligomerization, 639
  NK cell expressing receptors of, 900f
Oligosaccharide, *E. coli*, 1132f
O-linked sugars, 183, 183f, 184f
OMP (Orotidine monophosphate), 266f
OMP decarboxylase. *See* Orotidine 5′-phosphate decarboxylase
OmpC porin
  channel, 1126
  trimer, 1129f
OmpF porin mutant D744, 1123f
Oncogene
  ovarian cancer, 587
  ras as, 236
1-(5′phosphoribosyl)-5-amino-4-imidazole carboxamide (AICAR), 255f
One-carbon transfer reactions, 830
OOH + H⁺ radical, 762
Oophorectomy, 588
OPG. *See* Osteoprotegerin
Opiate, 968
Opiate receptors, 969f
Opioid(s)
  definition of, 968
  endogenous, 955, 957
  morphine and, 968–974, 968t, 969f
  peptides of, 968
    AA sequences of, 968t
    affinities of, 970t

Opioid receptors, 446, 968, 970t
  δ-receptors, 969, 971f
  G protein linking of, 970
  κ-receptors, 969–970, 972f, 973f
  major, 969, 969f
  μ-receptors, 969, 970t, 971f
  orphanin as not binding to, 446
Opium, 968
Opsin, 837, 837f
Opsonization, 916f, 917, 942f
Oral contraceptives, 853
Organ systems, 2–5, 3f, 4f
  alimentary, 3, 4f
  blood, 3
  cardiovascular, 3, 4f
  diseases of central nervous system with, 239–244, 240f–244f
  endocrine, 3
  integumentary, 3
  lymphatic, 3, 4f
  nervous system, 3
  reproductive, 4
  respiratory, 3, 4f
  skeletal-muscular, 3, 4f
  urinary, 3
Ork II, 650
Ornithine, 497, 536, 538f, 543f
  metabolism of arginine and, 542f
  proline/glutamate conversion of, 541, 545
Ornithine decarboxylase, 541, 542f, 543f
Ornithine transcarbamylase, 500, 501f, 502f, 503t
Orotic acid, 245, 248f, 249f, 713–714
Orotic aciduria, pyrimidine metabolism with, 288t
Orotidine 5′-phosphate decarboxylase (OMP decarboxylase)
  structure of, 264, 267f
  UMP as competitive inhibitor of, 268
Orotidine monophosphate. *See* OMP
Orphan receptors, 339, 488–491, 491f
  PPAR, 488
Orphanin (OFQ), 446
  structure of, 970
Orphanin FQ receptor (OFQR), 968
  agonist for, 970

Osmolality, 698
Osmoreceptor, 380
Osmosis, 697–698
  balance in, 706
Osmotic lysis, 901–902, 903f
Osmotic pressure, 704
Osteoclast differentiation factor, 632, 636f
Osteoclasts, 632
  in collagen formation, 769f
  CSF role in, 632, 636f, 637f
  motility of, 632
Osteomalacia, 840
Osteoprotegerin (OPG), 591, 592f
OT. *See* Oxytocin
Ovalbumin upstream promoter transcription factor (COUP-TF), 479
Ovarian cancer
  metastases of, 590f
  new approaches to, 587–594
  oncogenes increasing risk of, 587
  stages of, 588, 589f, 590, 590f
  symptoms of, 588
Ovarian carcinoma, 588, 589f
Ovarian cycle, 406, 413, 415f, 416f
  follicular phase of, 413
  LH/progesterone levels in, 413, 415f
  luteal phase of, 413, 414f
Ovaries, 588, 589f
  enlarged, 588
Overhanging ends. *See* Sticky ends
Overt cystic fibrosis, 688
Ovulation, 413
Oxalic acid, 832
Oxaloacetate, 546, 549
  in fatty acid degradation, 561
Oxidation, 843
  amino acid, 514, 514f, 515f
  depression of aerobic, 500
  fatty acid, 202–205, 202f–205f, 554
  β-oxidation, 204–205, 205f, 207, 208f, 489
  fatty acid, 204–205, 205f, 207, 208f, 489, 566
  pathway, 560, 560f
Oxidative deamination, 52, 53f
Oxidative pentose phosphate pathway, 278

1230 INDEX

Oxidative stress, 673f
Oxidoreductases, 107
Oxygen
  availability of, 877
  binding of, 871, 872f, 873, 873f
    affinity for, 873, 873f, 879
  consumption, T_3 increase of, 800
  dissociation curves, 878, 878f
  saturation curve, 877
Oxygen deficiency, 758. *See also* Hypoxia
Oxygen transport, 743, 871–877, 872f–876f, 877t
  carbon dioxide in, 878–879, 878f
  hemoglobin responsible for, 871
Oxygenase mixed function, 522
Oxyhemoglobin, 874, 874f, 877
Oxy-hemoglobin, structure of, 875f
Oxytocin (OT), 380
  gene structure of, 386f
  release of, 380, 432
    suckling response, 383f
  structure of, 381f
Oxytocin receptor, 380
  human, 384f, 386f
Oysters, 740, 776

P35, 467
P38, 986f
P50, 325*f*
  homodimer of, 326
P53, 587
P75, 626, 628f
P90-ribosomal protein S6 kinase (p90RSK), 716f
P90RSK. *See* p90-ribosomal protein S6 kinase
P160 proteins, 342
P160 proteins/coactivators, 342, 346f
P300, 326, 356
P450, 567f, 784, 785f, 786f
P450$_{17\alpha}$, 453f
P450$_{c11\beta}$, 453f
P450$_{c21}$, 453f
PABA. *See* Para-aminobenzoic acid
PAH. *See* Phenylalanine hydroxylase

Pain, 955–962, 960f, 988
  acute v. chronic, 955
  chemical messengers sensing, 957
  endorphins inhibiting, 955, 957
  extent of, 968–969
    neuronal plasticity altering, 972
  nerve ending sensing, 955
  noxious stimuli in, 955
  opioids/ morphine altering, 968–974, 968t, 969f
  ovarian cancer, symptoms of, 588
  pathways of, 956f, 972, 973f, 974, 974f
  signaling, 955, 956f, 957, 959f, 965, 969f, 972
  transmission of, substance P, 962, 963f–966f, 965, 988
  untreated, 955
Palmitic acid, 200f
Palmitoyl protein anchors, 236f
Palmitoylation, 475, 723
  sites of, 892f
Palmitoylcarnitine matrix-binding protein, 726f
Palmitoyl-ethanolamide, 977f
Pancreas
  acinar cells of, 441
  alpha cells of, 133, 135f, 173f, 559
  alpha-amylase with, 93
  beta cells of, 133, 135f, 138, 142–145, 144f–145f, 173f, 559
    destruction of, diabetes type 1 with, 930
  delta cells of, 133, 134, 135f, 173f, 559
  depolarization in, 142
  digestive enzymes from, 441
  ductal cells of, 441
  endocrine, 133, 135f
  exocrine, 133
  exocytosis of, 142
  HCO$_3$ homeostasis in, 693f
  immune attack in, 932
  insufficiency of, 685
  islets of Langerhans in, 133, 135f, 144f, 173f
  proteases, 439–441
Pancreatic islet cell tumor, amyloids with, 172t

Pancreatic lipase, 202, 203f
Pancreatic phospholipase, 202, 203f
Panhypopituitarism, 365–367
  hypothalamus stalk and, 365
  symptoms of, appearance, 366
Panic attacks, 1022
Pantoic acid, pantothenic acid formed from, 818, 819f
Pantothenate, 818
Pantothenic acid (vitamin B_5), 119t, 818, 819f
  formation of, 818, 819f
  pantothenate deficiency with, 818
Para-aminobenzoic acid (PABA), 828, 829f
Paracellular diffusion, 737f
Paracrine, 402
Paraprotein, abnormal, 888
Parasite(s), 881
  viruses as, 1069
Parasitic symbionts, 1069
Parasympathetic nervous system, 439, 1027f
  acetylcholine and, 1027–1028, 1027f
Parathyroid hormone (PTH), 840
Parietal cells, 436
  secretion of, 439, 440f
    intrinsic factor with, 825
Parkinson's disease, 1031
Pars intermedia cells, 352, 366f, 466
Pars intermedia-like cells (PILC), 368, 378f
Pars reticulata, 1031
Passive diffusion, 697–698, 701f, 705
Passive transport, 697
PAT proteins, 561
  domain
    pf, 730f
  structure of, 564f
Pathogens
  BBB preventing, 735f
  E. coli, 1147–1148, 1150t
  streptococcus pyogenes, 943t, 946
Pathways. *See also* Signal transduction pathway(s)
  adrenergic receptor, 1041
  anandamide, 977f
  bacterial cell biosynthesis, 1139–1140, 1142f

INDEX 1231

Pathways (Continued)
  blood coagulation, 866f, 868f, 869
  complement system
    alternative, 934–935, 935f, 937f, 938f, 949, 950f
    classical, 934, 935, 935f, 937f, 946
    factor members of, 942
    overview of, 935f
    regulators of, 942–949
  constitutive secretory, 697f
  creatine/creatinine, 538f
  exocytosis, 697f
  extrinsic, 866f
    apoptosis, 492, 493, 494f, 495f
    triggering of, 865
  Fas, 904f
  GABA/glutamate metabolism, 999f
  glutamine, 500f, 507f
  gp130-JAK-STAT, 638f
  intrinsic, apoptosis, 492, 493, 494f, 495f
  JAK/STAT, 632, 638f, 658, 661f, 669
  L-arginine-nitric oxide, 967f
  MAPK, 609, 626, 637f, 650, 666, 710, 713f
    adrenergic receptor, 1042f, 1043f
    iron/p38 kinases with, 666
  microviridae, 1098f
  mitochondrial cell-death, 901f
  NMDA receptor, 995f
  pain, 956f, 972, 973f, 974, 974f
  pentose phosphate, 801, 802, 805f
  phospholipase C, 707, 709
  polyamine, 541, 542f
  protein, E. coli, 1126, 1130f
  protein tyrosine kinase-dependent, 899
  purine synthesis, 263f
  Rab4-mediated recycling, 982f
  Rab5, 981
  retinoid metabolism, 836f
  salvage pathway, 271
Payer's patches, 33, 34f
pBR322, 305, 308f
  restriction map of, 308f
PC. See Phosphocholine
p/CAF. See CBP-associated protein
PDGF. See Platelet-derived growth factor

PDGF β-receptor, 601
PDGF-associated proteins, 601, 604
PDGF-BB, 601, 602f
PDGF-C, 604
PDGF-CC, dimeric growth factor domains of, 604, 605f
PDGFRα, 600, 603f, 604, 604t
PDGFRβ, 600, 604t
PDZ domain, 694f, 789f, 992, 995f
PE. See Phosphatidylethanolamine
Pellagra, 119t, 814
Pentapeptide
  E. coli, 1113, 1114f, 1121f
  NAG-NAM, 1113
Pentose phosphate isomerase, 281, 281f
Pentose phosphate pathway, 801, 802, 805f
  enzymes of, 284f–285f, 474
  glycolysis' connection with, 281, 282f
Pentose-containing trinucleotides, cellular concentrations of, 278
Pentraxins, 951
Pepsin, 439
Pepsinogen, 439
PEPT1. See Proton-dependent dipeptide transporter
Peptide(s)
  ANP, 454
  CLP, 368
  foreign, 929–930, 929f, 931f
  GLP-1, 558f
  MHC molecules with fragments of, type 1 diabetes with, 929f, 930
  MHC protein binding, 923f
  MHC-1, 906f
  opioid
    AA sequences of, 968t
    affinities of, 970t
  PRL-releasing, 430
  viral/TAX, HLA-A0201/human T cell with, 930f
Peptide bond, 54, 54f
  amino acids moved to ribosomes for, 67–73, 71f–75f, 80f
  elongation factor for, 70, 74–77, 78f, 79f
  initiation factor for, 68, 73–74, 75f–77f

Peptidoglycan, 1107, 1111f, 1113, 1114f
  -associated lipoproteins, 1132f
  binding of, 1119f, 1123f
  of gram-negative E. coli, 1114f, 1118
  monomer, 1117f
    bactoprenol with, 1120f
  porins in, 1123f
  synthesis, 1107, 1121f
Peptidyl site (P-site), 68, 70, 74f
Peptidyllysine residues, 765
Per-Arnt-Sm homology domain, 640f
Perforin, 901f, 903
  action mechanism of, 903f
  C2 domain, 905f
  NK cell granule containing, 907f
  pores, 901f
Perforin-granzyme-mediated necrosis, 901f
Pericytes, 735f
Perilipin, 561, 562
Peripheral cannabinoid receptor (CB-2), 975, 975f, 980, 980f, 981f
  cyclic AMP inhibited by, 981, 981f
Peripheral leukocytes, 880
Peripheral membrane protein (FABPpm), 727f
Peripheral nociceptors, 956f, 974f
Periplasmic space, 1126
Periportal cells, 506, 508f
Permeases, 697
Pernicious anemia, 828
Peroxidase(s), 740, 773, 796
  active site, 979f
  eosinophil release of, 881
  thyroid, 426f, 922
Peroxidation, fatty acid, 843
Peroxide, 762
Peroxisome(s), 25–26, 26f, 514
  proliferator-activate receptor-α, 982
  size/number of, 488–489
Peroxisome proliferator-activated receptor (PPAR), 488, 490–491, 490f
  versions of, 491
Peroxisome proliferators, 488–491, 489f, 490f
Peroxy free radicals, 843

Peroxy-BH4, 522
Peroxyhydrobiopterin, 522
Peroxypterin iron, 522
PEST sequence, 82
PG. See Prostaglandin(s)
PG 2. See Prostaglandin J2
PGA1, 491
PGD$_2$-G, 978f
PGE$_2$, 955, 956f
PGE$_2$ receptor, 957f
PGES. See Prostaglandin E$_2$ synthase
PGF$_{2\alpha}$G, 978f
PGF$_2$G, 978f
PGH$_2$-G, 978f
PGI$_2$-G. See Prostacyclin glycerol
PGIS. See Prostaglandin synthase
PGs. See Prostaglandins
pH, 22, 30–32
   buffering system, 871
   gastric lumen, 435
   Henderson-Hasselbalch equation for, 48
   titration of glycine with, 48f
Phage(s), 1089, 1092. See also Bacteriophage(s)
   DNA, 1103
   dormant, 1091
   infection
      E. coli, 1091, 1092, 1095, 1101, 1103
      sequence of, 1101, 1103
   lysogenic cycle of, 1106
   shape of, 1092, 1095
   T, 1089, 1090f, 1091, 1093t, 1094f
Phage λ, 1101f
   integrase, 1105f
   life cycle of, 1092f
   protein gpD, 1104f
Phage λ receptor, 1107f
Phagocytes
   chemotaxis of, 600–601, 951
   FC receptors on, 916f
   lymphatic system and, 894
   mature, 880, 881f
   mononuclear, 632, 637f
   opsonization of, 917
Phagocytosis, 695, 764, 937f
   pinocytosis compared with, 695t, 696f
Phencyclidine (PCP), 991
   administration of, 995
   structure of, 992f
Phenobarbital, 996

Phenol, pK values for, 47t
Phenotype. See Blood type(s)
Phenylalanine, 770
   metabolite conversion of, 527f
   serum concentration of, 523
   transamination to phenylpyruvate of, 524, 527f, 531
   tyrosine and, 521–524, 523f, 531f
Phenylalanine decarboxylase, 516t
Phenylalanine hydroxylase (PAH), 522, 523f
   deficiency of, 527f
   tetrameric, 522, 524f
Phenylalanine hydroxylation, 521, 523f
Phenylethanolamine N-methyl transferase (PNMT), 448, 451f, 526
Phenylketonuria (PKU), 54, 527f, 531, 531f
   in children, 523, 527f
   classic, 523
Phenylpyruvate, phenylalanine transanimated to, 524, 527f, 531
Pheomelanin, 527
PhoE porin trimer, 1126, 1129f
Phosducin, 838f
Phosphatase, 551
   acid/alkaline, 93
   glucose-6-, 161, 161f, 169, 171t, 288t
   phosphoprotein, 1039, 1041
   protein, 164, 166f, 556f
      CFTR with, 688
   protein tyrosine, 723, 899
   serine conversion of, 545, 545f
Phosphate, 1126
Phosphate-dependent glutaminase, 500f
   kidney, 499
Phosphatidic acid, 222f, 975
   phospholipase D releasing, 981f
Phosphatidylcholine, 220, 221f, 518. See also Lecithin
Phosphatidylethanolamine (PE), 222f, 977f, 981f
Phosphatidylglycerol, 222f
Phosphatidylinositol, 222f, 609
   lipid, 222f, 609
   lipid metabolism, 725f
   muscarinic receptors with, 1017

Phosphatidylinositol 4,5-bisphosphate (PIP$_2$), 716f
Phosphatidylinositol (3,4,5)-triphosphate (PIP$_3$), 713f
Phosphatidylinositol-3-kinase (PI3K), 405, 405f, 645, 653f, 654, 661f, 670f, 713f
Phosphatidylserine, 222f
Phosphocholine (PC), 951, 952f
Phosphocholine-binding site, 951
Phosphocreatine, 536, 538f, 540f
Phosphodiesterase, 540, 837
5-phospho-D-ribosyl-1-pyrophosphate (PRPP) conversion, 252, 252f
Phosphoglucomutase, 156
   chains with glucose-1,6-bisphosphate, 160f
   mechanism, 159f
6-phosphogluconate dehydrogenase, 282f, 284f
6-phosphogluconolactonase, 282f, 284f
3-phosphoglycerate, 545, 545f
Phosphoinositol-bis phosphate (PIP$_2$), 709
Phospholipase A$_2$ (PLA$_2$), 467, 521f
   anandamide with, 977f
   arachidonic acid with, 965, 975
   cell membrane interaction with, 470, 470f
   structure of, 469f
Phospholipase C (PLC), 420, 626. See also PLC-1
   gamma pathway, signal transduction through, 624f
   hydrolysis, 418
   pathway of, DAG production through, 707, 709
   stimulation of, 622, 1028
Phospholipase C gamma pathway, signal transduction through, 624f
Phospholipase D, 881
   anandamide from, 975, 977f
   phosphatidic acid from, 981f
Phospholipase-β, 962
Phospholipases, 554
Phospholipid turnover, 430
Phospholipids, 220–226, 221f–225f, 226t
   glycosphingolipids as, 220, 222f, 223f, 226–232, 227f–233f

INDEX 1233

Phospholipids *(Continued)*
  sphingomyelin as, 220–221, 222f
  substrate digestion of, by PLA_2, 470f
  triacylglycerides v., 220
Phosphonate-inhibited D-Ala, 1114f
Phosphopantetheine, 214, 214f
Phosphoprotein phosphatase, 1039, 1041
Phosphoribosyl aminoimidazole carboxamide formyltransferase (AICAR transformylase), 256f
Phosphoribosyl aminoimidazole-succinocarboxamide synthase (SAICAR synthase), 255f
  structure of, 260f
Phosphoribosyl formylglycinamidine synthase (FGAM synthase), 254f
  structure of, 259f
Phosphoribosyl glycinamide formyltransferase (GAR transformylase), 254f
Phosphoribosyl-aminoimidazole carboxylase (AIR carboxylase), 254f
  crystal structure of, 260f
Phosphoribosyl-aminoimidazole synthase (AIR synthase), 254f
Phosphoribosylation, 271
Phosphoribosyl-pyrophosphate synthetase, crystal structure of, 257f
Phosphorus, calcium and, 531, 837
Phosphorylase, 156, 159f, 163, 166f, 169, 169f, 170f, 171t
  glycogen storage diseases affecting, 171t
  structure of, 159f
Phosphorylation, 478
  ACC inhibited by, 554
  cAMP-dependent, 689f
  estrogen receptor, 486, 486f
  I_kB, 326, 658
  PDGFRα, 603f
  PKA sites for, 564f
  progesterone, 478–479, 482f
  protein kinase A, 1027, 1033
  protein kinase C, 1027
  selenide, 776
  STAT, 661f, 988f
  3-phosphoserine, 545, 545f

Photophobia, 808
Phototherapy, 808
Phytic acid, 740, 741f
Phytyl side chain, 841
Phytylmenaquinone, 843, 844f
PI3K. *See* Phosphatidy-linositol-3-kinase
PIC. *See* Preinitiation complex
Pica, 740
PIF. *See* Prolactin release-inhibiting factor
Pili, 1110f
Pilocarpine, 685, 686f
Pineal body, 1019, 1021
Pineal gland, 532, 532f, 533f
  functions of, 535f
  neural connections between eyes and, 533
Pineal methoxyindoles biosynthesis, 537f
Pinealocytes, 532f, 533
Pinocytosis, 695, 695t, 696f
  phagocytosis compared with, 695t, 696f
PIP_2. *See* Phosphatidylinositol 4,5-bisphosphate
PIP_2 hydrolysis, 438
PIP_3. *See* Phosphatidylinositol (3,4,5)-triphosphate
Pituitary. *See also* Anterior pituitary cells
  pars intermedia, 352, 366f, 466
    PILC (*pars intermedia*-like), 368, 378f
  hormones, 378, 380
  human, 366f
Pituitary stalk, 367f
  human, 366f
pK values, ionizable groups in protein, 47f, 47t
PKA phosphorylation sites, 564f
PKC. *See* Protein kinase C
PKU. *See* Phenylketonuria
PLA_2. *See* Phospholipase A_2
Placenta, 413
Plasma, antibodies in, blood groups with, 890t
Plasma cells, antibody derivation from, 890t, 910, 911, 914
Plasma membrane. *See* Cell membrane
Plasma osmolarity (Na^+concentration), 430, 693f

Plasma volume. *See* Hypovolemia
Plasmalogen, 222f
Plasmid, foreign DNA introduced into, 305, 309f
Plasmid therapy, gene therapy with, 315t
Plasminogen, 867, 868f, 869
  catalytic domain of, 867f
Plasticizers, 488
Platelet-derived growth factor (PDGF), 600–601, 602f, 603f, 653f
Platelets, 600, 879, 883, 884, 898
  aggregation of, 949
  diameter of, 882
  fibrin formation on, 865f
  megakaryocyte transition to, 630f
PLC. *See* Phospholipase C
PLC-1, 713f
PLC-δ, 905f
Pleural mesothelial cells, 327
PLK AMP-PCP-pyridoxamine complex, substrates of, 821f
PLP. *See* Pyridoxal phosphate; Pyridoxine 5′-phosphate
Plummer-Vinson syndrome, 740
Pluripotent stem cells, 879–880, 880f
PMCA1b. *See* Calcium-ATPase
PMP. *See* Pyridoxamine phosphate
PMP oxidase, 818
Pneumococcal cell wall, 951
PNMT. *See* Phenylethanolamine N-methyl transferase
Point mutation, 239, 465–466
Pol β. *See* DNA polymerase β
Pol II. *See* DNA polymerase II
Pol III. *See* DNA polymerase III
POL polypeptide, HIV, 1062f
Polarized cell membrane, 961
Polio virus, 1071f
Polyamine(s), 991
  biosynthesis of, 543f
  structure of, 541, 542f
Polyamine pathway, 541, 542f
Polyclonal antibodies, 915
Polycyclic phencyclidine, 991
Polydeoxyribonucleotide synthase. *See* DNA ligase
Polyglutamate, 828–829, 829f, 830
Polyglutamine, expansions of, 240

Polyglutamine tracts, 240
Polymorphonuclear leukocytes, 880, 881f
Polypeptide hormone chain, *E. coli*, 1122f
Polypeptide hormone receptors, 486
Polypeptide hormones, 365–444, 367f. *See also* Glucagon; Substance P
  actions of releasing, 367f, 370t, 371t, 372t, 373t, 387–406, 391f
  anterior pituitary, 368, 373t–377t, 378, 379f
  GAG-POL, 1056
  gastrointestinal, 435–444, 436t
  gonadotropins, 406–415
  half-life of, 430
  humoral mechanism of, 367–378
  MHC encoded, 675f
  neurofascin, 730f
  panhypopituitarism, 365–367
  posterior pituitary, 380–385, 381f, 382f, 383f, 384f, 385f, 386f
  PRL, 368, 419, 430–435
  prolactin, 430–435
  thyrotropin, 417–428
Polypeptide trace, TG-FαisEGFR50 complex, 617f
Polyprotein(s)
  cleavage of viral, 1078
  ENV, 1062f
  Gag, 1056
  POL, 1062f
Polysaccharides, *E. coli*, 1127f, 1130, 1131f
Polysomes, ribosomes grouped as, 65
Polyunsaturated fats, 200f, 218
Polyunsaturated fatty acids, 200f, 218
Pompe's disease, 171t
Pore
  formation of, 902, 904f
  invading pathogenic cell, 934f
Porin(s), *E. coli*, 1124, 1125f–1127f, 1126
  number per cell of, 1125
  proteins in, 1125
  slow, 1123f
  sucrose-specific, 1125, 1129f
  trimers for, 1129f

Porin Omp32, electrostatic potential of, 1124f
Porin OmpA-PptF, family, 1123f
Porin OmpF, 1125, 1125f, 1126f, 1127f
  mutant D744, 1123f
Porin ScrY, 1125, 1126, 1129f
Pork, 776
Porphobilinogen, 753
Porphyrias, 756, 758t
Porphyrin synthesis, 756
Porphyrins, 756, 760, 879
Posterior pituitary, 380–385, 381f, 382f, 383f, 384f, 385f, 386f
Posttranscriptional gene silencing (PTGS), 311
Posttranslational modifications, 236
Posttranslational palmitoylation, 540–541
POT. *See* Proton-dependent oligopeptide transporters
Potassium
  aldosterone synthesis and, 458
  channels
    adenylate cyclase as opening, 446
    PDZ domains with, 992
Potassium ion, 984, 986f, 1023f
  channels, 731, 970, 1017
PPA. *See* Preproenkephalin A
PPAR. *See* Peroxisome proliferator-activated receptor
PPARα, 342
  crystal structures of human, 343f
PPARγ
  crystal structure of human, 489, 490f
  ligand-binding domain of, 489, 490f
Pravastatin, HMG-CoA reductase inhibition with, 195f
Pre-A-NK cells, 903
Precipitin, 911
Precursor(s)
  catecholamine, 521, 524
  chymotrypsinogen, 441
  creatinine/creatine, 536
  HIV, gag/pol/env genes as, 1054f
  IL, 675
  melanin/metabolite, 521–522

Precursor(s) *(Continued)*
  molybdopterin formation, 788, 791f
  procaspase, 493f
  protein C, 857, 859f
  T cell, 896
  trypsinogen, 441
Precursor Z, 788, 791f
Prediabetes, 131
Pregnancy
  autoimmune diseases during, 920
  dietary metal intake during, zinc, 778
  hypopituitarism and, 367
  iron loss during, 739
  milk synthesis during, 430
  progesterone in, agonistic, 465–466
Pregnanediol, 568
Δ5-pregnenolone, 452
Preinitiation complex (PIC), 330, 331f
Preintegration complex, HIV viral DNA, 1056
Premature gray hair, 818
Preponderant circulating form, 418
Preproenkephalin A (PPA), 1031, 1033f
Preprogastrin, 438f
PreproGIP, 443–444
Preproinsulin, 138, 139f
Preproopiomelanocortin, 368
  precursor protein, 378f
Preprosomatostatin, 398, 401f
Preprotachykinin, 962
Preprotein, 78
Preprotein convertase enzymes, 173
Preprothrombin, 846
Pre-RNA, 320
Presynaptic neurons, 968
PRF. *See* Prolactin-releasing factor
Primase, 294, 296f
Primer synthesis, 1094f
Primosome, 294
Prion disease, 33–35, 34f, 734
  Alpers syndrome, 33, 35
  blood transfusions with, 33
  bovine form of, 33
  corneal transplants with, 35
  Creutzfeldt-Jacob disease, 33, 35

Prion disease (Continued)
   familial, 242
   genetic (autosomal) mode of transmission of, 35
   Gerstmann-Sträussler-Scheinker syndrome, 33
   growth hormone injections with, 35
   kuru, 33
   meat ingestion causing, 33
   Peyer's patches with, 33, 34f
PRL. See Prolactin
Proapoptotic Bid, 591–592, 593f
Proapoptotic factor Bax, 224
Procaspase precursor, 493f
Procaspase-3 activation, 592
Procaspase-8, 902f
Procaspase-9, 592, 902f
Procollagen, collagen maturation process from, 768, 768f, 769f
Proenkephalin A (PPA), 1031
Proerythroblasts production, 641f
Progesterone, 430
   as agonist, 465–466
   FAAH activated by, 984
   inactivation of, 568
   ovarian cycle levels of LH and, 413, 415f
   phosphorylation of, 478–479, 482f
Progesterone receptor, 478–479, 482f
Programmed cell death. See Apoptosis
Progressive spastic quadriplegia, 503t
Proinflammatory cytokine(s), 654
Proinsulin, 644, 648f
   mini-, 649f
Prokaryotic ribosomes, 67, 67f
Prolactin (PRL), 368, 419, 430–435
   dopamine inhibiting, 432
   melatonin increasing, 535
   secretion of, 433f
   signal transduction, 431f
   suckling response release of, 383f
   TRH releasing, 430, 432
Prolactin receptor, 430
   x-ray structure of, 432f
Prolactin release-inhibiting factor (PIF), 368, 432

Prolactin-releasing factor (PRF), 368, 372t, 432
Prolactin-releasing hormone, 430
Proliferator-activated receptor-α peroxisome, 982
Proline, 40f, 44t, 536, 538f, 549, 550f, 1003
   α-helix structure disrupted with, 58, 81
   hydroxylation, 832, 833f
   ornithine conversion to, 541, 545
   transporters, 721
Proline oxidase, 549
Proline-4-hydroxylase, 833f
Prometaphase, 14
Promoter(s), 329
   bacteriophage, 1097
   CRH, 387
      regulation of, 390f
   eukaryotic, basal, 329f
   FAAH, 984, 988f
   ferrochelatase, 758
   HRE on, 341
   LHβ, repression of, 412f
   LTRs as, 1052
Properdin, 938f, 949, 950f
Prophage, 1091
Prophase, 14
β-propiolactone, 301f
Propionic acid, 499
Propionyl CoA, 518, 550
Propionyl CoA carboxylase, 550, 820
Proposed stoichiometry, 460f
Propreproopiomelanocortin, 378f
Proproteinase E, 858f
Propylthiouracil (PTU), 797f
Prostacyclin, 978
Prostacyclin glycerol (PGI$_2$-G), 978f
Prostacyclin H$_2$-glycerol (PGH$_2$-G), 978f
Prostaglandin(s), 489–490, 491, 955, 956f, 978f
   arachidonic acid as substrate for, 201
   cortisol and, 470
   cox1 responsible for, 120
   ethanolamides of, 977f
   pain signals of, 955, 956f, 957
   PLA$_2$ as substrate for, 467
   receptors for, 957f
   relatives of, 468f

Prostaglandin D synthase, 978f
Prostaglandin D$_2$-glycerol (PGD$_2$-G), 978f
Prostaglandin E$_2$ synthase (PGES), 978f
Prostaglandin F$_{2\alpha}$ glycerol (PGF$_2$G), 978f
Prostaglandin F$_{2\alpha}$ synthase (PGF$_{2\alpha}$G), 978f
Prostaglandin J2 (PG)2, 491
Prostaglandin synthase (PGIS), 978f
ProstaglandinE$_2$-glycerol (PGE$_2$G), 978f
Prostaglandins (PGs), 468f
Prostate cancer, 486
Prostate cancer cells (LNCaP), 788f
Prostate gland, acid phosphatase with, 93
Prosthetic groups, 117f–125f, 119–125
   coenzymes v., 117f–118f, 119–120
   defined, 119
   metalloenzymes, 120–125, 120f–125f
      cox1 as, 120, 121f
      cox2 as, 120–123, 121f–122f
      cyclooxygenase as, 120–123, 121f–122f
      SOD as, 122f, 123–125, 124f
Protachykinin, 962
Protease(s)
   aspartate-specific cysteine, 492
   in blood clotting mechanism, 864
   C3, 935
   complement system, 933–934
   coronavirus main, 1089f
   HIV, 1063f
      AIDS treatment based on, 1067
      inhibitor of, 1059, 1064f
      mechanism of, 1062f
   lysosomal, 916f
   pancreas, 439–441
   serine, 857, 858f, 864, 864f
      inhibitor of, 864f
Proteasome, 184, 654
   26S, 83, 84f
Protein(s), 33–90. See also Binding proteins; Glycoprotein(s); Growth factor(s); Lipoproteins; Protein kinase(s); specific proteins

Protein(s) (Continued)
  aggregation of, type 2 diabetes, 170–173, 171f–173f, 172t
  amino acid, 39–54, 40f–43f, 44t, 45f–53f, 47t
    sequences of, 56–57, 57f
  amino acids moved to ribosomes for, 67–73, 71f–75f
    A-site with, 68, 70, 74f
    E-site with, 70, 74f
    P-site with, 68, 70, 74f
  antiviral, 642–643
  ATP-synthesizing, 780
  bacteriophage, 1095–1096, 1102f, 1104f
  BAX, 224, 493, 494f, 902f, 995, 995f
  blood, 867, 884, 886f, 887f, 888
  blood group, 185, 187f
  bone morphogenic, 612
  calcium-transporting, 530–531
  capsid-stabilizing, 1104f
  catalytic, 96–97, 97f, 859f
  chaperone/chaperone-like, 80, 760
  chaperonin, 81
  chylomicrons with, 202, 203f
  classification of, 86, 86f–89f
    CATH, 86, 88f–89f
    collagen structure in, 86, 87f
    fibrous, 86
    globular, 58, 86, 86f
    solubility in, 86, 87f
  coding of, 857
    mRNA, 54, 54f
  complement system of, 934f, 935
  complexity, four levels in, 56, 63f
  C-reactive, 951–953, 952f
  degradation of, 81–83, 82f–84f
    ligase in, 82, 83f
    proteasome in, 184
    ubiquitin in, 81, 82, 83f
  denatured, 61
  elongation factor for, 70, 74–77, 78f, 79f
    EF-1α-GTP as, 74–77, 78f, 79f
    EF-2-GTP as, 74–77, 78f, 79f
  eNOS surrounding, 734, 734f
  envelope, 1076
    HIV, 1053f
  Fas-associated, 902f
  fibrous, 86

Protein(s) (Continued)
  flavo-, 514, 514f, 515f, 806
  folding, 81
  G, 430, 432, 677, 713f, 1021
    activation of, 1034f
    inhibited release of, 968
    inhibitory, 436
  G protein-coupled receptor, 408, 438, 957f, 970
  globular class of, 58, 86, 86f
  glycation of, 876f
  glycophosphatidylinositol anchoring for, 238, 723
  heat shock, 80–81, 340, 349, 478, 1006
  histone-like, 1131–1134, 1134f
  HIV, 1048, 1049f, 1050–1052, 1050f, 1053f, 1054f
  inhibitors to synthesis of, 83–85, 85f
  inhibitory, 612, 902f
  initiation factor for, 68, 73–74, 75f–77f
    eIF-4F as, 74, 75f–77f
    eukaryotic initiation factor-2 as, 73–74, 75f–77f
    phosphorylation with, 74, 76f
  INR-binding, 335
  kinase, 486, 658
  milk, 430
  myeloma, 888
  neocapsid, 1085, 1087, 1087f, 1088f
  nuclear, CBP, 326, 342, 344, 346f
  nucleocapsid, 1050
  pancreas proteases, 439–441
  peptide bond with, 54, 54f
  plasma, CRP, 951–953, 952f
  poly-
    ENV, 1062f
    GAG, 1056, 1062f
    POL, 1056
    viral, 1078
  posttranslational modifications of, 236
  preproopiomelanocortin precursor, 378f
  preprosomatostatin, 398
  primary structure of, 56–57, 57f, 63f
  prion, 35–54, 36f–43f, 44t, 45f–53f, 47t

Protein(s) (Continued)
  prion disease with, 33–35, 34f, 242
  repeating, 1069f
  ribosomes in structuring, 63–67, 65f–71f, 65t
  scaffold, shank, 992
  scaffolding, 604, 1095–1096, 1097, 1100f
  secondary structure of, 57–62, 58f–63f
    disulfide bond in, 60, 60f
    electrostatic bonding in, 59, 60f
    -helical form of, 58, 58f
    hydrogen bonding in, 58, 59f, 60f
    β-sheet form of, 58, 58f
  streptococcal M, 946, 948f
  sugars of
    N-linked, 183, 183f, 184f
    O-linked, 183, 183f, 184f
  surfactant, 466, 467f
  synapse-associated, 995, 997f
  synthesis, 62–63, 64f
  termination factor for, 68, 77, 80f
    release factor in, 77
  transfer to mitochondrion of completed, 78–81
  transmembrane, 970
  viral, 638, 1051
Protein C, 853, 856, 857, 857t. See also Activated protein C
  catalytic domains of, 859f
  cleavage of, 884
  gene encoding for, 857
  gla domain of, 857
  structure of human, 858f
Protein F, 1096
Protein J, 1095–1096, 1097f
Protein kinase(s), 486. See also specific kinases
  activity of, 565f
  JAK, 661f
  receptors, 600
  serines in sequence of, 486f, 612–613
Protein kinase A, 105, 106f, 163, 166f, 168f, 390, 411f, 561
  aquaporin mechanism with, 698, 699f
  calcium ion channels opened by, 411–413, 1001

Protein kinase A *(Continued)*
  cyclic AMP activation of, 410, 411f, 419, 432
  cyclic AMP-dependent, 688
  gene expression as culmination of, 973f
  IP$_3$/DAG activation of, 1021
  phosphorylation by, 554, 1027, 1033
  sodium ion channels opened by, 970, 1001
Protein kinase C (PKC), 163, 167, 168f, 418, 420, 430, 438, 609
  activation of
    DAG, 710, 713f
      phospholipase C stimulation of, 622, 626
    aldosterone circulation and, 452, 454
    DAG with, 965
  phosphorylation site for, 1027
Protein nutrition, 849–850, 851t
Protein phosphatase(s)
  CFTR with, 688
  triglyceride acted on by, 556f
Protein phosphatase-1, 167, 168f
  insulin stimulation of, 164
  tumor promoter okadaic acid bound to, 166f
Protein phosphotyrosines, 723
Protein S, 853, 856, 856f, 857, 857t
  modular arrangement of, 862f
  SHBG-like region of, 862f
Protein sequence analysis (PSA), 61f
Protein tyrosine kinase
  activities, 613
  -dependent pathways, 899
Protein tyrosine kinase receptors. *See* PDGFRα
Protein tyrosine, phosphatases of, 723, 899
Proteinaceous channel, facilitated diffusion, 698
Proteoglycans, 188
Proteolytic enzymes, renin, 452, 455f
Proteolytic processing, 492
Proteome, 90, 90f, 91f, 313
Proteomics, 591
Prothrombin (factor II), 846, 867
  conversion to thrombin of, 884, 888
  time determinations, 854

Prothrombinase, 858
Protoheme IX, 756
Proton(s) (H$^+$)
  antiport pf Sodium-, 729
  membrane transport of, 714–719, 715f–719f
  Na$^+$ antiport and, 714, 715f
  cytoskeletal network with, 716f
  in oxygen release, 878
Proton-dependent dipeptide transporter (PEPT1), 718–719
Proton-dependent oligopeptide transporters (POT), 719
Proton-linked channels, MCT example of, 715, 717f
Protoporphyrinogen IX, 753, 753f
Protoporphyrinogen IX oxidase, 756f
Proximal box (P-box), 461
Proximal zinc finger, 463, 464f
Prozac, 533, 1022, 1027
  structure of, 1024f
PrP^c. *See* Cellular prion protein
PRPP. *See* 5-phospho-D-ribosyl-1-pyrophosphate
PRPP synthetase, purine metabolism disorder with, 288t
PrP^{Sc}
  conversion diagram for, 38f
  propagation in cell of, 35–54, 36f–43f, 44t, 45f–53f, 47t
  three-dimensional structure of, 37f
PSA. *See* Protein sequence analysis
PSD-95 group, 995, 997f
Pseudoknot, 317, 318, 318f
Pseudopodia, 695, 696f
P-site. *See* Peptidyl site
Pteridine ring, 828, 829f
Pterins, molybdenum as form of, 788
Pteroyl-γ-L-glutamic acid structure, 829f
PTGS. *See* Posttranscriptional gene silencing
PTH. *See* Parathyroid hormone
PTU. *See* Propylthiouracil
Puberty, 486
Pulmonary artery, 871, 872f
Pulmonary embolism, 853, 854, 855f, 856
Pulmonary fibrosis, plural thickening in, 325*f*

Pulmonary vein, 871, 872f
Pulsatile pattern, GH secretion, 395
Purine(s)
  biosynthesis, 252–268, 252f–269f
    regulation of, 252, 263f
  interconversions, 268–270, 270f, 272f–273f
    IMP to AMP, 268–270, 270f, 272f
    IMP to GMP, 268–270, 270f, 273f
  metabolism disorders of, 283, 288t
  nucleotide catabolism of, 271, 274f
  nucleotide cycle of, 275f
Purine nucleotide cycle, 275f
Purine nucleotide phosphorylase, 271, 274f
Purine ring, carbon/nitrogen sources of, 252, 252f
Purines synthetic pathway regulation, 263f
Purines/pyrimidines, 244–288, 245f–270f, 272f–287f, 288t
  adenine as, 244, 247f, 248f, 250f, 271, 274f
  base-pairing with, 249–251, 250f, 251f
  biosynthesis of, 252–268, 252f–269f
  cytosine as, 244, 247f, 248f, 250f, 268, 297f
  disorders of, 283, 288t
  guanine as, 244, 247f, 248f, 250f, 271, 274f
  hypoxanthine as, 245, 248f, 249f, 271, 274f, 287f
  models of, 247f
  not occurring in nucleic acids, 249f
  nucleic acids as, 244–288, 245f–270f, 272f–287f, 288t
  nucleotide catabolism, 271–282, 274f–287f
  numbering of atoms in rings of, 246f
  orotic acid as, 245, 248f, 249f
  structures of, 245f
  thymine, 244, 247f, 248f
  uracil as, 244, 247f, 248f, 271, 297f
  xanthine as, 245, 248f, 249f, 271, 274f

Puromycin, 85, 85f
Putamen, 1032f
Putrescine, 536, 538f, 541
Pyk2, 670f
Pyridoxal, 818
Pyridoxal kinase, 818, 820f
  catalysis, 822f
Pyridoxal phosphate (PLP), 51, 52f, 112–118, 113f, 119t, 508, 509, 509f, 511, 511f, 818
  in amino acid decarboxylation, 516, 517f
  cystathionine synthase containing, 518, 520
  in serine deamination, 513, 513f
  structure of, 548f
  vitamin B_6 form, 818
    coenzyme conversion from, 818, 820f, 822f
Pyridoxamine (vitamin B_6), 818, 820f, 821f
  deficiency of, 1004
Pyridoxamine kinase, 818, 820f
Pyridoxamine phosphate (PMP), 508, 510f, 511, 511f, 818, 820f
Pyridoxine 5'-phosphate (PLP), 818, 820f, 822f
Pyridoxine kinase, 818, 822f
Pyridoxine phosphate, 818, 820f
Pyridoxine (PN) [Vitamin B_6], 113f, 119t, 511f, 818, 820f
  as cystathionine synthase treatment, 520
  deficiency in, 518, 1004
Pyrimidine. See also Purines/pyrimidines
  E. coli synthesis of, 1139–1140, 1142f
  metabolism disorders of, 283, 288t
  nucleotide catabolism, 271, 277f
  structure of, derivation carbon/nitrogen in, 253, 263f
  synthesis of, 253
    carbamoyl phosphate in, 265f
Pyrogenic exotoxins (SPE), 948f
Pyroglutamate, 371t
Pyroglutamyl peptidase II, 430
Pyroglutamyl-histidine bond, 430, 542f
Pyrroline 5-carboxylate, 549

Pyruvate, 513f, 546, 547f, 549, 561
  carbon dioxide with, 879
  cofactor, 508
Pyruvate carboxylase, 500f
  active site cleft of, 821, 825f
  biotin carboxylase subunit of, 824f
  tetrahedron-like arrangement of subunits of, 825f
Pyruvate dehydrogenase, 175f, 176, 176f, 801, 802f, 803f
  E. coli, 1138, 1138f
Pyruvate kinase, 174, 174f
Pyruvic acid, 18, 20f

QFQ4 opioid receptor, 968
Quaternary structure, 61, 63f
Quinolinic acid, 990f
Quisqualic acid, 990f

R domain, 689f
R form hemoglobin, 874
RA. See Retinoic acid
Rab4-mediated recycling pathways, 982f
Rab5 pathway, 981
Racemase, 46, 509–510
  alanine conversion reaction of, 1118f
Radioactivity, DNA mutations/damage from, 298–299
Raf-1, 604, 626
Raft aggregation, 725f
Ranitidine (Zantac), 542
RANKL. See NF-kB ligand
Ranvier, nodes of, 731, 731f
RAR. See Retinoic acid receptor
Ras, 236–238, 237f, 238f, 609, 637f
  CaaX motif of, 238
  GAP regulation of, 713f
  signaling of, 626, 650
Ras-Raf-1, 606f
RCA. See Complement activation
RCA proteins, microorganism interactions with, 943t
RDA. See Recommended daily allowance
Reactive oxygen species, 326, 740, 995

Receptor(s), 7. See also Baroreceptor
  acetylcholine, 448, 450f, 923f, 1004f, 1011, 1013f, 1016f
    muscarinic, 1017, 1017f
    nicotinic, 1013f, 1014, 1014f, 1016f
    nicotinicoid superfamily of, 1003, 1019
  adrenergic, 163, 165f, 167, 168f, 448, 451t, 1031–1032, 1034f, 1036f, 1038f, 1041, 1042f, 1043f
  agouti, 418
  androgen, 486, 487f, 488f
  angiotensin, 447, 452, 454, 456f
  antibody, 896f, 916f, 927, 943
  arionatriuretic, 455f
  βARK kinase of, 1033, 1039
  bradykinin/PGE2, 957f
  capsaicin, 961, 961f, 982–983, 984f, 985f
  cellular locations of, 7, 26–27, 27f
  class 1, 349
  clustering of, 1007
  colony-stimulating factor, 601, 629, 632f, 634f, 637f
  conformation of, 344, 483–488
  cytoplasmic, 339f, 340, 348f
  death, 591
  decoy, 591, 593f, 594
  desensitization of, 1033
  docking sites for, 562
  dopamine, 432, 1028, 1029f–1031f, 1031
  E. coli BtuB, 1139, 1141f
  EAA, 500, 988, 989t, 990–991, 990f, 992, 995, 999
  Fas, 902f, 904f, 924, 925f
  ferric citrate, 1125, 1129f
  folic acid intestinal, 829
  G protein, 677
  G protein-coupled, 408, 438, 957f, 970, 980, 980f, 981f, 990, 991f, 1028
  GABA, 998–999
  gastrin/CCK_B, 438
  ghrelin, 398f
  GLP-1, 559
  glucocorticoid, 349–356, 353f, 354f, 355f, 462f, 470, 474
    homodimeric, 464f

INDEX 1239

Receptor(s) (Continued)
  glutamate, 559, 967f, 990
  glycerine, 1007
  glycine, 1003
  growth factor, 593f, 601, 602f, 603f, 604, 607f, 609, 612f, 613, 620t, 621f, 622, 710, 713f
    insulin-like, 645, 653f
    NGF, 625
    PDGF β, 602f
    trkA, 624, 625, 627f
    TSH, 419
  histamine, 436, 437t, 542, 543f
  hormone, 374t, 380, 384f, 385f, 410, 413f
    activation mechanism for, 478, 481f
    estrogen/estrogenic, 349, 478, 485f, 566
    growth, 395, 397f, 402, 404f, 405
    human CRH, 389f
    leuteinizing hCG, 407f
    melatonin, 533, 535, 536f
    nuclear, 198, 481f
    oxytocin, human, 384f, 386f
    polypeptide, 430, 432f, 486
    progesterone, 478–479, 482f
    serotonin, 1003, 1018f, 1019, 1020f, 1021, 1021t, 1022f, 1024f, 1025f
    steroid, 490–491, 490f
    testosterone, 478
    thyroid, 417f, 419, 430, 800
    vasopressin, 380, 384f, 385f
  immunoglobulin-like, 903
  Inositol triphosphate, 707, 708f, 709, 711, 713f, 970, 992, 1021
  insulin, 133–138, 134f, 136f–138f
  interferon, 639, 647f, 654
  interleukin, 632f, 639f, 654, 661f–663f, 665–666, 670f, 669, 677, 680, 680f, 903
  ion channels associated with, 1014
  KIR, 903, 906f
  LDL, 189–191, 190f–192f, 1074f
  ligand/liganded, 478
    vanilloid, 961, 961f, 981f, 982, 984f

Receptor(s) (Continued)
  liver X, 198
  -mediated endocytosis, 695
  MR, 458, 458f, 461f, 462f, 463, 478
  neuro-, 788, 794f
  NK cell, 909f
  nociceptin, 446, 446f
  nuclear
    bound with corepressor, 348f
    superfamily classes of, 351f, 488
  oligomeric, 639, 900f
  opiate, 969f
  opioid, 446, 968, 969–970, 969f, 970t, 971f, 972f, 973f
  orphan, 488–491, 491f
  phosphorylations of, 478
  protein kinase, 600
  RAR, 339, 530
  retinoid, 198, 349
  scavenger, 233
  seven membrane, 408, 438, 558f
  somatostatin, 398, 400f
  steroid, 350f, 461, 461f
  substance P, 964f, 965f
  T cell, 420, 425f, 929, 929f, 930f, 948f
  tachykinin, 962
  toll-like, 654
  trafficking of, 981, 982f
  transferrin, 747, 752f, 753, 753f
  tumor necrosis factor, 926f, 927f
  unliganded forms of, 478–482
  V1a, 385f
  virus CD46, ICAM-1, CD4, 947f
  vitamin D, 527, 530f, 784, 785f
  α_2 receptor, 1032
  β-receptor
    agonist levels of, 1033
    desensitization of, 1040f
  δ-receptors, 969, 971f
    structure of, 972f
  κ-receptors, 969–970, 972f, 973f
  μ-receptors, 969, 970t, 971f
Receptor-receptor interaction, 604
Recognition sites, 302f, 306f
Recombinant DNA, 305, 309f

Recommended daily allowance (RDA). See also Intake
  folic acid, 830
  iron, 742t
  magnesium, 782, 782t
  micronutrients, 850t
  molybdenum, 795t
  niacin, 812
Recoverin, 1037f
Red blood cell(s) (erythrocytes), 637, 871, 874, 879, 880f, 893. See also Cell membrane; Erythro-BH4; Erythropoiesis
  appearance of, 880f
  carbon dioxide reactions in, 879f
  CBC for, 885t–886t
  colony-forming units of, 636f
  creatine kinase with, 95
  deglycosylation mechanism in, 877
  diameter of, 880
  EPO increase of, 644f
  erythroid cells differentiating into mature, 760
  hemoglobin formation in, 743, 760, 762, 763f
  immature, hemoglobin production in, 743, 760, 764
  life span of, 739–740, 760
  macromolecules on surface of, 889
  normal, hemoglobin in, 762
  reactions occurring in, 879f
  vitamin E deficiency and, 843
Red blood cell count
  CBC for, 883, 885t–886t
  iron deficiency reduction of, 739
  normalizing system for, 641f
Regulator of virion (REV), 1050
RelA, 323, 325f
Relaxation response, 445
Release factor, 77
Relenza, 1076, 1077f
Renal artery, 871, 872f
Renal lithiasis, purine metabolism disorder with, 288t
renal tubular cell, 458f, 459f
Renin
  enzyme, 452, 455f
  secretion of, 457–458
  structure of, 455f

Renin-angiotensin system, 452, 454f, 457–458
Renzapride, 1021t
Repressor genes
　cI, 1106f
　　E. coli with, 1101, 1103f
　cII, 1101, 1106, 1106f
　　E. coli with, 1101
　lac, 1135–1139, 1139f
Reproductive system, 4
RER. See Rough endoplasmic reticulum
Resinferatoxin (RTX), model of capsaicin and, 985f
Resorption
　bone, 840
　of neutral/cationic AAs, 817f
Respiratory chain, 177, 178f, 180f, 770, 772f
Respiratory failure, urea cycle disorder, 497
Respiratory system, 3, 4f
Responsive element, 490
　estrogen, 485f, 486
Restriction endonucleases, 300, 302f. See also Restriction enzymes
　action of E. coli with, 302f
Restriction enzymes, 300–305, 302f–305f, 306t, 309f
　blunt-ended products with, 302, 302f
　Eco RI as, 302–304, 302f, 303f, 305f, 306t
　Eco RV as, 303, 304f, 306t
　Hae III as, 302, 302f, 306t
　Hind III as, 302, 302f, 306t
　recognition sites with, 302f, 306f
　restriction sites for, 304
　Sma I as, 302, 302f, 306t
　sticky ends with, 302–303, 302f, 309f
Restriction map, pBR322 in, 308f
Retention, sodium, 448
Retinal, 835–836, 835f–836f
　trans-isomer, 1034f
Retinoic acid (RA), 835f
　transport of, 836
Retinoic acid receptor (RAR), 339, 530, 835, 835f, 836, 836f
Retinoid, metabolism pathways for, 836f
Retinoid receptor, 198, 349

Retinoid X receptor (RXR), 491, 530, 800, 836
Retinol, 836
Retinol dehydrogenase, 835, 836f
Retinol palmitate, 835
Retroviral infections, 925, 927
Retroviruses
　gene therapy with, 315t
　HIV as, 1048
　retroviral integration in, DNA cutting/joining steps, 1058f
Reuptake mechanism, 990, 1020f, 1023f
REV. See Regulator of virion
REV response element (RRE), 1052f
Reverse transcriptase, 307, 311f, 1069
　HIV, 1050, 1053f, 1055f, 1056f, 1064
　structure of virus, 311f
Reverse transcription, HIV, 1052, 1053f, 1055f, 1056f, 1057f, 1066f
Reward, 1001
RFC-1 transporter orientation, 830f
Rh antigens, 891, 892f
Rh immunoglobulin (RhoGAM), 891
Rh negative blood type, 891
Rh positive blood type, 891
RhAG antigen, 892f
RHCE antigen, 892f
RHCE-related genes, 891, 891f
r/hCRF, 388t
RhD antigen, 892f
RHd-related genes, 891, 891f
Rheumatoid arthritis, 920, 922
Rhinoviruses, human, 947f, 1069, 1071f, 1072
Rho D antigen, 891
Rho kinase 1 (ROCK1), 716f
Rhodopsin, 837, 837f, 1032, 1034f, 1037
　light sensing and, 1037f
　signaling, 1034f
Rhodopsin kinase Grk1, 1037f
RhoGAM, 891
Riboflavin (Vitamin B_2), 119t, 806–808, 806f, 807f, 809f
　deficiency of, 808
Riboflavin kinase, 806, 807f
Riboflavin-5′-phosphate (FMN), 806, 807f

Ribonuclease, 441
Ribonucleic acid (RNA), 316–321
　antisense, 311
　bacteriophage, 1102f, 1103
　double-stranded, 311
　heterogeneous nuclear, 319
　heteronuclear, 919
　HIV, 1048, 1049f, 1050–1052, 1052f, 1054f
　　LTRs in, 1052, 1054f
　hybrid, 1056, 1056f
　information transfer from gene through, 316f
　in lambda genome, 1102f
　major types of, 316, 316f
　messenger, 9, 54, 54f, 62–63, 67, 70, 71f, 74, 316–321, 317f, 320f
　Polio virus, 1071f
　polymerase II, subunits of, 331t
　positive single-stranded, 1085f
　pre, 320
　ribose with, 244
　ribosomal, 11, 316, 316f
　SARS, 1084f–1085f, 1085–1087
　secondary structures of, 317
　　bulge loops as, 317, 318f
　　bulges as, 317, 318f
　　duplexes as, 316, 318f
　　hairpins as, 316, 318f, 320
　　internal loops as, 317, 318f
　　junctions as, 317, 318f
　　pseudoknot as, 317, 318f, 319
　　single-stranded regions as, 316, 318f
　short inhibitory, 311
　7SL, 319
　small nuclear, 318, 321f
　small nucleolar, 318, 320f
　transfer, 63, 65, 65f, 67–73, 71f–75f, 316–321, 316f, 317f
　uracil partly released with catabolism of, 271
　viral, 1048, 1049f, 1050, 1064
Ribonucleotide reductase, 281, 740
　NDP to dNDP conversion by, 286f
Ribose, 244
　atomic model of, 286f
Ribose-5-phosphate, from glucose-6-phosphate, 278, 281, 281f–283f

Ribose-5-phosphate isomerase, crystal structure of *E. coli*, 282f
Ribosomal ribonucleic acid (rRNA), 11, 316, 316f
Ribosomes, 21
  amino acids moved to, 67–73, 71f–75f
    A-site with, 68, 70, 74f
    elongation factor for, 70, 74–77, 78f, 79f
    E-site with, 70, 74f
    initiation factor for, 68, 73–74, 75f–77f
    P-site with, 68, 70, 74f
    termination factor for, 68, 77, 80f
  bound, 65
  crystal structures of, 65, 66f
  eukaryotic, 65
  free, 65
  mitochondria with, 65
  prokaryotic, 67, 67f
  rRNA, 11, 316, 316f
  subunits of, 63–67, 65f–71f, 65t
  tRNA with, 65, 65t
Ribozyme, 319, 321
  A2486 location of, 70
  metals coordinated in, 125, 125f
Ribulose-5-phosphate, 182f, 281f
Rice, polished/unpolished, 806
Rickets, 119t
Right lymph duct, 894f
Ring structure
  chair/boat configurations with, 147, 148f, 150f
  sugars with, 147, 147f–150f
RNA. *See* Ribonucleic acid
RNA interference (RNAi), 311
RNA pol II
  architecture of, 334f
  structure of, 334f, 336f
  subunits of, 331t
  transcription by, 329f, 354f
RNA polymerase (RNAP)
  bacterial cell, 1131
  as bound to DNA, 337f
  INR-box binding of, 335
  transcription by, N protein allowing, 1099, 1101
RNA polymerase II, clamp mechanism of, 335
RNA/DNA hybrid, 1056, 1056f

RNAi. *See* RNA interference
RNAP. *See* RNA polymerase; RNAP elongation complex
RNAP elongation complex, 338f
RNase H, 294, 312
ROCK1, 716f
Rod cell(s), 837, 837f
  cGMP-gated channels in, 1037f
Rod cell membrane, 837
Rofecoxib, 127f
ROS, 834f
Rosiglitazone, 490
Rough endoplasmic reticulum (RER), 16, 21, 22f, 24f
RRE. *See* REV response element
rRNA. *See* Ribosomal ribonucleic acid
RSAs, 945
R-Smad stoichiometry, 615f
R-state ATCase, 1146f
RTX. *See* Resinferatoxin
RXR. *See* Retinoid X receptor
RXRα, 342
  crystal structures of human, 343f, 490f
  ligand-binding domains of, 489, 490f
Ryanodine receptors (RYRs), 707f, 710
  model of, 711f
RYRs. *See* Ryanodine receptors

26S proteasome, 83, 84f
Saccharin, 149, 152f
S-adenosyl methionine (SAM), 518, 519f
S-adenosylhomocysteine, 831f
S-adenosylmethionine-dependent methylation, reactions of, 831f
SAICAR. *See* ADP + Pi 1-(5′phosphoribosyl)-4-(N-succinocarboxamide)-5-aminoimidazole
SAICAR synthase. *See* Phosphoribosyl aminoimidazole-succinocarboxamide synthase
Salivary glands, 796
Salivation, 439
Salmon, 776
Salrole (Sassafras), 301f
Salvage pathway, purine bases recovered through, 271

SAM. *See* S-adenosyl methionine
SAP90 group, 995, 997f
SAP97, 995
SAPs. *See* Synapse-associated proteins
Saquinavir, 1059, 1064f
Sarcolemma, 724
Sarcoplasmic reticulum Ca^{2+} ATPase, 712f
SARS-CoV genome, hypothesized, 1087f
Satins, 193
  HMG-CoA reductase inhibitors as, 195f
Saturated fats, 198, 199f, 200f, 491
  lauric acid as, 200f
  myristic acid as, 200f
  palmitic acid as, 200f
  stearic acid as, 199f, 200f
Sauvagine, 387, 387f, 388t
Saxitoxin, 731f
Scaffold, 610f, 992
Scaffolding proteins, 604, 1095–1096, 1097, 1100f
Scanning confocal microscope, 313
Scavenger receptor, 233
SCF. *See* Stem cell factor
Schiff bases, 118, 119t, 508, 510, 510f, 511f, 876f
Schizophrenia, 1031
  subtype, 717
Schwann cells, 998
SCID
  gene therapy for, 314, 315
  purine metabolism disorder with, 288t
SCR domains, 949f
SCR1, 944f
SCR2, 944f
Scrapie, 33. *See also* PrP^{Sc}
ScrY, 1125
Scurvy, 119t
Seborrhea, 808
Sec synthase, 776, 776f
Secretin, 439
  family, 444
Secretion
  acid/histamine, 439
  aldosterone, 454f
  ANP, 454
  anterior pituitary GH, 392
  bicarbonate, 441

Secretion (Continued)
  bile, 506f
  constitutive, 695
  cytokine, 475, 899
  gastric acid, 435–436
  GH, 394f, 395f
    preformed, 395
  GHRH, 392, 394f
  HCl, 439f
  hormone, 454f
  IGF-I, 394f
  insulin, 142, 144f, 447, 559, 559f
  macrophage IL-1, 654
  milk, 380
  mineralocorticoids, 452
  parietal cells, 439, 440f
    intrinsic factor from, 825
  PRL, 419
  progesterone, 413
  renin, 457–458
  signal for
    hGH, 695
    human fibrillin-1, 695
  T_4, 796f
  TGFα, 616
  thyroid hormone, 428f, 921f
  TRH, 418
Secretory granules, 899
Sec-tRNA^{Sec}, 776, 776f
SeCys. See Selenium-cysteine
Segmented neutrophils (polymorphonuclear leukocytes), 880, 881f
Seizures, 1004
Selective serotonin reuptake inhibitors (SSRIs), 1022–1023
Selenide phosphorylation, 776
Selenium, 773–776, 774f–776f
  dietary sources of, 776
Selenium-containing deiodinases (D1, D2, D3), 422, 429f
Selenium-cysteine (SeCys), 773, 774f, 775f, 776f
Selenium-cysteine synthase. See Sec synthase
Selenocysteine residue, 773
Selenodeiodinases, 774f
Selenophosphate (SeP), 775–776, 776f
Selenophosphate synthetase (SePS), 773, 775
Semiconservative model, 295, 298f

Semienzymic reaction, antibodies, 910
Seminiferous tubules, 409, 410
Sense strand, 62–63
Sensitivity, light/dark, 532–533, 1032
SeP. See Selenophosphate
SePS. See Selenophosphate synthetase
Septum pellucidum, 366
Sequenase, 1091, 1093f, 1094f
Sequestration, 1040f
SER. See Smooth endoplasmic reticulum
Serine, 40f, 44t, 518
  deamination of, 513, 513f
  glycine conversion of, 545, 546f, 549, 547f
  metabolism of, 545, 546f
  in protein kinase sequences, 486f, 612–613
Serine dehydratase, 513, 513f, 549, 548f
Serine hydroxymethyl-transferase, 545, 546f, 549f
  structure of, 549f
Serine protease
  inhibitor of, 867
  thrombin, 846, 857, 864
    conversion of factor XIII to factor XIIIa by, 847
  VKD, 857, 858f
Serine protease 1 cofactors, 548f, 943, 1003,
Serine-threonine kinase Akt, 653f
Sermorelin, 392
Seroconversion, HIV negative to HIV positive, 1067
Serotonergic system, 1019, 1020f, 1027
Serotonin, 352, 448, 533, 533f, 1003, 1019, 1023f, 1028
  amino acid decarboxylation with, 516
  development and, 1022
  excess, 814
  malignant carcinoid syndrome/excess, 814, 816
  pineal methoxyindoles biosynthesis from, 537f
  structure of, 534f, 1019f
  synaptic cleft and, 533
  synthesis of, 532, 532f, 535, 1019f, 1020f
  translocation, 1027

Serotonin receptor(s), 1018f, 1019, 1020f, 1021, 1022f, 1024f
  characteristics of, 1021t
  classes of, 1025f
Serotonin transporter (SERT), 1023f, 1025f
Serotonin type 3 receptor (5-HT$_3$R), 1003
Serotonin-N-acetyltransferase (SNAT), 537f
Serotypes
  E. coli, 1148
  HRV binding of, 1072, 1073f
SERT. See Serotonin transporter
Sertoli cell, 409, 410
Ser-tRNA^{Sec}, 776, 776f
Serum, 523, 879
  composition of, 1045t
  human, 751f
Serum albumin, 884
  crystal structure of human, 886f
Serum ferritin, 741
Serum glutamate-oxaloacetate transaminase. See Aspartate aminotransferase
Serum glutamate-pyruvate transaminase. See Alanine aminotransferase
Serum iron level, 741
Serum-binding proteins, IGF, 648
Serum/glucocorticoid-inducible kinase (Sgk), 457, 458f
Severe Acute Respiratory Syndrome (SARS), 1078, 1081, 1083f, 1085–1086, 1085f
  genome of, 1085f, 1087f
  replication concepts for, 1086f
Sex hormones, 483–488, 483f–488f. See also specific sex hormones
  female, 566
Sexual conjugation, pili in, 1107
Sexual maturation, 533
Sgk. See Serum/glucocorticoid-inducible kinase
s-GOT, 508, 509f
SH2 domain, 606f
SH3 domain, 997f
Shank scaffold proteins, 992
SHBG-like region, protein S, 862f
β-sheet, 58, 58f, 81
Shock, bacterial M protein inducing, 948f

Short inhibitory RNA (siRNA), 312
Short tandem repeat polymorphism (STRP), 239
Short-term memory loss, 446
Shrimp, 776
S-HT$_3$R. *See* Serotonin type 3 receptor
Sialic acid structure, 229
Sickle cell disease, gene therapy for, 314
Side chain cleavage enzyme system, 453f
sIFN-γ complex, 646f
Signal transducer and activator of transcription (STAT), 405, 632. *See also* STAT1; STAT3; STAT5
  activation of, 682f
  increased expression of, 588
  ovarian cancer with, 587–588
  pathways of, 632, 638f, 658, 661f, 664, 666
Signal transduction, 368, 406f, 695. *See also* Substance P
  angiotensin I receptor, 456f
  arg-vasopressin, 382f
  ascorbic acid, 834f
  chemokine-mediated, 677
  DAP10 transmembrane adaptor for, 899
  dopamine receptor, 1031f
  ERK, 604, 658
  glucocorticoid, 479f
  gp130-JAK-STAT, 638f
  growth factor, 406f
  hGH, 695
  HIV-1, 1051f
  insulin, 137f
  muscarinic, 1017
  myeloid/CSF-1R, 637f
  NFκB, 834f, 951
  NMDAR-PSD-95 complex for, 995f
  nuclear translocation, 352
  pain, 955, 956f, 957, 965, 969f, 972
    neurotransmission of, 959f, 988
    through PLC gamma pathway, 624f
  prolactin, 431f
  ras, 626, 650
  rhodopsin, 1034f

Signal transduction *(Continued)*
  somatotropic cell, 396f
  Substance P, 966f
  TRH, 418
  visual, 1037f
Signal transduction pathway(s)
  adrenergic receptor, 1042f, 1043f
  CSF-1R regulation of, myeloid cell, 637f
  EGF-receptor, 612f
  FSH receptor, 414f
  gp130-JAK-STAT, 638f
  insulin, 137f
  insulin-like growth factor, 648, 653f
  interferon, 647f
  interleukin, 658, 659f, 661f, 673f, 674f, 678f, 679f, 682f
  MAPK, 606f
  NK cell, 908f
  p75, 628f
  phospholipase C gamma, 624f
  PLC gamma, 624f
  TGFβ, 615f
  TSH receptor, 424f
Signature Cys-loop proteins, 1003
Sildenafil (Viagra), 540
Sildenafil citrate (Viagra), 540
Simple diffusion, 701
Single-stranded binding proteins (SSB proteins), 294, 295f
Single-stranded DNA, 1094f
  circular, 1099f
Single-stranded RNA, 316, 317f, 1085f
siRNA. *See* Short inhibitory RNA
Skeletal tissues, IGF-I/IGF-II stimulation of, 644
Skeletal-muscular system, 3, 4f
  contraction in, 894
  creatine as contained in, 540
Skeleton, magnesium % in, 780
Skin. *See also* Albinism
  color of, 527
  development of, 677
  immune reactions of, 678f
  melanocyte in, 527, 528f
  ultraviolet light in, 513f
Skin inflammation, 119t
7SL RNA, 319

SLE. *See* Systemic *lupus erythematosus*
Sma I, 302, 302f, 306t
Smac, 592
Smad complex, 615f
Smad2, 613, 615f
Smad3, 613, 615f
Smad4, 613, 615f
  stoichiometry of R-smad complex with, 615f
Smad7, 613
Small intestine
  brush border membrane of, 783
  iron regulation in, 739
Small nuclear ribonucleic acid (snRNA), 318, 321f
Small nucleolar ribonucleic acid (snoRNA), 318, 320f
  diagram of, 319f
  structure of, 320f, 321f
sMltA, 1113, 1115f
Smooth endoplasmic reticulum (SER), 21, 22f
Smooth muscle cells, 654
SMRT, 342
  binding of, 344f
sn3 position, 518
SNAT. *See* Serotonin-N-acetyltransferase
S-nitrosoglutathione (GS-NO), 983f
snoRNA. *See* Small nucleolar ribonucleic acid
snRNA. *See* Small nuclear ribonucleic acid
SOD. *See* Superoxide dismutase
Sodium
  conductance, 729–731
  depletion of, 452
  in glucose transport, 704–705, 705f
  homeostasis, 1028
  retention of, 448
Sodium benzoate, 503t
Sodium channels
  activation of, 729
  opening of, 1037f
  subunits of, 728f
  voltage-gated, 728f, 729–731, 729f
Sodium conductance channel (HSP), 458f
Sodium ion, 454f
  aldosterone action on, 456–457

Sodium ion *(Continued)*
  channels, protein kinase A opening, 970, 1001
  -dependent reuptake system, 990
  in membrane transport, 704
  -proton exchanger, 454f, 1028
Sodium Phenylacetate, 503t
Sodium-dependent amino acid transporters (symports), 442
Sodium-dependent L-glutamate transporters, 722
Sodium/iodide symporter (NIS), 796
  structure model/mechanism of, 798f
Sodium/potassium ATPase (Na$^+$/K$^+$ ATPase), 702, 702f, 706, 800
Sodium/potassium gradient, nerve impulses through, thiamine involved in, 803
Sodium-proton antiport, 729
Sodium-proton exchanger, CFTR increase of, 688, 691, 693f
Solute uptake kinetics, 701
Somatic recombination, antibody formation, 917, 919
Somatocrinin, 370t, 392
Somatoliberin, 392
Somatomedin C, 644
Somatostatin(s) (SS), 368, 432, 436. *See also* Preprosomatostatin
  actions of, 405
    acid secretion inhibiting, 439
    cell proliferation and, 406f
    insulin secretion, 559
  delta cells producing, 133, 173f, 559
  fasting and, 419f
  negative control of GH by, 392, 398
  structure of, 400f
Somatostatin 5 receptor, 398, 400f
Somatostatin receptors, 398, 400f
Somatostatinergic neuron, 392
Somatotrope, 392, 395, 396f
Somatrophs, 368
Sorbitol, 149, 152f
SP. *See* Substance P
S-palmitoylation, 236

Spark chamber, 39
SPE. *See* Pyrogenic exotoxins
Specific response element (SRE), 478
Spermatogenesis, 408, 409, 410
Spermidine, 541, 543f, 991
Spermine, 541, 991
S-phase progression, cell cycle, 614
Spherical melanosomes, 527
Sphingolipids, 186, 220, 222f
  ceramide of, 220
  cerebrosides of, 220
  gangliosides of, 220
  globosides of, 220
  glycosphingolipids as, 220, 222f, 223f, 226–232, 227f–233f
  sphingomyelin as, 220–221, 222f
  sulfatides of, 220
Sphingomyelin, 220–221, 222f, 224f, 225f
  biosynthesis from ceramide of, 220–221, 224f
  sphingomyelinase in breakdown of, 221, 224, 225f
Sphingomyelin lipids, 1045
Sphingomyelinase, 221, 224, 225f, 226, 226t
  acid, 221, 226
  Niemann-Pick disease with deficiency in, 226t
  sphingomyelin breakdown with, 221, 224, 225f
Sphingosine kinase, 224
Spikes, virus surface, 1048, 1049f, 1066f, 1086
Spinach, 740
Spinal central canal astrocytes, 998
Spinal cord, dorsal horn of, 955, 956f
Spinal neurons, 974f
Spinobulbar muscular atrophy, 240, 242f
Spiny GABAergic neurons, 1001
Spironolactone, 466, 466f
Spleen, 896, 915
Splenda™. *See* Sucralose
Sponge cells, 189
Spooling. *See* Elongation
S-prenylation, 236
S-protein, 1088f

Squalene, cholesterol synthesis with, 193, 194f
Src homology domains, 604
Src kinase, 670f
SRC-1 coactivator, crystal structure of, 342t
SRE. *See* Steroid receptor enhancer
SREBP-1. *See* Sterol response element-binding protein
SRP, human, crystal structure of, 319f
SS. *See* Somatostatin(s)
SSB proteins. *See* Single-stranded binding proteins
SSRIs. *See* Selective serotonin reuptake inhibitors
StAR. *See* Steroid acute response
Starch, 153–154
  amylopectin as, 153, 153f, 154f
  amylose as, 153, 153f
  anhydroglucose with, 153, 153f
  defined, 153
  digestion of, 442–443
Start codon, 67, 71f
STAT. *See* Signal transducer and activator of transcription
STAT1, 609, 647f, 988f
STAT3, 342, 405, 430, 431f, 988f
STAT5, 638f, 670f, 988f
  phosphorylation of, 661f
Stearic acid, 199f, 200f
Stem cell factor (SCF), 601
Stem cell precursors, 630f
Stem cell progenitors, 629
Stereoisomers, 46
Steroid acute response (StAR), 390, 409
  action of, 410
Steroid hormone(s), 409, 445–496. *See also* Cortisol
  adrenal cortex, 449f, 452, 453f
  adrenal medulla, 446, 448–451
  aldosterone, 387, 447, 452–466
  11β HSD2 not being a substrate of, 463, 465f
  acetal grouping of, 477f
  ACTH and, 452
  action of, 456–457, 458f, 459f
  cortisol v., 452, 463, 478
  ENaC synthesis with, 729
  fluctuations of, 452
  formation of, 453f
  hyperaldosteronism with, 465

INDEX 1245

Steroid hormone(s) *(Continued)*
    inactivation of, 568
    Na$^+$/K$^+$-ATPase and, 458, 458f
    rapid effects of, 460f
    second receptor of, 454
    secretion, 454f
    DHEA, 387, 452, 453f, 474–476, 475f, 476t
    intermediate required for synthesis of, 452
    metabolism of, 566–570, 567f–569f
    orphan receptors and, 488–491, 491f
    peroxisome proliferators, 488–491, 489f, 490f
    reactions leading to, 453f
    sex hormones, 483–488, 483f–488f
    synthesis of, ascorbate involvement in, 832
    x-ray structures of, 476–478, 477f
        ball-and-stick representations of, 477f
Steroid hormone receptors
    activities, locations of, 461f
    activities model for, 350f
    cytoplasmic, 339f, 340
    superfamily
        PPAR member of, 490–491, 490f
        retinoic acid with, 836
        zinc fingers in, 461
Steroid receptor enhancer (SRE), 458f
Steroid synthetic mechanism, 409
Steroidogenic cells, 561
Sterol, 554
Sterol response element-binding protein (SREBP-1), 218
Stevia, 152f
Sticky ends, 302–303, 302f, 309f
Stoichiometry
    proposed, 460f
    R-Smad/smad4 complex, 615f
Stomach, GH release ghrelin in, 392
Stop codon, 67, 70, 71f, 77
Straight line equation, 99
S-transferase, GSH, 521
Streptavidin, 820, 825

Streptococcus pyogenes, 943t, 946, 948f
    M protein of, 946, 948f
Streptomycin, 83, 85f
Stress
    acute, 445, 446
    adrenal medulla and, 446, 448–451
    anxiolytic substance modulating, 446
    chronic, 445–446
    cortisol and, 445–450, 446f, 449f, 451f, 466, 467
    cortisol release during, 445–446, 446f, 449f
    endothelial cell oxidant, 476
        nitric oxide bioavailability and, 476
    GSH efflux in, 691
    hormonal responses to, 447f
        CRH/ACTH, 447, 447f
    oxidative, 673f
    response
        catecholamines in, 450, 450f, 451f, 526f
        sympathetic nervous system, 447f
Stress-induced diabetes, 445, 447–448
    cortisol as factor in, 133
Stroke, 853
STRP. *See* Short tandem repeat polymorphism
Strychnine, 1003–1004, 1014
    structure of, 1005f
Stuart factor, 865
Subclavian artery, 871, 872f
Subclavian vein, 871, 872f, 894
Substance P (NK1) (SP), 962, 963f–966f, 965, 1031, 1033f
    chemical formula of, 963f
    EAAs with, 988, 989f
    NK1R with, 964f
    signaling mechanism for, 966f
    symptoms regulated by, 962
    synapse crossing of, 969f
Substance P receptor (NK1R), 964f, 989f
    binding of, 966f
    signaling/effector systems stimulated by activation of, 966f

Substance P-Neurokinin-1 receptor complex, 965f
Substantia nigra, 1031, 1032f, 1033f
Subthalamic nucleus, 1032f
α-subunit, 419, 422
β-1-subunit, sugar moiety, glycated hemoglobin β-1-subunit, 875
β-subunit, 419
Succinate dehydrogenase, 740, 768, 772f, 808f, 809f
    flavins as, 806
Succinate, GABA degraded to, 999, 999f
Succinic semialdehyde, 999f
Succinyl CoA, 546, 550, 551f, 753, 753f, 802, 804f, 820, 825
Suckling, 419, 430
Suckling response, 380, 383f
Sucralose (Splenda™), 149, 152f
Sucrose, 149, 150f–152f
    -specific porin, 1125, 1129f
Sugar moiety, of glycated hemoglobin β-1-subunit, 875
Sugars, simple, 146–152, 147f–152f
    aspartame as, 54, 149, 152f
    atomic models of, 149, 151f
    disaccharides as, 147–149, 149f, 150f
    Fischer projection of, 147f, 148f
    fructose as, 149f, 150f, 152f
    galactose as, 149f, 150f
    glycyrrhizin as, 152f
    Haworth projection of, 148f
    lactose as, 149f, 150f
    maltose as, 149f, 150f
    monosaccharides as, 146–147, 147f, 150f
    naturally occurring, 149, 152f
    ring structure of, 147, 147f–150f
    saccharin, 149, 152f
    sorbitol as, 149, 152f
    stevia as, 152f
    stick structure of, 147f
    sucralose as, 149, 152f
    sucrose as, 149, 150f–152f
    synthetic, 149, 152f
    thaumatin as, 152f
    xylitol as, 149, 152f
Sulfatide lipodosis, 226t, 227

Sulfatides, 220, 226, 227, 227f
Sulfation, 395
Sulfite oxidase, 788, 793f
　reaction catalyzed by, 793f
Sulfur, 771
Sumatriptan, 1021t
Sunlight, 530, 837
Supercoiling
　E. coli DNA, 1131–1134, 1134f
　negative DNA, 1131
Superoxide dismutase (SOD), 122f, 123–125, 124f, 764, 766f, 767f
　structure of copper-zinc, 124f, 125
Superoxide radicals, 764
Surfactant protein, 466, 467f
Surgery, Cushing's disease, 474
Susumu Tonegawa, 917
Swallowing, problems in, 740
SWI/SNF. See Switching/sucrose non-fermenting
Switching/sucrose non-fermenting (SWI/SNF) complex, 356, 356f
　human 2 mating type, 357f, 359
Sympathetic nervous system, 1027f
　stress responses of, 447f, 448
　　norepinephrine derived from, 448
Symport, 442, 704
Synapse, 957, 959, 989f
　dopaminergic, 1033f
　magnification of, 959f
　substance P across, 969f
Synapse-associated proteins (SAPs), 995, 997f
Synaptic cleft, 533, 959
Synaptic gap, 959f
Synthetic sweeteners, 149, 152f
Systemic circulation, human, 872f
Systemic *lupus erythematosus* (SLE), 920, 925
　genetic susceptibility to, 925

T cell(s), 654, 658, 665, 677
　absence of, 670f
　antibody production from, 911
　antigen, 725f
　CD4
　　autoreactive, 930, 932f, 933f
　　HIV levels of, 1048f
　　pathogenic, 932

T cell(s) *(Continued)*
　CD8, 671, 675, 932
　　autoreactive, 930, 932f, 933f
　　cytotoxic, perforin with, 901f
　　cytotoxicity involvement of, 923–924
　　Graves' disease with, 923
　helper, 675
　killer, 904f
　leptin in, 984
　maturation of, 894
　natural killer, 639, 899, 901–903, 904f, 906f–909f
　　oligomeric receptors expressed on, 900f
　　tumor cell destruction by, 901, 901f, 907f, 908f
　pathogenic, 922
　precursor, 896
　proliferation, 661f
　TAX interaction with human, 930f
T cell receptor, 927
　complex among TAX/HLA-A0201 and human, 930f
　MHC protein interactions with, 929, 929f
T cell-mediated effects/damage, 923
T form hemoglobin, 874
T lymphocytes. See T cell(s)
T phages, 1089, 1090f, 1091
　characteristics of, 1093t
　even, 1091f, 1092
T1R domain, 654, 657f
T_3. See Triiodothyronine
T_3 receptors, 403f, 425f
　structure of, 426f
T_4. See Thyroxine
T4 lysozyme, 1120f
T4 phage, 1090f, 1093f
T_4, thyroxine. See Tetraiodothyronine
T7 DNA polymerase, 1091, 1094f
　structure of, 1093f
T7 phage, 1091, 1093t, 1094f
Tachykinin(s), 962t
　receptors recognizing, 962
Tachykinin neuropeptide family, 962t
TAFs. See Transcriptional activator factors
Tagamet. See Cimetidine

Talc, 588
Tamiflu, 1076, 1077f
TAR. See Transcriptional activation region
Target cells, 959f
　dendrites of receiving, 959f
　lysis of, 942f
Target end organs, for anterior pituitary hormones, 379f
Tarui's disease, 171t
TAT. See Transactivator of transcription
TAT phosphorylates, 1050
TAT proteins, 326, 1050, 1050f, 1051–1052
TATA box, 329, 329f, 330, 390f
TATA motif region, 1131
TATAAA consensus sequence, 330
TATA-binding protein, 330, 330f
　binding of, 331f, 333
Taurine, 1003
TAX. See Viral peptide
Taxol, 588, 591
Tay-Sachs disease, 226t
TBG. See Thyroid-binding globulin; Thyroxine-binding globulin
tBID, 591–592, 593f, 902f
TBPA. See Thyroid-binding prealbumin (transthyretin)
TCA cycle. See Tricarboxylic acid cycle
T-cell cultures, 594
TCR β-chain, 931f
TCRα-chain, 931f
Teeth, 783
Tegaserod, 1021t
Teichoic acid, 951
Telophase, 14
Termination, 329
Termination factor, ribosome/peptide bond, 68, 77, 80f
Testosterone, 566
　binding of, 478
　blood level of, 486
　formation of, 408, 568
　LH synthesis of, 409–410
　negative feedback by, 410
　release of, 430
　structure of, 410, 412f
Testosterone receptor, 478
Tetracycline, 85, 85f

Tetrahydrobiopterin (BH4), 475, 522, 526, 540
Tetrahydrocannabinol (THC), 980, 980f
structure of, 980f
Tetrahydrofolate (THF), 830, 831f
Tetrahydrofolic acid, 115f, 118, 119
Tetraiodothyronine (T_4, thyroxine), 366, 418, 422, 428f, 775f
Tetrameric hemoglobin, 762, 763f
Tetrameric PAH partial structure, 522, 524f
Tetrapyrroles, 756
Tetrodotoxin, 731f
Tf. See Transferrin
TFIIB recognition element (BRE), 329, 329f
TFIID, 360
TGF. See Transforming growth factor
TGFα, 612, 614, 616, 616f
secretion of, 616
TGFα–EGFR complexes, 616
TGFαisEGFR50 complex polypeptide trace, 617f
TGFβ, 612, 613, 613f, 614
signal transduction pathway of, 615f
TGFβ gene family, 613, 613f
TGFβ type II receptor, 613
extracellular domain of, 614f
TGFβ1, 613
crystal structure of, 614f
TGFβ1 dimer, 615f
TGFβ2, 613
TGFβ4, 613
TGFβ5, 613
TH. See Tyrosine hydroxylase
Th1 cells, 669
Th2 cells, 658, 664f, 671, 675
Thalamus
pain in
pathway in, 974f
signals involving, 955
serotonergic system with, 1020f
α-thalassemia-1, 762
α-thalassemia-2, 762
α-thalassemias, 762, 763t
β-thalassemias, 760, 762, 763t
Thalassemia minor, 762
Thalassemias, characteristics of, 763t

Thaumatin, 152f
THC. See Tetrahydrocannabinol
Thermoregulation, 425
THF. See Tetrahydrofolate
THF-like-2, 594
Thiamine (Vitamin B_1), 119t, 801–806
conversion to TPP of, 801, 802f
deficiency, 803, 804
Thiamine diphosphate, 804f
Thiamine diphosphokinase, 801, 802f
Thiamine diphosphotransferase, 801, 802f
Thiamine monophosphate, 804
Thiamine pyrophosphate (TPP), 118, 119t, 801
disease of, 804
mechanism of, 804f
thiamine conversion to, 801, 802f
Thiamine triphosphate, 804
Thiol, pK values for, 47t
Thioredoxin reductase, 281, 773, 774f
structure of, 287f
Thioredoxin-binding domain, 1093f
ThO, 671, 675
Thoracid duct, 545, 545f, 894
Threonine, 40f, 44t, 547f, 549, 612–613, 810f
Threonine dehydratase, 513, 513f
Threonine dehydrogenase (NAD+), 549, 549f, 551, 560, 812
Thrombin (factor IIa), 846, 857, 864
conversion of factor XIII to factor XIIIa by, 847
prothrombin conversion of, 884, 888
Thrombokinase, 884, 888f
Thrombomodulin-dependent generation, APC, 857
Thrombophilia, 856
Thromboplastin, 884, 888f
Thrombosis, deep vein, 853–867
Thrombospondin, 949
repeat type 1 domains of, properdin containing, 949
Thromboxane A_2-glycerol (TXA_2-G), 978f
Thromboxanes, 467, 472f, 978f

Thrombus, 855f
deep vein, 865
Thymidine kinase
cell cycle and, 278
reaction, 271, 276
structure of, 279f
Thymidine phosphorylase, reaction, 276
Thymidylate synthase, 278
dUMP to dTMP conversion catalyzed by, 278f
structure of, 279f
Thymine, 244, 247f, 248f
Thymine dimer
repair of, 300f
ultraviolet light causing, 299, 299f
Thymosin, 896
Thymosin β4, 896, 897f
blood clotting with, 898f
release of, 898
Thymosin β9, 897f
Thymus, 894
T cells of, 883
anti-insulin antibodies in, 930
pathogenic, 922
Thyrocyte, 419, 422, 922
TSH and, 421
Thyroglobulin, 418, 422, 426f, 796–797
binding to thyroid peroxidase of, 922
T4 residues on, 799f
Thyroglobulin-like structure, 421, 426f
Thyroid
epithelial cells, 418, 799f
hormone precursor, 418f
human, 429f
iron-copper ratio in function of, 740
TRH and, 795, 796f
Thyroid follicle, 418f, 799f, 920. See also Thyrocyte
cell, 417, 418, 419
thyroid hormone produced by, 920–922
Thyroid gland hypertrophy, 796f
Thyroid hormone(s)
biosynthesis of, 797f
DNA-binding domain of, 420
iodine in formation of, 795

1248 INDEX

Thyroid hormone(s) *(Continued)*
  pathway of, 417f
  secretion of, 428f, 921f
  structure of, 800f
  thyroid follicle cell production of, 920–922
  tissue targets for, 423
Thyroid hormone receptor (TR), 417f, 461f, 530, 800, 836
  human, 800
Thyroid peroxidase, 426f, 922
Thyroid receptor-binding protein (TRBP), 342, 345f
  binding of, to ERα/TRβ, 345f
Thyroid response element, 425
Thyroid-binding globulin (TBG), 418
Thyroid-binding prealbumin (transthyretin) (TBPA), 797, 797f
Thyroid-releasing hormone (TRH), thyroid gland control by, 795, 796f
Thyroid-stimulating hormone (TSH), 366, 375t, 418, 419. *See also* hTSH; TSH receptor
  complex of TSHR and, 423
  deficiency of, 366
  human (hTSH), 376t
  iodine with, 795, 796f, 797f
  molecule of, carbohydrate substitutions in, 423f
  thyrocytes and, 421
  thyroid follicular cells under influence of, 797f
Thyroliberinase, 430
Thyronines, 796, 799f
Thyrotoxicosis, 921, 921f
Thyrotrophic cell, 417–418, 419, 921f
Thyrotrophs, 368, 795
Thyrotropin, 417–428
  release of, 417
Thyrotropin-releasing hormone (TRH), 368, 372t, 417, 419. *See also* TRH receptor
  fasting/SS pathways/leptin and, 419f
  PRL release by, 430, 432
  secretion of, 418
  signal transduction of, 418
  as tripeptide/composition of, 418

Thyroxine (T4), 797f
  deiodination of, 799f
  synthesis/release of, 795–796, 796f, 799f, 921f
  thyroglobulin containing residues of, 799f
Thyroxine-binding globulin (TBG), 797f
TIDA. *See* Tuberoinfundibular dopaminergic system
TIM complex. *See* Translocator Inner Membrane
TIP47, 561, 564f
Tissue
  cells, diabetic, 134f
  connective, 832
  injury to, 880, 923
    macrophages leading to, 924–925
  lymphoid, 883
Tissue factor, 658, 884
  extracellular domain of, 888f
Tissue targets, thyroid hormone, 423
TL2, 594
T-lymphocyte receptors, 948f
TNF. *See* Tumor necrosis factor
TNF family, 595t
TNF superfamily, 594–600, 595t–598t, 599f, 600f
TNF trimers, 600f
TNFα, 224, 323, 470, 673f, 675, 926f
  structure of, 926f
TNFa mutant, ribbon drawing of, 599f
TNFα receptor (TNFαR1), 926f, 927f
TNFαR1. *See* TNFα receptor
TNFR1, 928f
TNFRSF. *See* Tumor necrosis factor receptor superfamily
TNFSF. *See* Tumor necrosis factor superfamily
α-tocopherol, 841, 843
  structures of, 841f
β-tocopherol, 841
γ-tocopherol, 841
α-tocotrienol, 841
  structure of, 841f
δ-tocotrienol, 841
Toll-like receptor 1, human, 654, 657f
Toll-like receptors, 654, 657f

Toll-like/IL-1R domain, 654
TOM complex. *See* Translocator outer membrane
Tonicity, membrane transport and, 698
Tonsils, 883
Topoisomerase I, DNA replication with, 291
Topoisomerase II, DNA replication with, 292
Toxicity, 837
TPP. *See* Thiamine pyrophosphate
TR. *See* Thyroid hormone receptor
Tra2.10, 947f
Trace metals, important dietary, 764
TRAIL. *See* Tumor necrosis factor-related apoptosis-inducing ligand
TRAIL-DR complex, crystal structure of, 599f
TRAIL-R3 trimers, 592f
Trans configuration, 199, 201f
Trans fats, 199, 201f
11-trans configuration, 837, 837f
Transacetylase, *E. coli*, 1138, 1138f
Transactivation factor activity, 349
Transactivator of transcription (TAT), 1050
Transaminases, 507, 508
Transamination, 50, 503, 508–512
  decarboxylation and, 509
  glutamate, 545
  leucine, 552
  mechanism of, 510f
  phenylalanine, 524
  phenylalanine to phenylpyruvate, 524, 527f, 531
Transcarbamylase, 497
Transcellular diffusion, 737f
Transcobalamin II, 825
Transcortin, 352, 466
Transcription
  asbestosis as aberrant, 323–327
  chromatin and, 357–359, 358f, 360f
  coactivators/corepressors, 337–344, 341f, 343f
  enhancers, 337

INDEX 1249

Transcription *(Continued)*
  fatty acid synthase, 218
  histone acetylation with decrease in, 243
  HIV reverse, 1052, 1053f, 1055f, 1056f, 1057f, 1066f
    mechanism of, 1057f
  IL-8/IL-6, 327
  initiation, multiple sites for, 461f
  mRNA, phage, 1103
  phases/definition of, 327
  progesterone receptor, 479
  RNA pol II, 329f, 354f
  start site of, 335
Transcription complex
  pre-, 341
  transcription factors and, 329–338, 332f, 333f
Transcription factor(s), 337. *See also* Signal transducer and activator of transcription
  calcium ion and, 411–413
  COUP-TF, 479
  glucocorticoid receptors as model, 349–356, 353f, 354f, 355f
  structures of, 332f, 333f
  transcription complex and, 329–338
Transcriptional activation region (TAR), 730f
Transcriptional activator factors (TAFs), 330
  chromatin and, 359
Transcriptional regulatory sequences (TRSs), 1086
Transducin, 837, 1037f
Transducin beta gamma, 838f
Transfer ribonucleic acid (tRNA), 63, 316–321, 316f, 317f
  amino acids moved to ribosomes with, 67–73, 71f–75f
  features of, 317f
  ribosomes with, 65, 65t
  tertiary structure of, 776f
Transferase, 107, 108f
  malonyl CoA, 726f
Transferrin (Tf), 743, 764
  cycle, 743
  human serum, 751f
Transferrin receptor, 747, 752f, 753

Transforming growth factor (TGF), 612–616, 613f–616f
  relatives of, 612–613, 613f
Transformylase and cyclohydrolase enzyme, crystal structure of, 261f
Transfusability, 889, 890t, 893, 893t
Transfusion. *See* Blood transfusion
Transgene, 240, 312
  type 1 diabetes with, 931–932
Transglucosylase, 156
Transglutaminase, 898
Transglycosidase, *E. coli*, 1120f
Transglycosylase, *E. coli*, 1113, 1115f, 1116f, 1118
Trans-golgi network, 697f
Trans-isomer, retinal, 1034f
Transketolase, 282f, 285f, 805f
  crystal structure of, 285f
  human, 285f
  reaction mechanism of, 805f
  yeast, 285f
Translocase, 170, 724
  CACT, 726f
  fatty acid, 727f
Translocation, serotonin, 1027
Translocator Inner Membrane (TIM complex), 80
Translocator Outer Membrane (TOM complex), 80
Transmembrane, 899, 908f
Transmembrane domain 4 (TM4), 743, 747f, 996. *See also* Cystic fibrosis transmembrane regulator
  GluR6, 998f
  structures of DMT1-, 747f
Transmembrane proteins, 8f
  opioid receptor, 970t
Transmembrane topology
  human vasopressin receptor, 385f
  ligand binding and, 456f
Transpeptidase, *E. coli*, 1114f
Transport. *See* Membrane transport; *specific transport channels*
Transretinoic acid, 835, 835f
Transsulfuration pathway, 518
Transthyretin, 418

Trauma
  brain injury
    deaths from, 366
    frequency of, 365–366
    hypothalamus stalk severed by, 365
TR_β, 800
  binding of TRBP to ERα and, 345f
TRBP. *See* Thyroid receptor-binding protein
β-trefoil motif, 617
Tremolite fibers, 323, 324f
TR_g ligand-binding domain, 800f
TRH. *See* Thyroid-releasing hormone; Thyrotropin-releasing hormone
TRH receptor, 419, 430
  forms of, 419
TRHR1, 419
TRHR2, 419
Triacylglycerides
  breakdown by lipases of, 211f
  glycerol from breakdown of, 182
  hydrolysis of, 207
  lipoproteins with, 233, 233f
  phospholipids v., 220
  synthesis of, 556f
  utilization of, 207, 208f
Triacylglycerol, 562f, 565f
Triacylglycerol lipase, 211f, 557
Triamcinolone, 351f
Tricarboxylic acid cycle (TCA cycle), 18, 20f, 50, 51f, 53f, 500f
Triglycerides
  digestion of, 202, 202f, 443
  fat absorption in chylomicrons of, 561
  structure of, 201, 201f
Triiodothyronine (T_3), 366, 418, 420, 422, 428f, 775f, 795–796, 797f, 921f. *See also* T_3 receptors
  formation of, 797, 797f, 799f
  Graves' disease and, 920
  oxygen consumption increased by, 800
  reverse, 800
Trimerization
  central domain of, 610f
  E. coli, 1122f, 1127f
  HIV, 1065f

Trimers
  E. coli porin, 1129f
  TNF, 599f
  TRAIL-R3, 592f
5,7,8-trimethyltocol, 841
5,8-trimethyltocol, 841
7,8-trimethyltocol, 841
Trinucleotide, 245, 248f. *See also specific nucleotide*
  structure of, 248f
Triosephosphate isomerase, 110f
Triskelion structure, 610f
trkA, NGF receptor, 625
tRNA. *See* Transfer ribonucleic acid
Tropocollagen
  fiber triplex, 768, 768f
  formation of, 769f
TRP channel, 708f
TRPV family, 783
TRPV5, 783, 783f
TRPV6, 783. *See also* Calcium transport channel
TRSs. *See* Transcriptional regulatory sequences
Tryglycerides
  breakdown of, 557
  protein phosphatases acting on, 556f
Trypsin, 441
Tryptophan, 40f, 44t, 537f, 816, 818
  catabolism of, 554f
  codon for, 518
  melatonin/serotonin synthesis from, 532, 532f
  metabolism of, 532–536, 532f–536f
  niacin conversion of, 812, 816f
Tryptophan hydroxylase, 533, 534f, 537f, 1019
TSH. *See* Thyroid-stimulating hormone
TSH receptor, 419
  activation of, 921f
  complex, 419, 421f
  in Graves' disease, 920, 923
  signal pathway, 424f
TSHR, 921f
  complex of TSH-, 423f
  in Graves' disease, 920
T-state, ATCase in, 1146f
Tuberoinfundibular dopaminergic system (TIDA), 432, 434f

Tubulin, 1007, 1009f
Tumor(s). *See also* Anti-tumor activity; Ovarian cancer
  carcinoid, 814
  formation of, TGFα in, 616
  ovarian, 588, 589
  secondary, 588
Tumor cells, NK destruction of, 901, 901f, 907f
Tumor necrosis factor (TNF). *See also* TNFα
  family of, 591
  ligands in, 901, 902f
  perforin-granzyme-mediated necrosis with, 901, 901f
Tumor necrosis factor receptor superfamily (TNFRSF), 594, 632f
Tumor necrosis factor superfamily (TNFSF), 594
Tumor necrosis factor-related apoptosis-inducing ligand (TRAIL), 591
  extended loop of, 594
  homotrimers of, 592f
  receptor pathways of, 593f
  resistance to, 594
  subunits of, 599f
Tumor suppressor genes, 587–588
Tumorigenesis, 617
Turnover number (K_{cat}), 100
TXA_2-G. *See* Thromboxane A_2-glycerol
Type 1 diabetes
  defined, 131
  development theory for, 930–933, 932f, 933f
  genetic background for, 932
  insulin-dependent, 145, 922
  tissue cell characteristics with, 134f
  transgene in, 931–932
Type 1 Hyperammonemia, 503t
  treatment for, 503t
Type 2 diabetes
  defined, 131
  genetic factors for, 131, 132f, 132t
  insulin independence of, 145
  protein aggregation with, 170–173, 171f–173f, 172t
  tissue cell characteristics with, 134f

Type 2 Hyperammonemia, 503t
  treatment for, 503t
Tyrosinase, 527
Tyrosine(s), 42f, 44t, 47f, 531, 531f, 770–771, 773f. *See also* L-tyrosine
  hydroxylase conversion to L-DOPA of, 448, 526
  hydroxylase step conversion of, 448, 526
  melanin synthesis from, 527, 527f
  phenylalanine and, 521–524, 523f, 531f
  residues, 422
Tyrosine hydroxylase (TH), 1033, 1039f
Tyrosine kinase (TK), 134, 136f, 404, 404f, 405, 617
  binding of, 992
  domains of, 623, 624f, 644, 650f
  families of, 624f
  by G proteins, 965
  -linked receptors, 713f
  receptors for, 622, 625, 629, 634f, 710
  -Src-ERK/P13K, 645
  type III receptors for, 629, 634f
Tyrosine kinase phosphorylation, 134, 136f

U1A protein crystal structure, 319f
Ubiquitin, 81–83
  pathway, 81
  structure of, 81–82, 82f
Ubiquitin monophosphate. *See* UMP
Ubiquitin triphosphate. *See* UTP
Ubiquitin-activating enzyme, 82, 83f
Ubiquitin-carrier protein, 82, 83f
Ubiquitin-conjugating enzyme, 83, 84f
Ubiquitin-proteasome system, 84f
Ubiquitin-protein ligase complex, VHL, 640f
Ucephan, 499
UDP. *See* Uridine diphosphate
UDP glucuronyl transferase, 743

UDP-glucose, glycogen formation from, 154, 155f, 157f, 162–163, 163f
UDP-N-acetylmuramoyl: L-alanine synthetase, 1111, 1112f, 1113
UGG codon, 518
Ultraviolet light
  DNA mutations/damage from, 298–299, 299f
  in skin, 513f
UMP (Ubiquitin monophosphate), 266f
  OMP decarboxylase inhibition with, 268
UN coating ATPase, 609
Underphosphorylated adipocyte HSL, 561
Uniport, 704, 705
Unsaturated fats, 198–199, 199f, 200f
  arachidonic acid as, 199–201, 200f
  linoleic acid as, 199, 200f, 208, 208f
  oleic acid as, 199, 199f, 200f, 208, 208f
Upstream stimulatory factors (USF), 218
Uptake
  Ca^{2+}, 789f
  calcium, 783–784, 783f
  chylomicron remnant, 843
  dietary metal, 743–753
  fatty acid, 723–728, 724f–728f
  Iodine, 427f
  iron, 743–753, 749f
  LDL, 191f
  solute, 701
  vasopressinergic neuron causing, 380
  zinc, 778
Uracil, 244, 247f, 248f
  deamination of cytosine at 4-position to form, 297f
  RNA catabolism with, 271
Urea, 53, 871
  from ammonia, 499
Urea cycle, 53, 500f, 501f, 536. See also Hyperammonemia
  disruptions in, 497–503
    infantile, rate of, 497
  enzymes, 499, 501f, 502f
  partial inhibition of, 500

Uric acid, purine nucleotide catabolism producing, 271, 274f
Uridine diphosphate (UDP). See also UDP glucuronyl transferase
  acetylglucosamine with, 512, 512f, 1111
  N-acetylmuramic acid-, 1107–1108
  peptidoglycan synthesis with, 1107–1108
  pyrimidine synthesis and, 1142f
Uridine kinase, 271, 276
Uridine phosphorylase, 271, 277f
Urinary system, 3, 1148
  excretion products, 568, 568f
Urine, cystathionine excretion in, 520
Urocanate, 513, 549, 550, 550f
Urocanate hydratase, 550
Urocanic acid, 513f
Urocortin, 388t
  human, 387, 387f
Urogenital-associated tissue, 883
Uroporphyrinogen decarboxylase, 753, 753f, 755f
Uroporphyrinogen I, 753, 755f
Uroporphyrinogen I synthase, 753, 755f
Uroporphyrinogen II cosynthase, 753, 755f
Uroporphyrinogen III, 753, 755f
USF. See Upstream stimulatory factors
UTP (Ubiquitin triphosphate)
  carbamoyl phosphate synthetase inhibition by, 268
  conversion to CTP of, 269f
  E. coli affinities for, 1140–1144
  synthesis route of, 269f

V gene segments, 919
V Leiden factor, 856, 856f
  venous Thromboembolic disease with, 857t
V1a, 385f
Vaginal bleeding, 588
Vagus nerve, 439
Vagus reflex, 439
Valine, 44f, 44t, 877
  catabolism of, 550
Valine decarboxylase, 516t

Valium, 101, 1008f
Van der Walls forces, 59, 60f
Van der Walls forces/interactions, 250, 913f
Vanilloid receptor (VR1), 961, 961f, 981f, 982, 984f
  CB-1 with, 983
  regulation of, 985f
Vanilloid receptors, 982
Vanillyl moiety, 985f
Variable domain fragment of antibody (HyHEL-10), 912f
Variable heavy chain (VH), 910, 917
Variable light chain (VL), 910, 917, 918f, 919
Vascular cell adhesion molecules (VCAMs), 470
Vascular circulatory system, 871
Vascular endothelial cell, 861f
  membrane, APC molecules in, 858, 861f
Vascular endothelial growth factor (VEGF) receptor 2, 601
Vasoactive compounds, 881
Vasoconstriction, 456
Vasodilation, 540, 542
Vasopressin (VP), 367, 380, 698, 699f, 1003. See also Antidiuretic hormone
  gene structure of, 386f
  interaction with membrane-spanning domains of, 385f
  release of, 380
    controls for, 380, 382f
    nicotine and, 380
  structure of, 381f
VAT. See Vesicular transporter
VCAMs. See Vascular cell adhesion molecules
VDJ recombination, 670f
VDLs, 843
VDR. See Vitamin D receptor
VDR ligand-binding domain, 784
Vegan, 850
Vegetarians, 850
  iron deficiency in, 740
VEGF. See Vascular endothelial growth factor
Veins
  deep, 853, 854f
  hepatic portal, 871, 872f
  pulmonary, 871, 872f

Veins *(Continued)*
  subclavian, 871, 872f, 894
  vascular circulatory system
    overview of, 871, 872f
Venous clots, 856
Venous Thromboembolic disease,
    Leiden factor in, 857t
Ventral striatum, 1001, 1002f
Ventricle, left, 871, 872f
Venule, 506f
Vertebrates
  intestinal tracts of, *E. coli* in,
    1107
  light sensing of, 1037f
Very long-chain fatty acid (VL-
    CFA), 723
Very low-density lipoproteins
    (VLDL), 145, 191f, 203, 204f
  functions of, 235f
Vesicular antiporter, 722
Vesicular monoamine transporter
    (VMAT), 722
Vesicular transporter (VAT),
    1033f
Vessels, lymphatic, 894
VH. *See* Variable heavy chain
VHL. *See* Von Hippel-Lindau
Viagra. *See* Sildenafil
VIF. *See* Viral infectivity factor
Vinyl chloride, 301f
Viral DNA, 1056, 1056f
  retroviral integration in, 1058f
Viral genome
  HIV-1/HIV-2, 1051–1052,
    1053f, 1054f
  DNA, 1059f
  packaging of, 1092, 1095
Viral infection, 925, 927
  cells with, apoptosis in, 904f
  IFNs with, 638
Viral infectivity factor (VIF),
    1050–1051
Viral integrase, 1056, 1060f,
    1061f
  phage λ, 1105f
Viral neuraminidase, 1072, 1076
  inhibitors of, 1076, 1077f
Viral nucleic acids, 1089, 1091f
Viral peptide (TAX), complex
    among human T cell/HLA-
    A0201 and, 930f
Viral polyprotein cleavage, 1078
Viral protein(s), 638
  HIV, 1051

Viral protein R (VPR), 1051
Viral protein U (VPU), 1051
Viral RNA
  HIV, 1048, 1049f, 1050, 1052,
    1053f, 1054f
  West Nile Virus, 1082f
Virion, 1069
  adenovirus, 1070f
  mature, 1097
Virus(es). *See also* Human immu-
    nodeficiency virus
  adeno-associated, 314t, 1074t
  bacteriophage, 1089–1107
    structure of, 1090f
  basic biology of, 1068–1069
  coronavirus, 1089f
  current, 1068–1089
  flaviviruses, 1076
  genetics of, 1069
  human rhino-, 947f
  hybrid, 1047–1048
  icosahedral symmetry of, 1069,
    1069f, 1090f
  influenza/bird flu, 1072, 1076,
    1076f
  measles, epitopes for binding
    antibodies and, 947f
  parasitic nature of, 1069
  replication, 1103
  SARS, 1078, 1081, 1083f, 1085–
    1086, 1085f, 1086f, 1087f
  shape of human, 1069
  spikes on surface of, 1048,
    1049f, 1066f, 1086
  VIF in, 1050–1051
  West Nile, 1076, 1078,
    1081f–1083f
Virus particles
  adenovirus, 1070f
  binding of, 1064, 1066f
  budding of, 1064, 1069
  Influenza, 108f
  neuraminic acid on surface of,
    1076, 1076f
  SARS, 1078, 1083f
Virus receptors, CD46/ICAM-1/
    CD4, 947f
Virus-binding site, 943, 946f
Vision
  opsin in process of, 837, 837f
  vitamin A deficiency and, 837
Vitamin(s), 113f, 119t, 801–847.
    See also Water-soluble vita-
    mins

Vitamin(s) *(Continued)*
  coenzymes with, 119t
  fat-soluble, 835–847, 835f–838f,
    839t, 840f–847f, 841t, 843t,
    847t
  water-soluble, list of major,
    801
Vitamin A, 835–837, 835f–837f
  deficiency of, 837
  food sources of, 839t
  toxicity, 837
Vitamin B deficiency, 818, 820
Vitamin B_1. *See* Thiamine
Vitamin B_2. *See* Riboflavin
Vitamin B_3. *See* Niacin
Vitamin B_5. *See* Pantothenic acid
Vitamin B_6. *See* Pyridoxamine;
    Pyridoxine
Vitamin B_{12}, 518, 756, 818, 1126
Vitamin C. *See* Ascorbic acid
Vitamin D, 837–841
  activated, 783
  -binding protein, 530
  deficiency of, 840
  degradation of, 784
  food sources of, 841t
  formation of, 527, 529f
  ligand-binding domain of, 530
Vitamin D receptor (VDR), 527,
    530f, 784, 785f, 836, 837
Vitamin E, 841–843, 841f–847f
  compounds of, 841
  deficiency of, 843
  food sources of, 847f
Vitamin H. *See* Biotin
Vitamin K, 843–847, 844f–847f
  deficiency of, 847
  food sources of, 847f
  forms of naturally-occurring,
    843
  structure of, 844f
Vitamin K_2, 843, 844f
Vitamin K-dependent (VKD)
  proteins, 846, 847f
  serine protease, 857, 858f
Vitamin K-dependent γ-carboxyl-
    ase, 844
Vitruvian man, 2, 3f
V-J combinations, in B-cell differ-
    entiation, 919
VKD. *See* Vitamin K-dependent
VKOR. *See* 2,3-epoxide reductase
VL. *See* Variable light chain

V_L light chain domain, 917
VLCFA. *See* Very long-chain fatty acid
VLDL. *See* Very low-density lipoproteins
VMAT. *See* Vesicular monoamine transporter
Voltage-gated channels, 1003
Von Gierke's disease, 169, 171t
  purine metabolism disorder with, 288t
Von Hippel-Lindau (VHL), ubiquitin-protein ligase complex, 640f
Von Willebrand factor, 865f
VP. *See* Vasopressin
VP receptors, 380, 384f, 385f
  V1a, 385f
VPR. *See* Viral protein R
VPU. *See* Viral protein U
VR1. *See* Vanilloid receptor

Warfarin, 846, 854
  structure of, 846f
Water
  apical membrane permeated by, 699f
  biological roles of, 27–28, 28f, 29f
  blood carrying, 871
  movement of, 698
Water-soluble vitamins, 801–838
  ascorbic acid, 770, 771, 818, 832–838, 833f, 834f
  biotin, 114f, 118, 119t, 820–821, 824f, 825f, 826f
  cobalamin, 825, 827f, 828, 828f
  folic acid, 119t, 818, 828–832, 829f
  Niacin, 119t, 808–818, 809f–817f
  pantothenic acid, 119t, 818, 819f
  pyridoxal, 818

Water-soluble vitamins (*Continued*)
  pyridoxamine, 818, 820f, 821f, 1004
  pyridoxine, 113f, 119t, 511f, 520, 818, 820f
  riboflavin, 119t, 806–808, 806f, 807f
  thiamine, 119t, 801–806, 802f, 804f
  conversion to TPP of, 801
Watson, James, 289
Watson-Crick base-pairing, 251f
Wernicke-Korsakoff syndrome, 806
West Nile Virus, 1076, 1078, 1081f–1083f
Whey acidic protein, 430
White blood cells. *See* Leukocytes
Wound healing, 832
  blood clotting as major part of, 853

X chromosome, HGPRT gene with, 283
Xa. *See* Activated factor X
Xanthine, 245, 248f, 249f
  guanine deaminated to form, 271, 274f
  uric acid conversion from, 271, 274f
Xanthine dehydrogenase (XMP), 806, 809f
Xanthine monophosphate (XMP), uric acid conversion from, 271, 274f
Xanthine oxidase, 128f, 129, 788
  allopurinol in inhabitation of, 283
  crystal structure of, 271, 274f
  purine metabolism disorder with, 288t
Xanthinuria, purine metabolism disorder with, 288t

Xanthosine monophosphate-IMP conversion, 268–270, 270f
x-chain, 919
xCT transporter, 721f
Xenobiotics, GSH, 521
XMP. *See* Xanthine dehydrogenase (XMP); Xanthine monophosphate
Xylitol, 149, 152f
Xylulose-5-phosphate, 281f

Y_2 receptor, 1033
yCCS chaperone, 766f
Yeast, 285f, 776
YINM-containing adaptor, 909f

Zantac. *See* Ranitidine
ZBD, 1094f
Z-DNA, 289, 291f
Zidovudine (AZT), 1067, 1068t
Zinc, 713, 776–780, 777f–780f
  coordination site, 777f
  RDA for, 778
Zinc family transporters (Zip), 778, 780f
Zinc finger, 349, 354f, 461, 490
  distal, 463, 464f
  motifs, 779f
  proximal, 463, 464f
Zip. *See* Zinc family transporters
Zip4, 780f
Zip5, 780f
Zn^{2+}, 991, 1026f
Zona fasciculata, 387, 449f, 452, 453f, 454, 456
Zona glomerulosa, 449f, 452, 453f, 454
Zona granulosa, 387
Zona reticularis, 449f, 452, 453f
Zona reticulosa, 387, 474
Zoster, 925, 927
Zwitterions, 45f–50f, 46–50, 47t, 1007f
Zymogen, 857